Handbook of Postharvest Technology

BOOKS IN SOILS, PLANTS, AND THE ENVIRONMENT

Soil Biochemistry, Volume 1, edited by A. D. McLaren and G. H. Peterson
Soil Biochemistry, Volume 2, edited by A. D. McLaren and J. Skujiņš
Soil Biochemistry, Volume 3, edited by E. A. Paul and A. D. McLaren
Soil Biochemistry, Volume 4, edited by E. A. Paul and A. D. McLaren
Soil Biochemistry, Volume 5, edited by E. A. Paul and J. N. Ladd
Soil Biochemistry, Volume 6, edited by Jean-Marc Bollag and G. Stotzky
Soil Biochemistry, Volume 7, edited by G. Stotzky and Jean-Marc Bollag
Soil Biochemistry, Volume 8, edited by Jean-Marc Bollag and G. Stotzky
Soil Biochemistry, Volume 9, edited by G. Stotzky and Jean-Marc Bollag

Organic Chemicals in the Soil Environment, Volumes 1 and 2, edited by C. A. I. Goring and J. W. Hamaker
Humic Substances in the Environment, M. Schnitzer and S. U. Khan
Microbial Life in the Soil: An Introduction, T. Hattori
Principles of Soil Chemistry, Kim H. Tan
Soil Analysis: Instrumental Techniques and Related Procedures, edited by Keith A. Smith
Soil Reclamation Processes: Microbiological Analyses and Applications, edited by Robert L. Tate III and Donald A. Klein
Symbiotic Nitrogen Fixation Technology, edited by Gerald H. Elkan
Soil–Water Interactions: Mechanisms and Applications, Shingo Iwata and Toshio Tabuchi with Benno P. Warkentin
Soil Analysis: Modern Instrumental Techniques, Second Edition, edited by Keith A. Smith
Soil Analysis: Physical Methods, edited by Keith A. Smith and Chris E. Mullins
Growth and Mineral Nutrition of Field Crops, N. K. Fageria, V. C. Baligar, and Charles Allan Jones
Semiarid Lands and Deserts: Soil Resource and Reclamation, edited by J. Skujiņš
Plant Roots: The Hidden Half, edited by Yoav Waisel, Amram Eshel, and Uzi Kafkafi
Plant Biochemical Regulators, edited by Harold W. Gausman

Maximizing Crop Yields, N. K. Fageria
Transgenic Plants: Fundamentals and Applications, edited by Andrew Hiatt
Soil Microbial Ecology: Applications in Agricultural and Environmental Management, edited by F. Blaine Metting, Jr.
Principles of Soil Chemistry: Second Edition, Kim H. Tan
Water Flow in Soils, edited by Tsuyoshi Miyazaki
Handbook of Plant and Crop Stress, edited by Mohammad Pessarakli
Genetic Improvement of Field Crops, edited by Gustavo A. Slafer
Agricultural Field Experiments: Design and Analysis, Roger G. Petersen
Environmental Soil Science, Kim H. Tan
Mechanisms of Plant Growth and Improved Productivity: Modern Approaches, edited by Amarjit S. Basra
Selenium in the Environment, edited by W. T. Frankenberger, Jr., and Sally Benson
Plant–Environment Interactions, edited by Robert E. Wilkinson
Handbook of Plant and Crop Physiology, edited by Mohammad Pessarakli
Handbook of Phytoalexin Metabolism and Action, edited by M. Daniel and R. P. Purkayastha
Soil–Water Interactions: Mechanisms and Applications, Second Edition, Revised and Expanded, Shingo Iwata, Toshio Tabuchi, and Benno P. Warkentin
Stored-Grain Ecosystems, edited by Digvir S. Jayas, Noel D. G. White, and William E. Muir
Agrochemicals from Natural Products, edited by C. R. A. Godfrey
Seed Development and Germination, edited by Jaime Kigel and Gad Galili
Nitrogen Fertilization in the Environment, edited by Peter Edward Bacon
Phytohormones in Soils: Microbial Production and Function, William T. Frankenberger, Jr., and Muhammad Arshad
Handbook of Weed Management Systems, edited by Albert E. Smith
Soil Sampling, Preparation, and Analysis, Kim H. Tan
Soil Erosion, Conservation, and Rehabilitation, edited by Menachem Agassi
Plant Roots: The Hidden Half, Second Edition, Revised and Expanded, edited by Yoav Waisel, Amram Eshel, and Uzi Kafkafi
Photoassimilate Distribution in Plants and Crops: Source–Sink Relationships, edited by Eli Zamski and Arthur A. Schaffer
Mass Spectrometry of Soils, edited by Thomas W. Boutton and Shinichi Yamasaki
Handbook of Photosynthesis, edited by Mohammad Pessarakli
Chemical and Isotopic Groundwater Hydrology: The Applied Approach, Second Edition, Revised and Expanded, Emanuel Mazor
Fauna in Soil Ecosystems: Recycling Processes, Nutrient Fluxes, and Agricultural Production, edited by Gero Benckiser
Soil and Plant Analysis in Sustainable Agriculture and Environment, edited by Teresa Hood and J. Benton Jones, Jr.
Seeds Handbook: Biology, Production, Processing, and Storage: B. B. Desai, P. M. Kotecha, and D. K. Salunkhe
Modern Soil Microbiology, edited by J. D. van Elsas, J. T. Trevors, and E. M. H. Wellington
Growth and Mineral Nutrition of Field Crops: Second Edition, N. K. Fageria, V. C. Baligar, and Charles Allan Jones
Fungal Pathogenesis in Plants and Crops: Molecular Biology and Host Defense Mechanisms, P. Vidhyasekaran
Plant Pathogen Detection and Disease Diagnosis, P. Narayanasamy
Agricultural Systems Modeling and Simulation, edited by Robert M. Peart and R. Bruce Curry
Agricultural Biotechnology, edited by Arie Altman
Plant–Microbe Interactions and Biological Control, edited by Greg J. Boland and L. David Kuykendall

Handbook of Soil Conditioners: Substances That Enhance the Physical Properties of Soil, edited by Arthur Wallace and Richard E. Terry
Environmental Chemistry of Selenium, edited by William T. Frankenberger, Jr., and Richard A. Engberg
Principles of Soil Chemistry: Third Edition, Revised and Expanded, Kim H. Tan
Sulfur in the Environment, edited by Douglas G. Maynard
Soil–Machine Interactions: A Finite Element Perspective, edited by Jie Shen and Radhey Lal Kushwaha
Mycotoxins in Agriculture and Food Safety, edited by Kaushal K. Sinha and Deepak Bhatnagar
Plant Amino Acids: Biochemistry and Biotechnology, edited by Bijay K. Singh
Handbook of Functional Plant Ecology, edited by Francisco I. Pugnaire and Fernando Valladares
Handbook of Plant and Crop Stress: Second Edition, Revised and Expanded, edited by Mohammad Pessarakli
Plant Responses to Environmental Stresses: From Phytohormones to Genome Reorganization, edited by H. R. Lerner
Handbook of Pest Management, edited by John R. Ruberson
Environmental Soil Science: Second Edition, Revised and Expanded, Kim H. Tan
Microbial Endophytes, edited by Charles W. Bacon and James F. White, Jr.
Plant–Environment Interactions: Second Edition, edited by Robert E. Wilkinson
Microbial Pest Control, Sushil K. Khetan
Soil and Environmental Analysis: Physical Methods, Second Edition, Revised and Expanded, edited by Keith A. Smith and Chris E. Mullins
The Rhizosphere: Biochemistry and Organic Substances at the Soil–Plant Interface, Roberto Pinton, Zeno Varanini, and Paolo Nannipieri
Woody Plants and Woody Plant Management: Ecology, Safety, and Environmental Impact, Rodney W. Bovey
Metals in the Environment, M. N. V. Prasad
Plant Pathogen Detection and Disease Diagnosis: Second Edition, Revised and Expanded, P. Narayanasamy
Handbook of Plant and Crop Physiology: Second Edition, Revised and Expanded, edited by Mohammad Pessarakli
Environmental Chemistry of Arsenic, edited by William T. Frankenberger, Jr.
Enzymes in the Environment: Activity, Ecology, and Applications, edited by Richard G. Burns and Richard P. Dick
Plant Roots: The Hidden Half, Third Edition, Revised and Expanded, edited by Yoav Waisel, Amram Eshel, and Uzi Kafkafi
Handbook of Plant Growth: pH as the Master Variable, edited by Zdenko Rengel
Biological Control of Major Crop Plant Diseases, edited by Samuel S. Gnanamanickam
Pesticides in Agriculture and the Environment, edited by Willis B. Wheeler
Mathematical Models of Crop Growth and Yield, Allen R. Overman and Richard Scholtz
Plant Biotechnology and Transgenic Plants, edited by Kirsi-Marja Oksman-Caldentey and Wolfgang Barz
Handbook of Postharvest Technology: Cereals, Fruits, Vegetables, Tea, and Spices, edited by Amalendu Chakraverty, Arun S. Mujumdar, G. S. Vijaya Raghavan, and Hosahalli S. Ramaswamy
Handbook of Soil Acidity, edited by Zdenko Rengel

Additional Volumes in Preparation

Humic Matter: Issues and Controversies in Soil and Environmental Science, Kim H. Tan

Molecular Host Resistance to Pests, S. Sadasivam and B. Thayumanavan

Handbook of Postharvest Technology

Cereals, Fruits, Vegetables, Tea, and Spices

edited by

Amalendu Chakraverty
Indian Institute of Technology
Kharagpur, India

Arun S. Mujumdar
National University of Singapore
Singapore

G. S. Vijaya Raghavan
Hosahalli S. Ramaswamy
McGill University
Sainte-Anne-de-Bellevue
Quebec, Canada

MARCEL DEKKER, INC. NEW YORK • BASEL

Library of Congress Cataloging-in-Publication Data
A catalog record for this book is available from the Library of Congress.

ISBN: 0-8247-0514-9

This book is printed on acid-free paper.

Headquarters
Marcel Dekker, Inc.
270 Madison Avenue, New York, NY 10016
tel: 212-696-9000; fax: 212-685-4540

Eastern Hemisphere Distribution
Marcel Dekker AG
Hutgasse 4, Postfach 812, CH-4001 Basel, Switzerland
tel: 41-61-260-6300; fax: 41-61-260-6333

World Wide Web
http://www.dekker.com

The publisher offers discounts on this book when ordered in bulk quantities. For more information, write to Special Sales/Professional Marketing at the headquarters address above.

Current printing (last digit):
10 9 8 7 6 5 4 3 2 1

PRINTED IN THE UNITED STATES OF AMERICA

Preface

In order to prevent huge quantitative as well as qualitative losses of fruits, vegetables, cereals, pulses, spices, and plantation crops, all steps of improved postharvest technology (PHT) must be carefully designed and implemented, beginning with harvesting and ending with consumption and utilization of their products and by-products.

To derive optimal benefit from production techniques, the engineering principles and practice of harvesting and threshing and their effects on grain yield have been outlined in this book. Drying is one of the most important operations in PHT. Hence, the theory, principles, methods, and commercial dryers associated with grain-drying systems have been narrated systematically. A chapter on the drying of fruits, vegetables, and spices is included as well. This book also deals with the principles of grain storage, infestation control and pesticide applications, warehouses, silos, and special storage methods. Present milling technologies of grains, especially processing and milling of rice and pulses, are illustrated and described. Rice husks and other agro-industrial by-products pose a serious disposal problem. Therefore, a chapter is devoted to the conversion and utilization of biomass, with an emphasis on combustion and furnaces, gasification and gasifiers, and chemical processing of biomass and by-products. Moreover, utilization of fruit and vegetable by-products is incorporated. Importance has also been placed on the structure, composition, properties, and grades of food grains. Postharvest technology of tea, coffee, cocoa, and spices has been included as well. Postharvest technology of fruits and vegetables is discussed, covering in detail postharvest physiology, maturity, quality, grades, cooling, storage, disease detection, packaging, transportation, handling, and irradiation.

Although PHT has been introduced as a field of study at various agricultural universities and food technological institutes all over the world in the last few decades, practically no attempt has been made to develop a comprehensive handbook of PHT that deals with engineering principles and modern technologies.

Thus, a comprehensive handbook covering both fundamentals and present practice of PHT of grains, fruits, and vegetables for the production of food, feed, chemicals, and energy should serve as a valuable source of information to a worldwide audience concerned with agricultural sciences and engineering, food technology, and other allied subjects.

Postharvest technology is an interdisciplinary subject. Therefore, the contributing authors of this book are specialists recognized in their respective disciplines.

We take this opportunity to express our heartfelt thanks to the chapter authors for their timely and valuable contributions. We wish to pay homage to the contributors Dr. A. C. Datta and Dr. R. S. Satake, who are no longer with us in person. Sincere thanks are due to the editorial staff of Marcel Dekker, Inc., and all the other people who assisted us directly and indirectly. The wholehearted cooperation of our families is also deeply appreciated.

Amalendu Chakraverty
Arun S. Mujumdar
G. S. Vijaya Raghavan
Hosahalli S. Ramaswamy

Introduction: Production, Trade, Losses, Causes, and Preservation

The need to increase food production to meet the requirements of a rapidly growing world population is widely recognized. Cereals, pulses, fruits, and vegetables are the important food crops in the world as these are the bulk sources of calories, proteins, and nutrients, and spices and plantation crops play an important role in the economies of many countries. To supply an adequate quantity of grains and other food to the expanding world population is a challenge to mankind. Rice and wheat have an added importance in national and international trade with political and social implications.

The supply of grains and other food crops can be augmented by increasing production as well as by reducing postharvest losses. The production of food has increased significantly during the last few decades due to successful research and development efforts in both areas. The use of recently developed high-yielding crops has created a high yield potential when it is supplemented with suitable application of fertilizer and modern management practice. The term ''green revolution'' is used to reveal the impact of high-yielding cultivars on the world of agriculture.

1 PRODUCTION AND TRADE

Worldwide wheat production has increased remarkably since the 1960s, as has the worldwide production of rice. In the period from 1950 to 1971, the world grain production nearly doubled. This dramatic increase is strongly due to the higher use of fertilizers and improved cultivars. Wheat and rough rice production in different countries is shown in Table 1.

Table 2 shows the major grain exporting and importing countries in 1997. The production of pulses and some fruits and vegetables in 1996 is shown in Tables 3 and 4,

Table 1 Wheat and Paddy Production (1000 MT) in Some Countries

Country	Wheat production, 1996	Paddy production, 1996
India	62620.0	120012.0
China	109005.0	190100.0
Russian Fed.	87000.0	2100.0
U.S.A.	62099.0	7771.0
Canada	30495.0	—
France	35946.0	116.0
Australia	23497.0	951.0
Pakistan	16907.0	5551.0
Argentina	5200.0	974.0
World	584870.0	562260.0

Source: *FAO Production Year Book*, Vol. 50, FAO, Rome, 1996.

Table 2 Grain Export and Import (million tonnes) Through Major Seaports of the World, 1997

Country	Export	Country	Import
3 Canadian ports	27.202	3 Egyptian ports	2.759
7 U.S. ports	94.804	7 Chinese ports	1.466
4 European ports	13.401	3 S. Korean ports	12.000
4 Australian ports*	12.509	7 Japanese ports*	8.674

* 1997–1998.
Source: *World Grain*, Nov. 1998.

Table 3 Pulse Production (1000 MT) in Some Continents/Countries

Continent/country	Production, 1996
Asia	28222
Africa	7651
Europe	9380
N. America	5541
S. America	3770
Australia	2186
India	14820
China	4979
Brazil	2862
France	2636
World	56774

Source: India—*FAO Production Year Book*, Vol. 49, FAO, Rome, 1995; others—FAO Production Year Book, Vol. 50, FAO, Rome, 1996.

Table 4 Fruit and Vegetable Production (million tonnes) in Selected Countries

	Production, 1996			
Country	Apple	Orange	Mango	Potato
China	16.00	2.26	1.21	46.03
India	1.20	2.00	10.00	17.94
Russian Fed.	1.80	—	—	38.53
Poland	1.70	—	—	22.50
Brazil	0.65	21.81	0.44	2.70
Mexico	0.65	3.56	—	1.20
France	2.46	—	—	6.46
Germany	1.59	—	—	13.60
U.S.A.	4.73	10.64	—	22.55
World	53.67	59.56	19.22	294.82

Source: *FAO Production Year Book*, Vol. 50, FAO, Rome, 1996.

respectively. The world supply, demand, and stock (1997–1998) of some important grains—wheat, rice, maize, and barley—along with their cultivation area and yield are presented in Table 5. In addition, Figures 1 and 2 represent the world prices of wheat from 1970–1971 to 1997–1998 and maize from 1981–1982 to 1997–1998, respectively. These reveal the international status of grains, fruits, and vegetables.

As regards the world trade activity (1996–1997) of food and feed grains—wheat, maize, barley, soybean, rice, and sorghum—it is interesting to note that some countries are perennial powerhouses in grain exports, such as the United States, Australia, Canada, the European Union countries, and Argentina, whereas Egypt, Japan, China, and Mexico nearly always rank among the top grain importers (*World Grain*, Nov. 1998).

2 LOSSES AND CAUSES

Hunger and malnutrition can exist in spite of adequate food production. These can be the result of uneven distribution, losses, and deterioration of available food resources. Hence, maximum utilization of available food and minimization of postharvest food losses are absolutely essential.

Losses vary by crop variety, year, pest, storage period, methods of threshing, drying, handling, storage, processing, transportation, and distribution according to both the climate and the culture in which the food is produced and consumed. With such an enormous variability, it is not surprising that reliable statistics of postharvest food losses are not available. It is also very difficult to determine the exact magnitude of losses. Fortunately, research and development and education activities related to postharvest technology of crops have been growing. For each postharvest operation there is a possibility of some losses either in quantity or in quality of crop product. For cereals, the overall postharvest losses are usually estimated to be in the range of 5–20%, whereas for fruits and vegetables it may vary from 20% to 50%. If these losses can be minimized, many countries of the world may become self-sufficient in food.

The major purpose of food processing is to protect food against deterioration. All food materials are subject to spoilage. The rate of spoilage of raw food commodities may

Table 5 World Grain Position (in million tonnes and hectares), 1997–1998

		WHEAT			
Supply		Demand		Ending stocks 132.0	
Beginning stocks	109.7	Food use	420.5		
Production	609.3	Feed use	99.9	Wheat area	230.8
Total	719.0	Other	66.6	Yield (tonnes/ha)	2.65
		Total	587.0		
		RICE (milled)			
Supply		Demand		Ending stocks 52.1	
Beginning Stocks	51.2	All uses			
Production	384.6			Rice area	148.2
Total	435.8			Yield (tonnes/ha)	2.6
		MAIZE			
Supply		Demand		Ending stocks 87.7	
Beginning Stocks	91.3	Feed	405.6		
Production	578.6	Other	176.6	Maize area	136.9
Total	669.9	Total	582.2	Yield (tonnes/ha)	4.23
		BARLEY			
Supply		Demand		Ending stocks 31.3	
Beginning Stocks	91.3	All uses	147.2		
Production	154.5			Barley area	65.3
Total	178.5			Yield (tonnes/ha)	2.37

Source: World Grain, Nov. 1998

be very high for fruits and vegetables and not as rapid in the case of cereals and pulses. The spoilage of food is due to three main causes: 1) microbial, 2) enzymatic, and 3) chemical.

All foods during storage are more or less infected with microbes, which cause decomposition of the food constituents, often with the production of evil-smelling and toxic substances. Hence, prevention of microbiological spoilage is essential in any preservation method.

Enzymes, being normal constituents of food, can break down its proteins, lipids, carbohydrates, etc., into smaller molecules and are also responsible for enzymatic browning or discoloration of food. Hence, no food can be preserved properly if its enzymes are not inactivated.

The different chemical constituents of food also react with one another or with the ambient oxygen, causing alteration in color, flavor, or nutrients.

3 PRESERVATION

Ideally, any method of food preservation should prevent all the above three types of spoilage, but none of the present industrial methods fulfills the requirements completely. All

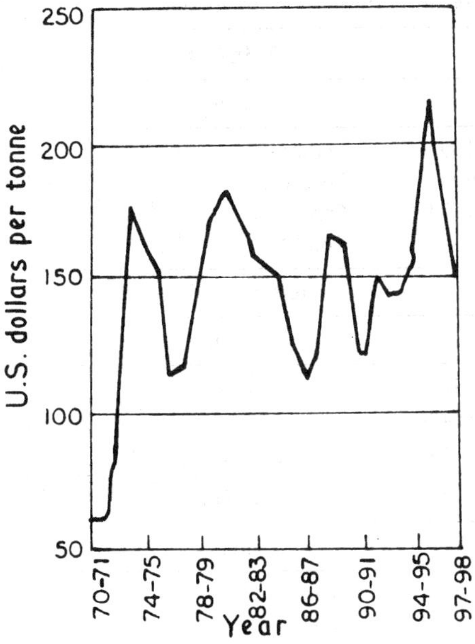

Fig. 1 Export wheat prices, 1970–1971 through 1997–1998 (July–June). (From *World Grain*, Nov. 1998.)

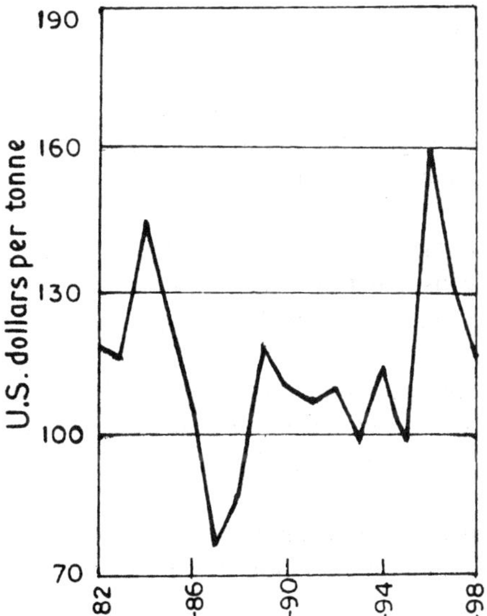

Fig. 2 Export maize prices, 1981–1982 through 1997–1998 (July–June). (From *World Grain*, Nov. 1998.)

these methods must prevent microbial spoilage, but they may be effective to varying degrees in preventing enzymatic and chemical spoilage.

Leaving aside potential innovative preservation techniques such as ohmic heating, pulsed electric field, edible coating, and encapsulation, generally, industrial methods of food preservation include:

Removal of moisture—drying/dehydration, concentration, etc.
Removal of heat—refrigeration/cold-storage, freezing, etc.
Addition of heat—canning, pasteurization, etc.
Addition of chemicals/preservatives
Fermentation
Other methods—application of high-frequency current, irradiation, etc.

Apart from these, various other technologies such as pyrolysis, gasification, combustion, and chemical and biochemical processing are also used for conversion of biomass and grain by-products to chemicals, energy, and other value-added products.

Amalendu Chakraverty

BIBLIOGRAPHY

A Chakraverty. Postharvest Technology. Enfield, NH: Science Publishers, 2001.
JG Ponte, K Kulp, eds. Handbook of Cereal Science and Technology, Second Edition, Revised and Expanded. New York: Marcel Dekker, 2000.

Contents

Contributors

Peter Alvo, M.Sc. Department of Agricultural and Biosystems Engineering, McGill University, Sainte-Anne-de-Bellevue, Quebec, Canada

János Beke, Ph.D. Department of Automotive and Thermal Technology, Faculty of Mechanical Engineering, Szent István University, Gödöllő, Hungary

Thomas H. J. Beveridge, Ph.D. Pacific Agri-Food Research Centre, Agriculture and Agri-Food Canada, Summerland, British Columbia, Canada

Sriman K. Bhattacharyya, Ph.D. Department of Civil Engineering, Indian Institute of Technology, Kharagpur, India

Amalendu Chakraverty, Ph.D. Post Harvest Technology Centre, Department of Agricultural and Food Engineering, Indian Institute of Technology, Kharagpur, India

Susanta Kumar Das, M.Tech, Ph.D. Post Harvest Technology Centre, Department of Agricultural and Food Engineering, Indian Institute of Technology, Kharagpur, India

Adhir C. Datta, Ph.D.† Department of Agricultural and Food Engineering, Indian Institute of Technology, Kharagpur, India

† Deceased.

Jennifer R. DeEll, Ph.D, P.Ag. Fresh Market Quality Program, Ontario Ministry of Agriculture and Food, Vineland Station, Ontario, Canada

Georges Dodds Department of Agricultural and Biosystems Engineering, McGill University, Sainte-Anne-de-Bellevue, Quebec, Canada

Bernie L. Dronzek, Ph.D. Department of Plant Science, University of Manitoba, Winnipeg, Manitoba, Canada

Waliaveetil E. Eipeson, Ph.D.* Department of Fruit and Vegetable Technology, Central Food Technological Research Institute, Mysore, India

István Farkas, D.Sc. Department of Physics and Process Control, Szent István University, Gödöllő, Hungary

Stefan Grabowski, Ph.D. Food Research and Development Centre, Agriculture and Agri-Food Canada, Saint-Hyacinthe, Quebec, Canada

Ananada P. Gupta Department of Civil Engineering, Indian Institute of Technology, Kharagpur, India

Catherine K. P. Hui Horticultural Research and Development Centre, Agriculture and Agri-Food Canada, Saint-Jean-sur-Richelieu, Quebec, Canada

Rajshekhar B. Hulasare, Ph.D. Department of Biosystems Engineering, University of Manitoba, Winnipeg, Manitoba, Canada

Digvir S. Jayas, Ph.D., P.Eng., P.Ag. Department of Biosystems Engineering, University of Manitoba, Winnipeg, Manitoba, Canada

Kamaruddin Abdullah, Dr. Department of Agricultural Engineering, Institut Pertanian Bogor, Bogor, Indonesia

Ajjamada C. Kushalappa, Ph.D. Department of Plant Science, McGill University, Sainte-Anne-de-Bellevue, Quebec, Canada

Monique Lacroix, Ph.D. Research Centre in Applied Microbiology and Biotechnology, Canadian Irradiation Centre and INRS–Institute Armand-Frappier, University of Quebec, Laval, Quebec, Canada

Denyse I. LeBlanc, M.Sc. Atlantic Food and Horticulture Research Centre, Agriculture and Agri-Food Canada, Kentville, Nova Scotia, Canada

---

* Retired.

Michèle Marcotte, Ph.D. Food Research and Development Centre, Agriculture and Agri-Food Canada, Saint-Hyacinthe, Quebec, Canada

Arun S. Mujumdar, Ph.D. Department of Mechanical Engineering, National University of Singapore, Singapore and Department of Chemical Engineering, McGill University, Quebec, Canada

Srikantayya Nagalakshmi, M.Sc.* Department of Plantation Products, Spices, and Flavour Technology, Central Food Technological Research Institute, Mysore, India

Hampapura V. Narasimha, M.Sc., Ph.D. Department of Grain Science and Technology, Central Food Technological Research Institute, Mysore, India

Rhambo T. Patil, Ph.D. Central Institute for Agricultural Engineering, Bhopal, India

Herman W. Peppelenbos, Dr. Postharvest Quality of Fresh Products, Agrotechnological Research Institute (ATO-DLO), Wageningen, The Netherlands

Robert K. Prange, Ph.D, P.Ag. Atlantic Food and Horticulture Research Centre, Agriculture and Agri-Food Canada, Kentville, Nova Scotia, Canada

V. M. Pratape Department of Grain Science and Technology, Central Food Technological Research Institute, Mysore, India

Bashyam Raghavan, M.Sc. Department of Plantation Products, Spices, and Flavour Technology, Central Food Technological Research Institute, Mysore, India

G. S. Vijaya Raghavan, B.E., M.Sc., Ph.D. Department of Agricultural and Biosystems Engineering, McGill University, Sainte-Anne-de-Bellevue, Quebec, Canada

Somiahnadar Rajendran, Ph.D. Department of Food Protectants and Infestation Control, Central Food Technological Research Institute, Mysore, India

N. Ramakrishnaiah Department of Grain Science and Technology, Central Food Technological Research Institute, Mysore, India

Kulathooran Ramalakshmi, M.Sc. Department of Plantation Products, Spices, and Flavour Technology, Central Food Technological Research Institute, Mysore, India

Hosahalli S. Ramaswamy, Ph.D. Department of Food Science and Agricultural Chemistry, McGill University, Sainte-Anne-de-Bellevue, Quebec, Canada

Ramesh S. Ramteke, Ph.D. Department of Fruit and Vegetable Technology, Central Food Technological Research Institute, Mysore, India

---

* Retired.

Byrappa Ranganna, Ph.D. Division of Agricultural Engineering, University of Agricultural Sciences, Bangalore, India

Cristina Ratti Département des Sols et de Génie Agroalimentaire, Université Laval, Sainte-Foy, Quebec, Canada

Timothy J. Rennie, M.Sc. Horticultural Research and Development Centre, Agriculture and Agri-Food Canada, Saint-Jean-sur-Richelieu, Quebec, Canada

Esmaeil Riahi, Ph.D. Department of Food Science and Agricultural Chemistry, McGill University, Sainte-Anne-de-Bellevue, Quebec, Canada

Shyam S. Sablani, Ph.D. Department of Bioresource and Agricultural Engineering, Sultan Qaboos University, Al-Khod, Muscat, Oman

Ashok K. Sarkar Milling Technology and Quality Control, Department of Food Technology, Canadian International Grains Institute, Winnipeg, Manitoba, Canada

Robert S. Satake, D.Eng.† Satake Corp., Hiroshima, Japan

James P. Smith, Ph.D. Department of Food Science and Agricultural Chemistry, McGill University, Sainte-Anne-de-Bellevue, Quebec, Canada

Shahab Sokhansanj, Ph.D. Department of Agriculture and Bioresource Engineering, University of Saskatchewan, Saskatoon, Saskatchewan, Canada

Samson A. Sotocinal, Ph.D. Department of Agricultural and Biosystems Engineering, McGill University, Sainte-Anne-de-Bellevue, Quebec, Canada

Clément Vigneault, Ph.D. Horticultural Research and Development Centre, Agriculture and Agri-Food Canada, Saint-Jean-sur-Richelieu, Quebec, Canada

Noel D. G. White, Ph.D. Cereal Research Centre, Agriculture and Agri-Food Canada, Winnipeg, Manitoba, Canada

† Deceased.

Handbook of Postharvest Technology

1

Structure and Composition of Cereal Grains and Legumes

ESMAEIL RIAHI and HOSAHALLI S. RAMASWAMY

McGill University, Sainte-Anne-de-Bellevue, Quebec, Canada

1 INTRODUCTION

Cereals are monocotyledonous plants that belong to the grass family. Based on botanists' approximation, there are about 350,000 plant species, of which about 195,000 species are economically important flowering plants. Nearly 50 species are cultivated worldwide and as few as 17 species provide 90% of human food supply and occupy about 75% of the total tilled land on earth. They consist of wheat, rice, corn, potato, barley, sweet potato, cassava, soybean, oat, sorghum, millet, rye, peanut, field bean, pea, banana, and coconut. The cereal grains such as wheat, rice, corn, barley, oat, rye, sorghum, and millet provide 50% of the food energy and 50% of the protein consumed on earth. Wheat, rice, and corn together make up three-fourths of the world's grain production. In general, cereal grains have been considered as the source of carbohydrates to supply food energy to the diet. Cereal grains, especially rice and wheat, provide the bulk of energy consumed on earth (Stoskopf, 1985).

The cereal crops that are grown for their edible fruit are generally called *grain*, but botanically referred to as *caryopsis*. The cereal seed consists of two major components, the *endosperm* and *embryo* or *germ*. The endosperm encompass the bulk of the seed and is the energy source of stored food. An outer wall called the *pericarp* that develops from the ovary wall encases the endosperm. A semipermeable layer under the pericarp, which is called *testa*, surrounds the embryo and is derived from the inner ovary wall. The testa is permeable to water, but not to dissolved salts, and is important for germination. The third layer, which is called *aleurone*, contains thick-walled cells that are free of starch. The pericarp, testa, and aleurone layer are collectively called the *bran*.

The legumes such as chickpea, black gram, mung bean, and pigeon pea, have an estimated 16,000–19,000 species in 750 genera. Asia ranks first both in area harvested and production. India accounts for 75 and 96% of the total global production of the chickpea and pigeon pea, respectively. The term *legume* originated from the Latin *legumen*, which means seed harvested in pods. The expression *food legumes* usually means the immature pods and seeds as well as mature dry seeds used as food by humans. Based on Food and Agricultural Organization (FAO) practice, the term *legume* is used for all leguminous plants. Legumes such as French bean, lima bean, or others, that contain a small amount of fat are termed *pulses*, and legumes that contain a higher amount of fat, such as soybean and peanuts, are termed *leguminous oilseeds*. Legumes are important sources of food in developing countries. Soybean, groundnut, dry bean, pea, broad bean, chickpea, and lentil are the common legumes in the most countries. In some countries, depending on the climatic condition and food habits, other legumes are grown. Legumes are next to cereals in terms of their economic and nutritional importance as human food sources. They are cultivated not only for their protein and carbohydrate content, but also because of the oil content of oilseed legumes such as soybeans and peanuts.

Legumes are reasonably priced sources of protein, generally about double that of most cereals, and have a high food value; also, they are fair sources of some vitamins and minerals. Legumes have almost the same caloric value per unit weight as cereals. Legumes are a better source of calcium than cereals and contain 100–200 mg of calcium per 100 g. Legumes, when compared with cereals, are a better source of iron, thiamine, riboflavin, and nicotinic acid. The utilization of legumes is highest in India and Latin America owing to religious restriction and food attitude. Legumes also contain some antinutritional factors, such as trypsin and chymotrypsin, phytate, lectins, polyphenols, flatulence-provoking and cyanogenic compounds, lathyrogens, estrogens, goitrogens, saponins, antivitamins, and allergens. However heat treatment is known to destroy the antinutrients, such as protease inhibitors and lectins, although it also destroys the vitamins and amino acids. Legumes are a good source of dietary fiber; the crude fiber, protein, and lipid components have a hypocholesterolemic effect.

The following is a brief account of the structure and composition of the major cereal crops and legumes.

2 CEREAL CROPS

2.1 Structure

2.1.1 Wheat

Wheat is a single-seeded fruit, 4- to 10-mm long, consisting of a germ and endosperm enclosed by an epidermis and a seed coat. The fruit coat or pericarp (45- to 50-μm thick) surrounds the seed and adheres closely to the seed coat. The wheat color, depending on the species and other factors, is red to white, and is due to material present in the seed coat. Wheat also is classified based on physical characteristics such as red, white, soft, hard, spring, or winter. The wheat kernel structure is shown in Fig. 1. The outer pericarp is composed of the epidermis and hypodermis. The epidermis consists of a single layer of cells that form the outer surface of the kernel. On the outer walls of the epidermal cells is the water-impervious cuticle. Some epidermal cells at the apex of the kernel are modified to form hairs. The hypodermis is composed of one to two layers of cells. The inner pericarp is composed of intermediate cells and cross-cells inward from the hypodermis. Long and cylindrical tube cells constitute the inner epidermis of the pericarp. In the crease, the seed coat joins the pigment strand, and together they form a complete coat about the endosperm

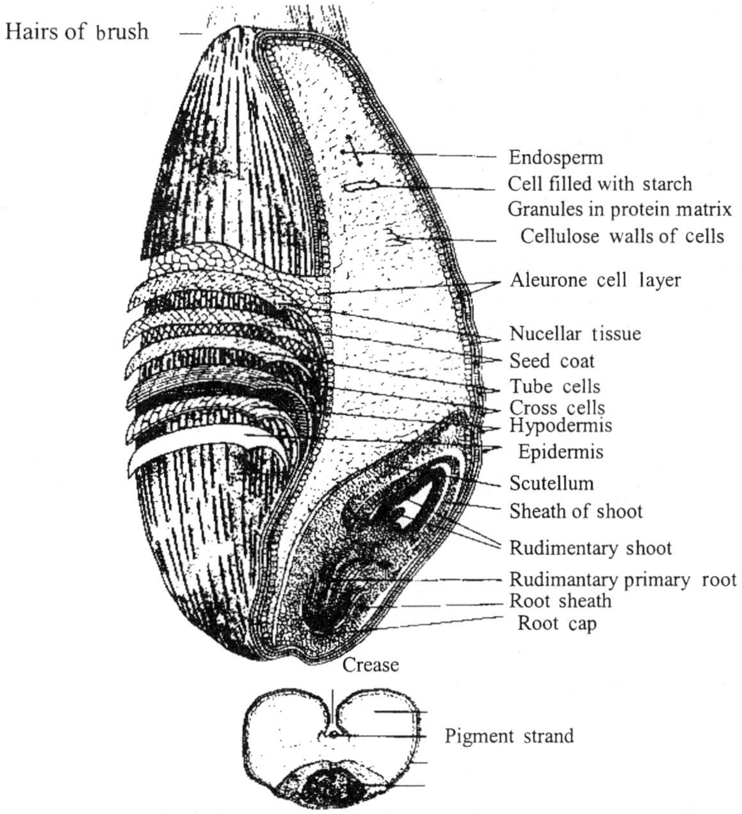

Fig. 1 Diagrammatic illustrations of wheat structure. (From Lasztity, 1999.)

and germ. Three layers can be distinguished in the seed coat: a thick outer cuticle, a ''color layer'' that contains pigment, and a very thin inner cuticle. The bran comprises all outer structures of the kernel inward to, and including, the aleurone layer. This layer is the outer layer of the endosperm, but is considered as part of the bran by millers. The aleurone layer is usually one cell thick and almost completely surrounds the kernel over the starchy endosperm and germ. The endosperm is composed of peripheral, prismatic, and central cells that are different in shape, size, and position within the kernel. The endosperm cells are packed with starch granules, which lie embedded in a matrix that is largely protein. Additional details on the wheat structure can be found in Lasztity (1999).

2.1.2 Corn

Corn or maize (*Zea mays* L.) is an important cereal crop in North America. Maize within a few weeks, develops from a small seed to a plant, typically 2- to 3.5-m tall. Corn apparently originated in Mexico and spread northward to Canada and southward to Argentina. The corn seed is a single fruit called the kernel. It includes an embryo, endosperm, aleurone, and pericarp. The pericarp is a thin outer layer that has a protection role for the endosperm and embryo. Pericarp thickness ranges from 25 to 140 μm among genotypes. Pericarp adheres tightly to the outer surface of the aleurone layer and is thought to impart semipermeable properties to the corn kernel. All parts of the pericarp are composed of

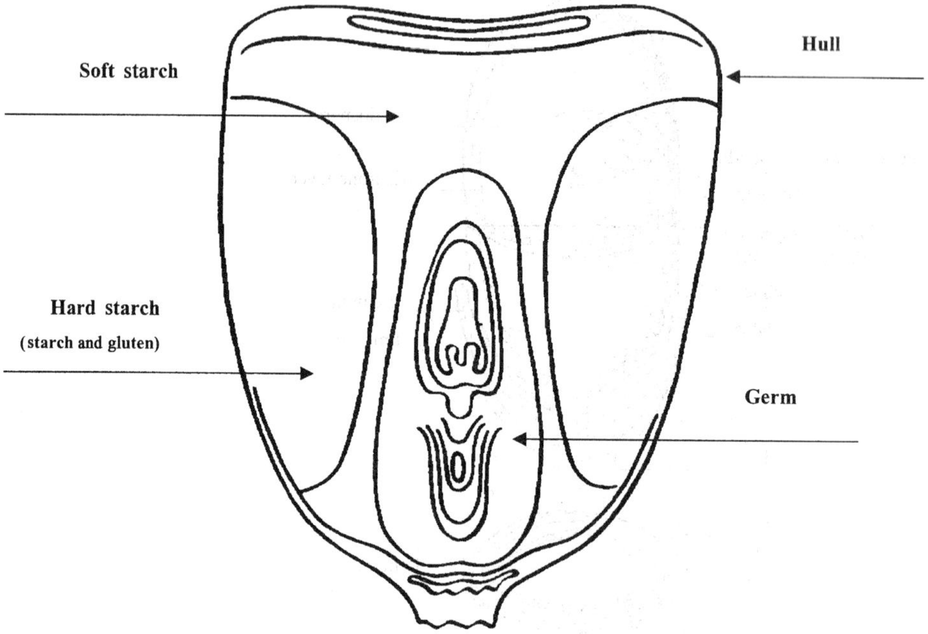

Fig. 2 Diagram of a corn kernel. (From Potter, 1986.)

dead cells that are cellulosic tubes. The innermost tube-cell layer is a row of longitudinal tubes pressed tightly against the aleurone layer. This layer is covered by a thick and rather compact layer, known as the mesocarp, composed of closely packed, empty, elongated cells with numerous pits. A waxy cutin layer that retards moisture exchange covers an outer layer of cells, the epidermis. The endosperm usually comprises 82–84% of the kernel dry weight and 86–89% starch by weight. The outer layer of endosperm or the aleurone layer is a single layer of cells of an entirely different appearance. This layer covers the entire starchy endosperm. The germ is composed of the embryo and the scutellum. The scutellum acts as the nutritive organ for the embryo, and the germ stores nutrients and hormones that are necessary for the initial stage of germination. A typical longitudinal section of a kernel of corn is shown in Fig. 2 and additional details can be found in Potter (1986).

2.1.3 Rice

Rice (*Oryza sativa* L.) is one of the major food staples in the world. The ripe rice is harvested as a covered grain (rough rice or paddy), in which the caryopsis is enclosed in a tough hull or husk composed mostly from silica. The pericarp is fused to the seed and comprises seed coat, nucellus, endosperm, and embryo. The caryopsis is covered by hull, composed of two modified leaves: the palea and larger lemma. The hull provides protection for the rice caryopsis. The hull also protects the grain from insect infestation and fungal damage. The hull consist of four structural layers: (a) an outer epidermis of highly silicified cells; (b) sclerenchyma or hypoderm fibers two- or three-cell–layers thick; (c) crushed, spongy parenchyma cells; and (d) inner epidermis of generally isodiametric cells. The

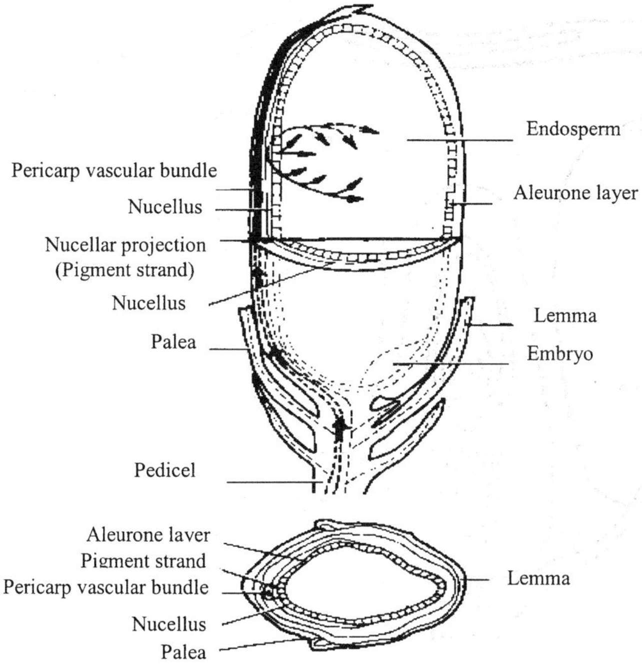

Fig. 3 Structure of the rice grain. (From Juliano, 1985.)

embryo or germ is very small and is located on the central side at the base of the grain. The typical structure of the rice grain is shown in Fig. 3; additional details can be found in Juliano (1985).

2.1.4 Barley

Barley (*Hordeum vulgare* L.) also belongs to the grass family and is one of the major ancient world's crops. It contributes to the human food, malt products, ranks the top ten crops, and is fourth among the cereals. In the commercial barley, the flowering glumes or husk is attached to the grain, whereas some varieties are hull-less and the grain is separate from the husk. The husk is usually pale yellow or buff and is made up of four types of cells, which are dead at maturity. The caryopsis is located in the husk and the pericarp is fused to the seed coat or testa. Within the seed coat the largest tissue is the starchy endosperm that is bonded to the aleurone layer. The embryo is located at the base of the grain. The longitudinal section of the mature barley is shown in Fig. 4, and further details can be found in MacGregor and Bhatty (1993).

2.1.5 Oat

Oat is grown for both grain and forage needs. The hull contributes to about 30% of the total kernel weight. It consists of leaf-like structures that tightly enclose the groat and provide protection during seed growth. At the early stage of growth, the hull assists in nutrient transport and contributes significantly to groat nutrition. Contribution of hulls to the total dietary fiber content of oat is considerable; the hemicellulose content of the oat

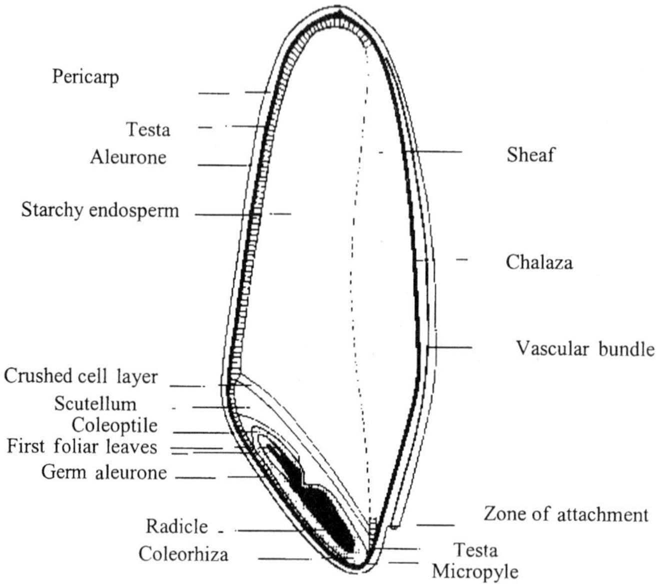

Fig. 4 Structure of the mature barley. (From MacGregor and Bhatty, 1993.)

hull is between 30 and 50%. After removing the hulls, the morphology of remaining groat is not unlike other common cereals. The groat is longer and more slender than wheat and barley and, mostly, is covered extensively with hairs. The groat consists of three morphological and chemically distinct components: bran, germ, and starchy endosperm. These components are traditional descriptions of commercial fractions and do not accurately reflect the genetic, chemical, or fractional characteristics of each fraction. The structure of the oat kernel is shown in Fig. 5 (Webster, 1986).

2.1.6 Rye

Rye (*Secale cereale* L.), another member of the grass family, has two species: *S. fragile* and *S. cereale*. Rye is used mostly in bread making. The mature rye grain is a caryopsis, dry, one-seeded fruit, grayish yellow, ranging from 6 to 8 mm in length and 2 to 3 mm in width. The ripe grain is free-threshing and normally grayish yellow. The seed consists of an embryo attached-through a scutellum to the endosperm and aleurone tissues. These are enclosed by the remnants of the nucellar epidermis, the testa or seed coat, and the pericarp or fruit coat. The aleurone is botanically the outer layer of the endosperm and, in rye, is generally one-cell thick. The aleurone layer surrounds the starchy endosperm and merges into the scutellum located between the endosperm and embryo. In the mature grain, the aleurone is characterized by the presence of numerous intensely staining aleurone granules. The starchy endosperm represents the bulk of the kernel and is composed of three types of cells: peripheral or subaleurone, prismatic, and central, which differ in shape, size, and location within the kernel. Figure 6 is a schematic of the longitudinal section of a rye cell (Kulp and Ponte, 2000).

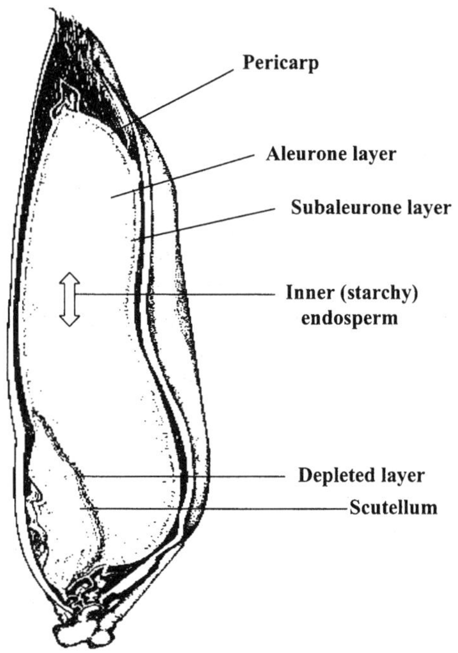

Fig. 5 Oat kernel structure. (From Webster, 1986.)

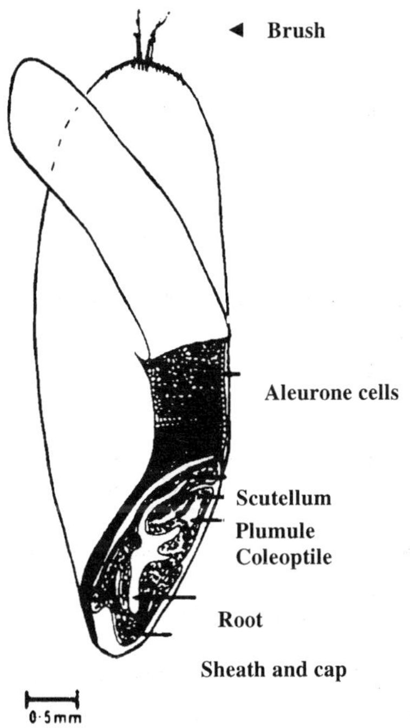

Fig. 6 Diagrammatic view of longitudinal section of rye grain. (From Bushuk and Campbell, 1976.)

2.1.7 Sorghum

Sorghum (*Sorghum bicolor* L.) is a major source of energy and protein in developing countries, especially in Africa and Asia. The sorghum kernel is roughly spherical and is composed of three main components: the seed coat, embryo, and endosperm. The seed coats consist of the fused pericarp and testa. The extreme outer layer is pericarp that is surrounded by a waxy cuticle. Some sorghums contain a complete testa that may or may not contain spots of pigment. The embryo consists of a large scutellum, an embryonic axis, a plumule, and a primary root. The embryo is relatively firmly embedded and difficult to remove by dry milling. The endosperm is the largest proportion of the kernel and consists of an aleurone layer. The peripheral layer is made up of cells containing a high proportion of protein. The layer after the peripheral layer, called the corneous layer, contains less protein and a higher proportion of starch than the peripheral layer. Figure 7 shows the structure of a sorghum grain; additional details can be found in Hulse et al. (1980) and Kulp and Ponte (2000). The mature sorghum grain comprises about 10% em-

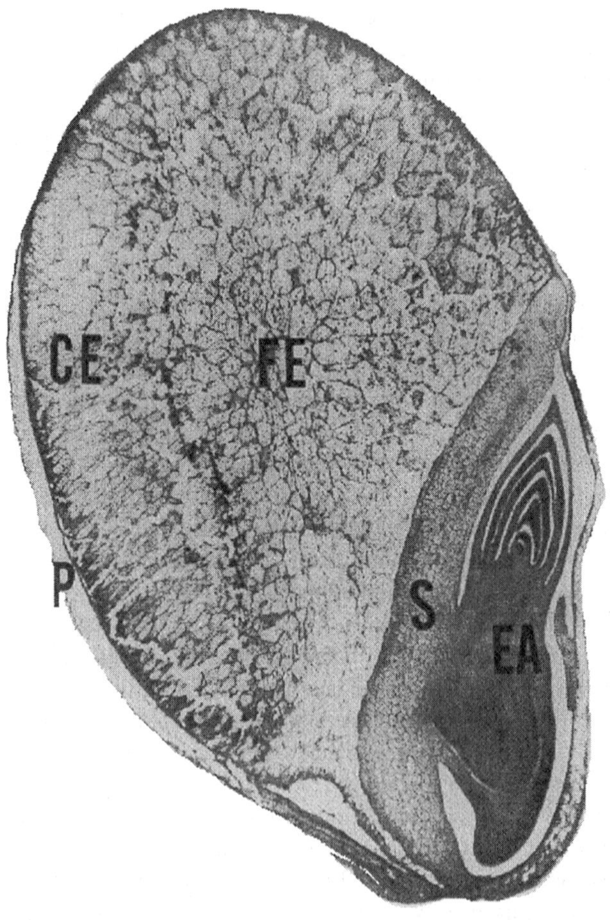

Fig. 7 Cross section of the sorghum grain: P, pericarp; CE, corneous endosperm; FE, floury endosperm; SA, stylar area; S, Scutellum; EA, embryonic axis. (From Hulse et al., 1980.)

bryo, 8% pericarp or bran layers, and 80% endosperm. These proportions may vary with variety, environmental condition, and degree of maturity. The embryo is rich in protein, lipid, minerals, and B vitamin groups.

2.2 Composition

The chemical composition of the cereals varies widely and depends on the environmental conditions, soil, variety, and fertilizer. The proximate composition of selected cereals is given in Table 1 along with typical values of vitamin and mineral composition. The distribution of different amino acids in cereal grains is tabulated in Table 2.

Wheat has a higher protein content than other cereals: The protein content varies from 7 to 22% depending on the variety. However, because of low availability of some essential amino acids in wheat (see Table 2), its biological value requires addition or supplementation with other amino acids. Several research efforts have focused on producing different wheat varieties with higher protein and essential fatty acids content.

Carbohydrates are the major chemical composition of the corn. However, the maize corn kernel is more than a rich source of carbohydrates, it is a source of enzymes for the study of biosynthesis, and genetic markers for genetic, biochemical, and genetic engineering studies. The starch granule is formed inside an amyloplast and arranged in an insoluble granule. Starch is the major carbohydrate in the kernel and comprises close to 72% of its dry weight. Starch also is found in the embryo, bran, and tip cap. Amylose makes up 25–30% of the starch, whereas amylopectin composes 70–75% of the starch. Monosaccharides, such as fructose and glucose, are found in equal proportions in the endosperm. Among the disaccharides, sucrose is the major sugar in kernels that comprise only 4–8% of kernel dry weight; maltose is also found at less than 0.4% of the kernel dry weight. The corn bran consists of 70% hemicelluloses, 23% cellulose, and 0.1% lignin on a dry weight basis. The protein content of the corn shows that it is poor in essential amino acids, such as tryptophan, lysine, and threonine, valine, and sulfur amino acids (see Table 2). The corn has only 4.4% oil (dry basis), but the amount of corn oil production is enormous, even though it is not considered as an oil seed crop. Triglycerides are the major composition (98.8%) of the refined commercial corn oil. Other types of lipids, such as phospholipids, glycolipids, hydrocarbons, sterols, free fatty acids, carotenoids, and waxes, are also found in corn oil. Corn has a high level of unsaturated fatty acid (linoleic acid). Corn oil is very stable compared with other seed oils owing to its low level of linolenic acid and the presence of natural antioxidant.

The composition of rice and its fraction depends on the cultivars, environmental conditions, and processing. The rice components distributed differently in aleurone, embryo, and other parts of the grain. The average brown rice protein content ranges from 4.3 to 18.2% with a mean value of 9.2%. Protein is the second most important rice component after carbohydrates. The outer tissues of the rice grain are rich in water-soluble proteins (albumin) and also salt-soluble proteins (globulin), but the endosperm is rich in glutelin. The milling fraction of the rice grain has a limited prolamin (alcohol-soluble proteins), and the nonprotein nitrogen (NPN) of the rice is about 2–4%. Rice starch is composed of a linear fraction—amylose—and branched fraction—amylopectin—that is a major factor in the eating and cooking quality of the rice. The waxy rice starch has approximately 0.8–1.3% amylose, whereas nonwaxy milled rice contain 7–33% amylose. The amylose content of the rice can be classified as 1–2% (waxy), 7–20% (intermediate), 20–25% (high), and more than 25% (very high). Waxy rice has a higher free-sugar level,

Table 1 Proximate, Vitamin, and Mineral Composition of Different Cereal Grains

		Wheat	Rye	Corn	Barley	Oats	Rice	Sorghum
Proximate composition	Moisture[a]	10	10.5	15	10.6	9.8	11.4	10.6
	Protein[a]	14.3	13.4	10.2	13	12	9.2	12.5
	Fat[a]	1.9	1.8	4.3	2.1	5.1	1.3	3.4
	Fiber[a]	3.4	2.2	2.3	5.6	12.4	2.2	2.2
	Ash[a]	1.8	1.9	1.2	2.7	3.6	1.6	2
Vitamin composition	Retinol	—	—	2.5[d]	—	—	0.0–0.08[b]	—
	Thiamine	9.9[b]	1.45[d]	3.8[d]	—	0.67[d]	2.6–3.3[b]	4.62[b]
	Riboflavin	3.1[b]	2.90[d]	1.4[d]	—	0.11[d]	0.6–1.1[b]	1.54[b]
	Niacine	48.3[b]	—	28[d]	—	0.80[d]	29–56[b]	48.4[b]
	Pyridoxin	4.7[b]	—	5.3[d]	—	0.21[d]	4–7[b]	5.94[b]
	Pantothenic acid	9.1[b]	—	6.6[d]	—	—	7–12[b]	12.54[b]
	Biotin	0.056[b]	—	0.08[d]	—	13[d]	0.04–0.08[b]	2.9[b]
	Folic acid	—	—	0.3[d]	—	104[d]	—	0.20[b]
Mineral composition	Calcium	—	31.5[f]	0.01–0.1[g]	406[b]	50[g]	0.1–0.8[g]	0.05[a]
	Magnesium	3740[a]	92[f]	0.09–1[g]	410[b]	141[g]	0.6–1.5[g]	0.19[a]
	Phosphorus	—	—	0.26–0.75[g]	5630[b]	450[g]	1.7–3.9[g]	0.35[a]
	Potassium	—	412[f]	0.32–0.72[g]	5070[b]	370[g]	1.5–3.7[g]	0.38[a]
	Sulfur	—	—	0.01–0.02[g]		—	0.4–1.6[g]	—
	Sodium	—	—	0–0.01	254[b]	4[g]	—	0.05[a]
	Chlorine	—	—	0.05[b]	—	—	500–800[b]	—
	Cobalt	4.4[e]	—	0.003–0.34[b]	—	—	—	3.10[a]
	Iodine	—	—	73–810[b]	—	4[g]	—	—
	Manganese	—	—	0.7–54[b]	18.9[b]	—	17–94[b]	10.80[a]
	Selenium	28.1[a]	—	0.01–1[b]	—	—	53–810[b]	—
	Zinc	2610[e]	—	12–30[b]	23.6[b]	3[g]	1.7–31[b]	15.4[a]
	Iron	5410[a]	2.7[f]	1–100[b]	36.7[b]	3.81[g]	—	50[a]

[a]Percentage.
[b]Range mean content (μg/g).
[c]U/g.
[d]mg/kg (dry basis).
[e]μg/100 g.
[f]mg/100 g.
[g]mg/g at 14% moisture.
Source: Bushuk and Campbell, 1976; Kulp and Ponte, 2000; Juliano, 1985; Pomeranz, 1971; Watson, 1987.

Table 2 Amino Acid Composition of Cereals

Amino acid (%)	Wheat	Rye	Corn	Barley	Oats	Rice	Sorghum
Arginine	0.80	0.53	0.51	0.60	0.80	0.51	0.40
Cystine	0.20	0.18	0.10	0.20	0.20	0.10	0.20
Histidine	0.30	0.27	0.20	0.30	0.20	0.10	0.03
Isoleucine	0.60	0.53	0.51	0.60	0.60	0.40	0.60
Leucine	1.00	0.71	0.11	0.90	1.00	0.60	1.60
Lysine	0.50	0.51	0.20	0.60	0.40	0.30	0.30
Methionine	0.20	0.50	0.10	0.20	0.20	0.20	0.10
Phenylalanine	0.70	0.70	0.51	0.70	0.70	0.40	0.51
Threonine	0.40	0.40	0.40	0.40	0.40	0.30	0.30
Tryptophan	0.20	0.10	0.10	0.20	0.20	0.10	0.10
Tyrosine	0.51	0.30	0.50	0.40	0.60	0.70	0.40
Valine	0.60	0.70	0.40	0.70	0.70	0.51	0.60

Source: Pomeranz, 1971.

especially maltodextrine. Lipids are in the form of spherosomes or lipid droplets, with a different size in the aleurone layer, subaleurone, and embryo. Rice has about 0.4% non-starch lipid at 14% moisture. Starch lipid from brown rice is close to 0.6–0.7% for non-waxy rice and 0.2% for waxy rice. Rice is also a source of several vitamins and minerals.

Carbohydrates represent the major source of energy in barley, accounting for about 80% of grain weight. Starch is the most abundant single component, accounting for up to 65%, and is composed mainly of amylose and amylopectin. Cellulosic microfibrils are found in the cell walls that reinforce a matrix that mainly comprises arabinoxylans and β-glucans. Proteins are the minor components in barley and consist 8–15% of the dry weight of the mature grain. Albumins and globulins represent 2.8 and 18.1% of the total proteins in barley, whereas glutelins account for 7–38% of the total. In barley, lipids are stored in oil droplets or spherosomes bounded by a simple membrane. The principal core lipid is triacylglycerol and small amounts of other nonpolar lipids. Total lipid content of whole barley grain among different varieties has been reported from 2.4 to 3.9%. The total lipids comprise 67–78% nonpolar lipids, 8–13% glycolipids, and 14–21% phospholipids.

The starch content of the oats has a higher lipid content (1.2%) than other cereals. The floury endosperm of the high-protein species contains a smaller proportion of starch granules than the lower-protein species. Oat starch is more like rice starch in both size and shape and is highly aggregated. Physical properties of the oat starch are influenced by the climate, genetic, and agronomic condition. The amylose content of the oat ranges between 16 and 27%. Oat has superior nutritional value when compared with other cereals owing to its higher protein quality and concentration (15–20%). Oats also have higher lipid concentrations than found among other cereal grain. The free-lipid percentage of oats is 5.0–9.0%, whereas wheat, rice, maize, barley, and rye have 2.1–3.8, 0.8–3.1, 3.9–5.8, 3.3–4.6, and 3.3–4.6% lipid, respectively. The lipids are distributed in different parts of the grain, but over 50% of the lipids are stored in the endosperm.

The starch composition of the rye flour is mostly the same as wheat flour. However, the individual granules are different in size and also the α-amylase activity in the mature kernel is high and, therefore, the viscosity of the dough is lower than wheat. Rye proteins are considered superior to other cereals in biological value; rye has a higher proportion of water- and salt-soluble proteins compared with the other cereals that have an effect in improving the amount of the essential amino acid lysine (see Table 2). The principal component of the rye is starch that is inversely related to the protein content. The crude fat content of rye is similar to that of wheat and barley, ranging from 1.5 to 2.0%. The unsaturated fatty acid content of rye is very high and is characterized by a higher linolenic acid content than found in other cereal grain.

In sorghum (nonwaxy), amylose is reported to range from 20 to 30% and, in some varieties, from 12 to 13% of the starch. Glucose appears to be the principal free sugar in sorghum. Compared with other cereal grains, cellulose and pentosan contents are high. Pentosans are located mostly in the pericarp and cell walls. Removal of the outer pericarp significantly reduces crude fiber of the sorghum. The mineral content of the sorghum, such as calcium, iron, and phosphorus, and also B vitamins, is reduced by removal of the outer pericarp and is also affected by phytic acid that affects their availability.

3 LEGUMES

3.1 Structure

Usually the food legumes are classified in two categories: the legumes in which energy is stored as fat (such as peanut and soybean), and those in which energy is stored as starch (chickpeas). The structure of the food leguminous plants is generally similar. Mature legume seeds have three major components: the seed coat, the cotyledons, and the embryo axis, which constitute 8, 90, and 2% of the seed, respectively. The structure of typical legume seeds and their various anatomical parts of the seed are shown in Figs. 8 and 9. The outer layer of the seed is the testa or seed coat. In most legumes, the endosperm is short-lived and, at maturity, it is reduced tȯ a thin layer surrounding the cotyledons or embryo. The external structure of the seed includes helium, micropyle, and raphe. After soaking and removing the seed coat of a bean, the endosperm comes off, and the reminder is composed of embryonic structure. The embryonic structure includes the shoot, which consists of two cotyledons, and a short axis above and below the cotyledons, which has

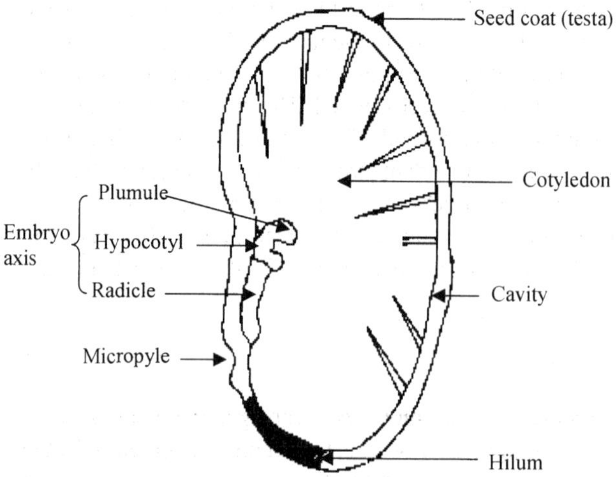

Fig. 8 Cross sections of a mature broad seed with one cotyledon removed. (From Kadam et al., 1989.)

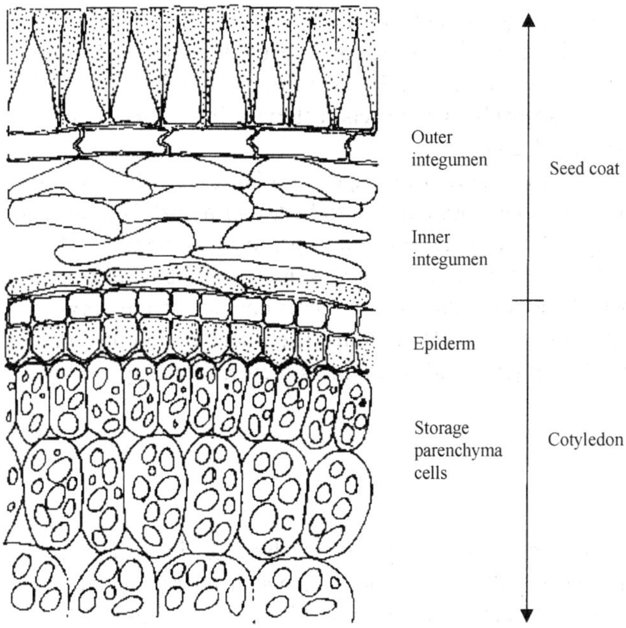

Fig. 9 Mung bean seed coat and cotyledon. (From Kadam et al., 1989.)

several foliage leaves, and terminates in the short tip. The embryonic stem and plumule are fairly well developed in the resting seed and lie between two cotyledons or seed leaves. The radicle or embryonic root has almost no protection except that provided by the seed coat. Therefore, the seed is unusually breakable, especially when it is dried and roughly treated.

Usually, legumes have a moderately thick seed coat. Legume seeds having thick seed coats have higher amounts of lipids. The outermost layer of the seed coat is usually known as the *cuticle*. The other two important features in the external topography of the seed include the *hilum* and *micropyle*. The hilum of the legume is different from others in shape and size, ranging from round to oblong, oval, or elliptical. The micropyle shows variation, ranging from circular and triangular to fork-shaped. The cotyledons of legumes seeds are composed of numerous oarenchymatous cells. The size of the parenchymatous cells ranges from 70 to 100 μm, and the most abundant structures in this region are starch. The distribution of nutrients in different seed fractions, calculated on a percentage of the whole seed, shows that the major portion of protein, ether extract, phosphorus, and iron is present in cotyledons, whereas 80–90% of crude fiber and 32–50% of calcium are present in seed coat.

3.2 Composition

The chemical composition of the legumes depends on the cultivars, geographical location, and growth condition. The proximate composition of selected legumes is given in Table 3, along with typical values of vitamin and mineral composition. Legumes usually contain a large amount of carbohydrates ranging from 24 to 68%. The carbohydrates include mono- and oilgosaccharides. Starch is the main carbohydrate in legumes similar to pinto

Table 3 Proximate, Vitamin, and Mineral Composition of Different Raw Legumes

		Peanut	Pigeon	Chickpea	Soybean	Lentils	Large	Black	Green	Fava	Mug	Cowpea
Proximate composition (%)	Moisture	5.6	10.6	10.7	8.6	10.5	10.9	10.6	10.7	10.6	9.7	11.7
	Protein	22.7	19.8	19.5	34.3	24.7	21.2	21.8	23.9	24.8	23.6	22
	Fat	44.5	1.3	5.7	18.7	1	1.1	1.4	1.3	1.4	1.4	1.3
	Carbohydrate	25.5	65.2	61.7	31.6	61.2	62.7	63.5	62.4	60.4	61.6	63.4
	Crude fiber	2.9	5.5	4	3.8	4.1	5.3	—	3.4	7	4.4	4.5
	Neutral fiber	5.5	13.6	6.1	12	10.4	11.3	13.3	5.7	14.9	9.2	7.7
	Ash	2.2	3.7	2.7	5.1	2.6	4.2	3.4	2.5	3.3	3.3	3.3
Vitamin composition (mg, %)	Thiamine	0.90	0.60	0.51	0.87	0.54	0.64	0.99	0.79	0.52	0.61	0.94
	Riboflavin	0.183	0.166	0.228	0.330	0.238	0.180	0.201	0.254	0.286	0.245	0.227
	Niacin	15.44	2.94	1.72	2.35	2.3	1.61	1.93	2.94	2.52	2.46	2.36
	Vitamin B_6	0.582	0.264	0.560	0.627	0.549	0.601	0.285	0.153	0.374	0.410	0.440
	Total folacin	0.401	0.343	0.481	0.250	0.432	0.308	0.47	0.322	0.431	0.490	0.545
	Pantothenic acid	2.92	1.35	1.32	1.73	1.78	1.32	0.99	1.91	1	1.71	1.39
	β-Carotene	—	36.2	29.1	46.3	34.9	—	—	160.2	47.4	54.1	28.0
Mineral composition (mg, %)	Phosphorus	460.4	317.4	365.7	477	408.5	366.5	380.3	332.9	373.3	348.8	426.5
	Potassium	786.6	1200.4	1044.2	1820	970	2017.3	1424.3	1049.5	1503.1	1192.2	1450.3
	Sodium	34.4	26.1	22.7	6.9	16.6	6.6	5.2	18.4	11.6	5.6	23
	Calcium	66.0	129.1	165	223.1	59.3	98.6	92.3	49.1	97.8	124.8	80.3
	Magnesium	268.3	171	202.7	284.5	180.7	236.2	195.6	157.2	214.7	243.6	250.2
	Zinc	5.28	2.87	3.54	4.48	3.51	2.97	3.96	2.73	3.35	2.62	3.77
	Manganese	2.99	1.79	2.14	5.43	1.31	1.67	1.17	1.13	4.59	1.06	1.28
	Copper	3.15	1.09	0.81	1.43	0.77	0.96	0.77	0.76	0.82	1.05	0.94
	Iron	5.92	8.26	6.23	8.66	8.07	7.43	4.82	5.02	6.66	8.80	7.54

Source: Matthews, 1989.

beans, but in soybean and lupine seeds starch content ranges from 0.2 to 3.5%. The oligosaccharides, such as those of the raffinose family (raffinose, stachyose, and verbascose), are the most predominant in legumes and are 31.1–76% of total soluble sugars. In some legumes, the percentage of the oligosaccaride is higher.

Consumption of large amount of beans causes flatulence in humans and animals that results in discomfort, abdominal rumblings, cramps, pain, and diarrhea, which is caused by the raffinose family. Legumes also contain large amounts of crude fiber, ranging from 1.2–13.5%. Cellulose is a major component of crude fiber in pink beans, whereas in other legumes, such as red gram and lentil, hemicelluloses are the major components of fiber. The legume starch usually contains more amylose than amylopectin. The starch structure is both crystalline and amorphous.

Legumes contain an appreciable amount of protein. The protein content depends on the species of legume and ranges from 15 to 45%. The crude protein content is a mixture of nitrogenous compounds, such as free amino acids, amines, lipids, purine and pyramidine bases, nucleic acid, and alkaloids. Globulins constitute most of the storage proteins in most legume seeds. The biological value of legume proteins is low owing to their content of sulfur amino acids. However, addition of methionine in the diet has a beneficial effect on the protein efficiency ratio in all legumes.

Some legumes, such as peanuts, soybeans, and winged beans, have a considerable amount of lipid, 50, 21, and 17%, respectively. However, lipid content of other species varies from 1 to 7.2%. The lipid content of the legumes depends on variety, origin, location, climate, soil, and environmental conditions. The major fatty acid contents of the legume are oleic, linoleic, and linolenic acid; the legume lipids consist of natural lipid, phospholipids, and glycolipids. The natural lipids are the predominant lipids in most legumes. Total lipid content of the legumes, especially natural lipids, increase to more than 20% during seed maturation. These lipids are highly sensitive to enzymatic and nonenzymatic oxidation, which results in the aldehydes, ketones, esters, and acids.

Legumes are a good source of minerals, such as calcium, iron, copper, zinc, potassium, and magnesium. Potassium is the main mineral of the legume and comprises about 25–30% of the total mineral content of the food legumes. Legumes also are a moderate source of iron and also have a sufficient amount of phosphorous, which exists as phytic acid. Legumes are a good source of thiamin, riboflavin, and niacin, whereas the carotene content of most species is very limited. Drying and storage diminish most of the vitamins. Legumes contain more vitamin E than cereal grains. Legumes also contain a considerable amount of folic acid; a high quantity of polysaccharides and lignin reduce the availability of B vitamins, especially B_6.

REFERENCES

Bushuk W, Campbell WP (1976). Morphology and chemistry of the rye grain. In Drews E, Evans LE, Rozsa TA, Scoles GJ, Seibel W, Simmonds DH, Starzycki S, eds. Rye: Production, Chemistry, and Technology. American Association of Cereal Chemists, St. Paul, MN, p. 63–105.

Egli DB (1998). Seed Biology and Yield of Grain Crops. Cab International, Wallingford, England p. 178.

Hulse JH, Laing EM, Pearson OE (1980). Sorghum and the Millets: Their Composition and Nutritive Value. Academic Press, New York, p. 997.

Juliano BO (1985). Rice: Chemistry and Technology. American Association of Cereal Chemists, St. Paul, MN, p. 774.

Kulp K, Ponte JG (2000). Handbook of Cereal Science and Technology, 2nd ed. Marcel Dekker, New York, p. 790.
Lasztity R (1999). Cereal Chemistry. Akademiai Kiado, Budapest. p. 308.
MacGregor AW, Bhatty RS (1993). Barley Chemistry and Technology. American Association of Cereal Chemists, St. Paul, MN, p. 486.
Matthews HR (1989). Legumes: Chemistry, Technology and Human Nutrition. Marcel Dekker, New York, p. 389.
Pomeranz Y (1971). Wheat Chemistry and Technology. American Association of Cereal Chemists, St. Paul, MN, p. 821.
Potter NN (1986). Food Science. AVI New York, p. 735.
Reddy NR, Pearson MD, Sathe SK, Salunkhe DK (1984). Chemical, nutritional and physiological aspects of dry bean carbohydrates: a review. Food Chem 13:25–68.
Kadam SS, Deshpande SS, Jambhale ND (1982). Seed structure and composition, In: Handbook of World Food Legumes: Nutritional Chemistry, Processing Technology and Utilization, vol 1. Salunkhe DK, Sathe SK, Reddy NR, CRC Press, Boca Raton, FL, pp. 23–45.
Stoskopf NC (1985). Cereal Grain Crops. Reston Publishing Reston, VA. p. 516.
Watson SA (1987). Structure and composition. In: Watson SA, Ramstad PE, eds. Corn: Chemistry and Technology. American Association of Cereal Chemists, St. Paul, MN, pp. 53–82.
Webster FH (1986). Oats: Chemistry and Technology. American Association of Cereal Chemists, St. Paul, MN, p. 433.

2

Physical and Thermal Properties of Cereal Grains

SHYAM S. SABLANI

Sultan Qaboos University, Al-Khod, Muscat, Oman

HOSAHALLI S. RAMASWAMY

McGill University, Sainte-Anne-de-Bellevue, Quebec, Canada

1 INTRODUCTION

Data on physical properties of grain are essential for the design of equipment for handling, aeration, and storage, as well as processing cereal grains and other agricultural materials. Basic thermal and moisture transport properties are also required for simulating heat and moisture transport phenomena during drying and storage. The most important such properties are the grain weight, sphericity, roundness, size, volume, shape, surface area, bulk density, kernel density, fractional porosity, static coefficient of friction against different materials and angle of repose, heat capacity, thermal conductivity, thermal diffusivity, moisture diffusivity, equilibrium moisture content, and latent heat of vaporization. These properties vary widely, depending on moisture content, temperature, and density of cereal grains. The experimental measurement of the physical and thermal properties of cereal grains is the concern of postharvest technologists and researchers.

Substantial research has been carried out, over the years, on gathering data for material property evaluation, and some excellent review articles have been published in various scientific journals on the physical, thermal, and moisture transport properties of plant and animal food materials (Nelson, 1973; Polley et al., 1981; Miles et al., 1983; Sweat, 1974). Several excellent books have also been published highlighting data on physical, thermal, chemical, and electromagnetic radiation properties of food and agricultural products covering a broad range of plant and animal food products (Mohsenin, 1980, 1981, 1984, 1986; Okos, 1986; Rahman, 1995; Rao and Rizvi, 1995). In most published books and reviews, the importance of these properties and the fundamentals involved in their measurement are highlighted, but compilation of some properties are limited mostly to fruits, vegetables, and other food products of plant or animal origin. Property data for grains is generally scarce and scattered. This chapter is designed to present relevant data for cereal grains and also a brief review of the methods commonly used for their estimation. It is hoped that this review will be helpful to postharvest technologists, as well as equipment design and process engineers, who are interested in the storage and handling of cereal grains.

2 PHYSICAL PROPERTIES

2.1 1000-Grain Weight

In handling and processing of grains, it is customary to know the weight of 1000 grain kernels. The 1000-grain weight is a good indicator of the grain size, which can vary relative to growing conditions and maturity, even for the same variety of a given crop. When compared with other crops at the same moisture level, the 1000-kernel weight will also provide an idea of relative size of the kernel for handling purposes. Pabis (1967) used 1000-grain weight and kernel density to determine the effective diameter of a kernel. Generally, this is measured directly by taking the weight of 1000 grain kernels. Fraser et al. (1978) measured the weight of 1000 kernels of fababeans (also known as fava beans) and presented a correlation of grain weight as a function of moisture content. Dutta et al. (1988a) found that the 1000-kernel weight of gram increased linearly with increasing moisture content from 10.9 to 28.4%. Bala and Woods (1991) measured the weight of randomly selected 100 malt grains and multiplied by 10 to give a 1000-grain weight. They also presented a linear regression equation for the 1000-kernel weight as a function of moisture content. Shepherd and Bharadwaj (1986a) assumed a linear relation between the 1000-kernel weight of pigeon pea and the moisture content, and measured the 1000-kernel weight at 0 and 14.7% moisture content to evaluate regression parameters. Deshpande et al. (1993) also observed a linear relation between 1000-kernel weight of soybean and its moisture content in the range 8.0–20.0%. Data on the 1000-grain weight of selected cereal grains are given in Table 1.

2.2 Sphericity and Roundness

Accurate estimation of shape-related parameters are important for determination of terminal velocity, drag coefficient, and Reynolds number. It is also important to know the shape before any heat or moisture transport analysis can be performed. *Sphericity* is defined as the ratio of the surface area of a sphere, which has the same volume as that of the solid, to the surface area of the solid. *Roundness* of a solid is a measure of the sharpness of its corners and is defined as the ratio of the largest projected area of an object in its natural rest position to the area of the smallest circumscribing circle (Curray, 1951). Higher values of sphericity and roundness indicate that the object's shape is closer being spherical. Curray (1951) suggested the following relation for calculation of sphericity and roundness of the grain:

$$\text{Sphericity} = \frac{d_i}{d_c} \tag{1}$$

Table 1 1000-Grain Weight, Sphericity, and Roundness of Grains

Product	Moisture content (% w.b.)	1000-grain weight (kg)	Sphericity	Roundness	Refs.
Faba[fava]beans	8.5–34.9	$0.371 + 3.7 \times 10^{-3}m$	—	—	Fraser et al., 1978
Gram	12.4–32.4	$0.156 + 1.562 \times 10^{-3}m$	0.735	0.697	Dutta et al., 1988a
			0.810[a]	0.813[a]	
Malt	4.89–41.66	$0.0352 + 0.627 \times 10^{-3}m$	—	—	Bala and Woods, 1991
Oilbean seed	4.3		0.605 ± 0.277	0.398 ± 0.357	Oje and Ugbor, 1991
Pigeon pea	12.8	0.076	0.822	0.818	Shepherd and Bhardwaj, 1986a
Soybean	8.0–20	$0.101 + 0.105\ m$	$0.803 + 0.06\ m$	—	Deshpande et al., 1993

[a]Without root: *m*, moisture content.

$$\text{Roundness} = \frac{A_p}{A_c} \tag{2}$$

where d_c and A_c represent the diameter and area of the smallest circumscribing circle, respectively, d_i denotes the diameter of the largest inscribing circle. A_p is the projected area of the grain.

Dutta et al. (1988a) used shadowgraphs of a grain in three mutually perpendicular positions to determine the sphericity and roundness of gram, with and without roots. If the root of gram is ignored, the sphericity and roundness values were higher. Oje and Ugbor (1991) found that 95% of oilbean seeds had a roundness of less than 0.55 and sphericity of less than 0.75, which explained the difficulty in getting the seeds to roll. They suggested that this property should help in the design of hoppers and dehulling equipment for the seed. Shepherd and Bharadwaj (1986a) approximated the shape of pigeon pea as a prolate spheroid and estimated the sphericity and roundness of pigeon pea at 12.8% moisture level using 20 shadowgraphs. They reported the sphericity and roundness values of pigeon pea as 0.822 and 0.818, respectively, with a standard deviation of less than 0.056. Deshpande et al. (1993) measured the linear dimensions of grains with a micrometer (reading to 0.01 mm) and used the relation [sphericity = $(LWT)^{0.33}/L$, where L = length, W = width, T = thickness], as proposed in Mohsenin (1981). They observed that the sphericity of soybean grain increased linearly with increasing moisture content from 8.0 to 20.0%. The sphericity and roundness values for selected grains are also included in Table 1.

2.3 Bulk Density

The bulk density of cereal grains is determined by measuring the weight of a grain sample of known volume. The grain sample is placed in a container of regular shape, and the excess on the top of the container is removed by sliding a string or stick along the top edge of the container. After the excess is removed completely the weight of the grain sample is measured. The bulk density of the grain sample is obtained simply by dividing the weight of the sample by the volume of the container. The bulk density gives a good idea of the storage space required for a known quantity of particular grain. Bulk density also influences the effective conductivity and other transport properties. From the storage point of view, it is important to determine the effect of moisture content on the bulk density of grains because the bulk density of some grains increase with an increasing moisture content, whereas it decreases for some other grains. The bulk densities of rough rice (Wratten et al., 1969) and short rice (Morita and Singh, 1979) were reported to increase linearly with an increasing moisture content between 12 and 18%, whereas the bulk densities of canola (Sokhansanj and Lang, 1996), fababeans (Fraser et al., 1978; Irvine et al., 1992), flaxseed (Irvine et al., 1992), gram (Dutta et al., 1988a), lentils (Irvine et al., 1992; Carman, 1996), malt (Bala and Woods, 1991), and soybean (Deshpande et al., 1993) decreased linearly with an increasing moisture content. Shephard and Bhardwaj (1986a) reported that the bulk density of pigeon pea is higher than that of soybean and grain sorghum but lower than that of fababean in the same moisture range. The bulk densities of rice bran (Jones et al., 1992; Tao et al., 1994), pigeon pea (Shepherd and Bhardwaj, 1986b; rice (Jones et al., 1992; Chandrasekhar and Chattopadhyay, 1989), and wheat (Jones et al., 1992) have also been determined as a function of moisture content for the different grains, which along with their associated moisture contents, are given in (Table 2).

2.4 Kernel Density

The *kernel* (true) *density* of a grain is defined as the ratio of the mass of a grain sample to the solid volume occupied by the sample. For the determination of kernel density of an average grain, two methods have been suggested: one involved the displacement of a gas, whereas the other used displacement of a liquid. In both methods, Archimedes' principle of fluid displacement is used to determine the volume. Wratten et al. (1969) determined the density of medium- and long-grain rice using an air-comparison pycnometer. The kernel density varied from 1324 to 1372 kg/m^3 for medium-grain rice (Saturn) and from 1362 to 1385 kg/m^3 for long-grain rice (Bluebonnet-50). The true density of pigeon pea grain was determined by measuring the grain volume using the water displacement method (Shepherd and Bhardwaj, 1986a). They found that the true density of pigeon pea decreased linearly with increasing moisture content in the range 5.9–22.0%. On comparing the true density of pigeon pea with that of other cereal grains, it was higher than that of soybean, but lower than that of corn in the moisture range selected for this study. Dutta et al. (1988a) used water displacement method and determined the true density of gram in the range of moisture content 8.8–23.7%. The true density of gram decreased linearly with increasing moisture content. Oje and Ugbor (1991) used the water displacement method and found the mean true density of oilbean seed was 1120 kg/m^3 at a moisture content level of 4.3%. Chang (1988) used a gas pycnometer with helium to determine the true density of corn, wheat, and sorghum kernels. Irvine et al. (1992) used an air comparison pycnometer to measure the grain volume of preweighed samples and determined the densities of flaxseed, lentils, and fababeans at different moisture content. They found that the true densities of Eston lentils, Laird lentils, and fababeans decreased by 1.3, 2.8, and 1.6%, respectively, when moisture content of these seeds increased between 10 and 18%. The kernel density of flaxseed remained relatively constant at approximately 1149 kg/m^3 between 8.2 and 12.6% moisture content. Deshpande et al. (1993) measured the kernel density of soybean using a liquid displacement method. The grains were coated with a very thin layer of epoxy resin adhesive (Araldite) to avoid any absorption of water during the experiment as used earlier by Dutta et al. (1988a). The adhesive was insoluble in water, resistant to heat, humidity, solvents, and acids. The increase in weight of the grain owing to adhesive coating was less than 1.5%, and there was no change in weight of the grain even if it was kept submerged in distilled water for 2 h. The effect of moisture content on kernel density of soybean grain showed a linear decrease with moisture content. Tao et al. (1994) used an indirect method to determine the volume of preweighed rice bran. They obtained the volume of the rice bran samples from porosity determination. The true densities of bran of long and medium rice were reported to be 1080 kg/m^3, and 1000 kg/m^3, respectively. The densities of wheat and canola kernels have also been determined using liquid displacement method (Sokhansanj and Lang, 1996). Kernel density values for different grains are summarized in Table 2.

2.5 Porosity

The porosity of grain is an important parameter that affects the kernel hardness, breakage susceptibility, milling, drying rate, and resistance to fungal development (Chang, 1988). Porosity is a property of grain that depends on its bulk and kernel densities. The grain porosity can be measured with the help of an air comparison pycnometer or by the mercury displacement method. The values of grain porosity of various grains, as measured with the former method, tend to be 5–10% higher (Thompson and Issacs, 1967). *Fractional*

Table 2 Bulk Density, Kernel Density, and Porosity of Grains

Product	Moisture content (%wb)	Bulk density (kg/m³)	Kernel density (kg/m³)	Porosity	Refs.
Bran	13.6	235	—	—	Jones et al., 1992
Canola[a]	5.0–19.0	672–661	1122–1118	0.401–0.409	Sokhansanj and Lang, 1996
Canola[b]		678–661	1126–1118	0.403–0.409	
Corn (BoJac)	11.8		1452 (0.53)		Chang, 1988
Corn (Stauffer)	12.0		1450 (1.29)		Chang, 1988
Faba[fava]beans	8.5–34.9	883–4.44 m	—	—	Freaser et al., 1978
Faba[fava]beans	12.6–21.9	815–761	1393–1373	—	Irvine et al., 1992
Flaxseed	7.0–15.1	634–574	1148–1143	—	Irvine et al., 1992
Gram	8.8–23.7	814.8–3.45 m	1340–2.73 m	39.1 + 0.145 m	Dutta et al., 1988a
Lentil	6.1–24.6	1212–8.95 m	—	27.3 + 0.159 m	Carman, 1996
Lentils (Eston)	11.4–18.0	825–783	1410–1392	—	Irvine et al., 1992
Lentils (Laird)	11.7–17.7	804–767	1431–1393	—	Irvine et al., 1992
Malt	4.89–32.4	527–4.48 m	—	—	Bala and Woods, 1991
Oilbean seed	4.3		1120 ± 34		Oje and Ugbor, 1991
Pigeon pea	12.8	782	1283		Shepherd and Bhardwaj, 1986a
Rice bran[b]	—	290	1080	0.73	Tao et al., 1994
Rice bran[d]	—	280	1000	0.72	Tao et al., 1994
Rice (rough)[c]	12–18	598–648	1324–1372	58.5–53.1	Wratten et al., 1969
Rice (rough)[b]	12–18	586–615	1362–1383	59.6–56.9	

Rice (short)	11.24–19.45	632–664	—	—	Morita and Singh, 1979
Rice (paddy)	8.9	639			Jones et al., 1992
Rice (white)[a]	12.9	851			Jones et al., 1992
Rice (white)[a]	13.3	823			Jones et al., 1992
Rice (white)[d]	13.2	839			Jones et al., 1992
Rice (Panoli)	10.5	780	1465	0.47	Chandrasekhar and Chattopadhyay, 1989
Sorghum (Ferry Morse)	12.7		1471 (0.68)		Chang, 1988
Sorghum (Funk)	11.2		1448 (0.60)		Chang, 1988
Soybean	8.0–20.0	748.9–166.3 m	1254.8–525.8 m	0.405–0.136 m	Deshpande et al, 1993
Wheat[a]	8.0–22.0	790–686	1374–1332	0.425–0.485	Sokhansanj and Lang, 1996
Wheat[b]		766–686	360–332	0.419–0.485	
Wheat	9.1	827			Jones et al., 1992
Whole meal	9.3	542			Jones et al., 1992
Wheat (hard) (Newton)	13.8		1476 (0.42)		Chang, 1988
Wheat (hard) (Centurk)	13.8		1469 (0.54)		Chang, 1988
Wheat (soft) (Hart)	13.6		1478 (1.13)		Chang, 1988
Wheat (soft) (Pike)	13.5		1463 (0.44)		Chang, 1988

[a]Adsorption.
[b]Desorption.
[c]Samples milled at different places.
[d]Long-grain.
[e]Medium-grain.
The values in parentheses are coefficients of variation (%).

porosity is defined as that fraction of the space in the bulk grain that is not occupied by the grain (Thompson and Issacs, 1967). The percentage fractional porosity was evaluated from the following relation:

$$P_f = \left(1 - \frac{\rho_b}{\rho_t}\right) \times 100 \quad (3)$$

where ρ_b = bulk density of grain (kg/m^3) and ρ_t = kernel density of grain (kg/m^3). The porosity of grains varies with moisture content. The porosity of canola (Sokhansanj and Lang, 1996), gram (Dutta et al., 1988a), lentil (Carman, 1996), and wheat (Sokhansanj and Lang, 1996) increase with an increasing moisture content, whereas Deshpande et al. (1993) observed that the porosity of soybean decreased with an increasing moisture between 8.0 and 20.0% (db). Porosity values for selected grains are also included in Table 2.

2.6 Coefficient of Friction

Static and dynamic coefficients of friction of grains on metals, wood, and other materials are needed for the design and prediction of grain motion in harvesting and handling equipment. These parameters are also important in determining the pressure of grain and silage against bin walls and silos. The static coefficients of various cereal grains have been determined relative to different structural materials. For the measurement of the coefficient of friction, a topless and bottomless container is placed on an adjustable tilting surface and filled with the grain, while the tilting surface is maintained in a horizontal orientation. The tilting surface with a box on its top is raised gradually at one end, with a screw device, until the box just starts to slide down, and the angle of tilt (in radians, coefficient of friction) is read from a graduated scale (Dutta et al., 1988a). The static coefficients of friction on sheet metal, plywood, and rubber were determined for red kedney beans, unshelled peanuts and soybeans (Chung and Verma, 1989), and lentils (Carman, 1996). Their study showed that the friction coefficient increased linearly with an increasing moisture content. Carman (1996) reported that static and dynamic coefficients of friction were greatest for lentil seeds against rubber and least for galvanized sheet metal, with plywood in between. They also reported that the effect of moisture content was greater on the dynamic coefficient because of the increased adhesion between the seed and the material surface at higher moisture values. When compared with other seeds, the coefficient of friction for lentil seed was higher than that of soybean seed and gram. The friction coefficient has also been estimated for gram (Dutta et al., 1988a), fababeans (Fraser et al., 1978; Irvine et al., 1992), flaxseed and lentils (Irvine et al., 1992), and oilbean seed (Oje and Ugbor, 1991). The values of static coefficient of friction for oilbean seed were highest for plywood, with grain perpendicular to the direction of motion, and lowest for glass. This coefficient was lower than values of gram, but higher than fababeans (Oje and Ugbor, 1991). Dutta et al. (1988a) found that the static coefficient of friction for gram was maximum for the plywood with grains perpendicular to the direction of slide and minimum against the galvanized steel. They also reported that the static coefficient of friction for gram was higher than those of barley, corn, soybean, wheat, fababeans, and pigeon pea. In most of the studies, the friction coefficient increased with an increasing moisture content. Table 3 summarizes some published data on the friction coefficients of selected grains.

Table 3 Coefficient of Friction of Grains

Product	Moisture content (% wb)	Coefficient of friction: Sheet metal	Plywood	Rubber	Refs.
Gram	9.95–31.9	0.348–0.489	0.384–0.559	—	Dutta et al., 1988a
	9.95–31.9		(0.409–0.582 m)[a]		
	9.95–31.9		(0.455–0.651 m)[b]		
Kidney beans (red)	10.4–15.1	0.250 + 0.0070 m	0.245 + 0.0081 m	0.266 + 0.0106 m	Chung and Verma, 1989
		0.203 + 0.0068 m	0.220 + 0.0063 m	0.239 + 0.0078 m	
Faba[fava]beans (Diana)	8.5–21.6	0.32–0.38	0.28–0.46[a]	—	Fraser et al., 1978
			0.32–0.55[b]		
Faba[fava]beans	12.6–21.9	0.29–0.36	0.28–0.37[a]	0.31–0.39[e]	Irvine et al. 1992
		0.52–0.61[d]	0.32–0.41[b]	0.29–0.40[f]	
Flaxseed	7.0–15.1	0.27–0.66	0.33–0.56[a]	0.42–0.62[e]	Irvine et al., 1992
		0.56–0.76[d]	0.43–0.60[b]	0.44–0.63[f]	
Lentils (Estons)	11.4–18.0	0.24–0.33	0.17–0.24[a]	0.32–0.41[e]	Irvine et al., 1992
		0.45–0.49[d]	0.25–0.31[b]	0.31–0.38[f]	
Lentils (Laird)	11.7–17.7	0.20–0.30	0.15–0.24[a]	0.27–0.34[c]	Irvine et al., 1992
		0.37–0.41[d]	0.21–0.29[b]	0.26–0.33[f]	
Lentil	6.1–24.6	0.307 + 0.0063 m	0.356 + 0.0051 m	0.428 + 0.0032 m	Carman, 1996
		0.253 + 0.0046 m	0.289 + 0.0043 m	0.348 + 0.0035 m	
Oilbean seed	4.3	0.331 ± 0.0162	(0.374 ± 0.0122)[a]	(0.299 ± 0.0089)[c]	Oje and Ugbor, 1991
			(0.418 ± 0.0182)[b]		
Peanuts (unshelled)	2.50–15.2	0.306 + 0.0122 m	0.366 + 0.0085 m	0.441 + 0.0049 m	Chung and Verma, 1989
		0.272 + 0.0065 m	0.310 + 0.0057 m	0.376 + 0.0034 m	
Soybean	8.00–16.5	0.291 + 0.0036 m	0.301 + 0.0054 m	0.319 + 0.0058 m	Chung and Verma, 1989
		0.256 + 0.0031 m	0.256 + 0.0044 m	0.255 + 0.0054 m	

[a]Grain parallel to wood surface.
[b]Grain perpendicular to wood surface.
[c]For glass.
[d]Corrugated steel.
[e]Steel-trowelled concrete.
[f]Wood-floated concrete.

Table 4 Angle of Repose (Emptying) of Cereal Grain

Product	Moisture (% wb)	Angle of repose degree (°)	Refs.
Faba[fava]beans (Diana)	8.5–20.9	20.6–23.5	Fraser et al., 1978
Faba[fava]beans	12.6–21.9	28.1–29.3	Irvine et al., 1992
Flaxseed	7.0–15.1	30.3–38.0	Irvine et al., 1992
Gram (*Cicer arietinum* L.)	12.4–32.4	$17.1 + 1.21\ \text{m}–0.0265\ \text{m}^2$	Dutta et al., 1988a
Lentils (Eston)	11.4–18.0	25.7–28.6	Irvine et al., 1992
Lentils (Laird)	11.7–17.7	21.6–23.9	Irvine et al., 1992
Oilbean seed	4.3	17	Oje and Ugbor, 1991

2.7 Angle of Repose

Various investigators (Dutta et al., 1988a; Fraser et al., 1978; Irvine et al., 1992; Oje and Ugbor, 1991) have studied the emptying angle of repose by using a specially constructed box with a removable front panel. The box is filled with grain, then the front panel is quickly removed. This allows the grain to flow to its natural slope. This slope is taken as a measure of the angle of repose. The emptying angle of repose has been measured for fababeans (Fraser et al., 1978, Irvine et al., 1992), gram (Dutta et al., 1988a), oilbean seed (Oje and Ugbor, 1991), and flaxseed and lentils (Irvine et al., 1992). Oje and Ugbor (1991) found that the emptying angle of response of oilbean seed was considerably higher than for soybean, corn, and wheat. This was probably because the smooth surface of oilbean seed which makes it easy for the seeds to slide on each other. Dutta et al. (1988a) showed that the emptying angle of repose increased with an increase in moisture content of gram which could be characterized by a second-order equation (Table 4).

2.8 Bulk Volume Shrinkage

Data are required on bulk volume shrinkage of grain for the mathematical solution of drying processes (moisture desorption). The bulk volume shrinkage is determined by placing a grain sample in a cylindrical container, exposing it to an air stream, and recording continuous changes in mass and depth of grain bed (Lang and Sokhansanj, 1993). Clark and Lamond (1988) determined the bulk volume shrinkage of wheat during air drying in a 2-ft bed and presented the following relationship:

$$\Delta X = 0.85\ \Delta M \tag{4}$$

where ΔX is bed shrinkage and ΔM is bed mean moisture reduction. Boyce (1965) presented a linear equation relating percentage shrinkage of barley bed to the average moisture content. The moisture of 0.31-m^2–bed area of barley decreased from 34 to 14% during through-flow drying, and the following relation was developed:

$$\Delta X = 25.21 - 0.66\ M \tag{5}$$

where M is mean bed moisture. Bala and Woods (1984) simulated deep-bed drying of malt in a cylindrical container with a bed height of 0.3 m and a diameter of 0.15 m. The barley sample was dried at 50°C with 12% relative humidity, and an exponential model was presented:

$$\Delta X = 15.91\,(1 - e^{-0.97\,\Delta M}) \tag{6}$$

Lang and Sokhansanj (1993) proposed shrinkage eqs (7) and (8) for wheat and canola dried from their initial moisture content of 9.1 and 6.4%, respectively:

$$\text{Wheat} \quad \Delta X = 1.0 + 1.3\,\Delta M \tag{7}$$

$$\text{Canola} \quad \Delta X = 0.8 + 0.9\,\Delta M \tag{8}$$

where

$$\Delta X = \frac{X_i - X}{X_i} \times 100\% \qquad \text{and} \qquad \Delta M = M_i - M\% \tag{9}$$

They attributed the difference in bulk volume shrinkage to the differences in constituents of the two seeds. The canola seed contained 42% oil on a mass basis; on the other hand, wheat kernel contained 65% carbohydrates of starch origin: starch is known to have higher affinity for water than oil. Lang and Sokhansanj (1993) also studied the effect of relative humidity (20–90%) and air temperature (20–90°C) and related shrinkage coefficient, $\lambda_m (= \Delta X/\Delta M)$ with relative humidity (RH) and temperature (T):

$$\text{Wheat} \quad \lambda_m = 1.3 + 0.4 \times 10^{-2}\,\text{RH} + 0.2 \times 10^{-3}\,T \tag{10}$$

$$\text{Canola} \quad \lambda_m = 0.9 + 0.7 \times 10^{-2}\,\text{RH} + 0.3 \times 10^{-3}\,T \tag{11}$$

The effects of relative humidity on shrinkage coefficient for wheat and canola were reported to be predominant, and temperature effects were statistically insignificant.

3 THERMAL PROPERTIES

Thermal and moisture-transport properties are required for simulating heat and moisture transfer phenomena during drying and storage of grains. Variability in composition and physical characteristics is typical for cereal grains as well as vegetables, fruits, and meat products. Moreover, a given grain product may have different thermal properties, depending on its origin, concentration, and previous history. The property values also vary with moisture content and temperature during drying. Although early analyses used constant uniform values of thermal properties, present-day numeric methods, such as finite-element analysis, can account for nonuniform thermal properties (Sweat, 1995). Hence, it is important to know the thermal property changes during drying or storage. Most studies have reported thermal properties of cereal grains as a function of moisture content. Some studies have evaluated the influence of temperature on thermal properties of grains (Dutta et al., 1988b; Shepherd and Bhardwaj, 1986b). Apart from using thermal property data for simulating drying and storage conditions, the data are also useful in the modeling of heat treatment to wheat, corn, and legumes, as it has shown some promise in stimulating germination (Mohsenin, 1986).

3.1 Specific Heat

Specific heat indicates amount of heat required to change the temperature of a material of unit mass by 1°. In mathematical form, specific heat C_p, is written as

$$C_p = \frac{Q}{m\,\Delta T} \tag{12}$$

where Q is the amount of heat, m is the mass of a material, and ΔT is the change in temperature. In the literature, many methods have been reported for the determination of specific heats of a variety of food products. In most published studies, a method of mixtures, also known as calorimetric method, was employed to determine specific heat. In this method, a material (usually water) of known specific heat C_{p1}, mass m_1, and temperature T_1 is allowed to exchange the heat in an adiabatic environment with another material for which the specific heat (C_{p2}) is to be determined, and its mass (m_2) and temperature (T_2; generally, lower or higher than T_1) are also known. Once mixture reaches the equilibrium temperature of T_e, the specific heat can be calculated from a heat-balance equation between the heat gained (or lost) by one material and heat lost (or gained) by the liquid, using the following expression:

$$C_{p2} = \frac{m_1 C_{p1}(T_e - T_1)}{m_2(T_2 - T_e)} \tag{13}$$

Great care must be taken to avoid heat exchange with the exterior. Encapsulation of the sample in a copper cylinder or plastic pouches can be used to avoid moisture migration. Rahaman (1995) has detailed the sources of errors in the mixture method.

Mohsenin (1980) presented a method of a guarded plate to measure the specific heat of agricultural materials. In this method, the specimen is surrounded by electrically heated thermal guards. These thermal guards are maintained at the same temperature as the specimen, which is also being heated electrically; thus, there is no heat loss. The specific heat of the sample can be calculated by a heat balance as Eq. (14).

$$C_p = \frac{VIt}{m(T_f - T_i)} \tag{14}$$

where V and I are the voltage (volt) and current (amp) supplied to the heater for the time t(s); T_f and T_i are the final and initial temperature of the sample of mass m(kg). Mohsenin also described an adiabatic chamber method to measure the specific heat of agricultural products. The adiabatic chamber is designed to prevent heat or moisture loss through the chamber. A measured quantity of heat is added by means of a heating wire immersed in the bulk material in a container placed in the test chamber. The specific heat of the material can be calculated from the heat balance as

$$Q = (m\, C_p\Delta T)_{\text{sample}} + (m\, C_p\Delta T)_{\text{container}} + (m\, C_p\Delta T)_{\text{chamber}} \tag{15}$$

Wright and Porterfield (1970) used a calorimeter to measure the specific heat of a single peanut. The calorimeter consisted of a small aluminum cup surrounded by a resistance heating coil and foam insulation. The temperatures of peanut and calorimeter were measured using two thermocouples. A constant power was supplied for specific time, and transient temperature data were gathered during heating and cooling of the calorimeter, with and without the sample. The specific heat was calculated by the heat balance, based on total heat supplied and absorbed, and heat desorption of water from the sample. The details of these and several other methods are discussed by Mohsenin (1980) and more recently by Rahman (1995).

Although these methods are simple, inexpensive, and direct, they only estimate the specific heat over large temperature ranges. Also the accuracy of the measurement is highly dependent on the accuracy with which the temperature rise of the calorimeter can be recorded (Wallapapan et al., 1986). The method of differential-scanning calorimetry

(DSC) is more advanced and sophisticated, but has been used to determine the specific heat of various food products. It requires very small (milligram) amounts of sample and has the capability of measuring very small thermal effects during thermal processes. The temperature dependence of specific heat can also be evaluated. However, the DSC equipment is expensive and requires careful sample preparation. For the measurement of specific heat, the rate of change of temperature ($\mathrm{d}T/\mathrm{d}t$) of the sample is measured as it is being heated. The heat flow rate ($\mathrm{d}Q/\mathrm{d}t$) into the sample of known mass (m) is measured and the specific heat of the sample is determined as:

$$C_\mathrm{p} = \frac{(\mathrm{d}Q/\mathrm{d}t)}{m(\mathrm{d}T/\mathrm{d}t)} \tag{16}$$

The details on application of DSC are presented by Lund (1983).

The method of mixture has been used to determine the specific heat of pea, rice (Ordinanz, 1946), rough and fully furnished rice, shelled rice, oat (Haswell, 1954), yellow dent corn, and soft white wheat (Kazarian and Hall, 1965), medium-grain rough rice (Wratten et al., 1969), beans, corn, rapeseed (Pabis et al., 1970), soybean (Watts and Bilanski, 1970), Spanish peanut (Wright and Porterfield, 1970), hard, red spring wheat (Muir and Viravanichai, 1972), sorghum (Sharma and Thompson, 1973), soybean (Alam and Shove, 1973), rice (Oshita et al., 1978), short-grain rice (Morita and Singh, 1979), gram (Dutta et al., 1988b), pigeon pea (Shepherd and Bharadwaj, 1986b), malt (Bala and Woods, 1991), oilbean seed (Oje and Ugbor, 1991), and four different varieties of rough and hulled rice (Oshita, 1992). In most studies, the influence of moisture content on specific heat has also been determined. Specific heat of cereal grains generally increases linearly with an increasing moisture content. Some studies have also evaluated the temperature effect on specific heat and expressed the specific heat as a function of moisture content and temperature by a second-order polynomial (Dutta et al., 1988b; Shepherd and Bharadwaj, 1986b). A plot of specific heat against the mass of seed revealed that lighter seeds tend to have high specific heat capacities, whereas heavier ones have lower specific heat capacities (Oje and Ugbor, 1991). Specific heat data along with any model developed for the prediction are given in Table 5.

3.2 Thermal Conductivity

Thermal conductivity of food materials depends on its chemical composition as well as the structure of the sample, and because of this the measurement of thermal conductivity of food materials is difficult and more challanging than specific heat measurement (Sweat, 1995). In general, two principal methods have been used for the measurement of thermal conductivity: a steady-state method and a trasient method (Wallapapan et al., 1986). The steady-state method, as based on Fourier's law of heat conduction, is given by Eq. 17:

$$q = -kA\frac{\mathrm{d}T}{\mathrm{d}t} \tag{17}$$

The most common steady-state techniques are the guarded hot-plate method and concentric cylinders method. A guarded hot-plate apparatus consists of a heat source surrounded by samples and then by a heat sink. Insulations are provided at both axial ends to avoid heat loss. The one-dimensional steady-state equation of heat conduction for homo-

Table 5 Specific Heat Data or Estimation Models for Grains

Product	Moisture (% wb)	Temperature (°C)	Specific heat (kJ/kg K)	Refs.
Bean	0–23.1	—	1.210–2.466	Pabis et al., 1970
Bean (broad)	0–23.1	—	1.344–2.600	Pabis et al., 1970
Corn	10.5	>0	1.170	Handbook, 1967
Corn	0–23.1	—	1.212–2.468	Pabis et al., 1970
Corn (yellow dent)	0.90–30.2	12–29	1.461 + 0.0355 m	Kazarian and Hall, 1965
Faba[fava]beans (Diana)	11.5–31.7	−32–0	0.65 + 0.048 m	Fraser et al., 1978
		0–20	1.23 + 0.027 m	
		20–40	1.54 + 0.026 m	
		40–60	1.62 + 0.020 m	
Gram (*Cicer arietinum* L.)	12.4–32.4	10–50	$-4190 + 11.9\,T + 0.0215\,T^2 - 373\,m - 0.165\,m^2 + 1.38\,m\,T$	Dutta et al., 1988b
Malt (Triumph, Sonja)	4.89–41.66	—	1.651 + 0.0412 m	Bala and Woods, 1991
Oat	11.7–17.8	—	1.658–1.859	Haswell, 1954
Oilbean seed	4.3	—	3.67 ± 0.84	Oje and Ugbor, 1991
Pea	14.0	0.1	1.840	Ordinanz, 1946
Pigeon pea	8.0–26.0	10–40	$[537.61 \times 10^{-3} + 653.17 \times 10^{-4}\,T - 84.74 \times 10^{-5}\,T^2 + 103.42 \times 10^{-5}\,m + 35.34 \times 10^{-5}\,m^2 + 0.67\,m\,T]$[a]	Shepherd and Bhardwaj, 1986b
Peanut (Spanish)	10.0	18–29	1.980–1.876	Wright and Porterfield, 1970
	20.0	18–29	2.240–2.077	
	30	18–29	2.458–2.240	
Rapeseed	0–23.1	—	2.244–3.500	Pabis et al., 1970
Rapeseed (Torch)	0.75–19.6	2	1.305–1.807	Moysey et al., 1977
	0.66–15.5	19	1.385–1.812	

Rice	11.73–15.61	0.1	1.760–1.840	Ordinanz, 1946
Rice bran (long, Lemont)	—	—	1.717	Tao et al., 1994
Rice bran (short, Nato)	—	—	1.717	Tao et al., 1994
Rice (rough)	10.2–17.0	—	1.566–1.872	Hswell, 1954
Rice (fully furnished)	10.8–17.4	20–24	1.181 + 0.0377 m	Haswell, 1954
Rice (paddy, medium-grain)	12.0–20.0	—	0.921 + 0.544 m	Wratten et al., 1969
Rice (rough, short-grain)	11.0–24.0	—	1.269 + 0.0349 m	Morita and Singh, 1979
Rice (rough, koshihikari)	11.0–23.0	—	1.27 + 0.0414 m	Oshita, 1992
Rice (rough, Ozora)	09.0–19.0	—	1.18 + 0.0437 m	Oshita, 1992
Rice (rough, Akibare)	12.0–23.0	—	1.24 + 0.0268 m	Oshita, 1992
Rice (rough, Ukonnoshiki)	12.0–22.0	—	1.18 + 0.0306 m	Oshita, 1992
Rice (hulled, koshihikari)	10.0–20.0	—	1.33 + 0.0328 m	Oshita, 1992
Rice (hulled, Ozora)	10.0–18.0	—	1.19 + 0.0450 m	Oshita, 1992
Rice (hulled, Akibare)	11.0–22.0	—	1.19 + 0.0306 m	Oshita, 1992
Rice (hulled, Ukonnoshiki)	11.0–23.0	—	1.17 + 0.0302 m	Oshita, 1992
Rice (Sasanishiki)	09–19.0	—	1.18 + 0.0411 m	Oshita et. al., 1978
Rice (shelled)	9.80–17.6	20–24	1.202 + 0.0381 m	Haswell, 1954
Sorghum	2.0–29.0	24	1.394 + 0.0322 m	Sharma and Thompson, 1973
Soybean	0.0	12–28	1.578	Alam and Shove, 1973
Soybean	8.7		1.855	
Soybean	15.1		2.010	
Soybean	17.1		2.061	
Soybean	21.2		2.169	
Soybean	27.5		2.345	
Soybean	7.4	30–128	1.876	Watts and Bilanski, 1970
Wheat (hard red spring)	1.0–19.0	0.6–21.1	1.093 + 0.0452 m	Muir and Viravanichai, 1972
Wheat (soft white)	0.7–20.3	10.7–32.2	1.394 + 0.0408 m	Kazarian and Hall, 1965

geneous isotropic materials without internal heat generation is used to calculate thermal conductivity (k) (Rahman, 1995):

$$Q = \frac{kA(T_o - T_i)}{l} \tag{18}$$

where Q is heat quantity; T_o and T_i are the outer and inner temperatures; and l is the sample thickness. In concentric cylinder method, the sample is placed between insulated ends of two concentric cylinders. The coolant is passed though the inner cylinder while the outer cylinder is being heated by a heater. The heat absorbed by coolant is assumed to be the heat that goes through the sample, and the thermal conductivity is calculated from the one-dimensional radial heat-transfer Eq. (19):

$$Q = kA_o\left(\frac{T_o - T_i}{r_o \ln(r_i/r_o)}\right) \tag{19}$$

where Q is heat quantity (J); A_o is area of the outer cylinder (m^2); T_o and T_i are the temperatures of outer and inner cylinder (°C); r_o and r_i are the radii of outer and inner cylinder m, respectively. These measurement methods and several other steady-state ones have been discussed by Mohsenin (1980) and by Wallapapan et al. (1986). More recently, Rahman (1995) also presented an excellent review of steady-state and quasi–steady-state techniques. The steady-state techniques, however, are not suitable for foods because of long temperature equilibration times, the possibility of moisture migration, and the need for a large sample size.

The line heat source method and its modified version, the thermal conductivity probe method, are the transient methods of thermal conductivity measurement. These methods are more convenient and simple, require only a short time, and use relatively small samples for measurement. The methods require measurement of transient temperature resulting from the applied heat flux, at some point in the material, using a line heat source or a probe at the same point. The thermal conductivity is then computed from Eq. (20):

$$k = \frac{q}{4\pi L(T_1 - T_2)} \ln\left(\frac{t_2}{t_1}\right) \tag{20}$$

where q is the power generated by a heater (W); T_1 and T_2 are the temperature of probe thermocouple (°C) at time t_1 and t_2 (s); and L is the length of probe thermocouple (m). The construction details of the thermal conductivity probe and an analysis of the method are given in Sweat (1995). Thermal conductivities of yellow dent corn and soft white wheat were obtained using a transient line heat source method (Kazarian and Hall, 1965). Bakke and Stiles (1935), Babbitt (1945), Moote (1953), Ojha et al. (1967), and Dua and Ojah (1969) used a steady-state heat-transfer equation in cylindrical coordinate to obtain thermal conductivity of grains. The transient methods have also been used to determine thermal conductivity of medium-grain, rough rice (Wratten et al., 1969), beans, corn, rapeseed (Pabis et al., 1970), short-grain, rough rice (Morita and Singh, 1979), sorghum (Miller, 1963; Sharma and Thompson, 1973), hard, red, spring wheat (Chandra and Muir, 1971), whole soybean, crushed soybean, powdered soybean (Jasonski and Bilanski, 1973), pigeon pea (Shephard and Bhardwaj, 1986b), gram (Dutta et al., 1988b). Dutta et al. (1988b) reported that the thermal conductivity of gram increased with both increase in temperature and moisture content, and it was higher than that of pigeon pea, sorghum, or wheat, but lower than that of bean or corn. Thermal conductivity data for selected grains are summarized in Table 6.

3.3 Thermal Diffusivity

Thermal diffusivity indicates how fast heat can propagate through the material under transient conduction of heat-transfer conditions. Physically it relates the ability of a material to conduct heat with its ability to store heat. The thermal diffusivity can be calculated by dividing the thermal conductivity by the product of specific heat and mass density:

$$\alpha = \frac{k}{(\rho C_p)} \tag{21}$$

where α is thermal diffusivity (m^2/s); k is thermal conductivity (W/m K); ρ is density (kg/m^3), and C_p is specific heat (J/kg K). Thermal diffusivity can be determined either by direct experiment or estimated from the thermal conductivity, specific heat, and density data. In the direct measurement, thermal diffusivity is based on the measured transient temperatures at any location in the material being heated under known boundary conditions. The analytical solution to the Fourier's heat conduction equation in appropriate geometry or Heisler charts can be used to calculate thermal diffusivity for infite slab, infinite cylinder, or sphere. For a finite-shaped object, Newman's rule is applied to combine the solution of infinite-shaped geometry.

Dickerson (1965) described an apparatus to measure thermal diffusivity based on transient heat transfer. The apparatus consists of a metal cylinder sample holder which is placed in an agitated water bath. Transient temperatures are measured at the surface and at the center of the sample. For this purpose, one thermocouple is soldered at the outer surface of the cylinder to monitor the surface temperature of the sample and another thermocouple probe is inserted at the center of the sample. The sample holder contains two Teflon caps, one at the bottom and another at the top. The cylinder is then placed in the agitated water bath and the transient temperature data are recorded until a constant rate of temperature rise is obtained for both inner and outer thermocouples. Thermal diffusivity of the sample is estimated from Eq. (22):

$$\alpha = \frac{\Phi r^2}{4(T_s - T_c)} \tag{22}$$

where Φ is the constant rate of temperature rise at the surface (°C/s); r is the radius of sample (m); $(T_s - T_c)$ is the maximum temperature difference (°C) between surface and center (at the steady-state heat-transfer condition). This method and several other methods for the measurement of thermal diffusivity are detailed in Rahman (1995).

Thermal diffusivity data for grains reported in the literature are summarized in Table 7. Kazarian and Hall (1965) directly measured thermal diffusivity of yellow dent corn and soft white wheat. Jones et al. (1992) made direct measurement of thermal diffusivity of white rice, paddy rice, wholemeal, bran, and wheat. The thermal diffusivity of rice bran was also measured by Tao et al. 1994). The thermal diffusivities of medium-grain, rough rice (Wratten et al., 1969) and short-grain, rough rice (Morita and Singh, 1979), pigeon pea (Shepherd and Bhardwaj, 1986b), gram (Dutta et al., 1988b) were calculated from thermal conductivity, specific heat, and density. Dutta et al. (1988b) observed that thermal diffusivity of gram increased with increased in temperature. The plot of thermal

Table 6 Thermal Conductivity Data or Estimation Models of Grains

Product	Moisture (% wb)	Temperature (°C)	Thermal conductivity (W/m K)	Refs.
Barley	10.0	10–30	0.139–0.143	Kustermann et al., 1981
Barley	30.0	10–30	0.165–0.169	Kustermann et al., 1981
Bean	0–23.1	—	0.136–0.245	Pabis et al., 1970
Bean (broad)	0–23.1	—	0.140–0.252	Pabis et al., 1970
Corn	0–23.1	—	0.158–0.284	Pabis et al., 1970
Corn	2.0–6.3	22–55	0.110–0.180	Kustermannn et al., 1981
Corn (yellow dent)	0.9–30.2	20.8–52.6	0.141 + 0.0011 m	Kazarian and Hall, 1965
Faba[fava]beans (JV-2)	13	76.0	0.23	Vinod and Bera, 1995
	20	71.56	0.28	
	23.1	69.42	0.30	
	26	66.66	0.33	
Gram (*Cicer arietinum* L.)	12.4–32.4	10–50	$-0.507 + 0.00255\,T - 2.13E\text{-}06\,T^2 + 4.24E\text{-}3\,T\,m - 6.56\text{E-}6\,m^2 + 6.48E\text{-}6mT$	Dutta et al., 1988b
Oat (white English)	12.5	27	0.130	Oxley, 1944
Oat	9.1–27.7	—	0.064–0.093	Bekke, 1935
Oat	10.0	10–30	0.128–132	Kustermann et al., 1981
Oat	30.0	10–30	0.170–176	Kustermann et al., 1981
Pea	10.0–30.0	—	0.129–0.232	Pabis et al., 1970
Pigeon pea	11–28	27	$9.91 \times 10^{-2} + 3.1 \times 10^{-3}$ m	Shepherd and Bhardwaj, 1986b
Peanuts (Spanish)	5.0–43.0	25	0.104–173	Suter et al., 1975
Rapeseed	0–23.1	—	0.160–0.288	Pabis et al., 1970
	10.0	10–30	0.133–0.136	Kustermann et al., 1981
	30.0	10–30	0.176–0.179	Kustermann et al., 1981
	6.3–12.8	4.4	0.108–0.137	Bilanski and Fisher, 1976
(Torch)	0.75–19.6	2	0.093–0.113	Moysey et al., 1977
	0.75–15.5	19	0.097–0.120	Moysey et al., 1977
Rice (kalingpong, brown)	12.0	52–69	0.820–0.901	Dua and Ojha, 1969
Rice (kalingpong, white)	12.0	52–69	0.919–1.023	Dua and Ojha, 1969

Rice bran (long, Lemont)	—	—	0.067	Tao et al., 1994
Rice bran (short, Nato)	—	—	0.064	Tao et al., 1994
Rice (paddy, medium-grain)	12.0–20.0	—	0.0865 + 0.0013 m	Wratten et al., 1969
Rice (rough, short-grain)	11.0–24.0	—	0.1000 + 0.0111 m	Morita and Singh, 1979
Rice (panoli)	10.5	—	0.225	Chandrasekhar and Chattopadhyay, 1992
Sorghum (hybrids Rs610)	13.0–22.0	5	0.1020 + 0.0022 m	Miller, 1963
Sorghum (variety NC + RS66)	1.0–23.0	24–49		
Sorghum	2–29.0	4–24	0.100–0.141	Sharma and Thompson, 1973
Soybean (whole)	11.2	10	0.095	Jasansky and Bilanski, 1973
		21	0.106	
		32	0.112	
		43	0.118	
		54	0.126	
		66	0.133	
Soybean (crushed)	10.1	10	0.085	Jasansky and Bilanski, 1973
		21	0.088	
		32	0.093	
		43	0.100	
		54	0.111	
		66	0.126	
Soybean (powder)	9.4	10	0.066	Jasansky and Bilanski, 1973
		21	0.074	
		32	0.078	
		43	0.087	
		54	0.095	
		66	0.104	
Wheat (hard red spring)	4.4–22.5	−27–20	0.139 + 0.0012 m	Chandra and Muir, 1971
Wheat (soft white)	0.7–20.3	9–23	0.117 + 0.00113 m	Kazarian and Hall, 1965
Wheat	10.0	10–30	0.152–0.155	Kustermann et al., 1981
Wheat	30.0	10–30	0.186–0.189	Kustermann et al., 1981

Table 7 Thermal Diffusivity Data and Estimation Models of Grains

Product	Moisture (% wb)	Temperature (°C)	Thermal diffusivity (m^2/s)	Refs.
Barley	10.0	10–30	9.43×10^{-8}–9.54×10^{-8}	Kustermann et al., 1981
Barley	30.0	10–30	11.0×10^{-8}–11.11×10^{-8}	Kustermann et al., 1981
Bean	0–23.1	—	8.54×10^{-8}–7.55×10^{-8}	Pabis et al., 1970
Bean (broad)	0–23.1	—	7.98×10^{-8}–7.43×10^{-8}	Pabis et al., 1970
Bran	13.6	30–60	1.72×10^{-7}	Jones et al., 1992
Corn	0–23.1	—	10.45×10^{-8}–9.23×10^{-8}	Pabis et al., 1970
Corn	2–40	22–45	8.89×10^{-8}–8.33×10^{-8}	Kustermann et al., 1981
Corn (yellow dent)	0.9	8.7–23.3	0.001020	Kazarian and Hall, 1965
	5.1		0.000983	
	9.8		0.000940	
	14.7		0.000906	
	20.1		0.000867	
	24.7		0.000888	
	30.2		0.000924	
Faba[fava]beans (JV-2)	13.0	76.0	1.00×10^{-7}	Vinod and Bera, 1995
	20.0	71.56	1.07×10^{-7}	
	23.1	69.42	1.10×10^{-7}	
	26.0	66.66	1.14×10^{-7}	
Gram (*Cicer arietinum* L.)	12.4–32.4	10–50	1.33E-6 − 8.45E-9 T + 1.40E-11 T^2 + 3.79E-8 m − 3.83 m^2 − 1.13E-10 mT	Dutta et al., 1988b
Oat	10.0	10–30	10×10^{-8}–10.15×10^{-8}	Kustermann et al., 1981
Oat	30.0	10–30	10.9×10^{-8}–11.0×10^{-8}	Kustermann et al., 1981
Pea	10.0–30.0	—	7.93×10^{-8}–6.82×10^{-8}	Pabis et al., 1970

Pigeon pea	11–28	27	$7.565 \times 10^{-8} + 1.714 \times 10^{-9}$ m $- 4.579 \times$ 10–11 $m^2 + 4.246 \times 10^{-13}$ m^3	Shepherd and Bhardwaj, 1986b
Rapeseed	0–23.1	—	7.1×10^{-8}–8.2×10^{-8}	Pabis et al., 1970
	10.0	—	7.29×10^{-8}–7.44×10^{-8}	Kustermann et al., 1981
	30.0	—	11.21×10^{-8}–11.36×10^{-8}	Kustermann et al., 1981
(Torch)	0.75–19.6	2	10.1×10^{-8}–9.42×10^{-8}	Moysey et al., 1977
(Torch)	0.75–19.6	19	$9.96 \times 10^{-8} \times 10^{-8}$	Moysey et al. 1977
Rice bran (long, Lamont)	—	—	1.33×10^{-7}	Tao et al., 1994
Rice bran (short, Nato)	—	—	1.36×10^{-7}	Tao et al., 1994
Rice (paddy)	8.9	30–60	1.07×10^{-7}	Jones et al., 1992
Rice (paddy, medium-grain)	12–20	Room temp	0.00135 + 2.49e-5 M	Wratten et al., 1969
Rice (rough, short-grain)	11–24	Room temp	0.00125 − 1.63e − 5 m	Morita and Singh, 1979
Rice (panoli)	10.5	—	8.65×10^{-8}	Chandrasekhar and Chattopadhyay, 1992
Rice (white[a])	12.9	30–60	1.00×10^{-7}	Jones et al., 1992
Rice (white[a])	13.3	30–60	1.00×10^{-7}	Jones et al., 1992
Rice (white[a])	13.2	30–60	1.01×10^{-7}	Jones et al., 1992
Sorghum	2–29.0	4–24	9.29×10^{-8}–8.2×10^{-8}	Sharma and Thompson, 1973
Soybean	8.0–40.0	20–50	4.5×10^{-8}–7.7×10^{-8}	Watts and Bilanski, 1973
Whole meal	9.3	30–60	1.02×10^{-7}	Jones et al., 1992
Wheat	10.0	10–30	9.07×10^{-8}–9.18×10^{-8}	Kustermann et al., 1981
	30.0	10–30	10.4×10^{-8}–10.5×10^{-8}	Kustermann et al., 1981
Wheat (soft white)	0.7	9.1–23.2	0.000927	Kazarian and Hall, 1965
	5.5		0.000896	
	10.3		0.000854	
	14.4		0.000801	
	20.3		0.000800	
Wheat	9.1	30–60	0.91×10^{-7}	Jones et al., 1992

[a]Samples milled at different places.

diffusivity with temperature at different levels of moisture content showed that the constant moisture curves converged with increase in temperature.

4 CONCLUSION

To design equipment and facilities for handling, processing, drying, and storing cereal grains, their physical and thermal properties must be known. The physical and thermal properties of cereal grains were collected from the literature, reviewed, and tabulated. In most, property values were reported as a linear function of moisture content. Some studies have also reported the influence of temperature on property values. A brief discussion on experimental techniques widely used for the measurement of properties was provided. It is hoped that this chapter will provide useful property data for postharvest technologists and engineers interested in design of handling equipment for grains.

REFERENCES

Alam A, Shove GC. 1973. Hygroscopicity and thermal properties of soybeans. Trans ASAE 16: 707–709.

Babbitt JD. 1945. The thermal properties of wheat in bulk. Can J Res F 23:388–401.

Bakke AL, Stiles H. 1935. Thermal conductivity of stored oats with different moisture content. Plant Physiol 10:521–524.

Bala BK, Woods JL. 1991. Physical and thermal properties of malt. Drying Technol 9:1091–1104.

Bala BK, Woods JL. 1984. Simulation of deep bed malt drying. J Agric Eng Res 30:235.

Bilanski WK, Fisher DR. 1976. Thermal conductivity of rapeseed. Trans ASAE 19:788.

Boyce DS. 1965. Grain moisture and temperature changes with position and time during thorough drying. J Agric Eng Res 10:333.

Carman K. 1996. Some physical properties of lentil seeds. J Agic Eng Res 63:83–92.

Chandra S, Muir WE. 1971. Thermal conductivity of spring wheat of low temperatures. Trans ASAE 14:644–646.

Chandeasekhar PR, Chattopadhyay PK. 1989. Heat transfer during fluidized bed puffing of rice grains. J Food Proc Eng 11:147–157.

Chang CS. 1988. Measuring density and porosity of grain kernels using a gas pycnometer. Cereal Chem 65:13–15.

Chung JH, Verma LR. 1989. Determination of friction coefficient of beans and peanuts. Trans ASAE 32:745–750.

Clark RG, Lamond WJ. 1968. Drying wheat in 2 ft bed. J Agric Eng Res 13:141.

Curray JK. 1951. Analysis of sphericity and roundness of quartz grain. MS dissertation, Pennsylvania State University, University Park, PA.

Deshpande SD, Bal S, Ojha TP. 1993. Physical properties of soybean. J Agric Eng Res 56:89–98.

Dickerson RW. 1965. An apparatus for measurement of thermal diffusivity of foods. Food Technol 19:198–204.

Dua KK, Ojha TP. 1969. Measurement of thermal conductivity of paddy grains and its by-products. J Agric Eng Res 14:11.

Dutta SK, Nema VK, Bhardwaj RK. 1988a. Physical properties of gram. J Agric Eng Res 39:259–268.

Dutta SK, Nema VK, Bhardwaj RK. 1988b. Thermal properties of gram. J Agric Eng Res 39:269–275.

Fraser BM, Verma SS, Muir WE. 1978. Some physical properties of fababeans. J Agric Eng Res 23:53–57.

Handbook of Fundamentals. 1967. New York, NY: ASHRAE.

Haswell QA. 1954. A note on the specific heat of rice, oats and their products. Cereal Chem 31: 431–442.
Irvine DA, Jayas DS, White NDG, Britton MG. 1992. Physical properties of flaxseed, lentils, and fababeans. Can Agric Eng 34:75–81.
Jasansky A, Bilanski WK. 1973. Thermal conductivity of whole and ground soybeans. Trans ASAE 16:100–103.
Jones JC, Wootton M, Green S. 1992. Measured thermal diffusivities of cereal products. Food Aust 44:501–503.
Kazarian EA, Hall CW. 1965. Thermal properties of grain. Trans ASAE 8:33–48.
Kunze OR. 1991. Moisture adsorption in cereal grain technology—a review with emphasis on rice. Appl Agric Eng 7:717–723.
Kustermann M, Scherer R, Kutzbach HD. 1981. Thermal conductivity and diffusivity of shelled corn and grain. J Food Processing Eng 4:137.
Lang W, Sokhansanj S. 1993. Bulk volume shrinkage during drying of wheat and canola. J Food Processing Eng 16:305–314.
Lund DB. (1983). Applications of differential scanning calorimetry in food. In: M Peleg, EB Bagley, eds. Physical Properties of Food. Westport, CT: AVI.
Miles CA, Van Beek G, Veerkamp CH. 1983. Calculation of thermo-physical properties of foods. In: R Jowitt et al., eds. Physical Properties of Foods. New York: Applied Science Publishers, 269–312.
Miller CF. 1963. Thermal conductivity and specific heat of sorghum grain. MS dissertation. Texas A&M University College Station TX.
Mohsenin NN. 1980. Thermal Properties of Foods and Agricultural Materials. New York: Gordon & Breach Science.
Mohsenin NN. 1981. Physical Properties of Food and Agricultural Materials. New York: Gordon & Breach Science.
Mohsenin NN. 1984. Electromagnetic Radiation Properties of Food and Agricultural Products. New York: Gordon & Breach Science.
Mohsenin NN. 1986. Physical Properties of Plant and Animal Materials. Structure, Physical Characteristics and Mechanical Properties, 2nd updated and revised ed. New York: Gordon & Breach Science.
Moote I. 1953. The effect of moisture on thermal properties of wheat. Can J Technol 29:57–69.
Morita T, Singh RP. 1979. Physical and thermal properties of short-grain rough rice. Trans ASAE 22:630–636.
Moysey EB, Shaw JT, Lampaman WP. 1977. The effect of temperature and moisture on the thermal properties of rapeseed. Trans ASAE 20:461.
Muir WE, Viravanichai A. 1972. Specific heat of wheat. J Agric Eng Res 17:338–342.
Nelson SO. 1973. Electrical properties of agricultural products—a critical review. Tran ASAE 16: 384–400.
Oje K, Ugbor EC. 1991. Some physical properties of oilbean seed. J Agri Eng Res 50:305–313.
Ojha TP, Farrall AW, Dhanak AM, Stine CM. 1967. Determination of heat transfer through powdered food products. Tran ASAE 10:543–544.
Okos MR. 1986. Physical and Chemical Properties of Food. St. Joseph, MI: ASAE Publication.
Ordinanz WD. Specific heats of foods in cooling. Food Ind 18:101.
Oshita S. 1992. Thermodynamical estimation of the specific heat of rice. In: Advances in Food Engineering. Boca Raton, FL: CRC Press, 125–135.
Oshita et al. 1978. (Cited by Oshita, 1992). Thermal properties of rough rice in bulk. Bull Fac Agric Mic Univ Japan, 57:115–125.
Oxley TA. 1944. The properties of grain in bulk. Soc Chem Ind Trans 63:53–57.
Pabis S. (1967). Grain drying in thin layers. Presented at the Agricultural Engineering Symposium of National Institute of Agricultural Engineering, Silsoe, paper 1/C/4.

Pabis S, Bilovitska E, Gadai SP. 1970. Thermal conductivity and thermal diffusivity in grain layers of some agricultural produces. J Eng Phys 19:1150.

Polley SL, Snyder OP, Kotnour P. 1980. A compilation of thermal properties of foods. Food Technol 34:76–80, 82–84, 86–88, 90–92, 94.

Rahman S. 1995. Food Properties Handbook. Boca Raton, FL: CRC Press.

Rao MA, Rizvi SSH. 1995. Engineering Properties of foods. 2nd ed, New York: Marcel Dekkar.

Sharma DK, Thompson TL. 1973. Specific heat and thermal conductivity of sorghum. Trans ASAE 16:114–117.

Shepherd H, Bhardwaj RK. 1986a. Thermal properties of pigeon pea. Cereal Foods World 31:466–470.

Shephard H, Bhardwaj RK. 1986b. Moisture dependent physical properties of pigeon pea. J Agric Eng Res 35:227–234.

Sokhansanj S, Lang W. 1996. Prediction of kernel and bulk volume of wheat and canoloa during adsorption and desorption. J Agric Eng Res 63:129–136.

Suter DA, Agrawal KK, Clary L. 1975. Thermal properties of peanut pods, hulls and kernels. Trans ASAE 18:370.

Sweat VE. 1995. Thermal properties of foods. In: MA Rao, SSH Rizvi, eds. Engineering Properties of Foods. 2nd ed. New York: Marcel Dekkar.

Sweat VE. 1974. Experimental values of thermal conductivity of selected fruits and vegetables. J Food Sci 39:1080–1083.

Tao J, Rao RM, Liuzzo JA. 1994. Selected thermo-physical properties of rice bran. Appl Agric Eng 10:709–711.

Thompson RA, Isaacs GW. 1967. Porosity determination of grains and seeds with air comparison pycnometer. Trans ASAE 10:693–696.

Vinod CH, Bera MB. 1995. Surface heat transfer coefficient of faba bean (*Vicia faba*, L) puffed with sand. J Food Sci Technol 32:94–97.

Wallapapan K, Sweat VE, Diehl KC, Engler CR. 1986. Thermal properties of porous foods. In: MR Okos, ed. Physical and Chemical Properties of Food. St. Joseph, MO: ASAE.

Watts KC, Bilanski WK. 1970. Calorimetric determination of the specific heat of soybeans. Can Agric Eng 12:45–56.

Wratten FT, Poole WD, Chesness JL, Bal S, Ramarao V. 1969. Physical and thermal properties of rough rice. Trans ASAE 12:801–803.

Wright ME, Porterfield JG. 1970. Specific heat of Spanish peanuts. Trans ASAE 13:508.

3

Grain-Grading Systems

RAJSHEKHAR B. HULASARE, DIGVIR S. JAYAS, and BERNIE L. DRONZEK

University of Manitoba, Winnipeg, Manitoba, Canada

1 INTRODUCTION

Grading systems for grain are necessary as an aid in marketing. Grain is more effectively marketed with some form of grading system in place. A grading system for grains significantly influences the efficiency of other operations, such as handling, transportation, processing, treatment, and storage. Hence, handling and transportation are considered a part of the grading system within the overall marketing context. Each of the major grain-exporting countries have a grain-grading system to facilitate grain marketing both in domestic and export markets. Grading systems facilitate trade in markets by striving to meet requirements of the buyers and, at the same time, provide impetus to producers to produce desired quality for equitable returns.

Grain that is produced and harvested in the field is physically heterogeneous and is a mixture of constituents, such as weeds, inert matter, stones, insects, and other debris. Variation in physical appearance of the grain arises from the environmental conditions under which the grain is produced, which may vary even within a farm. The grain may be ready for harvest in one portion of a farm, whereas it may still be at an immature stage in another part of the farm. Other characteristics of the grain, such as moisture content, fungal infection, or discoloration, although not obvious, also affect the quality and storage life of grains.

Grade, therefore, should represent a mixture having characteristics that conform to the parameters established for that grade. *Grading* may be defined as ''the segregation of heterogeneous material into a series of grades reflecting different quality characteristics of significance to the users'' (Canada Grains Council, 1982).

The quality of grain is an important factor that affects the price customers or end users are willing to pay. The requirements differ widely based on end use, such as high-protein wheat for baking pan breads, or low-protein wheat for noodles and flat breads. Other factors such as foreign material content, uniformity within and between shipments, dockage, and moisture content also determine the quality and the price. Ideally, a grading system should take into account the customer preference (end users) and also ensure equitable returns to the producers. Grading systems of exporting countries are revised regularly to adapt to changing global trade in grains (trade without borders or various international trade treaties), customer preferences, and logistics of grain movement over distances.

Grading is the segregation of grain into parcels of defined quality to facilitate price determination (Canadian Grain Commission [CGC], 1993). A successful grain-grading system must achieve the following results:

1. Assure the producer an equitable grain price relative to its quality.
2. Facilitate grain handling by encouraging the collective storage of bulk lots of similar quality.
3. Establish a method of relating prices to quality to simplify trading.
4. Enable the buyer to consistently obtain the same quality of grain.
5. Separate grain into a sufficient number of quality divisions so that the buyers have a choice of grades, but at the same time, limit the number of grades to facilitate handling and storing in an efficient bulk-handling system.

In this chapter, a brief history of the development of grading systems is presented. The Canadian-grading system is described in detail, using wheat as an example, and is compared with the United States- and Australian-grading systems.

2 HISTORICAL PERSPECTIVE

Grain trading has always been between areas of deficit and surplus production. The areas may be adjacent, as between a country farm and a nearby city, or far apart, as between Australia and the United Kingdom. In trading between adjacent areas, the seller and the buyer generally confronted each other in the presence of the grain and bargained over the grains' value until an agreement was reached. The bargaining involved estimates of the overall quality of the grain, the amount of dockage present, and the needs of the seller to sell and the buyer to buy. For distant areas, at least one entrepreneur (or more often several) was involved between the seller and the buyer. Sometimes the contract to purchase included provisions for discounts should the delivered grain not equal the sample, or for arbitrators to intervene if a dispute arose over the quality of the grain.

Grain markets essentially operated in this manner for centuries, whether in the Mediterranean region 2300 years ago, or in Europe in the late 1800s (Irvine, 1983). This system had certain limitations because the "quality" specification was only the statement of origin. The buyer was not certain about the quality of grain that he would receive. If the grain was not up to the standard, the buyer collected a penalty (or paid less than) agreed on in the contract. With the grain received being inferior, the buyer was unable to use the grain for its original intended purpose. For example, grain intended for use in bread making may have to be used for feed if it is of inferior quality.

Grading systems evolved in different countries in the world over many years. Social, political, topographical, technological, and market constraints determined the pace of development of grading systems for grains. In the United States, an initial step was taken by the Chicago Board of Trade in 1856 by establishing grades for wheat. The step was taken to market the surpluses in the sparsely populated Midwest to the eastern seaboard and the overseas market. Similar conditions existed in southern Russia at the start of the 19th century, but the grading or handling system did not evolve, owing to social and technological factors. Wealthy landlords were major producers, and the railways had not been introduced. Similar conditions existed in Argentina and Australia in the latter half of the 19th century, but failed to produce any new grain-handling methods. Canada, because of increased production of wheat, evolved its grading system with legislation in 1863 for grain-grading and handling.

2.1 Beginning of the United States-Grading System

As the grain fields of the U. S. Midwest opened up, surplus production was moved by river, rail, or wagon to Chicago and Milwaukee for onward movement to Canada or the eastern seaboard of the U. S. and later to Europe. In 1858 (Irvine, 1983), the Chicago Board of Trade, followed by the Milwaukee Chamber of Commerce, established a series of grades for wheat, based on cleanliness and bulk density to regulate the quality of wheat on the market.

Club wheat: clean, 60 lb per Winchester bushel (a cylinder with a diameter of 18.5 in. and a depth of 8.0 in.) (780 kg/m^3)
No. 1 Spring: clean, 56 lb per Winchester bushel (725 kg/m^3)
No. 2 Spring: reasonably clean, 50 lb per Winchester bushel (650 kg/m^3)
Rejected: 40 lb per Winchester bushel (520 kg/m^3)

The proposed grades were for the trading community and there was no idea of compulsory grading of grains. The merchants paid a fee to have their grain graded, and this applied to grain coming into the Chicago market from other markets and also grain moving to markets outside of Chicago.

2.2 Origin of the Canadian-Grading System

Wheat production in Ontario increased from 87,090 t to 1.11 Mt annually from 1842 to 1882 (Irvine, 1983). Most of the increase was in winter wheat and was due to many factors, such as an advancing frontier, large inflow of immigrants to work the land and become producers, high prices, and good markets. The Crimean War of 1853–1856 improved markets and prices in the United Kingdom and with the reciprocity agreement between Canada and the United States, from 1854 to 1866, there was a considerable increase in two-way flow of wheat and flour across the U. S.–Canadian border. During the American Civil War of 1861–1865, considerable trade across the border was based on grades established by the Chicago Board of Trade. Although wheats grown on either side of the border were much the same, the economically stronger U. S. merchants usually had an advantage owing to their own grading system. In 1863, the Province of Canada enacted legislation (Irvine, 1983) for grain grading and inspection, with the grades being defined along the lines established by the Chicago Board of Trade. The legislation defined six grades for wheat and allowed the Boards of Trade (in Quebec, Montreal, Toronto, Kingston, Hamilton, and London) to establish Boards of Examiners who would recommend the appointment of inspectors, approve standard samples, and arbitrate disputes.

3 TYPES OF GRADING SYSTEMS

Two types of grading systems (CGC, 1993), are in use throughout the world: the fair average quality system and the numeric system.

3.1 Fair Average Quality System

In this system, samples are prepared to represent the average quality of the current crop and are distributed to purchasers with guarantees that shipments will be near the average quality. Normally, the buyer arranges to have shipments sampled by an independent agency on its arrival at port or final destination. The buyer who believes the shipment is not of fair average quality may use these samples as evidence for negotiations or for arbitration of the dispute. Only a few countries use this system (CGC, 1993).

3.2 Numeric System

The numeric system separates grain into divisions of quality that are defined by grading factors. Each division is identified by a grade name or number, and grain is bought and sold on the basis of these grades. Buyers obtain the specific quality desired by selecting the grade. In some grading systems, such as Canada's, a certificate of grade is acceptable to an importer, and no additional samples are required as evidence of quality. In other systems, importers demand additional safeguards, such as a sample, together with the analysis of the actual grain being shipped (CGC, 1993). The grading systems of the United States, Canada, and Australia are numeric.

4 CANADIAN-GRADING SYSTEM

4.1 Establishment of Grades

The Canadian grading system for grains is a numeric system defined by the Canada Grain Act. The act empowers the Canadian Grain Commission (CGC), as the regulating body, to frame, administer, and implement the grading system. The system relies heavily on the kernel characteristics and visual distinguishability of the kernels for assigning grades. For this purpose, the act has laid out procedures, standard equipment, personnel training for inspection, and control measures for effective implementation of the system.

The Canada Grain Act established two committees, for Eastern and Western regions of Canada, called the Eastern Grain Standards Committee and the Western Grain Standards Committee, respectively. The committees meet independently to review primary and export standard samples and recommend the grade definitions to the CGC. The committees are also responsible for recommending grade specifications to the CGC. Statutory grades are established, amended, or deleted according to the Schedule III of the Canada Grain Regulations under the Canada Grain Act. These statutory grades are the basis of the Canadian-grading system for grains. The grades are recognized and accepted by all (domestic and foreign) purchasers of Canadian grain.

The CGC is empowered to establish grades in cases where no grade is established by regulations. For example, the CGC is empowered to establish grades for grain that is out of condition, especially binned grain marketed on specifications, and screenings.

4.2 Grade Definitions

The Canadian-grading system for grains (wheat) is represented in Fig. 1. The three most important criteria for classification are based on kernel texture (hard or soft), bran color (red or white), and growth habit (spring or winter). In the Canadian-grading system, a registered variety of any class of wheat must be visually distinguishable from registered varieties of any other class of wheat. Accordingly, each Canadian wheat class has evolved unique properties through selective breeding that makes it suited for specific end uses.

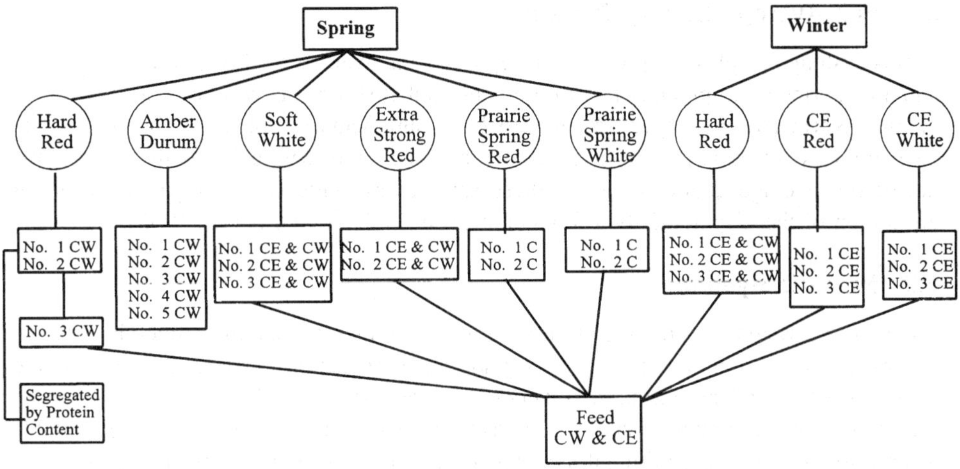

Fig. 1 The Canadian-grading system: circles, class; rectangles, Canada feed; CW, Canada Western; C, Canada; CE, Canada Eastern. (From Williams, 1993.)

For example, durum wheat is very hard, making it suitable for semolina milling and has properties ideally suited to manufacture high-quality pasta; whereas Amber durum adulterated with red spring wheat would yield less semolina and produce low-quality pasta. The various Canadian wheat classes each have unique hardness, specific kernel shape, and intrinsic protein properties to meet specific end-product requirement. The Canadian-grading system, through CGC, employs a rigid policy on licensing (or ''release'') of new varieties. New varieties must conform to exacting kernel characteristics and functional quality specifications. The functional quality specifications relate to end use and consist of milling and baking tests that each variety is subjected to for 3 years in field trials in different locations. Sufficient grades are established for each wheat class to allow buyers to decide according to their needs. Four classes or grades have been established by the Canada Grain Act:

Class 1: includes all the statutory grades set out in Schedule III of the Canada Grain Regulations. The bulk of Canadian grain grading falls into this class of grades.

Class 2: includes grades established for new and for special crops, and for grain demonstrating particular qualities, or limited production, or both.

Class 3: comprises grades for grain that do not qualify for statutory grades because of their condition or admixture.

Class 4: includes all grades established for by-products of grain, or for screenings removed during cleaning.

The designations, Canada, Canada Western, and Canada Eastern, included in the various grade names, indicate the geographic areas in which the grain is grown. Class 5 is provided for amber durum wheat (Canada Western), which includes grades that are inferior to Class 4, but are superior to Feed, and limits and tolerances are also specified for this class.

4.3 Standard Samples

Most factors considered in grading must be determined by visual examination. To provide a comparison to assist in grading, or to set grading levels, standard samples are prepared annually for all grain grades defined in Schedule III of the regulations and for any other special grades (under Classes 2 and 3 of grade definitions) established by the CGC. All standard samples are prepared by the CGC's inspection division and are submitted to Standard Committees for review. The committees recommend specifications for grades of grain, and select and recommend standard samples to the CGC for consideration and approval. To prepare the standard samples for each grade, quantities of grain from the current crop are obtained from primary elevators and from railcars unloading at terminal elevators. The samples represent characteristics of the major grading factors of that growing season and are a cross section from all production areas. The term *standard* in the Canadian grading system, thus means minimum visual quality (CGC, 1993).

There are two distinctly different types of samples prepared each year: the primary standard samples and the export standard samples. Because the two samples are used for different purposes, their preparation procedures are different.

4.3.1 Primary Standard Sample

A primary standard sample is prepared for each statutory grade of grain listed in Schedule III of the Canada Grain Regulations and for any other grades the CGC considers appropriate. The sample is used as a visual guide in grading grain at the domestic level and represents, as nearly as possible, the minimum visual quality for the grade, as defined in the Canada Grain Regulations.

After harvest, grain samples from the new crop are sent to the inspection division of CGC. Portions of the previous year's standards are included because of the carryover stocks from year to year. The samples are graded, analyzed, physically tested, and blended into samples representing minima in visual quality, by statutory definition, for each grade. Portions of the samples are then sent to CGC's Grain Research Laboratory (GRL) for quality testing. After the analysis, the sample and all the pertinent data are submitted to the appropriate Standard Committee (Eastern or Western) and, if acceptable, are recommended for adoption by the CGC.

On approval of the proposed samples, they become the official primary standard samples. They are supplied to CGC and private-sector grain inspectors to be used as guides in grading grain at the domestic level and at terminal elevators. They may also be supplied to importers and overseas agencies when an export standard sample has not been established for the grade. In a conflict in interpretation between the primary standard sample and the official definition of the grade, the official definition takes precedence.

4.3.2 Export Standard Sample

Export standard samples are prepared for grades of hard red spring wheat, amber durum wheat, and any other grain that, in the opinion of the CGC, is likely to be sold for export. They provide a guarantee to buyers that shipments during the crop year will be close to the visual quality of the export standard sample of the grade for which the shipment is certified. They represent the average quality of all grain of that grade likely to be exported during the crop year and, in effect, represent the minimum visual quality the buyer will receive for a specified grade.

Export standard samples include grain from both the current harvest and existing stocks, because existing grain stocks are generally available in market in the first part of the crop year. The Inspection Division grades, analyzes, and blends individual samples into composite samples representing the average quality of all grain of that grade likely to be sold for export. The preliminary samples are then submitted to the GRL. The samples are subjected to full-scale tests to determine grading factors, among other criteria of quality: test weight, varietal purity, vitreousness, soundness, and limits of foreign material; and also their milling and baking qualities. The samples with pertinent data are submitted to the Western Grain Standards Committee for review, and follow the same procedure as that for the primary standard samples to be eligible for official export standard samples.

Export standard samples are used by CGC inspectors as visual guides in the grading of export grain. They are also supplied to buyers around the world as representative samples of the visual quality of graded shipments. Export standard samples take precedence over the official definition of the grade if there is any conflict in interpretation. Establishing export standard samples ensures that the quality of exports is consistently maintained near the average quality of the grade.

4.4 Grading Factors

Grades of grain provide the means of rapidly and reliably characterizing quality to meet the demands of a handling and marketing system. Definitions of ''degree of soundness'' provide grain inspectors with a means of visually evaluating factors that cannot be precisely measured. Factors such as maturity and weather damage in cereal grains, color in oilseeds, and kernel size uniformity of pulse crops are judgmental visual determinations made in comparison with standard or guide samples. Measurable factors, such as admixtures of foreign material and severe damage from improper handling or weathering, are assessed using equipment and procedures standard to the inspection process. Grading factors are commonly referred to as degrading factors, because a failure to meet the requisite quality level in each factor will result in a lower grade.

Commonly, five key-grading factors are considered for grading grains. Not all may be applicable for all the crops; for example, test weight is not applied to canola (rapeseed) as the correlation of test weight with oil content is unclear. The various grading factors are described in the following. Minimum and maximum limits, tolerance levels, equipment and procedures for testing for various grading factors, and measures such as moisture content, dockage, and protein content are given in detail in the *Grain Grading Handbook for Western Canada*, a manual published by the CGC.

4.4.1 Test Weight

Test weight is the bulk density of the grain expressed in kilograms per hectoliter (kg/hL) for both the domestic and export grades and is determined by specific procedures using approved equipment (CGC, 1993a). For example, No. 1 Canada Western Red Spring wheat has a minimum test weight of 75 kg/hL.

4.4.2 Varietal Purity

Varietal purity is determined by the percentage of unregistered and deregistered varieties of the same class. By restricting varieties of inferior quality, varietal purity standards ensure that a class of crop (e.g., wheat) maintains its intrinsically high quality. Any varieties not equal to the standard will qualify only for lower grades. For example, Neepawa is the varietal standard of quality for red spring wheat, and Hercules the varietal standard of quality for amber durum wheat.

4.4.3 Vitreousness

Vitreousness is the glossy or shiny appearance that indicates hardness. Generally, the more vitreous a kernel, the higher its protein content. Nonvitreous kernels are, by definition, those that are broken, damaged, show signs of starchiness, or are severely bleached.

4.4.4 Soundness

Soundness refers to the extent and degree of overall physical damage and in most circumstances is the single most important grading factor. Sound kernels are well-developed, mature, and physically undamaged. Damage includes such things as frost, immaturity, weathering during harvest, diseases, and the effects of unfavorable storage conditions.

4.4.5 Maximum Limits of Foreign Material

Foreign material refers to that material, other than grain, of the same class remaining in the sample after the dockage has been removed. These may include other cereal grains, inseparable seeds, thistle heads, heated kernels, severe moldy grain, and pieces of stems.

4.4.6 Other Factors Affecting Grading

Apart from the foregoing five factors, three other measures: dockage, moisture content, and protein content, also affect the monetary value of the grain.

4.4.6.1 Dockage

Dockage consists of all the readily removable material that can be removed from a sample of grain using approved methods and equipment before assigning an official grade.

4.4.6.2 Moisture Content

Moisture content is defined as the percentage of moisture in cleaned grain. The moisture content of cleaned grain must be within specified limits to qualify for categories in grades within class definitions. For example, for wheat, the limits are

Straight grade: less than 14.5% (wet mass basis)
Tough: 14.6–17.0%
Damp: greater than 17.0%

Moisture content is tested on standardized equipment that is operated in strict accordance with instructions and routinely checked for accuracy (CGC, 1993b). The moisture content of grains can be reduced by mixing with drier grain, or by drying grain using ambient or heated air.

4.4.6.3 Protein Content

Protein content determination was introduced for wheat destined for export. All No. 1 and No. 2 Canada Western Red Spring wheat is binned at export terminals on the basis of protein content determined by near-infrared analyzers. The main objectives of the protein segregation system for wheat are as follows:

1. To improve Canada's precision and flexibility in meeting world market demands for wheat of milling quality
2. To reflect the market value of the protein of high-grade wheat directly back to the producer

It is important that any sample receives the same grade at every inspection under every circumstance. Because grading is dependent on visual examination and judgment of individual inspectors, various controls and procedures have been imposed to ensure consistency of grading among inspectors. There are standardized-grading procedures, sampling methods, standardized equipment, guide and average samples that are followed scrupulously to ensure effective grading of grain. Inspectors are carefully selected based on their aptitude and educational qualifications and are trained in all aspects of grading grains.

5 AUSTRALIAN-GRADING SYSTEM

Grain production in Australia is mostly in the coastal areas receiving sufficient rainfall to support grain production. Rainfall restriction along with soil capabilities and topography restrict grain production to particular coastal areas. Wheat is a major crop suited to these areas and is grown in five states: Western Australia, South Australia, Victoria, New South Wales, and Queensland. The individual states produce different classes of grain, particularly wheat. The warm climatic conditions during harvest and storage make grain more prone to storage problems. Hence, use of a central storage (warehouse) is more common than on-farm storage. All of the wheat-growing states have their own state-controlled system of elevators, and storage and transportation are handled by bulk-handling authorities (BHA) of state (Anonymous, 1998). The states have a high degree of autonomy, extending to grain production and marketing.

The wheat classification is summarized in Fig. 2. There are four milling classes and two nonmilling classes. Australian wheat is marketed on a state basis, with the state name incorporated in the description. For example, South Australian Standard White (ASW) would be offered as ASW(SA). The Australian Wheat Board (AWB) is the apex authority empowered by the Wheat Industry Stabilization Act, that sets the receival standards and also supervises grain storage and transportation through BHAs. The AWB assumes a supervisory role over the inspection of wheat in storage to ensure that the grain is maintained in a sound and insect-free condition.

5.1 Receival Standards

The classification of wheat into the various quality classes and grades begins when the farmers deliver their wheat into the central bulk-handling system. Before the deliveries can be received, the wheat must conform to a strict receival standard. In the receival standard system, the producer signs and hands over an affidavit stating the variety of wheat that the producer is delivering, and he or she is liable for punitive action if the variety is incorrectly declared. The system demonstrates that a nonvisual varietal identification sys-

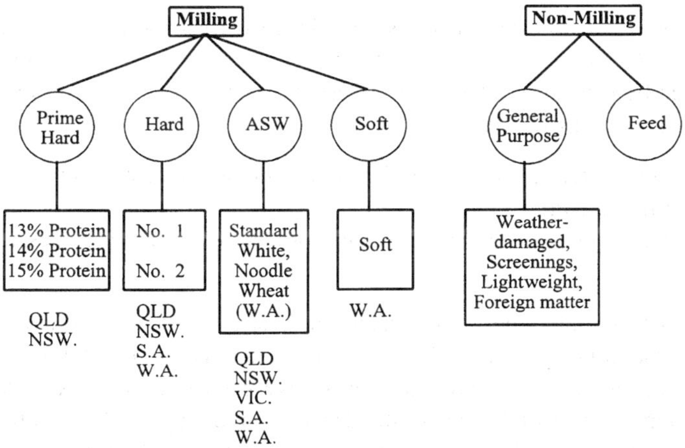

Fig. 2 Australian wheat classification: circle, class; square, grade; VIC, Victoria; QLD, Queensland; S.A., South Australia; NSW, New South Wales. (From Williams, 1993.)

tem is possible. Only few samples are taken if there is little doubt about variety, and the overall experience of the AWB has been that producers generally declare correct varieties.

Although applied by the BHAs in each state, the receival standards are determined by the AWB. These standards are collectively referred to as Australian Standard White (ASW) specification. Broadly, they refer to moisture content (12.5% maximum), test weight (74 kg/hL, minimum), no live insects, a range of strict tolerances for unmillable material, foreign material, foreign seed, weather damaged grain, and a nil tolerance for sprouted kernels. Wheat that does not conform with these basic specifications is downgraded, depending on the severity of the defect. Wheat grown in specific regions that meets the basic specifications and also meets additional standards for protein content and variety type is segregated separately and is sold at a premium.

5.2 Assessing Quality

Grain quality is generally defined in terms of simple physical criteria, such as test weight, moisture content, and the percentage of foreign material. The basic criteria used to evaluate wheat quality are: protein content, grain hardness, dough properties, and milling quality (Anonymous, 1998).

Protein content has a greater influence on overall processing quality than any other single factor, and although wheat varieties differ in their capacity to accumulate protein, the environment is the major determinant of the protein level of a particular wheat grade. The other three quality factors are related to the wheat variety, although there is some environmental influence. These criteria form the basis of the Australian-grading system. As a result, the wheat in each class is relatively homogeneous and clearly shows identifiable quality characteristics. The various grading factors, tolerances, and limits for ASW are given in Table 1.

Table 1 Comparison of Standard of Quality and Tolerances for the United States, Australian, and Canadian-Grading Systems for No. 1 (Top Grade) Wheat

Standards of quality/maximum limits	United States (U. S. No. 1 Hard Red Spring wheat)	Canadian (No. 1CWRS)	Australian (Prime hard)
Standard of quality			
Minimum test weight (kg/hL)	76[a]	75	74
Variety	NA	Any varieties of red spring wheat equal to or better than Neepawa	NA
Minimum hard vitreous kernels	75%	65.0%	NA
Degree of soundness	NA	Reasonably well-matured and free from damaged kernels.	NA
Maximum limits			
Foreign material			
Other than cereal grains	NA	About 0.2%	
Total (including cereal grains)	0.4%	About 0.75%	1.0%
Unmillable material			
Screenings below the screen	NA	NA	5.0%
Screenings above the screen	NA	NA	3.0%
Wheats of other classes or varieties			
Contrasting classes	1.0%	1.0%	NA
Total	3.0%[b]	3.0%[b]	
Damaged kernels			
Heat (part of total percentage)	0.2%		Nil
Total	2.0%	[c]	Nil

Table 1 Continued

Standards of quality/maximum limits	United States (U. S. No. 1 Hard Red Spring wheat)	Canadian (No. 1CWRS)	Australian (Prime hard)
Sprouted	[d]	0.5%	Nil
Binburnt, severely mildewed	[d]	2 K[e]	Nil
Rotted, moldy	[d]		
Heated (incl. binburnt)	Presence	0.10%	Nil[f]
Fireburnt	[d]	Nil	Nil
Stones	0.1%	3 K	Nil
Earth and sand	[g]	[g]	20, 1[h]
Ergot	0.05%	3 K	1 K[i]
Sclerotia	[d]	3 K	NA
Smudge	[d]	30 K	Nil[j]
Total smudge and blackpoint	[d]	10.0%	NA
Shrunken and broken	3.0%		[k]
Shrunken		6.0%	
Broken		6.0%	
Total	3.0%[l]	7.0%	
Degermed	[d]	4.0%[m]	NA
Grass-green	[d]	0.75%[n]	1%
Pink	[d]	1.5%	2%
Artificial stain, no residue	[d]	Nil	NA
Natural stain	[d]	0.5%	5%[o]
Insect damage	31 K/100 g		1%
Sawfly, midge	NA	2.0%[p]	NA
Grasshopper, army worm	NA	1.0%[p]	NA
Dark, immature	NA	1.0%	[q]
Insect infestation (live)	2 lw, or 1 lw + 1 oli, or 2 oli[r]	Nil	Nil
(dead)	NA	NA	5
Pickling compounds, chemicals not approved for stored grain	NA	NA	Nil

Apart from the foregoing general standards, the countries have various tolerances for other material found in grain samples. The Australian system has limits for field insects, earcockle, pea weevil, snails, loose smut, and various types of seed contaminants (classed into 7 types). The U.S. system also has count limits for animal filth, castor beans, crotalaria seeds, glass, stones, and unknown foreign substances.

[a]Test weight is in lb/bu in the U. S. system and is converted into kg/hL using the formula: $[1.292 \times \text{lb/bu}] + 1.419 = \text{kg/hL}$, for all wheats except durum wheat.

[b]Includes contrasting classes.

[c]Included in damaged kernels of other types, such as binburnt, severely mildewed, rotted, moldy, heated (including binburnt), or fireburnt.

[d]Included in damaged kernels of wheat of various types such as black-tip fungus, heat-damaged, blight or scab, frost-damaged (blistered), frost-damaged (candied), frost-damaged (flaked), frost-damaged (discolored black or brown), germ damaged (mold), green damage (immature), mold-like, other damage (cracks, breaks, or chews), sprout-damaged, insect-bored, and germ-damaged (sick).

[e]The letter K refers to kernel-size pieces in 500 g.

[f]Applicable for heat damaged, binburnt, or moldy kernels and is maximum percentage by count.

[g]Included in stones criterion.

[h]The tolerance refers to wheat ergot (i.e., number of pieces per 0.5 L and the tolerance is 2 cm for ryegrass ergot and refers to maximum length that the pieces are not to exceed when aligned end to end.

[i]Maximum tolerance of 20 grains of sand and 1 pea size piece of earth per sample.

[j]Applicable for kernels infested with ball smut (stinking smut) in percentage by count.

[k]Included in unmillable and foreign material.

[l]Includes damaged kernels (total), foreign material, and shrunken and broken kernels.

[m]Tolerances apply to kernels not classed as sprouted.

[n]Tolerances are given as a general guide and may be increased or reduced in the judgment of the inspectors after consideration of thc overall quality of a sample.

[o]Maximum percentage by count in a 300-grain sample, including 2% for pink fungal stained.

[p]Tolerances are not absolute maximums. Inspectors must consider the degree of damage in conjunction with the overall quality of the sample.

[q]Included in natural stain grain.

[r]lw, live weevils; oli, other live insects injurious to stored grain.

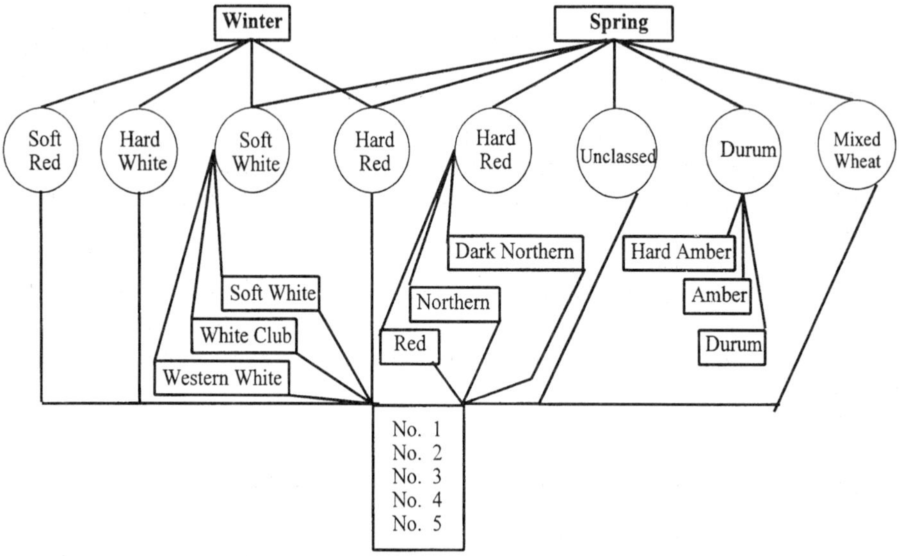

Fig. 3 Grades of U. S. wheats: circle, class; rectangle, subclass; square, grade. (From Williams, 1993.)

6 UNITED STATES-GRADING SYSTEM

The U. S.-grading system is summarized in Fig. 3. The U. S. Grain Standards Act provides the basis for grading and inspection of grains and makes Federal Grain Inspection Service (FGIS) responsible for the grading and inspection of grain. The grading is a single-standard system, applying equally to both domestic and export sales. Varietal restrictions are not imposed, and licensing of new varieties is without rigid control. Grades are based on factors such as test weight, damaged kernels, and foreign material. The measure of test weight (i.e., pounds) differs from the Canadian and Australian systems (kg/hL). After the grain is graded by grading factors, the following guidelines (USDA, 1995) are followed to designate grades:

1. The letters ''U.S.''
2. The abbreviation ''No.'' and the number of the grade, or the words ''Sample grade.''
3. The words ''or better'' when applicable.
4. The subclass or, in the case of Hard Red Winter wheat, Hard White wheat, Mixed wheat, Soft Red Winter wheat, and Unclassed wheat, the class.
5. The applicable special grade in alphabetical order, except for treated wheat. The grade designation for treated wheat includes, following the class or subclass and any special grade designations, the word ''treated'' followed by a statement indicating the kind of treatment (e.g., scoured, limed, washed, and such); and the word ''dockage'' and the percentage thereof.

Additionally, remarks for a particular class of wheat also needs to be included in the certificate. For example, for Western White wheat, the name and percentage of White Club wheat or percentage of protein, if applicable. Special grades for the unusual conditions of grains are also made part of grade designations. For example, U. S. No. 2 Dark Northern Spring wheat containing more than 0.05% of ergot with dockage of 0.1% would be designated as U. S. No. 2 Dark Northern Spring wheat, dockage 0.1%. Similarly, U. S. No. 1 Amber Durum wheat with 0.2% dockage and treated with lime would be designated as U. S. No. 1 Amber Durum wheat, treated (limed), dockage 0.2%. Optional grade designation such as ''or better'' is designated on request of the farmer and is assigned to U. S. No. 2 and 3, but not to U. S. No. 1. An example of designation would be ''U. S. No. 2, or better, Hard Red Winter wheat.''

Unlike the Canadian- and Australian-grading systems (zero tolerance for insect infestation), the U. S.-grading system has tolerance limits for the insect infestation. For example, for wheat, samples meeting or exceeding two live weevils, or one live weevil and one live stored-grain insect, or two live stored-grain insects, are considered infested (see Table 1).

7 ADVANTAGES OF A GRADING SYSTEM FOR GRAINS

The advantages of having a grading system are numerous, the primary objective being maximum net return from the grain to be extracted from the market. In ''globalization of trade'' or ''trade without borders,'' with new trade agreements and treaties, this becomes an important criteria. The various advantages of grading system follow:

1. The producer is assured of equitable grain price relative to its quality, and this provides an impetus to farmers to produce, harvest, and store grain of high quality.
2. The system allows for ''right of appeal'' against unfair grading so that producer gets correct or consistent grade for grain and subsequent returns (in Australian system arbitration of grade dispute is not provided, but disputes are settled at receival point by the supervisors of BHA).
3. Product selection is easier at the user level owing to various levels of grades. Similarly, producer can offer grain at a price based on grade and reflecting its end use. End user selection and offer by the producer on the basis of grade, facilitates efficient handling, storage, transportation, and distribution of grains.
4. Grades facilitate trading, because price information on grade basis is easier to transmit, and quoted prices for grades connote a grain with a particular set of characteristics. This also makes it easier for transactions in the absence of the grain itself, and this procedure has become highly sophisticated, with reliance on grades as a measure of quality.
5. Transporting and handling have become more efficient, as the grain with similar characteristics (grade) can be bulked into larger lots, reducing the cost of segregation of lots with the same grade. The bulking, in turn, reduces variability within a particular lot. Storage and maintenance are also facilitated, as bulked lots of same grade are easier to store and maintain for quality than several segregated lots.
6. Grading and cleaning reduce the volume of grain to be moved. The inferior grades may not have to be moved if a local market exists. These, in turn, reduce the overall transportation cost, especially when transportation capacity is a limiting factor. Grades also facilitate determination of value of losses in transit. Quality deterioration can be measured by an effective grading system.

7. Grading has directly resulted in an increased demand for grain because grade signifies the associated characteristics that are understood by all buyers. Given a choice of grades for end use, quantities purchased increase and, at the same time, encourage competition between buyers and sellers and also between the buyers, regardless of size (quantity).

8 DISADVANTAGES OF A GRADING SYSTEM FOR GRAINS

The grading systems for grains were developed and improved in various countries with variations according to geographic locations, climatic factors, political environment, and the technological changes taking place in farming, storage, and preservation. The local demand and global trade also influence the grading system. Owing to these various factors, the grading systems undergo changes from time to time and can have the following disadvantages, such as.

1. The number of grades provided for each grain is a debatable issue. An excessive number of grades means a smaller quantity would fall in each grade. Increased number of segregations (grades) effectively reduces the throughput capacity of the handling and transporting system. For each grade that requires physical segregation, separate binning at the elevator represents a cost to the system and a constraint to the elevator capacity and space. In practice, only a few grades may be handled whereas many would remain ''on the shelf'' to be used in only unique situations.
2. Moisture content, although not a grading factor, decides the monetary returns to the producer. For example, in the Canadian system wheat is classed as straight (14.5% wet basis [wb]), tough (14.6–17.0% wb), and damp (greater than 17.0% wb). The grain that is tough or damp has to be dried and segregated. This means additional handling, drying, and operating cost and can interfere with throughput of the elevator if not carefully planned. The real value of the grain lies in its dry matter content. Existing grading systems allocate the same price for a range of moisture content in a particular class, and determine moisture content on a wet basis. Hence, for dry wheat, the same price prevails up to 14.5% moisture. The producer, therefore, does not receive the real value for his grain, and this has also resulted in producers increasing their returns by rewetting the grain and delivering it near the maximum of the allowable level. Grain with high-moisture content reduces the throughput, in addition to greater risk of spoilage in storage, or the additional cost of drying. Efforts at selling grain on a dry-matter basis have not found favor with the grain industry (Hill, 1990). For example, in the United States, as far back as 1909 (Hill, 1990), a proposal for selling grain on a dry-matter basis did not receive any response from the grain industry, and when the proposal surfaced again, in 1982, the industry strongly opposed it, and continues to do so.
3. The variations in measurement of specific criteria for grains in various countries also creates confusion in the minds of buyers. For example, use of 2.0-mm screen instead of 1.68-mm screen may result in differences of over 5% in the determination of shrunken and broken wheat kernels in a shipment. International standardization procedure are not likely to evolve, mainly owing to diversity in perception of what constitutes quality, and differences in definitions of factors

used to determine grade. The Codex Alimentarius Commission was created in 1962, under charter of the United Nations (Hill, 1990) to establish uniform standards for many products and commodities, but little progress was made on adopting international standards for grains. Agreement has not been achieved, even on fundamental issues, such as measurement of moisture content, or its definition for setting grain standards.

9 CONCLUSION

The major grading factors, such as test weight, moisture content, and foreign material, are common to all of the three grading systems, with variation in tolerance limits based on the class of grain and its characteristics. The various other factors, such as heat-damaged kernels, treated grain, burnt, moldy, smudge, and such, have varying limits of tolerances. All three grading systems are numeric, but the Canadian and Australian systems have different standards for the domestic and export markets, whereas the United States has a single standard system for both domestic and export purposes. The limits and tolerances for various factors vary from country to country owing to variations in climate, production, technological considerations, and marketing constraints.

REFERENCES

Anonymous. 1998. The Australian grain industry. URL address: *http://www.awb.com.au/agindust.htm*, accessed on March 23, 1998.

Canada Grains Council (CGC). 1982. Philosophy of grading. In: Grain Grading for Efficiency and Profit. Winnipeg, MB: Canada Grain Council, 6 p.

CGC. 1993a. Appendix B—Test weight determination. In: Grain Handling Handbook for Western Canada. Winnipeg, MB: Canadian Grain Commission, pp. 19–22.

CGC. 1993b. Appendix C—Moisture testing. In: Grain Handling Handbook for Western Canada. Winnipeg, MB: Canadian Grain Commission, pp. 23–31.

CGC. 1993. Quality control of Canadian grain. In: Grains and Oilseeds Handling, Marketing, Processing. 4th ed. Vol. 1. Winnipeg, MB. Canadian International Grain Institute, pp. 289–308.

Hill LD. 1990. The persistent issues in search for equitable grades. In: Grain Grades and Standards: Historical Issues Shaping the Future. Urbana, IL: University of Illinois Press, pp. 269–315.

Irvine GN. 1983. The history and evolution of the Western Canadian Wheat grading and handling system. In: Wheat Grading in Western Canada (1883–1983). Winnipeg, MB: Canadian Grain Commission, 131 p.

USDA. 1995. Official United States Standards for Grain, 7 CFR Part 810. Grain Inspection and Stockyards Administration. Washington, DC: FGIS, United States Department of Agriculture.

Williams PC. 1993. The world of wheat. In: Grains and Oilseeds Handling, Marketing, Processing, 4th ed. Vol. 2. Winnipeg, MB: Canadian International Grain Institute, pp. 578–590.

4

Harvesting and Threshing

ADHIR C. DATTA[†]

Indian Institute of Technology, Kharagpur, India

1 INTRODUCTION

In preparing quality grains and seeds from cereal and pulse crops for consumers, it is desirable that these be separated from the matured crop plants. All grains from paddy and wheat must be effectively stripped from the whole plant stem; gram, mustard, pigeon pea, lentil, and soybean must be threshed for seed from the plants, peanuts must be threshed or picked from the vines; and corn must be shelled from the cob. It will not be possible to separate the seeds and grains from different crops with a single machine. In removing seeds or grains from the plant stem effectively, a series of operations are performed by the individual process elements as the material passes through it. These functional elements are put together to form a successful machine called a *combine harvester*. A *combine* is, therefore, a combined harvester–thresher put together to harvest all the small grains (such as cereals, pulses, oilseeds, and millets), sorghums, soybeans, and many other crops. Such machines are quite large, requiring a high-capacity power source (30–112 kW) besides requiring large fields, highly skilled operators, and technicians to use and maintain them. On the other hand, a large-category farmer, growing more than one crop cannot afford a separate combine for each crop and, also, maintain it over the year for next season. High-overhead cost, lack of technical know-how of functional elements, availability of genuine spares, periodic maintenance, loss of grain, and adaptability of a combine as a multiharvest machine are the prime hurdles in making the decision to purchase a combine. In spite of these difficulties, cereals and pulse crop combines as harvester and thresher are becoming popular in all the countries where labor is scarce at the peak period of harvest, and where manual harvesting and threshing operations are costly.

Present-day mechanization in harvesting and threshing of crops has undergone many changes. Much of the grains and seed crops in the field are harvested, threshed, and sepa-

[†] Deceased.

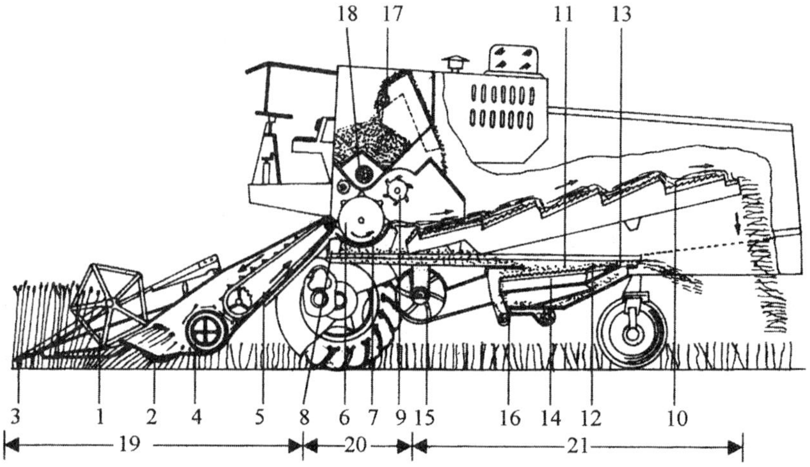

Fig. 1 Line diagram of a grain combine showing basic functional components:

1. Reel (pickup type)
2. Reciprocating cutter-bar
3. Stalk divider
4. Conveyor auger
5. Crop feeder–conveyor
6. Threshing cylinder
7. Concave
8. Stone trap
9. Straw guide drum
10. Straw walker
11. Flap sieve
12. Twin-nose sieve
13. Rake
14. Bottom sieve
15. Fan blower
16. Grain auger and grain elevator
17. Tank-filling auger
18. Tank auger and unloading auger
19. Header unit
20. Threshing system
21. Winnowing and cleaning systems

rated from foreign materials with combined harvester–threshers called grain combines (Fig. 1). According to their power source, combines are classified as tractor-operated, trailed-type, and engine-operated, self-propelled type. The prime operational functions in sequence of a combine are (a) header or crop-gathering unit, (b) cutting the standing crops, (c) feeding the cut crops to the cylinder, (d) threshing the grains from the stalk or stem, (e) separating the grains or seeds from the straw, (f) cleaning the grains of chaff and other foreign materials, (g) elevating the clean and sound grains to a tank, and (h) transferring grains from tank to truck. Although these functions can also be achieved by the individual machines separately, all such operations in the field are done in one pass of the combine, the efficiency of which in terms of sound grain recovery and effective separation will depend on the performance of each unit.

In areas where conditions are not suited to combine operations and fields are small, people carry out hand-cutting of crop plants with a sickle, scythe, or cradle, and also with rotary or reciprocating action cutters; the cut crops are left in the field for a few days for further drying, and, thereafter, make crop bundles and bring them to a threshing yard where grains are separated by the treading of animals or beating on a log of wood. The grains are then separated from the straw and chaff against natural wind or by using a winnower; a stationary threshing machine in which grains or seeds are threshed and separated with a blower. In the Philippines and in a few European countries, grain tillers are stripped off, using a stripper harvester, and tillers are then threshed separately using an axial-flow power thresher in the field, and the leftover plants are burnt, whereas in the United States, machines are taken to the field where crops are harvested and threshed, and the grains are brought back in a tank. Indian farmers prefer to leave the cut crops in the field for 3–5 days for drying of unevenly ripened seeds or kernels before they are threshed with the stationary thresher.

The most appropriate methods of mechanization of grain crop harvesting would be direct combining, or reaping and binding, followed by stationary threshing. In the former case, the plant stems as crop residue are left in the field to be converted into manure forgoing its economic value. But in countries such as India, Pakistan, China, and Bangladesh, the plant stem is extensively used in industries to manufacture *Khas khas* (a handmade mat of paddy straw to keep rooms cool), ropes, straw boards, parchment paper, wall mats, and it is also used as animal feed. The reaper–binder on the other hand, uses the precision-built binding machine. It has been reported that the cost of imported twine for binding the bundles is more than the manual gathering and tying in the conventional method (1–3). Additionally, owing to high initial investment and operational cost, and high cost of imported twines, the tractor-mounted or self-propelled cutter–binders are not becoming popular in many countries.

Ideally, sound grains from agricultural crops are required to be separated from materials, such as straw and chaff, small debris, and remaining unthreshed materials. All these are performed in a harvester–thresher combination or in a combine. Therefore, it is essential to describe the mechanisms involved in the different machine elements for the function of harvesting, threshing, separating, cleaning, and loading grains and seed crops individually or in combination. Additionally, emphasis is also laid on describing the principles, processes, and design requirements in harvesting and threshing operations of cereal and pulse crops.

2 HARVESTING

Harvesting, an important field operation for any food grain crops, is the cutting process carried out when the crop attains physiological maturity such that a maximum recovery of quality product is obtained. Harvesting a crop at an appropriate stage of maturity minimizes the field losses, thereby, increasing the total yield as well as head yield. Consequently, to draw maximum benefit from the crop, harvesting must be done at the right time. Farmers generally judge the maturity by visual observation of color of leaf and panicle. Frequently, the observation of this index leads to delay in harvesting the crop owing to visual error and, as a result, the grain moisture becomes reduced to a very low level. Recent investigations have shown that the delayed harvesting at low grain moisture led to considerable increase in field losses and deterioration in the grain quality, resulting in reduction in milling outcome. Timely harvest, on the other hand, can prevent such a loss. It has been reported (4) that, in South India, a total of 22% field loss and 2% reduction in milling yield occurred because of improper and delayed harvesting of an IR-8 variety of paddy. Exact harvest time for each crop is, therefore, crucially essential for maximum recovery.

2.1 Harvest Time

At field level, grain loss caused by shattering, lodging, and damage by birds and rats will occur if it is not harvested in time. For milling, timely harvest of crops ensures less breakage of milled grain, quality grain, and better markets of consumer acceptance. Undeveloped grains yield low test weight, starch content, and market value. Grains cease growing and add less to their dry matter content when they reach the harder stage toward maturity. Accordingly, ripening of grain is not uniform among the different earheads and the grains within an earhead. Harvest time of most cereal and pulse crops is conventionally identified from (a) the color of the leaf (straw color) and the panicle, which is often judged as golden yellow, and (b) the grain moisture content (wb) at harvest. Truly matured grain at harvest will have low grain moisture and most of the spikelets at the upper part change from yellow to golden yellow, whereas, the kernels have passed from the dough stage into a hard, matured condition. Time decision on harvesting for maximum production is not well-defined by researchers. For the cereal and pulse crops, there is a specific time span (in days) called the harvesting period during which the crop can be harvested without any significant loss in head yield and when the grain loss is minimum. This is counted on the basis of the number of days either after heading or flowering (DAF) or after sowing (DAS), or is based on the grain moisture content (wb) at harvest. The optimum period of harvest (OPH) or critical period of harvest (CPH) is the time span when the crop could be harvested without any significant loss (at the 5% level) in total yield and head yield of a crop. The duration of harvest is, therefore, bounded by the CPH of a crop. However, the climatic condition, crop variety, sowing date, and moisture content of grain at harvest influence the grain loss, yield, and duration of harvest.

Grain loss and yield of crops are interrelated. Grain losses of cereals and pulses occur primarily during preharvest and at delayed-harvesting operations when crop overdrying results in shattered grains. Reduction in head yield under delayed harvest is due to the development of sun checks in grains (5,6), which develops by alternate wetting at night and drying during the day time. Ranganath et al. (7) observed that maximum head rice was obtained between 24 and 34 DAF and between 32 and 38 DAF during dry and wet seasons, respectively, when the moisture content was bounded between 19 and 25% in the former season and between 18 and 21% in the latter season. Any harvesting beyond these optimum periods resulted in significant loss in head rice yield, either due to underripening or overripening of the grains. Grain losses are influenced by the timings of harvest and the days counted from the sowing date (8–11), or the day counted after flowering date (7,12–14) for most of the crops. It has been reported (6) that wheat loss with manual harvest ranges between 3 and 7% and increases with the increased delay in harvest. Least loss occurs with moisture ranging between 19 and 22% (5). Therefore, the moisture content of grain is considered an important factor in deciding when to harvest a crop.

Dutt (9) and Datta and Dutt (11) for gram (chickpea), Saxena (10) for soybean, and Ranganath et al. (7), Mukundan (12) Rafey (13), and Sajwan (14) for different varieties of paddy, conducted field trials with a view to establishing the effect of the date of harvest on grain yield and head yield. The best-fit yield curves of short (JS-2 and Vijaya) and long duration (JS-7244 and B-108) pulse crops, and Jaya variety (wet and dry season) of paddy crop are illustrated in Fig. 2. Depending on the type of crop, the reduction in yield was observed when harvested either early or later than the targeted-harvesting period. The shaded zone and the line AB of each curve as determined statistically at a 5% level, represent the critical period of harvest (CPH) and the harvest duration counted either from the DAS or DAF. The moisture content and the harvest duration corresponding to line AB of the yield curve represent the grain moisture at harvest and the harvest period, respectively. The studies conducted by Bhole et al. (21) show that the moisture content at harvest affects the head and milling yields (Fig. 3), and each declines rapidly after a

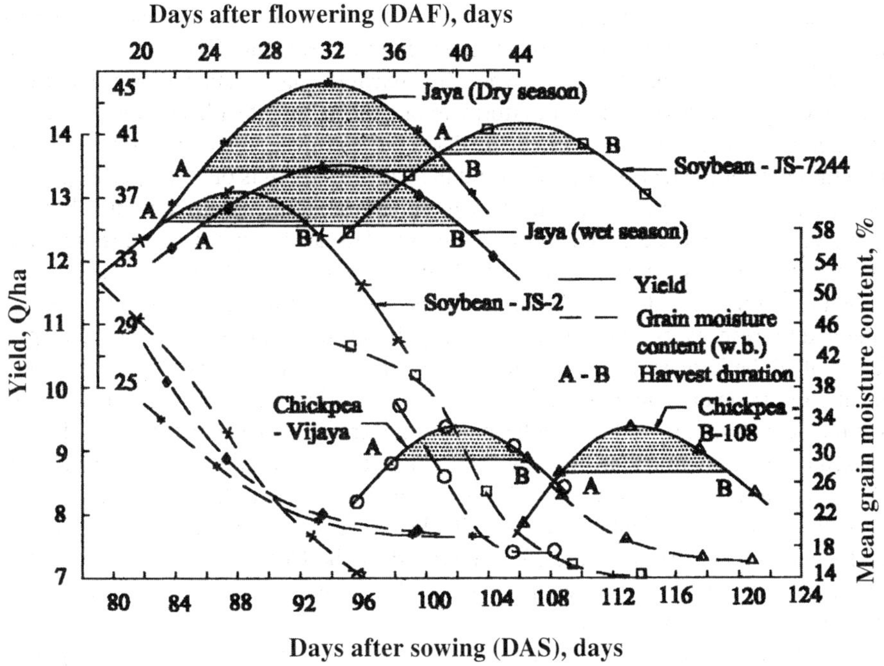

Fig. 2 Effect of DAS or DAF on yield and grain moisture for paddy, chickpea, and soybean. (From Refs. 9, 10, 14.)

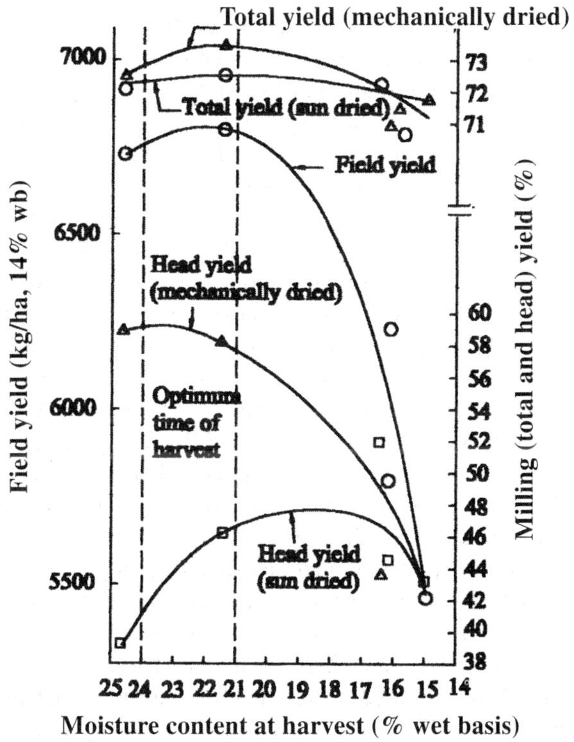

Fig. 3 Effect of moisture content at harvest on head yield and milling yield of IR-8 variety of paddy. (From Ref. 21.)

specific range of grain moisture. Therefore, delayed harvesting beyond the critical range of moisture content (24–21% for IR-8 variety of paddy) will reduce net recovery of grain. As suggested by the researchers, the recommended range of the CPH, the grain moisture, and the harvest duration for cereal and pulse crops are presented in Table 1. It serves as the guideline for the harvesting of common crops. The sowing date, crop variety, and climate will bring considerable variation in the CPH, harvest duration, and the grain moisture. Any unscheduled rain of short duration will not only delay the CPH by 2–5 days, but will also result in the shattering of grain as well as loss and deterioration of grain quality.

In light of the foregoing discussion, it is apparent that the crops must be harvested within a limited time span (harvest duration) for a specific variety. Use of mechanical harvesters—fast cutting of crops, with reduced human drudgery—is therefore essential and preferred so that the harvesting is over within the specified period when the optimum grain moisture range prevails. Delayed harvesting, either because of unpredictable weather and pilferage or nonavailability of labor, especially during peak months of harvest, will

Table 1 Critical Period of Harvest and Range of Grain Moisture for Cereal and Pulse Crops

Crop	Critical period of harvest (DAF)	Grain moisture (%)	Harvest duration (days)	Refs.
Paddy	26–33	27.0–22.0	6–10	15,16,17
	25–29 (Ratna, Jaya, Pankaj)	24.0–20.0	5–9	18
	20–28 (Jaya)	25.0–19.0	8–10	13
	24–40 (Jaya, dry season)	26.0–20.0	7–9	14
	24–40 (Jaya, wet season)	29.0–21.0	7–10	14
	16–28 (Ratna)	25.0–19.0	9–12	13,14
	18–28 (Pankaj)	25.0–19.0	9–11	13
	18–28 (Pusa 2–21)	25.0–19.0	9–11	13,14
Wheat	27–31	20.0–14.0	4–7	5,6,19
Chickpea	98–107[a] (Short-duration, Vijaya)	37.0–17.5	9–11	9,11
	109–119[a] (Long-duration, B-108)	23.5–15.5	9–12	9,11
Soybean	84–92[a] (Short-duration, JS-2)	32.4–18.5	7–10	10
	101–112[a] (Long-duration, JS-7244)	36.0–14.5	10–13	10
Jowar	104–106 (CSH–4)	21.0–24.0	3–6	20

[a]Counted after DAS.

cause excessive drying of grain, development of sun checks, and increase in shattering losses (4,6,8,9) resulting in decreased head yield and reduced milling recovery.

2.2 Cutting Principles

For cutting biological materials, operation of a sharp or serrated-edged knife is employed. It causes failure, primarily in shear. Shear failure is achieved either by using a single-cutting element, such as a sickle, scythe, rotary cutters, or flail-type cutters or by employing double-cutting elements, as in scissors or reciprocating-type cutters. Single-element cutters, either moving with high velocity (linear or rotary motion) or stationary relative to the machine, are enough when the material being cut is thick and relatively strong in bending, or is supported over a fixed surface such as wood or ground to act as one of the other shearing elements. In this case, the crop material is supported well with the ground and is strong in bending, similar to sugarcane, pigeon pea, or maize, and the material would be able to resist the cutting force exerted by operation of a single-cutting element. In all cutting processes, failure in shear or impact or both is possible when a system of forces acts on the material. Before shear failure, the material is invariably first compressed, then bends (deformation), which increases the work required in a cutting operation (22,23). Plant deformation, ∂ along the applied force, F is functionally expressed as,

$$\partial = K\frac{h}{\sigma A}F \tag{1}$$

where;

A	=	effective cross-sectional area
h	=	cutting height or stubble height
σ	=	shear stress of material
K	=	constant of proportionality that depends on knife characteristics and the material being cut

In cutting crop stems of cereal and pulses that are weak in bending, single-element cutters moving at high velocity (linear or rotary motion) will require a low-cutting force for close operation (minimum stubble height) of the cutting element. On the contrary, in the cutting of a group of stalks at an increased stubble height, greater cutting force will be experienced. Most cereal and pulse crops are harvested at ground level using two sharp shearing elements that meet and pass each other with little or no clearance. For failure in shear, either or both the elements may be moving, and the motion may be linear with uniform velocity (as in scissors), reciprocating (oscillating shear-bar type), or rotary (as in rotary cutters). Cutting of crop stems on a plane perpendicular to the fiber direction is achieved by the fast moving of sharp shearing elements against the slow or stationary element called a ledger plate, which is also beveled and frequently serrated. Shearing occurs at the intersecting point of the elements that maintain almost no clearance. Increased clearance between them will either result in the crop bending instead of cutting, or clogging the cutting unit. More often, in an oscillating shear-bar–type cutting unit, design emphasis is given related to cutting effectiveness, drive to reciprocating mechanism, and safety in operation, instead of force and power requirements.

Cutting of crop plants either by a serrated-edged sickle following linear motion or by a high-speed serrated/sawtooth-edged rotary disk cutter follows the theory suggested

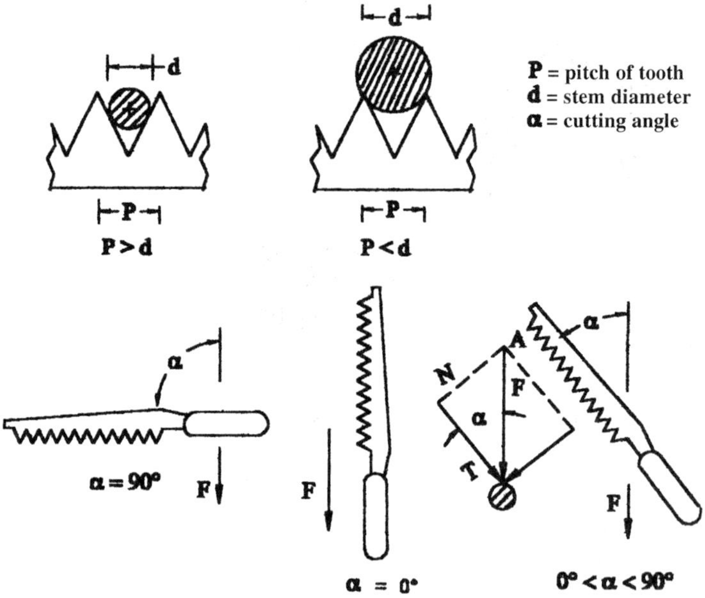

Fig. 4 Illustration on theory of cutting. (From Refs. 24, 25.)

by Kawasthima (24). According to him, straight cutting occurs along the edge of the unit teeth when the pitch of the teeth is larger than the plant diameter. For plants having a larger diameter than the pitch of the teeth (Fig. 4), continuous sliding of the individual teeth digs into the plant and cutting occurs. In the latter case, cutting resistance is considered to be much less than the former. That is why most cereal crop harvesters (rotary disk cutters or oscillating shear-bar–type cutting units), including sickles, are provided with a serrated-edged blade having a pitch less than the plant diameter. A similar principle is said to be applicable to plain blades for which teeth, according to Ivashko (25), are considered to be continuous microscopic chisels that help in smooth digging into the stem for frictional cutting. The shape of these chisels depends on the material, heat treatment, and way of grinding. The cutting edge becomes blunt once the microscopic chisels are broken.

Ivashko (25) explained the principal procedure for determining the component distribution of the cutting edge and the parameters responsible for shear failure. According to him, for a blade edge making an angle α (cutting angle) with the direction of applied force, the cutting force will have two components, normal force $N = F \sin \alpha$, which pushes the tip of the tooth into the material, making the perforations in it, and tangential force, $T = F \cos \alpha$, which continuously slides the teeth along the material for complete cutting. Cutting is difficult when $\alpha = 0°$ or $90°$ because in either case, the combinations of utilization of the bottom of the teeth and sliding motion during the process of cutting are present. For a reciprocating-bar–type cutter with cutting angle α and angle of action γ (Fig. 5), the cutting resistance F can be expressed as

$$F = N_1(\sin \gamma + \mu \cos \gamma + \mu) \tag{2}$$

in which the following approximation has been accounted

$$R_1 \approx R_2 \approx \mu N_1 \approx \mu N_2 \tag{3}$$

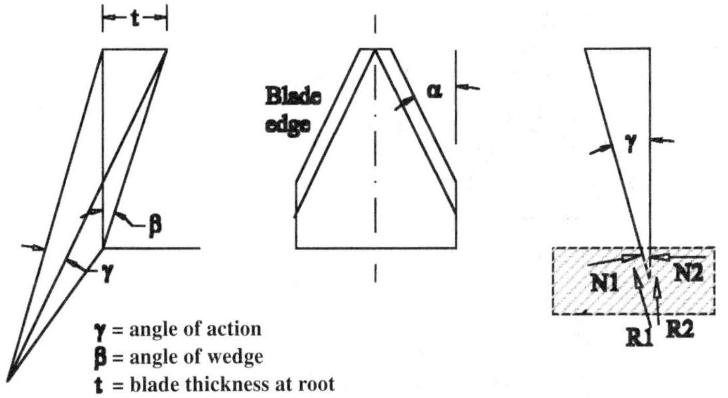

Fig. 5 Blade edge configuration and resistance to cutting. (From Ref. 26.)

where,

R_1, R_2 = frictional resistances,
N_1, N_2 = normal reactions,
μ = coefficient of friction between the blade and material.

If β is the angle of wedge, then its relation with other angles will be

$$\tan \gamma = \tan \beta \cdot \tan \alpha \tag{4}$$

Under the condition that N_1 and μ are constants, the force required to cut a straw is proportional to the angle of action, γ. Hence, for the fixed value of γ, a minimum cutting force would be required either by decreasing the angle of wedge β or cutting angle α.

2.3 Cutting Force and Energy

Comparative performance of cutting elements used in harvester design can be judged by their cutting energy requirements, cutting force, and stress applied. Cutting of cereal and forage crops by harvesters using counteredged cutting, as in a reciprocating cutter-bar, is primarily due to shear which, according to Chancellor (27), accounts for 65% of the total energy, 25% of which is spent in compression, and the rest in bending the plants. Various research workers have investigated the effects of different process variables in cutting agricultural materials and have established that the force and energy parameters to cut the materials tend to lie within fairly constrained limits. These variables were primarily influenced by the characteristics of the measuring system (28) with which the measurements were made. However, the cutting energy values observed in the laboratory were slightly lower than those obtained for cereal and pulse crop harvesters operated in the field. O'Dogherty and Gale (29) conducted laboratory experiments for cutting a single grass stem using a moving blade or blades attached to the counterweighted rotating arm along with a quartz piezoelectric force transducer. They showed that there is a critical cutting speed for sharp cutting, and that when cutting above this speed, the specific energy required is practically constant. For cutting speeds less than the critical value, there is generally a rapid increase in specific energy, except at speeds as low as 5–10 m/s, at which there is a reduction in specific energy when stems remain uncut. Efficient cutting at speeds higher than the critical value will have a specific energy as low as 10–20 MJ/mm^2 whereas for inefficient cutting, values will have a range as high as 400 MJ/mm^2. In general, the critical cutting speed of grass stem was in the range of 15–30 m/s. However, over the speed range of 5–40 m/s, there was no significant variation in the specific peak force, which ranged up to about 70 N/mm. The specific cutting energy and specific peak force were reduced to as low as 15 MJ/mm^2 and 2–15 N/mm, respectively, when a sharp blade at higher rake angle (30°) was used to cut the grass stems. The double-shear arrangement of blades for cutting grass stem was most effective, as it not only reduced the critical speed value to 10–20 m/s, but also resulted in relatively low specific energy and specific peak force values (29).

McRandal and McNulty (30) developed theoretical models of impact-cutting behavior of forage crops, but did not establish a well-defined critical speed in their studies on the cutting of assemblages of stems. However, their results showed a linear reduction of about 25% in specific energy when the cutting velocity was increased from 20–60 m/s. Similar laboratory studies have been conducted by many researchers for the measurement of cutting energy and the critical cutting speed for different field crops and fodders. Some such common-measuring devices used by them are presented in Table 2.

The pendulum displacement method is the most popular technique to measure accurately the dynamic cutting energy (E_i) in joules of cereal, pulse, and fodder crops, and the standard expression (9,31–33) used is given by,

$$E_i = WR(\cos\theta_u - \cos\theta_o) \times 10^{-7} \tag{5}$$

where,

W = mass of pendulum with cutting element (g).
R = distance of C.G. of the pendulum bar from the axis of rotation (cm).

Table 2 Energy-Measuring Devices Used by Different Researchers

Sl. no.	Measuring devices (methods)	Type of rig	Crop material tested	Form of stem	Maximum cutting velocity (m/s)	Ref.
1	Pendulum displacement	Pendulum	Maize	Single stem	3.95	32
2	Pendulum displacement	Rotating blade	Alfalfa	Single stem	60.00	35
3	Pendulum displacement	Pendulum	Paddy, alfalfa	Single stem	2.77	31
				Single stalk	3.25	33
4	Pendulum displacement	Pendulum	Alfalfa, Sudan grass	Group of stems	9.75	36
5	Semiconductor strain gauges on blade holder	Rotating blade	Grass	Group of stems	60.00	30
6	Pendulum displacement	Pendulum	Corn	Corn kernel	1.26	37
7	Strain gauge on blade and straw support	Rotating blade	Hemp	Single stem	18.90	38
8	Transducer on blade and stem support	Linear motion of blade	Alfalfa, sunflower	Single stem	26.00	39
9	Transducer on blade and stem support	Linear motion of blade	corn	Single stem	60.00	27
10	Pendulum displacement	Pendulum	Bengal gram	Single pedicel	0.75	9
11	Flywheel-type ensilage cutter	Rotating blade	Paddy, Jowar	Group of stems	3.72	40
				Group of stems	3.95	40
12	Quartz piezoelectric force transducer attached to blade	Moving blade	grass stem	Single stem	15.00	29
				Group of stems	30.00	29

θ_u = upswing angle of pendulum arm with vertical after cutting the specimen (°);
θ_o = upswing angle of pendulum arm with vertical at the end of a free swing (°).

Fisher et al. (34) also used a pendulum knife to measure energy required to cut a single stalk at a time of alfalfa, blue grass, orchard grass and others, at different moisture contents (72–87%) and plant diameters. At each moisture content, the energy requirement followed a second-order polynomial relation with their diameter, and for each type of grasses. The pendulum displacement technique has also been applied to detach pedicel or to break pod (9) when the specimen is placed at the equilibrium position and in the path of a pendulum swing. Dynamic shear force and, hence, the stress required to cut a specimen as well as to detach the pedicel or to break a pod can be determined along with the energy measurement if the pendulum arm is mounted with a set of strain gauges at selected locations. The output of the strain gauges recorded with a strain-recorder, such as a dynograph, will simultaneously provide force data for a precalibrated strain-sensitive transducer. Rajput and Bhole (31) used the pendulum displacement method to measure dynamic properties of paddy stem at different crop heights, using a single-blade operated at 2.77 m/s. It was reported that the cutting stress (70–140 kg/cm^2) increased along its height, whereas the force and energy required varied inversely with moisture content and directly with its sectional area. Furthermore, cross-sectional area showed a significant effect on cutting force and energy, and 30°-bevel angle blade provided the minimum force and energy required to cut at the 10-cm height of a plant. Halyk and Hurlbut (41) observed the shear strength of alfalfa stem internodes between 4.08 and 183.5 kg/cm^2, and it decreased with the increase in moisture content at all heights. Liljedahl et al. (42) established the effect of moisture content of forage stalks from 8 to 75% (wb) on specific energy and observed the highest cutting energy at a moisture content of 30–35%. Similar investigations by Chancellor (27) and Datta et al. (40) also found the maximum cutting force at a moisture content of 35–40% for forage stalks. Chancellor (27) found that at the cutting speed of 2.54 cm/s, the shear blade with a 24°-bevel angle were the most efficient in cutting alfalfa stem, and the energy required for cutting a group of stems ranged between 0.4 and 2.4 hp-h per ton of dry matter to cut into 12.5-mm lengths. Prasad and Gupta (32) also observed the minimum-cutting energy per unit area and cutting stress of maize stalks (73.6%) for a shear blade with 23°-bevel angle. Datta et al. (40) determined the dynamic shear stress of different forage crops, including paddy straw using flywheel type ensilage cutter operated by an induction motor and utilized the relation given by Eq. (6):

$$S_d = \frac{0.102(P_l - P_n)}{\omega R_c A_s} \tag{6}$$

where,

S_d = dynamic shear stress (kg/cm^2);
P_l and P_n = average instantaneous power at load and at no-load respectively (W);
R_c = distance of the center of throat from the center of rotation (m);
A_s = cross-sectional area of stalks under shear (cm^2);
ω = angular velocity of knife (rad/s).

It was reported (40) that the dynamic shear stress ranged between 0.12 and 0.254 kg/cm^2 (8.5% m.c. db) for paddy and 0.10 and 0.155 kg/cm^2 (74% m.c. db) for jowar, and decreased linearly with an increase in knife velocity beyond 2.7 m/s. Furthermore, increase in moisture content increased dynamic shear stress exponentially for paddy straw. Therefore, it was recommended to cut green fodder at a high-moisture content and at an angular velocity of 2.7 m/s or more.

2.4 Harvesting Hand Tools

Harvesting of field crop is considered a labor-intensive operation and takes about 185–340 man-h/ha to cut and bundle paddy or wheat crops (16,43,44), and 170–200 man-h/ha (2,4) for cutting paddy crop. It requires nearly 20% of the total cost of operation (16). This operation as per the availability of labor continues for many more days when leftover crops overdry, causing an overall loss of 7–14%. The occurrence of rain with wind will increase the labor requirement and loss of grain, even up to 22% (4). Although there is a continuous effort to mechanize different farming operations at the national level, harvesting operation of cereal, pulse, and fodder crops among small, marginal, and medium farmers is carried out manually using a sickle.

For many centuries, the harvesting of cereal, pulse, and fodder crops involved hand-cutting with sickle, scythe, or cradle, either cut at ground level or stripped at earhead level (stripping). The sickle, a locally made hand tool, is conventionally operated by women laborers and has a curved, but serrated or plain-edged blade segment with one wooden handle. The blade shape closely resembles the segment of an ellipse, and the handle is conveniently designed to match the palm configuration of an operator for better grip and to reduce tool slippage. Blacksmiths make blades with spring steel, high carbon steel, and temper to a Brinnel hardness of 250 H_B to maintain sharpness and increased life. Serrations with five to ten teeth per centimeter at the inner edge of the blade are made under red-hot conditions. Although blade serration is most common in sickles, the studies conducted on smooth- and serrated-edged sickles indicate that there is no significant difference in either effective cutting or in the energy required to harvest crops.

According to the agroecological conditions, palm configurations, and style of operation, there are about 59 different sickles (45) used in the country. Their shape, size, and weight, as well as the handle geometry vary widely. The field capacity of these sickles varies between 0.008 and 0.090 ha/day, while their weight range is between 60 and 459 g. The Sikkim sickle is the lightest one, and the Simla sickle is the heaviest that provided the maximum field capacity (45). Pandey and Devnani (44) developed an improved CIAE sickle for the Bhopal region and compared its performance with the Maharashtra, Punjab, and local sickles for harvesting wheat crops of variety *Kalyan Sona*, *Narmada–4*, and *Sonalika*. The improved CIAE sickle performed better relative to time required to harvest 1 ha, harvest losses, and labor requirement. It was indicated that the sickles available with the farmers of all the states are not designed scientifically. So, the Indian Council of Agricultural Research (ICAR) entrusted Farm Implements and Machinery (FIM) centers located in areas of different agroclimatic conditions of the country to develop suitable sickles for their region. Based on ergonomic considerations and operation style, a total of eight different sickles were developed and recommended for farmers' use. The construction and geometric details of some such sickles that are gradually becoming popular among the farmers are presented in Table 3 in reference to the design dimensions shown in Fig. 6.

2.5 Power-Operated Harvesters

Present-day harvesters and reapers have been developed, after several modifications, to suit one's need for a timely harvest of the cereal and pulse crops, and to reduce human

Table 3 Physical Dimensions of Some Sickles of India

Sl. no.	Sickle	Thickness, t (mm)	Weight, (g)	Dimensions (mm)										θ (°)
				L_1	L_2	L_3	L_4	L_5	L_6	L_7	L_8	L_9	L_{10}	
1	Bhopal	5.0	350	190	58	45	225	75	144	110	125	28	70	110
2	Coimbatore	2.0	125	150	47	27	160	113	50	18	100	30	55	150
3	CIAE[a]	5.5	260	225	36	28	220	167	94	25	175	—	46	150
4	Dapoli[a]	1.3	175	168	37	26	168	176	140	35	136	35	13	125
5	Kharagpur	1.5	120	144	24	32	250	130	45	134	128	35	32	145
6	Maharasthra Agro.	1.3	100	140	22	25	130	134	101	7	188	23	15	140
7	Pune	4.0	250	168	45	25	195	155	107	80	95	30	41	123
8	Punjab[a]	4.5	250	235	27	31	225	167	104	30	175	—	52	145
9	Sikkim	1.0	60	180	35	20	210	70	25	30	105	22	46	170
10	Simla[a]	2.0	450	190	90	38	203	210	140	50	140	35	92	115

[a]Raised handle; others are in-line handle.
Source: Ref. 45.

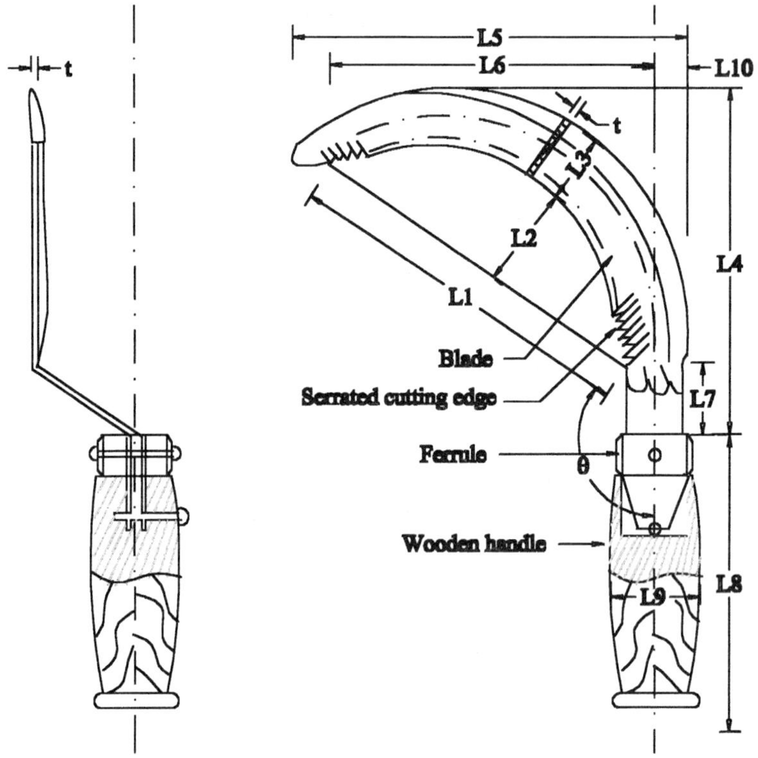

Fig. 6 Major dimensions of a sickle.

drudgery during the harvest period. These are, no doubt, faster methods of harvesting crops. A range of harvesters operated by 1.5 to several horsepower engines (gasoline or diesel, including tractor–power tiller) are manufactured commercially. These harvesters and forage choppers adopt one of the following cutting mechanisms:

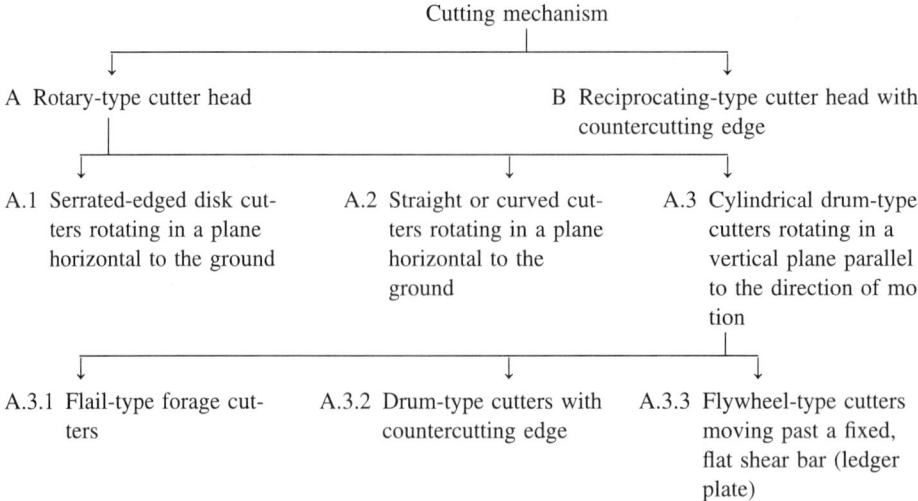

Cutting mechanism

- A Rotary-type cutter head
  - A.1 Serrated-edged disk cutters rotating in a plane horizontal to the ground
  - A.2 Straight or curved cutters rotating in a plane horizontal to the ground
  - A.3 Cylindrical drum-type cutters rotating in a vertical plane parallel to the direction of motion
    - A.3.1 Flail-type forage cutters
    - A.3.2 Drum-type cutters with countercutting edge
    - A.3.3 Flywheel-type cutters moving past a fixed, flat shear bar (ledger plate)
- B Reciprocating-type cutter head with countercutting edge

Based on the types of cutting mechanisms for biological materials, rotary-type cutter heads with serrated-edged disk (A.1), or straight or curved cutters (A.2), and reciprocating-type cutter head with countercutting edge (B) are most common in cereal crop harvesters or reapers, while the remaining cutting mechanisms are employed either for cutting grasses and fodders or in forage choppers. Engine-operated, back-mounted, walking-type disk cutters; engine-operated, walking-type, rotary disk cutters; engine-operated, front-mounted walking-type, reciprocating action harvesters; tractor-operated, front- or side-mounted reciprocating action harvesters; and engine-operated riding-type, stripper harvester, are the most common forms of cereal crop harvesters. The constructional and operational details along with the performance characteristics of each are discussed hereunder.

2.5.1 Rotary cutters

Circular blade, with sharp- or serrated-edged cutting units are extensively used in knapsack-type rotary cutters as well as in engine-driven walking-type (self-propelled) harvesters that use a small-capacity, light-weight gas engine of 0.45–1.88 kW. Drive to the disk blade is given from the engine mounted at the operator's back through a flexible cable or rigid shaft, whereas large-capacity, engine-driven rotary cutters derive power through suitable V-belt and pulley drive because of safety in operation. Each type of harvesters uses a rotary disk cutter of 22–45 cm in diameter, having 35–60 teeth at 5°–10°, angle of the teeth and are operated at 2500–4000 rpm (peripheral velocity 30–35 m/s) for cutting cereals, grass, bush, hedge, fodder stalks, and the like. Such cutters at high speed generate a continuous wind blast, resulting in random scatter of cut crop plants. Therefore, crop collection requires added labor and energy, up to a limit of 55–75 man-h/ha if rotary cutters are not provided with a proper crop guide assembly and deflectors. Furthermore, it is essential to operate the cutter as close to the ground as possible for clean cutting of cereal crops that are weak in bending. Bianer et al. (22) indicated that a rotor speed of 40–50 m/s is required for clean cutting of cereals. Operating speed beyond this range will increase power consumption, with no definite improvement in cutting performance.

This type of cutters uses two specific designs of drive shafts: flexible and hollow shafts for knapsack-type and shoulder-type, respectively. As each is a body-borne reaper, a heavy work load is imposed on the operator, who becomes tired even after a short run of about 1 h (46). Of the two, hollow, rigid shafts are conventionally capable of transmitting relatively higher torque for cutting biological plants, such as pigeon pea, maize, soybean, or sugarcane, that have a higher-bending strength. Mazumdar (47) evaluated the performance of both types of shafts based on power consumption. Flexible shafts required significantly lower running power, compared with the rigid shaft, and the reduction was of the order of 126–207% in the blade speed range of 30–40 m/s.

Mazumdar (47) made an intensive study on the knapsack-type rotary-cutting machine, with a view to developing a single-row rotary paddy harvester relative to mechanical and ergonomic considerations. The schematic diagram of the prototype, with crop guide, is shown in Fig. 7. It was recommended that in such cutters, a crop guide assembly should be provided with a provision to adjust the deflector plate angle and blade-shielding plate for effective windrowing. Blade speed, shear angle, blade clearance, crop moisture, crop density, and variety showed significant effect on efficient cutting and windrowing operations. A closed-tooth rotary blade with countercutting edge significantly reduced shatter loss in the speed range of 30.3–46.1 m/s, when compared with a blade without a countercutting edge. However, blade speed beyond 36.1 m/s reduced shatter loss significantly. Singh and Verma (46) reported that a rotary cutter without a crop guide assembly would

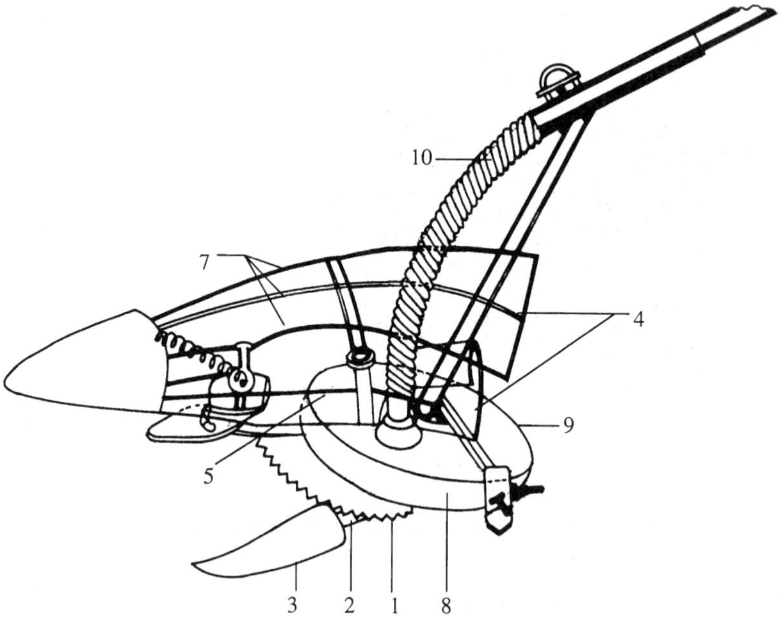

Fig. 7 Prototype of knapsack-type reaper with rotary cutter; observe plant deflector and crop divider:

1. Rotary cutter
2. Counter-edge plate
3. Left crop divider
4. Crop guide assembly
5. Main plant deflector
6. Central crop divider
7. Auxiliary plant flender
8. Blade-shielding plate
9. Safety shield
10. Flexible drive shaft inside flexible rubber cover

(From Ref. 47.)

require three persons for 19.8–26.0 h to harvest 1 ha of cereal crop, and for lodged crops more time would be required, as it is operated against the direction of lodging and operated idle while traveling in the direction of lodging (48).

Figure 8 shows the sectional view of a Japanese model (49) lightweight, portable, 1.0-m Mametora reaper (Model: SM–100) that uses mainly a single planetary rotary saw as blade, dividing rods, guide rod, straw-gathering, and disposing levers for paddy crops. This unit is for mounting at the front of a 3.5 hp two-wheel gasoline tractor (power tiller, model Mametora UK–13). The reaper input shaft receives the drive from the tractor pulley through a V-belt and pulley. Before the reaping operation, it required manual harvesting to clear a 2 × 2-m space at the corner for turning and to cut one or two rows close to the bunds on which to place the ejected cut crops. The performance of this machine was evaluated and modified as needed by developing countries, such as India, Indonesia, Iran, Pakistan, the Philippines, and Thailand, for different varieties of paddy crops. The overall performance data of the reaper are presented in Table 4. It was reported in general, that the circular saw blade required sharpening after approximately every 10 h of use. Furthermore, the dwarf variety crops could not be collected successfully and ejected by the machine owing to clogging. The shattering loss was primarily due to the impact of the plants on the collecting and ejecting plates by the dividing and gathering mechanism, and this

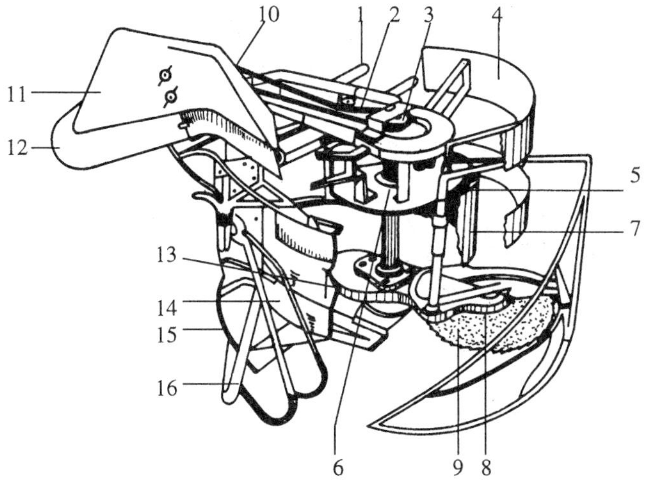

Fig. 8 Sectional view of a Mametora reaper showing different components:

1. Disposing arm
2. Main belt
3. V-pulley of input shaft
4. Gathering lever
5. Gathering roll
6. Gathering cam
7. Cover (middle)
8. Drum case
9. Circular saw
10. Engine pulley
11. Straw-disposing lever (upper)
12. Straw-disposing lever (lower)
13. Base holder
14. Cover (lower)
15. Saucer
16. Guide rod

(From Ref. 49.)

loss was considered to be moderately high. The overall grain loss of up to about 13% in India, 16% in Indonesia, 10.5% in Pakistan, 12% in the Philippines, and 8% in Thailand was reported. This loss, however, was affected by the crop variety, stage of harvest (maturity), moisture content, and machine adjustments.

2.5.2 Reciprocating Action Cutters

All cereal and forage crop harvesters are provided with reciprocating action-type cutting units of different designs. Based on the construction and distance between the adjacent guards (also called fingers) three types of cutter-bar are employed in harvesting machines: (a) standard or conventional type, (b) low-cut type, and (c) medium-cut type cutter-bar. The typical cross view of all three types is shown in Fig. 9. Because of the variations in the distance of the guards, the knife edges in extreme positions may or may not coincide with guards, but generally the knife edge stroke length(s) for all types of cutters will lie between 75 and 110 mm. The knife edge stroke length, s in the standard-type cutter-bar, is equal to the distance between the adjacent guards, and the knives in their extreme positions lie in the fingers. The standard cutters with s = 76.2-mm stroke or pitch are used in mowers and reapers (grass, cereal, and commercial crops). The 90-mm–pitch cutters are used to cut thick-stalk commercial crops, corn, sunflower, and the like. The distance between the axes of symmetry of successive guards in low-cut–type cutter-bar

Table 4 Performance Range of Mametora Reaper in Developing Countries

Sl. no.	Particulars	India	Indonesia	Pakistan	Philippines	Thailand
1	Crop condition					
	Crop	Paddy	Paddy	Wheat	Paddy	Paddy
	Variety	Jagannath	IR-36	Sandal	IR-36,42,46	RD-7
	Susceptibility to shattering	Yes	Yes	Yes	Yes, no, no	Yes
	Plant height (cm)	48–50	64–67	—	84–105	93–95
	Grain yield (t/ha)	2.2–2.8	3.5–4.4	3.9	4.1–6.9	3.3–4.0
2	Field performance					
	Forward velocity, (km/h)	0.804–0.828	0.924–1.044	1.08	0.756–1.176	1.002–1.500
	Working width (cm)	89–90	82–90	95	91–97	90
	Field capacity (ha/h)	0.06–0.07	0.05–0.07	0.08	0.06–0.11	0.09–0.11
	Field efficiency (%)	85–88	62–81	70	71–98	68–86
	Fuel (gasoline) consumption (L/ha)	10.9–11.8	11.4–20.0	13.0	9.9–19.5	12.2–14.9
3	Total machine loss (shattering) (%)	8.8–13.0	12.0–16.2	10.5	2.7–12.3	0.5–8.0
4	Labor requirement (man-h/ha)					
	Unskilled					
	Preparing field before harvest	6.8–8.2	0	0	13.3–45.3	10.0–12.0
	Gathering and others after harvest	30–33	113–200	57	28–71	20–25
	Total	40	163	57	44–116	30–37
	Skilled	15–17	12–51	13	9–17	9–12

Source: Ref. 49.

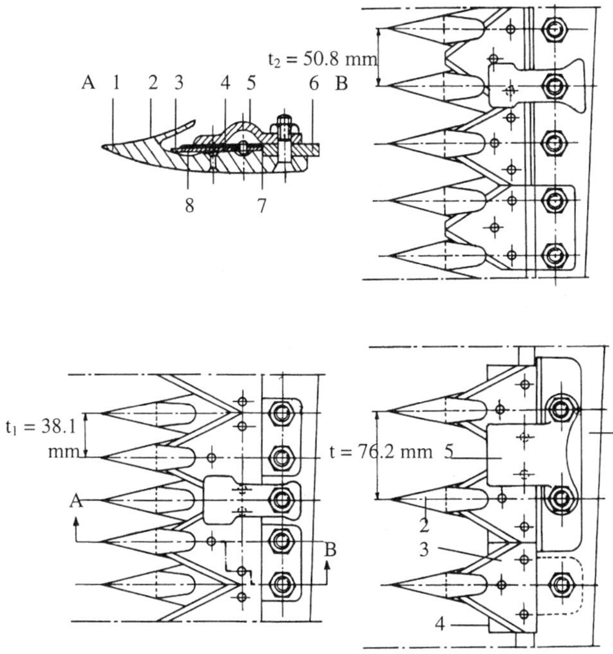

Fig. 9 Cutter-bar geometry showing different reciprocating-action cutters and major components:

1. Guard point
2. Knife guard (finger)
3. Reciprocating knife
4. Knife bar
5. Knife clip
6. Main bar
7. Knife back
8. Ledger plate

is half the knife edge stroke length. Hence, the knife edge stroke length, s for low-cut–type cutter, is twice the distance between the adjacent guards and lies in the first and third fingers at extreme positions of the knife. Such cutters with $s = 76.2$-mm–stroke are used in mowers, whereas those with $s = 101.6$-mm–stroke are used in trailed combines. But in the medium-cut–type cutter-bar, the distance between the adjacent guards is more than that of low-cut type, but less than that of standard type. This type of cutter with $s = 76.2$-mm–stroke is used in mowers of foreign manufacture.

The reciprocating-action cutters with stationary countercutting edges make to-and-fro motions between the successive guards when the cutting of crop stems occurs. The included angle between the cutting edges is near 38°, and this angle is less than normal for wheat crop and increases for serrated edges (33,50). The knife edges of such a cutter-bar derive reciprocating motion through a crank mechanism and make forward movement during field operation. Relative to ground level, any point on the knife edge will, therefore, make a curvilinear path. It is possible to obtain the knife edge's path of any selected point on the edge graphically if the crank radius r or the knife edge stroke s and distance l traveled by the machine in a half revolution of crank, are known. The curvilinear path traced by the knife edges is steep for standard type, followed by low-cut and medium-cut–type cutter-bar because of the wide spacings of consecutive knife edges of the latter. This phenomenon suggests use of the low value of forward travel per cut (feed rate) to stroke length ratio for standard type to reduce the uncut zone.

2.5.2.1 Cutter-Bar Components and Operation

Harvesting machines employing any of the three types of cutter-bar; namely, conventional, low-cut, and medium-cut–type with countercutting edge, are provided with the following essential components.

1. Reciprocating motion
   a. Trapezoidal-shaped knife/slider
   b. Knife back riveted with knife section
   c. Part length of pitman
2. Rotary motion
   a. Crank and crankpin
   b. Balancing counterweight
   c. Telescopic shaft and PTO
   d. Part length of pitman
3. Stationary
   a. Main bar (beam)
   b. Wearing plate
   c. Knife guards (fingers)
   d. Guard lip
   e. Knife clip or holder
   f. Adjustable inner and outer shoes
   g. Ledger plate
   h. Crop dividers
   i. Lift linkage and drag bar
   j. Support and adjustable frames

2.5.2.1.1 MAIN BAR. The main bar made of high-grade steel is the strong component of cutter-bar to which all other parts of cutting mechanism are attached directly or indirectly. It provides the main support to all the elements and acts as a cantilever bar or beam.

2.5.2.1.2 THE GUARDS. Typically shaped cast iron or malleable cast iron guards placed at an interval equal to the stroke length provide a place for the ledger plates and protect the cutting units. Their shapes helps in dividing the crop materials being cut and at the same time pushes them to the cutting edge.

2.5.2.1.3 THE CUTTER/SLIDER. It consists of a series of trapezoidal-shaped, serrated–smooth-edged knives and each is riveted in line with the knife back at regular intervals equal to the stroke length. It reciprocates between the centers of the successive guards during operation and is supported on ledger plate attached to the guard. The cutter is conventionally made of medium- to high-carbon steel (0.6–0.8% carbon) or spring steel.

2.5.2.1.4 THE LEDGER PLATES. Ledger plates, made of spring steel of medium- to high-carbon steel, are the stationary cutting element forming one-half of the cutting mechanism and are riveted with the guard. The edges of the ledger plates are serrated on the underside. Care should be taken to maintain its sharpness and serrations for effective cutting of crop stems.

2.5.2.1.5 Wearing Plates. The wearing plates, placed at the back of the cutter/slider, provide necessary support to the knife as well as absorb the rearward thrust of the knife. Because of the continuous-reciprocating action of the knife, they wear out fast, causing the rear side of the knife to drop down and make poor contact with the ledger plate. Ultimately, this results in uneven and poor cutting, requiring excessive draft during operation. If regular attention is not given to the wearing plate, increased clearance between the knife and ledger plate will cause increased deformation in bending and a wedging effect will result. They are normally made of high-carbon steel or alloy steel.

2.5.2.1.6 The Knife Back. The knife back is a long iron strip of medium-carbon steel to which each knife section is riveted. It thus forms the slider/cutter.

2.5.2.1.7 The Knife Clip or Holder. As the name implies, the knife clip or holder helps in holding the knife section close to the ledger plates. Such knife clips are placed three or four guards apart so that minimum clearance is maintained. They are nut-bolted to the main bar and the guard extension. The alternating vertical reaction F_v in the cutter-bar, results in vibration and increased clearance. This makes the knife play up and down, and causes poor cutting. In such a situation, they should be hammered down. They are generally made of malleable iron or spring steel.

2.5.2.1.8 Inner and Outer Shoes. These are the large shoe-like runners, one at each end of the cutter-bar that support the entire cutter-bar during operation. The detachable sole, provided underneath each shoe, is to adjust the stubble height, while the pointed and projected front part of the shoes demarcates the cutting width.

The constructional details of most components as well as the cutter-bar cross-section are shown in Fig. 10. The adjustable inner and outer shoes of 0.5- to 3.5-m–long cutter demarcate the cutting width as well as give support, and gauge the stubble height of crop during operation for mowers and side-mounted reapers. The cutter-bar of such machines essentially acts as a cantilever and, therefore, is supported by a pull-bar, drag link, and

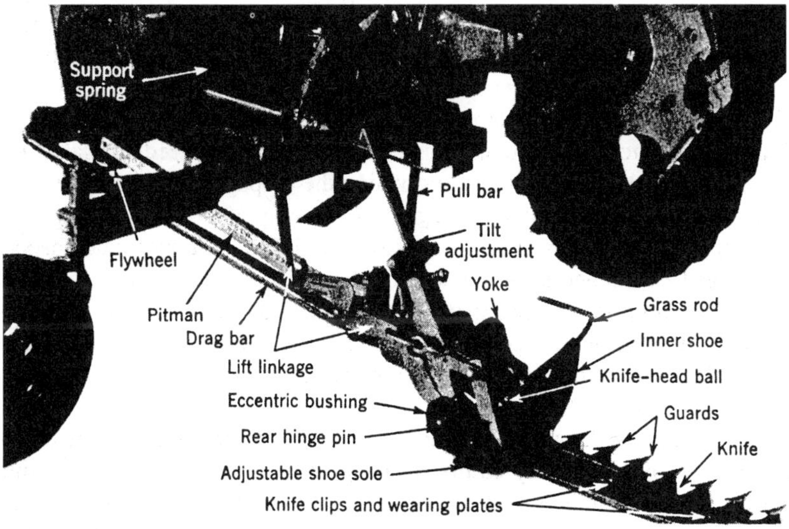

Fig. 10 A typical view of a semimounted mower with PTO drive (JI Case Co.). (From Ref. 22.)

linkages to maintain level and floating condition, and also to prevent rearward movement of the free end. The lift linkage is made adjustable to vary the weights supported by the inner and outer shoes to prevent bouncing of the cutter-bar. Therefore, these components are made strong and adjustable for proper functioning of the cutter-bar.

The reciprocating-action cutter-bars are powered from tractor PTO through telescopic shaft for rear-mounted and semimounted mowers; tractor PTO through a pair of V-belts for front-mounted reapers, side pulley through the V-belt for side-mounted reapers, and stationary engine through a gear box and V-belts for a self-propelled reaper. The cutter-bar speed of a mower is kept double that of the reaper and lies between 800 and 1000 rpm or 1600 and 2000 cutting strokes per minute (22) or knife velocity of 2.85 and 3.75 m/s for an 8-cm–stroke length. This speed is more than enough for scissoring action of the knife for crops to fail in pure shear. A correctly registered cutter-bar with properly adjusted knife clearance will require a little lower value of knife velocity than normal for cutting stems that are weak in bending.

Although the side-mounted reapers attached between the front and rear tractor wheels provide better visibility and respond directly to steering, because of complexity in mounting and power transmission system, front-mounted reapers either self-propelled or tractor-operated are more common for harvesting cereal crops. The semimounted and mounted-type cutter-bar attached at the rear of a tractor are generally preferred for mowing operation.

2.5.2.2 *Cutter-Bar Adjustments*

Smooth or underserrated cutting edges of the cutter-bar are often resharpened when needed by grinding the beveled top surfaces. As discussed in an earlier section, the underserrated knives perform better for thick-stemmed crops, and the smooth-edged knife or fine saw-tooth (higher teeth per inch) knife will function efficiently in cutting grasses. The countercutting edges (ledger plates) serrated on the underside are never sharpened, but are replaced if they do not function properly. Occasional checks of the scissoring elements, therefore, are very essential.

The wearing plates mounted with knife clips, essentially absorb the rearward thrust of the knife and give vertical support for the rear of the knife section. The knife clips, spaced three or four guards apart, prevent vertical movement of knife owing to unbalanced vertical action, F_v. At the same time, for proper functioning of the cutter-bar, there should be little or no clearance between the knife section and the ledger plate. This is maintained by keeping the knife clips close to the knife so that there is enough clearance for knife movement. If they become worn and allow the knife to play up and down, making less contact with ledger plates, they are required to be hammered down. Under this situation, occasional use of lubricant in moving parts will reduce wear loss and improve cutting performance. Increased clearance will allow deformation in bending, resulting in wedging effect and increased power consumption. Therefore, it is obvious that the knife may not cut the crop stems if the clearance between the knife clip and knife section is equal or more than the thickness of stem being cut.

The knife is required to be registered for smooth cutting of the crop stem. In a properly registered cutter-bar, the knife sections are centered in the guards at each end of the stroke when the knife touches and crosses two, three, or four guards, respectively, by the standard, low-cut–and medium-cut–type cutters. Such adjustments are primarily made either by adjusting the pitman length or by moving the entire cutter-bar in or out relative to the crankpin.

A cutter-bar is said to be in proper alignment if, during operation, the pitman and the knives are in one line and in vertical plane perpendicular to the direction of motion. This adjustment is very crucial for side-mounted reapers, and rear-mounted and semi-mounted mowers that act as cantilevers. The free end of such cutter-bars at rest, as is customary, is given a lead of about 2 cm for every 1 m length of cutter-bar so that the crop resistance during operation and friction between the cutter-bar and the ground will bring the cutter-bar in line with the pitman. Such an adjustment is not required for front-mounted (tractor) reapers and self-propelled reaper windrower

2.5.2.3 *Kinematics of the Cutter-Bar*

Conventionally, all three types of reciprocating cutter-bar use a plane, asymmetric crank system in mowers, reaper windrowers, and combine harvesters for which the axis of rotation of the crank is perpendicular to the plane of oscillation. Such cutter-bars are required to be operated at some prefixed level equal to stubble height; therefore, the crank center is located at a height m above the cutter-bar level (Fig. 11). The ratio m/L, where L is the pitman length, called eccentricity (ξ), plays an important role in deciding crank radius (r) and stroke length (s).

If a and b represent the lateral distances of knife bar at the end of a stroke, then

$$s + b = a \tag{7}$$

or

$$s + \sqrt{[(L - r)^2 - m^2]} = \sqrt{[(L + r)^2 - m^2]}$$

Squaring twice and rearranging yields,

$$2r = s\sqrt{\left[1 - \frac{4m^2}{(4L^2 - s^2)}\right]}$$

Neglecting s^2 term before $4L^2$, then

$$r = \frac{s}{2}\sqrt{(1 - \xi^2)} \tag{8}$$

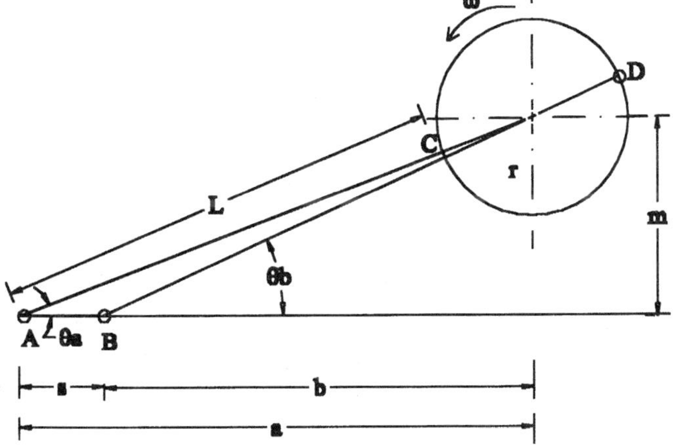

Fig. 11 Pitman at extreme positions.

Since, ξ in Eq. (8) is less than unity, $s > 2r$ and for fixed stroke length, increase in ξ will reduce r. Furthermore, from Fig. 11, the angles made by the pitman with horizontal at each end of the stroke will be

$$\theta_a = \sin^{-1}\left(\frac{m}{L + r}\right) \quad \text{and} \quad \theta_b = \sin^{-1}\left(\frac{m}{L - r}\right) \tag{9}$$

2.5.2.4 Balancing of Reciprocating Cutters

Reciprocating-action cutters used in reapers and mowers are often operated at high forward speeds and on uneven fields. The reciprocating and rotary components result in unbalanced forces at the crankpin end as well as at cutter-bar end. Balancing of the rotating and reciprocating components is absolutely necessary for smooth cutting and to reduce wear. Periodic alternating vertical reaction at the knife end tends to increase clearance between the slider (knife) and ledger plates.

Figure 12 depicts the force relation of a cutter-bar–crank combination (22). It is assumed that the crank rotates at a uniform angular velocity ω, and the proportions of pitman mass m_1 and m_2 are concentrated, respectively, at the slider end, making a reciprocating motion, and at the crankpin, making a rotary motion. At time $t = 0$, the crankpin is at D and pitman is at A. After a lapse of time t, the crank rotates through θ_c when the pitman end moves to B, causing a displacement x. Now from geometry,

$$\begin{aligned} x &= CB - CA \\ &= [L \cos \theta_p - r \cos \theta_c] - [\sqrt{L^2 - m^2} - r] \end{aligned} \tag{10}$$

where, θ_p = pitman angle with the ground.

Furthermore;

$$\sin \theta_p = \frac{m + r \sin \theta_c}{L}$$

$$\therefore \cos \theta_p = \frac{1}{L}\left(L^2 - m^2 - r^2 \sin^2 \theta_c - 2\, mr \sin \theta_c\right)^{\frac{1}{2}}$$

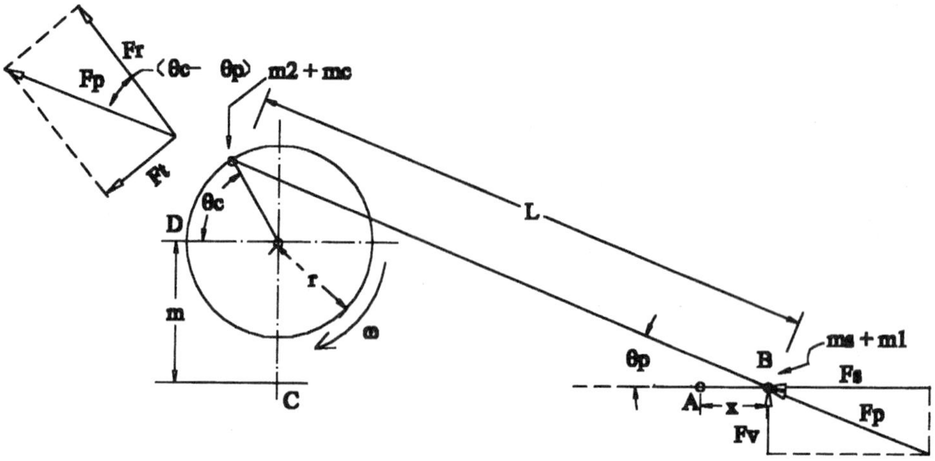

Fig. 12 Forces on reciprocating parts of crank and slider arrangement.

Substituting $\cos \theta_p$ in Eq. (10)

$$x = (L^2 - m^2 - r^2 \sin^2 \theta_c - 2\,mr \sin \theta_c)^{1/2} - r \cos \theta_c - (L^2 - m^2)^{1/2} + r \tag{11}$$

Because Eq. (11) is a function of θ_c, the first derivative of x with respect to time will give instantaneous velocity v, whereas the second derivative will provide acceleration f of the slider. Hence,

$$\begin{aligned} v &= \frac{dx}{dt} = \omega \frac{dx}{d\theta_c} \\ &= \omega \left[r \sin \theta_c - \frac{(r^2/2)(\sin 2\theta_c) + mr \cos \theta_c}{(L^2 - m^2 - r^2 \sin^2 \theta_c - 2\,mr \sin \theta_c)^{1/2}} \right] \end{aligned}$$

Neglecting the terms $r^2 \sin^2 \theta_c$ and $2\,mr \sin \theta_c$ before L^2,

$$v = \omega \left[r \sin \theta_c - \frac{r^2 \sin 2\theta_c}{2(L^2 - m^2)^{1/2}} - \frac{mr \cos \theta_c}{(L^2 - m^2)^{1/2}} \right] \tag{12}$$

Differentiating Eq. (12) for instantaneous acceleration f

$$\begin{aligned} f &= \frac{dv}{dt} \\ &= \omega^2 \frac{dv}{d\theta_c} \\ &= \omega^2\, r \left[\cos \theta_c + \frac{m \sin \theta_c}{(L^2 - m^2)^{1/2}} - \frac{r \cos 2\theta_c}{(L^2 - m^2)^{1/2}} \right] \end{aligned} \tag{13}$$

substituting $K = m/(L^2 - m^2)^{1/2}$; Eq. (13) becomes

$$f = \omega^2\, r \left[\cos \theta_c + K \sin \theta_c - \frac{Kr}{m} \cos 2\theta_c \right] \tag{14}$$

Hence, the inertia force F_s of the sliding components in a horizontal plane can be represented by

$$F_s = (m_s + m_1) f \tag{15}$$

where m_s represents the mass of slider (reciprocating knife). The same expression was proposed by Bainer et al. (22), and Lal and Datta (51) proved it. The balancing of reciprocating parts can be done by providing the second reciprocating mass at 180° out-of-phase and in-line with the slider. Because the slider and the part mass of the pitman are under continuous oscillation and the acceleration is a function of θ_c, an unbalanced, periodic, alternating vertical reaction F_v at knife end is introduced (see Fig. 12). The pitman force, F_p resultant of F_v and F_s acts on the crankpin E such that,

$$F_v = F_p \sin \theta_p \tag{16}$$

and

$$F_p = \frac{F_s}{\cos \theta_p} \tag{17}$$

The pitman in this situation is under compression and the force F_p at the crankpin will have two components: tangential force F_t, balanced by the flywheel, and a power source, while the radial force F_r is conventionally balanced by a proportionate counterweight placed on the flywheel at a predecided location, r_1 from the crank center and exactly opposite the crankpin. Therefore, from Fig. 12

$$F_r = F_p \cos(\theta_c - \theta_p) = F_s \frac{\cos(\theta_c - \theta_p)}{\cos \theta_p} \tag{18}$$

The force F_r, being the function of θ_c (θ_p replaced by θ_c), cannot be fully balanced by a rotating counterweight at all crank positions. However, it is proposed to place a counterweight on the flywheel, as stated equivalent to $\frac{1}{2}(F_r)_{max}$ to balance F_r. Furthermore, the rotating components, such as the crankpin mass (m_c) and part mass of pitman (m_2) result in unbalanced centrifugal force, F_c.

$$F_c = (m_c + m_2)\omega^2 r \tag{19}$$

which is independent of crank angle and can be balanced for all crank positions by placing a suitable mass opposite the crankpin at a predetermined distance. Therefore, the total force of counterweight F_{cw}

$$F_{cw} = F_c + \tfrac{1}{2}(F_r)_{max} \tag{20}$$

The counterbalance mass corresponding to F_{cw} is determined as follows.

2.5.2.5 The $(F_r)_{max}$ and Counterbalance Mass

In balancing the reciprocating-action cutters, it is necessary to determine $(F_r)_{max}$. This is

$$\frac{dF_r}{d\theta_c} = 0 \tag{21}$$

Now,

$$F_r = F_s \frac{\cos(\theta_c - \theta_p)}{\cos \theta_p}$$

From Eqs. (14) and (15)

$$F_r = (m_s + m_1)\omega^2 r \left[\cos \theta_c + K \sin \theta_c - \frac{Kr}{m} \cos 2\theta_c\right][\cos \theta_c + \sin \theta_c \tan \theta_p]$$

Substituting

$$\tan \theta_p = \frac{r \sin \theta_p + m}{(L^2 - m^2)^{1/2}} = \frac{Kr}{m} \sin \theta_c + K \tag{22}$$

We obtain,

$$\begin{aligned}
F_r &= (m_s + m_1)\omega^2 r\left[\cos\theta_c + K\sin\theta_c - \frac{Kr}{m}\cos 2\theta_c\right]\\
&\quad\times\left[\cos\theta_c + \sin\theta_c\left(\frac{Kr}{m}\sin\theta_c + K\right)\right]\\
&= (m_s + m_1)\omega^2 r\left[\cos^2\theta_c + K\sin\theta_c\cos\theta_c - \frac{Kr}{m}\cos\theta_c\cos 2\theta_c\right]\\
&\quad + (m_s + m_1)\omega^2 r\left[\sin\theta_c\cos\theta_c + K\sin^2\theta_c - \frac{Kr}{m}\sin\theta_c\cos 2\theta_c\right]\\
&\quad\times\left(\frac{Kr}{m}\sin\theta_c + K\right)\\
&= (m_s + m_1)\omega^2 r\left[\cos^2\theta_c + K\sin\theta_c\cos\theta_c - \frac{Kr}{m}\cos\theta_c\cos 2\theta_c\right]\\
&\quad + (m_s + m_1)\omega^2 rK\left[\sin\theta_c\cos\theta_c + K\sin^2\theta_c - \frac{Kr}{m}\sin\theta_c\cos 2\theta_c\right]\\
&\quad + (m_s + m_1)\frac{\omega^2 r^2 K}{m}\left[\sin^2\theta_c\cos\theta_c + K\sin^3\theta_c - \frac{Kr}{m}\sin^2\theta_c\cos 2\theta_c\right]
\end{aligned}\tag{23}$$

Differentiating Eq. (23) w.r.t θ_c and putting it equal to zero

$$\begin{aligned}
\frac{dF_r}{d\theta_c} &= -\sin 2\theta_c + K\cos 2\theta_c + 2\frac{Kr}{m}\cos\theta_c\sin 2\theta_c + \frac{Kr}{m}\sin\theta_c\cos 2\theta_c\\
&\quad + K\cos 2\theta_c + K^2\sin 2\theta_c + 2\frac{K^2 r}{m}\sin\theta_c\sin 2\theta_c - \frac{K^2 r}{m}\cos\theta_c\cos 2\theta_c\\
&\quad - \frac{Kr}{m}\sin^3\theta_c + \frac{Kr}{m}\sin 2\theta_c\cos\theta_c + 3\frac{K^2 r}{m}\sin^2\theta_c\cos\theta_c\\
&\quad + 2\frac{K^2 r^2}{m^2}\sin^2\theta_c\sin 2\theta_c - \frac{K^2 r^2}{m^2}\sin 2\theta_c\cos 2\theta_c = 0
\end{aligned}$$

On simplification and rearranging the terms,

$$\begin{aligned}
&\left[\frac{Kr}{m}\sin\theta_c - \frac{K^2 r}{m}\cos\theta_c - \frac{K^2 r^2}{m^2}\sin 2\theta_c + 2K\right]\cos 2\theta_c\\
&+\left[3\frac{Kr}{m}\cos\theta_c + 2\frac{K^2 r}{m}\sin\theta_c + 2\frac{K^2 r^2}{m^2}\sin^2\theta_c + K^2 - 1\right]\sin 2\theta_c\\
&+\frac{Kr}{m}\left[3K\cos\theta_c - \sin\theta_c\right]\sin^2\theta_c = 0
\end{aligned}\tag{24}$$

The minimum and maximum values of crank angles (θ_{min} and θ_{max}) satisfying Eq. (24) have been obtained by developing a suitable program and setting the specific values of *r*, *m*, and *L*. These are summarized in Table 5. Once we know the values of crank angles, θ_{min} and θ_{max} from Table 5, two values of $(F_r)_{max}$ can be determined from Eq. (23) and the maximum value of F_r is selected for balancing the rotating component.

Table 5 Minimum and Maximum Crank Angles for $(F_r)_{max}$

Crank radius r (cm)	Level of cutter-bar from crank center, m, (cm)	Minimum and maximum crank angles (°) — Pitman length, L (cm): 50		60		70		80		90		100	
		θ_{min} (°)	θ_{max} (°)	θ_{min} (°)	θ_{max} (°)	θ_{min} (°)	θ_{max} (°)	θ_{min} (°)	θ_{max} (°)	θ_{min} (°)	θ_{max} (°)	θ_{min} (°)	θ_{max} (°)
3.8	30	26.802	187.926	21.694	183.541	18.134	180.726	15.539	178.746	13.576	177.268	12.046	176.120
		(0.750)[a]		(0.577)[a]		(0.474)[a]		(0.405)[a]		(0.354)[a]		(0.314)[a]	
	35	31.377	193.181	25.364	187.371	21.207	183.772	18.175	181.290	15.879	179.461	14.087	178.051
		(0.980)[a]		(0.718)[a]		(0.577)[a]		(0.487)[a]		(0.422)		(0.374)[a]	
	40	36.353	199.242	29.159	191.535	24.340	186.996	20.848	183.942	18.208	181.725	16.148	180.031
		(1.333)[a]		(0.894)[a]		(0.696)[a]		(0.577)[a]		(0.496)[a]		(0.436)[a]	
4.1	30	27.129	187.806	21.341	183.411	18.352	180.602	15.712	178.631	13.714	177.162	12.158	176.022
	35	31.678	193.115	25.638	187.268	21.437	183.663	18.363	181.185	16.032	179.957	14.213	177.957
	40	36.607	199.238	29.422	191.465	24.573	186.907	21.046	183.849	18.373	181.633	16.286	176.943
4.4	30	27.456	187.684	22.242	183.279	18.572	180.476	15.887	178.514	13.855	177.055	12.272	175.924
	35	31.975	193.047	25.913	187.164	21.668	183.553	18.553	181.078	16.187	179.261	14.341	177.864
	40	36.857	199.235	29.684	191.395	24.808	186.816	21.245	183.754	18.540	181.541	16.425	179.855
4.7	30	27.782	187.559	22.519	183.146	18.796	180.349	16.065	178.397	13.997	176.948	10.314	175.825
	35	32.269	192.978	26.188	187.058	21.902	183.442	18.745	180.971	16.345	179.160	14.470	177.770
	40	37.103	199.233	29.944	191.324	25.043	186.724	21.446	183.659	18.708	181.448	16.566	179.767
5.0	30	28.107	187.432	22.799	183.011	19.022	180.222	16.245	178.280	14.142	176.840	12.505	175.726
	35	32.560	192.909	26.463	186.951	22.137	183.330	18.939	180.862	16.504	179.058	14.602	177.675
	40	37.344	199.232	30.203	191.252	25.279	186.632	21.649	183.563	18.879	181.354	16.709	179.678
5.3	30	28.430	187.303	23.080	182.874	19.251	180.093	16.249	178.161	14.290	176.732	12.625	175.627
	35	32.847	192.838	26.738	186.842	22.374	183.216	19.136	180.753	16.666	178.956	14.735	177.580
	40	37.581	199.231	30.461	191.179	25.516	186.538	21.853	183.465	19.051	181.259	16.854	179.588
5.6	30	28.752	187.172	23.363	182.735	19.483	179.962	16.615	178.042	14.439	176.623	12.747	175.527
	35	33.131	192.767	27.012	186.732	22.612	183.101	19.334	180.643	16.830	178.853	14.870	177.484
	40	37.813	199.231	30.716	191.105	25.752	186.443	22.058	183.367	19.225	181.164	17.000	179.498

[a] $k = \dfrac{m}{(L^2 - m^2)^{1/2}}$; $\omega = 94.25$ rad/s; $m_s = 4$ kg; $m_1 = 1.4$ kg.

Because F_{cw} in Eq. (20) would be a fixed quantity and independent of crank angle, the remaining unbalanced force F_{uf}, cannot be balanced at any instant and is given by Eq. (25).

$$\begin{aligned} F_{uf} &= (F_c + F_r) - [F_c + {}^1\!/_2\,(F_r)_{max}] \\ &= (F_c + F_r) - F_{cw} \end{aligned} \tag{25}$$

The balancing of rotating components at the crankpin end, described by Raney (52), is given in Fig. 13. The designer, therefore, is required to determine the countermass, m_3 to balance F_{cw}, and this is calculated using Eq. (26):

$$m_3 = \frac{(m_c + m_2)\omega^2 r + {}^1\!/_2\,(F_r)_{max}}{\omega^2 r_1} \tag{26}$$

where, r_1 refers to the location of a counterbalance mass (m_3) for its placement opposite the crankpin.

2.5.3 Vertical Conveyor–Reaper–Windrower

Reaper normally refers to a machine for harvesting the grain crop, and a *windrower* is for fodder crops. The windrower normally cuts the crop and leaves the stocks in a windrow for drying and curing. The term reaper–windrower is used to emphasize that the machine cuts the grain crops and lays them in the form of a windrow with panicles on one side for easy manual picking and binding.

The reaping unit of a cereal crop harvester is conventionally equipped with reciprocating-action cutters that are mounted either in front of a tractor and powered from PTO through V-belts and telescopic shaft, or attached in front of an engine and powered through a gear box and V-belts. Either type of harvester conveys the cut crops in erect condition during operation and shifts them vertically along the side to lay them uniformly at the right

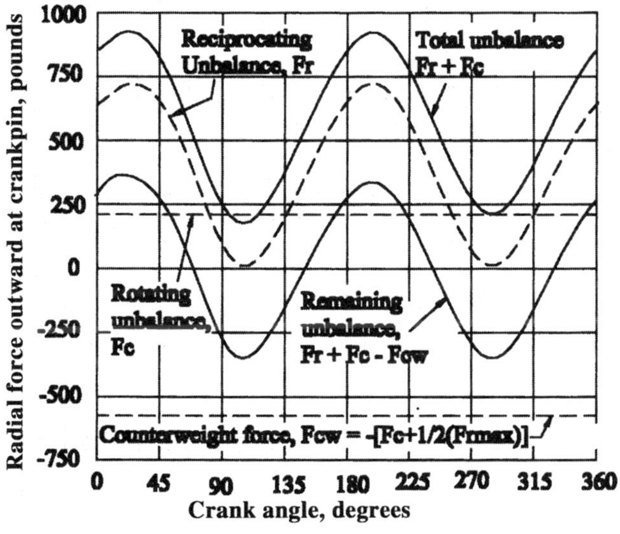

Fig. 13 Unbalance radial forces at crankpin of 7 ft (2.1 m) reciprocating-type mower. (From Ref. 52.)

side for manual picking and binding. Tractor front-mounted vertical conveyor–reaper–windrowers are commonly found in large farms of developed and underdeveloped countries, including India, because of better visibility, saving of time and energy, and ease of maneuverability with a hydraulic control system. On the other hand, a self-propelled, vertical conveyor–reaper, an addition in the farming community, was been introduced about 15 years ago, and it is now popular among the semimedium- and medium-class farmers. The cutting and conveying mechanisms of either type of reaper windrower are identical in principle, and are the outcome of joint effort of Chinese Academy of Agricultural Mechanization Sciences (CAAMS) and the Agricultural Engineering Department, International Rice Research Institute (IRRI), the Philippines.

In February 1982, the CAAMS–IRRI, in the Philippines, developed a 1.0-m reaper as an attachment to 2.2-kW (3-hp) gasoline engine-powered tiller for harvesting paddy crop. The CAAMS–IRRI 1.0-m reaper–harvester (Fig. 14) is a lightweight (95-kg total, 40-kg reaper), single-speed, walking-type, self-propelled, compact machine with an average capacity of 2.4 ha/day. It was developed to fill the need for a faster and more efficient method of harvesting the rice crop (and is also used for wheat). It is simple in operation and convenient to service and maintain. The CAAMS–IRRI reaper consists of a frame, front sheet, cutter-bar assembly, gathering header assemblies that include a star wheel, two flat belts with lugs, vertical conveyors, hitch frame attachment with an adjustable skid, V-belt, and roller chain drives. The reaper is driven from the PTO pulley of the tiller

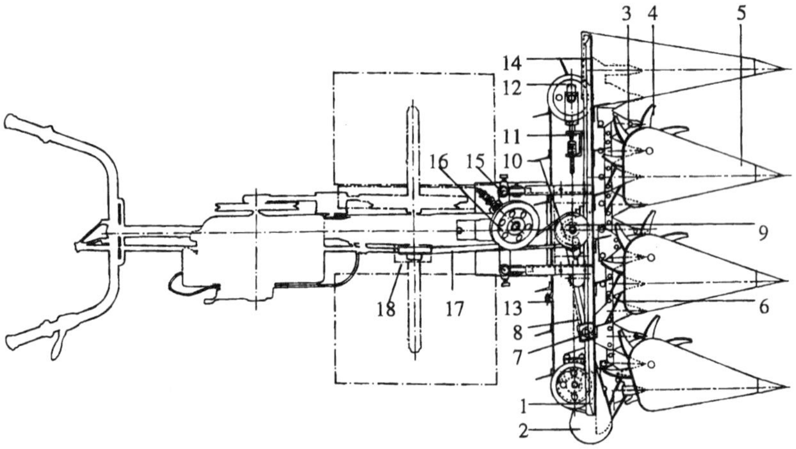

Fig. 14 Schematic drawing of a CAAMS–IRRI 1.0-m reaper windrower:

1. Main frame
2. Discharge plate
3. Cutter-bar assembly
4. Star wheel
5. Gathering header
6. Pressure spring
7. Eccentric crank and pitman
8. Roller chain drive
9. Main-drive shaft
10. Reaper input pulley
11. Flat-belt tensioner
12. Flat-belt pulley
13. Flat-belt conveyor
14. Front sheet metal
15. Hitch frame attachment
16. Inclined spring-loaded idler
17. Quarter-turn V-belt drive
18. PTO pulley

(From Operator's Manual of the Reaper. Issue 1; Aug 1982.)

to the reaper input pulley through a quarter-turn V-belt drive. A springs loaded belt idler serves as tensioner and guides the V-belt to track smoothly during operation. The input pulley drives the cutter-bar assembly through the pitman rod and eccentric mechanism. The main shaft also drives the conveyor lugged belts by means of sprockets and a roller chain. The lugs on the flat-belt conveyor mesh with the star wheels and cause them to rotate. During the reaping operation, the gathering headers guide the crop to the star wheels that hold the crop while it is cut, then guide the crop into the double vertical flat-belt conveyor. The pressure springs hold the crop against the conveyor until it is discharged vertically into a new windrow at the right side of the reaper for manual picking and tying of crops with straw bunch. During operation, the vertically conveyed crops are laid on the ground in a clear windrow maintaining the direction of tillers perpendicular to the direction of travel, and none of the panicles come in contact with the moving parts of the machine. It is thus apparent that the operator always has to take a left turn with the reaper during harvesting.

The CAAMS–IRRI 1.0-m reaper has a pair of 450-mm–diameter, and 250-mm–wide cage wheels, but it is not provided with differential and brake systems. This reaper performs the cutting and windrowing operations very successfully during its lengthwise travel in high-density paddy and wheat fields. The only difficulty faced with this type of reaper is that it is difficult to make a proper turn and, therefore, reaping at the corners is never perfect. Research workers of various institutions, entrepreneurs, and private organizations have tested the reaper intensively and modified it accordingly to suit the local requirements without altering the cutting and windrowing mechanisms. Such reapers, with different designs of chassis and power transmission systems, powered by a 2.2- to 3.8-kW diesel/gasoline engine are manufactured by numerous companies in India and abroad.

The front-mounted, power–tiller-operated riding-type Chinese vertical conveyor reaper (49) was probably the first successful reaper that used the reciprocating-type cutter-bar of 1.6-m width. The design and construction of each component was the same as that of the CAAMS–IRRI 1.0-m reaper. This machine was supplied to Indonesia, the Philippines, and Thailand for tests and feasibility trials under different field and agroclimatic conditions, such as dry field transplanted, wet field transplanted, and upland field direct seeded. The performance of the machine was very good in dry field and in upland field conditions where the crop was broadcasted, directly seeded, or transplanted. The average field capacity was 0.26 ha/h, little less than the expected value of 0.33 ha/h, probably because of more time loss in turning and positioning the machine (49). The machine capacity was increased with plot size, and the recommended minimum plot size was stated as 0.13 ha, below which the capacity fell very abruptly. In wet fields, the performance of the machine was not very satisfactory. The control and mobility of the machine were affected by the irregularity of the hard pan. The machine, even with the cage wheels, used to bog down and trampled over the crop area. Hence, except for the difficulty of operation in a wet field, there was no specific problem in cutting and windrowing of paddy or wheat crops.

In paddy-growing countries, power tillers of 2.2- to 9.0-kW capacity are very common with the farmers for rototilling and puddling operations. To make the power tiller a more versatile machine, the use of a front-mounted vertical conveyor–reaper–windrower based on the CAAMS–IRRI design has been reported (1,2,53,54) as very encouraging both in paddy as well as in wheat fields. The constructional details of a power tiller front-mounted vertical conveyor–reaper–windrower is shown in Fig. 15. In such a reaper–windrower, the drive to the crank mechanism; hence, to the cutter-bar, is provided from

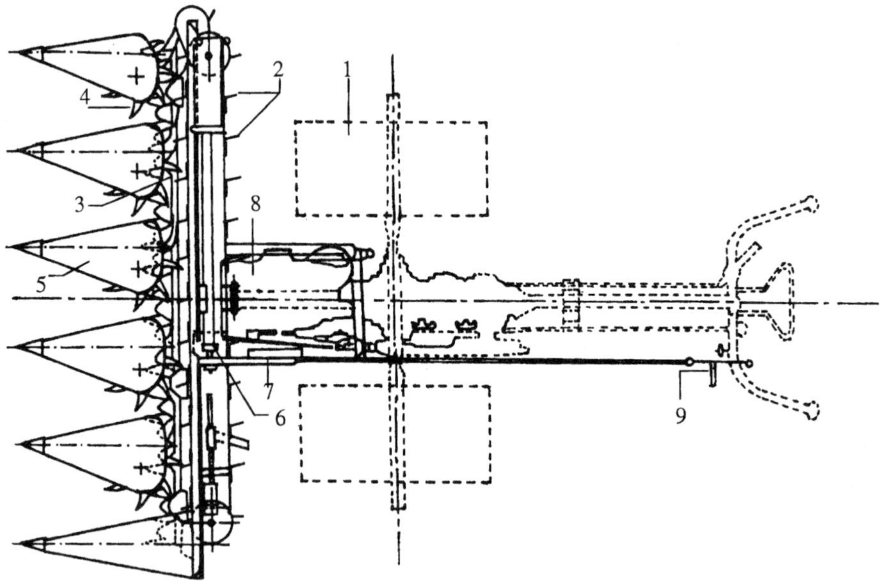

Fig. 15 Plan view of power tiller front-mounted vertical conveyor reaper windrower:

1. Pneumatic wheel
2. Belt lugs
3. Pressure spring
4. Star wheel
5. Gathering header
6. Main-drive pulley
7. Engine-drive pulley
8. Engine
9. Clutch lever

the engine pulley by V-belts and pulleys through a manually controlled clutch arrangement. The crop stubble height is adjusted by regulating the height of a gauge wheel placed between the power tiller engine and the back of the cutter head. Such an attachment to the power tiller has been commercialized in most of the countries, and the machines are being used by the farmers in harvesting cereal crops. The overall performance of the reaper, as reported by the research workers, is presented in Table 6.

The construction and design details of the 1.5- 2.4-m cutter-bar of a tractor front-mounted, vertical conveyor–reaper–windrower (Fig. 16) are similar to those of the self-propelled reaper, with an exception that it is a large-capacity machine and has been designed for heavy-duty work. Furthermore, the rotational motion to the input shaft of the crank mechanism is given from the tractor PTO through a double V-pulley connecting a long drive shaft, telescopic shaft, and bevel gear. The input shaft also drives two canvas conveyor lugged belts, one closed to the cutter-bar and the other 30 cm above the ground, by a heavy-duty roller chain and sprockets. The lifting and lowering of the header unit is controlled with the help of the tractor's hydraulic system through a pair of steel ropes, one end of which is connected to the each lower link and the other end to the bracket hook attached with the header unit, and each is guided through two separate grooved pulleys. It has been reported that the vertical conveyor–reaper–windrower with complete attachment and accessories for a 26-kW tractor is more convenient and economical in harvesting paddy or wheat crops than either owning a self-propelled combine harvester or manual harvesting and picking of crop followed by threshing with a power thresher. The

Table 6 Performance of Three Common Types of Vertical Conveyor–Reaper–Windrower

Sl. no.	Type of machine	Suitable for crops	Wt. of machine without power source, (kg)	Cutterbar speed (strokes/min)	Designed forward velocity (km/h)	Range of observed parameters					Ref.
						Forward speed (km/h)	Field efficiency (%)	Labor for reaping collecting, and binding (man-h/ha)	Fuel consumption (L/ha)	Total gram loss (shattering and windrowing) (%)	
1	Self-propelled 1.0-m–cutter-bar with 4-row crop dividers	Unlodged paddy and wheat	84	1200	3.4	3.06	60–64	32–35	1.3–1.4	2.1–4.5	—
2	Power tiller front-mounted 1.6-m Chinese vertical conveyor reaper with 6-rowcrop dividers	Paddy and wheat	88	1200	3.6	3.15–3.84	50–63	30–37	1.5–2.8	0.5–2.0	49
3	Power tiller front-mounted 1.6-m–cutter-bar with 6-row crop dividers	Unlodged paddy and wheat	85	1224	3.6	2.7–3.2	51–64	25–38	1.6–3.0	0.5–1.8	2
4	Power tiller front-mounted 1.6-m–cutter-bar with 7-row crop dividers	Barley and unlodged paddy, wheat	87	1224	2.9	2.2–2.9	67–85	38–48	1.4–1.8	3.8–8.0	1
5	Tractor front-mounted 2.05-m–cutter-bar with 9-row crop dividers	Barley and unlodged paddy, wheat	164	1550	3.6	2.2–2.9	56–73	42–49	3.7–4.8	3.9–7.2	1
6	Tractor front-mounted 2.01-m–cutter-bar with 8-row crop dividers (Farm Machinery Institute, Pakistan)	Unlodged paddy and wheat	160	1500	3.6	3.3	59	43	3.5	3.31	55

Fig. 16 Tractor front-mounted vertical conveyor–reaper–windrower: observe the crop windrow. (Courtesy of Punjab Agricultural University, Ludhiana, India.)

functioning of the tractor front-mounted, reaper–windrower reported by various workers (3,55,56) is most encouraging. Such types of reaper–windrower for attachment to a 26-kW tractor are manufactured by many companies in India and other developing countries. It is thus apparent that the various forms of vertical conveyor–reaper–windrower would contribute to the mechanization of cereal crop harvesting under the prevalent sociotechno-economic conditions and help in increasing the production and the productivity of the crop.

3 THRESHING

The prime objective of the threshing process is to detach the sound or undamaged grain kernels from the plants. In some crops, it also involves the removal of kernels from the protective cover, called husk or pod. It is achieved by striking, treading, squeezing, tearing, and rubbing actions or by combinations of these methods. Traditionally, threshing is performed by treading the grain under the hooves of animals, striking the grains with sticks, beating them over a log of wood or bamboo grating, or over a ladder, or using a stationary power thresher of specific design for each crop, or between the rasp-bar and concave of a combine. The machine used for the purpose of grain detachment and separation is called a thresher and was introduced in India about 1960.

Timely preparation of seedbed and sowing of seasonal crops immediately after harvest are the prime desire of all the farmers. To meet these requirements, threshing either needs to be delayed or the farmers are required to use a stationary thresher. Delayed threshing will cause not only spoilage of grain, but will also increase the breakage percentage during milling. Hence, the use of stationary thresher, a faster method of grain detachment and separation for cereal and pulse crops has become very common among all groups of farmers. Stationary threshers are conventionally used when the fields are small and conditions are not favorable for combine operation. For speedy and effective threshing, pedal- and power-operated threshers are used, respectively, by the small to marginal- and medium-class farmers. With the use of these threshers, threshing of most crops and separation of grains and seeds from unthreshed debris has become mechanized in all developed and underdeveloped countries.

3.1 Traditional Threshing

Conventionally, harvested paddy and wheat crops are either dried in the field or on a cemented floor for 3–5 days to bring down the moisture content from 27–40% to 15–20% when threshing operations are carried out. Threshing of immatured or moist grain would result not only in more breakage, but would also require higher impact force for grain detachment, and cleaning and grain separation from leftovers (broken stalks and chaffs) would become difficult, requiring more impact power. An average laborer can thresh 15–22 kg/h of grain by hand-beating, or 110–140 kg/h by treading the grain under the feet of animals. In either method, sound grains are separated using a winnower operated by human or by a power unit or dropping the mix against the medium to high natural wind. Thus, the separation of sound grain and manual bagging require additional labor, making the entire process tedious, time-consuming, and labor-intensive. It thus brings down the overall threshing ability of a laborer to 12–18 kg/h in manual threshing and 80–120 kg/h by treading the grain under the feet of animals. Therefore, traditional methods of threshing are considered a slow process of grain detachment. Small and marginal farmers of South and Southeast Asia knowingly follow these methods because of capital constraint and limited production of cereal and pulse crops.

3.2 Pedal Threshing

Threshers of different designs and capacities are being manufactured by various manufacturers in all the countries. They are either throw-in or hold-on type. The Japanese-type rotary drum thresher, a hold-on type is the first of its kind developed in Japan for threshing paddy crops. Such a thresher is common with small farmers of paddy-growing countries. It is cheap, compact, and simple in construction. It consists of a threshing cylinder (42 cm in diameter and 40–70 cm long) with wire loops, driving mechanism, and supporting frame. The rotary motion to the drum is given by a crank mechanism from a treadle and two cast iron gears (80 and 20 teeth) with 1:4-speed gain to achieve a cylinder speed between 300 and 375 rpm (6–7.8 m/s). The threshing drum mounted with a large number of wireloops at its periphery and at regular intervals shatters grain by impact and combing actions. Most grains from paddy are detached when the tiller ends of crop bundle is held over the rotating cylinder along the motion direction. One man can thresh about 1.5–2.0 q/day but additional labor is required to separate grain from chaff and other debris. To increase the capacity, farmers engage four to six laborers for threshing over a 1.5- to 2.4-m–long threshing cylinder made locally and operated by a 3.75-kW diesel engine through a flat belt. An additional two- to four men are engaged for the supply of crop bundles near the thresher and grain separation. This method in contrast with manual threshing, therefore, saves time and energy, and at the same time reduces human drudgery to a large extent.

3.3 Power Threshing

The prime functions of threshing units are to detach sound grain kernels and kernels from husk or pod; separate them from broken stalks, leftovers, and chaffs; convey and deliver the sound and unbroken grains or seeds, free from foreign materials, to a delivery outlet and bag the grains. Because operations such as detachment, separation, conveying, and elevating of grains or kernels are done simultaneously and in sequence, a prime-mover with a capacity in the 3.5- to 30.0-kW range is necessary, depending on the crop parameters, type, and size of threshing cylinder and feed rate. Power-driven threshers are very similar to the threshing units used in a combine harvester. The large-sized threshers are mostly throw-in type (Fig. 17) and are provided with a belt conveyor as self-feeder to the threshing cylinder on which the workmen, uniformly and evenly, place the crop material. Throw-in types are conventionally axial-flow threshers that not only detach cereal and pulse grain or seeds by impact and combing actions, but also break the leftover straw during their flow axially against the louvers provided on the inner surface of the cylinder cover. Effective removal of the grain from tillers and seed separation from husk occur in an axial-flow thresher, rather than in low-capacity hold-on–type threshers (Fig. 18) in which crop panicles are held manually for some time against the rotating threshing drum. The leftovers in the form of *bhusa* (broken straw and chaffs) obtained from axial-flow thresher are used as cattle feed or left in the field as fertilizer, whereas the whole crop straw obtained from the hold-on threshers is used as the raw material for industries manufacturing straw boards, *Khas khas*, ropes, and are also used to thatch roofs, and as animal feed.

Detachment of seeds and breaking of pods are primarily achieved by impact action, wherein squeezing, rubbing, combing, and tearing actions are also associated between the

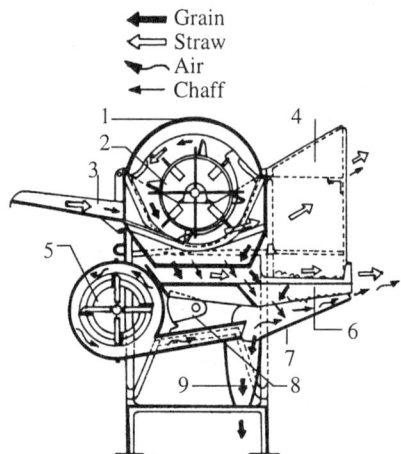

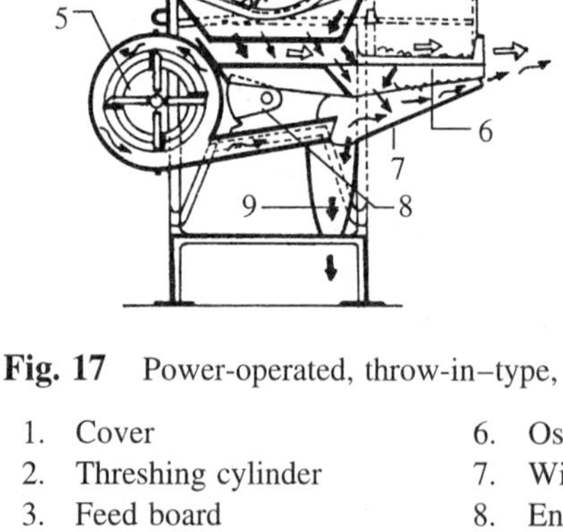

Fig. 17 Power-operated, throw-in–type, stationary thresher:

1. Cover
2. Threshing cylinder
3. Feed board
4. Straw discharge guard
5. Blower
6. Oscillating screen
7. Windboard
8. Engine
9. Grain discharge chute

(From Ref. 57.)

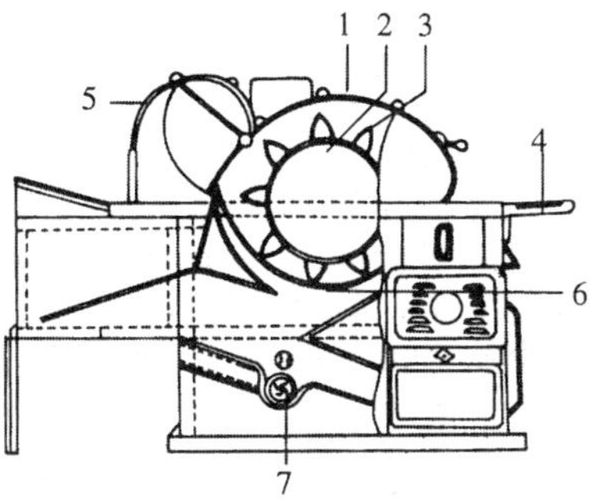

Fig. 18 Power-operated, hold-on–type thresher:

1. Cover
2. Threshing cylinder
3. Threshing element
4. Feeding table
5. Chaff outlet
6. Concave grate
7. Grain auger

(From Ref. 57.)

threshing elements and concave grating. Unlike manual threshing, the impact actions in threshers are simulated by rotating the cylinder(s) having threshing elements of different designs at their periphery and along the drum length. When such cylinders are rotated at high speed (7.5–30.0 m/s), most grains or pods from the straw, which is moving relatively slowly, are shattered by the repeated impacts. The remaining grains are threshed by the combing, rubbing, squeezing, and tearing actions as the crops accelerate and pass through the narrow passages between the cylinder element tip and the concave, as well as with the louvers. During the separation process, it is assumed that, theoretically, a sheet of crops giving cushioning effect to the grains rolls over the cylinder when impact shatters most grains. Increased crop thickness or high feed rate will, therefore, relatively reduce grain damage and separation from panicle, and vice versa. Proper selection and adjustment of design and system parameters (discussed later) for effective threshing will primarily depend on the moisture content and feed rate of the selected crop.

3.4 Threshing Cylinder and Concave

The threshing cylinder and the matching concave are the heart of a combine harvester and stationary power thresher. Most commonly used threshing cylinders are provided with wire-loop, spiked-tooth, flat-bar, rasp-bar, or angle iron bar as elements called the threshing elements of the cylinder. Hammer mill- and serrated flywheel-types are also used in axial-flow threshers. The constructional details as well as the operating principles of each type of threshing cylinder are discussed in the following.

3.4.1 Wire-Loop and Spike-Toothed Cylinders

The elements, such as wire-loops and spike-tooth or studs, are placed at regular intervals along the drum length, but staggered in subsequent rows around its periphery to provide impact, rubbing, and combined actions to the crop during its flow through the restricted passages. The concave grating of steel rods and flats covers one-third to five-twelfth of drum periphery. It is pivoted at the rear, whereas the front is made adjustable vertically to control clearance between them. A typical 3.75-kW thresher may have 650 × 480-mm–diameter threshing cylinder with 40–50 beaters (6-mm rod for wire-loop or 18- to 20-mm side square/diameter bar as a spike or stud), designed to be operated at 20.5 and 24.0 m/s for paddy and wheat crops, respectively. The total number of rows of wire-loop (for paddy) or spike-tooth (for wheat) on the drum periphery depends on the crop, moisture content, feed rate, and peripheral velocity. Rubber or plastic-lined loops, spike-tooth, and corrugated spikes, or studs as beaters are also used for difficult threshing conditions.

3.4.2 Rasp-Bar Cylinders

The rasp-bar-type threshing cylinders are provided with left and right rasps, or a raised configuration alternately along its length and at regular intervals over the drum periphery to balance the side thrust-with numbers varying between six and ten (even number). The wrap angle of the stationary rasps of the concave grate is between 110° and 125° of the cylinder. Grain separation is obtained between the corrugated bars and concave grate. Sometimes, a flat-bar grate, in place of corrugated grates, is used as the concave. Rubbing and tearing actions are primarily responsible for grain detachment and breakage of straw. Apart from groundnut, rasp-bar cylinders are adopted for a wide variety of crops and threshing conditions (22) because the actions of rasps on crops are mild. It is recommended for pulse crop threshing in stationary threshers and cereal crop threshing at high moisture in a straight-through combine harvester.

3.4.3 Flat-Bar Cylinders

The construction of a threshing cylinder, with or without corrugations in flat-bars, is similar to that of a spike-tooth cylinder of a stationary thresher, but is made more rigid and strong for threshing wheat crops. Besides the use of standard concave grating of steel rods and flat-bars, as in other threshers, the louvers are placed on the inner surface of the cover for axial-flow of crop materials, effective straw breakage, and grain separation. The grain detachment and straw breakage are primarily due to impact, and successive grain separation is due to rubbing and tearing actions against the concave grate and louvers.

3.4.4 Angle-Iron-Bar Cylinders

A total of six to ten angle-iron bars in place of rasp-bars are mounted rigidly and regularly over the drum periphery to form the cylinder, and a concave grating of steel rods and flats is provided to separate grains from panicles, primarily by impact. Subsequent grain detachment during its flow through a restricted passage occurs by rubbing and tearing actions in axial-flow threshers. This type of threshing cylinder is easily adaptable to a wide variety of crop and threshing conditions, and has less tendency to break up the straw. The angle-bar cylinders, therefore, are recommended for threshing millets, barley, and other small seeds, and also in straight-through combines.

3.4.5 Hammer-Mill- or Beater-Type Cylinders

Threshers using this type of cylinder are called drummy, hammer-mill-type, or beater-type threshers. They are normally recommended for threshing crops such as wheat, barley, sorghum, and pearl millet. The threshing drum of such threshers consists of beaters mounted on a shaft that rotates inside a concave closed casing that is provided with a cleaning system. Additionally, angle-iron ribs, parallel to the drum axis, are provided at the upper casing for effective grain separation. A typical 5.6-kW drummy thresher may have 650-mm–diameter and 460-mm–wide threshing cylinder with 10–16 beaters designed to rotate at a peripheral speed of 16–20 m/s. The cylinder is rotated counter-clockwise when viewed from the feeding end. Grains are primarily detached by impact action of the beaters, and subsequent separation is obtained by rubbing and tearing actions against the concave grate and angle-iron ribs. The cleaning sieves and an aspirator–blower are provided on a separate shaft. The oscillating sieves placed below the concave receive the threshed material (grains, chaff, and broken straw). The lighter materials during their fall are sucked by the aspirator–blower and thrown at a greater distance. The upper sieve receives thick straw (mostly nodes for wheat) and overflows, while the clean grains are received by the bottom sieve. The lowermost sieve separates broken and fine materials from clean, sound grains of wheat. This type of thresher requires more power and can provide fine bruised wheat straw (*bhusa*).

3.4.6 Serrated Flywheel- or Chaff-Cutter-Type Cylinder

Threshers provided with this type of cylinder are also called serrated flywheel-type, chaff-cutter-type, syndicator- or toka-type threshers that are primarily used to thresh wheat crop that may become wet owing to the postharvest monsoon. Unlike beaters in a hammer-mill-type thresher, the syndicator-type thresher is provided with a flywheel having serrations on its periphery and sides, closed casing, and concave, and ensilage cutter-type chopping blades fitted at regular intervals on the flywheel rim. Through the feeding chute, the whole crop is fed axially into the threshing cylinder. The grain separation is obtained by rubbing actions between the serrated surfaces of the flywheel and stationary concave grate. Additional beaters impart impact. The material makes three-quarter turns axially during which threshing occurs before encountering the chopping knife which cuts the straw into small pieces. A typical 7.5-kW toka thresher may have 900 × 580-mm–sized threshing drum with closed concave and two to four chopping knives. They are designed to operate at an optimum peripheral velocity of 18.2–20.7 m/s. Such threshers are popular among the farmers because they can thresh moist crop and consume less power, provide chopped straw as animal feed, and deliver clean grain separately.

3.5 Factors Influencing Threshing Performance

The prime functions of a thresher is the detachment of sound grain kernels and seeds from their protective cover, free from foreign materials. The performance criteria of a thresher will thus include the percentage of grain threshed (E_t), the percentage of grain damaged (E_d), the percentage of grain unthreshed (E_u), cleaning efficiency (E_c), the percentage of blown grain (E_b), the percentage of grain loss (E_l), and the percentage of recovery (E_r). These are outlined in the Bureau of Indian Standards (BIS–IS: 6284-1975) and RNAM Test Codes (57), defined as follows:

$$E_t = \frac{W_t}{W_A} \times 100 \tag{27}$$

$$= \left[1 - \frac{W_u}{W_A}\right] \times 100$$

$$= (1 - E_u) \times 100 \tag{28}$$

$$E_d = \frac{W_d}{W_A} \times 100 \tag{29}$$

$$E_c = \frac{W_c}{W_m} \times 100 \tag{30}$$

$$E_b = \frac{W_b}{W_A} \times 100 \tag{31}$$

$$E_l = \frac{W_l}{W_A} \times 100 \tag{32}$$

$$E_r = \frac{W_r}{W_A} \times 100 \tag{33}$$

where;

W_t = quantity of threshed grain obtained per unit time from all outlets.
W_A = total grain input per unit time by weight.
W_u = quantity of unthreshed grain per unit time obtained from all outlets.
W_d = quantity of damaged grain collected at all outlets per unit time.
W_c = quantity of whole grain per unit time obtained from the main grain outlet.
W_m = quantity of whole material per unit time at the main grain outlet.
W_b = quantity of whole grain collected at chaffed-straw outlet per unit time.
W_l = quantity of whole grain, damaged grain, and unthreshed grain per unit time at chaff outlet and straw outlet (hold-on) or chaffed-straw outlet (throw-in) and scattered grain per unit time.
W_r = quantity of threshed grain (whole or damaged) per unit time collected at the main grain outlet.

All threshers are required to be evaluated for their performance under the conditions recommended by the manufacturer and at the specified speed of the thresher. Three different feeding rates, including optimum feed rate, as specified by the manufacturers for each crop, must be considered. Each test is required to last for at least 30 min and should be replicated thrice. At the end of each test, various weights of the samples at different outlets, running time, and feed rate are taken for computation of thresher performance using Eqs. (27)–(33).

The performance of a thresher is affected by several factors. These factors may be divided into three categories: those related to crop (crop type and its moisture content); those related to design (type of elements, number of rows of elements, concave length, and cylinder diameter); and those related to the system (feed rate, peripheral velocity, and cylinder–concave clearance). Most of these factors show a significant effect on the threshing performance. Based on the work of Arnold et al. (4,58,59) for a rasp-bar-type cylinder with an open grate concave, Wieneke (60) developed a generalized graphic representation of performance characteristics (Fig. 19) of a thresher involving nine variables. It may be observed that Wieneke (60) in place of percentage grain threshed (E_t), considered percentage of grain separated through concave (E_s), which is defined as,

$$E_s = \frac{W_s}{W_A} \times 100 \tag{34}$$

where W_s represents grain quantity collected after flow through the concave and, therefore, includes unthreshed grain and blown grain. So, for a given crop condition and thresher design parameters, system variables (Wieneke called them adjustments and operational

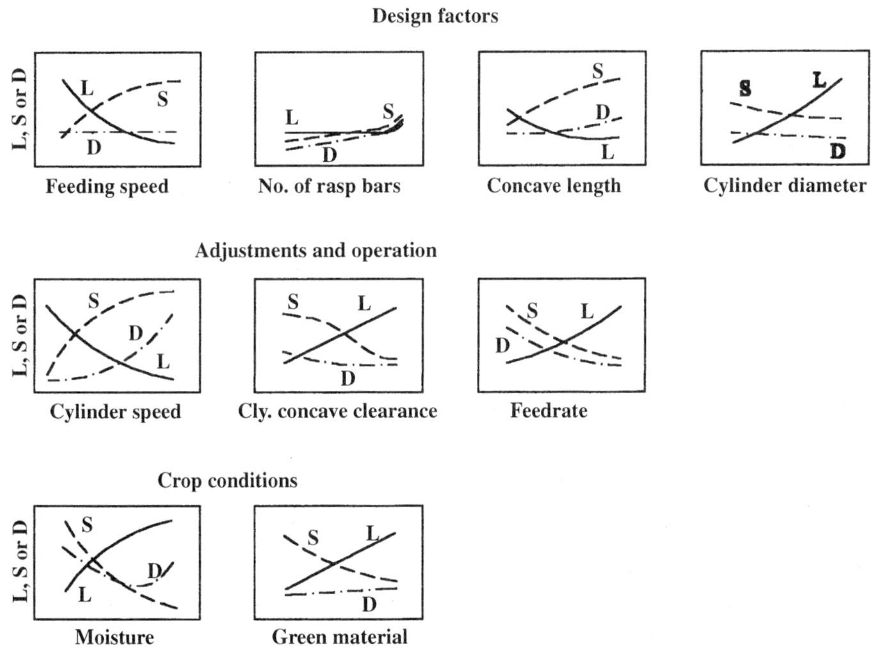

Fig. 19 General functional characteristics of a rasp-bar cylinder and open concave: L, unthreshed grain loss; D, grain damage; S, percentage grain separated through concave. (From Ref. 60.)

parameters) need to be negotiated to obtain maximum grain separation (E_s), minimum grain damage (E_d), and percentage grain unthreshed (E_u). In general, for threshing a crop with a low moisture content, increased peripheral velocity and decreased concave clearance will result in a thoroughness of threshing that has higher values of E_s, and E_t, and E_d, and lower values of E_u will be observed at a reduced feed rate.

Many researchers have evaluated the performance of threshers for a wide range of crop–cylinder–system variables. The optimum and near optimum range of cylinder–concave parameters for cereal and pulse crops are summarized in Table 7. It is apparent that there is no specific cylinder–concave configuration that can be recommended for a crop for the best threshing effectiveness. Conversely, for the same cylinder–concave configuration, more than one crop can be effectively threshed following necessary adjustments within the threshing system. Therefore, for a thresher, proper adjustments of peripheral velocity, concave clearance, and number of rows of threshing elements will be necessary to thresh a wide variety of dried crops. In evaluating a thresher, energy input and peak power requirements are given less weight than its characteristic performance for obtaining a maximum quantity of sound clean grains or seeds from cereal and pulse crops.

3.6 Grain or Seed Separation

Separating sound grain or seed from chaff, brokens, and straw is the next important function of a thresher. The process begins as the grains, chaff, and straw first leave the concave openings. During operation, a large portion of threshed grain or seed along with some chaff and straw is separated from the straw by the threshing unit. Rotary separators and oscillatory straw carriers are the common forms of grain separators adopted in all stationary

Table 7 Range of Cylinder–Concave Parameters in Threshers for Cereal and Pulse Crops

Crop	Peripheral velocity (m/s)	Cylinder configuration	Concave configuration	Wrap angle degree	Concave clearance (mm)	Ref.
Paddy	20.0	Spike-tooth	Rod and flat-bar grating	202	6.0–10.0	46
	11.7–13.0	Spike-tooth	Rod and flat-bar grating and louvers	170	10.0–13.0	61
	18.6–22.4	Wire-loop	Wire-loop	105	3.0–9.0	62
	18.6–22.4	Rasp-bar	Rasp-bar	105	3.0–9.0	62
	20.0–25.0	Rasp-bar	Rasp-bar	—	5.0–9.0	22
Wheat	10.5	Spike-tooth	Rod and flat-bar grating and louvers	170	9.0–12.0	61
	17.0–18.4	Spike-tooth	Rod and flat-bar grating	170	13.0–19.0	63
	20.0–24.0	Rasp-bar (double cylinder, one above the other)	Rod and flat-bar grating	200 and 160	14.0–16.0	64
	25.0–30.0	Rasp-bar	Rasp-bar	—	5.0–12.0	22
Barley	22.5–27.5	Rasp-bar	Rasp-bar	—	6.0–12.5	22
Peas	10.0–15.0	Rasp-bar	Rasp-bar	—	7.5–18.0	22
Pigeon pea	9.4–10.5	Spike-tooth	Rod and flat-bar grating and louvers	170	7.0–9.0	61
Chickpea	8.5–10.0	Spike-tooth	Rod and flat-bar grating and louvers	185	30.0–35.0	65
	13.0–14.0	Rasp-bar	Rod and square-bar grating	110	10.0–12.5	8
	13.0–15.7	Spike-tooth	Rod and flat-bar grating and louvers	170	5.0–8.0	61
	13.5–14.3	Plastic-covered peg-tooth	Rod and square-bar grating	110	10.0–12.5	9
	13.5–14.3	Plastic-covered 45°, wire-loop	Rod and square-bar grating	110	10.0–13.0	9
Soybean	11.0–12.0	Rasp-bar	Rod and square-bar grating	180	8.0–13.0	10
	12.5–17.5	Rasp-bar	Rasp-bar	—	7.5–21.0	22
	11.0–12.0	Plastic-covered spike-tooth	Rod and square-bar grating	180	10.0–15.0	10
	5.5–6.5	Spike-tooth	Angle-iron ribs parallel to cylinder axis for upper concave	135 (upper concave)	35.0 (upper concave)	66
			Square-bar grating for lower concave	165 (lower concave)	15.0 (lower concave)	

threshers and combine harvesters. Primary separation of most grains is due to the impeller blower (rotary separator) provided at the other end of the threshing drum or centrifugal blower that blows off the light materials through the chaff outlet. Subsequent separation of the remaining free seeds and unthreshed seeds occur as they are agitated and moved over the oscillating straw carrier. The oscillating straw rack, a one-piece straw carrier, and multiple sections (pieces) of straw walker are commonly used for the separation process and have been accepted widely for grain harvesting and threshing machines.

3.6.1 Straw Walkers

A multiple piece, oscillating straw walker, having three or four narrow sections placed side by side in the machine, is commonly used in recent threshers for grain separation. Multiple throw crankshafts, one at the front and the other at the rear are attached to the individual section. The crank throw of each section is spaced at equiangular spacing ($\pi/2$ or $2\pi/3$ radians) around the circle of motion. The recommended crank speed of the oscillating straw racks or walkers is between 215 and 260 rpm. Either a too high or too low speed of the crank will increase seed losses (22) as spilled grain. During its motion, the straw rack accelerates the straw in rearward and upward directions in one and one-half cycle of rotation whereas in the return stroke, the rack tends to leave the straw in midair for a moment. As a result, the grains heavier than straw fall onto the section near to the discharge end. In the next stroke of the crank, the grains move one step toward the rear. This process continues and separates the grain from straw as they walk over the straw rack or walker.

3.7 Cleaning

Seeds or grains after separation from straw still contain a sufficient quantity of chaff and broken plant residue that passed through the oscillating straw rack or straw walker with the seed. The functions of the cleaning unit are separation of the sound and whole grains, return of unthreshed and partially threshed earhead to the cylinder, and removal of leftovers. Cleaning of grains from such a mixture is either assisted by the oscillating sieving mechanism or by the pneumatic method. The latter utilizes the aerodynamic characteristics of particle mixtures. Cleaning of grain by winnowing, an ancient method, is not really effective and seldom finds its place in threshing and harvesting machines. In all cleaning mechanisms, however, the assistance of air is mainly utilized (a) to remove light materials, (b) to assist in orienting the particles on the perforated sieves, and (c) to move the particles caught in the opening of perforated sieves during the oscillation process.

Further removal of chaff and other debris from grain, and separation of brokens are simultaneously carried out by the cleaning unit, which comprises one or more oscillating sieves having circular perforations of uniform size, a pedal-type conventional fan delivering air at the front and up through the sieves, an extension at the rear of the upper sieve, and augers for conveying the tailings and clean grains to the elevators. For effective removal of light materials, an adjustable air deflector is provided between the sieves and the fan. In the process of sieving, the mixture moves rearward over the perforated surface, horizontally across it, and the seeds or brokens that are smaller than the size and shape of the opening fall down to the next sieve by gravity, when the air blast carries away most of the light materials. Unthreshed and partially threshed earheads, and other materials that pass through the upper sieve and not through the lower sieve (that has smaller openings) are discharged to the tailing auger. The process of cleaning is accelerated by oscillating the sieve(s) continuously and using a fan to deliver an air blast through the sieve(s).

Besides the pneumatic force created by the fan, the efficiency of sieving and, hence, the removal of chaff, is affected by (a) seed parameters—size, shape, and weight of each seed; (b) sieve parameters—size, shape, and spacing of perforations; and (c) system parameters—rated frequency, amplitude, and direction of oscillation of the sieve(s) and their combination. Because of the improper selection of system variables for a selected set of sieves, the seeds are often caught in the openings, thereby reducing the efficiency of sieves. Physical properties of the grain, chaff, and straw accordingly play an important role in the cleaning process by sieves and fan. Feller (67) observed that the clogging rates generally increase more for particles that are irregular and undersized than for the oversized ones. In such a situation, sieving efficiency can be improved if its acceleration is made higher than that from gravity. MacAulay and Lee (68) conducted grain separation studies on oscillating combine sieves at the University of Guelph, Ontario. At low feed rates of wheat, aerodynamic separation of grain from straw occurred over the sieve and the best separation was achieved at a 36° sieve lip angle and at an oscillation frequency between 5.5 and 6.4 c/s. Moreover, increase in feed rate caused mat formation on the sieve that decreased the sieving efficiency and increased the grain loss.

Gorial and O'Callaghan (69) applied the pneumatic technique to analyze the separation of grain from straw and chaff in a vertical air stream. They studied the influence of particle characteristics, air velocity, speed, and the direction of the injection of material on the separation process. A mathematical model describing particle motion was developed to simulate the paths of grain and straw particles moving in an air stream with a view to determining the effectiveness of separation and the level of grain losses. Considering the net force on a particle moving vertically in a fluid as the difference between the drag force and the gravitational force, and the relative velocities of particle in the x and y direction for air velocity v_a, the following mathematical model (69) describing particle motion was developed.

1. The displacement (S_{yf}) for a falling particle in the y-direction;

$$S_{yf} = \frac{v_t^2}{g} \ln \cosh(gt/v_t) \tag{35}$$

2. The displacement (S_{yr}) for a rising particle in the y-direction:

$$S_{yr} = -\frac{v_t^2}{g} \ln \sinh(gt/v_t) \tag{36}$$

3. The displacement (S_x) in the x-direction,

$$S_x = v_a t + \frac{v_t^2}{g} \ln \frac{c}{(c + gt/v_t)} \tag{37}$$

where,

v_t = suspension velocity or terminal velocity of particle
t = time elapsed
g = gravitational acceleration
c = integration constant.

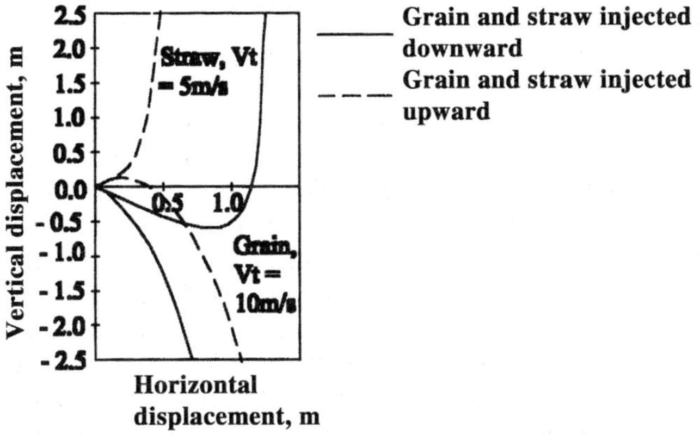

Fig. 20 Trajectory of grain and straw in vertical airstream. (From Ref. 69.)

Gorial and O'Callaghan (69) used the aerodynamic and physical properties of grain and straw to calculate the net force acting on a particle and obtained their trajectories in a vertical airstream (Fig. 20). It was stated that for the downward injection, the particles continue on a downward path, and the heavier grain particles fall nearer the wall than the lighter straw particles, which will move over 1 m from the point of injection before they are entrained in the airstream and carried upward. For an upward injection of grain and straw, on the other hand, the straw always moves upward, requiring less width of column to initiate the separation. The trajectories in a vertical airstream of a mixture of wheat particles, typical of the range observed in a threshed sample, are shown in Fig. 21 for horizontal injection.

On a comparative basis, it is apparent from the foregoing figures that the upward injection of grain and straw mixture in a vertical airstream is best, followed by horizontal and downward injections, for effective separation and minimum grain loss. However,

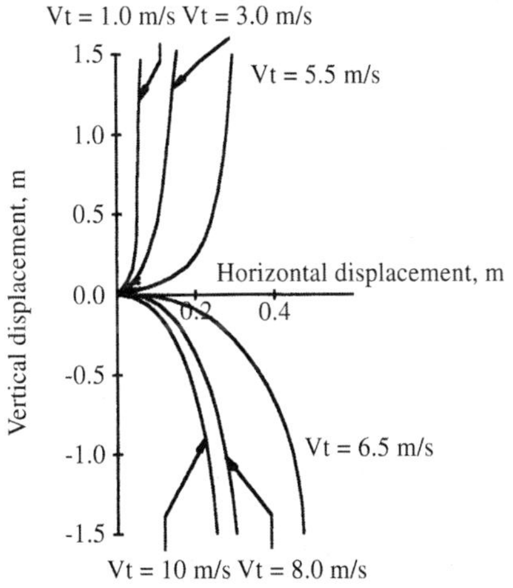

Fig. 21 Trajectory of threshed mixture in a vertical airstream: Injection velocity of particle, 0.5 m/s; air velocity 6 m/s; direction of injection was horizontal. (From Ref. 69.)

Table 8 Aerodynamic and Physical Properties of Grains

Grain	Mass (mg)	Terminal velocity (m/s)	Geom. dia. (mm)	Eqiv. dia. (mm)	Drag coeff. (C_d)	R_e	Sphericity	Shape factor Z_m measured	Z_p predicted
Adzuki	115.60	12.6	5.40	—	0.57	4200	0.79	0.40	0.39
Barley	37.60	7.5	4.08	3.70	0.98	1802	0.49	0.23	0.19
Beans	444.00	13.3	8.35	—	0.80	6970	0.55	0.29	0.30
Black eye	136.40	12.7	5.90	—	0.79	3993	0.73	0.29	0.34
Buckwheat	27.60	8.9	3.60	—	0.61	1983	0.84	0.38	0.41
Chickpeas	391.40	12.7	8.50	—	0.81	6823	0.82	0.28	0.36
Green lentil	78.70	8.3	5.00	—	1.00	2068	0.71	0.23	0.35
Maize	321.20	11.6	7.50	7.62	0.81	5721	0.70	0.29	0.38
Marrowfat	397.20	15.2	8.10	—	0.60	7714	0.83	0.38	0.40
Millet	6.00	7.6	2.04	—	0.56	971	0.95	0.41	0.47
Mung	75.50	12.0	4.60	—	0.55	3466	0.90	0.42	0.44
Pinto	360.30	14.0	8.10	—	0.63	7120	0.65	0.37	0.32
Rapeseed	4.46	6.7	2.05	1.94	0.57	828	0.93	0.40	0.41
Rice	16.50	7.2	2.75	2.75	0.84	1269	0.45	0.27	0.24
Rye	33.40	7.8	3.70	—	0.87	1804	0.46	0.26	0.23
Sesame	2.36	4.4	1.66	1.62	0.76	450	0.64	0.30	0.31
Sorghum	30.90	9.7	3.61	3.50	0.59	2156	0.80	0.39	0.38
Soybean	197.00	14.0	6.22	6.14	0.50	5646	0.93	0.46	0.47
Wheat	30.20	7.8	3.36	3.29	0.85	1692	0.61	0.27	0.30
Whole lentil	33.40	9.9	3.50	—	0.62	2178	0.85	0.37	0.42

Source: Ref. 70.

this is affected by the aerodynamic and physical properties of grain and straw, their orientation, airspeed, and direction of injection. Some important aerodynamic and physical properties of common grains associated with cleaning and separation are presented in Table 8.

4 COMBINE HARVESTER

The combine is the most modern and effective farm machine adapted to mechanize labor-intensive operations, such as harvesting and threshing simultaneously, at the field level. Combine harvesting of grain and seed crops are carried out to some extent by the large-area farmers who grow single crops in a field and in fields where combine operation favors the conditions. The grain combine, being the harvester–thresher put together, is required to perform a series of functions. The stationary thresher employs the same principle and possesses the same basic components as in a combine, but it differs in the feeding arrangement of the crops during field operation, and inclusion of a straw stacker. Hence, most emphasis in the following paragraphs has been applied to these components, rather than describing the operation principles, power requirements, and seed losses at different stages.

Fig. 22 Self-propelled grain combine harvester. (Courtesy of Punjab Tractor Ltd., Chandigarh, India.)

4.1 Combine Types and Their Merits

In harvesting grain and seed crops, the two most common systems—namely, direct combining and windrowing and combining—are followed. In each system, major operations performed in obtaining sound seeds or grains are (a) cutting of field crops and windrowing them; (b) picking up from the windrow; (c) conveying and feeding the cut crops into the threshing mechanism; (d) threshing or detachment of sound grain kernels and kernels from husk and pod; (e) separating the grains or kernels from broken stalks (or husk), chaff, and straw; (f) cleaning the grains or kernels from chaff and other foreign leftovers; and (g) loading the grains in a trailer or wagon. In a direct-combining or straight-through combine harvester (Fig. 22), the cut materials are directly conveyed into the threshing cylinder during operation in the field, whereas the windrowing-and-combining system allows sufficient time to evenly dry the plant straw and unevenly ripened grains or kernels before the threshing operation. Therefore, the reaping-and-windrowing operations are carried out several days (3–5 days) earlier than with the direct-combining method. In a windrowing-and-combining system, the reel and cutter-bar header is replaced by a pick-up attachment in the combine. The windrow is picked up gently by the pick-up header and then taken into the combine through a crop-feeder conveyor (see Fig. 1) for subsequent operations. This system, however, reduces the weather hazards to the standing crop. Hence, the windrow–combine system requires an extra operation over direct combining, but each system requires the minimum labor input. The windrow–combine system is better followed for heavy vegetative crops, such as alfalfa, grown for seed purpose and leftovers for animal feed. Bainer et al. (22) reported that the combine harvesting of 85–95% of barley, soybean, and wheat crops in the United States have been achieved since 1950, because other methods of harvesting and threshing are costly, and require more labor.

Besides the two foregoing types, combines are also classified as self-propelled flatland combine, straight-through combine, and trailed or self-propelled hillside combine.

In developed countries, self-propelled combines with 2.5- to 4.5-m–cutting widths are common, compared with the trailed machines because of merits such as, (a) ease of operation and better maneuverability, (b) better visibility and control, and (c) minimum grain loss at the opening up of fields. The high cost of repair and maintenance, and nonavailability of most spare parts and accessories are the prime hurdles faced by the farmers in owning a self-propelled grain combine. Although the initial cost of a self-propelled combine is nearly twice that of the trailed machine, the foregoing constraints are often borne before the merits of the former.

The trailed or self-propelled hillside combine is a specially designed combine to harvest grain and seed crops along the contours on steep slopes of up to 35%. The important design requirements of a hillside combine are to keep the threshing and separating assembly level and make the long header unit work parallel to the ground. For this purpose, provisions are made for shifting one or both the main wheels vertically to meet the first requirement. In some modern self-propelled combines, automatic leveling is achieved using a heavy, damped pendulum that controls the pressure switch which, in turn, actuates the solenoid valves in the hydraulic system. On the other hand, the header unit, to remain floated and to work parallel to the field surface, is hinged rigidly to the main frame of the combine to give support at either end. The leveling of the header unit of a trailed hillside combine is done through manually controlled, power-driven mechanisms. The principal use of a hillside combine is very much limited because of the high initial investment, complexity in design, nonavailability of spare parts, and need for skilled technicians to maintain it.

4.2 Header Unit of a Combine Harvester

The header unit refers to the cutting and conveying mechanism of a combine harvester; therefore, it includes the reel, the cutter-bar, an inclined canvas conveyor or a platform for receiving the cut crops, and feeding them into the cylinder–concave assembly where threshing and separation take place. The entire header unit of a combine is normally hinged at the front to enable adjustment of the stubble height between 4 and 6 cm during harvesting and up to 1 m during transport or moving from one plot to another.

The conventional reel in self-propelled or trailed combines is either ground-driven or power-driven with interchangeable sprockets to vary its peripheral velocity. It is provided with four to eight slats (bats) fixed rigidly on radial arms. For effective laying of cut crops on the conveyor and to meet the different crop requirements, the reel position is made adjustable both horizontally and vertically. This is because the reel position relative to the cutter-bar level, and its speed in reference to the forward velocity, affect header loss (shatter and cutter-bar losses) beside the crop condition and cutting height. For optimum combine operation, the reel axis is kept to a level such that the bottom edges of the slats at the lowest point of their travel are either in the same level or a little lower than the cutter-bar level, and this is adjusted slightly ahead of the cutter-bar. During operation, the peripheral speed of the reel is adjusted between 20 and 45% in excess of the forward speed of the combine to meet the crop maturity level and to evenly spread the cut crops on the conveyor, with heads pointing inward. In other words, the reel index, the ratio of the peripheral velocity of reel tip to the forward velocity of the combine, should be between 1.20 and 1.45. Lower reel speeds will result in uneven laying of crops on the conveyor and less shattering losses, and vice versa.

In a windrow–combine system, the design and construction of the pickup reel on the header unit are different from that used either in straight-through, flatland, or hillside combine. The pickup reel collects the cut crops and passes the material either to the inclined canvas conveyor or auger for feeding them to the cylinder–concave assembly. Such a pickup reel has proved effective even to lift the lodged crops before cutting. Unlike the conventional reel in a straight-through combine, each slat in the pickup reel is replaced with a large number of closely spaced curved fingers attached with the square bar that follows mechanisms such as, eccentric spider control, cam control, or planetary gear control, so that they always remain parallel to the ground. This arrangement causes the fingers to enter the windrow or the lodged material and lift it gently above the cutter-bar that cuts the crop during its forward motion. Therefore, the pickup-type reel in a combine is always preferred over the conventionally used fixed-bat-type reel that results in much higher header loss than using the former.

The cutter-bar adopted in all types of combine is a reciprocating type the construction and design of which are similar to those of a mower. The cutter speed of a combine is kept half the speed of a mower, and it lies in the range between 400 and 450 rpm (800–900 strokes/min). Second, the knife sections are serrated on the upper surface and are sharpened when needed. On the other hand, the ledger plates are made smooth, and the guards are tempered and hardened following the special methods. Beside these, the straight-through self-propelled combines (see Figs. 1 and 22) will have additional and essential components such as, feeding fingers, stone trap, and the straw guide drum, for the proper functioning of the machine.

4.3 Initial Settings, Design Parameters, Troubles, and Remedies of Combines

In the interest of the designers, research engineers, technicians, and field men, a range of design parameters associated with the operations, such as harvesting, threshing, separating, and cleaning of selected grain combines, compiled from the manufacturer's literature, initial basic settings, troubles, and remedies during field operations are presented in Tables 9–11. These will additionally help achieve a better understanding, control, and design of grain-harvesting equipment.

Table 9 Initial Basic Settings of Machine Components for Important Crops[a]

Sl. no.	Crop	Drum speed (rpm)	Drum speed (m/s)	Concave clearance Front (mm)	Concave clearance Rear (mm)	Sieves Upper (mm)	Sieves Lower (mm)	Speed of straw walker (rpm)
1	Gram	450–700	14.1–22.0	15	11	16–19	7–10	190
2	Mustard	450–700	14.1–22.0	10	5	16–19	4–5	190
3	Paddy	600–800	18.8–25.1	17	14	16–19	5–6	190
4	Soybean	250–600	7.8–18.8	15	11	16–19	8–10	190
5	Wheat	900–1000	28.3–31.4	15	7	16–19	6–8	190

[a]Final adjustments are made after inspection of grain quality, crop condition, and cleanliness of delivered material.

Source: Kartar Agro. Industries (P) Ltd., Patiala, India.

Table 10 Design Particulars of Selected Self-Propelled Grain Combines

Sl. no.	Particulars	Combine–harvester specifications				
1	Combine–harvester model no. and type	Kartar[a] 3500, self-propelled	Kartar[a] 4000, self-propelled	Kartar[a] 4500, self-propelled	Swaraj[b] 8100, self-propelled	Axia S 514[c], self-propelled
2	Make	Kartar Agro. Industries	Kartar Agro. Industries	Kartar Agro. Industries	Punjab Tractors, Ltd.[b]	Axia Eng. Co.[c]
3	Crop	Paddy, wheat, soybean, safflower, and sunflower	Paddy, wheat, soybean, safflower, and sunflower	Paddy, wheat, soybean, safflower, and sunflower	Paddy, wheat, barley, gram, mustard, peas, and beans	Paddy, wheat, barley, oats, gram, mustard, rape, linseed, clover, peas, beans, unicorn, poppy
4	Engine	P 4 (Perkins)	Ashok Leyland ALU 370	Ashok Leyland ALU 400	Ashok Leyland ALU 400	Ashok Leyland ALU 400
	Number of cylinders	Four	Six	Six	Six	Six
	BHP of engine	55 hp at 2400 rpm	91 hp at 2000 rpm	105 hp at 2200 rpm	105 hp at 2200 rpm	105 hp at 2200 rpm
5	Cutter-bar type	Reciprocating	Reciprocating	Reciprocating	Reciprocating	Reciprocating
	Width (cm)	350	400	450	420	426
	Height adjustment	Hydraulic	Hydraulic	Hydraulic	Hydraulic	Hydraulic
	Min. cutting ht (cm)	10	10	10	7	—
	Max. cutting ht., (cm)	100	100	100	120	—
	Speed (m/s)	—	—	—	—	1.45
	Stroke length (mm)	—	—	—	—	90
6	Reel					
	Type	Pickup	Pickup	Pickup	Pickup	Pickup
	Diameter (cm)	–	–	—	—	100
	Speed adjustments	Mechanical	Mechanical	Mechanical	Mechanical	Mechanical
	Height adjustment	Mechanical	Hydraulic	Hydraulic	Hydraulic	Hydraulic
7	Threshing drum-type	Rasp-bar	Rasp-bar	Rasp-bar/stud	Rasp-bar	Rasp-bar
	Diameter (cm)	60	60	60	61.5–118.5	60
	Length (cm)	76.5	100	126	127.4	130
	Peripheral velocity (m/s)	16.8–37.7	16.8–37.7	16.8–37.7	19.3–80.7	18.9–40.8
	Speed (rpm)	535–1200	535–1200	535–1200	600–1300	603–1300
	Speed adjustment	Mechanical	Mechanical	Mechanical	Mechanical	Mechanical

8	Concave					
	Gap between concave and thresher					
	At inlet 1 (mm)	16–39	16–39	16–39	7–15	—
	At outlet (mm)	3–16	3–16	3–16	—	—
	Gap adjustment	Mechanical	Mechanical	Mechanical	Mechanical	—
9	Straw walker					
	Number	Three	Four	Five	Five	Five
	Area	3(320 × 25)	4(330 × 25)	5(330 × 25)	5 (5.16 m^2)	5 (5.2 m^2)
	Speed (rpm)	160	160	160	—	—
10	Cleaning					
	Area (cm × cm)	170 × 71	174 × 97	174 × 122	1.85 m^2 (upper sieve) 1.77 m^2 (lower sieve)	1.39 m^2 (upper sieve) 1.17 m^2 (lower sieve)
	Adjustment	Mechanical	Mechanical	Mechanical	Mechanical	Mechanical
11	Ground speed					
	Forward (km/h)	1.2–20.9 (three)	1.5–21.0 (three)	1.5–21.0 (three)	1.5–20.2 (three)	1.4–20.0 (three)
	Reverse (km/h)	3.5–8.7 (one)	3.5–9.0 (one)	3.5–9.0 (one)	3.9–9.1 (one)	3.4–8.5 (one)
12	Fan					
	Diameter (cm)	52.5	54.0	54.0	—	—
	No. of blades	Four	Five	Five	—	—
	Adjustments	Mechanical	Mechanical	Mechanical	—	—
13	Steering system	Hydraulic	Hydraulic	Hydraulic	Hydraulic	Hydraulic
14	Brake system	—	—	—	Disk type mechanical	—
15	Grain tank capacity (m^3)	1.25	1.5	3.0	2.0	2.3
16	Weight (dry) (kg)	4300	5500	6000	7500	6800
17	Ground clearance (cm)	—	—	—	36.5 (min.)	—
18	Field capacity (ha/h)					
	Wheat	1.01	1.42	1.82	1.74	1.22–1.62
	Paddy	0.81	1.22	1.42	1.40	0.81–1.22

Source: Service/operator's manuals of three leading combine manufacturers in India, namely:
[a]Kartar Agro. Industries (P) Ltd., Patiala—147202, Punjab, India.
[b]Punjab Tractors Ltd., Near Chandigarh—160055, Punjab, India.
[c]Axia Eng. Co., Patiala—147001, Punjab, India.

Table 11 Probable Troubles and Remedies in Combine Harvesting

Sl. no.	Trouble	Probable reason	Remedies
1	Crops falling ahead	a. Low reel speed	a. Increase reel speed
		b. Improper placement of cutter-bar and reel	b. Adjust reel location vertically and horizontally
2	Uneven cutting of crops	a. Cutter-bar at higher level	a. Adjust the cutter-bar height
		b. Knife section not close to the ledger plate	b. Reduce gap between knife clip and knife section (hammer knife clip tip)
		c. All ledger plates attached to guard not level	c. Level all ledger plates
		d. Knife sections not registered	d. Register the knife section
		e. Damaged knife, guards, and wearing plates	e. Replace them as needed
3	Falling of heads of cut crops	a. Reel height too low or too high	a. Adjust reel height
4	High shattering loss	a. Higher ground-to-reel speed	a. Decrease ground speed to reduce bating of crop by the reel
		b. Improper reel speed	b. Adjust the reel speed to match ground speed
		c. Clogging of straw walkers	c. Clean the openings of the straw walker and increase frequency of oscillation
		d. High airspeed developed by the blower	d. Reduce blower speed and change direction of wind board
		e. Overmatured crop	e. Reduce pickup reel and ground speeds proportionately
		f. Excessive crop feed	f. Reduce forward speed
		g. Inadequate openings of adjustable sieve	g. Adjust the openings of the adjustable sieve
		h. Choking of sieves	h. Clean the sieves

5 Wrapping crop in front of the auger	a. Pickup reel height higher b. High reel speed c. Pickup lines pitched high	a. Adjust pickup reel height b. Reduce reel speed in accordance with forward speed c. Reduce the pitch of lines
6 Pickup reel does not pull crop	a. Header unit much above the ground b. Pickup reel height is more	a. Lower the header unit height b. Lower pick-up reel height
7 Clogging of cutter-bar	a. More clearance between knife section and ledger plate b. High crop feed rate c. More cutting height d. Knife section not close to ledger plate	a. Provide minimum clearance between knife section and ledger plate all through the cutter-bar b. Adjust the auger–platform gap depending on the crop c. Adjust cutter-bar height close to ground d. Reduce gap between knife clip and knife section (hammer knife clip tip)
8 Unsatisfactory cutting	a. Improper knife registration b. Bent fingers c. Excessive clearance between knife and ledger plate or knife and knife clip d. Blunt out knife section e. Excessive vibration of cutter-bar assembly	a. Register the knife section b. Make proper alignment of the fingers c. Adjust the clearance d. Replace or resharpen the worn-out blades e. Excessive play in cutter-bar and knife drive be removed
9 Hydraulic cylinder not lifting cutter-bar properly	a. Drive belt to gear type pump loose b. Leakage in hydraulic pump	a. Adjust the tension of V-belt of hydraulic hand pump b. Repair the hydraulic pump

Table 11 Continued

Sl. no.	Trouble	Probable reason	Remedies
10	Overloading of threshing drum	a. Too low a speed for threshing drum b. Low cylinder–concave clearance	a. Increase threshing drum speed b. Increase gap between cylinder and concave
11	Uneven crop feeding to the thresher	a. Feeding chain too tight b. Dents in feeding auger c. Auger shaft not aligned	a. Adjust chain tension b. Repair auger dents c. Align auger shaft
12	Clogging of threshing drum	a. Cylinder speed low b. Drive belt of straw walker is loose c. Clogging of straw walkers d. High crop feed rate to threshing cylinder	a. Increase cylinder speed b. Adjust belt tension c. Clean straw walker d. Reduce forward speed of combine
13	Ineffective threshing	a. More cylinder–concave clearance b. Less peripheral velocity of threshing drum c. Excessive feed rate entering the drum	a. Reduce the clearance b. Increase drum speed c. Increase drum or reduce forward speed
14	Excessive grain damage	a. High speed of threshing drum b. Low cylinder–concave clearance c. Recirculating clean grains are cracked when re-threshed	a. Reduce peripheral velocity of threshing drum b. Increase cylinder–concave clearance c. Open sieve openings to reduce the trailings

	d. Uneven or low feed rate of crop	d. Adjust pickup reel and forward speed to reduce damage
	e. Auger shaft not aligned or dented auger	e. Align auger shaft properly and repair auger dents
	f. Inadequate crop feed to the cylinder–concave assembly	f. Increase the ground speed or reduce pickup reel speed
15 Unthreshed earheads in the straw	a. Low peripheral velocity of threshing cylinder	a. Increase the speed of threshing drum
	b. Higher cylinder–concave clearance	b. Reduce the cylinder–concave clearance
	c. Low peripheral velocity of cleaning unit	c. Increase the speed of engine driving the cleaning unit
	d. High blower speed	d. Reduce blower rpm
16 Short straw and chaff in the grain tank	a. Inadequate blower speed	a. Increase blower speed a little or adjust wind board for more air
17 Cutter-bar clutch slipping	a. Excessive crop feed to cutter-bar auger	a. Reduce the machine speed
	b. Drive chain to cutter-bar jam	b. Adjust the chain tension
18 Clogging elevator	a. Grain tank full	a. Drain grain from grain tank
	b. Slackness of elevator–conveyor chain	b. Adjust the elevator–conveyor chain tension
	c. Slippage of V-belt	c. Adjust V-belt tension
	d. Low air velocity	d. Adjust the direction of air blast

Source: Service/operator's manuals of the following leading companies in India:
Kartar Agro. Industries (P) Ltd., Patiala—147202, Punjab, India.
Punjab Tractors Ltd., Near Chandigarh—160055, Punjab, India.
Axia Eng. Co., Patiala—147001, Punjab, India.

APPENDIX

Abbreviations

AMA	Agricultural Mechanization in Asia, Africa, and Latin America
ASAE	American Society of Agricultural Engineers
BIS	Bureau of Indian Standard
CIAE	Central Institute of Agricultural Engineering
Conn	Colonel
FIM	Farm Implements and Machinery
ICAR	Indian Council of Agricultural Research
IIT	Indian Institute of Technology
IS	Indian Standard
PAU	Punjab Agricultural University
PTO	Power take off
qty	Quantity
RNAM	Regional Network for Agricultural Machinery
UNDP	United Nations Development Program
WB	West Bengal
wt	Weight

Symbols

A	effective cross-sectional area, m^2
A_s	cross-sectional area of stalks under shear, m^2
a, b	lateral distances of knife bar at the end of a stroke, m
C_d	drag coefficient
CPH	critical period of harvest, days
DAF	days after heading or flowering, days
DAS	days after sowing, days
E_b	percentage blown grain, %
E_c	cleaning efficiency, %
E_d	percentage grain damaged, %
E_i	dynamic cutting energy, J
E_l	percentage grain loss, %
E_r	percentage grain recovery, %
E_s	percentage grain separated through concave, %
E_t	percentage grain threshed, %
E_u	percentage grain unthreshed, %
F	cutting resistance, N
f	instantaneous acceleration of slider, m/s^2
F_c	centrifugal force at crank pin, N
F_{cw}	total force of counterweight, N
F_p	force along the pitman at crank pin, N
F_r	radial component of F_p at crank pin, N
$(F_r)_{max}$	maximum value of force F_r, N
F_s	inertia force of sliding components in horizontal plane, N
F_{uf}	unbalanced radial force at the crankpin, N
F_v	alternating vertical reaction at knife end, N
h	cutting height or stubble height, m

Symbols (cont'd)

K	constant
L	pitman length, m
$L_1 \ldots L_{10}$	dimensions of sickle, m
m	height of crank center above cutter-bar level, m
m_1	part mass of pitman acting at slider, kg
m_2	part mass of pitman acting at crankpin, kg
m_3	counterbalance mass opposite the crankpin at radius r_1, kg.
m_s	mass of slider, kg
N	normal load, N
N_1, N_2	normal reactions, N
OPH	optimum period of harvest, days
P_1, P_n	average instantaneous power at load and at no-load respectively, W
R	distance of CG of pendulum bar from the axis of rotation, m
r	crank radius, m
r_1	location of counterbalance mass opposite the crankpin, m
R_1, R_2	frictional resistance, N
R_c	distance of the center of throat from the center of rotation, m
R_e	Reynold's number
S_d	dynamic shear stress, N/m^2
S_x	displacement of particle in x direction, m.
S_{yf}	displacement of falling particle in y direction, m
S_{yr}	displacement of rising particle in y direction, m
s	stroke length, m
T	tangential force, N
t	thickness, m
v	instantaneous velocity, m/s
v_a	air velocity, m/s
v_t	suspension velocity or terminal velocity of particle, m/s
W	mass of pendulum with cutting element, kg
W_A	total grain input per unit time by weight, kg/h
W_b	quantity of whole grain collected at chaffed-straw outlet per unit time, kg/h
W_c	quantity of whole grain per unit time obtained from the main grain outlet, kg/h
W_d	quantity of damaged grain collected at all outlets per unit time, kg/h
W_l	quantity of whole grain, damaged grain, and unthreshed grain per unit time at chaff outlet and straw outlet (hold-on) or chaffed–straw outlet (throw-in) and scattered grain per unit time, kg/h
W_m	quantity of whole material per unit time at the main grain outlet, kg/h
W_r	quantity of threshed grain (whole or damaged) per unit time collected at the main grain outlet, kg/h
W_s	quantity of grain per unit time collected after flow through concave, kg/h
W_t	quantity of threshed grain obtained per unit time from all outlets, kg/h
W_u	quantity of unthreshed grain per unit time obtained from all outlets, kg/h
Z_m	measured shape factor
Z_p	predicted shape factor
α	cutting angle, degree

Symbols (cont'd)

β	angle of wedge, degree
γ	angle of action, degree
θ_0	upswing angle of pendulum arm with vertical at the end of free swing, degree
θ_a, θ_b	pitman angle with horizontal at the ends of a stroke, degree
θ_c	crank angle after a lapse of time *t*, degree
θ_{ma}	maximum crank angle, degree
θ_{mi}	minimum crank angle, degree
θ_p	pitman angle with horizontal after a lapse of time *t*, degree
θ_u	upswing angle of pendulum arm with vertical after cutting the specimen, degree
ξ	eccentricity
μ	coefficient of friction between the blade and the material
σ	shear stress of material, N/m^2
ω	angular velocity, rad/s

REFERENCES

1. Devnani RS, Pandey MM. 1985. Design development and field evaluation of vertical conveyor reaper windrower. AMA 16:41–52.
2. Garg IK, Sharma VK, Singh S. 1984. A power tiller-mounted vertical conveyor reaper windrower. AMA 15:40–44.
3. Verma SR, Garg RL. 1969. Development and performance of tractor mounted PTO operated self-raking reaper. J Agric Eng 14:29–37.
4. Michael AM, Ojha TP 1987. Principles of Agricultural Engineering. vol. 1. New Delhi. ML Jain for Jain Brothers.
5. Arnold RE, Calldwell F, Davies ACW. 1958. The effect of moisture content on the grain and drum setting of the combine harvester on the quality of oats. J Agric Eng Res 3:336.
6. Iqbal M, Sheikh GS, Sial JK. 1980. Harvesting and threshing losses of wheat with mechanical and conventional methods. AMA, 11:66–70.
7. Ranganath KA, Bhashyam MK, Bhaskar T Rao, Desikachar HSR. 1970. Influence of time of harvest and environmental factors on grain yield and milling breakage of paddy. J Food Sci Technol 7:144–147.
8. Bukhari S, Ibupoto KA, Jamro GH, Khohro GA. 1991. Influence of timing and date of harvest on wheat grain losses. AMA 22:56–58.
9. Dutt B. 1993. Influence of crop-cylinder parameters on selection and evaluation of gram thresher. PhD Dissertation Dept of Agric Food Eng, IIT, Kharagpur, India.
10. Saxena BB. 1987. Effect of harvesting stage and threshing variables on yield and grain quality of soybean. PhD Dissertation Dept of Agric Food Eng, IIT Kharagpur, India.
11. Datta AC, Dutt B. 1995. Optimum harvest periods and yield potential of gram. Presented at the 31st Annual Convention of ISAE. Kerala Agric University, Trichur, India, Dec 28–30.
12. Mukundan K. 1976. Optimization of agronomic resources for maximizing grain and mill yield of rice. PhD Dissertation, Dept Agric Food Eng IIT Kharagpur, India.
13. Rafey, A. 1980. Studies on agronomic practices influencing yield, milling out-turn and quality of rice. PhD Dissertation Dept Agric Food Eng, IIT Kharagpur, India.
14. Sajwan KS. 1980. Influence of field environment and post-harvest operations on milling out-turn and quality of rice. PhD Dissertation, Dept Agric Food Eng, IIT Kharagpur, India.
15. U. S. Dept. of Agriculture. 1973. Rice in the United States: varieties and production. Agric Handbook 289. Agric Res Serv.

16. Agric Food Eng Dept. 1979. Annual Report. Coordinated Research Scheme on Energy Requirement in Intensive Agricultural Production. IIT Kharagpur, India.
17. Bhattacharya KK, Chatterge, BN. 1977. Studies on maturity of rice grain. Rice Proc Eng Center Rep, IIT Kharagpur, India.
18. Bose SP, Chattopadyay PK. 1976. Harvesting and drying of high moisture paddy. Rice Proc Eng Center Rep 2:33–37.
19. Hunt, D. 1977. Farm power and machinery management. In: Laboratory Manual and Work Book. 7th ed. Ames, IO: Iowa State University Press, 1–365.
20. Dahatonde BN, Adhao SH. 1978. Studies on optimum harvesting period of jowar variety CSH-4. J Maharashtra Agric Univ 3:184–186.
21. Bhole NG, et al. 1970. Paddy harvesting and drying studies. Rice Proc Eng Center IIT Kharagpur, India.
22. Bainer R, Kepner RA, Berger EL. 1963. Principles of Farm Machinery. New York: John Wiley & Sons.
23. Kepner RA. 1952. Analysis of cutting action of a mower. Agric Eng 33:693.
24. Kawasthima M. 1951. Survey of shape and cutting force of serrated sickle on the market. Memoir, Coll Agric Utsunomiya Univ, 3, Japan.
25. Ivashko AA. 1958. [Problems of cutting theory of organic matters by knife]. (Voprositeopii Vezanila Organitecheskikh Materlalov Lezvieum, Tractori-selkhomashini, 2.
26. Ezaki H. 1959. Theoretical studies on fundamental elements for the design of small harvester. J Kanto-Tosan Natl Exp Sta, Ser 2.
27. Chancellor WJ. 1958. Energy requirements for cutting forage. Agric Eng 39:633–636.
28. Chancellor WJ. 1988. Cutting of biological materials. In: Brown RH ed. Handbook of Engineering in Agriculture. Vol 1. CRC Series in Agriculture, Boca Raton, FL: CRC Press, pp. 35–63.
29. O'Dogherty MJ, Gal GE. 1986. Laboratory studies of the cutting of grain stems. J Agric Eng Res 33:115–129.
30. McRandal DM, McNulty PB. 1978. Impact cutting behavior of forage crops. II. Field test. J Agric Eng Res, 23:329.
31. Rajput DS, Bhole NG. 1973. Static and dynamic shear properties of paddy stem. Harvester. (Agric Eng Dept, IIT Kharagpur), 15:17–21.
32. Prasad J, Gupta OP. 1975. Mechanical properties of maize stalk as related to harvesting. J Agric Eng Res 20:79–87.
33. Kepner RA. 1952. Analysis of the cutting action of a mower. Agric Eng 33:693–704.
34. Fisher OA, Kolega JJ, Wheeler WC. 1957. An evaluation of the energy requirements to cut forage, grasses and legumes. Univ Conn Storrs Agric Exp Progr Rep 17.
35. Prince RP, Wheeler WC, Fish DA. 1958. Discussion of energy requirements for cutting forage. Agric Eng 39:638–639, 652.
36. Feller R. 1959. Effects of knife angles and velocities on the cutting of stalks without a counter-edge. J Agric Eng Res 4:277–293.
37. Srivastava AK, Herum FL, Steve, KK. 1976. Impact parameters related to physical damage to corn kernel. Trans ASAE 19:1147–1151.
38. Novikov J. 1957. Theory and calculations of rotary cutting apparatus with working blades, Sel'khozmashina 8(1).
39. Dobler K. 1973. [Use of quartz crystal transducer in measuring force components in moving forage crops] Melrkomponenten-Schnittkraft—measuring mit Quartz-Kristallab fuechmern beim Mahen ben Halmgut. ATM Messtech Praxis 4:61.
40. Datta AC, Chakraborty AK, Gupta OP. 1969. Dynamic shear stress of different crops. Harvester, Agric Eng Dept IIT Kharagpur, India, 11:99–103.
41. Halyk RM, Hurlbut LW. 1968. Tensile and shear strength characteristics of alfalfa stems. Trans ASAE 11:256–257.
42. Liljedahl JB, Jackson GL, De Graff RP, Schroeder, ME. 1961. Measurement of shearing energy. Agric Eng 42:298–301.

43. Annual Report. 1977–78. Harvest and Post Harvest Technology Scheme. Agric Food Eng Dep, IIT Kharagpur, India.
44. Pandey MM, Devnani RS. 1981. Development of improved CIAE sickle for harvesting cereals. AMA 12:54–58.
45. Pandey MM. 1981. Harvesting hand tools of India. Tech Bull CIAE/81/25, CIAE, Nabibagh, Bhopal, India.
46. Singh S, Verma SR. 1978. Field test report on Beaver knapsack type power reaping machine. Punjab Agric Univ Ludhiana, India.
47. Mazumdar M. 1982. A knapsack type cereal harvester with a rotary cutting element from engineering and ergonomic considerations. PhD Dissertation Dept Agric Food Eng IIT Kharagpur, India.
48. Verma SR. 1978. Reapers and windrowers for cereal harvesting. Proc Summer Institute on Harvesting Equipment. PAU, Ludhiana, India.
49. Anon. 1983. Testing, evaluation of cereal harvesters. Technical Series 4. Economic and Social Commission for Asia and Pacific. RNAM, C/O UNDP, P.O. Box 7285 ADC, Pasay City, Philippines, pp 19–45.
50. Kanafojski CZ, Karwowski T. 1976. Agricultural Machines Theory and Construction. vol 2. Crop Harvesting Machines, Springfield, VA: U. S. Dept Commerce National Technical Information Service. 22161.
51. Lal R, Datta AC. 1971. Agricultural Engineering Through Worked Examples. Saroj Prakashan, 646 Katra, Allahabad-2, India.
52. Raney RR. 1946. Vibration control in farm machinery. Unpublished Paper, International Harvester Co.
53. Chauhan AM. 1983. Power tiller front-mounted reaper. J Agric Eng 10:65–68.
54. Devnani RS, Nag KN. 1970. The power tiller harvester. J Agric Eng 7:61–64.
55. Singh, Choudhury AP, Gee–Clough D. 1988. Performance evaluation of mechanical reapers in Pakistan. AMA 19:47–52.
56. Verma SR, Garg RL. 1971. Design and development of tractor front mounted reaper. J Agric Eng 8:55–59.
57. RNAM 1983. RNAM Test Codes and Procedures for Farm Machinery. Technology Series 12. Economic and Social Commission for Asia and Pacific, RNAM, C/O UN–ESCAP, UN Building, Rajadamnern Nok Avenue, Bangkok 10200, Thailand, pp. 207–246.
58. Arnold RE. 1964. Experiments with rasp bar threshing drum. I. Some factors affecting performance. J Agric Eng Res 9:99.
59. Arnold RE, Lake JR. 1964. Experiments with rasp bar threshing drums. II. Comparison of open and closed concaves. J Agric Eng Res 9:250.
60. Wieneke F. 1964. Performance characteristics of the rasp-bar thresher. Grundlagen Landtechn 21–33. In: Brown RH, ed. Handbook of Engineering in Agriculture. Vol I. CRC Series in Agriculture, Boca Raton, FL: CRC Press, 1988; 65–84.
61. Singhal OP, Thierstein GE. 1987. Development of an axial-flow thresher with multi-crop potential. AMA 18:57–65.
62. Sarwar JG, Khan AU. 1987. Comparative performance of rasp-bar and wire-loop cylinders for threshing rice crop. AMA 18:37–42.
63. Behera BK, Dash SK, Das DK. 1990. Development and testing of a power operated wheat thresher. AMA 21:15–21.
64. Sewell AJ. 1990. Some effects of concave to drum clearance and concave design on small grain threshing drum performance. J Agric Eng Res 33:115–217.
65. Anwar MT, Gupta CP. 1990. Performance evaluation of chickpea thresher in Pakistan. AMA 21:23–28.
66. Mazumdar KL. 1994. Study on design and operational parameters of spike-tooth threshing system for soybean crop. PhD Dissertation Dept Agric Food Eng, IIT Kharagpur, India.
67. Feller R. 1977. Clogging rate of screens as affected by particle size. Trans ASAE 20:758.

68. MacAulay JT, Lee JHA. 1969. Grain separation on oscillating combine sieves as affected by material entrance condition. Trans ASAE 12:648.
69. Gorial BY, O'Callaghan JR. 1991. Separation of grain from straw in a vertical air stream. J Agric Eng Res 48:111–122.
70. Gorial BY, O'Callaghan JR. 1990. Aerodynamic properties of grains/straw materials. J Agric Eng Res 46:275–290.

5

Grain Drying: Basic Principles

ARUN S. MUJUMDAR

National University of Singapore, Singapore

JÁNOS BEKE

Szent István University, Gödöllő, Hungary

1 INTRODUCTION: BASIC PRINCIPLES AND TERMINOLOGY

Drying is one of the most important technologies for the preservation of grains. Grains must be dried to the levels of moisture suitable for storage, without spoilage; the levels depend on the specific grains. Sun and solar drying has been practiced extensively since ancient times. However, the technique is not suited for the large quantities of various grains that, today, must be dried commercially. This chapter is aimed at providing only the basic principles and termiology as far as convective drying is concerned. Subsequent chapters will discuss aspects of grain dryers used in practice.

Drying is a complex operation involving transient transfer of heat and mass along with several rate processes, such as physical or chemical transformations that, in turn, may cause changes in product quality as well as the mechanisms of heat and mass transfer. Physical changes that may occur include shrinkage, puffing, crystallization, and glass transitions. In some cases, desirable or undesirable chemical or biochemical reactions may occur, leading to changes in color, texture, odor, or properties of the solid product. In the manufacture of catalysts, for example, drying conditions can yield significant differences in the activity of the catalyst by changing the internal surface area.

Drying occurs by effecting vaporization of the liquid by supplying heat to the wet feedstock. As noted earlier, heat may be supplied by convection (direct dryers), by conduction (contact or indirect dryers), radiation, or volumetrically by placing the wet material in a microwave or radio frequency electromagnetic field. Over 85% of industrial dryers are of the convective type with hot air or direct combustion gases as the drying medium.

Over 99% of the applications involve removal of water. All modes except the dielectric (microwave and radio frequency) supply heat at the boundaries of the drying object so that the heat must diffuse into the solid primarily by conduction. The liquid must travel to the boundary of the material before it is transported away by the carrier gas (or by application of vacuum for nonconvective dryers).

Transport of moisture within the solid may occur by any one or more of the following mechanisms of mass transfer:

Liquid diffusion, if the wet solid is at a temperature below the boiling point of the liquid
Vapor diffusion, if the liquid vaporizes within material
Knudsen diffusion, if drying takes place at very low temperatures and pressures (e.g., in freeze-drying)
Surface diffusion (possible although not proved)
Hydrostatic pressure differences, when internal vaporization rates exceed the rate of vapor transport through the solid to the surroundings
Combinations of the foregoing mechanisms

Note that since the physical structure of the drying solid is subject to change during drying, the mechanism of moisture transfer may also change with elapsed time of drying.

1.1 Thermodynamic Properties of Air–Water Mixtures and Moist Solids

1.1.1 Psychrometry

Most dryers are of the direct (or convective) type. In other words, hot air is used both to supply the heat for evaporation and to carry away the evaporated moisture from the product. Notable exceptions are freeze and vacuum dryers, which are used almost exclusively for drying heat-sensitive products because they tend to be significantly more expensive than dryers that operate near atmospheric pressure. Another exception is the emerging technology of superheated steam drying (Mujumdar, 1995). In certain cases, such as the drum-drying of pasty foods, some or all of the heat is supplied indirectly by conduction.

Drying with heated air implies humidification and cooling of the air in a well-insulated (adiabatic) dryer. Thus, hygrothermal properties of humid air are required for the design calculations of such dryers. Table 1 summarizes the essential thermodynamic and

Table 1 Thermodynamic and Transport Properties of Air–Water Systems

Property	Expression
P_v	$P_v = 100\exp[27.0214 - (6887/T_{abs}) - 5.32\ \ln(T_{abs}/273.16)]$
Y	$Y = 0.622\ RH\ P_v/(P - RH\ P_v)$
c_{pg}	$c_{pg} = 1.00926 \times 10^3 - 4.0403 \times 10^{-2}\ T + 6.1759 \times 10^{-4}\ T^2 - 4.097 \times 10^{-7}\ T^3$
k_g	$k_s = 2.425 \times 10^{-2} - 7.889 \times 10^{-5}\ T - 1.790 \times 10^{-8}\ T^2 - 8.570 \times 10^{-12}\ T^3$
ρ_g	$\rho_s = PM_g/(RT_{abs})$
μ_g	$\mu_g = 1.691 \times 10^{-5} + 4.984 \times 10^{-8}\ T - 3.187 \times 10^{-11}\ T^2 + 1.319 \times 10^{-14}\ T^3$
c_{pv}	$c_{pv} = 1.883 - 1.6737 \times 10^{-4}\ T + 8.4386 \times 10^{-7}\ T^2 - 2.6966 \times 10^{-10}\ T^3$
c_{pw}	$c_{pw} = 2.8223 + 1.1828 \times 10^{-2}\ T - 3.5043 \times 10^{-5}\ T^2 + 3.601 \times 10^{-8}\ T^3$

Source: Mujumdar, 1995; Pakowski et al., 1991.

Table 2 Definitions of Commonly Encountered Terms in Psychrometry and Drying

Term/symbol	Meaning
Adiabatic saturation temperature, T_{as}	Equilibrium gas temperature reached by unsaturated gas and vaporizing liquid under adiabatic conditions. (Note: For air–water system only, it is equal to the wet-bulb temperature; T_{wb}).
Bound moisture	Liquid physically or chemically bound to solid matrix such that it exerts a vapor pressure lower than that of pure liquid at the same temperature.
Constant-rate–drying period	Under constant-drying conditions, it is the drying period when evaporation rate per unit drying area is constant (when surface moisture is removed)
Dew point	Temperature at which a given unsaturated air–vapor mixture becomes saturated
Dry-bulb temperature	Temperature measured by a (dry) thermometer immersed in vapor–gas mixture
Equilibrium moisture content, X^*	At a given temperature and pressure, the moisture content of moist solid in equilibirum with the gas–vapor mixture (zero for nonhygroscopic solids)
Critical moisture content, X_c	Moisture content at which the drying rate first begins to drop (under constant-drying conditions)
Humid heat	Heat required to raise the temperature of unit mass of dry air and its associated vapor through 1° (J kg^{-1} K^{-1} or Btu lb^{-1} °F^{-1})
Humidity, absolute	Mass of water vapor per unit mass of dry gas (kg kg^{-1} or lb lb^{-1})
Humidity, relative	Ratio of partial pressure of water vapor in gas–vapor mixture to equilibrium vapor pressure at the same temperature
Unbound moisture	Moisture in solid that exerts vapor pressure equal to that of pure liquid at the same temperature
Water activity, a_w	Ratio of vapor pressure exerted by water in solid to that of pure water at the same temperature
Wet-bulb temperature, T_{wb}	Liquid temperature attained when large amounts of air–vapor mixture is contacted with the surface. In purely convective drying, drying surface reaches T_{wb} during the constant-rate period

transport properties of the air–water system. In Table 2, a listing of brief definitions of various terms encountered in drying and psychrometry is given. It also includes several terms not explicitly discussed in the text.

Figure 1 is a psychrometric chart for the air–water system. It shows the relation between the temperature (abscissa) and absolute humidity (ordinate; in kilogram water per kilogram dry air) of humid air at 1 atm absolute pressure from 0° to 130°C. Lines representing the percentage of humidity and adiabatic saturation are drawn according to the thermodynamic definitions of these terms. Equations for the adiabatic saturation and wet-bulb temperature lines on the chart are as follows (Geankopolis, 1993):

$$\frac{Y - Y_{as}}{T - T_{as}} = -\frac{c_s}{\lambda_{as}} = -\frac{1.005 + 1.88Y}{\lambda_{as}} \tag{1}$$

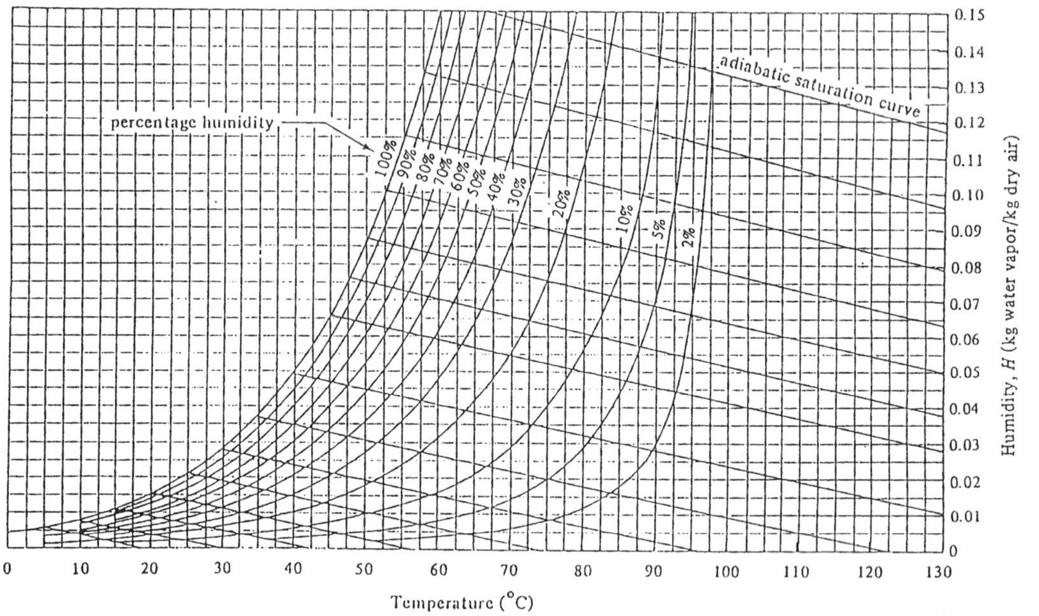

Fig. 1 Psychrometric chart for the air–water system.

and

$$\frac{Y - Y_{wb}}{T - T_{wb}} = -\frac{h/M_{air}k_y}{\lambda_{wb}} \tag{2}$$

The ratio ($h/M_{air}k_y$), the psychrometric ratio, lies between 0.96 and 1.005 for air–water vapor mixtures; thus it is nearly equal to the value of humid heat c_s. If the effect of humidity is neglected, the adiabatic saturation and wet-bulb temperatures (T_{as} and T_{wb}, respectively) are almost equal for the air–water system. Note, however, that T_{as} and T_w are conceptually quite different. The adiabatic saturation temperature is a gas temperature and a thermodynamic entity, whereas the wet-bulb temperature is a heat and mass transfer rate-based entity and refers to the temperature of the liquid phase. Under constant drying conditions, the surface of the drying material attains the wet-bulb temperature if heat transfer is by pure convection. The wet-bulb temperature is independent of surface geometry of the analogy between heat and mass transfer.

Most handbooks of engineering provide more detailed psychrometric charts, including additional information and extended temperature ranges. Mujumdar (1995) also includes numerous psychrometric charts for several gas–organic vapor systems.

1.1.2 Equilibrium Moisture Content

The moisture content of a wet solid in equilibrium with air of given humidity and temperature is termed the *equilibrium moisture content* (EMC). A plot of EMC at a given temperature versus the relative humidity is termed *sorption isotherm.* An isotherm obtained by exposing the solid to air of increasing humidity gives the *adsorption isotherm.* That obtained by exposing the solid to air of decreasing humidity is known as the *desorption*

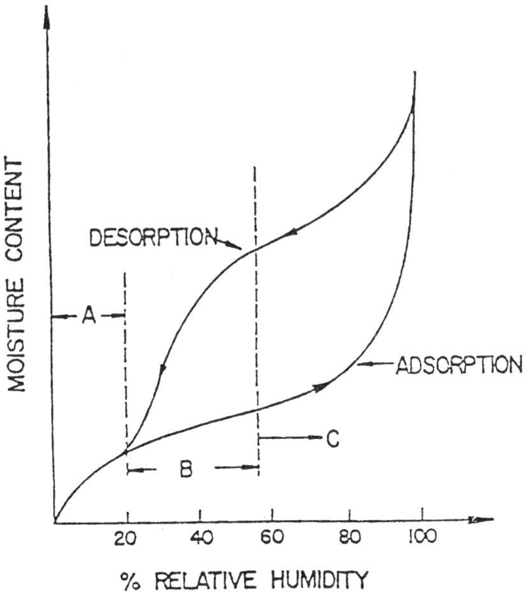

Fig. 2 Typical sorption isotherms.

isotherm. Clearly, the latter is of interest in drying as the moisture content of the solids progressively decreases. Most drying materials display *hysteresis* in that the two isotherms are not identical.

Figure 2 shows the general shape of the typical sorption isotherms. They are characterized by three distinct zones: A, B, and C, which are indicative of different water-binding mechanisms at individual sites on the solid matrix. In region A, water is tightly bound to the sites and is unavailable for reaction. In this region, there is essential monolayer adsorption of water vapor and no distinction exists between the adsorption and desorption isotherms. In region B, the water is more loosely bound. The vapor pressure depression below the equilibrium vapor pressure of water at the same temperature is due to its confinement in smaller capillaries. Water in region C is even more loosely held in larger capillaries. It is available for reactions and as a solvent.

Numerous hypotheses have been proposed to explain the hysteresis. The reader is referred elsewhere (Bruin and Luyben, 1980; Fortes and Okos, 1980; Bruin, 1988) for more information on the topic.

Figure 3 shows schematically the shapes of the equilibrium moisture curves for various types of solids. Figure 4 shows the various types of moisture defined in Table 2. Desorption isotherms are also independent on external pressure. However, in all practical cases of interest, this effect may be neglected.

According to Keey (1978), the temperature dependence of the equilibrium moisture content can be correlated by:

$$\left[\frac{\Delta X^*}{\Delta T}\right]_{\Psi=constant} = -\alpha X^* \tag{3}$$

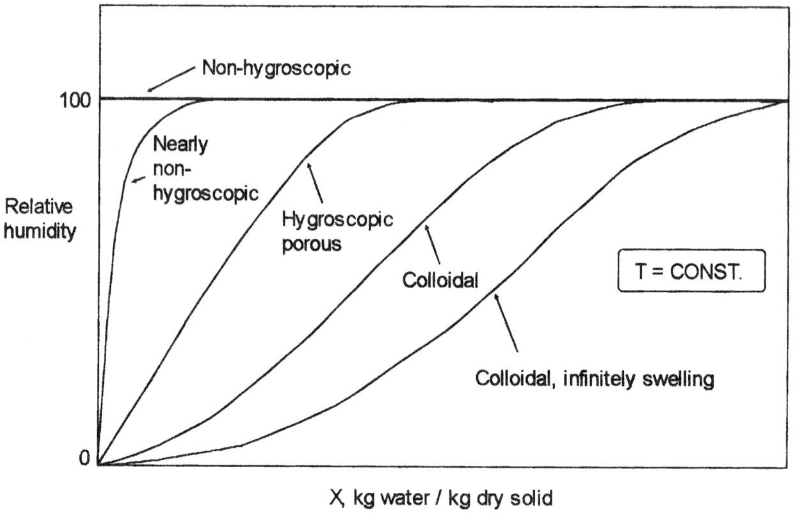

Fig. 3 Equilibrium moisture content curves for various types of solids.

where X^* is the dry-basis of equilibrium moisture content, T is the temperature and Ψ is the relative humidity of air. The parameter α ranges from 0.005 to 0.01 K^{-1}. This correlation may be used to estimate the temperature dependence of X^* if no data are available.

For hygroscopic solids, the enthalpy of the attached moisture is less than that of pure liquid by an amount equal to the binding energy, which is also termed the *enthalpy of wetting*, ΔH_w (Keey, 1978). It includes the heat of sorption, hydration, and solution and may be estimated from the following equation:

$$\left.\frac{d(\ln \Psi)}{d(1/T)}\right|_{X=constant} = -\frac{\Delta H_w}{R_g T} \tag{4}$$

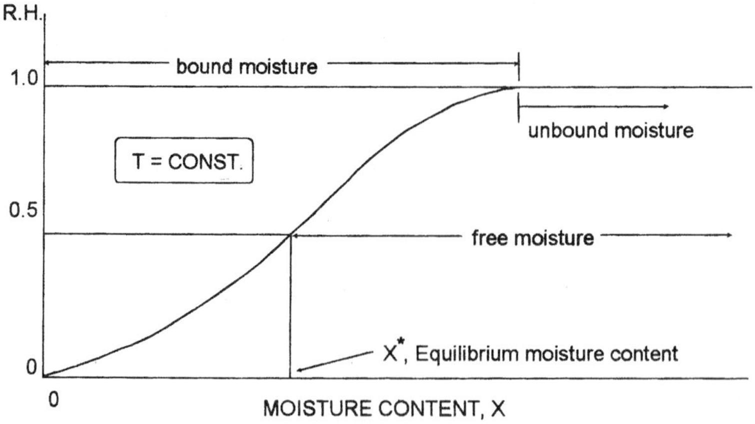

Fig. 4 Various types of moisture content.

A plot of ln (Ψ) against $1/T$ is linear with a slope of $\Delta H_w/R_g$ where R_g is the universal gas constant (R_g = 8.314 $\times$ 10^3 kg kgmol^{-1} K^{-1}). Note that the total heat required to evaporate bound water is the sum of the latent heat of vaporization and the heat of wetting; the latter is a function of the moisture content X. The heat of wetting is 0 for unbound water and increases with decreasing X. Since ΔH_w is responsible for lowering the vapor pressure of bound water, at the same relative humidity, ΔH_w is almost the same for all materials (Keey, 1978).

For most materials, the moisture-binding energy is positive; generally it is a monotonically decreasing function of the moisture content, with a value of zero for unbound moisture. For hydrophobic food materials (e.g., peanut oil or starches at lower temperatures) the binding energy can, however, be negative.

In general, water sorption data must be determined experimentally. Some 80 correlations, ranging from those based on theory to those that are purely empirical, have appeared in the literature. Two of the most extensive compilations are developed by Wolf et al. (1985) and Iglesias and Chirife (1982). Aside from temperature, the physical structure as well as composition of the material also affect water sorption. The pore structure and size as well as physical or chemical transformations during drying can cause significant variations in the moisture-binding ability of the solid.

1.1.3 Water Activity

In drying some materials that require careful hygiene attention (e.g., food), the availability of water for growth of microorganisms, germination of spores, and participation in several types of chemical reaction becomes an important issue. This availability, which depends on relative pressure or water activity, a_w, is defined as the ratio of the partial pressure, p, of water over the wet solid system to the equilibrium vapor pressure p_w, of water at the same temperature. Thus a_w, which is also equal to the relative humidity of the surrounding humid air, is defined as:

$$a_w = \frac{p}{p_w} \tag{5}$$

Different Shapes of the X versus a_w curves are observed, depending on the type of material (e.g., high-, medium-, or low-hygroscopicity solids).

Table 3 lists the measured minimum a_w values for microbial growth or spore germination. If a_w is reduced below these values by dehydration or by adding water-binding agents, such as sugars, glycerol, or salt, microbial growth is inhibited. Such additives should not affect the flavor, taste, or other quality criteria, however. As the amounts of soluble additives needed to depress a_w even by 0.1 are quite large, dehydration becomes particularly attractive for high-moisture foods as a way to reduce a_w. Figure 5 shows schematically the water activity versus moisture content curve for different types of food. Rockland and Beuchat (1987) provide an extensive compilation of results on water activity and its applications.

Figure 6 shows the general nature of the deterioration reaction rates as a function of a_w for food systems. Aside from microbial damage, which typically occurs for $a_w >$ 0.70, oxidation, nonenzymatic browning (Maillard reactions), and enzymatic reactions can occur even at very low a_w levels during drying. Laboratory or pilot testing is essential to ascertain than no damage occurs in the selected drying process, for this cannot, in general, be predicted.

Table 3 Minimum Water Activity a_w for Microbial Growth and Spore Germination

Microorganism	Water activity
Pseudomonas, Bacillus cereus spores	0.97
B. subtilis, C. botulinum spores	0.95
C. botulinum, Salmonella	0.93
Most bacteria	0.91
Most yeast	0.88
Aspergillus niger	0.85
Most molds	0.80
Halophilic bacteria	0.75
Xerophilic fungi	0.65
Osmophilic yeast	0.62

Source: Brockmann, 1973.

1.1.4 Equilibrium Moisture Content for Grains

Numerous theoretical, semiempirical and empirical models have been proposed in the literature to fit the equilibrium moisture content isotherms for grains. The theoretical models are based on capillary condensation or adsorption (Brunaner, Emmett, Teller [BET], Guggenheim, Anderson, de Boer, [GAB]). Henderson and Chung proposed a purely empirical model to fit measured data. In recent years, these have become more popular to fit the empirical data. Brooker et al. (1992) have discussed details of various models. Sun and Woods (1993) have provided a useful review of the equilibrium moisture data and models for wheat alone. A modified form of the Henderson equation had been adapted by the American Society of Agricultural Engineers (ASAE) as standard equipment for

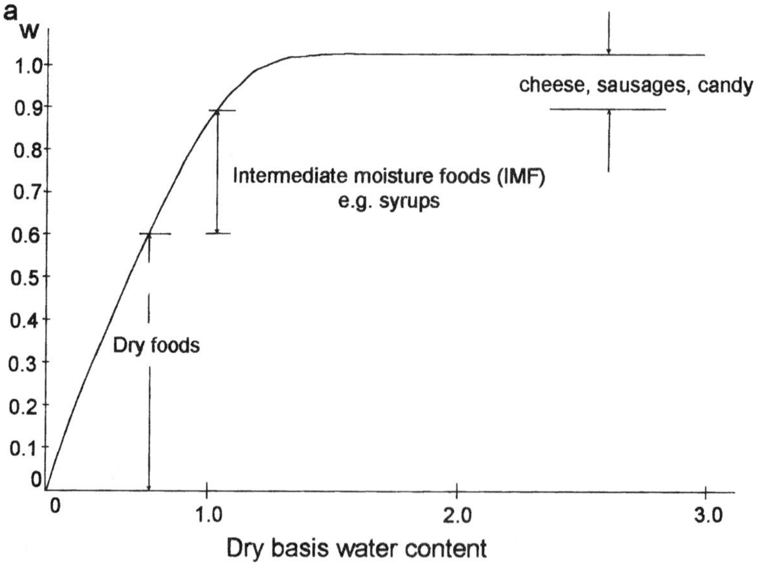

Fig. 5 Water activity versus moisture content plot for different types of food.

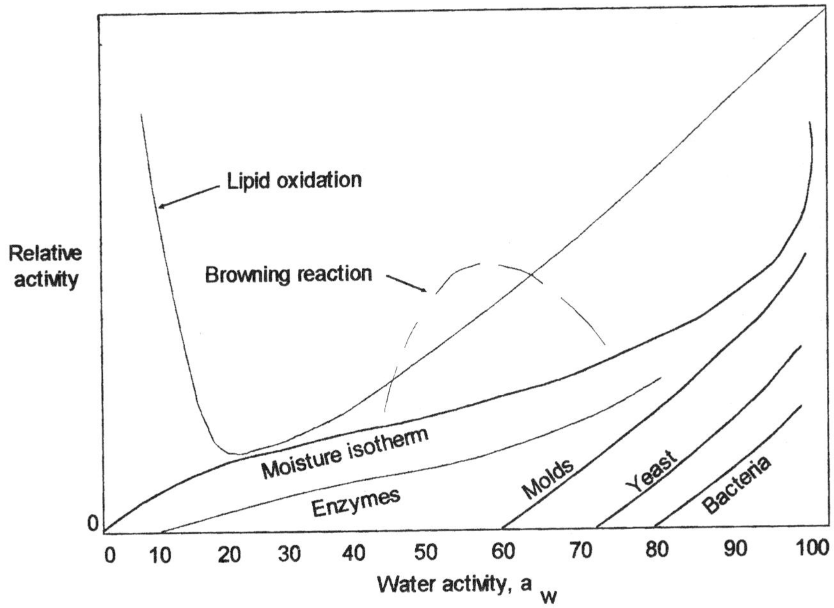

Fig. 6 Deterioration rates as a function of water activity for food systems.

various grains. For various grains and over specific relative humidity ranges, different models have been reported to be better than the modified Henderson equation. For the purpose of design calculations, however, most models are within the range of engineering accuracy.

The modified Henderson equation has the following form:

$$1 - RH = \exp[- k\,(T + C)\,(100\,X^n)]$$

where

RH = relative humidity
X = dry basis moisture content (fractional)
T = temperature, in Kelvin
k, n, C = experimental constants for grain

Table 4 is a list of values of k, n, and C for selected grains.

Table 4 Values of Constants in Henderson Equation

Grain	$k \times 10^5$	n	c
Barley	2.2919	2.0123	195.267
Beans	2.0899	1.8812	254.23
Corn	8.6541	1.8634	49.81
Rice	1.9187	2.4451	51.161
Sorghum	0.8532	2.4751	113.725
Wheat, durum	2.5738	2.2110	70.318

Table 5 Constants in Chung and Pfost (1967) Model

Grain	A	B	C
Barley	0.33363	0.05028	91.323
Corn	0.33872	0.05897	30.205
Rice	0.29394	0.046015	35.703
Wheat	0.37761	0.055318	112.35

Another empirical model in common use, proposed by Chung and Pfost (1967), is described by

$$X = A - B \ln[-(T + C) \ln(RH)]$$

where A, B, and C are empirical constants for a given grain. Table 5 lists values of these constants for selected grains.

Note that some grains display a hysteresis phenomenon. For drying calculations the desorption isotherm is the entity of interest.

From the moisture description isotherms at various temperatures the heat of vaporization h_v of moisture from grains at different moisture contents and temperatures can be computed using the Clapeyron equation for an isothermal phase change (see Brooker et al., 1992). Table 6 gives values of the heat of vaporization for water from several grains as functions of temperature (10° and 38°C) and moisture content (5, 10, and 20%, d.b.).

Note that the pure water has a heat of vaporization of 2.477×10^3 and 2.453×10^3 at 10° and 38°C, respectively.

1.2 Drying Kinetics

Consider the drying of a wet solid under fixed drying conditions. In the most general cases, after an initial period of adjustment, the dry-basis moisture content X decreases linearly with time t following the start of the evaporation. This is followed by a nonlinear

Table 6 Heat of Vaporization of Moisture in Selected Grains

Moisture content d.b.	T = 10°C kJ/kg 10^{-3}	T = 38°C kJ/kg 10^3
Wheat		
5	2.897	2.821
10	2.835	2.760
20	2.632	2.563
Corn		
5	3.007	2.928
10	2.921	2.844
20	2.695	2.625
Sorghum		
5	2.767	2.697
10	2.728	2.656
20	2.574	2.507

decrease in X with t until, after an prolonged time, the solid reaches its equilibrium moisture content X^*, and drying stops. In terms of free moisture content, defined as:

$$X_f = (X - X^*) \tag{6}$$

the drying rate drops to 0 at $X_f = 0$.

By convention, the drying rate N is defined as:

$$N = -\frac{M_s}{A}\frac{dX}{dt} \text{ or } -\frac{M_s}{A}\frac{dX_f}{dt} \tag{7}$$

under constant-drying conditions. Here, N(kg m^{-2} h^{-1}) is the rate of water evaporation; A is the evaporation area (may be different from heat transfer area); and M_s is the mass of bone-dry solid. If A is not known, then the drying rate may be expressed in kilograms of water evaporated per hour.

A plot of N versus X (or X_f) is the so-called drying rate curve. This curve must always be obtained under constant-drying conditions. In actual dryers, the drying material is generally exposed to varying drying conditions (e.g., different relative gas–solid velocities, different gas temperatures and humidities, and different flow orientations). Thus, it is necessary to develop a method to interpolate or extrapolate the limited-drying rate data over a range of operating conditions.

Figure 7 shows a typical ''textbook'' drying rate curve displaying an initial constant rate period where $N = N_c$ = constant. The constant rate period is fully governed by the rates of external heat and mass transfer because a film of free water is always available at the evaporating surface. This drying period is nearly independent of the material being dried. Many foods and agricultural products, however, do not display the constant rate period at all because internal heat and mass transfer rates determine the rate at which water becomes available at the exposed evaporating surface.

At the so-called critical moisture content X_c, N begins to fall with further decrease in X, as water cannot migrate at the rate N_c to the surface owing to internal transport limitations. The mechanism underlying this phenomenon depends on both the material and drying conditions. The drying surface first becomes partially unsaturated and then

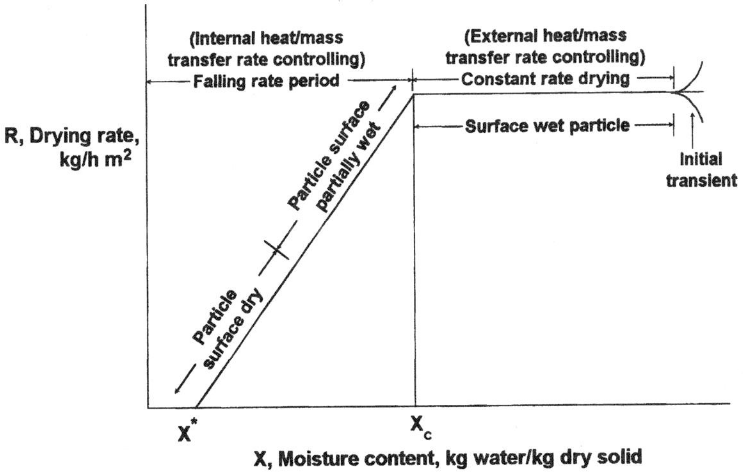

Fig. 7 Typical textbook batch-drying rate curve under constant-drying conditions.

fully unsaturated until it reaches the equilibrium moisture content X^*. Detailed discussions of drying rate curves are given by Keey (1991), and Mujumdar and Menon (1995).

A material may display more than one critical moisture content at which the drying rate curve shows a sharp change of shape. This is generally associated with changes in the underlying mechanism of drying owing to structural or chemical changes. It is also important to note that X_c is not solely a material property. It depends on the drying rate under otherwise similar conditions, and it must be determined experimentally.

It is easy to see that N_c can be calculated using empirical or analytical techniques to estimate the external heat/mass transfer rates (Keey, 1978; Geankopolis, 1993). Thus,

$$N_c = \frac{\sum q}{\lambda_s} \tag{8}$$

where $\sum q$ represents the sum of heat fluxes due to convection, conduction, or radiation, and λ_s is the latent heat of vaporization at the solid temperature. In purely convective drying, the drying surface is always saturated with water in the constant rate period; thus, the liquid film attains the wet-bulb temperature. The wet-bulb temperature is independent of the geometry of the drying owing to the analogy between heat and mass transfer.

The drying rate in the falling rate period(s) is a function of X (or X_f) and must be determined experimentally for a given material being dried in a given type of dryer.

If the drying rate curve (N vs. X) is known, the total drying time required to reduce the solid moisture content from X_1 to X_2 can be simply calculated by:

$$t_d = -\int_{X_1}^{X_2} \frac{M_s}{A} \frac{dX}{N} \tag{9}$$

Table 7 lists expressions for the drying time for constant rate, linear falling rates, and a falling rate controlled by liquid diffusion of water in a thin slab. The subscripts c and f refer to the constant and falling rate periods, respectively. The total drying time is, of course, a sum of drying times in two succeeding periods. Different analytical expressions are obtained for the drying times t_f depending on the functional form of N or the model used to describe the falling rate (e.g., liquid diffusion, capillary, or evaporation-condensation). For some solids, a receding front model (wherein the evaporating surface recedes into the drying solid) yielods a good agreement with experimental observations. The principal goal of all falling rate drying models is to allow reliable extrapolation of drying kinetic data over various operating conditions and product geometries.

The expression for t_f in Table 7 using the liquid diffusion model (Fick's second law of diffusion form applied to diffusion in solids with no real fundamental basis) is obtained by analytically solving the following partial differential equation:

$$\frac{\partial X_f}{\partial t} = D_L \frac{\partial^2 X_f}{\partial x^2} \tag{10}$$

subject to the following initial and boundary conditions:

$$X_f = X_i, \text{ everywhere in the slab at } t = 0$$

$$X_f = 0, \text{ at } x = a \text{ (top, evaporating surface), and} \tag{11}$$

$$\frac{\partial X_f}{\partial x} = 0, \text{ at } x = 0 \text{ (bottom, nonevaporating surface)}$$

Table 7 Drying Times for Various Drying Rate Models

Model	Drying time
Kinetic model, $N = -\frac{M_s}{A}\frac{dX}{dt}$	t_d = Drying time to reach final moisture content X_2 from initial moisture content X_1
$N = N(X)$ (general)	$t_d = \frac{M_s}{A}\int_{X_2}^{X_1}\frac{dX}{N}$
$N = N_c$ (constant-rate period)	$t_c = -\frac{M_s}{A}\frac{(X_2 - X_1)}{N_c}$
$N = aX + b$ (falling-rate period)	$t_f = \frac{M_s}{A}\frac{(X_1 - X_2)}{(N_1 - N_2)}\ln\frac{N_1}{N_2}$
$N = Ax \quad X^* \leq X_2 \leq X_c$	$t_f = \frac{M_s X_c}{AN_c}\ln\frac{X_c}{X_2}$
Liquid diffusion model DL = *constant*, $X_2 = X_c$ Slab; one-dimensional diffusion, evaporating surface at X^*	$t_f = \frac{a^2}{\pi D_L}\ln\frac{8}{\pi^2}\frac{X_1}{X_2}$ X = average free moisture content a = half-thickness of slab

Source: Mujumdar, 1997.

The model assumes one-dimensional liquid diffusion with constant effective diffusivity, D_L, and no heat (Soret) effects. X_2 is the average free moisture content at $t = t_f$ obtained by integrating the analytical solution $X_f(x, t_f)$ over the thickness of the slab, a. The expression in Table 7 is only applicable for long drying times because it is obtained only by retaining the first term in the infinite series solution of the partial differential equation.

The moisture diffusivity in solids is a function of both temperature and moisture content. For strongly shrinking materials, the mathematical model used to define D_L must also account for the changes in diffusion path. The temperature dependence of diffusivity is adequately described by the Arrhenius equation as follows:

$$D_L = D_{L0}\exp[-E_a / R\, T_{abs}] \tag{12}$$

where D_L is the diffusivity, E_a is the activation energy and T_{abs} is the absolute temperature. Okos et al. (1992) have given an extensive compilation of D_L and E_a values for various food materials. Zogzas et al. (1996) provide methods of moisture diffusivity measurement and an extensive bibliography on the topic. Empirical correlations for the effective moisture diffusivity for some selected grains are given in Table 8.

1.3 Moisture Diffusivity in Grains

Zogzas and co-workers (1996) have made an extensive compilation of experimental data on moisture diffusivity for grains and various foodstuffs. Effects of moisture content and

Table 8 Various Empirical Models for Temperature and Moisture Dependence of Moisture Diffusivity in Selected Grains

Grain	Model formulation
Corn	$D(X, T) = a_0 \exp(a_1 X) \exp\left(-\frac{a_2}{T}\right)$
	$a_0 = 7.817 \cdot 10\text{-}5$ (or $4.067 \cdot 10\ -5$), $a^1 = 5.5$, $a^2 = 4850$ $0.05 \leq X \leq 0.4$, $283 \leq T \leq 300$, $R_{eq} = 4$ mm
Corn	$D(X, T) = a_0 \exp(a_1\ T - a_2)\ X \exp\left(\frac{-a_3}{T}\right)$
	$a_0 = 2.54.10^{-8}$, $a_1 = 1.2343.10^{-1}$, $a_2 = 45.47$, $a_3 = 1183.3$ $0.1 \leq X \leq 0.4$, $293 \leq T \leq 338$.
	$a_0 = 101.41$, $a_1 = 4.71$, $a_2 = 49.5$, $a_3 = 6400$, $0.1 \leq X \leq 0.25$, $303 \leq T \leq 323$.
Rice	$D(X, T) = a_0 \exp(a_1\ T - a_2)\ X \exp\left(\frac{-a_3}{T}\right)$
	$a_0 = 2.744.10^{-6}$, $a_1 = 1.589.10^{-3}$, $a_2 = 0.379$, $a_3 = 4294.8$, $0.26 \leq X \leq 0.32$, $321.8 \leq T \leq 355.2$
Wheat	$D(X, T) = (a_0 + a_1\ X + a_2\ X^2) \exp(-\ a_3\ T_c)$ $a_0 = -32.10^{-9}$, $a_1 = 674.10^{-9}$, $a_2 = -2327.10^{-9}$, $a_3 = 0.033$ T_c is in °C

temperature are discussed in detail. Over 92% of reported data lie in the range 10^{-12} and 10^{-8} m^2/s. Significant variation in reported data can be ascribed to variation in the experimental as well as analytical technique employed. The Arrhenius-type relation is often used to empirically fit the data on effect of temperature on diffusivity with reasonable success, that is:

$$D_L = D_{L0} \exp\left(-\frac{E_a}{RT}\right)$$

where D_{L0} is the Arrhenius factor m^2/s; E_a is the activation energy for moisture diffusion, (kJ/mol); R is the ideal gas constant (kJ/mol K); and T is the absolute temperature K. E_a for most grains is close to 30 kJ/mol. Note that D_{L0} and E_a are strongly correlated, so it is difficult to compare accuracy of various data for E_a from the calculated values of D_L alone.

Different empirical formulations have been used to combine the effects of moisture and temperature for the calculations of D_L. The effect of moisture content can be introduced into D_{L0} or E_a, or both. Also, polynomial factors have been fitted empirically. There is no theoretical basis for choosing one formulation over the other, however.

Table 9 Approximate Ranges of Effective Moisture Diffusivity Values for Some Grains

Material	Moisture content (kg/kg, d.b.)	Temperature (°C)	Diffusivity (m^2/s)
Rice	0.10–0.25	30–50	3.8×10^{-8} –2.5×10^{-7}
Soybeans	0.07	30	7.5×10^{-13}–5.4×10^{-12}
Wheat	0.12–0.30	21–80	6.9×10^{-12}–2.8×10^{-10}

Source: Zogzas et al., 1996; Marinos-Kouris and Marouris, 1995.

D_L is not a true material property, and care should be taken in applying effective diffusivity correlations obtained with simple geometric shapes (e.g., slab, cylinder, or sphere) to the more complex shapes actually encountered in practice, as this may lead to incorrect calculated results (Gong et al., 1997).

Table 9 is a list of the ranges of the liquid water diffusivity for rice, soybeans, and wheat over temperature ranges of interest in drying. Note that one to two orders-of-magnitude variation in the diffusivity is not uncommon during a single drying run.

Keey (1978) and Geankopolis (1993), among others, have provided analytical expressions for liquid diffusion and capillarity models of falling rate drying. Table 10 gives solution of the one-dimensional transient partial differential equations for cartesian, cylindrical, and spherical coordinate systems. These results can be used to estimate the diffusivity from the falling rate drying data or to estimate the drying rate and drying time if the diffusivity value is known.

The diffusivity D_L is a strong function of X_f as well as temperature and must be determined experimentally. Thus, the liquid diffusion model should be considered purely as an empirical representation of drying in the falling rate period. More advanced models are available, but their widespread use in the design of dryers is hampered by the need for the extensive empirical information required to solve the governing equations. Turner and Mujumdar (1997) provide a wide assortment of mathematical models of drying and

Table 10 Solution to Fick's Second Law for Some Simple Geometries

Geometry	Boundary conditions	Dimensionless average free MC
Flat plate of thickness $2b$	$t = 0; -b < z < b; X = X_0$ $t > 0; z = \pm b; X = X^*$	$X = \frac{8}{\pi^2} \sum_{n=1}^{\infty} \frac{1}{(2n-1)} \exp\left[-(2n-1)^2 \frac{\pi^2}{4b}\left(\frac{D_L t}{b}\right)\right]$
Infinitely long cylinder of radius R	$t = 0; 0 < r < R; X = X_0$ $t > 0; r = R; X = X^*$	$X = 4 \sum_{n=1}^{\infty} \frac{1}{R^2 a_n^2} \exp(-D_L a_n^2 t)$ where a_n are positive roots of the equation $J_0(Ra_n) = 0$
Sphere of radius R	$t = 0; 0 < r < R; X = X_0$ $t > 0; r = R; X = X^*$	$X = \frac{6}{\pi^2} \sum_{n=1}^{\infty} \frac{1}{n^2} \exp\left[\frac{-n^2 \pi^2}{R}\left(\frac{D_L t}{R}\right)\right]$

Source: Pakowski and Mujumdar, 1995.

dryers, and also discuss the application of various techniques for the numerical solution of the complex governing equations.

One simple approach to interpolating a given falling rate curve over a relatively narrow range of operating conditions is that first proposed by van Meel (1958). The plot of normalized drying rate $\nu = N/N_c$ versus normalized free moisture content $\eta = (X - X^*)/(X_c - X^*)$ was nearly independent of the drying conditions. This plot, called the characteristic drying rate curve, is illustrated in Fig. 8. Thus, if the constant rate-drying rate N_c can be estimated and the equilibrium moisture content data are available, then the falling rate curve can be estimated using this highly simplified approach. Extrapolation over wide ranges is not recommended, however.

Waananen et al. (1993) have provided an extensive bioliography for over 200 references dealing with models for drying porous solids. Such models are useful to describe drying processes for engineering design, analysis, and optimization. A mathematical description of the process is based on the physical mechanisms of internal heat and mass transfer that control the process resistances, as well as the structural and thermodynamic assumptions made to formulate the model. In the constant rate period, the overall drying rate is determined solely by the heat and mass transfer conditions external to the material being dried, such as the temperature, gas velocity, total pressure, and partial pressure of the vapor. In the falling rate period, the rates of internal heat and mass transfer determine the drying rate. Modeling of drying becomes complicated because more than one mechanism may contribute to the total mass transfer rate and the contributions from different mechanisms may even change during the drying process.

Diffusional mass transfer of the liquid phase is the most commonly assumed mechanism of moisture transfer used in modeling the drying that takes place, under locally applied pressure, at temperatures below the boiling point of the liquid. At higher temperatures, the pore pressure may rise substantially and cause a hydrodynamically driven flow of vapor that, in turn, may cause a pressure-driven flow of liquid in the porous material.

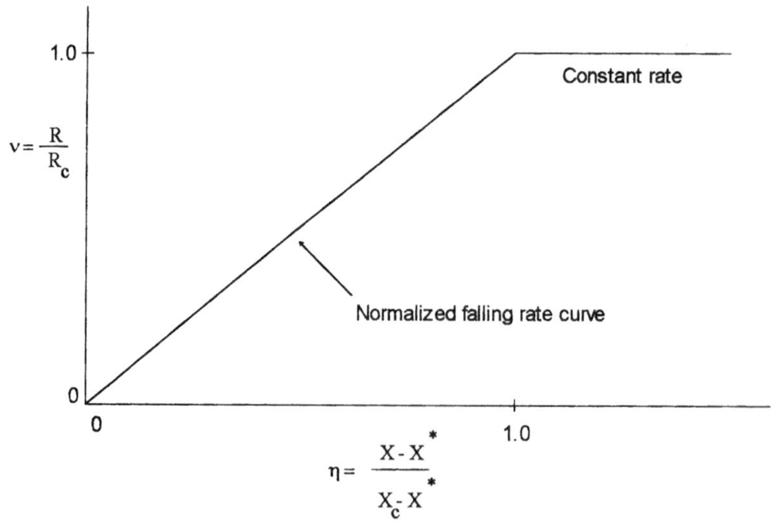

Fig. 8 Characteristic drying rate curve.

For solids with continuous pores, a surface–tension-driven flow (capillary flow) may occur as a result of capillary forces caused by the interfacial tension between the water and the solid. In the simplest model, a modified form of the Poiseuille flow can be used in conjunction with the capillary force equation to estimate the rate of drying. Geankopolis (1993) has shown that such a model predicts that the drying rate in the falling rate period is proportional to the free-moisture content in the solid. At low, solid moisture contents, however, the diffusion model may be more appropriate.

The moisture flux caused by capillarity can be expressed in terms of the product of a liquid conductivity parameter and moisture gradient. Here, the governing equation has the same form as the diffusion equation.

For certain materials and under conditions such as those encountered in freeze-drying, a ''receding-front'' model involving a moving boundary between ''dry'' and ''wet'' zones often describes the mechanism of drying much more realistically than does the simple liquid diffusion or capillarity model. Examination of the freeze-drying of a thin slab indicates that the rate of drying is dependent on the rate of heat transfer to the ''dry–wet'' interface and the mass transfer resistance offered by the porous dry layer to permeation of the vapor that sublimes from the interface. Because of the low pressure encountered in freeze-drying, Knudsen diffusion may be significant. Liapis and Marchello (1984) have discussed models of freeze-drying involving both unbound and bound moisture.

When drying materials are under intense-drying conditions, diffusion or capillarity models generally do not apply. If evaporation can occur within the material, there is a danger of the so-called vapor-lock that occurs within the capillary structure causing breaks in liquid-filled capillaries. This phenomenon can cause departure from the classic drying curve [e.g., no constant-rate drying may appear under intense-drying conditions, but may do so under milder-drying conditions (Zaharchuk, 1993)].

1.4 Drying Kinetics of Grains

Without exception grains are dried in convection (direct-type) dryers using heated air as the drying medium. Use of conduction, radiation, or dielectric heating is negligible in practice. Most dryers are thus basically packed-bed fluid-type, with heated drying air percolating through the bed, which may be stationary or moving under gravity. Use of fluidization, spouting, vibration, agitation, or other, is possible; but not often useful in practice. The direction of the airflow may be upward, downward, or in the cross-flow direction relative to a stationary or moving bed. Often internals are placed within the bed of grains to ensure better flow of the grains and more uniform air distribution to avoid wet and hot spots.

Except when harvested in wet season, most of the drying of grains occurs in the following rate period. The internal moisture diffusion model adequately describes the falling rate. Newton's cooling–law-type model is also applied, for example:

$$\frac{d\bar{X}}{dt} = k\,(\bar{X} - X^*)$$

or, separating variables and integrating

$$\frac{\bar{X}(t) - X^*}{\bar{X}_i - X^*} = \exp(k\,t)$$

where $\overline{X}_i$ is the initial dry-basis moisture content of the grain. A modification of the foregoing equation that gives an additional fitting parameter n is the so-called Page equation:

$$\frac{\overline{X}(t) - X^*}{X_i - X^*} = \exp(-k\, t^n)$$

where k and n are determined empirically.

The foregoing models are often referred to as ''thin-layer'' models, as they assume no external resistance to moisture transfer. In practice, deep beds are used in which the drying conditions (e.g., temperature of grain and air as well as air humidity) change continuously. Hence, in modeling deep beds it is necessary to account for changes in external conditions even if the ''thin-layer'' kinetics are internally controlled. Brooker et al. (1992) have discussed various deep–bed-drying models for grains; the reader is referred to this reference and others for details.

2 CLOSING REMARKS

Dryer calculations require knowledge of various thermophysical properties and quality parameters; these are integral parts of other chapters in this handbook. The reader is referred to the website www.geocities.com/drying_guru/ for information about current sources and resources useful in the selection, design, and operation of dryers.

NOMENCLATURE

A	evaporation area, m^2
a_w	water activity, –
B	empirical constant
c_p	specific heat, J kg^{-1} K^{-1}
c_s	humid heat, J kg^{-1} K^{-1}
C	empirical constant
D_L	effective diffusivity, m^2 s^{-1}
D_{LO}	effective diffusivity at reference temperature, m^2 s^{-1}
E_a	activation energy, J
ΔH_w	enthalpy of wetting, J kg^{-1}
h	convective heat transfer coefficient, W m^{-2} K^{-1}
k	empirical constant
k_g	thermal conductivity, W m^{-1} K^{-1}
k_y	convective mass transfer coefficient, kg mol s^{-1} m^{-2} mol $frac^{-1}$
M_{air}	molar mass of air, kg mol^{-1}
M_s	mass of bone dry solid, kg
n, N	empirical constant
N	drying rate, kg m^{-2} h^{-1}
P_v	vapor pressure of pure water, Pa
p	partial pressure, Pa
p_w	equilibrium vapor pressure of water, Pa
RH	relative humidity, decimal fraction, –
R_g	universal gas constant, 8.314 J mol^{-1} K^{-1}
T	temperature, °C

T_{abs}	absolute temperature, K
T_{wb}	wet-bulb temperature, °C (or K)
t	time, s (or h)
X	total moisture content, kg water/kg dry solid, –
X_c	critical moisture content, kg water/kg dry solid, –
$\overline{X}$	average moisture content, kg water/kg dry solid, –
X_f	free moisture content, kg waer/kg dry solid, –
X^*	equilibrium moisture content, kg water/kg dry solid, –
Y	absolute air humidity, kg water vapor/kg dry air

Greek letters

η	normalized drying rate, –
λ_s	latent heat of vaporization, J kg^{-1}
μ_g	dynamic viscosity, kg m^{-1} s^{-1}
ν	normalized drying rate, –
ρ_g	density, kg m^{-3}

Subscripts

c	constant rate period
f	falling rate period
g	gas
i	initial
s	solid
ν	vapor
w	water
wb	wet-bulb

REFERENCES

Brockmann MC, 1973. Intermediate moisture foods. In: WB van Arsdel, MJ Copley, AI Morgan, eds. Food Dehydration. AVI, Westport.

Brooker DB, Bakker–Arkema FW, Hall CW, 1992. Drying and Storage of Grains and Oilseeds. AVI, New York.

Bruin S, 1988. Preconcentration and drying of food materials: Thijssen Memorial Symposium. Proceedings of the International Symposium on Preconcentration and Drying of Foods. Eindhoven, The Netherlands.

Bruin S, Luyben KCAM, 1980. Drying of food materials: a review of recent developments. In: AS Mujumdar, ed. Advances in Drying, Vol. 1. Hemisphere, Washington, pp 155–216.

Chung DS, Pfost HB, 1967. Adsorption and desorption of water vapor by cereal grains and their products. Trans ASAE 10:552–575.

Fortes M, Okos MR, 1980. Drying theories: their bases and limitations as applied to foods and grains. In: AS Mujumdar, ed. Advances in Drying, Vol. 1. Hemisphere, Washington, pp. 119–154.

Geankopolis CJ, 1993. Transport Processes and Unit Operations, 3rd ed. Prentice Hall, Englewood Cliffs, NJ.

Gong Z–X, Devahastin S, Mujumdar AS, 1997. A two-dimensional finite element model for wheat drying in a novel rotating jet spouted bed. Drying Technol Int J 15:575–592.

Henderson SM, 1952. A Basic concept of equilibrium moisture. Agric Eng 33:29–31.

Iglesias HA, Chirife J, 1982. Handbook of Food Isotherms: Water Sorption Parameters for Food and Food Components. Academic, New York.
Keey RB, 1978. Introduction to Industrial Drying Operations. Pergamon, Oxford.
Keey RB, 1992. Drying of Loose and Particulate Materials. Hemisphere, Washington.
Liapis A, Marchello JM, 1984. Advances in modeling and control of freeze drying. In: AS Mujumdar, ed. Advances in Drying. Vol. 3. Hemisphere, Washington, pp 217–244.
Marinos–Kouris D, Maroulis ZB, 1995. Transport properties in the drying of solids, In: AS Mujumdar, ed. Handbook of Industrial Drying. 2nd ed. Marcel Dekker, New York, pp 113–159.
Mujumdar AS, ed, 1995. Handbook of Industrial Drying, 2nd ed. Marcel Dekker, New York.
Mujumdar AS, 1995. Superheated steam drying. In: AS Mujumdar, ed. Handbook of Industrial Drying, 2nd ed., Marcel Dekker, New York, pp 1071–1086.
Mujumdar AS, 1997. Drying fundamentals. In: CGJ Baker, ed. Industrial Drying of Foods. Blackie Academic, London, pp 7–30.
Mujumdar AS, Menon AS, 1995. Drying of solids. In: AS Mujumdar, ed. Handbook of Industrial Drying, 2nd ed. Marcel Dekker, New York, pp 1–46.
Okos MR, Narsimhan G, Singh RK, Weitnauer AC, 1992. Food dehydration. In: DR Heldman, DB Lund, eds. Handbook of Food Engineering. Marcel Dekker, New York, pp. 437–562.
Pakowski Z, Bartczak Z, Strumillo C, Stenstrom S, 1991. Evaluation of equations approximating thermodynamic and transport properties of water, steam, and air for use in CAD of drying processes. Drying Technol 9:753–773.
Pakowski Z, Mujumdar AS, 1995. Basic process calculations in drying. In: AS Mujumdar, ed. Handbook of Industrial Drying, 2nd ed. Marcel Dekker, New York, pp 71–112.
Rockland LB, Beuchat LR, 1987. Water Activity: Theory and Applications to Food. Marcel Dekker, New York.
Sun DW, Woods JL, 1993. The moisture content/relative humidity equilibrium relationships for wheat, a review, 1993. Drying Technol 11:1523–1551.
Turner I, Mujumdar AS, eds. 1997. Mathematical Modeling and Numerical Techniques in Drying Technology. Marcel Dekker, New York.
van Meel DA, 1958. Adiabatic convection batch drying with recirculation of air. Chem Eng Sci 9: 36–44.
Waananen KM, Litchfield JB, Okos MR, 1993. Classification of drying models for porous solids. Drying Technology 11:1–40.
Wolf W, Spiess WEL, Jung G, 1985. Sorption Isotherms and Water Activity of Food Materials. Elsevier, Amsterdam.
Zaharchuk DJ, 1993. Intense drying and spalling of agglomerate spheres (iron ore pellets). PhD Dissertation, Chemical Engineering Department, University of Waterloo, Waterloo, Canada.
Zogzas NP, Maroulis ZB, Marinos–Kouris D, 1996. Moisture diffusivity data compilation in foodstuffs. Drying Technol J 14:pp.2225–2253.

Supplemental References on Grain Drying

Bala BK, 1997. Drying and Storage of Cereal Grains. Science Publishers, Enfield, USA.
Brooker DB, Bakker–Arkema FW, Hall CW, 1992. Drying and storage of grains and oilseeds. van Nostrand Reinhold, New York.
Brooker DB, Bakker–Arkema FW, Hall CW, 1974. Drying Cereal Grains, AVI Publishing, Westport, CT.
Hall CW, 1980. Drying and storage of agricultural crops. AVI, Westport, CT.
Pabis S, Jayas DS, Cenkowski S, 1998. Grain Drying: Theory and Practice. John Wiley & Sons, New York.

6

Grain-Drying Systems

SUSANTA KUMAR DAS and AMALENDU CHAKRAVERTY

Indian Institute of Technology, Kharagpur, India

1 INTRODUCTION

Moisture content in the cereal grain plays a vital role in the chain of handling and storage. Germination, microbial growth, insect infestation, deterioration of color, development of off-flavor, and lowering of nutritive value are some common quality factors associated with storage of high-moisture grain that render the commodity unfit for human consumption. Thus, removal of moisture becomes a crucial step to provide extended storage life, facility of handling, retention or enhancement of quality, and new products for further processing.

Commonly, the process of thermally removing small amounts of moisture from materials is referred to as drying, which can be accomplished in various ways—some of which are specific to the commodity, location, or the volume of the material to be handled. Nevertheless, removal of moisture from the grain is a very energy-intensive process; thus, the efficiency of a drying operation in terms of energy and time has economic consequences for commercial viability. In addition to these, maintenance of hygienic condition, prevention or control of quantitative and qualitative losses of materials, and management of proper space utilization should be adhered to.

The final moisture content of the product to be achieved is largely decided by the storage environment and its storability or set tolerance limits on the quality attributes. This, in turn, dictates the selection of a drying system, drying time, and the range of operating parameter values.

2 CLASSIFICATION OF GRAIN-DRYING SYSTEMS

Classification of grain-drying systems is generally based on the mode of heat transfer to the material (i.e., conduction, convection, and radiation). The conduction mode of drying, also termed nonadiabatic, or indirect mode of drying, is accomplished by direct contact of the product with a heated surface or heated particles. In convection drying, the air is heated to the desired level of temperature and is used as a medium of convective heat transfer. In the radiation-heating mode, the material being dried absorbs energy from some source that emits electromagnetic radiation. This absorbed energy is converted to heat that vaporizes the water molecules inside the food material. Ideally, all these modes of heat transfer act together; it is the quantitative proportional contribution of heat transfer that determines the mode of heat transfer occurring during drying of material.

3 CONDUCTION DRYING

When heat for drying flows from the surface of the retaining wall to the wet solid, this phenomenon is known as conduction drying. This mode of drying is generally more appropriate for very wet solids. Conduction drying of materials can also be accomplished by direct contact with the heated particles.

3.1 Drying by Direct Contact with Heated Surface

The materials to be dried are placed over a heated bed (stationary, vibratory, or moving) that supplies heat for vaporization of moisture. The evolved vapor is removed under subatmospheric pressure (vacuum), or by a stream of gas that acts as a carrier of moisture. In spite of high thermal efficiency compared with a convective dryer (1), conduction dryers are expensive to build and to operate. Consequently, these are commercially used only for low temperature and an atmosphere of low-level oxygen for drying of valuable materials that are essentially thermolabile and easily oxidizable.

In this category, the most important one is the steam-tube rotary dryer that can handle free-flowing granular materials, such as cereal grains or the like. It consists of a cylinder with a length/diameter ratio generally varying from 4 to 10. The cylinder is installed on a suitable bearing assembly at a small angle (2°–4°) to the horizontal. Feed enters at one end of the cylinder that progresses forward by virtue of rotation (4–6 rpm) and the slope, and the dried product is discharged at the other end. The inside of the rotary dryer is provided with one or more rows of metal tubes, installed longitudinally in the shell, through which steam flows. This kind of indirect dryer is reported to be suitable for drying cattle feed, brewer grain, parboiled paddy, corn, wheat, and such.

The performance of conduction drying with this type of rotary dryer can be improved by the combination of conduction and convection modes of heat transfer. The hot gas in addition to transferring heat by convection also acts as an aid for removal of moisture from the materials. A typical rotary dryer has been reported for drying parboiled paddy (2). It comprises a cylindrical shell, 30 ft long and 4 ft in diameter, with 48 pairs of 2-in. and $1^1/_2$-in. (nominal) size steam tubes in two rows inside the shell. These steam pipes are suitable for using up to 50 psig steam in combination with common steam inlet and condensation outlet ports. Longitudinal flights are fixed at the shell for lifting and forward movement of the paddy inside the dryer. While the feed is introduced through a hopper with an arrangement of a feed rate control mechanism by a screw feeder and sliding plate, hot air is blown at the entrance of the feed end breeching box. The rotary shell is rotated at 2–6 rpm by a motor through a speed regulation gear and pulley, and belt-drive mechanisms.

The major advantages of this category of dryers are maintaining the uniformity of drying and for handling high-moisture grains and kernels having a tendency to stick together. However, these types of dryers are more popular for drying chemicals than for agricultural products (3).

3.2 Drying by Direct Contact with Heated Particles

Removal of moisture as well as puffing of cereal grains is traditionally practiced in southeast Asian countries using heated particles in direct contact with the grain. However, no large-scale commercial unit has yet been available for drying grains by this technique.

A heated-sand–drying system has been reported to be very effective for drying paddy (20–29% moisture content). One part of paddy is mixed with two parts of sand (120°–150°C) in a rotating drum for 5–6 min (4). The discharged mixture of sand and grain is separated by a screen, and the sand is recycled to the drum. While the grains are parboiled in situ, there is simultaneous removal of moisture from the grains. Moisture removal is very rapid if the sand temperature is increased to 190°–240°C and mixed with paddy for 15–60 s. A significant increase in yield of head rice is noted compared with that of a shed-dried sample. Removal of grain moisture from 24% to 15% (db) is possible with either a small sand/grain ratio with moderately high sand temperature, or with a large sand/grain ratio with low initial sand temperature (5). From an energy point of view, a larger sand/grain mass ratio with higher initial sand temperature is favored.

4 CONVECTION DRYING

The most popular grain-drying technique is convection drying. In this system, the heat transfer for drying is accomplished by direct contact between the wet material and a stream of hot gas (generally air). The vaporized moisture is carried away from the grain by the heating medium. Both continuous and batchwise convection-drying systems are practiced for drying wet grains. Convection drying can be further subdivided, as given in Fig. 1.

4.1 Natural Air Drying

Natural air drying is essentially in situ drying of grain that is contained in bins commonly used for grain storage. The bin is equipped with a fan or blower, a duct system, a perforated

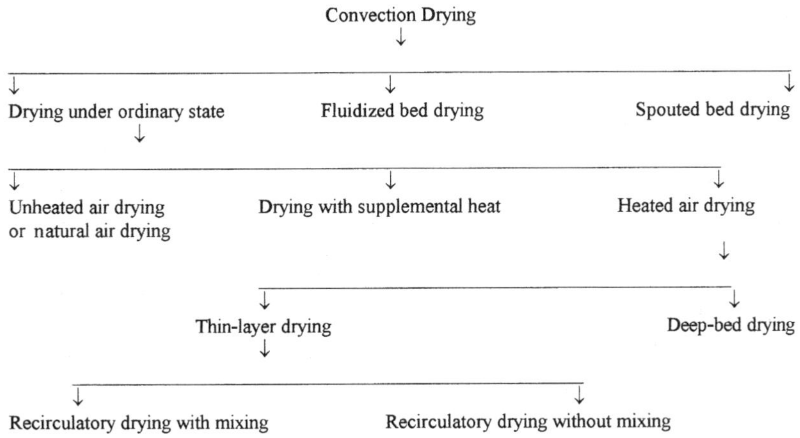

Fig. 1 Classification of convection grain-drying system.

floor, and exhaust vents for the moist air to escape. The primary need in this drying system is the estimation of the required specific airflow rate (i.e., volume of air per unit volume of grain per unit time, [cu ft/bu h^{-1} or m^3/m^3 min^{-1}]). The air might be blown upward or sucked downward for the moisture migration, and the arrangement is entirely a matter of choice, considering advantages or disadvantages of the system. The recommended airflow rates for a 1.2 to 1.8-m depth of grain, are 150–250 m^3/m^3 h^{-1} and 250–500 m^3/m^3 h^{-1}, respectively, for small grains and others, such as peas, beans, and shelled corn and ear corn (6). The approximate airflow rate can be calculated using Eq. (1), on the assumption that the air will be unheated and a negligible amount of sensible heat transfer will occur (7).

$$\frac{Q}{v} C_p(t_a - t_g)\theta = L_{fg}(W_d)\,(x_o - x_f) \tag{1}$$

where Q is the airflow rate (m^3/s), v is specific volume of air (m^3/kg), C_p is specific heat of air (kJ/kgK^{-1}), t_a and t_g are temperatures (°C) of air at the inlet and when leaving the grain, respectively, L_{fg} is the latent heat of vaporization (kJ/kg), W_d is dry matter in the material (kg), x_o and x_f are the grain initial and final moisture contents (decimal, dry basis), and θ is the drying time (s). The rates of drying depend on the local weather conditions, which must be taken into consideration in this equation. The minimum airflow rate for natural air drying depends on the nature of grain and the initial moisture of the grain. For wheat with moisture content of 20% (db), the flow rate of air should be about 2.4 m^3/m^3 min^{-1}. The corresponding values for oat, shelled corn, and sorghum with similar moisture content should be 1.6, 2.4, and 2.4 m^3/m^3 min^{-1} (8).

One important aspect for natural air drying of wet grain is the specific airflow rate. If the wet grain is placed on the top of the dry grain, the air that emerges through the dry grain is almost in equilibrium condition and virtually does not alter the drying potential. Correlation among the depth of wet grain, bin diameter, and airflow rate per unit volume of wet grain placed above the bed of dry grain in a fan-bin system has been suggested (7). Low initial investment and maintenance cost, no fuel, minimal or near-absence of mold growth, and need of least supervision are some of the favorable features of the natural–air-drying system for grains. However, a long drying time, weather dependency, large drying space, and very low efficiency in humid areas are considered to be unfavorable criteria for this drying system (9).

4.2 Drying with Supplemental Heat

Drying with supplemental heat involves drying of grain in the storage bin using a constant heat source to raise the air temperature a few degrees (4°–12°C) above ambient. The heating source could be preferentially an electric-heating system. However, a running blower or motor can raise the air temperature to a small extent. This method of drying is advantageous for its low-cost operation and design simplicity, besides its usefulness for places where the weather remains moist for a prolonged period. Two adverse effects are generally encountered in this process of drying with supplemental heat. First, the grain may have a chance of overdrying. Second, the higher temperature of the exhaust air from the drying zone will heat the underdried grain, which may induce rapid mold growth.

Drying with this method usually requires a continuous ventilation system for about 1–3 weeks. Chances of overdying become minimal if the air temperature is increased by 5°C or less. Also the management is relatively simple, with a minimum-handling opera-

Table 1 Advantages and Disadvantages of Supplemental Heated Air Drying of Grain

Advantages	Disadvantages
Low cost of equipment and maintenance	Fire hazards to a certain extent
Independent of weather	Danger of accelerating mold growth
Low fuel cost	Rate of drying is still slow
Requires less supervision	Useful particularly for short-term storage on the farm
Most efficient use of bin capacity	

tion. This low-temperature air drying induces a minimum of stress in the kernel. Thus, the development of cracks in the grain is also less. Recommended airflow rates vary from 80 to 165 m^3/h $tonne^{-1}$ of grain (6). Table 1 shows the advantages and disadvantages of the supplemental heated air-drying system (9).

4.3 Heated Air Drying

The first two methods of drying systems, namely natural air drying and drying with supplemental heat, require 1–3 weeks, or even longer, to reach the final safe moisture level of the grain. In heated air drying of grains, the air is heated considerably. This is usually favored or practiced when the grains need to be marketed immediately or contain very high moisture (e.g., parboiled paddy). However, it is not uncommon to use heated air drying for grains that are to be stored for a certain length of time before distribution or consumption.

The heated air-drying system for grains has certain advantages over the natural air-drying and supplemental-heated air-drying systems. These are mainly a faster rate of drying, high drying capacity, independence from climatic conditions, usefulness for both long- and short-term storage of grains, and suitability for bulk drying. Higher initial investment and maintenance cost, high fuel consumption, danger of fire hazards, and the necessity of skilled or trained personnel for control of drying are some of the unfavorable features for this drying system. Heated air grain-drying systems are again subdivided into two distinct heads: namely, thin-layer drying and deep-bed drying.

4.3.1 Thin-Layer Grain-Drying System

Thin-layer grain-drying refers to the drying process in which all grains are uniformly exposed to identical conditions of temperature, relative humidity, and flow rate of air throughout the entire period of drying. This ensures better quality of the dried product. Generally, a grain bed depth of up to 0.2 m thick is considered as thin-layer. All commercial flow dryers are designed based on thin-layer–drying principles.

Depending on the direction of movement of grain and air movement inside the dryer, this group of flow dryers is classified as cross-flow, concurrent-flow, countercurrent-flow, and mixed-flow systems. The preferred high-temperature continuous grain-drying system in the United States is the cross-flow dryer (7), in which the flow of air is perpendicular to the downward movement of grains. The mixed-flow grain dryers are the most common. The countercurrent-flow drying system where the movements of air and grain are opposite

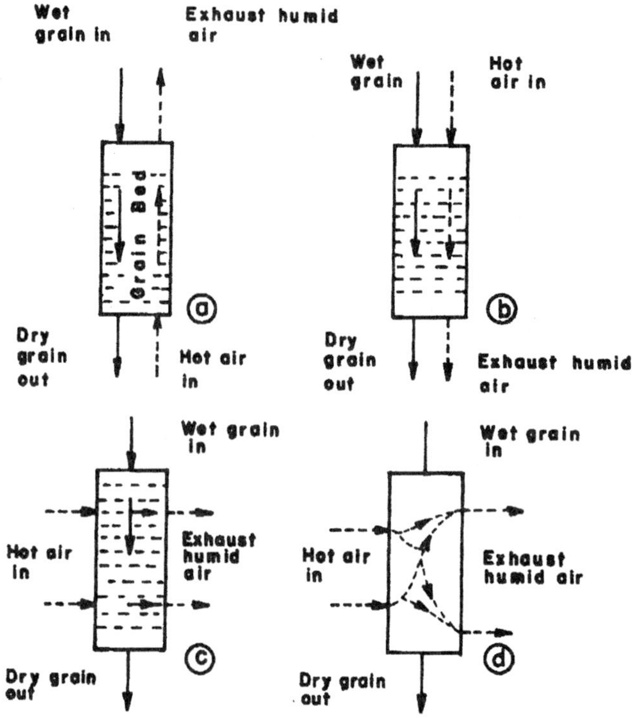

Fig. 2 Different modes of airflow in a thin-layer grain-drying system: (a) countercurrent-flow, (b) concurrent-flow, (c) cross-flow, and (d) mixed-flow.

is used mainly for cooling grain and drying with in-bin continuous-flow dryers. The system with a parallel flow of grain and air (concurrent-flow) is a recent design. Figure 2 shows different modes of airflow.

4.3.2 Nonmixing Thin-Layer Grain Drying

The design of this kind of thin-layer–drying system is very simple where the drying takes place in a column of grain held between two screens, and the grain bed moves downward slowly in a plug-flow manner (9). The airflow is perpendicular to the grain flow. A high air velocity can be used, without any chance of blowing out grain. Owing to nonmixing in the drying zone, the grain layer adjacent to the air inlet side is dried by hotter, but less humid, air than that of the exhaust side. However, the mixing occurs at the outlet ports for the grains. These grains are then either recirculated to the feed hopper of the dryer or conveyed to the tempering bin or storage house. Figure 3 is a schematic of a nonmixing thin-layer–drying system.

4.3.3 Mixing-Type Thin-Layer Grain Drying

The baffle-type and LSU dryers are examples of the mixing-type thin-layer–drying system. The baffle-type dryer design is quite similar to the nonmixing type (see Fig. 3), but a number of inclined sheet metal plate baffles are used in the path of grain movement in the downward direction. The grains are mixed during their downward movement, while the hot air flows in a crosswise manner. Figure 4 shows the schematic arrangement.

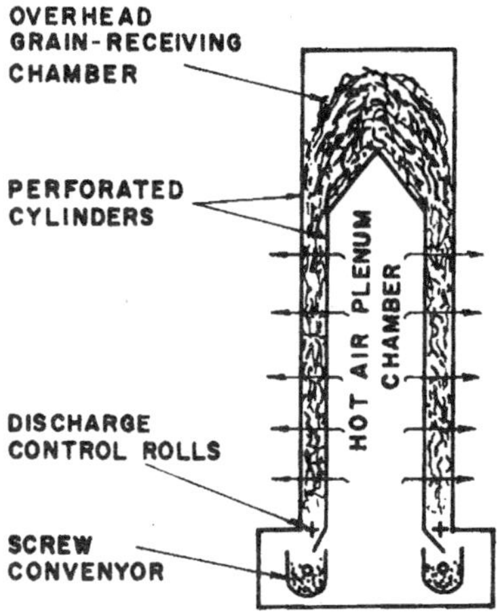

Fig. 3 Nonmixing-type columnar grain-drying system.

In a Louisiana State University (LSU) dryer, the air flows through a row of inverted "V-shaped" channels. The grain moves downward in between the space of the inverted V-ports (distance ~20 cm), and mixing of grains occurs by the staggered position of the alternate rows of air channels. The channels in a particular row are open to the incoming hot air from the blower, but these are closed at the opposite end. In the adjacent row, the channels are closed to the air inlet side but open at the exhaust side. This alternate open-and-closed-ends arrangement acts as air inlet and exhaust ports, respectively. Figure 5 shows the schematic of this kind of drying system.

In the mixing-type of grain dryers, generally the lighter particles, such as chaffs, straw, and immature grains, are blown out with the exhaust air. Thus, the quality of the products becomes improved. In general, a high air temperature is used for mixing-type thin-layer grain-drying systems (10).

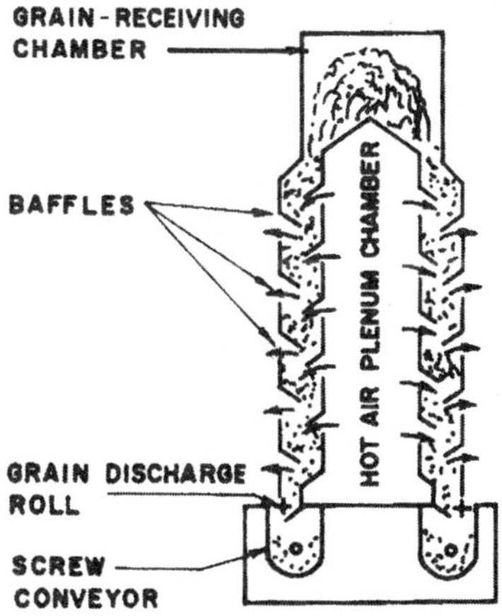

Fig. 4 A mixing-type baffle dryer.

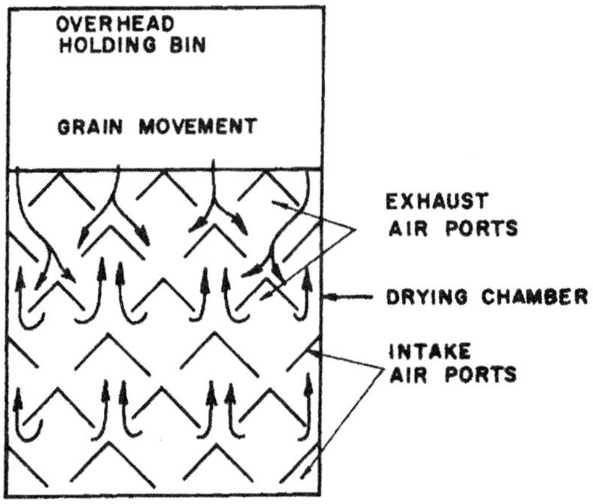

Fig. 5 Schematic of the drying chamber of a mixing-type LSU grain dryer.

The empirical Eq. (2) can be used for evaluating the time of drying in a thin-layer of grain (11):

$$MR = a \exp(-k\theta) \tag{2}$$

where a is considered the grain shape factor, and k is the drying constant, s^{-1}. The moisture ratio (MR) in this equation is defined as the ratio of $(m - m_e)/(m_0 - m_e)$. The terms m_0 and m_e are the initial and equilibrium moisture contents (% db), and m is the moisture content at any time θ (s). The variation in drying rates among long, medium, and bold (short) varieties of paddy is mainly attributed to the dimension and shape of the grain. However, the diffusivity coefficient (D; m^2/s) for all these three types is considered to be the same as that expressed by Eq. (3):

$$D = A \exp\left(-\frac{B}{T}\right) \tag{3}$$

where A and B are constants (7).

4.3.4 Deep-Bed Drying

Deep-bed drying is generally practiced for on-farm drying of grains. In this drying system, all grains in the dryer are not exposed to the same conditions of drying. The grains are placed on a false perforated floor that is raised to a certain height above the ground. The space below this floor acts as an air plenum chamber. Natural air or hot air is blown through the bed above it. Figure 6 shows a deep-bed grain-drying system.

The airflow through the grain mass in the bed carries the moisture. The moisture transfer from the grain to the outgoing air takes place in a finite depth or zone, which is also termed *drying front.* This drying front moves upward in the bed in the direction of air movement. To avoid mold growth at the upper layer of the bed, this drying front should move quickly. Considering the large pressure drop and, consequently, the higher energy consumption for airflow across the bed, the volumetric airflow rate per unit volume of grain is kept to a minimum level that would be just sufficient to prevent the spoilage of grains in the bed.

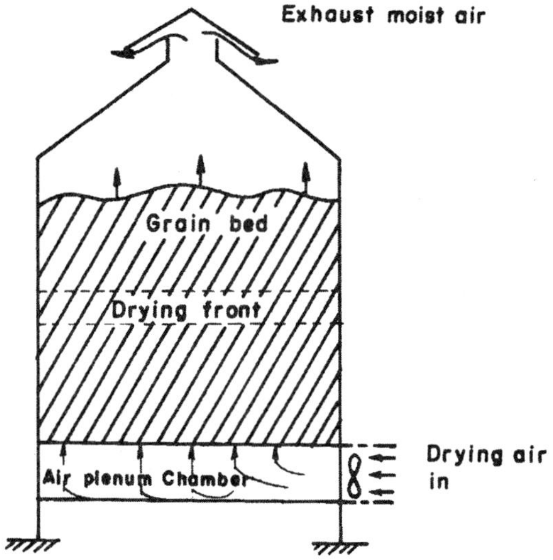

Fig. 6 Schematic of a deep-bed grain-drying system.

The grain at the upper layer of the drying zone remains at its initial moisture level (m_0), and the exhaust air just above the drying zone is in equilibrium with the moist grains. The grain exchanges the moisture in the drying zone by evaporation of water from it; the supply of energy from the air for such evaporation processes causes cooling of the escaping air, as manifested by a drop in air temperature. The depth of the drying zone, however, depends on the air velocity or height of the bed. If the level of air velocity is high or bed height is less, the drying zone may extend entirely throughout the bed. In contrast to thin-layer drying, the level of air velocity is considered critical to deep-bed drying.

In deep-bed drying, two distinct drying rates can be observed; the first period corresponds to the maximum rate of drying and can be expressed (12) as Eq. (4):

$$\frac{\Delta m}{\Delta \theta} = \alpha G(H_e - H_i) = \frac{W_d(m_o - m_m)}{100\ \theta_m} \tag{4}$$

where α is the cross-sectional area of the dryer (m^2); G is mass velocity of dry air (kg/m^2 s^{-1}); H_i and H_e are the absolute humidities of air (kg/kg) at the entrance and exhaust point of the drying front, respectively; m_m and m_o are average moisture contents of the grain (% db) at the end of a maximum rate period and at the initial condition, respectively; θ_m is the drying time (s) for this maximum drying rate period, and W_d is the weight of bone-dry grain (kg). During the falling rate period, the drying time (θ_f) can be obtained from Eq. (5):

$$\theta_f = \frac{1}{k} \ln \frac{m_m - m_e}{m_f - m_e} \tag{5}$$

where m_f and m_e are the final and equilibrium moisture contents (% db), respectively.

All batch-type static-bed dryers are designed on the deep-bed–drying principle. An example of a deep-bed grain–drying system is in-bin drying of grain, which is commonly used for seed drying and grain storage. Different capacities and sizes of bin dryers are used to obtain various drying rates. Because they use low airflow rates, their energy utilization factor is higher. The size selection of this kind of deep-bed grain-drying system (bin size) is generally based on the quantity of grain available per day that should be dried within 24 h (6).

4.3.5 Fluidized Bed Drying

When the fluid velocity through a bed of solid particles steadily increases, then both the pressure drop in the system and the drag on the individual particles increase. Ultimately, the particles start to move and become suspended in the fluid. The condition of fully suspended particles that behaves as a dense fluid is called fluidization, and the bed is termed a fluidized bed. Dryers in which the solids are fluidized by the drying gas are called fluidized bed dryers. These kinds of dryers are useful in various drying applications (13).

Food particles suitable for fluidized bed drying should be in the size range of 20 μm–10 mm. Cereal grains can be easily fluidized over a wide range of moisture contents. The air velocity should be as high as possible to promote rapid drying. The minimum or incipient fluidization velocity (V_f) for nonspherical particles and entrainment or terminal velocity (V_e) for spherical particles (m/s) can be calculated (13,14) from Eqs. (6) and (7):

$$V_f = \frac{(\rho_s - \rho_a)g}{\mu} \frac{d_p^2 \varepsilon^3 \phi}{150(1 - \varepsilon)} \tag{6}$$

and

$$V_e = \left(\frac{4d_p(\rho_s - \rho_a)g}{3C_d\rho_a} \right)^{1/2} \tag{7}$$

where ρ_s and ρ_a are the densities (kg/m^3) of solid particles and gas, respectively; d_p is the diameter of the particle (m); ε is the void fraction of the bed; μ is the viscosity of air (kg/m s^{-1}); C_d is the drag coefficient (which is 0.44 for Reynolds number of the air in the range of 500–200,000); g is acceleration due to gravity (m/s^2); and ϕ is sphericity of the particle. For low a Reynolds number, both V_e and V_f vary with d_p^2, $(\rho_s - \rho_a)$, and $1/\mu$. Thus, the ratio of V_e/V_f depends on mainly the void fraction at minimum fluidization (13); thus,

$$\frac{V_e}{V_f} = \frac{8.33(1 - \varepsilon)}{\phi^2 \varepsilon^3} \tag{8}$$

For nonspherical particles ϕ is less than 1, and ε is greater than 0.45 (approximate value for spherical particles is 0.45). For large-shaped spherical grains with a Reynolds number higher than 10^3, the ratio of velocities is

$$\frac{V_e}{V_f} = \frac{2.32}{\varepsilon^{3/2}} \tag{9}$$

The pressure drop (Δp) per unit bed height (L) (Pa/m) at minimum fluidization velocity can be expressed as

$$\frac{\Delta p}{L} = g(1 - \varepsilon)(\rho_s - \rho_a) \tag{10}$$

The upper limit of gas flow rate is often determined by the loss of fine product with the exhaust gas (15).

For drying of grain in a fluidized bed, perforated plate bed-support is sufficient, and the system is usually operated at atmospheric pressure, although operations above or below atmospheric pressures are possible (14). Better management of drying condition can be accomplished in a batch-operated fluidized bed dryer. However, batch units are suitable for small-scale operation.

Two types of continuous fluidized bed drying are practiced (i.e., plug-flow fluidized bed dryer and well-mixed fluidized bed dryer). In the former, the feed is introduced at one end (wet end) and the product moves forward in a plug-flow pattern until they are discharged at the other end (dry end) over an adjustable weir (Fig. 7). A uniform drying is accomplished because of relatively narrow distribution of residence time of the particles in the bed. However, the system has limited applications for easily fluidizable materials. In the latter type, the particles are vigorously mixed in the bed, maintaining uniform temperature, and the inlet particles are dried up quickly under dispersed conditions. The ratio of bed length to its width for a well-mixed–type fluidized bed dryer is equal to or less than 1, and this is generally suitable for those particles that are difficult to fluidize (14). Because of broader distribution of residence time, drying results are seldom uniform, compared with a plug-flow type. A plug-flow fluidized bed dryer is more commonly used for grain drying than is the well-mixed type.

4.3.6 Spouted Bed Drying

The spouted bed dryer is a particular design of a fluidized bed dryer. Coarse solids too large to fluidize can be handled in a spouted bed. Its applicability has been well recognized for granular materials with diameters that are generally above 5 mm (e.g., grains). In this type of dryer, high-velocity air is allowed to enter through a centrally located nozzle in a conical-base, vertical chamber (Fig. 8). The zone of fast-moving particles at the center creates a spout, and the grains (or granular materials) move downward at a slower rate surrounding the central spout. Thus, the downward-moving grains generally receive the countercurrent-moving air at the annular zone. The initial drying takes place at a higher rate in the central spout, but the later drying in the annular bed is relatively mild.

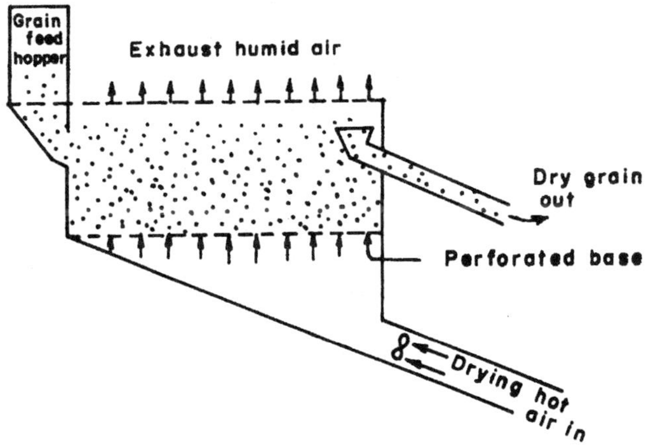

Fig. 7 Schematic of a fluidized bed grain-drying system.

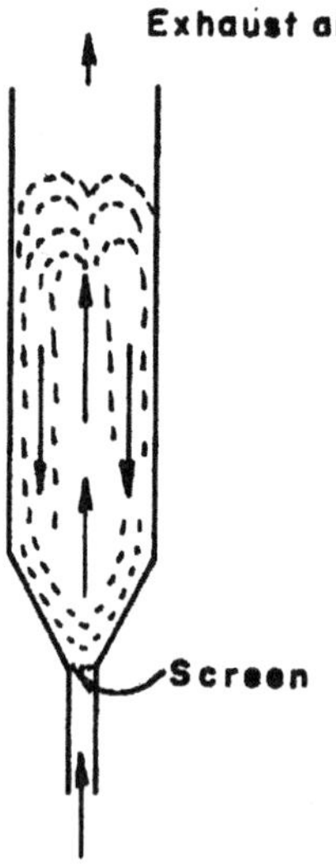

Fig. 8 Schematic of a spouted bed.

An empirical correlation between the minimum superficial fluid velocity V_s(m/s) through the bed necessary for spouting in a 3- to 12-in.–diameter column can be obtained as follows (16):

$$V_s = \left(\frac{d_p}{d_c}\right)\left(\frac{d_o}{d_c}\right)^{0.33}\left[\frac{2gl(\rho_s - \rho_a)}{\rho_a}\right]^{1/2} \tag{11}$$

where d_p is the particle diameter (m); d_c is the column (bed) diameter (m); d_0 is the fluid inlet nozzle diameter (m); l is the bed height (m); ρ_s is the density of solid (kg/m^3); and ρ_a is the density of fluid (kg/m^3).

The classic spouted bed dryer as described earlier has some limitations. The capacity is limited because of maximum attainable height in the bed. Also, scale-up of the system is difficult beyond a 1-m–column diameter. A significant improvement of performance has been observed by introduction of a hollow, tall vertical tube (draft tube) some distance above the nozzle (17). The draft tube may be impermeable, porous, or partially porous.

The central nozzle system causes a significant pressure loss, and this can be reduced by another design of spouted bed grain dryer by introducing air tangentially with the help of a horizontal slit and swirling ring. This configuration is reported to ensure uniform particle circulation in the bed, and to favor formation of a spouted channel. A very recently developed tangential airfed, spouted bed dryer unit, equipped with an inner screw, ensures nondependence on the airflow rate and handling of smaller particulate ($d_c/d_p > 500$). This is not possible with a conventional spouted bed (17).

Removal of moisture from wheat for its use as a seed, from 22 to 25% (db) to about 17.5–19% can be attained in a spouted bed dryer. The temperature of air should be in the range between 50° and 70°C, with an operating gas velocity of 3.14 m/s (minimum fluidizing gas velocity of 1.22 m/s and terminal velocity of 10.2 m/s). Drying of wheat for processing mills can be accomplished in a spouted bed dryer by air at 100–140°C without any damage to its composition. Wheat can also be dried in a spouted bed dryer within 20 min by air at 100–102°C from its initial moisture content of 17% without any damage to the embryo. Continuous drying of shelled corn (13 kg; 14% db) at an air temperature of 160°C and flow rate of 120 m^3/h has been tried with a spouted bed dryer. The system consists of an arrangement of air feeding through a swirling ring, a draft tube, and a screw conveyer inside the bed. The drying results give water removal efficiency of 376 $kg/m^2\ h^{-1}$, specific energy consumption of 3535 kJ/kg water and reduction in bed pressure loss from 9.8–13.7 to 2.9–3.9 kPa. The mean residence time of particles in the bed is 20 min (17).

5 RADIATION DRYING

The electromagnetic radiation, with wavelengths ranging between 0.25 μm and 0.2 m, is used for the drying of different materials. The radiation sources include solar, infrared, dielectric, and microwave. The solar spectrum of radiation (0.25–2.5 μm) does not penetrate the skin of the materials. The magnitude of absorbed radiation depends primarily on the average radiative properties of the materials, direction of incidence, and the wavelength of the radiation. Infrared radiation (4- 8-μm band) can penetrate to a certain depth in the material, and the drying of coatings, thin sheet, and films is generally accomplished by this spectrum of radiation. The radio frequencies are used to heat some moist solids volumetrically by reducing the internal resistance to heat transfer, and the energy is absorbed selectively by the water (for its high dielectric value) molecules within the materials. Because of involvement of high cost, this method of drying is useful for drying high-unit–value products or removal of final moisture content of the products, which is generally not otherwise possible with other drying systems.

5.1 Sun Drying

A major quantity of grain is still dried under the sun in most of the Third World countries. This drying system can be classified into three categories: drying of standing crops, drying of grains in stalk, and drying of threshed grains, as described in the following sections.

5.1.1 Drying of Standing Crops

The grains are dried on the standing crops until they attain proper moisture content. They are harvested and threshed thereafter. Besides a considerable quantity of shattering loss of grains, this slow-drying process takes about 2–3 weeks after the grains attain maturity. Development of sun checks is a common phenomenon that is attributed to fluctuation of temperature and relative humidity of the ambient air during day- and nighttime. Thus, a high yield of broken product is obtained during milling of grains.

5.1.2 Drying Grains on the Stalk

The crop harvested at high moisture content is left either as such or in small bundles in the field until it is dried to proper moisture content. Although the losses from shattering, birds, and rodents are unavoidable in this system, there are fewer sun checks in the grain, compared to drying threshed grains in the sun. In Japan, the harvested rice plants are tied in small bundles with the panicle downward and are hung on racks or poles to avoid direct exposure of the grain to the sun. In this drying process, however, the drying rate is somewhat slow. It requires 7–20 days, depending on weather conditions.

5.1.3 Drying of Threshed Grains

The method of open-air sun drying is a most common practice when the grains are harvested at a relatively high moisture content (to avoid shattering loss), and threshed. In this drying method, the grains are spread in a thin layer (3- 5-cm thick) on the hard soil, a concrete floor, or straw mats and exposed to the sun. Intermittent stirring of the grains improves the rate as well as uniformity of the drying process. In the tropics, drying of harvested paddy usually takes about 2–3 days (dependent on the weather condition and solar insolation) to bring down the moisture content from 20–22% to 13 to 14% (db).

An energy balance for an open-air sun drying of grain can be visualized as,

Solar energy absorbed on the grain surface =

[(energy required for moisture evaporation from the grain)

+ (energy lost by convection to the surrounding

air from the upper surface of the grain)

+ (energy lost by conduction to the ground from the bottom layer of the grain)].

The solution to this heat and mass balance equation is complicated by several factors. These are diurnal variations in solar radiation (both beam and diffuse components), wind speed (and, thereby, both heat and mass transfer coefficients), ambient temperature, relative humidity, ground temperature, and different physical characteristics of the materials (7).

This simple method of drying requires no fuel; preserves viability, germinability, and baking quality of the dried grains; reduces microbial activity and insect infestation; and does not induce discoloration of the product. However, it is highly labor intensive, weather dependent, prone to unhygienic contamination (dust, dirt, debris, etc.), unsuitable for bulk handling, and requires a specially-constructed large floor space.

5.2 Solar Drying

Basically, a solar dryer consists of a heat-transfer unit (absorber) that receives and absorbs the solar spectrum of radiation. The absorber converts the radiation energy into thermal energy that can be transferred to the working fluid flowing in contact with the absorber. Because there are many options to increase the overall efficiency of the system, there are large variations in the solar dryer design.

Based on available designs, solar dryers may be classified into several categories that depend on the mode of heating or the operational mode of heat derived from solar radiation. Among many classifications suggested, the classification based on the operating mode (12) appears to be more logical: natural-convection dryers and forced-convection dryers. The former does not require any mechanical or electrical power to run a fan or blower for the air movement that is necessary to accomplish the drying operation. The forced-convection dryers, on the other hand, require the fan or blower for maintaining a current of hot air through or over the wet materials. These can be further classified as direct- and indirect-mode dryers.

In the direct mode, solar energy is received directly on a small batch of materials placed inside the dryer. A moderate airflow rate at 50°–60°C is provided by natural convection to accomplish the drying. This solar-drying system is commonly used in developing countries or tropical areas. In the indirect-mode solar dryers, the materials inside the dryer do not receive the energy directly from the sun. Generally, the drying air is preheated by sucking the air through a roof-integrated solar collector (with a provision for a supplemented heating system) before it enters the bed of material inside a bin or storage house.

5.2.1 Natural Convection Solar Dryers

Such dryers appear to be more attractive for use in developing countries because they do not use a fan or blower for air circulation inside the dryer. Moreover, they are inexpensive and easy to operate by unskilled workers. However, there are some limitations in this category of dryers: slow drying, not much control of temperature and humidity, smaller capacity; and change in color and flavor of some products owing to direct exposure to the sun. In its simplest form, they consist of some kind of enclosure and a transparent cover. The materials to be dried receive direct sunlight and are heated; also, the high temperature in the enclosure facilitates the removal of moisture through natural ventilation of air.

Among direct-mode natural convection solar dryers the solar cabinet dryer is the simplest form first reported by Lof (18). It consists of a small box (the length is generally three times the width) with insulated sides and perforated bottom, and the interior surface is painted black. The top cover is glazed with a single- or double-layered transparent cover and tilted at a certain angle to reduce the reflective losses (thereby, more transmission) of solar spectrum of radiation incident on it. The materials to be dried are placed on perforated trays (generally made of wire mesh), and these are kept inside the cabinet. The hot air escapes through the upper ventilation holes by natural convection, and this creates a partial vacuum inside the cabinet. Thus, it sucks ambient air through the bottom holes. An air temperature as high as 90°C has been recorded under no-load conditions. In general, a cabinet solar dryer can reduce the drying time by at least one-third to half, compared with open-field sun drying, and is reported to be better than other dryers for small-scale operations in tropical regions.

In indirect-mode solar dryers, the hot air while passing through the bed of material carries the moisture and ultimately escapes out through a duct or chimney. The decrease in density of air on heating inside the chimney further increases the bouncy force and thus induces higher linear velocity of the exhaust air (19,20).

A typical unit of this type of solar-drying system consists of an inclined (20°) solar collector (0.9 times the latitude), a batch dryer, and an identical collector positioned vertically (which acts as a chimney). All these are joined in series and face due south. Besides receiving solar energy on the inclined collector to preheat the air before it enters the dryer, the exhaust air from the dryer is further heated in the vertical chimney, which causes the air to escape through the chimney at a higher velocity. Because of this, an adequate airflow through the product ensures an increased drying rate and reduced drying time.

The drying time with this system depends on the depth of the bed, initial moisture content, humidity of the ambient air, the level of solar energy, and other meteorological parameters. The drying rates for the products decrease with the increase in bed thickness. This makes the design unsuitable for a packed bed configuration in which air circulation by natural convection would be greatly hindered. However, in some modified designs, installation of a wind-driven fan at the exhaust end of the chimney shows some improvement in the efficiency of the system (21).

5.2.2 Forced Convection Solar Dryers

In this category of solar dryers, operating blowers ensure adequate air circulation through the product. Thus, a larger mass of materials can be handled with much reduction in drying time. Higher thermal efficiency is possible with some designs. Bin-type forced convection solar dryers are quite common. In all drying bins, there is a false or perforated floor with as much as 15% or more open area. Since low temperature and slow drying is preferred in this kind of bin drying, warm air with a temperature rise of 5–10°C can be generated by the solar air-heating system operated under forced convection mode.

A typical collector is the bin-wall collector that becomes an integral part of the bin itself; Figure 9 shows such a system. The south-side wall (half of the perimeter of a circular bin) is painted black and covered with transparent UV-stabilized plastic film. A suitable gap between the wall and the plastic cover is maintained to facilitate airflow through this channel. In the rectangular bin, the south-facing roof is used as a solar collector. Air is sucked in by a fan through the space between the cover and the wall, and the heated air thus obtained is forced through the grain bed. Permanent covers such as translucent Fiberglass sheet is used for better durability, but its removal becomes imperative during the nondrying period; otherwise, excessive heat development inside the bin will spoil the lot through undesirable moisture migration. Nevertheless, the bin-wall collector performance increases in the winter months owing to the incidence of higher solar radiation on the vertical surface. A reduction in drying time, as much as 50–75%, has been observed for drying crops using this type of solar bin dryers when the rise in air temperature varied between 5° and 12°C. An airflow rate of 2 $m^3/min\ m^{-3}$ of corn and a collector area of 0.5 m^2/m^3 of corn have been reported to accomplish the drying with this type of forced convection solar dryer (12).

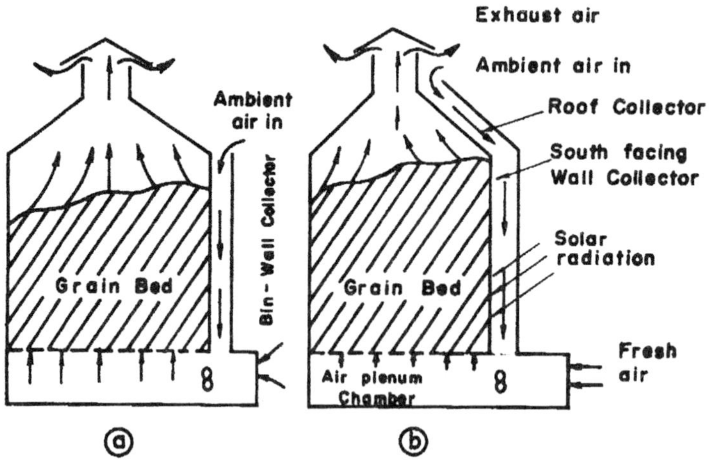

Fig. 9 Forced convection bin–wall collectors for grain drying: (a) circular bin, (b) rectangular bin.

5.2.3 Integrated Array of Solar Collectors

The application of solar air heaters for grain drying requires a large collector area. Often, the required area is arrived at simply by scaling-up from laboratory data on the thermal efficiency of a representative small solar collector unit after taking a suitable allowance as a safe factor. However, the thermal efficiency of an array of these small collectors is lower than that of a single unit (22). Thus, it leads to an erroneous design. The thermal performance of the total collector system also depends on the configuration of the small collector units in the array. Use of separate units to constitute the total system gives large thermal losses from the edges that could be minimized to a large extent using an integrated array of collector modules.

In general, the total useful energy gain of a collector is expressed as follows (24):

$$Q_u = 1000\ wC_p(t_o - t_a) = A_c F_R[(\tau\alpha)_e I_T - U_L(t_i - t_a)] \tag{12}$$

where Q_u is the total useful energy available from the collector (W/m^2), w is the mass flow rate of air (kg/s), C_p is the specific heat of air (kJ/kg K^{-1}), t_o and t_a are the collector outlet and ambient temperatures (°C), A_C is the collector area (m^2), F_R is the collector heat removal factor, $(\tau\alpha)_e$ is the effective transmittance–absorptance product, I_T is the insolation on the tilted collector (W/m^2), U_L is the overall heat transfer coefficient (W/m^2 K^{-1}), and t_i is the temperature of air at the collector inlet (°C).

Figure 10 shows a schematic of an integrated array of collectors (20 m^2) that comprises ten collectors having an area of 2 m^2 for each. The thermal efficiency (η) of this system can be defined as follows:

$$\eta = \frac{wC_p(t_o - t_i)}{A_c I_T} \times 10^5 \tag{13}$$

The efficiency of the system can be improved using a suitable airflow (Fig. 11). The prediction of outlet air temperature under various organizations of airflow can be made by an iterative procedure (23). The success or accuracy of the predictive model, however,

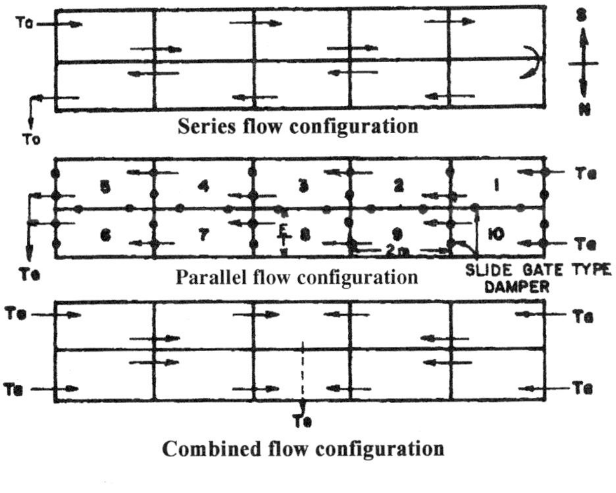

Fig. 10 Schematic of different airflow configurations with the integrated array of collectors.

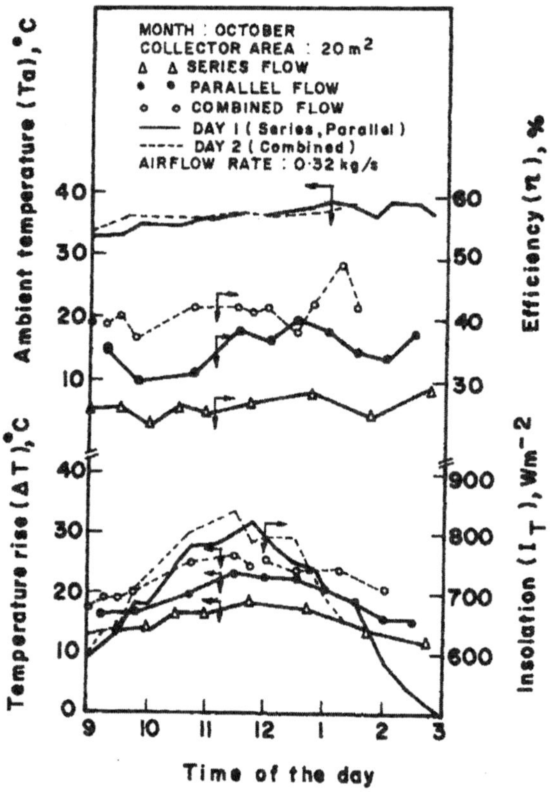

Fig. 11 Performance of an integrated array of collectors for different airflow configurations. (From Ref. 22.)

depends on the realistic estimation of the heat-transfer coefficients, which, in turn, are functions of the absorber geometry, airflow geometry, and tilt angle of the collector.

The following multiple regression equation has been developed to correlate rise in air temperature (Δt) with total collector area (A_c) and mass flow rate of air (w) for the foregoing integrated system having a collector area of 10 m^2 (25):

$$\Delta t = (t_o - t_i) = 7.88 - 78.6\,w + 176.4\,w^2 + 4.69\,A_c - 0.093\,A_c^2 - 0.292\,A_c w \qquad (14)$$

for $0.1 \le w \le 0.283$ kg/s, $2 \le A_c \le 10$ m^2, $771 \le I_T \le 850$ W/m^2.

The collector tilt is an important factor in receiving a reasonably high insolation throughout the year and, thereby, a better annual performance of the system. The optimum tilt generally depends on the latitude of the place. In winter months, the average insolation on a surface facing the equator is maximum if the collector is inclined at an angle greater than the latitude (24,25).

5.2.4 Hybrid-Type Solar Grain-Drying Systems

Because solar radiation is intermittently available, some auxiliary air-heating system should be provided for continuous operation of the drying process during the night as well as on cloudy or rainy days. This auxiliary energy source may be an electrical heating unit, an oil/gas-fired system, or furnaces utilizing biomass as a fuel. A provisional storage unit is optional, but helpful, to store excess solar energy from peak radiation hours and then deliver it to the ambient air during low or no radiation hours. The system generally used for this is rock bed storage, thermal storage, or a phase change material (PCM) storage; the last is a cost-prohibitive operation.

A typical example of this hybrid system is an integrated solar–husk-fired flue gas grain dryer of 1 tonne/day drying capacity (26). It comprises an inclined roof–solar flat plate collector for heating air, a blower for sucking heated air from either collector or furnace, a husk-fired furnace and a two-directional airflow batch dryer of 0.5-tonne–holding capacity (Fig. 12).

The collector is made of 25-mm–thick wood planks, and the total collector area (20 m^2) is divided into ten small units having a frontal area of 2000 mm × 1000 mm each. The collectors are organized in an array of two rows and five columns. The whole assembly is installed on eight rigid mild steel angle supports at a certain height above the ground. The collector is inclined at 30° with the horizontal—this has been estimated to be optimum for the testing place (for 22.3° N; 87.2° E). The absorber plates are made of corrugated G.I. sheet-painted matte-black, with a special nonselective black paint. The top cover is single-window glass. The air is sucked through the collector by passing air over the absorber and below the glazing surface. This system can be operated with a series, parallel, or combined airflow configuration, with the help of slide-gate–type dampers (see Fig. 10). The inclined step-grate husk-fired furnace consists of two chambers—a combustion chamber and a precipitation chamber. The area and inclination of the step grate are 3 m^2 and 45°, respectively. The two-directional airflow batch dryer has one air inlet at the bottom plenum and another at the top. The hot air flows into the dryer in either upward or downward direction to facilitate uniform drying of paddy (27). A perforated sheet is

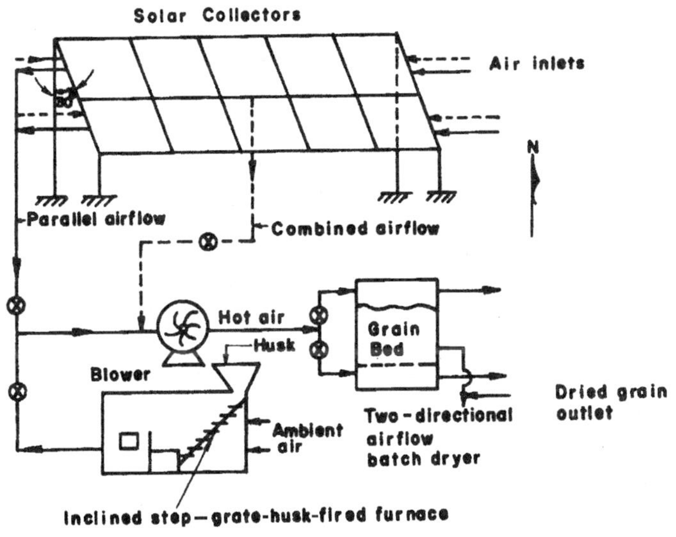

Fig. 12 Schematic of solar–husk-fired grain-drying system: solid lines, parallel flow; dashed lines, combined flow.

fixed at the bottom part of the dryer to hold paddy on it. A suitable capacity blower blows heated air.

The parboiled paddy of about 40% (db) initial moisture content has been dried to 17.3% (db) in two passes. In the first pass, solar-heated air at 58.0 ± 2°C for 3.0 h has been used. In the second pass, the flue gas at 110°C from the husk-fired furnace has been used for about 1.0 h with a tempering period of 2–4 h between the two passes. The milling quality of the dried parboiled paddy has been good. The seed and raw paddy dried intermittently at 40° and 50°C with tempering between the passes has shown significant quality improvement.

5.2.5 Continuous Solar Grain Dryers

The continuous solar grain dryer also has three basic units, similar to that of a batch dryer. However, in this system, the grain, instead of remaining stagnant, moves downward while facing the hot air in a cross-flow, counterflow, or concurrent-flow manner (28). The main component of this dryer is the vertical column bin, which holds the grain. This grain moves, under gravity flow, between two perforated sheets. This is dried by the hot air flowing across the bed. The collector consists of a matrix-type absorber comprising several layers of black-painted wire-mesh (mild steel) netting. The collector and the air duct (flexible) are mounted on a cart, and the collector tilt angles can be adjusted for receiving maximum solar radiation and to obtain higher overall thermal efficiency of the system. A centrifugal electrical blower is coupled to the system for drawing ambient air through the collector absorber and delivers the hot air to the moving grain column.

5.3 Drying by Electromagnetic Radiation

5.3.1 Dielectric and Microwave Drying

Drying by dielectric heating covers both radio frequency (RF) and microwave systems. This method of drying has been used industrially for years because of its advantages (e.g., volumetric dissipation of energy throughout the product, ability to maintain uniformity in moisture content within the material, and such). However, the main limitation in its application is the economy of the whole process. In both systems, heating is accomplished by absorption of energy by the material, which is essentially a dielectric insulator when it is placed in a high-frequency electric field.

There are certain frequency ranges that are agreed on internationally for use in heating. This frequency is called ISM (industrial, scientific, and medical) bands. At radio frequency these are 13.56 ± 0.00678 MHz, 27.12 ± 0.16272 MHz, and 40.68 ± 0.02034 MHz, whereas microwave frequencies, the ISM bands are 915 MHz and 2450 ± 50 MHz (29).

The production and transferring of power (from RF) to drying applicators occurs through the RF applicator. It is a part of the whole system that applies the high-frequency field to the product. It also modulates the amount of RF power supplied by the RF generator circuit. The microwave-heating system is quite different from RF-heating systems; however, it consists of a high-frequency power source (magnetron), a power transmission medium, a tuning system, and an applicator.

Dielectric drying relies on the principle that energy is absorbed by the water present in the material undergoing drying. The absorbed energy causes the temperature to rise, evaporates water vapor, and reduces the moisture level. Calculation of the actual amount of energy (or power) absorbed by a dielectric body is essential for a full understanding of the drying process by the use of radio frequency and microwave energy. The amount of microwave energy dissipated as heat in a determined volume of material can be expressed (30) by Eq. (15).

$$\frac{P}{V} = 2\Pi\gamma\varepsilon_l E_{\mathrm{loc}}^2 \qquad (15)$$

where P is microwave power dissipated as heat (W), V is the volume of material (m^3), γ is the frequency of the electromagnetic field (Hz), ε_l is the loss factor of the material, and E_{loc} is the local electric field strength within the material (V/m).

In the drying context, for which the moisture is removed continuously, the dielectric loss decreases and, consequently, the magnitude of heating reduces. It is interesting that in some materials, this decrease in loss factor is so critical that below a certain moisture content (critical moisture content), they apparently become transparent to this spectrum of electromagnetic radiation. For highly hygroscopic materials, this occurs between 10 and 40% db (31).

Unlike removal of moisture in conventional drying with hot air, the drying by dielectric and microwave systems is quite fast because of rapid generation of vapor within the material, and the total pressure gradient thus established controls the drying process. Although the systems can moderate the drying rate, rapid heating may lead to scorching or burning of the products. In some cases, the hindrance to escape of the generated steam inside may cause rupture of the material into pieces or even an explosion. Combined methods of conventional drying with hot air and dielectric or microwave heating will be more cost-effective. Removal of water vapor from the surface by hot air is relatively efficient. On the other hand, electromagnetic heating most efficiently expedites the vapor diffusion or pumping action of moisture from the interior. The efficiency of this combined system can be improved or maintained by diverse methods, such as preheating, booster drying, and finish drying (31).

The industrial applications of microwave or dielectric heating are numerous, but either form of energy is suitable on a case-to-case basis, as no general rule is available in the selection of one over the other. However, the selection of either a microwave or dielectric-drying system is based on some factor, such as size of load, watt density, power, geometry, system compatibility, self-regulation, self-limitation, or other. Dielectric drying has several applications in the food sectors, which include drying of breakfast cereals, crackers, and biscuits. Several industrial systems utilize microwave drying. In food applications, drying pasta using microwave and hot air of a controlled humidity, results in a substantial reduction in drying time compared with conventional drying. Besides the microwave-drying system's favorable features of low energy consumption, operation, and maintenance, it also provides reduction in microbial and insect infestation in the dried products (31). However, applications of these dielectric and microwave processes in grain drying have not been reported on a commercial scale.

5.3.2 Infrared Drying of Grains

Application of infrared drying is most common in dehydration of coated films and in finish drying of paper and board. However, its applications to other products, such as foodstuffs, are still not common in the industrial sectors. The infrared heating of foods for moisture vaporization is accomplished by absorption of thermal spectrum of electromagnetic radiation (0.75–100 μm) by the materials. This includes near infrared (0.75–3.0 μm), medium infrared (3.0–25 μm), and far infrared (25–100 μm). The effectiveness of heating and drying by infrared, however, depends on the intensity of radiation heat transfer (between the source and the receiver). In the drying process, this intensity of heat exchange depends mainly on specific factors. These are the spectrum of the wavelength (temperature of the radiating source), optical and thermophysical characteristics of the products being dried, intensity and uniformity of radiation (relative position of the source and the irradiated object—called geometric parameter), and physical characteristics of the intervening medium in the drying chamber.

The knowledge of radiative properties such as reflectivity, absorptivity, emmisivity, and transmissivity of the materials is essential in understanding the infrared-drying system. Materials with low reflectivity in the infrared spectrum of radiation (and high or medium absorptivity, depending on the drying process) are preferred to minimize power consumption for heating. Low absorptivity, and consequently high transmissivity (or penetration depth of the radiation into the material) would be desirable for drying thick moist foods to avoid excessive thermal damage of the surface through intensive heating. The heat flux on the surface of the irradiated materials is about 20–100 times greater than that of convective drying (32). The radiative properties of the materials depend on the wavelength of radiation, its thickness, and its moisture content. However, their variations are quite difficult to estimate for foodstuffs because they are a complex mixture of different biomolecules, salts, and water (33). Transmissivity of many materials is higher at lower wavelengths (32). Besides the transmissivity and thickness, the temperature field of an irradiated material is dependent on the temperature and flow rate of the gaseous medium in the drying chamber. The effectiveness of the drying process with the intermittent-drying technique and combined treatment of infrared and convective methods increases in terms of reduction in drying time, less energy consumption, and product quality improvement.

The main component of an infrared dryer is the radiator (or called emitter). Depending on the heating mechanism, this is classified as two types. It may be either an electrically heated radiator (metal sheath radiant rod, quartz tube, or quartz lamp), or a gas-fired radiator. In the last type, one side of a metal or ceramic perforated plate is heated by a gas flame, whereas the other side of this heated body is exposed to the irradiated material.

Infrared drying of foodstuffs is quite uncommon; however, laboratory-level studies on infrared drying of grains reveal the possibility of its applications on a larger scale. The infrared (thermal) treatment applied to a thick layer of freshly harvested moist grains in conjunction with vigorous mixing of the layer shows considerable reduction in drying time. While intensive irradiation accelerates the dehydration, mixing prevents overheating the grain beyond the permissible limit. Drying of grains (e.g., wheat) by infrared radiation shows a constant rate period similar to convection drying of a layer of grain. However, the reasons for such are different from each other. In infrared drying, the constant-rate period depends on the sufficiently intensive delivery of the moisture from the inside layer to the drying surface. On the other hand, in convective drying of a substantially thick layer (0.1–0.2 m) of sample at a relatively low airflow rate, it arises from low moisture transfer from the surface (32). A low flow rate (0.01 m/s) of cold air (near 20°C) causes more lowering in temperature of the surface than that of the center layer, which helps in movement of inside moisture toward the evaporating surface. Thus, the rate of drying is increased.

Selections of radiation parameters, such as the voltage of the lamps and distance between the source and the layer of grain, depend on the temperature of the grain during

Table 2 Advantages and Disadvantages of Infrared Drying of Grains

Advantages	Disadvantages
Reduction in drying time	Scaling up of this process, which is not always straightforward
Penetration of radiation directly into the product without heating the surroundings	It is essentially a surface dryer, and laboratory equipment must be tested in the plant to ensure the design's effectiveness
High efficiency to convert electrical energy into heat (for electrically heated radiators)	
Uniform product heating with directional characteristics for selecting part of a large object	
Easy to program and manipulate the heating cycle	
Low product deterioration	
Inexpensive compared with dielectric and microwave drying	

the drying process. The rate of drying increases considerably by increasing the voltage, but excessive increase in the same will cause the grain temperature to increase beyond the permissible limit.

The advantages and disadvantages of infrared drying (33) are shown in Table 2.

6 OTHER METHODS OF GRAIN DRYING

Chemical and physical dehydrating agents have long been used in various process industries. Various chemical treatments have been tried for dewatering and preservation of foods (34). Drying of grains by using relatively chemically inert hygroscopic materials (desiccants) shows some encouraging results. However, both of these methods of dewatering–desorption of water remain of merely scientific interest, and large-scale commercial implementation of this process is yet to be achieved.

6.1 Drying by Chemicals

Wet paddy at a moisture content of 30% is treated with 0.05–0.25% acetic acid and calcium propionate in the bagged condition and stored for about a month. This treatment neither brings down the moisture content nor arrests the growth of yeast and microbial activity of the paddy. Neither discoloration nor bacterial action can be stopped when the same moist paddy is given a treatment with urea at 0.5–1.0% levels. Treatment with ferrous sulfate, either alone or in combination with common salt, aggravates the rate of discoloration and development of heat. Moreover, neither of these two chemicals are considered to be safe food additives, nor are they are able to bring down the moisture of the wet paddy (34).

Common salt is generally considered to be a safe food additive and is inexpensive. Its use for preserving monsoon-harvested paddy has been successful in both laboratory and large-scale trials. In addition to controlling infection, it also retards sprouting and causes dewatering of rice kernels to a safe storage level, by an osmosis mechanism. In a large-scale trial with this process, freshly harvested Indian ADT-27 variety of paddy of 24% moisture content is treated with 4–5% (w/w) dry common salt. The grain is bagged after dewatering for 24–36 h and is finally stored in bulk for about 2–3 months. No increase in temperature or in microbial infection has been observed in the paddy when treated with more than 4.0% salt. The paddy attains a moisture level of 18% just after dewatering and that level comes down to 15% at the end of 1 month. The overall quality of the paddy is quite good, although the grain surface appears to remain wet, which is attributed to the presence of a dry salt layer on the husk. The paddy treated with about 5.0% common salt gives nearly 90–95% viability after about 6 months of storage. The solution leached out from the moist paddy after this treatment does not contain any dissolved sugars or amino acids, and thus ensures that the nutritive quality of the kernel is well preserved.

A reduction in drying time has been observed by treating the parboiled paddy with brine solution (35). By circulating 15% brine solution for 10–20 min to the hot soaked (at 65°C for 4.5 h) paddy containing 28–30% moisture, the grain loses its moisture and the kernel moisture comes down to a level of 24%. After steaming, it takes only 2 h to dry the parboiled paddy to 15% moisture content by mechanical drying at an air temperature of 90°C. The salt does not enter the paddy kernel while withdrawing moisture from it by osmosis, and the percentage of salt content on the rice is negligible. Nonopening of the paddy husk is a distinct advantage of this method, although to a small extent it may affect the efficiency of the mill's shelling unit.

Drying of grains with desiccant materials has some advantages. The drying can be performed at a much lower temperature that facilitates not only increased energy conservation, but also preserves the quality for some commodities, compared with high-temperature convective air drying (36). The dewatering process proceeds through a physical phenomenon: surface adsorption and capillary condensation. A desirable desiccant material in this aspect should be easily reactivated at a moderately elevated temperature. It should also remain free-flowing, even under saturated condition, and possess a high-adsorption ratio (weight of moisture adsorbed per unit weight of material when exposed to an environment of 100% relative humidity). Materials such as silica gel, sodium and calcium bentonite, and anhydrous calcium chloride have been tried for drying cereal grains such as corn, pecan or others (36–39).

Generally, two desiccant-drying systems have been tried: one involves an intimate mixture of the grain and desiccant material in a closed environment (static-desiccant drying), and the other employs air circulation through separate beds of reactivated desiccant material and grain in sequence. A typical static–desiccant-drying system at an optimum ratio of corn/desiccant (bentonite) of 1:1 reduces the moisture level of grains sharply from 32.5% to 17.4% within 24 h and is followed by a slow process thereafter. A higher ratio of 3.1 (corn/silica gel) decreases the moisture content of the grain from 24.9% to 14.6% within the same contact time (37).

7 CONCLUSIONS

So far, thin-layer and deep-bed grain-drying systems have been the common practice worldwide. The deep-bed–drying system is advantageous for a small-capacity grain-drying unit. In this drying system, a batch dryer operating with heated or unheated air could be easily employed to remove a small percentage of moisture from the grain. However, this requires a longer drying time compared with a thin-layer–drying system. The thin-layer, continuous-flow grain-drying system is generally practiced commercially for large-scale drying. Inspite of some advantages, such as short drying time, uniform drying and quality, weather-independence, and so on. High capital investment and specific energy consumption must be considered in adopting this drying system.

In most of the developing and underdeveloped countries in the world, wherever solar energy is abundantly available, a major quantity of food grain is still dried under the sun. However, this inexhaustible solar energy can be harnessed efficiently for batch drying of cereal grains at low to moderate air temperatures using a relatively simple and inexpensive flat plate collector system. Particularly, the solar collectors are suitable for drying the seed grains and in situ drying of grains in the storage bins without creating any environmental pollution. Because of the intermittent nature of solar energy, crop drying using a combination of both biomass and solar energy sources offers a good alternative option for a continuous drying system.

There have been many research and development efforts to use electromagnetic radiation (i.e., solar, infrared, dielectric, and microwave) in the field of drying or dehydration of crops. Particularly in the recent past, the development of microwave–vacuum drying was one such effort for the drying of both raw and parboiled paddy and other cereal grains. The application of microwave for drying parboiled paddy reveals that, under optimum conditions of microwave input power, pressure in the chamber, and drying time, the drying could be accomplished in a single pass with good milling and other qualities of the dried product. The microwave-based system is generally technologically superior to commercial-drying systems, but it requires a different set of operational and safety standards. However, a commercial grain-drying plant based on any one of the electromagnetic radiation technologies is yet to develop.

Although application of common salt for preserving and drying freshly harvested paddy has been made on large-scale trials, this technique still suffers from corrosion and other inherent problems that are quite detrimental to commercial exploitation.

NOMENCLATURE

A	constant in Eq. (3)
A_c	solar collector area in Eqs. (12), m^2
a	grain shape factor in Eq. (2), dimensionless
B	constant in Eq. (3)
C_d	drag coefficient in Eq. (7), dimensionless
C_p	specific heat of air, kJ/kg K^{-1}
D	diffusivity coefficient for moisture in Eq. (3), m^2/s
d_c	column diameter in spouted bed dryer in Eq. (11), m
d_o	nozzle diameter in spouted bed dryer in Eq. (11), m
d_p	solid-particle diameter, m
E_{loc}	local electric field strength in Eq. (15), V/m
F_R	collector heat removal factor in Eq. (12), dimensionless
G	mass velocity of dry air in Eq. (4), $kg/m^2\ s^{-1}$
g	acceleration due to gravity, m/s^2
H_e	absolute humidity of exhaust air in Eq. (4), kg/kg

H_i	absolute humidity of inlet air in Eq. (4), kg/kg
I_T	solar insolation on the tilted collector in Eq. (12), W/m^2
k	drying rate constant, s^{-1}
L	height of fluidized bed in Eq. (10), m
L_{fg}	latent heat of vaporization of water, kJ/kg
l	bed height in spouted bed dryer in Eq. (11), m
MR	moisture ratio for grain in Eq. (2), dimensionless
m	moisture content of grain at any time, % db
m_m	average moisture content of the grain at the end of maximum rate period, % db
m_e	equilibrium moisture content of grain, % db
m_f	final moisture content of grain at the end of drying in Eq. (5), % db
m_0	initial moisture content of the grain at the dryer inlet, % db
P	microwave power dissipated as heat in Eq. (15), W
Δp	pressure drop in fluidized bed dryer in Eq. (10), Pa
Q	volumetric flow rate of air, m^3/s
Q_u	useful energy available from a collector in Eq. (12), W/m^2
T	absolute temperature of the air, K
t_a	temperature of air at the dryer inlet in Eq. (1), °C; ambient temperature in Eq. (12), °C
t_e	temperature of air at the dryer outlet, °C
t_g	temperature of air leaving the grain, °C
t_i	air temperature at the collector inlet in Eq. (12), °C
t_o	air temperature at the collector outlet in Eq. (12), °C
U_L	overall heat transfer coefficient in Eq. (12), W/m^{2}K^{-1}
V	volume of material receiving microwave energy in Eq. (15), m^3
V_e	terminal velocity of particle in Eq. (7), m/s
V_f	minimum fluidization velocity in Eq. (6), m/s
V_s	minimum superficial fluid velocity in Eq. (11), m/s
v	specific volume of air, m^3/kg
W_d	dry matter in grain, kg
w	mass flow rate of air, kg/s
x_f	final moisture content of the grain in Eq. (1), decimal, dry basis
x_0	initial moisture content of grain in Eq. (1), decimal, dry basis

Greek Symbols

α	cross-sectional area of the dryer in Eq. (4), m^2
γ	frequency of electromagnetic field in Eq. (15), Hz
η	thermal efficiency of solar air-heating system, dimensionless
ε_l	loss factor of material receiving microwave energy in Eq. (15), dimensionless
ε	void fraction of grain bed, dimensionless
θ	drying time, s
θ_m	drying time for maximum rate period in Eq. [4], s
θ_f	drying time for falling rate period in Eq. [5], s
μ	viscosity of air, kg/ms^{-1}
ρ_a	density of air, kg/m^3
ρ_s	density of solid particle, kg/m^3
φ	spehericity of solid particle, dimensionless
$(\tau\alpha)_e$	effective transmittance–absorptance product in Eq. (12), dimensionless

REFERENCES

1. AS Mujumdar, AS Menon. Drying of solids: principles, classification and selection of dryers. In: AS Mujumdar, ed. Handbook of Industrial Drying. Vol. 1. 2nd ed. New York: Marcel Dekker, 1995, pp. 1–39.
2. A Chakraverty. Post Harvest Technology of Cereals, Pulses and Oilseeds, 3rd. ed., New Delhi: Oxford & IBH Publishing, 1995.
3. CGJ Baker. The design of flights in cascade rotary dryers. Drying Technol 6:631, 1988.
4. NGC Ienger, R Bhaskar, P Dharmarajan. Studies on sand parboiling and drying of paddy. J Agric Eng 8:51, 1971.
5. MV Ramakumar, GSV Raghavan, S Javaregowda. Operating conditions for drying of paddy in a rotating drum with sand medium. Drying Technol 14:595, 1996.
6. VGS Raghavan. Drying of agricultural products. In: AS Mujumder ed. Handbook of Industrial Drying. Vol. 1. 2nd ed., New York: Marcel Dekker, 1995, pp. 627–642.
7. DB Brooker, FW Bakker–Arkema, CW Hall. Drying and Storage of Grains and Oilseeds. New York: Van Nortrand Reinhold, 1992.
8. USDA. Drying of shelled corn and small grains. Farmer Bull 2214, 1965.
9. A Chakraverty. Bulletin on Paddy and Other Grain Drying Systems. Kharagpur, India: Post Harvest Technology Centre, Indian Institute of Technology, 1987.
10. P Pillair. Rice Postproduction Mannual. New Delhi: Wiley Eastern, 1988.
11. Y Wang, RP Singh. A single layer drying equation for rough rice. Presented at American Society of Agricultural Engineers, St. Joseph, MI; 1978, paper no. 78–3001.
12. MS Sodha, NK Bansal, A Kumar, PK Bansal, MAS Malik. Solar Crop Drying. Vol. 1. Boca Raton, FL: CRC Press, 1987.
13. WL McCabe, JC Smith, P Harriot. Unit Operations of Chemical Engineering. 5th ed. Singapore: McGraw Hill International, 1993.
14. JC Brennan. Fluidised bed drying. In: R Macrae, RK Robinson, MJ Sadler eds. Encyclopaedia of Food Science and Food Technology and Nutrition. San Diego: Academic Press 1993, pp. 1463–1469.
15. S Hovmand. Fluidized bed drying. In: AS Mujumdar, ed. Handbook of Industrial Drying, Vol. 1. 2nd ed. New York: Marcel Dekker, 1995, pp. 195–248.
16. KB Mathur, PE Gishler. A technique for contacting gases with coarse solid particles. AICE J 1:157, 1955.
17. E Pallai, T Szentmarjay, AS Mujumdar. Spouted bed drying, In: AS Mujumdar. ed. Handbook of Industrial Drying. Vol. 1. 2nd ed. New York: Marcel Dekker, 1995, pp. 453–488.
18. GOG Lof. Recent investigations in the use of solar energy for drying of solids. Solar Energy, 6:122, 1962.
19. RHB Excell. Basic design theory for a suitable solar rice dryer. Renewable Energy Rev J 1: 1, 1980.
20. SK Das, Y Kumar. Design and performance of a solar dryer with vertical chimney suitable for rural application. Energy Conver Manage 29:129, 1989.
21. B Kilkis. Solar energy assisted crop and fruit drying systems: theory and applications. In: Energy Conservation and Use of Renewable Energies in Bioindustries. Oxford: Pergamon Press, 1981, pp. 307.
22. A Chakraverty, SK Das. Performance of an integrated array of solar collector modules for different air flow rates and flow organisations. Energy Conver Manage 30:29, 1990.
23. SK Das. Development of an integrated array of solar air heating collectors for a paddy dryer of 0.5 tonne holding capacity. PhD dissertation, Indian Institute of Technology, Kharagpur, India, 1989.
24. JA Duffie, WA Beckman. Solar Engineering of Thermal Processes. New York: John Wiley & Sons, 1980.
25. SK Das, A Chakraverty. Annual performance of integrated collector modules at a suitable tilt for air heating system. Energy Conver Manage 30:251, 1990.

26. A Chakraverty, SK Das, Design and testing of an integrated solar-cum-husk fired paddy dryer of one tonne per day capacity. In: AS Mujumder ed. Drying 87. Washington, DC: Hemisphere 1987.
27. A Chakraverty, SK Das. Development of a two directional air flow paddy dryer coupled with an integrated array of solar air heating modules. Energy Convers Manage 33:183, 1992.
28. S Pattanayak, P Sengupta, BC Roychoudhury. Continuous solar grain dryer. In: Proceeding of International Solar Energy Society Congress, New Delhi, 1978, pp. 1964–1966.
29. PL Jones, AT Rowley. Dielectric drying. Drying Technol 14:1063, 1996.
30. A Garcia, JL Bueno. Improving energy efficiency in combined microwave convective drying. Drying Technol 16:123, 1998.
31. RF Schiffmann. Microwave and dielectric drying. In: AS Mujumdar, ed. Handbook of Industrial Drying. Vol. 1. 2nd ed. New York: Marcel Dekker, 1995, pp. 345–372.
32. AS Ginzburg. Application of Infrared Radiation in Food Processing. Chem Process Eng Ser. London: Leonard Hill, 1969.
33. C Ratti, AS Mujumdar. Infrared drying. In AS Mujumdar, ed. Handbook of Industrial Drying. Vol. 1. 2nd ed. New York: Marcel Dekker, 1995, pp. 567–588.
34. CS Shivanna, B Kudarathullah. Preservation of freshly harvested moist paddy with chemicals. Bull Grain Technol 9:269, 1971.
35. CS Shivanna. Traditional and modern methods of parboiling and drying of paddy. Food Ind J 4:1, 1971.
36. MT Danziger, MP Steinberg, AI Nelson. Drying of field corn with silica gel, Trans Am Soc Agric Eng 15:1071, 1972.
37. AC Srivastava, WK Bilanski. Use of desiccant for drying of maize. AMA, Autumn, 49:1981.
38. KC Watt, WK Bilanski, DR Menzies. Comparison of drying of corn using sodium and calcium bentonite, Can J Agric Eng 28:35, 1986.
39. SR Ghate, MS Chhinnan. Pecan drying with silica gel. Energy Agric 2:11, 1983.

7

Commercial Grain Dryers

G. S. VIJAYA RAGHAVAN

McGill University, Sainte-Anne-de-Bellevue, Quebec, Canada

1 INTRODUCTION

For centuries, cereals have been the primary source of food for the human population. Maize, rice, and wheat together account for over 80% of the world cereal production of 2 billion metric tons (MT) in 1997. Other important cereal productions include barley, rye, oats, millet, and sorghum. Although the cultivated land base for cereals dropped slightly from 716 to 701 million ha^{-1}, from 1980 to 1997, FAO statistics show that productivity increased by over 30% to 2.97 MT ha^{-1} during this period (1). Although this is a clear indication of technological improvements in production, crop production energy has been estimated to be but 20–25% of the total used (2). Postharvest energy consumption is primarily in processing (including drying), transportation, and cooking. This breaks down to roughly 15, 5, and 60% of the total, respectively, depending on the commodity and in which region it is produced. In the developing nations, very little commercial energy is used for drying because the natural-drying potential of the climate is high and usually sufficient to bring moisture content to safe storage levels. However, about 34% of the world's cereal crop is produced in nations where artificial drying of certain species is a necessity, and commercial energy inputs for reducing moisture in crops, such as grain corn; can be as high as 10% of the energy input (3). Thus, substantial energy inputs are used in drying cereal crops from typical harvest moisture contents of 20–30% to safe levels for storage (10–13%).

About 126 kg of water must be removed per MT of coarse grain harvested for use in the long term, and the corresponding energy requirements range from 0.38 to over 0.63 GJ/MT of grain, depending on the type of commercial dryer used, the required moisture reduction, and the operating conditions. One can estimate that the over 1 billion L of diesel fuel are used each year for this operation. Although the energy and investment costs of grain drying are significant, they are offset by several advantages. First, the crop may be harvested earlier and closer to its ideal moisture content, thus reducing losses in the field owing to inclement weather, pathogenic microorganisms, birds, and wildlife. This also permits autumn field operations to be performed while the soil is relatively dry, reducing damage to the physical structure of the soil, and in some regions, may permit a second crop. Last, proper drying and aeration of tough or damp grain reduces or eliminates spoilage problems in storage caused of hot spots and insect infestations. However, when high—temperature-drying systems are used, a fraction of the load becomes overdried, and losses in marketable weight and quality are incurred.

Among convective-drying technologies, there are literally hundreds of variants; yet only certain types of dryer are presently used in the grain industry. Not all grain dryers are suited to a given geographic area or farm. The choice of system depends on the annual volume produced, the marketing pattern, the type of farm, the material to be dried, and the capacity and nature of existing facilities. Presently, most commercially available dryers are based on heated air (convection dryers); however, there has been substantial research aimed at adapting other technologies to grain drying. Two general areas are involved: (a) improvement of heated air dryers to improve energy efficiency or to eliminate overdrying (e.g., fluidized and spouted beds concepts), and (b) alternative modes of heat or mass transfer. Among the alternative modes are conduction drying, conduction drying using particulate media, infrared drying, microwave drying, and heat pump drying. Although improving fuel efficiency is one of the main motivations for continued research, several other advantages may be sought. Among them are faster drying, better quality of dried grain, and incorporation of renewable energies; however, advances in one direction often require a trade-off somewhere else. For example, faster drying is often associated with poorer quality; improved fuel efficiency requires trade-offs in drying time or system capacity. This chapter is intended to provide an introduction of the various types of grain dryers presently available on the market so that the reader may understand how a given dryer is selected for a given farming operation.

2 CROP CONDITIONING

The term *crop conditioning* refers to moisture-reduction methods that are implemented before heated-air drying or, in some circumstances, replace heated-air drying altogether. Crop conditioning includes aeration, natural air drying, in-storage drying with supplemental heat, and multistage drying. Aeration, natural air drying, and natural air drying with supplemental heat are in a category referred to as low-temperature drying, which was the norm before the 1940s. Since then, the enormous increases in per hectare productivity and in production levels of individual farms have led to the implementation of high temperature dryers, fuelled by cheap fossil fuels, to streamline operations at the end of the cropping season. With increases in fuel prices and concern about the use of nonrenewable energies, there is a renewed tendency to implement low-temperature drying techniques when possible.

2.1 Aeration

Aeration consists essentially of moving small amounts of unheated air through a pile of grain to equalize the grain temperature and prevent moisture migration in bins exposed to significant changes in ambient temperature. Aeration is also used to cool grain after drying, to keep damp grain cool until it can be dried, to remove storage odors, or to distribute fumigants in the grain mass. This operation is usually carried out in a storage bin that is equipped with a fan, duct system, and perforated floor, along with exhaust vents to provide escape for moist air. Whether or not the ventilating air is blown upward or sucked down through the grain is largely a matter of choice. Upward ventilation is more commonly used, although there are advantages and disadvantages to each of these methods. An important advantage of using upward ventilation is that it allows storage temperatures to be measured easily because the warmest air is always at the top of the pile. The recommended airflow rate for normal aeration of shelled corn, soybeans, and small grains at 125 Pa (0.5 in H_2O) is 5 $m^3h^{-1}m^{-3}$ of grain (0.1 cfm bu^{-1}) (4). Aeration of damp grain at 500–750 Pa (2–3 in H_2O) requires higher flow rates of the order of 50 $m^3h^{-1}m^{-3}$ of grain (1 cfm bu^{-1}) (5).

It is important that the aeration fan should not be run when the relative humidity of the ambient air is too high. For example, during fall and winter, the operator should select days when the average relative humidity is less than or equal to 70% (6), and the air temperature is more than −1.1°C (30°F). Furthermore, bins of 40 m^3 (1000 bu) or less generally do not require aeration if the grain is loaded dry.

2.2 Natural Air Drying

Natural air drying requires higher airflow rates than those used for aeration, but is conducted in the same type of setup, as just described. Airflow rates are typically 150–250 $m^3h^{-1}m^{-3}$ (3–5 cfm/bu) and 250–500 $m^3h^{-1}m^{-3}$ (5–10 cfm/bu) for bed depths of 1.2–1.8 m (4–6 ft), respectively (4). These figures apply to small grains, peas, and beans, and shelled and ear corn (6).

2.3 In-Storage Drying with Supplemental Heat

Drying can be accomplished in the storage bin by ventilating with air heated to 4–12°C (7–22°F) above ambient through a duct system or through one centrally placed cylinder, as with batch drying. This method is adequate for bins having capacities of up to 100 ton (7). This method usually requires continuous operation of the ventilation system over a period of 1–3 weeks, but has the advantage of permitting the handling of huge amounts in one fill.

In-storage drying may also be carried out on a bar floor provided with a fan capable of providing airflow rates of 80–165 $m^3h^{-1}ton^{-1}$ of grain and an appropriate system of floor and lateral ducts. The advantages of this method are low-cost and simplicity.

2.4 Multistage Drying

The term *multistage drying* refers to any process that uses high-temperature drying in combination with aeration or natural air drying. An outline of two such processes, dryeration and combination drying, follows.

2.4.1 Dryeration

Dryeration is a two-stage process by which grain is dried in a heated air dryer to within about 2% of its ''dry'' moisture content and then moved to an aerating bin where it is left to steep without ventilation for about 10 h (5). This allows time for moisture to migrate to the surface of the kernel. The grain is then aerated for about 12 h at airflow rates on the order of 25–50 $m^3h^{-1}m^{-3}$ of grain (0.5–1 cfm/bu).

The advantages of this process are

1. Higher drying temperatures can be used during the removal of most of the moisture, as the residence time is shorter, which improves the speed of the initial drying.
2. The drying system capacity can increase by up to 60% because the initial drying is faster, and there is no residence time for cooling.
3. The last few moisture percentage points, which are the most difficult to remove efficiently, require only energy for the blower because the heat already contained within the grain is sufficient to drive the moisture to the surface, resulting in fuel savings of 20% or more.
4. The grain quality is improved by slower cooling, which results in fewer heat stress cracks.

If the air is blown up through the grain, there is often considerable condensation on the roof and walls of the bin. Therefore, the grain must be moved to another bin for storage. The amount of condensation on the roof can be reduced by pulling the air down through the grain or by cooling the grain immediately after it comes out of the dryer.

2.4.2 Combination Drying

Combination drying is an extension of the dryeration process, and it is primarily used for grain with very high harvest moisture >25% (8). This method involves reducing the moisture content to 19–23% in a high-temperature dryer, then moving the grain to a bin dryer in which drying is completed using natural air or supplemental heat. The use of high temperatures when the grain is relatively moist reduces the risk of stress cracking. With this method the output of the high-temperature dryer is increased to two or three times that obtained when it is used for complete drying. In addition, energy requirements may be reduced by as much as 50%. Airflows for the bin-drying portion of the process are between 45 and 90 $m^3h^{-1}m^{-3}$ of grain (0.9 and 1.8 cfm/bu). Combination drying is an excellent compromise between fuel efficiency and processing time.

The choice between dryeration and combination drying depends on the amount of grain to be dried, its typical initial moisture content at harvest, and the cost of energy and capital investment involved. If small amounts of grain at relatively low-moisture contents are to be dried, the purchase of equipment for combination drying would not be warranted. Combination drying is more suited to high moisture contents and large volumes of grain. In all cases, for bins 100 m^3 or larger, aeration ducts large enough for airflows of at least 36 $m^3h^{-1}m^{-3}$ of grain (0.7 cfm/bu) should be provided. Because fully perforated bin floors allow the greatest number of options, they should be considered for installation on all new, large storage bins.

3 ARTIFICIALLY HEATED AIR DRYING

The drying temperature is an important consideration in any heated air process, for energy efficiency, fuel costs, processing time, and output product quality depend to a large extent on this factor. Suggested ranges for drying temperatures vary, depending on the intended end-use of the grain. A few recommendations for natural- and heated-air drying and the maximum drying temperatures to be used on grain for seed, commercial use, and animal feed are listed in Table 1. Drying time and airflow rate are also important. However, these vary according to the drying temperature and type of dryer used. Finally, significant energy

Table 1 Recommendations for Drying Grain with Natural Air and Heated Air

Parameter	Ear corn	Shelled corn	Wheat	Oats	Barley	Sorghum	Soy	Rice	Peanuts
Max. moisture content of crop at harvesting for satisfactory drying:									
With natural air (%)[a]	30	25	20	20	20	20	20	25	45–50
With heated air (%)[a]	35	35	25	25	25	25	25	25	45–50
Max. moisture content of crop for safe storage in a tight structure (%)[b]	13	13	13 (12%)[a]	13 (12%)[c]	13	12	11	12	13
Max. relative humidity for natural air drying to safe storage level (%)	60	60	60	60	60	60	65	60	75
Max. temp. of heated air when crop is:									
Used for seed (°C)	43	43	43	43	41	43	43	43	32
Sold for commercial use (°C)[a]	54	54	60	60	41	60	49	43	32
Animal feed (°C)[a]	82	82	82	82	82	82			

[a] Moisture contents on wet basis: (a) higher temperatures than those listed may be used when the corn is dried under carefully controlled conditions; temperature of the kernels does not exceed 54°C at any time: (b) if there is any possibility that the crop may be sold, use the lower temperature as listed for commercial use.

[b] If the products are to be stored for long periods, the moisture content should be 1–2% lower than shown in this tabulation.

[c] Seed.

reductions may be obtained by matching the final moisture content to the required storage time and the anticipated storage temperature. For example, if a storage temperature of 15°C is maintained, corn can be safely stored for nearly 2 months at 18% moisture, whereas it can be stored for only about 10 days at 24% moisture. At these moisture contents, corn can be stored for up to 9 months and 1 month, respectively, at a storage temperature of 4°C (9).

The two major types of heated air grain dryers are bin dryers and portable dryers. Bin dryers are available in batch, recirculating, and continuous categories, whereas portable dryers are commonly available in nonrecirculating and recirculating types.

3.1 Bin Dryers

Bin dryers are usually operated at lower airflow rates than other types and are generally more energy-efficient, although slower than most other types of dryer. The rule of thumb in the selection of bin dryer size and capacity is that the grain harvested in a day should be dried to safe storage level within 24 h to permit loading of the next day's harvest.

3.1.1 Batch Dryers

''Batch-in-bin'' systems are the least expensive for drying grain. The main components are a bin with a perforated floor, a grain spreader, a fan and heater unit, a sweep auger, and an underfloor unloading auger (Fig. 1). The heater fan starts when the first load of grain is put in and continues to operate as long as is required to lower the average grain moisture content to the desired level. The drying rate depends on grain depth, temperature of the heated air, and airflow rate. As a general rule, an airflow of 450 $m^3h^{-1}m^{-3}$ of grain (9 cfm/bu) leads to efficient drying. This is the airflow on the exhaust side of the grain bed and is a function of the initial flow rate, the grain bed depth, and the grain's packing characteristics. The pressure drop across the grain bed is usually measured with a manometer and used to determine the required fan capacity from charts usually supplied by the fan manufacturer.

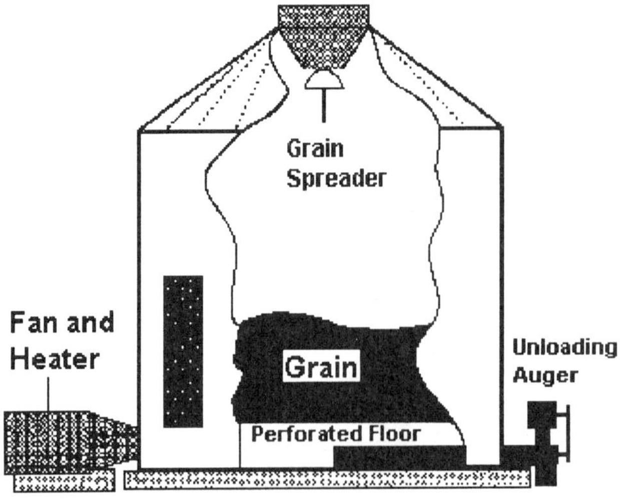

Fig. 1 Typical batch bin dryer.

For a given grain depth, the drying rate can be increased by raising the air temperature, but this increases the chance of overdrying near the floor. An adequate air temperature for the crop being dried can be determined from the initial moisture content (see Table 1). The grain must be cooled after drying and before storage. This is done by using the dryer fan to blow cool air over the bed, or by transferring the grain to an aerated storage bin.

Alternate heating and cooling cycles are sometimes invoked to reduce the moisture differential between the drier grain near the perforated floor and the damper grain near the top of the grain column.

Some bin dryers have overhead, perforated, cone-shaped drying floors supported about 1 m below the roof (Fig. 2). A heater fan unit is installed below the perforated floor and blows warm air up through the grain.

When one batch of dry grain is dropped to a perforated floor at the bottom of the bin where it is cooled by an aeration fan, the next batch is loaded and dried on the dryer floor above. Cool, dry grain is transferred to another storage bin by an underfloor auger. The advantage of this system is that drying can continue while the grain is being cooled and transferred.

Vertical stirring augers may be installed in bin dryers to promote more uniform drying and permit a higher airflow rate, thus increasing the drying rate for a given crop. Although the use of stirring augers may result in slightly lower fuel efficiencies, the larger possible batch size, the increased drying rate, and the reduction in quality losses from overdrying at the bottom outweigh this disadvantage.

3.1.2 Recirculating Dryers

In recirculating dryers, grain is constantly mixed during heating. One example of a recirculating dryer is shown in Fig. 3. A slanted floor causes the grain to move toward a vertical auger situated in the center of the dryer. The auger picks up the grain and delivers it to the top of the grain bin. The result is a more uniformly dried crop than that obtained using nonrecirculating types.

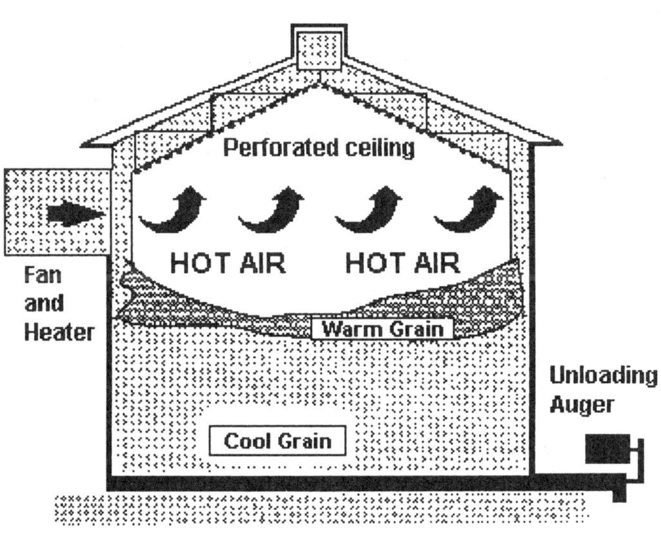

Fig. 2 A bin dryer with overhead drying floor.

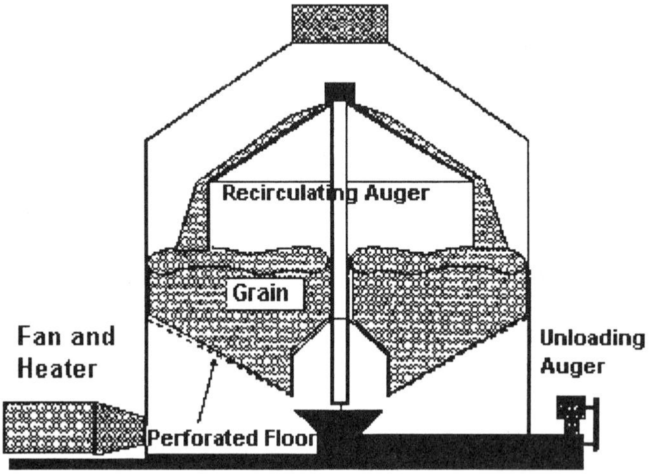

Fig. 3 Recirculating dryer.

The dryer shown in Fig. 4 is used as a recirculating batch or continuous-flow dryer. When being used as a recirculator, an ''under grain'' sweep-auger moves grain to the center of the perforated bin floor where it is picked up by a vertical auger and delivered to a grain spreader. When the dryer is operated as a continuous-flow dryer, the grain traveling up the vertical auger is transferred to an aeration bin by an inclined auger.

3.1.3 Continuous-Flow Dryers

Of the many types of continuous-flow dryers, the cross-flow dryer is the most commonly used for grain drying (Fig. 5).

The grain is loaded at the top and passed down both sides of the hot and cold plenums before entering the unloading augers. Grain flow rate is controlled manually or by a thermostat near the outside of the grain column. As fan capacity is decreased or column width increased, more efficient use of heat results; however, the moisture differential between grain on the inside and outside layers increases.

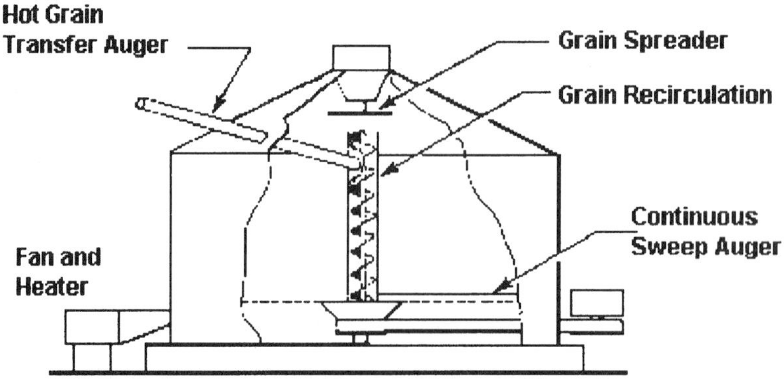

Fig. 4 Recirculating bin dryer.

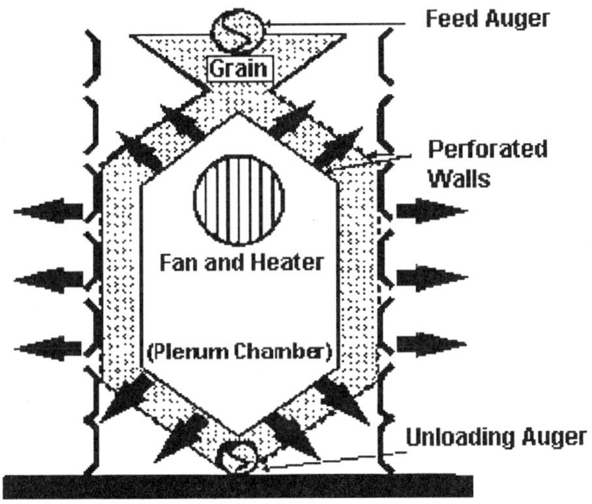

Fig. 5 Cross-flow dryer.

Some continuous-flow dryers use three fans and three plenums, each with individual temperature controls. These may be run with two heating sections and one cooling section, or with three heating sections. In the latter case the grain must be cooled in an aerated bin (see Figs. 6a, 6b).

Farm Fans (Indianapolis, IN) has a series of dryers of this type that they term continuous multistage dryers, ranging in capacity from about 5 to 27 ton/h (265–1220 bu/h) based on drying and cooling corn from 25 to 15% moisture.

Several companies recycle drying or cooling air. Two common techniques of accomplishing this are shown in Fig. 6. Some manufacturers use the system shown in Fig. 6a. Here, ambient temperature air is drawn through the grain in the cooling section and then passed through the fan heater unit of the midsection. This system results in more energy saving than the system shown in Fig. 6b because air from the first heating section

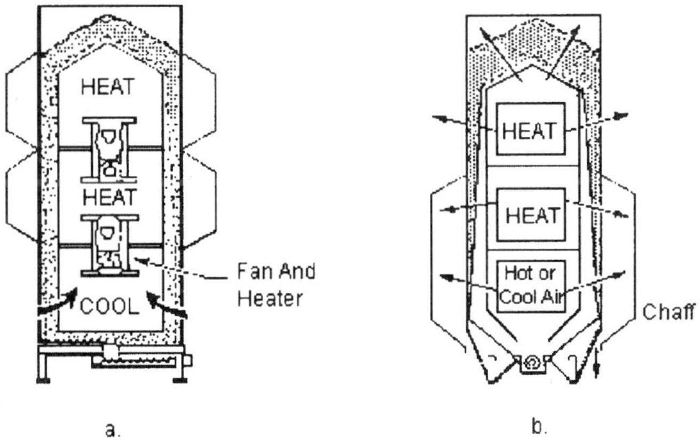

Fig. 6 Heat recovery systems: (a) reverse cooling; (b) one-way airflow.

is recycled. Its disadvantage is that chaff and fine material may be drawn into the midsection hot air plenum, necessitating frequent cleaning.

Most continuous-flow dryers are of the stationary type, although some of the smaller-sized dryers are portable. For example, Gilmore and Tatge Manufacturing Company, Inc. (Clay Center, KS) make a concentric cylinder-type portable dryer that handles 7.8 ton/h (350 bu/h) based on moisture removal from 20.5 to 15.5%. Grain column width on many of these dryers is 0.30 m as compared to the 0.45 m found on the GT-Tox-o-Wik recirculating batch dryers. Note that the moisture differential across the grain column is lowered as its width is decreased. A thinner column, therefore, means that, for a given average moisture content, the inner layer is overdried less. Thus, using a continuous-flow dryer might be of some benefit when drying heat-sensitive small grains, such as wheat, oats, and barley.

Another type of continuous-flow dryer is the parallel-flow dryer in which the grain moves in the same direction as the hot airflow. This results in more uniform drying and reduces the danger of heat damage. Furthermore, because no screens are used in parallel-flow dryers, small seed crops can be dried without leakage.

Continuous-flow dryers are not well suited for the drying of small quantities of different types of grain because start-up and emptying of these dryers is inefficient. Accurate moisture control is difficult to achieve until a uniform flow is established. Continuous-flow dryers are best in situations in which large quantities of grain must be dried without frequent changes from one type to another.

3.2 Rotary Dryers

Rotary dryers are now employed in the grain industry to a limited extent. The major components of the system are a long, inclined cylindrical shell, a fan and heater unit, loading and unloading augers, and a variable-speed drive. The operating principle of this type of dryer is to repeatedly lift the grain using a set of flights along the perimeter of the cylindrical shell and drop them into a stream of heated air. Most rotary grain dryers are of the concurrent-flow type in which grain and heated air are introduced at one end of the shell and dried grain and moist air exit at the other end. The grain is moved through the entire length of the shell by cascading a certain distance along the periphery of the inclined shell, such that in each fall, the grain is moved closer to the exit.

A typical commercial rotary dryer has a shell diameter of 1–2 m, a length of 15–30 m, and a slope 2–4° from the horizontal. The shell rotates at 4–8 rpm, and the drying air temperature is 121–288°C (10). Rotary dryers are increasing in popularity for the parboiling of rice because they are particularly suited to drying high moisture particles that tend to stick together and cannot be suitably dried in bin or column dryers. The specific energy consumption of the system cannot be fairly compared with that of other dryers, for the parboiled rice is at a much higher moisture content than those materials dried in other types of units.

3.3 Portable Dryers

Portable dryers generally appeal to the farmer who has grain bins in various locations, or who does custom drying off the farm. Portable dryers may be used without a proper grain-handling system to fill an immediate need in an emergency situation; however, they are normally not used when drying is beneficial, but not necessary, owing to the inconvenience of setup and dismantling of the system. The two types of portable batch dryers are nonrecirculating and recirculating.

3.3.1 Nonrecirculating Dryers

Most nonrecirculating dryers have a fully enclosed concentric cylinder configuration. They are loaded from the top, and drying is accomplished by blowing hot air radially through a column of grain. Although the inside grain layer (the layer near the hot air plenum) tends to overdry while the outside layer remains underdried as occurs in the batch-in-bin dryer, the damp and dry grain are mixed as the grain is removed from the dryer so that a satisfactory product results.

Other types of portable nonrecirculating batch dryers exist, such as wagon or truck box dryers. These use a heater fan unit, similar to that used for bin drying, that is connected to smaller air ducts suspended at midheight of the box or located on its floor. If suspended air ducts are used, exhaust ducts on the floor are a necessity.

These types of automatic dryers are equipped with thermostats or timers to control heating and unloading cycles and can be completely automated if a mechanized grain-handling system is used.

3.3.2 Recirculating Dryers

Portable recirculating batch dryers are essentially the same as nonrecirculating models except that they have a central auger that picks up grain near the bottom of the column and deposits it at the top (Fig. 7). A complete recirculation of grain occurs roughly every 15 min. Most common dryers of this type come in sizes ranging from 10 to 18.5 m^3 (300–525 bu) bin capacity. These dryers are often used by medium-sized of farms in eastern

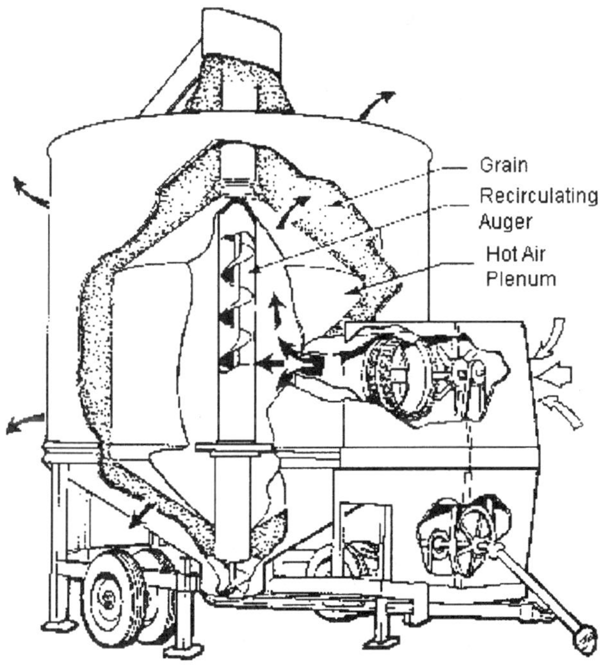

Fig. 7 Typical portable dryer.

Table 2 Recommended Drying System Based on the Annual Farm Production at Harvest

Annual production (ton)	Type of drying system
22–60	Natural air drying
60–445	Natural air drying with supplemental heat
445–1556	Batch-in-bin dryers
Above 1556	Portable and continuous flow dryers

North America that cannot afford a more expensive continuous-flow model. The dryers may be used for virtually any crop, if the maximum safe drying temperature is not exceeded. However, their disadvantage is that constant augering can cause damage to certain seeds, such as beans, peas, and malting barley, especially when they are nearly dry.

3.4 Dryer Selection

Selection of a continuous-flow, batch, or batch-in-bin dryer depends largely on the amount of grain to be dried and the facilities a farmer already has available when the dryer is purchased. For example, a farmer who already has a good-sized storage bin and only a small volume of grain to store would likely use an in-storage dryer, rather than purchase a portable dryer and ''wet grain'' holding bin. In-storage dyers usually are suitable only for farms harvesting less than 160 ha.

The recommendations concerning the type of system to be used can be made based on the annual production of a given farm, as illustrated in Table 2. Although the capacity range presented here is for corn, it can be extended to other grains and cereals. The recommendations are based on harvest moisture conditions in the central United States region. The crops most often dried in Canada and the United States by artificial means are corn (maize) and beans. Wheat, oats, and barley are harvested in the dry season and usually come off the field at a low enough moisture content for safe storage. If need be, the grain may be dried with natural air on sunny, warm days.

Most of the dryers in Canada are found in Quebec and Ontario (11), many of these being portable batch types. Larger, continuous-flow models may be found at cooperatives across the country or on the larger farms in southwestern Ontario where farmers are growing 320–360 ha of their own crop.

4 ARTIFICIAL DRYING IN DEVELOPING COUNTRIES

Where humidity is too high to allow grain to be adequately dried by natural means, it is necessary to supply heat to the drying crop. The most popular forms of artificial drying may be categorized according to the depth of grain being dried. These are (a) deep-layer drying, (b) in-sack drying, and (c) shallow-layer drying.

Deep-layer dryers consist of silo bins (rectangular warehouses) fitted with ducting or false floors through which air is forced. Depths of up to 3.5 m of grain may be dried at one time (12).

An in-sack dryer is made of a platform that contains holes just large enough to hold jute sacks full of grain. Heated air is blown up through the holes (and grain) by a heater and fan unit. The platform may be constructed from locally available material. A typical oil-fired unit that handles 2–5 ton of grain is equipped with a fan that delivers 9700 m^3h^{-1} of air heated to 14°C above ambient temperature and consumes about 4.5 L h^{-1} of oil (11). For 2-ton loading, the moisture removal rate is about 1%/h.

Shallow-layer dryers are those consisting of trays, cascades, or columns in which a thin layer of grain is exposed to hot air. In these dryers, the hot airstream is at the highest safe temperature and the amount of drying is determined by the length of time the grain is allowed to remain in the dryer, either as a stationary batch or as a slow-moving stream. Because the layer of grain being dried is thin (less than about 0.20 m), no significant moisture gradient develops through the grain (11). This means that the drying temperature is limited only by the possibility of heat damage to the grain.

Another simple, but effective, type of artificial dryer utilizes a locally built platform dryer in which the products of combustion of local fuel are not allowed to pass through the grain. The heated air passes through the produce by means of natural air movement or convection currents. One such dryer built at Mokwa, Nigeria, uses a pit (which became the hot air plenum) covered by the drying floor, the firebox being located outside the plenum chamber.

Yet another type of dryer is the horizontal dryer, which contains a number of chambers, each being divided by horizontal, equidistant, screen-bottomed trays placed on horizontal pivots. Damp grain is placed on the top tray in a layer 0.16 to 0.18-m deep and is tipped to the next set of trays after an initial drying period. As this type of dryer is normally operated as a batch dryer, it is an advantage to have two cooling chambers per unit so that one batch may be loaded into the dryer while the other is being removed from the machine. A typical setup of this type would include a double drying chamber, a cleaning unit, and augers or elevating units for tilting the dryer and elevating the grain to storage.

5 SOLAR ENERGY IN DRYING

An alternative that is being encouraged in hot, dry countries of Asia and Africa is solar drying. Solar heat is trapped with a solar collector constructed from an aluminum sheet painted black. The collector may be fixed to the drying bin in such a way that an air space exists between it and the bin wall. Energy absorbed by the collector heats the ventilating air by a few degrees as it is forced through the air space. In North America, these types of dryers have been known to operate satisfactorily with grain moisture contents up to 25%, even on cloudy days (7). The reason for this is that solar energy is about half visible light and half infrared rays, the latter being able to penetrate clouds. On rainy days and nights supplemental heat may be supplied electrically.

In countries where harvesting time occurs at the beginning of the dry season, the most popular method of drying is exposure to the sun. Crops are often left to dry in the field before harvesting. In some countries various crops are dried on scaffolds or inverted latticework cones. Another method is to lay paddy, maize, cobs, and other crops on heaps of stubble and then to cover them with stubble. At the village level, probably the most common practice is to spread the harvested threshed or shelled crop on the ground or on a specially prepared area (e.g., matting, sacking, mud/cow dung mixture, or concrete) exposed to the sun.

In humid countries, initial crop drying may take place as just outlined; however, further drying is accomplished by placing the crop in a ventilating storage area. A more effective type of drying than sun drying is shallow layer drying. This form of drying may be achieved by spreading the produce in a layer on the ground or on wire bottom trays that are supported above the ground. Cribs may also be constructed for drying maize on the cob or unthreshed legumes and cereals. These are usually oriented so that the long axis is facing the prevailing wind. They often have roofs or wide overhangs to protect the drying crop from rain.

In most warm countries, a commercial dryer is too expensive and not essential enough for a single farmer to consider its purchase. China, India, and countries on the continent of Africa are examples of places where solar drying by direct exposure or by a cheaply constructed collector is employed.

Solar energy can be and is used by some North American farmers for low-temperature drying situations with unsophisticated collectors (12).

6 NONCONVENTIONAL METHODS

Increases in the price of fossil fuels have prompted researchers to investigate and develop more energy efficient dryers (13–16). One attempt at reducing fuel cost was to pass unheated air through large beds of absorbent material such as silica gel before passing it through the grain. The problem was that the gel itself had to be dried at high temperatures, making the operation expensive.

Other methods related to enhancement of heat-transfer techniques are also being studied. Particle–particle heat transfer is one such technique that has led to the design of many experimental dryers. Although the heat-transfer rates and efficiency have been higher than for convection systems, the moisture removal was low owing to restriction of mass transfer of moisture. If the moisture to be removed is limited, conduction dryers may be recommended. Hybrid systems consisting of a initial heating by particle-to-particle conduction, followed by heated-air drying have been contemplated. Richard and Raghavan have dealt with this topic extensively (17). They discuss the theoretical aspects, experimental data, and demonstrate the potential of this method. The main advantage of this type of dryer is its rapidity. Following this concept, a continuous-flow conduction grain processor/dryer was developed in the late 1980s at McGill University, Quebec, Canada; it is shown in Fig. 8 and fully described in a paper and two patents by Pannu and Raghavan

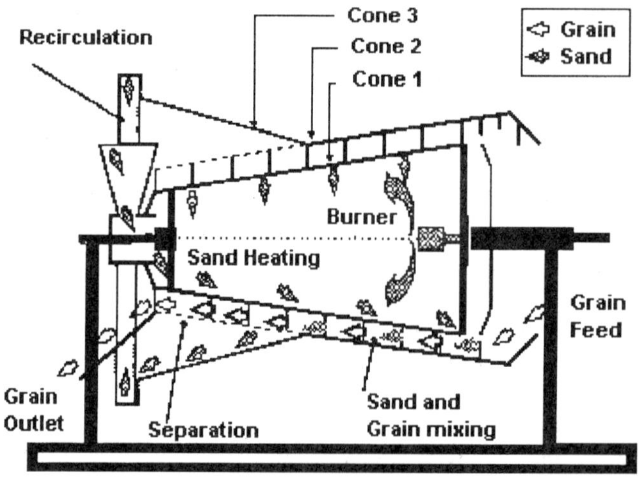

Fig. 8 Particulate medium dryer developed at Macdonald Campus.

Fig. 9 Particulate medium thermal processor, Macdonald Campus.

(18–20). It is based on particle–particle heat transfer and was designed to control mixing and heating time and provide ease of separation of the grain and the particulates. As a dryer, however, the most serious limitation was resistance to mass transfer owing to rapid saturation of the air in the conical drum. Some work was performed using zeolites as the particulate media in an attempt to improve removal of moisture from the air. This improved the mass transfer (moisture removal) by 50–130%, and although fine zeolite powders tended to stick to the finished product, nutritional studies on ruminants did not indicate deleterious effects (21,22).

The design was extensively modified in the 1990s (23), leading to the construction of a 1 $ton^{-1}h^{-1}$ thermal processor for grain to efficiently accomplish tasks such as roasting, popping, disinfestation, and degermination (Fig. 9). A 10 $ton^{-1}h^{-1}$ commercial unit is now being developed.

In regions where hydroelectric power is available, it may be reasonable to adapt microwave-heating technology for grain drying applications. One motivation for this approach is that hydroelectric power is considered to be nonpolluting. The main disadvantage is the initial capital cost of equipment. Nevertheless, there may be substantial savings in drying time owing to rapid internal heat generation and pressure-driven expulsion of internal moisture. Some recent work has shown that it is possible to dry grains to seed quality in a combined microwave–convective environment (24). As with particulate medium heating, microwaves may be more practical as an initial heating method integrated with heated air drying.

ACKNOWLEDGMENTS

The author thanks the following companies for providing information on different types of dryers: Beard Industries, Frankfort, IN; Caldwell Manufacturing Company, Kearney, NE; Farm Fans, Indianapolis, IN; Gilmore and Tatge Manufacturing Company, Inc., Clay Center, KS; Long Manufacturing N.C. Inc., Tarboro, NC; Martin Steel Corp., Mansfield, OH; and Mathews Company, Crystal Lake, IL. The help of Mr. P. Alvo in revising the original version of this chapter which appeared in the *Handbook of Industrial Drying*. Vol. 1, 1995 (AS Mujumdar, ed.), Marcel Dekker, Inc. is also appreciated.

REFERENCES

1. FAO Statistics Database on Internet. 1998.
2. Parikh JK, Syed S. Energy use in the post-harvest (PHF) system of developing countries. Energy Agric 6:325–351, 1988.
3. Smil V, Nachman P, Long TV II. Technological changes and the energy cost of U.S. grain corn. Energy Agric 2:177–192, 1983.
4. Agriculture Canada. Drying and conditioning. In: Agricultural Materials Handling Manual, Part 3. Ottawa: The Queen's Printer, 1962, pp 1–31.
5. Foster GH. Drying cereal grains. In: Storage of Cereal Grains and Their Products. St. Paul, Min: Am Soc Cereal Chem 1984, pp 79–116.
6. Brooker DB, Bakker–Arkema FW, Hall CW. Grain drying systems. In: Drying Cereal Grains. Westport CT: AVI Publishing, 1974, pp 145–184.
7. Nash MJ. Cereal grains, legume grains and oil seeds. In: Crop Conservation and Storage, New York: Pergamon Press, 1978, pp 27–79.
8. Friesent OH. Heated-Air Grain Dryers, Ottawa Information Services Agriculture Canada Publication 1700, 1981, pp 8–25.
9. Ritchie JD. Sourcebook for Farm Energy Alternatives. NY: McGraw-Hill, 1983.
10. Brooker D, Bakker-Arkema FW, Hall CW. Drying and Storage of Grains and Oilseeds. Westport, CT: AVI, 1992.
11. Otten L, Brown R, Anderson K. A study of a commercial cross-flow grain dryer. Can Agric Eng 22:163–170, 1980.
12. Hall DW. Handling and Storage of Food Grains in Tropical and Subtropical Areas. Food and Agriculture Organization of the United Nations, 1970, pp 1–198.
13. Meiring A, Daynard TB, Brown R, Otten L. Dryer performance and energy use in corn drying. Can Agric Eng 19:49–54, 1977.
14. Mittal S, Otten L. Evaluation of various fan and heater management schemes for low temperature corn drying. Can Agric Eng 23:97–100, 1981.
15. Mujumdar AS, Raghavan GSV. Canadian research and development in drying—A survey. In: Drying '84 New York: Hemisphere/McGraw-Hill, 1984.
16. Sturton SL, Bilanski WK, Menzie DR. Drying of cereal grains with the dessicant Bentonite. Can Agric Eng 23:101–104, 1981.
17. Richard P, Raghavan GSV. Drying and processing by immersion in a heated particulate medium. In: Advances in Drying. Vol. 3. New York: Hemisphere, pp 39–70, 1984.
18. Pannu K, Raghavan GSV. A continuous flow particulate medium grain processor. Can Agric Eng 29:39–43, 1987.
19. Raghavan GSV, Pannu KS. Méthode et appareil de sèchage et de traitement à la chaleur d'un matériau a l'état granulaire. Canada Patent 1254381, 1989.
20. Raghavan GSV, Pannu KS. Method and apparatus for drying granular material. US. Patent 4597737, July 1, 1986.
21. Raghavan GSV, Alikhan Z, Fanous M, Block E. Enhanced grain drying by conduction heating using molecular sieves. Trans ASAE 31:1289–1294, 1988.
22. Alikhani Z, Raghavan GSV, Block F. Effect of particulate medium drying on nutritive quality of corn. Can Agric Eng 33:79–84, 1990.
23. Sotocinal S. Design fabrication and testing of a particulate medium thermal processor. PhD dissertation McGill University, 1997.
24. Raghavan GSV, Alvo P, Shivhare US. Microwave drying of cereal grain: advantages and limitations. CAB Postharvest News Inform 4:79N–83N, 1993.

8

Grain Storage: Perspectives and Problems

SOMIAHNADAR RAJENDRAN

Central Food Technological Research Institute, Mysore, India

1 INTRODUCTION

1.1 Necessity for Grain Storage

Food security—''access by all people at all times to enough food for an active, healthy life''—is an important commitment for every nation (1). Increased agricultural production particularly the cereals, the staple food of people and the livestock, should match with scientific storage and preservation of the produce to ensure an uninterrupted supply to the world's ever-increasing population. Although grain production, or more specifically the harvest period, is relatively short, the consumption extends throughout the year. Grains are stored by farmers for their consumption or for seed purposes. Traders and marketing agencies store grain for financial gain. Government organizations play a role in domestic security, price stabilization, and in earning valuable foreign exchange through export of grains. The overall objective of grain storage is to preserve the quality, including nutritive value, and to keep the grains in good condition for marketing and processing, thereby reducing product and financial losses. The role of storage has become more significant, as the grain quality standards with reference to pests, pesticides, and other contaminants in national and international markets have become increasingly rigorous. Storage of food grains, a component of postharvest operation, is an ongoing challenge for both industrialized and developing countries. Aspects of grain storage have been reviewed earlier (2–4).

1.2 Grain Properties

Cereal grains are basically single-seeded fruits in which the starchy endosperm is the predominant portion (50–70%), whereas the germ, the most nutritive part of the grain that contains proteins, lipids, minerals, and vitamins, constitutes a small part. The endosperm is surrounded by the bran that comprises the aleurone layer and other layers of highly compressed cells, nucellus, seed coat, and pericarp. In naked grains (e.g., wheat and maize), the pericarp protects the grain from fungal invasion and transfer of moisture. Grains such as barley and paddy are covered with husks or hulls comprising lemma and palea. The hull protects the grain from mechanical damage and against fungal and insect attack. Grain being a living organism, it respires. This results in the oxidative breakdown of plant storage materials into carbon dioxide, water, and energy. The energy liberated during this aerobic respiration is utilized by the cells for metabolic processes, resulting in the release of heat (5).

The granular structure of the grains has many advantages for their handling and storage. It facilitates easy flow through pipes or conveyers, to pour or aspirate, or to fluidize. The interstitial spaces in grains allow air to be blown through the grain mass for removal of excess heat (cooling), moisture (aeration), and to change the composition of the surrounding air (controlled atmosphere storage) (6). The intergranular air serves as an excellent insulation and is responsible for temperature gradients in grains in tall silos.

The water or moisture content is an important component of the grain and is the crucial factor affecting the storage of grain. The moisture content is taken into consideration during procurement and marketing. The growth and multiplication of insect pests and fungi, the major spoilage organisms of the grains, are dependent on the available water in the grain. Grains at the time of arrival to a storage site contain dockage which includes broken grains, foreign materials, and dust. The presence of dockage is conducive for the development of insect pests and it impedes effective aeration of the grain in a silo or bin (7). Grain dust is composed of fine particles of 15- to 120-μm starch granules, bran flakes, or pieces of chaff formed by the abrasion of grains during handling. Beyond certain limits, the dusts form an explosive mixture with air in the presence of a spark.

Grain has low thermal conductivity and specific heat. Hence, in bulk storage, in the absence of forced air circulation, the heat produced by the metabolic processes of the grain (and grain pests) does not dissipate readily, resulting in ''hot spots.'' Variations in grain hardness between cereal grains and between cultivars of the same species (e.g., wheat) have been reported. Grain hardness is an important property, as it can influence the grain's ability to resist damage during handling and insect attack. Bulk density or mass per hectoliter i.e., weight of a given volume of grain including the voids, expressed as pounds (lb)/bushel in the United States and kilogram (kg)/hectoliter in most other countries, is a widely used grain property in commercial contracts. Low bulk density means the grain is infested or infected, or that physiologically immature grains and impurities are present. Grains are hygroscopic and gain or loose water depending on the water vapor present in them and in ambient air to attain equilibrium. Once the equilibrium moisture content is reached the grain no longer absorbs moisture from its surrounding air. The equilibrium moisture content (ERH) values are different for various grains. For a specific grain, the values increase with increasing relative humidity and decrease slightly with increasing temperatures.

1.3 Environmental Factors Influencing Grain Quality

The quality of stored grain depends on four important factors: (a) initial condition of the grain; (b) environmental conditions during the period of storage; (c) biotic factors, such as insects, rodents, and microorganisms; and (d) various treatments applied on the grain during the storage period (e.g., aeration, drying, fumigation, controlled atmospheres, grain protectants).

1.3.1 Initial Condition of the Grain

The storage properties of the grain are likely to be affected even before reaching the storage premise. Environmental conditions during growth and maturation of grains, the degree of maturity at the time of harvest, harvesting methods, and the method of handling grain in the farms influence the storage quality. Fungal and insect pest activity can start in the field itself. High temperatures and humidity during maturation of corn are conducive to aflatoxin formation. The level of infection by the storage fungi has been linked to the amount of kernel damage as a result of combined harvesting and handling. Insects such as the larger grain borer *Prostephanus truncatus* and the maize weevil *Sitophilus zeamais*, which are active flyers, infest the grain in the field itself. It is not uncommon for the maize and paddy crops to be harvested at high moisture levels of more than 15%, which is not suitable for storage as such. It needs drying artificially or by sunlight (8). The sanitary condition or physical state of the newly harvested grain (i.e., moisture content, cleanliness, and specific weight) will affect the storage quality. Therefore, it is necessary to conduct quality tests on the grain before binning or taking into storage. These tests include determination of moisture content, bulk density, germination tests (glutamic acid decarboxylase activity or tetrazolium tests), amylase activity (Hegberge test), free fatty acids, nonreducing sugar contents, detection of insect infestation, and mold counts. Indices of fungal load can also be assessed by chitin and ergosterol assays. These details will help assess the shelf life potential of the grain; hence, assessment of grain quality is an important storage strategy (6). Among the quality parameters the water activity or moisture content of the grain is the most significant because it has an influential effect on the multiplication of insect pests and growth of other spoilage organisms. The moisture content will not be uniform in bulk storage and is likely to vary between farms, between truck loads or grain lots, and in bag storage between peripheral and inner bags, and in bulk storage between top layer and periphery or in depth. The highest moisture prevailing in any part of the grain bulk has more practical importance than the average moisture content.

1.3.2 Physical Factors

Because pest activity is dependent on temperature, the latter plays an important role during grain storage. The optimum temperatures for the growth and multiplication of insects, fungi, and mites in stored grains are 25–32, 30, and 25°C, respectively. Development of insects and mites is decreased, respectively, at temperatures less than 15° and 5°C. The mold growth, however, is affected only at 0°C or below. As the temperature rises, the rate of respiration of the grain and the pests in it increases. Furthermore, the enzymatic activity of grains goes up. This enhanced biological activity leads to rapid quality deterioration at higher temperatures. Another interrelated factor is the moisture or water activity of the grain. Moisture content in the range of 12–14% is favorable for insect development. When the water activity is 0.9 or more, mold and other microorganisms thrive. When the water activity is low, the pest activity is automatically reduced. Temperature and moisture together largely determine the length of safe storage life. Respiration of grain and the pests leads to the consumption of oxygen and release of carbon dioxide during storage. Composition of the intergranular atmosphere influences the type of metabolism of the microorganisms and the grain. It also affects nonenzymatic reactions and certain enzymatic reactions. Oxygen and carbon dioxide levels also affect insect population and mold growth.

Interactions of the physical factors along with the biological process in the grain storage ecosystem lead to changes in the grain composition and its functional properties. Free fatty acids and glycerol are formed when grain lipids are broken down by lipases. The free fatty acid content of grain is a sensitive indicator of deterioration during storage (9,10). Changes in carbohydrates include decreases in the amounts of nonreducing sugars and total sugars. Considerable loss in vitamin A activity in yellow corn and sorghum, decreases in water-soluble B vitamins, and an increases in the level of phosphorus compounds in grains during storage have been reported (10).

Functional properties such as bread-making quality of wheat, milling and cooking quality of rice, malting quality of barley, and starch separation in corn are also likely to be affected during storage (9,10). Storage of rice leads to aging and aged rice is preferred in some regions in Asia. Decreased levels of 2-acetyl-1-pyrroline, an aroma principle in aromatic rice following storage of rice, have been observed. Reductions in the content of free amino acids, nonreducing sugars, and albumin have been noted in ordinary rice (9).

1.3.3 Biotic Factors and Treatment Effects

Biotic factors such as insects, mites, rodents, birds, and microorganisms are responsible for quantitative and qualitative losses of the stored grains. They are also responsible for the contamination, heating, and associated storage problems, and in extreme cases may pose health hazards. The biotic factors are discussed in detail in Sec. 3, and the effects of treatments, such as aeration, drying, fumigation, and protectants, are discussed in Sec. 4.

2 STORAGE SYSTEMS

2.1 Farm Level Storage

For safe storage of food grains a storage facility, whether indoors or outdoors, small or large, should ensure protection against physical factors such as adverse weather, high temperatures, rain and snow, and keep out biotic factors including insects, mites, rodents, birds, and microorganisms. The structure should be pilferage-proof and must be designed for easy grain handling and to facilitate effective fumigations. A major part of the grain produced is stored at the farms in most of the countries for their own consumption and to fetch a higher price at a later period (8,11). The storage loss at farm level generally exceeds 4%, which is relatively more than in centralized storage (11). The storage structures are of varying capacities, ranging from 100 kg to few metric tons, and are made of locally available building materials to withstand climatic conditions. Some of the traditional structures used are earthen pots, underground pits, mud bins, and (maize) cribs. Suitability of bins made of plywood, ferrocement, metal, and high-molecular weight, high-density polyethylene (HMHDPE) of 0.5- to 3-metric–ton capacity was examined by Krishnamurthy and Majumder (12) for farm level storage in developing countries. The plywood bin was the most suitable among them. Traditional underground structures in various shapes and sizes are used in African countries and to a limited extent in Argentina, India, and United States. In principle they are like hermetic storages preserving the grain quality by the natural buildup of high carbon dioxide levels and low-level oxygen atmospheres that are lethal to grain pests and microbes (13). The traditional structures are simple and cheaper, but do not adequately protect against pests and adverse weather conditions. Merits and drawbacks of various traditional storages have been discussed (11). At farm level, metal bins or silos of various sizes have also been used (8). In countries such as Australia and the United States, the bin capacity exceeds 50 metric ton and is also equipped with facilities for aeration.

2.2 Bagged Storage

In developed countries, grains are stored in bulk in silos, flat storages, and in grain elevators, whereas in the developing countries, they are generally, stored in gunny or woven polypropylene bags in the conventional warehouses (8,14). Bag storage is labor-intensive and thus incurs higher operating costs and losses to pests, and results in more spillage. In some of the warehouses, if the flooring is not properly constructed, water seepage occurs. This increases the humidity in the warehouse, favoring multiplication of *Cryptolestes* spp.; also, it damages the bottom layer of bags. The advantages of bagged storage system are the lower capital costs without any need for sophisticated aeration and fumigant circulation facilities. However, much less research has focused on the possible loss reduction in bag storage systems in the developing countries (8). The existing system may continue in these countries because of small farm sizes, handling of many grain varieties, and cheaper manual labor. Moreover, it will be uneconomical to change over to a capital intensive bulk storage system.

2.3 Bulk Storage

The system of bulk handling and storage in silos or bins and elevators has been adopted by the industrialized countries as a result of mechanized harvesting and postharvest operations. Bins and silos of varying capacities, built of masonry, wood, reinforced concrete, or metal, with hopper or a flat bottom are the modern grain storage structures in the developed nations. In hoppered bins grains flow out by gravity; hence, they are self-cleaning and do not require shoveling. Flat-bottomed bins are cheaper to construct and provide more storage spaces but cause some delay in outloading the grains as the shoveling is slow. Bins in round forms, rather than oval or hexagonal, are preferred because of the greatest strength (15). Silos are generally provided with facilities for monitoring and recording the temperature of the grain at various depths. They are also equipped for aeration and recirculation of fumigant gases. Unlike in wooden and concrete silos, temperature gradients, resulting in large moisture transfers are common in metal silos because of a high thermal conductivity of the metal (6). Spoutlines are a problem for concern in bulk storages. During binning, fine particles, brokens, smaller grains, and weeds remain at the center of the pile, whereas whole-grain kernels flow away from the slope. This core of high-dockage in the center of the pile, known as spoutline, is the source of pest proliferation and heat development in a silo, and it impedes air circulation, thereby affecting the storage period of grain (15). Of late, importance has been given to make silos sufficiently gas-tight to adopt controlled atmosphere storage and for effective phosphine fumigations. Techniques to seal silos with flexible polyurethane, acrylic-based, or elastic adhesive sealants, have been developed, particularly in Australia (16). The bulk storage system is being improved further by installing pest-monitoring system (acoustic detection) and automation for aeration, grain cooling, and pest-control measures.

2.4 Hermetic Storage

Storage of grains under airtight conditions is an ancient method in which insect population and mold growth are checked by natural buildup of carbon dioxide and depletion of oxygen by the respiration of grains and organisms. The same principle has been applied in underground storages, ''volcani cubes,'' flexible silos supported by a welded mesh frame of 50–1,000 metric ton and pad or bunker storage of 10,000- to 50,000 metric ton capacity. The merits and suitability of the technique for the storage of grains in tropical and subtropical climates have been discussed by Navarro et al. (17). However, Annis and Banks (18) claimed that the hermetic storage will not be effective when the grain is dry, the infestation is less than ten insects per kilogram of grain, and under the air ingress rate of 1%/day in underground and semiunderground stores; 2%/day in bunkers and sealed stacks; and 5%/day in aboveground storage. The hermetic storage system has to be supplemented with fumigation or controlled atmosphere treatment. Storage studies of bagged maize in flexible liners (''volcani cubes'') with and without introduced carbon dioxide, supports the foregoing concept (19).

In semiunderground bins in Cyprus and Kenya, grain has been stored for 3 years under sealed storage wherein condensation of moisture was observed (17). In airtight underground storages in Argentina about 0.5% loss of grain was recorded in a storage period of 2–3 years (8). Underground structures are economical for small holdings. However, in large-scale storages there is a problem in grain handling. Moreover, the quality of grain stored underground does not meet the current quality standards of zero-tolerance for insects (11). Airtight storage or sealed storage of grain stacks with PVC sheets has been standardized in Australia. The technique is cost-effective when it is intended to store grains indoors for not less than 3 months. Unlike hermetic storage, the atmosphere inside the enclosure is enriched with carbon dioxide or phosphine gas to control the initial infestation in the stack (20,21). Long-term storage of milled rice stacks under high carbon dioxide atmosphere is being routinely carried out in Indonesia and successful field trials have been conducted in India (22), and also; in other Asian countries. Hermetic storage offers a residue-free storage system, yet some pest control method has to be integrated into the system for its complete success.

2.5 Outdoor Storage

Sometimes need arises for temporary storages both in developing and developed countries. Temporary outdoor structures are often used when (a) there is lack of space in permanent storages, (b) the cost of construction of new structures are prohibitive, and (c) there is a surplus production in the localized areas where proper transport facilities are not available to move the stocks. In India, whenever the conventional godowns and silos are full, bag-stacks are built outdoor on a raised area (plinth), and the stacks are covered with 250-μm–thick, tailor-made polyethylene covers (Fig. 1). The stacks are aerated at least once-a-week by raising the covers to the seventh or eighth layer. This storage technique, called cover and plinth (CAP) is currently used for wheat and paddy. The drawbacks of CAP are that the food grains cannot be effectively fumigated and the covers are likely to be damaged by wind and during rains (8). In China, about 10% of grains produced is stored outdoors. Maize, wheat, and paddy grains are stored (a) in rectangular bag-stacks built on raised plinths and the top and sides of the stacks covered with either PVC sheets or locally made mats; and (b) in mat silos built by making a circular wall with bamboo mats on a raised platform. The grain-filled silos are then covered with rice straw or PVC or metal sheeting. The storage period may go up to 3 years, depending on the necessity (23).

Volcani cubes made of PVC liners of 830-μm thickness, with a grain holding capacity ranging from 5 to 50 metric tons have been developed in Israel for outdoor storage

Fig. 1 Outdoor storage of wheat in India.

(24). This acts as hermetic storage structure and has been tested under field conditions in some of the countries in Africa and Asia. In a recent study in the Philippines, it has been reported that insects in grain inside the volcani cube was controlled only when carbon dioxide was added (19). On usage, over a period, the liners are known to lose their plasticity, but their gas-retention property is significantly increased. Rodent damage to the liners has been rarely encountered (17). Pad or bunker is a temporary storage structure for storing wheat in Australia (25). They are built as rectangular structures on well-drained waterproofed sites with concrete- or steel-framed side walls. Inloading and outloading are done by specially designed machines. After filling, the grains are covered with a high-quality PVC fabric. As much as 50,000 metric tons of grains are stored in a bunker, which is fumigable.

The necessity for outdoor storages is likely to increase further because it is cheaper than permanent structures. Structures suitable for storage of high-moisture grains are the immediate need in the countries where maize or paddy is harvested at higher moisture levels and cannot afford any energy-intensive drying operation. Cross infestation by active fliers (e.g., the grain borers, *Rhyzopertha dominica* and *Prostephanus truncatus*) poses an additional problem for outdoor storages in the tropical and subtropical regions.

3 LOSSES DURING STORAGE

Pest organisms, comprising insects and mites among invertebrates; birds and rodents among vertebrates; and microorganisms including fungi, yeast, and bacteria, cause depredation of grains in storage. The interaction of all the pests in the grain ecosystem has a cumulative effect on the grain quality and thus leads to qualitative, quantitative, and nutritive losses.

3.1 Insects

Insects are the first among the invasive forces to start the interaction with the grain, and they are one of the major threats to the maintenance of grain quality during storage. They consume, contaminate, and disseminate microflora. There are over 100 species of insects infesting stored grains, the majority of them are beetles, some are moths, and the rest are psocids, which are primitive insects (Table 1). In moth pests, only the larvae are destructive, as the adults are nonfeeding and short-living. In the case of beetles, in addition to larvae, the adults are usually long-living and cause damage to the grains by way of feeding and by probing for oviposition. The insects have four life stages—egg, larva (with four to six instars), pupa, and adult. In certain species the preadult stages develop inside the grain. These include angoumois grain moth, *Sitotroga cerealella*, and the grain borers *R. dominica* and *P. truncatus* and the grain weevils, *Sitophilus* spp.

Grain weevils are the predominant grain pests in the world. The granary weevil *S. granarius* is restricted to temperate regions; the maize weevil *S. zeamais* to warm and humid regions, particularly in the maize-growing areas; and the rice weevil, *S. oryzae* mainly in the tropical countries. The grain borers, *R. dominica* and *P. truncatus* cause more damage to the grains by their boring activity, rather than by feeding itself. The adults are strong fliers; hence, they cause field infestation and cross-infestations. *R. dominica* is a serious pest on wheat and paddy rice in the warm and drier areas. *Prostephanus truncatus*, originally from Central and South America, has now become a serious pest of corn, particularly unhusked corn, in East and West Africa (26). Among dermestids the khapra beetle, *Trogoderma granarium* is the most dreaded pest. The destructive khapra larvae under adverse conditions enter diapause and hide in cracks and crevices of the warehouses. They are tolerant to most of the pesticides. Flour beetles, *Tribolium* spp. are secondary pests, yet they thrive in grain storages because broken grains and dockage will normally be

Table 1 Important Stored Grain Insect Pests

Common name	Scientific name	Major commodity
Beetles		
Rice weevil	*Sitophilus oryzae*	Wheat, rice
Maize weevil	*S. zeamais*	Maize
Granary weevil	*S. granarius*	Wheat
Khapra beetle	*Trogoderma granarium*	Wheat, rice
Lesser grain borer	*Rhyzopertha dominica*	Wheat, paddy
Larger grain borer	*Prostephanus truncatus*	Maize
Red flour beetle	*Tribolium castaneum*	Rice
Confused flour beetle	*T. confusum*	Rice
Long-headed flour beetle	*Latheticus oryzae*	Rice
Saw-toothed grain beetle	*Oryzaephilus surinamensis*	Rice
Grain beetles	*Crytolestes* spp.	Rice
Moths		
Angoumois grain moth	*Sitotroga cerealella*	Paddy
Almond moth	*Ephestia cautella*	Wheat, rice
Rice moth	*Corcyra cephalonica*	Rice
Indian meal moth	*Plodia interpunctella*	Wheat
Psocids		
Booklice	*Liposcelis* spp.	Rice

present along with sound kernels. Psocids, commonly encountered in grain storage premises, contribute to the contamination of grains as filth. They were once believed to be only scavengers and mold feeders. However, it has now been established that they do cause visible damage and loss of grain of about 3% in a storage period of 6 months. Psocid abundance is an annoying problem to the godown managers and for the laborers who carry grain in bags (27). The pest status of psocids has gained prominence in recent years. Among moths, *S. cerealella* attacks newly harvested corn, sorghum, and paddy rice. Infestations are largely restricted to the peripheral layers. The other pests, *Plodia interpunctella* (Indian meal moth), *Ephestia cautella* (tropical warehouse moth), and *Corcyra cephalonica* (rice moth) complete their life cycle outside the grain itself. The significance of grain insect pests and their biology have been discussed elsewhere (28).

Stored grain insects cause quantitative losses primarily by way of feeding and, in species such as *Sitophilus* spp., damage is done while depositing their eggs. Adults of *R. dominica* and *P. truncatus* create a lot of grain dust because of their innate behavior of boring. Insects such as *P. interpunctella* and *T. granarium* preferentially feed on the grain embryo and cause germination losses. There is a considerable loss of nonreducing, reducing, and total sugars in the infested grains. Insects contaminate the grains with their excreta containing uric acid, exuviae, fragments of dead insects, and webbing (by pyralid moths). Insect contamination decreases the market value of the grains and frequently results in the rejection of the material. Insects also play an active role in dissemination of storage fungi. The insect activity, notably by internal infestors, results in the generation of heat in the local areas. The hot spots cause moisture and temperature gradients in the grain mass that favor mold growth leading to deterioration of grains' quality.

The level of insect infestation in grain is one of the criteria for grain quality assessment. There are several methods to detect the insect infestation in grains (Table 2) and in grain storage structures. Sampling and sieving, the oldest method, is not accurate for

Table 2 Methods of Detection and Monitoring of Stored Grain Insects

Method	Remarks
Nontrapping methods	
ELISA	Species-specific; commercially used
Aural or acoustic	For hidden infestation; can be automated
NMR	Still at experimental stage
NIR	For detecting mites
Radiography	For hidden infestation
Sampling and sieving	Widely adopted; not accurate at low-level infestation
Specific gravity	For hidden infestation; qualitative
Cracking and floatation	For hidden infestation
Staining techniques	Specific for *Sitophilus* spp.
Ninhydrin reaction	Egg and early larval stages not detected
Uric acid analysis	Officially accepted; time-consuming
Carbon dioxide analysis	Slow, not applicable for high-moisture grain
Trapping methods	
Physical traps	For adult beetles
Pheromone traps	Species-specific for moths and beetles
Food-based attractant traps	Attracts several species; simple and cheaper

detecting low-level infestations, yet it is commonly practiced in all developing nations. Specific gravity method or the floatation technique is a simple qualitative method (not applicable for hulled grains and for corn) for detecting internal infestation. In recent years, more attention has been focused on aural or acoustic techniques (29) and enzyme-linked immunosorbent assay (ELISA)-based detection (30). By use of the acoustic method infestation by *Sitophilus* spp. and *R. dominica* can be detected. Automation in insect monitoring in grain silos by the acoustic method and identifying the species by the type of noise produced will be practiced soon. The ELISA test, nuclear magnetic resonance (NMR) and near infrared (NIR) techniques, radiography, floatation method, staining, and ninhydrin reaction methods are specific for detecting the internal infestors, which contribute fragments, the major contaminants in processed foods. Pheromone (sex and aggregation) traps that help detect or locate the infestations at an early stage, are used in grain storage facilities. The traps have also been deployed for controlling pest populations in grain storage.

3.2 Mites

Mites, the tiny (<1-mm size) arthropods, occur in grains both in tropical and temperate regions. There are about 54 species recorded in stored grains and flour. Although they are abundant, they have not been studied in detail. Also, there is a lack of information on the quantitative losses in stored grains caused by mites (28). The moisture content of the grain is the critical factor for their survival. They multiply rapidly in damp and moldy grain at low temperatures. In heavy infestation, mites impart a characteristic odor owing to their lipid secretions. They eat the germ part of the grains and disseminate storage fungi and bacteria (31); they cause dermatitis and allergies in humans. Mites in stored grain may be either saprophytic (e.g., *Acarus siro* and *Typophagus putrescentiae*), parasitic, or predatory (*Pyemotes* spp., *Cheyletus* spp). The life cycle consists of egg, larva, nymph (one to three instars), and adult. During the nymphal stage, in some species under certain environmental conditions, there is a hypopus stage that is tolerant of adverse conditions.

3.3 Storage Fungi

Field and storage fungi, yeast, and bacteria, present either internally or externally in the cereals, have been implicated in the deterioration of grains during storage. Yeast and yeast-like fungi are predominant in sealed silos when the oxygen level is low and grain moisture is high. Not much work has been carried out on yeasts. The role of bacteria has been noted in grain only at a water activity exceeding 0.9 when complete degradation of the grain has already taken place. Both field and storage fungi originate from soil and decaying debris, but they may also be present in harvesting and grain-handling equipments. Important field fungi that infect the grain in the standing crop or in the farm before threshing are *Alternaria*, *Cladosporium*, *Fusarium*, and *Drechsclora* spp. They infect the grains having more than 20% moisture. A typical example is infestation of ear corn in a growing plant by *Fusarium* spp., which persists in crib storage. Similarly, *Alternaria* spp. infect high-moisture, shelled corn awaiting drying or when it is slowly dried. The damaging effects of field fungi are stopped as the grain moisture comes down during storage and the field fungi either die or remain as dormant mycelium in the grain.

Aspergillus and *Penicillium* spp. are the important storage fungi. However, if climatic conditions are conducive, these fungi are also found infecting the grain in the field. In high-moisture or wet grains, species of *Rhizopus*, *Mucor*, and *Nigrospora* are encountered. *Wallemia sebi* has been reported in grains heavily infested with mites. Different species of *Aspergillus*, the dominant group of fungi, can proliferate in the absence of free water, and each has specific limits of grain moisture levels. These limits are 13.8–14.3%, 14.0–14.5%, 14.5–15.0, 15.5–16.0, and 17–18% for *A. halophilicus*, *A. restrictus*, *A. glaucus*, *A. candidus*, *A. ochraceus*, and *A. flavus*, respectively. All the species kill the germ and cause discoloration (32). *A. candidus*, *A. ochraceus*, and *A. flavus* generate heat, resulting in caking and a musty odor of the grain. *Penicillium* spp. develop fast in grains with moisture content in the range of 16.5–20%.

The development of storage fungi in the grain is influenced by such factors as the level of infection before storage, grain temperature, moisture content, or more appropriately the water activity, dockage content, and degree of insect and mite infestation. The optimum temperatures for the growth of *Penicillium* spp. range between 20° and 25°C; for *A. restrictus* and *A. glaucus*, 30°–35°C; and for *A. flavus* 40°–45°C. Most fungi require at least an a_w of 0.7 to grow, and they cause spoilage at or above an a_w of 0.8. Composition of the intergranular air also influences the growth of fungi. Oxygen at less than 1% level retards the development (33). Stored grain insects, particularly the secondary pests and scavengers, moving inside the grain mass disseminate fungi and facilitate inoculation. The grains damaged by insects help the mold invade.

Grain germ is the most susceptible part for the storage fungi for invasion and subsequent development. The germ is killed and discoloration takes place following infection, and this lowers the grade of the grain as ''sick'' or ''damaged'', thereby, reducing the market value. The baking quality of wheat is also affected (6). Besides dry matter loss, biochemical changes such as hydrolysis of glycerides of grain lipids giving rise to increased levels of free fatty acids have been observed (5). Furthermore, the commodity becomes rancid with a musty odor caused by volatiles developed from the oxidation of free unsaturated fatty acids. Any malodorous or discolored grain is not acceptable to food processors, as it will have repercussions in the end-use products. Active respiration of fungi and other microorganisms leads to heating of the grain (5). When the temperature rises to 55°C, grains become caked and black, with a burned appearance (Fig. 2). At this stage, the grains become unacceptable and unfit for human as well as animal consumption.

The early stages of infection by storage fungi can be detected by enumeration techniques after plating surface-disinfected kernels on a suitable agar medium (direct-plating method) or by a dilution-plating method. Qualitative tests, based on the estimation of constituents of fungi (chitin and ergosterol), or a breakdown product in the grain (free fatty acid), are also followed. The levels of free fatty acids, which are directly related to the degree of infection by mold are also estimated (10). Certain volatile compounds released following the biochemical changes in grain brought out by fungi, can be identified and monitored to indicate the level of mold activity in the stored grain. Other reported methods include detection of fungal biomass by measuring immunofluorescence, spectrofluorimetry, and species-specific ELISA technique for *Aspergillus* spp., *Penicillium islandicum*, and an inihibition radioimmunoassay for *A. repens* (32).

Mycotoxins are one of the secondary metabolites produced by fungi at certain stages of their development. The production of mycotoxins by the fungi in grains has major implications for human health and livestock production. The incidence of mycotoxins in grains and extent of losses owing to the presence of toxins vary from country to country. *Aspergillus* spp. and *Penicillium* spp. are the predominant fungi implicated in mycotoxin production and (besides groundnut) maize, followed by rice, are the major grains affected. Recently, Lubulwa and Davis (34) have reported that the total annual and social costs from aflatoxin in maize in Indonesia, Philippines, and Thailand in 1991 have been 319

Fig. 2 Bags containing milled rice damaged by storage fungi.

million U.S. dollars. Indonesia has been reported to be the country most affected by aflatoxin in maize.

Important mycotoxins and the causative agents are listed in Table 3. Aflatoxin, particularly aflatoxin-B_1 is the most notorious toxin present in food commodities, including cereal grains and feeds, in most parts of the world. International tolerance limits for aflatoxin in grains range from 10–30 ppb for total aflatoxin and 5–20 ppb for aflatoxin B_1 (34). The mycotoxins in grains are detected and determined by thin-layer chromatography (TLC), high-performance liquid chromatography (HPLC), or gas chromatography (GC) methods after extraction and cleanup (35). Possible synergism of different mycotoxins and interrelation between the presence of mycotoxins and human diseases have been the topics of considerable interest and concern in recent years (33). Suspected relations are between (a) incidence of malaria and aflatoxin, and (b) between cancer and immunotoxicity of fumonisins and aflatoxin (34,36).

Any increase in fungal activity in grain is reflected in a temperature rise in the storage. Hence, temperature monitoring is essential during the storage period besides periodical testing of the grain. Moisture being one of the critical factors for fungal development, it should be brought down to safe levels. At 27°C, for example, the safe limits for milled rice, wheat, and shelled corn as well as paddy rice are 13.0, 13.5, and 15%, respectively, although minor variations are present between the cultivars. Under practical situations, however, moisture content of the harvested grain varies. Corn and paddy are normally harvested at high-moisture levels. In all of these examples, grain moisture should be uniformly and rapidly reduced to safe moisture level. Storing the grain at low temperatures and chilled aeration of the grain will also aid in reducing the growth of mold. Organic acids and their salts, such as propionic acid, sorbic acid, acetic acid, and isobutyric acid either alone or in different combinations, are used. Besides the cost, acid treatment cor-

Table 3 Mycotoxins and the Important Toxigenic Fungi in Cereals

Mycotoxin	Fungi	Toxicosis in humans or livestock
Aflatoxins	*Aspergillus flavus* *A. parasiticus* *A. nomius*	Carcinogen, teratogen, immunosuppressive
Citreoviridin	*Penicillium citreonigrum* *Eupenicillium ochrosalmoneum* *A. terreus*	Neurotoxin
Citrinin	*P. citrinum* *A. terreus* *P. verrucosum*	Nephropathy (in swine)
Fumonisins	*Fusarium proliferatum* *F. nygamai* *F. napiforme* *F. verticillioides* (= *F. moniliforme*)	Leukoencephalomalacia in horses and pulmonary edema in swine; ''spiking syndrome'' in poultry; Suspected carcinogen in humans
Deoxynivalenol	*F. graminearum* *F. culmorum* *F. crookwellense*	Immunosuppressive
Ochratoxin A	*P. verrucosum* *A. ochraceous*	Nephrotoxic, teratogenic, and carcinogenic
Sterigmatocystin	*A. versicolor* *Emericella nidulans*	Carcinogen
Zearalenone	*F. sporotrichioides* *F. pallidoroseum* *F. graminearum* *F. culmorum*	Reproductive disorders in swine; possible human carcinogen

rodes metal parts in storages, affects seed viability, and acid-treated grain is generally used for animal feed. Fungicides, such as iprodione and thiobendazole at low levels, have also been tested for grain preservation. Application of ammonia (0.05%) or sulfur dioxide as gas in the grain-drying system inhibits fungal development. The effect of fumigation and controlled-atmosphere storage on growth of fungi has been reviewed (33).

3.4 Rodents

Rats and mice are the predominant and notorious vertebrate pests causing direct and indirect losses in grain storage. They cause damage to the crop in field-harvested grain before and during storage and also during processing and utilization. Quantitative loss of grains by direct consumption by rodents has been estimated to be 1% whereas overall loss by spillage and damage may exceed 10%. Contamination of cereal grains by urine, fecal pellets, saliva, hair, and body fragments is a major sanitary problem contributing to public health hazard. They are involved in the spread of diseases such as plague, typhus, leptrospiral jaundice, rat-bite fever, salmonellosis, and trichinosis in human beings. Rodents carry insects, mites, and microflora on their body and cause cross-contamination of grain stocks. They also cause extensive damage to the storage structures, electrical installations, and water pipes by gnawing (37). The extent of damage to food grains and nonfood items vary from place to place, year to year, and also depends on the level of rodent populations.

Field (wild) and commensal rodents belong to the family Muridae. Among commensal rats three important rats are *Mus musculus* the house mouse; *Rattus norvegicus* the Norway rat, sewer rat, or Brown rat; and *R. rattus* the black or granary rat. Some field rats invade rural and urban grain storages and also cause damage to stored grains, and they are of local importance. These include *Arvicanthis nitoticus* and *Praomys natalensis* in sub-Saharan Africa; *Oryzomys* spp. the rice rats in South and Central America; *Bandicota bengalensis*; Indian mole rat in India, and *B. exulans*, the Polynesian rat in Asia. Voles, a group of rodents belonging to the family Cricetidae also cause damage to stored grain in rural areas. The other rodents that feed on stored grains are the ground and tree squirrels (e.g.; *Citellus* spp., *Tamias* spp., *Xerus* spp., *Funiscurus* spp., and *Halosciurus* spp.) belonging to the family Sciuridae (37). The biology, habits, and habitats of commensal rats have been extensively studied (38). Rats and mice live for less than 2 years on an average and they live in colonies. They are nocturnal showing peak activity shortly after dusk. However, when hungry they also move around in search of food during other times. They follow regular paths called runways along walls or other objects that presents a vertical plane. They nest in a safe spot where the supply of food and water is assured. While feeding on dry grains rats drink water at about 15–30 mL/day, whereas the requirement of water for mice is much less, 1 mL/day. Nearly 70–80 fecal pellets are dropped daily, and about 40–50-mL urine excreted and hundreds of hairs are shed by rats. Rats can live without food for 3 days and resist thirst up to 4–6 h. An adult mouse eats about 3 g of food per day, but discards partially eaten food causing spillage and a significant quantity of contaminated grain. In bag-stacks heavy mice infestation may lead to collapse of the stacks. Rats are highly adaptable to any environment and have a high fecundity rate.

Surveillance for the presence of rats and mice in the storage premises and the level of infestation and the type of infestation form part of the pest management operations. Their presence and identification can be accomplished by making observations on the droppings, urine stains, hairs, foot prints, greasy paths, characteristic odor in the premises, and damaged structures and grain bags by gnawing and by live rat movements (39). The size and lengths of the fecal pellets vary among species. Because rats move or run on the same pathway close to the walls, the runways can be located by dusting the area to a depth of 4 mm. The infestations can also be confirmed by the presence of damaged metal and plastic pipes, electrical cable, and wooden structures.

Knowledge about the type and degree of rodent infestation in the premises is essential for understanding suitable control measures. Various control methods are available for rodent control (Table 4). They are not equally effective against all species. The basic approach in rodent control is to prevent the rodents from reaching the food grains by arranging suitable barriers in building structures and by rodent-proofing. Unnecessary

Table 4 Methods of Rodent Control

Lethal methods	Nonlethal methods
Trapping	Hygiene/sanitation
Chemical	Barriers/exclusion
Predators	Rodent-proofing
Pathogens	Chemosterilants
	Repellants

openings that are accessible to rats must be blocked with sheet metal, expanded metal to concrete so that the rats do not gnaw to enter. The ventilators and windows should be covered with wire of 1- 3-cm^2–opening size for rats and 0.6-cm^2 size for mice. Exclusion of mice needs greater care as they can squeeze in through smaller apertures. Sanitation within and around the storage facilities should be given priority in control strategies. Spillage should not be allowed and routine dusting, sweeping and disposal of trash and refuse are important. Dead storage areas in godowns, flat storages, and other storage structures need special attention. Areas surrounding the storage facilities extending up to 20 m all around must be cleared of weeds and harborages.

Despite precautions, rodents are likely to invade a new or an uninfested storage structure when trapping or chemical control methods have to be undertaken. Repellents based on ultrasonic, electrical, chemical, and electromagnetic sources have not been particularly effective under practical conditions (37). Use of chemosterilant can form one of the components of integrated rodent management. A reduction in population can be brought out by using antifertility compounds that are specific for either sex in the form of bait formulations. Formulations based on α-chlorohydrin, a male sterilant and toxicant, are marketed in some countries (40). In the natural environment, wild animals such as snakes, owls, hawks, ferrets, and mongooses check the rodent population. However, in grain storage facilities, the use of these predatory animals including dogs and cats poses practical problems. Predators can play a significant role in on-farm storage areas.

Traps are useful where (a) there is low level of infestation in the premises, (b) accidental poisoning is likely to take place and the use of poisons is to be avoided, (c) when rats are bait-shy, and (d) to avoid any dead-rat problem. Trapping is not an efficient method, as only adults and surplus populations are caught, whereas young ones and trap-shy rats are not affected. Various types of traps used with or without food baits are snap traps, steel traps, and wooden or metal box traps with single or multiple entry. Traps need regular inspection and setting and the trapped rats and dead rats need to be recovered from the storages promptly.

Chemical control methods involve fumigation and use of contact or stomach poisons. For rapid control, poisoning in a suitable bait formulation is adopted. Precautions such as safe handling of the rodenticides, prompt disposal of poisoned rats, and removal and destruction of all uneaten baits at the end of the control program are to be followed. Poison baiting is the most popular, economical, and quick method adopted all over the world. Bait, which is compatible with the poison in terms of effectiveness and acceptable to the pest in suitable form is kept inside bait stations placed in the rodent runways. Baits must be fresh and, therefore, need replacements often. When some acute poisons are used, prebaiting will be required to avoid bait-shyness and to increase the acceptance of baits by the rodents. Bait formulations may be used in the form of powder, cake, or block, or in bags, the choice determined by the behavior of the particular species. Bait in bags has the advantage of being protected from spillage and moisture. Rodenticides are also used as a tracking powder or poison (coumachlor) to overcome the bait shyness or bait nonacceptance problem. When dusted in rat runways and at the entrance of burrows, the dust will stick to its body or feet from which the poison is taken inside the system when the rat cleans it by licking. The choice of chemical poison depends on the need and the situation. The cost, local availability, acceptability of baits by rats, toxicity to the target pest, lack of toxicity to nontarget species, and regulatory restrictions are the factors considered in selecting the rodenticide. They are used in different concentrations in baits and they differ in their speed of lethal action against rats (Table 5). Among the fast-acting or acute

Table 5 Rodenticides and Their Toxicity

Rodenticide	Concentration in bait (% w/w)	Time to death (days)
Zinc phosphide	1–2.5	<0.1
Sodium fluoroacetate	0.2–0.3	<3
Warfarin	0.025	5–10
Calciferol	0.1	3–5
Brodifacoum	0.005	3–7
Bromadiolone	0.005	4–8

poisons, zinc phosphide is cheaper, produces rapid kill, and is widely used as a single-dose poison for rodent control. Anticoagulants, which are either hydroxycoumarins or indanediones, disrupt the mechanism that controls blood-clotting (vitamin K cycle) and cause fatal internal hemorrhages. Their action is cumulative, and they need to be ingested over a period of a few days or even several days to be effective. Rodents readily accept anticoagulants in bait formulations and the problem of primary and secondary poisoning is considerably less.

Vitamin K is an effective antidote for anticoagulants. The first generation anticoagulants such as warfarin, coumachlor, and diphacinone are known for delayed effect and need ingestion by rats for several days; therefore they are expensive in terms of bait and labor. Resistance problem is another disadvantage with these anticoagulants. Quick-acting second-generation anticoagulants were introduced in 1970s to overcome the resistance problem. They are the halogenated hydroxycoumarin derivatives, such as difenacoum and bromadiolone. Some of the anticoagulants such as brodifacoum or flocoumafen are effective in a single dose at 20–50 mg/kg. In some situations, fumigants, such as hydrocyanic acid gas and phosphine, are used for control. Burrows are treated with calcium cyanide powder releasing hydrocyanic acid or aluminium and magnesium phosphide pellets or ablets releasing phosphine gas. Fumigation of storage facilities will also eliminate rodent infestation in addition to grain insect pests. Controlled atmospheres particularly carbon dioxide-rich atmospheres are also effective although the technique is seldom practiced against rodents (41).

To sum up, an integrated approach involving good housekeeping, maintenance of hygienic conditions in and around, physical means of trapping, use of poisoning by bait formulations, and burrow fumigations is essential to contain the rodent menace. As they invade the storage structures and migrate to new areas in search of food, a continuous management program is required.

3.5 Birds

The extent of losses to stored grains by birds has been less studied than that of rats. Losses result from direct feeding and from contamination by their droppings and feathers. Their activity starts immediately after harvest, during threshing and drying and during storage, particularly in bag-storage in developing countries. The birds also have a role in bringing into grain stores grain insects and microflora (37). Major pests are the pigeons *Columba livia* and the sparrow *Passer domesticus*. Other opportunistic grain feeders such as parrots, doves, and weavers are of local importance. Pigeons feed in flocks and travel up to 8 km for foraging. They are so much accustomed, that they have little fear of humans. In India,

Fig. 3 Bird menace, a common problem in grain storage.

due to daily transaction of grain bags in the grain storage depots, spillages are often inevitable. Pigeons feed on the spillage (Fig. 3). Sparrows are smaller and feed in flocks covering distances up to 2–3 km. Preventive measures are very important in reducing losses by birds. Making the premises inaccessible to birds by netting or grills or hanging plastic strips is necessary. All ventilators and windows need wire netting to prevent the birds from invading, nesting, or roosting. Scaring devices based on vision, noise, or recorded alarm calls will become ineffective after some period. Chemical (4-aminopyridine)-based baits when eaten by few pigeons in a flock induce them to make distress calls with tremors or convulsions that will scare away other birds. In some countries, trapping or sedation is used and the trapped or sedated birds are shifted far away from the premises. Chemical-based lethal method or scaring techniques against birds are generally considered inhuman and are not acceptable everywhere.

4 GRAIN PROTECTION

4.1 Physical Methods

Protection of grain from the ravages of spoilage organisms such as insects, mites, and microorganisms, by physical means such as (a) manipulation of temperature, water activity, and composition of the atmosphere; (b) application of inert dusts; and (c) mechanical separation and removal of insect pests, have been used even before protective chemicals and fumigants were introduced. Most of the physical treatments are residue-free processes and the grain quality generally remains unaffected. However, as compared with chemical methods, physical methods are normally expensive, but some of them are already in commercial use, whereas others are yet to be applied judiciously (42).

4.1.1 Sanitation

Before binning or before building bag-stacks, the storage facilities should be cleaned thoroughly to eliminate insects breeding on sweepings, spillage, and residual grains. As infestation is likely to be carried in harvesting machinery, the combines should also be cleaned. Sanitation of the premises is an integral part of the protection strategies. It will be advantageous to give a residual spray treatment with approved contact insecticides on the floor and wall surfaces and fumigate empty storages. In Australia, a slurry of an inert dust (Dryacide—a diatomacious earth coated with silica aerogels) is used for treating the inside structures and is claimed to have given protection for 12 months (43). The cleanliness around storage facilities is also important. Weeds and unwanted materials around the stores should be removed or destroyed, as they harbor the pests particularly the vertebrates. The sanitary condition of the grain matters a lot in affecting its storage life. The impurities, dust and insects present in the grain must be removed by mechanical sieving or by aspiration.

4.1.2 Drying

For a longer storage life, the moisture content of the grain should be within the safe limit. However, in newly harvested grain the moisture content will exceed the limit, particularly in paddy and maize during wet seasons it will be 25–30% in Southeast Asia and Asia, and in wheat in Canada (44,45). Before storing, grain has to be dried either by natural or artificial means to bring down the moisture to an optimum level to avert stack burning and spoilage by the microbial activity and mycotoxin production. In paddy any delay in drying affects germination, producing discolored grains owing to nonenzymatic browning and reduces head rice yield. Drying is performed to bring down the grain moisture to an acceptable level for storage, marketing, or processing. Natural sun-drying is still widely practiced at rural level in many developing countries (45). However, this ancient technique of exploiting free energy has not been improved when compared with mechanical or other forms of drying (46). In artificial drying, grain is subjected to the action of a generally hot-air flow in a dryer produced by a heat energy source. It is an endothermic process involving heat and mechanical energy. Drying by artificial means is expensive, and it is not widely used in the developing countries where there are no incentives for dried grain (9).

4.1.3 Aeration

Temperature is a major factor involved in regulating the level of pest activity in stored grain. When grain temperature is 15°C or more, the insect pests develop quickly and cause heating and other related problems affecting the storage life of the grain. Aeration is a part of the preventive management system, and the primary objective is to reduce the temperature of the grain in bulk storage (44). It involves forced movement of atmospheric air through stored grain by means of fan, and it has very little effect on moisture content of the grain. The effectiveness of aeration is affected by the dockage, brokens, and other impurities present in the grain. Aeration also helps eliminate abnormal odors associated with mold growth, to achieve uniform distribution of a fumigant for insect control, and to desorb the fumigant on termination of the treatment. Aeration is a common practice in all industrialized countries. Lately, an interlinked ''Smart Aeration'' and microprocessor controlled ''PMCAM system'' have been developed in Australia for automation of aeration of grain in bulk storage units (47).

4.1.4 Grain Monitoring

Regular inspection of grain in storage is essential to make timely pest management decisions such as fumigation, aeration, or top-dressing with protectants. The grain is likely to develop hot spots owing to increase in insect population and mold growth. It is also likely that moisture condensation takes place owing to moisture and temperature gradients and consequent air movements in the bulk storage. Grain temperatures inside bins in various depths are to be recorded regularly and the data checked for any changes. In bulk storage, physical traps, such as probe traps and cone traps, are useful in trapping active and mobile adult beetles. Bait bags containing a mixture of broken wheat, peanut, and kibbled carobs or brown rice are useful for detecting infestation in godowns (48,49). Pheromone traps are used for the detection of moth pests and grain borers. Acoustic monitoring of pests, mentioned in Sec. 3.1., can be done by keeping sensors parallel to the temperature sensors in silos and bins. There has been intensive research on the automation of pest monitoring in bulk storage in industrialized countries such as France and United States (50,51).

4.1.5 Inert Dusts

Inert dusts, such as wood ash, paddy husk ash, koalins, lime, and clay materials; have been traditionally used in the developing countries for grain preservation (52). They act as a desiccant, absorbing water from the insect body and may also have an abrasive action. Some of the dusts such as silica aerogels absorb the waxy layer of the exoskeleton of the insects. They act slowly and take 20 or more days to cause insect mortality. They are relatively safer than conventional insecticides. The disadvantages are that they affect grain bulk density, flowability, and grain-handling properties, and the dusts containing crystalline silica may cause silicosis and other respiratory diseases. Inert dusts are generally less effective at higher relative humidity. The particle size also matters in its effectiveness. There are five types of inert dusts (52): (a) nonsilica dusts (e.g., limestone, lime, katelsons); (b) ash, clays, sand; (c) diatomaceous earths, which are the fossilized diatoms of marine or freshwater origin and are composed mainly of amorphous hydrated silica; (d) synthetic silicates and precipitated silicas; and (e) silica aerogels obtained by drying aqueous solutions of sodium silicate. The first two items are being used traditionally in the developing countries. They are effective, usually at higher application rates exceeding 5% by weight. Diatomaceous earths in commercial formulations are used at an application rate of 0.1% w/w, whereas silica aerogels with finer-particle sizes are used at lower rates (52). Aqueous slurries of ''Dryacide,'' containing mainly amorphous hydrated silica, have been effective for fabric treatment (43). Commercial formulations of dust containing 80–90% silicon dioxide include Dri-Die, Insecto, and Perma-Gaurd (US), Dryacide (Australia), Insectigone (Canada), and Aerosil R974 (Germany). Diatomaceous earths have been registered for use in United States as a food additive.

4.1.6 Temperature

Insects being poikilotherms are sensitive to large temperature changes in the environment. Increase or decrease in temperatures outside the optimum range of 25–32°C results in developmental delay, drop in reproduction, and mortality at the end. Insects are killed rapidly by heat rather than by cold treatment. Tolerance to heat treatment varies depending on the insect species, stage, and age of the insect, and its physiological state. Late larva of *S. oryzae* and pupa of *R. dominica* are the most tolerant. During heat disinfestation, the tolerance of grain to heat should also be considered. As the limits for insect mortality and safety limits for grain quality are very narrow, it is important that the temperature and the period of treatment be monitored closely, lest grain may become discolored, germinability impaired, and gluten protein affected. Moreover, grain after heating should be cooled immediately for safe storage. The sources for heat treatment may be microwave, dielectric heating, hot air, fluidized bed, or infrared. Among the systems studied so far, only the fluidized bed heating has been developed to a commercial prototype level (53). Apart from fluidized bed, dielectric heating is promising, as it can selectively act on insects. A major impediment for heat disinfestation is the capital cost. However, among the physical methods, heat disinfestation technique offers promise to fit into modern grain storage (42).

In temperate countries during summer and in tropical climatic regions in any season, grain chilling or grain aeration with refrigerated air is carried out. Ambient air is passed through refrigerated coils to reduce its temperature and the cooled air is reheated slightly to reduce the humidity to 60–75%. Chilled air is blown through the grain to bring down the temperature to prevent spoilage and pest activity. A mobile refrigeration system that can control the temperature as well as the humidity of the aeration air is used. Currently, grain chilling is used for wheat, maize, and rice, and there are about seven commercial manufacturers (the majority in Europe) of mobile grain chillers in the world (44). Unlike ambient air aeration, grain cooling by refrigerated air requires only a few days of treatment, but it is relatively more expensive. Normal aeration is dependent on the presence of cold ambient air, whereas chilled aeration can be carried out at any time, and both temperature and humidity of the air going inside the bin can be adjusted. Grain chilling must be repeated at intervals, otherwise grain temperature may increase shortly after chilling. At low temperatures the degradation of insecticides is less. Therefore, the amount of pesticides used and the level of pesticide residues formed are reduced considerably in grain aerated with refrigerated air. This is an important advantage of chilled aeration (54).

4.1.7 Irradiation

Ionizing radiation for insect control in stored grains has been studied since the 1950s. Most investigations have focused on the sterilizing potential of gamma-radiation emitted from cobalt-60 and cesium-137, or accelerated electrons of less than 10-V energy from a cathode. The treatment causes mortality as well as sterility in grain insect pests and the effect occurs at all temperatures. Irradiation has been permitted for use in food commodities up to an overall average dose of 10 kGy by the FAO/IAEO/WHO Joint Expert Committee on wholesomeness of irradiated food (55), and many countries have permitted the use of irradiation for disinfesting grains. The major advantage in grain protection by irradiation treatment is that it leaves no chemical residues. The treatment is a highly technical job, and the public is still wary about irradiated foods. There are only four pilot-plant irradiation facilities in the world (56). Among the developing nations, commercial application of the technique in Indonesia for controlling *S. oryzae* in milled rice in polyethylene–polyester copolymer bags has been reported (57).

4.1.8 Controlled Atmospheres

Normal grain atmosphere consisting of 78% nitrogen, 21% oxygen, 0.03% carbon dioxide, and the balance argon and other gases can be altered either naturally, as in airtight (hermetic) storage, or by artificial means by introducing carbon dioxide or nitrogen and burner gas in sealed storage. In hermetic storage, oxygen depletion and carbon dioxide increase are brought out by the respiratory activities of the grain as well as the pests in the ecosystem (see Sec. 2.4). The degree of control depends on the retention of insecticidal atmosphere, which is dependent on the storage structure and its airtightness. A certain level of grain damage is inevitable in hermetic storage. However, in storage of grain in sealed structures in an atmosphere wherein the gaseous composition is artificially changed to become toxic to the pests, is variously called controlled atmosphere or modified atmosphere storage. The system protects the grain quality, controls insect and mite pests, and prevents mold growth. This residue-free control technique, applicable in gas-tight rigid structures and flexible enclosures, needs technical expertise to carry out the treatment and is relatively costly when compared with conventional fumigations. The controlled atmospheres applied may be of either low-oxygen (0.5% oxygen and 99.5% nitrogen) or high-carbon dioxide atmosphere (40–80% carbon dioxide, balance air) or burner gas (0.5% oxygen, 13–21% carbon dioxide, balance mainly nitrogen). The choice between the three depends on local availability of the gases, suitability of the structures, and affordability. The required gases for producing the controlled atmospheres may be supplied from industrial or off-site sources, on it may also be generated on-site (41,58).

The technique of controlled atmosphere storage of grain is well advanced over the past 20 years mainly in the industrialized countries. Leaky storage structures and subsequent costs are the major constraints for the adoption of the technique in other countries. Nevertheless, the application of technique is likely to expand in the near future owing to factors such as development of insect resistance to conventional fumigants, and contact insecticides and residue problems. Moreover, the system is considered as ''organic'' and is exempt from residues (41).

4.2 Chemical Methods

4.2.1 Fumigants

Fumigants are chemicals available as gases, liquids, and in solid formulations, but act on the insect and other pests in gaseous state. Fumigation plays a key role in grain preservation as it controls insects developing inside and outside the grain and crawling and hidden pests. Unlike spray treatments with contact insecticides, fumigation is a curative treatment with no lingering or residual effect. Fumigated stacks or bulk grains are readily reinfested, if not protected properly, or in the absence of an effective prophylaxis. Another advantage with fumigants is that they can be applied in different types of storage situations. Some of the factors influencing fumigation include temperature, relative humidity, grain types and their conditions (size, moisture content, impurities present), storage structure, and type of insect pests and their resistance status. Fumigants enter the insect mainly through the respiratory system and the poisoning of insects is influenced by the rate of respiration. The rate of respiration of insects increases in response to the rise in temperature. Thus, fumigations are generally more effective at more than 20°C, assuming other favorable conditions exist. At lower temperatures, sorption of the fumigant by the commodity is increased; therefore, some allowance in dosage will be needed. For phosphine, fumigation at temperatures less than 15°C is not recommended, as the fumigant is less effective at lower temperatures, and some additional requirement such as a vaporizer is required during the application of methyl bromide. Fumigants are sorbed by grains during the exposure period. The rate of sorption depends on the characteristics of the fumigant and the grain and the prevailing temperature (60). The fumigant may react with the grain constituents forming fixed residues (irreversible sorption) or is desorbed on termination of the treatment (reversible sorption). Fine materials and other impurities impede gas distribution in bulk storage and absorb the fumigant more than that by the grain itself, resulting in inadequate gas concentration. Relative humidity is important in phosphine fumigation when the rate of decomposition of aluminium phosphide formulations to liberate phosphine gas is determined by the humidity. To facilitate decomposition, a minimum of 30% humidity is necessary, when aluminium phosphide preparations are used for grain fumigations. The type of storage, particularly the standard of gas tightness of structure, matters in successful treatments. In rigid structures such as silos and whole godowns, sealing should be done perfectly to prevent the gas loss through leakage and to achieve target concentrations or the (concentration × time) *Ct* product. In bag-stack fumigations under gas-proof sheets, the type and quality of the sheet used influence the retention of gas concentrations. The sheet or cover should be properly weighted down to the floor using sand snakes, loose sand, or adhesive tape. If mud is used for sealing during 7-day fumigation with phosphine, it should be checked for cracks and rectified (Fig. 4). The toxicity of fumigants on insects depends on the species, its life stage, metabolic state, and resistance status. Sedentary stages such as egg and pupa are generally less susceptible to fumigants. Additionally, diapausing larvae of *Trogoderma granarium* and pyralid moths (e.g., *Ephestia cautella*) and hypopi of mites are tolerant. Grain insect pests notably *R. dominica* and *Tribolium castaneum* have developed a high degree of resistance to phosphine fumigant in many countries. Dosage regimens for currently used fumigants, such as methyl bromide and phosphine, take into consideration the variables such as temperature, gas leakage, insect tolerance and resistance, and grain types for effective treatment (60,61). Gas monitoring is important during fumigation to predict its success and to supplement dosage or to extend the exposure period if necessary. For phosphine fumigations the success is ascertained based on the final-day concentration of 100 ppm and higher (for nonresistant strains only), and for methyl bromide treatments the prescribed *Ct* product at 25°C is 150 g h/m^3. If the target concentration or *Ct* has not been achieved it must be considered that the fumigation has failed and the reasons for this should be investigated and rectified accordingly. Residues are formed in fumigated grains and the residue levels should not exceed the national and international tolerances.

Currently there are only two major fumigants for grain disinfestation (i.e., phosphine [hydrogen phosphide] and methyl bromide) (Table 6). Phosphine is the most widely used fumigant in the world. It is the preferred chemical for routine grain disinfestations in the developing countries where other alternative techniques, such as controlled atmosphere storage, are expensive or cannot be readily adopted. Phosphine has many desirable properties of a fumigant; however, it is also likely to induce quality changes when applied repeatedly, or when grain moisture is high and aeration has not been carried out adequately (62). Pure phosphine is odorless and is flammable. However, commercial preparations of aluminum and magnesium phosphides—the active ingredients for phosphine—contain substances (e.g., ammonium carbamate) that reduce fire hazard or restrict the rapid release of the gas. Metal phosphide formulations are available as tablets, pellets, sachet, and as "plates" to suit different storage situations. They release phosphine gas on exposure to atmospheric moisture; the rate of release depends on the type of formulation, ambient temperature, and humidity. Magnesium phosphide formulations release phosphine completely and rapidly; therefore, they are more suitable for temperate climates. In developed countries, phosphine is also available in cylinders mixed with liquid carbon dioxide at 2–3% w/w. The patented cylinder-based formulation, Phosphume is used for treating grains

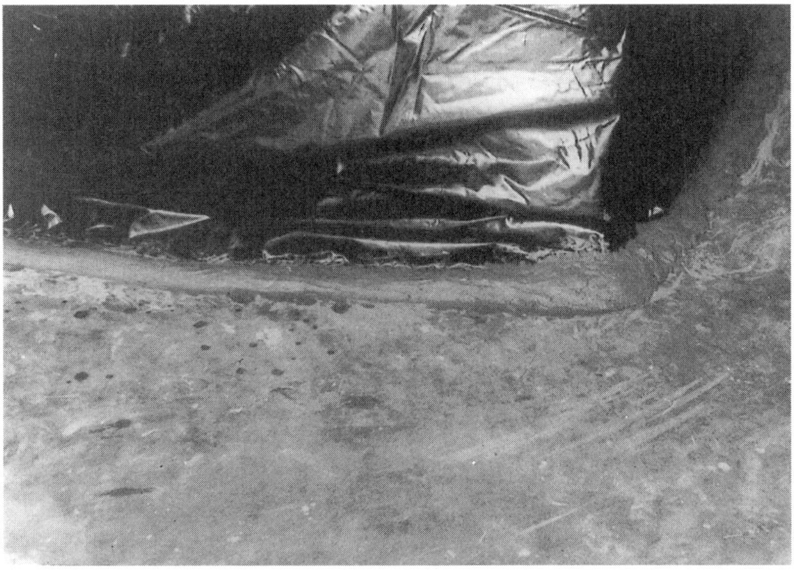

(A)

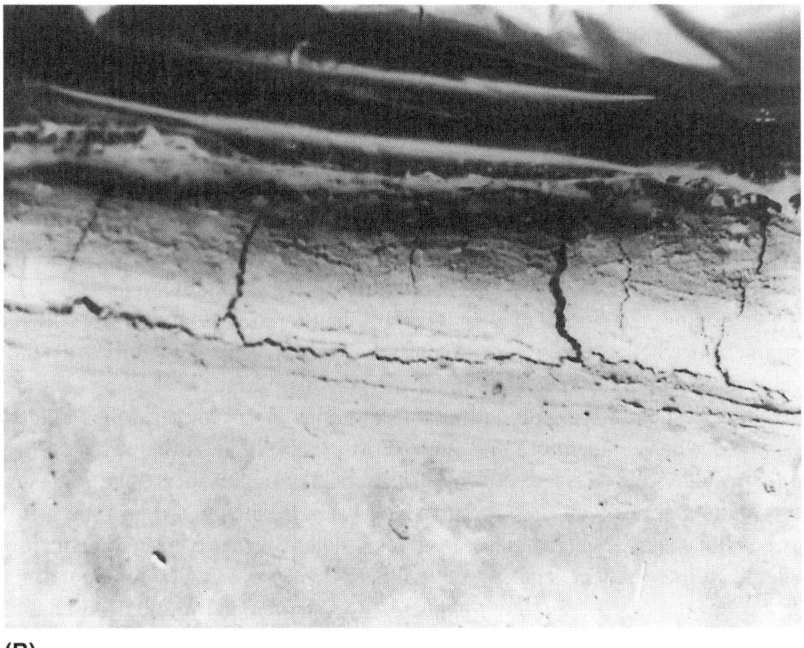

(B)

Fig. 4 During bag-stack fumigations floor sealing with mud leads to gas leakage, as the mud dries up and cracks.

Table 6 Properties and Other Data on Major Grain Fumigants

	Phosphine	Methyl bromide
Formula	PH_3	CH_3Br
Molecular weight	34	95
Boiling point (°C)	−87.4	3.6
Vapor pressure at 25°C (atm)	38.5	2.4
Flammability limits (% vol)	1.8	10–16
Solubility in water at 25°C (g/L)	0.3	13.4
TLV–TWA (ppm)	0.3	5.0
Availability	Solid preparations; cylinder formulations	Cylinder; cans
Dosage (g/m^3)	1–3	16–48
Exposure period (d)	5–28	1
Residue limits (ppm)[a]	0.10	50
Major drawback	Insect resistance	Ozone depletor

[a] ppm of phosphine and bromide ion, respectively, for phosphine and methyl bromide in whole grains.

in vertical silos, pads, or other bulk storage and farm bins by a pressurized distribution system called Siroflo, or without any pressurized distribution (63,64). The Siroflo technique has been claimed to be applicable to vertical silos, even though they are not completely gastight, and it is known to be effective also against phosphine-resistant strains. There are other specialized phosphine application methods such as Sirofume an automatic topping-up process, Phyto Explo system, and the J system or closed loop fumigation. These patented techniques facilitate uniform gas distribution for effective disinfestation in different storage situation (65).

Methyl bromide a colorless and odorless gas is available in pressurized steel cylinders or in 450- and 680-g cans, with or without 2% w/w chloropicrin as the warning gas. The fumigant plays a key role in quarantine treatments and is the preferred fumigant when there is a time constraint and when phosphine-resistant strains are present. It is sorbed by the commodities relatively more than that of phosphine. Materials meant for seed or malting purposes are avoided, as methyl bromide will impair germinability at higher doses, longer exposure periods, and when temperature and grain moisture are high. Moreover it forms inorganic bromide residues and may affect cooking and baking quality of treated grains.

Methyl bromide is applied on the top surface of grain in silos or on bag-stacks to allow it to distribute by gravity penetration. It requires gas delivery tube arrangements for dosing the fumigant from cylinders. The compound has been listed as an ozone-depleting substance, and the developed nations have planned to phase it out by 2005. Although developing countries have been currently exempted on this deadline and allowed to retain it for quarantine and preshipment uses, pressure for total phaseout in all countries is already there. At present, research priorities have focused on the development of methods to minimize atmospheric emission of methyl bromide and to find out suitable substitutes. Some of the candidate fumigants investigated or reconsidered as alternatives are ethyl formate, carbon disulfide, and carbonyl sulfide. Ethyl formate has been tested under field conditions in India, and it proved effective at a dosage of 300–400 g/m^3 with 72-h–exposure period (66). Carbonyl sulfide has been extensively studied in Australia for disinfesting grains and other stored products (67). Fumigation technology is facing constraints from biological to political sources, and the challenges are to be encountered despite budget cuts for research and development and the declining number of experts.

4.2.2 Contact Insecticides

Synthetic insecticides play a significant role in grain preservation along with other control measures and regular hygiene and sanitation measures. Insecticides kill insects already present and prevent cross-infestation and reinfestation of insect-free grain. The insecticides are used at all levels of storage and have versatile applicability (68). Before binning or bringing cleaned grain into a store, floor and wall surfaces and other likely places are sprayed to eliminate the resident insects. Insecticides used for this purpose are malathion, methoxychlor, chlorpyriphos methyl, synergized pyrethrins, pirimiphos-methyl, bromophos, and iodofenphos, the choice depends on local availability, regulatory restrictions, and cost (56). In the developing countries, besides floor areas and wall surfaces in the grain storage depots,/godowns and the bag-stacks are routinely sprayed with insecticides such as fenitrothion, deltamethrin, and synergized pyrethrins. Studies in Africa and Asian countries indicate that such repeated treatments in the tropical climates are not cost-effective and facilitate resistance development (69). Under the tropical climatic conditions the insecticides are degraded quickly, they are lost by oxidation, volatilization, and lose their effectiveness as they become absorbed by dust and by abrasion during cleaning operations (70). To control the active flyers, a space treatment with dichlorvos (DDVP) is applied. In some countries diclorovos resin strips, which slowly emit insecticide vapor, are used in grain stores. Diclorovos has both contact and fumigant action and adult moths are quickly knocked down by the chemical. To reach inaccessible places in flat storage and warehouses, an aerosol treatment or fogging will be needed. In the tropical countries direct mixing of insecticides with the grain is generally not adopted. However, in the industrialized countries admixture of approved insecticides with the food grains at specified levels are still permitted so that the bulk-stored grain may remain protected from insect attack during the entire storage period. Malathion (8 g a.i./t), pirimiphos-methyl (4 g/t), chlorpyriphos-methyl (2.5 g/t), biorosmethrin (1.5 g/t), and dichlorvos (10 g/t) are some of the insecticides used in different countries. Malathion has been used for more than 30 years as a grain protectant and for prophylactic spray on storage structures. Insect resistance to malathion is a common problem in most parts of the world; therefore, it has been largely replaced by more potent compounds (71). Chlorpyrifos-methyl and pirimiphos-methyl are also organophosphorus insecticides, with broad-spectrum activity. Some strains of *R. dominica* are, however, tolerant to them. Both compounds are more persistent than malathion. Methacriphos is a potent organophosphorus insecticide showing very good effect under temperature climatic conditions. Unlike other organophosphorus compounds, it has appreciable toxicity against *R. dominica*. Among the pyrethroids, deltamethrin and bioresmethrin are important. Both are effective against *R. dominica*. Deltamethrin remains effective on treated surfaces for a longer period even under highly humid conditions. *Sitophilus* spp. are generally tolerant to deltamethrin. A comprehensive review on the properties and use of various grain protectants has been published (68).

The efficacy of the contact insecticides is modified by ambient temperature and humidity, grain moisture, and the type of surface treated. Higher temperatures and grain moisture lead to quicker breakdown of the insecticides and thus the efficacy is reduced. In fact, under hot and humid conditions, higher dosages of the insecticides are recommended. With some protectants the effectiveness is likely to be influenced by the type of commodity, as well as by its moisture content. Reduced insecticidal potency has been observed in fenitrothion when the moisture content of maize is higher. When used as a spray on bag-stacks or wall and floor surfaces, the type of surface area (jute or polypropylene sacks or cement or concrete surfaces) and the formulation (wettable or emulsifiable concentrate) influence the persistence and toxicity of insecticides. Jute surfaces are highly sorptive; hence, insecticides are less persistent on jute than on polypropylene or multipaper-walled bags. Insecticides are generally more persistent on plywood or galvanized iron than on concrete surfaces, and wettable powders persist for a longer period than emulcifiable concentrates. Potential insecticides tested recently, as grain protectants are cyfluthrin and prallethrin (71). Insecticides in combinations have also been used to exploit their individual potencies against grain insect pests. The combinations may be between two insecticides of the same (malathion and dichlorvos) or different groups (bioresmethrin and fenitrothion) or between two compounds of entirely different modes of action (bioresmethrin and methoprene) (72).

Insecticide residue and insect resistance are serious problems for concern in grain storage. Widespread resistance to grain protectant chemicals and residual insecticides has been reported. There is a growing concern about pesticide residues in food commodities by the public; therefore, restrictions on the use of insecticides by regulatory bodies have been increasing. Use of contact insecticides as protectants by direct admixture with grain has been decreasing in recent years. However, as with residual sprays, fogging, and aerosol applications, the use of insecticides is likely to continue in grain storage. In some situations, inert dusts can be used as substitutes for conventional insecticides.

4.2.3 Insect Growth Regulators

Insect growth regulators that specifically interfere with biochemical processes in insects and prevent the early stages from reaching the adult stage have been investigated for controlling stored grain insects. Chitin synthesis inhibitors affecting the molting process during the development of insects belong to benzoylphenylurea group of chemicals (e.g., diflubenzuron and fenoxycarb). Although they have been evaluated in the laboratory, they are yet to be cleared for use in grain protection. Methoprene, a juvenile hormone analogue, has been registered for use in the United States and some other industrialized countries for mixing with the grains. The juvenile hormone analogues are not highly effective against internal infestors such as *Sitophilus* spp. and against the very early stages of insects. The compound acts by ingestion and through contact. Methoprene has already been used as a grain protectant. It is more effective when used in mixture with contact insecticides such as chlorpyriphos-methyl, malathion, and methacrifos (72).

4.3 Biological Control

There has been a growing tendency in favor of nonchemical methods. In this context, the biological control method has been reexamined as a part of integrated stored grain pest management. There are situations in which biological control, which is residue-free but has species-specific action, will play an active role. Control may be manifested by the action of predators, parasitoids, or pathogens. The warehouse pirate bug *Xylocoris flavipes* and the polyphagus predatory bug *Lyctocoris campestris* are the important predators attacking the immature stages of stored grain insect pests. Parasitoids are very small hymenopteran insects, parasitizing egg or larva. Important parasitoids are *Bracon hebetor* on *Ephestia cautella*, *Trichogramma evanescens* on *Ephestia kuehniella*, *Cephalonomia waterstoni* on *Cryptolestes ferrugineus* and *Choetospila elegans* on *Ryzopertha dominica*.

The adult wasps are very small (1–2 mm in size) and can be easily removed by sieving. There have been intensive field studies in the African countries on the control of *Prostephanus truncatus* with its natural enemy, *Teretriosoma nigrescens* belonging to the family Histeridae. In the United States, the Environmental Protection Agency (EPA) has recently exempted tolerances for predators and parasites in stored products (57). Among the pathogens, various strains of *Bacillus thuringiensis* that produce insecticidal crystal proteins have been studied, particularly for the control of pyralid moths. *B. thuringiensis* formulations have been used for top dressing of grains in bins or other bulk storages for controlling *Ephestia cautella* and *Plodia interpunctella* (56). The drawbacks with the use of *B. thuringiensis* are development of insect resistance to the biocide, and the beetles are highly tolerant. Thus, there are avenues for utilization of biocontrol agents under certain situations and for controlling a particular pest.

4.4 Expert Systems

Stored grain ecosystem is a complex niche and it involves several variables such as temperature, water activity, pest development and dispersion, biological activity, and pesticide degradations. A detailed knowledge about the events in the storage system is essential to predict pest outbreaks and advise suitable control measures or other management action. In this context, computer programs, called Expert Systems or Decision Support Systems, have been developed to assume the role of an experienced storage manager or a storage expert in offering solutions with reference to storage problems under a given set of conditions. It can also forecast any possible outbreak of pests during the storage period. It is possible to interlink an expert system with remote sensors recording the temperature, grain moisture content, as well as signals for acoustic detection of insect pests in bulk storages and, in turn, it enables automated control measures (73). It can also play a role in training (74). A decision support system for stored grain pest management has been developed for the first time in the United Kingdom as a training device (75). Most of the expert systems developed so far are in prototype form, whereas a well-advanced system has been introduced by Longstaff and Cornish (74) for application in Australia and probably other countries. An expert system will not be able to advise or predict suitably, unless the inputs are thorough. Database on growth of microorganisms under various storage conditions, on population dynamics of mites in grain storage, and about cost–benefit analysis of different management options are still needed for incorporation into expert systems.

5 CONCLUSION

Grain storage has dual functions: (a) to stop grain degradation, and (b) to maintain quality to the satisfaction of consumers and to the national and international market demands. Grain management strategies have two opposite goals: they are zero-tolerance of insects and zero residues. To achieve this, there is now more emphasis on control techniques involving physical processes and even biological agents. Industrialized countries are well ahead in the technology by adopting nonchemical methods, such as aeration, chilling of grain, and controlled-atmosphere storage. Insect pest-trapping, methods for monitoring and detection of infestation, and automation in infestation detection in bulk storages are now well developed. Insect traps, which are already in use in the developed countries, will play a larger role in the management strategy. Expert systems that simulate storage have been introduced in some countries. However, there is a wide gap in the technology of storage between the industrialized and developing nations. This has been especially so in tropical and humid regions, where the pest problem is acute, control measures are improperly performed, either owing to lack of knowledge or to cost factors, and insects' resistance to pesticides is common. To bridge the gap, technical tie-ups or collaborative research programs between advanced and developing nations are needed. There are certain major issues common to all countries, which can only be resolved by a joint venture. These include mycotoxins in grains (76) and the incidence of insecticide and fumigant resistance in insects (77). In addition to incentives for grain at safe moisture levels, development of a grain-drying system that is affordable in a developing country, will help reduce the mycotoxin problem to some extent. Large expenditures are involved in development of newer insecticides and to generate data to meet registration requirements. The absence of newer chemicals and regulatory restrictions or reregistration requirements for the chemicals already in use exacerbate the situation. There has been a change in the pest scenario such that some of the insect pests have gained prominence probably owing to elimination or absence of their natural enemies. Psocids in storage facilities in hot and humid regions in Asia and the larger grain borer *Prostephanus truncatus* in African countries are typical examples.

Physical control methods have been one focus, albeit that they are costlier and total insect kill may not always occur when used alone. Heat treatment is promising, for it acts quite rapidly, but one needs to know more about its mechanism and the sensitivity system in insects to potentiate the action. A combination of two physical methods or supplementation of a physical method with a chemical method at reduced dosages often may be economical and efficient. Top dressing with Dryacide in a silo while applying Siroflo (63), or aeration and application of protectants plus grain cooling, and admixture of insect growth regulator with contact insecticides, are some of the combinations already tested and found effective. In view of the priority for nonchemical control measures, monitoring of grain conditions and pest activity during the storage period has become essential to predict any depredation or pest development and to plan appropriate protection measures. Automation in monitoring and the linking of monitoring system to expert systems has gained prominence in recent years. The expert systems need more inputs about the grain storage ecosystem from tropical or developing countries where the pest situation is acute and grain deteriorating factors are more favorable.

Problems such as cuts in research grants and a decline in number of storage entomologists are affecting the progress of research in grain storage. Likely phaseout of methyl bromide and widespread phosphine resistance are major impediments for the existing fumigation technology. The repercussions of these constraints will be reflected more in the developing nations, which may lack either the expertise to implement advanced control system or the resources to change over from conventional control methods, or both.

ACKNOWLEDGMENTS

I thank Mr. S. K. Majumder and Dr. A. N. Raghunathan for valuable comments and suggestions for improving the manuscript and Mr. V. A. Daniel for checking the minor mistakes in the manuscript.

REFERENCES

1. World Bank. Poverty and Hunger: Issues and Options for Food Security in Developing Countries. A World Bank Policy Study. Washington, DC, 1986.
2. Jayas DS, White NDG, Muir WE. Stored Grain Ecosystems. Marcel Dekker, New York, 1995.

3. Gewinner J, Harnisch R, Muck O. Manual on the Preservation of Post-Harvest Grain Losses. GTZ, Eschborn. 1996.
4. Sauer DB, ed. Storage of Cereal Grains and Their Products. American Association of Cereal Chemistry, St. Paul, MN, 1992.
5. Mills JT. Spoilage and Heating of Stored Agricultural Products: Prevention Detection and Control. Agriculture Canada Publication 1823E, Canadian Government Publishing Centre, Ottawa, 1989.
6. Multon JL. Spoilage mechanisms of grains and seeds in the post-harvest ecosystem, the resulting losses and strategies for the defence of stocks. In: Preservation and Storage of Grains, Seeds and Their By-Products: Cereals, Oilseeds, Pulses and Animal Feed. Lavoisier, New York, 1988; pp 3–63.
7. Flinn PW, McGaughey WH, Burkholder WE. Effects of fine material on insect infestation: a review. In: Fine Material in Grain. North Central Regional Research Publication 332. Ohio Agricultural Research and Development Center, 1992; pp 24–30.
8. Semple RL. Post harvest technology: storage development and application in developing Asian countries. In: Grain Storage Systems in Selected Asian Countries. eds. Semple RL, Hicks PA, Lozare JV, Castermans A, eds. REAPASIA, 1992; pp 9–59.
9. Juliano BO. Concerns for quality maintenance during storage of cereals and cereal products. In: Stored Product Protection: Proceedings of the 6th International Working Conference on Stored-Product Protection. Highley E, Wright EJ, Banks HJ, Champ BR, eds. CAB International, Wallingford, 1994; pp 663–665.
10. Pomeranz Y. Biochemical, functional and nutritive changes during storage. In: Storage of Cereal Grains and Their Products. Sauer DB, ed. American Association of Cereal Chemistry, St. Paul, MN, 1992; pp 55–141.
11. McFarlane JA. Storage methods in relation to post-harvest losses in cereals. Insect Sci Appl 1988; 9:747–754.
12. Krishnamurthy TS, Majumder SK. A comparative evaluation of some storage bins for rural grain storage. J Ind Acad Wood Sci 1978; 9:1–16.
13. Shejbal J, de Bioslambert JN. Modified atmosphere storage of grains. In: Preservation and Storage of Grains, Seeds and Their By-Products: Cereals, Oilseeds, Pulses, and Animal Feed. Multon JL, ed. Lavoisier, New York, 1988; pp 749–777.
14. Kennedy L, Devereau AD. Observations of large-scale outdoor maize storage in jute and woven polypropylene sacks in Zimbabwe. In: Stored Product Protection: Proceedings of the 6th International Working Conference on Stored-Product Protection. Highley E, Wright EJ, Banks HJ, Champ BR, eds. CAB International, Wallingford, 1994; pp 290–295.
15. Bailey JE. Whole grain storage. In: Storage of Cereal Grains and Their Products. Sauer DB, ed. American Association of Cereal Chemistry, St. Paul, MN, 1992; pp 157–182.
16. Newman CJE. Specification and design of enclosures for gas treatment. In: Fumigation and Controlled Atmosphere Storage of Grain. Champ BR, Highley E, Banks HJ, eds. ACIAR Proceedings 25. 1990; pp 108–130.
17. Navarro S, Donahaye JE, Fishman S. The future of hermetic storage of dry grains in tropical and subtropical climates. In: Stored Product Protection: Proceedings of the 6th International Working Conference on Stored-Product Protection. Highley E, Wright EJ, Banks HJ, Champ BR, eds. CAB International, Wallingford, 1994; pp 130–138.
18. Annis PC, Banks HJ. Is hermetic storage of grains feasible in modern agricultural systems? In: Pest Control and Sustainable Agriculture. Corey SA, Dall DJ, Milne WM, eds. CSIRO, Melbourne, 1993; pp 479–482.
19. Alvindia DG, Caliboso FM, Sabio GC, Regpala AR. Modified atmosphere storage of bagged maize outdoors using flexible liners: a preliminary report. In: Stored Product Protection: Proceedings of the 6th International Working Conference on Stored-Product Protection. Highley E, Wright EJ, Banks HJ, Champ BR, eds. CAB International, Wallingford, 1994; pp 22–26.
20. Annis PC, Graver J van S. Suggested Recommendations for the Fumigation of Grain in the

ASEAN Region, Part 2, Carbon Dioxide Fumigation of Bag-Stacks Sealed in Plastic Enclosures: an Operations Manual. Kuala Lumpur, AFHB/Canberra, ACIAR, 1991.
21. Graver J van S, Annis PC. Suggested Recommendations for the Fumigation of Grain in the ASEAN Region. Part 3. Phosphine Fumigation of Bag-Stacks Sealed in Plastic Enclosures: an Operations Manual AFHB/ Canberra, ACIAR, Kuala Lumpur, 1994.
22. Narasimhan KS, Rajendran S, Krishnamurthy TS, Chandy Z, Khedkar PM, Hakeem MA. Carbon dioxide fumigation trials in India. In: Proceedings of an International Conference on Controlled Atmosphere and Fumigation in Grain Storages, Winnipeg, Canada, June 1992. Navarro S, Donahaye E, eds. Caspit Press, Jerusalem, 1993; pp 37–49.
23. Ren Yonglin. Outdoor storage of grain in China. GASGA Newslett 1991; 15:16–17.
24. Donahaye E, Navarro S. Volcani cubes—a storage system for bagged or crated commodities. GASGA Newslett 1989; 13:3.
25. Graver J van S, Annis PC. Outdoor storage systems: their role and research requirements. In Grain Postharvest Research and Development: Priorities of the Nineties. Proceedings of the Twelth ASEAN Seminar on Grain Postharvest Technology, Surabaya, Indonesia, August 1989. Naewbanij EJO, ed. AGPP, Bangkok, 1991; pp 37–42.
26. GASGA (Group for Assistance on Systems Relating to Grains After Harvest). Larger Grain Borer. Technical Leaflet No. 1, NRI, Kent, 1987.
27. Kleith U, Pike V. Economic assessment of psocid infestations in rice storage. Trop Sci 1995; 35:280–289.
28. Haines CP. Insects and Arachnids of Tropical Stored Products: Their Biology and Identification. Natural Resources Institute, Kent, 1991.
29. Hagstrum DW, Flinn PW, Shuman D. Acoustical monitoring of stored-grain insects: an automated system. In: Stored Product Protection: Proceedings of the 6th International Working Conference on Stored-Product Protection. Highley E, Wright EJ, Banks HJ, Champ BR, eds. CAB International, Wallingford, 1994; pp 403–405.
30. Kitto GB, Quinn FA, Burkholder WE. Development of immunoassays for quantitative detection of insects in stored products. In: Stored Product Protection: Proceedings of the 6th International Working Conference on Stored-Product Protection. Highley E, Wright EJ, Banks HJ, Champ BR, eds. CAB International, Wallingford, 1994; pp 415–420.
31. Fleurat–Lessard F. Grain mites, general characteristics and consequences of their presence in stocks. In: Preservation and Storage of Grains, Seeds and Their By-Products: Cereals, Oilseeds, Pulses and Animal Feed. Multon JL, ed. Lavoisier, New York, 1988; pp 409–416.
32. Sauer DB, Meronuck RA, Christensen CM. Microflora. In: Storage of Cereal Grains and Their Products. Sauer DB, ed. American Association of Cereal Chemistry, St. Paul, MN, 1992; pp 313–340.
33. Hocking AD. Effects of fumigation and modified atmosphere storage on growth of fungi and production of mycotoxins. In: Stored Products: Proceedings of an International Conference, Bangkok. Champ BR, Highley E, Hocking AD, Pitt JL, eds. ACIAR Proceedings 36, 1991; pp 145–156.
34. Lubulwa ASG, Davis JS. Estimating the social costs of the impacts of fungi and aflatoxins in maize and peanuts. In: Stored Product Protection: Proceedings of the 6th International Working Conference on Stored-Product Protection. Highley E, Wright EJ, Banks HJ, Champ BR, eds. CAB International, Wallingford, 1994; pp 1017–1042.
35. Frayssinet C, Cahagnier B. Detection and assay of toxins in grains and seeds. In: Preservation and Storage of Grains, Seeds and Their By-Products: Cereals, Oilseeds, Pulses and Animal Feed. Multon JL, ed. Lavoisier, New York, 1988; pp 527–549.
36. ACIAR. Mycotoxins in food. ACIAR Postharvest Newslett 1995; 35:8.
37. Smith RH. Rodents and birds as invaders of stored-grain ecosystems. In: Stored-Grain Ecosystems. Jayas DS, White NDG, Muir WE, eds. Marcel Dekker, New York, 1995; pp 289–323.
38. Lund M. Rodents as commensal pests. In: Rodent Pests and Their Control. Buckle AP, Smith RH, eds. CAB International, Wallingford, 1994; pp 23–43.

39. Harris KL, Baur FJ. Rodents. In: Storage of Cereal Grains and Their Products. Sauer DB, ed. American Association of Cereal Chemistry, St. Paul, MN, 1992; pp 393–434.
40. Marsh RE. Chemosterilants for rodent control. In: Rodent Pest Management. Prakash I, ed. CRC Press, Boca Raton, 1988; pp 101–169.
41. Banks HJ, Annis PC. Comparative advantages of high CO_2 and low O_2 types of controlled atmospheres for grain storage. In: Food Preservation by Modified Atmospheres. Calderon M, Barkai–Golan R, eds. CRC Press, Boca Raton, 1990; pp 94–122.
42. Banks HJ, Field P. Physical methods for insect control in stored-grain ecosystems. In: Stored-Grain Ecosystems. Jayas DS, White NDG, Muir WE eds. Marcel Dekker, New York, 1995; pp 353–409.
43. Longstaff BC. The management of stored product pests by non-chemical means: an Australian perspective. J Stored Prod Res 1994; 30:179–185.
44. Maier DE. Preservation of grain with aeration. In: Grain Drying in Asia. Champ BR, Highley E, Johnson GI eds. ACIAR Publication 71, 1996; pp 379–397.
45. Ren Yonglin, Graver J van S. Grain drying in China: problems and priorities. In: Grain Drying in Asia. Champ BR Highley E, Johnson GI eds. ACIAR Proceedings 71, 1996; pp 23–29.
46. Lantin RM, Paita BL, Manaligod HT. Revisiting sun drying of grain: widely adopted but technologically neglected. In: Grain Drying in Asia. Champ BR Highley E, Johnson GI, eds. ACIAR Proceedings 71, 1996; pp. 302–307.
47. Banks HJ. New technology breaths life into aeration. Stored Grain Austr Aug 1997.
48. Hodges RJ, Halid H, Rees DP, Meik J, Sarjono J. Insect traps tested as an aid to pest management in milled rice stores. J Stored Prod Res 1995; 21:215–229.
49. Haines CP, Rees DP, Ryder K, Sistyanto S, Cahyana Y. Brown-rice bait-bags for monitoring insect pest populations in bag stacks of milled rice as an aid to pest control decision-making. In: Proceedings of 5th International Working Conference Stored-Product Protection. Fleurat–Lessard F, Ducom P, eds. Bordeaux, Fr, 1991; pp 1351–1358.
50. Hagstrum DW. Field monitoring and prediction of stored grain insect populations. Postharvest News Inform 1994; 5:39N–45N.
51. Fleurat–Lessard F, Andrieu AJ, Wilkin DR. New trends in stored grain infestation inside storage bins for permanent infestation risk monitoring. In: Stored Product Protection: Proceedings of the 6th International Working Conference on Stored-Product Protection. Highley E, Wright EJ, Banks, HJ, Champ BR, eds. CAB International, Wallingford, 1994; pp 397–402.
52. Golob P. Current status and future perspectives of inert dusts for control of stored product insects. J Stored Prod Res 1997; 33:69–79.
53. Evans DE, Thorpe GR, Dermott T. The disinfestation of wheat in a continuous-flow fluidized bed. J Stored Prod Res 1983; 19:125–137.
54. Longstaff BC. Temperature manupulation and the management of insecticide resistance in stored grain pests: a simulation study for the rice weevil, *Sitophilus oryzae*. Ecol Modelling 1988; 42:303–313.
55. Ahmed M. Irradiation disinfestation of stored foods. In: Proceedings of the 5th International Working Conference on Stored Product Protection. Fleurat–Lessard F, Ducom P, eds. Bordeaux, Fr, 1990; pp 1105–1116.
56. Harein PK, Davis R. Control of stored-grain insects. In: Storage of Cereal Grains and Their Products. Sauer DB, eds. American Association of Cereal Chemistry, St. Paul, MN, 1992; pp 491–534.
57. UNEP (United Nations Environmental Protection). Montreal Protocol on Substances that Deplete the Ozone Layer: 1994. Report of the Methyl Bromide Technical Options Committee—1995 Assessment, 1995.
58. Paster N, Calderon M, Menesherov M, Barak V, Mora M. Application of biogenerated modified atmospheres for insect control in small grain bins. Trop Sci 1991; 32:355–358.
59. Banks HJ. Uptake and release of fumigants by grain: sorption/desorption phenomena. In: Proceedings of an International Conference on Controlled Atmosphere and Fumigation in Grain Storages. Navarro S, Donahaye EJ, eds. Caspit Press., Jerusalem, 1993; pp 241–260.

60. EPPO. EPPO recommendations on fumigation standards. EPPO Bull 1984; 589–609.
61. AFHB/ACIAR. Suggested Recommendations for the Fumigation of Grain in the ASEAN Region. Part 1. Principles and General Practice. AFHB, Kuala Lumpur/ACIAR, Canberra, 1989.
62. Rajendran S, Gunasekaran N. Effects of phosphine fumigant on stored products. Pesticide Outlook 1995; 6:10–12.
63. Winks RG, Russel, GF. Effectiveness of SIROFLO in vertical silos. In: Stored Product Protection: Proceedings of the 6th International Working Conference on Stored-product Protection. Highley E, Wright EJ, Banks HJ, Champ BR, eds. CAB International, Wallingford, 1994; pp 22–26.
64. Chakrabarti B, Mills KA, Bell CH, Wonter–Smith TJ, Clifton AL. The use of a cylinder-based formulation of 2–3% phosphine in liquid carbon dioxide. In: Procedings of the 5th International Working Conference on Stored-Product Protection. Highley E, Wright EJ, Banks HJ, Champ BR, eds. CAB International, Wallingford, 1991; pp 775–783.
65. Highley E, Wright EJ, Banks HJ, Champ BR, eds. Stored Product Protection. Proceedings of the 6th International Working Conference on Stored-Product Protection. CAB International, Wallingford, 1994.
66. Muthu M, Rajendran S, Krishnamurthy TS, Narasimhan KS, Rangaswamy JR, Jayaram M, Majumder SK. Ethyl formate as a safe general fumigant. In: Controlled Atmosphere and Fumigation in Grain Storages. Ripp BE, Banks HJ, Bond EJ, Calverley DJ, Jay EG, Navarro S, eds. Elsevier, Amsterdem, 1984; pp 369–393.
67. Desmarchelier JM. Carbonyl sulphide as a fumigant for control of insects and mites. In: Proceedings of the 6th International Working Conference on Stored-Product Protection. Highley E, Wright EJ, Banks HJ, Champ BR, eds. CAB International, Wallingford, 1994; 78–82.
68. Snelson JT. Grain Protectants. ACIAR Monogr 3. ACIAR, Canberra, 1987.
69. Hodges RJ, Sidik M, Halid H, Conway JA. The cost efficiency of routine applicaitons of contact insecticide for protecting bagged milled rice under typical ASEAN storage conditions. In: Proceedings of the 13th ASEAN Seminar on Grain Postharvest Technology. Naewbanij JO, ed. AGPP, Bangkok, 1992; pp 295–309.
70. Gudrups I, Harris A, Dales M. Are residual insecticide applications to store surfaces worth using? In: Proceedings of the 6th International Working Conference on Stored-Product Protection. Highley E, Wright EJ, Banks HJ, Champ BR, eds. CAB International, Wallingford, 1994; pp 785–789.
71. Arthur FH. Grain protectant chemicals: present status and future trends. In: Stored Product Protection: Proceedings of the 6th International Working Conference on Stored-Product Protection. Highley E, Wright EJ, Banks HJ, Champ BR, eds. CAB International, Wallingford, 1994; pp 719–721.
72. Bengston M, Strange AC. Recent developments in grain protectants for use in Australia. In: Stored Product Protection: Proceedings of the 6th International Working Conference on Stored-Product Protection. Highley E, Wright EJ, Banks HJ, Champ BR, eds. CAB International, Wallingford, 1994; pp 751–754.
73. Wilkin DR, Mumford JD, Norton GA. The role of expert systems in current and future grain protection. In: Proceedings of 5th International Working Conference on Stored-Product Protection. Fleurat–Lessard F, Ducom P, eds. Bordeaux, Fr, 1990; pp 2039–2046.
74. Longstaff BC, Cornish P. ''Pest man''—a decision support system for pest management in the Australian grain-industry. AI Appl Nat Resource Manage 1994; 8:13–23.
75. Dennis T. An expert system for stored grain pest management. PhD dissertation. London University, 1988.
76. Highley E, Johnson GI. Mycotoxin Contamination in Grains. Presented in the 17th ASEAN Technical Seminar on Grain Postharvest Technology, Lumut, Malaysia, 25–27 July 1995. ACIAR Tech Rep 1996; 37:145 p.
77. Subramanyam B, Hagstrum DW. Resistance measurement and management. In: Integrated Management of Insects in Stored Products. Subramanyam B, Hagstrum DW, eds. Marcel Dekker, New York 1996; pp 331–397.

9

Structural Considerations: Warehouse and Silo

ANANADA P. GUPTA and SRIMAN K. BHATTACHARYYA

Indian Institute of Technology, Kharagpur, India

1 INTRODUCTION

1.1 Grain Storage Principles

Grains are generally stored in warehouses or in storage bins. These range from small units to very large complexes. Indeed, the type of storage facilities are related to transportation modes, handling systems, and need for fumigation.

The *warehouse* structures—typically single-storied buildings and commonly referred to as *mill buildings*—are characterized by the use of trusses to support the roof, large spans between columns, few if any interior walls or partitions, and the existence of industrial equipment—such as cranes—within the building. Many times, as an alternate form, the warehouse structure is constructed as a rigid-frame type, eliminating the roof trusses. The exterior walls of warehouses may be of precast concrete panels, brick masonry, or metal sheathing. If metal sheathing is used, it is supported by horizontal girts attached to the columns. The walls are generally nonload-bearing, but must be strong enough to resist the lateral forces from wind or earthquake. However, whatever material is selected for covering, it should be weather-tight, corrosion-resistant, fireproof, and economical.

The design of single-storied buildings for lateral forces caused by wind or earthquake is not usually critical. However, provision of a proper bracing system to resist these lateral forces is extremely important. Lateral forces in both the transverse and the longitudinal directions must be considered.

To facilitate handling of materials, usually few walls and partitions within the building are provided. Floor slabs are laid directly on the ground, and whenever special mechanical equipment is required, it is placed directly on a special foundation to eliminate possible vibration of the structure caused by operating machinery.

Crane ways in industrial buildings pose special design problems related to the crane columns. The stiffness of the portion of columns supporting the crane is usually increased to carry the combined loads from crane and roof.

The rigid-framed warehouse structures are mostly conventional linear developments of plane frames. A rigid frame resists external loads essentially by virtue of bending moments developed in the ends of members. Thus the connections in a rigid frame must transmit moment as well as thrust and shear. Several variations of plane frames are possible. In general, rigid frames may be classified as single-storied or multistoried frames and single-span and multispan frames. For larger spans, open-web systems are adopted. It is beyond the scope of this chapter to give design principles and computational aids for all variations of plane frames. Widely accepted single-bay, pitched rigid frames are considered for providing guidelines for analysis and design. However, for the analysis of frames with multiple bays or with multiple stories, designers may refer to structural analysis books for further details (10). Because adaptation of frames with multiple bays or with multiple stories is very rare in practice, and because these kinds of frames involve several possible combinations, these are not dealt with in this chapter.

Before proceeding with the analysis of forces and moments in the frame, the restraints at the column bases must be established. The forces that may be developed at a column base are shear, thrust, and moment reactions. When a base is designed to resist moment, it is called a *fixed base*, and when no moment is allowed to be transmitted it is called a *hinged base*. There are various other forms of grain storage facilities, including those that are axisymmetrical or suspended systems. These are not in common use.

For bulk storage common modes adopted are *bins*, *bunkers*, or *silos*, in isolation or in a battery. Here, again, the most common form is deep silo. The design of these containers has evolved over time, in the sense that the dynamic effects of loading and unloading are now taken into consideration, as are the effects of moisture variation in the grains stored.

Warehouse structures may or may not incorporate a gantry crane. The most common features are the vertical long walls and the heaved roof. Essentially, they are developed around basic pitched-portal frames.

2 DESIGN PRINCIPLES OF A WAREHOUSE

The warehouse may or may not have facilities for a gantry crane. If provision of gantry cranes is required in a warehouse, the frames should be designed considering the additional load that will be developed by the movements of the cranes and possible combinations thereof.

2.1 Warehouse Without Facilities for Cranes

Warehouses without facilities for cranes normally have prismatic columns and a roof slope of close to 1:2.5. The pitched-portal frames are symmetrical, with the moment of inertia (I) of rafter K times that of column; K can vary between 0.5 and 1.5. These are rigid-jointed frames placed on foundations with hinged or fixed bases. Figures 1A–D give computational aids for hinged-base frames and Figs. 2A–D give those of fixed-base frames.

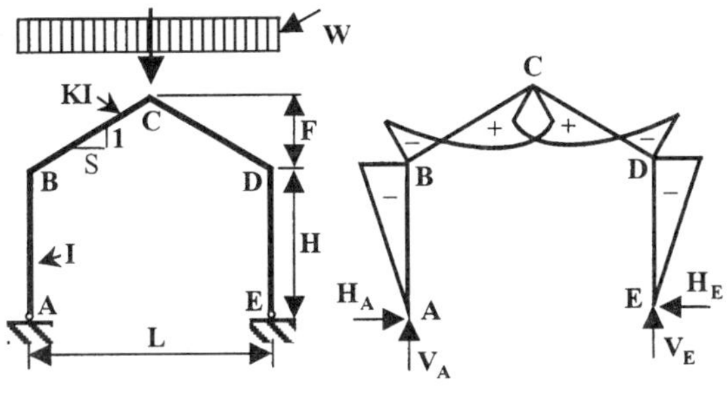

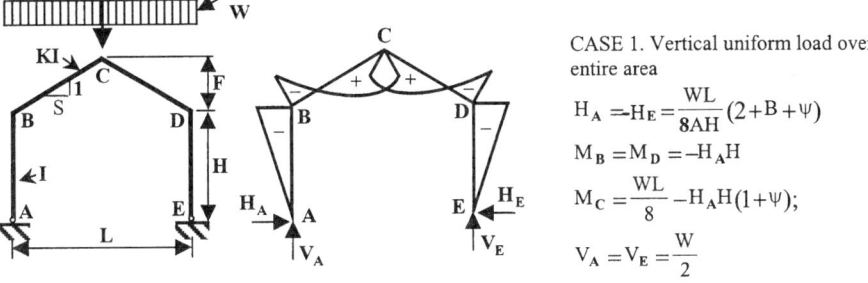

CASE 1. Vertical uniform load over entire area

$$H_A = -H_E = \frac{WL}{8AH}(2+B+\psi)$$

$$M_B = M_D = -H_A H$$

$$M_C = \frac{WL}{8} - H_A H(1+\psi);$$

$$V_A = V_E = \frac{W}{2}$$

Fig. 1A Hinged-base gable frame with vertical UDL over the whole frame.

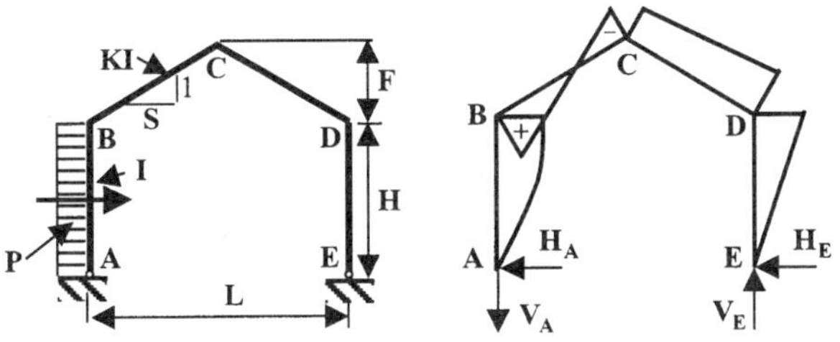

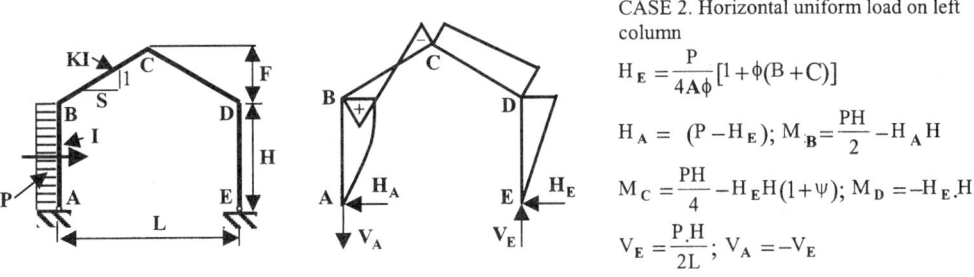

CASE 2. Horizontal uniform load on left column

$$H_E = \frac{P}{4A\phi}[1+\phi(B+C)]$$

$$H_A = (P - H_E);\ M_B = \frac{PH}{2} - H_A H$$

$$M_C = \frac{PH}{4} - H_E H(1+\psi);\ M_D = -H_E.H$$

$$V_E = \frac{P.H}{2L};\ V_A = -V_E$$

Fig. 1B Hinged-base gable frame with horizontal UDL applied to left column.

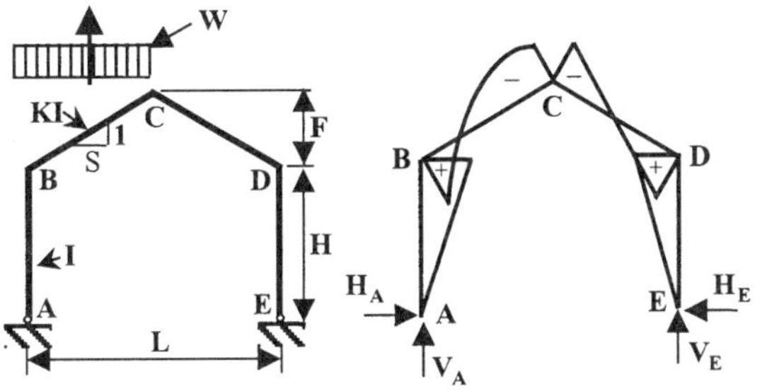

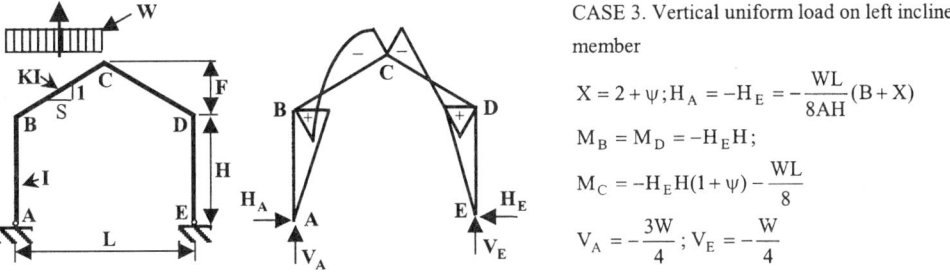

CASE 3. Vertical uniform load on left inclined member

$$X = 2+\psi;\ H_A = -H_E = -\frac{WL}{8AH}(B+X)$$

$$M_B = M_D = -H_E H;$$

$$M_C = -H_E H(1+\psi) - \frac{WL}{8}$$

$$V_A = -\frac{3W}{4};\ V_E = -\frac{W}{4}$$

Fig. 1C Hinged-base gable frame with v ertical UDL over the left rafter.

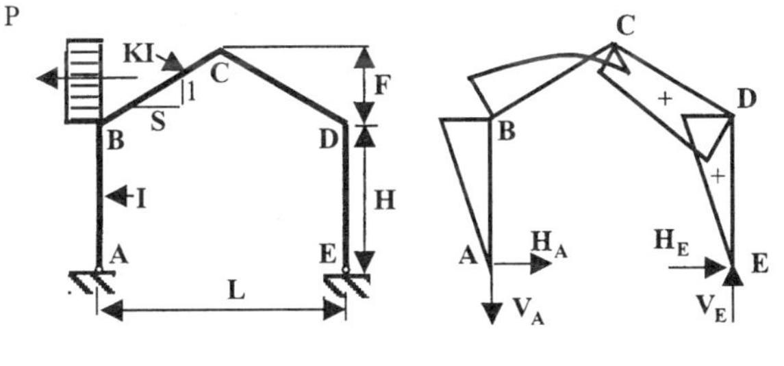

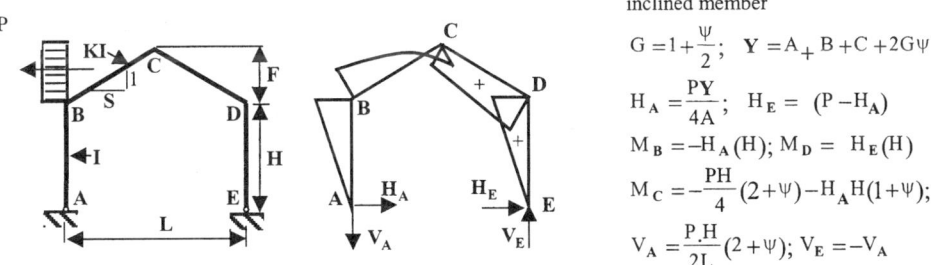

CASE 4. Horizontal uniform load on left inclined member

$$G = 1+\frac{\psi}{2};\quad Y = A_{+} B + C + 2G\psi$$

$$H_A = \frac{PY}{4A};\quad H_E = (P - H_A)$$

$$M_B = -H_A(H);\ M_D = H_E(H)$$

$$M_C = -\frac{PH}{4}(2+\psi) - H_A H(1+\psi);$$

$$V_A = \frac{P.H}{2L}(2+\psi);\ V_E = -V_A$$

Fig. 1D Hinged-base gable frame with horizontal UDL applied to left rafter.

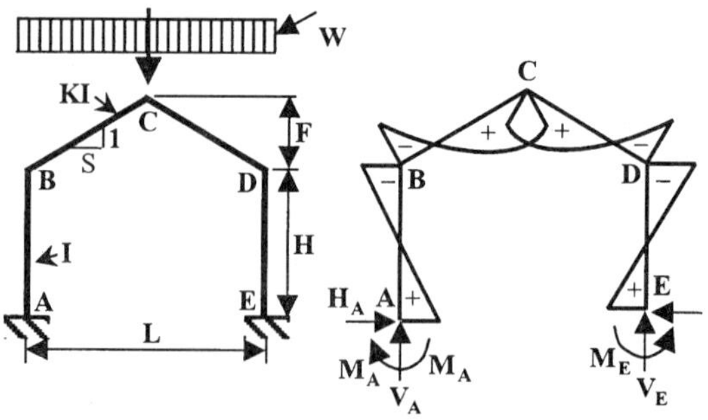

CASE 1. Vertical uniform load over entire girder

$$G = 2 + \frac{5A\psi}{4};\ J = 2A + \frac{5B\psi}{8}$$

$$M_B = M_D = -\frac{WLG\phi}{E};$$

$$M_A = M_E = \frac{WL\phi}{E}(J - G)$$

$$H_A = H_E = \frac{WLJ\phi}{EH};\ V_A = V_E = \frac{W}{2}$$

$$M_C = \frac{-WLG\phi}{E} + \frac{WL}{8} - H_E F$$

Fig. 2A Fixed-base gable frame with vertical UDL over the whole frame.

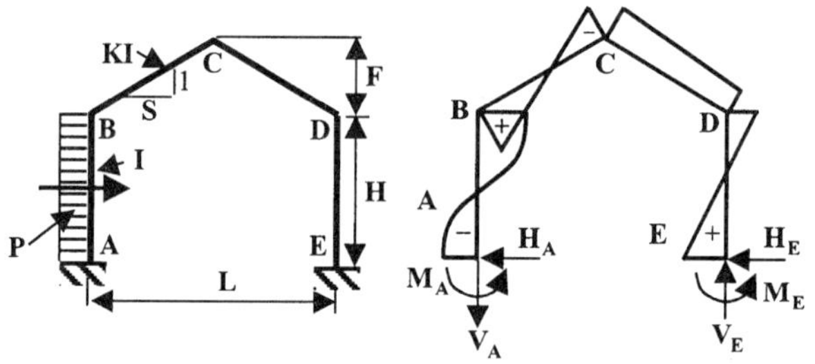

CASE 2. Horizontal uniform load on left column

$$\left.\begin{matrix} M_B \\ M_D \end{matrix}\right\rangle = \frac{PH}{2E}(4 - 3A) \pm \frac{4PH}{D};$$

$$H_E = \frac{P}{4E}(3B - 8A);\ H_A = -(P - H_E);$$

$$M_A = M_B + H_E H - \frac{PH}{2};\ M_E = M_D + H_E \cdot H$$

$$M_C = \frac{PH}{2E}(4 - 3A) - H_E F$$

$$V_E = \frac{8PH}{DL};\ V_A = -V_E$$

Fig. 2B Fixed-base gable frame with horizontal UDL applied to left column.

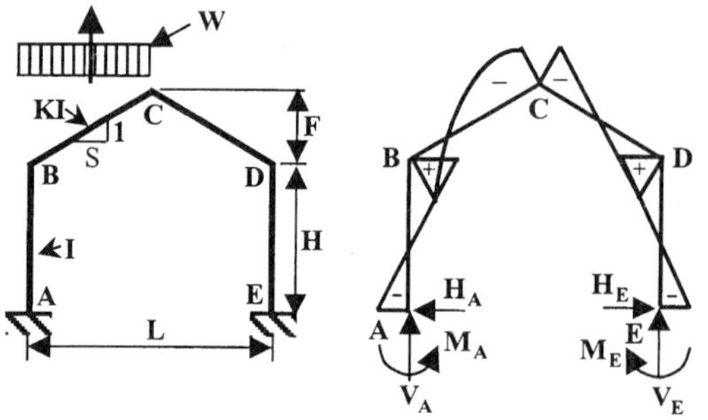

CASE 3. Vertical uniform load on left inclined member

$$G = 2 + \frac{5A\psi}{2};\ J = 2A + \frac{5B\psi}{8}$$

$$\left.\begin{matrix} M_B \\ M_D \end{matrix}\right\rangle = WL\phi\left(\frac{G}{E} \pm \frac{1}{2D}\right);\ M_E = M_D + H_E H$$

$$H_A = H_E = -\frac{WLJ\phi}{EH};\ M_A = M_B + H_A H$$

$$M_C = \frac{WLG\phi}{E} - \frac{WL}{8} - H_E F$$

$$V_E = -\frac{W}{4D}(D - 4\phi);\ V_A = -(W + V_E)$$

Fig. 2C Fixed-base gable frame with vertical UDL over the left rafter.

In such analysis of indeterminate structures, two sets of data are to be arranged as inputs. The first set defines the frame shape, size, support conditions, and relative stiffness of the members. The second set defines the loading to which the frame is subjected. The loads are due to dead weight, live load, wind effect, earthquake forces, or other. Much of this loading is guided by codal provisions and is dependent on the type of exposure, location, and such. These provisions can be finally presented in four loading schemes, as shown in Figs. 1 and 2, A–D.

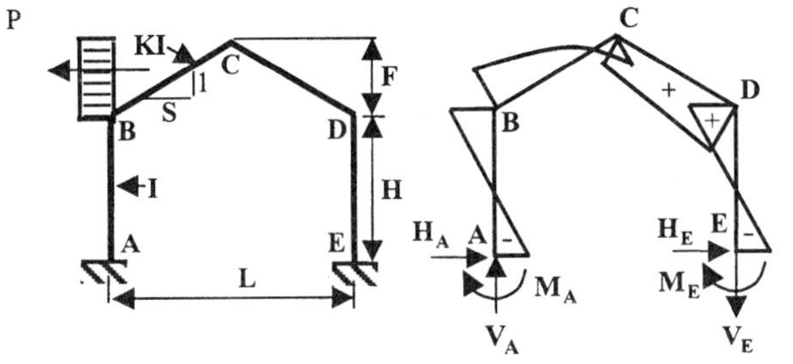

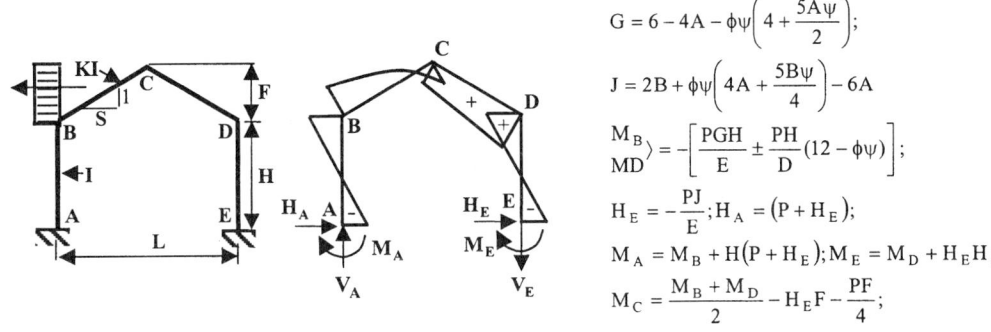

CASE 4. Horizontal uniform load on left inclined member

$$G = 6 - 4A - \phi\psi\left(4 + \frac{5A\psi}{2}\right);$$

$$J = 2B + \phi\psi\left(4A + \frac{5B\psi}{4}\right) - 6A$$

$$\left.\begin{matrix} M_B \\ M_D \end{matrix}\right\rangle = -\left[\frac{PGH}{E} \pm \frac{PH}{D}(12 - \phi\psi)\right];$$

$$H_E = -\frac{PJ}{E}; H_A = (P + H_E);$$

$$M_A = M_B + H(P + H_E); M_E = M_D + H_E H;$$

$$M_C = \frac{M_B + M_D}{2} - H_E F - \frac{PF}{4};$$

Fig. 2D Fixed-base gable frame with horizontal UDL applied to left rafter.

All other possibilities can be obtained and solved using superposition.

The attempt here is to suggest formulas for solving the frames for two kinds of support conditions under four different kinds of loading. It would appear that a set of tables may be developed, using these formulas, as design aids.

2.1.1 Hinged Bases

Some basic parameters are first computed based on frame geometry and sectional properties (see Figs. 1A–D). Constant sections for columns and also for rafters are assumed.

$$F = \frac{L}{2S}; \qquad Q = \left(\frac{L}{2}\right)\sqrt{1 + \frac{1}{S^2}}$$

$$\phi = \frac{I_{AB}}{I_{BC}}\frac{Q}{H}; \qquad \psi = \frac{F}{H}$$

$$A = 4\left(3 + 3\psi + \psi^2 + \frac{1}{\phi}\right)$$

$$B = 2(3 + 2\psi)$$

$$C = 2\left(3 + \psi + \frac{2}{\phi}\right)$$

Parameters F, Q, ϕ, φ, A, B, C are to be used along with Figs. 1A–D. The parameter span L, slope of the rafter S, and height H are indicated in the figure.

An example of evaluating the reactions and moments of a single pitched-portal frame with hinged bases for two load cases follows:

Example: Data given are $H/L = 0.5$; $K = 1.25$ and $S = 2.5$

The basic parameters evaluated as based on these data (see Sec. 2.1.1), are

$F = L/5$; $Q = 0.54L$; $\phi = 0.864$; $\Psi = 0.4$; $A = 22.07$;

$B = 7.6$; $C = 11.43$

Referring to Fig. 1A:

Load case 1: $M_B = M_D = (-)\ 0.0566\ WL$

$M_C = (+)\ 0.0457\ WL$

$H_A = H_E = (+)\ 0.1132\ W$

$V_A = V_E = (+)\ 0.5000\ W$

Referring to Fig 1B:

Load case 2: $M_B = (+)\ 0.2712$ PH

$M_C = (-)\ 0.0702$ PH

$M_D = (-)\ 0.2287$ PH

$H_A = (-)\ 0.7712$ P

$H_E = (-)\ 0.2287$ P

$V_A = (-)\ 0.2500$ P

$V_E = (+)\ 0.2500$ P

2.1.2 Fixed Bases

As in hinged bases, some basic parameters are also first computed based on frame geometry (see Figs. 2A–D) for fixed bases. Similar to hinged bases in the preceding section, these are framed structures without provision for cranes.

$$F = \frac{L}{2S}; \qquad Q = \frac{L}{2}\sqrt{1 + \frac{1}{S^2}}$$

$$\phi = \frac{I_{\mathrm{AB}}}{I_{\mathrm{BC}}} \cdot \frac{Q}{H}; \qquad \psi = \frac{F}{H}$$

$$A = \frac{3(1 - \phi\psi)}{2(1 + \phi\psi^2)}; \qquad B = \frac{6(1 + \phi)}{1 + \phi\psi^2}$$

$$D = 16(3 + \phi); \qquad E = 12[2 + 2\phi - A(1 - \phi\psi)]$$

Parameters F, Q, ϕ, ψ, A, B, D, and E are to be used along with Figs. 2A–2D. Parameters such as span L, height of frame H, slope of rafter, S are to be known for a frame.

2.2 Warehouse with Facilities for Cranes

These frames have fixed bases. The columns have a larger cross-section up to the level of the gantry (Fig. 3). The computational aids suggested are quite general. Some assumptions, such as axial deformations, are neglected without significantly affecting the results. Using principles of minimization of energy, the following set of equations may be written:

$$M_G \cdot a_{11} - H_G \cdot a_{12} + V_G \cdot a_{13} = f_1(q) \qquad (1)$$

$$- M_G \cdot a_{21} + H_G \cdot a_{22} - V_G \cdot a_{23} = f_2(q) \qquad (2)$$

$$M_G \cdot a_{31} - H_G \cdot a_{32} + V_G \cdot a_{33} = f_3(q) \qquad (3)$$

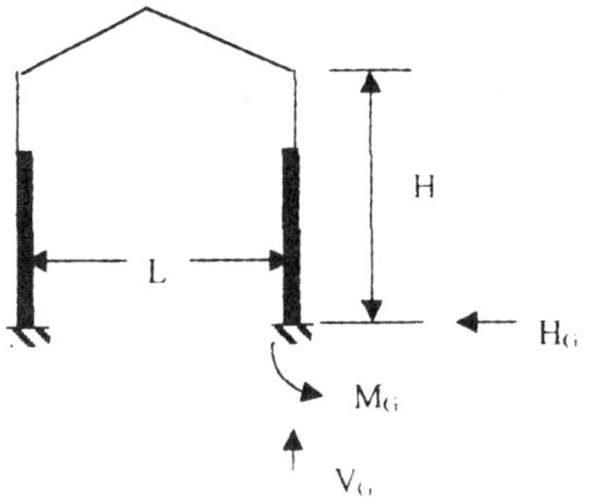

Fig. 3 Gable frame with stepped column showing redundance.

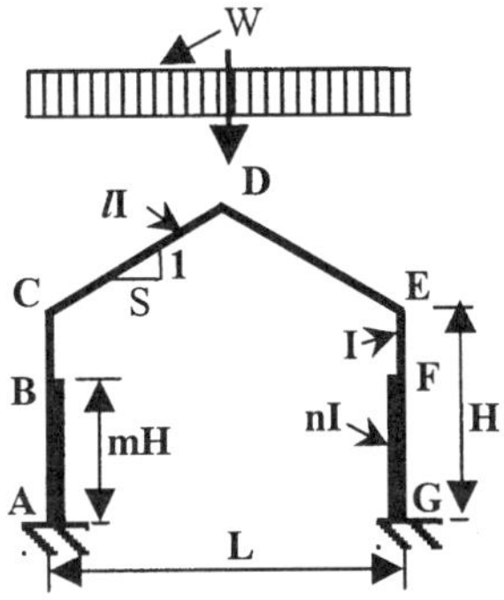

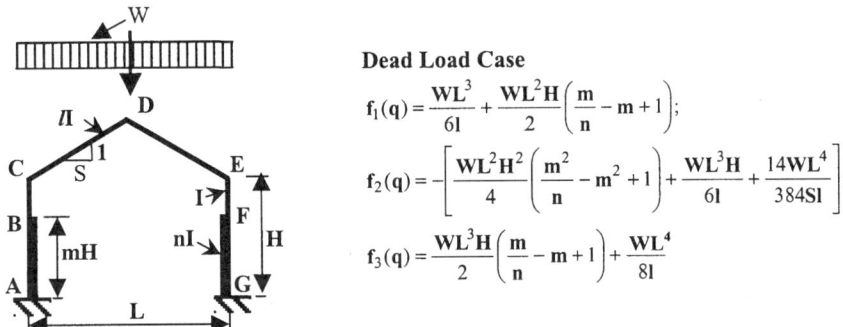

Dead Load Case

$$f_1(q) = \frac{WL^3}{6I} + \frac{WL^2H}{2}\left(\frac{m}{n} - m + 1\right);$$

$$f_2(q) = -\left[\frac{WL^2H^2}{4}\left(\frac{m^2}{n} - m^2 + 1\right) + \frac{WL^3H}{6I} + \frac{14WL^4}{384SI}\right];$$

$$f_3(q) = \frac{WL^3H}{2}\left(\frac{m}{n} - m + 1\right) + \frac{WL^4}{8I}$$

Fig. 4A Gable frame having stepped columns with vertical UDL over the whole frame.

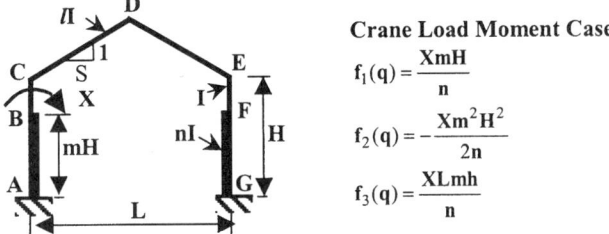

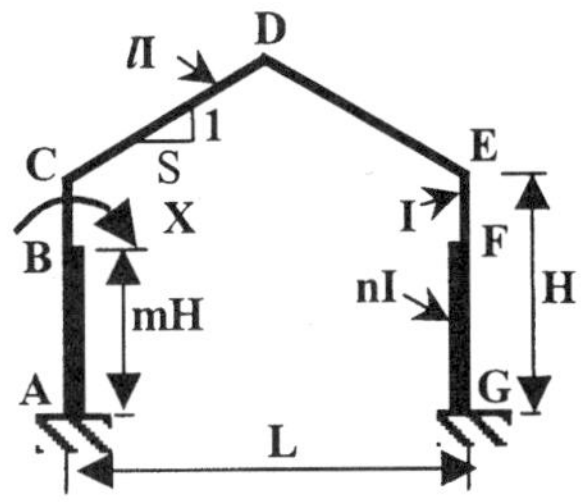

Crane Load Moment Case

$$f_1(q) = \frac{XmH}{n}$$

$$f_2(q) = -\frac{Xm^2H^2}{2n}$$

$$f_3(q) = \frac{XLmh}{n}$$

Fig. 4B Gable frame having stepped columns with vertical crane load applied on left column.

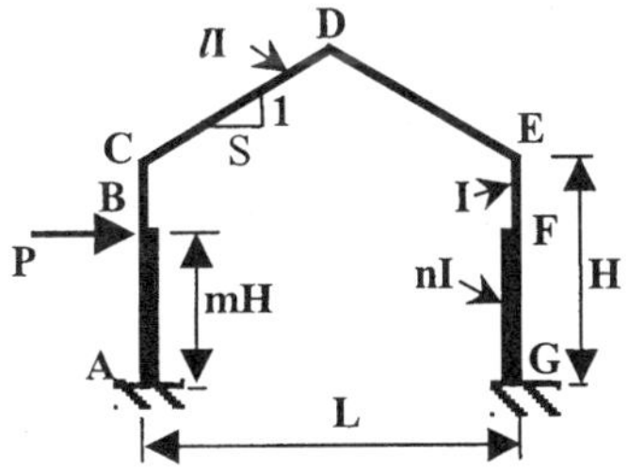

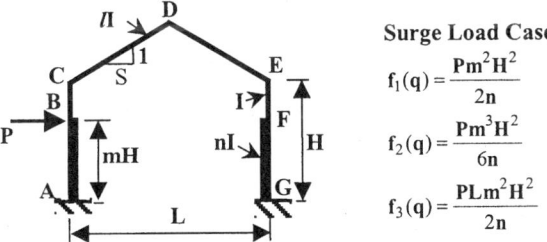

Surge Load Case

$$f_1(q) = \frac{Pm^2H^2}{2n}$$

$$f_2(q) = \frac{Pm^3H^2}{6n}$$

$$f_3(q) = \frac{PLm^2H^2}{2n}$$

Fig. 4C Gable frame having stepped columns with surge load on left column.

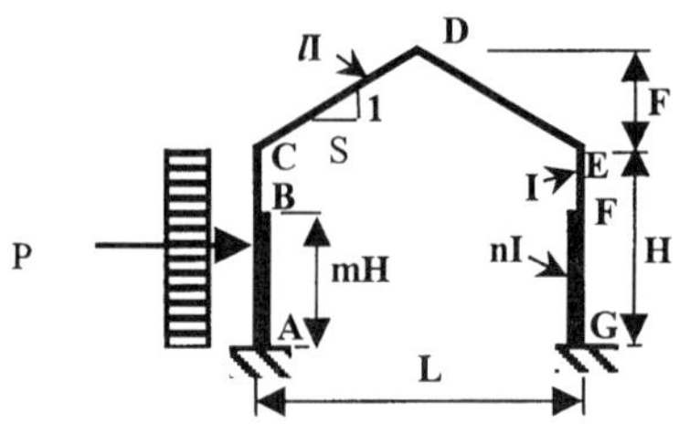

Wind Load on Column

$$f_1(q) = \frac{PH^3}{6}\left[\left(\frac{m^3}{n} - m^3 + 1\right) - 3\left(\frac{m^2}{n} - m^2 + 1\right) + 3\left(\frac{m}{n} - m + 1\right)\right]$$

$$f_2(q) = -\frac{PH^4}{24}\left[3\left(\frac{m^4}{n} - m^4 + 1\right) - 8\left(\frac{m^3}{n} - m^3 + 1\right) + 6\left(\frac{m^2}{n} - m^2 + 1\right)\right]$$

$$f_3(q) = \frac{PLH^3}{6}\left[\left(\frac{m^3}{n} - m^3 + 1\right) - 3\left(\frac{m^2}{n} - m^2 + 1\right) + 3\left(\frac{m}{n} - m + 1\right)\right]$$

Fig. 4D Gable frame having stepped columns with horizontal UDL over the whole frame.

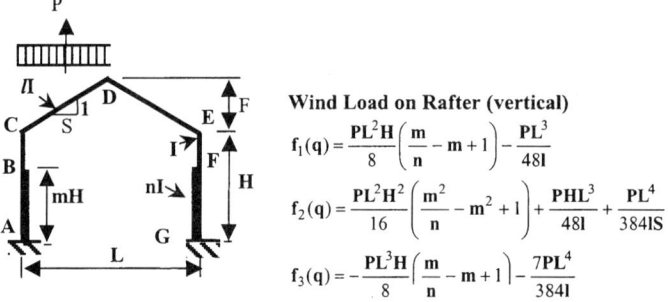

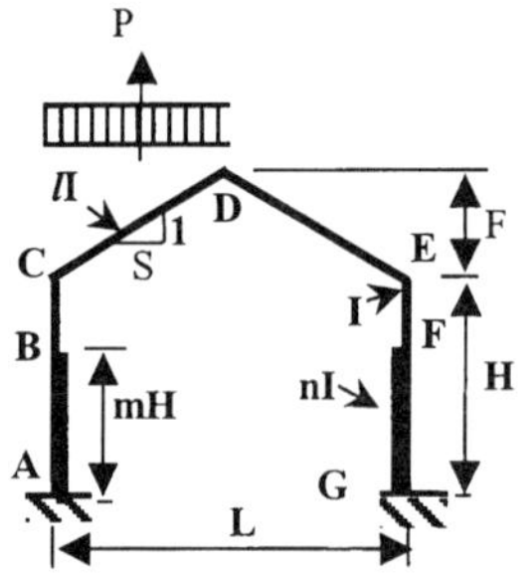

Wind Load on Rafter (vertical)

$$f_1(q) = \frac{PL^2H}{8}\left(\frac{m}{n} - m + 1\right) - \frac{PL^3}{48I}$$

$$f_2(q) = \frac{PL^2H^2}{16}\left(\frac{m^2}{n} - m^2 + 1\right) + \frac{PHL^3}{48I} + \frac{PL^4}{384IS}$$

$$f_3(q) = -\frac{PL^3H}{8}\left(\frac{m}{n} - m + 1\right) - \frac{7PL^4}{384I}$$

Fig. 4E Gable frame having stepped columns with vertical crane load applied on left column.

Wind Load on Rafter (Horizontal)

$$f_1(q) = \frac{PLH^2}{4S}\left(\frac{m^2}{n} - m^2 + 1\right) - \frac{PLH^2}{2S}\left(\frac{m}{n} - m + 1\right)$$

$$- \frac{PL^2H}{8S^2}\left(\frac{m}{n} - m + 1\right) - \frac{PL^3}{48IS^2}$$

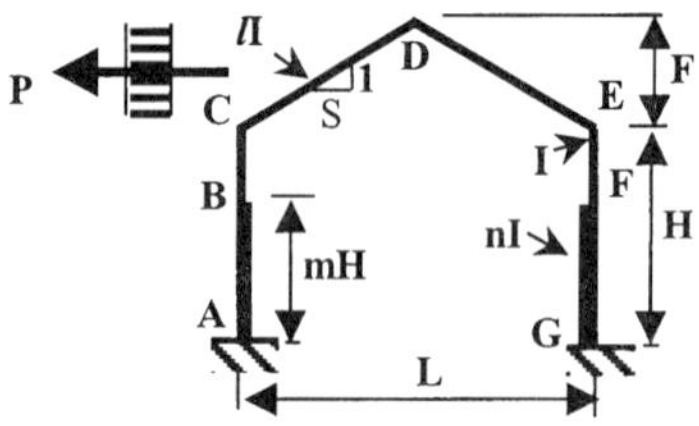

$$f_2(q) = \frac{PL^4}{384IS^3} + \frac{PL^3H}{48IS^2} + \frac{PL^2H^2}{16S^2}\left(\frac{m^2}{n} - m^2 + 1\right)$$

$$+ \frac{PLH^2}{4S}\left(\frac{m^2}{n} - m^2 + 1\right) - \frac{PLH^3}{6S}\left(\frac{m^3}{n} - m^3 + 1\right)$$

$$f_3(q) = \frac{PL^2H^2}{4S}\left(\frac{m^2}{n} - m^2 + 1\right) - \frac{PL^2H^2}{2S}\left(\frac{m}{n} - m + 1\right)$$

$$- \frac{PL^3H}{8S^2}\left(\frac{m}{n} - m + 1\right) - \frac{7PL^4}{384IS^2}$$

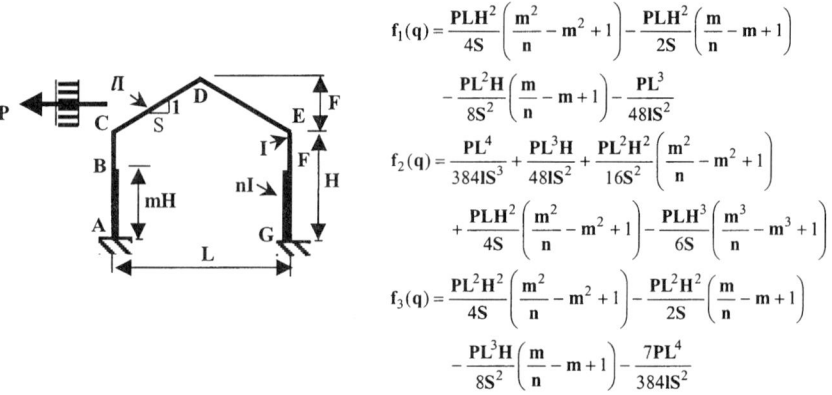

Fig. 4F Gable frame having stepped columns with horizontal UDL applied on left rafter.

in matrix form as

$$\{\mathbf{R}\}[\mathbf{A}] = \{\mathbf{q}\} \tag{4}$$

where, $\{\mathbf{R}\}$ is the reaction matrix, $\{\mathbf{q}\}$ is the load matrix and different elements of $[\mathbf{A}]$ are as follows. The parameters m, n, and ℓ are as shown in Fig. 4A–F.

$$\mathbf{a}_{11} = 2H\left(\frac{m}{n} - m + 1\right) + L\left(\frac{1}{\ell}\right)$$

$$\mathbf{a}_{12} = H^2\left(\frac{m^2}{n} - m^2 + 1\right) + \frac{L^2}{4}\left(\frac{1}{\ell S}\right) + HL\left(\frac{1}{\ell}\right) = a_{21}$$

$$\mathbf{a}_{13} = \frac{L^2}{2}\left(\frac{1}{\ell}\right) + HL\left(\frac{m}{n} - m + 1\right) = \mathbf{a}_{31}$$

$$\mathbf{a}_{22} = \frac{2H^3}{3}\left(\frac{m^3}{n} - m^3 + 1\right) + H^2L\left(\frac{1}{\ell}\right) + \frac{HL^2}{2}\left(\frac{1}{\ell S}\right) + \frac{L^3}{12}\left(\frac{1}{\ell S^2}\right)$$

$$\mathbf{a}_{23} = \frac{H^2L}{2}\left(\frac{m^2}{n} - m^2 + 1\right) + \frac{HL^2}{2}\left(\frac{1}{\ell}\right) + \frac{L^3}{8}\left(\frac{1}{\ell S}\right) = \mathbf{a}_{32}$$

$$\mathbf{a}_{33} = L^2H\left(\frac{m}{n} - m + 1\right) + \frac{L^3}{3}\left(\frac{1}{\ell}\right)$$

Solving the set of simultaneous equations involving M_G, H_G, and V_G

$$M_G = \frac{(a_{23}a_{32} - a_{22}a_{33})[a_{23}f_1(q) + a_{13}f_2(q)] + (a_{13}a_{22} - a_{12}a_{23})[a_{33}f_2(q) + a_{23}f_3(q)]}{(a_{23}a_{32} - a_{22}a_{33})(a_{11}a_{23} - a_{13}a_{21}) + (a_{13}a_{22} - a_{12}a_{23})(a_{23}a_{31} - a_{21}a_{33})}$$

$$H_G = \frac{[a_{23}f_1(q) + a_{13}f_2(q)] - (a_{11}a_{23} - a_{13}a_{21})M_G}{(a_{13}a_{22} - a_{12}a_{23})}$$

and

$$V_G = \frac{f_1(q) - a_{11}M_G + a_{12}H_G}{a_{13}}$$

The expressions for $f_1(q)$, $f_2(q)$, and $f_3(q)$ for different loading cases are given in Fig. 4A–F.

3 DESIGN PRINCIPLES OF SILOS

3.1 Introduction

Silos are large-sized, deep containers to store granular or powdery materials. There are several ways to define silos. Ordinarily, a storage bin may be called a silo if its depth is greater than twice its width. However, from the structural point of view, if the plane of rupture of the material stored—drawn from the bottom edge of the bin—does not intersect the free surface of the material stored, the bin is known as a *silo*. Conversely, if the rupture plane—drawn from the bottom edge of the bin—intersects the free surface of the material stored, the bin is termed a *bunker*. However, the terms, *bin*, *silo*, and *bunker* have different meanings in different parts of the world. In some countries the term *bin* is generally assumed to include both silos and bunkers. According to the Eurocode 1, the geometrical limits applicable for silos are $h_1/d < 10$, $h_1 < 100$ m, $d < 50$ m (where h_1 = height; d = diameter). Given these limitations, *silo* is defined as a *slender silo* if $h_1/d > 1.5$ or a *squat silo* if $h_1/d < 1.5$.

3.2 Common Forms, Materials, and Structural Responses

Silos are generally circular in cross section, although different forms, such as square or rectangular cross sections, are also adopted. However, square or rectangular cross sections are commonly adopted for shallow bins or bunkers. When calculating the size of a silo of a specified capacity, the unit weight of the material should not be overestimated and too small a value should not be assumed for the angle of internal friction. As a guideline, Table 1 shows the unit weight and the angle of internal friction for some commonly used materials (4).

The materials generally used for the construction of silos are structural steel and reinforced concrete. Both materials have their own merits and demerits. A silo having a circular cross section may also be formed with thin-walled plates without any stiffener.

High strength, durability, workability, long life, and resistance to fire may be considered as the main properties of the concrete silos. On the other hand, because the steel members have high strength, the steel silos can resist high loads with comparatively lighter weight and smaller-sized members. The steel silos are gas- and water-tight because of higher density. Fabrication, erection, and handling or dismantling are easier. However, the main drawback with steel is that it is susceptible to corrosion.

For the design of their bottoms, silos may or may not have premises underneath the bottom. The choice of the type of silo depends first on the properties of the loose material and in the unloading equipment selected.

Some of the materials that require storage in silos are hygroscopic and, in contact with moist air, cohesion builds up to such a degree that the material can be loosened only by mechanical means. Raw sugar, salt, and some granular materials fall in this category. Flour is also hygroscopic, but in relatively dry storage the cohesion is weak, and the material can be easily loosened by compressed air so that it flows out of the cells by gravity. Nonhygroscopic or less hygroscopic materials are stored most economically in

Table 1 Unit Weight and Angle of Internal Friction of Some Commonly Used Materials

Materials	Unit weight (kN/m^3)	Angle of internal friction (Φ) in degrees
Wheat	8.50	28°
Paddy	5.75	36°
Rice	9.00	33°
Maize	8.00	30°
Barley	6.00	27°
Corn	8.00	27°
Sugar	8.20	35°
Wheat flour	7.00	30°

relatively high cellular silos the individual cells of which are emptied by gravity flow. Hygroscopic materials are stored in large-spaced silos in which the material can be broken loose mechanically before being taken out by appropriate conveyor systems. Also some of the highly hygroscopic materials (sugar, salt, etc.) may attack concrete, so that a protective coating for the inside surfaces of walls is required unless special cement and dense, closed surfaces are used. If hygroscopic sticky materials are not dried before storage, considerable cohesion and arching can be expected; consequently, the cell walls are subjected to unsymmetrical loading during emptying and have to resist considerably higher horizontal pressure than in the fully-filled condition. Sometimes silos are required to be provided with driers and equipment for infected grain-treatment facilities.

3.3 Assessment of Forces: Pressure Theories

The stored materials in the storage bins exert horizontal pressure to the side walls of the bins in addition to the vertical forces from weight of the material. The horizontal pressure varies under different conditions and for different configurations of the bins. Pressure along the depth of the silo varies during the filling and emptying processes of the materials (Fig. 5).

Location of discharge opening may also substantially change the pressure distribution on the bin walls. Discharge through eccentrically loaded openings causes change in lateral pressure. The pressure increases on one side and decreases on the other. The silo walls should have adequate horizontal bending strength and stiffness to resist the unbalanced loading that occurs during eccentric draw off. The equilibrium of the contained material is changed while unloading the silo. If the silo is unloaded from the top, the frictional load on the wall may change owing to reexpansion of the material; however, the lateral pressures remain the same as those that occur during filling.

When unloading a free-flowing material from the center of a hopper at the bottom, one of two different modes of flow may occur. The mode of flow depends primarily on the nature of the contained material, and the size of the silo and the hopper. The modes are generally termed *core flow* and *mass flow* (Fig. 6). Core flow generally gives rise to some increase in lateral pressure from the filled condition. Mass flow occurs in silos with steep-sided hoppers. To ensure the convergent downward movement of the entire mass

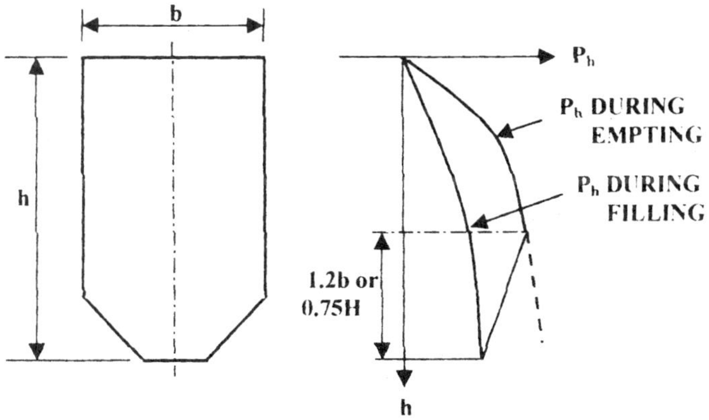

Fig. 5 Variation of pressure along the depth of a silo.

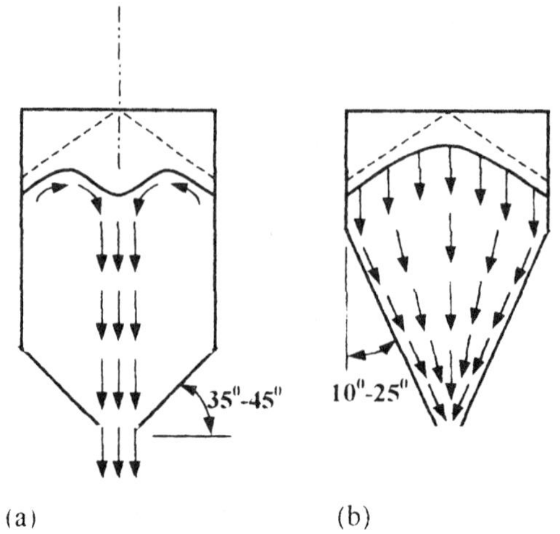

Fig. 6 Condition of flow of material: (a) core flow, (b) mass flow.

of the material as a whole, steep-sided hoppers are provided. However, this action produces substantial local increase in lateral pressure at the intersection between the vertical walls and the hopper bottom. Sometimes a situation may arise in which the flow may be a borderline between core flow and mass flow and, in the process, the stored material is intermittently in core or mass flow. This occurs when the hopper is almost—but not quite—steep and smooth enough for mass flow and leads to jerky flow. At high withdrawal rates, these jerks can be destructive.

The volumetric change of the stored material owing to increase in moisture content and temperature fluctuations can cause the development of high pressures.

In general, the factors that affect the pressure distribution are moisture content of the stored material, particle size gradation, angularity of particles of the stored material, temperature of the material, rate of filling, amount of aeration during filling, aeration during withdrawal, and so on.

Different theories are in vogue for the determination of pressure distribution on the walls. The horizontal pressures exerted by the material on the walls are calculated using Janssen's theory, Airy's theory, or Coulomb's theory. In this *Handbook*, Janssen's theory

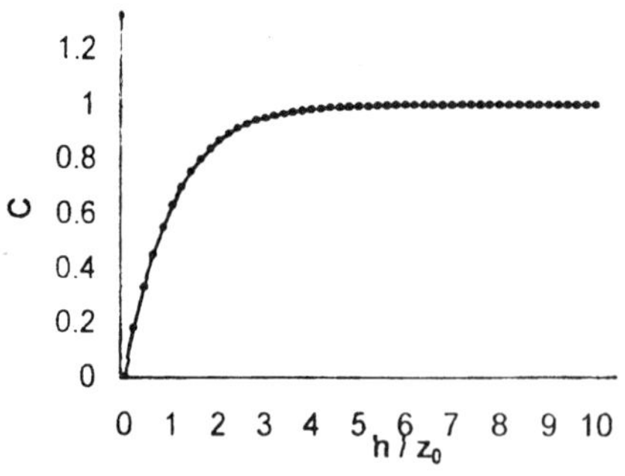

Fig. 7 Variation of C with h/z_0.

Table 2 Values of Pressure Ratio (K) and Angle of Wall Friction (Φ') of Different Material

Sl. No	Material	Pressure ratio (K) during		Angle wall friction (Φ') during	
		Filling (K_f)	Emptying (K_e)	Filling (Φ'_f)	Emptying (Φ_e)
1	Granular material $s \geq 0.2$ mm	0.5	1.0	0.75	0.6
2	Powdery material $s \leq 0.06$ mm	0.5	0.7	1.0	1.0

is considered for the calculations of pressure. The variation of vertical and horizontal pressure may be represented as

$$p_i(h) = p_i(\max)[1 - e^{-h/z_0}]$$

$$p_i(h) = p_i(\max)C$$

Where $p_i(h)$ is the pressure at any height h and $p_i(\max)$ is the maximum pressure developed in the wall; $z_0 = R/(\mu' K)$; $C = [1 - e^{-h/z_0}]$; R is a parameter known as hydraulic mean radius and is obtained by dividing the plan area (A) of the silo by the plan perimeter (P); μ' is the coefficient of wall friction. The variation of parameter C with h/z_0 is as shown in Fig. 7.

K is the pressure ratio, which generally lies between $(1 - \sin \Phi)/(1 + \sin \Phi)$ and $(1 + \sin \Phi)/(1 - \sin \Phi)$. However, the exact value of K is obtained experimentally. The acceptable values of K in different codes for broadly classified material according to the particle size (s) are as indicated in Table 2 (4).

Values of the maximum pressures (p_v, p_h, p_w) during filling and emptying conditions are obtained in terms of unit weight, weighted perimeter, pressure ratios, and coefficient of wall frictions (Table. 3)

3.4 Guidelines for Analysis

The walls of silos are designed to resist bending moments and tension caused by the developed pressure of the stored material. The development of bending moment or tension depends on the type of cross section chosen for the silos. In silos with circular cross sections, the circular wall is designed for the hoop tension, which is equal to ($p_h \times d/2.0$); where p_h is the horizontal pressure exerted by the material and d is the diameter of

Table 3 Values of Maximum Pressures During Filling and Emptying Conditions

Types of pressures	Maximum values of pressures during	
	Filling	Emptying
P_v	$\omega R/K_i\mu'_f$	$\omega R/K_e\mu'_e$
P_h	$\omega R/\mu'_f$	$\omega R/\mu'_e$
P_w	ωR	ωR

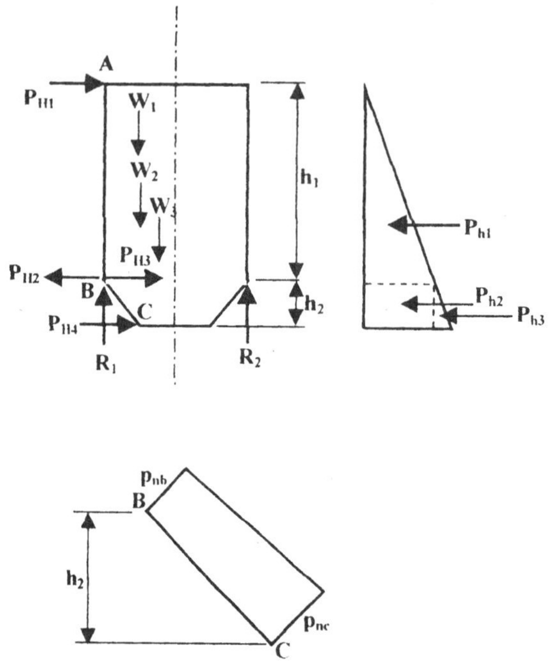

Fig. 8 Distribution of resultant pressure on walls.

the silo. The longitudinal tension develops in the wall if the silo is supported at places other than the waste level.

For silos of rectangular or square cross sections, the possibility of spanning may be of two types. Combined bending and direct tension will occur if the walls are spanning horizontally; conversely, if the walls are spanning vertically, bending is taken care of in the vertical direction, whereas direct tension is taken care of in the horizontal direction. Figure 8 shows the distribution of resultant pressure on the walls.

The design of sloping hopper bottoms in the form of inverted truncated pyramids on each sloping side, consists of finding the center of pressure, the intensity of pressure normal to the slope at this point, and the mean span. The bending moments at the center and edge of each slope are then calculated. Figure 9 shows the distribution of resultant pressure on the hopper bottom.

Many times the bins may have to be adopted attached together in series. The designer should be aware of the possible bending moments in silos of any shape in such situations. Such bending moments will be present in walls of silo groups especially when some cells are full and some are empty. They may occur as the flowpath changes, and their magnitude may increase with the rate of discharge.

A typical silo with a circular cross section consists of a cylindrical wall, top ring beam, bottom ring beam, and conical hopper. The capacity of the silo may be expressed in parametric form as

$$Q = \frac{1}{4}\pi d^3 Y$$

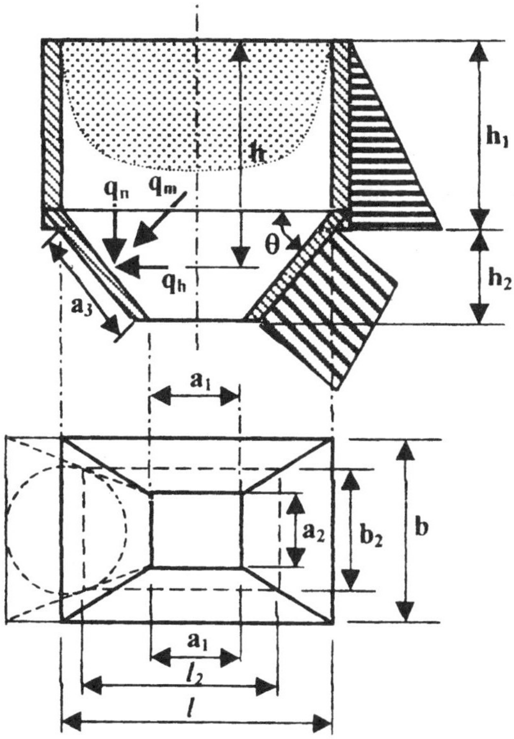

Fig. 9 Resultant pressure on hopper bottom.

where

$$Y = (K_1 + 0.167 \tan \theta)$$

Hence,

$$d = \left(\frac{4Q}{\pi Y}\right)^{1/3}$$

Q is the storage capacity of silo in cubic meters (m^3); d is the diameter of the silo in meters, and K_1 is a factor defined as the ratio of height to diameter (h_1/d). θ is the inclination of the hopper bottom with the horizontal. The recommended inside diameters of silos are (7):

1. For heavy grain (such as wheat): 6.0 m
2. For light grain (such as sunflower): 12.0–18.0 m

The ring beams provided at the top and at the junction of a cylindrical wall and hopper part are primarily designed for the radial forces, which introduce hoop tension or compression. The vertical component of the loads at those levels are transferred to the wall or to the supporting system.

Conical hoppers are essentially subjected to only meridional and hoop tensions.

3.5 Optimization Scheme

Several optimization methods have been developed for solving different types of problems. The optimization techniques are useful in finding the minimum of a function of several variables under a prescribed set of constraints. The mathematical programming basically involves the approach of a general method for expressing constraints of many types and the explicit use of an objective function, such as minimum weight, minimum cost, or other. While stipulating constraints, the aspects assumed are that every member should be optimally stressed and deformations will remain within a specified threshold. The problem of optimization of a silo is related to various parameters, as has been observed in the foregoing discussions.

4 DESIGN AND DESIGN AIDS FOR SILOS

4.1 Form and Capacity

A few examples are worked out to demonstrate the usage of the *Handbook.* A few design details are also provided for ready reference.

Example 1. Select the diameter and height of a silo to store grains with a storage capacity of 500 m^3. Assume the inclination of hopper bottom with horizontal as 30 degrees. As has been mentioned before

$$Q = \frac{1}{4}\pi d^3 Y$$

where

$$Y = (K_1 + 0.167 \tan \theta)$$

Assuming K_1 (h_1/d) as 4 and for $\theta = 30°$, the value of Y is 4.0964. Hence, the value of the diameter d is 5.375. If we adopt a diameter of 5.5 m, the height of the silo, excluding the hopper part, is 4×5.5 m $= 22.0$ m.

Example 2. Select the diameter and height of a silo to store maize, with a storage capacity of 200 metric ton. Assume the inclination of hopper bottom with horizontal as 30 degrees. If we refer to Table 1, the unit weight of maize is 8 kN/m^3. Hence, the volume of the silo required to store maize is (2000/8)m^3 = 250 m^3. As noted before,

$$Q = \frac{1}{4}\pi d^3 Y$$

where

$$Y = (K_1 + 0.167 \tan \theta).$$

If we assume the value of $K_1(h_1/d)$ as 3 and for $\theta = 30°$, the value of Y is 3.096. Hence, the value of the diameter is 4.68 m. If we assume a diameter of 5.0 m, the height of the silo, excluding the hopper part, is $3 \times 5.0 = 15.0$ m.

Example 3. Design a circular silo of internal diameter 4.0 m and height 15.0 m to store wheat. The depth of the hopper part is 3.0 m.

Refer to Table 1, the unit weight of the material is 8.5 kN/m^3 and the angle of internal friction $(\Phi) = 28°$.

Refer to Table 2 (grain size of material > 0.2 mm), $K_f = 0.5$; $K_e = 1.0$; $\Phi_t = 0.75$; $\Phi = 21°$; $\Phi_e = 0.6$; $\Phi = 16.8°$.

Angle of wall friction: during filling ($\mu_f = \tan \Phi_f$) = 0.3838; during emptying ($\mu_e = \tan \Phi_e$) = 0.302

$$\text{Hydraulic mean radius } (R) = A/P = \frac{\pi d^2/4}{\pi d}$$

$$= d/4 = 1.0 \text{ (for the present problem).}$$

$$\text{Hence } z_{0_e} = \frac{R}{\mu_c K_c} = 3.113$$

$$\text{And } z_{0f} = \frac{R}{\mu_f K_f} = 5.2110$$

It is apparent from the foregoing calculations that the ratio h/Z_0 is more during emptying Horizontal and vertical pressures at different depths are calculated as follows: Refer to Fig. 7.

At 3.0 m from top

$$\frac{h}{z_{0_e}} = \frac{3.0}{3.113} = 0.906$$

Hence $1 - e^{-h/z_{0_e}} = 0.596$

$$p_h = \frac{8.50}{0.302} \times 0.596 = 16.775 \text{ kN/m}^2$$

$$p_v = \frac{p_h}{K_e} = \frac{16.775}{1.0} = 16.775 \text{ kN/m}^2$$

Table 4 shows the foregoing values at different heights of the silo.

The horizontal pressure is maximum at a depth of 15.0 m from top.
Hoop tension = (27.84 × 4.0/2.0) = 55.68 kN/m.

4.2 Design of Component Parts

4.2.1 Design of Silo Wall

For a steel silo, the thickness of plate is determined from consideration of the compressive and tensile stresses to which the wall will be subjected by the total vertical weight. The vertical weight includes weight of the material stored above that level, self-weight of plate,

Table 4 Computed Values of Pressures at Different Depths of a Silo

Height from top of silo	3.0 m	6.0 m	9.0 m	12.0 m	15.0 m
h/z_{0_e}	0.906	1.812	2.718	3.624	4.53
$1 - e^{-h/z_0}$	0.596	0.837	0.934	0.973	0.989
p_h (kN/m^2)	16.775	23.56	26.29	27.786	27.84
p_v (kN/m^2)	16.775	23.56	26.29	27.786	27.84

and lining material. The total weight over the cross-sectional area yields compressive stress. The Poisson effect of compressive stress along with the hoop stress yields total tensile stress.

For a concrete silo, the horizontal ring reinforcements are provided to take care of the tensile stresses, whereas vertical bars are provided to take care of compressive or bending stresses.

For silos with rectangular cross sections, if the wall spans vertically, horizontal reinforcement is provided to resist the direct tension, and vertical reinforcement to resist the bending moments. The horizontal bending moments owing to the continuity at corners should be considered.

4.2.2 Design of the Hopper

The total vertical load transferred at the hopper is calculated. The longitudinal stress and the hoop stresses are calculated for a particular thickness of plate provided. If the stresses are within acceptable limits, the assumed thickness of the plate is adopted.

The amount of horizontal reinforcement required for the hopper bottoms, provided in the form of an inverted truncated pyramid, is based on the horizontal direct tension combined with the bending moment. At the top of the slope, the bending moment and the component of the hanging-up force are combined to determine the reinforcement required in the upper face at the top of the slope. Although the horizontal span of the sloping side is considerably reduced toward the outlet, the amount of reinforcement should not be reduced below that determined for the center of pressure.

4.2.3 Design of the Ring Beam

The ring beam is provided at waist level and may be supported on column members. The ring beam supports the load of material stored, the weight of the silo, and the weight of the platform, if any. While designing a steel silo, a builtup section—either in the form of a ''I'' or a box section—may be adopted for the beam, and the bending, shear, and torsional stresses should be checked.

The horizontal component of the hanging-up force tends to produce horizontal bending moment in the ring beam. The ring beam should withstand this horizontal moment owing to the horizontal component of this hanging-up force and the outward pressure from the contained material, whichever is critical.

4.3 Reinforcement Detailing

A few typical reinforcement details at the corner of the concrete silo are shown in Fig. 10.

NOTATIONS

A	Cross-sectional area of silo
a_1, a_2	Dimensions of hopper opening
b	Side dimension of rectangular silo
d	Diameter of silo
h	Total height of silo
h_1	Height of silo
h_2	Height of hopper
K	Pressure ratio in silo; coefficient of moment of inertia for rafter of frames

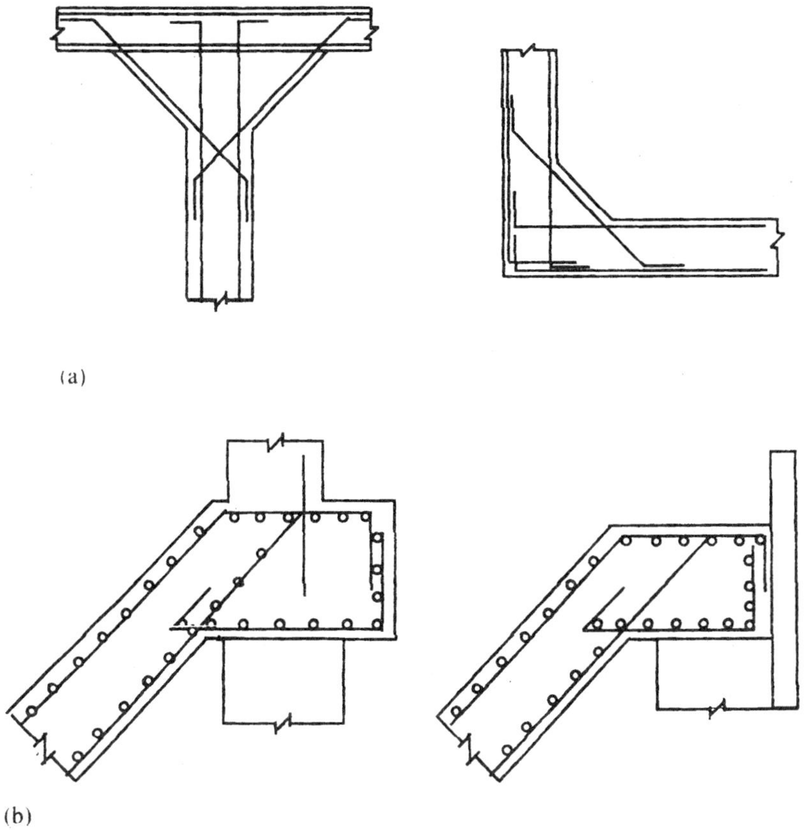

Fig. 10 Typical details for concrete silo: (a) reinforcement position at intersecting walls, (b) typical details of hopper-supporting beams.

l	Side dimension of rectangular silo; coefficient of moment of inertia for rafter of frames
p	Pressure
p_h	Horizontal pressure distribution
p_v	Vertical pressure distribution
p_w	Vertical pressure from friction
P	Perimeter of cross section; lateral load on frame
P_h	Horizontal pressure on side wall
P_H	Resultant thrust on side wall
Q	Storage capacity in m^3
q_h	Horizontal thrust in the hopper
q_m	Resultant thrust in the hopper
q_n	Vertical thrust in the hopper
R_1, R_2	Reactive forces of silo
s	Particle size of grains
S	Slope of pitched roof of frames
W	Load
μ	Coefficient of wall friction

θ	Inclination of hopper with horizontal
ϕ	Angle of internal friction for silo; ratio of moment of inertias for warehouse
ω	Unit weight of material
σ	Dimension ratio for warehouse structures
M_A	Moment at node A
H_A	Horizontal reaction at A
V_A	Vertical reaction at A
UDL	Uniformly distributed load

REFERENCES

1. Baikov VN. Reinforced Concrete Structures. MIR Publishers, Moscow, 1978; pp 273–283.
2. Billig K. Structural Concrete. McMillan & Co. London, 1960; pp 754–767.
3. Clarke JL, Garas IK, Armer GST. Design of Concrete Structures—The Use of Model Analysis. Elsevier Applied Science London, 1995; pp 320–326.
4. I.S.:4995 (part 1). Criteria for Design of Reinforced Concrete Bins for the Storage of Granular and Powdery Materials. Bureau of Indian Standards, 1974.
5. Komendant AE. Contemporary Concrete Structures. McGraw—Hill, 1972; pp 614–623.
6. Kounddis AN. Steel Structures—Eurosteel '95.' 1995.
7. Mukhanov K. Design of Metal Structures. MIR Publishers, Moscow, 1980.
8. Nash MJ. Crop Conservation and Storage. Pergamon Press, London, 1978; pp 128–135.
9. Reynolds CE, Steedman JC. Reinforced Concrete Designer's Handbook. 10th ed. Rupa & Co., New Delhi, India, 1988.
10. Steel Designers' Manual. 4th ed. ELBS Publications. England, 1983; pp 332–344.
11. Mallick SK, Gupta AP. Reinforced Concrete. 6th ed. Oxford & IBH Publishing, New Delhi, India, 1990.
12. Bresler B, Lin TY, Scalzi JB. Design of Steel Structures. 2nd ed. Wiley Eastern New Delhi, India.
13. Meng Q, et al. Finite element analysis of bulk solids Flow: I. Development of a model based on a secant constitutive relationship. Agric Eng Res 1997; 67:151.
14. Meng Q, et al. Finite element analysis of bulk solids flow: II. Application to a parametric study. Agric Eng Res 1997; 67:151.

10

Controlled Atmosphere Storage of Grain

NOEL D. G. WHITE

Agriculture and Agri-Food Canada, Winnipeg, Manitoba, Canada

DIGVIR S. JAYAS

University of Manitoba, Winnipeg, Manitoba, Canada

1 INTRODUCTION

Controlled atmospheres (CA) are used as periodic treatments to control pests (insects and mites) in stored grain or, less frequently, as long-term storage environments to prevent pest occurrence (insects, mites, and molds) or spontaneous combustion, as in poppy seeds that must be stored in nitrogen gas (Mills, 1989). The CAs contain either high levels of carbon dioxide (CO_2), or high levels of nitrogen (N_2), with virtual elimination of oxygen (O_2).

Grain in bulk storage is an immature ecosystem (Sinha, 1995) consisting of the living grain itself, numerous species of insects, mites, and molds, both on and in the seed. The nonliving part of the system includes humidity and grain moisture, temperature, and intergranular gases. When the grain is warm (>25° C) or becomes moist, owing to bin leakage or to moisture migration by convection currents, the biological organisms begin to respire. Oxygen is consumed and heat, moisture, and carbon dioxide are produced.

Canada has a legally defined zero-tolerance for insect pests in stored grain; hence, they must be controlled whenever detected. Dozens of insects are found in stored grain (Sinha and Watters, 1985) in low numbers, and many are fungus feeders that do not directly feed on the grain (White et al., 1995a). The most common grain-feeding insects in western Canada are the rusty grain beetle, *Cryptolestes ferrugineus* (Stephens) and the red flour beetle, *Tribolium castaneum* (Herbst), often being detected in up to 46% of farm granaries (Madrid et al., 1990). The whole-seed feeders—the rice weevil *Sitophilus oryzae*

L. and the lesser grain borer *Rhyzopertha dominica* F.—are present (Madrid et al., 1990; Fields et al., 1993), but rarely become established in farm-stored grain because they multiply relatively slowly (Fields and White, 1997), and the grain generally cools down after harvest.

Few contact insecticides are registered for use on grain because of residue problems (White and Leesch, 1996). The fumigant methyl bromide is not used directly on stored grain in Canada, although it is used widely throughout the world. Methyl bromide is currently being banned because of the damage it does to atmospheric ozone. The only fumigant widely used is phosphine gas, and insect resistance to it is becoming more widespread (Mills, 1983). In many circumstances, in relatively airtight structures, the use of CA is economically competitive with phosphine and is safe to apply. Although CO_2 is a ''greenhouse'' gas for the earth and efforts are being made to minimize its production, CO_2 used in CA is a by-product of industrial processes or natural processes, such as tapping natural gas wells. The gas that would normally be directly vented to the atmosphere is compressed and put to a useful purpose.

In the past two decades an abundant literature has accumulated and six major conferences have been held to discuss the uses and effects of modified atmospheres in protecting the quality of commodities (Donahaye et al. 1997, 2001; Navarro and Donahaye, 1993; Champ et al., 1990; Ripp, 1984; Shejbal, 1980). An excellent review on CA technology and theory is also given by Bell and Armitage (1992).

This chapter gives a brief review of literature related to the CA technology and discusses its usefulness from a biological and engineering viewpoint.

2 HERMETIC STORAGE

In many parts of the world, grain is stored in airtight pits or containers. The natural respiration of damp grain and fungi consumes the O_2, which falls from near 21% in air to 1 or 2% and produces CO_2, raising it from about 0.035% in air to near 20%. This nearly anaerobic environment kills insects and mites and prevents aerobic fungi from growing. Occasionally, if the grain is too moist, yeasts and lactic acid-forming bacteria will multiply. Generally high levels of CO_2 and depleted O_2 maintain the quality of stored grain for long periods (White and Jayas, 1992; Banks, 1981).

3 CONTROLLED ATMOSPHERES

Nitrogen gas atmospheres have been used successfully in numerous fumigations, notably in Italy (Shejbal et al., 1973; Shejbal, 1980; Tranchino, 1980; Tranchino et al., 1980). Initial guidelines for the use of N_2 on stored cereals were developed in Australia in the early 1970s (Banks and Annis, 1977). The nitrogen is removed from air with large commercial membrane technology units or applied from compressed gas tanks. Nitrogen is purified by cryogenic separation of liquified air or from compressed air by pressure-swing absorption (carbon acting as a molecular sieve) (Bell and Armitage, 1992). The main drawback to N_2 fumigation is that structures must be completely airtight. Any gas leakage allows O_2 to enter the structure, causing the fumigation to fail.

The production of combustion gases from portable internal combustion engines burning clean propane or methane can be an effective fumigation technique. The introduced gas contains less than 1% O_2, close to 20% CO_2, and the balance N_2 (Storey, 1980; McGaughey and Akins, 1989). These gases can be produced as needed to flush the bulk grain and maintain low O_2 levels, even in leaky bolted grain bins if a plastic sheet is placed over the surface of the grain bulk (McGaughey and Akins, 1989). The gas generator may operate on the principle of catalytic conversion (Navarro et al., 1979) or open-flame combustion (Storey, 1973).

Carbon dioxide fumigation is generally favored because moderate levels of CO_2 will kill insects even at relatively high O_2 levels because of (a) complex physiological processes in the insects (Nicolas and Sillans, 1989), and (b) dessication because spiracles remain open and water loss cannot be regulated (Mellanby, 1934; Jay et al., 1971; Navarro and Calderon, 1974; Jay and Cuff, 1981).

Unlike a standard phosphine fumigation, which is a rapid neurotoxin, for which about 2000 ppm are initially generated from solid tablets, gradually decreasing over several days through bin leakage (levels must remain above 100 ppm, 0.01% for 3 days), CO_2 must be maintained at high levels. About 60% CO_2 seems near-optimal but lower levels (~20%) are effective if exposure time is long enough (weeks). The product of gas concentration and time is critical in all fumigations (Bond, 1984).

CAs can be introduced into a granary from the top or bottom, but because CO_2 is 1.3 times heavier than air it is often best to apply it from the top of the bin, as a compressed gas (snow) or dry ice. The CO_2 levels can be monitored by gas detection tubes (e.g., Draeger tubes), gas chromatography (White et al., 1990), or with portable infrared analyzers (e.g., Gow-Mac Instrument Co., Bound Brook, NJ).

Sheeted bag stacks (rice or paddy) have been successfully treated with CO_2 in southeast Asia (Annis and van Someren Greve, 1984; Annis, 1990). To be successful, a thorough glue seal must be made between ground sheet and canopy. Portable flexible fumigation enclosures are also now available (Donahaye and Navarro, 1990).

4 FACTORS AFFECTING CA

Increases in CO_2 concentration at a constant relative humidity and temperature usually produce increased insect mortality with decreased exposure time. The effectiveness of CO_2 is reduced with decreasing temperature (Spratt et al., 1985; White et al., 1988) and increasing relative humidity (Pearman and Jay, 1970). Responses to high CO_2 (Table 1) and low O_2 (Table 2) among insect species vary considerably (Krishnamurthy et al., 1986).

The toxicity of CO_2 for a particular insect species depends on the life stage present (i.e., its rate of respiration and surface area/volume ratio), behavior of the insect in response to gas gradients, and slow sublethal effects of intoxication.

5 INSECT BEHAVIOR

Various stored-product beetles move in response to gradients of CO_2 or O_2 (Willis and Roth, 1954; Shejbal et al., 1973; Navarro et al., 1981; White et al., 1993). Such locomotory behavior could have a major effect on pest control in nonairtight structures if some insects were able to move to areas where they could survive the fumigation (Navarro et al., 1981).

The rusty grain beetle is relatively susceptible to CO_2 toxicity; adults are the most tolerant life stage (Shunmugam et al., 1993). Adult *C. ferrugineus* are also attracted to moist grain (Loschiavo, 1983) or rotting grain (White and Loschiavo, 1988), partially because they are facultative fungivores (Loschiavo and Sinha, 1966).

Adults placed in the center of horizontal columns move equally in both directions in controls, but are attracted to the higher levels of CO_2 under gas concentration gradients

Table 1 Time in Days to Kill Stored-Product Insects Using Different Concentrations of Carbon Dioxide in Air, Temperatures 20–29°C

			CO_2 (%)				
Family	Species	Stage[a]	20	40	60	80	100
Cucujidae (beetle)	*Cryptolestes ferrugineus*	e	—	2	1	—	—
		l	—	1	1	—	—
		p	—	3	2	—	—
		a	—	6–13	2–4	<3	<2
	Oryzaephilus mercator	e	—	—	—	—	2.5
		l	—	—	—	—	1.5
		p	—	—	—	—	2.5
		a	—	—	—	—	0.5
	O. surinamensis	a	<14	<14	<3	<3	>1
Bostrichidae (beetle)	*Rhyzopertha dominica*	e	>4	—	—	—	—
		a	7.0	4.0	1.5	1.5	2.0
Curculionidae (beetle)	*Sitophilus granarius*	e	—	—	—	—	9.5
		l	—	—	—	—	9.0
		p	—	—	—	—	17.8
		a	>22.0	6.5	6.5	2.5	5.0
Curculionidae (beetle)	*S. oryzae*	e	15.5	4.5	3.5	3.5	7.5
		l	>14.0	>7.0	3.0	2.0	7.0
		p	>24.0	8.5	6.0	8.5	20.0
		a	7.0	3.0	1.5	1.0	4.5
Tenebrionidae (beetle)	*Tribolium castaneum*	e	>4.0	2.0	<2.0	<2.0	2.5
		l	>16.5	16.5	<7.0	<7.0	2.0
		p	—	—	3.0	<5.0	4.0
		a	>5.0	>14.0	1.5	3.0	3.0
	T. confusum	e	—	—	—	—	2.5
		l	—	—	—	—	1.5
		p	—	—	—	—	6.5
		a	9.0	3.0	3.5	2.0	1.0
Dermestidae (beetle)	*Trogoderma granarium*	e	—	14.0	3.0	—	—
		l	>25.0	>25.0	>25.0	16.0	11.5
		p	—	—	5.5	5.0	3.5
		a	—	13.0	2.5	—	—
Phycitidae (moth)	*Plodia interpunctella*	e	2.5	3.0	—	—	—
		l	>2.0	>1.5	—	—	1.5
		p	—	<3.0	<3.0	<3.0	<3.0
		a	—	<7.0	<7.0	<7.0	—
	Ephestia cautella	e	—	—	—	—	2.2
		l	—	—	<5.0	—	1.5
		p	—	—	<3.0	—	2.0
		a	—	—	<2.0	—	0.5
Anobiidae (beetle)	*Lasioderma serricorne*	e	—	<2.0	<1.0	5.5	—
		l	—	5.0	3.0	3.0	—
		p	—	6.0	4.0	3.5	—
		a	—	2.0	1.5	2.0	—
Phycitidae (moth)	*Ephestia kuehniella*	e	—	—	—	—	<1.5
		l	—	—	—	—	<6.5
Acaridae (mite)	*Tyrophagus putrescentiae*	e	—	—	—	—	>5.0
		l	—	—	—	—	0.5
		n	—	—	—	—	0.5

[a] Stages: e, egg; l, larva; p, pupa; a, adult; n, nymph.
Source: Annis, 1986.

Table 2 Time in Days to Kill Stored-Product Insects at Various Low Concentrations of O_2, Temperatures 20–29°C

			O_2 (%)			
Family	Species	Stage[a]	0.0	1.0	2.0	3.0
Cucujidae (beetle)	*Cryptolestes ferrugineus*	a	<2.0	—	—	—
	Oryzaephilus mercator	e	2.0	—	—	—
		l	1.0	—	—	—
		p	2.0	—	—	—
		a	0.5	—	—	—
	O. surinamensis	a	<1.0	—	—	—
Bostrichidae (beetle)	*Rhyzopertha dominica*	e	—	>4.0	—	—
		a	2.0	>4.0	>4.0	>4.0
Curculionidae (beetle)	*Sitophilus granarius*	e	9.0	—	—	—
		l	4.5	—	—	—
		p	8.5	—	—	—
		a	5.0	16.0	17.0	>17.0
Curculionidae (beetle)	*S. oryzae*	e	9.0	<7.0	<7.0	14.0
		p	20.0	>14.0	>14.0	>14.0
		a	4.5	8.5	>21.0	>21.0
Tenebrionidae (beetle)	*Tribolium castaneum*	e	2.5	1.5	3.0	4.0
		l	1.5	6.5	>14.0	>14.0
		p	4.0	>3.0	—	—
		a	1.5	6.0	>14.0	>14.0
	T. confusum	e	2.5	—	—	—
		l	1.0	—	—	—
		p	6.5	—	—	—
		a	4.5	>7.0	>7.0	>7.0
Dermestidae (beetle)	*Trogoderma granarium*	l	12.0	—	—	—
		p	4.0	—	—	—
Phycitidae (moth)	*Plodia interpunctella*	e	1.5	3.0	3.0	>4.5
		l	1.5	>4.0	>4.0	>4.0
		p	3.0	3.0	6.0	>7.0
		a	1.0	<7.0	12.5	>14.0
	Ephestia cautella	e	1.5	1.5	—	—
		l	1.0	0.5	—	4.0
		p	<2.0	1.0	—	—
		a	0.5	0.5	—	—
Gelechiidae (moth)	*Sitotroga cerealella*	l	—	3.5	—	—
		p	—	4.5	—	—
		a	<1.0	1.0	—	—
Tenebrionidae (beetle)	*Tenebrio molitor*	l	<1.0	—	—	—

[a] Stages: e, egg; l, larva; p, pupa; a, adult.
Source: Annis, 1986.

ranging from 3 to 37% (White et al., 1993). More than half the insects were dead at 72 h in CO_2 levels above 31%. Because added CO_2 accumulates in lower regions of grain bulks (Alagusundaram et al., 1993), storage operators may be able to use the positive geotropism of adult *C. ferrugineus* and the tendency of the insect to be attracted to high-CO_2 levels to assist in the control of this pest with controlled atmospheres.

6 SUBLETHAL EFFECTS ON INSECTS

Insects are occasionally exposed to elevated, but nonlethal, concentrations of CO_2 either because of CO_2 fumigation in nonairtight grain bins, which can take 3 weeks or more and may not be effective (Alagusundarum et al., 1993), or because of the natural increase in CO_2 from the metabolism of insects, molds, and grain in hot spots. Natural CO_2 levels between 2 and 3% by volume in air (which is 100-fold higher than the ambient air levels) have been found throughout infested nonairtight granaries (Sinha et al., 1986). Levels ranging from 5 to 18% have been reported in localized areas because of insect and microfloral respiration in wet and warm grain (White and Sinha, 1980; White et al., 1982).

Little is known about the effects of CO_2 levels in the range of 1–20% on stored-product beetles. Maximizing the speed of pest control with higher CO_2 levels (>60%) at 20–29°C has been the aim of most researchers in the past (Annis, 1986). Oxygen levels of 3% are not lethal to *T. castaneum* in 30 days, indicating a high tolerance for low O_2 concentrations (Annis and Dowsett, 1993).

Stored-grain insects and their various life stages are affected differentially by CO_2. These effects can be both physiological (Annis, 1986; Krishnamurthy et al., 1986) and behaviorial (Navarro et al., 1982; White et al., 1993).

The effect of concentrations of CO_2 that can be produced by biological respiration (7.5–19.2%) on oviposition of adult *Tribolium castaneum*, flat grain beetles, *Cryptolestes pusillus* (Schonherr), or *C. ferrugineus* was determined. Relative to controls, *T. castaneum*, *C. pusillus*, and *C. ferrugineus*, exposed to 7.5% CO_2 for 1 week, had numbers of offspring reduced by 43, 94, and 50%, respectively, and the total population at 6 weeks was reduced 53, 84, and 19%, respectively. With CO_2 levels of 17.1%, or higher, for 1 week, no offspring were produced, and exposed adults had a high mortality. Eggs and subsequent immatures of the confused flour beetle, *Tribolium confusum* J. du Val, *T. castaneum*, or *C. ferrugineus* were exposed for 3 weeks to elevated levels of CO_2 at 22°C. Insect development was similar to 7.5 and 8.6% CO_2, with a mean mortality of 43, 62, and 30% greater than controls for *T. confusum*, *T. castaneum*, and *C. ferrugineus*, respectively. Also, mean levels of 5.8–8.3% CO_2 for 7 weeks reduced, on all sampling dates, populations of *T. confusum* by 85%, *T. castaneum* by 99%, *C. pusillus* by 68%, and *C. ferrugineus* by 54%. Although *T. castaneum* had a greater oviposition rate than *C. pusillus* at 7.5% CO_2, immature mortality was greater for *T. castaneum* (White et al., 1995b).

7 DEVELOPMENT OF TOLERANCE OR RESISTANCE TO CA

Bond and Buckland (1979) investigated the response of the granary weevil, *Sitophilus granarius* (L.) to selection under high CO_2 atmospheres. The exposure of adults to 42% CO_2 for seven successive generations, and to 75% CO_2 for four successive generations produced insects with 3.3- and 1.8-fold increases in their resistance to the elevated CO_2. Navarro et al. (1985) reported on the development of resistance to CO_2-rich atmospheres among the adults of *S. oryzae*. Two groups of insects were exposed to 40% CO_2 in air for seven successive generations, and to 75% CO_2 in air for ten successive generations, respectively, at 26°C and 99% relative humidity (RH). The generations of insects subjected to selection pressure were compared with unexposed insects for their tolerance factor (LT_{95} selected generation/LT_{95} nonselected generation; LT_{95} = lethal time in hours for 95% mortality). The results indicated that *S. oryzae* has the genetic potential to develop resistance to CO_2-rich atmospheres. The tolerance factor at the seventh generation (under 40% CO_2) was 2.15 and that at the tenth generation (under 75% CO_2) 3.34. Reduction of relative humidity to 60% and augmentation of O_2 concentration to 21% at these CO_2 levels did not markedly alter the tolerance factor, indicating that the tolerance in these insects was largely due to the action of CO_2. Removal of selection pressure for five generations in the 40% CO_2 group and for four generations in the 75% CO_2 group resulted in a significant reduction in their tolerance to the CO_2-rich atmosphere, indicating that the strains obtained were not completely isogenic. Donahaye (1990a) exposed 40 generations of *T. castaneum* to atmospheres containing 65% CO_2, 20% O_2, and 15% N_2 at 95% RH, and found a resistance factor of 9.2 to CO_2. Removal from controlled atmosphere environments from the 13th to 21st generations resulted in only a 17% decrease in resistance to CO_2. Although resistance can occur to some extent, it is unlikely to occur under normal storage conditions when O_2 levels are reduced, relative humidity is lower than 95%, and successive generations of isolated populations would not be exposed to CO_2. Selection also resulted in slight resistance to low-O_2 environments (Donahaye, 1990b).

8 PHYSICAL ASPECTS OF CO_2 FUMIGATION

8.1 Diffusion of CO_2 Through Bulk Grain

The determination of the diffusion coefficient of one gas through another gas or one gas through interspaces of granular materials is frequently required in many engineering applications. A new transient method to determine a diffusion coefficient was developed. The method has been applied to determine the diffusion coefficient of CO_2 through wheat bulks. The newly determined diffusion coefficient, 4.11 mm^2/s, compares well with the previously published value, 4.15 mm^2/s, considering the possible differences in this biologically and physically variable material (Singh, Jayas et al., 1985).

The diffusion coefficient of carbon dioxide through barley bulks was determined using a transient method at three grain moisture contents (12.5, 15.0, and 18% wet mass basis), three temperatures (15°, 25°, and 40°C), two porosities (48 and 54%), two directions of gas flow (upward and downward), two kernel orientations (horizontal and vertical), and three initial gas concentrations (20, 40, and 60%). The diffusion equation was solved using Crank–Nicolson's implicit finite difference method. The diffusion coefficient of CO_2 through barley bulks was in the range of 5.06–8.36 mm^2/s. The diffusion coefficient decreased with increases in temperature and porosity. The diffusion coefficient was greater for downward flow than for upward flow and was greater for flow parallel to the major axes of the kernels (horizontal) than flow perpendicular to the major axes of the kernels (vertical). The diffusion coefficient did not change with initial gas concentration (Shunmugam et al., 1993).

8.2 Convective Pore Velocity of CO_2

The phenomena that affect transport of CO_2 include natural convection, diffusion, sorption (adsorption and absorption) or desorption, and forced distribution (Banks, 1990). These phenomena are influenced by many factors, including external environmental factors, properties of the grain through which the gas is moving, and properties of the gases introduced into the storage. Interactions that are complex and not yet fully understood occur among these factors.

A mathematical model used for predicting the gas flow rates, owing to natural transport phenomena, must incorporate the contributions made by each individual transport process involved. Alagusundaram et al. (1996a,b) developed a three-dimensional diffusion model for CO_2 in grain bins. This model predicted lower concentrations of CO_2 than the concentrations measured in experimental bins. The noninclusion of the movement of interstitial air by convection currents within the grain bulk or leaks in the storage structure might have caused these deviations.

The movement of carbon dioxide gas through a variety of grain bulks subjected to controlled temperature differences was observed. The CO_2 concentration, as it varied with time, was used in a finite-difference, diffusion–convection model to determine an average convective-pore velocity of CO_2 through grain bulks. The average convective-pore velocity increased as the temperature difference increased when the CO_2 was introduced into the grain bulk from the bottom surface. The average convective pore-velocity was not affected by the temperature difference when the CO_2 was introduced into the grain bulk from the top surface. The average convective-pore velocity was smaller for canola than for two cereal grains (barley and wheat) or lentils when the CO_2 was introduced into the grain bulk from the top and bottom surfaces. The lowest average convective-pore velocity was 4.30×10^{-6} m/s for canola at a temperature difference of 20°C when CO_2 was introduced into the grain bulk from the bottom surface. The largest average convective-pore velocity was 2.42×10^{-3} m/s for wheat at a temperature difference of 40°C when CO_2 was introduced into the grain bulk from the top surface (Bundus et al., 1996).

9 CO_2 SORPTION

Grains and oilseeds in storage adsorb CO_2 at varying rates. This adsorption can cause partial vacuums in sealed bins and affect levels of intergranular CO_2 available to kill insects (Cofie–Agblor et al., 1995).

Carbon dioxide uptake by wheat, barley (hulled, hulless), oats (hulled, hulless), and canola increased asymptotically with increasing exposure time at all test temperatures and initial CO_2 concentrations (Fig. 1). The calculated amount of CO_2 sorbed at equilibrium decreased linearly with increasing temperature from 20 to 30°C at both initial concentrations. The sorption of CO_2 was most rapid in canola at 20°C, attaining sorption equilibrium in 24 h compared with 98% of the equilibrium amount sorbed in 144 h in oats at 20°C. At similar moisture contents, CO_2 sorption was greatest in hulless oats at all temperatures and initial concentrations and the maximum amount of CO_2 sorbed was 492 mg/kg hulless oats at 12% moisture content and 20°C, whereas sorption was least in hulless barley and hulless oats at 18% moisture content and 30°C. For wheat and hulless oats, sorption decreased with increasing moisture content from 12 to 18% (Fig. 2), whereas for hulless barley it increased initially from 12 to 14.8% moisture content and then decreased as moisture content increased. At 25°C, the amounts of CO_2 sorbed at equilibrium by both hulless barley and barley were not significantly different, but the amounts sorbed at equilibrium by hulless oats and oats were significantly different (Cofie–Agblor et al., 1998).

10 GAS LOSS FROM BOLTED METAL BINS

The walls of bolted bins are not gastight even if they are initially caulked. Also, roofs are not airtight. To fumigate these bins, a plastic sheet of polyvinylidene chloride (PVDC)

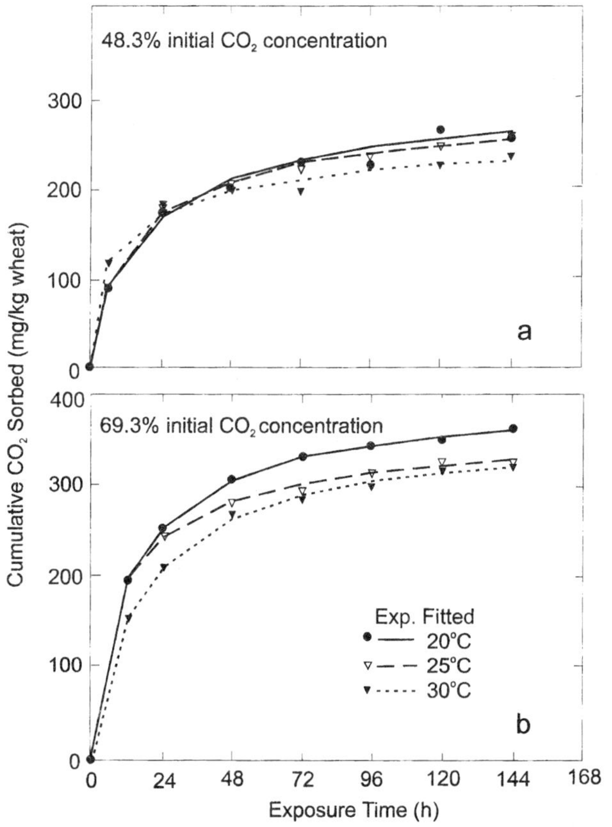

Fig. 1 The effect of initial gas concentration and grain temperature on sorption of CO_2 by hard red spring wheat at 14% moisture content. Experimental values are symbols, fitted curves are lines. (From Cofie-Agblor et al., 1998.)

that is impervious to CO_2 diffusion must be placed over the top of the grain bulk (Alagusundaram et al., 1995a). Using two published mathematical models (The Lawrence Berkeley Laboratory model; Sherman and Grimsrud, 1980) and the Banks and Annis (1984) model, the rate of CA gas loss from a granary can be very crudely estimated (Peck et al. 1994). There are many variables that must be subjectively estimated that make the models unacceptable without further refinement.

By using pressure tests and CO_2 gas loss measurements, Lukasiewicz et al. (1999) demonstrated that the seams between steel wall plates could be effectively sealed with caulking compound. Pressure decay tests were based on the recommendations of Banks and Annis (1984) that a structure needs a gas-loss half-life of at least 5 min for CO_2 (i.e., 2500–1250 Pa in 5 min).

11 SEALING BINS

Even when efforts are made to seal a bolted-metal granary and CO_2 is added daily, incomplete insect control is obtained (Alagusundaram et al., 1995b) (Fig. 3).

Experiments were conducted in two 5.56-m–diameter farm bins to determine the mortality of caged adult rusty grain beetles under elevated CO_2 concentrations. The bins

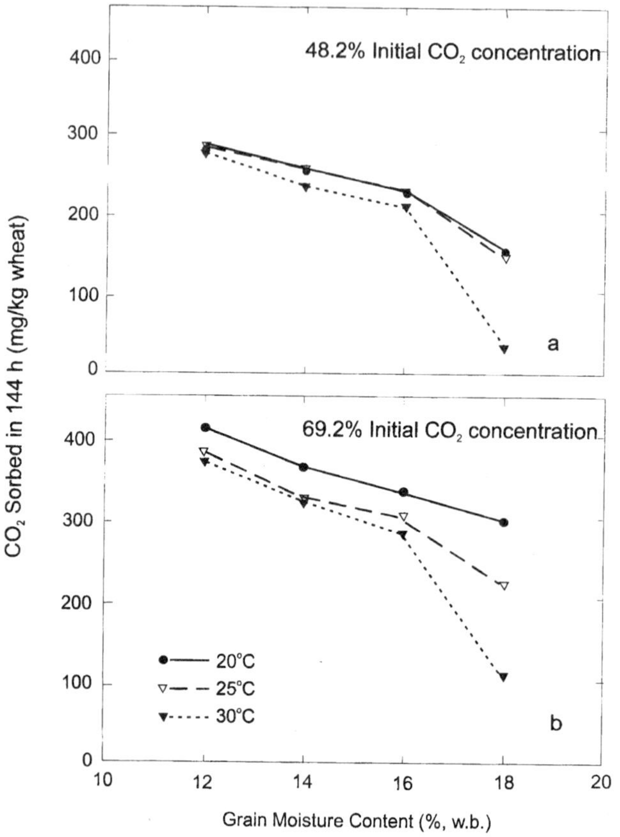

Fig. 2 Combined effects of grain temperature and moisture content on the CO_2 sorption by wheat exposed to two initial CO_2 concentrations for 144 h. Experimental values are symbols, fitted curves are lines. (From Cofie–Agblor et al., 1997.)

were filled with wheat to a depth of 2.5 m. Dry ice was used to create high CO_2 concentrations in the wheat bulks. Two different modes of application of dry ice were used: (a) pellets on the grain surface and in the aeration duct, and (b) pellets on the grain surface and blocks in insulated boxes on the grain surface. The pellets exposed to the ambient conditions on the grain surface and in the aeration duct sublimated quickly and had to be replenished at frequent intervals. Dry ice blocks in insulated boxes, however, maintained high CO_2 concentrations without replenishment for longer than 15 days. In both modes of application, the observed CO_2 concentrations in the intergranular gas were about 15 and 30% at 2.05 m and 0.55 m above the floor, respectively. At 0.55 m above the floor, the mortality of rusty grain beetle adults was more than 90%, whereas in the top portions of the bulk (2.05 m above the floor) the mortality was only 30% (see Fig. 3). On average about two-thirds of the insects were killed (see Fig. 3).

Mann et al. (1997a) demonstrated that welded steel hopper bins could be effectively made airtight, with a pressure-relief valve, and recirculating pump to keep CO_2 levels uniform. Similar sealing techniques were used at the bottom and top of the bin (Figs. 4 and 5).

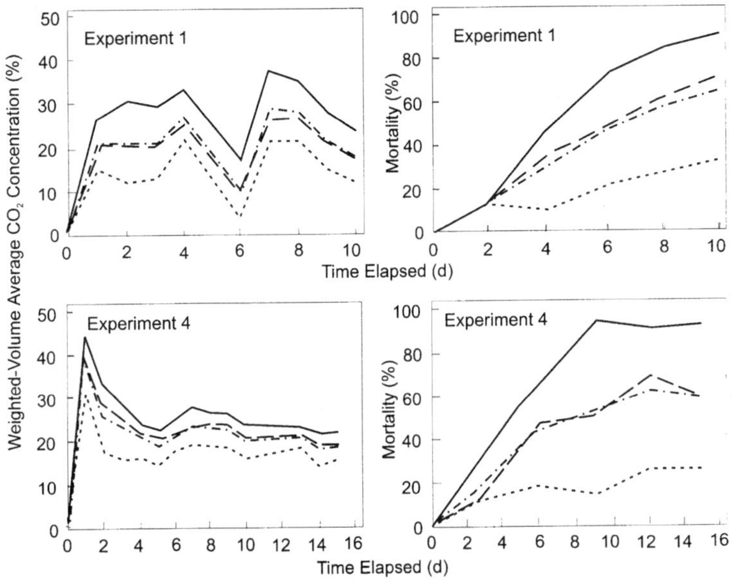

Fig. 3 Changes in weighted–volume average CO_2 concentration with time (left) and mortality of *C. ferrugineus* with time (right) when dry ice pellets were introduced on the grain surface and in the aeration duct (experiment 1) and when dry ice pellets were introduced on the grain surface and dry ice blocks in insulated boxes were placed on the grain surface (experiment 4): solid line, 0.55 m; dashed line, 1.3 m; periods, 2.05 m above the floor; and dashed–dotted lines, average for whole bin. (From Alagusundaram et al., 1995a,b.)

Fumigation of grain with carbon dioxide requires the maintenance of high CO_2 concentrations (20–60% by volume in air) for extended periods (>4 days). Grain storage structures currently used on Canadian farms are not airtight and should be sealed if they are to be used for fumigation with CO_2. Various sealing techniques were applied to two welded-steel hopper bins to improve their gastightness. Each sealing technique was tested by adding 136 kg of dry ice pellets to a steel box connected to the empty bin by ABS

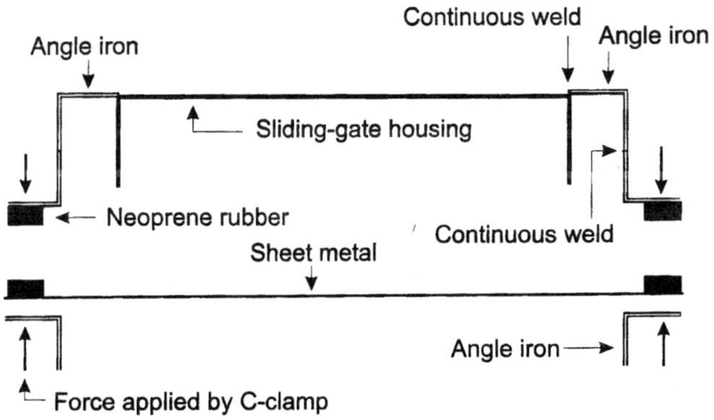

Fig. 4 Schematic of the sealing method used to seal the opening in the bottom cone of welded-steel hopper bins. (From Mann et al., 1997a.)

Fig. 5 A view of the bottom cone opening of the welded-steel hopper bin after being sealed. (From Mann et al., 1997a.)

piping. If the CO_2 displaced air, this would have created a CO_2 concentration of approximately 65% inside the empty bin. Carbon dioxide concentrations continually increased as the sealing improved, but never reached 65%. Uniformity of CO_2 within the bin in both the radial and vertical directions also improved as the sealing improved. Fumigating with CO_2 using this type of bin will be possible if either (a) the length of exposure at the lower CO_2 concentration is increased, or (b) additional dry ice is added to the bin to compensate for leakage.

By using this sealing technique, rusty grain beetles in 80 tons of wheat were controlled in 10 days at 13, 16, or 20°C with CO_2 levels remaining near 40% (Mann et al., 1997b). The cost of retrofitting the bin was about 320 dollars Can. The cost of dry ice was about 85 dollars, which is equivalent to the cost of phosphine gas that would be used on a grain bulk of this size.

Controlled atmospheres have been used successfully in concrete primary elevators. Manitoba Pool Elevators in Starbuck, Manitoba with Praxair Ltd. supplying CO_2, fumigated 209 metric ton bins (280 m^3). Initially there was a 25% gas loss from the bin. Then a gas impermeable plastic was used to seal the bottom gate, which provided an effective seal, but was time-consuming to install; it also created the problem of removal in filled bins, where the plastic could become entangled in the horizontal-unloading auger. To overcome this, a new gate design was implemented, which proved to be gastight and was operated with an external lever. The new design reduced the CO_2 makeup portion to 5%.

Typically less than 1 metric ton of CO_2 was used for each batch fumigation which took 4 h to attain 80–100% concentrations at the top gas-sampling location. The cost of this procedure is very competitive with traditional methods of grain fumigation excluding initial equipment and capital; a week of CO_2 application costs less than a penny-per-bushel (Chekerda, 1997).

The only drawback to using CO_2 in concrete elevators is that extended use can cause significant carbonation with gradual exposure of reinforcing steel, which may rust and weaken (Banks and McCabe, 1988), although this problem was considered minor by the authors.

12 RAILROAD HOPPER CARS

Extensive efforts to make railroad hopper cars airtight for fumigation of grain with CO_2 were unsuccessful even in a stationary car. The top fiberglass hatches and bottom hoppers require design modifications to make the cars suitable for CA (Mann et al., 1997c).

13 COMMERCIAL USE OF CA

Australians were the original pioneers in the use of CA for grain treatment. Large percentages of farm bins are now constructed airtight (rolled steel), with pressure-relief valves for either standard phosphine or CO_2 fumigation. At one time in the mid-1980s a warehouse holding 300,000 metric ton of wheat was regularly treated with CO_2 piped from a nearby industrial park (Kwinana, Western Australia). About 350 metric tons of CO_2 were placed in the 300,000-metric–ton shed and recirculated; 5 extra tonnes were added per day. Insects were controlled in 10-days (Mills and White, 1984; Ripp, 1984).

Currently, bins with CA capability are being installed in a newly constructed grain terminal elevator run by Cargill Ltd. and Alberta Wheat Pool in Vancouver, British Columbia, Canada.

The use of CA in grain storage can be practical and cost-effective if supplies of CO_2 are readily available and essentially airtight structures are available. The most feasible structure present on farms or at primary elevators is a welded steel hopper bin that has been retrofit with appropriate seals, pressure-relief valve, and recirculation pump.

ACKNOWLEDGMENT

We thank R. Sims, Cereal Research Centre, AAFC, Winnipeg, Manitoba, for preparing the figures.

REFERENCES

Alagusundaram K, Jayas DS, Muir WE, White NDG, Sinha RN (1993). Distribution of introduced carbon dioxide in farm granaries: an experimental investigation. Am Soc Agric Eng, St. Joseph, MI. Paper 93-6019. 20 pp.

Alagusundaram K, Jayas DS, Muir WE, White NDG, Sinha RN (1995a). Distribution of introduced carbon dioxide through wheat bulks contained in bolted metal bins. Trans Am Soc Agric Eng 38:895–901.

Alagusundaram K, Jayas DS, Muir WE, White NDG, Sinha RN (1995b). Controlling *Cryptolestes ferrugineus* (Stephens) adults in stored wheat in bolted-metal bins using elevated carbon dioxide. Can Agric Eng 37:217–223.

Alagusundaram K, Jayas DS, Muir WE, White NDG (1996a). The convective diffusive transfer of carbon dioxide through stored grain bulks—a finite element model. Trans Am Soc Agric Eng 39:1505–1510.

Alagusundaram K, Jayas DS, Muir WE, White NDG, Sinha RN (1996b). Finite element model of three-dimensional carbon dioxide movement in grain bins. Can Agric Eng 38:75–82.

Annis PC (1986). Towards rational controlled atmosphere dosage schedules: a review of current knowledge. In: Proceedings of an International Working Conference on Stored Product Protection. 4th Donahaye E, Navarro S, eds. Maor-Wallach, Jerusalem, pp 128–148.

Annis PC (1990). Sealed storage of bag stacks: status of the technology. In: Fumigation and Controlled Atmosphere Storage of Grain. Champ BR, Highley E, Banks HJ, eds. Australian Cent Intern Agricultural Research Procedures. No. 25.

Annis PC, Dowsett HA (1993). Low oxygen disinfestation of grain: exposure periods needed for high morality. In: Proceedings of an International Conference on Controlled Atmospheres and Fumigation in Grain Storages. Navarro S, Donahaye E, eds. Caspit Press, Jerusalem, pp 71–83.

Annis PC, van Someren Greve J (1984). The use of carbon dioxide for quality preservation in small sheeted bag stacks. In: Controlled Atmospheres and Fumigation in Grain Storages. Ripp BE, et al, eds. Developmental Agricultural Engineering, vol 5. Elsevier, Amsterdam.

Banks HJ (1981). Effects of controlled atmosphere storage on grain quality: a review. Food Technol Aust 33:335–340.

Banks HJ (1990). Behaviour of gases in grain storages. In: Fumigation and Controlled Atmosphere Storage of Grain. Champ BR, Highley E, Banks HJ, eds. Proc 25, Austr Cent Intern Agric Res Canberra, Australia.

Banks HJ, Annis PC (1977). Suggested procedures for controlled atmosphere storage of dry grain. CSIRO Aust Div Entomol Tech Paper 13. 23 pp.

Banks HJ, Annis PC (1984). Importance of processes of natural ventilation to fumigation and controlled atmosphere storage. In: Controlled Atmospheres and Fumigation in Grain Storages. Ripp BE, ed. Elsevier, Amsterdam.

Banks HJ, McCabe JB (1988). Uptake of carbon dioxide by concrete and implications of this process for grain storage. J Stored Prod Res 24:183–192.

Bell CH, Armitage DM (1992). Alternative storage practices. In: Storage of Cereal Grains and Their Products. Sauer DB, ed. American Association of Cereal Chemistry, St. Paul, MN, pp 249–311.

Bond EJ (1984). Manual of Fumigation for Insect Control. FAO Plant Product Protein Publ 54. 432 pp.

Bond EJ, Buckland CT (1979). Development of resistance of carbon dioxide in the granary weevil. J Econ Entomol 72:770–771.

Bundus CL, Jayas DS, Muir WE, White NDG, Ruth D (1996). Average convective-pore velocity of carbon dioxide gas through grain bulks. Can Agric Eng 38:91–98.

Champ BR, Highley E, Banks HJ, eds. (1990). Fumigation and Controlled Atmospheres. Storage of Grain. Proc 25, Austr Center Int Agric Res Canberra, Australia.

Chekerda C (1997). Combined effort tests carbon dioxide for grain fumigation. Forum on Stored Grain Products. No. 21, Vol 12 Demianyk CJ, ed. Agr Agri-Food Can, Winnipeg, MB.

Cofie–Agblor R, Muir WE Simicio R, Cenkowski S, Jayas, DS (1995). Characteristic of carbon dioxide sorption by stored wheat. J Stored Prod Res 31:317–324.

Cofie–Agblor R, Muir WE, Jayas DS, White NDG (1998). Carbon dioxide sorption by grains and canola at two CO_2 concentrations. J Stored Prod Res 34:159–170.

Donahaye E (1990a). Laboratory selection of resistance by the red flour beetle, *Tribolium castaneum* (Herbst), to a carbon dioxide-enriched atmosphere. Phytoparasitica 18:299–308.

Donahaye E (1990b). Laboratory selection of resistance by the red flour beetle, *Tribolium castaneum* (Herbst), to an atmosphere of low oxygen concentration. Phytoparasitica 18:189–202.

Donahaye E, Navarro S (1990). Flexible PVC liners for hermetic or modified atmosphere storage of stacked commodities. In: Fumigation and Controlled Atmosphere Storage of Grain. Champ BR, Highley E, Banks HJ, eds. Proc. 25 Aust Center Intern Agric Res, Canberra, Australia.

Donahaye EJ, Navarro S, Vanarva A. (1997). Proceedings of the International Conference on Controlled Atmosphere and Fumigation in Grain Storages. Nicosia, Cyprus, Printco Ltd., Nicosia, Cyprus.

Donahaye EJ, Navarro S, Leesch JG. 2001. Proceedings of the International Conference on Controlled Atmosphere and Fumigation in Grain Storages. Fresno, CA, Executive Printing Services, Clovis, CA.

Fields PG, White NDG (1997). Survival and multiplication of stored-product beetles at simulated and actual winter temperatures. Can Entomol 129:887–898.

Fields PG, van Loon J, Dolinski, MG, Harris JL, Burkholder WE (1993). The distribution of *Rhyzopertha dominica* in western Canada. Can Entomol 125:317–328.

Jay EG, Cuff W (1981). Weight loss and mortality of three life stages of *Tribolium castaneum* (Herbst) when exposed to four modified atmospheres. J Stored Prod Res 17:117–124.

Jay EG, Arbogast RT, Pearman GC (1971). Relative humidity: its importance in the control of stored-product insects with modified atmospheric gas concentrations. J Stored Prod Res 6:325–329.

Krishnamurthy TS, Spratt EC, Bell CH (1986). The toxicity of carbon dioxide to adult beetles in low oxygen atmospheres. J Stored Prod Res 22:145–151.

Loschiavo SR (1983). Distribution of the rusty grain beetle (Coleoptera: Cucujidae) in columns of wheat stored dry with localized high moisture content. J Econ Entomol 76:881–884.

Loschiavo SR, Sinha RN (1966). Feeding, oviposition, and aggregation by the rusty grain beetle, *Cryptolestes ferrugineus* (Coleoptera: Cucujidae) on seed-borne fungi. Ann Entomol Soc Am 59:578–585.

Lukasiewicz M, Jayas DS, Muir WE, White NDG (1999). Gas leakage through wall seams of bolted-metal bins. Can Agric Eng 41:65–71.

Madrid FJ, White NDG, Loschiavo SR (1990). Insects in stored cereals, and their association with farming practices in southern Manitoba. Can Entomol 122:515–523.

Mann DD, Jayas DS, White NDG, Muir WE (1997a). Sealing of welded steel hopper bins for carbon dioxide fumigation of stored grain. Can Agric Eng 39:91–97.

Mann DD, Jayas DS, Muir WE, White NDG (1997b) Conducting a successful CO_2 fumigation in a welded-steel hopper bin. Presented at the American Society of Agricultural Engineering, St. Joseph, MI, paper 97-6064.

Mann DD, Waplak S, Jayas DS, White NDG (1997c). Potential of controlled atmospheres for insect control in hopper cars. Presented at American Society of Agricultural Engineering, St. Joseph, MI, paper RRV97-304.

McGaughey WH, Akins RG (1989). Application of modified atmospheres in farm grain storage bins. J Stored Prod Res 25:201–210.

Mellanby K (1934). The site of water loss from insects. Proc R Soc Ser B 116:139–149.

Mills JT (1989). Spoilage and heating of stored agricultural products. Prevention, detection and control. Agric Agri-Food Canada Publ 1823E, Canadian Government Publication Centre, Ottawa, ON.

Mills JT, White NDG (1984). Improved procedures for controlling insects in stored grains. Foreign Visit Report on Science and Technology. Australia. Agric Agri-Food Canada Publ IST84-13.

Mills KA (1983). Resistance to the fumigant hydrogen phosphide in some stored product species associated with repeated inadequate treatments. Ges Allg Angew Entomol 4:98–101.

Navarro S, Calderon M (1974). Exposure of *Ephestia cautella* (Wlk.) pupae to carbon dioxide concentrations at different relative humidities: the effect on adult emergence and loss in weight. J Stored Prod Res 10:237–241.

Navarro S, Donahaye E, eds. (1993). Proc Int Conf Controlled Atmospheres and Fumigation in Grain Storages. Caspit, Jerusalem.

Navarro S, Gonen M, Schwartz A. (1979). Large-scale trials on the use of controlled atmospheres for the control of stored grain insects. In Proc 2nd Int Work Conf Stored Prod Entomol, Manhattan, KS, pp 260–270.

Navarro S, Amos TG, Williams P (1981). The effect of oxygen and carbon dioxide gradients on the vertical dispersion of grain insects in wheat. J Stored Prod Res 17:101–107.

Navarro S, Dias R, Donahaye E (1985). Induced tolerance of *Sitophilus oryzae* adults to carbon dioxide. J Stored Prod Res 21:207–213.

Nicolas G, Sillans D (1989). Immediate and latent effects of carbon dioxide on insects. Annu Rev Entomol 34:97–116.

Pearman GC, Jay EG (1970). The effect of relative humidity on the toxicity of carbon dioxide to *Tribolium castaneum* in peanuts. J Ga Entomol Soc 5:61–64.

Peck M, Jayas DS, White NDG (1994). Prediction of gas loss from bolted metal bins caused by changing environmental conditions. Presented at the American Society of Agricultural Engineering, St. Joseph, MI. Paper 94-6039.

Ripp BE (1984). Modification of a very large grain store for controlled atmosphere use. In: Controlled Atmosphere and Fumigation in Grain Storages. Ripp BE, ed. Elsevier, New York, pp 281–292.

Ripp BE, Banks HJ, Bond EJ, Calverly DJ, Jay EG, Navarro S, eds. (1984). Controlled Atmosphere and Fumigation in Grain Storages. Elsevier, Amsterdam.

Shejbal J, ed. (1980). Controlled Atmosphere Storage of Grains. Elsevier, Amsterdam.

Shejbal J, Tonolo A, Careri G (1973). Conservation of wheat in silos under nitrogen. Ann Technol Agric 122:773–785.

Sherman MH, Grimsrud DT (1980). Measurement of infiltration using fan pressurization and weather data. In: 1st AIC Conference, Air Infiltration Instrumentation and Measuring Techniques, Berkshire, UK, pp 279–322.

Shunmugam G, Jayas DS, White NDG (1993). Effects of controlled atmospheres on all life stages of the rusty grain beetle. J Appl Zool Res 4:114–117.

Singh (Jayas) DS, Muir WE, Sinha RN (1985). Transient method to determine the diffusion coefficient of gases. Can Agric Eng 27:67–72.

Sinha RN (1995). The stored grain ecosystem. In: Stored-Grain Ecosystems. Jayas DS, White NDG, Muir WE, eds. Marcel Dekker, New York.

Sinha RN, Waterer D, Muir WE (1986). Carbon dioxide concentrations associated with insect infestations of stored grain. I. Natural infestation of corn, barley, and wheat in farm granaries. Sci Aliments 6:91–98.

Spratt E, Dignan G, Banks HJ (1985). The effect of high concentrations of carbon dioxide in air on *Trogoderma granarium* Everts (Coleoptera: Dermestidae). J Stored Prod Res 21:41–46.

Storey CL (1973). Exothermic inert-atmosphere generators for control of insects in stored wheat. J Econ Entomol 66:511–514.

Storey CL (1980). Mortality of various stored product insects in low oxygen atmospheres produced by an exothermic inert atmosphere generator. In: Controlled Atmosphere Storage of Grain. Shejbal J, ed. Elsevier, Amsterdam.

Tranchino L (1980). Economic aspects of nitrogen storage of grains. In: Controlled Atmosphere Storage of Grains. Shejbal J, ed. Elsevier, Amsterdam.

Tranchino L, Agostinelli P, Constantini A, Shejbal J (1980). The first Italian large scale facilities for the storage of cereal grains in nitrogen. In: Controlled Atmosphere Storage of Grains. Shejbal J, ed. Elsevier, Amsterdam.

White NDG, Jayas DS (1993). Quality changes in grain under controlled atmosphere storage. In: Proc Int Conf Controlled Atmosphere and Fumigation in Grain Storages. Navarro S, Donahaye E, eds. Caspit, Jerusalem, pp 205–214.

White NDG, Leesch JG (1996). Chemical control. In: Integrated Management of Insects in Stored Products. Subramanyam Bh, Hagstrum DW, eds. Marcel Dekker, New York, pp 287–330.

White NDG, Loschiavo SR (1988). Effects of localized regions of high moisture grain on efficiency of insect traps capturing adult *Tribolium castaneum* and *Cryptolestes ferrugineus* in stored wheat. Tribolium Inf Bull 28:97–100.

White NDG, Sinha RN (1980). Changes in stored wheat ecosystems infested with two combinations of insect species. Can J Zool 58:1524–1534.

White NDG, Barker PS, Demianyk CJ (1995a). Beetles associated with grain captured in flight by suction traps in southern Manitoba. Proc Entomol Soc Man 51:1–11.

White NDG, Jayas DS, Muir WE (1995b). Toxicity of carbon dioxide at biologically producible levels to stored product beetles. Environ Entomol 24:640–647.

White NDG, Jayas DS, Sinha RN (1988). Interaction of carbon dioxide and oxygen levels, and temperature on adult survival and multiplication of *Cryptolestes ferrugineus* (Coleoptera: Cucujidae) in stored wheat. Phytoprotection 69:31–39.

White NDG, Jayas DS, Sinha RN (1990). Carbon dioxide as a control agent for the rusty grain beetle (Coleoptera: Cucujidae) in stored wheat. J Econ Entomol 83:277–288.

White NDG, Sinha RN, Muir WE (1982). Intergranular carbon dioxide as an indicator of biological activity associated with the spoilage of stored wheat. Can Agric Eng 24:35–42.

White NDG, Sinha RN, Jayas DS, Muir WE (1993). Movement of *Cryptolestes ferrugineus* (Coleoptera: Cucujidae) through carbon dioxide gradients in stored wheat. J Econ Entomol 86: 1846–1851.

Willis ER, Roth LM (1954). Reactions of flour beetles of the genus *Tribolium* to carbon dioxide and air. J Exp Zool 127:117–152.

11

Grain-Milling Operations

ASHOK K. SARKAR

Canadian International Grains Institute, Winnipeg, Manitoba, Canada

1 EVOLUTION OF GRAIN MILLING

The first step between postharvest of cereals and finished, processed cereal-based foods is the milling operation. The history of milling operations is very rich and as old as human civilization. Historians have considered the level of milling technology as an indicator of the development of a given civilization at that point in time.

Figure 1 provides a chronological illustration of the development of all forms of grain-milling equipment and the time periods during which they were developed and were operational. Although the milling tools and the processes have changed radically, the basic purpose of milling remains essentially the same—to improve palatability and digestibility of the grain.

The development of the grain-milling operation in its present form is primarily because of the following reasons:

1. Economies of scale
2. Need for improved product quality
3. Wider product range

Most grain-milling facilities built today would include a high degree of automation allowing process monitoring and control through the application of programmable logic controllers with computer interface. On-line quality monitoring, a futuristic proposition only 10–15 years ago, is now more commonplace.

2 OVERVIEW OF GRAIN-MILLING PROCESS

There are a multitude of cereal-based processed food products available in the marketplace nowadays. The nature of these food products partly influences the milling requirements

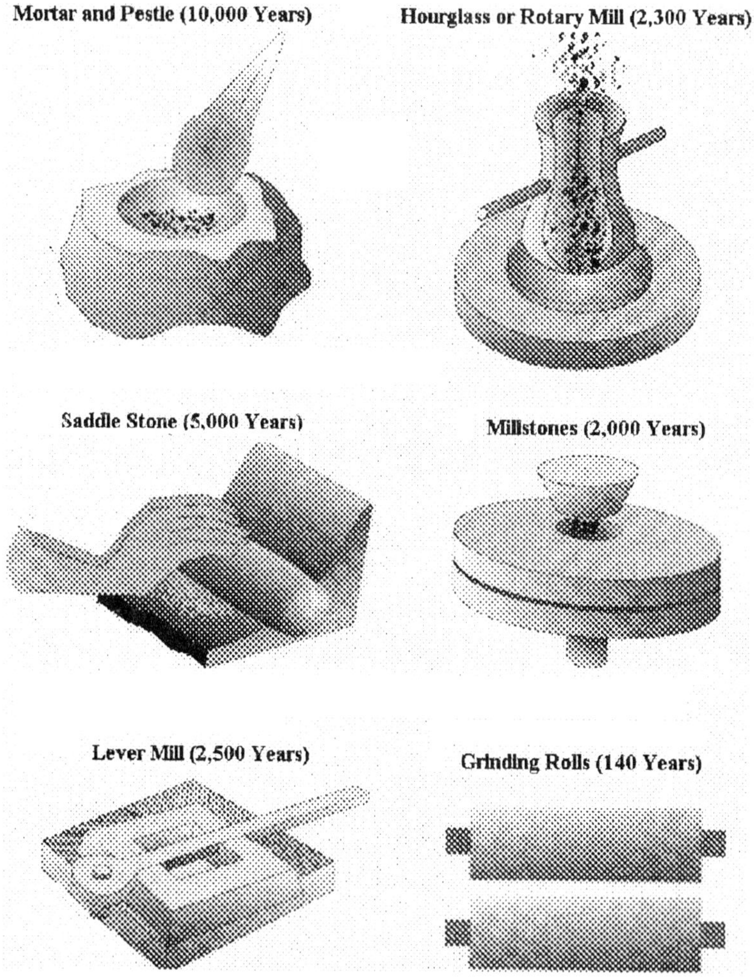

Fig. 1 Chronologic development of grain-milling equipment.

specific to the product quality. Different cereal grains pose unique challenges to the miller and each possess unique quality attributes in meeting requirements to produce widely differing end products. Thus, the milling process for each of these ranges between being somewhat different, to barely having any resemblance to each other. Nevertheless, despite this there are certain processing elements that are common to all, and these follow:

1. All the incoming grain is cleaned, inspected, and stored according to quality before milling.
2. Preparation of grain by tempering or conditioning to facilitate removal of hulls, bran, and germ from the endosperm.
3. If it is received unhulled, sometimes this preparation of the grain before further milling or processing may require dehulling.
4. Careful removal of the hulls, bran, or germ from the endosperm in the actual milling process, either by dehulling or pearling, and polishing the grain (oats,

Table 1 Basic Processing Steps and End Products in Grain Milling

Cereal grains	Wheat	Corn	Oats	Barley	Rice
Storage bin assignment and inventory control	Grain/intake and storage	Grain/intake and storage	Grain/intake and storage	Grain/intake and storage	Grain/intake and storage
Grain preparation	Cleaning, tempering, or conditioning	Cleaning, tempering, or conditioning	Cleaning, hulling; aspiration steaming and drying	Cleaning	Cleaning, drying Steeping, steaming, drying
Processing	Break system grading, purification, sizing system, reduction system, flour dressing	Degermination, drying, cooling, aspiration, roller milling, sifting and aspirations	Cutting, steaming, and rolling screening and cooling	Pearling, aspiration, grading, grinding, sifting, and parsifying	Shelling, paddy separation, hulling, polishing, aspiration, grading, coating
End products	Straight, grade flour, patent flour Clear flour Whole wheat flour	Cereal grits Brewers' grits Cornmeal Corn flour	Oat groats, regular rolled oats Oat bran Oat flour	Pot barley Pearled barley Barley semolina Barley flour	Coated milled rice, second heads rice Screenings rice Brewers' rice

barley, and rice), or by a series of gradual grinding and sifting to release the pure endosperm in either granular form (corn and durum wheat) or fine powder form as in meal and flour (rye and wheat).
5. Physical or chemical treatment to enhance product quality.
6. Packaging, storage, and handling of finished products.

Table 1 highlights the basic steps involved in the processing of the various types of grain and their end products.

3 WHEAT FLOUR MILLING

Common wheat is milled into flour for the production of various types of leavened and unleavened bread products; oriental noodles, steam buns, and various other types of Asian foods; cakes, cookies, crackers, and a host of other food products, in addition to various industrial processes including starch and gluten production, and so on. Common wheat flour and granular semolina, also referred to as *farina* in North America, are also used for the production of pasta products in some parts of Latin America. Generally speaking, for such purposes durum wheat is the most suitable and, therefore, is the most commonly milled. From milling and end product quality requirement perspectives, all types of wheat may be categorized into three groups:

Hard wheat
Soft wheat
Durum wheat

The modern wheat flour-milling operation can be divided into six steps:

1. Wheat reception and storage
2. Preparation of wheat for milling
3. The milling process
4. Flour collection and treatment
5. Generating finished products and their handling
6. By-products of milling and their utilization

The schematic flowchart (Fig. 2) shows the entire operation from beginning to end. These steps are essentially common to milling of all types of wheat; however, there are variations in milling that are specific to each of the three groups of wheat. Therefore, the milling of soft wheat and durum wheats are covered separately in the latter section of this chapter. The description of the six steps here primarily apply to the milling of hard wheat.

3.1 Wheat Reception and Storage

The primary function at this stage is to weigh, preclean, and store a sufficient quantity of wheat according to quality so that the mill will have a dependable supply of suitable quality of wheat or wheat mix necessary for flour production of the required quality. The total storage capacity, the range of finished flour quality requirements, and sourcing of wheat domestically as well as from overseas will determine the size and number of bins required. For example, mills producing only one or two types of flour and with an abundance of domestic wheat supply easily accessible would obviously require fewer bins and a small storage capacity, as compared with those mills that are dependent on wheat from

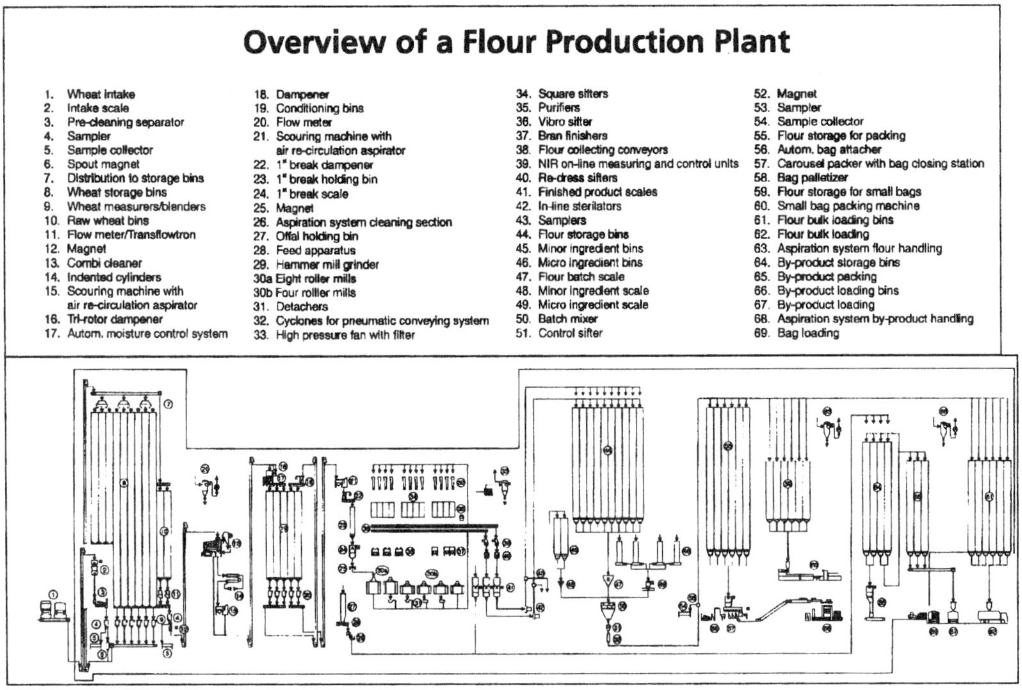

Fig. 2 Schematic flow diagram of a flour mill. (Courtesy of Bühler, Inc.)

overseas and that also are required to produce a variety of flours for a wide range of applications.

Each mill assesses its optimum storage capacity based on its location and unique requirements to ensure a constant supply of suitable wheats to maintain production while avoiding storage of excessive quantities to prevent carrying charges.

The level of inspection and evaluation of quality of incoming wheat is again dependent on the source of wheat and how rigid the quality requirement is. For instance, a flour mill in the United Kingdom receiving wheat from various local sources would have wheat samples from every truck quickly assessed for basic quality with analyses for protein, the falling number (a measure for α-amylase activity), moisture, test weight, and other tests appropriate to ascertain wheat soundness and freedom from kernel damage (e.g., heat damage). The wheat is then discharged and directed to appropriate bins to ensure that wheats of similar quality are stored together and that there is no undesirable mixing of wheat varieties suitable for bread with varieties meant for cakes and cookies. A rapid method of analysis using near-infrared reflectance spectroscopy (NIR) permits very prompt testing for contents of protein and moisture. Larger mills may use more quality parameters to identify the quality of the wheat being preserved. This practice enables the miller to be more precise in formulating wheat grists (wheat mixes) that will yield flour of the required quality at the least cost. Mills in the United Kingdom also blend wheat from other European Union (EU) countries with wheats from other countries, such as Canada and the United States with their own domestic wheat varieties. Therefore, the wheat blending can be a very complex part of wheat preparation for milling because frequently there are several wheat grists with which a mill has to work, and different wheats that can vary considerably in terms of quality and in price. Therefore, proper segregation of wheat during storage is a very critical first step in the profitable operation of a flour mill.

A flour mill in Canada by contrast would not have to conduct extensive tests before unloading the wheat on arrival at the mill. Some basic tests will be carried out before the wheat is used, but the tests required for identity preservation before binning is rendered redundant because grain is bought on a guaranteed basis of class, grade, and protein segregation wherever applicable.

Looking now at the process of grain intake, precleaning, and storage, the objectives here are as follows:

- High throughput
- Accurately weighing the wheat and thus ensure the recording of any discrepancy between the quantity ordered and that received
- Basic cleaning to remove:
  - Large objects that may block the bin outlets or damage downstream equipment
  - Fine material and dust that could hamper the flow of wheat, may harbor infestation, cause respiratory problems, cause malfunctioning of equipment components, such as bearings, and possibly contribute to dust explosions
  - Identity preservation of all wheat received

The process of wheat intake begins as the grain is dumped into a pit that has a metal grate on top to prevent any large chunks or oversized foreign material, such as wood, metal, stones, or other), from entering the precleaning system. The hopper of the pit channels the wheat through a chute to a conveying element that brings the wheat to a weigh scale. The chute between the hopper and the conveying element, which most commonly is a bucket elevator, is usually a slide that may be adjusted to regulate the flow of grain through the intake process. Also provided at this point is a magnet that will prevent any tramp ferrous metal to pass through. Underneath the weighing scale there is generally a surge bin equipped with low- and high-level indicators. As the level in the bin goes down below the low-level, the process starts, and when it goes above the high-level the process stops. This ensures a constant flow of grain through the system. The grain is next fed into a grain separator which removes coarse, large-sized foreign material, as well as fine material such as dirt, sand, and lighter foreign material, such as chaff, husk, and such. The grain separator consists of two screens: the one on top has a larger opening than the wheat size, whereas the one at the bottom has finer perforations to allow only the very fine material, such as dust, sand, and the like, to pass through. The grain falls through the top screen and goes over the bottom screen into an aspiration channel. A feeding mechanism ensures that the grain is fed uniformly and evenly across the width of the channel in a thin layer to allow effective exposure of the entire material to the airstream for exceptionally searching removal of the lighter foreign material from the heavier sound wheat kernels. Often the separator is preceded by a magnet. The emphasis here is more on throughput and overall basic cleaning. The grain after being precleaned is sent to another bucket elevator that transports the grain to the top of the silo bins where it is dropped on a distributing chain conveyor for feeding into the individual bins. Air-operated slide gates are installed for specific bin selection from a remote point.

The schematic diagram of a modern intake system is shown in Fig. 3. The diagram shows a bucket elevator, which receives wheat from a rail car or truck, and takes it to a weighing scale. The scale is followed by a separator with an aspiration channel.

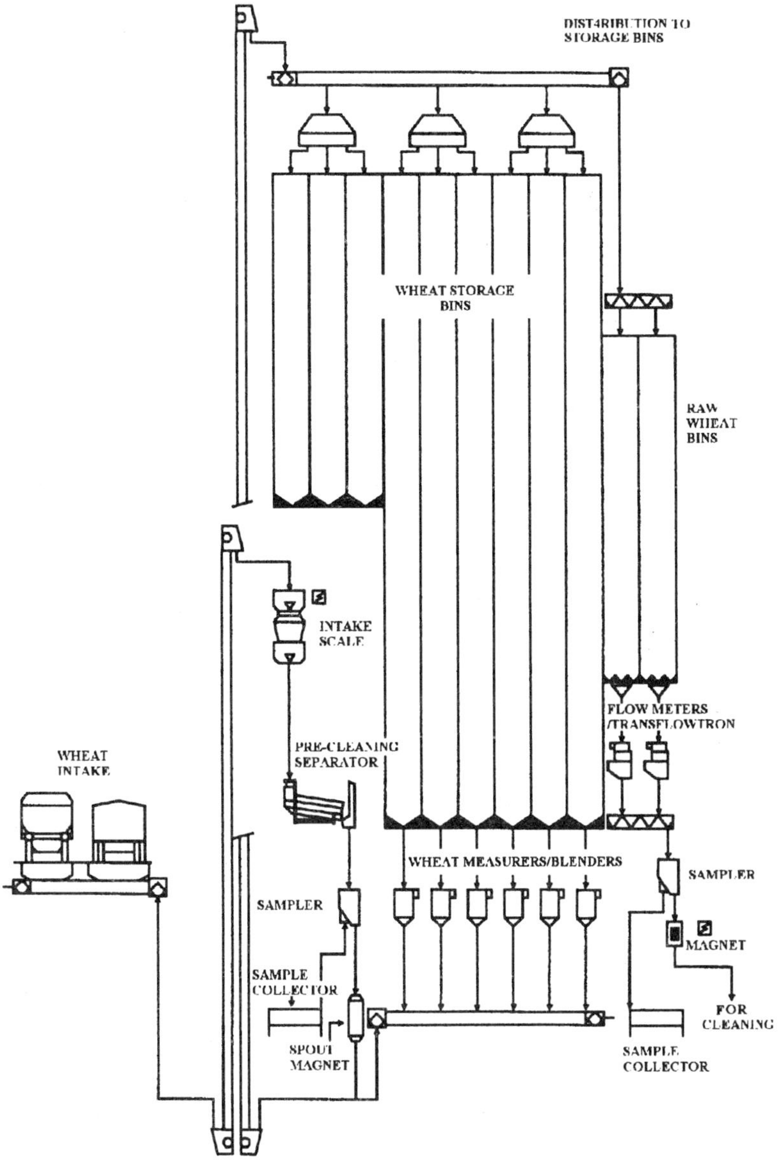

Fig. 3 Wheat intake and storage. (Courtesy of Bühler, Inc.)

To ensure even flow of the grain in a manner of first-in and first-out of the bin there is usually a multiple outlet under each bin. Generally, at this point, bins are discharged for one of the following reasons:

1. To supply a single type of wheat or to blend various types of wheat for preparation of wheat grists–mixes of desired quality for the cleaning house or screen-room as it might be referred to by some millers.
2. To turn the wheat over from one bin to another to aerate, cool, and dry the

grain for protecting quality and safeguarding against any potential infestation, and subsequent heating of the grain, either because of infestation or because of the presence of moisture higher than what is considered safe for storage for a prolonged period. A maximum grain moisture content of 14% is considered safe for storage in most tropical and subtropical conditions.

3.1.1 Metering Devices

There are two types of metering devices that can be used to draw wheat out of the bins and they are as follows:

1. Devices that measure according to volume
2. Devices that measure according to weight

The first category includes a simple graduated slide and volumetric measurers, whereas the second category would have devices such as automated weighing scales and flow balancers that work on the basis of weight.

3.1.1.1 Volumetric Devices

3.1.1.1.1 SLIDES. Simple graduated slides are most commonly used in the milling plants in North America. The purpose of using such slides is to regulate the quantity of wheat that may be allowed to flow through the system without choking or overloading the equipment and conveying elements. Second, to regulate the quantity of wheat from various bins in varying proportions to arrive at a desired blended wheat quality. Generally the slides work satisfactorily in the North American situation as milling plants here in most situations blend similar wheat types not only in terms of hardness, but also in terms of kernel shape and size.

When blending wheats of widely varying moisture, kernel size and test weight (density) for equal opening of the slides would not necessarily guarantee equal quantities of grain being discharged. Varying levels of moisture content within a given type of wheat would have an influence on its volume and density and thus would affect its flow rate through a given opening. Therefore, in such situations, the use of slides could lead to inaccurate blending proportions.

3.1.1.1.2 VOLUMETRIC MEASURERS. Such measurers are quite commonly used by flour mills around the world. Even though they regulate the quantity according to volume, they are certainly better than slides in terms of adjustment control.

Figure 4 shows a volumetric measurer. The wheat flows in from the top into the

Fig. 4 Volumetric measurer. (Courtesy of Bühler, Inc.)

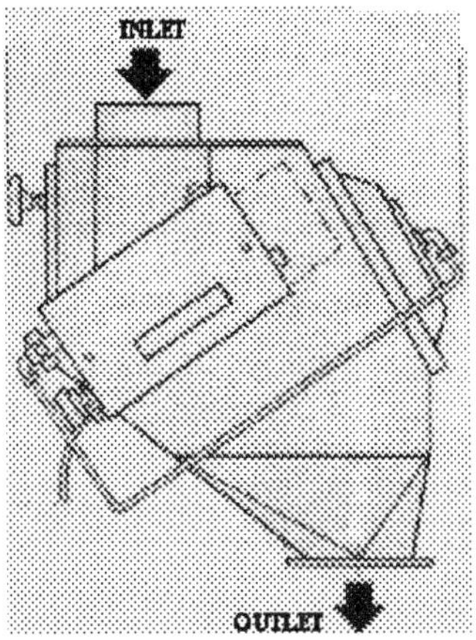

Fig. 5 Flow balancers. (Courtesy of Bühler, Inc.)

back end of the machine. There are rotating pockets, inside the housing that carry the wheat from the back of the machine and drop it down as the pockets are facing down in their rotation. The width of the pockets can be adjusted on the run by turning the wheel that is provided on both sides of the measurer.

These measurers can be split into two parts, each handling a different wheat; therefore, the pocket width adjustment is necessary for both sides. Also provided on each side is a scale that will provide a reading that corresponds to the pocket width. Thus, the ruler scale adjustments on various measurers would reflect its opening, which can be translated in the percentage format to ascertain whether the desired quantities are being added. Since the width of an opening can be precisely monitored by reading the scale, openings can be calibrated for various types of wheat for a relatively higher degree of accuracy.

3.1.1.2 Flow Balancers

This device allows continuous flow of wheat at a certain rate of flow in terms of metric tons per hour. This rate of flow can be controlled by setting the rate of flow required in the weight control section. The scale where the adjustment is made is connected to a balance beam. The wheat is fed through a feed gate, which is set in equilibrium with the controlling weight. Proportioning of the wheat through the feed gate needs to be accomplished to maintain constant throughput as set on the scale. This feed is controlled through the air-operated gate by closing or opening it to maintain the equilibrium. The wheat as it enters through the feed gate hits a plate with a force that is directly proportional to the rate of flow (Fig. 5).

Flow balancers are most commonly used to regulate wheat flow and blending on the basis of weight. They are popular because they are accurate and continuous in operation, unlike older automated weighing systems consisting of a number of synchronized weighing scales, under a corresponding number of bins.

Generally, a battery of flow balancers would have each unit connected to individual bins. Thus, an appropriate wheat mix may be prepared and sent to another set of bins that would have prepared wheat mixes ready for cleaning and tempering operations.

3.2 Preparation of Wheat for Milling

This step of the process can rightfully be split into three parts:

1. Wheat blending to formulate grists or mixes
2. Cleaning the wheat to remove all foreign materials and adhering dirt
3. Tempering or conditioning the wheat to bring it to its optimum milling condition

These operations may not necessarily follow the same order as just outlined. For instance, wheat blending might have been carried out before the cleaning process even begins. This practice is quite common in many parts of the world, including North America. The primary reason for this is that in most situations wheat varieties that are blended have similar degrees of hardness; therefore, these wheats can be cleaned and tempered together. However, most mills in Europe would be milling mixed wheat grists of hard, medium hard, and even soft wheat. In such situations the hard, medium hard, and soft wheat would be cleaned and tempered separately and then blended to obtain optimum performance. The reasons for this will be clarified in the tempering and conditioning section (see Sec. 3.2.8).

3.2.1 Wheat Blending to Formulate Grists or Mixes

Appropriate quality attributes would obviously be essential in the wheat for production of finished flours of required specification. Sometimes all the required properties may not be present in a single wheat type. On the other hand, a given wheat may possess all the necessary quality requirements, but its cost may be prohibitive. Wheat cost is the largest single cost in a bag of flour. Even modest savings made in wheat costs translate into major savings over a long period.

In practical situations, there are times when certain types of wheat are not even available for purchase. There are also many countries (including developed countries such as Switzerland and United Kingdom) that have regulations restricting or imposing tariffs on the importation of wheat to promote or protect the domestic wheat crop industry.

For any of the foregoing reasons it is very common for flour mills to generate several wheat blends to meet their specific quality requirements within given constraints. Therefore, the best solutions are usually unique, with wheat-blending decisions being influenced by the following factors:

Quality
Price
Availability
Policy

Optimization of raw material purchasing and usage is achieved through (a) accurate wheat evaluation, (b) cost minimization of wheat mixes and flour blends using linear programming, (c) projection of future wheat requirements.

Formulation of a grist or wheat mix may be anywhere from being very simple in some cases, to being very complex in others, that require computer application software to perform wheat blend cost minimization using linear programming. The procedure essentially involves evaluation of the quality of all the wheats relative to price and then making the selection that would meet the target quality at least cost. This exercise would obviously take availability and policy—the two other important factors—into account.

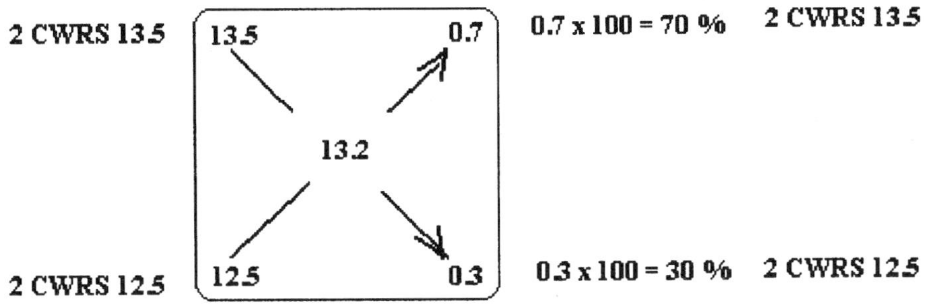

Fig. 6 Pearson square.

The following criteria are important when considering flour quality for a wide range of applications:

Protein content
Protein quality
Starch quality and condition
Presence of enzymes

Most of these criteria can, to some extent, be influenced by the milling process. However, the judicious selection of wheat will have a direct bearing on the overall flour quality relative to these quality criteria. Some of these criteria have been explained in more detail in the latter part of this chapter under ''Functional Properties'' (see Sec. 3.3.10). The following example will best illustrate a simple wheat grist formulation:

A bakers patent flour is to be milled to supply a large mechanized bakery in Canada that requires:

Protein: 12.5%
Ash: 0.50%
Flour with strong protein quality
Farinograph water absorption: 65.0% minimum

Millers in Canada are aware that for flour with strong protein quality and a higher level of water absorption, as indicated in the foregoing, the only class of wheat they need to consider is the CWRS class. This class of wheat is available in three grades and, in addition, the top two grades are segregated according to four levels of protein; for example, if there are only two wheats available for blending (i.e., 2CWRS 13.5 and 2CWRS 12.5).

Let us now look at how these two wheats will be blended to provide flour of the required quality. Use of CWRS wheat only would ensure that strong protein quality and high water absorption requirements will be met. Historical milling data also confirm that flour ash content of 0.50% can be readily achieved when milling the no. 2 grade of CWRS at an approximate flour extraction rate of 75.0%. We also know that flour protein required is 12.5%, and that there is a loss of protein of approximately 0.7% from wheat to flour; therefore, it is clear that the wheat blend protein should be 13.2%.

If we use this information in a Pearson Square, we can calculate the percentage of each of the two wheats required in the blend as shown in Fig. 6.

This method of calculation is very simple. The difference in protein between the targeted wheat blend and individual wheat protein are placed in diagonally opposite corners and multiplied by 100 to obtain a percentage. Thus the solution is that 70% of 2CWRS 13.5 and 30% of 2CWRS 12.5 must be blended to obtain a grist that when milled will produce flour with quality according to specifications. If a cleaning house has a capacity of 10 tons/h, this would mean that the bin-discharging devices such as slides, volumetric measurers or flow balancers must be set to draw 3 and 7 tons/h under the 2CWRS 12.5 and 2CWRS 13.5 bins, respectively.

3.2.2 Cleaning House or Screen Room

During harvesting and postharvest handling and storage of wheat at the farm, it becomes exposed to a wide range of foreign matter comprised of various types of seeds such as oats, barley, wild oats, corn, pea, flax, rapeseed, mustard, wild buckwheat, and such.

During the harvesting and threshing process some soil dirt, stones, mudballs, chaff, husk, straws, and unthreshed wheat also become mixed with the wheat mass. Broken, shrunken, thin and immature wheat kernels (green kernels) are also considered undesirable, as they pose various kinds of problems in quality and processing. There are also various types of grain defects, damages due to the environment, microbial contamination, and infestation that may cause concern to the miller and, in worst case, may even render the wheat unsuitable for processing. With a proper grading system in place, as in major grain-exporting countries, tolerance levels for all of the factors listed as part of the grading factors are established. This is the only way that quality can be protected, for these defects cannot be separated from the wheat. Major exporting countries also have limits set on total foreign material that a particular class and grade of wheat may contain.

Wheat from various sources may contain as little as 0.5% to as high as more than 5% of foreign material. These levels may not include shrunken and broken, or diseased and damaged kernels.

The objective of the cleaning house is very simple: to perform a meticulous removal of all foreign material that comes along with the wheat, including shrunken, broken, and immature kernels, before the wheat is sent for tempering or conditioning.

3.2.3 Principles of Cleaning

When we examine the physical properties of all of the foreign material and compare them individually with the wheat, we would find certain clear distinctions among them. The machine-manufacturing companies have taken advantage of these differences in their physical behavior to isolate these foreign materials from the wheat mass. Table 2 shows a grouping of the foreign material relative to common characteristics for their removal from wheat. The table also shows the principle that has been applied in the machine for removal and as well as the types of equipment that are used.

3.2.4 Cleaning House Equipment

3.2.4.1 Separator

The separator is used for cleaning wheat and in many different types of grain- and seed-cleaning operations. They are available in many different makes and designs. The basic equipment consists of two shaking screens. The top screen, with larger perforations than the wheat size, removes larger impurities while allowing wheat kernels to drop through. Wheat kernels along with finer impurities are then spread on the bottom screen, which has finer perforations than the wheat size, allowing only finer impurities to pass through the screen and be removed separately. Wheat kernels, thus, tail over the bottom screen

Table 2 Cleaning Equipment, Principles of Separation, and Impurities Removed

Properties	Types of foreign material	Equipment in use
Size		
Larger	Unhulled kernels, corn, soybean as well as any material larger than the wheat size.	Grain separators, scalperator, and any other machines that use screens or sieves for separation.
Smaller	Rapeseed, flax, mustard, sand, dust and any material smaller than the wheat size.	
Specific gravity		
Lighter	Barley, oats, wild oats, and ergot	Table separators, gravity selector, and combinator
Denser	Stones, metals, and mudballs	Destoner
Shape		
Shorter	Broken wheat, cockle, and wild buckwheat	Disc separator, indented cylinders and spiral seed separator
Longer	Oats, barley, wild oats, and ergot	
Round	Cockle	
Magnet properties	Ferrous and ferrous alloy objects	Electrical and permanent magnets
Differences in terminal velocity	Chaff, husk, straws, dust, and shriveled kernels	Aspiration channels, air recycling chambers
Friction or abrasion	Outermost bran layer, crease and surface contamination	Horizontal and vertical scourers

and before leaving the machine usually pass through an aspiration channel for the removal of lighter material. One such aspiration channel with an air-recycling feature will be described later. A modern separator "classifier" is shown in Fig. 7.

This machine can be used for a variety of grain, coffee bean, and cocoa bean cleaning and grading operations. This is helped by the fact that the screen slope for a specific application can be selected, also, the stroke and angle of throw of the screen box are adjustable. The equipment is sturdy in design with a very low maintenance and power requirement.

Fig. 7 Grain separator. (Courtesy of Bühler, Inc.)

3.2.4.2 Air-Recycling Aspirator

Aspiration channels are used with many cleaning equipment designs for the removal of dust that comes with the wheat and dust that is created by attrition each time the grain is handled. Additionally, it also removes chaff, husk, lighter seeds, and immature kernels. Aspiration channels are used with most cleaning equipment because no single aspirator can remove all of the material in one single pass. Even if it were possible to do so, there is always some dust created by the continuous handling of the grain. This means that total air requirement for all cleaning equipment and related filtration in the dust collectors will be high; thus, through recycling the fresh air requirement is much less and with the following benefits:

Smaller aspiration trunking (lines)
Reduced filtration area
Low power consumption

3.2.4.2.1 DESIGN AND OPERATION. This type of separator can be used for a wide range of grain, such as, wheat, barley, oats, corn, and rye. The grain enters a vibratory feeder through a feed inlet (Fig. 8). The feeder helps spread the grain across the full width of the aspiration channel and allows its feeding into the channel as a thin layer. As the low-density material (chaff, husk, and dust) and the grain enter the aspiration channel, with a double adjustable wall for higher efficiency of separation, they become loosened as they begin to float. The suspended low-density materials are drawn up through the channel space along with the air currents. The light material and air end up in a separating zone that consists of an expanded chamber where the air loses its velocity and the lighter materials begin to settle down at the bottom. The bottom is narrowed down and has a screw conveyor that collects the settled light material from across the width of this separator and discharges it into an outlet through a retarding gate. The relatively dust-free air is recycled back by a radial fan into the aspiration channel. Because the grain is heavy, it is not lifted by the air currents and begins to fall down and exits the machine through finger valves.

All of the air can be recycled back. The inlet feed hopper to the machine, however, should be kept under exhaust through connecting it to a central exhaust system. This machine can be used on its own or in conjunction with a separator or scourer.

3.2.4.3 Destoner

A destoner removes heavier impurities, such as stones, from relatively lighter material to be cleaned, such as wheat kernels, taking advantage of the differences in their floating characteristics as the principle of separation. In a similar way it can also be applied to clean different types of grain, oilseeds, and beans for removal of heavier impurities, such as, stones, metals, and mud balls (Fig. 9). The grain falls through an inlet on one end of the machine onto an inclined oscillating screen deck. Air currents are uniformly drawn up through the deck to create a fluidized bed of material. Stones being heavier remain in contact with the screen deck and continue to ascend owing to the oscillating movement until they drop out at the top end into a rubber sleeve. After sufficient accumulation of stones, the rubber sleeve opens up and the stones are dropped into a box. The grain to be cleaned, being lighter, is supported on a cushion of air and begins to move up a little bit before falling back toward the material outlet at the lower end.

3.2.4.4 Gravity Tables

Gravity tables can be used for different types of grain cleaning and sorting on the basis of relative differences among densities of the material being handled. The separation prin-

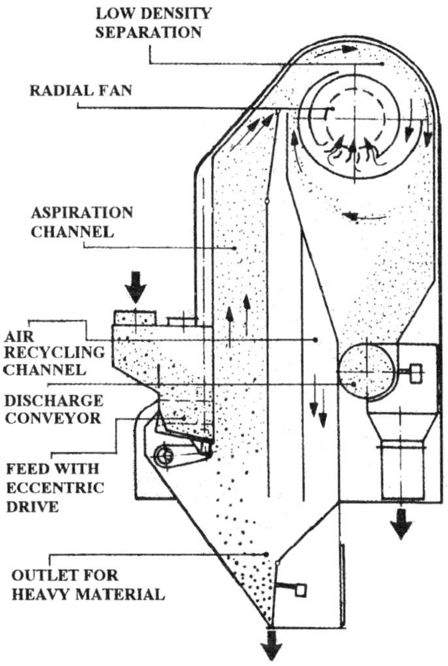

Fig. 8 Air recycling separator. (Courtesy of Bühler, Inc.)

ciple (Fig. 10) is very similar to a destoner. The main difference is that it makes more than two separations on the basis of density: light, mixed, and heavy.

The woven-wired surface deck vibrates at an angle laterally to the gravity table, rather than longitudinally, as in a destoner. The lighter fraction may include barley, oats, ergot, and such, when cleaning wheat. A mixed product may consist primarily of wheat and some lighter material, and the heavy fraction consist of stones. If a destoner has

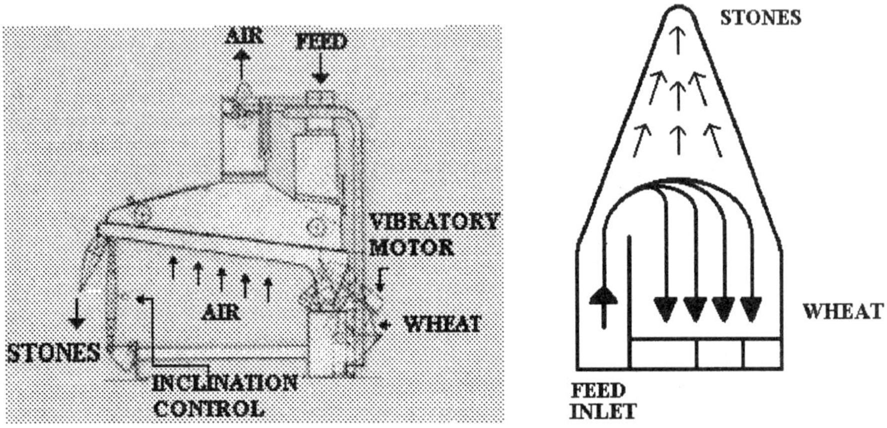

Fig. 9 Destoner. (Courtesy of Bühler, Inc.)

already been used for stone removal, then the machine is adjusted such that the heavier fraction is clean wheat.

3.2.4.5 Disc Separator

This machine consists of a number of discs that are mounted on a shaft that rotates on a horizontal axis (Fig. 11a). These discs rotate at close to 60 rpm. Both sides of the disc surface consists of indented pockets of a certain shape to do a specific separation. For example, when shorter seeds and broken wheat are meant to be lifted from the mass of wheat, the pockets are shorter and smaller than when wheat is supposed to be lifted out of longer seeds, such as barley and oats (see Fig. 11b).

The wheat to be cleaned is fed through an inlet at one end (head end) of the machine at the top. As the wheat falls through the inlet hopper, its flow can be regulated by a slide to guide it to the first few discs as desired. The disc pockets are cut to provide a slight undercut to help with the lifting and hold the lifted material better. The discs pick up the material that fit the shape and size of the pocket, leaving the longer material behind. Those materials that are not lifted are conveyed to the next set of discs through the conveying action of the inclined plates mounted on the spokes of each disc. This way the unlifted

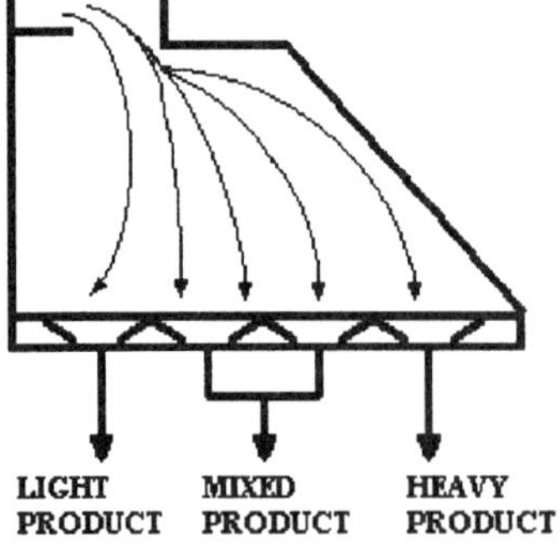

Fig. 10 Operation of a gravity table.

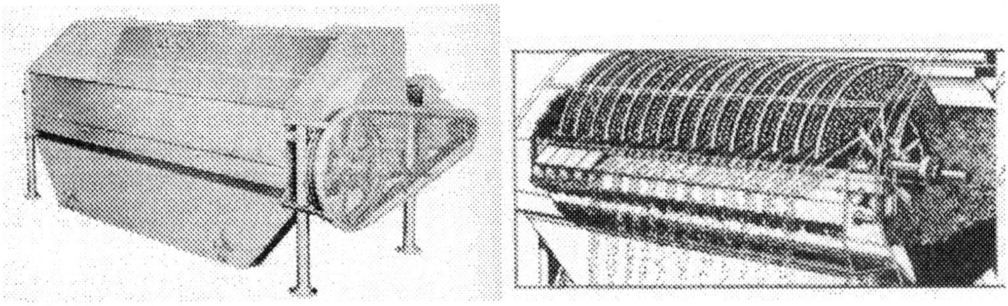

(a)

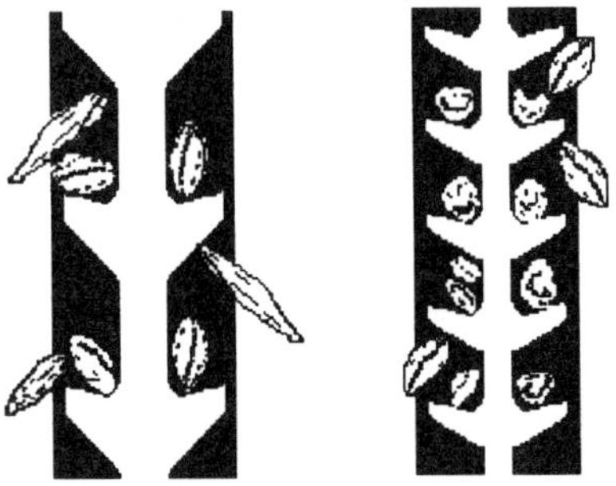

(b)

Fig. 11 (a) Disc separators; (b) longer pockets, shorter pockets. [Courtesy of Carter Day Industries (Canada) Ltd.]

material moves to the very end (tail end) of the machine where there is an opening to allow the material to exit the machine and discharge into a spout. The opening can be controlled by means of an adjustable slide to ensure the level of grain mass in the machine is optimum for efficient operation.

The lifted material is discharged, as the pockets face down in their rotation, into small troughs located between the discs. The lifted material from the troughs can all be allowed to collect in a hopper below, or some of the lifted material, that may require further treatment, can be directed to a small conveyor. This conveyor top is covered with small flaps. The conveyor is located on the edge of the passage of lifted material from the troughs before they roll over and drop in to the hopper below. The last few discs generally have liftings that may not be pure and, therefore, by opening the little flaps corresponding to these liftings allows the material to be conveyed back to the head end of the machine once again for retreatment.

3.2.4.6 Indented Cylinders

Indented cylinders are also used for length separation, as is the disc separator. It removes shorter seeds from the mass of wheat as well as picks up wheat kernels from the mass of longer seeds, such as oats and barley. Instead of using several rotating discs with pockets it uses a long rotating cylinder with indents. The design and the operation of this can be described as follows.

The wheat stream enters the head end of the machine (Fig. 12a) and reaches the indented surface which picks up the materials that fit the shape of the indentations. The material is picked up by the indentations in their upward motion and carried up to a point beyond which the material falls out of the edge of the indented pockets (see Fig. 12b). The material discharged is collected into a built-in trough (see Fig. 12c). The trough runs along the length of the cylinder and remains stationary. The position of the trough, in terms of inclination, can be adjusted from outside the cylinder through a hand wheel and its position of inclination can be observed on a connected indicator. The inclination must be adjusted to obtain an optimum separation. Inside the trough there is a conveyor that conveys the lifted material to the other end of the machine, where it is channeled to exit through a spout. The material that is not lifted and remains on the inside surface of the cylinder also is moved gradually to the other end (tail end) of the machine where it exits through a separate spout. To ensure that the unlifted material is properly spread evenly across the surface for better and effective operation, a spreader is used to spread the material to allow more effective material and surface contact (see Fig. 12c).

Usually, there are two or more cylinders that run together to remove shorter seeds, longer seeds, and for retreatment.

3.2.4.7 Scourer

Cleaning the surface and crease dirt from the grain and removing the outer layer of bran that becomes loosened during the tempering process is carried out by frictional cleaning

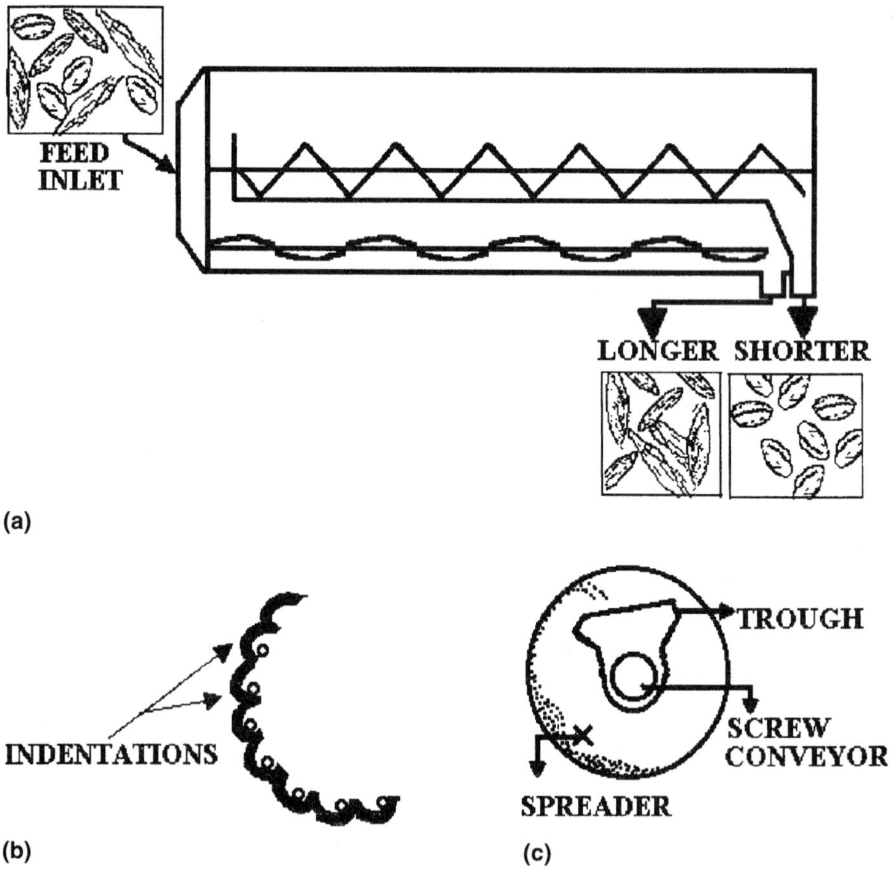

Fig. 12 (a) Indented cylinder; (b) indented pockets; (c) material discharged into trough.

using a scourer. Horizontal scourers are commonly used, although sometime ago vertical scourers were also available. The friction is generated as the grain passes through a wired screen jacket inside which rotating beaters throw and impact the grain against it causing loosened bran layers and surface dirt to detach. The scourings go through the openings of the wire jacket while clean scoured grain passes through an outlet at the other end of the machine. The scoured grain is always aspirated for removal of any remaining finer particles as a result of scouring.

The scourer shown in Fig. 13a, provides a much more comprehensive and efficient scouring action by employing a rather unique approach to scouring. It consists of a segmented rotor with conveying screw and spikes. This rotates within a stationary jacket. The jacket has a number of cast iron spike plates and screen as segments. A thorough scouring function in this machine is accomplished by the following:

Interaction between the rotor spikes and the screen jacket
Interaction of the rotating and stationary spikes
Interaction between the rotating conveying screw and stationary screw

3.2.4.8 The Cleaning House Operation and Flow Diagrams

The flow of a cleaning house is shown in Fig. 14. Figure 14a shows a more conventional type of cleaning house, whereas Fig. 14b shows a modern cleaning house flow.

The sequence of flow of the uncleaned wheat is based on the following reasoning:

1. Remove any foreign material that can damage or cause obstruction in the subsequent equipment in the flow. Remove most of the foreign material as soon as possible so that the downstream equipment will have a lighter load and operate more efficiently. Invariably, the first machine used in the sequence is the grain separator. However, generally speaking, it may be preceded by a magnet. The

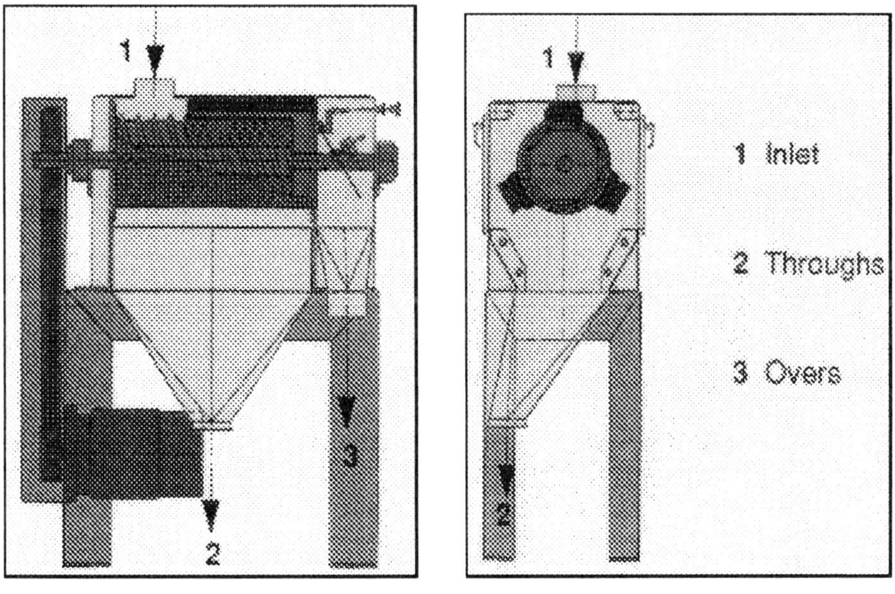

Fig. 13 (a) Scourer; (b) scourer: end view. (Courtesy of Bühler, Inc.)

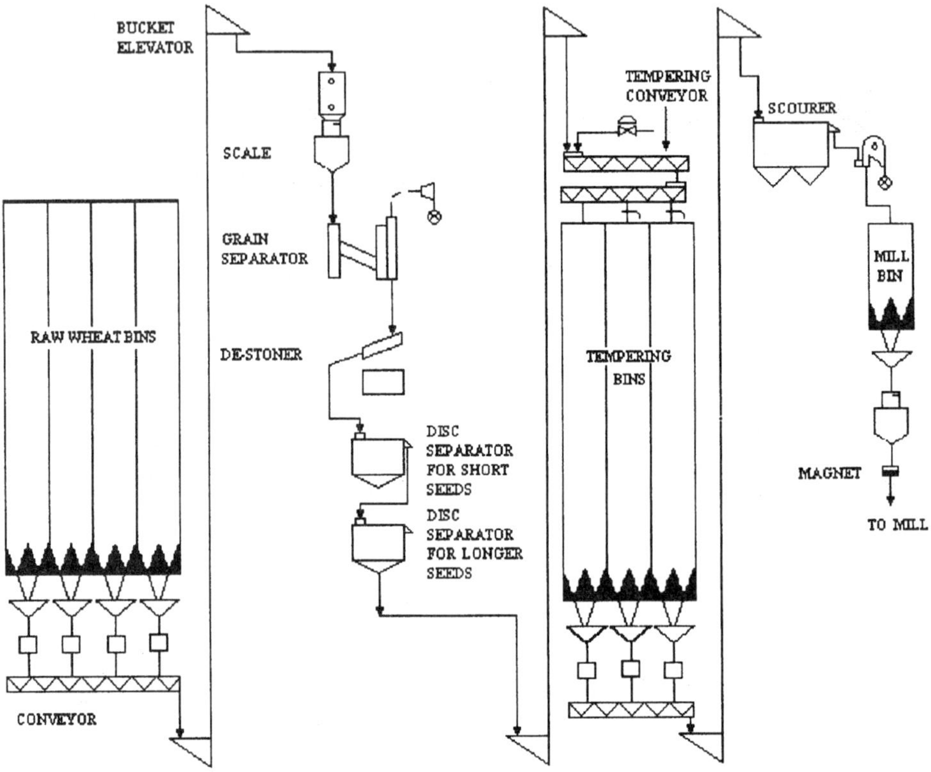

(a)

Fig. 14 (a) A conventional cleaning house;

separator removes the bulk of the impurities (i.e., all the materials that are larger and smaller as well as lighter than wheat).

2. Remove any remaining stones and heavy material such as metal particles (ferrous and nonferrous), of the same size or shape, but heavier than the wheat kernel. If these heavy materials are not removed at this stage, their potential chances to cause equipment damage or wear is much greater. Any large and small heavy material such as big stones and small grit would be removed by the grain separator that precedes the destoner.
3. The next machines that are used in the sequence remove seeds that are both longer and shorter than the length of the wheat kernels. The equipment used here is either disk separators, indented cylinders, or a combination of both. These seeds have cross sections similar to the wheat kernel; therefore, they manage to remain in the wheat mass. Shorter seeds that are removed may contain a significant quantity of broken wheat kernels. Therefore, in some of the cleaning house flow diagrams instead of sending this material to the screenings bin, it is sometimes sent to a spiral seed separator. This permits the broken wheat to be recovered from the other undesirable material such as dark round seeds (cockle) which tend to spiral down the outer groove as they gather speed.

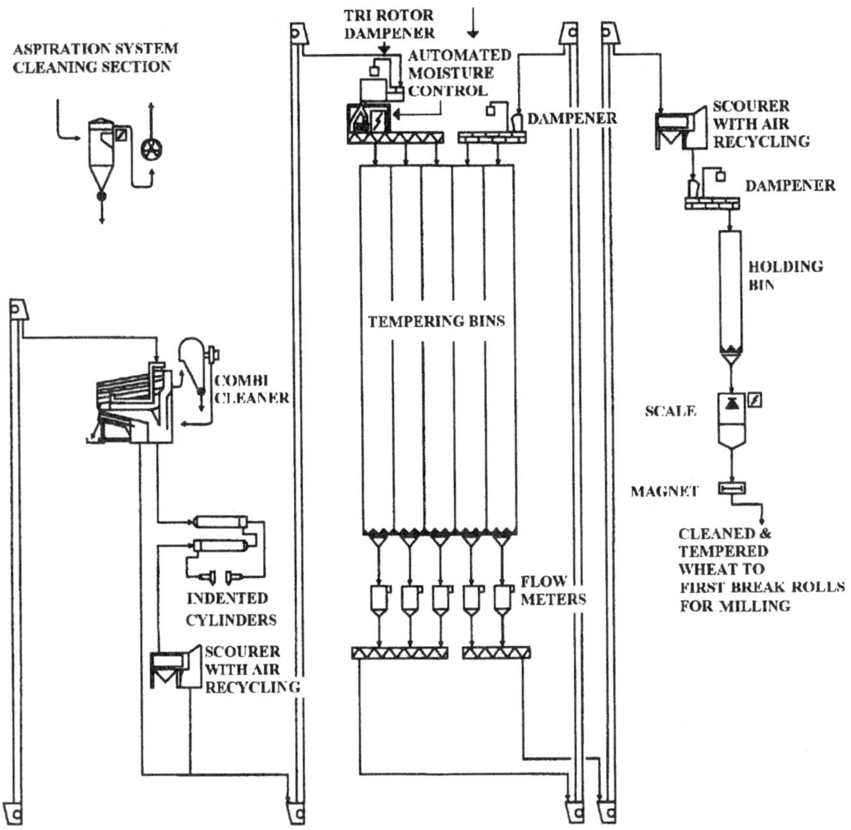

(b)

Fig. 14 (b) A more modern cleaning system. (Courtesy of Bühler, Inc.)

Round seeds tend to move out toward the outer edge of the spiraling slide, which has multiple grooves to allow the material with differing flowing properties to be separated into specific channels.

4. Generally, at this point, the wheat is practically free of all foreign material, and it is ready for the next process, referred to as *tempering* or *conditioning*. However, if the wheat has a lot of surface dirt at this point a scourer before the tempering process would be considered appropriate. Otherwise, in the tempering process when water is added to the wheat, it would tend to become muddy.
5. After the tempering process, which involves addition of water, the moistened wheat is allowed to sit in the bin for a specified time to allow the moisture to be absorbed by the wheat kernels. As the moisture is first added to the wheat kernels they have a tendency to swell. When the moisture is fully absorbed and the wheat kernels seem to become somewhat dry, the outermost layer of the bran tends to become loosened. Therefore, at this point, a frictional or abrasive action helps to remove this outer bran layer, which being very brittle would otherwise become rather easily shredded in the milling process. Such frictional

cleaning is performed at the very end of the cleaning sequence because hard objects such as stones and nonferrous metal in the wheat can cause severe damage to the equipment. Magnets always precede scourers or any other equipment that runs at high speed. In addition, magnets are generally installed at several key points throughout the milling process and flour-handling systems.

6. All items of wheat-cleaning equipment are also connected through duct work to a central dust control system. All the dust cannot be removed at one given point, no matter how efficient the design may be, and each time the grain is handled some dust is created; therefore, it is collected at several points beginning with the first cleaning machine—the grain separator—which normally is equipped with an efficient aspiration channel at the back to remove the bulk of the dust at the very first step of cleaning. Because all this cleaning equipment is exhausted centrally through this dust collection system, it keeps all the equipment under a slightly negative pressure, thus preventing any escaping of the dust into the atmosphere.

Even though the reasoning for the sequence of the flow holds equally well for both the conventional- and as well as modern-cleaning systems, there is a fundamental departure from the conventional-cleaning to the modern-cleaning approach. The items of cleaning equipment used in most milling plants have not actually changed much in their fundamental design. The change that has taken place in this area has been the introduction of equipment separating the heavy from light fraction within the wheat mass. The early generation of this type of equipment was introduced approximately 29 years ago by Bühler, Inc. of Uzwil in Switzerland. Table 3 lists the chronologic development of these types of equipment and their functions.

As can be observed when comparing Fig. 14a with 14b. In the modern cleaning flow immediately following the passage through the first cleaning equipment (combi-cleaner, see Fig. 15), the wheat is divided into a heavier and lighter stream. This equipment is versatile enough to perform three foregoing functions as well as removes lighter material through an air-recycling aspiration channel.

Incorporating such a machine in the flow dramatically simplifies the cleaning-house flow and operation. The heavier fraction of wheat stream that generally is regulated at 80% consists primarily of sound and heavy kernels, with no foreign material, and thus can be sent directly to the tempering system. The lighter stream contains most of the foreign material such as lighter seeds. Therefore, this stream has to go through the indented cylinders or disk separators for the removal of shorter and longer seeds as well as it has to go through the scourer and aspiration. Because the lighter stream is only 20% of the

Table 3 Chronologic Development of Equipment Separating Heavy from Light Fractions

Equipment name	Function	Year (Approx.)
Concentrator	Separation of heavy stream from light	early 1970s
Combinator	Separation of heavy stream from light Function of a destoner	early 1980s
Combi-cleaner	Separation of heavy stream from light Function of a grain separator with an aspiration channel Function of a destoner	early 1990s

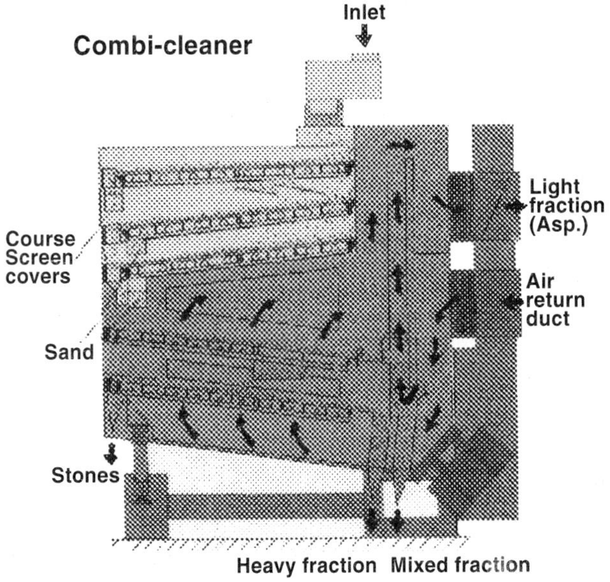

Fig. 15 Combi-cleaner. (Courtesy of Bühler, Inc.)

total wheat inflow, the capacities required for these items of equipment are only 20% of the capacity of the total system. This means not only savings in terms of equipment costs, but also in terms of operating costs, as less air would be required to vent or exhaust these machines and smaller motor size would be adequate enough to power them.

3.2.5 Tempering or Conditioning Process

This is an integral part of the process that immediately follows the cleaning operation. The terms *tempering* or *conditioning* may be defined as, "To bring the wheat to its optimum condition for milling to obtain optimal milling performance in terms of production and quality."

It involves the addition of water to achieve the following:

1. Raise the moisture level of wheat to a predetermined level considered optimum for milling under the prevailing conditions.
2. Allow a predetermined time, considered optimum, under the prevailing conditions, for the moist wheat to rest in the bin

   Eliminate any differences in the moisture content of individual wheat kernels among the tempered wheat following the moisture addition, by allowing time for moisture transference from kernel to kernel to achieve uniformity.

   Complete penetration of moisture inside the wheat kernel.

The optimum milling moisture and tempering time are both required to obtain optimum milling performance relative to the following objectives:

Toughen up the bran so that it will resist powdering in the milling process

Enhance easier separation of the bran from the endosperm

Mellow the endosperm sufficiently that it can be rather easily ground into flour
Allow the ground material to be sifted easily and efficiently

When analyzing these objectives, it is clear that drier wheat when ground will sift easily, as well as that the bran separation from endosperm in drier wheat can be accomplished easily, although the bran will have a tendency to chip or shred. On the other hand, moist wheat will have tougher bran that will resist powdering in the grinding process and the endosperm will be mellower. Therefore, a delicate balance between the two, which may be further influenced by several other factors, would provide the optimum conditions. Although all the foregoing objectives are influenced by the milling moisture level, the third point relating to the mellowness will also be heavily influenced by tempering time, which must be sufficient to allow a complete penetration of the desired moisture level inside the kernel. Optimum milling moisture and tempering time will be determined by

Kernel texture (hardness)
Ambient temperature and humidity
End use quality requirements
Required flour moisture

Milling performance for production would be measured by the following functions:

Throughput (milling capacity)
Flour extraction
Milling moisture

Milling performance relative to quality would be measured by the following attributes:

Flour moisture
Flour granulation
Physical starch damage level in the flour
Flour ash content or color grade value as a measure of flour refinement

3.2.5.1 Influence of Kernel Texture (Hardness) on Tempering

Harder kernel textures require longer tempering times and higher milling moisture levels for good milling performance. Hardness can be measured in various ways. Experienced millers may tend to simply bite a few kernels, when the tempered wheat is about to be ground, to determine if it is mellow enough to be ground satisfactorily. Hardness can be measured using NIR, single-kernel characterization test, and a few other methods of testing, including the particle size index test or PSI. For wheat that has never been milled before ascertaining its hardness value, this would help obtain a near-perfect optimum tempering requirement the very first time it is milled. Table 4 provides a list of all types (classes) of Canadian wheat and their typical hardness value range as measured by PSI tests.

3.2.5.1.1 EFFECTS OF THE SUBOPTIMUM TEMPERING ON THE MILLING PERFORMANCE. The bran will tend to shred a lot more if the wheat is too dry for milling during the grinding process, causing bran specks to appear in the flour. This causes poorer flour color and higher flour ash content owing to the presence of higher levels of bran parts in the flour. The flour extraction rate however is going to increase, as the bran cleans up rather easily. On the other hand, if the wheat is at a higher level of milling moisture than what will be considered optimum, then the ground material will be difficult to sift; consequently, the throughput is reduced, and flour extraction is also reduced as it becomes

Table 4 Typical Hardness Value Range for Canadian Wheat Classes

Canadian wheat classes	PSI (%)	Hard
Canada western amber durum	35–44	↓
Canada western extra strong	48–55	
Canada western red spring	50–56	
Canada western red winter	55–60	
Canada prairie spring red	58–64	
Canada prairie spring white	60–68	
Canada western soft white spring	65–70	Soft

difficult to clean the bran. In other words, the higher the milling moisture is above the optimum level, the lower will be the throughput. The same inverse relation would also exist with the flour extraction rate as shown below:

Milling moisture over optimum is inversely proportional to throughput

Milling moisture over optimum is inversely proportional to flour extraction

3.2.5.2 *Calculation of Water Addition*

The quantity of water added to raise the moisture level to the targeted percentage will be dependent on the following:

Natural moisture content of the wheat	(M_1) in %
Desired moisture level of the wheat	(M_2) in %
Rate of flow of wheat	(W) in kg/h

The following formula may be used for calculation of water in liters per hour (L/h):

$$\text{Water addition rate (L/h)} = W\left(\frac{(M_2 - M_1)}{(100 - M_2)}\right)$$

3.2.6 Equipment for Tempering

3.2.6.1 *Conventional System*

The conventional tempering system involves a slow-moving screw conveyor into which wheat enters at one end. An atomized spray of water is applied on the wheat mass in conveyance with the help of compressed air. The quantity of water is regulated by a water flowmeter. The water is set manually at a predetermined level using the foregoing formula. As the wheat enters the spout that directs it to the tempering conveyor it hits a plate connected to a pressure switch. This switch is connected through a control circuit with a solenoid valve that cuts off the water line in the event there is no wheat flow, owing to interruption in the process. Even though the movement of the conveying blades is slow, there are generally short paddle blades that retard the forward motion of the wheat and thus creating more mixing action. In such a system up to 3% of water may be added in one stage. If more water needs to be added it is done over two stages.

3.2.6.2 Intensive-Mixing System

Such systems have also been in use for many years and have proved their usefulness in terms of the following:

1. Ability to add more water in one stage from 5 to 7% depending on the equipment type.
2. Ability to reduce the tempering time by the following:
   Vigorous mixing action enhances initial water absorption (superficial).
   Uniform distribution of the moisture; therefore, kernel-to-kernel moisture transference is not required.
   Some systems also use steam application to effect a reduction in tempering time.

The influence of temperature on the tempering time is quite significant. In fact, during the 1960s, it was quite common to choose between what was known as a hot-tempering system and a cold-tempering system. Cold tempering was the regular tempering, with ambient water temperature, as we have today, except that back then the method and control of water addition and as well as the tempering equipment was quite different. Hot tempering referred to the passing of the moist wheat through a tower with hot radiators inside so that during the passage of wheat from the top inlet to the bottom outlet it would be exposed to the hot radiators and, in general, the surrounding warm temperature. This would reduce the tempering time to a fraction (as low as 6 h for hard wheat) of what it would take with cold tempering (as high as 72 h for hard wheat). The hot conditioning also had an added benefit of improving or modifying the gluten properties, which made the flour more suitable for baking. However, the use of hot-conditioning systems gradually declined for the following reasons:

Higher capital investment, upkeep, and attention were required to ensure efficient operation
Additional energy cost to heat the radiators
Danger of overheating the moist wheat, especially, if the wheat flow halted somewhere within the column

In Canada, during winter, wheat may be received at the flour mill at very low temperatures. This is certainly advantageous as a natural safeguard against the development and propagation of the insect infestation problems. However, low temperatures are not good for water penetration during tempering, as the process is very slow; second, wheat is a bad conductor of heat; therefore, during the milling process, as the cold wheat passes through the pairs of corrugated rolls, the bran tends to chip, rather than curl up, as it does when the wheat is warm and moist.

If the incoming wheat temperature is very low, it is sometimes heated to raise its temperature to the ambient room temperature before water is added in the tempering process. In Canada, it is a common practice to add hot water to the wheat in the tempering process, which does help in the initial absorption of the water.

3.2.6.2.1 EQUIPMENT USING INTENSIVE MIXING SYSTEMS. There are different types of equipment available for mixing wheat and water uniformly up to a level of between 5 and 7% over a short conveying length. Even though these types of equipment vary in terms of design and other technical specifications, they all have one common objective and that is to provide vigorous mixing.

Intensive dampener and technovator, both incorporating aggressive mixing action, were introduced to accomplish the goals.

3.2.6.2.2 TECHNOVATOR. Technovator, a very popular tempering equipment in North America, consists of a large grain mixer on an inclined plane. Inside the grain mixer trough a high-speed rotor with paddle blades configured in such a manner that it provides more of an agitating action and less of a conveying one. Therefore, the grain and moisture become mixed more thoroughly as it is conveyed up on the inclined plane to its outlet. A provision for adding hot water as well as steam, if required, are made available.

3.2.6.2.3 INTENSIVE DAMPENERS. Figure 16 shows an intensive dampener. It has a stainless steel casing of either 1 or 2 m in length and usually 30 cm in diameter. Inside the casing rotates a high-speed rotor of standard steel. Beaters attached to the rotor are also of stainless steel. The wheat enters at one end, and a measured quantity of water is sprayed on it. The high speed rotor ensures uniform and rapid moisture distribution. The percentage of water that can be added in one step, therefore, is as high as 5%. Stainless steel construction, coupled with a complete self-emptying feature, enhances the degree of sanitation that is required by the flour mills.

A further improvement in the dampening system has become available. It allows gentler, but more efficient mixing action to enable absorbtion of the required quantity of water by the wheat undergoing tempering, leaving very little of superficial moisture. The system is called a trirotor dampening system (Fig. 17).

3.2.6.2.4 TRIROTOR DAMPENING SYSTEM. This dampening system consists of a three-sectional jacketed housing that can be easily taken apart for maintenance. Inside the housing there are three rotors (see Fig. 17b) that allow the whirling wheat mass an intimate exposure for a very uniform and effective dispersion of water.

The action of the three rotors helps diffuse the moisture quickly without being too aggressive in terms of causing friction or abrasion. Unnecessary abrasion creates breakage of wheat and premature separation of outermost layer of bran. This system allows a moisture addition of up to 7% in one passage; thereby, eliminating the need for a two-stage tempering process for most situations.

The measured moisture content that is added in the tempering process is calculated and controlled by moisture control systems, as described in the following section.

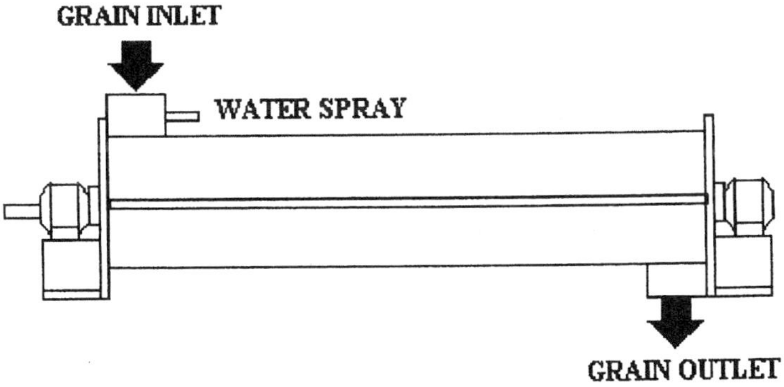

Fig. 16 An intensive dampener. (Courtesy of Bühler, Inc.)

(a)

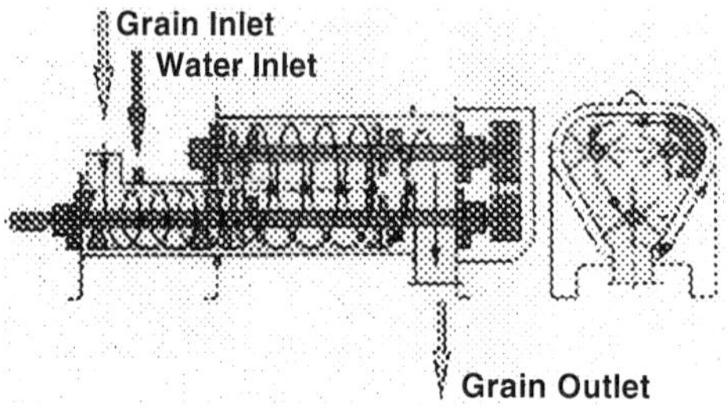

(b)

Fig. 17 (a) Trirotor dampener; (b) mixing action of the rotors. (Courtesy of Bühler, Inc.)

3.2.7 Moisture Control Systems

To maintain uniform target moisture in the tempered wheat, it is essential to have a moisture control unit that will detect the moisture of incoming wheat and will adjust the quantity of water accordingly to maintain the targeted final desired moisture content. Many years ago this function was performed manually by controlling water in a flowmeter. The quantity of water to be added was calculated on the basis of the flow rate of wheat, natural moisture of the wheat, and desired moisture level of the wheat. The only automated feature was the shutting off of the water when a flow detector in the wheat spout did not detect any wheat and activated a solenoid valve to shut the water flow. Various systems have been developed to control the moisture addition in a tempering system.

One such moisture control unit that works in conjunction with a trirotor system will be described.

It consists of four components. They are as follows:

1. Flow rate measurer and controller
2. Moisture content measurer for the continuous measurement of the wheat moisture
3. Water flowmeter for metering the calculated quantity of water
4. Processor unit that calculates and automatically controls the moisture to be added relative to the preset final desired moisture and the actual measured moisture

On-line moisture measurement is done on a continuous basis, which allows the moisture addition to be continuously regulated until the desired moisture is achieved. Determination of moisture is done by using safe microwave technology, rather than gamma rays

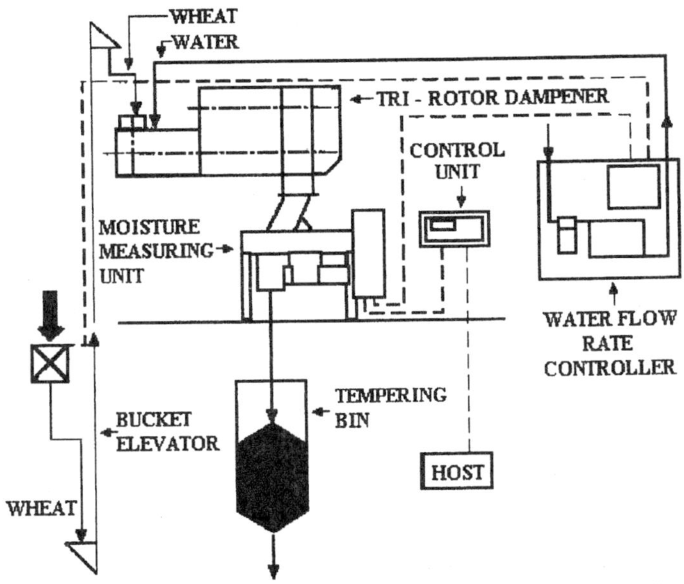

Fig. 18 Moisture addition using automated moisture control unit on feedback system. (Courtesy of Bühler, Inc.)

as used by some systems. The range of moisture measured successfully by the moisture content measurer can be as low as less than 9%, to as high as freshly dampened wheat following a short rest time.

This system operates in a feedback mode, as shown in Fig. 18. This means moisture is measured after water addition on freshly dampened wheat and correction is made based on a preset desired final moisture. The system can also be setup to operate on a feed-forward basis for which moisture is measured before water addition and necessary water addition is made on the basis of preset final desired moisture.

3.2.7.1 Chlorination of the Tempering Water

Tempering water is often chlorinated to keep the total microbial activity plate count numbers as low as possible. In the beginning of the process as the plant is being sanitized, it is usual to begin with a higher level of chlorine treatment of 300 ppm in the water. Following several weeks of this application the level can be brought down to about 100 ppm for regular application.

3.2.8 Optimum Tempering Conditions

Optimum tempering in terms of moisture and time is dependent on prevailing circumstances and conditions, including the ambient temperature and humidity at the location. Therefore, it is difficult to suggest an optimum milling moisture and tempering time that will satisfy the specific requirements of all situations. Table 5 provides a general guide of how different the milling moistures and tempering times can be for CWRS class of wheat in different countries.

The tempering conditions are altered, depending on wheat hardness as well as on the end-use quality requirements of the flour. Table 6 provides a range of the wheat-

Table 5 Optimum Tempering Conditions for CWRS Class Wheat

Factors	North America	South Korea	Switzerland	Mexico
Milling moisture (%)	15.5–16.5	15.5–16.5	17.0–17.5	15.5–16.0
Tempering time (h)	8–24	40–48	36	50–60

tempering requirements for hard, soft, and durum wheats as applied by flour mills in Canada.

When comparing the tempering requirements of the hard and soft wheat from Table 6, it is quite clear that the milling moisture as well as tempering time requirement for soft wheat are less than those for hard wheat owing to the differences in kernel hardness. Therefore, when milling a blend of hard and soft wheat, they are cleaned and tempered separately and then blended. Hardness alone, however, does not determine the ultimate tempering requirements, as can be seen from Table 6 when comparing the tempering requirements of hard and durum wheat. Although durum wheat is the harder of the two, its tempering time requirement is shorter than for hard wheat. The reason is that the end-use quality requirements are quite different. Hard wheat is milled to produce a white and fine powdery material as finished product, whereas durum wheat is milled to produce a bright golden yellow granular material. Therefore even though it is harder than hard wheat, its tempering time is usually shorter to allow a higher percentage of granular material production to meet its required objectives.

3.3 The Milling Process and Principal Milling Equipment

Milling of wheat into flour essentially involves a series of grinding and sifting operations with in-between purification of the granular endosperm particles. Grinding, sifting, and purification operations are carried out by the rollermill, plansifter, and purifier, respectively. To have a clear understanding of the process, it is important to understand the operation of each of the three principal types of equipment. These equipment types are manufactured by several equipment-manufacturing companies. Therefore, it is natural to expect some differences in terms of durability, design, operating efficiency, and price. However, the basic principle of operation is common for all. They are described in the following section.

3.3.1 Rollermills

3.3.1.1 Design

A rollermill has two sides (Fig. 19). These sides are separated, with separate drives, product feeding, and ground product pickup. Both sides may be allowed to grind the same or different stocks, depending on the quantity of a given stock to be ground. Each side consists of a pair of grinding rolls on a horizontal plane. Some of the older models of

Table 6 Tempering Requirements of Hard, Soft, and Durum Wheats

	Hard wheat	Soft wheat	Durum wheat
Milling moisture (%)	15.5–16.5	14.0–15.0	15.5–16.0
Tempering time (h)	8–24	4–6	4–10

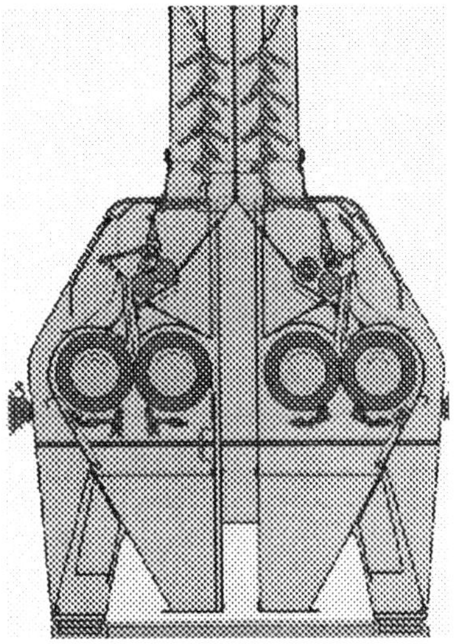

(a)

(b)

Fig. 19 (a) Cutaway section of a rollermill showing two sides; (b) series of rollermills. (Courtesy of Bühler, Inc.)

rollermills had rolls placed diagonally. Both pairs of rolls are housed in a heavy cast-iron frame. The type of grinding action required by the rolls would determine what type of surface these grinding rolls should have. For example, the grinding of wheat and subsequent break stocks requires corrugated rolls that become progressively finer going from first break to the last break passage. The grinding of pure endosperm particles in the reduction rolls involves application of smooth rolls. The grinding of composite particles of bran and endosperm requires finely corrugated rolls. Therefore, all types of grinding action in the milling process are performed by rollermills that appear to be the same from their outer appearance (see Fig. 19b) but may be different in terms of surface of the rolls. Grinding objectives also determine the differential speed of the front and back roll of the pair. The front roll within the pair runs at higher speed than the back roll. The differential speed for break rolls is 2.5:1, whereas that of a reduction passage is 1.25:1. Generally, fast rolls run at between 500 and 550 rpm. Therefore, the slow roll speed of a break passage would be between 200 and 220 rpm.

Feed rolls, that allow the stock to be ground to be fed properly to the ''nip'' of the grinding rolls, are located above the grinding rolls. Feed rolls are much smaller in diameter, but are of the same length as the grinding rolls, and they run at much slower speed. Feed roll speeds can be varied to control the rate of feed. For most granular products a pair of feed rolls is used, but for bulky break stocks the rear feed roll is replaced by a paddle-bladed distributing (stock) conveyor. Just above the feed rolls there is a feed gate that holds the feed in the feed hopper. When the feed gate is slightly open and feed rolls turn, the product flows downward on to the ''nip'' of the grinding rolls. Many rollermills use a balanced feed gate that allows automatic opening or closing of the feed gate corresponding to the level or weight of the feed present in the hopper. On one hand, this ensures that the product does not back up in the feed hopper and, on the other hand, that the feed is spread across the entire width of the grinding rolls so that metal does not grind against metal.

Grinding rolls are heavy and cylindrical. They are made up of close-grained cast iron, with a hard surface referred to as ''chill.'' The composition of chill is very hard, to resist wear during grinding. The depth of the chill is approximately 12 mm, and once this depth is worn out through repeated resurfacing of the grinding rolls they become unfit for reuse. The diameter of the grinding rolls are either 225 mm (North American) or 250 mm (European). They are commonly available in lengths of 800 and 1000 mm. A small shaft extends from each end of the roll, at which a self-aligned roller bearing is mounted. The housing of the bearing is anchored to the frame of the rollermill using bolts.

Each pair of rolls within the rollermill has its own drive. The fast roll is belt-driven, either from an individual motor drive or from a transmission belt from a counter line shaft in a group drive. The slow roll is driven either from a gear transmission or from a belt transmission from the fast roll. Feed rolls in all conventional rollermills are driven by a small transmission belt from the main roll drive. In very modern rollermills they are driven by individual motors. The adjustment of the roll gap between front and back roll is accomplished by turning a handwheel located in front of the machine. Turning the handwheel moves the slow (back) roll forward or backward allowing alteration of the roll gap, as desired. The fast (front) roll does not shift.

Corrugated rolls are provided with brushes that are located underneath the rolls for removing any moist ground material sticking to its surface, whereas smooth rolls are provided with metal scrapers to accomplish the same thing.

3.3.1.2 Operation

Product enters through a large cylindrical sightglass at the top. This allows an operator to observe the level of material in a rollermill from a distance. In most modern rollermills, there is a sensor located within the sightglass hopper that allows the rolls to disengage if the sensor does not detect product. This prevents any accidental wearing out of roll surface. The material to be ground enters the feed hopper where the combination of feed rolls' forward rotation and the opening of the balanced feed gate allows the product to be spread evenly across the entire width of the feed gate. Thus an uniformly thick curtain of product feed is introduced to the nip of the grinding rolls.

The rolls rotate inward to provide the grinding action. The grinding gap is adjusted by turning the handwheel on both ends of the machine to ensure the gap is even. The desired gap is achieved by examining the ground stock underneath the rolls. The ground material is dropped into the hopper below, where it is collected and picked up by a pneumatic lift that carries the ground material vertically up two or three floors higher to an air cyclone. Within the cyclone the material loses its velocity and begins to settle down at the bottom where it is discharged out through an airlock running underneath the cyclone.

It is obvious that when intensive grinding is required the roll gap must be reduced. There are, however, other factors that intimately affect the grinding results and, therefore, the subsequent finished product quality. The following section on corrugation specifications provides some insight into the various technical aspects of the corrugations and their relation to the grinding results.

3.3.2 Corrugation Specifications

The following are the key factors that affect the grinding results by the corrugation properties:

Number of corrugation per centimeter
Spiral of corrugation
Corrugation profile
Disposition of corrugation

3.3.2.1 Number of Corrugations

The number of corrugations per centimeter increases when going down from first break to the last break passage. This is because material to be ground becomes thinner as the progression is made. If, for example, finer corrugations were used for grinding wheat in the first break, the combination of roll gap pressure required for grinding and higher number of corrugations would create a lot of fine material, including cutting of the bran. On the other hand, finer corrugations are required for the latter breaks: for example, in the last break passage there is no longer much endosperm attached to the bran; therefore, with fine corrugations, it is possible to grind relatively closely and be effective in detaching any remaining endosperm in the bran.

Likewise, for sizing stocks, the composite particle size is smaller, and a light grind with finely corrugated rolls helps clip off the bran chip attached to the endosperm particles. Corrugations for first break can be as low as 3.9/cm, and for the sizing about 12/cm.

3.3.2.2 Spiral

When more cutting is required with corrugated rolls, reducing the grinding gap will accomplish that, but it will also generate a lot of fine material. Increasing the spiral (Fig. 20) of the corrugation provides cutting, without creating fine material.

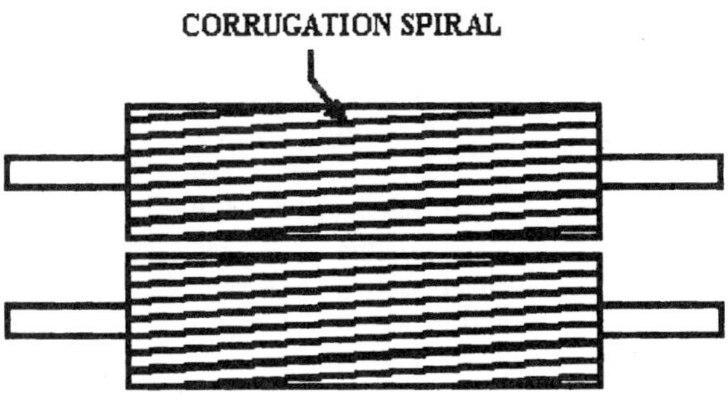

Fig. 20 Roll corrugation spiral. (Courtesy of Bühler, Inc.)

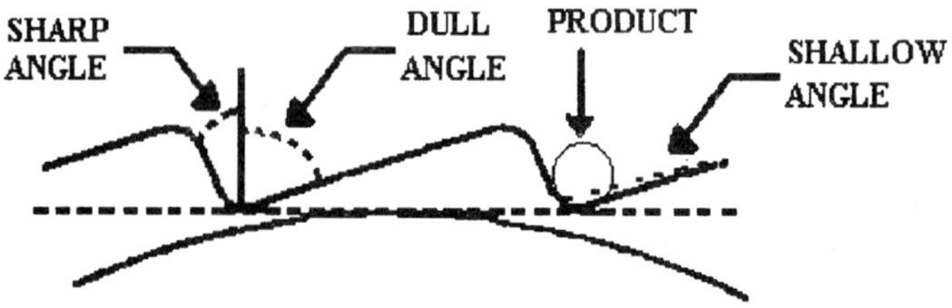

Fig. 21 Corrugation profile.

Usually, spiral is represented as a percentage. Corrugation spirals range from 2 to 10%, or even higher. In most North American flour mills, spirals may begin for first break at 2 or 4% and remains at 4% for rest of the breaks or may go up to 6%.

3.3.2.3 Profile

Corrugations come with many different types of profile. In most the angle of the cut is unequal, providing a sharp and a dull angle. Profiles are chosen on the basis of grinding objectives. For example, in first, second, and third breaks when milling hard wheat, the desire is to create as much granular endosperm material as possible. This necessitates the application of corrugations that provide depth, meaning the sharp and dull angle of the corrugation (Fig. 21) is not too wide.

Usually, angles of 30–35 degrees for sharp and 60–65 degrees for a dull angles are preferred. On the other hand, wider angles are preferred for lower break passages, as the thin layer of bran with very little endosperm attached can be removed only with flatter angles (40 or 70 degrees) and finer (more) corrugations.

North American corrugation profiles are designated by names (Table 7), and their specifications seem quite complex to the uninitiated.

3.3.2.4 Disposition

Everything else being constant the position of the corrugation on the fast and slow roll alone can determine the severity of grinding. For example, when the sharp angle of the

Table 7 Typical Corrugation Specifications in a Hard Wheat Mill

Grinding passage	No. of corrugations per centimeter	Spiral (%)	Profile name	Fast- and slow-roll differential
1st break	3.9–4.7	2–4	Modified Dawson or Getchell	2.5:1
2nd break C*	4.7–5.5	2–4	Modified Dawson or Getchell	2.5:1
2nd break F**	5.5–6.3	2–4	Modified Dawson or Getchell	2.5:1
3rd break C*	5.5–6.3	2–4	Modified Dawson or Getchell	2.5:1
3rd break F**	6.3–7.1	2–4	Modified Dawson or Getchell	2.5:1
4th break C*	6.3–8.7	4	Modified Dawson or Getchell	2.5:1
4th break F**	7.9–9.4	4	Modified Dawson or Getchell	2.5:1
5th break C*	7.9–9.4	4	Modified Dawson or Getchell	2.5:1
5th break F**	9.4–11.0	4	Modified Dawson or Getchell	2.5:1

* Coarse.
** Fine.

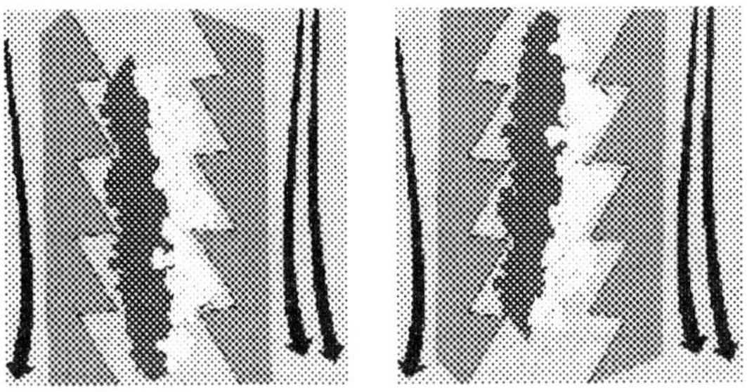
(a) (b)

Fig. 22 (a) Dull-to-dull; (b) sharp-to-sharp.

fast roll is facing down and the slow roll is facing up (Fig. 22b) the grinding action is severe and it provides a lot of cutting. This particular disposition of the corrugation is referred to as ''sharp-to-sharp.'' The two arrows in the figures represents fast roll, and the single arrow represents the slow roll.

In a similar way the opposite disposition would provide the least amount of cutting and more of producing finer material as the rolls are pressed closer. This disposition is referred to as ''dull-to-dull'' (see Fig. 22a). In flour milling the objective is to release the endosperm without cutting the bran, therefore, almost always break rolls are dispositioned as ''dull-to-dull.'' Table 7 shows a typical corrugation specifications for a hard wheat mill.

3.3.3 Sifter (Plansifter)

Ground material from a grinding passage in a rollermill is always sent to a sifting section for sifting. Sifting operation is provided by a ''plansifter'' also commonly referred to as sifter. A sifter consists of a number of sifting compartments or sections. These sections are all independent of each other. Each section has its own set of sieves. These sieves are stacked up, one above the other, and the number of sieves in a given stack may vary, depending on the sifter type, anywhere from 14 to 30. A sifter generally comes with four, six, or a maximum of eight sections (Fig. 23a). These sections are located within a box and there are two boxes in a sifter containing an even number of sections. The two boxes are joined together with a driving mechanism in the center. The boxes are suspended from girders above by four sets of canes at each end of the corner. Canes are used for suspension as they are strong and flexible to allow free swinging of the sifter. The drive consists of a vertical countershaft that has weights attached to it in the center. As the shaft is rotated by a motor-driven pulley from the top the sifter with the counterbalanced weights begin to swing on a horizontal plane. A pencil fastened to the bottom end of the sifter would draw a perfect circle. This indicates that the sifter is in balance. The diameter of the circle for European sifter is 75 mm, and speed ranges from 210 to 225 rpm. North American sifters have a rotational diameter of 100 mm, with a slower revolution of 180–190 rpm.

A sieve comprises a wooden frame, over which screen cloth is stretched and secured either by means of staples or glue, and an outer frame consisting of a collecting tray (see Fig. 23c). The sieve frame is inserted on this outer frame. The wooden sieve frame has appropriate screen cloth on top, and on the bottom side a coarse wire screen is in place to hold plastic or cotton pad cleaners. As the sifter rotates the cleaners move around keeping the underside of the sieve clean. Since the sieve frame is inserted on the outer

(a)

(b)

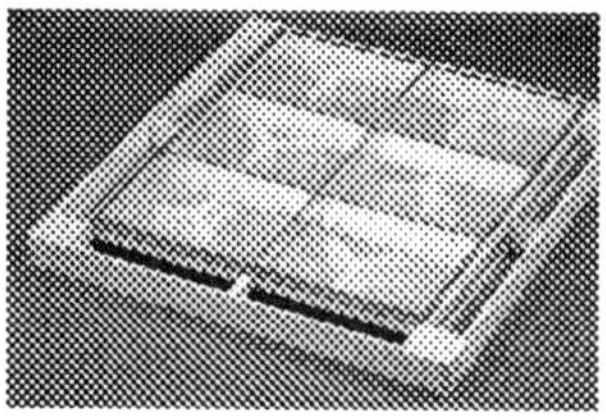

(c)

Fig. 23 (a) Plansifter; (b) series of plansifters; (c) sifter sieve. (Courtesy of Bühler, Inc.)

frame consisting of a collecting tray, material that passes through the sieve is collected in this tray and is guided out of it through an opening along one side of the frame. The material that passes over the sieve exits the sieve through another opening in the outer frame.

Material feed inlet for each section is located at the top of the section, whereas the outlets are at the bottom. There is generally one inlet per section as each section handles a given type of material. However, sometimes a section may be divided to handle two different types of material when the quantities being handled are much less and, then, there will be two inlets located at the top at two opposite ends. The ground material is sifted into at least two, and as high as seven different fractions, depending on the material; therefore, there are anywhere from two to seven outlets per section. These inlets and outlets are part of the sifter and thus swing with the sifter as the sifter is rotating. Their connection with the corresponding spouts, which are fixed, is done by cotton sleeves (stockings).

3.3.3.1 Operation

The material for sifting is allowed to enter the sifter section through the feed inlet at the top. As the material reaches the top of the sieve stack the sifting operation begins by separating the material into a number of fractions according to size. Channels within the sifter section and within the sieve frames and the outlet frame allow guiding of the separated fractions out of the sifter section. The outlets from the outlet frame are connected to spouts at the bottom through which the sifted fractions are appropriately directed to their destinations.

3.3.4 Purifiers

Sifters are used to sift ground material into several fractions that are graded according to size, for it uses sieves of varying screen sizes as a separating medium. Each graded fraction, especially, the granular fraction following sifting, consists of similar-sized particles of

Pure endosperm
Composite particles of endosperm and bran–germ attached to each other
Bran

Purifiers facilitate a further separation of the sifted graded stock of similar size into pure endosperm, composite, and bran particles.

A purifier comprises two sections (Fig. 24a). There are three decks of sieves in each section. These decks (layers) of sieves in each section oscillate back and forth as a result of an eccentric drive. Each layer of sieves has four individual sieves. The sieves are progressively coarse going from head to the tail end within the layer; however, going from the top layer to the bottom, the sieves are progressively finer. To keep the sieve apertures clean, each sieve has a pair of guide rails (see Fig. 24c) mounted on the frame under the sieve cloth. A brush assembly moves up and down these guide rails, assisted by the oscillation, keeping the underside of the sieve clean. The sieve layers slope slightly downward from head to the tail end. The materials that still remains on top of each of the three layers of sieves are channeled out into three spouts. The material, that falls through all three layers of sieves to the bottom, is collected altogether into either one of the two shaking channels provided at the bottom or split into two separate fractions and collected separately. There are several small flaps just above and along the length of the shaking channels. Deflecting the flaps allows directing the falling material from one channel to the other. The shaking channels act as conveyors in moving the material collected to its outlet.

Air is drawn up through the material layers spread over the sieves (see Fig. 24b) to create a stratification of the material. Therefore, provision is made to carefully regulate this air that is being drawn up through the shaking sieves by providing a compartmented hood over the top layer of sieves. This hood is connected to an air trunk providing the aspiration air. Each compartment can be individually regulated to optimize airflow across the length and width of the sieve layer.

3.3.4.1 Operation

The graded material from a sifter enters the purifier through a feed inlet just ahead of the sieve layers. The feeding device ensures that the material is spread evenly across the full

(a)

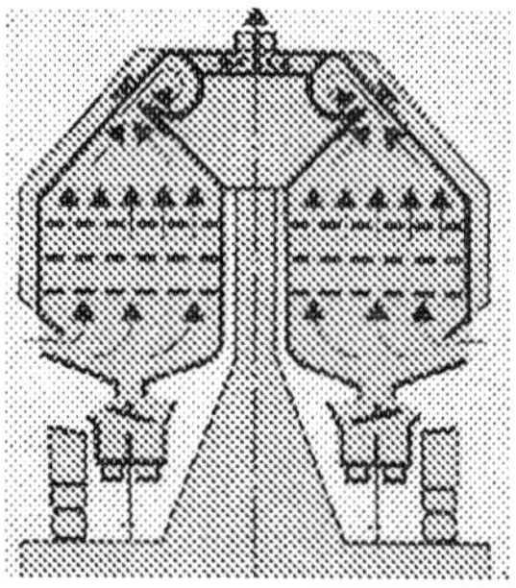

(b)

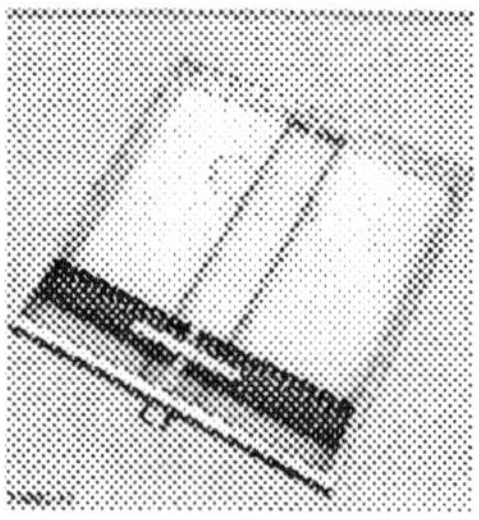

(c)

Fig. 24 (a) A modern purifier; (b) air being drawn up through the sieves; (c) underside of a purifier sieve showing brush. (Courtesy of Bühler, Inc.)

width of the sieve layer. The fluid bed of material begins to stratify as a result of the combination of controlled air that is drawn up through the sieves and oscillation of the sieve layers. Stratification of material according to density causes the lowest-density material to rise to the top and where it is removed by air under suction. The lighter branny material and germ particles with endosperm attached is carried all the way to the end of the sieve layers and exit the machine from spouts at the tail end. Pure granular endosperm particles being dense fall through the sieve layers. All the material dropping through the sieve layers, beginning from head end of the machine to the tail end, can be collected into one shaking channel at the bottom. However, near the tail end of the machine, the material starts to thin out; consequently, impure material of bran and germ with some endosperm attached may begin to drop through. When this becomes evident, while examining the material under the purifier, flaps over the channels can be deflected to collect the impure material in the second channel.

3.3.5 Flour-Milling Equipment Layout

The most common configuration of milling equipment layout involves placement of rollermills one floor above the ground floor (Fig. 25). Material, once ground in a rollermill, is carried by pneumatic conveying (see Fig. 25), to the sifter, which is generally located three floors above the ground level. Purifiers and bran finishers or dusters, which are used to remove the last traces of endosperm from the bran, are usually located two floors above the ground level. The purpose of having a multistory building for flour mills is to take full advantage of the gravity to feed various equipment in sequential operation. If two pieces of equipment are located on floors above each other, but are wide apart from each other so that a spout connecting them would have a very shallow angle, then a horizontal screw conveyor is used for discharging the material. Sometimes an additional floor is incorporated to permit such feeding using a spout.

3.3.6 Pneumatic Conveying

The ground material from the rollermill is collected in a hopper located within the rollermill. The outlet of the hopper is connected to a spout below that allows material to be discharged into a pneumatic lift. Air under negative pressure (suction) in the pneumatic lift allows the ground material to be carried up to the sifter floor. A pneumatic lift consists of a narrow metal tube, primarily comprising a long vertical section, with a small horizontal section at the bottom and at the top (see Fig. 25). The bottom end is connected to the spout that allows the ground material, discharged from the rollermill, to enter the pneumatic line. The top end is connected to a metal cyclone tangentially (see Fig. 25).

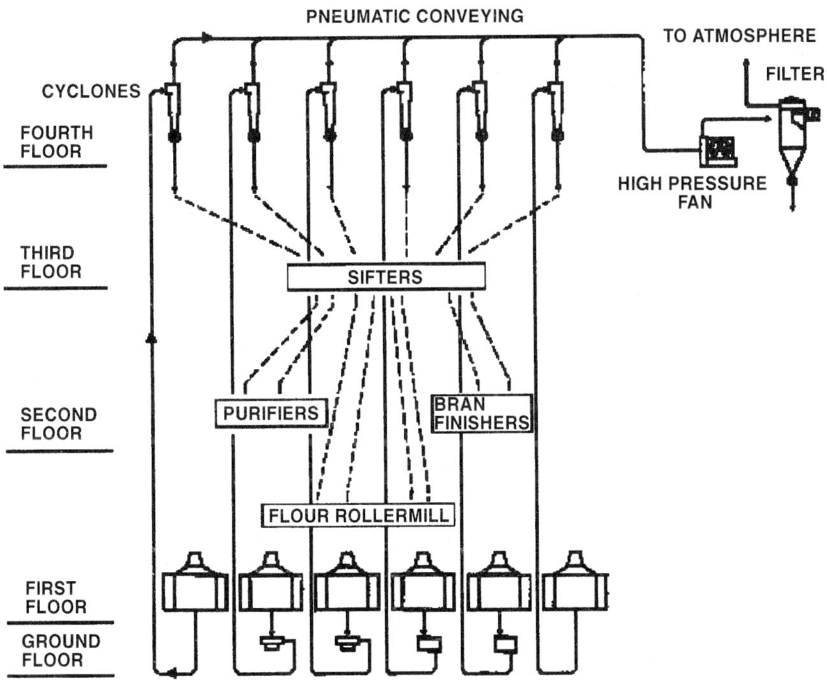

Fig. 25 Flour milling equipment layout and pneumatic conveying.

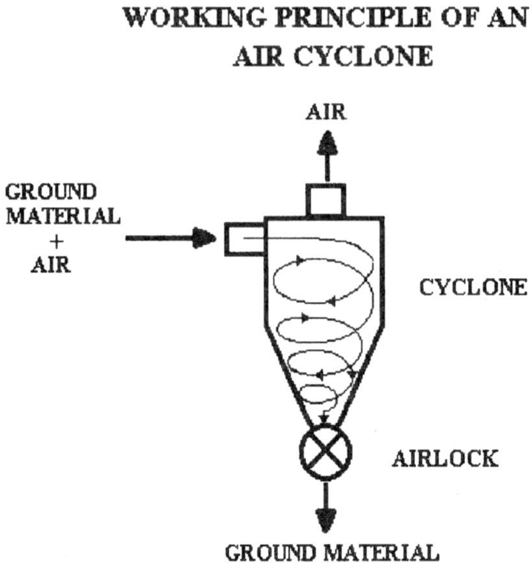

Fig. 26 Air cyclone.

The function of the cyclone is to allow the separation of the conveyed material from the conveying air. Figure 26 shows the working principle of a cyclone.

The air and the material under conveyance enter the cyclone tangentially. The ground material at entry strikes the inside wall of the cyclone and loses its velocity and begins to settle down. The clean air along with a very insignificant quantity of light material, which does not settle down, leaves through the top of the cyclone into a long common trunking that has several cyclones connected to it. This trunking is connected to an inlet of a centrifugal high-pressure fan that provides the suction air. The outlet of this fan is connected by trunking to blow air with fine material that did not settle down, into a filter dust collector. This dust collector consists of long filter sleeves which help entrap virtually all fine material before the clean air is vented out in to the atmosphere.

3.3.7 The Process of Flour Milling

When the wheat comes to the mill for processing, it must be free of all types of foreign material, and it should be in its optimum condition for milling. Any inconsistencies and errors made earlier in the process are very difficult to counteract during actual milling. Proper wheat blending, cleaning, and tempering processes are just as important as the actual milling process.

The milling process involves a series of grinding and sifting operations with in-between purification of the granular endosperm particles. This process of grinding, purification and sifting can be fairly complex depending on the flour quality (refinement) and rate of flour extraction required. *The most common objective in flour milling is to produce as much as possible of white refined flour that is free of bran and germ contamination.* The basic reason for this is financial return, which is much higher for white flour than it is for bran. The obvious question that arises at this point is that how much flour could be produced from a wheat after it has been clean and tempered. The answer to that will depend on several factors, such as

Type of wheat, wheat quality (e.g., test weight, presence of immature kernels, and other)
Quality of equipment and mill flowsheet
Processing factors
Required flour quality (e.g., ash content or color value) which is a measure of flour refinement

However, in general, the percentage of white flour of acceptable quality, produced from good quality wheat is about 75–76%, even though the endosperm content of wheat is approximately 83.0% (Fig. 27).

The constituents of a kernel of wheat are:

Endosperm	83.0%
Bran	14.5%
Germ	2.5%

In an ideal milling process a miller thus would be able to obtain 83.0% of flour. However, that never happens because bran adheres strongly to the endosperm, and it is not very cleanly separated. Looking at the cross-sectional view of the wheat kernel (see Fig. 27), it seems obvious that if the wheat kernel did not have a crease, perhaps the kernel

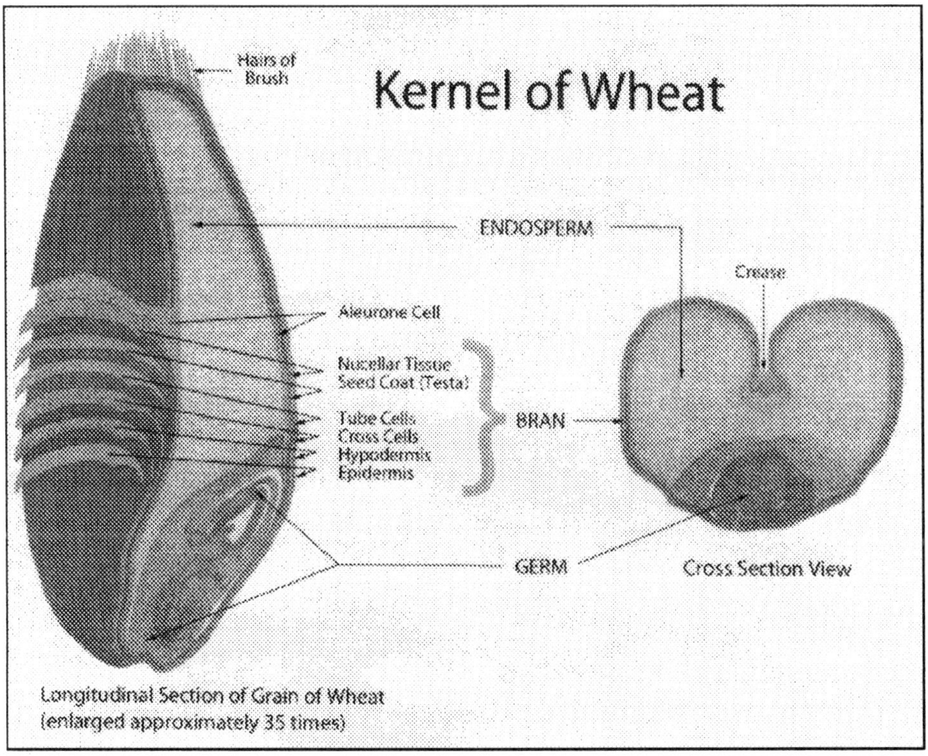

Fig. 27 A longitudinal and cross section of a kernel of wheat. (Courtesy of the North American Millers Assn.)

could then be pearled and thus provide a higher percentage of endosperm for suitable flour production.

Because pearling action through abrasion or friction would still leave some bran in the crease area, a different approach to bran removal must be applied. The crease therefore poses a technological challenge to the miller, making the separation of the bran more complex.

The *gradual break and reduction system* was developed to meet this challenge. This processing system essentially involves the releasing of the endosperm from the wheat kernels on a gradual basis and then the reduction of these endosperm particles into fine flour particles over a number of stages.

The milling process thus can be divided into the following three parts:

1. Breaking the wheat kernels and gradually releasing the endosperm in granular form
2. Grading of all the released granular endosperm, according to particle size, and its purification
3. Reduction of the purified granular endosperm particles into fine flour particle size

This sequence of the process has been graphically illustrated in Fig. 28. Even though grinding and sifting functions are repeated throughout the milling process, the specific objectives are not the same in all parts of the process. This has been explained earlier.

Let us now look at the milling process, beginning with the very first grinding step.

3.3.7.1 Breaking and Releasing the Endosperm

Breaking and releasing the endosperm from wheat kernel begins when the cleaned and tempered wheat is initially ground using corrugated rolls that are carefully adjusted to provide the appropriate gap between the grinding rolls to obtain satisfactory grinding results. Satisfactory grinding results are defined by the following requirements:

1. Opening the wheat kernels sufficiently to allow the release of some endosperm subject to the following constraints:
   Not to cut up the wheat too much.
   Not to produce any flour if possible.

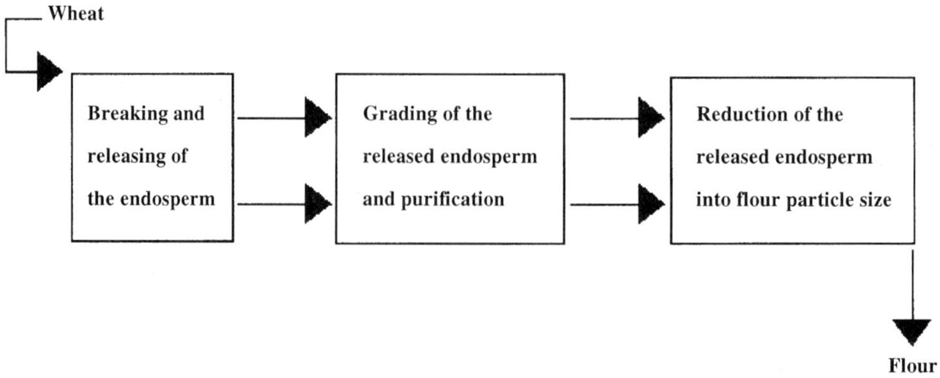

Fig. 28 Simplified block flow diagram of the milling process.

2. The ground stock would contain the following material listed in order of decreasing particle size and the degree of bran presence or contamination as well as quantity.

   The bulk of ground material consists of the coarsest material, comprising large chunks of endosperm with large pieces of bran attached. This is removed first by the top set of sieves in the plansifter. This material requires further processing for releasing of more endosperm from it.

   The next group of particles consists of the following:
   a. Coarse granular pure endosperm particles, also referred to as semolina in British terminology and farina in North American terminology.
   b. Particles containing endosperm, bran or germ attached to each other, also referred to as composite particles.
   c. Some branny material of size similar to a and b. The mixture of stock is later segregated according to material as described in items a–c.
3. The next group of material is a mixture of similar types of material, as listed in the foregoing, except it has a higher percentage of endosperm and much less bran and germ, and particle size will be finer. In British terminology it is referred to as fine semolina and in North America it may be referred to as middlings.
4. The fine particles of endosperm with very little bran specks, known as fine middlings or dunst, are removed next.
5. A small quantity of flour is also produced even though every attempt is made to minimize its production at this stage of the process.

The grinding of wheat is carried out in first break corrugated rolls and the ground stock consists of the mixture of the foregoing listed material. The physical separation of the material takes place in a plansifter. A break sifter while removing the coarsest material comprises bran and endosperm, also referred to as break stocks, grades the remainder of the released material in a number of groups according to the particle size as described previously. This grading part will be described very shortly in more detail.

The break stock is fed to the next break-grinding passage in sequence through a spout connected from the sifter outlet to the inlet of the second break roll-grinding passage. In the second break passage the same function is repeated as more endosperm is released gradually and more of the released material is graded according to particle size to create segregation, as described in the foregoing list. The break stock from second break reaches the third break-grinding passage where similar action is repeated and more endosperm is released and graded. The break stock that reaches the fourth and fifth break, if the latter are used, would have very little endosperm remaining on the bran flakes as most of it would already have been extracted and graded. Therefore, the purpose of the last break-grinding passage might be described as cleaning up the bran before it moves on to a machine commonly known as bran duster or finisher. This particular machine removes the last traces of the endosperm that still adheres to the bran. There are two different sizes of bran that are produced. The coarse bran is referred to as bran, while the fine bran is referred to as shorts. Both of these are passed through separate dusters: namely, bran duster and shorts duster. Thus, although a very gradual breaking and releasing of the endosperm from wheat and subsequent break stocks is desired, at the same time a miller has to ensure that all the endosperm is released and the bran is cleaned up, using the available four or five break passages and a bran duster, before dispatching the bran to the bran bin.

3.3.7.2 Grading of the Released Endosperm and Purification

3.3.7.2.1 GRADING. After the endosperm particles are released from individual break-grinding passages, they are graded according to particle size in the corresponding sifter sections as already explained. There are several graded groups of material or stocks that are sifted: namely, coarse semolina, fine semolina, and middlings. This is because the grading of the material within narrow ranges of particle size is important, as it helps the purifiers to effectively segregate the pure endosperm particles, composite, and branny particles of the same size. Additional sifter sections are often required to complete the grading function, and these sections are then referred to as graders, or redressers, or dusters.

3.3.7.2.2 PURIFICATION. A modern purifier has been described earlier. Purification helps create a stratification of material according to density. This means the pure endosperm particles being heavier than the similar size composite particles and branny material would fall through the layer to the bottom while the lightest branny material of the same size floats up on top. The composite particles of endosperm and bran–germ attached to each other remain somewhere in the middle of the layer.

The pure endosperm particles thus, fall through the sieve openings first and are collected from the front part of the machine. The composite particles are carried toward the end of the machine where the layer starts to thin out. This type of material is collected from the back section of the machine. The branny material with very little endosperm attached is allowed to go over the top of the end sieve layers and is collected from the outlets located at the rear of the machine. This stratification of density layers can be effectively attained if the particle sizes of the pure and impure endosperm are fairly close to each other. If the particle sizes are quite different, the integrity of the purification action will be compromised. Thus, grading is very important to create necessary bands of graded stocks falling within a narrow range of particle size that would allow effective purification. Therefore, in the milling process, there are usually three or four purification passages that are allowed to obtain a satisfactory level of segregation between pure and contaminated stocks.

3.3.7.3 Reduction of Purified Endosperm Particles into Flour

Purification passages segregate the stocks into pure endosperm, composite particles, and branny material. The branny material ends up with the lower break stocks where the remaining endosperm is removed and bran is separated. Composite materials go to finely corrugated rolls where they are lightly ground to clip the endosperm and scratch off the bran pieces attached to them. These ground materials are sifted to allow the bran to go to the lower breaks and the retrieved endosperm particles, after being purified, are sent with other purified endosperm particles to be ground into fine flour particle size in the reduction rolls. This entire system of grinding the composite particles followed by sifting and purification is referred to as the *scratch* or *sizing* system.

Reduction of purified endosperm particles, from semolina, middlings, and sizing purifiers, into flour is the ultimate goal in the production of refined white flour. The objective in this part of the process is to grind pure endosperm particles into flour as much as possible. Because this material to be ground is virtually free of any bran contamination, there is no risk of shredding or powdering the bran along with endosperm as it was in the previous parts of the operation.

Grinding is done on smooth rolls, for the aim here is size reduction. This process of size reduction is achieved on a gradual basis involving several grinding passages, each of which is followed by sifting. The number of grinding and sifting steps could be at least six and as high as ten or more.

The purified endosperm particles are directed to the first reduction grinding passage where roll gap is narrowed to yield a substantial quantity of flour. The ground material is sifted and the flour is removed by sifting through the fine apertures of nylon cloth sieves. The material that remains on top of the sieves is then directed to the second reduction-grinding passage for further size reduction. Even the most purified endosperm particles do contain a few bran specks and parts of wheat germ. These parts of germ are flattened when they pass through a pair of smooth rolls because they are tough and naturally pliable unlike the endosperm particles, which are brittle. Thus, when ground, endosperm particles, being brittle, break down into finer particles, whereas bran and germ parts either remain the same size or slightly increase in size through flattening. Therefore, small quantities of bran and germ fragments can be easily sifted out on the basis of the size differential. The purity of the mainstream product is protected in this manner by first eliminating the bran fragments through purification and, subsequently, by avoiding contamination with any remaining bran and germ fragments by sifting and collecting similar material from three or four reduction passages together and directing them to a reduction passage that is either referred to as *coarse reduction* or *collecting passage*. There are usually two coarse reduction passages in the reduction system. Each one of these is placed after two or three reduction passages which would feed the bran–germ fragments to it. Therefore, in a conventional flow diagram germ is removed from a coarse reduction passage.

As we go down from the first to the last reduction passage, the percentage of flour extracted decreases in each of these stages. Generally speaking, the flour generated from the combined first four reduction passages is approximately 60% of the total flour that is produced. As each reduction passage extracts white flour from the material that is fed to it, the subsequent reduction passages downstream continue to receive less and less material. In addition to that, the residual material for grinding as it progresses through the reduction system also becomes darker, and it appears to be quite contaminated with fine branny material that cannot be further separated from it. This is because the residual materials on the lower reduction passages are largely devoid of the white flour, which has already been extracted. Also the branny contaminated materials obtained from the purification and lower break passages are directed to the lower reduction-grinding passages. For this reason the flour sieves used in the sifter sections become progressively finer down the reduction system. Since this helps in preventing the flour sifted from unnecessary contamination with bran specks, particularly from passages that are relatively high in bran contamination. From the very last reduction passage what still remains over the flour nylon sieves, is referred to as *middlings* or *mill run*, and this may be sold separately or added to the shorts and sold as such.

3.3.8 Flowsheet Concepts

The flowsheet of a flour mill graphically describes the origin and destination of all the ground material and their fractions throughout the milling process. As can be seen from the simplified flowsheet (Fig. 29), all the endosperm particles released from the first, second, and third break, being of good quality and primarily from central part of the wheat kernel, are flowed together. These endosperm particles are further graded according to particle size into coarse semolina, fine semolina, and middlings. Even though they are all virtually free of any bran contamination, they are sent to the first, second, and third reduction grinding passages, respectively, because of the differences in their particle size. This allows optimizing of the grinding roll gap setting relative to the particle size of the material. When a mixture of large and small particles is ground, the grinding performance is never optimum. For instance, if the setting were appropriate for larger particles, the smaller

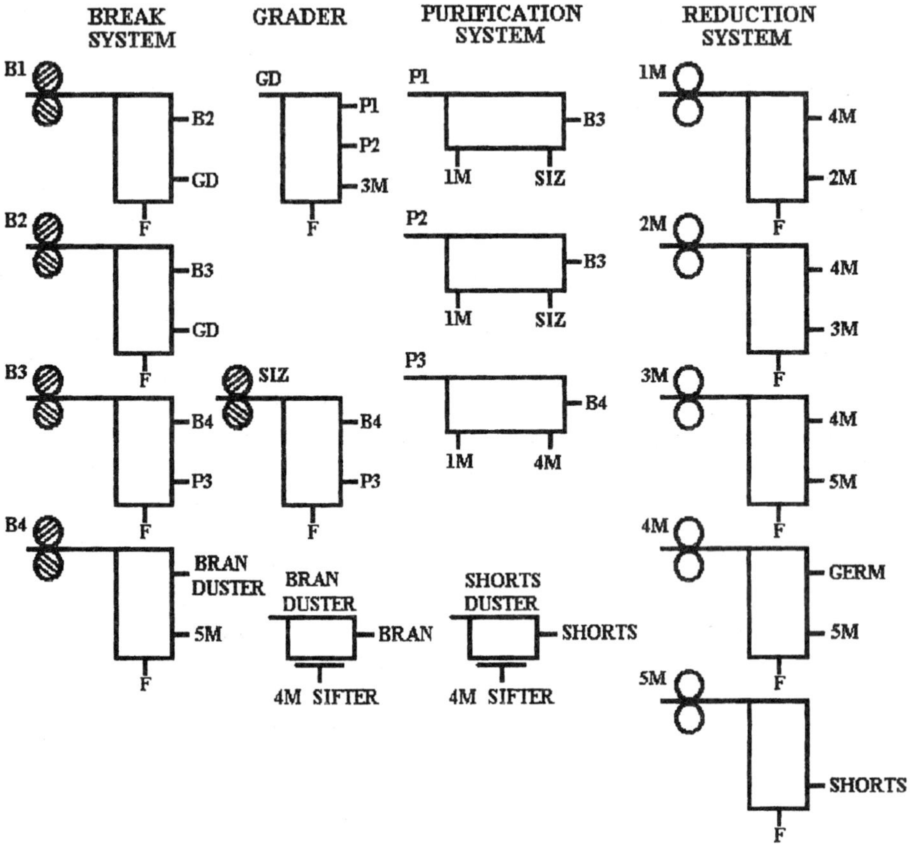

Fig. 29 A simple flowsheet.

particles remain unaffected or undertreated. If the setting is targeted for smaller particles, the larger particles are overground or overtreated.

Therefore, it is quite clear from the foregoing that various fractions of material are best grouped together for further processing on the basis of their degree of contamination with bran and germ (purity) and also on the basis of their particle size. It is also clear that although the reason for grouping materials on the basis of particle size is to optimize grinding and purification performance, the reason for grouping on the basis of bran contamination is to achieve an optimum yield of good quality flour from sifting. Relatively coarser nylon sieves are used for ground materials that are free of any contamination thus allowing a high percentage of refined flour to be extracted. Relatively finer nylon sieves are used for material contaminated with bran and germ particles. The fineness of the aperture size of the nylons is proportional to the degree of contamination as well as the mean particle size of the material to be sifted. While segregating material on its basis of degree of purity and particle size, the quantity of the material segregated must also be carefully considered. Each machine passage, such as grinding rolls, purification passage, and plansifter section, requires a certain quantity for optimal operation. Although on one hand, selection of numerous fractions within narrow bands for purity and particle size allows one to be discriminating; on the other hand, it may leave much less quantity in each of these fractions, resulting in less than optimum performance. Therefore, when designing a flowsheet these two aspects must be balanced, and often the integrity of quality (purity) and particle size is required to be compromised to maintain balance with quantity. Since different mills operate under different conditions, the consideration given to flowsheet design of a flour mill relative to the foregoing factors warrants much careful planning and thought.

Therefore, in a flowsheet design, flow of material considerations are based on three factors:

1. Quality
2. Particle size
3. Quantity

Other important considerations when designing a flowsheet are as follows:

1. Types of wheat to be ground: for example, milling durum wheat requires more purifiers whereas milling soft wheat requires more sifters with either less purifiers or no purifiers.
2. Flour refinement required: higher degree of refinement requires a more comprehensive flowsheet with more purifiers.
3. Flour end-use requirements.

There are some important rules for flowsheet design, which follow:

1. Fractions of material should not be routed back to the same machine passage or to a preceding part of the system as it creates a closed loop.
2. Refined flour should be produced as soon as practically possible, and as soon as a flour stream is produced it should be routed to the flour collection system. Otherwise the chance of contaminating the flour is greater.
3. Generally speaking, once a stock has been ground on a pair of smooth rolls it should not be passed through a pair of corrugated rolls.

The flowsheet of a flour mill (Fig. 29) can be considered the blueprint of its milling process, and it is generally regarded as confidential information by millers.

This flowsheet also shows various points in the diagram where flour is extracted. Although, in general, these flours may appear to be similar, a close visual examination of all of their flour streams would reveal visual differences in terms of color and, occasionally, fine specks of bran count.

3.3.9 Milling Performance Evaluation

Flour sells at a much higher price than bran; therefore, the measurement of flour production is a routine function in controlling the process. This important measurement of the performance is commonly referred to as the *flour extraction rate* and is calculated as a quantity of flour produced as a percentage of the wheat used. Variations of this calculation are applied to provide such information as is critical to the operation over a wider range of situations, as given in Table 8.

The objective is to produce as much flour as possible. However, the resultant flour should be free of any bran and germ contamination. Although, a flour contaminated with bran can easily be distinguished visually, the evaluation of minute differences sometimes can be quite subjective. Thus, for an objective evaluation either an ash test or a color measurement of the flour sample is performed. A higher presence of bran parts increases

Table 8 Various Methods for Calculation of Flour Extraction Rate

No.	Formula	Description of the method	Application
1.	Weight of flour sold × 100/ weight of wheat as received	This method takes into account the foreign material and natural moisture content of the wheat, which have a bearing on the flour extraction rate. The higher the percentage of these two factors the lower is the flour extraction rate and vice versa.	Used by management/mill accountants
2.	Hourly weight of flour × 100/hourly weight of wheat	This method takes into account the hourly reading of the wheat scale just before milling (after cleaning and tempering) and the hourly reading of the flour scale in the mill. The flour extraction thus calculated is sensitive to moisture evaporation loss from grinding.	Used by shift operators to enable them to adjust the milling equipment to obtain target extraction. Because the scale readings are easily accessible and they are monitored.
3.	Weight of flour × 100/ weight of total products	In this method the weight of flour is calculated as a percentage of the total weight of all the finished products (i.e., flour and all the by-products). This method of calculation is not influenced by moisture evaporation loss, presence of foreign material and the natural moisture content of the wheat.	Used in situations where flour and wheat scales are not available while the flour and all the other by-products are constantly bagged as they are produced
4.	Same calculation as in method 2 except wheat and flour weights are calculated on a constant-moisture basis.	This method uses weight correction for flour and wheat at generally 14% moisture thus eliminating any positive or negative influence on the flour extraction due to moisture.	This method may be used when precisely comparing the flour extraction rates of various wheats. Used in experimental studies.
5.	Using a calculation method similar to 1 except wheat and flour moistures are corrected to a standard constant moisture.	This method provides the most accurate flour extraction rate potential of a wheat or wheat mix.	Used where accurate extraction rates are necessary for comparison of various types of wheat, such as in laboratory milling comparisons.

the ash content of the flour, as pure endosperm has a much lower ash content than bran. Similarly, whiter flour free of branny material will be much brighter in color compared with the bran-contaminated flour which will be darker.

It is customary for millers to express flour extraction rates at a given ash content or a color value. Flour extraction rate expressed without an ash or color value does not help in the assessment of its potential flour extraction at an acceptable quality level.

As is indicated in the flowsheet, flour streams are drawn from each sifter section. Because these flour streams differ from each other in terms of bran contamination, they are visually somewhat different from one another. Thus, these flour streams would differ in ash content and color values. They would also differ in protein content and many functional properties such as protein quality and enzymatic activity. This is because in a kernel of wheat the protein content increases from the center of the kernel toward the bran. Therefore, flour streams from the central part (first reduction) of the wheat kernel are lower in protein content relative to the flour streams that come from closer to the bran (fourth break). Also, flour streams coming from the bran and germ end are higher in α-amylase activity.

Another important tool for measuring milling performance is plotting a cumulative ash curve. If the curve initially remains flat, indicating no or little increase in ash as the percentage of flour increases, and toward the higher end the flour ash begins to ascend then this will be considered as a good-milling performance as compared with a curve that begins to ascend at an earlier point (Figs. 30 and 31).

Cumulative protein curves may also be drawn, and they follow a similar pattern. To obtain a cumulative ash curve the individual flour streams are arranged in ascending order of ash content. The percentage of each flour stream is also listed beside its ash content. By applying the weighted average calculation method, the cumulative ash content for the cumulative flour percentage is calculated. Table 9 provides such data for both ash and protein calculation.

Because the white flour commands a much higher price than the by-products, it is easy to understand why so much importance is attached to achieving a high flour extraction rate. As might be expected; however, the extraction rate is not a complete parameter per se. On one hand, a miller would attempt to extract as much white flour as possible by

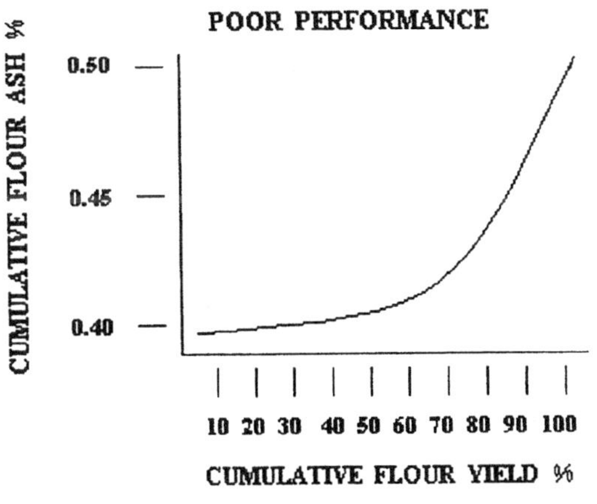

Fig. 30 Ash curve—poor performance.

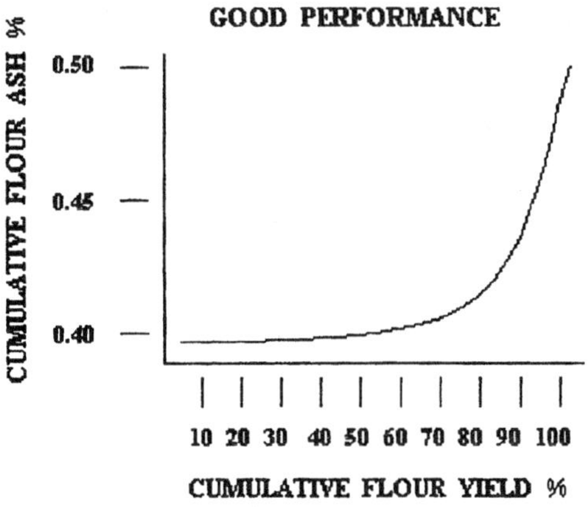

Fig. 31 Ash curve—good performance.

carefully removing the last traces of endosperm from the bran flakes. On the other hand, in the process, the final trace of flour that is extracted would increase the total ash content of the flour and make the overall flour a darker color. (Ash content provides a measure of the extent of contamination with bran.) Thus, flour extraction rate is meaningful only in conjunction with its ash content or color.

3.3.10 Functional Properties of Flour

Functional properties of flour are measurable characteristics that are related, quite predictably, to the secondary processing of flour. This section deals briefly with some of the more common properties of concern to the miller.

Table 9 Cumulative Ash and Protein of Flour Streams

Flour stream	Proportion of total flour (%)	Cumulative flour (%)	Ash (%)	Cumulative ash (%)	Protein (%)	Cumulative protein (%)
1 Reduction	23	23.00	0.37	0.37	11.7	11.7
2 Reduction	16	39.00	0.38	0.37	11.6	11.7
3 Reduction	12	51.00	0.39	0.38	11.4	11.6
Sizing	5	56.00	0.46	0.39	12.0	11.6
2 Break	7	63.00	0.48	0.40	14.8	12.0
1 Break	4.2	67.20	0.50	0.40	14.6	12.1
4 Reduction	7.2	74.40	0.51	0.41	12.3	12.2
1 Coarse reduction	4.3	78.70	0.54	0.42	12.2	12.2
Grader	5.2	83.90	0.59	0.43	14.4	12.3
3 Break	4.7	88.60	0.64	0.44	17.8	12.6
5 Reduction	4.4	93.00	0.69	0.45	12.8	12.6
Filter	1.6	94.60	0.75	0.46	14.9	12.6
2 Coarse reduction	1	95.60	0.85	0.46	12.5	12.6
4 Break	1.5	97.10	0.93	0.47	18.3	12.7
Bran duster	0.5	97.60	1.40	0.47	18.6	12.8
6 Reduction	1	98.60	1.47	0.48	13.2	12.8
Shorts duster	0.7	99.30	1.56	0.49	16.3	12.8
7 Reduction	0.7	100.00	1.62	0.50	13.2	12.8

3.3.10.1 Ash Content and Color

Ash content is the proportion of incombustible mineral matter that remains after a measured quantity of flour is completely incinerated. There is a direct relation between ash content and degree of bran contamination (as explained in the next paragraph). Consequently, ash content provides a useful measure of flour purity and, therefore, of milling efficiency.

Within a kernel of wheat, ash content increases from the center toward the bran, as might be expected because mineral matter is concentrated in the outer (bran) layers of the kernel. Thus, the flour that comes from the center of the kernel is the whitest, whereas the flour scraped off the bran is the darkest. Therefore, the flour streams from head reduction passages are the whitest and the lowest in ash content, whereas flour streams from the last break and reduction passage are darkest and highest in ash content.

3.3.10.2 Protein

Similar to ash content, protein content also increases from the center of the endosperm toward the bran skin within a wheat kernel. Therefore, flour streams from head reduction passages are lowest in protein content and flour streams from lower break passages are highest. Wheat flour protein when hydrated forms gluten. Gluten provides the viscoelastic properties to the dough, which is very important for baking. Wheat flour protein, with the exception of rye, to some extent, has this unique property among all cereal grains.

3.3.10.3 Flour Strength and Flour Stream Properties

As flour and water are mixed together to form a dough, protein hydrates and gluten is formed. The viscoelastic properties of the dough is characterized by gluten properties. Gluten properties and consequently dough properties vary from wheat flour to wheat flour. Some gluten can be too weak and extensible, whereas other gluten can be too strong to stretch and, therefore, inextensible. The dough properties would, therefore, correspond to gluten properties. Flour with strong gluten properties would exhibit strong physical dough properties. This characteristic of flour is referred to as *flour strength.* Flour strength is an important property for the purposes of bread baking. Flour streams from final break and lower reduction passages exhibit weaker properties and lower baking tolerance. This is due to their relatively higher degree of contamination with bran and germ. Flour streams from head break and head reduction passages show relatively stronger physical dough properties and higher baking tolerance. The flour stream coming off the germ end, with a higher degree of germ contamination, is relatively high in enzymatic activity. Flour streams from break passages are much lower in starch damage, whereas starch damage is much higher in flour streams from lower reduction passages. All of these properties play an important role in the end-use quality of the flour.

3.4 Flour Collection and Treatment

Individual flour streams are produced at various stages of the milling process. These flour streams are collected together in a group of one, two, or three, or even more. The groupings of flour streams are based on their ash content or flour color. The whitest flour streams, which have low ash, are put together in flour conveyor no. 1 (Fig. 32). This flour is also referred to as ''patent flour.'' After the removal of the *patent flour* the remainder flour is called ''clears flour.'' *Clears* flour is sometimes separated into ''first clears'' and ''second clears.'' This is shown in Fig. 32 where *first clears* and *second clears* are going into flour conveyors no. 2 and no. 3, respectively. When all of the flour streams are collected together to produce one flour, it is referred to as *straight run* flour. When various flour streams

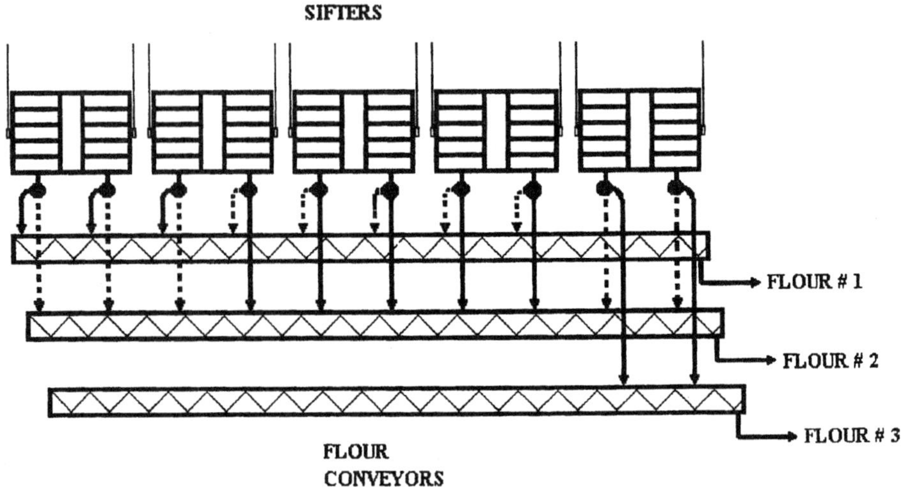

Fig. 32 Flour streams collection.

are split two or more ways, the practice is commonly known as *split run* milling. As can be seen in the Fig. 32, after the flour streams are separated by sifting, they are directed to flour-collecting conveyors through appropriate spouting. Generally, there is more than one conveyor. The conveyors are located beneath the sifter floor. The flour spouts are arranged so that they can swivel at the elbow where the spout bends just above the flour conveyor. Flour spouts, thus, can be swung from one conveyor to the other. The alternative arrangement for switching flour streams involves the use of diverter valves, which allows a given flour stream to be diverted to any of the flour conveyors of choice.

Table 10 shows collection of flour streams in three groups. The ash contents and protein contents together with the proportions of each flour stream for this example are taken from Table 9. In practice, flour streams with similar properties are combined into corresponding flour conveyors. In this example, three flour conveyors are used, providing three combinations of the original 18 flour streams. Table 10, thus, shows flour proportion, ash, and protein content of the three combinations, designated 1, 2, and 3. If, instead, the flour streams in Table 9 were combined in sequence, the ash content and protein content of the cumulative flour at each step would be shown by the ''cumulative ash'' and ''cumulative protein.'' Thus, the straight grade flour (last line) would have an ash content of 0.50% and a protein content of 12.8%.

In Canada, flour millers use a device called a ''divide board,'' that provides an additional level of control whereby flours leaving the conveyors may be further divided or recombined as desired. This process is referred to as *divide milling.*

A divide board is equipped with diverter valves. The number of inlets as well as outlets on a divide board is same as the number of conveyors. If all the diverter valves are set in the neutral position (Fig. 33), the flour from each conveyor passes through the divide board unaltered. Depending on the flour quality requirement these valves are adjusted to further combine or split flours coming from various conveyors (Figs. 34 and 35).

Figure 34 illustrates a divide board set to produce flour for large-scale bakeries, which requires slightly lower protein content. This flour is used in a fully mechanized bakery to produce pan breads. It is generally a straight run flour.

Table 10 Analysis of Split Run Flours

Flour stream	Proportion of total flour (%)	Cumulative flour (%)	Ash (%)	Cumulative ash (%)	Protein (%)	Cumulative protein (%)
1 Reduction	23	23.00	0.37	0.37	11.7	11.7
2 Reduction	16	39.00	0.38	0.37	11.6	11.7
3 Reduction	12	51.00	0.39	0.38	11.4	11.6
Sizing	5	56.00	0.46	0.39	12.0	11.6
2 Break	7	63.00	0.48	0.40	14.8	12.0
Flour no. 1		63.00		0.40		12.0
1 Break	4.2	4.20	0.50	0.50	14.6	14.6
4 Reduction	7.2	11.40	0.51	0.51	12.3	13.1
1 Coarse reduction	4.3	15.70	0.54	0.52	12.2	12.9
Grader	5.2	20.90	0.59	0.53	14.4	13.3
3 Break	4.7	25.60	0.64	0.55	17.8	14.1
5 Reduction	4.4	30.00	0.69	0.57	12.8	13.9
Flour no. 2		30.00		0.57		13.9
Filter	1.6	1.60	0.75	0.75	14.9	14.9
2 Coarse reduction	1	2.60	0.85	0.79	12.5	14.0
4 Break	1.5	4.10	0.93	0.84	18.3	15.6
Bran duster	0.5	4.60	1.40	0.90	18.6	15.9
6 Reduction	1	5.60	1.47	1.00	13.2	15.4
Shorts duster	0.7	6.30	1.56	1.06	16.3	15.5
7 Reduction	0.7	7.00	1.62	1.12	13.2	15.3
Flour no. 3		7.00		1.12		15.3

Figure 35 illustrates how the divide board is set to divert 48% of flour from no. 1 into no. 2 to produce a strong bakers flour. This setting would then also produce 15% all-purpose or household–family flour.

3.4.1 Flour Treatment

The practice of flour treatment is an integral part of the milling process in many countries. A miller may be required to treat the flour to accomplish the following objectives:

1. Enrich the flour with vitamins and minerals to fulfill the nutritional requirements
2. Enhance the appearance of flour color by bleaching
3. Modify or improve the flour properties to tailor it for a specific application

There are some countries in the world where finished flour does not undergo any treatment. This would most likely be true where quality is not considered particularly critical. This situation may exist in many underdeveloped regions of the world. Even in such cases, flour imported or received as part of an aid package may contain vitamins and minerals. The following section describes each of these objectives in more detail.

3.4.1.1 Enrichment

As shown in Fig. 27, endosperm makes up about 83% of a wheat kernel; however, it contains, relative to the whole kernel of wheat, only a small portion of the important vitamins (Table 11).

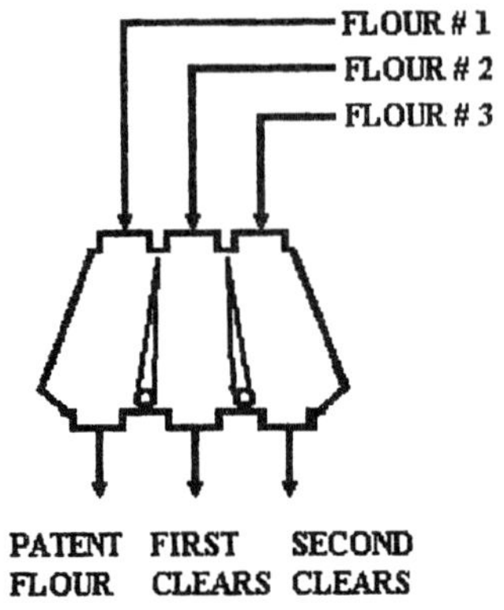

Fig. 33 Neutral position.

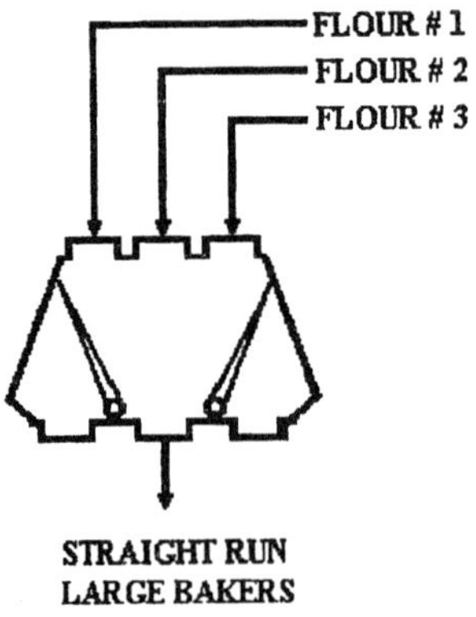

Fig. 34 Straight run.

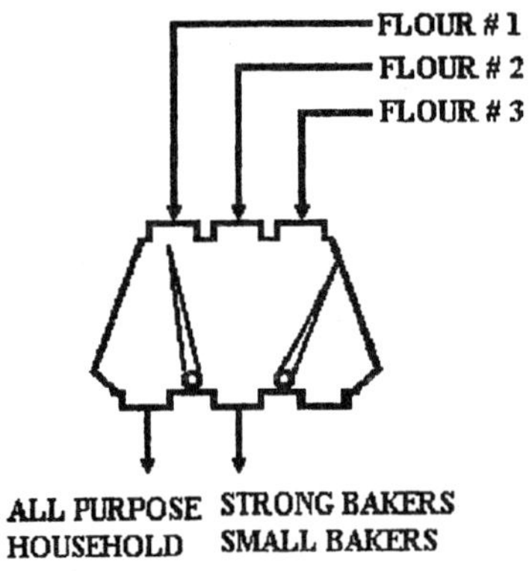

Fig. 35 All purpose and strong bakers.

Table 11 Percentage of Vitamins in Endosperm[a]

Vitamins	%
Riboflavin	32
Niacin	12
Thiamin	3

[a] a kernel of wheat.

In the pursuit of the production of the white refined flour, many of the nutrients are lost with the removed bran and germ fractions. In North America, it is a common practice, therefore, to enrich the flour with these vitamins and some minerals. In Canada, however, flour must be enriched to a prescribed level according to the federal law (Table 12).

There are a number of countries that would require the addition of the foregoing enrichments, however, to varying degrees there are certain special needs that may require the use of specific vitamins or minerals. For example, in the United Kingdom, the supplementation of calcium is required by law. This necessitates the addition of calcium carbonate in the flour.

3.4.1.2 Enhancement of Flour Appearance

The presence of fine bran particles and flour streams from the bran end (low-grade) have not only a deleterious effect on the functionality of the flour, but also have an adverse effect on the flour appearance in terms of *dress* (bran specks) and color. Furthermore, within a given variety of wheat, the higher the flour extraction rate, the greater will be polyphenol oxidase enzyme levels (Hatcher and Kruger, 1993). Polyphenol oxidase has a darkening effect on flour doughs. The color progressively darkens with time. This means further deterioration of the color and color stability of the dough and resultant products as the flour is processed in the secondary-processing industries. Additionally, freshly milled flours generally have a *creamy* color.

From the foregoing, it is obvious that much of the flour appearance problem (color and dress) can be resolved through appropriate flour dressing in the sifters and by controlling the flour extraction rate to a point where the problems related to the bran specks, color of the flour, color stability of the dough, and subsequently its processed products,

Table 12 Minimum Levels of Nutrient Requirement of Flour in Canada[a]

Nutrient	Minimum (mg/100 g)
Thiamin	0.64
Riboflavin	0.40
Niacin	5.3
Iron	4.4
Folic acid	0.15

[a] *Source*: FDA requirements.

cease to exist. Flour color whether dark (high extraction) or creamy/yellowish (freshly milled) has a good correlation with finished product color. Therefore, to obtain a whiter crumb in bread the flour also has to become whiter. The creamy color of the freshly milled flour is due to the presence of carotenoid pigments. There are two ways to reduce or eliminate this: (a) natural aging of flour, (b) whitening of flour through additives.

3.4.1.2.1 NATURAL AGING. A few weeks of storage following the production of flour tends to whiten the flour by reducing the pigmentation. This is referred to as *natural aging* of flour. In addition to the improvement in flour color, the aging process also has beneficial effects on the dough characteristics and the finished product quality. There are a number of countries where flour is not allowed to be treated chemically. In such situations flour is kept in the storage bins for about 1 week to almost 4 weeks before bagging and shipping. During this time the flour may be homogenized by turning it over. It is obviously an expensive proposition to carry the flour for a number of weeks after its production. Economical solutions are available in which flour is allowed to be chemically treated, as described next.

3.4.1.2.2 WHITENING WITH ADDITIVES. Flour may be bleached with benzoyl peroxide to whiten. It is a very common practice throughout the world to bleach flour using this chemical. Chlorine would also bleach, but it also has other undesirable influences.

Bread produced from flour bleached with benzoyl peroxide clearly has a whiter crumb when compared with bread crumb color from unbleached flour. There are other products, such as enzyme-based and flour that is extracted from legumes that can also be used for whitening the bread crumb. Products such as soy flour are added at the bakery level for such purposes. On dry flour, however, benzoyl peroxide is the principal bleaching agent that is used. It is added in powder form with a common application rate of 50 ppm. The bleaching action is affected by temperature, time, and rate of application.

At an average-operating temperature of 30°C it would take about 1–2 days to complete the action, as the benzoyl peroxide breaks down to benzoic acid. Addition of higher levels of benzoyl peroxide would hasten the bleaching action; however, there is a risk of some residue remaining in the flour.

3.4.1.3 Additives to Improve Functional Properties

The finished bread product quality is a function of gas production and gas retention. These two factors must be in a proper balance to yield an optimum finished product. Changes in wheat quality, milling variables, and other conditions would have an influence on these factors. Thus, maintaining a perfect balance can be quite challenging and almost impossible without the use of any additives.

Gas production is a function of α- and β amylase activity, starch damage, and yeast. Gas retention is a function of protein content and protein quality.

3.4.1.3.1 ADDITIVES FOR GAS PRODUCTION. Although additives for gas retention properties will be discussed later, let us first look at the diastatic supplements that affect the gas production.

3.4.1.3.2 DIASTATIC SUPPLEMENT. The starch component in the flour is broken down to form dextrins and fermentable sugars, which are subsequently broken down to release carbon dioxide. The α- and β-amylase, naturally present in the flour, actually work on the damaged starch to start the process. While physical damage of the starch is primarily influenced by milling variables, such as grinding pressure, tempering, ratio of break and reduction flours, and wheat hardness, the α-amylase activity is dependent on the wheat quality. If the wheat is too sound then there is very little activity and as the sprout damage level increases in wheat the activity intensifies. If the wheat has very little α-amylase activity, obviously, there will be insufficient gas production at the end; therefore, to ensure optimum gas production it is necessary to supplement the α-amylase in the form of an additive. The α-amylase can be obtained from three different sources: fungal, cereal-based, and bacterial.

The use of fungal α-amylase became more popular in Canada a little over 20 years ago. Before that mills were using malt flour (cereal-based). The advantages of using the fungal α-amylase as compared with the addition of malt flour are as follows:

Better standardization.
Its thermal death point is lower; therefore, there is a built-in protection against over-treatment.
It can be concentrated in its production; therefore this allows addition of only smaller quantities.

In Canada a popular product for fungal α-amylase addition is referred to as DOH TONE. A similar product is also used in the United States.

3.4.1.4 Additives for Gas Retention

The dough must have appropriate gas retention properties whereby it will be expanded easily by the gas that is produced, without collapsing or causing a big rupture. The gas retention properties are dependent on the protein content as well as protein quality. When a flour is deficient in protein content, adding *gluten* in powder form is a common practice to overcome the deficiency.

Although good protein quality is an attribute of wheat quality, the protein quality can be further improved through the oxidation process of the flour. The effect of the flour oxidation process is described next.

3.4.1.5 Flour Oxidation

The oxidation process strengthens the flour by increasing its resistance. Increased resistance with a good balance of extensibility improves the thin–film-forming properties of the gluten when stretched, thereby, allowing it to retain the expanding gas bubbles in the dough. Some of the common oxidizing agents used as flour additives will be discussed.

3.4.1.5.1 POTASSIUM BROMATE. Potassium bromate has been recognized as the most popular and effective oxidant in treating flour to improve its baking properties. Its effectiveness has been thought to stem from its being a slow-acting improver, and its economical cost makes it even more attractive. Its action extends throughout the baking process, including in the oven where a desirable increase in volume, referred to as *oven jump*, is obtained. There are some other oxidizing agents for which action is considered to be limited only up to the mixing stage.

Potassium bromate, however, is banned in many countries, including Canada, and the list is growing because it is a carcinogen. Some studies have indicated that it does not break down completely to form potassium bromide; thus, there is a residual potassium bromate that can be detected in bread.

3.4.1.5.2 AZODICARBONAMIDE. Many years ago, it was a common practice to keep flour in storage to naturally age or mature it to whiten and to improve its functional properties. Flour is still matured that way in some countries; however, it is quite common nowadays to mature flour artificially by using additives such as azodicarbonamide (ADA). It is not economically viable to hold the flour in the bin for 3–4 weeks. Also flour stored in the bin for a longer period is likely to become infested. The common application rate for ADA is 4–5 ppm. At this level it acts as a maturing agent and helps maintain uniformity in flour quality relative to what is referred to as *green* flour.

Flour milled from freshly harvested wheat would require a higher level of ADA. Higher levels would also be required for strong bakers and clears flours that are typically higher in ash. Maximum allowable limit of ADA is 45 ppm in North America, and when it is used as an oxidizing agent, such as a bromate replacer, then the application rate is much higher than that used for maturing purposes.

ADA is fast-acting and its action is fully completed at the end of the mixing stage in the baking process. In some processes involving longer fermentation, it may actually convert into biurea before the mixing stage, for it is highly reactive. One way to solve that problem is by coating ADA and thus delaying its action. Coating, however, adds to the cost. There are two concentrations at which ADA is available in North America. The first one is Maturox (a trade name) at 10% concentration premix and the second one is a premix of ADA concentrate at 23% concentration. The flour mills in Canada routinely use this additive.

3.4.1.5.2 ASCORBIC ACID. Similar to other additives, ascorbic acid is also sold as a premix and is available under the trade names AA-10 and AA-25W. Ascorbic acid in the past has been of only limited use, primarily owing to its cost. Since the removal of potassium bromate, it is quite commonly used by millers as an oxidizing agent. The typical application rate is anywhere from 25 to 75 ppm. Its effect on increasing loaf volume may not be as dramatic as would be expected from flour treated with potassium bromate, but it provides tighter and finer grain in the bread crumb. It also helps extend the mix time in the baking process. The ascorbic acid requires enzymes in flour to oxidize it to form dehydroascorbic acid. It is this dehydroascorbic acid that oxidizes the flour. Therefore, the effect of ascorbic acid varies with flour owing to the varying levels of enzymes present. Dehydroascorbic acid can not be used by itself, for it is highly unstable. As a replacer for potassium bromate, ascorbic acid in various combination with ADA has been tried with some success. The maximum allowable limit for ascorbic acid is 200 ppm and, in some situations, it may be added very close to its upper limit as a bromate replacer.

3.4.1.5.3 REDUCING AGENT. Oxidation is particularly helpful when gluten is much more extensible in relation to its resistance. Some flours have strong gluten with much higher resistance and low extensibility. To improve gas retention properties in these flours, gluten has to be made more extensible (mellow) to enable appropriate stretching of gluten without breaking. In such a case, flour is treated with a reducing agent. L-Cysteine hydrochloride is commonly used for such purposes. Proteolytic enzymes, such as protease may also be used for such application.

L-Cysteine hydrochloride is also added at the mill level for specific flours. It is highly corrosive in pure form and, therefore, would be diluted to form a stabilized, noncorrosive premix. It helps reduce mix time and mellows the dough by making it more extensible. In Canada its addition is allowed at the mill level, whereas in the United States, it is allowed only at the bakery level. Although the maximum allowable limit is 90 ppm, the typical addition rate would be 20–30 ppm. Extensibility in pizza and tortilla doughs is important; therefore, it is added at higher levels. In bread flours, it is used primarily in a specific baking application referred to as *no-time dough*.

Table 13 Flour Additives Commonly Used in Canada[a]

Flour additive	Maximum permitted level	Physical state	Application rate (ppm)	Purpose
Azodicarbonamide	45 ppm	Powder	2–20	Maturing and oxidation
Chlorine	Sufficient for bleaching	Gas	1000–1400	Bleaching, maturing
Benzoyl peroxide	150 ppm	Powder	50	Bleaching
Ascorbic acid	200 ppm	Powder	70	Improver, oxidant
L-Cysteine (hydrochloride)	90 ppm	Powder	30	Mix time reducer/relaxer
Fungal α-amylase	As required	Powder	As required	Diastatic supplement

[a] *Source*: Flour Treatment Agents table published by AIC Canada as per FDA requirements.

3.4.2 Treatment of Soft Wheat Flours

3.4.2.1 Chlorine

Chlorine is added in gaseous form to cake flours. In addition to the bleaching of flour chlorine helps improve the volume, grain, and texture of the finished product. This is particularly applicable to high-ratio cake. Cake flour is generally chlorinated to bring the pH level of the flour down to about 4.8. Generally, a treatment level of 1250 ppm helps realize the required pH level of 4.8.

Table 13 lists commonly used additives. It also provides various other information, such as permitted levels and common application rates.

3.4.2.2 Dispensing Additives into Flour

The addition of flour additives are in very small quantities; therefore, to be evenly distributed in the flour their volumes are increased by diluting with inert free-flowing material. It is important after the additives are dispensed that they be thoroughly mixed with the flour in a flour conveyor. It is quite common to have the flour additive feeders directly located above the flour conveyors in North America. The powder feeders used in North American flour mills are very simple in design and very effective in operation.

3.5 Generating Finished Products and Their Handling

As described earlier in Sec. 3.4, flour streams may be collected into more than one flour conveyor. It was also mentioned that these flours can be further recombined or divided, using *divide boards*, to produce finished flours of targeted quality. Divide boards have their limitations. In situations when numerous finished flours of varying quality are required, a flour-blending system becomes a necessity.

A comprehensive flour-blending system can accomplish the objective very precisely. Such facilities, additionally, allow a flour production plant to manufacture *flour* mixes that need to be formulated according to a given recipe that requires blending of macro- and microingredients together with the flour in a batch-blending environment. In such systems flours from all the conveyors are sent to individual bins, rather than through the divide boards. There are a sufficient number of flour bins to accommodate a large number

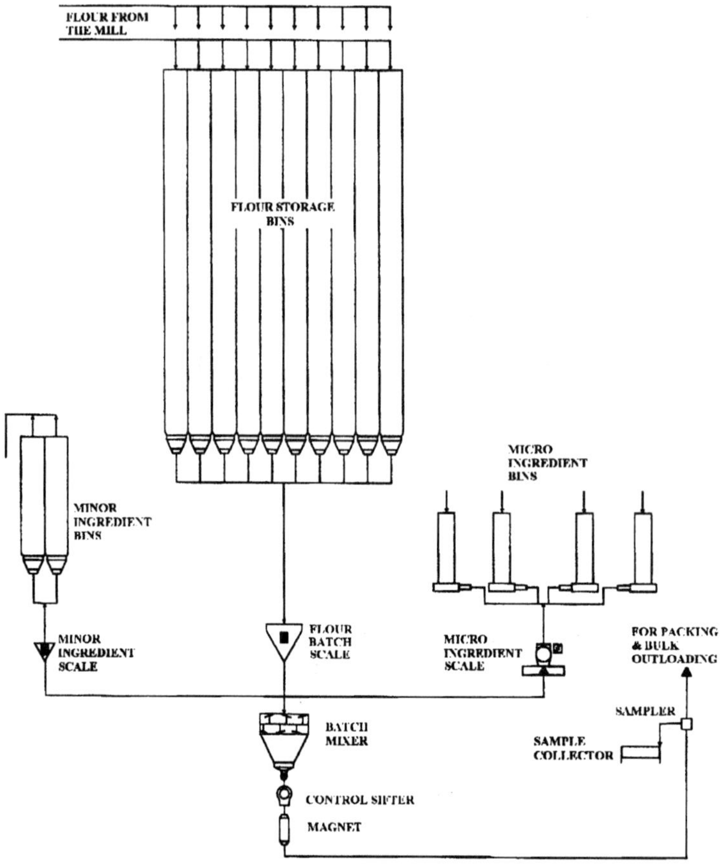

Fig. 36 A comprehensive flour-blending system. (Courtesy of Bühler, Inc.)

of base flours. Such a system is shown in the Fig. 36. These systems are seen in Europe as well as in those facilities around the world where smaller lots of wide ranges of specific finished flours are required.

Base flour bins may consist of patent (flour no. 1), first clears (flour no. 2) and second clears (flour no. 3), as described earlier; for example, from milling of a hard red spring wheat with a strong gluten. Some base flour bins may comprise patent and clears flours from medium hard, semistrong wheat, and also from weak and soft wheat. There may be more base flour bins dedicated for high- and low-protein hard wheat flours. The following examples indicate how various base flours may be blended to generate a wide range of suitable finished flours.

3.5.1 Different Types of Flour

1. Household, family, or all-purpose flour: only the patent flour (no. 1) of low-protein (11.5–12.0%) and low-ash content (0.38–0.40%) from hard red spring wheat is drawn from the bin, and there may not be any blending required.
2. Large bakers flour may require blending of all three flours (no. 1, no. 2, and no. 3) from hard red spring wheat. If a straight-run flour from this wheat is

binned, then this bin may be drawn without requiring any further blending. Flour protein and ash content requirements for such flours commonly range from 12.5 to 13.0% and 0.50 to 0.52%, respectively.

3. Strong bakers flour may be produced from a straight-run milling using higher level of wheat protein. Alternatively, it can be generated by blending the patent flour and clears flour 50% each from hard red spring wheat. The protein and ash content range for such flours would be 13.0–13.5% and 0.52–0.56%.
4. In flour for starch and gluten separation: first and second clears from hard red spring wheat may be blended in the ratio of 80–85% to 15–20%, respectively.
5. Whole wheat flour may be produced by grinding the whole grain into a meal form of required particle size. However, more commonly, flour mills blend 85–88% of strong bakers flour base with 12–15% of fine bran of appropriate particles size. In Canada the wheat germ is left out, thereby increasing the shelf life of the whole wheat flour.
6. Pastry flour: patent and clears from soft wheat flours may be blended to produce pastry flour. If straight-run flour from soft wheat is being binned, then there is no blending required. Protein content for such flour is approximately 8.5–9.5% with an ash content varying from 0.48–0.52%. This flour can also be used for cookies.
7. High-ratio cake flour: patent flour from soft wheat with very low protein and low ash of approximately 8.0 and 0.35%, respectively is used. The cake flours usually have very fine particle sizes.
8. Cookie flour may be a straight-run flour or a blend of some patent and more clears flour. For example, 20% of patent flour may be blended with 80% of clears.
9. Cracker flour requires a little bit of strength; therefore, a blend of 50% medium strength straight-run flour with straight-run soft wheat flour may produce desirable results. The protein content in cracker flour may be about 9.5–10.5%.

3.5.2 Finished Products Handling

Handling of finished products involves storage, handling, and dispatching finished flours and by-products. Such a facility is shown in Fig. 37. As can be seen from Fig. 37, flour is stored, handled, and dispatched for three principal forms of utilization, as follows:

1. Storage and bagging in large bags using a high-speed carousel packer. This is primarily for use in small bakeries.
2. Storage and bagging in small bags of various sizes to be sold in supermarkets as consumer (household) flour.
3. Storage and loading in bulk for deliveries to the large-scale mechanized bakeries in bulk tankers or trucks.

Figure 37 also shows the bagging as well as bulk-loading of by-products.

3.6 Finished By-Products and Their Utilization

Most of the endosperm of a wheat kernel is ground into fine flour and sometimes a small portion of it is additionally extracted as granular refined product referred to as semolina or farina. These are considered the main products of a flour mill. There is a small portion of endosperm that becomes highly contaminated with bran during the milling process and

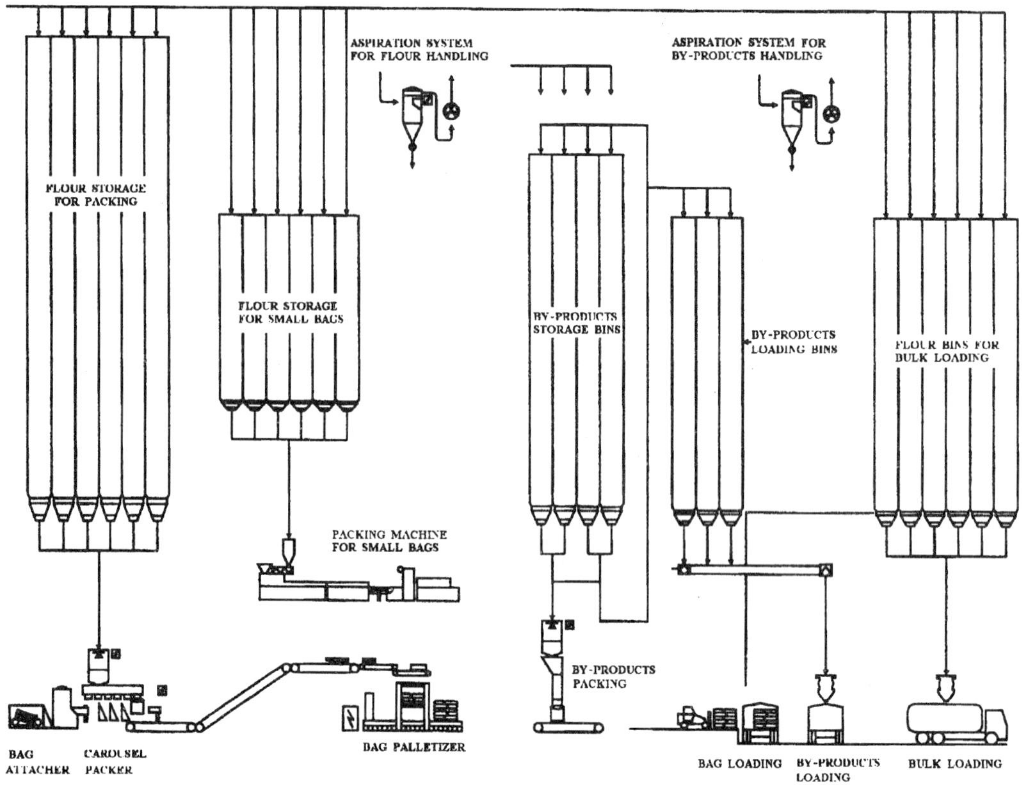

Fig. 37 Finished flour and by-products storage, handling, bagging, and bulk loading. (Courtesy of Bühler, Inc.)

thus cannot be recovered. This material, commonly referred to as middlings in North America, along with bran, shorts, and germ form the by-products of the milling industry.

3.6.1 Bran

Bran, as described earlier in the milling process, is removed from the break system. After cleaning up and sifting in the last break passage, the coarsest material that is removed is the bran. The next coarsest material removed in the same passage following the removal of the bran is the fine bran or shorts which will be described next. Bran from the sifter is sent to the bran duster for a final clean up before being sent to the storage bin.

Bran is sold in bags or bulk, and sometimes it is pelletized. Bran has a very low bulk density; therefore, converting it into pellets reduces its volume substantially. This is particularly beneficial when bran has to be transported over a long distance. The crude protein and fiber levels are high in bran, making it particularly suitable for animal feed ration.

Some bran is also used for human consumption, such as in whole wheat breads, bran muffins, and breakfast cereals.

3.6.1.1 Shorts

Shorts is also passed through a shorts duster after it is removed in the break system. It essentially consists of relatively finer pieces of bran particles. Its main application would be in poultry rations. In small plants, middlings are sometimes mixed with shorts and marketed as one product.

Clean shorts without any middlings in it, may also be blended with flour directly on line for the production of whole wheat flour.

3.6.1.2 Middlings

Middlings are the fibrous residual branny or endosperm material from the lower reduction rolls that cannot be processed further to produce any more flour of acceptable quality. Further attempts of extracting any more flour from it would result in inferior quality flour with poor functionality. Therefore, it is sold as animal feed.

3.6.1.3 Germ

Sometimes germ may not be separately removed and, in that case, it goes with bran. Germ in most flour mills is usually removed from the coarse-reduction sifting passage as the coarsest fraction. Recovery of germ in this manner is relatively small compared with those systems where additional equipment is installed to obtain a high recovery of food-grade germ. Germ contains very high protein and is also a good natural source for vitamin E. The utilization of germ as breakfast cereal is quite common. Since germ contains oil, it needs to be stabilized before it can be safely stored for a reasonable time. Germ is toasted for that purpose. Lower grades of germ may also be used for animal feed.

4 SOFT WHEAT MILLING

Hard wheat with strong gluten quality and high gluten content is used for milling flour for bread. Suitable flour quality for the production of cakes, cookies, pastries, and crackers require use of weak flour quality made from milling of soft wheat with low gluten content and weak gluten properties. The milling process described before relates primarily to the milling of hard wheat. The milling process for hard wheat can be well adapted for milling of soft wheat, as they are quite similar, except for the differences in the physical properties of the endosperm of the two types of wheat. The bran in soft wheat adheres more strongly to the endosperm than in hard wheat; therefore, it becomes more difficult to separate.

Figure 38 shows a cross section of a kernel of hard, soft, and durum wheat. As can be seen the soft wheat endosperm looks starchy and opaque whereas endosperm of hard wheat looks hard, crystalline, and translucent. The cell structure of soft wheat endosperm is very weak and it breaks readily, whereas that of hard wheat is very strong.

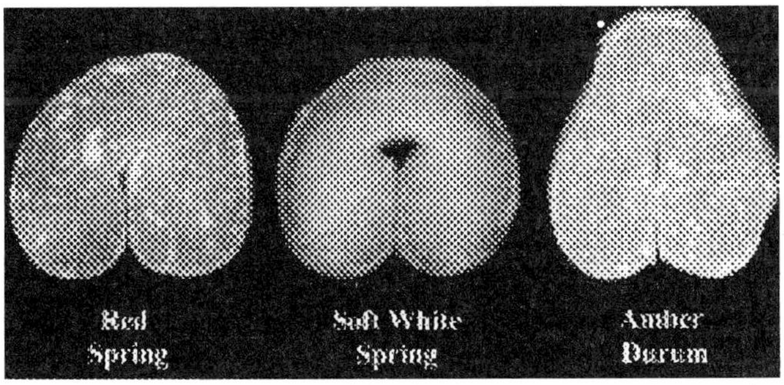

Fig. 38 Hard wheat, soft wheat, and durum wheat. (Courtesy of the Canadian Grain Commission.)

When viewed under a microscope, soft wheat flour appears woolly, whereas hard wheat flour appears to be hard and crystalline. These differences require a somewhat different approach for the preparation and milling of the soft wheat than that for hard wheat.

4.1 Preparation of Soft Wheat for Milling

The principles of separation of all of the cleaning equipment is based on kernel size, shape, and density. Because soft wheat kernels are similar to hard wheat kernels relative to all of these characteristics, they are cleaned in the same manner as hard wheats.

Tempering requirements for soft wheat are quite different and critical for achieving acceptable milling results. Soft wheat requires a relatively lower milling moisture than hard wheat for easier and cleaner separation of bran. The texture of a soft wheat kernel is soft relative to the texture of a hard wheat kernel. Also the bran is relatively more pliable. Therefore, there is little or no risk of shredding the bran when milling soft wheat at a lower moisture content. Additionally, the tempering time required for soft wheat is also lower than hard wheat, for longer tempering time is required for mellowing the endosperm. The cell structure of the soft wheat being weak allows much quicker moisture penetration relative to the hard wheat. Lower levels of moisture addition in soft wheat also is a factor in reduced tempering time requirement. The average milling moisture of soft wheats is 14.0–15.0%. Tempering time varies from 4 to 10 h.

4.2 Milling

Milling of soft wheat requires a lot more attention to detail than does the milling of hard wheat. This is because, being sticky in nature, the soft wheat ground material is hard to sift and separate. Thus, the milled products are not free-flowing causing spouts, lifts, and other equipment to become easily plugged.

In addition, when milling soft wheat, attention must be paid particularly in the following areas:

1. Break rolls are adjusted in such a manner that excessive production of break flour is avoided. In general, the percentage of break flour production in the milling of soft wheat is much higher than in the milling of hard wheat.
2. Since bran adheres more strongly to the endosperm in soft wheat, it is not easy to clean bran only by grinding and sifting of break stocks. Additional bran dusters are incorporated much earlier in the flow. For example, the ground material from third break before going to the sifter is passed through a bran duster. This process is also repeated for fourth and fifth break.
3. The reduction rolls must be carefully adjusted so that they do not grind any harder than they should, otherwise endosperm particles, instead of being reduced to flour-particle size, flatten out forming endosperm flakes. These flakes then are unable to fall through the fine apertures of the nylon sieve; thus, they are carried over unnecessarily to the next passage.
4. Flake disruptors or detachers are used to break up the flakes following their passage through a reduction roll and before sifting. In the absence of these detachers, especially on critical passages, some extraction loss would be experienced.

5. Throughput should be closely monitored, as any slight increase of throughput may cause the mill to plug up or may cause a loss of flour extraction.

The purification system does not play a critical role in soft wheat milling, because rather than remaining in coarse granular form, the endosperm readily breaks down into a fine-particle size, which cannot be properly purified. Some soft wheat mills do not use any purifiers at all.

Sticky material require more sifting area; therefore, soft wheat mills are designed with additional sifting surfaces when compared with hard wheat mills. When milling soft wheat on the same mill as hard wheat, the throughput for soft wheat may have to be reduced by as much as 30% to accommodate the additional sifting area required.

4.3 Soft Wheat Flours: Production and Utilization

There are two types of flours commonly produced from soft wheat: (a) pastry flour and (b) cracker flour. In addition to these some mills also produce special cake flour by removing between 10 and 25% of the patent flour. When higher percentage of patent flour such as 25% is removed, the remainder of the flour is sold to cookie manufacturers. The ash content of this flour would be about 0.52–0.55%. If only 10% of flour is removed then the remaining flour would have the same application as pastry flour (Table 14).

Cake flours are sifted to a fine granulation. This is accomplished by using fine nylon sieve covers in a sifter. In some facilities, fine grinding (pin milling) and air classification systems may also be used to obtain a low-protein fraction that otherwise is sometimes hard to achieve in a conventional manner. Figure 39 describes the process flow.

A portion of the straight-grade flour with 10.8% protein from the mill is sent to a separate system of fine grinding in a pin mill followed by air classification. Air classification essentially involves drawing the finely ground flour particles through a series of cyclones under suction. This allows coarser particles to settle first in the cyclone. Medium and fine fractions are removed, in that order, through the subsequent cyclones. The medium fraction with the lowest protein (approximately 6%) is used as cake flour.

Cake flours are sometimes dried for preparation of cake mixes. These flours may be dried to as low as 8% moisture to achieve (a) better shelf life and (b) greater stability.

Some leavening agents that are added in cake mixes sometimes may react with moisture. This is no longer a problem, as the products that are used today are much more

Table 14 Soft Wheat Flour Types

	Pastry flour	High-ratio cake flour	Cracker flour
Wheat mix (%)	Low protein soft wheat mix	Lowest protein flour stream combination from a low-protein soft wheat mix	Medium strength 50% + soft (weak) 50%
Flour extraction (%)	73.0–74.0 (straight grade)	10–20 (top pat.)	74.0–75.0
Flour protein (%)	8.5–9.5	7.0–8.0	9.0–10.0
Flour ash (%)	0.48	0.35–0.40	0.50
Chlorination (pH)		4.4–4.8	—

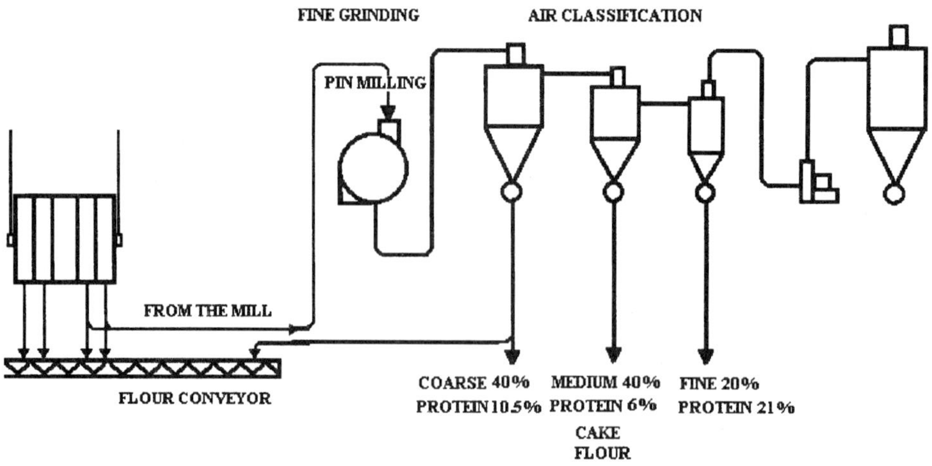

Fig. 39 Cake flour production using fine grinding and air classification.

stable and therefore drying flour is no longer a common practice. Flour-drying is also cost-prohibitive.

Dry flour enhances baking performance in terms of (a) volume, (b) crumb, and (c) texture.

5 DURUM WHEAT MILLING

Durum wheat milling requires a very different approach from common wheat (hard and soft) milling. Whereas the objective in common wheat milling is to produce a maximum quantity of white speck-free flour, the objective in durum wheat milling is to produce a maximum quantity of bright golden yellow semolina (a granular product) free from specks of bran, germ, or foreign material. Durum semolina is used primarily for pasta products and couscous (preferred in North Africa).

Durum wheat is quite different from most common wheats in kernel characteristics. Amber durum wheat kernels are bolder, longer, and have a lighter seed coat color. The endosperm is extremely hard and yellow. Thus, amber durum wheat is intrinsically suitable for the production of attractive bright golden yellow semolina.

The major quality criteria required by most processors are as follows:

A bright golden yellow semolina
A very low tolerance for black and brown specks
Freedom from pieces of grit
Strict compliance of semolina particle size to the specified granulation
Sufficient protein content and strong gluten to provide firmness to the "bite" in pasta products and tolerance to overcooking
Compliance for bacteria, yeast, and mold counts

These differences from flour in end product requirements and in kernel characteristics are responsible for the significant differences required in the approach to milling durum wheat.

5.1 Cleaning Durum Wheat

Special attention to cleaning is required for durum wheat. The reasons for this follow:

All dark material, such as stones and mud balls and dark seeds, such as ergot contamination or wild buckwheat, must be completely removed, as the tolerance for grit and black specks is very low.

The extraneous foreign material can be removed; however, if the discoloration is present within the wheat kernel, such as black point and in a severe situation smudge, it obviously becomes very difficult to eliminate the problem.

Separation of longer seeds such as, oats, barley, wild oats, and those infected with ergot is particularly difficult, as durum wheat kernels are also fairly similar in length; therefore, they cannot be easily removed by machines that cause separation on the basis of length, such as disk separators and indented cylinders.

If not completely removed, these and other foreign materials are likely to be fragmented in the milling process and end up as small dark specks in the finished semolina, in spite of all efforts to remove them by purification and sifting. Pasta products made from specky semolina have unattractive spots clearly visible through the transparent wrappers or windows that characterize pasta packages.

Ergot is a major contaminant of durum wheat. It is a fungus that grows in place of a kernel on the wheat floret. It is black, toxic in large amounts, and difficult to separate from durum wheat, because it is very similar in size and shape, but is somewhat less dense. One single ergot grain may shatter into a multitude of fine dark particles if permitted to reach the break system. Barley, oats, and wild oats are difficult to remove because they are also very similar to durum wheat in size and shape, but are also slightly less dense. Stones and mudballs must be removed completely, as stones produce gritty particles and mudballs crumble and adversely affect color of the semolina. Consequently, cleaning on the basis of differences in specific gravity provides a major focus in durum wheat cleaning.

Blackpoint is a disease that can affect durum wheat. It leaves black dots on the seed coat. If this problem is present it can be controlled, to some extent, by putting the wheat through intensive scouring, thereby removing the outer layer of the bran and with that the problem of discoloration. Scouring would additionally remove the germ and also help reduce the microbial contamination if any were present. However, if this discoloration extends through the crease of wheat it becomes difficult to control. Smudge causes severe discoloration of the portion of endosperm as well; therefore, avoidance of discoloring specks in semolina becomes virtually impossible.

The only way to control this problem, as well as many others, would be to process superior grades of durum wheat, which would severely limit or eliminate many of the degrading factors. Figure 40 shows a typical durum wheat-cleaning facility.

Here, the grain separator removes foreign material that is larger, smaller, and lighter than wheat. Next the wheat passes through a machine called a *combinator*. This removes stones of the same size as wheat. It also segregates the wheat mass into a lighter and a heavier stream. The heavier stream which may be close to 80% of the total stream is almost free of any foreign material. It may or may not be passed through an indented cylinder for removal of some round and short seeds before being sent for tempering. The lighter fraction consists of most of the foreign material that is lighter than sound wheat. The lighter fraction is then passed through a series of cleaning equipment for removal of the various impurities.

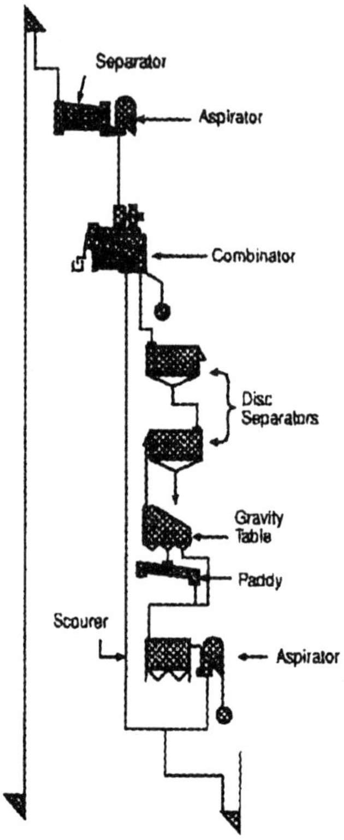

Fig. 40 A durum wheat-cleaning diagram. (Courtesy of Bühler, Inc.)

5.2 Tempering Durum Wheat

Durum wheat is much harder than common wheat, yet it is tempered for a much shorter time, generally between 4 and 10 h. The reason for this is obvious. Because the objective in durum wheat milling is to produce as large a quantity of semolina as possible, keeping the endosperm hard would help realize this goal. A longer period of tempering makes the endosperm mellow and would cause production of more fine and flour particles. Longer periods of tempering also have a bleaching effect on the endosperm color, affecting the semolina color, which subsequently affects the color of the finished pasta product.

In general, over the years, the semolina particle size requirement for pasta is becoming progressively finer. Most recent Polimatik pasta-processing technology permits a very short mixing time of less than 1 min. The semolina particle size requirement for such process is generally much finer than conventional granulation. In such cases the tempering time is extended to reduce the particle size without increasing any undue levels of starch damage. Higher levels of starch damage would affect the texture of cooked pasta and also contribute to cooking loss.

Durum wheat is usually milled at about 16% moisture content. The moisture of semolina would be dependent on the milling process and relative humidity. Generally speaking, semolina moisture will be in the range 14.0–14.5%. A fine spray of water approximately 1 h before milling to add 0.5–1.0% moisture, has a beneficial effect. It is after this moisture addition that the final milling moisture reaches 16.0%. This helps toughening up the bran to resist shredding during the milling process.

5.3 Particle Size

The milling process objectives for durum wheat are derived from the quality needs of the finished semolina and the level of its extraction requirement. Its particle size is a quality factor that is affected through milling process, and this requirement may vary depending on the process, and on the end product requirement. For example, homemade couscous requires much coarser granulation than pasta products. In general, whatever the particle size requirement may be for a given product or process, it should have as narrow a size range as possible. This provides a uniform-sized product. Large particles often do not hydrate fully during the mixing cycle of the pasta manufacturing, and this shows up as white specks in the finished pasta products. Table 15 provides particle size granulation of a conventional semolina and a semolina of fine particle size required for Polimatik systems.

5.4 Semolina Extraction

The extraction rate of semolina is an important consideration. Semolina extraction of 66–68% is considered acceptable by most millers. Extraction rates exceeding 70% are considered good. There is some low-grade flour that is also produced in small quantities and is of low commercial value. The total extraction of the semolina together with the flour adds up to 76–78% or higher.

5.5 Milling of Durum Wheat

A durum wheat milling flowchart is depicted in block form in Fig. 41.

As can be seen from the diagram there is a secondary purification system that does not exist in flour milling. There are various differences in the approach to semolina milling when compared with flour milling:

More break passages to allow gradual releasing of granular endosperm and a minimum quantity of flour

Table 15 Semolina Granulations Required for Conventional and Polimatik Systems

Conventional semolina		Semolina for Polimatik system	
(μ)	(%)	(μ)	(%)
Over 500	8–9		
Over 350	60–62		
Over 250	15–18	355–0	100
Over 150	4–5		
Through 150	2–3		

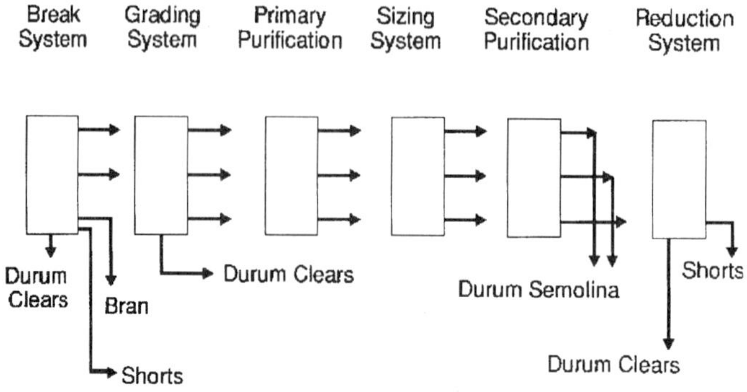

Fig. 41 Durum wheat-milling block flow diagram. (Courtesy of the Canadian International Grains Institute.)

More comprehensive grading system to allow grading of released endosperm into several narrow ranges of particle size

An elaborate purification system to purify the various narrow ranges of particle size

5.5.1 Break System

Durum wheat milling requires, as much as possible, a longer break system (Fig. 42) to enable the release of the endosperm in large granular form and minimize the production of any finer particles and flour.

In some situations there may be as many as seven or eight break passages incorporated. Most commonly a six-passage break system is used. This would allow the break releases to be lower on the corresponding break passage when compared with that obtained in flour milling. Because, first, the emphasis is more on the production of granular material during the break system grinding, more cutting, rather than keeping the bran flakes large, is intended. This is accomplished through positioning the grinding rolls in sharp-to-sharp disposition.

5.5.2 Grading System

The longer break system is designed to generate a relatively higher percentage of granular material which, obviously, would have to be graded next. Therefore, the requirement for elaborate grading of released endosperm particles is logical. The grading operation facilitates their separation into numerous narrow-range bands of particle size. This is essential for effective purification action. Effective purification (stratification) can be attained only if material fed to the purifier falls within a narrow particle size range.

5.5.3 Purification System

The purification system is very comprehensive in all durum wheat mills. In most mills the purification function is divided into two stages: primary purification and secondary purification. In the primary purification, the coarse granular endosperm particles from the break system after being graded, are purified. Purified streams would contain pure endosperm particles and composite particles of endosperm with bran, germ, or both attached to it. These streams would be then directed to the sizing system. The secondary purification system is fed by the granular endosperm material after it has been gently ground for size reduction and to detach the bran and germ from the endosperm by the sizing system. The purified semolina of appropriate size, and free of any bran specks, leaves the secondary purification system and is the finished product. The overall purification system can be considered the most important feature in the durum-milling process.

5.5.4 Sizing System

The purpose of the sizing system is to grind lightly to obtain appropriate granulation of the semolina as well as to scratch off bran and germ fragments attached to the endosperm in the composite particles. Finely corrugated rolls ranging from 8 to 16 corrugations per centimeter for the first through sixth sizing passages, respectively, may be used. In addition to choosing proper corrugations, a correct roll gap setting to obtain a light grind is quite critical, as it has a direct effect on the granulation.

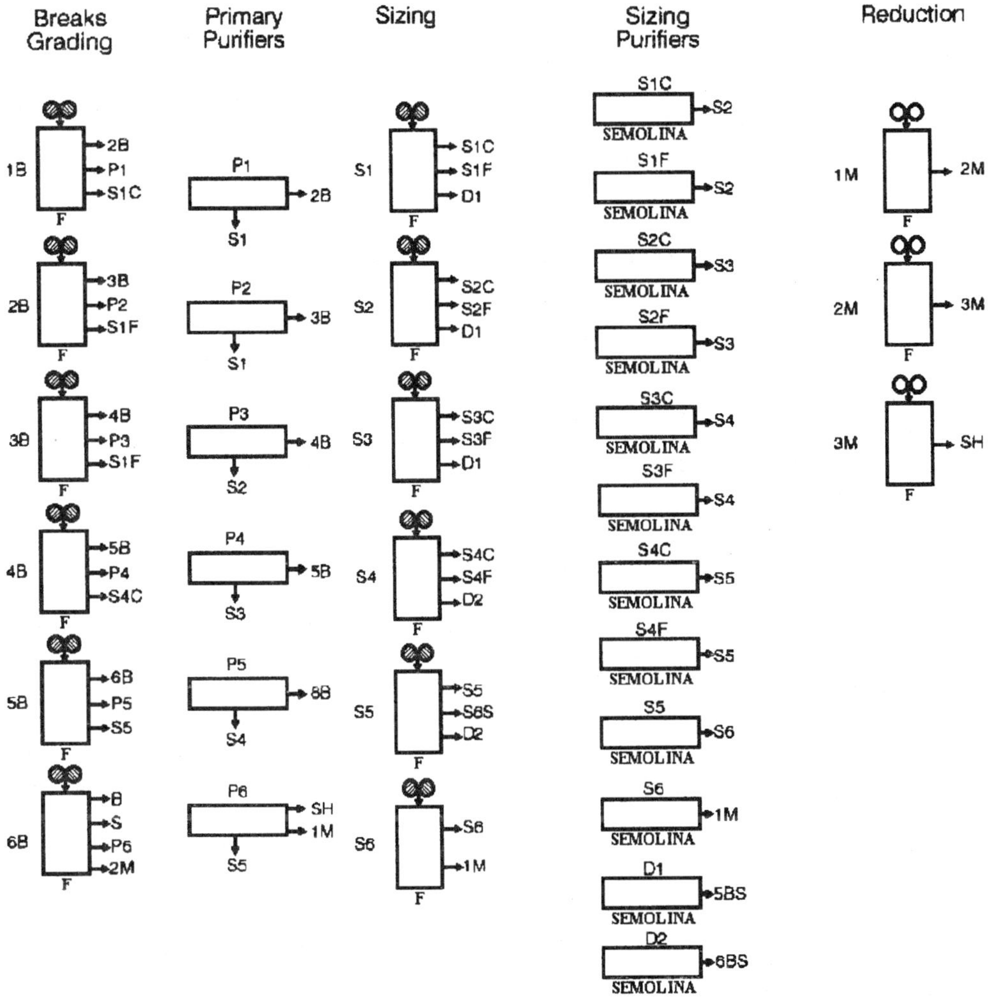

Fig. 42 Durum wheat-milling flow diagram. (Courtesy of the Canadian International Grains Institute.)

Table 16 Semolina Grades and Their Extraction Rates

Semolina grade	Durum wheat grade	Targeted extraction (%)
1	2 Durum wheat	70
2	3 Durum wheat	67
3	3 Durum wheat	69 + pasta reused

5.5.5 Reduction System

The reduction system in durum milling has a very limited function. Ideally, the entire endosperm of durum wheat would be reduced to semolina of desired granulation, completely eliminating the need for a reduction system. Unfortunately, in practice, there is always a small proportion of inferior granular stock that comes from the last break, the last sizing and tail end of the sizing purification. These stocks are too fine, too poor in color, and too contaminated with bran to be converted for inclusion in the mainstream semolina. Instead, they are routed to the reduction system where they are ground in smooth rolls to produce flour. A short reduction system of three to four passages is adequate for this purpose.

5.6 Blending of Semolina and Quality Optimization

Finished semolina quality is dependent on the quality of the durum wheat and the effectiveness of the milling process. Poor crop condition, especially low-protein content in the wheat makes the situation very difficult. This is because not much can be done in the mill to raise the protein content except to blend with wheat of higher grades and higher protein content, if it were available. Millers may remain vigilant not to make any unnecessary errors that may end up causing undue protein loss in semolina. If a semolina blending facility is available, this would provide opportunities for the adoption of production techniques that would allow a plant to optimize its milling results and product quality relatively more than a conventional approach.

As an example, there may be three different qualities of semolina produced from no. 2 and no. 3 grades of durum wheat in the absence of no. 1 grade of durum wheat in a poor crop year. Table 16 shows how three varying qualities of semolina are produced from these two wheats and their targeted extraction rates.

The blending of semolina instead of various grades of durum wheat obviously means greater control and flexibility in terms of corresponding desired quality of a multitude of products, ranging from the high-quality end to economy pasta products. For example, 80% of no. 1 semolina may be blended with 20% of no. 3 semolina for the production of premium pasta products, such as, rotini, rigatoni, or other. For the production of short and nonsensitive shapes where starch release is not critical, such as tubetti and elbows, 100% of no. 3 semolina may be used.

ACKNOWLEDGMENTS

Most of the cleaning, tempering, and milling equipment and process diagrams included in this chapter have been provided by Bühler, Inc., Uzwil, Switzerland. Kind permission from Bühler, Inc. for their use is greatly appreciated. The support extended by the following organizations in providing equipment diagrams and descriptions is also greatly appreciated: Carter Day Industries (Canada) Ltd., North American Millers Association, AIC, Canada, and the Canadian Grain Commission.

REFERENCES

1. Dexter JE, Sarkar AK. Roller milling operation, Flour. Encyclopedia of Food Sciences, Food Technology and Nutrition. Academic Press, New York, 1993.
2. Hatcher DW, Kruger JE. Distribution of polyphenol oxidase in flour millstreams of Canadian common wheat classes milled to three extraction rates. Cereal Chem 1993; 70 (1)51–55.
3. Sarkar A. Flour milling. In: Grains and Oilseeds—Handling, Marketing and Processing. Vol. 2, 4th ed. Canadian International Grains Institute, Winnipeg, MB, 1993.
4. Sarkar A. Optimizing wheat mixes for end-use. Assoc Oper Millers Bull. Jan, 1988, AOM, Leawood, Kansas.
5. Sarkar A. Application software in the milling business. Assoc Oper Millers Bull. Sept, 1990, AOM, Leawood, Kansas.
6. Sarkar A. Milling technology, Ble 2000: enjeux et strategie, Algiers, Algeria, Feb, 2000.
7. Quality Evaluation Methods for Red Spring Wheat, 2nd ed., revised 1992. Grain Research Laboratory, Canadian Grain Commission, Winnipeg, MB.

12

Specialty Milling

ASHOK K. SARKAR

Canadian International Grains Institute, Winnipeg, Manitoba, Canada

1 BARLEY MILLING

1.1 Introduction

The use of barley for human food application has a long history. It was consumed by earlier civilizations as a staple, along with other cereal grains. Its consumption as a staple thus dates back thousands of years, and it continued to remain dominant until the latter part of the 19th century. As the primary (milling) and secondary (bakery, pasta, noodles, and such) processing industries began to develop and processed food became more available, the consumption of barley and other grains in whole-grain form began to decline. Barley is still consumed as part of the staple diet in some countries, especially where consumption in the form of whole grain is still in limited practice.

The consumption of barley today, especially in developed countries, is for nutritional benefits.

Barley is used primarily for the following purposes:

1. Malting, brewing, and spirits
2. Food uses
3. Feed ingredient

1.2 Processing of Barley

Barley hulls are removed by shelling. Following the removal of the hulls, it is pearled and may be cut and pearled again. Pearled barley may also be flaked; also, it may be further reduced in size to form granular products of varying degrees of particle sizes that meet the required product specifications. The total yield of pearled barley products, including barley flour, may add up to about 70%. Generally speaking, pearled barley yields

Table 1 Milled Barley Products and Their End Uses

Country	Milled products	End products
North America and Europe	Pot barley	Stew, casseroles, barley water
	Pearled barley	Salads, puddings, and faster cooking
	Rolled barley	Cookies, granola, porridge
	Barley flour	Home baking
Japan	Whole barley roasted	Barley tea
	Pearled barley	Shochu liquor and miso (soup base)
	Pearled and pressed	Rice extender
	Pearled and cut	Rice extender
	Pearled, cut, and pressed	Rice extender
	Flour	Noodles and related items
Morocco and Tunisia	Pearled barley	Home cooking (sometimes roasted and ground)
	Coarse granular barley semolina	Couscous
	Medium semolina	Couscous
	Fine semolina	Soup
	Barley flour	Flat bread

would be lower than that. It is dependent on the quality of pearled barley required and the quality of barley that is being processed.

The very first step before beginning the processing of barley is the consideration of quality criteria for the selection of barley for a given end product. This is covered in the following section in detail. Although hulless barley is becoming more popular for food uses, barley with hulls, both two and six rows are quite commonly used for such applications. A wide variety of end products are produced from barley. Table 1 lists some of the applications.

1.2.1 Quality Criteria for Barley Selection

As can be seen from Table 1, for all applications except barley tea, barley kernels must be pearled. Therefore, when selecting barley for milling the following criteria are examined:

1. Pearling properties
2. Test weight
3. Moisture contents
4. Broken kernels
5. Free from damage and discoloration
6. Uniformity
7. Desired properties for secondary processing

Good pearling properties for barley would mean that the hulls are easily removed from the kernel, while at the same time maintaining a high percentage of pearled barley yield. Poor pearling properties would cause greater loss of endosperm when attempting to remove the hulls during the pearling operation. Plumper and bolder kernels allow them-

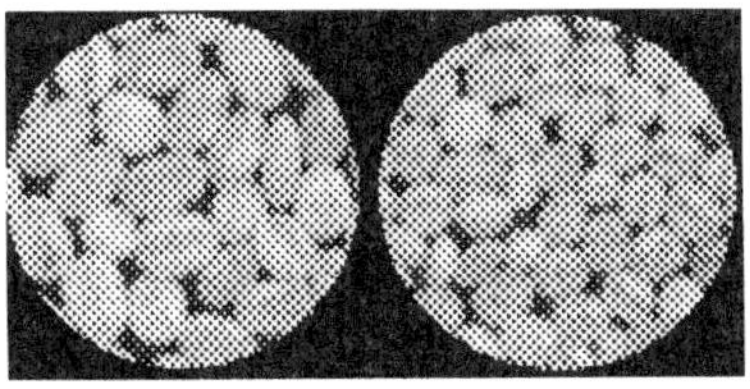

Fig. 1 Pearled barley of varying yield.

selves to be pearled more easily than smaller kernels. Figure 1 shows two samples of barley that are pearled to obtain satisfactory removal of hulls and comparable brightness of the pearled barley. The barley sample with larger kernels yields a higher percentage of pearled barley than the sample with smaller kernels.

The first six attributes of the seven listed adequately protect the quality requirements of most barley processors who process barley for food. Although all of these points are self-explanatory, the last one needs further elaboration. In North America the use of barley for food is primarily for nutritional benefits, and the milled barley products application in the secondary processing industry does not demand as discriminating functional properties requirements as in some of the other countries, such as Japan. For example, the presence of high levels of *steely* kernels, which are vitreous kernels with a gray hardness, is considered undesirable by some processors, especially in Japan and in North Africa.

The properties for secondary processing would obviously be dependent on the type of end product. For example, pearled and cut barley for use as a rice extender, for which the color and shape of pearled barley is desired to be similar to that of rice, requires bright pearled color and good endosperm texture, which means fewer broken and steely kernels. Plump kernels with shallow and narrow creases provide uniform and higher pearled barley yield and are therefore preferred. Six-row or hulless barley is preferred because their size and flavor are similar to rice.

For miso, for which soaking at a constant rate is more critical, uniformity in kernel quality in terms of size and texture is critical. Mealy endosperm texture is preferable, along with plumpness, uniform pearled color, and a narrow and shallow crease. Preference for barley types varies with the region within Japan. In the western region, hulless barley is preferred, and it is mixed with rice for supplementation. In the eastern region, two-row barley is used, and it is mixed with rice or soybean. A yellowish color of the barley is acceptable for miso to make the appearance of the barley brighter. For barley tea, six-row barley is more commonly used, as this has been the traditional choice. The smaller kernels from six-row barley facilitate more even roasting of the kernels than the larger kernels from two-row barley. In some regions hulless barley is used.

For shochu liquor, barley with a high carbohydrate content is required because it is converted to alcohol. Therefore, two-row barley is preferred.

1.2.2 Cleaning

The cleaning process for barley is dependent on how many impurities it contains. The cleaning begins with a magnet followed by a separator. The separator removes larger, smaller, and lighter foreign material by using screens that are larger and smaller than the size of the barley, and an aspiration channel, respectively, for the lighter material. This is carried out just as explained under wheat flour milling.

The barley is then directed to a destoner that removes stones and other heavy material of the same size as barley. Other heavy material that is removed may include nonferrous metal objects, glass, and such. Besides affecting the quality and being injurious if consumed, these are also particularly damaging to the pearlers that impart friction as a mode of its operation.

This equipment is a standard feature in most barley-cleaning houses. If barley is very dirty with a lot of surface chaff, husks, and seeds that are shorter, such as cockle, and longer, such as sticks, oats, and wild oats, then the cleaning house is equipped with additional cleaning equipment.

A battery of indented cylinders or disk separators are used for removal of short and long seeds that have a cross section similar to barley, but have different length, which is why they could not be removed by a separator earlier. Separation of oats may be difficult, for many barley kernels would be similar in length; therefore, long seed separation may be carried out following the removal of the hulls. For surface chaff and hull, a machine that uses a rotor with finger beaters running enclosed within a perforated sheet metal cover, can be quite helpful. This is followed by aspiration for removal of the loosened material.

At this point the barley is clean and ready for further processing.

1.2.3 Tempering or Conditioning

Barley is generally processed without any addition of water because it is naturally soft. There are, however, some varieties grown in different parts of the world, such as in North Africa, that are extremely hard and very low in moisture. It may be impossible to mill these varieties without any addition of water. Therefore, in these situations there may be a provision for light-tempering, following a light-pearling action.

The reason for pearling lightly before moisture addition is that moist barley will take more power for pearling and also that pearling performance will suffer. That light pearling may suffice in these situations is because the barley is further pearled or processed in the mill; thus, any remaining hulls may be removed there.

1.2.4 Barley Pearler

The principal equipment in the milling of barley is a barley pearler. Such a machine has been in use for many years. Its introduction made the grinding stones extinct in the barley-pearling operation and helped the development of the barley-milling process on an industrial scale, and thus, small-scale barley-milling operation evolved into an industry.

A barley huller–pearler essentially involves removal of the hulls through abrasion and friction between a grinding medium and a screen jacket. Although the grinding medium still consists of several stone (emery) rings attached to a central shaft that runs at a high speed (Fig. 2), the grain and hardness of the stone rings are designed to provide an effective hulling and pearling action. The intensity of such action can be controlled through four segmented outlet slides controlled by micrometer screws from outside to regulate the desired length of time the material is to be treated; thus, the hulling and pearling degree can be controlled.

The operation of this machine involves feeding the barley from the top through a sight glass (see Fig. 2). As the barley descends, filling the space between the emery rings

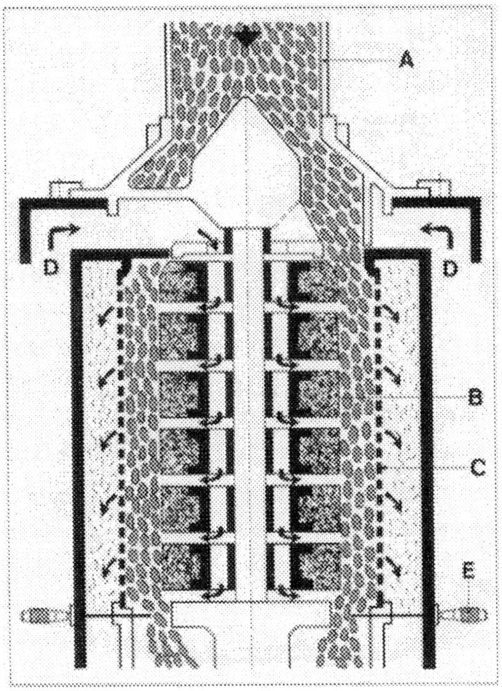

Fig. 2 Operation of a barley pearler. (Courtesy of Codema Inc.)

and the perforated screen jacket surrounding the rings, it is subjected to friction among the kernels, stones, and the screen. This friction and abrasion loosens and, ultimately, detaches the hulls. Such friction does generate heat; therefore, to control the temperature, air is allowed to enter through the rotor area at the top, which helps direct the detached hulls through the screen perforations. This suction air with hulls is carried over to a cyclone or filter (dust collector). The hulling and pearling degree is easily controlled while the machine is in operation. The power requirement for an efficient machine such as this would be much lower than poorly designed older machines that can take as much as 50% more power for the same load. Although this machine is vertical, there are also horizontal pearling machines available.

1.2.5 Barley Milling

1.2.5.1 North America and Europe

In most situations milling of barley primarily involves varying degree of pearling action, with or without in-between cutting followed by grading on screens using a shaker or sifter, to provide various grades of pearled barley. This is a common practice in North America and Europe. A simple version of such a facility is shown in Fig. 3.

In the United Kingdom, barley is passed through first pearling stage for the removal of the hulls. The product following the first pass is referred to as blocked barley, which is similar to pot barley. Blocked barley may then be passed through a second and third stage of pearling for removal of the remaining hulls and also the outer layer of the endosperm. The final yield of pearled barley is approximately 70%. Blocked barley cutting

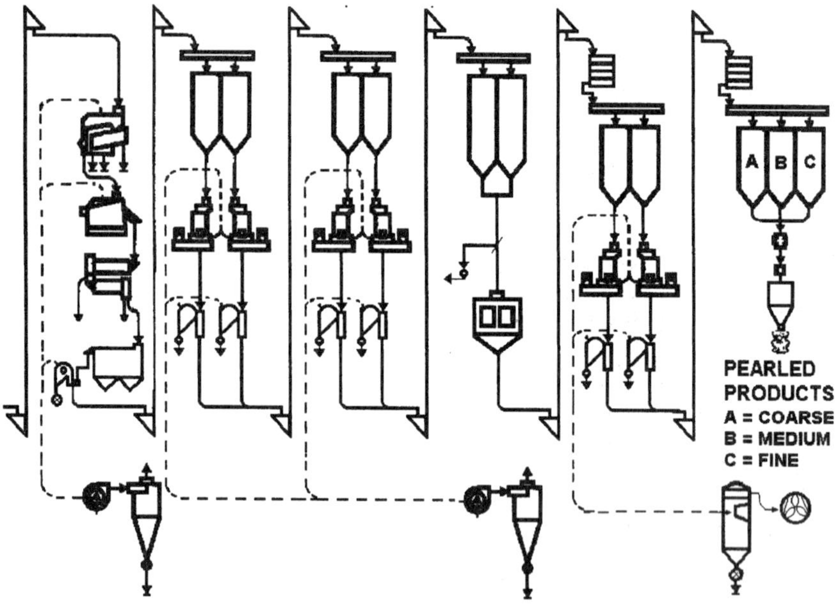

Fig. 3 Barley-milling process (pearled and cut).

can be accomplished using a groat cutter. Pearled barley is also used for producing flaked barley through steam application and use of a pair of large-diameter rolls.

Barley milling in Germany involves pearling cutting and pearling. A lot of barley is processed through various pearling stages with inbetween cutting. The final pearling stage allows the product to be pearled to almost a round-shaped finished, pearled barley product.

The pearled barley and the pearled and cut barley may also be rolled or flaked. This is a common practice in Japan.

1.2.5.2 Japan

In Japan, there are three principal finished barley milled products that are produced. These products, along with their percentage of production, are shown in Table 3. These products are produced as follows:

1.2.5.2.1 PEARLED AND ROLLED BARLEY. The milling process involves pearling of barley to about 40%. The pearled material is then graded by sifting to obtain uniform kernel size. The pearled product is then rolled by passing it through a pair of rolls. The product is shown in Fig. 4.

1.2.5.2.2 PEARLED AND HALF-SPLIT BARLEY. The barley is pearled to a degree of 20%. The pearled material is then graded for uniform kernel size. This product is then cut in half and pearled again to about 30%. This is shown in Fig. 5.

1.2.5.2.3 PEARLED, HALF-SPLIT, AND ROLLED BARLEY. This is produced using the same milling procedure as described for pearled and half-split barley. This product is further processed by passing it through a pair of rolls to flatten it, as shown in Fig. 6.

Table 2 provides a list of milled barley products and their percentage of production.

For application as rice extender all of these three milled barley products may be used. Table 3 provides an average pearled barley yield by application.

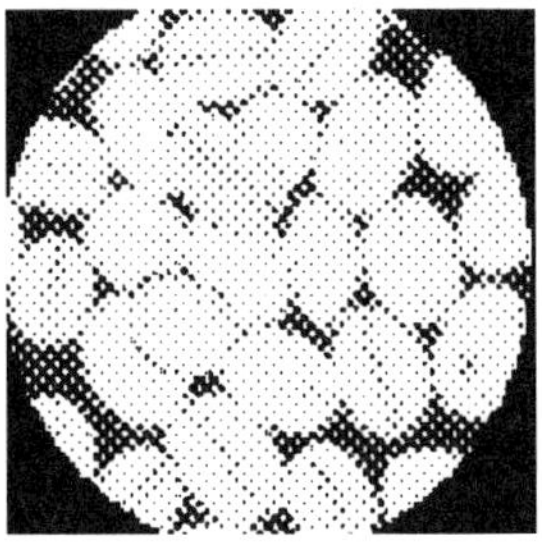

Fig. 4 Pearled and rolled barley.

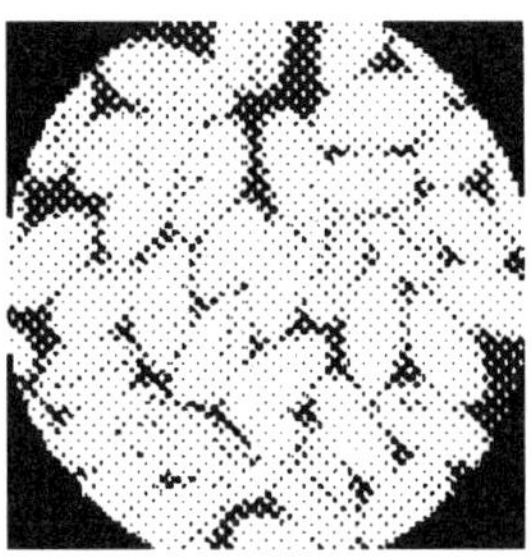

Fig. 5 Pearled and split barley.

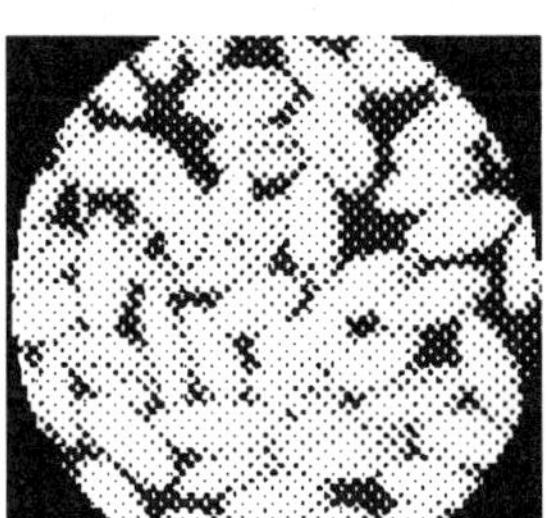

Fig. 6 Pearled, split, and rolled barley.

Table 2 Milled Barley Products and Their Percentage of Production

Products	Production (%)
Pearled and rolled barley	42
Pearled and half-split barley	30
Pearled, half-split, and rolled barley	28

Table 3 Pearled Barley Yield Percentage by Application

Application	Pearled barley yield (%)
Rice extender	52
Miso	72
Shochu alcohol	67

1.2.5.3 North Africa

In some parts of the world, such as Morocco and Tunisia, the hulled barley would serve as raw material to the milling section where rollermills, sifters, and purifiers may be used for further processing. The processing objectives would involve the following:

1. Size reduction to suit various product application requirements
2. Further purification of the product

Light pearling coupled with this additional processing helps boost the extraction rate while offering different granulation and refinement for a creative product mix (Fig. 7).

Table 4 is an example of the finished products, extraction rates, and application of the milled barley in that region. A barley mill would also normally produce some pearled barley in addition to the other granular finished products as given in Table 5.

The milling process involves cleaning the barley as it passes through intermediate storage bins to the barley pearlers placed in a series of first-, second- and third-pearling operations. In each stage of the pearling operation there may be one or more pearlers, as necessary, to split the load. The barley-milling capacity, for example, may be 1.0 t/h. If each pearler handled 500 kg/h, there would then be a requirement for two separate pearling lines. Each pearler may use a 25-hp motor. Therefore, it is too expensive to pearl

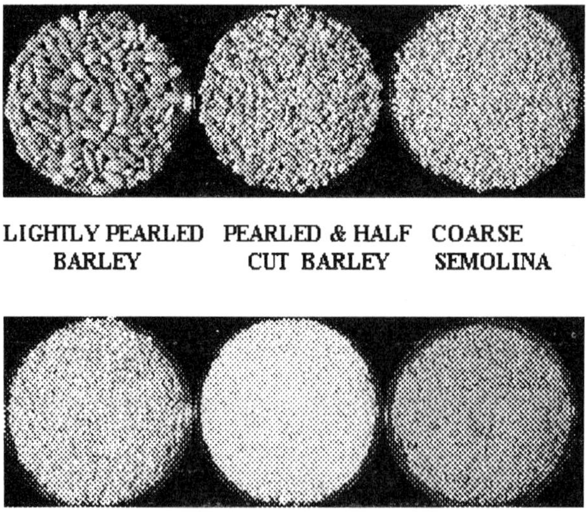

Fig. 7 Milled barley products in North Africa.

Table 4 Milled Barley Products, Extraction Rates, and Their Application

Products	Extraction rate (%)	End use application
Coarse semolina	45–50	Couscous
Medium semolina	45–50	Couscous
Fine semolina	45–50	Soup
Flour	23–25	Bread
Bran	25–30	Feed

from a power-consumption standpoint. In each stage of pearling approximately 8–12% of the hulled material is removed. Depending on the type of barley being processed, 24–30% of the material may be removed. For example, when using a good variety of barley, only two pearling stages are required, whereas for poor quality, three pearling stages may become essential if it is tough and hard. If the barley is damp it is more difficult to pearl, and the temperature of the product increases. If the barley is going to be milled only into flour, then it should be somehow softened through a tempering process which, in turn, may cause some processing problems. Pearled barley is warm as it comes out of the pearler. The pearling machine consists of stone wheels (emery/carborundum) mounted on a vertical shaft running within a perforated sheet metal, slotted screen.

The pearled barley then moves on to a bin, and may be lightly tempered if it is very hard, before being binned. From there it may be passed through a machine referred to as a brush machine. The basic difference between this machine and a scourer is that instead of steel beaters it uses brushes; thus, providing a polish to the material before it is milled.

The grinding and sifting of pearled barley may be carried out in a series of rollermills and sifter sections. Figure 8 shows such a facility. This particular flow has four passages of grinding, sifting, and purification.

1.3 Finished Products and End Uses

Finished, milled barley products are required to meet a wide range of secondary processing and even more diversified end use requirements. Some of the principal end use applications are described in the following four subsections.

Table 5 Barley Milled Products and Application

Finished products	Application
Pearled barley	Home cooking
Cut and pearled barley	Home cooking
Coarse semolina	Couscous
Fine semolina	Soup
Flour	Bread
Bran	Feed

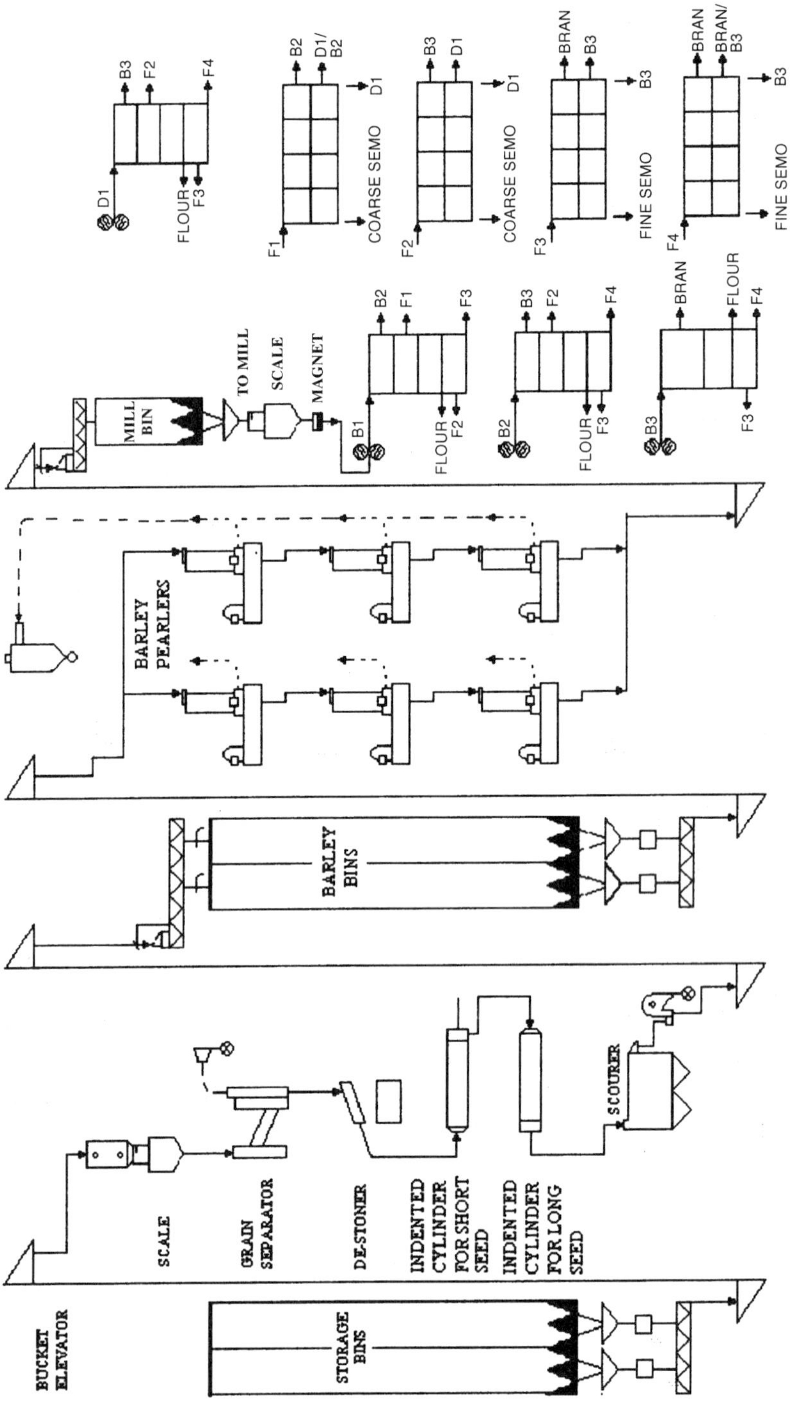

Fig. 8 Barley-milling process (semolina).

1.3.1 Whole Barley Grain

The barley is cleaned and graded for uniform kernel size. The whole barley grain is sent for roasting without any hulling or pearling process. A smaller-sized kernel is preferred because it needs to be evenly roasted. Roasted barley can then be ground to produce semolina, in some parts of North Africa (see Sec. 1.3.3 under semolina). In Japan, roasted barley is used for brewing tea (barley tea). It is a popular cold beverage, especially for summer. Roasted barley sometimes may be ground for filling tea bags.

1.3.2 Pot and Pearled Barley

Pot barley is somewhat larger than pearled barley because it is further refined (pearled) to produce pearled barley. The cooking time for pot barley, therefore, is slightly longer (approx. 40 min) than pearled barley. There applications have been listed in Table 1. They are primarily used in soups and salads. Pearled barley may be used as is or further processed (split or pressed) for following applications:

1. Rice extender: Pearled barley may be added to rice as rice extender in Japan. The addition level usually does not exceed 20%. The barley kernels following pearling and cutting followed by pearling action again end up having an appearance similar to rice kernels. Traditionally this practice was for economic reasons. It is still practiced because of tradition and as well as the awareness of its nutritional attributes.
2. Miso is a popular soup base in Japan that is made from soybean, rice, barley, and salt. The mixture ferments for a long time, up to 18 months. Consistent pearled barley kernels are used for good results.
3. Shochu is a liquor that is made from pearled barley, yeast, and *Koji*. The starch of barley is converted into sugar by *Koji* and the sugar is then converted into alcohol by the yeast. *Koji* is a bacteria used from ancient times. It is also used for soybean paste, soy sauce, and sake.

Pearled barley is also used for various types of ethnic home cooking application in North Africa and in many other parts of the world.

1.3.3 Barley Semolina

Coarse semolina is primarily used for homemade couscous in North Africa. Couscous is agglomerated semolina particles that are prepared by adding water, mixing, and drying followed by size reduction and then sifting to obtain similar-sized large particles. This product is cooked by steam and served with meat and vegetables.

In the rural areas of some parts of North Africa, barley is roasted before it is ground down to semolina particle size. This helps bring out a characteristic nutty flavor. Also, in this region a characteristic bluish color of the semolina of some of the barley varieties is preferred. In some European countries barley has been traditionally bleached to remove this color.

1.3.4 Barley Flour

Barley flour has been used in some countries along with wheat flour for the production of bread made from yeast-fermented dough. The percentage of barley flour addition in these situations may vary from 10 to 20% with little or no appreciable negative influence on the end product's quality. The reason for the addition of barley flour in these situations has been primarily because of the lack of domestic wheat supply, and the importing of wheat to make up the shortfall is uneconomical.

Because of the nutritional benefits that the barley flour has to offer, it may be added to that level (10–20%) to supplement the nutrients. Adding a higher percentage of barley flour would progressively affect the volume, crumb, and would yield a heavy bread. Barley flour is also used in small percentages in Japan for the production of certain types of noodles. It is also used in North Africa for the production of flat bread, for which it may be used up to 70%. This bread has a coarse texture and is a popular specialty in the rural areas.

1.4 Benefits of Barley Usage for Food

The utilization of barley for food purposes is rather limited, but it does have a potential for growth and offers a number of advantages:

1. Process. It is easier to process and has a simple flow diagram relative to other grains.
2. Economical. Its lower processing costs, combined with lower raw material costs, make it a more attractive and affordable substitute for other grain products.
3. Variety of finished products. Barley-milled products are available in many different shapes and forms making its application in various secondary processing and end uses rather convenient.
4. Functionality. Barley flour has good water retention capability. This property has an important implication for food application as it adds weight to the product and helps maintain its freshness. It also improves the texture of the product. Viscosity is the other property that is supplemented by barley components (β-glucans and starch). The utilization of these components as thickeners for soup, sauces, salad dressings, and a host of other products including beverages, has great potential.
5. Nutritional values of barley. The recently recognized nutritional benefits of barley (Table 6), has brought about renewed interest in its application for food purposes. Barley has both soluble and insoluble fiber. It has a high concentration of tocotrienols, which act as antioxidants. Barley has β-glucans which are believed to be responsible for improved blood glucose and lipid levels in diabetics (type II). Waxy barley varieties have higher levels of β-glucans.

Table 6 Nutritional Benefits of Barley

Nutrient component	Description	Action
Fiber	A good source of both soluble and insoluble fiber	Lowers blood cholesterol and important for healthy intestinal function
Tocotrienols and tocopherols	Act as antioxidants	Reduce LDL cholesterol or bad cholesterol
Vitamins and minerals	Good source of vitamin B and minerals	Thiamin, riboflavin, niacin, and chromium
β-Glucans	Higher levels are found in waxy-type barley	Improved blood glucose and lipid levels among diabetics

2 CORN MILLING

2.1 Introduction

Corn is an important cereal crop, serving as a staple to a large population that is spread over parts of Africa, Asia, and North and South America. Its consumption as breakfast cereal, snacks, a variety of foods, food ingredients, and beverages, such as beer and spirits, covers all the regions of the world. It is the most diversified of all cereal grains in terms of its application. Utilization of corn can be summarized as follows:

Human food and nutrition
Industrial processing and utilization
Feed ingredient for animal rations

Corn is grown under diversified conditions. Its production is generally large, and it is relatively economical. Abundant availability and economical price provides it a competitive edge as an important component in the foregoing applications.

Corn is available most commonly in two different colors for processing: yellow and white. Which color may be preferred would be strongly influenced by regional preferences and the end product. For example, the southern states of the United States prefer cooking grits milled from white corn for its light color and flavor. Flaking grits, on the other hand, are preferred milled from yellow corn with hard textured endosperm. There are many different types of corn that are available and its texture has a strong influence on the production of grits. On the basis of texture, flint corn has the hardest texture whereas floury corn is the softest. For wheat, the hard or soft wheat texture is attributable to the presence of the hard vitreous kernels in relation to the percentage of soft, mealy/starchy kernels. In corn, hardness is derived from the presence of hard, also referred to as *horny*, endosperm in relation to the soft or mealy (floury) endosperm within a kernel. A kernel of corn consists of the following:

Constituents	%
Endosperm	80–82
Germ	10–12
Bran	5–6
Tip cap	1–1.5

2.2 Dry Milling of Corn

As mentioned under wheat flour milling, the crease of the wheat poses a problem, and this weighs heavily when designing the approach to the milling process. Similarly, when milling corn the germ poses a problem if it is not completely removed. This is because the germ in corn contains a high percentage of fat, up to 30% on dry basis, and the germ itself makes up to about 10–12% of the whole kernel. Therefore, when making grits and cornmeal if some of the germ fragments remain in the product it is going to be detrimental to the shelf life of that product. Following is an approximate breakdown of the fat content throughout the kernel:

Component	%
Whole corn	4–6
Endosperm	0.5–1.0
Germ	25–30
Bran (hull)	5–6

2.2.1 Milling Process

In situations where corn grits and meal are going to be consumed within a short time from the time it is produced, the complete removal of germ is not as critical; therefore, a complicated degermination process is not necessary.

There are some similarities, just as there are many contrasts that exist when comparing corn-milling with wheat flour-milling processes. The main difference would be in tempering, for degermination and dehulling takes place before the corn is milled. The ground corn materials in process are coarse and abrasive in handling, thus requiring perforated sheet metal and wear-resistant heavier-duty wire meshes for sifting. The cleaning process is quite similar to wheat cleaning and is described next.

2.2.1.1 Cleaning

Cleaning corn is relatively simpler compared with wheat. It may require a lot of attention in the cleaning house if rodent pellet residue is found in the finished products. The rodent pellets are hard to remove and thus may pose a problem. The best way to overcome this problem is to demand higher quality grades of corn with very low or zero tolerance for these pellets. If the corn arrives at the mill with high natural moisture that may not be safe for storage, such as, over 15–16% under moderate temperature, then storage time in the bins feeding the cleaning house should be carefully monitored. If the corn is dried to the safe storage moisture levels then care should be taken not to dry excessively, as it may damage the kernel, causing fissuring or cracking of the endosperm, which in turn, affects the clean separation of germ and bran from the endosperm. Discoloration may also be caused through overdrying and this will detract from the visual attractiveness of the finished grits, as the end product is required to be free of all foreign material fragments and particularly free of any discoloring specks.

Figure 9 shows a cleaning house diagram of corn. After the corn is weighed in a scale it passes through a magnet for removal of tramp metal. It then passes on to a grain separator with an aspiration channel. This removes impurities that are larger, finer, and lighter than corn. This machine is also equipped with an air-recycling system that allows efficient aspiration of lighter foreign materials without requiring the energy as would be conventionally perceived. After the bulk of the foreign material is removed it moves on to a destoner for removal of stones the sizes of which are similar to that of corn.

Generally, this is the extent of cleaning that is required. The cleaned corn passes on to a tempering system where, depending on the degermination and the milling process applied, the addition of moisture and tempering time are determined. This particular cleaning diagram also shows a conditioning unit that provides moisture addition at high temperatures to help loosen the bran and germ before it is dehulled and degerminated in the next machine in sequence. This equipment is shown in Fig. 10 and described in detail later.

2.2.1.2 Tempering or Conditioning

The type of end products, quality, and the extraction level requirements would determine the level of degermination and the appropriate milling process necessary to accomplish

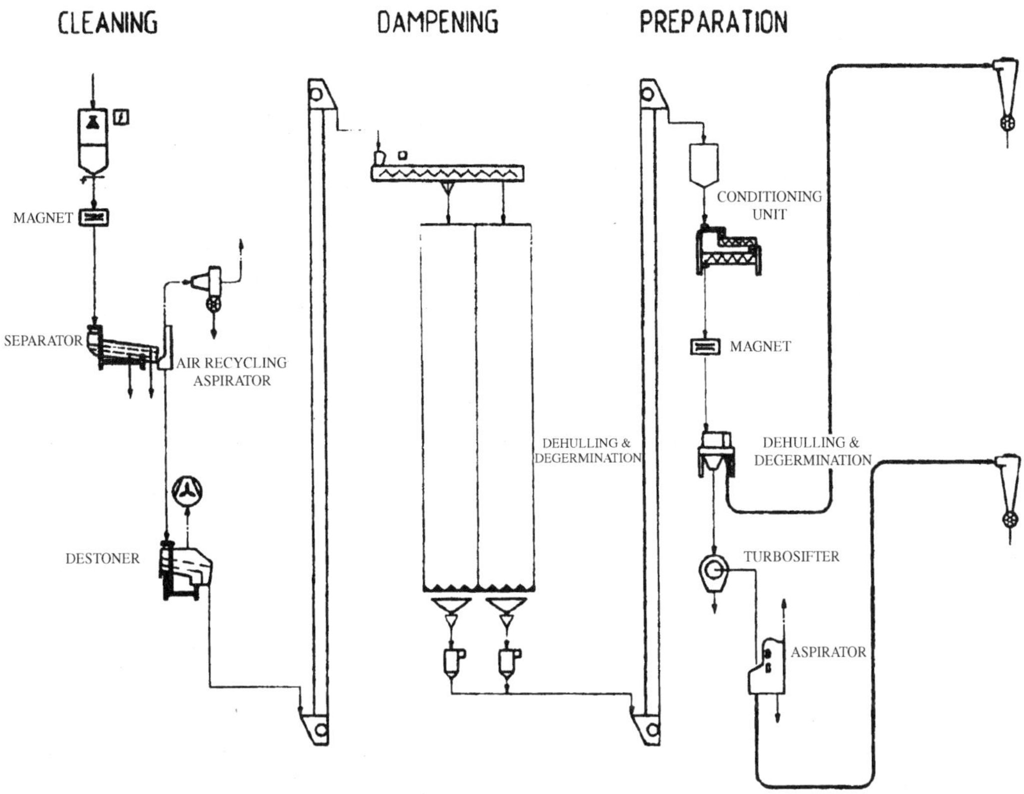

Fig. 9 Corn cleaning and degerming operation. (Courtesy of Bühler, Inc.)

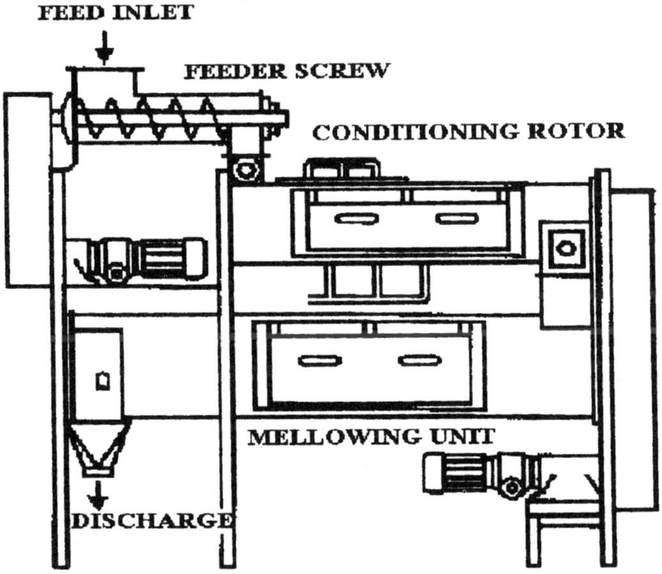

Fig. 10 Conditioning unit for corn. (Courtesy of Bühler, Inc.)

the objectives. Once the level of degermination and the milling requirements have been identified, the tempering requirements can then be determined. As in wheat flour milling, the optimum milling moisture and tempering time are targeted for removal of bran. In corn milling, these parameters focus on both the removal of the bran and germ. Because complete removal of germ may be considered critical to a varying degree, depending on the combination of the end products that are to be produced, tempering time and moisture levels would also vary.

Table 7 provides a guideline for moisture addition relative to degermination and milling process applied. The physical addition of the moisture, in a conventional system, takes place in a conveyor just as in wheat flour milling. The only difference is that since moisture penetration or absorption in corn is much slower than wheat, a longer conveyor is required relative to wheat tempering, and conveyor speed is critical. This process cannot be speeded up by simply replacing a paddle bladed conditioning conveyor by an intensive dampening system with high speed mixing screw as is done with wheat when higher than 3% moisture is required to be added in one stage. In corn, an intensive dampening system might have negative effect, as corn is much harder and thus could create a lot of broken kernels.

2.2.1.3 Conditioning Unit

A conditioning unit, specifically designed for corn, helps bring it to an optimum condition for easier removal of the hull and the germ. This unit is designed to apply heat and moisture, through a possible combination of steam and hot water, to allow loosening of the hull so that it can be readily removed in the dehulling and degermination process. Figure 10 shows a section of the machine.

It has a variable speed feeder screw feeding to the conditioning rotor located under the feeder. After the conditioning takes place in the unit with moisture, heat, and the thorough mixing by means of paddles, the conditioned corn is passed down to the mellowing unit. The mellowing unit consists of a rotor with paddles attached. It has a retarding gate at the discharge end of the unit that can be adjusted to ensure that the unit is running with full depth of the stock and that the uniform mellowing time in the unit is maintained. The maximum level of steam added is 3% and hot water 4% at 80°C.

Table 7 Tempering Conditions for Various Degermination and Milling Process

Process	Moisture (%)	Tempering time (h)	Main product
Germ removal through rollermill grinding	15.5–16.5	2–6	Grits for household consumption (fat 1.2–1.5%, approx.)
Degermination with Beall or similar equipment	20–24	2–6	Flaking and brewery grits (fat less than 1.0%)
Decorticating and use of rollermills	17–18	2–6	Brewery grits with less than 1.0% fat and cooking grits
Impacting or entoleting (dry milling)	14.5–15.5	2 or less	Germ and meal separation for feed or food uses

2.2.1.4 Corn Dehuller and Degerminator

The machine can be used for wet and dry degermination processes with two specific models allowing one for wet and the other for dry application. This machine provides a vigorous friction action that can be described as follows:

Friction between kernels
Friction between kernels and rotor jacket
Friction between kernels and beaters mounted on the rotor

Figure 11 shows the profile of the machine.

After the corn is conditioned in the conditioning unit with hot water and steam, it is fed through a gravity spout into the inlet of this machine. The warm, moist, and conditioned corn is then subjected to a thorough abrasive action. This action would help remove the hull and partial degermination of the corn. The hull, germ parts, and some broken kernels fall through the perforations of the sheet metal jacket. The treated kernels pass over to the end of the machine where they fall through an opening controlled by a retarding slide gate. The removal of the hulls can be controlled by adjusting the retarding slide gate.

In the dry application, dry corn is fed into the machine where the rotor action breaks up the kernel removing the germ. Parts of the broken kernel and the germ fall through the wire screen jacket; large pieces of the corn are carried and discharged through the end of the machine.

2.2.1.5 Beall Degerminator

The Beall degerminator is widely used in North America. It is particularly useful for the production of a higher level of grits with minimum of fat content. It helps dislodge the germ without unduly cutting or reducing its size. For the production of flaking grits it is critical that germ is removed cleanly while producing large chunks of grits.

The machine has a simple but very effective design that can be described as follows: The tempered corn at a high moisture level of 20–24% enters the machine. The germ and hull are removed through attrition as the moist corn is pushed through between the conical-shaped cast iron rotor running within a housing of similar shape. The scrubbing and scouring action between the cone rotor and housing not only loosens hull and removes

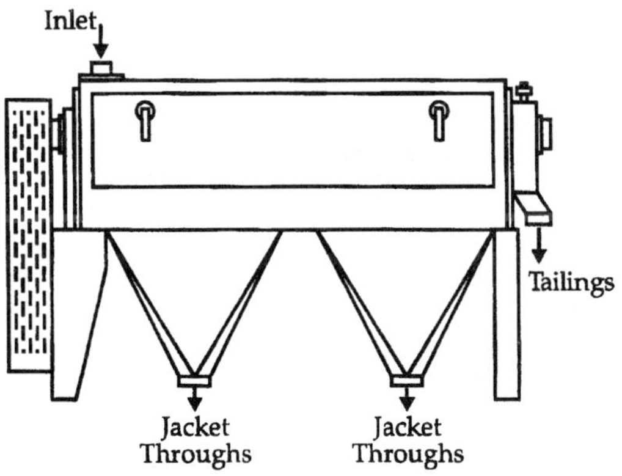

Fig. 11 Dehuller and degerminator. (Courtesy of Bühler, Inc.)

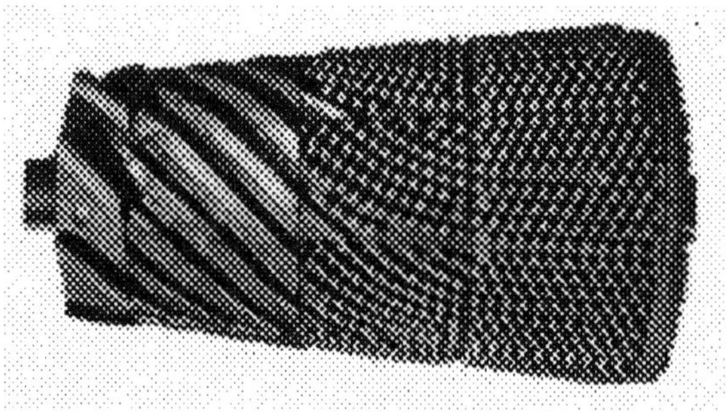

Fig. 12 Patterned cone rotor. (Courtesy of the Beall Degerminator Co.)

germ, but also breaks the corn into large pieces of two or more. The cone-shaped rotor surface is divided into three different surface configurations (Fig. 12).

The smaller end, being the front end, has an auger shape, whereas the middle part has a knob-shaped surface. The remainder surface leading up to the discharge end has studs. The housing or casing has two perforated screen plates on each side, but the rest of the inside surface area has knob-shaped plates. A side view of this machine is shown in Fig. 13.

The screen perforations allow the loosened hulls, germ, and some small parts of the kernel to fall out, while the larger grits are carried through to the tailend of the machine. A weighted gate is located near the tailend of the machine that allows the discharge of the larger grits. The loading of the tailgate can be varied by means of a weight. The loading affects the outflow of the discharged material; thus, its regulation will influence the severity of the action. Additionally, a shifting lever, located near the discharge end, allows varying the clearance between the rotor and the casing while the machine is in operation.

Thus, by manipulating the discharge gate loading and clearance between the rotor and casing the Beall degerminator can be adjusted to provide a high performance according to specific requirements for any size of corn. The rotor speed can also be adjusted to

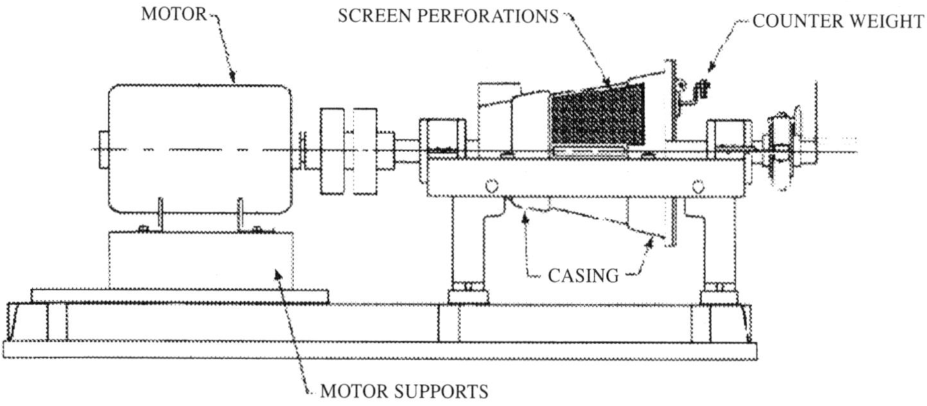

Fig. 13 Side view of a Beall degerminator. (Courtesy of the Beall Degerminator Co.)

desired level. The rotor is provided with chilled iron knob surface to resist wear. The rotor itself is made up of five sections; therefore, in case of wear rather than replacing the whole rotor, only the section with wear is replaced.

2.3 Common Degermination and Milling Practices

Dry milling of corn can be carried out in many different ways. Some of the commonly used commercial practices will be described next. All of these processes involve removal of germ and hull. If the germ is not required to be removed then the milling process becomes much simpler. The processes described are adaptations of practices in Europe and as well as in North America. One of the characteristic differences between the milling practices in the two continents is that in Europe there is much more use of purifiers in the mill, whereas in North America the use of purifiers is limited and the use of aspirators is more extensive.

2.3.1 Degermination and Milling with Rollermills

The milling equipment used in this process is similar to that used in wheat flour mills. After cleaning and short tempering of corn to approximately 15.5–16.0% moisture it is sent to first-break passage for grinding. There may be many different configurations of equipment that can be used with advantage to suit each specific situation. Generally, there are four break passages, two sizing or scratch system passages and four purification passages. Aspirators are used in many different locations in the milling section for removal of hulls, for they are created each time some processing takes place. The aspirators are located ahead of second-, third-, and fourth-break passages (Fig. 14).

Table 8 provides some technical specifications for the rolls and corrugations. Table 9 provides an indication of surface allocation required by this type of milling system.

The actual milling process involves grinding in the break system as would be done in the wheat flour-milling process. Because of the short tempering time and high moisture

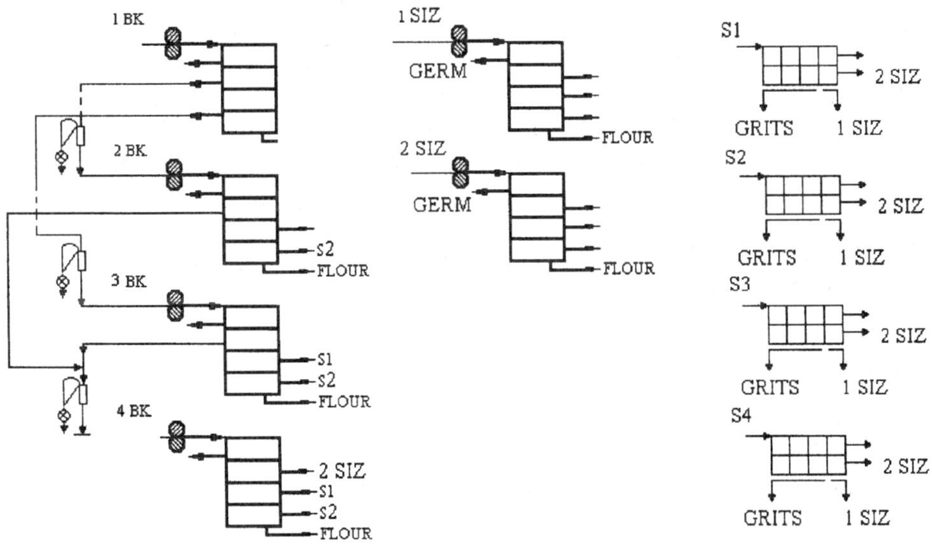

Fig. 14 Degerming with rollermills.

Table 8 Roll Corrugation Specification in Europe

Grinding passages	Roll surface (mm/100 kg/24 h)	Corrugations (cm)	Corrugation spiral (%)	Roll disposition	Roll differential
1 Break	1	3	8	Dull/dull	1.5:1
2 Break	1.5	4		Sharp/sharp	2.5:1
3 Break	1	5	to	Sharp/sharp	2.5:1
4 Break	1	6		Sharp/sharp	2.5:1
1 Sizing	1.5	8		Sharp/sharp	2.5:1
2 Sizing	1	9	12	Sharp/sharp	2.5:1

on the surface, germ becomes loosened and, through grinding, it is removed in the initial breaks. The top screen of the sifter, which is generally a perforated sheet metal cover, removes the germ-rich fraction. Wire meshes may become clogged, so they may not be used as top screens.

Germ removal is followed by the next coarsest material, which is usually passed through the aspirators before being fed to the second-break passage. The aspirator takes out any loosened hulls. Because generally grits and coarse meals are intended to be produced here, the granular material is passed through purifiers to purify the grits. Sometimes some markets also require production of corn flour; if this were a requirement then some reduction rolls which are finely corrugated would be required in the flow.

The extraction of grits and finer products is strongly correlated with the vitreousness of the corn. Generally speaking, the total extraction of grits and other finer granular products, including flour, which will pass through a 60-wire mesh, may be up to 65%. For household-cooking grits the fat content obtained under this process is between 1.2 and 1.5% on a moisture-free basis. If grits with less than 1% fat content is required, then the extraction of this particular product may be around 40%.

A more comprehensive and modern corn mill flowchart using rollermills for degermination, is shown in Fig. 15. The diagram shows four break passages, five reduction passages, and four purification passages. This system will provide a higher level of overall extraction and will also be able to handle a higher load and will have an improved finished product quality. This is because it is more comprehensive in terms of the number of processing passages as well as that the machines that are in use here are all better than the ones used in the previous diagram for performance and many other features.

2.3.2 Wet Degermination with Beall or Similar Equipment

This process is ideal for the extraction of high percentage of large grits or flaking grits with a lowest possible fat content. These flaking grits are used as raw material for the

Table 9 Milling Surface Allocation

Equipment	Value	Unit
Roll surface	7–10	mm/100 kg/24 h
Sifter surface	0.04–0.06	M^2/100 kg/24 h
Purifier surface	2–3	mm width of sieve/100 kg/24 h

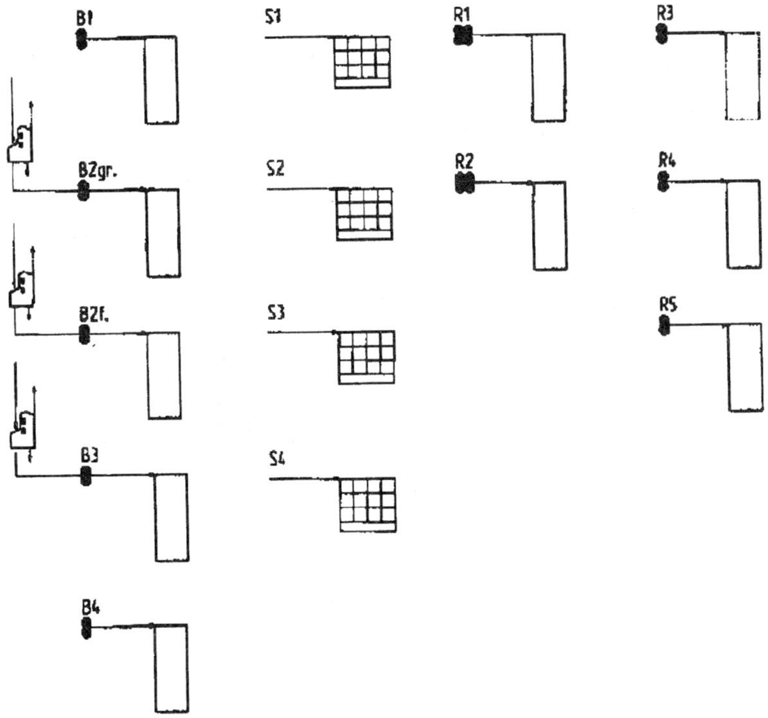

Fig. 15 Corn mill-flow diagram using rollermills for degermination. (Courtesy of Bühler, Inc.)

manufacturing of corn flakes. This process may also be used for the production of brewers' grits, which are smaller in particle size than flaking grits but have a similar stringent fat content requirement of less than 1% on a dry-moisture basis. For the production of brewers' grits, the particle size of the flaking grits is further reduced by grinding in the rollermills. This process yields a high percentage of brewers' grits at the required fat content specification.

A very high percentage of flaking grits in the United States is produced by this process using a ''Beall.'' The ''Beall'' removes the hull and the germ in one single passage through attrition, as was described earlier. The coarser material that is discharged through the tailgate and the material that falls through the screen have a very high moisture level. These streams may be combined or kept separately and dried and cooled before further processing, to a lower moisture level of about 16% or less. The cleaning, conditioning, and degerming processes are shown in Fig. 16, and a process flowchart of the mill is shown in Fig. 17.

Drying, as with any cereal grain, must be carried out carefully, as fast drying with high temperatures may cause heat stress resulting in fissures; thereby, developing cracks in flaking grits.

A similar machine in operation but different in design is the dehuller and degerminator (see Fig. 11). Regardless of the type of machine being in use, in this process, preparation of the corn for maximum removal of hulls and the germ requires addition of moisture

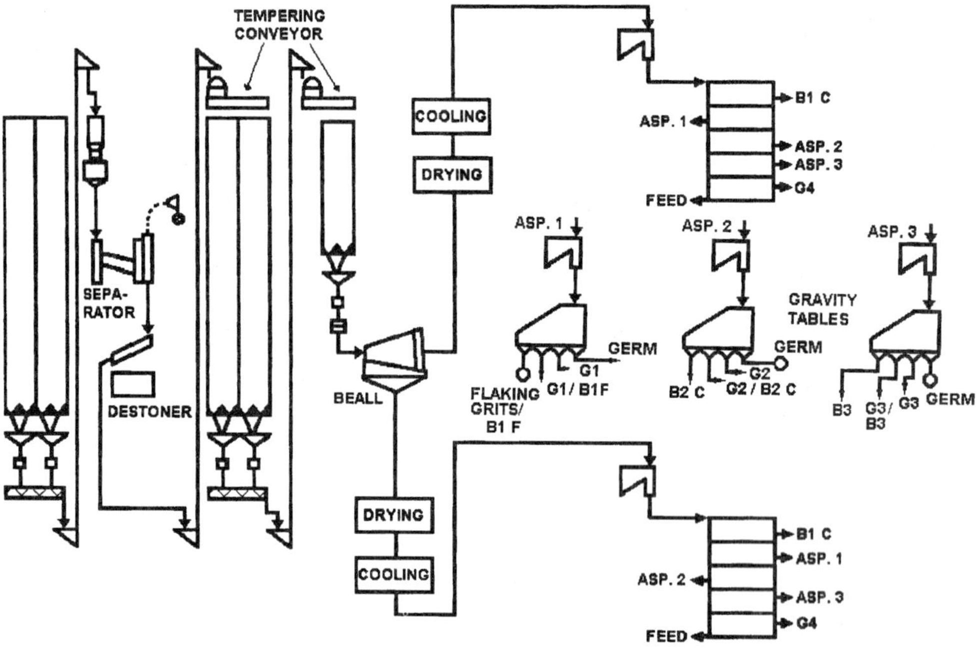

Fig. 16 Degerming with Beall degerminator.

and heat. This may be accomplished through the use of the conditioning unit as has been previously described (see Fig. 10).

The actual process using such a conditioning unit, along with dehuller and degerminator equipment, is shown in Fig. 18. It involves bringing the moisture of the corn up to 20–24% by application of hot water at 80°C and combining it with steam as necessary. Addition of such a high level of moisture also means that the finished products and the by products will also have a high level of moisture. In addition to that, handling the moist material coming out of the degerminating system for further processing, would be very difficult. Therefore, these materials are dried in one of two ways:

1. When the corn's moisture before degermination is near 20–22%, some drying of the material continues through its natural handling by means of aspirators and transportation, and such. However, the finished products and by-products would still have high enough moisture that they would require drying down to 14% or less before shipping.
2. When moisture of corn before degermination is near 25%, it becomes almost essential to dry the material. This can be accomplished by using rotary steam tube-type dryers. Following the drying cycle the material needs to be cooled down to about 30°C or less. The moisture of the material is reduced to about 16.0%.

In this drying procedure, the finer "throughs" and the coarser "overtails" from the degerminating equipment may be combined for drying purposes unless the mill has a large production capacity.

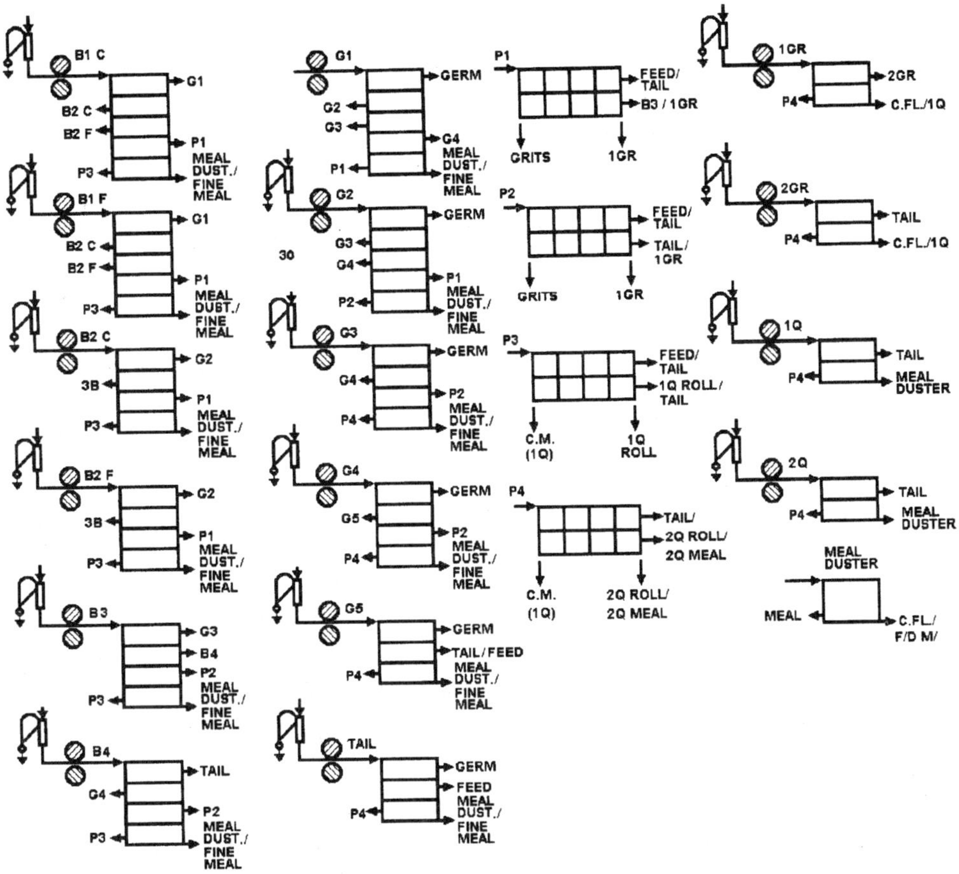

Fig. 17 Flow diagram of a corn mill.

However, when drying is occurring through the application of aspirators and through natural movement and handling of the material in process, the "overs" and "throughs" of the degerminating equipment are sent to their respective grading in the sifters. Figure 18 shows such a system. The system consists of two graders, three break passages, three sizing passages, and four purification passages. Out of the four purification passages two are actually using just aspirators for removing hulls, whereas the remaining two are purifier sections.

The grading is followed by further separations in their own respective gravity tables for the separation of germ and other fractions. The other fractions can be grouped into two types of stocks. One is pure large grits with very low fat content, whereas the other is rich in germ. The former can be designated as flaking grits and sent for packing. The germ-rich fraction is also coarse-break stock, with bran/hulls that need to be processed using a short milling system for removal of the hulls and germ while reducing the particle size of the grits. If brewers' grits is required more than large, flaking grits, then this milling system can adequately meet the objectives. This system involves three to four break passages, three sizing passages, and four purification passages. Break passages in combination

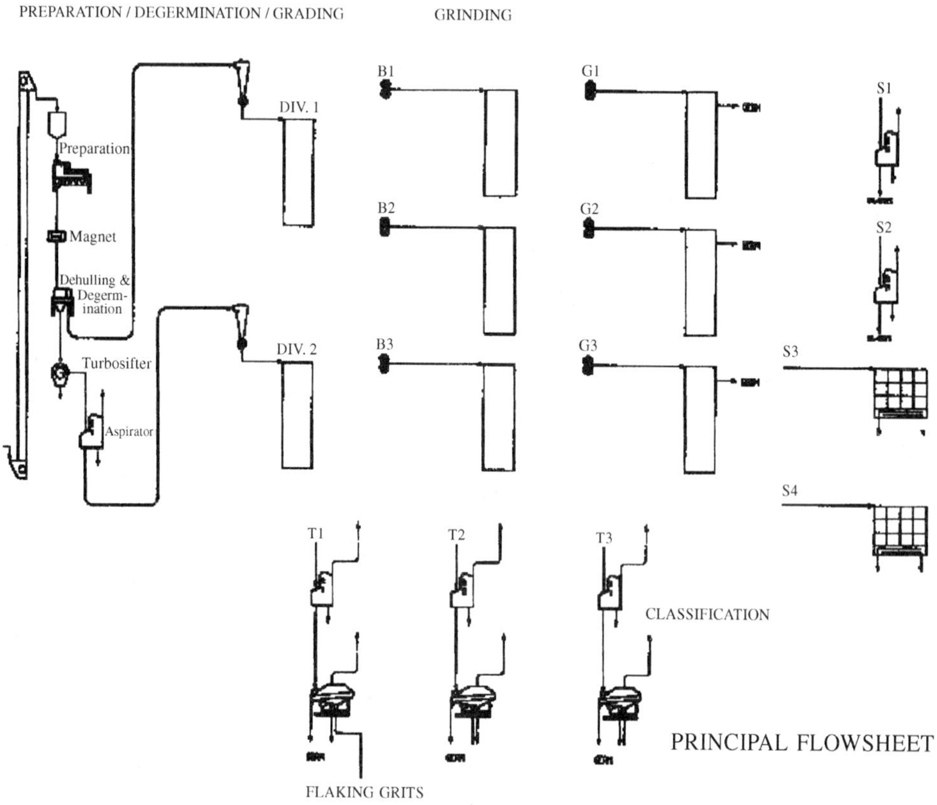

Fig. 18 Wet degermination for the production of flaking grits and brewers' grits. (Courtesy of Bühler, Inc.)

with sizing passages help reduce the particle size of the grits and remove the remaining germ pieces and hulls.

The total extraction of grits, which may include flaking grits, brewers' grits, cooking grits, or combinations thereof, may range from 60–65% with a fat content of less than 1%.

2.3.3 Dehulling, Degermination, and Use of Rollermills

The main difference in this type of process when compared with the previous one is that the moisture level used in the corn before degermination is lower. After the corn is cleaned its moisture is raised to about 17% through the application of hot water or steam. Because the moisture is not as high as the second process described for flaking grits, the percentage of grits production is slightly higher overall; but obviously, if flaking grits are desired to be produced with this system, the percentage of that product will be less. The conditioned corn at 17% or more is passed through dehulling and degerminating equipment. The rest of the milling system is very similar to the system described in the second process, with three breaks, two sizings, and four purification passages. Depending on the quality of the corn and on the end product quality requirements relative to fat content, fiber, and particle size requirements, the extraction of grits in this process would range from 50 to 60% under 1% of fat content. Since the moisture addition for germ preparation is not as high as that in the second process, drying of stocks is not necessary.

Table 10 Milling Systems and Extraction, Granulation, and Fat Values

	Degermination with rollermills			Degermination with dehuller and degerminator		
Finished products	Granulation (μm)	Extraction (%)	Fat %	Granulation (μm)	Extraction (%)	Fat (%)
Flaking grits				5600–3500	40–60	<0.6
Brewers' grits				1500–500	55–65	<1.0
Cooking grits	2500–1400	60–62	1.2–1.5			
Meal and flour	<500	6–8	2–6			
Germ		8–12	20–16		10–6	<22
Branny feed		20–24	9–7		25–20	7–12

2.3.4 Impacting or Entoleting

Such a process is used primarily for feed purposes. After cleaning the corn a very small percentage of moisture is added by way of preparation to bring the moisture of the corn to about 15%. After passing through the impact degerminators the material is graded in a sifter section where the coarsest fraction is sent to another passage of impact degermination. This material is once again graded in another sifter section. The remainders of the graded material from both sections are passed on to two gravity tables, one for each grading section. Adding two more grinding passages would help increase the recovery of germ even more. The germ thus obtained may be used for feed purposes or can be used for oil extraction. The rest of the materials, such as grits and hulls, are also used for feed purposes. If necessary, this process can also be used for the production of cooking grits and germ.

2.4 Milling Results

Table 10 provides a summarized milling performance evaluation, relative to extraction, granulation, and fat content, that may be expected from the two types of milling systems. Different processing facilities depending on their requirements apply the relevant milling system.

2.5 Corn Products and Their End Uses

Corn products are highly diversified in terms of their end uses. The end use requirements dictate the granulation or particle size of the milled corn products. Figure 19 shows various finished products.

Table 11 briefly describes the various milled corn products and their end use applications. Popcorn and sweet corn do not require any processing such as milling or similar type of processing for consumption. Popcorn is popped while sweet corn is boiled for consumption. Corn is also used for making syrup, dextrose, alcohol, modified starch, and cooking ingredients. The first step in manufacturing all of the foregoing requires the extraction of starch from corn through the wet-milling process. This process is described next.

2.6 Wet Milling of Corn

Wet milling of corn is performed to produce starch, gluten meal, sweeteners, ethanol, and others. The process involves steeping the corn in water under controlled conditions fol-

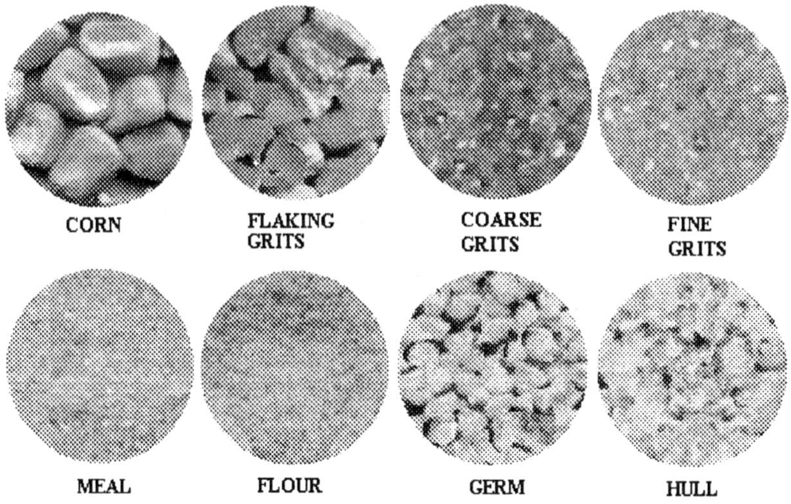

Fig. 19 Milled corn products.

lowed by milling. The components are then separated through screening, grinding, washing, centrifuging, and washing.

2.6.1 Milling Objectives

The milling objectives are significantly different for wet milling owing to the end product requirements, for finished products here are no longer grits, coarse and fine meal, and flour. The production of grits, meal, and flour requires particle size reduction of endosperm and refinement of these from bran, hulls, and germ. Therefore, this is achieved by the conventional dry-milling approach of cleaning, tempering, degerminating, drying, cooling, grading, aspiration, grinding, sifting, and purifying. Essentially grits, meals, and flour would all consist of primarily starch and protein, and the main difference among them would be the particle size.

Table 11 Milled Corn Products and Their Applications

Milled products	Applications	Characteristics
Flaking grits	Corn flakes	Coarse granulation (3.5–6 US mesh). yellow corn with nonfissured, vitreous endosperm to provide higher yield of grits
Grits	Table use such as polenta, porridge, snack foods, brewers' grits	Appropriate balance of granulation (14–28) to provide reasonable cooking time and homogeneous cooked texture without causing formation of lumps
Meal and flour	Tortillas, bread mix, pasta, snack foods, flat breads, and such	Finer particle size with wide variation. More specific particle size is dependent on the type of finished product

When corn is required to yield finished products on the basis of its constituents, such as a very high yield of refined starch, germ, gluten meal, or feed, a quite different approach to isolate these various components is required.

It is quite clear from the explanations provided in dry corn milling, that of the four different degerminating processes described, the best recovery of germ is accomplished in the process where highest level of tempering moisture is added. Germ and hull can be removed more effectively by tempering to a much higher level of moisture (24–25%) and by controlled attrition within the rotor and the knobbed jacket of the Beall degerminator. Following the dislodging of the germ and separation of some of the hull the physical removal of these eventually takes place in the gravity tables and aspirators. This separation of germ and hull from large chunks of pure and composite endosperm particles with some remaining germ and hulls, is primarily accomplished on the basis of density and size. Germ and the hull, having differing specific gravity and being larger (especially germ and germ-rich streams) than the chunks of pure and composite endosperm particles, are separated by gravity tables (germ) and aspirators (hull). The remaining germ and hull in the composite endosperm particles are separated by further careful grinding, aspirating (hull), and concentrating the germ-rich large fractions by sifting.

From the foregoing, the principal physical characteristics that are taken advantage of dry corn milling and the processing approach taken, for the separation of germ and hull from the endosperm, can be summarized as follows:

1. Differences in weight
2. Differences in particle size
3. The higher the tempering moisture, the better the germ recovery
4. Size reduction

2.6.2 Milling Approach in the Wet Process

A high yield of refined starch cannot be obtained from dry milling even with fine grinding and air classification, which still exists in limited practice primarily for soft wheat milling. Even best results obtained with such application may give a protein content of about 6% in the starch-rich fraction. Furthermore, good results are not obtained in terms of protein shift when hard-textured grain, such as corn, is processed. Even in wet milling harder varieties of corn are required to be steeped in water for a longer time to yield desirable level of refined starch (Fig. 20).

To meet the end product requirement objectives, the wet-milling process uses the following approach:

1. Degerm the corn under wet condition at near 45% moisture to obtain higher yield of germ. The wet condition also prevents unnecessary shredding of hull that can be removed later in the process in the form of fiber, on the basis of its size.
2. Make use of water as a medium for separation of lighter germ flowing over the top from heavier particles settling under at the bottom.
3. Make use of size for removing and washing of germ and fiber through wet steeping.

2.7 The Wet-Milling Process

Corn is cleaned in the same manner as in dry corn milling. The cleaned corn then moves on to the next step in the process, referred to as steeping and described in the following:

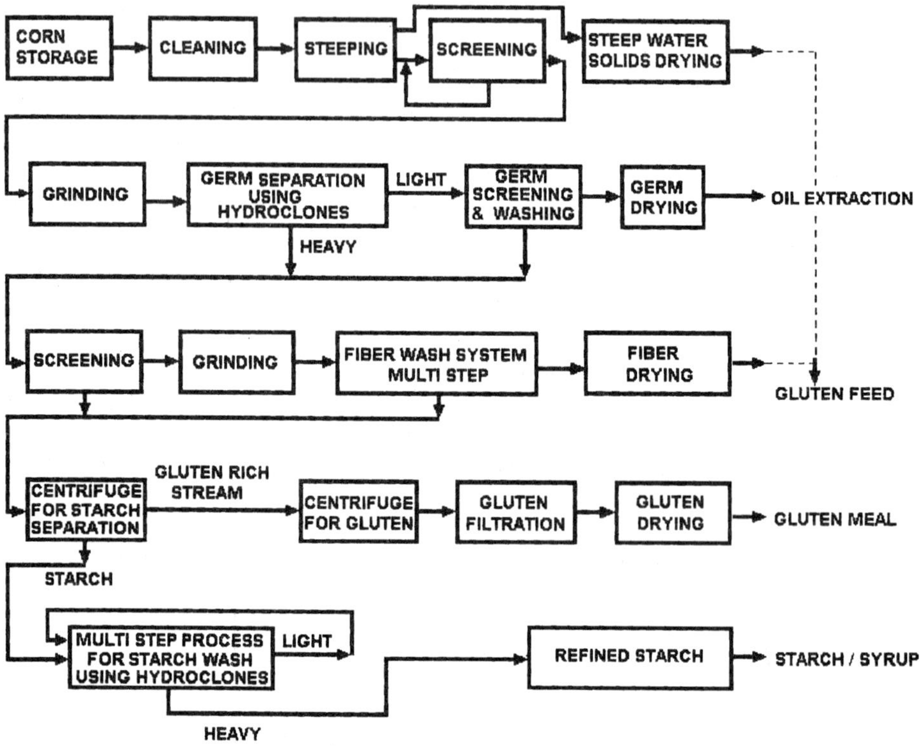

Fig. 20 Simplified block flow diagram of the wet-milling process.

2.7.1 Steeping

Steeping in wet milling replaces tempering in dry milling. In this process, therefore, softening of the kernels is accomplished. However, this is achieved by soaking the corn in water rather than adding water to it. As in tempering, temperature and time are critical for the penetration of moisture and, therefore, are controlled carefully. Additionally, treatment of water with sulfur dioxide is carried out, which may further hasten the moisture penetration and, more importantly, is thought responsible for breaking up the protein matrix surrounding the starch granules. This facilitates starch and protein separation further and helps in increasing the percentage of refined starch yield. Steeping is carried out in tanks, and steep water that ends up with some corn solubles is sent for concentration through evaporation. The residue obtained has nutritive value for feed rations.

2.7.2 Germ Separation

The steeped corn is ground in disk mills to release germ in as large a size as possible. Care is taken in grinding adjustment to prevent any severe crushing of the corn. Similar to the Beall degerminator, the grinding surface of the disks is shaped as knobs. Because of the presence of high moisture, the process of germ dislodging action is very gentle. The ground material slurry is taken to a separating medium for separation into light and heavy fractions. This involves the use of hydrocyclones (Fig. 21a), where the lighter material containing germ floats on top and exits as overflow, while the relatively heavier fraction that starts to settle underneath is removed from the bottom (see Fig. 21b). The

(a)

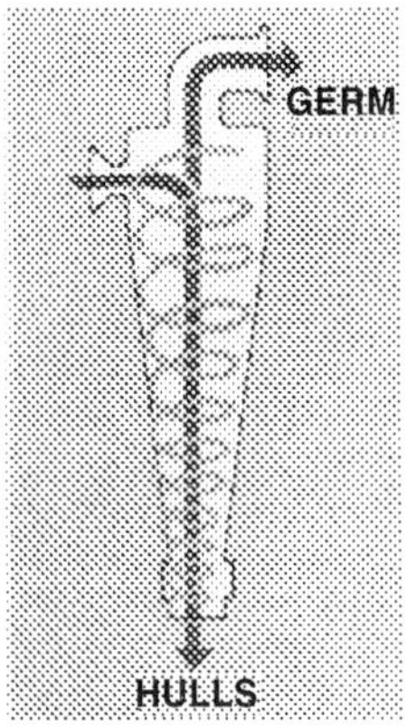

(b)

Fig. 21 (a) Series of hydrocyclones separating germ; (b) operating principle of hydrocyclone. (Reproduced from Bulletin No. Food-3, courtesy of Alfa Laval Separations, Inc.)

operating principle of the hydrocyclone is similar to that of the air cyclone used in suction pneumatic conveying of products described in Chap. 11. The main difference between the two being the former uses water, whereas the latter uses air as a carrying medium.

The material that begins to settle at the bottom may still contain germ that needs to be removed and recovered. Therefore, this material is passed through a second set of grinding mills for relatively close grinding to release the remaining germ. A schematic diagram of germ separation is shown in Fig. 22.

The released germ is separated from the rest, as described before, with the overflow going for germ wet screening. Wet screening allows the germ to be washed while being physically separated from the rest using appropriate screen size and type. The germ is then sent for drying and oil extraction. The heavier material settling at the bottom is moved to the next step of fiber removal.

2.7.3 Fiber Separation

The rest of the corn slurry is passed through finer screens to separate the fiber (mostly hull) from starch and gluten. A schematic drawing of fiber washing and drying is shown in Fig. 23.

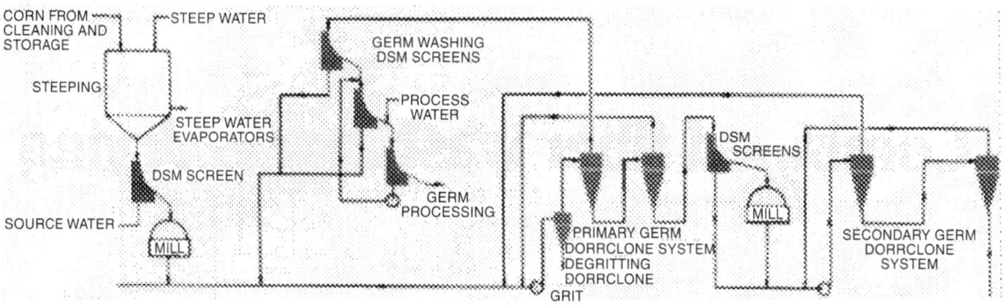

Fig. 22 A schematic diagram of germ separation in wet milling. (Reproduced from Bulletin No. Food-3, courtesy of Alfa Laval Separations, Inc.)

Larger-sized particles that are unable to pass through the screen size consist of primarily fiber and some particles of starch and fiber attached to each other. The remainder of the slurry, including gluten and much of the starch, passes through to the next step of starch and gluten separation, which will be described in the next section. Continuing on with the larger-sized fiber and starch particles attached to each other, these must be detached through some form of disintegration in an impact or attrition mill. Following the passage through the mill, the slurry consisting of the ground material is sent to a multistep fiber wash system (Fig. 23) consisting of several screens for washing the starch from the fiber. The fiber thus separated is sent for removal of water and concentration of solids (fiber). Screen centrifuges may be used for that purpose. Figures 24 and 25 show the operation of a fiber-separating screen and the screening centrifuge.

2.7.4 Starch and Gluten Separation

The remainder of the slurry from initial screening of the fiber along with the slurry that remains from the fiber wash system are directed to the starch and gluten separation stage. This consists of various centrifuges for clarification of process water and thickening of streams and to separate starch-rich slurry from high-protein gluten stream. Starch-rich slurry is directed to a multistep starch-washing system for further concentration and refinement, and this phase of the operation will be discussed in the next section. The gluten stream moves on to another centrifuge for further concentration of solids.

Because gluten is not soluble in water and is lower in density, relative to starch, helps capture it quite efficiently, while starch escapes along with the water. Gluten-rich material is then sent for filtration. Following filtration gluten is sent for drying. This is

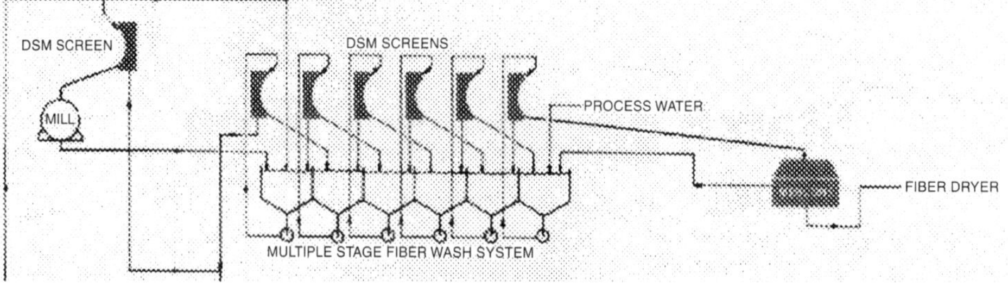

Fig. 23 A schematic diagram of fiber washing and drying. (Reproduced from Bulletin No. Food-3, courtesy of Alfa Laval Separations, Inc.)

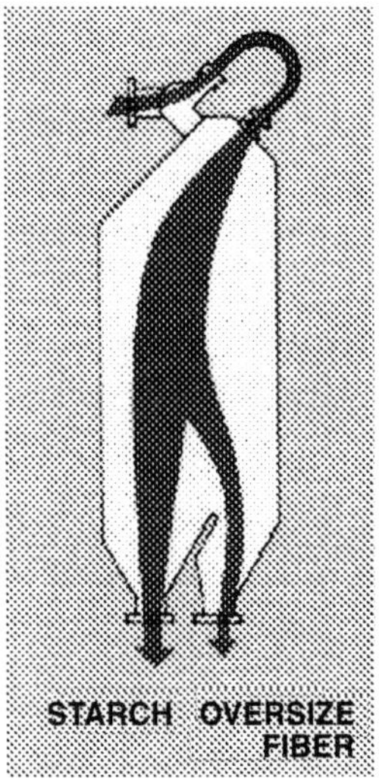

Fig. 24 Fiber-separating screens. (Reproduced from Bulletin No. Food-3, courtesy of Alfa Laval Separations, Inc.)

then sold as gluten meal for animal feed rations. The starch and gluten separation process is shown in Fig. 26. Economic viability of a wet-milling process depends largely on the effectiveness of the starch and gluten separation at this stage. The higher the recovery of high-protein gluten with minimum loss of starch, the greater is the profitability. Starch-rich slurry following the primary separation may contain only 3–5% gluten.

2.7.5 Starch Refinement

Starch-rich slurry following primary separation is sent to a multistep starch-washing process. In this process, diluted starch stream is washed several times until all the protein (gluten) is removed. This process uses a battery of hydroclones for separating the remaining protein from starch. When fine particles of starch and gluten are in suspension,

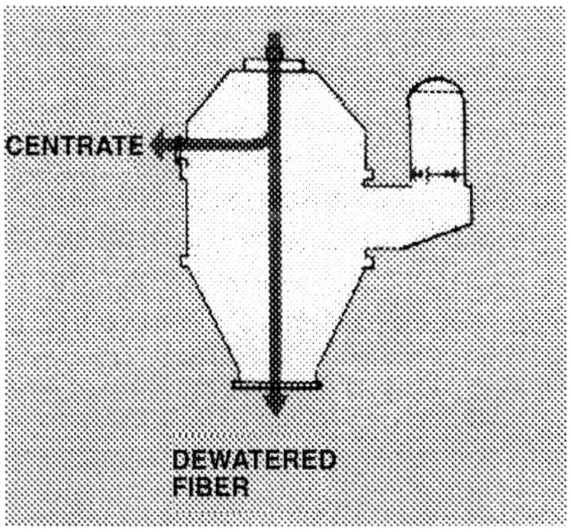

Fig. 25 Screening centrifuge. (Reproduced from Bulletin No. Food-3, courtesy of Alfa Laval Separations, Inc.)

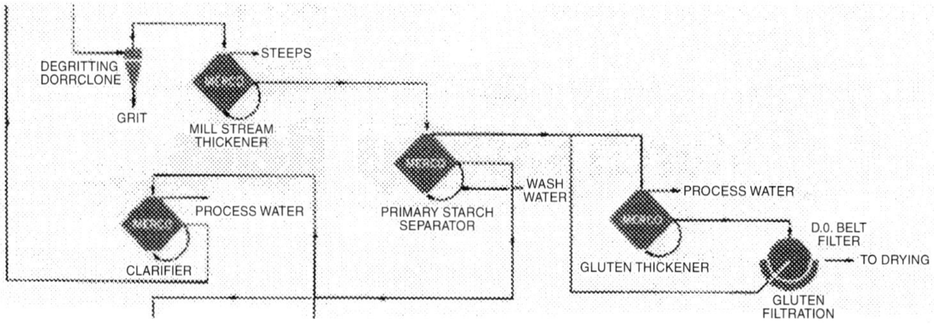

Fig. 26 A schematic diagram of starch and gluten separation. (Reproduced from Bulletin No. Food-3, courtesy of Alfa Laval Separations, Inc.)

the denser starch particles would tend to settle at the bottom, whereas lighter gluten particles would tend to float on top. The effectiveness of this separation is largely dependent on the force with which the water carrying the starch and gluten particles strike the inside of the cyclone. The necessary force required is provided by a series of pumps. Therefore, in a hydroclone, lighter gluten-rich fraction would exit from the top, while heavier starch continues to be refined as it passes from one set of hydroclones to the next. A starch-washing system is shown in Fig. 27.

Starch obtained this way is highly refined with a degree of purity of close to 99.5%. The refined starch may be converted to syrup and dextrose, or sold as is. Sometimes starch may be modified and sold as modified starch. For the production of alcohol, starch is first converted to dextrose, which is then fermented into alcohol.

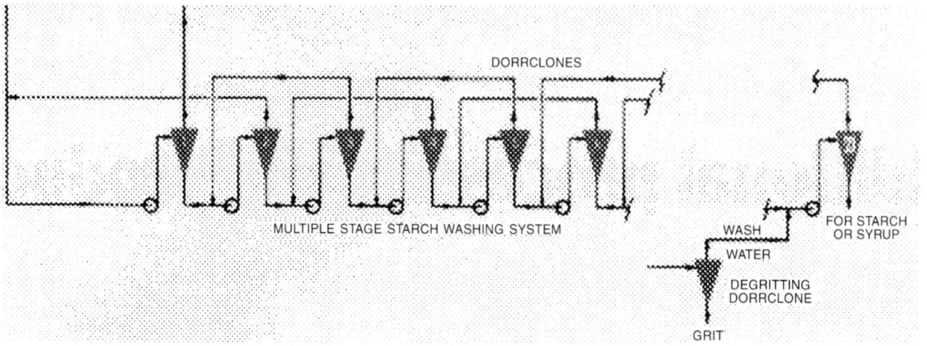

Fig. 27 A schematic diagram of starch-washing and refinement process. (Reproduced from Bulletin No. Food-3, courtesy of Alfa Laval Separations, Inc.)

3 OAT MILLING

3.1 Introduction

The utilization of oats, as with other cereal grains, is primarily in the area of animal feed, human consumption, and industrial uses. Although the proportion of oats used for human consumption is much less than for other uses; nevertheless, its role in the overall human diet is a significant one. Its high nutritional value combined with its characteristic flavor has helped it to secure an important place in the breakfast cereals group and other processed food.

The milling process of oats is very different from the rest of the cereal grains. This is primarily for three reasons:

1. The hulls strongly adhere to the kernel or groat.
2. Owing to their high fat content, the oat kernels need to be stabilized through heat treatment for prevention of rancidity.
3. The form of its consumption is quite different from the rest.

3.2 Quality of Oats

Selection criteria of quality of oats for milling requirements are as follows:

Low moisture content.
Clean as possible and, particularly, freedom from wild oats, barley, and other seeds of similar length and density that are hard to remove.
Low tolerance for discolored kernels as well as uniform kernel size are very important.
High test weight (bulk density) for thin seeds and higher percentage of hulls reduce yields and cause processing problems.
Low tolerance for immature, light oat kernels that would affect the yield and also would be hard to remove.

3.3 Milling Process

The milling process begins with cleaning the oats. Cleaning is followed by grading before dehulling. Dehulling operations remove the hulls, and oat kernels without the hulls (groats) are sent for conditioning and drying. The conditioned groats are then graded and subsequently sent for cutting or flaking. Some of this material may also be ground and sifted for oat flour and bran production, discussed later in some detail. The sequence of operations in modern oat-processing plants is generally very similar to what is shown in Fig. 28.

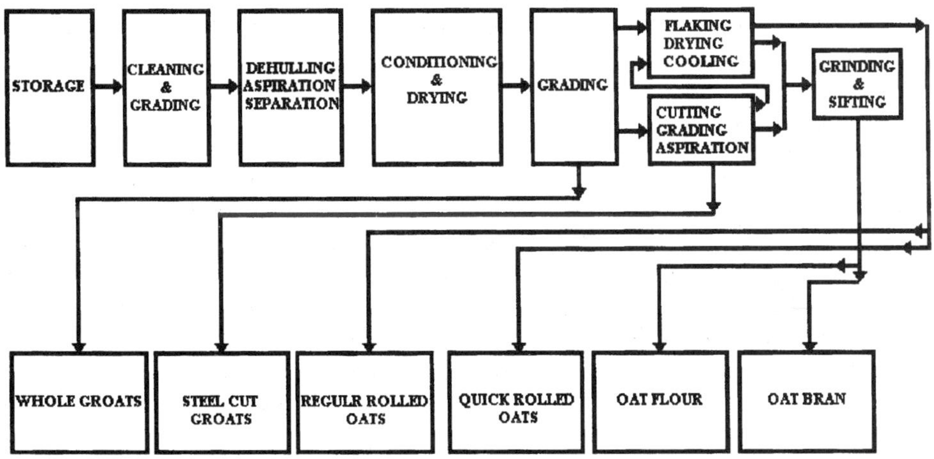

Fig. 28 A simplified block flow diagram of oat-milling process.

The cleaning system for oats uses equipment similar to that seen in the cleaning house of the other cereal grain-milling systems, except the clipper.

3.3.1 Cleaning

The cleaning of oats begins with a weighing scale, followed by a magnet for removal of any ferrous metal or its alloy from the oat stream. The first cleaning machine in sequence is a clipping machine. This separates the double oats that, if not separated here, would cause problems in the hulling operation. This machine also clips off the ends of oat kernels to improve the ''hullability'' of the oats during the hulling operation. A grain separator or classifier is used next, with an air-recycling aspiration system attached. This would help remove larger, finer, and lighter foreign material from oats. An air-recycling system helps conserve energy. The grain separator is followed by a destoner, where stones, mud balls, and other heavy material are removed. From there it is passed on to machines making separations on the basis of length, such as, disk separators or indented cylinders, for removal of longer and shorter seeds. A battery of these machines along with width graders may also be used for the removal of poor quality oats and other cereal grains while performing the important function of grading the oats. Choosing appropriate pocket shape, indentation, and width size for the disk separator, indented cylinder, and width grader, respectively, a highly selective grading of oats and separation of double oats, barley, wheat, and thin oats can be obtained.

Grading function can also be accomplished using drum graders which make separations on the basis of width. In the older oat-processing plants grading is also carried out using a grading sifter with various screen sizes to sort small, medium, and large oats.

3.3.2 Grading of Oats

Once the bulk of foreign material is removed, the next step is to grade the oats on the basis of size into large, medium, and short oats. Optimization of hulling efficiency is attained when oats are sorted out by size and sent to their respective hullers where the speed and load on the huller have been adjusted to suit the specific size of the oats. Small oats, for instance, interfere not only during hulling operation, they also pose a problem during separation from groats following the dehulling operation. This can be accomplished as shown in Fig. 29. As can be seen in the cleaning diagram, the clipping machine is not being used as double oats are being removed by disk separator later on in the grading process.

3.3.3 Hulling Operation

In this part of the process the inedible hulls that protect the kernel of oat are removed. If the natural moisture content of oats before the hulling process is higher it will be more difficult to hull. The ease with which hulls can be removed from the oat kernel is referred to as *hullability*. Therefore, the oats with higher natural moisture have relatively poor hullability. The converse is also true. In the cleaning section the use of a clipper machine, which cuts off the ends of the oats, helps to improve the hullability of the oats.

The *hulling efficiency*, also referred to as *hulling degree*, is the quantity of hulled kernels expressed as percentage of the total oats. This operation is considered very critical to oat millers. Just as in wheat flour milling a higher yield of white refined flour is important, similarly in oat processing higher yield of oat groats is critical. Oat *groats* is the kernel of oat after all the hulls and chaff are removed from the oat kernel.

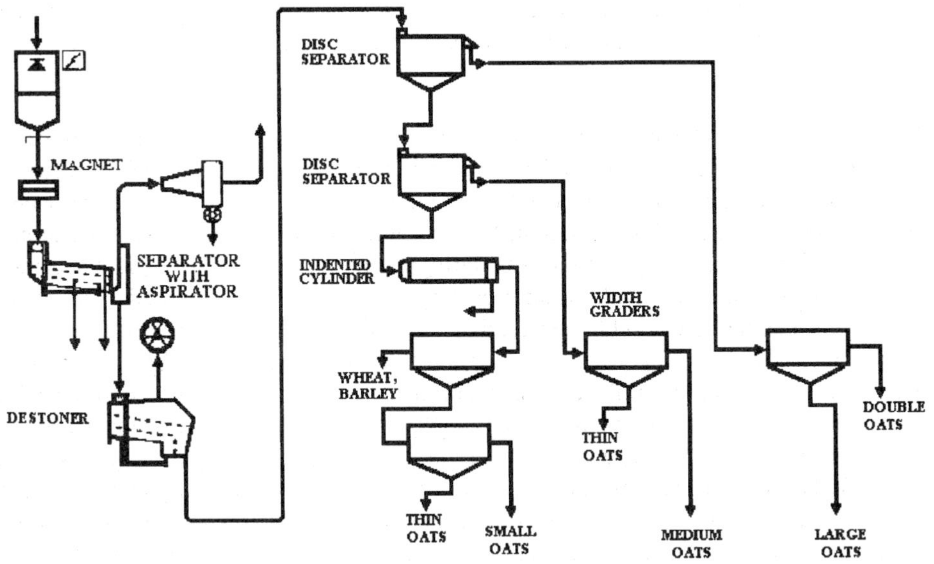

Fig. 29 Cleaning and grading of oats.

In the hulling operation, therefore, the following factors must be very carefully considered to ensure the highest degree of performance:

1. All the oat grain must be dehulled.
2. There is none or a minimum of broken material created.

Although these two points are attainable together up to a point, beyond that their relation becomes inversely proportional. Therefore, in any hulling operation there will always be a small percentage of unhulled grain as well as a small percentage of broken material that will remain following the hulling activity. The performance of a given hulling operation can thus be evaluated on the basis of how high the aforementioned point becomes, in terms of hulling efficiency, before the percentage of broken material begins to increase.

Hullability is influenced by the following factors:

Moisture content of the grain
Load on the impact huller
Peripheral speed of the rotor
Impact ring material
Size of the oats

1. Moisture. The higher the moisture content, the more difficult it is to remove the hulls; however, the risk of creating a lot of broken material is minimized.
2. Load. The higher the load, the lower will be the dehulling efficiency; that is there will be a higher percentage of unhulled grain remaining after the dehulling operation.
3. Peripheral speed. The higher the peripheral speed the greater is the dehulling efficiency; however, the percentage of the broken material would also be higher; also, the impact ring will wear off sooner.
4. Impact ring material. The composition of the material affects the hulling efficiency, percentage of broken material creation, and the useful life of the impact

ring. For instance, ceramic lasts much longer than most material, but its hulling efficiency is poor.
5. Size of the oats. The greater the uniformity of the oat grains the better will be the dehulling results.

Operationally, the following measures are taken to ensure that optimum performance can be attained at all times:

1. High-quality impact hullers are used with the provision of adjustable speed that can correspond to the hullability of varying types of oat kernel, thereby ensuring optimum performance at all times.
2. Varying sizes of oats can cause many problems as it will be hard to target proper adjustment, such as, speed and load, to obtain a high hulling efficiency. Therefore, in large processing plants the oats are graded into two or three sizes, using length separations and width separation, as explained in the earlier section, before being dehulled in separate lines. This allows optimum adjustment in each line. If small oats remain present with large oats, not only does that cause hindrance in terms of effective dehulling, but also this leads to problems in effective separation of the dehulled kernels (groats) from the hulls.
3. Steps are taken to remove the light oats as much as possible in the cleaning house. Usually strong aspiration helps lift light oats out of the main oat stream. A good cleaning operation helps lighten the load on the hullers. It is the light oats that do not hull easily and cause backing up of the feed on the hullers, reducing the capacity.

3.3.4 Hulling Equipment

Many years ago hulling of oats was carried out by stone hullers. These required low moistures for dehulling; therefore, kiln drying preceded the hulling operation. The modern hulling operation consists entirely of impact hullers. This is simply owing to the economics of the operation. The impact hullers require less horsepower. The percentage of broken material created is very small here, as oats in their natural moisture (12–14%) will be less susceptible to fragmentation. Lower energy costs of kiln drying follows the dehulling process and, therefore, there is much less material to dry. However, there are some oat-processing plants, although in minority, that use drying before impact hulling, as this makes the hull brittle and the hullability is much improved.

3.3.5 Impact Hullers

The hulls of the oat are separated from the oat kernel through impact using an impact huller. Figure 30 shows such a machine. It consists of a high-speed rotor, the speed of which is easily controlled to provide optimum setting relative to the hulling characteristics of the oats. The speed is adjustable through a frequency converter. The rotor consists of impeller flights that are grooved to facilitate the alignment of oat grains horizontally in the direction of the motion, permitting it to strike an impact ring. The impact ring is constructed to resist wear and tear, and its condition can be monitored through optical and acoustical wear detectors. The impact ring is mounted in the housing cover. Impact ring material selection is based on its performance (hulling efficiency/broken) and durability. Rubber rings are preferred for high-hulling efficiency and less broken material. Rubber rings wear out faster. The other popular material is Carborundum, which has good hulling efficiency and also longer useful life; however, it creates a slightly higher percentage of

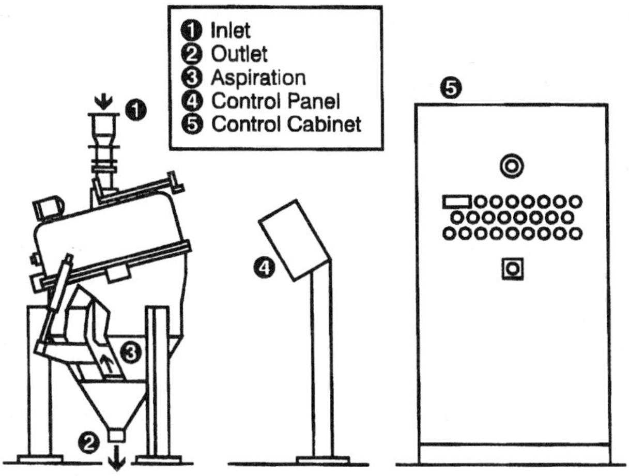

Fig. 30 An impact huller. (Courtesy of Bühler, Inc.)

broken material. If a kernel is only two-thirds of its original size then it may not be considered as broken.

As the oat kernels strike the impact ring the hulls are separated. The mixture of material consisting of groats, hulls, and unhulled kernels leaves through a hopper outlet for separation. Groats make up about 68–72% of the oats after the hulls are removed from it.

3.3.6 Aspiration and Separation

This mixture of product is then passed through an air-recycling aspirator for the physical removal of the hulls from the rest of the material. Following which the rest of the material is passed through a clipper, as used in the cleaning house, for clipping off the brush hairs and fuzz. Some of the hulls that are still remaining in this mixture, owing to its abrasive nature, are beneficial to this process of removing the fuzz and the brush hairs.

The groats with a small percentage of unhulled oats then move on to a table separator. A double-deck table separator may be used here to separate the lighter unhulled oats from the heavier denser groats. A table separator makes the separation between light and heavy material of similar size and shape on the basis of the following:

Stratification to create density layers through oscillating motion of the sieve decks

Followed by the application of triangular impact plates in staggered position on the deck where the lighter material that floats up on striking continues to move up and is discharged from the top. The heavy materials at the bottom, through similar impact action, continue to move downward and are separated

The inclination of the deck is adjusted to offer the best results

The mixture is fed at the center of the table. A double-deck would allow the lower deck to retreat the heavier material from the top deck. This would permit the high quality separation between the groats and unhulled oats that is necessary to meet the quality stipulations. Figure 31 shows the principle of operation.

The unhulled oats are returned back to the impact huller (Fig. 32). In some processes the unhulled oats are sent to a separate huller. For example, if there are three or four

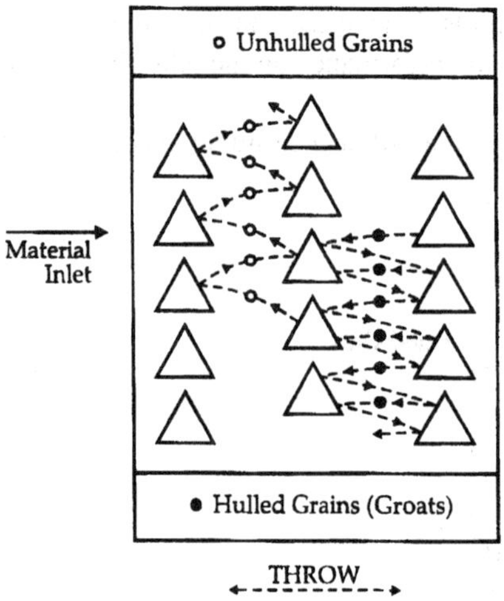

Fig. 31 Operating principle of a table separator. (Courtesy of Bühler, Inc.)

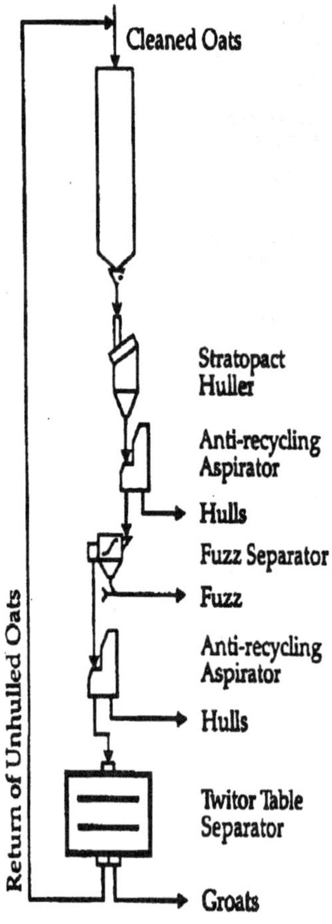

Fig. 32 Hulling operation, with unhulled oats returned to the same huller. (Courtesy of Bühler, Inc.)

primary hullers in operation, the unhulled oat kernels from these can be combined and sent to a separate huller adjusted to provide optimum performance on the rehulling operation (Fig. 33).

If the unhulled kernels are sent back to a separate huller then huller speed is set at a level higher than the previous huller.

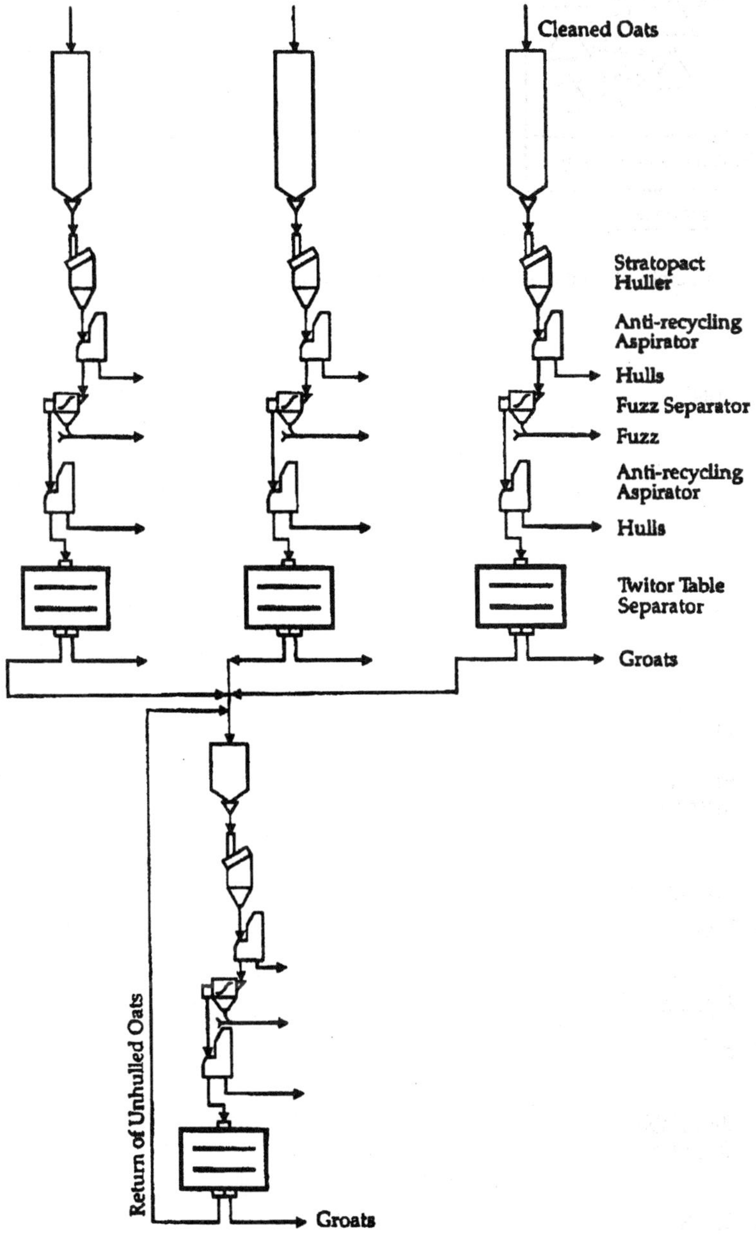

Fig. 33 Hulling operation, with unhulled oats returned to a separate huller.

3.3.7 Conditioning and Kiln Drying

This process involves conditioning with steam followed by drying; therefore, the treatment provided here is also referred to as *hydrothermal* treatment. There are still some oat-processing plants that practice hulling after the heat application. In such cases the oat grain with hulls intact are simply dried and there is no conditioning with steam. Even if the drying is carried out following the hulling operation, there are some plants that may exclude conditioning.

This is an important part of the total process, for the overall treatment provided here has far-reaching implications on the product quality in terms of taste, storage, and functionality. The goals of this hydrothermal treatment are as follows:

- To inactivate the complex enzyme systems that include primarily the fat-splitting enzyme (lipase) and oxidizing enzymes
- To develop the formation of characteristic oat flavor (toasted)
- To affect the functional and finished product quality
- To gelatinize some of the starch (outer starch layers)
- To lower the microbial count

3.3.7.1 Steam Conditioning

The purpose of the steamer is to hot-condition the groats by adding the required moisture and heat. This is accomplished through the controlled application of saturated steam which is applied directly. The groats are fed from a bin above to the top part of the steam conditioner. Steam is injected at several points in an agitator, where the groats are evenly exposed to the steam, and gives uniform treatment to raise the moisture and temperature level to a point that the desired objectives of inactivating the enzymes and partial gelatinization of the starch are met.

The rate of application of steam into the agitator is controlled according to the required temperature of the conditioned groats. A discharger at the bottom of the unit is used to control the output of the steam conditioner unit.

3.3.7.2 Kiln Drying

The objectives of drying the moist hot oat kernels are as follows:

1. To drive out much of the moisture that was added during steam conditioning, which is about 4–5%.
2. Additionally, to maintain the temperature of the oat kernels at levels (near 95°C) similar to those during steam conditioning, for an extended time through the drying cycle.

The parameters, listed in the second objective, of temperature and drying time or retention time as it may be referred to, are variables. These can be regulated as required to suit the specific needs of a given production facility.

Primarily there are two types of dryers that are used for this purpose: those with steam-heated radiators, and those that use several steam-jacketed pans arranged one above the other. The drying temperature in the pan dryers is about the same, approximately 95°C ± 5°C, as in the radiator dryer. The use of pan dryers is almost extinct.

Drying through contact heat from steam-heated radiators is an efficient form of drying offering the following advantages:

Energy efficient through the application of heat insulators around the radiators
Uniform heat treatment by an even flow through the column of radiators that are staggered to ensure a thorough contact and exposure to the heat source
High sanitary and safety standards as a result of a good combination of high-quality engineering and use of appropriate material for construction, such as, radiators that are made of either steel or stainless steel and are pressure tested
Electronic control of material flowing through

3.3.7.3 Operation of a Radiator Kiln Dryer

The kiln dryer described in this section consists of several heating sections, followed by cooling sections placed one on top of the other with louvers (Fig. 34). The heating section begins at the top of this column dryer, whereas the cooling section is near the bottom.

The heating section is made up of several radiators that are heated by saturated steam at 7 bar and 165°C (can be regulated). Insulators are placed around the periphery of the radiators to prevent heat loss. There are a number of heating sections, and each has a thermometer for product temperature measurement.

The product from the steamer enters the dryer at the top where the heating sections are located. The oat kernels flow through the louvers of the various heating sections and lose moisture on coming in contact with the radiators. The moisture is removed by the air passing through the louvers. The oat kernels gradually descend through the cooling section down the column where it is cooled by air currents that are independent of the ones in the heating sections. The oat kernels ultimately leave the cooling section through an outlet where a flow balancer (described in wheat flour milling) may be incorporated to control the discharge rate. Control of the discharge rate here allows the regulation of the time of the oat kernels in the dryer for heat treatment. The heating temperature is also adjustable as was indicated before.

The combination of steaming and controlled drying relative to temperature and time are critical-processing factors that allow the process to be optimized for the following targets:

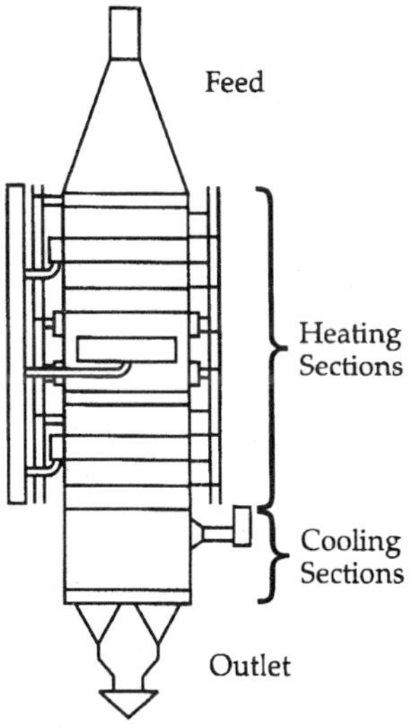

Fig. 34 Operation of a radiator kiln dryer.

1. Neutralize or inactivate the complex enzyme system that otherwise would have negative influence on product quality, such as lipolytic and oxidizing enzymes. Lipolytic enzymes are responsible for breaking the fat and causing the formation of free fatty acids which ultimately results in rancidity with associated bitter and soapy taste.
2. This process also stabilizes any free fatty acids already present.
3. Formation of flavors in the oat kernels which is subsequently passed on to oat products.
4. Functional properties and finished product quality are influenced.

3.3.8 Grading

Following the conditioning and drying, the groats may be sent to a polisher if the color is important; however, this step may not be essential. There is also a small loss of weight through polishing. The next step is grading during which small and large groats are separated. The small groats may be sent for steel cutting, while large groats are sent for flaking. Width graders are used for making separation between large groats and small groats. A double-deck–table separator may be used here to remove any remaining oats in the mass of groats. These oats are returned back to the hulling operation.

After the grading operation the groats are further processed to produce finished products for packaging. The market demand and economics will determine the product mixes that are required. A processing plant may be required to produce a range of finished products, such as

Thick rolled oats
Regular rolled oats
Quick rolled oats
Baby rolled oats
Instant rolled oats
Oat flour
Oat bran

3.3.9 Cutting of Groats

To produce the foregoing products one would require that some of the whole groats be processed into thick old-fashioned-type rolled oats, whereas some of the groats will be cut, also referred to as steel cut, for the production of quick rolled oats and other types. A drum-type groat cutter is used to cut the groats into four pieces. These materials, consisting of small and large pieces, are sorted in a small sifter as follows:

1. Large steel cuts
2. Regular steel cuts
3. Small or baby steel cuts

These steel cuts are passed through flaking rolls in the flaking system to produce a variety of flakes of different sizes and properties.

3.3.10 Flaking Operation

Flaking involves application of pressure or compression between two flaking rolls to flatten the large groats or steel-cut groats into flakes of a desired thickness and property.

The rolls are wear-resistant, and there is a provision for maintaining a desired optimum roll temperature through water cooling or preheating. The availability of roll pressure of 400–500 kN ensures formation of very thin flakes if so desired. Here, the roll gaps can be varied accurately to provide flakes of any desired thickness consistently. Thickness affects the cooking time.

Whole groats or any steel-cut–sized groats can be flaked. Before flaking, preparation of the material is essential. This preparation involves steaming in a steamer, similar to the one used before kiln drying. The addition of moisture and raising the temperature allows the groats to become more tough and pliable and thus improve flaking characteristics. This process also has beneficial influence on the finished product quality in terms of further gelatinization of the starch. This improves not only digestibility, but also has other positive influences on the functional and end use quality characteristics: namely, water absorption. The whole groats are flaked into a variety of old-fashioned rolled oats. Large steel cuts are flaked into quick rolled oats that require only 1 min to prepare. A thicker version of the same, when processed through roll gap adjustment, may require about 3 min.

After the flaking operation, the rolled oats are dried and cooled in a bed dryer. The drying target of the finished rolled oats is reduction to about 11.0% moisture. The products are packaged following this operation.

3.3.11 Production of Oat Flour and Oat Bran

The rollermills, as in wheat flour mills, may be used or even hammermills may be used to grind different-sized groats or finer products into one type of flour with all the bran and flour pulverized to a fine granulation. This flour may be referred to as whole oat flour. By being more selective in terms of grinding and then sifting to remove coarse material, oat flour is produced. This coarse material may be ground again for removal of some more flour from it. The flour obtained in this gradual manner would be a low ash oat flour, and the coarse material remaining now may be referred to as oat bran.

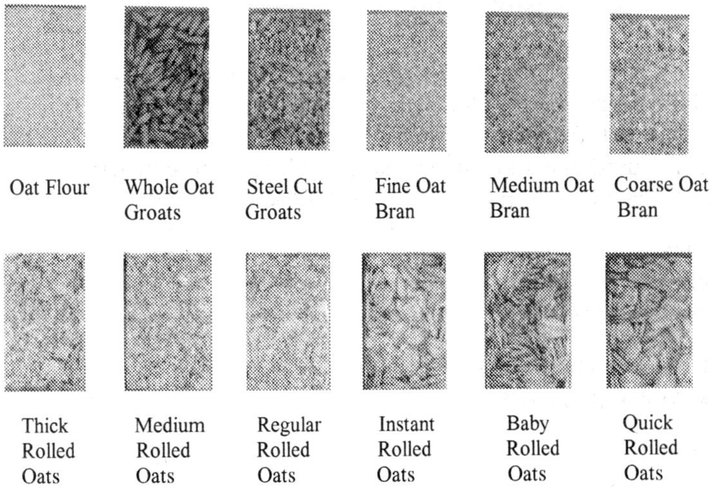

Fig. 35 Processed oat products. (Courtesy of Can-Oat Milling Products, Inc.)

3.4 Finished Oat Products' Quality Requirements and Specifications

There are many types of oat products that are produced in an oat-processing plant; that is, various types of rolled oats, oat flour, and oat bran (Fig. 35). These products are used as a major or minor ingredient in a wide range of end product application. These applications include manufacturing of cookies, crackers, cold and hot cereals, and the like. Apart from enriching the various end products with the nutritional benefits, these oat products are also required to have certain quality attributes that are necessitated by the processing requirements when these oat products are used as an ingredient. Some of these quality attributes are as follows:

1. Absorption rate and capacity
2. Flake durability/integrity
3. Neutralizing the enzyme activity
4. Flavor formation
5. Texture

The end use quality (functional properties) of wheat flour is strongly influenced by the quality of wheat from which it is milled and, to a much lesser extent, the milling process variables. For example, flours for cookies and cakes require soft wheat with low protein content and weak protein quality relative to the flour for white sandwich bread, which requires wheat of higher protein content and stronger protein quality.

Table 12 Critical Functions in Oat Processing That Influence Quality Attributes of Processed Oat Products

Type of oat product	Property	Critical oat-processing function	Application
Rolled oats, oat flour, and bran	Absorption rate and capacity	Flaking (flake thickness), grinding (granulation and bran content), conditioning (degree of gelatinization)	Cookies, extruded cereal products, and crackers
All products	Inactivating enzymes and stabilizing free fatty acids	Conditioning	All applications for shelf life
Various types of rolled oats	Flake durability	Steaming and flaking (tough and pliable flakes)	For texture or visual attractiveness in cookies
All products	Flavor characterization	Conditioning (toasting)	All applications, including cookies, crackers, and extruded products, for desirable characteristic oat flavor.
Steel cut groats, crushed oats, and oat flour	Texture	Groat cutting (size), grinding (granulation)	Crackers for more texture

Table 13 Quality Specifications of Oat Products

Finished products	Enzyme	Protein	Moisture	Free fatty acids	Granulation		Hulls
					Overs	Through	
Regular rolled oats	Negative	16.0 ± 2.0%	11.0 ± 1.0%	0.5 max	US no. 4, 4.75 mm, 10–30%	US no. 10, 2.0 mm, 5% max	4 max/50 g
Instant rolled oats	Negative	16.0 ± 2.0%	11.0 ± 1.0%	0.5 max	US no. 4, 4.75 mm, 5% max	US no. 14, 1.4 mm, 10% max	3 max/50 g
Quick rolled oats	Negative	16.0 ± 2.0%	11.0 ± 1.0%	0.5 max	US no. 8, 2.36 mm, 60% min	US no. 40, 425 μm, 5% max	3 max/50 g
Whole oat groats	Negative	16.0 ± 2.0%	11.0 ± 1.0%	0.5 max		US no. 20, 850 μm, 5% max	4 max/50 g
Oat flour	Negative	16.0 ± 2.0%	10.0 ± 1.0%	0.5 max	US no. 20, 850 μm, 1% max	US no. 35, 500 μm, 80% min	
Oat bran	Negative	17.0% min	11.0 ± 1.0%	0.5 max	US no. 14, 1.4 mm, 4–7%	US no. 40, 425 μm, 5% max	6 max/50 g
Total dietary fiber	min 18% in oat bran						

Source: Can-Oat Milling Products, Inc.

Finished oat products quality, on the other hand, is highly dependent on the oat-processing conditions. All types of finished oat products are processed from the same oats. For example, whole groats, steel cut groats, thick rolled oats, quick rolled oats, oat bran, and oat flour, all are processed somewhat differently to meet specific end use quality requirements but are milled from the same oats.

Although the required quality attributes of a specific oat product vary with the product, there are some quality requirements, such as flavor formation, absorption, and neutralization of enzymic activity, that are common to all products. Table 12 summarizes various quality attributes that are provided by various oat products and critical functions in the oat processing that influence these attributes. Table 13 provides a general guideline of the quality specifications of some oat products.

ACKNOWLEDGMENTS

Most of the cleaning, tempering, and milling equipment, and the process diagrams and photographs included in this chapter have been provided by the courtesy of Bühler, Inc., Uzwil, Switzerland. The corn wet-milling process schematic diagrams as well as the equipment photos and diagrams have been provided by the courtesy of Alfa Laval Separations, Inc., Oak Brook, IL. The following companies have also provided equipment or process diagrams and photos.

Codema Inc., Minneapolis, MN
The Beall Degerminator Co., Decatur, IL
Can-Oat Milling Products, Inc., Portage La Prairie, MB, Canada
Carter Day Industries (Canada), Ltd.

Kind permission from these companies and valuable information from many others in this respect is greatly appreciated.

REFERENCES

1. Bühler, Inc. Equipment brochures on corn and oat processing. Bühler, Inc., Uzwil, Switzerland.
2. Barley Medley. Spring 1997. Alberta Barley Commission, Calgary, AB, Canada.
3. Decomatic 2 brochure. Codema Inc., Minneapolis, MN.
4. Beall degerminator brochure. Decatur, IL.
5. Bulletin FOOD—3. 1996, Dorr–Oliver, Inc. (now Alfa Laval Separations, Inc.), Oak Brook, IL.
6. Sampling of Products. Can–Oat Milling, Portage La Prairie, MB, Canada.
7. Institute Images. May 1999. Canadian International Grains Institute, Winnipeg, MB, Canada.

13

Rice Milling and Processing

ROBERT S. SATAKE[†]

Satake Corp., Hiroshima, Japan

1 INTRODUCTION

The development of rice-milling machinery originally began approximately 140 years ago in the United Kingdom. Approximately 35 years later, the first rice-milling machines appeared in Japan. However, only within the past 40 years have machines resembling the modern rice-milling units of today come into being.

In rice-milling technology, the primary objective is to produce a maximum number of unbroken grains that have had their bran layers uniformly removed, resulting in an appearance that is of a desired color and luster.

The secondary technical objectives of rice milling are as follows:

Maximize practical throughput capacity
Minimize consumption of power
Minimize consumption of consumable parts
Maximize productivity of human resources

To achieve these objectives, there has been a steady and consistent effort to develop methods and machines that can optimize the entire process of rice milling.

2 MAIN MACHINES FOR RICE MILLING AND PROCESSING

2.1 Mechanical Cleaning

The objective of the mechanical-cleaning system, one of the steps in the milling process, is the efficient removal of any foreign material or objectionable seeds that may be present

[†] Deceased.

in the incoming paddy. Foreign materials include straw, dust, stones, metal, glass, and animal life. Objectionable seeds include any agricultural seed other than rice, such as red rice, morning glory, indigo, wheat, and soybeans.

Mechanical cleaning is a logical process containing several steps, and each step is dependent on the efficiency of preceding steps. This process is accomplished by using differences between the physical properties of rice and anything other than rice to remove objectionable materials.

2.1.1 Paddy Cleaners

Paddy cleaners provide screening and aspiration operations as shown in Fig. 1. A slotted perforated screen (upper screen) removes objects that are appreciably wider than rice (i.e., seed in pods, corn, beans, or other), whereas a round perforated screen (lower screen) removes objects that are significantly shorter than rice. The aspiration section removes objects that are significantly lighter than rice (i.e., dust, husks, and such).

Keys to the performance of paddy cleaners are (a) uniformity of feed, (b) perforation size and shape, and (c) air velocity. These determine the effectiveness of screening and aspiration and affect the throughput capacity of the machine.

2.1.2 Destoners

After paddy cleaners remove foreign materials and seeds that are (a) wider than rice, (b) shorter than rice, and (c) less dense than rice, a process is required to remove objects, such as stones, mud, and even dense glass, which are more dense than rice.

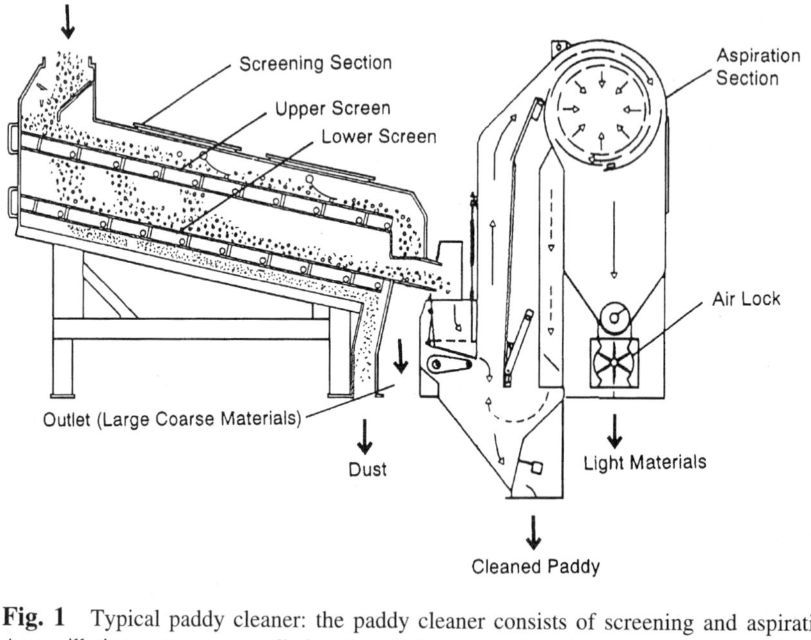

Fig. 1 Typical paddy cleaner: the paddy cleaner consists of screening and aspiration sections. An oscillating movement applied to the sloping screen by vibratory motors causes paddy to move downward across the screen surface. The aspiration section separates light materials from the paddy, allowing the cleaned paddy to drop through to a bottom outlet. (Courtesy of Satake Corp.)

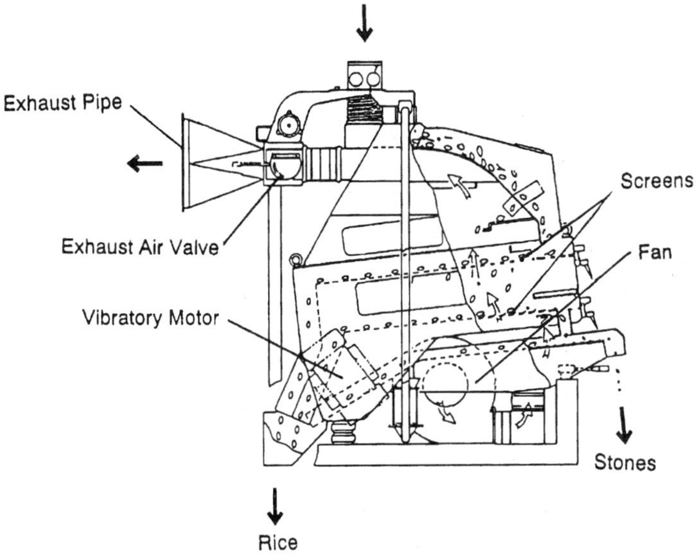

Fig. 2 Typical destoner: a screen assembly that is simply supported is caused to oscillate in a linear direction by a pair of contrarotating vibratory motors. (Courtesy of Satake Corp.)

Depending on the rice's variety and harvesting techniques, these foreign objects may be present in paddy. A simple gravity separator, called a destoner, fluidizes a bed of paddy and then conveys the heavier stones to the high side of a sloping screen. The lighter paddy is not in contact with the conveying surface and thus moves downward under the force of gravity. In this way stones are removed from the paddy stream. A typical destoner is shown in Fig. 2.

2.2 Mechanical Classification (Paddy Processing)

The objective of any mechanical classification system is the efficient separation of a combined stream of rice into two or more groups containing grains with similar physical characteristics.

The operations that involve mechanical classification include:

1. Separation of immature grains from paddy, either before the parboiling process, or to maximize the efficiency of the husking and bran removal processes that will follow.
2. Removal of objectionable seeds, such as red rice, that are thicker than a rice kernel.
3. Separation of broken grains from whole grains.

The removal of objectionable seeds and the classification of broken grains usually occur later in the rice-milling process. These processes will be discussed later in this chapter.

To remove immature grains, which are thinner than normal grains, a thickness grader equipped with slotted cylinders is used. Mature grains remain inside the cylinder, while

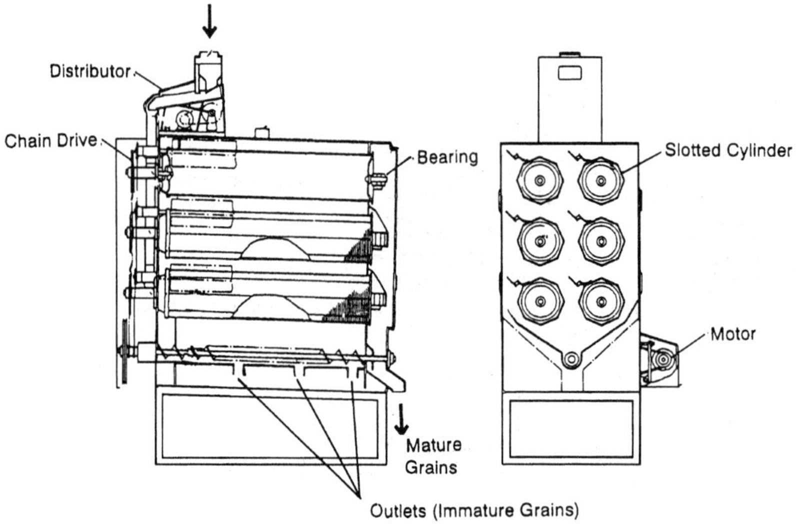

Fig. 3 Thickness grader equipped with six cylinders. (Courtesy of Satake Corp.)

immature grains pass through the slots. The cross-section of a thickness grader equipped with six slotted cylinders is shown in Fig. 3.

2.3 Husking and Husk Aspiration

The efficiencies of the husking, paddy separation, and husking of return paddy are so interdependent that it is difficult to consider them separately. Any analysis of problems in brown rice preparation must consider the system as a whole. Still, it is necessary to understand each individual operation.

2.3.1 Paddy Huskers and Husk Aspirators

Paddy is fed into the gap between two rubber rollers that are rotating at different speeds and in opposite directions. Compression provides the necessary normal force to cause friction between the paddy and the surface of the rubber rolls. Because of this friction, each side of the paddy grain attempts to travel at the same linear speed as its corresponding roll. Because this is impossible, the husk is sheared or ruptured and falls away as the grain is released from the rolls. A typical mechanism of the paddy husker is shown in Fig. 4.

After passing through the shelling rollers, the resulting mixture of ''husked'' rice or brown rice and husks that have been removed is introduced into the aspirator section. A cross section showing a paddy husker and aspirator is shown in Fig. 5.

In the husk aspirator, the brown rice and husks are distributed into a thin layer and drop into a column of air moving with an adjustable velocity. The lighter husks are then lifted from the rice. After the husks have been removed, the velocity of the air is decreased, allowing the husks to settle and be collected and removed from the machine.

The factors that influence the performance of the paddy husker and husk aspirator are as follows:

1. Moisture content of rice. In general, moist paddy does not shell easily.
2. Uniform feeding across the entire length of husking roll. In general, uniform

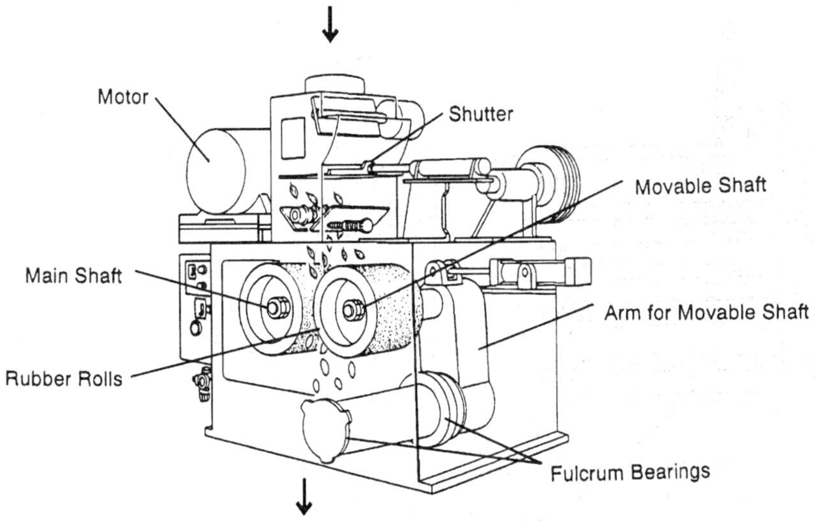

Fig. 4 Typical mechanism of paddy husker. (Courtesy of Satake Corp.)

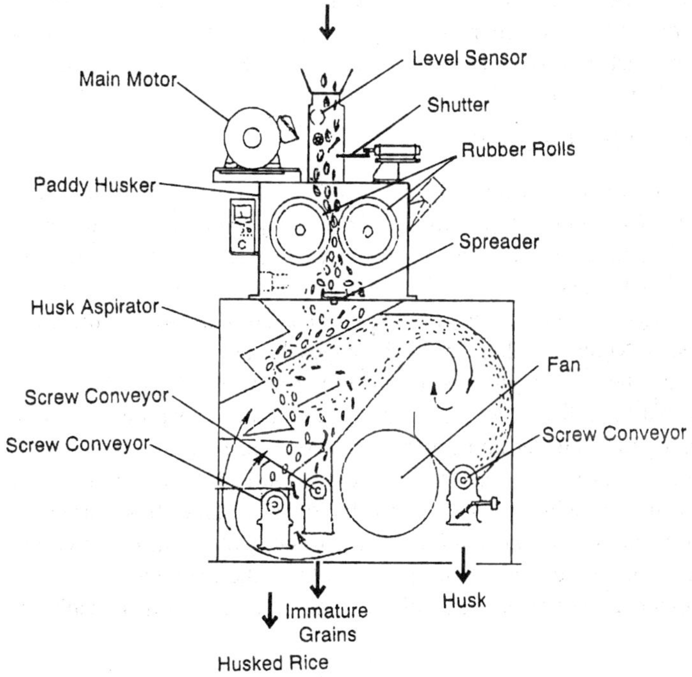

Fig. 5 Cross section of paddy husker and husk aspirator. (Courtesy of Satake Corp.)

feed across the entire length of the husking roll results in longer roll life and less breakage.

3. Speed ratio of husking rolls. It is the difference in peripheral speed of the two rolls that is the driving force behind the husking action. If this differential speed is significantly reduced, husking efficiency will suffer.
4. Husking pressure. Without compression there is no friction to cause each side of the grain to tend to travel at the same speed as its roll. This compressive force is essential for husking, but it is also the force that breaks rice and accelerates wear of the rubber rolls.
5. Uniformly distributed feed across face of aspirator. Uniform distribution results in uniform air distribution. Thus, there is better aspiration at high capacities, less contamination of husks in the rice stream, and less loss of rice grains with the husk stream.

2.3.2 Paddy Separators

Paddy separators are fundamentally ''density'' separators. Separation is achieved by fluidization, which causes denser brown rice grains to sink to the bottom of a bed of rice whereas lighter paddy grains rise to the top. The principle of paddy separation is shown in Fig. 6.

A typical-tray type paddy separator is shown in Fig. 7.

The factors that determine the efficiency of paddy separators are (a) feed rate, (b) angle of inclination of separating trays, (c) angle of inclination from feed to discharge ends of the machine, (d) speed of oscillation, and (e) position of the adjustable dividers.

2.3.3 Brown Rice Production System

A typical brown rice production system consisting of a paddy husker, husk aspirator, and paddy separator is shown in Fig. 8.

The efficiency of a brown rice production system is determined by the following factors:

1. Husking ratio
2. Percentage of brown rice in the mixture

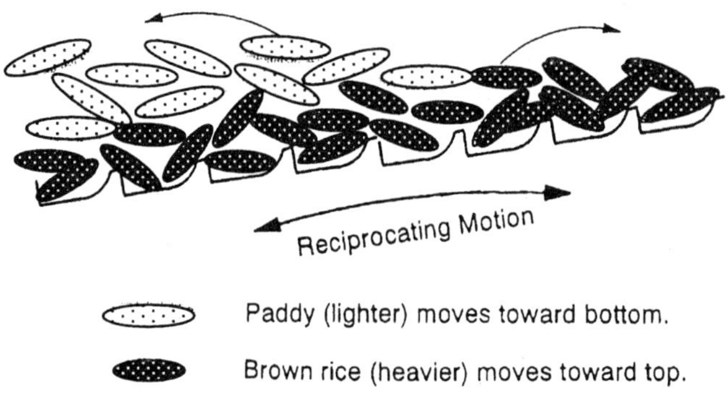
Reciprocating Motion
Paddy (lighter) moves toward bottom.
Brown rice (heavier) moves toward top.

Fig. 6 Principle of paddy separation: brown rice grains sink to the bottom of a bed of rice and move up the separating tray. Lighter paddy grains rise and slide down toward the lower end of the tray, leaving a mixture of brown rice and paddy in the center section.

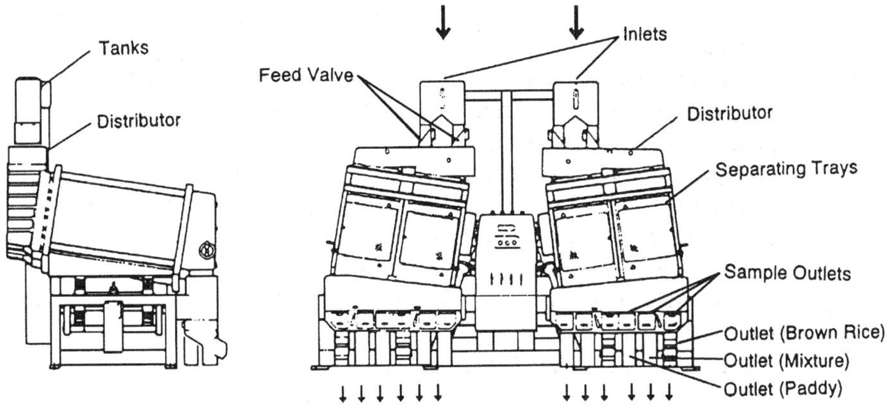

Fig. 7 Tray-type paddy separator. (Courtesy of Satake Corp.)

3. Rubber roll consumption
4. Power consumed during husking

Measurements of these factors should be made by sampling and by analysis of the flows before and after paddy husking and paddy separating.

2.4 Rice Milling (Bran Removal)

Bran removal is the process by which the germ and bran layers are selectively removed from the rice kernel, leaving the endosperm largely intact and with a desired surface appearance. The term *milling* is applied to this process, although it may be more appropriate to call this process *whitening* to distinguish it from the whole of processing steps (i.e., cleaning, husking, whitening, grading, and such), which are also commonly termed *rice milling*.

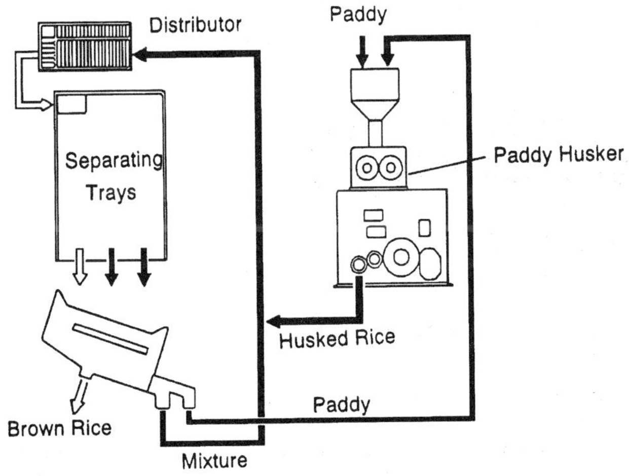

Fig. 8 Typical brown rice production system.

Rice milling can be conceptualized as the partial rupture of the bran layer and its separation from the internal endosperm of the grain. The rice grain structure consists of an inner hard, starchy endosperm, with outer layers of soft bran covering this hard core. The objective is to completely remove the bran layers, in the most cost-effective manner, without damage to the endosperm. Several mechanisms can be employed in the mechanical action. These include tearing by friction, cutting, grinding, and impact. Although all of these may be present in varying degrees in an actual milling system, modern rice-milling systems can be classified into two categories of milling equipment. Friction-type machines use primarily friction and cutting actions. These machines use relatively high pressure with low peripheral speed. Grinding-type machines employ mainly grinding and impact actions. They use low pressure with high peripheral speed.

2.4.1 Milling at Low Pressure with High Peripheral Speed

In some milling systems the bran layers of the rice grains are removed by a grinding action and an impact force, as grains come in contact with an abrasive roll. An abrasive roll rotates on a central shaft in a perforated metal cylinder. Such systems are termed *abrasive-type* rice milling. A typical high-speed abrasive–type-milling machine is shown in Fig. 9. Abrasive-type milling requires a peripheral speed of over 600 m/min, and an average milling pressure of less than 50 gf/cm^2 (5000 N/m^2).

2.4.2 Milling at High Pressure with Low Peripheral Speed

In other milling systems, frictional force and cutting force remove the bran layers. The grains are rubbed vigorously against each other and against the metallic surfaces of the milling chamber. The rice grains are enclosed in polygonal perforated metal enclosure in which a rotating eccentric-milling roll agitates the grains and provides the pressure to bring about the frictional forces.

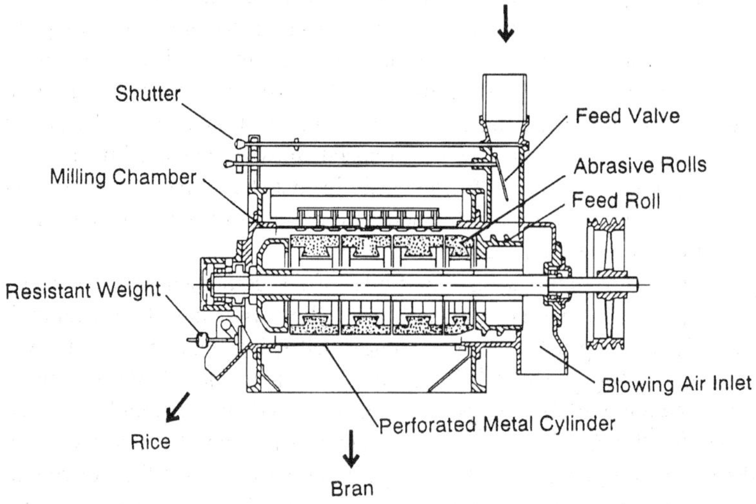

Fig. 9 High-speed, abrasive-type rice-milling machine: grinding against an abrasive medium provides the milling action.

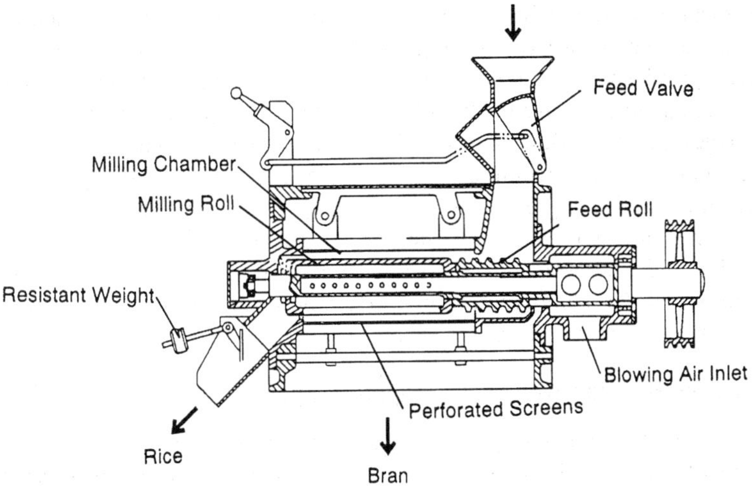

Fig. 10 Typical low-speed, friction-type rice-milling machine: here milling takes place as a result of frictional forces applied to the rice grains.

Because friction is the major force of milling, this is termed *friction-type* rice milling. A typical friction–type-milling machine is shown in Fig. 10. To be effective, friction–type-milling requires a peripheral speed of the milling roll of less than 600 m/min. and an average milling pressure of over 100 gf/cm^2 (10,000 N/m^2).

2.4.3 Combined Abrasive-Type and Friction-Type Milling System

In the past, different rice-milling machines have appeared, varying in form and function, but all falling into either the high-speed abrasive-type or low-speed friction-type categories. Although all of these machines could succeed to a greater or lesser extent, it could be seen that each type had its own strengths and weaknesses. Especially, the low-speed friction–type-milling machines had the advantages of providing a more polished, smoother grain surface than could be provided by the high-speed abrasive-type. On the other hand, the higher pressures needed in the friction-type create significant stresses on the grains and lead to the formation of broken grains. The high-speed abrasive-type machine is good for preserving milling yields, but there are drawbacks in that the surface of the milled rice will be rough and unattractive. It seems that neither the friction-type nor the abrasive-type, by itself, can achieve the desired results.

It was only with the development, in the 1960s, of the combined abrasive–type-and friction–type-milling system that the goals of high-milling efficiency along with high head rice yield could be achieved. This was accomplished using high-speed, abrasive-type machines, followed by low-speed, friction-type milling machines. Tests performed in Japan confirmed that the initial abrasive action of a combined system abrades the rice grain surface and thereby effectively raises the coefficient of friction of the rice surface. With an increased coefficient of friction, less-milling pressure is needed in the subsequent friction action. The lower-milling pressure results in a reduction in the generation of broken grains. There is also a benefit in finishing the milling process with low-speed friction action, so that the rice grain surface is smooth and polished.

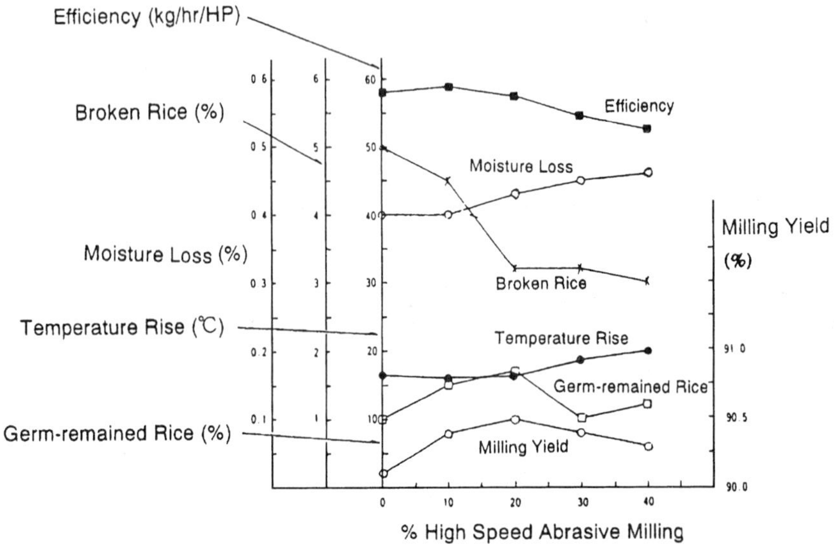

Fig. 11 The effect of combined abrasive-type and friction-type milling systems: for this test on Japonica rice, milling yield was maximized when high-speed, abrasive-type milling accounts for 20% of total milling. (Courtesy of Satake Corp.)

Figure 11 shows the characteristics of a combined abrasive–type- and friction–type-milling system. It may be noted that at 0%, milling by abrasion (bottom scale), which indicates 100% milling by friction, the highest percentage of broken grains is generated. In this particular data set, which was derived from tests on Japonica rice in Japan, the optimum combination of abrasive–type- and friction–type-milling was accomplished when the abrasive portion is the highest. In general, the high-speed abrasive portion of milling should be limited to removal of bran only and be sufficient to substantially increase the coefficient of friction on the surface of the rice grains for the subsequent milling by friction.

2.4.4 New Vertical-Milling System

Many modern-milling systems combine high-speed abrasive-type and low-speed friction-type milling machines, which are usually employed in a ''horizontal'' manner. The ''horizontal'' here designates the juxtaposition of the main shaft of the machine as well as the direction of flow of rice. Both are in a horizontal orientation. Figures 9 and 10 are examples of so-called horizontally aligned milling machines. There are several technical drawbacks that prevent the horizontal milling machines from reaching optimum performance. One drawback has been that density and pressure within the milling chamber tend to vary, especially from the bottom to the top of the milling chamber. The main cause of this is gravity. In addition, in most units, the suction air used to carry the loosened bran away from the milling chamber pulls the rice to the bottom of the milling chamber. Thus, it causes higher pressure and density at the bottom portion of the milling chamber. This difference of pressure and density creates conditions under which efficiency is less than optimum because the upper portion of the milling chamber is underutilized. Thus, uneven pressure creates conditions for increases in broken grains.

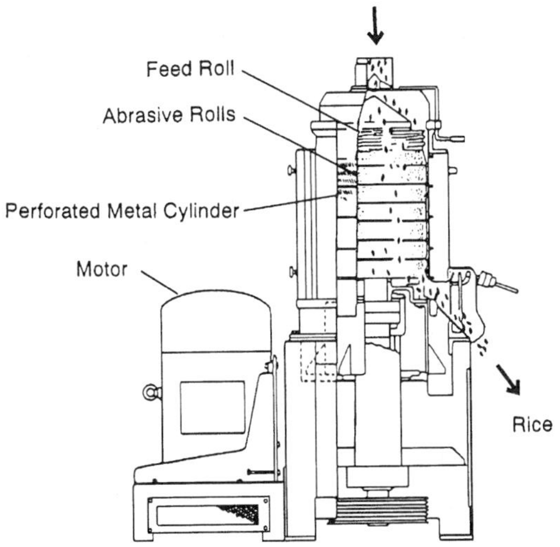

Fig. 12 Cross section of vertical (top-down), abrasive-type rice-milling machine. Rice enters from the top of the machine and falls into a milling chamber, which consists of abrasive rolls rotating within perforated metal screens. Milled rice is discharged from the bottom outlet. (Courtesy of Satake Corp.)

To overcome the aforementioned drawbacks, vertical-milling machines, such as shown in Figs. 12 and 13, have been developed. These are of two types—namely, the high-speed abrasive-type and the low-speed friction-type—and these are utilized in a combined abrasive-type and friction-type system. The main features of these machines are that the shaft and milling chamber are vertically oriented and the flow of rice is in a vertical

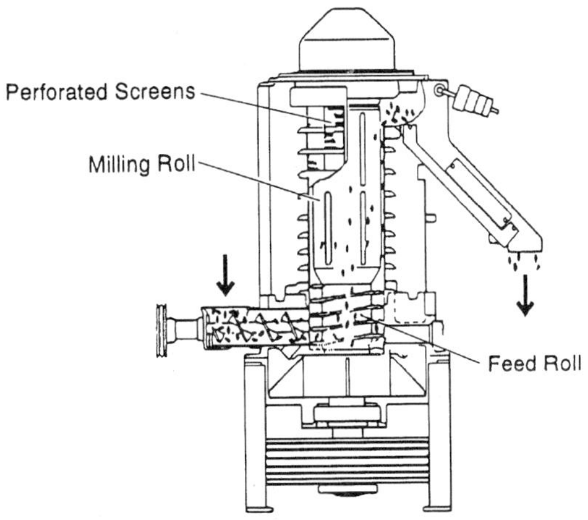

Fig. 13 Cross section of a vertical (bottom-up), friction-type rice-milling machine. Rice entering the inlet is conveyed horizontally by a feeding screw, where it comes into contact with a feed roll and is propelled upward into a milling chamber consisting of a steel milling roll rotating within perforated metal screens. Milled rice is forced from the top of the machine and is discharged through the outlet. (Courtesy of Satake Corp.)

direction. The interaction of pressure and weight through all layers of rice within the chamber achieves maximum utilization of the milling surfaces, thus enhancing efficiency as well as preventing uneven pressures and the excess pressure that causes broken grains. In this way, higher-milling efficiency and higher-milling yields with fewer broken grains can be achieved.

By combining the vertical high-speed abrasive-type with the vertical low-speed friction-type machines, an improved system can be evolved for almost any type of rice. For the Japonica short- and medium-grain rice, high germ removal and high polishing can be achieved without sacrificing whole grain yield. For Indica long grains, this vertical system can be useful for reducing the amount of broken grains, particularly for brittle rice varieties.

Another advantage of the combined vertical system is that a water mist polishing system can be added to the final-pass through the vertical friction–type-milling machine. This enhances the final bran removal process and helps produce milled rice without residual bran. This water mist also cools the rice. A more detailed discussion of the benefits of water mist polishing is undertaken in the following section.

Two additional benefits related to installation are as follows: Floor space required for installation is minimized, compared with horizontal-milling machines. Mills can be upgraded to higher efficiencies and higher capacities within limited floor space. The second benefit of the vertical-milling machines involves a reduction of conveying equipment. As each of the subsequent units uses bottom-up flow, rice can flow automatically from one machine to the next without any conveyor or elevating equipment. Figure 14 shows how the units can be connected in a multibreak installation.

2.5 Water Mist Polishing

As mentioned earlier, one of the goals in rice processing is enhancing the quality of the finished rice.

The "water mist" polishing machine was developed as a processing step to improve the finish or appearance of milled rice. As the name suggests, this machine uses a fine mist or spray of water to remove traces of dust, bran, and rice polish from the surfaces of milled rice grains. Moreover, bran and polish hidden deep inside the grooves of rice grains can be stripped away. The water mist and the subsequent rubbing action of the machine impart a high polish or luster to the milled rice, which substantially elevates its appearance. From a dull appearance, the rice grains can be transformed into shiny, almost translucent kernels, which have good visual appeal. This shiny appearance is even visible in photographs (Fig. 15).

2.5.1 Construction of Water Mist Polishing Machine

The water mist polishing machine is similar to a horizontal low-speed friction-type rice-milling machine. The main differences are that the milling action is much less severe, as suited to its polishing function, rather than a whitening one, and that the length of the actual milling chamber is almost twice than that of traditional friction-type machines. In this way, the relatively lower milling pressure of the water mist polishing machine is compensated by a longer product retention time within the milling chamber. This arrangement is ideal for producing a polishing effect at low pressure, without generation of broken grains.

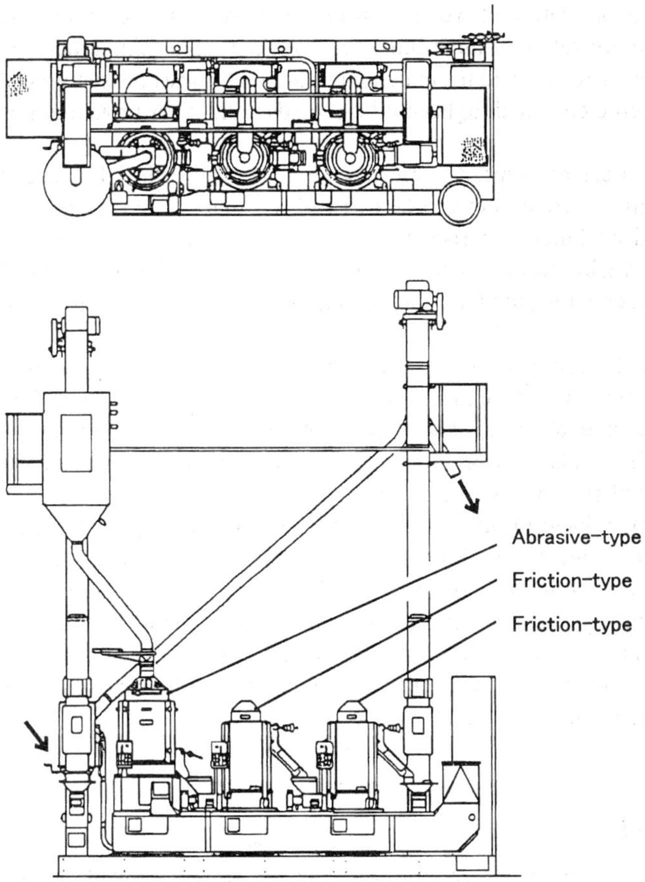

Fig. 14 A typical three-break vertical rice-milling system consists of one unit of abrasive-type and two units of friction-type machines: rice entering the abrasive-type unit is milled and discharged from the bottom of the unit where it is conveyed into the inlet of the adjoining machine. This rice continues to flow in the same manner to the subsequent friction-type unit and is discharged from the third-break friction-type unit. (Courtesy of Satake Corp.)

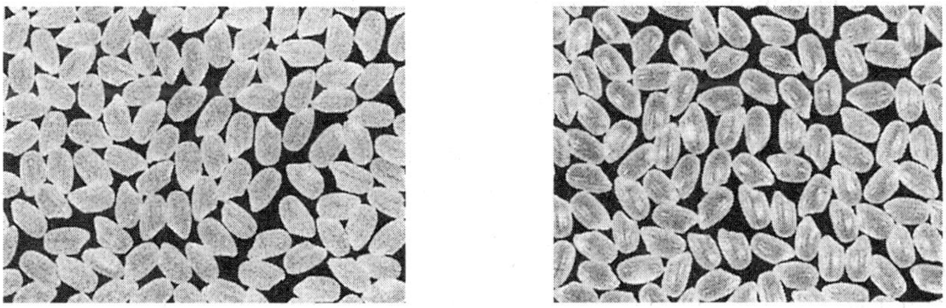

Fig. 15 Water polishing cleans the last traces of dust, bran, and rice polish from the surfaces of rice grains and, at the same time, imparts a high polish to the milled rice by rubbing the grains together vigorously.

Water and compressed air tubes are carried to the back of the milling chamber through the hollow main shaft. The tubes are connected to an atomizing device just inside the first portion of the milling chamber. Thus, a fine mist is applied to the rice as it begins the milling process. A bran suction line attached to the bottom of the milling chamber pulls air through the chamber and removes any loosened material.

The degree of cleaning and polishing depends mainly on the product flow rate and the amount of water introduced. For brilliant, highly polished rice, multiple passes through the machine may be necessary. Typical water addition is about 0.3–0.4% by weight. Insufficient water addition will decrease the polishing effect. Increasing the amount of water above the optimum level will not lead to further polishing, as the rice surface becomes overmoistened, which interferes with polishing.

2.5.2 Benefits of Water Mist Polishing

The addition of moisture softens the surface of partially milled rice grains so that it is easier to remove bran in the production of milled rice. As water addition facilitates bran removal, lower milling chamber pressures can be employed in the milling process, and this leads to the formation of fewer broken grains and higher total yields. Furthermore, the addition of water cools the rice during milling and also helps reduce grain moisture loss.

2.5.3 Enhancement of Storage Stability

The treatment of rice with water mist polishing enhances the storage life and stability of the milled rice. The oil in the residual bran on the surface of milled rice grains can quickly be oxidized to form free fatty acids, which impart a rancid smell and taste to rice. Removal of the residual bran with the water mist polishing machine may therefore extend the safe storage period for milled rice.

Data, shown in Fig. 16, suggests that the water-polished rice (''clean rice'') is also more resistant to mold growth. Water-polished rice resisted mold growth for 7–10 days longer than regular rice. This phenomenon may be due to the presence of residual bran on the surface and in the grooves of ordinary rice, which presents an ideal environment for the promotion of mold growth.

2.6 Rice Moisture Conditioning

2.6.1 Problems with Low-Moisture Rice

In commercial rice milling, moisture content can have a significant influence on the processing characteristics of the rice and the economical return to the miller. Rice containing low levels of moisture tends to be more rigid and inelastic than rice of higher moisture. Cooking and other qualities may also be affected.

The low-moisture problem usually originates in the field during hot, dry weather in late summer, or at the rice dryer owing to overdrying of some rice lots. A method is needed by which moisture can be safely added to rice grains so that the optimum moisture content can be restored. Because water addition increases weight, it also increases total yield of milled rice.

2.6.2 Rice Moisture Conditioner

For some time, a rate of moisture addition of 0.2%/h represented the maximum practical rate. Later it was discovered that improvements could be made if a water mist was applied

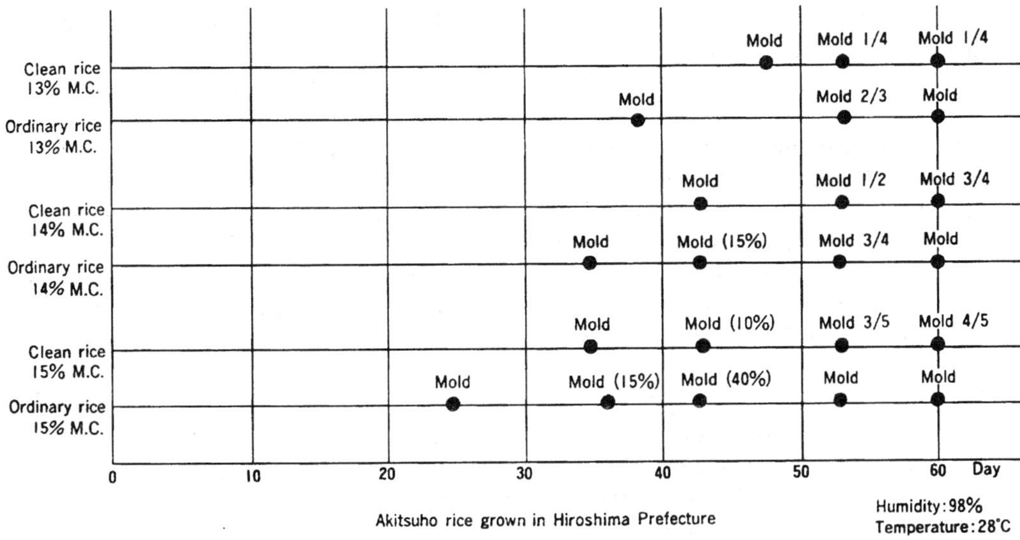

Fig. 16 Comparison of mold growth for ordinary and water-polished rice: tests indicate that, at a given moisture content, water-polished rice ("clean rice") resists mold growth for 7–10 days longer than ordinary rice. (Courtesy of Satake Corp.)

in a pressurized vessel. In previous rice conditioners, moisture had always been added under normal atmospheric pressure. The application of pressure to 0.025–0.030 kgf/cm^2 above atmospheric pressure in a chamber increases the safe penetration of moisture into the grain, thereby accelerating the moisture absorption rate from 0.2%/h up to 1.0% without damaging the rice grains. Figure 17 shows the device constructed to accomplish this. Basically this is a continuous-flow pressure vessel with air lock cutoff valves at the inlets and outlets. Pressurized moist air is injected into the vessel in a regulated fashion, and the pressure inside the chamber helps promote the penetration of the moisture into the

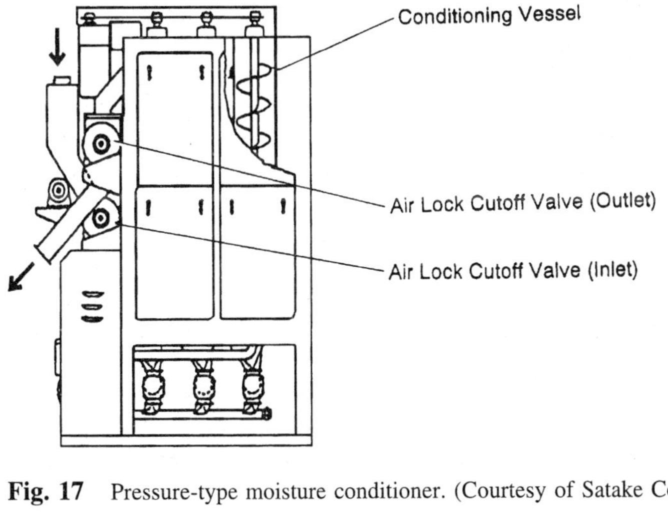

Fig. 17 Pressure-type moisture conditioner. (Courtesy of Satake Corp.)

kernels. Retention time inside the chamber is about 8 min. A moisture sensor on the interior of the chamber regulates the amount of moisture injected into the chamber.

2.6.3 Benefits of Rice Conditioning

Figure 18 shows that moisturizing rice can have a substantial positive effect on the cooking and eating quality of milled rice. Samples of rice at 13.0, 14.0, and 15.0% moisture content were soaked in water at 20°C for 40 min. Under these conditions, moisturized rice of 15% moisture content developed the least number of cracks. During tests, a distinctive "snap-crackle-pop" noise can be heard coming from low-moisture rice, indicating the severity of the stresses being built up within the rice grains as moisture begins to enter the kernel.

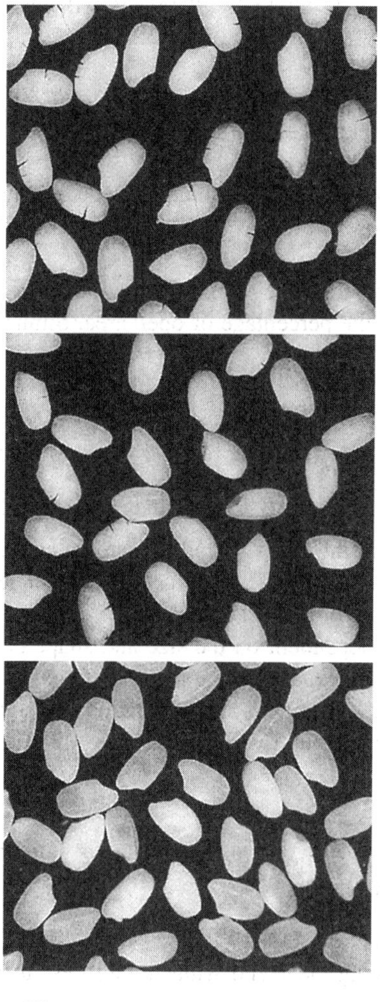

(a)

(b)

(c)

Fig. 18 Rice cracking during soaking (in water at 20°C for 40 min) moisturized to (a) 13.0, (b) 14.0, and (c) 15.0% moisture contents. Rice moisturized to 15.0% moisture content shows the least number of cracks compared with rice of lower moisture.

This cracking phenomenon during soaking is of great importance, particularly in those countries where cultural practices entail prolonged washing and soaking of the rice before cooking. Still, the same phenomenon also takes place during the initial phase of cooking rice, so this is a universal problem.

The potential benefit of moisture addition to rice is one of enhanced quality. Not only are the weight and volume of the rice increased, but the quality is also improved in terms of yielding better-tasting, better-cooking rice. It is expected that future milling technology will be expanded to include moisture conditioning as a standard processing step.

2.7 Separation of Broken Grains from Whole Grains

In most rice mills broken grains are removed from whole grains through a process of length-grading actions. Later these broken grains can either be marketed separately or blended with whole grains to produce specific products.

2.7.1 Sifters

The separation of milled rice into whole and broken grains typically requires a high-capacity, moderately efficient length grader.

The sifter typically separates the milled rice into three streams: whole grains, a mixture of whole and broken grains, and small broken grains with some larger broken grains.

The dimensions of the opening in the screen determine the efficiency of sifting. The capacity of the machine is largely determined by the percentage of open area. Similar to paddy cleaners, sifters require uniform feeding.

2.7.2 Indented Cylinder Graders

After a rough separation is obtained with a sifter, a series of more precise-length graders is used. This is done using indented cylinder graders (Fig. 19).

Broken grains are lifted by the indents and by the cylinder's rotary motion. At some point, depending on their size and weight, they will fall from the indents and be collected in an adjustable trough, which discharges at the front of the machine. The principle of this machine is shown in Fig. 20.

The size of the objects to be lifted will be dependent on the size and shape of the indent. The size of the lifted grains collected will depend on the speed of rotation of the cylinder and the angle of the collecting trough. The higher the speed of rotation, the greater the quantity and size of the objects collected in the trough will be.

2.8 Color Sorting

The objectives of the color sorting are to improve the quality of rice and the yield. The improvement of the quality of rice includes the following:

1. Removal of paddy and other foreign materials
2. Removal of inorganic materials including clear and white impurities such as glass fragments, plastics, and stones
3. Removal of stained rice
4. Removal of ''peck'' (dark spot) damage

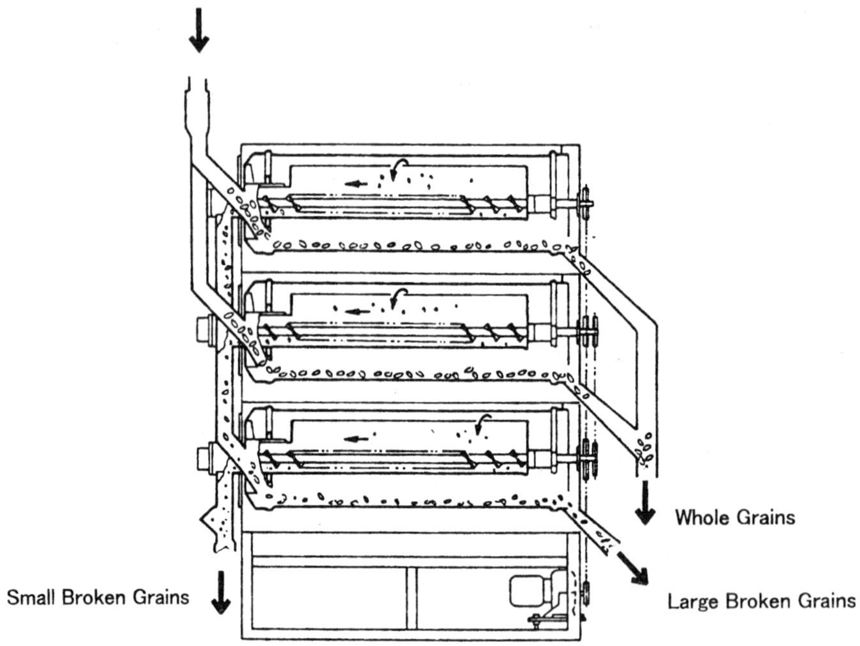

Fig. 19 Indented cylinder grader equipped with three cylinders. (Courtesy of Satake Corp.)

Sorters can easily and efficiently eliminate paddy from milled rice. Thus, they enable the modern mill to improve yields by avoiding the use of excessive pressure in their bran removal systems.

2.8.1 Color Sorters

A color sorter feeds a constant amount of product through a hopper and a vibratory feeder to an orienting device. This orienting device, such as a chute or slide, orients the product for proper presentation to a viewing section. As the product passes through the viewing section, its light intensity is measured and compared with a stored standard. Both reflective and transmissive methods can be used to determine the acceptability of the product. A

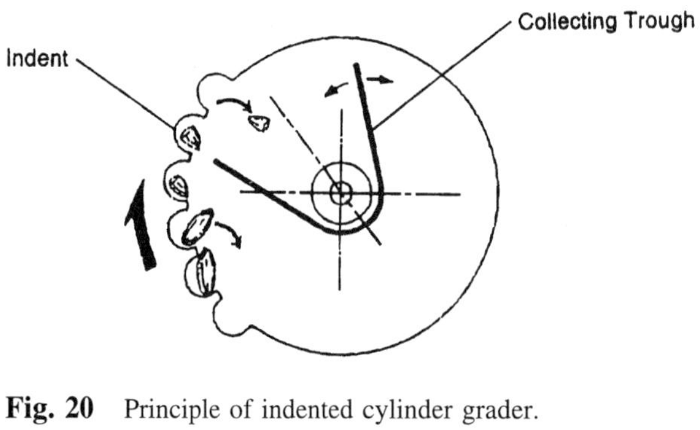

Fig. 20 Principle of indented cylinder grader.

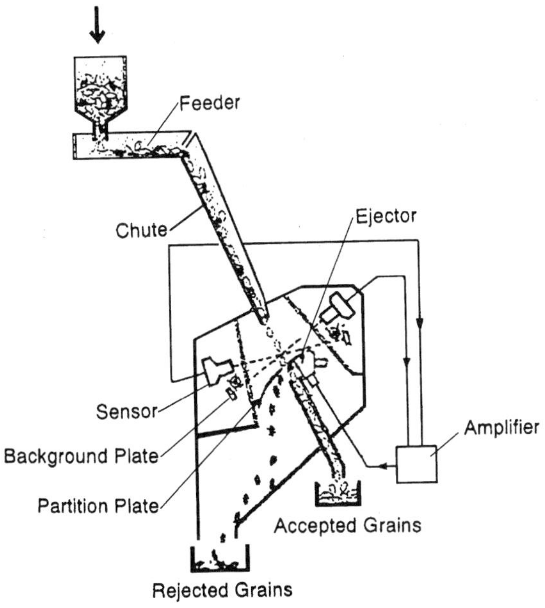

Fig. 21 Typical schematic of color sorter. (Courtesy of Satake Corp.)

momentary blast of air from a bank of ejectors directs unacceptable materials to the rejected grain hopper. A typical schematic of the color sorter is shown in Fig. 21.

2.9 Rice Washing

In Japan, much time and attention are devoted to the preparation of rice before cooking. The principal step in this preparation is washing the rice. In the United States, it is recommended that rice should not be washed in order to retain the required nutrients. But, in rice-centered cultures such as Japan, washing is seen as indispensable for the proper preparation of rice.

2.9.1 Development of Rice-Washing Machines

For institutional preparation (restaurants, hospitals, schools, and such), rice washing can be time-consuming, costly, and cumbersome, owing to the large volumes of rice involved. A large-scale rice washing operation poses some special problems. First, water addition must be handled carefully. Otherwise, extensive cracking of the grains can occur, and this can seriously degrade rice and completely negate any benefits from washing. Attention must also be paid to the drying step applied after washing. Too little drying will leave the rice moist, inviting mold and handling problems. Too much drying, however, could partially gelatinize the grains, leading to uneven cooking and poor flavor. Finally, because any washing process produces wastewater, there is the problem of wastewater disposal.

2.9.2 The "Jiff" Rice Process

To overcome the problems associated with commercial, large-scale washing of rice, a new washing system has been developed. The system consists of two separate parts: a washing and drying unit and a wastewater treatment unit. The entire system has been termed the

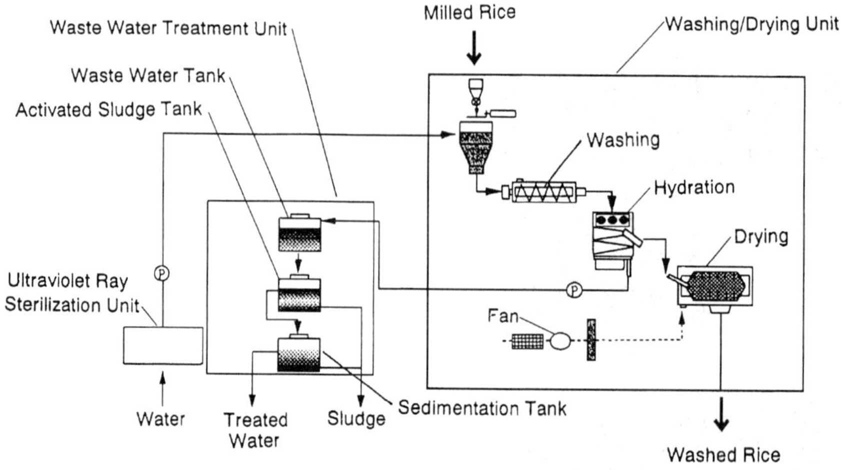

Fig. 22 Flow diagram of "Jiff" rice process. (Courtesy of Satake Corp.)

"Jiff" rice process in Japan to denote the fact that its "washed" rice product can be prepared quickly and easily.

A washing and drying unit adds water to the rice, agitates the moistened rice, and then dries the rice, in a continuous process flow. The moisture of the rice can be adjusted to remain constant with no absorption or desorption of moisture.

The second part of the process is the treatment of the wastewater. During the washing step organic solids become dissolved in the washwater, and those solids pose a significant disposal problem. Usually this type of wastewater is discharged into the municipal sewage system; however, owing to the high loading of suspended and dissolved organic solids, the costs charged by municipal sewage plants can be substantial. Therefore, a method has been developed to reduce this contaminant loading before it reaches the municipal system.

This system consists of an integrated wastewater treatment, which, by precipitating and settling action, removes almost all of the suspended and dissolved solids from the water. The solids and sediment can be discharged or dried and added to the bran fraction, because the nutritional quality of the dissolved solids is suitable for mixing with rice bran for animal feed. The treated water discharged from the system is acceptable for discharge into municipal sewage systems. The flow diagram of the "Jiff" rice process is shown in Fig. 22.

3 QUALITY CONTROL

In all countries where rice is handled, an inspector is always required to analyze the quality and grade of rice. The decision of the inspector is, however, sometimes subjective. To improve the efficiency and performance of rice inspection, various laboratory instruments have been developed.

3.1 New Laboratory Instruments for Quality Control

3.1.1 Rice Analyzer

In the conventional method of rice grading in Japan, a brown rice sample of approximately 1000 grains is visually inspected. This is done by an experienced rice inspector, who can

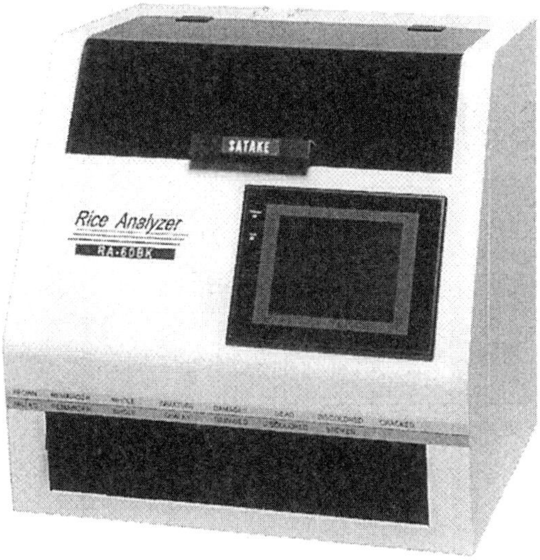

Fig. 23 Rice analyzer. (Courtesy of Satake Corp.)

identify good, immature, dead, damaged, cracked, and discolored grains. However, an automatic, easy-to-use system has been developed to inspect a 1000-grain sample of brown or milled rice with high accuracy and repeatability (Fig. 23).

3.1.2 Broken Rice Analyzer

A broken rice analyzer has been developed that is equipped with artificial intelligence and advanced image-processing capabilities. The broken rice analyzer examines a 25- to 50-g sample of milled rice and determines its broken percentages based on weight. The principle of this analysis is shown in Fig. 24.

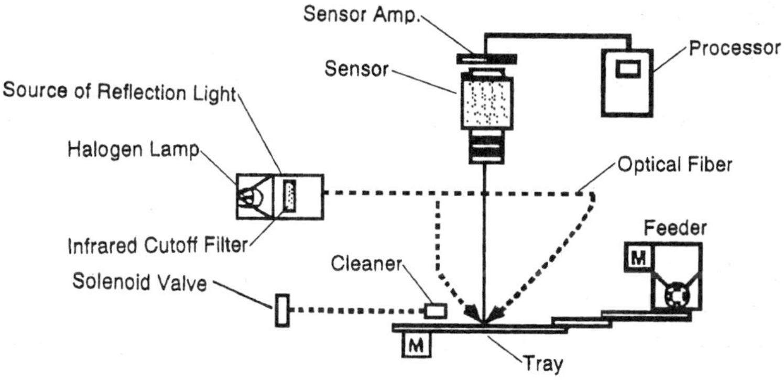

Fig. 24 Principle of broken rice analyzer: (1) rice is transferred to an analysis tray for scanning; (2) a camera sensor scans individual rice grains; (3) the kernel images are transformed into features—artificial intelligence algorithms analyze the individual grain features and determine whether or not the grain is broken; (4) weights of the individual grains are computed, and the percentage broken grains by weight is determined; and (5) the results of the analysis are reported on the analyzer's display and optically transferred to a computer or printer. (Courtesy of Satake Corp.)

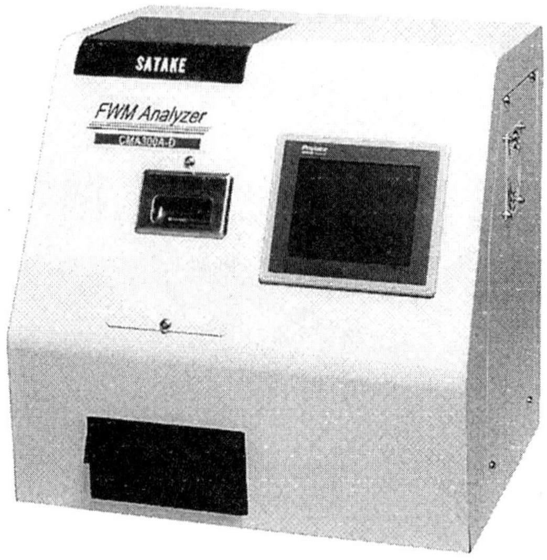

Fig. 25 Desktop type FWM analyzer. (Courtesy of Satake Corp.)

3.1.3 FWM Analyzer (Fat, Whiteness, and Degree of Milling)

The FWM analyzer is equipped with the near-infrared ray technology and is capable of simultaneously determining the fat content, degree of whiteness, and degree of milling of rice. Both desktop and on-line types are available. In the on-line type, the analyzer sends milling degree data to a computerized control panel which, in turn, sends signals to milling

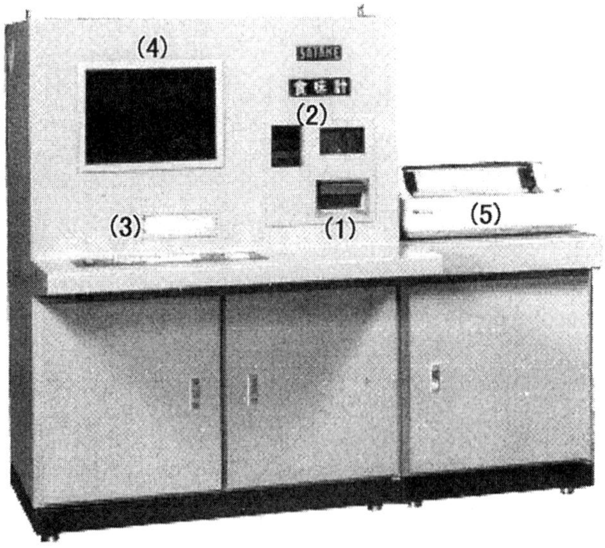

Fig. 26 The rice taste analyzer consists of a (1) sample drawer, (2) optics chamber, (3) microprocessor, (4) display, and (5) printer. (Courtesy of Satake Corp.)

machines to obtain the required milling degree. This system is considered a basic requirement of fully automated rice mills (Fig. 25).

3.1.4 Rice Taste Analyzer

A rice taste analyzer determines the taste of rice by detecting the amount of starches, proteins, degree of fat acidification, and moisture content of milled rice. A near-infrared analyzer is used for the quantitative assessment. The rice taste analyzer that is commonly used in Japan is shown in Fig. 26.

4 CONCLUSION

Quality standards for milled rice will continue to be improved. In addition, customers in the near future will demand more than merely rice with a specific appearance. Future milling systems may also require the capability to control, to one degree or another, such characteristics as water absorption rate, surface texture, and protein content. Because of this, rice mills will continually have to adopt advanced milling "technology."

In such an environment, it will be their knowledge of rice-milling technology that separates one company from another. Rice mill management will have to be familiar, to an increasingly greater degree, with the technological advancements that are being offered to the industry.

REFERENCES

1. Satake T (1990). Modern Rice-Milling Technology. University of Tokyo Press, Tokyo.
2. Bond N (1997). A Presentation to the Satake USA Milling College.
3. Kanemoto S (1997). Research on Rice Milling and Process Technologies.

14

Dehulling and Splitting Pulses

SHAHAB SOKHANSANJ

University of Saskatchewan, Saskatoon, Saskatchewan, Canada

RHAMBO T. PATIL

Central Institute for Agricultural Engineering, Bhopal, India

1 INTRODUCTION

Pulses are part of the legumes, a family of plants that have the unique characteristic of hosting symbiotic bacteria that fix atmospheric nitrogen in the plant roots. This not only reduces fertilizer requirements, but also improves soil structure. Pulses provide green pods for vegetables and nutritious fodder for cattle because they are rich in protein and in the essential amino acid *lysine*. Dry pulses are also a major food staple of populations in India, the Middle East, Africa, and South America.

The response to diversification of agriculture from conventional crops to pulses has been phenomenal. Canadian production of pulses was a record 1.89 million tonnes in 1994, up 43% from 1993. The production of lentils was 0.45 million tonnes which was higher by 29%, whereas field peas production was 1.44 million tonnes up by 48%. The acreage under pulses has increased to the level of 2.7 million acres in 1994 compared with 0.65 million acres in 1986, an almost four times increase in area and 4.6 times increase in production. The higher increase in production than area shows that production has been above average for these crops which was complemented with much improved crop quality. Canada exported 0.22 million tonnes of lentils and 0.69 million tonnes of field peas (including chickpea). Exports of field peas doubled during 1993 and 1994 owing to increased shipments to most major markets, notably Spain, India, and Central America (1).

Before cooking or other processing operations, it is often necessary to remove the fibrous seed coat (hull). In addition to reducing the fiber content, this improves the appearance, texture, quality, and palatability of the chickpea and reduces the cooking time (2). Dehulled seeds are easily digested and efficiently utilized by the body.

Pulse decortication and splitting is an age-old technology that has developed through the trial-and-error approach. This is the probable reason for nonuniformity in methods and machinery adopted by pulse processors in different parts of the world. Despite several research reports on the efficiency of some indigenous or mechanized dehulling systems there is little information in the literature on quantitative aspects. With the objective of studying the state of technology of pulse processing, the authors reviewed literature pertaining to pulse dehulling and then visited India, Iran, and Europe to examine processes and machinery used in dehulling pulses. This paper presents the review of research as well as a description of the technologies used in those countries. A partial list of terminologies used in pulse processing around the world is given in Appendix I.

2 PHYSICOCHEMICAL CHARACTERISTICS OF PULSES

The effectiveness of dehulling depends on the properties of the grain and the type of machine used. The yield of desired and undesired products obtained from dehulling depends on the following grain parameters (3):

1. The type of grain and its properties
2. Bond strength between kernel and hull, strength of kernel, and strength of hull
3. The proportion of hull in the kernel
4. Grain size and uniformity
5. Moisture content of the grain and the difference in moisture content between the hull and the rest of the kernel
6. Extent of hydrothermal treatment given to the grain
7. Proportion of dehulled kernel in the grain
8. The ease of classification of dehulled kernels and hulls

2.1 Seed Structure

Pulses all have a similar structure (4), but differ in color, shape, size, and thickness of the seed coat. Mature seeds have three major components: the seed coat, the cotyledons, and the embryo. Table 1 gives the approximate mass proportion of each component.

The seed coat, or hull accounts for 7–15% of the whole seed mass. Cotyledons are about 85% of the seed mass, and the embryo constitutes the remaining 1–4%. The external structures of the seed are the testa (i.e., seed coat), hilum, micropyle, and raphe (see Fig. 1 on page 402). The testa is the outer most part of the seed and covers almost all of the seed surface. The hilum is an oval scar on the seed coat where the seed was attached to the stalk. The micropyle is a small opening in the seed coat next to the hilum. The raphe is a ridge on the side of the hilum opposite the micropyle.

When the seed coat is removed from grain, the remaining part is the embryonic structure. The embryonic structure consists of two cotyledons (or seed leaves) and a short

Table 1 The Mass Proportion of Different Parts of the Seed

	Seed coat	Cotyledon	Embryo
Chickpea (*Cicer arietinum* L.)	15.0	84.0	1.0
Pea (*Pisum sativum*, L.)	10.0	88.7	1.3
Lentil (*Lens culinaris*, Medik.)	8.0	90.0	2.0

Source: Ref. 4.

Table 2 Characteristics of Field Pea Cultivars (*P. sativum* L.)

Cultivar	Hull (% of seed)	Hull thickness (mm)	Seed coat breakage (%, with TADD)	Cotyledon adhesion (N)
Bellevue	7.3	0.0559	1.1	8.8
Titan	7.0	0.0635	1.9	6.9
Trapper	7.6	0.0656	4.3	10.5
Tara	8.5	0.0720	6.1	7.9
Lenca	7.4	0.0665	3.7	10.8
Tipu	7.0	0.0635	2.9	9.6

Source: Ref. 5.

axis above and below them. The two cotyledons are not physically attached to each other except at the axis and a weak protection provided by the seed coat. Thus the seed is unusually vulnerable to breakage. Table 2 shows the thickness of the seed coat (or hull) and the force required to separate the two halves of the cotyledons in field peas (*Pisum satiyum*, L.), measured in newtons (N) using an Instron Universal Testing machine. The required force is in the range 7.9–10.8 N. The seed coat breakages, measured using a tangential abrasive dehulling device (TADD) (5), vary from 1.1% of the seed mass to 6.1%, which demonstrates the large variability in durability of the hull among varieties of peas.

The outermost layer of the seed coat is the cuticle, and it can be smooth or rough. Both the micropyle and hilum have been related to the permeability of the testa and to water absorption.

2.2 Seed Composition

Table 3 gives the approximate chemical and nutritional composition of the pulses. The proteins, ether extract (including lipids), phosphorus, and iron are contained in the cotyledons for the most part. About 80–90% of the crude fiber and 32–50% of calcium are concentrated in the seed coat. Although the embryo is rich in nutrients, it is the smallest part of the seed and holds only 3% of the total seed nutrients

The starch is found mostly in the cotyledon, and the granules are embedded in a dense proteinaceous matrix. The average size of a starch granule is on the order of 25–28 μm. The starch granules are a mixture of amylose and amylopectin. The pulses have a larger sugar content than cereals. The hull tightly envelopes the endosperm along its wrinkles and ridges and it is generally attached to a thin layer of the gum. The gum is reported to contain pentosans, hexosans, other polysaccarides, and uronic acids (6). The

Table 3 A General Composition (%) of Nutrients in Pulses

	Protein	Starches	Sugar	Fiber	Lipids
Chickpea	14.9–29.6	60.6	3.5–9.0	25.6	5.0
Pea	21.2–32.9	56.6	5.3–8.7	8.7	1.2
Lentil	20.4–30.5	59.7	4.2–6.1	—	1.2

Source: Ref. 4.

separation and removal of the hull depends on the nature of the gum and its hydration properties.

Pulses are categorized in two groups based on their milling characteristics: (a) easy to mill (e.g, lentil, chickpea, pea), and (b) difficult to mill (e.g., pigeon pea, black gram, green gram, beans, cowpea). The bond between the husk and the endosperm is weak in easy-to-mill pulses, whereas they are held firmly together by the gummy layer in difficult-to-mill pulses.

2.3 Seed Attributes

The physical characteristics of the following grains were measured; chickpeas (Kabuli and Desi), field peas, Laird lentils, and Eston lentils. The dimensions were measured using the image-processing technique. The mass of the grains was obtained using an analytical balance having accuracy of ±0.01 g. The 1000 seeds were first counted out in replicates of five samples before weighing. The bulk density was measured as the standard test weight method in five replicates. The particle density was obtained by using the volume (toluene) displacement method. The porosity of the bulk grain was obtained from bulk and kernel densities.

Table 4 lists data on the dimensions, weight, and densities of the pulses tested. The bulk grain porosity ranges from 28% for Kabuli chickpea to 47% for Laird lentil. The weight of 1000-seeds measured for these grains also shows a variation from about 33 g for Eston lentils to almost 600 g for Kabuli chickpea. The bulk density of the pulses ranges from 759 kg/m^3 for Laird lentils to 820 kg/m^3 for field peas. The dimensions measured are the major and minor diameters, roundness (r), and range of equivalent diameters. Some seeds, such as field peas, are quite round ($r = 0.99$), whereas Desi chickpea is irregularly shaped ($r = 0.88$), and lentils are lens-shaped ($r = 0.73$).

2.3.1 Chickpea

Chickpea (*Cicer arietinum* L.) is an important pulse crop used mainly for food. It is the major pulse crop in India where it is known as gram or Bengal gram. In Iran it is called *Nokhod.* Chickpeas are used either whole (fried, boiled), dehusked (roasted), split (dal in India and lappeh in Iran), or milled as flour. Husks and bits of the chickpea are used as feed for animals. The percentage of components of the chickpea varies with cultivars. Hulls usually account for 12–15% of the mass, but can be as low as 5–6% in some varieties and as high as 16–18% in others (2). The germ forms 2–4% of the grain, and the cotyledons represent up to 84–86% of the seed mass.

Chickpeas consist of 18–20% protein, 62% carbohydrates, 4% fat, and are a rich source of calcium, iron, and niacin. There are two main types of chickpea: (1) Desi or deshi (*Cicer arietinum* L.) for which the skin color ranges from brown to yellow; and (2) Kabuli or white chickpea (*Cicer kabulium*, L.) for which the skin color is white, the hull is thin, and the seed is larger and rounder than desi. The outer layers of the cotyledons contain larger concentrations of protein and more trypsin inhibitor than the inner layers. The ridged pyramid-shaped seed is likely to be scoured to spherical shape during abrasion in a dehulling machine. Scouring is greater when grain is moist and softer. Any grinding of the cotyledon causes more loss of protein and trypsin inhibitor because these materials are more concentrated in the outer part of the seed. One major problem with splitting is the loss of germ. In a mixture, the larger grains are scoured more than the smaller ones.

Table 4 Physical Characteristics of Commonly Grown Pulses in Canada

	1000-seed mass (g)	Particle density (kg/m^3)	Bulk density (kg/m^3)	Porosity fraction	Major diameter (mm)	Minor diameter (mm)	Spherical equivalent diameter (mm)	Sphericity, fraction
Desi chickpea	248.4	1255.6	773.0	0.38	8.73	7.01	6.31–9.72	0.88
SD[c]	1.2	4.1	1.1	0.01	0.75	0.54		0.03
Kabuli pea	600.7	1118.7	803.1	0.28	10.6	9.26	9.06–11.27	0.93
SD	1.4	3.7	1.0	0.01	0.27	0.05		0.02
Field pea	219.2	1268.0	820.5	0.35	7.55	6.84	6.49–7.9	0.99
SD	4	2.1	3.4	0.01	0.4	0.4		0.01
Laird lentil	82.9	1426.0	759.0	0.47	7.0	6.9 (2.77)[a]	6.26–7.36	0.73[b]
SD	0.4	0.4	2.7	—	0.39	0.42 (0.15)[a]		0.01
Eston lentil	33.1	1395.0	762.0	0.45	4.78	4.49 (2.28)[a]	4.14–4.92	0.78[b]
SD	0.2	0.8	0.6	—	0.28	0.27 (0.16)[a]		0.01

[a] Seed thickness.
[b] Sphericity calculated with thickness.
[c] SD, standard deviation.

2.3.2 Pea

The pea (*Pisum sativum* L.) is also an important pulse crop that is used both for food and feed. As food it is used either dehusked and roasted whole or as dal. Peas are classified into (1) green or garden peas (*Pisum sativum* var. Hortense) for vegetable, and (2) field peas (*Pisum sativum* var. Arvense) that are mostly used as split (dal) for food or for feed. Peas contain 22–24% protein, 56.6–62.1% carbocarbohydrates, 1.8% fat, and appreciable amounts of calcium, iron, phosphorus and vitamins B_1, B_2, and niacin.

2.3.3 Lentil

Lentils (*Lens culinaris*, Medik.) are another important pulse crop that are mainly used as splits in India (Masur dal) and Turkey. The whole seed is also used in different dishes, especially in Iran (mixed with rice, Adas polo). Lentil seeds are round and lens-shaped about 7-mm diameter and 3-mm thick, with seed coat color ranging from clear to yellow, green, or blotched purple. The seed surface is generally smooth. However, the seed coat of weathered lentils may be wrinkled. The cotyledon color may be orange, yellow, or green; the latter turning yellow after a period in storage. The three main components of a lentil seed, the seed coat, cotyledons, and the embryo, constitute respectively, 8, 90, 2% of the total seed weight.

Schematic views of a lentil seed are given in Fig. 1. The two most distinctive external structures of a lentil seed are the hilum and the micropyle. The hilum is an oval scar on the edge of the seed where the seed was connected to the stalk before harvest. A central opening about 2-mm-long runs its entire length. The micropyle is a small opening in the

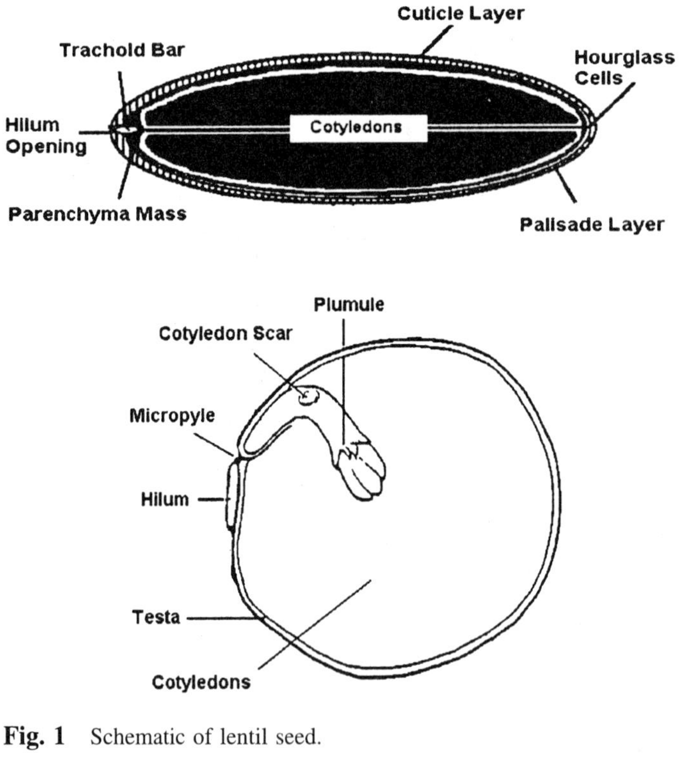

Fig. 1 Schematic of lentil seed.

seed coat near one end of the hilum, and is the site of entry for pollen. The seed coat of a lentil seed is 40- to 50-μm thick.

3 UNIT OPERATIONS IN PULSE PROCESSING

The sequence of operations in pulse processing is premilling treatment, dehusking or decortication, and splitting. Large variations exist in the steps followed in each operation.

In traditional operations, the dehulling is done either by the wet or dry method. In the wet method, the pulse is soaked in water for several hours before sun drying and milling. In the dry method, a small quantity of water or oil is applied onto the grain and conditioning time is usually shorter. Apparently, the wet method is no longer used in dal mills in India (7) In the modern methods, high-temperature–short-time heating is used for conditioning the grain for loosening the husk. It is claimed that this treatment leads to 5–10% higher dal yields. The general process flow chart for modern pulse processing is shown in Fig. 2.

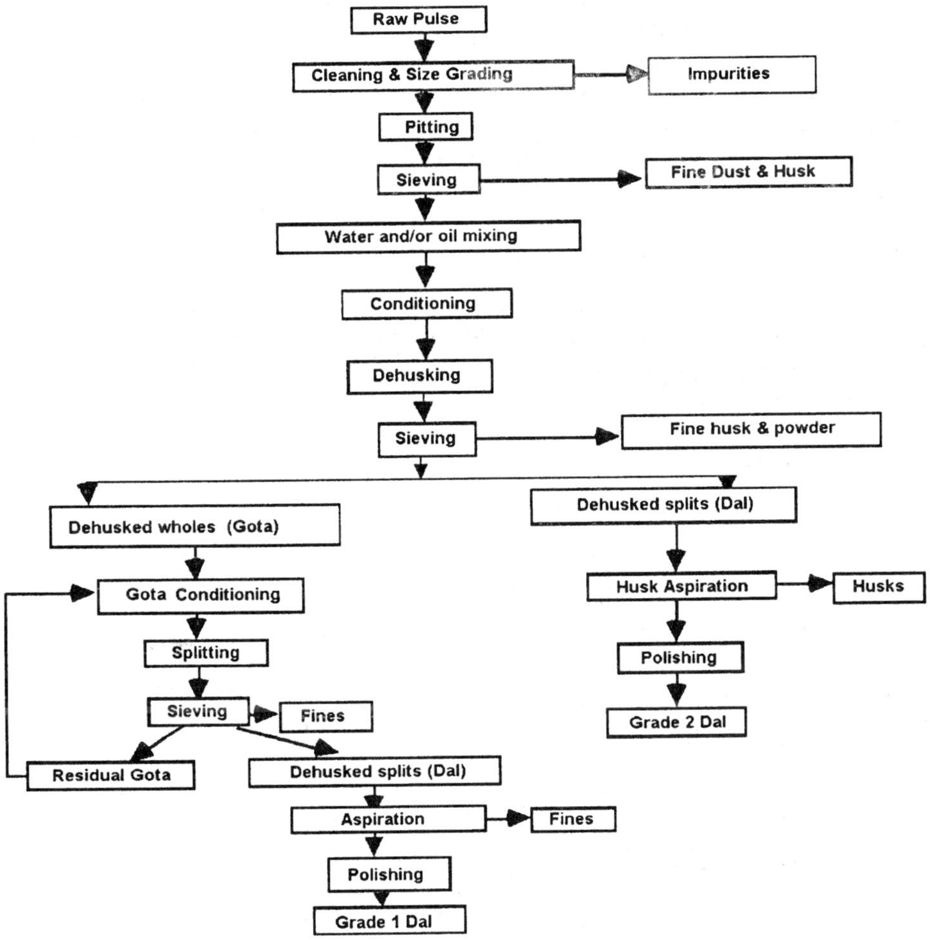

Fig. 2 General flowchart for pulse-milling process.

Table 5 Size, Capacity Range, and Operational Details of Dal Mill Machinery

Machine and size	Speed (rpm)	Power (kW)	Capacity (*t*/*h*)
Rotary cleaner (212 × 150 × 90–393 × 150 × 120 cm)	18–60	0.38–1.50	1.4–2.0
Roller (17.5 × 20 × 30–45 × 35 × 45 cm)	900–1100	2.25–3.75	0.4–3.0
Husk fan (40 × 20–120 × 75 cm)	500–1000	0.38–1.13	0.8–1.0
Flat screen grader (150 × 75–240 × 90 cm)	60–75 (stroke 2.5–3.75 cm)	0.38–1.13	0.5
URD sheller (45–60 cm)	250–400	2.25–3.75	1.0–2.0
Impact (patka machine)	90–250	0.19–0.75	1.4–3.0
Splitter (45 × 90–60 × 75 cm)			
Worm machine (120 × 20–600 × 20 cm)	90–250	0.19–0.75	1.4–3.0
Elevators (bucket) (240–900 cm)	125–250	0.19–1.50	2.0–3.0

Source: Ref. 7.

The description of each unit operation and equipment used are explained in the following. Table 5 lists specifications for typical series of machinery for dal milling operations. The specifications of commercial pulse-processing machinery are given in Appendix II.

3.1 Cleaning and Grading

Pulses must be cleaned during the process, because they may be delivered containing up to 20% impurities. Foreign materials include pod walls, broken branches, soil, cereals, oilseeds, weed seeds, diseased and deformed seeds, and stones. Initial cleaning of lentils takes place on scalpers and airscreens.

Raw material is cleaned by removing dust, dirt, foreign material, off-sized, immature, and infested grains. The cleaned grain is graded into uniform sizes. Air and rotary screens with round holes are used for cleaning. In rotary screens, however, the grain does not have equal opportunity to come in contact with the sieve before reaching the end of the separation zone, which leads to improper grading.

After the lentils are conditioned with water and tempered, the seeds are graded on round-hole rotary screens. The graded seeds are then dehulled, and the mixture of whole and split lentils, hulls, broken seed, and fines are passed through an inclined gravity separator to remove broken seed and fine particles. Whole seeds are screened from splits and are returned to the dehuller. The splits are fed into a horizontal gravity separator to separate dehulled from hulled splits.

Normally, two types of cleaners are available: rotary screen cleaners and reciprocating screen cleaners. The screens are used to remove the foreign materials on the basis of size difference. The suction fan removes lighter material such as dust particles. The separation of splits from whole grain is done by the screens (8).

The rotary screen consists of four compartments of different-sized screens fitted on 50-mm–diameter shaft. The screens clean different pulse crops. The machine is fitted on a sloped foundation (slope 50 mm for 1 m length) and is operated at a low speed (18–

30 rpm) for better performance. The body of a machine can be made of wood or iron (mild steel), depending on the manufacturer or customer choice.

The reciprocating flat screen cleaner has three screens and one dust (suction) fan. The foreign material, including dirt, dust, and such, is separated, and the pulse is graded for milling purposes. A stroke of 37–50 mm is provided for the reciprocating unit. Details of size, capacity, and other specifications are shown in Table 5.

The functions of various cleaning and grading equipment for pulses are described in the following (9).

Scalpers are used to remove large unwanted trash and fines. The separation is done on reciprocating or rotating screens.

Air-screen machines are used as scalpers for size separation as well as weight separation. The material is aspirated to remove light materials, then passed over screens to remove large materials as overs. The unders are passed on a second screen to separate fines from desired size seeds.

Disk and cylinder separators consist of series of cast iron disks mounted on a shaft revolving at a very precise speed relative to disk diameter within a cylindrical housing. The disks have precise undercut pockets with size and shape variations which allow the smaller seeds to be lifted and to reject larger seeds.

Indent cylinders are machines of choice for length separation. They are almost horizontal rotating cylinders lined with hemispherical depressions. Short seeds are picked up in the indent and lifted up and thrown into an auger to be carried out of the rotating cylinder.

Width and thickness separators are typically rotating, cylindrical, perforated shells. Larger seeds will not pass through and discharge from the end of the cylinder.

Gravity separators separate seeds based on a combination of shape, size, specific gravity, and surface characteristics. The seed mixture is fed onto an oscillating deck with a carefully controlled air movement to fluidize the material. The mixture is stratified—lighter up and heavier down; the seed layers along the deck in different directions toward discharge ends.

Spiral separators work on the basis of the shape of the seed to roll at different rates. The angle of the flights of the spiral has to be adjusted based on the type of crop.

Color sorters have proved to be very effective but need considerable adjustment, capital investment, and are slow, although it is expected that improvements in speed will be made.

Pulse cleaning can be accomplished by an air screen machine with the addition of a spiral, gravity separator, destoner, and perhaps color sorter. The indent cylinder can help for lentils. Cylinders for width and thickness separation should be used with care, because these machines tend to damage lentils. These machines are also seldom used for peas. Gravity tables are commonly used for lentils and some times for peas. Such tables are very useful for difficult separations. Disk and cylinder separation is seldom used for peas.

3.2 Seed Conditioning to Promote Dehulling

Conditioning is a general term applied to heating, cooling, wetting, drying, or any combination of these processes. The effects of conditioning on the seed could be to toughen the hull (bran), loosen the bond between the hull and the cotyledon, crack the seed coat, and harden the cotyledon to resist damage.

In a review of food legume processing, it was reported that in developing countries, whole grains are soaked in water for a short time before pounding in mortar and pestle fashion to remove the hulls (10). Heat treatment of moistened grains, or in some dry grains, makes the hull easier to remove as it becomes brittle and cracks. Also, the cotyledons tend to shrink more than the hull during this process, resulting in the hull being loosened from the cotyledon. Addition of moisture softens the grains and make them susceptible to scouring; whereas drying hardens the grains and increases their resistance to scouring. Pigeon peas are dried from a typical moisture content of 11 to 6–7% to provide increased resistance to peripheral scouring.

The tempering of small lentils before dehulling has been investigated (11). Combinations of the following factors were studied: (a) two lentil sizes, one lot 4.0–4.5 mm and another lot 4.5–5.0 mm; (b) soaking in water for 1, 5, 10, or 30 min; (c) drying the seed at 19° or 36°C for a durations of 0, 30, 60, and 120 min; (d) tempering the material for 0 or for 24 h. The dehulling efficiency was calculated as the sum of percentages (a) of split dehulled seed and (b) whole dehulled and hulled seeds. Hulled seeds were included as those seeds not considered as a loss by industry. The overall dehulling efficiency was 81.1%, consisting of 67.7% dehulled split, 10.1% hulled whole seed, and 3.2% dehulled whole.

Dehulling efficiencies were best for the following conditions (not inclusive): smaller-sized seed, shortest drying time, least immersion time, and longer tempering time. The drying temperature did not affect dehulling efficiency. The increased seed moisture content (longer immersion) had a negative correlation with the dehulling efficiency. There was a significant increase in the split hulled seeds and a slight increase in the number of hulled whole seed with increased moisture content.

3.2.1 Pitting

Whole pulses are passed through abrasive roller machine for scratching the seed coat to facilitate the entry of oil and water in the grain during premilling treatment. Some seeds (2–5%) are dehusked during pitting.

3.2.2 Oil and Water Treatment

Edible oil treatment is used to loosen the husk of difficult-to-mill pulses. The quantity of oil used varies from mill to mill. The quantity of oil required is estimated to be 3g/kg. Water treatment, which varies with crop, is used to expand seeds, which helps in loosening the husk by contraction of cotyledons during drying. Some dal millers apply water and oil simultaneously. This is done to reduce the process time.

Addition of large quantities of water during processing and nonremoval of the entire quantity during drying adds about 3–5% to the weight of the product and reduces the storage life of dal. Large variations in quantity of water added lead to increased energy costs for moisture removal and process time. This also reduces dal recovery.

The machine used for mixing in oil or water is a screw conveyor fitted with enclosed inlet and outlet with a full-flight or cut-flight screw. The screw is rotated slowly to achieve proper mixing of oil or water with the pulse grains. Cut-screw flights increase agitation and help to increase retention time for better mixing.

Treated grains are heaped and covered with gunny bags and left for 12–18 h. This helps penetration of the oil and water into the cotyledons after mixing and grain temperature equilibration after drying in the sun. Some wooden or cement tanks are used for storing the treated grains.

3.2.3 Kernel Temperature and Moisture Content

The effect of moisture content (6–20%) and temperature (−40° to 40°C) on seed coat durability of field peas was studied by using the rubberized aluminum disk in TADD (12). Seed coat breakage was most affected by seed moisture content, followed by temperature and cultivar. At all temperature levels, seed coat breakage increased linearly with decreasing moisture content. This may be due to changes in either tissue elasticity or the binding between the inner seed coat surface of the cotyledon. When the tissues begin to lose moisture, there is an increase in brittleness caused either by an increase in crystallization or by a change in cell orientation (13).

Conditioning buckwheat to water activities in the range of 0.11–0.98 influences the dehulling efficiency (14). Drier samples had a better dehulling efficiency, but the groats had more brokens than the moister sample. Treatment temperatures between 1° and 40°C did not affect dehulling efficiency.

The effect of temperature and moisture content on dehulling efficiency of three varieties of pigeon peas was studied in the laboratory (6). Differences between varieties exist. Figures 3 and 4 show the effect of the kernel moisture content and heat treatment on the dehulling efficiency of one variety of pigeon pea. The kernel moisture content was progressively reduced by drying at 50°C (from 11 down to approximately 4%), or progressively increased by wetting with addition of water (from 11 to about 18%). Figure 3 shows that reduced moisture (6%) gives the highest dehulling efficiency. When the seed was heat-treated by blowing hot air into the sample, the efficiency increased progressively leveling off at 100°–150°C. However, it was recommended that the temperature be carefully chosen so that the seed quality is not damaged.

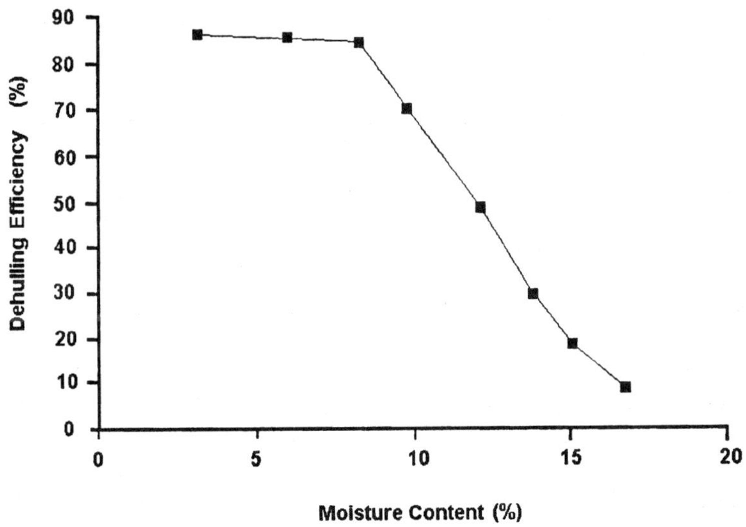

Fig. 3 Dehulling efficiency of pigeon peas vs. moisture.

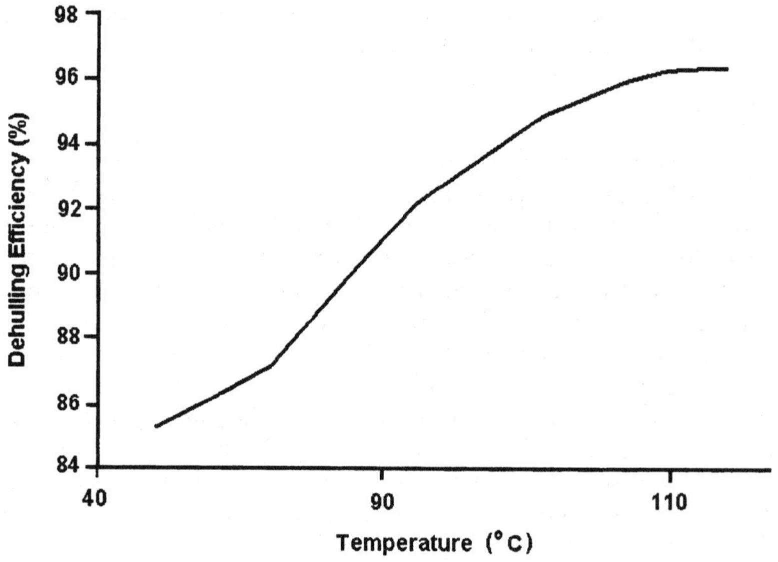

Fig. 4 Dehulling efficiency of pigeon pea vs. air temperature.

3.2.4 Drying

Normally, pulses are sun-dried as part of premilling treatment and for the loosening of husk. The size of a drying yard has no generalized correlation with the capacity of a mill, and drying time ranges between 3 and 8 days. Some times thick-layer, sun-drying is followed, which results in nonuniform drying and breakage during milling. This is a crucial part in premilling treatment and needs consideration. The premilling treatment, as such, varies from miller to miller and thereby results in repeated exposure of grains to milling machinery, adds to breakage and lowers the dal recovery. Standardization for sequence of operation, including pretreatment for each pulse crops, would help save these losses to a greater extent.

The hull of the matured soybean kernel is loosely attached to the cotyledons and, therefore, could be detached fairly easily in cracking of conditioned beans. A 1–2% reduction in bean moisture content promotes dehulling. Heat treatment of beans at 93°C for 15 min breaks the bond between hulls and cotyledon. The relative humidity of the drying air is the most significant factor affecting the level of seed coat cracks in soybean. Cracks increase with increases in drying air temperature, initial moisture content, and drying rate, and decrease with increased final moisture content and drying air relative humidity.

Beans should be dried to a proper moisture content and stored for days so that the hulls separate from the cotyledons easily after cracking (15). A drying air relative humidity below 40% results in cracks in the soybean seed coat and separation of the seed coat from the cotyledons.

In the conventional system of soybean dehulling, beans are dried to about 10% moisture content and cooled and tempered for moisture uniformity before dehulling. In fluidized-beds, beans (up to 15% moisture content) are heated rapidly to dry enough for cracking without needing to cool the beans.

Heated beans are cracked to halves in a single pair of cracking mills and then fed to an impacter to knock the hull loose from the bean. At higher temperature, short time and uniform heating result in easier dehulling.

3.2.4.1 Sun Drying

Normally, sun drying is practiced in drying yards. However, enterprises with limited space use roof tops for drying. Grains are spread in a thin layer (3–10 cm) and turned frequently with a rake for even drying. Some dal millers prefer thick layers (up to 15 cm) of grains in the drying yard for reasons of quality improvement. The drying period for grains varies from 1 to 5 days depending on weather conditions, thickness of layer, and pulse crop. The size of the drying yard depends on the capacity of the mill, availability of land, and recommendation of the manufacturer. Most of the dal millers employ sun-drying in open drying yards: 0–1 m^2/kg of pulse is sometimes used as a rule of thumb.

Some millers avoid drying or reduce drying time and mill the pulses after tempering for 8–12 h. This results in heavy breakage during milling. Some of the dal mills are equipped with dryers, but these are preferably used in the rainy season or in situations of unfavorable weather conditions. Some mechanical dal dryers have been introduced by different firms. Some technical data are reported in Table 6.

3.2.4.2 Losses in Conditioning

Some pulses such as green gram and pigeon pea have a tendency to split before they are completely dehusked. This leads to repeated exposure of grain to milling machinery for complete husk removal, and leads to high breakage. Insufficient premilling treatment also contributes to this problem. Another important aspect in pulse milling that leads to breakage is problems in separation of gota and whole grains. The unprocessed lot consisting of a mixture of gota and whole grains at the end of a pass is reprocessed for dal making.

Table 6 Details of Mechanical Dal Dryers of Various Makes

	Thermax Pune		S.R. Industries,	Ganesh Eng.,	Annapurna Industries,
Description	DD-10	DD-30	Madras	Madras	Patna
Holding capacity (t)	1.2	3	3.5	NA	0.8
Drying capacity (t/h)	1.2	3	1.5	1	0.8
Moisture removal (%)	5	5	4	NA	NA
Drying air temp. (°C)	60	80	60	150	150
Grain temp. (°C)	NA	NA	NA	75	NA
Heat exchanger	Yes	Yes	Yes	No	No
Type of dryer	Batch	Batch	Batch	Continuous	Batch
Power (hp)	8	11	17	5	NA
Fuel energy/h	7–81	10–121	51	45	80 kg
	LDO	LDO	Furnace oil	kW	Coal
Price (Can $)	3000	4600	3000	NA	8000
(Ref. year)	(1984)	(1984)	(1984)		(1980)
Extra yield claimed (%)	NA	NA	2	1.5–2	NA
Dryers in operation (h)	200[a]	27[a]	6	NA	1
(Ref. year)	(1984)	(1984)	(1980)	NA	(1980)

[a] Includes uses other than dal drying.
Source: Ref. 7.

Since commercial equipment available for gota separation is not effective owing to the same size of grain and gota, the gota in the mixture is again exposed to a premilling treatment, which renders it weak and leads to a high percentage of brokens during exposure to the dehusking machine.

3.3 Dehusking of Pulses

Dehusking or shelling of pulses is a preparatory operation for splitting. Dehusking is preferably achieved by subjecting the grains to abrasive force and splitting by attrition or impact. Generally, three to nine passes are required for milling various pulses, and this depends on the type of pulse crop, premilling treatment given, processing technique adopted, grain size, and variety. Splitting during milling is normally a disadvantage, as the splits scour at the edges causing powder loss. One of the causes of higher loss is more surface area when grain is split.

Carborundum emery-coated rollers are used for dehusking different pulses. The rollers are of two types: cylindrical and tapered. The foundation is horizontal for the tapered rollers, whereas it is sloped for the cylindrical roller. Normally, a slope of 15 cm/m in foundation for the entire length of the machine is recommended. The body of a roller is made of steel on which the mixture of Carborundum emery, chemical cement, and salt is applied and dried to make a uniform, circular abrasive surface. The granular size of emery varies from crop to crop and is recommended by manufacturers. The inlet and outlet of the roller can be adjusted for regulation of flow and retention time.

These rollers are available in different sizes and are specified as 23 $\times$ 53.5 $\times$ 4 cm, where 23 is the diameter of the roller, 53.5 is its length, and 4 is the diameter of the shaft. However, some manufacturers indicate only diameter and length of roller (e.g., 23 $\times$ 53.5 cm). For the tapered rollers, the specifications are given as 20 $\times$ 17.5 $\times$ 30 cm or 35 $\times$ 30 $\times$ 45 cm, where the first two figures denote diameters and the last is the length. The power requirement, capacity, roller size, and speed (rpm) varies from manufacturer to manufacturer.

3.3.1 TADD

Figure 5 shows the schematic drawing of the tangential abrasive dehulling device (TADD). This machine was developed at the Prairie Regional Laboratory (PRL) of the National Research Council of Canada in Saskatoon and is used for testing the dehulling properties of grains.

A resinoid, steel cutoff disk of 250-mm diameter and 3.12-mm thickness is mounted directly on a 0.0375-kW, 1725-rpm electric motor, which is supported on a stand. A spring-operated solenoid rubs two O rings on the disk mount to act as a break when current to the motor is discontinued. An aluminum head plate holds eight stainless steel sample cups (46.88-mm diameter and 17.19-mm deep) above the resinoid disk. The cups open at both ends, are mounted vertically with their centers equally spaced around a 184.37-mm–diameter circle. A rubber-faced aluminum cover plate closes the tops of the cups when the machine is in operation. With the machine assembled, the threaded head plate supports are adujusted so that the lower edges of the cups are lightly ground by the resonoid disk to assure minimum clearance between the cups and disk.

In operation, weighed samples of grain are placed in the cup, the cover plate fastened in position, and the resonoid disk rotated under the cups at 1725 rpm for a specified time. The abraded samples are then removed from the sample cups with the vaccum sample collector.

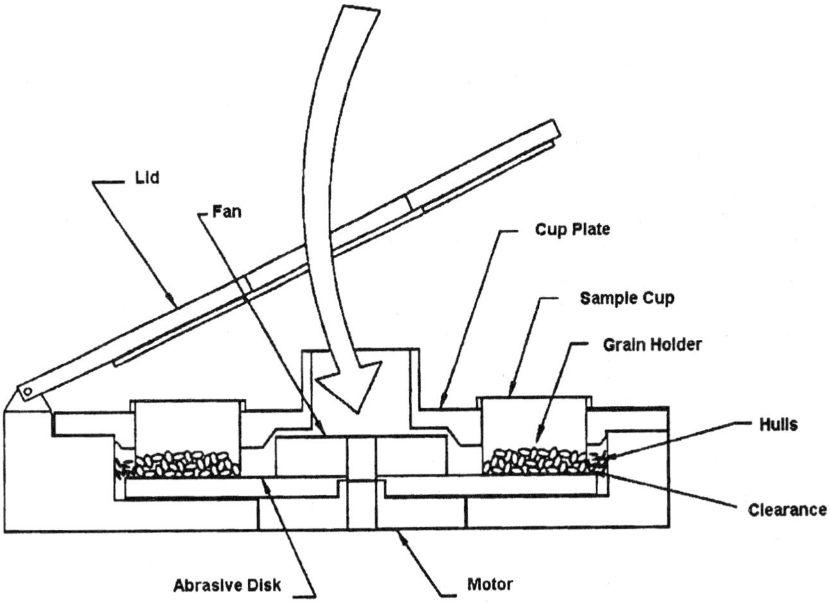

Fig. 5 Tangential Abrasive Dehulling Device (TADD).

3.3.1.1 CIAE Design

An abrasive emery roller cylinder mill to dehusk and split pulses to make dal (Fig. 6) was developed at the Pulse Processing Laboratory of the Central Institute of Agricultural Engineering, Bhopal, India (16). The mill consists of 250-mm–diameter cylinder coated with emery paste and an outer working layer of Carborundum. The clearance between the outer screen cage and inner abrasive roller was maintained throughout at 10 mm. The performance of the mill was evaluated evaluated for different pulses and at different speeds. The maximum milling efficiency was obtained at 850–900 rpm (13.5- to 14-m/s roller surface speed). Dal recovery was 74–75%.

Five degrees of Carborundum grit sizes were tested on dehulling of pigeon pea, green gram, and black gram (17). The grit or grade sizes ranged from 18 to 40, corresponding to 0.50 to 0.16 mm. The cylinder on which the grit size was formed was 300-mm in diameter and 400-mm long. It rotates at 900 rpm, corresponding to a tangential velocity of 14 m/s. The optimum size was 0.22–0.49 mm for pigeon pea and 0.16- to 0.30-mm range for the green gram. Finer or coarser grit size were found to produce either more broken grain and powder or insufficient abrasion.

3.3.1.2 PRL Design

An intermediate-sized, batch dehuller was developed at PRL (18). Grains are dehulled by abrasion provided by abrasive disks of 250-mm diameter mounted horizontally on an 18-mm–diameter shaft. The Carborundum stones, resinoid disks, or a combination of disk and stones can be used. The two end disks are angled to avoid accumulation of grain in the corners (Fig. 7). The case (300 × 300 mm) is mounted on pillow blocks and is free to rotate on the shaft. In operation the position of the cage is fixed in one or two positions by a spring-loaded–locking pin for loading or unloading. The fines are separated from

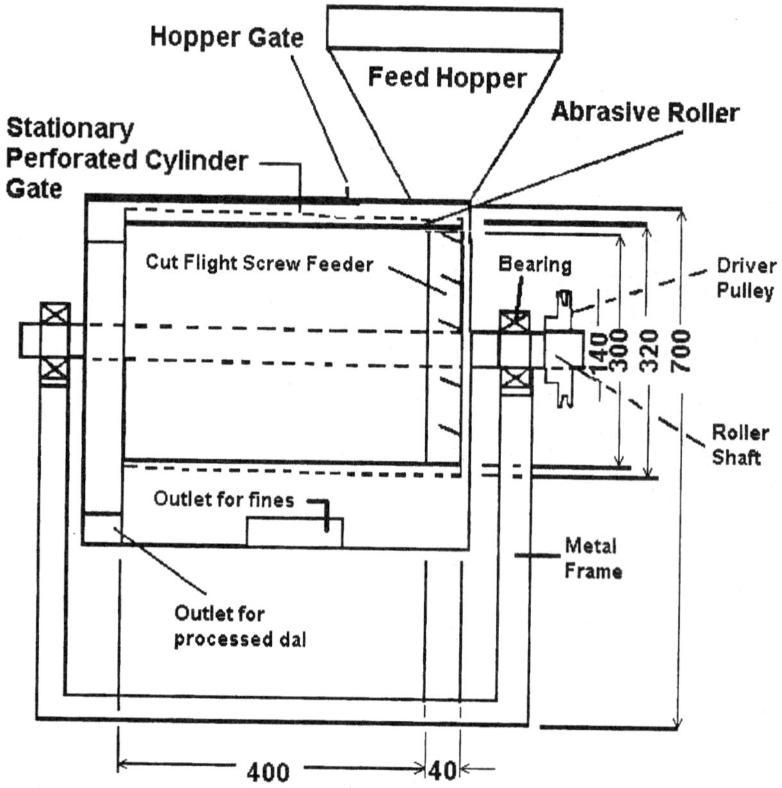

Fig. 6 Horizontal abrasive roller for pulse dehusking (CIAE design).

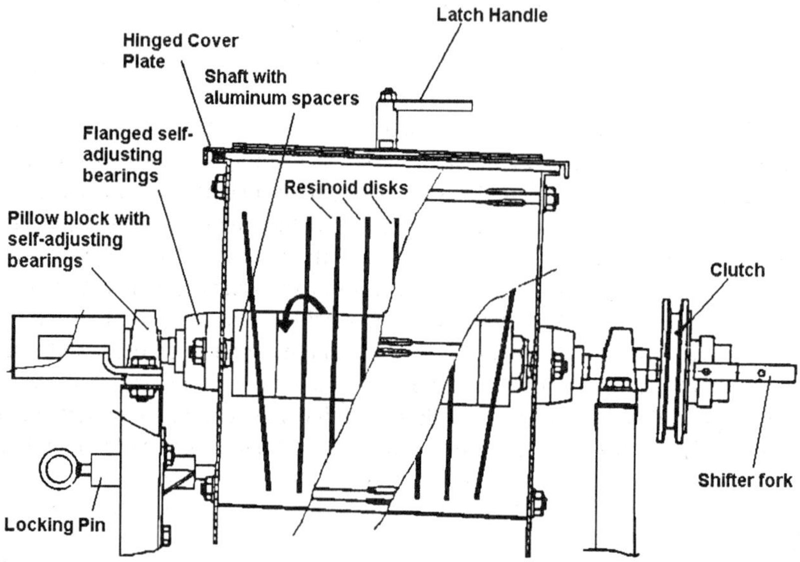

Fig. 7 Miniabrasive batch dehuller for pulse dehusking (PRL design).

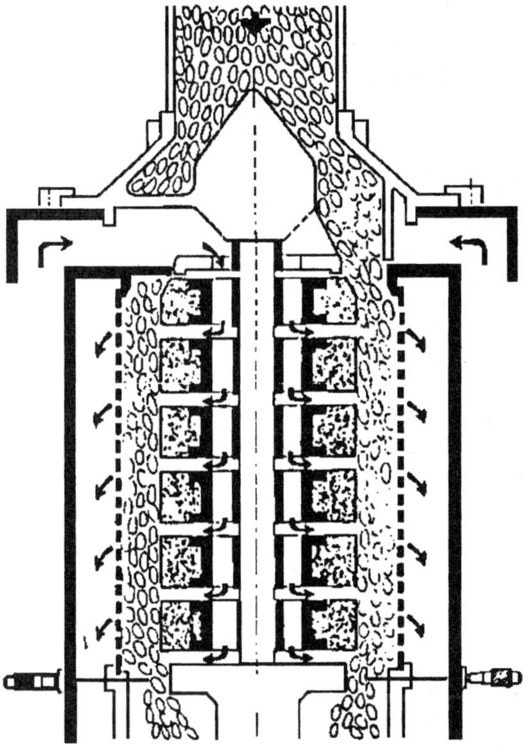

Fig. 8 Vertical abrasive dehuller for pulses (Schule design).

the dehulled grain by aspiration. The speed of the abrasive disks can be varied from 500 to 2000 rpm, depending on the type of grain being dehulled.

3.3.1.3 Schule Design

The vertical abrasive dehusking machines are used by European manufacturers (Fig. 8). The material enters from the top and completely fills the working chamber. Shelling is achieved by means of emery disks. These disks rotate within the material, which is moving downward. The weight of grain ensures adequate pressure of the material toward the emery disks. The shelling degree depends on residence time of the material. The residence time and throughput can be adjusted by the discharge gate. A separate fan is provided to aspirate and cool the unit by drawing air from the top of the machine. The aspiration air passes through the gaps between the emery disks and through the material to be dehusked, carrying with it the particles of husks through the perforations of the perforations of the cage. The proper use of possible adjustments of shelling machine and appropriate conditioning ensures splitting of pulses without undesired production of flour and particles removed by abrasion (19).

3.3.1.4 CFTRI Design

A minimill dehuller with a capacity of 150–200 kg/h has been developed by CFTRI, Mysore, India (Fig. 9). The mill consists of an emery-coated metal cone fixed to a vertical

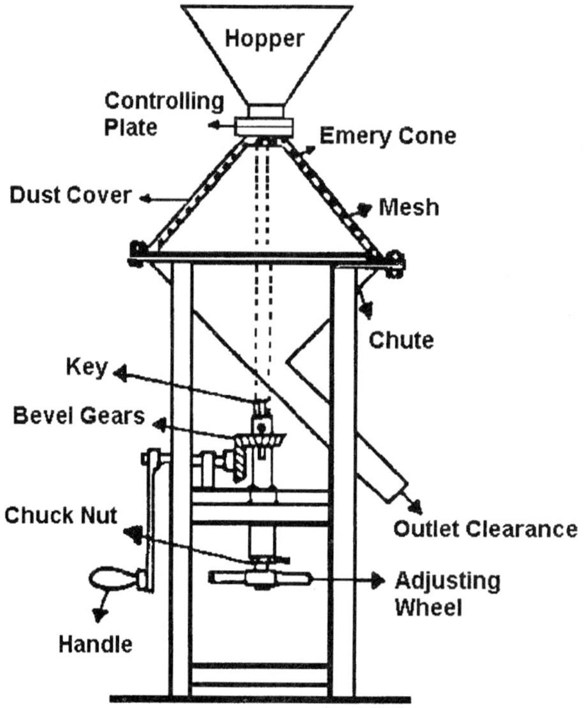

Fig. 9 CFTRI minimill for dehusking pulses.

shaft and rotating inside a fixed conical wire mesh screen. A screw at the base of the shaft can raise or lower the emery-coated cone. The screen and the cone are concentric, and their clearance is about the diameter of a grain. The grain stream is regulated into the machine such that it does not jam the machine (20). A dust cover, with a hopper at the top, envelopes the screen. Another hopper collects the mill stream. The seed must be preconditioned before dehulling. Chickpeas are dehulled in this unit giving 78–81% dal yield; 1–2% brokens; and 17–20% husks and powder (21).

A small laboratory-scale dehuller was developed for oat dehulling (22). The unit has been manufactured by Kustom Stainless Steel Equipment Inc. Winnipeg, Canada, and has been adapted by the Canadian Grain Commission for identification of tan-colored oat varieties such as Riel cultivar.

3.4 Comparing Vertical and Horizontal Rollers

The decortication efficiency [see Sec. 4.1., Eq. (7)] and seed coat removal effcienciency [see Sec. 4.1., Eq. (8)] of a vertical roller mill (Decomatic 2, Bernhard Keller AG Switzerland) and a horizontal mill (S. K. Engineering and Allied Works, India) were evaluated (23). Both machines use rotating stones inside a wire mesh cage. The Decomatic roller is a vertical cylinder 340-mm in diameter rotating at 2400 rpm; the S. K. Engineering roller is a truncated cone 250 mm at the entry and 350 mm at the exit, rotating at 1000 rpm. In both, the machines are full of grain and the flow rate is adjusted by adjusting the gate at the outlet. They tested a specific chickpea variety and broad bean. Chickpeas were tested in a commercial production without any control. The preconditioning of the samples was not reported. The results showed that the decortication efficiency of Decomatic was 83.7% and that of the S. K. Engineering was 75.9%. The seed coat removal efficiency of Decomatic was 84.4% and the S. K. Engineering was 86.7%. The results with broad bean showed excessive damage to the seed by both machines. There were also large variations among the efficiencies from sample to sample; thus, it was concluded that in these type of machines the treatment of the grain is not uniform.

3.5 Husk Separation and Grading

Husks are separated by aspiration and sold as animal feed. Some fine brokens are present in the husk and, if separated, can make available an extra quantity for human consumption. Grading adds to the quality of product by separating the dehusked unsplit grains (gota), whole grains, brokens, and dal in two grades. However, problems in separation of gota and whole grains owing to their same size and shape, leads to repeated exposure of grains to the machine and breakage during milling. A new method of separating whole and decorticated grain based on their bouncing properties has been reported (8). A succession of four hard surfaces is used to bounce the whole seed and the decorticated seed. A 61% efficiency of separation was established in the laboratory trials for pigeon peas.

The mixture of brokens and husk is separated from the dehusked pulses and splits by using a suction fan or blower and the mixture is used as animal feed. To separate the brokens for human consumption, the use of a specific gravity separator is advocated. This machine gives separation of about 5% brokens available in the mixture. The machine is supplied with three, four, and five fans, with a capacity ranging from 635 to 4500 kg/h.

3.5.1 Polishing

Whole pulses, such as pea, black gram, green gram, and splits (dal), are polished for value-adding. Some consumers prefer unpolished dal, whereas others need dal with attractive color (polished dal). Accordingly, dal is polished in different ways such as nylon polish, oil–water polish, color polish, and so on.

3.5.1.1 Removal of Powder and Dust

The cylindrical roller mounted with hard rubber, leather, or emery cone polisher, and roller mounted with brushes are used for the purpose. The powder particles are removed by the rubbing action. Speed and sizes of these types of polisher are similar to those of the cylindrical dehusking roller (see Table 5).

Another type of machinery provided for this purpose is a set of screw conveyors arranged in battery for repeated rubbings. The flights and shaft are covered with nylon rope or velvet cloth. The speed of each screw conveyor varies. The repeated rubbing adds to the luster of the dal, which makes it more attractive. These polishers are commonly known as nylon polisher or velvet polisher, depending on the material used, and are available in a set of 2, 3, 4, or 5 screw conveyors.

3.5.1.2 Oil and Water Polish

The screw conveyor similar to one supplied for oil and water mixing is provided for oil and water polish. The speed, size, and capacity, are similar to those of the oil and water mixing machine.

Table 7 Emery Sizes (No.) for Roller Machine

	Operation		
Crop	Pitting	Splitting	Polishing
Arhar (pigeon pea)	16	24	—
Mung (green gram)	24	30	35–40
Urd (black gram)	24	30	35–40
Chickpea	—	24	—

Source: Ref. 7.

3.5.2 Splitting of Pulses

Splitting of dehusked pulses and pulse seeds is one of the major operations in the dal mill. It is aimed at the production of perfect splits, with edges and without breakage. Different types of equipment are employed for the purpose: roller machine, under runner disk sheller, attrition mill (Chakki), elevator and hard surface, and impact sheller.

3.5.2.1 Roller Machine

The machine similar to one used for dehusking is used for splitting of different pulses. A course emery coating is required for the splitter roller (Table 7). The rollers are used for splitting green gram, pigeon pea, lentil, and others. Roller machines are based on the principle of abrasion. Revolutions and diameter of roller (peripheral speed), roughness of surface, length of roller (hold-up time), abrasion force, abrasion pressure, and clearance between the roller and the lower sieve are some of the factors that determine the extent of dehusking and scouring on this machine.

3.5.2.2 Under Runner Disk Sheller

The machine is simple in construction. It has two horizontal disks with emery coating of 12-mm thickness. The upper disk is stationary; the lower one rotates to cause splitting of dehusked pulse (gota). It is used for splitting dehusked black gram, chickpea, lentil, pigeon pea, and soybean. The capacity of a machine depends on its size and speed. The sheller machines cause breakage as high as 30–40% particularly if the grains are not thoroughly size-graded. Revolutions of the disks, peripheral speed as determined by the speed of rotation and diameter, roughness of contact surfaces and their parallelism, the distance and duration the grains roll (under pressure) between the revolving disks play important roles in splitting or breakage in these machines. At Kisan Krishi Yantra Udyog, Kanpur, India, a modified underrunner disk sheller has been developed. In this unit, the upper stationary disk is of rubber, with shore hardness of 40. This machine handles grain gently and results in less breakage.

3.5.2.3 Attrition Mill

Attrition mills of vertically or horizontally rotating stone disks or emery disks are also used to split pulses.

3.5.2.4 Elevator and Hard Surface

This combination is used by some dal millers, and it is believed that breakage is lower. Dehusked pulses are dropped from a height of about 10–15 ft on a hard surface and splits are obtained. This type of arrangement is normally used only for pigeon peas.

3.5.2.5 Impact Splitter

The machine consists of a mild steel blade impeller mounted on a shaft enclosed in a casing. The gota is fed to the machine and rotating blades throw the gota on the hard body (casing) of the unit, causing the splits. The gota is split by impact. The unit is manufactured in different sizes (see Table 5).

3.6 Dal Grading and Gota Separation Machinery

The machinery similar to the one used for cleaning and grading of pulses is used for gota separation and dal grading. A three- and four-compartment rotary sieve is used for gota separation and dal grading, respectively. Some manufacturers supply a reciprocating-type screen with 37- to 50-mm–stroke length for gota separation and dal grading.

A combination of rotary screen cleaner for raw material cleaning and reciprocating screens for gota separation and dal grading is also employed by some millers. The size of perforations vary depending on the type of grain to be processed. The specifications of the machine vary with the manufacturer and capacity of a particular unit (see Table 5). These units are built in a wooden or mild steel frame and are used for the separation of gota, splits, and brokens. The dal grader receives the product from the splitting unit and grades gota, dal in two grades, brokens, and powder.

3.7 Pulse Conditioning and Dehusking for Modern Mills

A conditioning unit consists of two sets of tempering bins, two pulse heating units, each having separate motorized blowers, air heat exchangers, grain-heating chambers, and feed hoppers. The temperature is controlled thermostatically. This entire unit is used for conditioning grain to loosen the husk.

3.8 Gota Conditioning and Splitting Unit for Modern Mills

A gota-conditioning unit consists of a motorized blower, air heat-exchanger, gota heating chamber, and thermostatic temperature control arrangement for gota heating and one blower for the aeration chamber, with a mechanized delivery outlet for aerating gota with moist air. The unit provides the conditions necessary for treating the gota for loosening the binding of the cotyledons and minimizing the breakage during splitting.

4 DEHULLING CHARACTERISTICS

The dehulling characteristic of any pulse is exhibited by the ease of dehusking and splitting. Dehusking is measured as the percentage of the original weight that has been dehulled. The splitting quality is reflected in the percentage of whole splits obtained after dehusking and splitting. The combined effect of both these factors is denoted as *dehulling effectiveness*. However, there are various definitions available for dehulling efficiency

Table 8 Dehulling Characteristics of Grain Legumes

Legume	Yield(%)[a]	Dehulling efficiency[b]	Intact seeds[c]
Soybean	88.7	0.72	91.3
Fababean	83.3	0.71	59.9
Field pea	87.3	0.61	47.6
Lentil	85.4	0.54	98.2
Kidney bean	84.2	0.51	0.0
Mung bean	74.2	0.33	18.1
Black-eyed cowpea	79.6	0.25	18.7
Brown cowpea	78.3	0.11	19.3

[a] The yield of dehulled grain when at least 90% of the hull has been removed from the seed.
[b] Dehulling efficiency = [hull removed (g/100 g seed)]/[100 − yield (100 g seed)].
[c] The weight percentage of seeds that have both cotyledons bound together after dehulling.
Source: Ref. 18.

based on the importance of the operation for a particular pulse, as well as its end use. The commonly adopted definitions follow. The dehulling characteristics of different pulses are given subsequently in Table 8.

4.1 Definitions of Dehulling Efficiency

The dehulling efficiency is defined by two expressions, dehulling effectiveness and quality of dehulling (16):

$$E_{\text{hulling}} = 100\left(1 - \frac{n_1}{n_2}\right) \tag{1}$$

where

E_{hulling} = effectiveness of dehulling
n_1, n_2 = mass of unhulled grains after and before hulling

The quality of dehulling is defined as

$$E_{\text{wk}} = \frac{k_2 - k_1}{(k_2 - k_1) + (d_2 - d_1) + (m_2 - m_1)} \tag{2}$$

where

E_{wk} = quality of dehulling
k_1, k_2 = mass of whole split kernels before and after hulling
d_1, d_2 = mass of crushed kernels before and after hulling
m_1, m_2 = mass of mealy waste before and after hulling

The overall dehulling efficiency (η) is then defined by Eq. (3):

$$\eta = E_{\text{hulling}} \times E_{\text{wk}} \tag{3}$$

The dehulling efficiency is given in terms of the following definitions (6):

1. The degree of dehusking (or dehulling), E_{dh}, more than 75% of the surface area of unsplit or split grain was without hull:

$$E_{dh} = \frac{m_1 + m_2}{M_1 + M_2} \tag{4}$$

where
m_1 and m_2 = mass of dehulled splits and unsplit grain
M_1 and M_2 = mass of total split and unsplit grain

2. Yield of dal, *Yd*, is considered in two definitions: (1) apparent dal yield, Y_{da}:

$$Y_{da} = \frac{M_1 + M_2}{M} \tag{5}$$

and true dal yield, Y_{dt}:

$$Y_{dt} = \frac{(M_1 + M_2) + (m_1 + m_2)}{M} \tag{6}$$

where m_1, m_2, M_1, and M_2 are defined as for Eq. (4); m_3 is the mass of the removed hulls, and M is the total original mass of grain.
The dehulling efficiencies are given in Ref. (23):

1. The decortication efficiency η_d, that quantifies the fraction of grains with a complete seed coat that have the seed coat have broken by the decorticator:

$$\eta_d = \frac{R_1 - R_2}{R_1} \tag{7}$$

where R_1 and R_2 are the percentage of undamaged grain with seed coat intact before and after decortication.

2. The coat removal efficiency quantifies the average amount of seed coat removed from each grain:

$$\eta_{cr} = \frac{C_1 - C_2}{C_g} \tag{8}$$

where C_1 and C_2 are mass of loose seed coat exiting and entering the decorticator, and C_g is mass of seed coat on grain entering the decorticator.

With pulses, best quality dal is produced when the milling efficiency is at its highest: that is, with minimum loss and highest quality. Scratching on abrasion of the cotyledon surface is not desirable in dal processing. Also, broken corners during splitting of grain, as sometimes happens in an under-runner disk sheller as well as a burr or plate mill are not desirable. The best quality dal is considered to be that which has no excessive scratching on the cotyledon surface and that in which the periphery of the split is sharp. The split should have a straight, flat surface that is not cupped.

As far as possible, one should obtain whole cotyledons, because even large brokens (three-fourth of cotyledons) are not liked by consumers. The optimum hydrothermal treatment is necessary to obtain good quality dal. Excessive water treatment has been reported to induce cupping in pulses, which lowers the quality drastically. The high-temperature–short time (HTST) treatment developed at CFTRI yields dal with sharp edges. However, if this process is not properly performed, it has been reported to give a charred smell to dal. Hence, proper hydrothermal treatment as well as critical adjustments of machines are very important in achieving good quality dal.

The size grading of the pulses is the most important operation in their processing. It is important because the large variations in the size of grain in commercial lots, especially in India. Hence, rotary screen graders are used to grade pulses to sizes 1 and 2. Generally, 80% of the finished product is of size 1.

ACKNOWLEDGMENTS

The authors wish to express their sincere thanks to Canadian Grain Select Co. of Eston, Bailey Brothers Seed Co. of Milden, and ADF-Saskatchewan Agriculture and Food for funding the study tour. Thanks are also due Dr. A. E. Slinkard, Dr. Bert Vandenberg, and Dr. R. T. Tyler for their advice and suggestions on survey during the tour. The help of Mr. Igor Solomon in taking measurements of physical properties of Canadian pulses is gratefully acknowledged. Our sincere thanks to all of the companies we visited in India, Iran, and Europe for their generosity in supplying useful information.

APPENDIX I

Terminologies Used in Pulse Processing

Bengal gram, Chana (India), Nokhod (Iran): chickpea
Chukki: attrition mill
Dal: split pulse grain (India)
Dehusking/decortication/dehulling: removal of the seed coat or husk from the seed
Gota: dehusked unsplit kernel
Gota machine: horizontal abrasive roller for pitting and dehusking
Kharif pulses: rainy season crop
Kori: unpolished dal
Lappeh: split pulse grain (Iran)
Masur: lentils
Moth: mothbean
Mung: mungbean
Rabi pulses: winter crop
Reel machine: rotary sieve cleaner
Splitting: splitting the whole dehusked seed into two halves (cotyledons)
Tur (Arhar): pigeon pea
Urd: black gram

APPENDIX II

Major Equipment Used in Pulse-Processing Industry

Machine: Dry destoner

Purpose: removing stones, magnetic and nonmagnetic metals, and mud balls. Both round particles and those with small diameter of up to 1.75 mm as well as heavy materials of hygroscopic structure are removed

Specifications

Capacity/h grain pulse (kg)	Overall dimensions Length (mm)	Width (mm)	Height (mm)	Air requirements (m^3/min)	Motor power vibrator (kW)
1000	1115	590	1420	20	1.1
3000	1115	1200	1675	55	2.2
6000	1115	2260	1675	110	4.4

Manufacturer: G. G. Dandekar Machine Works, Ltd. Dandekarwadi, Bhiwandi 412 302, India

Machine: Rotary flat sieves

Purpose: cleaning and grading granular material

Specifications

Capacity/h approx. (rice) (kg)	Overall dimensions[a] Length (mm)	Width (mm)	Height (mm)	Speed (rpm)	Motor power (kW)
1500	2450 (2000)	2050 (1000)	3250	140	0.75
3000	2450 (1500)	2550 (2000)	3250	140	1.10

[a] The values in parentheses are screen sizes.

Manufacturer: G. G. Dandekar Machine Works, Ltd. Dandekarwadi, Bhiwandi 412 302, India

Machine: Dal scourer (gota machine)

Purpose: pitting and dehusking of pulses

Specifications

Capacity/h approx. (dal) (kg)	Overall dimensions Length (mm)	Width (mm)	Height (mm)	Speed (rpm)	Motor power (kW)
750–950	1500	600	1200	900	3.70

Manufacturer: G. G. Dandekar Machine Works, Ltd. Dandekarwadi, Bhiwandi 412 302, India

Capacity/h approx. (dal) (kg)	Overall dimensions Length (mm)	Width (mm)	Height (mm)	Motor power (kW)
1000	1825	750	1250	5.60
1000	1400	750	1320	5.60
2000	1400	1600	1320	11.20
(Two units of 1 tonnes/h capacity are provided)				

Manufacturer: S. K. Engineering and Allied Works 5, Digiha, Station Road, Bahraich 271 801, India

Machine: Dal splitter

Purpose: splitting the dehusked pulses (gota)

Specifications

Capacity/h approx. (dal) (kg)	Overall dimensions Length (mm)	Width (mm)	Height (mm)	Speed (rpm)	Motor power (kW)
1000	1000	550	1500	300–475	0.70

Manufacturer: G. G. Dandekar Machine Works, Ltd. Dandekarwadi, Bhiwandi 412 302, India

Machine: Water–oil mixer

Purpose: application of oil–water to raw pulses

Specifications

Capacity/h approx (kg)	Overall dimensions Length (mm)	Width (mm)	Height (mm)	Speed (rpm)	Motor power (kW)
1500	2200	400	1000	100	0.70
3500	2200	600	1000	100	2.20

Manufacturer: G. G. Dandekar Machine Works, Ltd. Dandekarwadi, Bhiwandi 412 302, India

Machine: Under-runner disk sheller

Purpose: separating outer husk from pulses

Specifications

Capacity/h approx. (kg)	Overall dimensions Length (mm)	Width (mm)	Height (mm)	Speed (rpm)	Disk diam. (mm)	Motor power (kW)
1000	1220	1220	1600	250	990	1.90
2000	1475	1475	1752	176	1425	3.75

Manufacturer: G. G. Dandekar Machine Works, Ltd. Dandekarwadi, Bhiwandi 412 302, India

Machine: Batch dryer for dal

Purpose: drying moistened pulses

Specifications

Capacity approx. (kg)	Overall dimensions Length (mm)	Width (mm)	Height (mm)	Drying temp. (°C)	Fuel cons. (mm)	Motor power (kW)
1200	5500	2500	1800	60	7–8	6.00
2500	3750	3000	1900	80	10–12	8.25

Manufacturer: Thermax, Ltd. Chinchwad, Pune 411 019, India

Machine: Rotary cleaning and grading machine

Purpose: cleaning and size grading of pulses and dal

Specifications

Capacity approx. (kg)	Overall dimensions Length (mm)	Width (mm)	Height (mm)	Speed (rpm)	Motor power (kW)
[a]	2745	1015	1420	100	0.75
[a]	3655	1015	1420	100	1.10
[a]	4570	1015	1420	100	1.5
[a]	5485	1015	1420	100	1.9

[a] Capacity varies based on the application, and sieve-cleaning scrapers are also provided to avoid screen blinding.

Manufacturer: S. K. Engineering and Allied Works 5, Digiha, Station Road, Bahraich 271 801, India

Machine Gota aerator/dal dryer

Purpose: conditioning gota before splitting

Specifications

Capacity approx. (kg)	Overall dimensions Length (mm)	Width (mm)	Height (mm)	Motor power (kW)
[a]	5000	3000	6500	5.00
[a]	5000	3000	6500	10.00
[a]	5000	6000	6500	10.00

[a] Capacity varies based on the application (aerating or drying).

Manufacturer: S. K. Engineering and Allied Works 5, Digiha, Station Road, Bahraich 271 801, India

Machine: Pulse heating unit

Purpose: conditioning pulses to loosen the husk.

Specifications

Capacity approx. (kg)	Overall dimensions Length (mm)	Width (mm)	Height (mm)	Speed (rpm)	Motor power (kW)
2000	3000	900	2500	2880	5.00

Manufacturer: S. K. Engineering and Allied Works 5, Digiha, Station Road, Bahraich 271 801, India

Machine: Vertical pulse dehusking unit

Purpose: dehusking of pulses

Specifications

Capacity approx. (kg)	Overall dimensions Length (mm)	Width (mm)	Height (mm)	Motor power (kW)	Power used (kW)	Air (m^3/min)
1000–2000	1400	800	1940	15.0	9–14	15–25
2000–13000	1485	800	1930	22.5–37.0	15–35	25–30

Manufacturer: F. H. Schule Muhlenbau GmbH Dieselstrasse 5, D-21465, Reinbek, Hamburg, Germany

Machine: Splitting machine

Purpose: splitting of peas, beans, grams, and lentils

Specifications

Capacity approx. (kg)	Overall dimensions Length (mm)	Width (mm)	Height (mm)	Motor power (kW)
2000	1220	650	1110	—

Manufacturer: F. H. Schule Muhlenbau GmbH Dieselstrasse 5, D-21465, Reinbek, Hamburg, Germany

Machine: Impact dehuller

Purpose: dehulling seeds and separation of husks

Specifications

Capacity approx. (kg)	Overall dimensions Length (mm)	Width (mm)	Height (mm)	Motor power (kW)
750	1750	825	1425	5.60
1500	975	575	1050	0.75

Manufacturer: Forsberg PO Box 510, Airport Road, Thief River Falls, MN 56701

Machine: Dehuller/scarifier

Purpose: for dehulling and scarification of seeds

Specifications

Capacity approx. (kg)	Overall dimensions Length (mm)	Width (mm)	Height (mm)	Motor power (kW)
1500	1200	1050	–	3.75

Manufacturer: Forsberg PO Box 510, Airport Road, Thief River Falls, MN 56701

REFERENCES

1. Speciality crop report. Saskatchewan Agriculture and Food, Regina, Saskatchewan, 1994; pp 162.
2. Kurien PP. Postharvest technology of chickpea. 1990; 369–381.
3. Chakraverty A. Post harvest technology of cereals, pulses and oilseeds. Oxford and IBH Publishing, New Delhi, India, 1988.
4. Kadam SS, Deshpande SS, Jambhale ND. Seed structure and composition. In: CRC Handbook of World Food Legumes: Nutritional Chemistry, Processing Technology, and Utilization. Vol. 1. DK Salunkhe, SS Kadam, eds. 1989, 23–57.
5. Reichert RD, Ehiwe AOF. Variability, heritability and physicochemical studies of seed coat durability in field pea. Can J Plant Sci 1987; 67:667–674.
6. Ramakrishnaiah N, Kurien PP. Variabilities in the dehulling characteristics of pigeon pea (*Cajanus cajan* L.) cultivars. J Food Sci Technol 1983; 20:287–291.
7. Kulkarni SD. Food legume processing techniques in India. Productivity 1991; 32:247–251.
8. Kurien PP, Ramakrishnaial N, Pratap VM. A method for separation of pearled pigeon pea grains from whole grains based on differences in their bouncing properties. Res Ind 1993; 38: 77–82.

9. Milne G. Seed cleaning and grading. In: Pulse Cleaning and Processing Workshop. Sokhanasanj et al., eds. Extension Division. University of Saskatchewan. Saskatoon, 1995.
10. Siegel A, Fawcett B. Food legume processing and utilization. IRDC-TSI. International Development Research Centre, Ottawa, Canada, 1976.
11. Erskine WI, Williams PC, Nakkoul H. Splitting and dehulling lentil (*Lens culinaris*): effects of seed size and different pretreatments. J Sci Food Agric 1991, 57:77–84.
12. Ehiwe AOF, Reichert RD, Schwab DJ, Humbert ES, Mazza G. Effect of seed moisture content and temperature on the seed coat durability of field pea. Cereal Chem 1987; 64:237–239.
13. Dorrel DG. Seedcoat damage in navy beans (*Phaseolus vulgaris* L.) induced by mechanical abuse. Ph D dissertation. Michigan State University, East Lansing, MI, 1968.
14. Mazza G, Campbell CG. Influence of water activity and temperature on dehulling of buckwheat. Cereal Chem 1985; 62:31–34.
14a. Shyeh JB, Rodda ED, Nelson AI. Evaluation of new soybean dehuller. Trans ASAE 1980; 23:523–528.
15. Galloway JP. Cleaning, cracking dehulling decorticating and flaking of oil-bearing materials. J Am Oil Chem Soc 1976; 53:271–274.
16. Sahay KM, Bisht BS. Development of a small abrasive cylindrical mill for milling pulses. Int J Food Sci Technol 1988; 23:17–22.
17. Sahay KM, Bisht BS. Optimization of grades of Carborundum for best recovery of dhal in milling of different pulses. Agric Eng Today 1987; 11(6):10–11,22.
18. Reichert RD, Oomah BD, Youngs CG. Factors affecting the efficiency of abrasive-type dehulling of grain legumes investigated with a new intermediate-sized, batch dehuller. J Food Sci 1984 49:267–272.
19. Schule-Technical literature on verti sheller and splitter for pulses. 1995 FH Schule Muhlenbau GmbH, Dieselstrasse 5, D-21465, Reinbek, Hamburg, G.
20. Kurien PP, Patil BS. A hand operated small scale pulse dehusking machine for rural use. Res Ind 1989; 34:213–216.
21. Kurien PP, Shri PS, Balakrishnan. Dhal processing technology by CFTRI: a boon to pulse growers and processors. Fusion Asia 1991:36–38.
22. Separating the hulls from the groats. Res Dev Bull 1989; 99:1–4.
23. Moser DP, Fielke JM. Evaluation of grain legume decorticating machinery. Agricultural Machinery Research and Design Center, School of Manufacturing and Mechanical Engineering. University of South Australia, 1993.

15

Milling of Pulses

HAMPAPURA V. NARASIMHA, N. RAMAKRISHNAIAH, and V. M. PRATAPE

Central Food Technological Research Institute, Mysore, India

1 INTRODUCTION

Grain legumes belong to the family of Leguminosae. The word legume is derived from the word *legumen*, meaning seeds harvested in pods. *Pulses*, derived from the Latin word *pulse*, meaning pottage, is an alternative term for the edible dry, mature seeds of the leguminous plant. Leguminous seeds, generally containing low fat, are popularly referred to as ''pulses.'' Seeds having higher fat (normally termed oil seeds), although they are classified as leguminous seeds, are not included under pulses. The nomenclature of the scientific names of commonly grown pulses is given in the Glossary (see Sec. 6.1).

Pulses are a major source of proteins in the diets of populations who are dependent mainly on cereals as staple foods around the world. Economically, they provide reasonably good quality protein at only a fraction of the cost of animal proteins. In addition, they supplement certain deficient amino acids (such as lysine) of the cereal proteins. The intensive efforts that are being made to increase the agricultural production of pulses are not yielding the desired results. This has left a gap between the availability and requirements for these grains, particularly in the developing countries. Therefore, every effort is needed to conserve them during postharvest handling and processing. In most parts of the world, pulses are traditionally consumed either in the whole or in the dehulled split form as soft-cooked products. The dehulled splits from pulses are known as *dhal* [*dal*] on the Indian subcontinent. While more than 75% of the pulses produced in India are marketed and consumed as dhals (dehulled split form), in most African countries they are dehulled at home just before consumption. Dehulling, therefore, is an important primary processing activity in India and many other pulse-consuming countries. Proper understanding of pulse milling is essential for reducing both qualitative and quantitative losses.

Table 1 Production of Pulses (World)

	Area (1000 ha)				Production (1000 mt)			
	1993	1994	1995	1996	1993	1994	1995	1996
World	66,767	69,321	70,317	71,034	56,886	57,782	55,997	56,774
Africa	12,075	12,242	12,657	14,200	6,608	6,933	7,267	7,551
NC America	4,469	5,036	5,092	5,021	4,746	5,697	5,797	5,541
South America	4,851	6,594	6,097	6,085	3,311	4,266	3,878	3,770
Asia	37,269	37,543	38,890	38,394	36,150	26,977	27,481	28,222
Europe	2,598	2,448	2,381	5,267	7,345	6,976	5,811	9,380
Oceania	1,871	1,973	2,082	2,006	2,520	1,287	2,464	2,267
Ethiopia	976	1,077	1,242	1,242	872	954	1,108	1,108
Nigeria	2,070	2,120	2,220	3,420	1,626	1,721	1,850	1,700
Canada	892	1,175	1,248	1,120	1,460	2,077	2,090	1,852
Mexico	1,991	2,182	2,151	2,307	1,436	1,509	1,428	1,688
Brazil	3,978	5,566	5,060	5,082	2,505	3,396	2,941	2,862
China	4,538	4,553	4,553	3,153	5,937	6,113	5,511	4,979
India	23,605	24,045	24,925	25,604	13,517	14,317	14,820	15,414
Myanmar	1,428	1,614	1,864	1,853	894	1,063	1,228	1,289
Turkey	1,977	1,881	1,887	1,895	1,946	1,679	1,824	1,831
France	754	689	593	553	3,749	3,850	2,792	2,636
Ukraine	1,237	1,191	1,095	1,075	2,898	2,636	1,518	1,845
Australia	1,835	1,937	2,045	1,958	2,446	216	2,381	2,186

Source: Ref. 6.

1.1 Production of Pulses

About 56 million tonnes of pulses are produced per annum throughout the world, and the Asian continent accounts for nearly 50% of the total pulse production (Table 1). Almost all pulses are grown in Asia; however, the major ones are chickpea, pigeon pea, cowpea, lentil, beans, green gram, and black gram. The production of peas is higher in Europe, and the major pulses on the American continent are peas and beans. Cowpeas are grown mainly in Africa along with beans, whereas pigeon pea, chickpea, green gram, and black gram are common pulses of Asia. On the Australian continent chickpea is the major pulse grown. The type of pulse grown on the continent depends on the agroclimatic conditions there. Major pulse-growing areas around the world are shown in Fig. 1. The production data indicates that more than 70% of the total world production of pulses is accounted for by the developing countries. They are mostly grown as rain-fed crops on unirrigated lands, normally with lower inputs and management practices. The production of pulses has remained almost static for the last three decades, partly owing to lack of increase in acreage and productivity. As a result of continued population increases and the consequent growing demand, coupled with static pulse production, the per capita availability of pulses has fallen significantly. In addition, pulses suffer considerable losses during transportation, postharvest handling, and storage. These losses coupled with stagnated production of pulses in most developing countries suggest a very strong need for application of sound methods for every step in the postharvest handling and processing. With this in view, an attempt has been made here to focus on the developments in milling of pulses that could be used as guidelines for efficient processing.

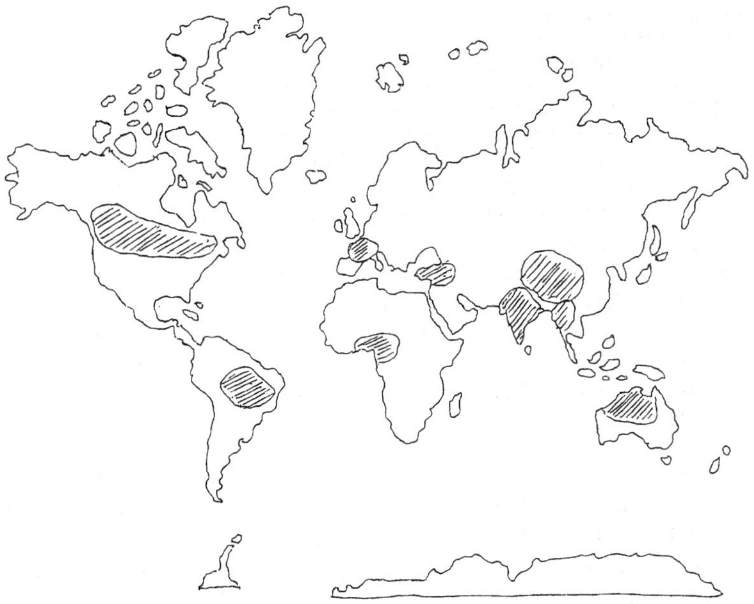

Fig. 1 Pulse production areas in the world.

2 MILLING METHODOLOGIES

Milling of pulses is practiced widely in Asia and Africa; it gives dehulled cotyledons with better appearance, texture, and cooking qualities. Pulse milling constitutes two major steps: loosening of the husk, followed by removal of the loosened husk in suitable milling machinery.

The first step is commonly referred to as *premilling treatment*, whereas the second is referred to as *milling* or *dehulling*. The first step of loosening the husk is achieved either by a wet or dry method. In the wet method, the grains are soaked in water for a few hours, drained, left in heaps (usually overnight), and very commonly dried in the sun. In the dry method, grains are mixed with a small amount of oil, usually after scarification of the husk. This scarification of the husk is commonly called *pitting* and is done to facilitate the oil penetration between the husk and the cotyledons. Oil-treated grains are heaped overnight and then dried in the sun for 2–5 days, with intermittent water spraying and mixing. In both of the premilling treatments, adherence of the husk to the cotyledon weakens and, consequently, its removal becomes easy.

The loosened seed coats of the pretreated pulses are removed in the subsequent operation of milling (Fig. 2). For this purpose different machines are used, depending on the type of pulse and scale of operation. Pulse milling is practiced at different levels: (a) home-scale, (b) cottage-scale, and (c) large-scale, and machines such as pestle and mortar or hand-driven disk mills (popularly known in India as *chakki*) are used in home-scale operation. In cottage-scale, a motorized plate mill, under-runner disk sheller, horizontal flour mill, or hullers (Figs. 3–6) are used for the milling; whereas emery-coated roller machines are used mainly in large-scale operations. Dry premilling treatment is preferred for larger-scale operation, although it is time-consuming and laborious.

The oldest and most common method of milling pulses at the home level is to pound

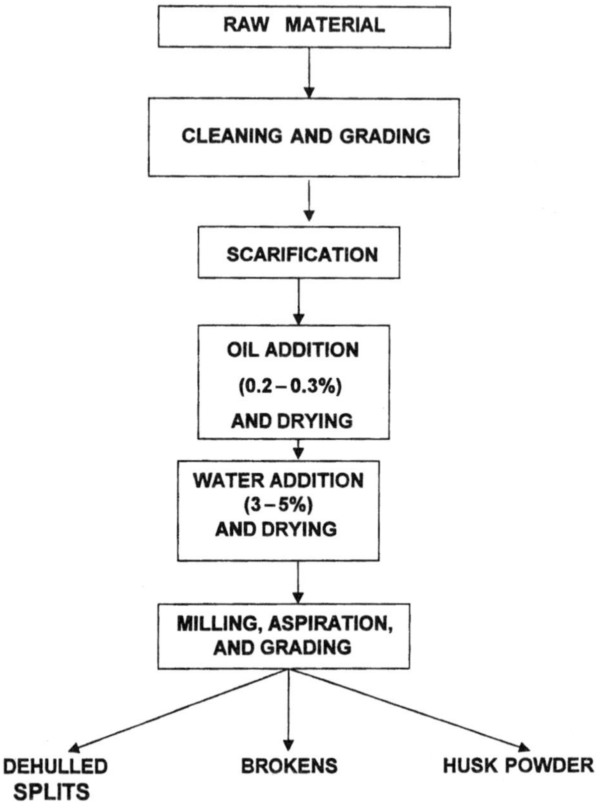

Fig. 2 Flowchart for processing of pulses.

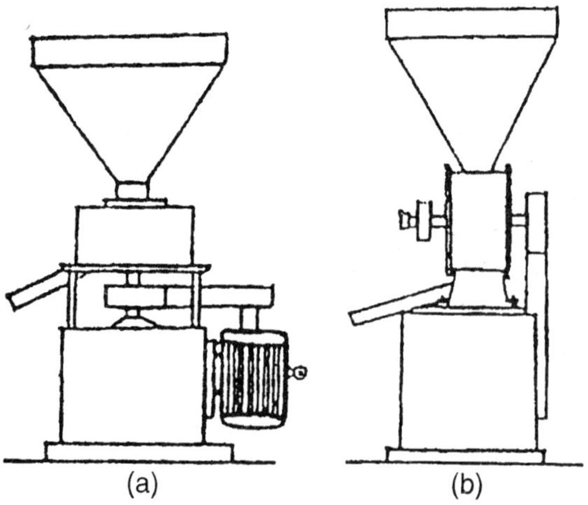

Fig. 3 Motorized stone grinders (chakkies): (a) horizontal, (b) vertical.

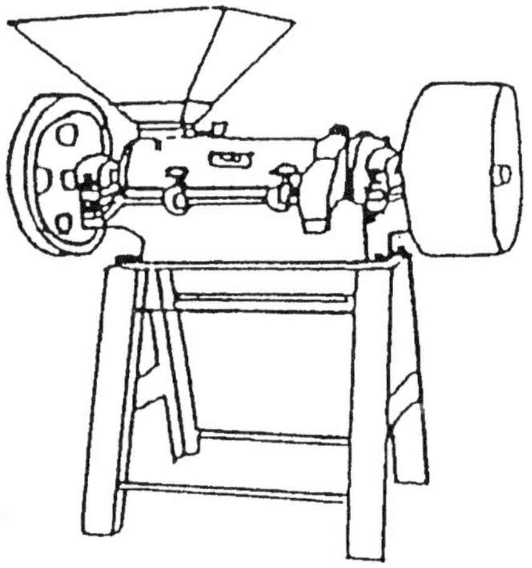

Fig. 4 Huller.

them in a pestle and mortar or in a stone grinder after drying the grains in the sun or after mixing with a small amount of water. The husk is then winnowed off to obtain clean cotyledons. This method is still in vogue in the rural sector (home level) in a few African countries. The method of dehulling cowpeas in West Africa depends on the variety and the dish for which it is required. Cowpea grains are soaked in water for a few hours and the husk is rubbed off or lightly pounded and separated by washing when it is required to make cooked gravy or a deep-fried snack product, popularly known as Akara in Nigeria. Whenever cowpea flour is required, dehulling is accomplished by the dry method. The steeping process is also practiced in several Southeast Asian countries.

Generally wet-processing methods are practiced in India, particularly in the small-scale and in rural sectors for producing small quantities of dhals to meet the household

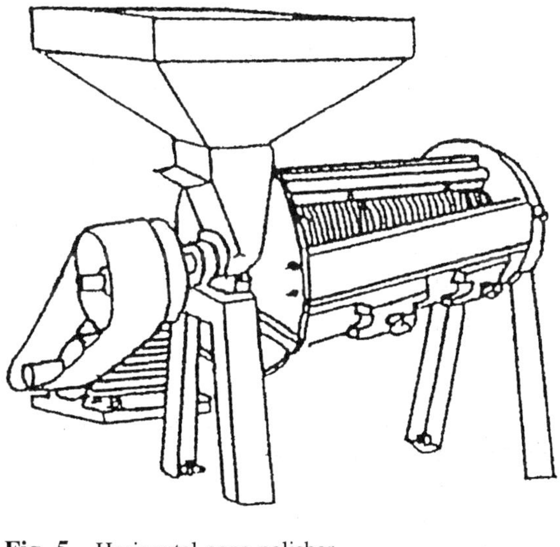

Fig. 5 Horizontal cone polisher.

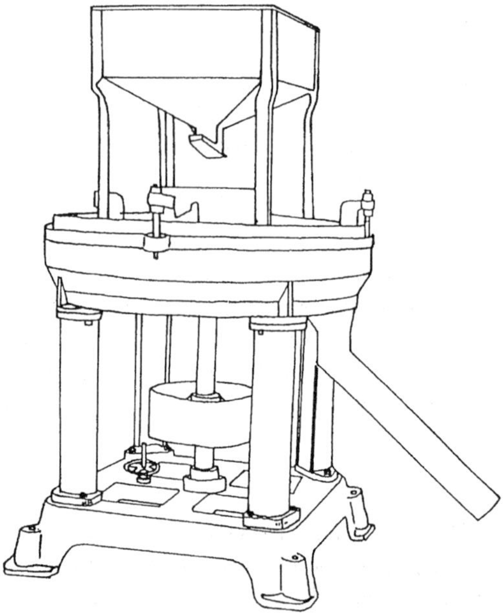

Fig. 6 Under-runner disk sheller.

requirements. Large-scale pulse mills in India, always prefer dry-processing methods. With the advent of the organized large-scale–milling systems in the last two decades, home-level operation and even small-scale pulse milling is becoming obsolete because it cannot compete with the product produced by the large-scale mills in the quality and yield of dhals.

Pulses can be grouped in two categories as ''easy-to-mill'' and ''difficult-to-mill'' depending on their milling characteristics. Because the seed coat (husk) of the pulses is tightly adhered to the cotyledons through a layer of gum, milling of pulses is different from that of milling of other grains such as paddy, for which husk loosely envelops the grain. Apart from this, the extent of husk adherence (i.e., the quantity and type of gum between the seed coat and cotyledons) determines whether the pulse is easy or difficult to mill. Generally easy-to-mill pulses such as chickpea, peas, or lentils are processed in small-scale mills, whereas all difficult-to-mill pulses such as red gram, green gram, black gram, and beans, are processed in large-scale pulse mills using dry premilling treatments. A general process for milling of pulses is indicated in the flowchart (see Fig. 2). However, variations exist in premilling treatments and milling machines used for different pulses, as each pulse has a specific characteristic end use. Hence, the milling schedules adopted for different pulses are outlined in the following subsections (1):

2.1.1 Pigeon Pea

This pulse poses the greatest difficulty in milling because the husk adhers tightly to the cotyledons. Generally, only the dry method is followed throughout the Indian subcontinent for milling of this pulse. Cleaned and size-graded grains are pitted in smooth roller machines, smeared with varying amounts (0.2–0.5%) of oil (any edible oil), tempered for about 12–24 h, sun-dried for 1–3 days, followed by spraying with water (2–6%), thoroughly mixed, heaped overnight, and then passed through the rollers for dehusking. This type of operation is repeated three or four times. After each dehusking operation, the husk, powder, and brokens are separated from dehusked split pulse (dhal). *Dehusked splits* obtained in this operation are considered as 'second grade' because their edges are not sharp and are usually rounded-off by scouring. The mixture of dehusked and unhusked grains obtained during processing (known as *Kappi*) is again mixed with water, as described earlier, equilibrated, and sun-dried. The sun-dried grains are either passed through the roller machine or split in a horizontal or vertical grinder or by using an impact-type machine. The dehusked splits thus obtained are considered as a ''first grade,'' because they would not have any chipped edges and would have a better consumer appeal. Quite often both first- and second-grade dehusked splits are mixed and marketed. The yield varies from 70 to 75% depending on the variety and the method followed. In large-scale mills, sun-drying is being replaced gradually with batch-type bin driers, as a result of which they are able to continue work throughout the year.

2.1.2 Chickpea

The chickpea is comparatively easy to mill. The cleaned and size-graded grains are pitted in smooth rollers at low peripheral speed. After pitting, the grains are mixed with about 5–10% water in a screw–conveyer-type mixer and heaped for a few hours to allow the water to seep in. The wetted grains are sun-dried for 1 or 2 days. The dried pulse is then passed through either a horizontal or vertical stone–emery grinder, where dehusking and splitting takes place simultaneously. The dehusked splits are separated from the husk and brokens with an appropriate aspirator and sifter; the remaining unhusked grains are dehulled by repeating the foregoing operation until all the grains are dehulled.

2.1.3 Black Gram

The cleaned and size-graded grains are scarified and pitted using emery rollers in two or three passes, so that complete scarification is effected. After each operation the husk and powder are separated. The scarified grains are then mixed with about 0.5% oil and heaped overnight for absorption. The grains are then sun-dried for 2 days. In some mills mechanical dryers are used. After drying, the grains are given a spray of water (2–5%), equilibrated, and passed through the rollers twice for dehusking. The splits obtained are termed second-grade dhal. The dehusked whole grains are either marketed as such at certain places in India or passed through a Burr mill for splitting. These splits are considered first-grade. The splits are ''polished'' with soapstone powder at the final stages. This is believed to give luster and enhance their market value.

2.1.4 Green Gram

The husk of green gram is thin, soft, and slippery. Although the husk tightly adhers to the grain surface, the two cotyledons are loosely attached and separate out easily. Hence, splitting occurs even before good dehusking can be effected. During the dehusking operation, there is also scouring of the cotyledons, resulting in large losses in the form of brokens and powder. The method generally followed is pitting, oiling (0.2–0.5%), and sun-drying, followed by dehulling and splitting in roller machines.

2.1.5 Peas and Lentil

The milling of these pulses is fairly easy as found for chickpea. General practice involves initial scouring, application of water, heaping, and sun-drying, followed by dehusking and

Fig. 7 Traditional dhal mill.

splitting in roller machines. After separating the splits, unhusked grains are treated again for a second time as in the first pass, and the process is repeated until all grains are dehusked and split.

2.2 Pulse-Milling Machinery (Traditional)

Pulse milling being an age-old industry in Asia, particularly in Bangladesh and the Indian subcontinent, simple mechanical devices such as pestle and mortar or a leg-operated pestle and mortar (*dhenki*) were in use about three decades ago, particularly for home-scale operation. In the community- or village-level operation, hand-operated wooden or stone grinders are still in use. With the progress in mechanical and electrical appliances, power-operated grinders (either horizontal or vertical) have come into existence for small commercial-level operations.

Large-scale commercial mills normally employ emery-coated horizontal rollers (tapered or cylindrical) along with appropriate separation systems (see Fig. 7). Many such mills are in operation in India and other Southeast Asian countries. Nearly 7000 such mills are working in India alone. However, manufacture of milling machinery is not an organized industry, but is of a localized nature. Normally, the mills are fabricated by local artisans in the areas. The power requirement, capacity, operating parameters, and specifications of the machinery vary from place to place, often decided by local artisans who have experience or expertise in this trade. But many fabricators depend on the trial-and-error approach, which leads to wastage of resources and lower yield of product.

2.3 Pulse-Milling Machinery: Current Status

Conventional pulse milling involves many unit operations, such as cleaning the raw material, size-grading, scarification of husk (pitting), oil mixing, water mixing, drying, dehulling, splitting, aspiration of husk, separation of brokens and splits, and finally, polishing. The different types of machines used for each unit operation are discussed here in.

2.3.1 Two Types of Mills Are in Vogue in India

One, which is very common in northern parts of India, is characterized by a two-storied setup similar to the roller flour mills. This type saves space and the number of elevators required. The second, which is more common in south India, is essentially a single-storied type, in which the material moves horizontally from one unit to the other through bucket elevators. In this system more elevators and space are required.

2.3.2 Cleaning and Size-Grading Systems

Pulses need to be cleaned of dust, dirt, immature grains, and other extraneous materials. They have to be graded according to size. These operations are carried out by two types of machines: rotary- or reel-type sieve-cleaners, and reciprocating sieve cleaners. The flat, reciprocating sieve-cleaners are extensively used both by large- and small-scale–milling units. However, this type is gradually being replaced by rotary-type cleaners in large mills.

2.3.3 Rotary-Type Screen Cleaner

The construction of a rotary-type screen cleaner is similar to one used in quarrying for grading of different-sized stones. It consists mainly of four compartments of screens fitted on a shaft. The machine is usually placed on a sloped platform and operates at low speed (18–30 rpm) for achieving better performance. The body of the machine is generally constructed of wood or mild steel. Advantages of this type of cleaners are lower dust and noise pollution and low-maintenance cost.

2.3.4 Reciprocating Screen Cleaner

The reciprocating-type cleaner consists of two or three compartments of different-sized screens (depending on the type of pulse being milled) and also a suction arrangement to remove dust and other light particles. Normally pulses are graded into two or three different sizes. Screens would usually be at an angle of about 5 degrees, with an amplitude of close to 15–20 mm; the speed is about 300–350 rpm. The size of the reciprocating cleaner varies from 150 × 75 cm to 240 × 90 cm depending on the required capacity.

2.3.5 Pitting and Dehulling Machines

These machines consist of emery-coated rollers, either cylindrical or slightly tapered, rotating in a cage of a perforated sieve (Fig. 8). Cylindrical rollers are mounted at an incline, whereas the tapered rollers are mounted perfectly horizontal. These arrangements help in easy movement of the pulse. The size of different cylindrical rollers varies from 75 × 25 cm to 90 × 35 cm. The tapered rollers come in different sizes, such as 17.5 × 20 × 60 cm to 35.0 × 45.0 × 90 cm. The annular gap between the roller and the wire mesh screen varies from 2 to 4 cm depending on the type and size of the pulse used. The power requirement, capacity, roller size, and speed of the rollers are all interrelated; hence, they vary from fabricator to fabricator. The inlet and outlet of the roller machine can be adjusted for regulation of flow rate and retention time.

The granular size of the emery grits used depends on the type of the pulse milled. Normally, it varies from 14 to 16 to 36 to 40 mesh (BSS). Generally, fine-grade emery is used for pitting of almost all pulses. Rough emery (14–16 mesh) is used for dehusking of pulses such as pigeon pea and chickpea, whereas fine emery (36–40 mesh) is used for

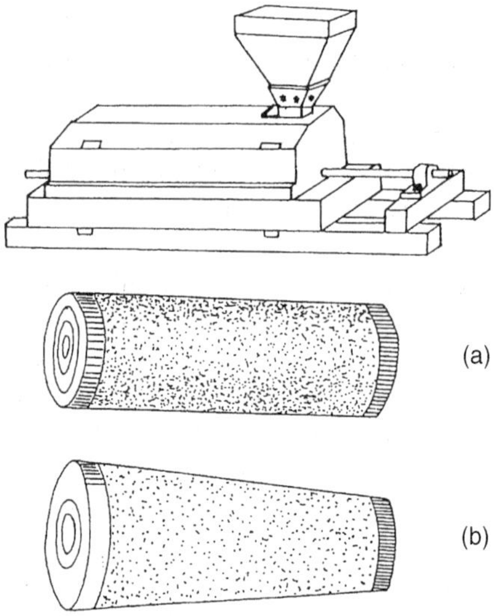

Fig. 8 Abrasive roller machine for milling of pulses with (a) cylindrical, and (b) tapered roller.

dehusking pulses such as black gram, green gram, and lentils. Normally, pitting is done at a lower peripheral speed (610–670 m/min), whereas for dehusking, particularly of pigeon pea, a high peripheral speed of 850–975 m/min is employed. Other pulses, such as black gram, green gram, and chickpea, are generally dehusked at a lower peripheral speed of 610–670 m/min.

2.3.6 Oiling and Watering systems

These systems are essentially screw- or paddle-type mixers with conveyor units. The screw is slowly rotated (50–70 rpm) to achieve proper mixing of oil–water with the grain. The length and width of the conveyors range between 1500 to 2500 and 200 to 300 mm, respectively.

2.3.7 Elevators

To convey grains and other millstreams from one unit operation to another, bucket elevators that are made of either wood or steel are normally employed. Capacity, number, and height of the elevators depend on the type of the mill.

2.3.8 Drying Equipment

Drying is an essential premilling step in the pulse-milling industry. Generally, sun-drying is performed when large drying yards are available, as in rural areas. When space is short, rooftops are conveniently used for the purpose. Because sun-drying depends on the climatic conditions, mills cannot run throughout the year. A few mills, therefore, have switched over to mechanical hot-air–drying systems. Most of the dryers in use are of the batch-type bin drier, with a holding capacity of 1–3 tonnes. These dryers use diesel, furnace oil, dry farm wastes, and groundnut husk as fuel.

2.3.9 Splitting Machines

The final product of the pulse-milling industry is the dehusked split cotyledon, popularly known as dhal in milling sector and consumers in the Indian subcontinent. During processing a mixture of dehusked and unhusked grains is obtained. Fully dehusked, unsplit grains are known in the industry as *gota* and are split in two after water treatment and sun-drying. In some pulses such as black gram, the gota is split without any treatment. For splitting, various machines similar to the under-runner disk (URD) shellers, vertical or horizontal attrition mills, roller machines, or impact machines are used. In a few mills, just dropping the dehusked whole grains on a hard surface from the discharge end of an elevator is also adapted for splitting.

The splitting roller is generally similar to the dehusking roller. However, the size of the emery grits (which reflects the smoothness or roughness of rollers) varies from pulse to pulse. An impact-type machine is very commonly used for splitting the dehusked grains although, in some places in India, only roller machines are used. Vertical or horizontal grinders or URD shelters are generally used for splitting chickpea, whereas gota from black gram is split using attrition mills.

2.3.10 Aspiration Systems

A box-type aspirator with a suction fan is generally used for separation of husk and powder. Brokens are separated by appropriate sieves, which are mixed later with husk and powder and sold as cattle feed. Cyclone-type separating systems are being introduced in a few pulse mills in India for improving efficiency and to reduce dust pollution.

2.3.11 Splits and Whole Grain Separators

These machines are similar to reciprocating-type screen graders and are generally employed for separation of splits from dehusked whole grains. An appropriate combination of perforated and slotted sieves is used for the separation of brokens, splits, and whole grains.

2.3.12 Polishing Systems

Polishing is the final treatment and is an optional operation in the pulse-milling industry before packing. This is done to provide a luster and improve the consumer appeal. Usually, a screw conveyor, with or without nylon ropes wound on to the shafts and flights, is used for this purpose. Depending on the consumer need, different polishing materials, such as water, oil, or soapstone powder are applied on to the split surface.

3 EVOLUTION OF MILLING MACHINERY AROUND THE WORLD IN THE LAST TWO DECADES

Traditional methods of milling pulses are generally laborious, wasteful, and dependent on climatic conditions. Hence, efforts have been made worldwide to improve them or develop better technologies. The Central Food Technological Research Institute (CFTRI), in Mysore, India, a pioneering research organization in postharvest food processing, has developed an improved method for milling almost all the pulses commonly consumed in the dehusked split form (11) (Fig. 9). This method is independent of the influence of climatic conditions and gives a higher yield in less time and cost of processing. The method essentially consists of a ''conditioning'' technique for effective loosening of the husk (premilling treatment) coupled with an improved dehusking roller machine (pearler) for efficient removal of the husk from the ''conditioned pulse.'' Cleaned and size-graded pulse is exposed to heated air at a specific temperature for a predetermined time in a specially designed conditioning chamber by a countercurrent through-flow technique. The

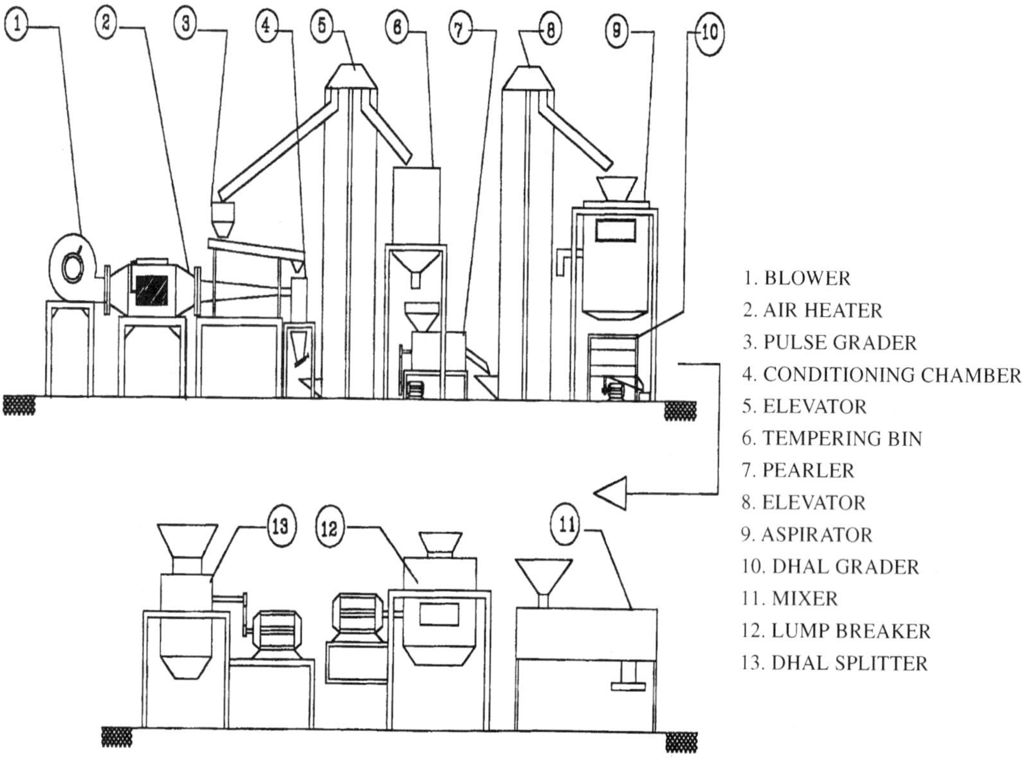

Fig. 9 Modern dhal mill developed by CFTRI, Mysore, India.

heated pulse is equilibrated to a critical moisture level (varying from pulse to pulse and also among different varieties of the same pulse) with gradual aeration in specially designed tempering bins. The husk becomes loosened by this technique and it is removed in an improved abrasion-type pearler. The pearler consists of an emery-coated roller, rotating inside a wire mesh cage. The clearance between the roller and the cage can be adjusted, depending on the size and shape of the pulse. Whole dehusked pulse (called *gota*) is humidified in an LSU-type aerator by passing humid air, and then split in an impact-type splitting machine. The process originally developed for red gram (Fig. 10) is suitably adapted for other common pulses. Several commercial mills are working in India and abroad based on the foregoing technology.

CFTRI, Mysore (India) has also developed (a) a hand-operated pulse dehusker (Fig. 11) and (b) a power-operated mini dhal mill (Fig. 12) to help the very small and rural-level processors (15). These machines work on the principle of abrasion and perform the operation of dehusking and splitting simultaneously. The units are suitable for dehulling bolder pulses such as chickpea, pigeon pea, peas, and soybeans. Traditional wet premilling treatment with suitable modifications could be adapted for milling of these pulses in these improved machines. These units have already started functioning in India and also have been included under various government schemes for countrywide propagation, with an objective of improvement in agrobased rural industries.

In Punjabrao Agriculture University at Akola (Maharashtra), India, a small mill has been developed for processing pigeon pea, green gram, and black gram. The unit is run

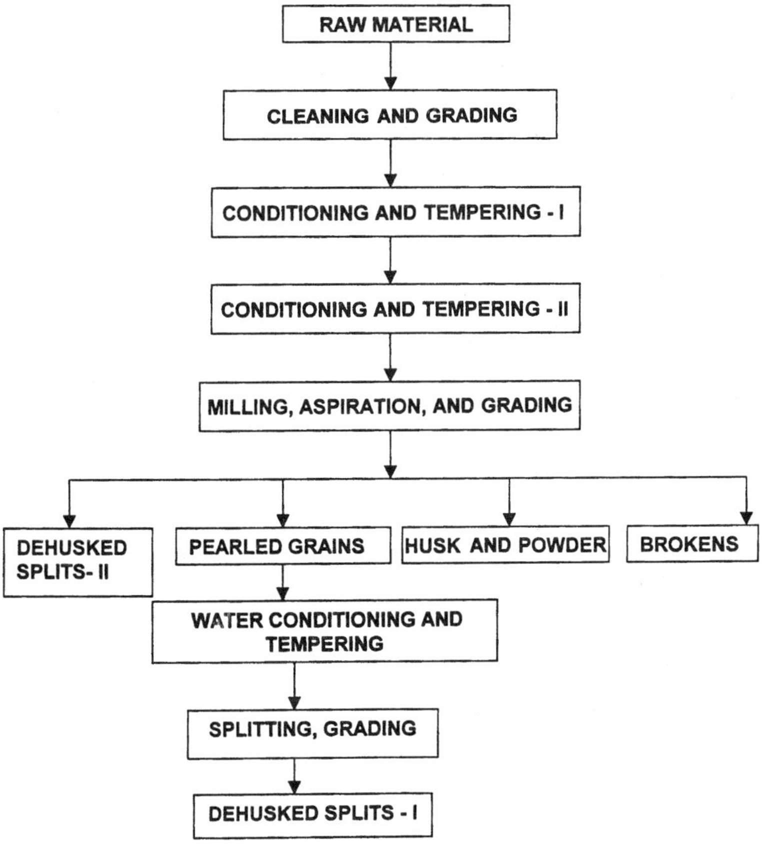

Fig. 10 Flowchart for milling of pigeon peas.

by a 2.5-horsepower (hp) single-phase motor. This mill consists of four components for carrying out unit operations such as dehulling, sieving, aspiration, and grading. The unit is reported to be suitable for rural areas because the capacity of the machine is about 125 kg/h (23).

Several machines developed for processing cereals have also been useful for milling pulses after some modifications. A few illustrations of these are (a) an under-runner disk sheller has been modified by a firm in India (Kissan Krishi Yantra Udyog, Kanpur, India) by replacing the top disk with rubber. The machine has a capacity of about 150 kg/h and is run by 3-hp motor. (b) One other firm (M/s. Tekade Industries, Amaravati, Maharashtra, India) has used a combination of cone- and roller-type units for milling pulses. (c) Indian Institute for Pulses Research at Kanpur, India has developed a disk-type vertical grinder–sheller for milling common pulses. (d) Small-scale pulse-dehulling machines have also been developed by many other Indian institutions such as Post Harvest Technology centers of the Indian Council of Agricultural Research, Delhi (India), and the Universities of Agriculture and Technology and Central Institute of Agriculture Engineering, Bhopal (India). However, these units are yet to become popular for milling of pulses across the country. Many of these developments may find appropriate use in a few Asian and African countries (30).

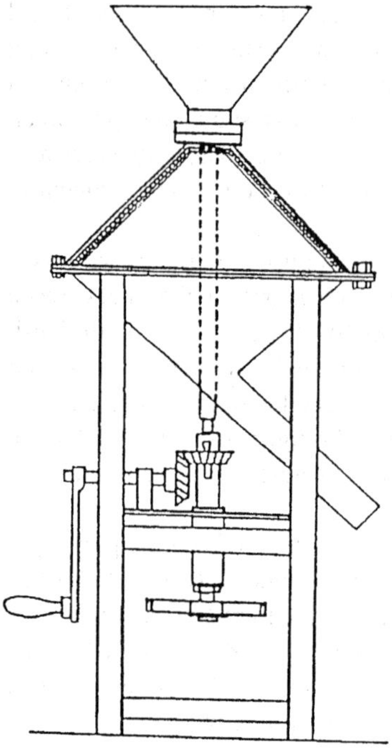

Fig. 11 Hand-operated pulse dehusker (CFTRI).

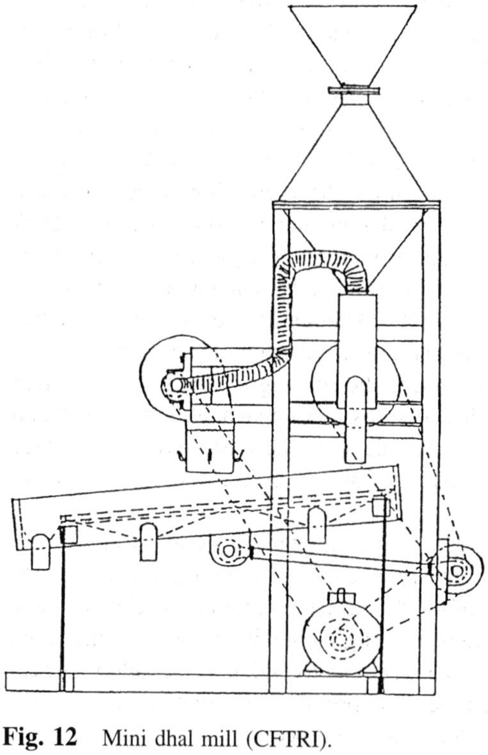

Fig. 12 Mini dhal mill (CFTRI).

An abrasive carborandum roller cylindrical mill has been developed to dehusk and split pulses (37,38). The capacity of this machine is 100 kg/h and is suitable for processing pigeon pea, green gram, and black gram. The mechanism consists of inner Carborundum grit-coated abrasive roller rotating inside an outer screen cage with a standard gap between the roller and the outer cage. The milling efficiency is reported to be 88%, with a recovery of 74–75%. The machine could be operated even by unskilled workers and is particularly suitable for rural areas.

A Hill mill unit has been successfully used in Prairie Regional Laboratory, Saskatoon, Saskachewan, Canada for dehulling of a few pulses such as brown Nigerian cowpea (34). The unit consists of a number of Carborundum disks fixed parallel on a shaft. Husk is successively removed by an abrasive action of the rotating disks, the amount removed being dependent on the throughput and retention time. An acceptable product would necessitate a removal of 14–27% polishing from cowpea at a throughput of 100–450 kg/h. This unit has been introduced for operation mainly in processing cowpea, particularly in the rural areas of some African countries (41).

A laboratory-type batch dehuller capable of dehulling 3–8 kg of legume grains per hour has been developed (35). This unit consists of abrasive wheels (10 in. [25 cm] in diameter) mounted on a horizontal shaft, and the grains are dehulled by abrasion provided by the wheels. A yield of 74–89% has been reported for at least 90%-dehulled grains when the unit was successfully applied for eight types of legumes. Abrasive-type disk hullers have been reported to be in use in Africa for milling various pulses.

3.1 Developments in Premilling Treatment

Pulse milling, as has already been indicated, usually consists of two operations (a) loosening of husk and (b) its removal. For loosening of the husk, pulses are treated in one of two ways before milling. It may be wet treatment (soaking in water for specified time followed by sun-drying), or dry treatment consisting of oil addition, water addition, and sun-drying. These treatments are traditionally followed to loosen the husk from the cotyledons. Various other pretreatments have been attempted by different research groups. For instance, vinegar (*sirka*) in place of vegetable oil in the dry premilling process and also use of other chemicals, such as sodium hydroxide and ammonia, for loosening of the husk have been reported to result in considerable improvements in the yield (7,44). Attempts have also been made to improve the dehulling of pigeon pea by soaking in sodium bicarbonate solutions of different concentrations (4–8%). Although the dehusking efficiency improved to 94% by this treatment, nutritional quality was reported to be reduced because of leaching of sugars, proteins, and electrolytes into the soaking medium (39,55). Apart from this, problems of discoloration of dehusked splits and nonavailability of husk and powder for cattle feed because of the residual chemicals in the husk may be envisaged after such pretreatments. These need to be closely examined before endorsing them for commercial application.

Recently, enzymatic pretreatment has also been attempted by Verma et al. (27) for dehulling of pigeon pea. Enzyme derived from the species *Aspergillus fumigatus* NCIM 902, has been used, and a hulling efficiency of 88.9% has been reported. A relatively new approach of steaming and equilibration of grains in airtight containers has been attempted for pretreatment of pigeon pea on laboratory scale (42). Although considerable improvement in the extent of dehulling has been reported, the problem of discoloration of a few oversteamed grains has been observed as a defect of the process.

3.2 Determination of Pulses' Milling Quality

Many times, several pulse samples have to be milled in the laboratory to determine their milling quality. This is especially required for breeders samples or for evaluation of a large number of market samples. Laboratory test-milling devices, generally used for cereals, such as (a) Carborundum disk rotating inside a metal sieve or wire mesh cage, or (b) a cone polisher used for milling cereals (e.g., Satake grain testing mill, Strong-Scott, Olmia Laboratory model cone polisher) are also being used for milling of pulse samples with certain modifications wherever required. A laboratory model barley pearler has been used by several research workers to determine the milling quality of pulse samples. A small-scale laboratory model of a tangential abrasive dehulling device (TADD) has also been developed (36). In this unit, abrasion is provided by a horizontally mounted rotating disk above which and in close proximity are vertically mounted, 10–12 bottomless sample cups. As the disk rotates, grains roll freely in the cups and are dehulled as they come in contact with the abrasive surface. Up to 12 samples can be milled simultaneously in this device.

The extent of milling is measured in terms of hulling efficiency (E_H) and milling efficiency (E_M). Whereas hulling efficiency is merely a measure of the extent of dehulling, milling efficiency (E_M) takes into account both dehulling and yield of finished product (dehulled split and unsplit grains) during milling. Several researchers have attempted to measure the hulling efficiency and milling efficiency taking various factors such as extent of husk removed, weight of unhulled and dehulled grains, extent of brokens and powder formation under defined milling conditions (27,36,37,39).

Ramakrishnaiah and Kurien (28) measured the milling efficiency (E_M) using the following expression:

$$E_M = \frac{\text{actual yield of dehulled grains}}{\text{theoretical yield}} \times \text{degree of dehulling}$$

Where the

actual yield is recovery of dehusked grains from the machine (actual) (%).
theoretical yield is maximum yield of dehusked grains that could be obtained (%) (i.e., wt. of whole grains − wt. of husk).

$$\text{Degree of dehulling (\%)} = \frac{\text{wt. of dehulled grains}}{\text{total wt. of sample milled}} \times 100$$

Whereas milling efficiency (E_M) is a measure of the performance of the dehulling unit, degree of dehulling (E_H) is more a grain character and is a measure of ease with which the grain could be dehulled under a given set of conditions. The objective of any dehulling unit is to achieve complete (100%) dehulling. In practice, it is difficult to achieve this, and generally maximum achievable milling efficiency is about 91%. However, the milling efficiency for a machine should be at least 77% to obtain an acceptable product. Dehusked splits with less than 97% degree of dehusking are normally unacceptable in the market.

Mandhyan and Jain (16) developed second-order regression equations and optimized clearance and speed of emery roller to increase the recovery of dehusked splits from pigeon peas. They reported 10.0 and 17.3% increase in the recovery from two varieties of pigeon pea. Phirke et al. (26) also studied dehulling index for two pigeon pea varieties using response surface modeling and TADD, with different pretreatments and conditions. They reported that hot milling (heat treatment after urea application) reduced the dehulling time by 50% and enhanced dehulling efficiency by 6–8% at optimum rotor speed of 12.6 m/s in the TADD having a dehulling disk with 30-mesh emery grade. More fundamental research is called for to understand the problems of variation in milling quality of different pulses and prediction of dehulling quality.

4 SEED CHARACTERISTICS THAT AFFECT MILLING

Several seed characteristics affect the dehulling efficiency (e.g., size and shape of the grains, husk content and its thickness, adherence of the husk to the cotyledons, and moisture content). Interaction of premilling treatments and seed characteristics play an important role in determining the dehulling quality. Selection of premilling treatment also depends on the seed characteristics of the grain.

4.1 Nature of Seed Coat

In pulses, a tough, single seed coat tightly envelopes the cotyledons. In some pulses, such as cowpea, green gram, and lentil, the seed coat is thin, forming about 5–10% of the grains, whereas in other pulses, such as chickpea and pigeon pea, it is thicker and constitute about 12–15% of the pulse; however one type of chickpea (*kabuli*) has a lower seed coat content.

As shown in Table 2, mean seed coat content ranges between 4.9 and 14.4% for different pulses, indicating a large variability. Variation occurs even within several varieties of the same pulse. This would significantly affect the expected yields of dehulled grains. The theoretical yields of dehulled grain, are determined by subtracting the seed coat content from the total seed mass.

4.2 Physical Characteristics of Grains

Dehulling characteristics are governed, to some extent, by the seed morphology, which varies immensely among legumes (35,37). Differences in the cell arrangements of the pulses' seed coats influence their dehulling characteristics. The seed coat in cowpea consists of a highly organized palisade cell structure. Sefa-Dedeh and Stanley (40) suggested

Table 2 Variability in Seed Coat Content of Different Pulses

Pulse	Number of genotypes	Range (%)	Mean (%)	Ref.
Chickpea (brown)	21	9.7–17.3	14.2	10
Chickpea (white)	19	3.7–7.0	4.9	10
Pigeon pea	22	12.6–17.2	14.4	40a
Green gram	24	7.4–11.4	8.8	3
Black gram	5	8.9–11.6	10.4	54
Lentil	6	7.0–8.0	7.2	56
Cowpea	3	9.0–11.5	10.5	11
Kidney beans	5	9.0–11.2	9.7	11
Horse gram	2	11.2–13.4	12.5	11

that cowpea varieties with thick, smooth seed coats (highly organized palisade cells) dehulled more satisfactorily than those with rough seed coats. In white-husked (*kabuli*) chickpeas, the outermost layer (epidermis) develops into a uniseriate palisade layer without thickening of the cell wall, whereas in brown-husked (*desi*) chickpeas, it develops into a multiseriate palisade layer that later becomes thick-walled sclereids, heavily stainable with toluidine blue (37). This would probably explain the easier-dehulling properties of brown (*desi*) chickpea varieties than those of the white-husked (*kabuli*) varieties.

Seed size is one of the factors affecting the dehulling process in pulses. It is a varietal character that is influenced by the growing season and location (4,57). Uniformity in size is also important for efficient dehulling. Dehulling efficiency is negatively and significantly correlated with seed size in green gram and cowpea (3). Yield obtained by TADD and barley pearler was negatively correlated with grain volume and seed size of pigeon pea, implying that yield would be reduce for bolder grains (50). The choice of the device for dehulling the pulse also depends on the size of the pulse. If the dehulling equipment, such as roller machine or stone grinder, is not properly set, large seeds are most likely to break, resulting in significant losses during dehulling (11,45). These workers further suggested that uniform and medium-sized pigeon pea seed would improve the efficiency of dehulling. Very small to small seeds are more difficult to dehull because they split even before husk removal and, hence, require several recycling steps and, therefore, are not generally preferred by millers. Williams et al. (56), however, reported that efficiency of dehulling and splitting of lentil is favored by large seed size, thin testa, a short storage period, and the correct wetting and drying practices. Furthermore, they reported that very bold seeds are also not accepted in pulse mills, because of heavy losses as broken seeds, indicating again the importance of grain size.

Similar to seed size, seed shape is a varietal characteristic in pulses (4). This characteristic is generally not affected by the growing environment. Pulses exist in various shapes such as spherical-shaped pea and pigeon pea, cylindrical-shaped green gram and black gram, pyramidal-shaped chickpea, flat oval-shaped field bean and horse gram, and kidney-shaped beans. This property plays a vital role in the selection of dehulling devices (11). Although round seeds are considered better for dehulling (46), very angular seeds lose excessive amounts of cotyledon material during dehulling. Sharper edges are preferentially lost from the angular seeds, whereas more seed mass is removed from flatter seeds. As a result, the flatter the seeds, the higher the amount of powder and brokens (i.e., small pieces of cotyledons). In addition, rounder seeds split more readily than flatter seeds, thus improving the efficiency of dehulling and splitting (13). According to these workers, the rounder the seeds, the better they are for dehulling.

The hardness of seeds is another important grain property affecting the milling quality. Kernels of some pulses such as green gram and cowpea are soft, whereas those of pigeon pea and kidney beans are hard at normal moisture levels. Chickpea and black gram kernels have medium hardness (11). Although the magnitude of correlations was low, grain hardness, has been reported to be negatively correlated with the yield of dehulled splits, which implies that hard grain genotypes of pigeon pea would produce lower yield (35,49). Variability in dehulling efficiency or yield of dehulled grain was significantly affected by hardness of the grain and resistance to splitting in individual cotyledons (35).

In addition, several environmental factors may influence the yield of dehulled splits from pulses. Variations in milling characteristics of pigeon pea, as influenced by variety and agroclimatic conditions, have been reported (28). Location and maturation of pigeon pea, which influence seed size, shape, and grain hardness, would directly affect the dhal yield in small- and large-scale–processing operations. Pigeon peas grown on light soils have better dehulling and cooking qualities (45). A few millers also have preferences for seed color, favoring white pigeon pea for two reasons: (a) yield is better when compared with other pigeon peas, (b) splits with a smaller degree of dehusking, but less visible white spots on leftover husk, can be sold in the market at a higher price than that obtained from colored seeds.

4.3 Effect of Varietal Differences on Dehulling Quality

A large variability has been reported in the dehulling quality of green gram, cowpea, chickpea, and pigeon pea cultivars, as determined by the tangential abrasive dehulling device (3,49). Among white-husked (*kabuli*) varieties of chickpea, the yield of dehulled grains varied from 84 to 90%, whereas the yield ranged from 73 to 83% among brown-husked (desi) varieties of chickpea (Table 3).

From the results of the studies in Table 3, it is apparent that a large variability exists in the dehulling quality of different pulses and even among different varieties of the same pulse. However, such a variability does not appear to have received much attention when breeding for high-yielding varieties, although dehulling quality of pigeon pea has been the subject of a few studies in the past. Pigeon pea varieties generally exhibit fewer variations in dehulling characteristics than cowpea varieties. The dehulling quality of the green gram cultivars is generally poor, because of their ease of splitting and long dehulling time (3). These workers suggested that resistance to seed splitting during dehulling and a loosely bound state of the seed coat to the cotyledons were the major seed quality requirements for a good dehulling property of these legumes. Ramakrishnaiah and Kurien (28) also observed wide variations in dehulling quality of pigeon pea cultivars (Table 4). Variations in the degree of dehusking obtained in this study are possibly the result of varying extents of loosening-of husk from the cotyledons after premilling treatments. These workers reported a wide range in the degree of dehusking (67–100%) of pigeon pea varieties. Furthermore; dehulling property was independent of seed size and husk content, but was greatly influenced by other varietal characteristics, such as quantity of gum between the husk and cotyledon, extent of husk adherence, and moisture content of the grain at the time of dehulling. In a subsequent study, Ramakrishnaiah and Kurien (29) reported an inverse relation between the content of nonstarchy polysaccharides and degree of dehusking in pigeon pea cultivars. In a recent study, variation in the yield of dehusked splits from pigeon pea genotypes ranging from 54.1 to 80.0% was reported (49). This study

Table 3 Variability in Dehusked Splits' Yield of Different Pulses

Pulses	No. of cultivars	Dehusked splits' yield (%)	Mean
Cowpea	6	60–78	69.0
Chickpea (brown-husked)	18	73–83	78.0
Chickpea (white-husked)	6	84–90	87.0
Pigeon pea	18	72–81	76.5
Green gram	12	64–73	68.5
Black gram	5	65–76	70.5

Source: Ref. 31.

Table 4 Yield of Dehulled Split Pulse and Degree of Dehusking of Different Cultivars of Pigeon Pea

Variety/cultivar	Husk content	Apparent yield	Degree of dehusking (%)	Calculated yield (%)
Commercial	12.5	80.2	67.1	76.8
BS-1	14.5	77.9	88.6	76.6
T-21	14.8	79.2	91.4	78.2
S-8	14.1	78.1	88.0	76.8
S-5	15.5	79.4	75.2	76.3
ICP-7221	12.3	80.7	81.3	78.8
ICP-7118	14.0	78.8	94.2	78.1
ICP-L	13.9	78.3	70.3	75.0
G-local	13.2	73.7	84.6	72.3
ICP-7120	14.1	79.4	82.3	77.4
HY-1	12.7	79.4	72.1	73.2
ICP-7182	13.5	78.6	90.8	77.6
Hy-4	14.1	78.5	85.8	76.9
Hyd-2	12.7	75.1	85.0	73.6
Hy-2	11.6	79.9	84.1	78.4
ICP-7119	13.0	78.1	85.2	76.6
Hy-3C	11.0	79.2	85.0	77.9
Hy-3A	10.5	79.8	87.6	78.8
S-1 41-31	14.8	82.0	100.0	82.0

Source: Ref. 28.

also reported that some newly developed varieties of pigeon pea showed very good dehulling quality. Kurien and Parpia (10) reported that smaller seeded varieties of pigeon pea grown in north India produced a lower yield because the seed coats were firmly attached to the cotyledons. Parpia (22) reported that *dehusked splits* yields of white pigeon pea were considerably higher than those of the red pigeon pea. However, later studies have not confirmed this observation; suggesting perhaps that the seed color may not significantly influence the dehulling quality of pigeon pea (49).

Swamy et al. (53) studied carbohydrates of pigeon pea in relation to its milling quality. They found that cold–water-soluble polysaccharides isolated from the interphase between the cotyledon and husk was mainly an arabinogalactan-type fraction in the easy-milling variety of pigeon pea, whereas the pectic-type fraction was higher in the difficult-to-mill variety of pigeon pea. Studies have also been carried out by Ramkrishnaiah and Kurien (32) to relate the nonstarchy polysaccharides with the milling quality of chickpea types. They observed a significant difference in the contents of hexoses and uronic acid of water-soluble nonstarchy polysaccharide (WS-NSP) fractions of husk between brown-husked (*desi*) and white-husked (*kabuli*) varieties (Table 5). The *kabuli* variety which is poor in milling quality contained more than double the amount of hexoses and uronic acid than those present in brown-husked (*desi*) variety which was easy-to-mill.

Phirke et al. (25) instrumentally measured seed coat adhesion to cotyledons in the pigeon pea varieties to study the effect of pretreatments. They observed that scarification, along with the application of heat to seeds treated with 8% urea solution reduced the seed

Table 5 Nonstarchy Polysaccharides (NSP) Content[a] of Brown-Husked (Desi) and White-Husked (Kabuli) Chickpea Cultivars[b]

Component	Type (unit)	Total NSP	Noncellulosic Water-soluble	Noncellulosic Water-insoluble	Cellulose
Whole grain	Desi (g)	202 ± 9	9.7 ± 0.3	99.3 ± 3.6	93.0 ± 5.3
	Kabuli (g)	124 ± 9	8.9 ± 1.5	79.2 ± 6.8	35.9 ± 1.4
Dehusked splits	Desi (g)	88 ± 5	5.2 ± 0.4	59.3 ± 2.4	23.5 ± 2.7
	Kabuli (g)	92 ± 8	4.8 ± 0.3	66.0 ± 5.3	21.2 ± 1.9
Husk	Desi (g)	900 ± 18	39.5 ± 4.0	344.5 ± 2.1	516.0 ± 16.0
	Kabuli (g)	714 ± 11	86.0 ± 11.4	322.0 ± 14.7	306.0 ± 13.3

[a] Values are means of duplicate determinations from each of three cultivars ± SD.
[b] Per kg dry weight.
Source: Ref. 32.

coat adhesion from 1.19 to 0.54 MPa. The instrument had a piezoelectric crystal cell transducer to measure the force needed for dehulling.

4.4 Dehulling Losses

The primary objective of dehulling is to remove only the seed coat from the cotyledons, but quite often noticeable amounts of cotyledon material and germ are also removed during the milling operation (2,41). As a result, considerable quantitative and qualitative losses occur during dehulling of pulses. The dehulling losses would depend primarily on the machinery employed for dehulling and the characteristics of the pulse being milled. The dehulling losses in terms of brokens were quite high (24.6%) in the stone grinder, and this might have been due to the attrition action of the stones employed for dehulling in this method (49). In commercial mills, product yields approach only 70%, which is much lower than the theoretical yield (20). The average dehusked splits yield from household and traditional commercial dehulling methods varied from 68 to 75% (22), which was 10–17% less than the theoretical average value close to 85%. Table 6 summarizes the data collected from commercial-milling systems in terms of yields of dehusked splits, powder, brokens, and husk fractions for different pulses milled in large- and small-scale–dehulling systems. This indicates that dehulling losses are significant and vary with the scale of operation and the pulse crop. The highest yield of dehusked splits was reported to be obtained from a modern dhal-making system developed at CFTRI (Mysore, India), where pulse was conditioned in a current of hot air before dehulling (12). This is further corroborated by the fact that a higher yield of dehusked splits was obtained in small-scale processing of pigeon pea when the pulse was heated in an open pan before dehulling using a stone grinder (45). Generally, losses in terms of brokens and powder fractions are higher in small-scale processing. Dehusked splits yields obtained by household-dehulling practice are noticeably lower than those obtained by the large-scale commercial dhal mills.

4.5 Effect of Dehulling on Nutrient Losses

Common methods of dehulling invariably remove most of the germ portion along with the husk, which results in losses of vitamins and proteins, the two important dietary constit-

Table 6 Dehulling Losses in Various Pulses

	Large-scale processing[a]			Small-scale processing[b]		
Pulse	Dehusked splits (%)	Brokens (%)	Husk + powder (%)	Dehusked splits (%)	Brokens (%)	Husk + powder (%)
Pigeon pea	72–78	2–8	18–25	55–70	15–20	15–20
Chickpea	75–80	1–6	14–23	65–70	10–15	15–20
Lentil	76–81	3–10	10–21	68–75	8–14	10–15
Green gram	70–78	5–10	17–24	60–75	10–15	8–15
Black gram	70–77	3–8	17–25	65–75	8–14	10–15

[a] Based on enquiry in 48 pulse mills in Maharashtra, Rajasthan, Madhya Pradesh, Tamil Nadu, and Uttar Pradesh, in India.
[b] Based on data collected from 90 small-scale pulse-milling units all over India.
Source: Ref. 30.

uents (2). In a laboratory study, a decrease in protein, calcium, and iron contents of the dehusked splits from chickpea and pigeon pea has been reported with an increase in dehulling time (48,50), as seen from the data presented in Table 7.

This was further corroborated by Narasimha (18), who found that the outermost layers (about 5%) of the cotyledon in pigeon pea are very rich in calcium (about 240 mg/100 g) and protein content (near 40%). Scouring of even 2–3% of the outer cotyledon material resulted in a significant loss of calcium and protein. However, influence of machine parameters on the milling yields and nutrient losses does not appear to have received the attention it deserves.

Table 7 Effect of Dehulling on the Chemical Constituents of Dehusked Splits and Powder Fractions of Chickpea (*cv.* Annigeri) and Pigeon Pea (*cv.* Cll)

Dehulling time (min)	Dhal			Powder		
	Protein (%)	Calcium (mg/100 g)	Iron (mg/100 g)	Protein (%)	Calcium (mg/100 g)	Iron (mg/100 g)
0	18.6[a]	43.0	5.7	—	—	—
	21.4[b]	64.9	5.7			
2	18.0	39.5	5.0	23.6	85.0	12.0
	20.8	51.7	4.1	31.2	167.8	17.3
4	17.5	38.0	4.8	21.8	65.5	10.5
	19.6	45.7	3.6	27.1	94.1	9.2
8	17.5	36.5	4.3	19.8	45.0	8.5
	19.6	45.7	3.6	27.1	94.1	9.2
12	16.4	35.0	3.8	18.9	45.0	7.0
	20.3	51.1	4.0	29.7	118.8	11.9
SEM	±0.18	±1.80	±0.21	±2.90	±0.30	
	±0.17	±2.83	±0.15	±2.00	±1.63	

[a] Chickpea.
[b] Pigeon pea.
All units are averages of two replicates, and are expressed on a moisture-free basis.
Source: Ref. 44.

Table 8 Effect of Dehusking on Cooking Time of Pigeon Pea Varieties in Comparison with Green Gram and Horse Gram

Variety	No. of cultivars	Cooking time (min) Whole seed	Dehusked splits
Pigeon pea: good-cooking cultivars	4	50–55	32–34
Pigeon pea: medium-cooking cultivars	9	52–62	40–50
Pigeon pea: poor-cooking cultivars	4	62–66	50–60
Green gram	1	60	25
Horse gram	1	130	100

Source: Ref. 18.

4.6 Influence of Dehulling and Premilling Treatments on Cooking Quality of Pulses and Dehusked Splits

Cooking time of whole pulses varies from 60 min (for pulses such as green gram and lentil) to 150–200 min (for pulses such as beans, horse gram, soybean, and such). Cooking times, which were reduced by almost 50% by mere dehusking, and cooking time of various *dhals* range from about 20 to 100 min, as seen from the data in Table 8 (18). Thus, one might surmize that perhaps dehulling evolved as a means to reduce the cooking time.

The cooking quality of dehusked splits is also influenced by the dehulling method, in particular, by the premilling treatments. However, there may not be a direct influence of the dehulling machines on the cooking time of dehusked splits. It is not the mechanical action of the roller machine or disk shellers that influence the cooking time, but the pretreatments given to seeds before dehulling that considerably influenced the cooking time, as observed in dehulling pigeon peas (43). Prolonged soaking of the seeds in water and subsequent sun or mechanical drying increased the cooking time of many grain legumes (17,18,21). Singh (44) observed that soaking and oven-drying (65°C, overnight) of pigeon pea seeds before dehulling considerably increased their cooking time (Table 9).

Table 9 The Cooking Time (min) of Pigeon Pea Dehusked Splits Obtained by Various Pretreatments Before Dehulling

	Pretreatments[a]				
Cultivar	None	Oil (1% w/w)	Water (1% w/v)[b]	NaCl (1% w/v)[b]	Na_2CO_3 (1% w/v)[b]
BDN-1	20	20	26	20	16
C-11	16	20	24	20	16
No. 148	16	16	20	14	12
LRG-30	16	18	22	18	14
LRG-36	14	18	24	14	12
SEM (±)	1.0	0.8	0.9	1.0	0.7

[a] After pretreatments, samples were dried at 65°C overnight before dehulling in TADD.
[b] Stored for 6 h.
Source: Ref. 44.

Furthermore, it was observed that although prolonged soaking of whole seeds in water increased the cooking times of respective dehusked splits, soaking in 1% solution of sodium carbonate generally decreased it considerably. According to this study, differences in the cooking time of dehusked splits were significant, owing to genotype and pretreatments. This implied that genotypes coupled with pretreatments would also play an important role in influencing the cooking time. Additional studies on the influence of pretreatments used in commercial dehusked splits mills on their cooking quality need to be performed.

4.7 Standards for Milling Quality

Several research workers have attempted to define milling quality of pulses. These have remained at only the laboratory level and, even today, the market value of the dehusked splits is largely decided by the traders. The price for the milled product is fixed, depending on number of grains with intact husk (partly or wholly) in the sample, chipping of edges of the cotyledons, extent of surface scouring of the grain, and the variety of the pulse. There is a substantial difference in the prices of first-, second-, and poor-quality dehusked splits, which is a matter of much concern to millers as well as to the consumers. Although limits have been set for pesticide residues, foreign matter, and other impurities for many pulses in several countries, unfortunately, no standards exist for milling quality of pulses. Perhaps market demand and supply has an edge over the price control, and the quality of dehusked splits is given a go-by. There is an urgent need for working out such specifications for ease of operation and to put pulse trading and processing on a firm scientific footing.

5 FUTURE RESEARCH NEEDS

Pulses are the most potential sources for providing the protein requirements to a large section of the population dependent on vegetarian diets, particularly in the developing countries. Most pulses are consumed in the dehusked, split form, popularly known as *dhals* [*dals*]. It is imperative that maximum recovery of good quality dehusked splits should be the main criterion in pulse milling. Recovery of dehusked splits from pulses depends on the proportion of the husk to cotyledon and the way it is attached to the cotyledons. Therefore, efforts must continue to evolve suitable genotypes, with lower husk contents, that can be easily separated from the cotyledons. Also it is desirable to develop varieties with uniformly sized grains, preferably round ones, to increase the dehusked splits yield. Identification and development of varieties with improved dehulling characteristics, therefore, should receive greater attention in the near future.

Pulse milling, as it is practiced today, is quite tedious, involving elaborate premilling treatments for loosening of the husk, and also a long-processing time. Continued efforts are needed to refine the premilling treatments such that processing time is shortened without affecting the product quality. Improvements are also needed in the milling machinery to reduce the loss of nutrients during the process of dehulling.

It is highly desirable to promote the efficient and low-cost–dehulling machines developed by various research and development institutions around the world, such as the CFTRI-designed mini dhal mill, so that many of the inefficient units such as stone grinders could be replaced for dehulling of pulses. Attention should be paid for educating the processors and consumers against certain undesirable and sometimes harmful practices (i.e., addition of colors and polishing aids for "improving" their appearance) so that the dehulled split pulses are rendered more safe for consumption. It is also desirable to develop complete-milling packages and schedules for different pulses and, quite often, for different varieties of the same pulse to improve the yields, retention of their nutritive value, and their acceptance for cooking and product-making qualities. There is also a need to develop small-capacity test-milling equipment to evaluate the dehulling quality of pulse cultivars, especially those for use in developing countries.

6 GLOSSARY

6.1 Pulses: Common and Botanical Names

Common name	Botanical name
Adzuki bean	*Vigna angularis*
Black gram, urad	*Vigna mungo* (*Phaseolus mungo*)
Broad bean Horse bean Field bean	*Vicia faba*
Cowpea	*Vigna unguilculata*
Chickpea Bengal gram Garbanzo	*Cicer arietinum*
Guar bean	*Cyamopsis tetragonoloba*
Green gram Mung bean	*Vigna radiata* (*Phaseolus aureus*)
Horse gram	*Macrotyloma uniflorum* (*Dolichos biflorus*)
Kidney bean, navy bean, pinto bean, Haricot bean	*Phaseolus vulgaris*
Khesari Lathyrus pea Grass pea	*Lathyrus sativus*
Lentil Masur	*Lens esculenta*
Moth bean	*Vigna aconitifolia* (*Phaseolus aconitifolius*)
Pea, garden pea	*Pisum sativum*
Pigeon pea Red gram, tur Congo pea	*Cajanus cajan*
Soybean	*Glycine max*
Velvet bean	*Mucuna pruriens*
Winged bean	*Psophocarpus tetragonolobus*

6.2 Common Indian Terms Used in Pulse Milling

Dhal (dal)	→	Dehusked split pulse or cotyledon
Kappi	→	Mixture of dehusked and unhusked (whole) grains, particularly used in milling of pigeon peas
Gota	→	Dehusked (pearled) unsplit grains of pigeon pea
Chakki	→	A form of plate mill (plates being made of either stone or emery or wood) also known as traditional stone grinder
Patka machine	→	Centrifugal-type dehusked splits splitter used in traditional dhal mills in India.
Desi chickpea	→	Chickpea cultivars grown traditionally in the Indian subcontinent, preferably with brown husk.
Kabuli chickpea	→	Generally large-sized chickpea cultivars with whitish thin husk used mainly as a table variety.

REFERENCES

1. Anon (1979): Grain legumes; processing and storage problems. Food Nutr Bull 1(2):1–7. Central Food Technological Research Institute, Mysore, India.
2. Aykroyd WR, Doughty J (1964). Legumes in Human Nutrition. FAO Nutr Stud 19, FAO, Rome, Italy.
3. Ehiwe AOF, Reichert RD (1987). Variability in dehulling quality of cow pea, pigeon pea and mung bean cultivars determined with TADD. Cereal Chem 64(2):86–90.
4. Erskine W, Williams PC, Nakkoul H (1985). Genetic and environmental variation in the seed size, protein, yield, and cooking quality of lentils. Field Crops Res 12:153–161.
5. Erskine W, Williams PC, Nakkou H (1991). Splitting and dehulling lentil: effects of seed size and different pre-treatments. J Sci Food Agric 57:77–84.
6. FAO Production Year Book (1996). 50:97–98.
7. Krishnamurthy K, Girish GK, Ramasivan T, Bose SK, Singh K, Tomer RPS (1972). A new process of removal of husk of red gram using *sirka*. Bull Grain Technol 10:181–186.
8. Kulkarni SD (1986). Pulse processing in India, a status report. Indian Agricultural Research Institute, New Delhi, India.
9. Kumar J, Singh U (1989). Seed coat thickness: variation and inheritance in a desi × kabuli cross. Indian J Genet 49:245–249.
10. Kurien PP, Parpia HAB (1968). Pulse milling in India: processing and milling of Tur, Arhar. J Food Sci Technol 5:203–207.
11. Kurien PP (1977). Grain legume milling technology. Presented at FAO Experts Consultants on Grain Legume Processing. Central Food Technological Research Institute, Mysore, India, November, 1977.
12. Kurien PP (1981). Advances in milling technology of pigeon pea. In: Proceedings of the International Workshop on Pigeon Pea. International Crops Research Institute for the Semi-Arid Tropics (ICRISAT). Patancheru, India, pp 321–328.
13. Kurien PP (1984). Dehulling technology of pulses. Res Ind 29:207–214.
14. Kurien PP (1987). Processing and utilisation of grain legumes, human nutrition. Food Sci Nutr 41(3–4):203–212.
15. Kurien PP, Patil BS (1989). A hand-operated small scale pulse dehusking machine for rural use. Res Ind 34:213–216.
16. Mandhyan BL, Jain SK (1993). Optimization of machine condition for milling of pigeon pea. J Food Eng 18:91–96.

17. Mattson S (1948). Cookability of yellow peas. Chem Abstr 42:4689.
18. Narasimha HV (1984). Factors affecting cooking quality of pulses. PhD dissertation, University of Mysore, India.
19. Narasimha HV (1997). Food grain situation in Uganda, present practices and future perspectives. Report submitted to FAO, Rome, p 21.
20. Natarajan CP, Shankar JV (1980). Nutritional consequences of primary processing of food grains in India. Indian Food Packer 34:44–52.
21. Paredes–Lopez O, Carabez–Treio A, Paimia–Tirado L, Reyes–Moreno C (1991). Influence of hardening procedure and soaking solution on cooking quality of common beans. Plant Foods Hum Nutr 41:155–164.
22. Parpia HAB (1973). Utilization problems in food legumes. In: Nutritional Improvement of Food Legumes by Breeding. Max Muller Protein Advisory Group of United Nations, New York, pp 281–295.
23. Phirke PS, Umbarkar SP, Tapre AB, Kubde AB (1992). Development and evaluation of mini dhal mill for villages. Agric Mech Asia, Afr Lat Am 23(4):62–64.
24. Phirke PS, Umbarkar SP, Kubde AB, Tapre AB (1993). Process development for pigeon pea milling at rural level. New Agric 4:47–56.
25. Phirke PS, Bhole NG, Adhaoo SH (1995). Shear forces for dehulling, splitting and breaking raw and pretreated pigeon pea. Int J Food Sci Technol 30:485–491.
26. Phirke PS, Bhole NG, Adhaoo SH (1996). Response surface modelling and optimization of dehulling of pigeon pea with different pre-treatments and conditions. J Food Sci Technol 33: 47–52.
27. Prasoon Verma, Saxena RP, Sarkar BC, Omre PK (1993). Enzymatic pretreatment of pigeon pea grains and its interaction with milling. J Food Sci Technol 30:368–370.
28. Ramakrishnaiah N, Kurien PP (1983). Variabilities in the dehulling characteristics of pigeon pea cultivars. J Food Sci Technol 20:287–291.
29. Ramakrishnaiah N, Kurien PP (1985). Nonstarchy polysaccharides of pigeon pea and their influence on dehulling characteristics. J Food Sci Technol 22:429–430.
30. Ramakrishnaiah N, Pratape VM, Narasimha HV (1993). National survey of pulse milling industry in India including rural processing, a status report, Central Food Technological Research Institute, Mysore, India.
31. Ramakrishnaiah N, Pratape VM, Kurien PP (1983). Central Food Technological Research Institute, Mysore, India. Unpublished data.
32. Ramakrishnaiah N, Kurien PP (1995). Non starchy polysaccharides content of desi and kabuli varieties of chickpea. Central Food Technological Research Institute, Mysore, India, Annual Report.
33. Reichert RD, Young CG (1976). Dehulling cereal grains and grain legumes for developing countries. 1. Quantitative comparison between attrition and abrasive type mills. Cereal Chem 53:829–835.
34. Reichert RD, Lorer EF, Youngs CG (1979). Village scale mechanical dehulling of cowpea. Cereal Chem 56:181–184.
35. Reichert RD, Ooman BD, Young CG (1984). Factors affecting the efficiency of abrasive type dehulling of grain legumes investigated with a new batch dehuller. J Food Sci Technol 49: 267–272.
36. Reichert RD, Tyler RT, York AE, Schwab DJ, Tatarynovich JE, M'wasaru MA (1986). Description of a production model or tangential abrasive dehulling device (TADD) and its application to breeders samples. Cereal Chem 63:201–207.
37. Sahay KM, Bisht BS (1988). Development of a small abrasive cylindrical mill for milling pulses. Int J Food Sci Technol 23:17–22.
38. Sahay KM (1990). Evaluation of a general purpose abrasive mill for dehulling of pulses. Int J Food Sci Technol 25:220–225.
39. Saxena RP, Laxmi Chand, Garg GK, Singh BPN (1989). Effect of bicarbonate soaking on

dehusking efficiency and composition of pigeon pea seeds and dhal. Int J Food Sci Technol 24:237–241.
40. Sefa–Dedeh S, Stanley DW (1979). The relationship of microstructure of cowpeas to water absorption and dehulling properties. Cereal Chem 56:379–386.
40a. Sharma YK, Tiwari AS, Rao KC, Mishra A (1977). Studies on chemical constituents and their influence on cookability in pegion pea. J Food Sci Technol 14:38–40.
41. Siegal A, Fawcett B (1976). Food legume processing and utilization with special emphasis on application in developing countries. IDRC-TSI, Ottawa, Canada.
42. Singh K, Saxena DC, Misra BK, Maheshwari PN (1993). Effect of moist heat treatment on milling of arhar (pigeon pea). Poster paper (GST—12) presented during IFCON—1993. Central Food Technological Research Institute, Mysore, India, September.
43. Singh U (1987). Cooking quality of some important Indian legumes. Presented at the Symposium on Present Status and Future Prospective in Technology of Food Grains. Feb 27–Mar 1. CFTRI, Mysore, India.
44. Singh U (1995). Methods of dehulling of pulses: a critical appraisal. J Food Sci Technol 32: 81–93.
45. Singh U, Jambunathan R (1981). A survey of methods of milling and consumer acceptance of pigeon pea in India. In: Proceedings of the International Workshop on Pigeon Pea. Vol. 8. International Crops Research Institute for the Semi-Arid Tropics (ICRISAT), Patancheru, India, pp 419–425.
46. Singh U, Jambunathan R (1990). Pigeon pea: post-harvest technology. In: The Pigeon Pea. CAB International, Wallingford, UK. pp 435–455.
47. Singh U, Manohar S, Singh AK (1984). The anatomical structure of desi and kabuli chickpea seed coats. Int Chickpea Newslett 10:26–27.
48. Singh U, Rao PV, Seetha R, Jambunathan R (1989). Nutrient losses due to scarification of pigeon pea cotyledons. J Food Sci 54:974–981.
49. Singh U, Santosa BAS, Rao PV (1992). Effect of dehulling methods and physical characteristics of grains on dhal yield of pigeon pea (*Cajanus cajan* L.) genotypes. J Food Sci Technol 29:350–353.
50. Singh U, Rao PV, Seetha R (1992a). Effect of dehulling on nutrient losses- in chickpea (*Cicer arietinum* L.). J Food Compos Anal 5:69–76.
51. Sosulsky FW, Dabrouski KJ (1984). Composition of phenolic acids in flours and hulls of 10 legume species. J Agric Food Chem 32:131–133.
52. Srivastava V, Mishra DP, Laxmi Chand, Gupta RK, Singh BPN (1988). Influence of soaking on various biochemical changes and dehusking efficiency in pigeon pea (*Cajanus cajan* L.) seeds. J Food Sci Technol 25:267–271.
53. Swamy NR, Ramakrishnaiah N, Kurien PP, Salimath PV (1991). Studies on carbohydrate of red gram in relation to milling. J Sci Food Agric 57:379–390.
54. Uma Chitra (1993). Effect of storage and processing on phytic acid levels in legumes and its interference with the utilization of protein and iron. PhD thesis, Andhra Pradesh Agricultural University, Hyderabad, India.
55. Vibha–Srivastava, Mishra DP, Laxmi Chand, Gupta RK, Singh BPN. (1988). Influence of soaking on various bio-chemical changes and dehusking efficiency in pigeon pea seeds. J Food Sci Technol 25:267–271.
56. Williams PC, Erskine W, Singh U (1993). Lentil processing, Lens–Newsletter 20:3–13.
57. Williams PC, Singh U (1987). The chickpeas—nutritional quality and evaluation of quality in breeding programmes. In: The Chickpeas—CAB International, Wallingford, Oxon, UK, pp 329–356.

16

Postharvest Physiology of Fresh Fruits and Vegetables

JENNIFER R. DeELL

Ontario Ministry of Agriculture and Food, Vineland Station, Ontario, Canada

ROBERT K. PRANGE

Agriculture and Agri-Food Canada, Kentville, Nova Scotia, Canada

HERMAN W. PEPPELENBOS

Agrotechnological Research Institute (ATO-DLO), Wageningen, The Netherlands

1 INTRODUCTION

Harvested fruits and vegetables continue to maintain physiological systems and sustain metabolic processes that were present before harvest. While attached to the plant, the losses from respiration and transpiration are replaced from the flow of sap, which contains water, photosynthates, and minerals (1); however, after harvest, the product is dependent entirely on its own food reserves and water content. Losses of water and substrates used in respiration can no longer be replaced and deterioration of the product begins.

Maturation, ripening, and senescence induce many changes in fruits and vegetables (2). Although a strict physiological distinction between fruit ripening and senescence is unclear, ripening hastens the onset of senescence and the probability of cell injury and death (3). Fruit ripening involves many complex changes, including seed maturation, color changes, abscission from the parent plant, tissue softening, volatile production, wax development on skin, and changes in respiration rate, ethylene production, tissue permeability, carbohydrate composition, organic acids and proteins (1).

Product respiration, transpiration, and ethylene production are major factors contributing to the deterioration of fresh fruits and vegetables. Reduction of these processes by technologies such as cooling and storage, enable the postharvest life of fresh produce to be prolonged. This chapter will concentrate primarily on respiratory gas exchange, transpiration, and the involvement of ethylene in postharvest physiology. Advancements in knowledge and technology will be the main focus within these subject areas, along with brief overviews of the physiological aspects. For a more general overview there are several excellent books dedicated completely to postharvest physiology (1,4,5).

2 GAS EXCHANGE

2.1 Respiration

Every plant tissue requires energy to remain alive and to support developmental changes. The energy is generated by respiration, which is the oxidative catabolism of carbohydrates. This process occurs in the mitochondria of living cells and mediates the release of energy and the formation of carbon skeletons necessary to the maintenance and synthetic reactions that occur after harvest (4). Respiration can be considered a series of enzymatic reactions, involving three pathways: glycolysis (glucose → pyruvate), the tricarboxylic acid cycle (pyruvate → CO_2), and oxidative phosphorylation (reduced nicotinamide adenine dinucleotide [NADH] + reduced flavin adenine dinucleotide [$FADH_2$] → adenosine triphosphate [ATP]). Oxygen (O_2) is consumed, whereas carbon dioxide (CO_2), some heat, and energy carriers are released. The process can be described by the simple equation:

$$C_6H_{12}O_6 + 6\ O_2 + 36\ ADP \rightarrow 6\ CO_2 + 6\ H_2O + 36\ ATP \quad (1)$$

Adenosine triphosphate (ATP) is the main energy carrier and is used in numerous cell functions. A reduction in respiration will result in decreased ATP production, since the vital role of respiration is the generation of ATP (energy). Consequently, less energy is available for processes associated with ripening, resulting in quality changes. The control of respiration by environmental factors such as temperature and gas composition is discussed in Sec. 2.4.

Accurate estimations of energy production and consumption are essential for understanding postharvest applications, such as altered gas conditions. On the energy production side, for instance, the belief that 36 ATP are released per glucose molecule is still under debate, with 32 or lower sometimes being reported. The contributions of alternative and residual respiration to O_2 consumption are important for the estimation of energy production (see Sec. 2.1.a). For energy use, it is important to know whether or not processes influencing the main quality attributes require a lot of energy, and if extreme treatments will cause an energy status that is unable to cover maintenance needs (see Sec. 2.2).

2.1.1 Alternative and Residual Respiration

A substantial part of the O_2 consumption might not be used for cytochrome respiration, but for alternative respiration or other O_2-consuming processes (residual respiration). Alternative respiration is an alternative to the cytochrome pathway of electron transport to O_2. It branches from the main respiratory chain at ubiquinone, and results in CO_2 and heat production but not ATP production. Alternative respiration can take place when cytochrome respiration is blocked; the importance of alternative respiration in vivo is still unclear (6).

Oxidative reactions are not solely due to respiratory activity in either fresh or minimally processed produce. Residual respiration is a term often used for O_2-consuming processes involving enzymes such as polyphenoloxidase (PPO) and peroxidase. Such enzymes are involved in wound repair reactions and in defense against intruding microorganisms. PPO, possibly the most studied enzyme in harvested fruits and vegetables, is responsible for the browning of plant tissues. In minimally processed produce, which is chopped, cut, sliced, or peeled, the level of tissue injury is much higher than in whole produce. Consequently, the level of metabolic activity and thus the respiration rate of minimally processed produce is often orders of magnitude higher than that of whole produce. Enzymes such as PPO will also be more active when present and may cause visible browning of cut surfaces.

Similar to alternative respiration, residual respiration does not result in ATP production. In avocado tissues, alternative and residual respiration can account for 60 to 70% of the total O_2 consumption (7). Although these numbers seem high, it is likely that both alternative and residual respiration contribute consistently to O_2 consumption. The three O_2-consuming pathways also respond differently to variations in gas concentrations (7,8), resulting in varying ratios between ATP-producing and non–ATP-producing O_2 consumption. These findings indicate that 36 molecules of ATP per O_2 molecule consumed is not consistently accurate for estimations of real-life ATP production. Therefore, when energy metabolism is quantified, a factor should be introduced to indicate the percentage of O_2 consumption resulting in ATP production.

2.1.2 Respiration Measurements

The rate at which fruits and vegetables ripen or senesce can be influenced and controlled by environmental factors such as temperature, humidity, and gas composition (Fig. 1). Both O_2 and CO_2 concentrations in the surrounding atmosphere can influence the quality and storage life of many fruits and vegetables. Extensive research on the use of altered gas conditions started in the 1920s with the work of Kidd and West (9), Thomas (10), and Blackman (11). Kidd and West's discovery of the climacteric and Blackman's studies of respiration in apples established the basis of modern postharvest physiology (12).

Respiratory gas exchange is often used as a general measure of the metabolic rate of tissues, since respiration has a central position in the overall metabolism of a plant (part). Measurements of gas exchange were first conducted in 1917 by Kidd on pea and mustard seeds (13). Blackman (11) was the first to measure the CO_2 production of apple fruits. Thornton (14) later measured O_2 consumption of harvested plant products other than apples. Since the mid-1940s gas exchange rates of fresh produce have been measured extensively under various O_2 and CO_2 atmospheres, temperature, and humidity conditions. These data are used for several postharvest applications, such as the determination of optimum conditions for controlled atmosphere (CA) storage rooms or modified atmosphere (MA) packages (see Sec. 2.4). There are various ways of measuring gas exchange, but the three methods used most often are:

1. *Static method*: a specific gas composition is generated around an object and the gas flow is closed for a specific period of time. Gas composition is measured at the beginning and end of the period (15).
2. *Flowthrough method*: a specific gas composition is generated around an object and the gas composition of the inward and outward flow is measured (16).
3. *MA method*: using a package with film of known O_2 and CO_2 permeability, the equilibrium concentrations that develop are measured. Gas exchange rates can then be calculated (17).

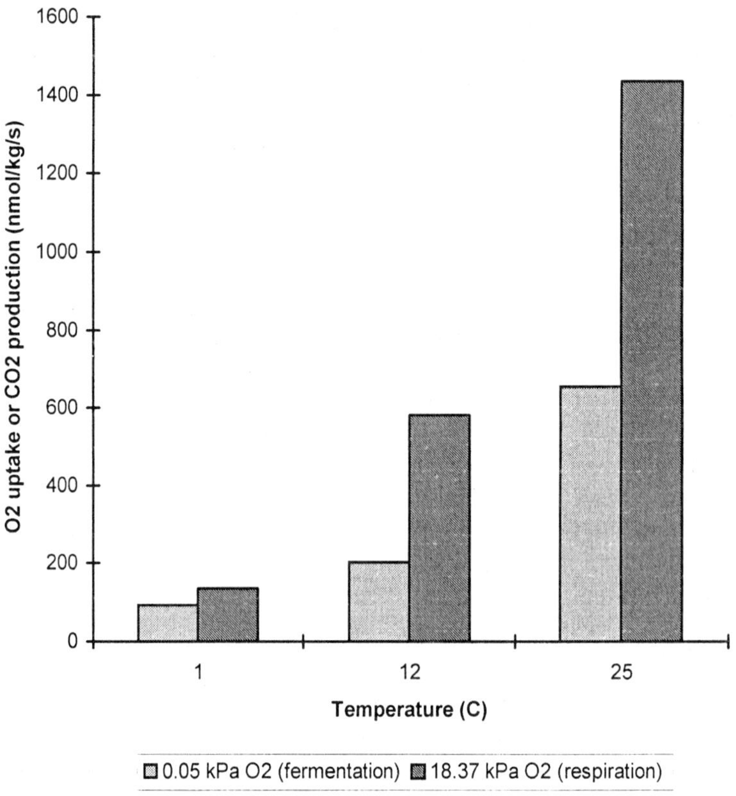

Fig. 1 The influence of temperature on respiration and fermentation rates of broccoli. Respiration is measured as O_2 uptake in ambient air and fermentation is measured as CO_2 production at 0 kPa O_2. (From Peppelenbos, unpublished data.)

Some experimental data utilizing gas exchange rates can be understood only if diffusion is taken into account (18). Generally, respiration rates are determined by measuring the O_2 and CO_2 concentrations in the atmosphere surrounding the product. It is known, however, that gas concentrations inside the product differ from the concentrations outside. Many diffusion barriers exist between the external atmosphere and the actual place where respiration occurs: the mitochondria. Thus bulky intact fruits respond to O_2 concentrations differently than cell cultures (16). Although the terminal oxidase of respiration is saturated at O_2 concentrations well below 5 kPa, apple respiration continues to increase at O_2 concentrations greater than 5 kPa. This behavior is often explained by the combination of respiratory responses with diffusion limitations within fruits (19,20).

2.1.3 Respiratory Quotient

Equation (1) represents the catabolism of glucose. The ratio of moles of CO_2 produced per mole of O_2 consumed is called the respiratory quotient (RQ) (21), and is 1 for glucose catabolism. When substances other than glucose are respired, the RQ is different than 1. For example, the oxidation of respiratory malic acid, a major substrate in apples (22), leads to the production of additional CO_2. The complete oxidation of malate by the tricarboxylic acid cycle results in a theoretical RQ of 1.6. The RQ can also exceed 1 when O_2

is not involved, such as in fermentation metabolism that becomes dominant when O_2 is limiting (see Sec. 2.2). RQ values below 1 can be expected when lipids or proteins, molecules often containing less oxygen than carbohydrates, are respired. Thus the RQ can be used as an indication of which substrates are being used in the respiratory pathway.

2.2 Fermentation

The main source of CO_2 production by plant tissues is respiration. At low O_2 or high CO_2 atmospheres, however, fermentation becomes increasingly important. The main fermentative metabolites found in plant tissues are ethanol, acetaldehyde, and lactic acid (23,24). Ethanol is the most abundant metabolite, especially under prolonged anoxia, resulting in additional CO_2 production (24,25). Ethanol fermentation is the catabolism of pyruvate to ethanol: the combination of glycolysis and fermentation can be expressed as:

$$C_6H_{12}O_6 + 2\ ADP \rightarrow 2\ CO_2 + 2\ C_2H_5OH + 2\ ATP \tag{2}$$

Physiological disorders of products stored in altered gas conditions are almost always found together with high concentrations of fermentative metabolites. Therefore, these metabolites are often considered to be the cause of storage disorders, such as necrotic or discolored tissues, off-flavors and off-odors (26).

One function of fermentation is the recycling of NADH to facilitate an increase in glycolysis. This enables the production of glycolytic ATP, although it is very inefficient compared to oxidative phosphorylation. An increase in fermentation helps the cell meet its ATP requirements (27), whereas the main injuries related to anoxia may eventually be due to changes in energy metabolism (25). There is no simple relation between survival and the rate of fermentation (24). The total amount of energy produced during respiration and fermentation should cover maintenance requirements. Disorders such as necrotic tissues are expected to develop when the gas conditions supplied result in lower energy production than maintenance energy (28).

2.2.1 Fermentative CO_2 Production

For many harvested plant products, such as apples, pears, and broccoli, an increase in CO_2 production can be observed when O_2 is decreased to very low concentrations (Fig. 2). One can assume that this additional CO_2 production is caused by increased fermentation, as the increase in CO_2 production is not accompanied by an increase in O_2 uptake. It is remarkable that for various products the CO_2 production at anoxia (fermentation) exceeds the CO_2 production in ambient air (respiration). CO_2 accumulation must often be avoided, or at least controlled, as elevated CO_2 can cause detrimental effects on product quality. For a thorough design of applications, such as CA rooms with dynamic control (see Sec. 2.5.1) where extremely low O_2 concentrations may be used, or MA packages in a nonoptimal transport chain where increased temperatures can temporarily cause low O_2 concentration in the package (see Sec. 2.5.2), a good estimation of fermentative CO_2 production is necessary.

An estimation of fermentative CO_2 production can be made using gas exchange measurements. The measurement of CO_2 production alone, however, is inadequate to estimate the increase in fermentation under low O_2 conditions. Instead, the measurements of both O_2 consumption and CO_2 production under various combinations of fixed O_2 and CO_2 concentrations are necessary. Such data are known for apple (29), tomato (30), and

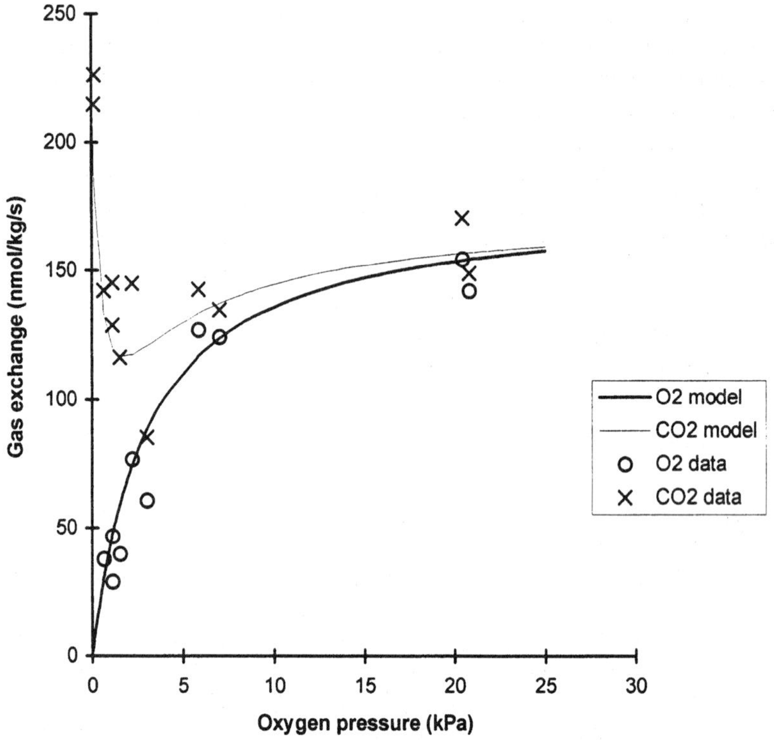

Fig. 2 Gas exchange rates of Cox's orange pippin apples stored for 3 days at 18°C in various O_2 concentrations. (Adapted from Ref. 28.)

mushrooms (31). Data are also known for tomato (20), blueberry (17), and raspberry (33) by the use of MA packages to derive various gas conditions.

2.3 Modeling Gas Exchange Rates

2.3.1 O_2 Uptake

Respiration and fermentation both result in mass fluxes of O_2 and CO_2. Several equations were developed to relate the fluxes to gas concentrations (30,34). The equations assuming a linear relation can cause serious prediction errors, especially at low O_2 concentrations at which they tend to overestimate O_2 uptake (20). Another drawback of empirical relations is that variables are often introduced that cannot be related to a specific physiological process, making it difficult to make extrapolations. Currently, the most widely used equation to describe respiration of a whole fruit is based on a mathematical description of the enzyme kinetics based on the Michaelis–Menten model. This approach has been used by several authors (18,35,36):

$$V_{O_2} = \frac{Vm_{O_2} * O_2}{Km_{O_2} + O_2} \tag{3}$$

where V_{O_2} is the O_2 consumption rate (nmol kg^{-1} s^{-1}), Vm_{O_2} is the maximum O_2 consumption rate (nmol kg^{-1} s^{-1}), O_2 is the O_2 concentration (kPa) and Km_{O_2} is the Michaelis constant for O_2 consumption (kPa O_2). In this equation it is assumed that the whole respiratory chain can be described by one enzyme-mediated reaction, with the substrate glucose considered as nonlimiting and the substrate O_2 as limiting. The Km_{O_2} in the equation refers to the O_2 concentration where the reaction rate (O_2 uptake rate) is half the maximum rate. Vm_{O_2} is the maximum reaction rate (nmol kg^{-1} s^{-1}), when O_2 is nonlimiting. Since not only low O_2 concentrations, but also high CO_2 concentrations, reduce respiration rates, Eq. (3) can be modified to include this type of inhibition. Three types of inhibition of enzyme functioning can be distinguished (37), but often the noncompetitive type is used (15,36):

$$V_{O_2} = \frac{Vm_{O_2} * O_2}{(Km_{O_2} + O_2) * (1 + CO_2/Km_{CO_2})} \tag{4}$$

where Km_{CO_2} is the Michaelis constant for CO_2 inhibition of O_2 consumption (kPa CO_2).

2.3.2 CO_2 Production

Equation (3) is also often applied to describe CO_2 production (36,38). This approach assumes no CO_2 production at 0 kPa O_2, which is correct if no fermentation takes place or if all of the fermentation that occurs leads to the formation of products such as lactate or alanine. However, ethanol is the primary fermentation product in most plant tissues (23, 24), leading to additional CO_2 production at low O_2 concentrations. Therefore, Eq. (3) is not suitable for products with increasing CO_2 production at low O_2 concentrations, such as asparagus and carrots (21), pear (16), cherry (32), and blueberry (17). To overcome this problem, models that make a distinction between CO_2 produced by oxidative metabolism and that produced by fermentative metabolism have been developed (31,39–41). Some prefer to base their models on an extension of Eq. (3) (31,40). Oxidative CO_2 production is calculated by multiplying the O_2 consumption by a specific RQ value, RQ_{ox}, which is the ratio between oxidative CO_2 production and O_2 uptake. RQ_{ox} is assumed to be independent of O_2 concentrations and O_2 is regarded as an inhibitor of fermentative CO_2 production. Total CO_2 production is described as (40):

$$V_{CO_2} = RQ_{ox} * V_{O_2} + \frac{RQ_{ox} * Vm_{O_2} * 10^{-10}}{(O_2 + a)^b} \tag{5}$$

where V_{CO_2} is the total CO_2 production rate (nmol kg^{-1} s^{-1}), and a and b are empirical constants, or as (15):

$$V_{CO_2} = RQ_{ox} * V_{O_2} + \frac{Vmf_{CO_2}}{1 + O_2/Kmf_{O_2}} \tag{6}$$

where Vmf_{CO_2} is the maximum fermentative CO_2 production rate (nmol kg^{-1} s^{-1}) and Kmf_{O_2} the Michaelis constant for the inhibition of fermentative CO_2 production by O_2. Specific parameters must be known for every product for the enzyme kinetics models. Accurate measurements of O_2 consumption and CO_2 production under various O_2 and CO_2 concentrations are needed to calculate these values. An example of Eqs. (1) and (6) fitted on gas exchange data is shown in Fig. 2.

2.4 Controlling Gas Exchange Rates

The reduction of respiration rate is often considered the process most affected by altered atmospheres (40,42). A direct relation between respiration rates and quality changes has been suggested (43,44). Although respiration is important, its reduction is not the only beneficial effect of altered atmospheres (26). Decreased O_2 and increased CO_2 levels also influence ethylene production and action (see Sec. 4: Ethylene) and suppress microbial growth. Nowadays, low O_2 and high CO_2 concentrations are involved in several techniques. The most common for fresh fruits and vegetables are CA storage and MA packaging. CA storage generally refers to decreased O_2 and increased CO_2 with monitoring and active adjustment of the gas composition, whereas MA refers to a difference in gas composition as compared with ambient air, without any active control of the gas composition.

2.4.1 CA Storage

Storage techniques based on altered gas conditions have a long history. Ancient Chinese writings report the transport of fruits in sealed clay pots with fresh leaves and grass added. The fruits generated a low O_2 and high CO_2 atmosphere which retarded their ripening (45,46). During the time of the Roman Empire, modified gas atmospheres were created by sealing underground pits filled with grain, and thereby protecting them from insects and rodents (4). In the beginning of the 19th century, Berard (47) demonstrated that fruits placed in closed containers did not ripen. Extensive research on the use of altered gas conditions to lengthen the postharvest life of produce started early in the 20th century, with the work of Kidd and West (9), Thomas (10), and Blackman (11).

Commercial storage under altered gas conditions started in England in 1929, when apples were stored in 10 kPa CO_2 and ambient O_2 (4). Reduced O_2 concentrations and increased CO_2 concentrations also proved to be beneficial for products other than apples. CA storage was adopted in the 1930s in Canada by C. A. Eaves and in the United States by R. E. Smock. After the Second World War, CA storage became increasingly important in many pome fruit growing countries. While regular CA facilities focus on maintaining O_2 concentrations between 3 and 4 kPa, the so-called ultra low oxygen (ULO) storage uses even lower concentrations. The advised O_2 concentrations for many apple cultivars are 1 to 2 kPa (48). It is now common for some apples to be stored for over a year in CA storage, e.g. Northern Spy apples in Nova Scotia (R. Prange, unpublished data).

2.4.2 Packaging

In contrast to CA, the gas composition within MA packaging is neither monitored nor adjusted. Depending on the O_2 sensitivity and metabolic activity of the product to be packaged, air or a predetermined gas mixture is used to flush the packages before closing. The use of ambient air as the packaging gas is obviously most economical, but is an option only when the respiration activity of the produce under the prevailing storage conditions is high enough to reduce the "in-pack" O_2 level fast enough not to cause physiological or microbial deterioration. For produce that has low levels of respiratory activity, flushing with a gas mixture composed of low O_2 and moderately high CO_2 is often used to shorten the time needed to reach the desired "in-pack" gas composition. After closing the package, the respiration of the product will result in a continued decrease in the O_2 content and a continued increase in the CO_2 content. Altered gas concentrations, however, cause a decrease in the respiration rate. Eventually an equilibrium concentration inside the package will be reached, which is the result of a balance between metabolic rates of the packed product and diffusion characteristics of the package materials. This explains the use of the term *equilibrium-modified atmosphere* (EMA) packaging. The package is often designed in such a way that the equilibrium concentrations resemble the optimal gas concentrations found in experiments where products are stored under a range of stable gas conditions. For a good application of MA packaging it is also essential to have insight into the respiratory characteristics of the product as they are affected by O_2 and CO_2 concentrations.

Often terms other than MA are used to refer to specific applications of packages. *Active packaging* is used for packages where specific compounds are added that absorb some of the gases in the package (O_2, CO_2, ethylene) or release specific compounds into the package atmosphere. *Modified humidity packaging* (MHP) is used for packages mainly designed to create an optimal humidity, instead of optimal O_2 or CO_2 concentrations, thereby preventing dehydration and microbial spoilage of the packaged product.

2.4.3 Product Characteristics

Some fruits and vegetables are well equipped to live detached from the plant, especially those designed for vegetative reproduction (carrot, potato, onion) and generative reproduction (seeds and fruits). These products often contain large amounts of carbohydrates that enable the maintenance of respiration and energy production. Other harvested products, such as leaves (spinach) or whole plants (lettuce, endive), do not contain much storage material and are susceptible to rapid senescence and wilting (see Sec. 3: Transpiration).

Large differences exist regarding the response of a plant product to CA storage and MA packaging. There are products, such as apples, whose storage life can be increased by months under low O_2 conditions. In contrast, others, such as carrots (49), do not respond positively to low O_2 or high CO_2 concentrations. In general, altered gas conditions are considered positive only within a certain range of concentrations, the so-called optimum concentrations. Much research has been directed toward determining the optimum concentrations, and these concentrations have been published for many products (48,50–52). However, one cannot generalize results because the optimum can depend on cultivar and even on the conditions under which a given cultivar has been produced, including soil type, climate, and other local factors (48). In addition, the optimal values for temperature, O_2, and CO_2 concentrations are often established separately, although interactions between temperature, O_2, and CO_2 concentrations (and probably also humidity) are known.

Currently, CA and MA are used commercially for a limited number of products. CA storage is important worldwide for apples and pears, and increasing in importance for cabbage (for salads), berries and currants. CA transport overseas is important for bananas and kiwis. MA packaging is increasingly used in supermarkets for products such as broccoli, corn, and lettuce, and especially for minimally processed products such as endive, spinach, vegetable mixes, and salads.

2.5 Practical Applications

2.5.1 Changes Over Time

Most harvested plant products show changes in metabolic activity after harvest. For climacteric fruits in particular, these changes are dramatic (see Sec. 4: Ethylene). Other products can also show such changes (Fig. 3).

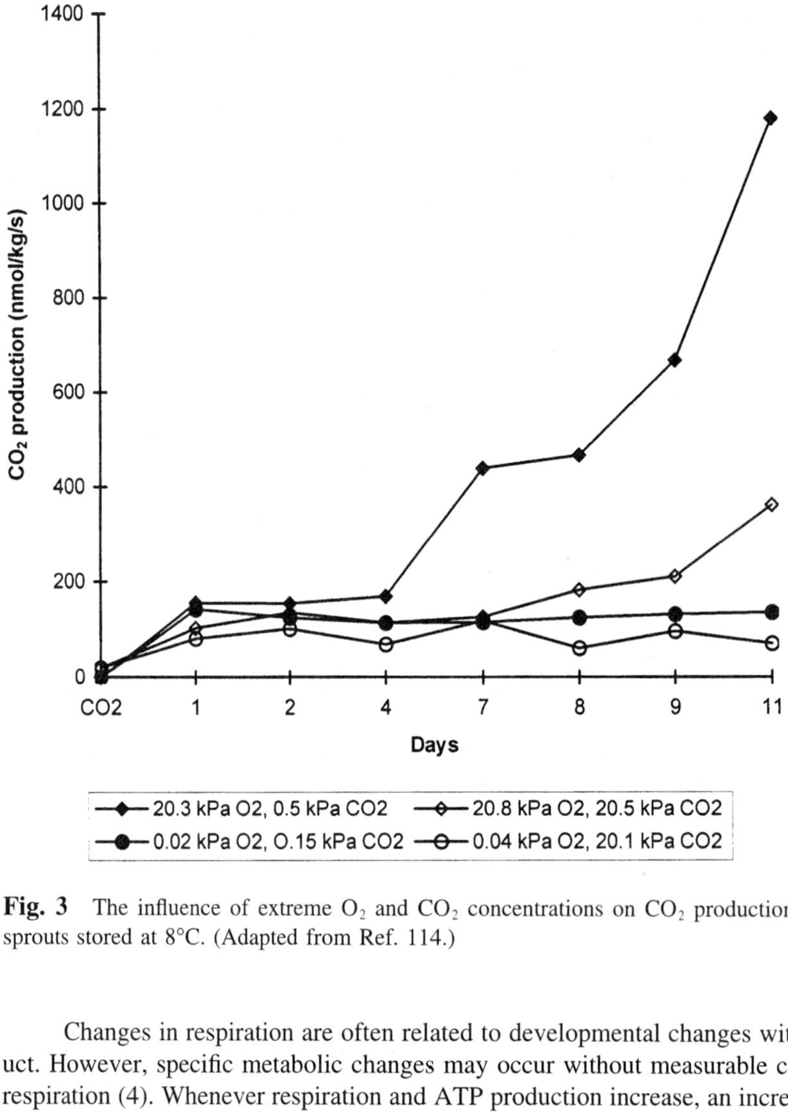

Fig. 3 The influence of extreme O_2 and CO_2 concentrations on CO_2 production of mungbean sprouts stored at 8°C. (Adapted from Ref. 114.)

Changes in respiration are often related to developmental changes within the product. However, specific metabolic changes may occur without measurable changes in net respiration (4). Whenever respiration and ATP production increase, an increase in energy requirements is expected. In such cases the storage of products at fixed low O_2 concentrations might cause problems, because energy needs are no longer met in these conditions. CA technology capable of responding to metabolic changes would be the alternative. Such a dynamic control of storage conditions (53,54) focuses on the O_2 concentration when CO_2 production is minimal (ACP; anaerobic compensation point). This, however, might not be the best procedure, since ACP tends to shift during storage, as observed in apple (28) and pear (Peppelenbos HW, unpublished data). An approach that focuses on the detection of fermentation metabolites, such as ethanol, in the storage atmosphere has also been suggested (55).

2.5.2 Temperature Control

Strict temperature control in the distribution chain is a prerequisite for optimal use of MA packages in practice, but in most countries the cool-chain between production, distribution, retail, and the consumer has many uncontrolled links. Changes in the permeabilities of most packaging films to gases in response to changes in temperature are generally less than changes in product respiration. Most of today's existing plastic films do not have the proper O_2/CO_2 permeability ratio to provide the ideal MA for many commodities at a given temperature. In view of all these variables and knowing that any change within or around the package will alter the dynamic equilibrium between the product and its environment, it is clear that knowledge about the limits of tolerance of a certain commodity is even more important for MA than it is for CA.

3 TRANSPIRATION

Fresh fruits and vegetables contain from 80 to 95% water, depending on the product. Within the fruit or vegetable, there is a continuum of intercellular spaces through which gases, including water vapor, move (56). It is assumed that this internal atmosphere is in a saturated condition (57,58). However, for most fresh fruits and vegetables, the presence of solutes and bound water in the liquid phase slightly reduces the equilibrium from 100% saturation to about 97% of the saturation vapor pressure of pure water (1). Transpiration is a mass-transfer process in which water vapor moves from the surface of a fruit or vegetable to the surrounding air. This process of moisture loss induces wilting, shrinkage, and loss of firmness and crispness of fruits and vegetables, and thus adversely affects the appearance, texture, flavor, and mass of produce. Most fruits and vegetables lose their freshness after 3 to 10% mass loss (59). Transpiration is considered to be the primary cause of postharvest losses and poor quality in leafy vegetables, such as lettuce, chard, spinach, cabbage, and green onion (60), and is considered the major cause of commercial and physiological deterioration in citrus fruits (61).

3.1 Theoretical Aspects

To understand transpiration, one must first understand Fick's law of diffusion. It states that the flux, which is the rate of movement of gas through a barrier (e.g., water vapor in and out of the plant tissue) is directly proportional to its partial pressure difference and inversely proportional to resistance, and can be viewed as:

$$\text{Flux} = \frac{P_s - P_a}{r} \tag{7}$$

where $P_s - P_a$ is the difference between the vapor pressure at the evaporating surface (P_s) and the surrounding air (P_a), and r is the resistance to vapor transfer across this barrier (s m^{-1}). This law clearly demonstrates that the driving force of transpiration is the gradient of water pressure between the plant tissue and the surrounding air. Water vapor pressure (kPa) is the pressure exerted by water vapor in the atmosphere and is a function of its density (D), mass of water per unit volume (gm^{-3}), and temperature (K), according to the ideal gas law:

$$P(\text{kPa}) = 4.62 \times 10^{-4} * D(\text{g m}^{-3}) * T(\text{K}) \tag{8}$$

By using Eq. (8), the vapor pressure values in Eq. (7) can be converted to vapor pressure density and flux of water vapor expressed as g m^{-2} s^{-1}.

The water vapor pressure of plant tissue depends on the temperature and the amount of solutes within the tissue. Since we are concerned with water movement out of fruits and vegetables, the difference in water vapor pressure between the interior of the produce and its surrounding atmosphere provides an estimate of product moisture loss.

Plants and their organs have a compromising structure in terms of diffusion. Although water moves relatively freely within the intercellular channels in many products (little resistance), a similar unrestricted exchange between the plant and its environment would result in rapid death due to water loss. Different sites along the pathway of diffusion provide varying levels of resistance to the movement of water molecules. Resistance to water and gas exchange is derived mainly from the cuticular layer (62), which is the interface between the plant and the atmosphere. Resistance from the fruit or vegetable surface thus includes that of the stomata, lenticels, cuticle, and epicuticular wax, trichomes, and hairs, and the periderm. Since the xylem vessels of harvested fruits and vegetables are probably occluded and their operation greatly impeded (59), water within a harvested commodity has to move through different routes via the continuum of the cell walls (56).

The transpiration coefficient (also referred to as conductance or permeance), which is an indication of the ease with which a fruit or vegetable surface gives off moisture, is expressed as the mass of moisture transpired per unit mass of commodity, per unit environmental water vapor pressure deficit per unit time (57,58). In some cases, the transpiration coefficient is expressed per unit of surface area of commodity, rather than per unit of mass. Free water surfaces have the highest transpiration coefficients and thus commodities such as lettuce and Brussels sprouts have very high transpiration coefficients (7400 and 6150 mg kg^{-1} s^{-1} MPa^{-1}, respectively), whereas products with thick, tough skins have very low transpiration coefficients (e.g., potatoes and apples [25 and 42 mg kg^{-1} s^{-1} MPa^{-1}, respectively) (56).

Both relative humidity and absolute humidity are commonly used to express the water content of air. Relative humidity (%) is defined as a ratio between the quantity of water vapor present and the maximum possible at that temperature and barometric pressure, whereas absolute humidity (g kg^{-1}) is a measure of the mass of water in a given mass of dry air, which is independent of temperature and pressure (63).

Psychrometric charts relate the various properties of moist air (Fig. 4). The scale along the horizontal axis represents dry bulb temperatures, based on a wet and dry bulb hygrometer. The vertical axis is the moisture content of the air, given as water vapor pressure (kPa). Dry air at all temperatures has zero water content and thus zero water vapor pressure. The maximum amount of water vapor that air can hold at a specific temperature is represented by the curved line at the top of the figure, equivalent to 100% relative humidity. Other curved lines illustrate constant relative humidity over a range of temperatures, indicating that air holds more water vapor at increasing temperatures. Generally, the maximum amount of water that air can hold doubles for every 11°C increase in temperature (64). This has important consequences on product cooling and illustrates the need to rapidly cool produce in order to minimize the vapor pressure difference between the produce and the air, and hence reduce transpiration. For example, even if saturated cool air is used for cooling, as long as the produce remains warmer it will lose water. Wet bulb temperatures are represented by lines that slope diagonally downward from left to right, whereas the horizontal lines indicate dewpoint temperatures (see Fig. 4). Dewpoint is the temperature to which moist air has to be lowered (at constant pressure) to initiate condensation (100% relative humidity) (63). Condensation caused from moving cooled packaged produce into warm moist air, can promote decay, weaken cardboard packages, and accelerate the warming of the produce. At low storage temperatures with high humidity, small tem-

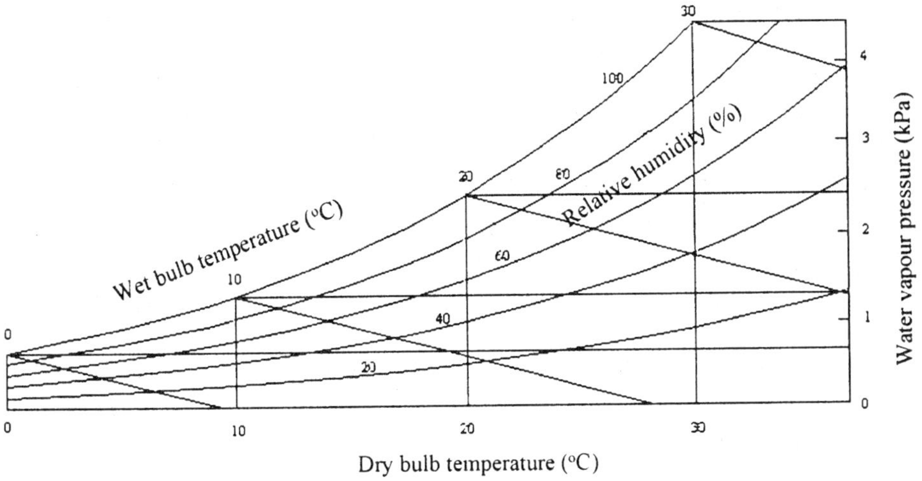

Fig. 4 Condensed psychrometric chart.

perature fluctuations can result in excessive condensation on cooling surfaces and increased water loss from the produce (1).

3.2 Factors Affecting Transpiration

3.2.1 Water Vapor Pressure Deficit

Transpiration results primarily from differences in water vapor pressure between the interior of the fruit or vegetable and the surrounding environment. The vapor pressure within a fresh commodity is thought to be solely dependent on the temperature of the commodity, whereas the vapor pressure of the surroundings is affected by temperature and relative humidity. Many researchers have found transpiration rates of fruits and vegetables to be linearly related to vapor pressure deficit (57,58). The relation commonly used is

$$\text{Transpiration rate} = \text{transpiration coefficient} \times \text{mass} \times \text{vapor pressure difference}$$

where the transpiration coefficient may be considered the reciprocal of the resistance and the vapor pressure difference is the vapor pressure within the product at the surface temperature of the product minus the vapor pressure in the surrounding environment.

In practice, the water vapor pressure deficit is often expressed in terms of relative humidity. Assuming the equilibrium water vapor pressure is equivalent to a relative humidity of 100%, evaporation is then proportional to the difference between the relative humidity in the air and at saturation (i.e., 100%). A change in relative humidity of 5%, for example, will have a much larger effect at high than at low humidity: a change from 98 to 93% increases evaporation by 250%, whereas a change from 85 to 80% would only increase evaporation by 33% (64).

3.2.2 Temperature

A difference in temperature between the product and the surrounding air can affect the equilibrium water vapor pressure, and thus transpiration. If the fruit or vegetable temperature is higher than that of the ambient air, the equilibrium water vapor pressure is increased substantially (64). If the difference between the product and air temperature is sufficiently large, such as during precooling, then the water vapor pressure of the air becomes very small relative to the equilibrium water vapor pressure. To reduce moisture loss under such conditions, it is more important to reduce the equilibrium water vapor pressure and thus the product temperature, than to cool the produce with humid air (64). For the same reason, non-uniform temperatures in a room with fruits or vegetables cause part of the load to be exposed to a high water vapor deficit, resulting in varying amounts of moisture loss and condensation.

Air temperature is also especially important when the moisture content of the air is expressed as relative humidity. The water vapor pressure deficit of air at 5°C and 95% relative humidity is 43% larger than the water vapor pressure deficit of air at 0°C and the same relative humidity (64).

3.2.3 Storage Environment

Moisture loss from fresh fruits and vegetables during storage is highly dependent on the relative humidity, air velocity, and heat of respiration, but also depends markedly on bulk density and the distance from where the air enters the mass of product (65). Models and experimental observations have shown the following: (a) An increase in air velocity (e.g., forced-air cooling) lowers the product temperature and the temperature gradient, thereby lowering the rate of moisture loss, except where dry air first contacts the fruit or vegetable. (b) For fruits or vegetables with a high transpiration coefficient, initial relative humidity has little effect on moisture loss, except where the air stream enters the load. In that situation, the effect of low relative humidity is severe. (c) For commodities with high transpiration coefficients, initial relative humidity has little effect on the relative humidity in the bulk of the produce load where the level approaches saturation. (d) Relative humidity is strongly affected by transpiration coefficients (defined earlier), but is relatively independent of the heat of respiration, air velocity, and initial relative humidity. (e) Increased heat of respiration markedly increases the gradient of moisture loss. Heat leakage into storage rooms will have a similar effect.

3.2.4 Product Respiration

Respiration produces water and heat, both of which have indirect effects on transpiration. The water produced remains within the fruit or vegetable tissue, whereas the heat is dissipated through direct heat transfer to the environment and through evaporation of water (66). The heat of respiration raises the product temperature and, therefore, increases transpiration. Apples lose small amounts of moisture when in a water-saturated environment, presumably at the same temperature as the fruit (67). This loss can be attributed to the heat of respiration, which raises the temperature of a commodity, creating a vapor pressure deficit and thus increasing evaporation.

3.2.5 Size, Shape, and Surface of a Commodity

Large fruits and vegetables possess a smaller surface-to-volume ratio than small commodities and thus tend to lose less moisture on a per unit mass basis (57,58). Larger fruits also generally possess thicker skins than smaller fruits, which could be another reason for the differences in transpiration rates (68).

The shape of a product can also affect the ratio of surface area to mass. For example, long, thin, cone-shaped carrots lose more mass than thick, cylindrical-shaped ones (69).

The longer, thinner carrots shrivel faster at the tips because of the greater surface area per unit mass. Relationships between transpiration rate and the surface-to-volume ratio have also been shown for apples, avocados, papayas, tomatoes, and watermelons (56).

Although the surface of some fruits and vegetables is almost entirely covered by an impervious waxy coating, many possible pathways for water loss exist, such as stem scars, lenticels, stomata, wounds, epidermal hairs, and cracks in the cuticle. The differences in surface structures account largely for the variation in transpiration rates between commodities and cultivars (57,58).

3.2.6 Maturity and Ripening

Transpiration rates of fruits and vegetables change with maturity. In some commodities, immature fruit transpire rapidly, decreasing until a certain stage of maturity is reached and then increasing again with over-maturity (70,71). This has been observed for apples (72) and plums (73). Other fruits may respond differently (74). Tomatoes appear to have a constant transpiration rate regardless of maturity, whereas tropical fruits, such as pawpaws, show an increase in transpiration with the appearance of color on the skin and the onset of the climacteric. Bananas and mangoes possess a steady and low initial rate of transpiration when green and unripe, followed by a rapid increase during the early stages of ripening. Ripe and overripe fruits then show a steady, but usually slowly rising rate.

Changes in the surface structure associated with maturity appear to be the reason for these variations in transpiration (57,58). In general, skin permeability of stored fruits tends to decrease with maturity, whereas respiration rate either varies or remains constant, depending on the nature of the product.

3.3 Controlling Transpiration

Transpiration of fresh fruits and vegetables can be reduced by minimizing the water vapor pressure difference between the produce and the air and/or by increasing the resistance of the product. Minimizing the water vapor pressure difference, which lowers the amount of water required to be evaporated from the produce before the air is saturated with water vapor, is generally achieved by lowering the temperature and/or raising the relative humidity of the air. The humidity of the surrounding atmosphere should be maintained at a level that produces a water vapor pressure close to that of the product. For example, very high relative humidities (95–99%) are generally required for succulent products, while much lower storage humidities (e.g., 60–70%) are needed for lower-moisture products, such as cured roots, tubers, and corms (4).

Although the relation between humidity and water exchange is relatively straightforward, the effects of temperature are more complex. Three thermal parameters have a clear effect on moisture exchange in storage (4): (a) the actual temperature, (b) the differential in temperature between the product and the atmosphere, and (c) the fluctuations in storage temperature. As illustrated by the psychrometric chart (Fig. 4), lowering the temperature decreases the amount of water that the air can hold. Therefore, at a given relative humidity the water vapor pressure difference between a product and the atmosphere will decrease with decreasing temperature, resulting in reduced transpiration. Thus, it is extremely important to cool harvested produce as quickly as possible to minimize water loss.

3.3.1 Storage Conditions

Increasing the cooling surface area has been widely used to control water vapor pressure (64). However, it is very difficult to increase the relative humidity over 90% with an increased cooling surface area alone. Direct humidification is another such method, which is the introduction of water or steam into the airstream after it leaves the cooling surface (75). Evaporation is limited by heat transfer when using a water spray, as the heat for evaporation has to be extracted from the water and from the air. An alternative is to inject steam, but the added energy has to be removed to cool this air again, adding considerably to the cost of refrigeration. Another modification tried successfully in Canada is spraying water directly on top of bulk-stored carrots with the refrigerated air moving down through the product (65). Evaporation removes some of the heat of respiration and the large surface area of the carrots ensures that the humidity of the air increases to saturation. Cooling with a wet coil is a modification of the direct humidification using a water spray in the air. In this case the water is sprayed onto the cooling surface. For operation near 0°C, the rate of airflow has to be very large and even then the air cannot be cooled below 1°C. Relative humidity levels higher than 97% are seldom obtained under these conditions and are usually less dependent on the amount of heat being removed (76).

Controlling water vapor pressure to control transpiration is used in three well-developed storage techniques: (a) jacketed storage (77), (b) the ''Humifresh'' system (76), and (c) hypobaric or low-pressure storage (78). However, detailed descriptions of these systems are beyond the scope of this introductory chapter on postharvest physiology.

Air movement is essential in refrigerated storage to remove the heat generated by stored fruits and vegetables. However, such air movement also tends to decrease the layer of moist air surrounding the product (the boundary layer). This increases the water vapor pressure difference near the surface of the product and thus increases transpiration. The faster the air movement, the greater the rate of water loss from the produce. Restricting the airflow around produce in cold storage can, therefore, reduce the rate of water loss. This can be achieved after initial cooling by decreasing the amount of air movement generated by the fans, either by running them at lower speeds or by reducing the length of time that they are operating (1). Open rooms with natural ventilation can also be modified to restrict airflow. However, there must be sufficient air movement to prevent large temperature gradients forming within the storage room.

3.3.2 Packaging

Transpiration can also be reduced by placing a physical barrier around the produce to reduce air movement across its surface. The simplest methods are to cover stacks of produce with tarpaulins, or to pack the produce into bags, boxes, or cartons (1). Close packing of produce restricts the passage of air around individual items and thus reduces transpiration. The degree to which the rate of transpiration is reduced is dependent on the permeability of the package to water vapor transfer, as well as the closeness of the containment. All materials commonly used are permeable to water to some extent. However, materials with relatively low rates of water transfer, such as polyethylene films, are considered to be good vapor barriers, whereas naked paper and fiberboard do not provide good control of transpiration because of their high permeability to water vapor. Packaging also reduces the rate of cooling by restricting air movement around individual items.

The use of plastic films, bags, wraps, liners, or bulk-box covers is an easy way to ensure high relative humidity around produce during storage, transport, and marketing (56). Such films increase resistance to water vapor transfer and thus produce a micro-atmosphere with higher relative humidity than the exterior. Since most films are not adequately permeable to be used as sealed packages, films usually have to be perforated or not closed tightly. Film packaging may also enhance decay, because of condensation within the package and the relatively high humidity.

The use of very thin plastic films for packaging individual products is a comparatively new technology. This technique is a special adaptation of film packaging, in which each product is sealed in plastic film and then passed through a hot-air tunnel to shrink the film (79,80). It greatly reduces shrinkage and mass loss, as well as improves the appearance, firmness, and overall storage life of some fruits, without any deleterious effect on flavor. For example, seal-packaging reduces mass loss, shriveling, and firmness loss in various citrus fruits (81–83), reduces mass loss in cucumbers (84), and may reduce decay in red bell peppers (85) and tomatoes (86).

Modified humidity packages (MHP) are designed for products in which dehydration causes the most important quality losses, and thus focus on controlling water vapor levels. When products such as leafy vegetables or bell peppers are not packed, quality losses such as wilting and shriveling can be observed very soon. In most ''closed'' packages (e.g., modified-atmosphere packaging), the relative humidity is close to saturation due to the water exchange between the product and surrounding atmosphere. This high humidity increases the probability for condensation and free water directly on the product, especially when the package is exposed to changing temperatures. Therefore, MHP systems are designed to control not only dehydration but condensation as well.

A package designed to have a high relative humidity at a high temperature will show condensation on the package surface or on the product if the temperature is decreased substantially. To counteract the effect of condensation, films have been developed that are coated with an antifog layer. Moisture then forms a continuous layer on the surface of such films, rather than separate droplets. This allows a clear view of the product and prevents water pooling at the bottom of a package.

Reducing water loss is one of the main aspects related to the packaging of minimally processed products, despite the usual emphasis on gas levels (20). However, for products such as onions, humidity should not be too high as it results in increased sprouting and decay. As with O_2 and CO_2, water vapor levels can be too high or too low, and an optimum level should be reached. For bell peppers this level has been estimated to be 92% relative humidity at 8°C (85). Lower relative humidity causes too much mass loss, whereas a higher relative humidity causes decay. MHP can effectively be used to minimize loss of quality in products for which water loss is the most predominant cause of quality changes, such as bell pepper and tomato. In such cases, the concentrations of O_2 and CO_2 in MHP are often close to that of ambient air.

3.3.3 Surface Coatings

Surface coatings have been applied to many fruits and vegetables to reduce transpiration losses, as well as to restore or enhance gloss and improve storage quality. For example, fruit surface coatings reduce the mass loss of papaya during storage (87); maintain better firmness, titratable acidity, and greener skin color in pears (88); decrease mass loss and the incidence of discolored stems and surface pitting in cherries during storage (89); retain flesh firmness and acidity in apples (90); and control mass loss and preserve gloss of citrus fruit (91).

Good coating formulations should have the following characteristics (56): nontoxic to humans, good permeability properties, stability in formulation, rapid drying, strong adherence to the fruit throughout the entire storage life, high gloss, and economical. However, it is possible that a specific coating may be acceptable for some produce but not for others, due to the properties of both the coating and the product.

All surface coatings offer some resistance to gas exchange, depending on their thickness and composition. Commercial carnauba waxes are more permeable to O_2, CO_2, and ethylene than waxes made from shellac and rosin (92). Hydrophilic coatings, such as those incorporating starch, carrageenan, or cellulose, retard the movement of these gases, but have less effect on water vapor (93). Since surface coatings can alter the O_2 and CO_2 levels within a product, they can also interfere with postharvest ripening, as observed in guavas coated with cellulose- or carnauba-based emulsions (94).

4 ETHYLENE

There are several recent publications (95–97) that review the general biology of ethylene (C_2H_4), including references to its role in postharvest storage of fruits and vegetables. Ethylene, which is produced naturally by many biological and nonbiological systems, is the sole known plant growth regulator that exists as a gas at normal biological temperatures. It can induce plant responses at very low concentrations (0.1 $\mu L\ L^{-1}$ or less), and has a wide variety of effects on virtually all stages of a plant's life cycle (e.g., germination, vegetative growth, flowering, fruiting, abscission, ripening, senescence, and dormancy). Ethylene is sometimes called the ''ripening gas'', because its most important postharvest effect is the acceleration of ripening and senescence. However, there are other postharvest effects of ethylene, including induction or suppression of potato sprout growth, loss of chlorophyll (degreening), and induction or suppression of disease resistance. Some of these effects are desirable and some are undesirable, depending on the intended use of the stored plant product. The presence and activity of ethylene in a storage system may go undetected if the storage operator does not appreciate its range of effects and/or does not have an ethylene detector.

4.1 Production of Ethylene

4.1.1 Biological Production

There is published evidence of ethylene production by animals, bacteria, fungi, and both nonflowering plants (e.g., algae and ferns) and flowering plants (e.g., gymnosperms and angiosperms). In flowering plants, all cells are capable of producing ethylene, but the rate of production may vary among cells and plant species. Horticultural commodities have been classified according to their ethylene production rates (97). In flowering plants, the formation of ethylene from methionine via the ACC pathway is now recognized as the major source of ethylene, whereas nonflowering plants do not possess ACC oxidase and produce ethylene from other unknown pathway(s) when they are stressed or damaged (99). In plant cells that are damaged or stressed, ethylene may be produced as a product of lipid peroxidation, but this has not been well-documented.

The formation of ethylene from carbons 3 and 4 of L-methionine involves two intermediary compounds, S-adenosylmethionine (AdoMet, formally known as SAM) and 1-aminocyclopropane-1-carboxylic acid (ACC) (Fig. 5). The three enzymes involved in this conversion are (a) AdoMet synthetase (EC 2.5.1.6) which catalyzes the conversion of L-methionine and ATP into AdoMet; (b) ACC synthase (EC 4.4.1.14) which catalyses the conversion of AdoMet to ACC. Another product of this reaction is 5′ methylthioadenosine

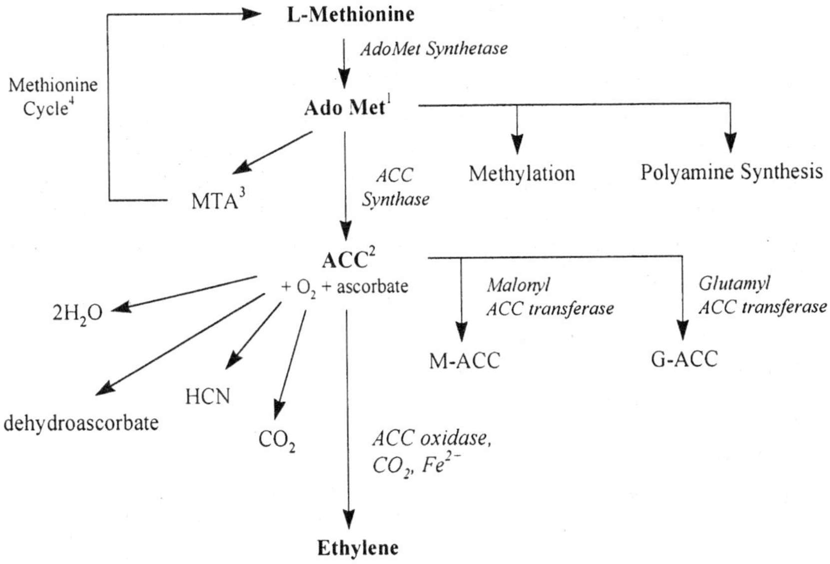

Fig. 5 Production of ethylene in higher plants: [1]S-adenosylmethionine (AdoMet); [2]1-amino-cyclopropane-1-carboxylic acid (ACC); [3]5′-methylthioadenosine (MTA); [4]also referred to as the Yang cycle (95). (Nomenclature adapted from Ref. 100.)

(MTA) that can be recycled to produce more methionine; and (c) ACC oxidase (formally known as ethylene-forming enzyme or EFE) which catalyzes the conversion of ACC to ethylene in the presence of CO_2 and Fe^{2+}. ACC can also be converted to malonyl-ACC or glutamyl-ACC in the presence of malonyl-ACC transferase or glutamyl-ACC transferase, respectively (Fig. 5).

All three enzymes are produced by multigene families (100,101). The rate of ethylene synthesis is controlled by the activity of ACC synthase, but it can sometimes be limited by ACC oxidase activity (95).

4.1.2 Non-Biological Production

Ethylene can be produced non-biologically in reactions where organic compounds are incompletely oxidized, such as incomplete combustion of organic fuels. Even before ethylene was identified, it was known that burning incense or exposing stored fruits and vegetables to smoke, kerosene heaters or illuminating gas produced responses now known to be ethylene-induced (95). Whenever the air surrounding stored fruits and vegetables contains a gaseous organic chemical or combustion gases, it is likely that ethylene-like responses will occur in the product, induced either by ethylene itself or by chemically similar compounds that mimic ethylene if present at concentrations higher than ethylene.

4.1.3 Ethylene, Respiration, and Fruit Ripening

In general, respiration rate is highest in an immature vegetable and declines with age. In fruits, there are two respiratory patterns. One group has a pattern similar to vegetables and are called non-climacteric fruits. The other group, called climacteric fruits, exhibits a temporary respiratory increase, which occurs when the fruit reaches full maturation and size and is entering the ripening stage. This respiration climacteric was first described in 1925 in apple (1) but it is now known to occur in other botanically unrelated fruit species (Table 1). Not all fruit have been examined to determine if they are climacteric or non-climacteric and, in some fruit species, there is disagreement on whether the fruit is climacteric or non-climacteric.

Climacteric fruits respond to ethylene differently than non-climacteric fruits and vegetables. In climacteric fruits, the respiration climacteric and subsequent ripening and senescence are associated with a coincident increase in endogenous ethylene production. Ethylene applied exogenously to climacteric fruits can also produce the same response, depending on the fruit species and its stage of maturity. Conversely, in non-climacteric fruits and vegetables, increasing the exogenous ethylene concentration increases the respiration and senescence rate, but both both rates decline if ethylene is removed. This response is repeatable compared with the single respiration increase in climacteric fruits.

Table 1 Classification of Some Edible Fruits According to Presence or Absence of a Respiration Climacteric During Ripening

Climacteric fruits	Non-climacteric fruits
Apple (*Malus domestica*)	Bell pepper (*Capsicum* spp.)
Apricot (*Prunus armeniaca*)	Carambola (*Averrhoa carambola*)
Asian pear (*Pyrus pyrifolia*)[a]	Cherry (*Prunus* spp.)
Avocado (*Persea americana*)	*Citrus* spp.[a]
Banana (*Musa* spp.)	Grape (*Vitis vinifera*)
Blackberry (*Rubus occidentalis*)	Lychee (*Litchi chinensis*)
Blueberry and Cranberry (*Vaccinium* spp.)[a]	Pineapple (*Ananas comosus*)
Cherimoya (*Annona cherimola*)	Pumpkin (*Cucurbita pepo*)
Eggplant (*Solanum melongena* var. *esculentum*)	Rambutan (*Nephelium lappaceum*)
	Strawberry (*Fragaria* spp.)
Feijoa (*Feijoa sellowiana*)	Tamarillo (tree tomato) (*Cyphomandra betacea*)
Fig (*Ficus carica*)	
Guava (*Psidium guajava*)	Watermelon (*Citrullus lanatus*)[a]
Jujube (*Ziziphus jujuba*)	
Kiwifruit (*Actinidia deliciosa*)	
Mango (*Mangifera indica*)	
Muskmelon (*Cucumis melo*)	Classification depends on other factors (e.g., attachment to plant, maturity at harvest)
Papaya (*Carica papaya*)	
Passionfruit (*Passiflora edulis*)	
Peach (*Prunus persica*)	
Pear (*Pyrus communis*)	Cucumber (*Cucumis sativus*)
Persimmon (*Diospyros kaki*)	Olive (*Olea europaea*)
Plum (*Prunus domestica, P. salicina*)	
Raspberry (*Rubus idaeus*)	
Sapote (*Calocarpum sapota*)	
Tomato (*Lycopersicon esculentum*)	

[a] Classification is not conclusive, or both types may be present.
Source: Refs. 1, 94, 114 and B. McGlasson (personal communication).

4.2 Ethylene Action (Perception and Signal Transduction) in Higher Plants

It is believed that ethylene exerts its effects by altering gene expression (102). There is evidence that ethylene can control both gene transcription and posttranscription processes (103). In the last 10 years considerable progress has been made in understanding how ethylene may be able to alter gene expression. The work has been reviewed by several authors (100,102,104,105). Using *Arabidopsis thaliana* (L.) Heynh., a pathway has been proposed comprising the genes (and associated protein products) that may be involved in ethylene perception, signal transduction, and control of gene transcription (Fig. 6).

There is considerable evidence that the ETR1 gene (and isoforms such as ERS) encodes for a membrane-bound protein that binds ethylene, whereas CTR1 encodes a protein that blocks the activity of EIN2 and subsequent downstream signal transduction. When present, ethylene may bind to and activate the ETR1 protein. The concentration of ethylene necessary to activate the ETR1 protein is quite low. In transgenic yeast containing the ETR1 gene, the K_d (dissociation constant) of ETR1 is 0.04 μL L^{-1} ethylene (gas phase) (104), which is close to the amount of ethylene required for a half-maximal response in the seedling growth assay. Binding of ethylene to ETR1 is reversible. The release of bound ethylene from the transgenic yeast has a half-life of 12 h, a rate similar to that observed with one class of binding activity reported from several plant sources (104). The ethylene–

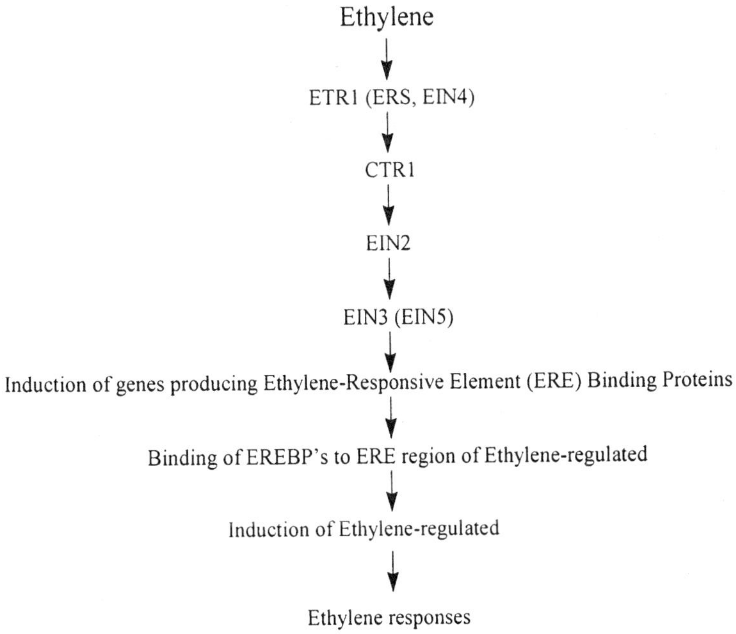

Fig. 6 Proposed linear sequence of gene action in the ethylene transduction pathway of *Arabidopsis thaliana* (L.) Heynh. In the absence of ethylene, CTR1 inactivates EIN2. In the presence of ethylene, an ethylene–ETR1 complex is formed that inactivates CTR1. Once CTR1 is inactive, the EIN2 gene is free to activate the rest of the sequence. (Adapted from Refs. 102, 104, and 105.)

ETR1 complex may then inactivate CTR1, either directly or indirectly (through an interaction with EIN4) and possibly through phosphorylation. With CTR1 inactivated, the EIN2 gene may then activate ethylene-regulated genes, possibly by acting on genes like EIN3 (or EIN5) and those producing EREBPs. More research in the future will verify the validity of this proposed sequence and how universal it may be in plants other than *A. thaliana*.

4.3 Inhibitors of Ethylene Production or Action

4.3.1 Ethylene Production

There are numerous inhibitors that endogenously or exogenously inhibit ethylene production (95). Many of these inhibitors vary in their capacity to inhibit ethylene production and occasionally may stimulate ethylene production (e.g., CO_2). The following are some of the best known and consistent inhibitors.

ACC synthase is inhibited by aminoethoxyvinylglycine (AVG) and aminooxyacetic acid (AOA). AOA has not been registered for commercial use, but AVG is now manufactured by Abbott Laboratories and registered for commercial use on apples and pears in the United States under the trade name ReTain (106,107; R Groen, Abbott Laboratories, personal communication).

ACC oxidase activity is inhibited by low O_2 concentrations or anaerobic conditions. Increasing temperature above 30°C progressively inactivates ACC oxidase (95). Cobalt (II) ions and nickel (II) ions inhibit ACC oxidase to a lesser extent.

4.3.2 Ethylene Action

Two types of inhibitors of ethylene action are recognized. The first group are mild toxicants that block ethylene action by slowing down cellular physiology. The second group of inhibitors of ethylene action act as competitive inhibitors, which means that they compete with ethylene by combining with the ethylene receptor proteins such as ETR1. In both types, the inhibition can be reversed by flushing the inhibitor out of the tissue. The action of the second group can be overcome by increasing the concentration of ethylene.

Examples of the first type of inhibitor are temperature (low or high), ethanol, and CO_2. Compounds that fit into the second group are silver ion [Ag (1)], 2,5-norbornadiene (NBD), diazocyclopentadiene (DACP), cyclopropene (CP), 3,3-dimethylcyclopropene (3,3-DMCP) and 1-methylcyclopropene (1-MCP) (108,109). None of these inhibitors are completely free of toxic or undesirable side effects, with the possible exception of ethanol and 1-MCP. The use of temperature, ethanol, and CO_2 on edible food products does not pose any problems, however, this is not true for Ag (I) and NBD. Silver ion is used commercially in the cut flower industry in the form of silver thiosulfate, but NBD will not likely be used commercially because it has an unpleasant odor and is a suspected carcinogen. The recently discovered inhibitor 1-MCP is nontoxic at concentrations that are active and has potential for commercial use (109).

4.3.3 CA Storage

The principles of CA storage and MA packaging have already been presented earlier in the respiration section of this chapter. With our current knowledge of ethylene physiology, it is apparent that CA storage provides the optimum conditions that reduce both ethylene production (low O_2) and ethylene action (high CO_2, low temperature). CA storage can reduce ethylene biosynthesis by its effects on both ACC synthase and ACC oxidase. Low O_2 and/or elevated CO_2 reduce ethylene biosynthesis in 'Golden Delicious' apples by delaying and suppressing expression of ACC synthase at the transcriptional level and by reducing the abundance of active ACC oxidase protein (110). Ethylene action can also be suppressed by low temperature and elevated CO_2 (95).

The possibility that CA storage would be more beneficial if ethylene was not present has been studied with mixed results. It appears that ethylene removal may be beneficial only when the ethylene concentration can be kept below physiological levels of 1 $\mu L\ L^{-1}$ or less (95). It has also been suggested that there is virtually no benefit to ethylene scrubbing in low-oxygen CA storage (1 kPa, rather than 3 to 5 kPa) due to the lower rate of ethylene production and ethylene concentration within the fruit in the presence of 1 kPa O_2 or less (95). However, ethylene scrubbing improves firmness retention in 'Elstar' apples stored in 1.2 kPa O_2 (ULO), compared with ULO with no ethylene scrubbing (110). Most of the research on ethylene scrubbing in CA conditions has been done on apples and it is not certain that other commodities would respond similarly.

4.4 Ethylene Addition

In some circumstances, ethylene is intentionally added to the stored product to accelerate some of the plant processes it controls. Various published summaries should be consulted for more details on treatment and safety procedures (see 112,113). The concentration of ethylene and treatment duration to achieve the desired effect will vary with the product. For most products, the threshold concentration is 0.1 $\mu L\ L^{-1}$ and the maximum effect is achieved at 10 $\mu L\ L^{-1}$ or less. However, some treatments may use up to 100 $\mu L\ L^{-1}$ to adjust for leaky rooms.

The source of ethylene can vary. Bottled ethylene gas is relatively inexpensive, but other sources may be preferred when bottled gas is not available or there is a safety concern (ethylene gas is explosive at concentrations between 2.7 and 36 kPa). Ethylene can be generated from commercial liquids that contain ethanol and possibly other proprietary products. Heating of the liquid breaks down the ethanol and ethylene is produced. Ethephon (2-chloroethane phosphonic acid), commercially available under various trade names, such as Ethrel, Florel, Cerone, Prep, and CEPA, is available in a liquid form. When the pH of the solution is raised above 5, the ethephon molecule is hydrolyzed, releasing several products, one of which is ethylene. The rate of hydrolysis increases as the pH increases. Calcium carbide (Ca_2C), which exists as a grayish solid, produces acetylene and trace amounts of ethylene when hydrolyzed (112). Acetylene has ethylene-like effects but it takes much higher concentrations, which increases the possibility of an explosion. This method is still popular in parts of the world where the other sources of ethylene are not available or are more expensive (e.g., India) (M Upadhaya, personal communication).

Perhaps the simplest and cheapest way to add ethylene is based on the fact that there is a wide range in the rate of ethylene production in plant products (98). Therefore, ethylene can be added to a storage by adding ripe plant products generating ethylene. This method may be appropriate in small-scale commercial operations or in the home.

4.5 Ethylene Removal

In many storage situations, it is important to remove ethylene to slow ripening and senescence. The simplest method to remove ethylene is to remove the source(s) and/or to ventilate the storage area with ethylene-free air. If this simple approach is not feasible there are alternative methods to remove unwanted ethylene. However, it is not certain with any of these methods that the ethylene concentration is constantly kept low enough to avoid induction of ethylene-based responses. The majority of the commercially available methods use one or more of the following processes (112,113).

Potassium permanganate ($KMnO_4$) oxidizes ethylene to CO_2 and H_2O. It is usually available commercially impregnated into a porous material that may also have some ethylene absorbing properties. Ethylene can be oxidized by ultraviolet (UV) lamps or ozone (O_3) generators (95,112). Although O_3 gas oxidizes ethylene, it is possible that some of the ethylene is directly oxidized by the UV radiation or the electric corona used in some O_3 generators. Ethylene is also oxidized if it is heated in the presence of a catalyst, such as platinized asbestos (95,112). There are commercial units available that successfully circulate the cold air from a storeroom, heat and oxidize the ethylene, and then remove the heat from the air before it is returned to the cold storeroom.

Ethylene can be adsorbed on activated or brominated charcoal, but this method is not as successful commercially as the other mentioned methods. Hypobaric (low-pressure) storage was promoted in the 1960s and 1970s as a storage method that could remove ethylene from the fruit. However, attempts to commercialize this technology have not been successful (95,112,113). Another interesting removal method that has been proposed is the use of certain soil bacteria that use ethylene as a biochemical substrate (112,113).

Two systems that remove O_2 from storage rooms, and may also remove ethylene, have been commercially available since the 1980s (95). One is the hollow fiber technology and the other is pressure swing adsorption (PSA). However, it has not been clearly demonstrated that the removal rate is sufficient to keep ethylene below 1 $\mu L\ L^{-1}$ in commercial storages. In 1997, a modification of PSA called vacuum swing adsorption (VSA) was commercially introduced, and is claimed to be more energy efficient (L Bakker, personal communication).

5 CONCLUDING REMARKS

This chapter has summarized the three primary physiological factors that affect the postharvest quality of fresh fruits and vegetables; namely, respiration, transpiration, and ethylene metabolism. Although each of these three factors were presented separately, it should be appreciated that they are not independent of each other. Not only do they respond similarly to certain environmental changes (e.g., lowering of storage temperature), but they influence each other. For example, lower respiration rates (and less heat of respiration) slows ethylene metabolism and transpiration rates, whereas high ethylene concentrations can increase respiration and transpiration rates. Therefore, disregarding the effects of any one of these three factors will almost certainly result in unacceptable quality deterioration. Furthermore, extremes of any of these factors may result in the development of specific physiological disorders, making the fresh produce unmarketable. On the other hand, physiological processes leading to enhanced quality (e.g., color development, softening, astringency loss, and aroma production) are also linked to, and strongly influenced by, respiration, transpiration, and ethylene metabolism.

REFERENCES

1. Wills R, McGlasson B, Graham D, Joyce D. Postharvest: An Introduction to the Physiology and Handling of Fruits, Vegetables and Ornamentals, 4th ed. CAB International, New York, 1998.

2. Watada AE, Herner RC, Kader AA, Romani RJ, Staby GL. Terminology for the description of developmental stages of horticultural crops. Hortscience 1984; 19:20–21.
3. Brady CJ. Fruit ripening. Annu Rev Plant Physiol 1987; 38:155–178.
4. Kays SJ. Postharvest Physiology of Perishable Plant Products. Van Nostrand Reinhold, New York, 1991.
5. Weichmann J, ed. Postharvest Physiology of Vegetables. Marcel Dekker, New York, 1987.
6. Lambers H, Atkin O. Carbon partitioning and source-sink interactions in plants. Curr Top Plant Physiol 1995; 13:226–238.
7. Lange DL, Kader AA. Changes in alternative pathway and mitochondrial respiration in avocado in response to elevated carbon dioxide levels. J Am Soc Hort Sci 1997; 122:245–252.
8. Hoefnagel M. Effects of nutrient limitations on respiration of plant cells in suspension culture. Dissertation, University of Leiden, 1993.
9. Kidd F, West C. Brown heart, a functional disease of apples and pears. Special report. Food Inv Board, Dept Sci Ind Res 1923; 12:1–54.
10. Thomas M. A quantative study of the production of ethyl alcohol and acetaldehyde by cells of the higher plants in relation to concentration of oxygen and carbon dioxide. Biochem J 1925; 19:927–947.
11. Blackman FF. Formulation of a catalytic system for the respiration of apples and its relation to oxygen. Proc R Soc Lond B 1928; 103:491–523.
12. Laties GG. [Review] Franklin Kidd, Charles West and FF Blackman: the start of modern postharvest physiology. Postharvest Biol Technol 1995; 5:1–10.
13. Kidd F. The retarding effect of carbon dioxide on respiration. Proc R Soc Lond B 1917; 87: 36–156.
14. Thornton NC. Carbon dioxide storage. III. The influence of carbon dioxide on the oxygen uptake by fruits and vegetables. Contr Boyce Thompson Inst 1933; 5:371–402.
15. Peppelenbos HW, Tijskens LMM, van 't Leven J, Wilkinson EC. Modelling oxidative and fermentative carbon dioxide production of fruits and vegetables. Postharvest Biol Technol 1996; 9:283–295.
16. Boersig MR, Kader AA, Romani RJ. Aerobic–anaerobic respiratory transition in pear fruit and cultured pear fruit cells. J Am Soc Hort Sci 1988; 113:869–873.
17. Beaudry RM, Cameron AC, Shirazi A, Dostal–Lange DL. Modified-atmosphere packaging of blueberry fruit: effect of temperature on package O_2 and CO_2. J Am Soc Hort Sci 1992; 117:436–441.
18. Chevillotte P. Relation between the reaction cytochrome oxidase–oxygen and oxygen uptake in cells in vivo. J Theor Bio 1973; 39:277–195.
19. Burton WG. Some biophysical principles underlying the controlled atmosphere storage of plant material. Ann Appl Biol 1974; 78:149–168.
20. Cameron AC, Talasila PC, Joles DW. Predicting film permeability needs for modified atmosphere packaging of lightly processed fruits and vegetables. Hortscience 1995; 30:25–34.
21. Platenius H. Effect of oxygen concentration on the respiration of some vegetables. Plant Physiol 1943; 18:671–684.
22. Hulme AC, Rhodes MJC. Pome fruits. In: Hulme AC, ed. The Biochemistry of Fruits and Their Products, Vol 2. Academic Press, London, 1972; 333–373.
23. Perata P, Alpi A. Plant responses to anaerobiosis. Plant Sci 1993; 93:1–17.
24. Ricard B, Couée I, Raymond P, Saglio PH, Saint-Ges V, Pradet A. Plant metabolism under anoxia. Plant Physiol Biochem 1994; 32:1–10.
25. Pfister-Sieber M, Brändle R. Aspects of plant behavior under anoxia and postanoxia. Proc R Soc Edinb, 1994; 102B:313–324.
26. Kader AA, Zagory D, Kerbel EL. Modified atmosphere packaging of fruits and vegetables. Crit Rev Food Sci Nutr 1989; 28:1–30.
27. Good AG, Muench DG. Long-term anaerobic metabolism in root tissue. Plant Physiol 1993; 101:1163–1168.

28. Peppelenbos HW, Rabbinge R. Respiratory characteristics and calculated ATP production of apple fruit in relation to tolerance to low O_2 concentrations. J Hort Sci 1996; 71:985–993.
29. Fidler JC, North CJ. The effect of conditions of storage on the respiration of apples. I. The effects of temperature and concentrations of carbon dioxide and oxygen on the production of carbon dioxide and uptake of oxygen. J Hort Sci 1967; 42:189–206.
30. Yang CC, Chinnan MS. Modeling the effect of O_2 and CO_2 on respiration and quality of stored tomatoes. Trans ASAE 1988; 31:920–925.
31. Peppelenbos HW, van 't Leven J, van Zwol BH, Tijskens LMM. The influence of O_2 and CO_2 on the quality of fresh mushrooms. Proc 6th International Controlled Atmosphere Research Conference, 1993; 746–758.
32. Cameron AC. Modified atmosphere packaging, a novel approach for optimizing package oxygen and carbon dioxide. Proc 5th International Controlled Atmosphere Research Conference, 1989; 2:197–208.
33. Joles DW, Cameron AC, Shirazi A, Petracek PD, Beaudry RM. Modified-atmosphere packaging of ''Heritage'' red raspberry fruit: respiratory response to reduced oxygen, enhanced carbon dioxide, and temperature. J Am Soc Hort Sci 1994; 119:540–545.
34. Hayakawa K, Henig YS, Gilbert SG. Formulae for predicting gas exchange of fresh produce in polymeric film. J Food Sci 1975; 40:186–191.
35. Banks NH, Hewett EW, Rajapakse NC, Cleland DJ, Austin PC, Stewart TM. Modelling fruit response to modified atmospheres. Proc 5th International Controlled Atmosphere Research Conference, 1989; 1:359–366.
36. Lee DS, Haggar PE, Lee J, Yam KL. Model for fresh produce respiration in modified atmospheres based on principles of enzyme kinetics. J Food Sci 1991; 56:1580–1585.
37. Chang R. Physical Chemistry with Applications to Biological Systems, 2nd ed. Macmillan Publishing, New York, 1981.
38. Song Y, Kim HK, Yam KL. Respiration rate of blueberry in modified atmosphere at various temperatures. J Am Soc Hort Sci 1992; 117:925–929.
39. Andrich G, Zinnai A, Balzini S, Silvestri S, Fiorentini R. The kinetic effect of pCO_2 on the respiration of rate of Golden Delicious apples. Acta Hort 1994; 368:374–381.
40. Banks NH, Dadzie BK, Cleland DJ. Reducing gas exchange of fruits with surface coatings. Postharvest Biol Technol 1993; 3:269–284.
41. Beaudry RM, Uyguanco ER, Lennington TM. Relationship between headspace and tissue ethanol levels of blueberry fruit and carrot roots sealed in LDPE packages. Proc 6th International Controlled Atmosphere Research Conference, 1993; 1:87–94.
42. Lee L, Arul J, Lencki R, Castaigne F. A review on modified atmosphere packaging and preservation of fresh fruits and vegetables: physiological basis and practical aspects—part I. Pack. Technol Sci 1995; 8:315–331.
43. Brash DW, Charles CM, Wright S, Bycroft BL. Shelf-life of stored asparagus is strongly related to postharvest respiratory activity. Postharvest Biol Technol 1995; 5:77–81.
44. Tijskens LMM. A model on the respiration of vegetable produce during postharvest treatments. In: Fenwick GR, Hedley C, Richards RL, Khokhar S, eds. Agri Food Quality. Royal Chemical Society, 1996; 179:322–327.
45. Floros JD. Controlled and modified atmospheres in food packaging and storage. Chem Eng Prog 1990; 6:25–32.
46. Jameson J. CA storage technology—recent developments and future potential. Proc COST94 Workshop, 1995; 1–12.
47. Berard JE. Memoire sur la maturation des fruits. Ann Chim Phys 1819; 16:152–251.
48. Kupferman E. Controlled atmosphere storage of apples. Proc 7th International Controlled Atmosphere Research Conference, 1997; 2:1–30.
49. Weichmann J. Physiological response of root crops to controlled atmospheres. Proc 2nd International Controlled Atmosphere Research Conference. Hort Rep 1977; 28:667–736.

50. Kader AA. A summary of CA requirements and recommendations for fruits other than apples and pears. Proc. 7th International Controlled Atmosphere Research Conference, 1997; 3:1–34.
51. Richardson DG, Kupferman E. Controlled atmosphere storage of pears. Proc. 7th International Controlled Atmosphere Research Conference, 1997; 2:31–35.
52. Saltveit ME. A summary of CA and MA requirements and recommendations for harvested vegetables. Proc 7th International Controlled Atmosphere Research Conference, 1997; 4:98–117.
53. Wollin AS, Little CR, Packer JS. Dynamic control of storage atmospheres. Proc. 4th International Controlled Atmosphere Research Conference, 1985; 308–315.
54. Wolfe GC, Black JL, Jordan RA. The dynamic control of storage atmospheres. Proc. 6th International Controlled Atmosphere Research Conference, 1993; 1:323–332.
55. Schouten SP, Prange RK, Verschoor J, Lammers TR, Oosterhaven J. Improvement of quality of Elstar apples by dynamic control of ULO conditions. Proc. 7th International Controlled Atmosphere Research Conference, 1997; 2:71–78.
56. Ben-Yehoshua S. Transpiration, water stress and gas exchange. In: Weichmann J, ed. Postharvest Physiology of Vegetables. Marcel Dekker, New York, 1987; 113–170.
57. Sastry SK, Baird CD, Buffington DE, Gaffney JJ. Transpiration rates of certain fruits and vegetables. Trans ASAE 1978; 84:237–254.
58. Sastry SK, Baird CD, Buffington DE, Gaffney JJ. Factors affecting rates of transpiration from stored products. ASAE Paper 1979; 79-4033.
59. Burton WG. Post-harvest Physiology of Food Crops. Longman Publishing, New York, 1982.
60. Kader AA. Postharvest quality maintenance of fruits and vegetables in developing countries. In: Lieberman M, ed. Plenum Press, New York, 1983; 455–470.
61. Ben-Yehoshua S. Gas exchange, transpiration and the commercial deterioration of stored orange fruit. J Am Soc Hort Sci 1969; 94:524–528.
62. Burg SP, Burg EA. Gas exchange in fruits. Physiol Plant 1965; 18:870–883.
63. Grierson W, Wardowski WF. Humidity in horticulture. Hortscience 1975; 10:356–360.
64. van den Berg L. Water vapor pressure. In: Weichmann J, ed. Postharvest Physiology of Vegetables. Marcel Dekker, New York, 1987; 203–230.
65. Thompson JF. Psychrometrics and perishable commodities. In: Kader AA, ed. Postharvest Technology of Horticultural Crops, 2nd ed. Publication 3311, Univ. of Calif., Div. of Agric. and Natural Resources, Oakland, CA, 1992; 79–84.
66. Meffert HF. Observations on weight loss of fruit during cold storage and transport. In: Annexe, 1970–3 Bulletin, International Institute of Refrigeration. 1970; 307–320.
67. Lentz CP, Rooke EA. Rates of moisture loss of apples under refrigerated storage conditions. Food Technol 1964; 18:119–121.
68. Karmarkar DV, Joshi BM. The relation of the size of the fruit to the loss of weight in storage. Indian J Agric Sci 1940; 10:1021–1029.
69. Apeland J, Baugerød H. Factors affecting weight loss in carrots. Acta Hort 1971; 20:92–97.
70. Christopher EP, Pianiazek SA. Visible shriveling of fruits. Rhode Island Agric Exp Stn Annu Rep 1943; 55:44–47.
71. Smock RM, Neubert AM. Apples and Apple Products. Interscience Publishers, New York, 1950.
72. Pieniazek SA. Maturity of apple fruits in relation to rate of transpiration. Proc Am Soc Hort Sci 1943; 42:231–237.
73. Smith AJM. Evaporation for plums. Great Britain Dept Sci Ind Res Food Invest Board Rpt 1935; 152–153.
74. Leonard ER. Studies in tropical fruits. X. Preliminary observations on transpiration during ripening. Ann Bot 1941; 5:89–119.
75. Lentz CP. Humidification of cold storages. Can J Technol 1954; 32:156–163.

76. Meredith D. The humifresh system: design and operating experience. In: Symposium on RH and the Storage of Fresh Fruits and Vegetables: Recent Research and Developments. Am Soc Htg Refrig Air Cond Eng Jan., 1973.
77. van den Berg L, Lentz CP. High humidity storage of vegetables and fruits. Hortscience 1978; 13:565–569.
78. Burg SR, Kosson R. Metabolism, heat transfer and water loss under hypobaric conditions. In: Lieberman M, ed. Postharvest Physiology and Crop Preservation. Plenum Press, New York, 1983; 399–424.
79. Ben-Yehoshua S. Extending the life of fruit by individual seal-packaging in plastic film; status and prospects. Plasticulture 1983; 58:45–57.
80. Ben-Yehoshua S. Individual seal-packaging of fruit and vegetables in plastic film—a new postharvest technique. Hortscience 1985; 20:32–37.
81. Ben-Yehoshua S, Kobiler I, Shapiro B. Some physiological effects of delaying deterioration of citrus fruits by individual seal packaging in high density polyethylene film. J Am Soc Hort Sci 1979; 104:868–872.
82. Ben-Yehoshua S, Kobiler I, Shapiro B. Effects of cooling versus seal-packaging with high-density polyethylene on keeping qualities of various citrus cultivars. J Am Soc Hort Sci 1981; 106:536–540.
83. Purvis AC. Effects of film thickness and storage temperature on water loss and internal quality of seal-packaged grapefruit. J Am Soc Hort Sci 1983; 108:562–566.
84. Risse LA, Chun D, McDonald RE, Miller WR. Volatile production and decay during storage of cucumbers waxed, imazalil-treated, and film-wrapped. Hortscience 1987; 22:274–276.
85. Rodov V, Ben-Yehoshua S, Fierman T, Fang D. Modified-humidity packaging reduces decay of harvested red bell pepper fruit. Hortscience 1995; 30:299–302.
86. Shirazi A, Cameron AC. Controlling relative humidity in modified atmosphere packages of tomato fruit. Hortscience 1992; 27:336–339.
87. Paull RE, Chen NJ. Waxing and plastic wraps influence water loss from papaya fruit during storage and ripening. J Am Soc Hort Sci 1989; 114:937–942.
88. Meheriuk M, Lau OL. Effect of two polymeric coatings on fruit quality of ''Bartlett'' and ''d'Anjou'' pears. J Am Soc Hort Sci 1988; 113:222–226.
89. Lidster PD. Some effects of emulsifiable coatings on weight loss, stem discoloration, and surface damage disorders in ''van'' sweet cherries. J Am Soc Hort Sci 1981; 106:478–480.
90. Lau OL, Meheriuk M. The effect of edible coatings on storage quality of McIntosh, Delicious and Spartan apples. Can J Plant Sci 1994; 74:847–852.
91. Hagenmaier RD, Baker RA. Layered coatings to control weight loss and preserve gloss of citrus fruit. Hortscience 1995; 30:296–298.
92. Hagenmaier RD, Shaw PE. Gas permeability of fruit coating waxes. J Am Soc Hort Sci 1992; 17:105–109.
93. Kester JJ, Fennema OR. Edible films and coatings: a review. Food Technol 1986; 40:47–59.
94. McGuire RG, Hallman GJ. Coating guavas with cellulose- or carnauba-based emulsions interferes with postharvest ripening. Hortscience 1995; 30:294–295.
95. Abeles PB, Morgan PW, Saltveit ME Jr. Ethylene in Plant Biology. 2nd ed. Academic Press, San Diego, 1992.
96. Kader AA, technical ed. Postharvest Technology of Horticultural Crops. 2nd ed. Publication 3311, Univ of Calif, Div of Agric and Natural Resources, Oakland, CA, 1992.
97. Mattoo AK, Suttle JC. The Plant Hormone Ethylene. CRC Press, Boca Raton, FL, 1991.
98. Kader AA. Postharvest biology and technology: an overview. In: Kader AA, Technical ed. Postharvest Technology of Horticultural Crops. 2nd ed. Publication 3311, Univ of Calif, Div of Agric and Natural Resources, Oakland, CA, 1992; 15–20.
99. John P. Ethylene biosynthesis: the role of 1-aminocyclopropane-1-carboxylate (ACC) oxidase, and its possible evolutionary origin. Physiol Plant 1997; 100:583–592.

100. Fluhr R, Mattoo AK. Ethylene—biosynthesis and perception. Crit Rev Plant Sci 1996; 15(5/6):479–523.
101. Morgan PW, Drew MC. Ethylene and plant responses to stress. Physiol Plant 1997; 100: 620–630.
102. Deikman J. Molecular mechanisms of ethylene regulation of gene transcription. Physiol Plant 1997; 100:561–566.
103. Lincoln JE, Fischer RL. Diverse mechanisms for the regulation of ethylene-inducible gene expression. Mol Gen Genet 1988; 212:71–75.
104. Bleecker AB, Schaller GE. The mechanism of ethylene perception. Plant Physiol 1996; 111: 653–660.
105. Ecker JR. The ethylene signal transduction pathway in plants. Science 1995; 268:667–675.
106. Good Fruit Grower. ReTain is registered for apples and pears. Good Fruit Grower 1997; 48(10):37.
107. Speth J. Realize potential with ReTain. Amer Fruit Grower 1997; June:25–27.
108. Sisler EC. Ethylene-binding components in plants. In: Mattoo AK, Suttle JC, eds. The Plant Hormone Ethylene. CRC Press, Boca Raton, FL, 1991; 81–99.
109. Sisler EC, Serek M. Inhibitors of ethylene responses in plants at the receptor level: recent developments. Physiol Plant 1997; 100:577–582.
110. Gorny JR, Kader AA. Controlled-atmosphere suppression of ACC synthase and ACC oxidase in ''Golden Delicious'' apples during long-term cold storage. J Am Soc Hort Sci 1996; 751–755.
111. van Schaik A. Ethyleenmanagement kan hardheidsverlies bij Elstar beperken. Fruitteelt 1996: 10–11.
112. Reid MS. Ethylene in postharvest technology. In: Kader AA, technical ed. Postharvest Technology of Horticultural Crops. Publication 3311, Univ of Calif, Div of Agric and Natural Resources, Oakland, CA, 1992:97–108.
113. Sherman M. Control of ethylene in the postharvest environment. Hortscience 1985; 20:57–60.
114. Peppelenbos HW, Brien L, Gorris LGM. The influence of carbon dioxide on gas exchange of mungbean sprouts at aerobic and anaerobic conditions. J Sci Food Agric 1998; 76:443–449.
115. Seymour GB, Taylor JE, Tucker GA. Biochemistry of Fruit Ripening. Chapman & Hall, London, 1993.

17

Maturity and Quality Grades for Fruits and Vegetables

THOMAS H. J. BEVERIDGE

Agriculture and Agri-Food Canada, Summerland, British Columbia, Canada

1 INTRODUCTION

Fruit and vegetable development starts with formation of an edible part—fruit setting, seedling emergence, tuber development, or stalk development—and ends with loss of edible character through physiological deterioration, development of fibrous character, or spoilage through microbiological intervention (Ryall and Lipton, 1972; Reid, 1992). The state of maturity of a fresh horticultural commodity is therefore a continuum along this development scale (Fig. 1). A time scale is associated with this development, but the exact length of the scale is variable and unique to the commodity. The term *mature* implies that point in the development scale when the horticultural commodity is in a state that is ready to use (process, store) or to eat. The breadth of the implications of the term *mature* may be appreciated by considering a commodity such as bean sprouts, which are mature at a very young physiological age, versus apples, in which maturity occurs toward the end of the development cycle. These two products span the maturity scale perceived from both a chronological and physiological development perspective. Both products are mature when ready to use or eat. Clearly, the term *mature* also depends upon context for interpretive precision in spite of attempts at term definitions (Watada et al., 1984). As a further complication, with a commodity like apples, which have serial customers such as a packing house/storage facility followed by retail outlets and in turn by the retail customer, maturity may carry different meanings. For the packing house, maturity means suitability for storage followed by sufficient shelf life to assure profitable supply to markets. On the other hand, the retail consumer may value color, flavor, or textural attributes. This implies two definitions of maturity and an appropriate compromise must be achieved through the marketing process.

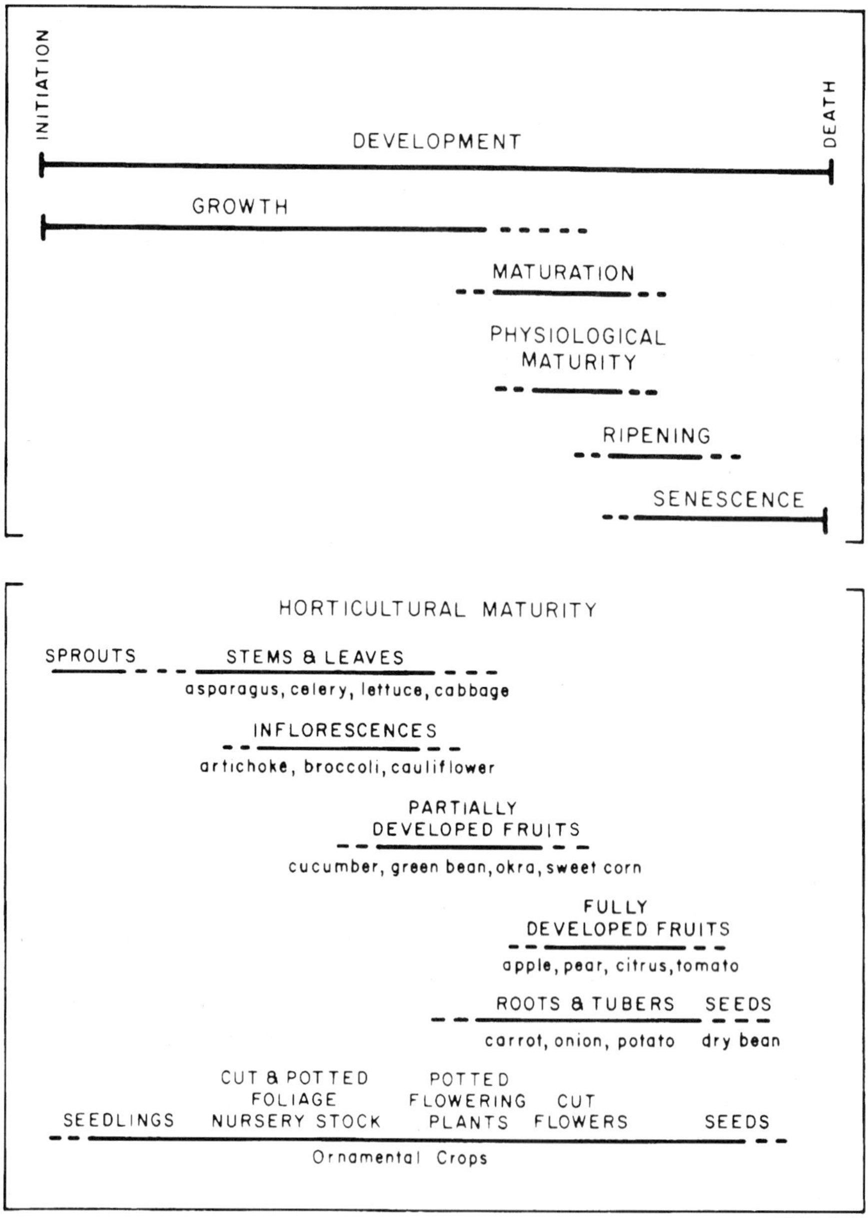

Fig. 1 Stages of development and senescence based on physiological processes and usage of horticultural crops. (From Watada et al., 1984.)

Quality, on the other hand, relates to the degree of satisfaction that the consumer or customer is likely to obtain through the use of the mature commodity. Clearly, this satisfaction is likely to be diminished if the commodity is over- or undermature, so that optimal quality is only obtainable at optimal maturity. Because of this equation, maturity and quality are interrelated and with market expectations or requirements. These expectations or requirements are expressed through regulations or guidelines published by grower groups, through contractual arrangements, or through legislative authority expressed as quality standards or grades and enforced through inspection agencies associated with government departments such as Agriculture and Agri-Food Canada (AAFC) or the United States Department of Agriculture (USDA).

2 GRADE STANDARDS

Countries generally legislatively regulate safety, inspection, and quality grades for agricultural commodities and food (Table 1; Kader, 1992; Canada Agricultural Products Act); so as to regulate trade and provide standardized coding for the various grade categories at least within a country. Import and export grades or categories are usually defined by the same legislation. Further regulations may be enacted within subfederal jurisdictions (state, provincial), which may have impact as well (Kader, 1992; Table 1). In Canada, the regulations have general provisions relating to the edible nature of the food, its health and safety attributes, and irradiation and packaging regulations. Packaging regulations pertain to branding, weights and measures, grade name displays, and display material generally. Further regulations relate to inspection procedures, administrative authority, fees, and provision for forfeiture and seizure.

These regulations can be obtained by writing to the appropriate government department or agency, but with the widespread use and availability of the Internet, governments are increasingly making the pertinent regulations available through this source. Table 1 is a listing of websites encountered during the preparation of this chapter. The list is not exhaustive but should begin to provide useful, quick access to government regulations concerning fruit and vegetables. Many websites are ''under construction'' and in a state of flux but clearly will be a major source of regulatory information in the future. Ideally others will add to the list as websites are developed by various governments and jurisdictions. Coverage from government to government is uneven and varies considerably; however, the sites published by the USDA and the Canadian Food Inspection Agency (CFIA) provide quite complete information. Others, such as the Australian and Californian websites, provide general information and addresses where more complete information may be obtained.

Tables of detailed Canadian grade standards for individual commodities are specified for fresh produce in the regulations. A synopsis of the Canadian lists outlining the parameters used in distinguishing the various grades is in Tables 2A, 2B, and 3. It should be remembered that issues of appropriate cleanliness and soundness are covered under the general regulations. In addition to the standards indicated, private arrangements between grower and processor or packer are possible. Under these arrangements, quality standards become contractual agreements with the requirements set by contract. California has established mandatory standards for fresh fruit and vegetables applicable within that state (Kader, 1992). The homepage for the California Department of Food and Agriculture is included in Table 1, but the department did not have detailed regulations online at the time of writing. California regulations are quite extensive for each commodity (Kader, 1992), and these Internet references can be utilized for direction to obtain regulation hardcopy. The international references allow access to the United Nations/Economic Commission for Europe (UN/ECE) regulations, but the site was under construction at the time of writing and was incomplete. However, the material available is full text and will provide extensive information on the UN/ECE regulations when completed. The foodnet reference

Table 1 Listing of Websites Carrying Information on Fresh and Processed Fruit and Vegetable Quality Standards and Regulations

Alberta
- Agriculture, Food and Rural Development Homepage
  - http://www.agric.gov.ab.ca
- Vegetables
  - http://www.agric.gov.ab.ca/ministry/acts/regs/ar370-90.html#3

Australia
- Home Page
  - http://www.dpie.gov.au/
- Quarantine and Inspection Service
  - http://www.dpie.gov.au/aqis/homepage/aqishome.htm

British Columbia
- Legislation Administered by the BC Ministry of Agriculture Fisheries and Food: summary of the Agricultural Produce Grading Act with linkages and addresses to the actual legislation through the Consolidated Statutes of British Columbia
  - http://www.agf.gov.bc.ca/legisltn.htm#NPMA

California
- California Department of Food and Agriculture Homepage
  - http://www.cdfa.ca.gov
- Fruit and Vegetable Inspection Programs
  - http://www.cdfa.ca.gov/inspection

Canada
- Canadian Food Inspection Agency Homepage
  - http://www.cfia-acia.agr.ca
- Fruit and Vegetable Regulations (Fresh and Processed)
  - http://www.cfia-acia.agr.ca/english/actsregs/freshfruit/home.html
  - http://www.cfia-acia.agr.ca/english/actsregs/process/grades.html
  - For regulations in French substitute /francais/ for /english/ in the address
- Statistical Information Concerning Fruit and Vegetables
  - 1996/97 Canadian Vegetable Situation and Trends
  - http://aceis.agr.ca/misb/hort/vegetabl.html
  - 1996/97 Canadian Fruit Situation and Trends
  - http://aceis.agr.ca/misb/hort/fruitsit.html

International
- United Nations/Economic Commission For Europe (UN/ECE) agricultural standards for fresh fruit and vegetables
  - http://www.unece.org/trade/agr/agr_ndxe.htm
  - or,
  - http://www.unicc.org/unecc/trade/agr/agr_ndxe.htm
    - Site under construction at time of writing
- Organization for Economic Co-operation and Development (OECD)
  - http://www.oecd.org/agr/activities/age_macs.htm

United States
- United States Department of Agriculture (USDA) Homepage
  - http://www.ams.usda.gov
- Agricultural Standards
  - http://www.ams.usda.gov/standards
- Fresh Fruit and Vegetable Quality Standards
  - http://www.ams.usda.gov/standards/stanfrfv.htm
- Processed Fruit and Vegetable Quality Standards
  - http:/www.ams.usda.gov/standards/standpfv.htm

Other
- http://footnet.fic.ca/regulat/regs.html

Table 2A Schedule of Canadian Grade Standards: Fruits

Commodity	Grade	Grading attributes
Apples	Canada Extra Fancy Canada Fancy Canada Commercial Canada Commercial Cookers Canada Hailed Canada No. 1 Peelers Canada No. 2 Peelers	Variety, color, striping, russeting, properly sized, appropriate to variety, shape, freedom from decay, insects, bruises. Properly packed. More leniency as grade decreases. Sizes and colors are specified in a table.
Apricots	Canada No. 1 Canada Domestic Canada Domestic Hailed	Varietal character, color, size, freedom from damage, insects, punctures. Properly packed. More leniency as grade decreases.
Blueberries	Canada No. 1	Typical coloration, size, freedom from insects. Properly packed.
Cantaloups	Canada No. 1	Typical of variety, mature, freedom from injury and insects.
Cherries	Canada No. 1 Canada Domestic Canada Orchard Run	Typical of variety, color, maturity, size, stems, freedom from damage.
Crabapples	Canada No. 1 Canada Domestic	Typical of variety. Color, maturity, size, freedom from damage, insects. Are properly packed.
Cranberries	Canada No. 1 Canada Domestic	Color, size, and maturity. Freedom from insect and other damage. Are properly packed.
Grapes	Canada No. 1 Canada Domestic	Typical of variety, color, size. Freedom from insect and other damage. Properly packed.
Peaches	Canada No. 1 Canada Domestic	Size, shape, color, maturity. Freedom from damage, insects. Properly packed.
Pears	Canada Extra Fancy Canada Fancy Canada Commercial	Typical of variety, size (defined for varieties), shape, color, maturity. Freedom from insect or other damage (bruises, hail, russeting), disease.
Plums, prunes	Canada No. 1 Canada Domestic	Size (defined for variety), shape, color, freedom from cracks, insects, physical damage. Properly packed.
Field rhubarb	Canada No. 1 Canada Domestic	Color, freshness (wilting), size, freedom from pests, disease, physical damage. Properly packed.
Strawberries	Canada No. 1	Color, shape, maturity, calyx attached. Freedom from disease, insects, or physical injury. Properly packed.

Source: Adapted from Canada Agricultural Products Act.

in the International listings provides access to the Canadian regulations and to the homepage of the USDA and should eventually allow access to the USDA regulations. However, the listings for the United States and Canada are much more direct. Some information on the Codex Alimentarius—Fresh Fruit and Vegetables is also available but is limited to a listing of a few commodities.

In Canada, both fruit and vegetable grade standards consist of three parameters

Table 2B Schedule of Canadian Grade Standards: Vegetables

Commodity	Grade	Grading attributes
Asparagus	Canada No. 1 Canada No. 1 slender Canada No. 2	Color, maturity, limited white stalk, diameter (specified), length, perception of freshness.
Beets	Canada No. 1 Canada No. 2	Varietal character, texture (lack of woodiness), size (defined), freedom from decay, disease.
Brussels sprouts	Canada No. 1 Canada No. 2	Texture (firmness), color, size, freedom from decay, disease.
Cabbages	Canada No. 1 Canada No. 2	Varietal character, texture (firmness), properly packed, freedom from decay, disease.
Carrots	Canada No. 1 Canada No. 2	Color, texture (lack of woodiness, softness), shape, not trimmed into crown, freedom from sunburn, decay, damage.
Cauliflowers	Canada No. 1 Canada No. 2	Size, color, "richness," maturity, freedom from decay, damage.
Celery	Canada No. 1 Canada No. 1 Heart	Freshness, maturity, size, properly packed, freedom from decay, disease, damage.
Sweet corn	Canada No. 1	Maturity, texture, size, freedom from decay, disease, damage.
Field cucumbers	Canada No. 1 Canada No. 2	Freshness, maturity, texture, color, size (specified), freedom from decay, disease, damage.
Greenhouse cucumbers	Canada No. 1 Canada No. 2	Texture and other attributes as for field cucumbers.
Head lettuce (iceberg)	Canada No. 1 Canada No. 2	Texture, varietal character, trimming, shape, size, freedom from decay, disease, damage.
Onions	Canada No. 1 Canada No. 1 Pickling	Varietal character, dried neck, texture (firmness), size, freedom from decay, disease, damage.
Parsnips	Canada No. 1 Canada No. 1 Cut Crowns Canada No. 2	Texture (woodiness), trimming, shape, maturity, size (specified), uniformity, freedom from decay, disease, damage.
Potatoes	Canada No. 1 Canada No. 1 Large Canada No. 2	Texture (firm), size, shape, cleanliness, skin loosening, freedom from decay, disease, damage.
Rutabagas	Canada No. 1	Texture (firm), maturity, size, trimming, cleanliness, freedom from decay, disease, damage.
Field tomatoes	Canada No. 1 Canada No. 2 Canada No. 1 Picklers Canada No. 2 Picklers	Maturity, size, shape, cracks, varietal character, freedom from decay, disease, damage.
Greenhouse tomatoes	Canada No. 1 Canada No. 1 Extra Large Canada Commercial Canada No.2	Maturity, size (specified), shape, varietal character, texture (firmness vs. softness), extent of ripening, freedom from decay, disease, damage.

Source: Adapted from Canada Agricultural Products Act.

Table 3 Selected Maturity Indicators for Several Fruit Commodities

Commodity	Maturity indicator
Grapes	Sugar accumulation, loss of acids, berry softening, skin coloration, ripening with varietal flavor/odor development.
Apple	Skin and flesh color, flesh firmness, sugars (soluble solids), acid, starch, days from full bloom to harvest.
Pear	Flesh firmness (important), ease of spur separation, days from full bloom (indices vary with variety). Harvested hard-green, ripened off the tree; color, texture, sweetness, typical varietal flavor development.
Citrus	Fruit color, taste (both termed unreliable), soluble solids. Soluble solids/acid ratio.
Cherries	Color change light red to black, soluble solids to 18%–22%, flesh firmness, light transmission properties.
Peach	Days after full bloom, size, firmness, ground color, sugars, acidity, starch.
Apricot	Dependent on intended purpose (fresh market, canning, etc.), soluble solids, flesh firmness, light transmission properties.
Blueberries	Freedom from injury and decay, plump, firm, uniformly blue color. Green or red indicator of immaturity (unripeness), soft texture may be indicator of overmaturity.
Strawberry	Fully red/at least 3/4 of berry surface red or pink.

Source: Adapted from Salunkhe and Kadam (1995).

(Tables 2A and 2B). These consist of the commodity name, a series of grade classes in decreasing order of perceived ''quality,'' and a series of attributes used to determine the grade standard. Tables 2A and 2B outlines the type of attribute used in the grade determination, but no attempt has been made to detail the actual regulations. The attributes relate to such properties as varietal character (should be in keeping with expectations), color, size, maturity, texture, and freedom from defects such as decay, disease, and physical damage. All of these attributes can be determined by trained inspectors in the field with minimal equipment. Attributes such as color and size may be quantified in the regulations that allow the use of rulers or comparators for inspection judgments. The ability of trained inspectors using minimal equipment to determine grade standards is important for the rapid assignment of grade to produce and is common to most grading systems (Kader, 1992; Table 1). However, the incorporation of quantitative measures such as color specification for apples in the Fresh Fruit and Vegetable Regulations of the Canada Agricultural Products Act is possible. California apparently defines apple maturity in terms of quantitative soluble solids content and a defined firmness test (Kader, 1992), and this application of quantitative measure extends to other commodities.

3 MATURITY

One of the most important parameters of the grade standard is the expressed or implied reference to the maturity of the commodity in question. In the first paragraph, maturity is defined as the state of being ''ready to use'' as defined by the customer. In some products, especially fruit products, a ripening process may be required to attain optimal readiness for consumption. This ripening process is usually denoted by changes in color, texture (usually softening), and flavor and provides a modifier for the idea of maturity. Thus commodities such as tomatoes or apples may be fully developed and mature at a green stage of development and are generally edible at that stage but are expected to ripen over time to provide optimal eating quality. The complexities of maturity definition have resulted in the development of a number of maturity indices that attempt to relate a chemical or physical measurement to the commercial maturity of the particular commodity. The intention is that these indices will reflect or predict the quality or quality grade of the commodity received by the customer.

4 MATURITY INDICES

Objective maturity standards are available for very few horticultural commodities, and most regulations and many research publications (Toivonen, 1997) rely on subjective measures of maturity. Commonly selected factors reflecting horticultural maturity include firmness, skin color, flesh color, sugar content, soluble solids, acid content, and pigments. Days from full bloom (Fig. 2) and heat unit accumulation (Table 4, Peas) during specific periods during the growing season are also used (Reid, 1992; Westwood, 1993). In other cases such as in bananas (Kotecha and Desai, 1995), a physical change is used, and the fruit is harvested when ridges on the surface of the skin change from angular to round. The development of maturity indices is an ongoing research issue (Richardson et al., 1997; Kappel et al., 1992; Abbott et al., 1997). In practice, any analytical or physical measurement that changes with physiological development and relates to customer acceptance has potential as a maturity index, and the use of sensory panels to demonstrate this relationship is a fairly recent development. In pears, using a 7-point hedonic scale in which 4 represents "just right," perceived firmness reflects an objective firmness measure whereas perceived juiciness is negatively correlated with objective firmness measurements (Kappel et al., 1995, Figs. 3A and 3B). In cherries the correlation exists with fruit weight, color, and firmness (Kappel et al., 1996). The maturity index should be easily determined, use simple equipment, and provide good correlation with efficacy in final use. A widely recognized good example is the starch-iodine test for starch in apple (Lau, 1988; Brookfield et al.,

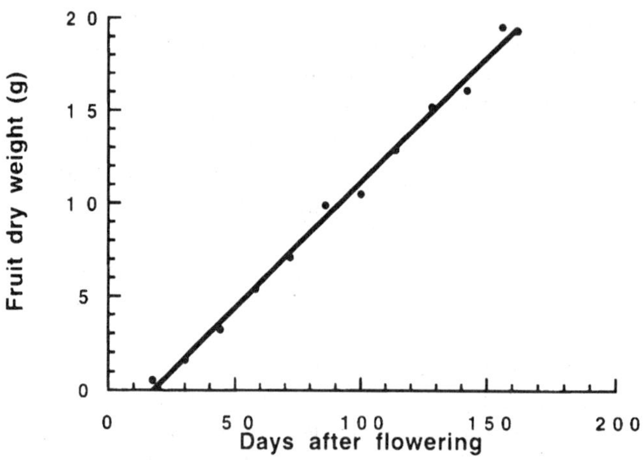

Fig. 2 Dry weight of kiwifruit from flowering until harvest. Fruit dry weight correlates well with soluble solids. (From Richardson et al., 1997.)

Table 4 Selected Maturity Indicators for Several Vegetable Commodities

Commodity	Maturity indicator
Potatoes	90–120 Days after planting
Tomatoes	Many possible maturities: mature green, pink, red, etc., see grade standards
Peppers	Color (green, red, yellow), smooth blocky shape, susceptible to chilling injury below 7°C
Peas	Accumulated heat units, alcohol insoluble solid (AIS), tenderness, size
Beans	14–18 Days from full bloom, size, fibrousness
Bean sprouts	Length, color
Onion	Top bending, size
Carrot	Size and color suitability for intended use, e.g., baby carrots
Radish	Size and cleanliness; suitability for use
Cabbage	Head firmness, appropriate color (green, purple), size
Cauliflower	Size, curds tight and compact, color
Broccoli	Firm head, closed florets, color, shape, lack of yellowing
Sweet corn	Sugar content (refractometer or equivalent)
Muskmelon	Distinct odor, ethylene development, soluble solids, skin ground color

Source: Adapted from Nonnecke (1989).

1997; Figs. 4 and 5), which is also available on the Internet (Priest and Lougheed, 1981). Detailed discussion of maturity and maturity indices for many common fruits are available in Salunkhe and Kadam (1995), and selected important attributes for several fruit commodities are summarized in Table 3. Maturity and ripeness indicators are particularly important for fruit commodities since their optimal harvesting or ripeness opportunity window tends to be relatively short. Similar data and discussion for vegetables are available in Nonnecke (1989), summarized in Table 4.

Firmness is a well recognized maturity index (Figs. 3A and 3B) and is commonly measured with a penetrometer, an instrument that determines the force required to push a probe of known diameter through the flesh of the fruit or vegetable. Commonly Magness-Taylor-type instruments are used, either hand operated or mounted in a device such as the Instron Universal Testing Machine (Bourne, 1982). These are discussed more fully in Sec. 4. The requirement that the index be easy and economical to measure makes the handheld or hand-operated instruments preferred. Color is measured instrumentally in terms of the Commission Internationale d'Eclairage (CIE) three-dimensional color solid or one of the derived color systems such as the L, a, b Hunter color-system (Clydesdale, 1984). Present instrumentation such as the Minolta Chroma Meter CR-300 or equivalent has made surface color measurements quick, easy, economical, and field-ready with battery operated models. Alternatively, matching to a color comparator such as the Ctifl (Centre technique interprofessional des fruits et legumes, 22 rue Bergere, 75009 PARIS) comparator for cherries may be used to define both color and size.

Maturation of fruit and vegetables causes profound changes in the biochemical characteristics of the commodity, which are reflected in changes of three easily measurable chemical components. Starch initially rises, then falls (Fig. 6), and this decrease is accompanied by rising soluble solids (Fig. 6). Acid levels decrease, and the result is an increased perception of sweetness as the sugar/acid ratio increases. Starch levels are easily assessed

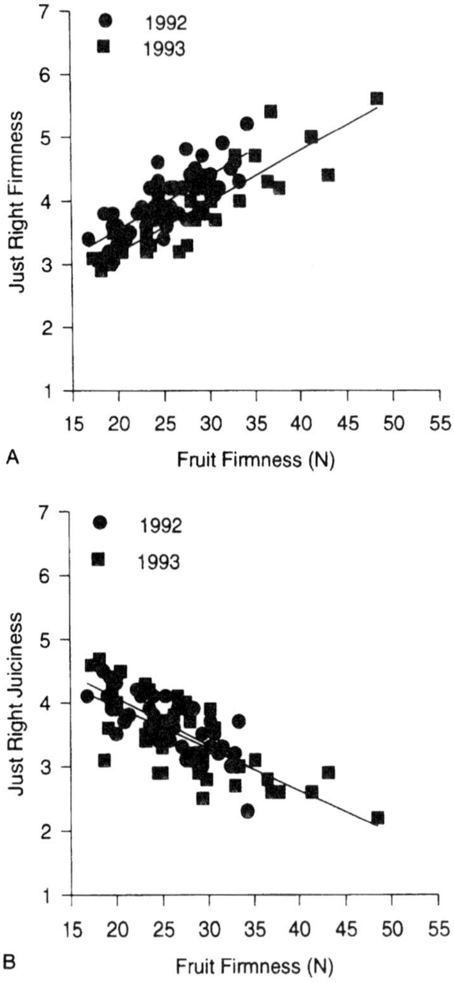

Fig. 3 Fruit firmness assessments of pear fruit in 1992 and 1993 for (A) firmness perception and (B) juiciness perception. (From Kappel et al., 1995.)

by the starch-iodine test (Fig. 4), and soluble solids are quickly and simply measured by refractive index calibrated as °Brix. Refractometers are widely available commercially as either hand held or bench mounted instruments and are usually temperature compensated internally. Titratable acidity assesses acid levels and is usually performed as a juice sample titrated to the phenolphthalein end point by sodium hydroxide. With colored samples, titration to pH 8.2–8.3 with a pH meter can substitute. Automatic equipment is widely available to allow titration of large numbers of samples. The results are commonly expressed in terms of the dominant organic acid expected in the sample—malic acid for apples or tartaric acid for grapes. For commodities expressing a respiration climacteric during maturation (e.g., apples, melons) ethylene measurements can be used as a measure of maturity (Reid, 1992). However, this procedure requires the use of a gas chromatograph and tends not to be rapid, simple, and economical. This restricts the use of ethylene as a maturity index to laboratory use.

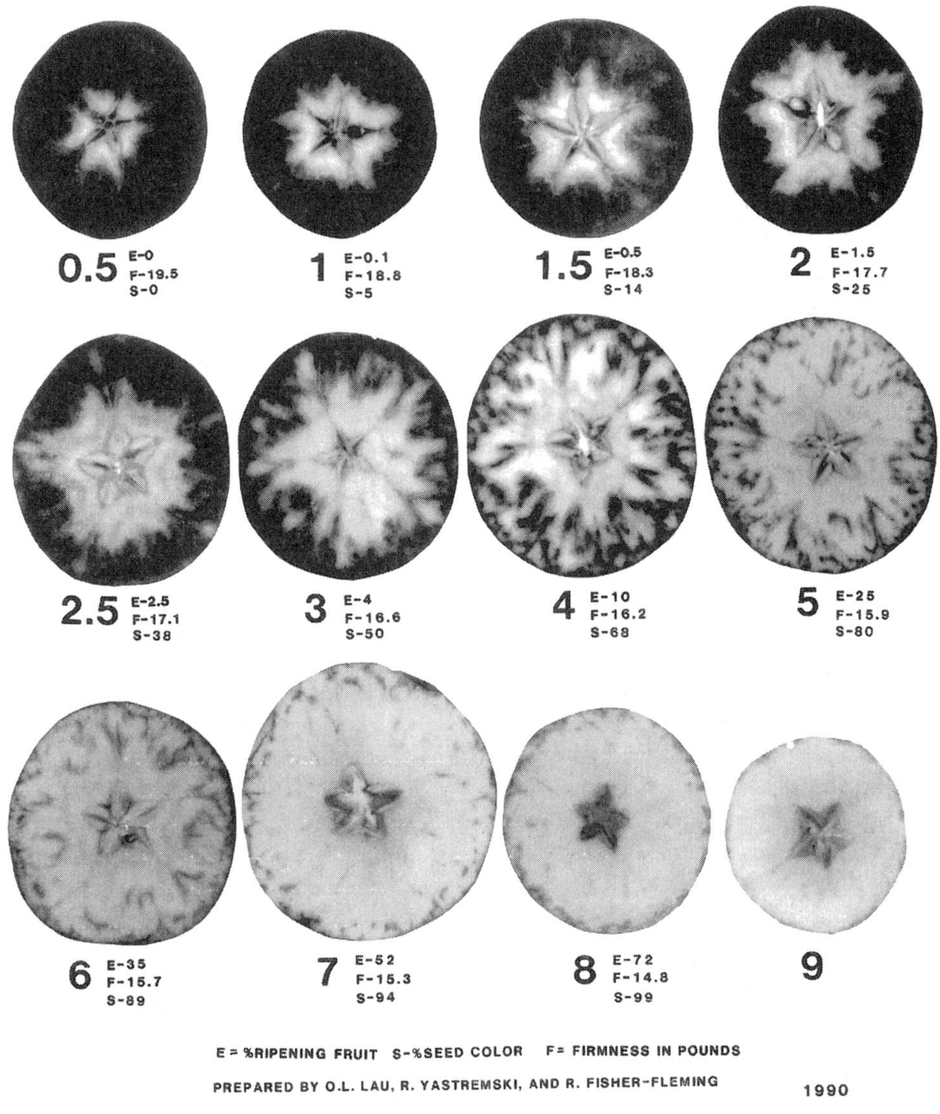

Fig. 4 Starch test guide for harvesting British Columbia McIntosh apples. (From Lau, 1995.)

5 PROCESSED FRUIT AND VEGETABLES

Because of the wide variety of and possibilities for processed fruit and vegetables, the quality grades and standards are much more extensive and complex. For this reason the discussion herein should be considered as general in nature and the detailed regulations consulted for specific answers on particular commodities or products. The Canadian grades and standards are discussed; the standards of both Canada and the United States are available on the Internet (Table 1). The Processed Products Regulations of the Canada Agricultural Products Act sets out grades, grade names, and standards for both canned and frozen fruit and vegetables. Also, the act provides for inspection and registration of agricultural

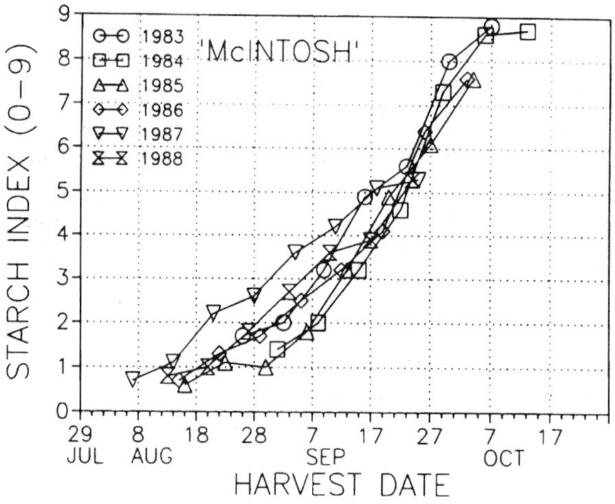

Fig. 5 Starch index values in McIntosh apples during maturation and ripening. (From Lau, 1995.)

product companies. A fruit or vegetable for which a standard is set out shall contain only the ingredients specified in the standard. There is provision for obtaining authorization to test market a food product that does not meet the regulations. In the United States, regulations pertaining to dehydrated, sugared, and miscellaneous products are included under this category.

Standard containers, can sizes, and volumes for canned material and net container weights for frozen materials (commonly 250 g, 375 g, or 500 g) are set out. Standards of container fill, usually in terms of drained weight for cans, depend on can size and packing media. Packing media may vary from fruit juice to heavy sugar syrup, and these clearly affect the drained weight. Size grades and label declarations, vegetable size grades, and particularly grade and fill declarations are regulated. Methods of determining drained weight are specified and inspection sampling plans and acceptance numbers for processed

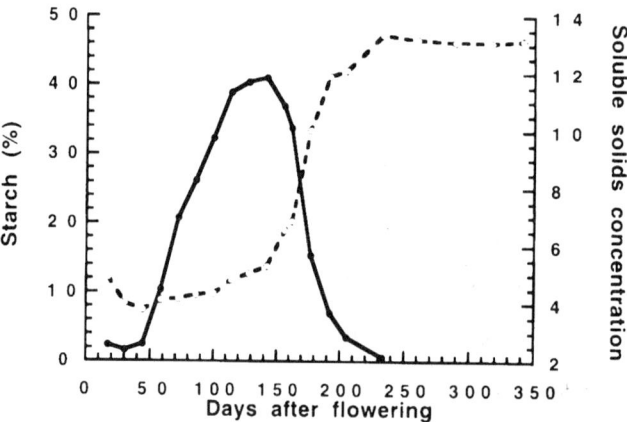

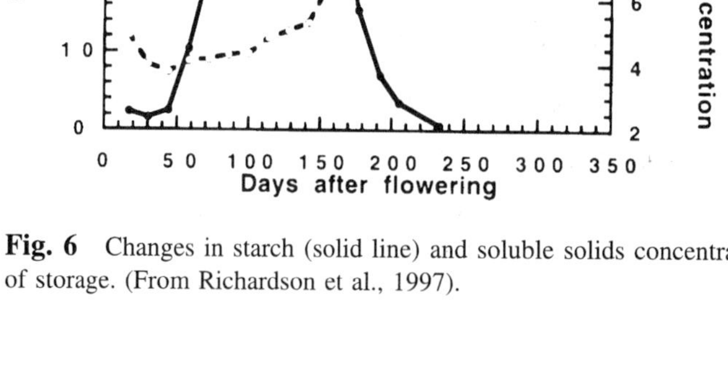

Fig. 6 Changes in starch (solid line) and soluble solids concentration from flowering until end of storage. (From Richardson et al., 1997).

commodity inspections are laid out in some detail. Regulations governing import, export, and interprovincial trade are delineated. The specific commodities covered under the Canada Agricultural Products Act are listed in Table 5 for canned fruit and vegetables and in Table 7 for frozen commodities. Types of factors differentiating grade standards are provided as example listings of apples as canned commodity and blueberries as frozen commodity (Tables 6 and 8). From these examples, it is clear the attributes considered are similar to those examined for fresh fruit and vegetables. Size, shape, color, varietal character, soundness, and suitability for consumption are important parameters. Further, appropriate fill weights, containers, labels, and other characteristics that make the pack suitable for commerce are also of concern. The examples discussed are provided for the reader's information but have no official status; therefore, the guidance of an office of the Canadian Food Inspection Agency should be sought in actual individual cases.

Table 5 Commodities and Grades for Canned Fruit, Fruit Products, and Vegetables[a]

Apples, sliced; apple juice; concentrated apple juice; apple juice from concentrate; applesauce
Asparagus, tips, cuts, or cuttings
Beans
Lima beans
Beets
Berries, small fruits
Blueberries
Strawberries
Carrots
Cherries
Corn
Fruit cocktail, fruit for salad, fruit salad
Mixed vegetables macedoina
Mixed vegetable juices
Mushrooms
Peaches
Pears
Plums
Peas
White potatoes
Sweet potatoes
Pumpkin and squash
Sauerkraut
Spinach
Tomatoes

[a] All commodities except those noted herein are graded Canada Fancy, Canada Choice, or Canada Standard. For indications of the differences between these grades see Table 6, which details the grades for apples. Exceptions are Apple Juice and Sauce (Canada Fancy, Canada Choice only), Sauerkraut (Canada Fancy, Canada Choice only), Mixed Vegetables Macedonia and Mixed Vegetable Juices (grade optional), Tomato Juice, Tomato Juice Concentrate and Puree (Canada Fancy, Canada Choice only). Other tomato product grades are optional.

Source: Adapted from Canada Agricultural Products Act.

Table 6 Apple Product Grades

Commodity[a]	Grade	Attributes[b]
Sliced Apples	Canada Fancy	Very good apple flavor, aroma; good color; uniform slices in size and color; free from core, peel, bruised tissue, insect injury, or other defect
	Canada Choice	Good flavor, aroma; fairly good color; slices practically uniform in size and shape; free from core peel, bruised tissue, insect injury, or other defects
	Canada Standard	Normal flavor, aroma; reasonably good color; reasonably free from core, peel, bruised tissue, insect injury, and other defects
Apple Juice	Canada Fancy	Possesses very good apple flavor; contains not less than 11.5% soluble solids; contains not less than 0.35% and not more than 0.70% malic acid (w/v); is practically free from defects
	Canada Choice	Possesses good apple flavor; contains not less than 10.5% soluble solids; contains not less than 0.30% and not more than 0.80% malic acid (w/v); fairly free from defects
Concentrated Apple Juice	Standards and Grades as for Apple Juice after dilution	Contains at least 68% soluble solids
Apple Juice from Concentrate	Standards and Grades as for Apple Juice after dilution	May contain apple juice, natural apple esters, ascorbic acid, carbon dioxide under pressure, sodium benzoate (1 g/kg maximum)
Applesauce	Canada Fancy	Very good apple flavor, aroma; good color, consistency, granular appearance; practically free from seed specs, skin, bruised tissue, carpel tissue, and other defects
	Canada Choice	Fairly good flavor, aroma, color, consistency; reasonably granular appearance; reasonably free from seed specs, skin, bruised tissue, carpel tissue, and other defects

[a] Each product contains a definition of the product, the expected styles, and definitions of the terms used to describe grade standards, for instance, the definition of apple juice and the styles as clarified or opalescent.
[b] Methods of soluble solids and malic acid determination are given and form part of the regulations.
Source: Canada Agricultural Products Act.

Table 7 Commodities and Grades for Frozen Fruits and Vegetables[a]

Apples, frozen apples (sliced); frozen apple juice concentrate
Apricots
Asparagus, frozen asparagus (tips, spears); frozen asparagus (cuts or cuttings)
Beans (green or wax)
Lima beans
Berries (small fruits)
Blueberries
Strawberries, frozen whole strawberries; frozen sliced strawberries
Broccoli
Brussels sprouts
Carrots (cut carrots, baby whole style; cut carrots, whole style; diced carrots; sliced carrots; whole carrots; whole baby carrots)
Cauliflower
Cherries, frozen cherries (red sour pitted); frozen sweet cherries
Corn, frozen whole kernel corn (whole grain); frozen cream style corn; frozen corn on cob
Fruit cocktail, frozen fruit cocktail, frozen fruits for salad; frozen fruit salad; frozen melon or cantaloupe balls
Mixed vegetables, frozen mixed vegetables; special blends of frozen vegetables
Concentrated orange juice
Peaches (halved, sliced, diced, or quartered)
Peas: frozen peas, frozen peas and carrots (diced, sliced, or whole)
French fried potatoes (straight cut or regular cut, shoestring or julienne, crinkle cut, crinkle cut shoestring, or crinkle cut julienne)
Squash: frozen squash (cooked); frozen uncooked squash (diced or cubed)
Rhubarb (cut)
Spinach (whole leaf, cut, or chopped)

[a] All commodities are graded Canada A or Canada B except blueberries and orange juice, which are graded Canada A, Canada B, or Canada C. For a sense of the differences among the grades see Table 8, where the grades for blueberries are detailed. In addition, for each commodity, the meaning of certain words (e.g., *practically* vs. *reasonably*) is defined for use in the grade standard.
Source: Canada Agricultural Products Act.

6 QUALITY MEASUREMENT OF PROCESSED PRODUCTS

The importance of subjective evaluation of quality or maturity should not be underestimated for processed fruit and vegetables (Jacobi et al., 1996; Toivonen, 1997). The technique clearly forms the basis for maturity and quality grades (Tables 6 and 8), and it is widely used to detect the onset of browning (enzymatic or nonenzymatic), yellowing, or other color changes that accompany processing or storage. Textural changes can also be assessed with these techniques. Equally, however, the instrumental techniques used to assess quality in fresh fruit and vegetables can be applied to processed material. Texture changes represent a major change that results from processing of fruit and vegetables and considerable effort has been directed toward measuring these changes by objective techniques (Bourne, 1982; Barbosa-Canovas et al., 1996). Generally these objective techniques make use of instrumentation such as the Instron Universal Testing Machine (or equivalent), currently available in sizes ranging from microsystems well adapted to measuring forces ranging from a few grams to several kilograms, to macrosystems measuring hundreds of kilograms force. The microsystems are especially useful for most foods. These

Table 8 Frozen Blueberry Grades

Commodity	Grade	Standard
Blueberries (from fresh sound ripe fruit of the blueberry bush): are cleaned and stemmed; may have been washed; properly packed, frozen, and maintained at temperatures necessary to preserve product Native or wild type; high bush or cultivated	Canada A	(a) Possess a good flavor characteristic of well-ripened blueberries; (b) possess a uniform color characteristic of well-ripened blueberries; (c) are in good condition; (d) are practically free of cap stems; (e) are practically free from defects including clusters, whole leaves, or any portion thereof, large stems, green berries, undeveloped berries, harmless extraneous material, or other defects
	Canada B	(a) Possess a good flavor characteristic of well-ripened blueberries; (b) possess a uniform color characteristic of well-ripened blueberries; (c) are in good condition; (d) are practically free of cap stems; (e) are fairly free from defects including clusters, whole leaves or any portion thereof, large stems, green berries, undeveloped berries, harmless extraneous material, or other defects
	Canada C	(a) Possess a flavor characteristic of well-ripened blueberries; (b) possess a fairly uniform color characteristic of well-ripened blueberries; (c) are in fairly good condition; (d) are reasonably free of cap stems; (e) are reasonably free from defects including clusters, whole leaves or any portion thereof, large stems, green berries, undeveloped berries, harmless extraneous material, or other defects

Source: Canada Agricultural Products Act.

systems can be equipped to perform texture analysis on many fresh and processed foods, including the attachments to allow traditional Warner-Bratzler shear (Sanchez et al., 1995), Kramer shear cell (Ng and Waldron, 1997), or puncture probe measurements (Bourne, 1982; Fig. 7). When processing tissues to form juices or purees, viscosity measurements (Collyer and Clegg; 1988, Fig. 8) or consistency measurements (Bourne, 1982) become appropriate. They include oscillatory rheometry (Marin, 1988) used to measure the dynamic elasticity (G') and dynamic viscosity (G'') parameters describing the viscoelastic behavior of materials (Tung and Paulson, 1995; Barbosa-Canovas et al., 1996). Because these latter measurement systems can utilize very small oscillations the structure of the tissue is minimally disturbed. Examination of unique features of processing such as the effects of cooking on texture (Fig. 9) is now becoming possible. These possibilities are

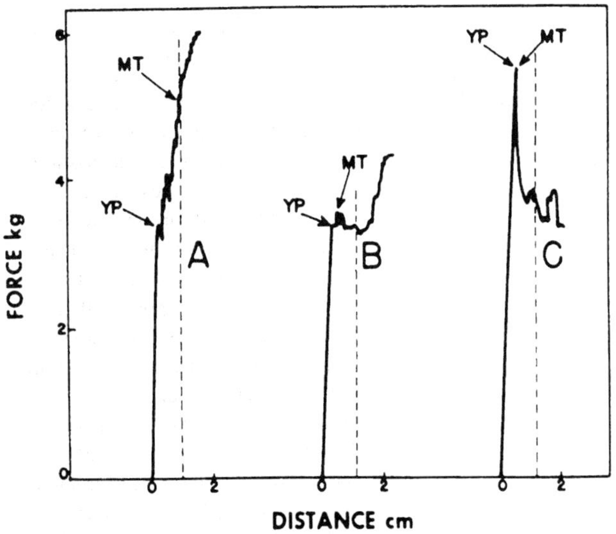

Fig. 7 Characteristic force-distance curves obtained by mounting the 5/16-in- (7.9-mm)-diameter tip of the Magness-Taylor Pressure Testor in the Instron. YP, yield point. MT, force reading that would have been obtained on a hand-operated pressure tester. Dotted lines represent the depth of penetration corresponding to the 5/16-in inscribed line on the pressure tester. Three types of product responses are displayed: A, products represented by products such as fresh apples; B, products represented by ripe pears and peaches; C, products such as raw vegetables. (From Bourne, 1980.)

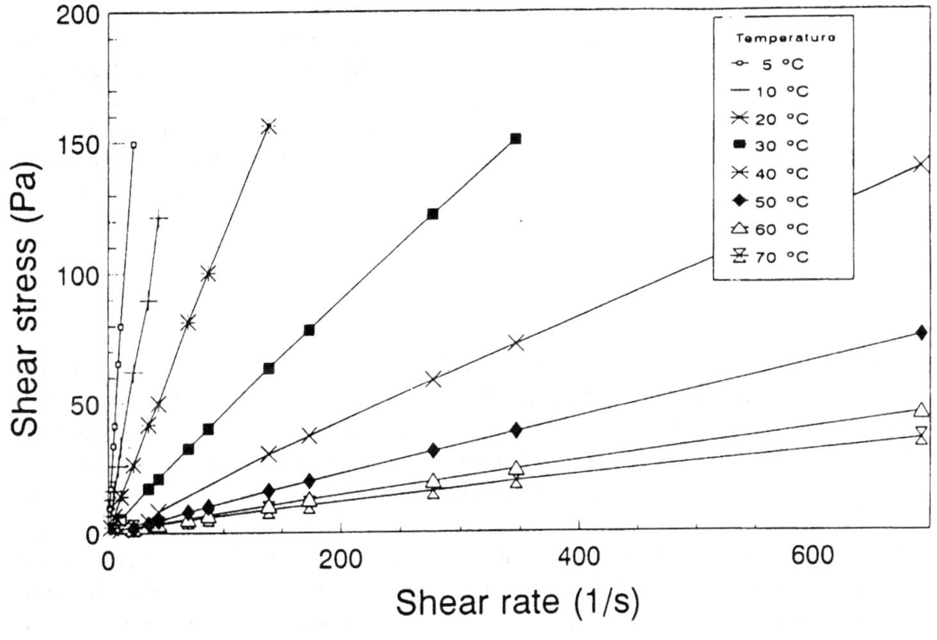

Fig. 8 Rheograms of cherry juice concentrate (74.0°Brix) at several temperatures. (From Giner et al., 1996.)

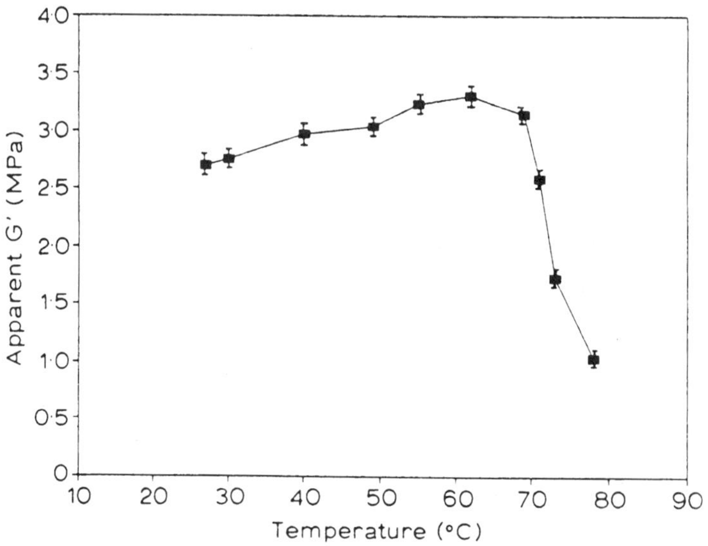

Fig. 9 Apparent G' during heating of raw carrot tissue. Sample 2 mm thick, 20.6 mm diameter; frequency 0.2 Hz; strain 1.12. (From Ramana and Taylor, 1992.)

only recently being recognized in research publications, and practical application awaits future developments.

7 FUTURE DEVELOPMENTS

The value and quantity of vegetables for processing have been decreasing steadily since 1991 (see 1996/97 Canadian Vegetable Situation and Trends in Table 1) as frozen vegetable consumption has increased and consumption of canned juice and vegetables has decreased. This situation seems to be paralleled in the fruit sector in which consumption of canned peaches and pears has declined 35%–40% since the early 1970s (see 1996/97 Canadian Fruit Situation and Trends in Table 1). The fresh market represents the growth market for both fruit and vegetables with wide availability of fresh produce produced locally or imported from around the globe. Of particular note is the fresh-cut market, which was expected to reach $19 billion and 25% of fresh produce sales by perhaps 2003–2004. Clearly, the immediate future belongs to the fresh market with storage of both fruit and vegetables combined with greenhouse production contributing to maintenance of fresh supplies over the marketing season. The importance of quality and grade standards for fresh produce will be emphasized by these trends.

The fresh-cut segment represents the processed portion of this fresh market, and in Canada there are no standards applicable to these products. Produce in plastic bags containing modified atmosphere packaging (MAP) are widely available, especially in the salad category but increasingly throughout the spectrum of fruit and vegetables. These products are perishable for both biochemical and microbiological reasons (Phillips, 1996), and the resulting concerns are driving consideration of quality standard regulations for these products. As research develops consensus on the regulatory requirements for these products, regulatory standards for both production and final product can be anticipated.

In commercial practice, fresh produce is passed through a packinghouse, where it is cleaned, graded, size/quality sorted, and packed for shipment to retail outlets. Sorting for size and color is largely automated, but most of the other quality/maturity attributes must be accomplished by visual examination. Automation of these later attributes will require precise, rapid examination of each horticultural commodity, individual by individual, nondestructively at very high speed. Potential methods for accomplishing this feat have been reviewed (Abbott et al., 1997; Clark et al., 1997). Detailed discussion is beyond the scope of this article, but as technical difficulties are overcome, one can anticipate the application of nondestructive techniques to sorting and quality/maturity selection processes. Sound waves can detect firmness or textural properties of individual fruit or vegetables. Fluorescence properties of individual fruits and vegetables change during maturation, ripening, and senescence (Song et al., 1997), and machine vision systems may allow definition of size and shape parameters (Abbott et al., 1997). Magnetic resonance imaging (MRI) has been demonstrated to detect internal quality defects, insect infestation of single fruit, and compositional differences between fruit structures (Clark et al., 1997), particularly with respect to oil bearing components and pits. The detection and quantitation of sugar and hence soluble solids by MRI are possible but presently difficult. As these techniques and others (Abbott et al., 1997) move from research and development to practical reality, they will create a revolution in the way fruit and vegetables are graded, sorted, and packed.

REFERENCES

Abbott, J. A., R. Lu, B. L. Upchurch, and R. L. Stroshine. 1997. Technologies for nondestructive quality evaluation of fruits and vegetables. Hortic. Rev. 20:1–120.

Barbosa-Canovas, G. V., J. L. Kokini, L. Ma, and A. Ibarz. 1996. The rheology of semiliquid foods. Adv. Food Nutr. Res. 39:1–69.

Bourne, M. C. 1982. Food Texture and Viscosity: Concept and Measurement. Academic Press, New York and London. pp. 120, 182.

Bourne, M. C. 1980. Texture evaluation of horticultural crops. HortScience 15:51–57.

Brookfield, P., P. Murphy, R. Harker, and E. MacRae. 1997. Starch degradation and starch pattern indices: Interpretation and relationship to maturity. Postharvest Biol. Technol. 11:23–30.

Canada Agricultural Products Act. Fresh Fruit and Vegetable Regulations. October 1991.

Clark, C. J., P. D. Hockings, D. C. Joyce, and R. A. Mazucco. 1997. Application of magnetic resonance imaging to pre- and post-harvest studies of fruits and vegetables. Postharvest Biol. Technol. 11:1–21.

Clydesdale, F. M. 1984. Color measurement. In Food Analysis Principles and Techniques. Vol. 1. Physical Characterization, ed. D. W. Gruenwedel and J. R. Whitaker. Marcel Dekker, New York and Basel.

Collyer, A. A., and D. W. Clegg. 1988. Rheological Measurement. Elsevier Applied Science, London and New York.

Giner, J., A. Ibarz, S. Garza, and S. Xhian-Quan. 1996. Rheology of clarified cherry juices. J. Food Eng. 30:147–154.

Jacobi, K. K., L. S. Wong, and J. E. Giles. 1996. Postharvest quality of zucchini (*Cucurbita pepo* L.): Following high humidity hot air disinfestation treatments and cool storage. Postharvest Biol. Technol. 7:309–316.

Kader, A. A. 1992. Standardization and inspection of fresh fruit and vegetables. In Postharvest Technology of Horticultural Crops, ed. A. A. Kader. Publication 3311, Division of Agriculture and Natural Resources, University of California. p. 191.

Kappel, F., M. Dever, and M. Bouthillier. 1992. Sensory evaluation of 'Gala' and 'Jonagold' strains. Fruit Varieties J. 46(1):37–43.

Kappel, F., R. Fisher-Fleming, and E. J. Hogue. 1995. Ideal pear sensory attributes and fruit characteristics. HortScience 30(5):988–993.

Kappel, F., R. Fisher-Fleming, and E. J. Hogue. 1996. Fruit characteristics and sensory attributes of an ideal sweet cherry. HortScience 31(3):443–446.

Kotecha, P. M., and B. B. Desai. 1995. Banana. In Handbook of Fruit Science and Technology, ed. Salunkhe, D. L., and S. S. Kadam. Marcel Dekker, New York, Basel, and Hong Kong.

Lau, O. L. 1988. Harvest indices, dessert quality, and storability of 'Jonagold' apples in air and controlled atmosphere storage. J. Am. Soc. Hortic. Sci. 113:564–569.

Lau, O. L. 1995. Harvest Handbook for B.C. 'McIntosh' Apples. Okanagan Federated Shippers Association, Summerland, Canada.

Marin G. 1988. Oscillatory rheometry. In Rheological Measurement, ed. A. A. Collyer and D. W. Clegg. Elsivier Applied Science, London and New York.

Ng, A., and K. W. Waldron. 1997. Effect of cooking and pre-cooking on cell wall chemistry in relation to firmness of carrot tissues. J. Sci. Food Agric. 73:503–512.

Nonnecke, I. L. 1989. Vegetable Production. Van Nostrand Reinhold, New York and London.

Phillips, C. A. 1996. Review: Modified atmosphere packaging and its effects on the microbiological quality and safety of produce. Int. J. Food Sci. Technol. 31:463–479.

Priest, K. L., and E. C. Lougheed. 1981. Evaluating apple maturity using the starch-iodine test. http://www.gov.on.ca/omafra/english/crops/facts/88-117.htm.

Ramana, S. V., and A. J. Taylor. 1992. Dynamic measurement of tissue rigidity during freezing and cooking of vegetables. J. Sci. Food Agric. 58:261–266.

Reid, M. S. 1992. Maturation and maturity indices. In Postharvest Technology of Horticultural Crops, ed. A. A. Kader. Publication 3311, Division of Agriculture and Natural Resources, University of California, p. 21.

Richardson, A., J. McAneney, and T. Dawson T. 1997. Kiwifruit quality issues. Orchardist 70(7): 49–51.

Ryall, L. A., and W. J. Lipton. 1972. Handling, Transportation, and Storage of Fruits and Vegetables. Vol. 1. Vegetables and Melons. AVI Publishing, Westport, Conn. p. 26.

Salunkhe, D. K., and S. S. Kadam. 1995. Handbook of Fruit Science and Technology, Production, Composition, Storage and Processing. Marcel Dekker, New York, Basel, and Hong Kong.

Sanchez, M. T., J. R. Hermida, G. Camo, and F. Torralbo. 1995. Texture of blanched and frozen asparagus. Alimentaria 261:49–52.

Song, J., W. Deng, R. M. Baudry, and P. R. Armstrong. 1997. Changes in chlorophyll fluorescence of apple fruit during maturation, ripening and senescence. HortScience 32:891–896.

Toivonen, P. M. A. 1997. The effects of storage temperature, storage duration, hydro-cooling and micro-perforated wrap on shelf life of broccoli (*Brassica olerccea* L., Italica group). Postharvest Biol. Technol. 10:59–65.

Tung, M. A. and A. T. Paulson. 1995. Rheological concepts for probing ingredient interactions in food systems. In Ingredient Interactions, Effects on Food Quality, ed. A. G. Gaonkar. Marcel Dekker, New York, Basel, and Hong Kong.

Watada, A. E., R. C. Herner, A. A. Kader, R. J. Romani, and G. L. Staby. 1984. Terminology for the description of developmental stages of horticultural crops. HortScience 19:20–21.

Westwood, M. N. 1993. Temperate-zone pomology. In Physiology and Culture, 3rd ed. Timber Press, Portland, Oreg., p. 303.

18

Cooling and Storage

TIMOTHY J. RENNIE and CLÉMENT VIGNEAULT

Agriculture and Agri-Food Canada, Saint-Jean-sur-Richelieu, Quebec, Canada

JENNIFER R. DeELL

Ontario Ministry of Agriculture and Food, Vineland Station, Ontario, Canada

G. S. VIJAYA RAGHAVAN

McGill University, Sainte-Anne-de-Bellevue, Quebec, Canada

1 INTRODUCTION

Precooling, refrigeration, proper relative humidity, and optimal atmospheric composition in storage facilities and packages are essential to reducing postharvest losses of commodities that are destined to reach the consumer in fresh condition. Besides the direct financial advantages for producers and distributors, elimination of losses through proper postharvest conditioning, storage, and handling can have an important influence on food availability and distribution and on environmental impacts of agriculture.

Proper storage is an important part of the marketing and distribution of horticultural commodities. The storage of produce has two main objectives: (a) to provide short-term storage to balance the daily fluctuations of supply and demand and (b) to provide long-term storage to extend the marketing season. Three factors involved in the deterioration of perishable commodities must be controlled to meet these objectives. The natural rate of respiration must be reduced as much as possible by controlling temperature and, for many commodities that respond positively to low O_2 and/or high CO_2 levels, the composition of the storage atmosphere. Moisture loss should also be minimized, and pathogenic microorganisms not be permitted to proliferate. Storage facilities may also be used for special treatments, such as the curing of potatoes or the degreening of oranges prior to shipment. There are many technologies available to create and maintain optimal temperature, relative humidity, and atmospheric composition for harvested horticultural products. This is partly due to the differences in recommended storage conditions over the range of products that are marketed as fresh, a partial list of which is given in Table 1. The choice of system is therefore a function of the range and type of products to be stored in a given facility, the volume handled, and a number of other factors that will be explained in the major sections of this chapter: precooling, cooling and refrigeration, and modification of atmospheric composition.

1.1 Background

1.1.1 Temperature

Temperature is the single most important factor affecting the deterioration rate of harvested commodities (1). The rate of deterioration is proportional to the respiration rate of the commodity, which is temperature-dependent. For each 10°C reduction in temperature, the respiration rate of a wide range of produce can be reduced by a factor of 2 to 4 (2). Moreover, the activity of postharvest pathogens and insects is also suppressed by low temperatures (3). Therefore, cooling and refrigeration are important to preserving the quality of fresh fruits and vegetables and to extending their storage lives.

The ideal storage temperature varies from product to product, and the temperature maintained in the storage area should be within 1°C (2°F) of that level (2). Lower temperatures may cause chilling injury and higher ones can reduce the storage life of the product. If the temperature is allowed to fluctuate beyond the desired range, the produce may experience increased water loss and condensation may develop on the product from the surrounding air (2), leading to the growth of microorganisms. Temperature fluctuations can be prevented by using the proper equipment to refrigerate the storage room.

The refrigeration unit must be able to handle the maximal refrigeration load that is expected. If the load is too great for the refrigeration unit, then the temperature rises. The refrigeration system should also be designed such that the air leaving the coils is as close to the desired room temperature as possible. This limits the frequency at which the system turns on and off. Use of coils with a large surface area can provide air leaving the coils at a temperature close to the room temperature while maintaining a large refrigeration capacity (2).

There should also be sufficient air circulation to keep the product at a uniform temperature and to prevent condensation, thus, uniformity of airflow around all of the produce is an important design consideration. Flow rates of 0.001 to 0.002 $m^3 \cdot s^{-1}$ of air per 1000 kg of product are generally adequate (2). The flow rate should be based on the maximal amount of produce that is expected to be stored in the room. Temperature may be controlled by placing thermostats in positions that represent the effective cooling temperature. They should not be placed near sources of heat or cold, such as doors, exterior walls, or air discharge areas of the cooling unit. The temperature measurement units are best placed about 1.5 m off the floor for easy reading (2).

The timing, degree, and type of cooling that can be used depend on the commodity to be stored and its end use since there are important differences in composition and physiological characteristics among the fruit and vegetables species used as foods. For example, leafy vegetables are cooled soon after harvest to maintain turgor, whereas tubers (potatoes, yams, sweet potatoes) are not. Potatoes, for instance, are held at 15°C to 25°C for up to 15 days to permit suberization (4), a process that promotes wound healing and reduces invasion by storage pathogens. Thereafter, potatoes are stored at 5°C to 10°C to

Table 1 Recommended Precooling Methods and Storage Conditions for Various Fruits and Vegetables

Produce	Precooling method[a]	Storage conditions
Apples (2, 14)	RC, FA, HC	0°C to 5°C, 1%–3% O_2, 1%–5% CO_2
Asparagus (14, 15)	HC, PI	0°C to 2°C, 95% to 100% RH
Apricots (2)	RC, FA	0°C to 5°C, 95% RH, 2% to 3% O_2, 2% to 3% CO_2
Artichokes (2, 49)	HC, FA, PI	0°C to 5°C, 90% to 95% RH, 2% to 3% O_2, 2% to 3% CO_2
Beans, snap (14, 49)	RC, FA, HC	8°C, 2% to 3% O_2, 4% to 7% CO_2
Beets (14, 33)	RC	0°C to 4°C, 95% RH
Blackberries (14, 33)	FA, RC	−0.5°C to 0°C, 90%–95% RH
Blueberry (20)	FA	Optimal at 1°C (3°C–4°C), 90% RH
Broccoli (2, 49)	FA, HC, PI, LI	Optimal at 0°C (0°C–5°C), 90%–95% RH, 1%–3% O_2, 5%–10% CO_2
Brussels sprouts (2, 15)	FA, HC, PI	0°C, 95%–100% RH
Cabbage (14, 33)	RC, FA	0°C, 92% RH
Cantaloupes, slip (2, 15)	HC, FA, PI	2°C–5°C, 95% RH
full slip (2, 15)	HC, FA, PI	0°C–2°C, 95% RH
Cauliflower (14, 15)	HC, VC	0°C, 95%–98% RH
Carrots (14, 33)	RC, PI	0°C to 2°C, 95% RH
Chinese cabbage (14, 15)	RC, FA, HC	0°C, 95%–100% RH
Celery (2, 49)	FA, HC, VC, WV	0°C–5°C, 90%–95% RH, 2%–4% O_2, 3%–5% CO_2
Cucumbers (2, 15)	RC, FA	10°C–13°C, 50%–55% RH
Eggplant (14, 15)	RC, FA	8°C–12°C, 90%–95% RH
Figs (2)	RC, FA, HC	0°C–5°C, 5%–10% O_2, 15%–20% CO_2
Garlic (2)	RC	0°C
Grapes (14, 33)	FA	−1°C to 0°C, 85% RH
Kiwifruit (2, 15)	FA, RC, HC	−0.5°C to 0°C, 90%–95% RH, 1%–2% O_2, 3%–5% CO_2; C_2H_4 must be below 20 ppb
Leeks (2, 15)	HC, PI	0°C, 95%–100% RH
Lettuce (14, 23)	HC, PI, VC	0°C, 95$^+$% RH
Mushrooms (2, 49)	FA, VC	Optimal at 0°C (0°C–5°C), normal O_2, 10%–25% CO_2
Nectarines (14, 15)	FA, HC	−0.5°C–0°C, 90%–95% RH
Okra (14, 49)	RC, FA	7°C–12°C, 90%–95% RH, normal O_2, 4%–10% CO_2
Onions (33)	No precooling	0°C, 75% RH
Peaches (14, 33)	FA, HC	−1°C to 0°C, 85% RH
Pears (14, 15)	FA, RC, HC	−1.5°C to −0.5°C, 90%–95% RH
Peas, green (14, 15)	FA, HC	0°C, 95%–98% RH
Peas, southern (14, 15)	FA, HC	4°C–5°C, 95% RH
Peppers, chili (dry) (2, 15)	RC, FA, VC	0°C–10°C, 32%–50% RH
Peppers, sweet (2, 15)	RC, FA, VC	7°C–13°C, 45%–55% RH
Plums (14, 15)	FA, HC	−0.5°C–0°C, 90%–95% RH
Potatoes (14, 33)	RC, FA	3°C–10°C, 90% RH
Pumpkins (33)	No precooling	10°C–13°C, 70% RH
Radish (14, 49)	PI	0°C, 90%–95% RH, 1%–2% O_2, 2%–3% CO_2
Raspberries (50)	FA	0°C to 0.5°C, 90% to 95% RH
Rutabagas (14, 15)	RC	0°C, 98%–100% RH
Spinach (14, 15)	HC, VC, PI	0°C, 95%–100% RH
Squash, summer (14, 15)	RC, FA	5°C–10°C, 95% RH
Squash, winter (15)	No precooling	10°C, 50%–70% RH
Strawberries (2, 11)	RC, FA	0°C, 95% RH, 5% to 10% O_2, 15% to 20% CO_2
Sweet Cherry (2, 51)	RC, FA, HC	0°C–5°C, 3%–10% O_2, 10%–15% CO_2
Sweet Corn (24, 52)	HC, VC, LI	0°C, 95% RH
Sweet Potatoes (33)	No precooling	10°C–15°C, 85% RH
Tamarillos (2, 15)	RC, FA	3°C–4°C, 85%–95% RH
Tomatoes (49)	RC, FA	Optimal at 12°C (12°C–20°C), 3%–5% O_2, 0%–3% CO_2
Turnip (14, 15)	RC, HC, VC, PI	0°C, 95% RH
Watermelons (11)	No precooling	4°C–10°C, 80%–85% RH

[a] RC, room cooling; FA, forced-air cooling; HC, hydrocooling; VC, vacuum cooling; PI, package icing; LI, liquid icing.

minimize conversion of starch to sugars. Many commodities are susceptible to chilling injury (e.g., banana, cranberry, cucumber, green pepper, and tomato), whereas others are sensitive to high CO_2 levels (pear, lettuce). Chilling injury can result in several sources of postharvest losses, including: surface lesions, water soaking of tissues, internal discoloration, breakdown of tissues, failure to ripen normally, accelerated senescence, greater susceptibility to decay, and compositional changes (5). Tropical and subtropical fruits are more likely to experience chilling injury than temperate climate produce; however, some apple cultivars and cranberries are particularly sensitive temperate types (6).

1.1.2 Relative Humidity

Relative humidity is another important factor for the long-term storage of perishable commodities. Low relative humidity allows excessive moisture loss from the produce, resulting in a loss of quality. The recommended humidity level for the storage of fresh fruits and vegetables is commodity specific; levels are generally in the range of 85% to 95%. Wilting and shriveling are likely to occur in most vegetable crops when stored at low humidities (7). However, relative humidities close to 100% can be ideal for the growth of microorganisms and can cause surface cracking in some produce (2). Low relative humidities are generally difficult to obtain in storage rooms. It is best to use large evaporator coils for the refrigeration system so that the coils operate at 3°C lower than the surrounding air temperature and thus reduce condensation (2). The smaller the temperature difference between the evaporator coils and the air, the less the moisture loss. Providing good insulation and minimizing air leaks are steps that can be taken to maintain high relative humidities (7). Mechanical humidifiers, spray nozzles, or steam systems may also be used to increase the humidity in the storage room (2). Pressure-atomized or heat-vaporized water can be added to the room to add moisture. A rate of 4 L water · h^{-1} per ton of refrigerant should be adequate to maintain a relative humidity of 95% (7). These systems are an added expense and require additional maintenance. Systems that have a lot of condensation on the evaporator coils need to be defrosted often, since the buildup of frost or ice on the coils results in decreased refrigeration capacity.

1.1.3 Atmospheric Composition

The deterioration of harvested produce can further be reduced by limiting the available oxygen in the storage atmosphere, usually in combination with maintaining higher carbon dioxide levels. This can be done by active means (controlled atmosphere storage [CA]) or passive means (modified atmosphere [MA] storage). It is generally accepted that limiting the O_2 supply reduces the respiration rate by approximately 50% at temperatures of 20°C to 25°C (3). If the product is refrigerated, the additional reduction due to low temperature is approximately 24%. These are equivalent to storage life increases of 100% and 33%, respectively, under the two temperature regimes (3). Higher CO_2 concentrations in the storage enclosure can also inhibit the respiration mechanism, however, some commodities are sensitive to higher-than-ambient CO_2 concentrations and can develop off-odors, discoloration, and other disorders (3). Therefore the type of system used to provide optimal conditions is dependent on the commodity. Finally, the control of ethylene (C_2H_2) is an important consideration in the storage of certain commodities (see Chapter 16 for a review of the role of this gas).

2 PRECOOLING

Cooling is the process of removing heat. The rate of cooling of a product is often referred to as its *1/2 cooling rate*. The 1/2 cooling rate is the time required to reduce the difference between the mean temperature of the product and the temperature of the cooling medium by one-half. It is expressed as a constant for a particular set of precooling conditions. Reduction of the temperature difference by 7/8 is usually three times that referred to as the 1/2 cooling rate and is often used as a commercial standard for the total time of cooling.

The sooner produce is cooled down, the better its chances at a relatively long storage life. Some fresh fruits and vegetables may deteriorate as much in 1 h at 26°C as in 1 week at 1°C (8), particularly if they have naturally high respiration rates. The distinction between precooling and cooling is that precooling is any method of removing field heat more rapidly than if the produce were simply placed in a storage chamber set at the desired temperature and allowed to cool.

Precooling can significantly extend the storage life of fresh fruits and vegetables, particularly when large quantities of produce harvested in warm temperatures are involved. In many situations, the bulk could take several days to cool to the final storage temperature if alternative means were not used. Precooling may involve a technology separate from that of normal cooling in a refrigerated environment or may be the result of modifications to the refrigerated storage (see Sec. 3.1).

2.1 Hydrocooling

Hydrocooling is the cooling of produce with cold water. This method is very effective for a wide range of products. Some hydrocooling facilities can handle up to 30,000 crates per day during the peak season (9). In general, the cooling times are in the range of 10 minutes to 1 hour (10). One of the advantages of hydrocooling over other methods is that the commodity does not lose moisture during the process (11).

There are two methods of hydrocooling: (a) immersion in a cold water bath and (b) shower cooling (2). Immersion systems are continuous flow systems and are most useful for products that have a higher density than water and therefore remain submerged (9). In continuous systems, the produce is conveyed through a tank of cold water and lifted out at the end of the tank by an inclined conveyor. The produce may be introduced in bulk or in containers. Since the conveyor speed is usually too slow to provide adequate water movement around the produce, circulating pumps or propellers are used to promote heat transfer away from the produce (12). A water velocity of $0.1\ m \cdot s^{-1}$ (20 ft/m) should be sufficient for rapid cooling (6). Lower-density produce such as cucumbers, squashes, and tomatoes is cooled by flotation in circulating water. However, when the produce is immersed in cool water, the air contained inside some commodities decreases in volume, creating a suction pressure that may facilitate the entry of pathogens into the tissues. It is therefore important to ensure that the circulating water is and remains clean.

Shower coolers involve overhead spraying of the produce with cold water and may be batch or continuous coolers (Fig. 1). The water is either pumped onto an overhead perforated pan and allowed to drip on to the produce or showered through spray nozzles. The pumping requirements for spray nozzles are much higher than for the perforated pan approach (10). When water is applied from overhead, the distance the water falls before hitting the produce should be kept below 15 to 20 cm (10). Drop heights that exceed this range can damage some produce. Products that are in pallet bins or field bins can be protected with a perforated cover (10).

In a batch operation, palletized containers are placed in a room and water is sprayed from the top. For shallow product depth, the water should be provided at a rate of 280 to 490 $L \cdot min^{-1} \cdot m^{-2}$ (liters per minute per meter squared) (10). Flow rates of 800 to

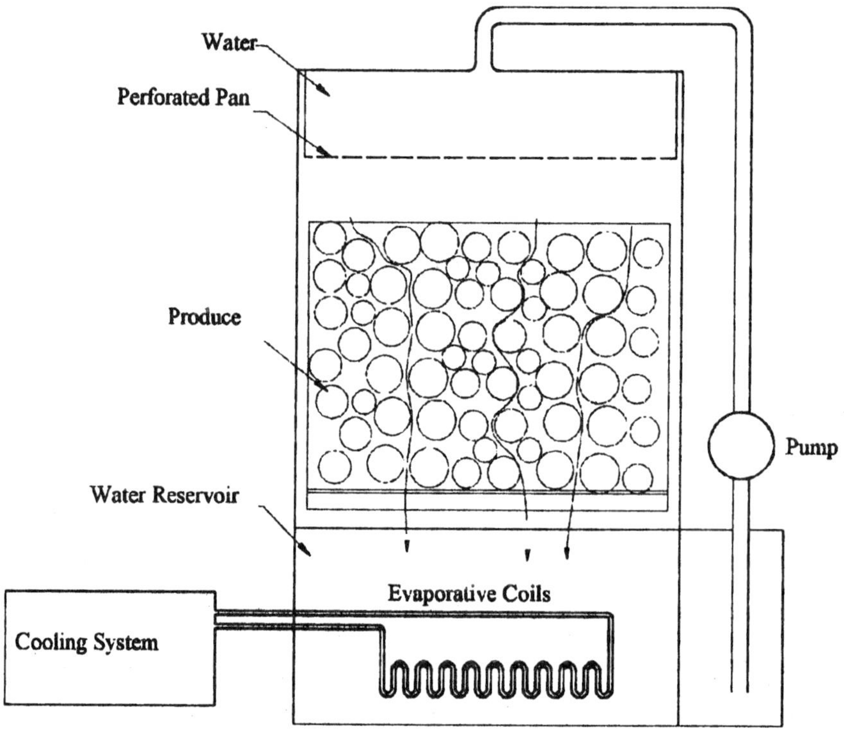

Fig. 1 Schematic view of a shower-type hydrocooler.

1000 L · min^{-1} · m^{-2} are used in commercial coolers for double stacked pallet bins (10). The water is collected in drains on the floor or in an underfloor reservoir, cooled, and then sprayed on the produce again. This cycle continues until the precooling is finished. If this is done in a refrigerated room, the produce may be stored for a short time in the same room after the precooling operation (9).

In a continuous flow shower system, the produce is transported on a conveyor much in the same way as in the immersion system. The time that the produce spends in the hydrocooler can be adjusted by changing the conveyor speed. If the produce is in bins or pallets, the containers may also be partly submerged as they travel along the conveyor. Water is sprayed on the produce from overhead.

The water must be sanitized, especially if it is reused. It should be taken from a clean source, either a well or a domestic supply (10) and should be disinfected by adding chlorine to the extent that the free chlorine level is 100 to 150 ppm (11). The water should be drained out of the hydrocooler at least daily and the system sanitized. Extremely dirty products should be washed before hydrocooling to decrease the amount of dirt that enters the system. Screens and filters serve to remove debris and dirt from the water before it is recycled.

Water used for hydrocooling is generally kept between 0°C and 0.5°C using mechanical refrigeration (10). Produce that is sensitive to chilling injury may be cooled in water at 0°C as long as the cooling time is limited (10). Hydrocoolers require a large refrigeration potential in a very short time. Ice is therefore often used to assist the mechanical refrigeration system (2) and is usually added to a water tank to cool the water or is used in an ice-accumulator refrigeration system. Financial costs for energy can be reduced by building the ice supply during off-peak hours and using the ice for cooling when the energy costs are high. The necessary refrigeration for a system can be decreased by insulating the hydrocooler. Up to half the refrigeration potential in some hydrocoolers can be lost as a result of insufficient insulation (12).

Hydrocooling is suitable for bulk and packaged produce. It is commonly used for melons, root vegetables, stem vegetables, and many types of tree fruits (10). Commodities that are hydrocooled must be tolerant to contact with water and to the levels of chlorine in the sanitized water. Commodities such as grapes and most berries must be ventilated after hydrocooling to remove surface water, which can otherwise encourage decay (10).

2.2 Contact Icing

One of the oldest and simplest cooling methods is contact icing, a method that is well suited to produce that is tolerant to long periods of cold (0°C) wet conditions (2). This involves filling packed containers or pallets with ice or covering pallets with ice. The contact between ice and produce causes rapid cooling. In general, reduction of product temperature from 35°C to 2°C requires a mass of ice equal to 38% of the product weight (13).

There are several different methods of filling the containers with ice. Individual package top icing is the simplest method. Ice is shoveled, raked, or blown on top of the product in the container. Cooling by this method is rather slow since the ice is only in contact with the top layer (11). It is not efficient for large operations because of the amount of labor involved in opening the containers, adding the ice, and then closing the containers. The coating of ice may block vent spaces, thereby restricting air movement and leaving the center of the load warm (11). Individual package top icing should be used only after precooling and before shipping, to assist in cooling and in maintaining high relative humidity (11).

An improvement to top icing is pallet box icing by layer. This method of icing is more labor intensive than top icing but the cooling is faster and more uniform (11). Crushed ice and produce are alternately layered in the pallet box. Thus, more ice surrounds the produce, which cools faster. It is recommended that all points in a bulk load of green leafy vegetables be within 150 mm (6 in) of the ice (14).

Liquid icing provides much faster cooling than individual package top icing. A slurry of cold water and ice is either drenched over the pallet of produce or pumped into the containers through the hand holds (Fig. 2). The water slurry causes the produce to float until the water drains out of the bottom of the container. As the water drains out, ice is distributed throughout the container. The produce settles and the ice falls into the spaces between individuals. This method creates very good contact between the ice and the product, resulting in good heat removal. The cold water of the slurry has a substantial effect on the cooling of the product. It has been shown that the cold water can contribute up to 40% of the cooling effect on broccoli (15). Liquid icing can reduce the pulp temperature to 0°C in a reasonable amount of time and maintain high relative humidity (11).

A greater financial investment is required for liquid icing than for the other methods of contact icing. Equipment includes an ice crusher, a slurry tank with mixer, a pump, and delivery hoses. With the manual equipment, two workers can liquid ice a 30-container pallet in 5 minutes (15). Automated systems can process five times more produce within

Fig. 2 Pallet exiting from a single-pallet liquid icing machine.

the same time, but they are expensive and require power. In the 1990s smaller-scale systems were developed to process 200 containers per hour with power requirements of only one-fifth of that required by automated systems (16). Full truck load top icing is a method that is used during the transport of produce. A 5- to 10-cm (2- to 4-in) layer of crushed ice is placed over a load of precooled produce prior to shipping. If it is used prior to shipment, the additional weight caused by the ice should be taken into account. Because a 12.2-m- (40-ft-) long trailer may need over 3600 kg (8000 lb) of top icing (15), the quantity of produce that can be shipped is limited (7).

Careful container selection is necessary if contact icing is to be used. The containers must be water resistant and large enough to accommodate the amount of ice required to cool the product. Waxed fiber board cartons are acceptable for contact icing since they have few openings and provide some insulation from the surrounding environment; however, they do not retain their strength over long periods when wet (15). Reusable plastic containers were specially designed for uniform ice distribution, ice retention, and resistance under very wet conditions (17). Wooden wire-bound crates, baskets, or hampers are also used.

An ice crusher may be needed to crush the ice if the available chip or flake ice is too large or if the ice is in the form of blocks. The ice that is to be used for any type of ice cooling should be no larger than 9.5 mm (3/8 in.) so that the ice particles can fill the voids between the produce (15). Small particles of ice are less likely to damage the produce than larger ones. Furthermore, ice particle sizes ranging from 4.5 to 5.1 mm (0.17 to 0.20 in.) used for liquid-ice systems provide the most uniform ice distribution and greater icing efficiency regardless of container type (16). The equilibrium temperature of a slurry of melting ice and water is 0°C (15), and 335 kJ (316.8 Btu) of heat is required to melt 1 kg (2.2 lb) of ice. Although the temperature of the slurry may be lowered by adding salt, the relatively small temperature reduction does not decrease the cooling time to any significant degree (15). For example, adding 18.2 kg of sodium chloride to 378.5 L of slurry reduces the temperature by only 2.8°C. Furthermore, such brine solutions can cause the produce to lose more water than desired (15). It is also important to keep the ambient air cold to reduce the melting rate of ice in the containers.

2.3 Vacuum Cooling

Vacuum cooling is one of the most rapid cooling methods. This method is suitable for produce that has a large surface area to mass ratio and readily releases water. Both these properties largely affect the rate of cooling and the final temperature of the product (9). Commodities such as lettuce, sweet corn, celery, green beans, and mushrooms can be cooled in 20 to 30 minutes. Vacuum cooling is recommended for leafy vegetables such as lettuce since the overlapping leaves create insulating air pockets that reduce the efficiency of other cooling methods. The main drawback of vacuum cooling is limited capacity, which is due to room and vacuum requirements. Vacuum cooler capacities range from 4 to 20 pallets per batch (2).

Produce is placed in an airtight chamber and the pressure in the chamber is decreased to the point that water boils at the desired cooling temperature (110) (Fig. 3). For example, a pressure of 0.610 kPa (4.6 mm Hg) permits water to boil at 0°C (9). Since the energy needed for the phase change of water from liquid to vapor is from the sensible heat of the produce, the produce cools close to the boiling temperature in a very short time. The evaporation of water also causes a weight loss in the produce. Generally, each 1% reduction in the weight of the produce due to moisture loss results in a temperature decrease of 6°C. Since this can have detrimental effects on product quality, the product can be wetted before vacuum cooling (9) or sprayed with a fine mist during cooling (2). The type of product container that is used has a negligible effect on the process of vacuum cooling (18). If the product is wrapped with a plastic film, the film must be perforated to obtain effective cooling (19). Airtight plastic film may be destroyed during the process as a result of the pressure differential that is created.

2.4 Forced-Air Cooling

Forced-air cooling is a method of precooling that can be used on a wide range of produce. It can cool produce 4 to 10 times faster than room cooling (12) and produces a more

Fig. 3 Portable vacuum cooler.

uniform temperature distribution in the pallet (20). This method works well for small-scale operations because of its cost effectiveness, compared to that of hydrocooling or vacuum cooling, and its high cooling rate (21). Cold air is forced through the product, rather than just surrounding the produce containers as is done in room cooling. Fans are used to create a static pressure difference across the two sides of the product containers (Fig. 4). This results in the air's being pulled through the containers, thus removing the warm air around the produce by convection. The static pressure that is created across the produce containers is in the range of 3 to 25 mm of water gauge, with a typical value of 12 mm (21). Cooling is faster and more uniform as the static pressure used is increased. An airflow rate of 0.5 to 3 L of air per second per kilogram ($L \cdot s^{-1} \cdot kg^{-1}$) of warm produce is needed for adequate heat removal (21). Fans are selected to operate with these criteria for static pressure and airflow. The quantity of air that does not pass through the

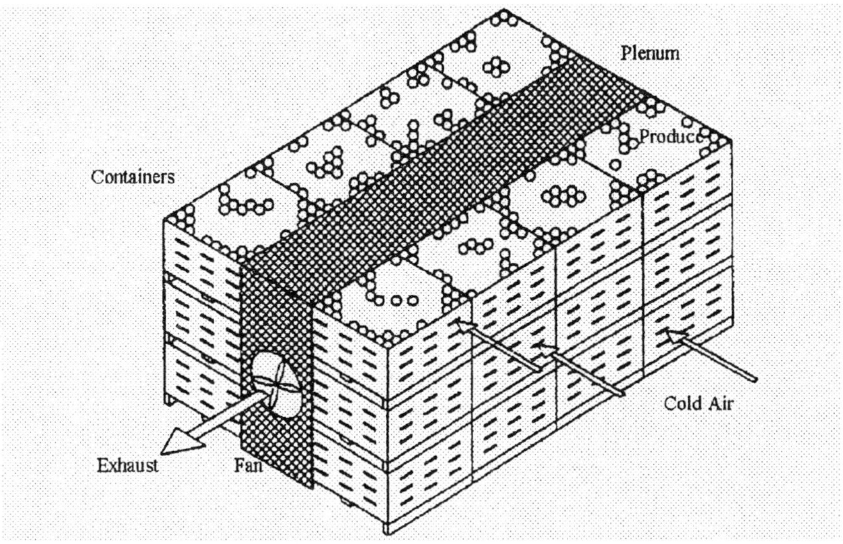

Fig. 4 Schematic of forced-air tunnel cooling.

produce should be minimized since it does not aid in cooling and results in a loss of efficiency.

Fan manufacturers usually present performance characteristics of their fans in graphic or tabulated form and indicate the optimal range of operation. It is recommended that a fan with a wide optimal flow rate range or a fan run by a variable-speed motor be purchased, if it is expected that the quantities of produce to cool will vary over the season. The fan speed and power requirements are indicated with the performance characteristics so that an economic analysis based on initial investment and operating costs may be done.

The position of containers must be such that the vents that are in adjacent containers are lined up in the direction of flow. These vents should be evenly distributed on the containers to allow for even air distribution throughout the produce (21). It is recommended that the total area of these vents be at least 5% to 10% of the surface area of the sides of the container (22) and that the produce inside the containers not be wrapped so that optimal heat transfer conditions can be achieved (23). It should be noted that increasing the amount of vent surface on a container by over 6% of the side surface results in a decrease in the strength of the container. Therefore, careful selection of containers is necessary (21). Reusable plastic containers allow up to 40% of the side surface of the containers to be used as vents. However, an opening percentage of 25% seems to be the optimum, taking into account cooling efficiency and physical support the container provides to the produce (17). Since the cooling rate is affected by the velocity of the air (20), proper fan selection is required to obtain the flow rates mentioned. This method of cooling requires greater refrigeration capacity than room cooling because of the higher rate of heat removal. It also requires fans that have the capacity to deliver the required flow rate at the ideal pressure.

Forced-air cooling is very attractive to small operations that harvest many types of produce, as it is suitable for a wide range of produce. It is believed that forced-air cooling can dry out some products, such as lettuce, spinach, mushrooms, and peaches. However, forced-air cooling of these products can be successful if the cooling is rapid and the relative humidity of the air is kept high (21).

Forced-air tunnel systems are the most used form of forced-air cooling. Two rows of palletized containers are placed parallel to each other with a space between them (Fig. 4). The aisle formed is covered on the top, along with one end to create a tunnel. A fan is placed at the free end to pull the air out of the air plenum tunnel, creating a static pressure difference across the containers. The cooled air around the outside of the pallets is pulled through the produce and into the plenum by the pressure difference. The costs associated with this type of forced-air system are minimal. The fan may be a portable unit that returns the air to the room or a permanent fan that directs the air back directly to the cooling coils. In the case of the portable fan, the warm air is allowed to return to the room before it is cooled. The position of the portable unit must be chosen to direct the warm air to the intake of the cooling system to prevent condensation on produce already cooled. In the case of permanent fans, the containers must also be rearranged after cooling to prevent possible dehydration.

Cold wall cooling systems are a more complex form of forced-air cooling but are very flexible with regard to the time that different types of produce require for adequate cooling and the time at which they arrive (Fig. 5). The main constraint is that pallets need to be moved as soon as they are cooled in order to prevent dehydration of the contents. A false wall is constructed, creating a permanent air plenum between the real wall and the false wall. This plenum is equipped with exhaust fans that circulate air across the cooling coils. Air inlets are located on the false wall but are not opened until a pallet load is placed up against the wall. Dampers control the air inlets so that they remain closed until a pallet load or a stack is placed up against the false wall. A good damper design allows different types of packages or partial pallets to be used. The false wall may be built with shelves so that several layers of pallets may be cooled (2).

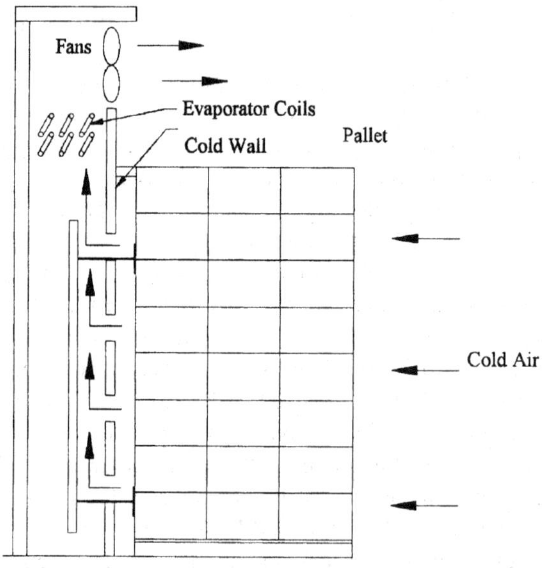

Fig. 5 Schematic of a cross-sectional view of a cold wall cooling system.

Serpentine cooling, a modification of the standard cold wall cooling method, is used for cooling pallet bins. Although the bins must have bottom ventilation, side ventilation is not necessary (2). Several rows of bins are pushed up against the false wall, each row having several layers. The number of rows that can be used is dependent on the desired cooling speed and the flow rate that is available. The openings on the pallet bins for the forklift are used as air supply and return plenums. Every second layer of the forklift openings that are facing the cold room is covered to prevent air from entering. Likewise, every second layer facing the false wall is covered. The coverings of the forklift openings are staggered, so that a layer that is covered when facing the room is not covered when facing the wall. The false wall must be designed so that the openings in the false wall face the forklift openings. When the fans are turned on, the fans pull air from the forklift openings that are against the false wall, causing a static pressure difference. Cold room air is forced into the forklift openings. The air then travels up or down through the produce to a forklift opening that returns it to the air plenum. This system has a few advantages over regular cold wall cooling. It provides very quick cooling because the air travels through a very shallow layer of produce, the rows are not limited in height (2), and large volumes of produce may be cooled at one time.

3 COOLING AND REFRIGERATION

3.1 Room Cooling

Room cooling is one of the most widely used forms of cooling, primarily because the room in which the produce is cooled also serves for longer-term storage, thus reducing overall handling requirements and costs. Since room cooling is quite slow compared to the methods described in the previous section, it has not been categorized as a precooling technique. It is not recommended for crops that have high respiration rates or are harvested in the warmer months, since significant quality deterioration may take place during the cooling period.

Produce that is room cooled must be tolerant to slow heat removal, as heat transfer is by conduction through the container walls rather than by convection. In general, crops that are harvested in the cool season or have low base respiration rates are suitable for this method. The time to cool the produce to the recommended storage temperature may range from a few hours to several days. The cooling time is shorter if the produce is unpacked because the produce is directly exposed to the cold air.

Field containers filled with fresh produce are placed into a refrigerated room. Cold air from evaporators near the ceiling travels across the top and between containers before returning to the evaporator. An airflow of at least 1 to 2 $m \cdot s^{-1}$ is needed to remove field heat effectively (2) and should be reduced to 0.05 to 0.1 $m \cdot s^{-1}$ for subsequent storage to prevent excessive moisture loss thereafter (12). It is possible to achieve better air distribution with several small evaporators evenly spaced along one wall than with one large evaporator (24). The pallet bins should be stacked in a manner that ensures that the cold air has good contact with all surfaces of the bins. Poor contact may not cool the produce in the center of the bin as rapidly as desired (3). Space should be left between all bins and the bins should be oriented so that the openings for forklifts run in the same direction as the airflow.

The type of packaging has an effect on the rate at which the produce cools. Cooling time can be reduced by providing ventilation in the containers. For example, a side venting area of 5% on boxes can reduce the cooling time by 25% compared to that of similar containers that are not vented (12). However, the addition of venting to containers decreases their stacking strength, depending on the size and location of the vents. Using 5% venting area reduces the fiberboard container strength by 2%–3% as long as the vents are not situated in the corners (12). The shape and size of the vents should not allow likely blocking of the vents by the produce within. It is more efficient to use a few large vents than numerous small vents (12). Plastic containers do not have this disadvantage of lost strength and can be designed to support the load of produce with up to 25% of the container surface area dedicated to venting (17).

Good room management can increase the rate of cooling of the produce. Gradually filling up the room with produce can result in a shorter cooling time than filling the room quickly. The lower cooling rate is a result of reduced field heat removal at any given time. For example, a room that is completely filled at one time requires a high refrigeration capacity to lower the temperature of the produce, because of the vast amounts of field and respiratory heat produced by products at high temperatures. If the room is only partially filled each day, then the necessary cooling load is less, and the highest load occurs on the final day, 10% of field heat plus 100% of respiration heat (12). This assumes that the crop can be harvested over a 10-day period.

Standard room cooling can be modified to improve performance. One modification is the use of ceiling jets to improve the rate of heat removal (2). This is accomplished by installing a false ceiling in the room. Cold air is forced into the space between the false ceiling and the roof. Cones in the ceiling conduct the air downward and pallets are arranged in a manner so that the air from the cones is forced down the corners of the pallets. The air then spreads into the channels between the stacked units. The floor should be marked to ensure that pallets can be properly placed.

Another useful modification is the addition of cooling bays. The cooling room is sectioned off into bays by installing partitions. A central aisle is left empty for easy access during forklift operations. The flow rate of air may be controlled for each bay, so warm produce can receive large airflows while it is being cooled and produce cooled already is not subjected to flow rates that can cause excessive moisture loss. The flow rate for cooled produce can be adjusted for optimal storage conditions. Cooled produce in one bay is not warmed by the addition of warm produce in another bay because of the partitions.

3.2 Mechanical Refrigeration

The vapor-recompression cycle is the most commonly used form of mechanical refrigeration (Fig. 6). Four major components are required: an expansion valve, an evaporator, a compressor, and a condenser (Fig. 7). High-pressure refrigerant passes through an expansion valve, where it is suddenly reduced in pressure. This results in *flashing*, a sudden drop in temperature caused by a sudden pressure drop. The flashed refrigerant should be at a lower temperature than the desired storage temperature, since it must be able to absorb heat from the storage room as it passes through the evaporator coils. The evaporator is a series of coils that are designed to promote heat transfer from the storage area to the refrigerant in the coils. The evaporation temperature of the refrigerant should also be below the desired storage temperature for maximal cooling capacity, because of the latent

Fig. 6 Refrigerated storage room cooled by mechanical refrigeration.

heat of vaporization. The refrigerant vaporizes as it collects heat during its flow through the evaporator, providing the cooling effect in the storage room. The vaporized refrigerant is then recompressed and its temperature increases. The high-pressure vapor thus enters the condensation coils at a vapor temperature higher than ambient temperature. As heat is transferred from the hot vapor to the air outside the storage room, the vapor condenses

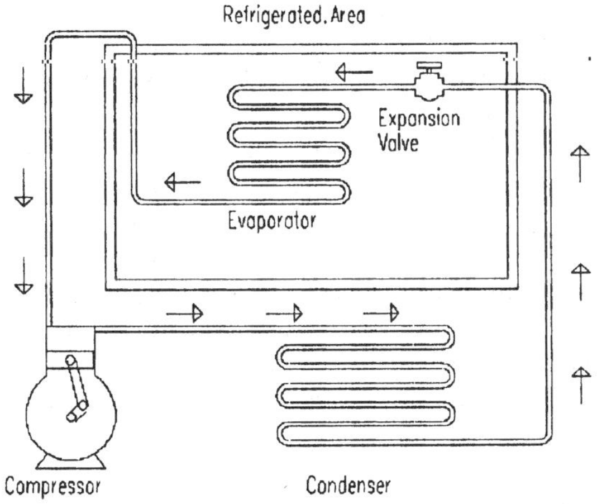

Fig. 7 Schematic of a vapor recompression refrigeration cycle.

back into a high-pressure liquid. This high-pressure liquid enters the expansion valve, and the cycle is repeated.

The expansion valve plays a major role in the vapor-recompression cycle, as it is used to control the amount of refrigerant that enters the evaporation coils and in smaller systems controls the pressure of the refrigerant. The valve aperture may be controlled manually or automatically by temperature or pressure sensors. There are many different types of expansion valves; the two most often used are capillary tubes and thermostatic expansion valves (2).

The evaporator is responsible for allowing heat to vaporize the refrigerant. The two designs most often used in storage systems for fruits and vegetables are the bare-pipe and the finned-tube evaporators. The bare-pipe evaporators are the simplest form and are easy to clean and defrost. The finned-tube evaporators have fins to increase the surface area of the evaporator and thus provide a higher heat transfer rate. Evaporators are also classified as direct-expansion or flooded types. In direct-expansion evaporators, there is no recirculation of the refrigerant in the coils before its return to the compressor. The flooded type recirculate any liquid refrigerant in the coils until it vaporizes before returning it to the compressor. Such a system has greater heat transfer efficiency. To maintain high efficiency, the evaporator should be defrosted periodically. Evaporators that are operating at 0°C are subjected to the accumulation of frost or ice (25). The frost may be removed by electric heaters, periodic flooding of the coils with water, directing of hot refrigerant into the coils, or defrosting with a brine or glycol solution (2).

Compressor types that are most commonly used are reciprocating, rotary, and centrifugal compressors. Compressors may be run by an electric motor or by an internal combustion engine. The reciprocating compressors can be used over a wide range of capacities but have high maintenance costs. Rotary compressors have lower maintenance costs but are not available in sizes lower than 23 kW (2).

There are three major types of condensers, water-cooled, air-cooled, and evaporative. In double-pipe condensers cold water is passed through a pipe that is encased by a larger pipe that carries the refrigerant. Heat is transferred by conduction from one pipe to the other. Shell and tube condensers are also used. These are similar to double-pipe condensers except that the water pipes enter a large shell that contains the refrigerant. The method of heat exchange is the same. Water-cooled condensers can be expensive, especially if water is limited or its disposal is difficult.

Air-cooled condensers are usually coils that are exposed to air. The refrigerant flows through the coils and the heat is conducted from the surface of the coils to the air by convection. Fans are often installed on air-cooled condensers to increase the rate of cooling by forcing air across the coils. Such condensers are cheaper and require less maintenance than water-cooled condensers, but they require a much larger surface area for the same amount of cooling.

Evaporative cooling condensers use a combination of the other two methods. Water is sprayed on the coils as air is drawn over them. The latent heat required to evaporate the water is extracted from the refrigerant. The water is collected in a pan below the coils and is used to spray the coils again. This method uses less water than the regular water-cooled condensers but units can be quite large.

The selection of a proper refrigerant is important for efficient use of a refrigeration system. Factors to consider include the cost of the refrigerant, the latent heat of vaporization, the condensing pressure, and the toxicity. Other considerations include the environmental impact and the nature of the refrigerant with respect to its chemical properties.

3.3 Evaporative Cooling

In some parts of the world, the climate is suitable for the use of evaporative cooling to provide some or all of the necessary refrigeration for cooling or storage. Evaporative cooling is a very economical and energy efficient technique. For evaporative cooling to be effective, the air used should have a relative humidity lower than 65% (7). An airstream is passed through a water spray or a membrane that is saturated with water. Because of the low relative humidity of the air passing through a zone of high moisture, the air acquires more water vapor. The added water vapor must evaporate from the surface of the membrane or from the spray, a process that requires energy. This energy arises from the water and the airstream, reducing the temperature of both. Theoretically, the lowest temperature that can be obtained from a single-stage evaporative cooler is the wet bulb temperature of the incoming air. Evaporative coolers can be designed to be 85% to 90% effective, or in some cases even more (26). A multiple-stage evaporator can be used to obtain temperatures that are below the wet bulb temperature of the incoming air. In a two-stage system, water is cooled to the wet bulb temperature of the air by an evaporative system. This water is then used in a water-to-air heat exchanger to reduce the temperature of the outside air without adding moisture to it. The dry bulb and wet bulb temperatures of the outside air subsequently decrease. The cool air can be passed through an evaporative cooler for further cooling if desired, down to its wet bulb temperature. The use of a two-stage system can reduce the temperature of the air by 6°C more than use of a single-stage cooler (2). The theoretical minimal temperature that can be obtained from a multiple-stage evaporator is the dew point temperature of the air (2). The system can use more than two stages, but the absolute reduction decreases for each stage. It is therefore not usually cost effective to include more than two stages.

3.4 Carbon Dioxide Cooling

Liquid CO_2 is a refrigerant that requires relatively low equipment investment as well as a high cooling capacity; however, operating costs are higher than for conventional vapor-recompression refrigeration systems. Liquid CO_2 can be bought in large canisters. When released through an orifice, it converts to CO_2 vapor. This process requires the addition of energy. The latent heat of CO_2 (2068 kPa and −17°C) is 265.2 $kJ \cdot kg^{-1}$ (27), at which point the CO_2 vapor is at a temperature of about −78.9°C. It absorbs more heat as it increases in temperature. The specific heat of the CO_2 vapor is 0.804 $kJ \cdot kg^{-1} \cdot °C^{-1}$. If the vapor temperature increased to 0°C, an additional 65.53 $kJ \cdot kg^{-1}$ of heat would be extracted from the surrounding environment, raising the total to 328.6 $kJ \cdot kg^{-1}$ of heat removed.

Some work using liquid CO_2 as a heat sink to cool produce has been done. In one study, CO_2 was forced over the top of strawberry flats at different speeds to determine the effect of velocity on the cooling rate (27). At 5 $m \cdot s^{-1}$, 20 kg of strawberries was cooled from 25°C to 2.5°C in 44 minutes. The rate of use of CO_2 for this test was 0.058 $kg\ CO_2 \cdot °C^{-1} \cdot kg^{-1}$ of strawberries, which was believed to be high because of the small scale of the tests. In similar tests, pallets with an average weight of 513 kg of blueberries used 0.022 $kg\ CO_2 \cdot °C^{-1} \cdot kg^{-1}$ of blueberries (28). It was estimated that 0.015 $kg\ CO_2 \cdot °C^{-1} \cdot kg^{-1}$ was used to cool the blueberries and the rest was lost to other sources (28), resulting in 68% system efficiency.

CO_2 can have further beneficial effects on produce by providing some postharvest disease control (29). Product moisture loss may be lower than during conventional refrigeration since the amount of moisture required to saturate the dry vapor is only a fraction of the amount of moisture that would condense on an evaporator (12).

3.5 Alternative Methods

3.5.1 Nighttime Cooling

In some parts of the world there is a large diurnal temperature swing, and, where the nighttime temperature is low enough, outside air may be used as a source of refrigeration. Nighttime cooling is especially useful on commodities that are stored at moderate temperatures (5°C–12°C) such as pumpkins, potatoes, onions, sweet potatoes, and hard-rind squash (2). Natural ventilation during the night is usually sufficient if the outside temperature is below the required range for 5 to 7 h each day (2). In such climates it is advantageous to harvest the product when its temperature is lowest (early in the morning or at night), thus reducing the refrigeration load.

The removal of heat by radiation to the nighttime sky is a promising method, although it is not often used. A clear nighttime sky is very cold and a large amount of heat can be exchanged through radiation. Simulations have shown that the air temperature around a black metal surface can cool to about 4°C below ambient nighttime temperatures (2).

3.5.2 Well Water

Groundwater temperature usually does not vary by more than 1°C during the year for any given location (30) if taken at a depth at which soil temperature is nearly constant. This depth varies with geographical region and can range from 2 to 9 m (31). The water can be used as a heat sink to cool produce if circulated through some type of heat exchanger.

3.5.3 High-Altitude Cooling

The temperature of air drops about 10°C for every 1000-m increase in altitude (2). This can be useful for cooling produce if the situation warrants. For example, produce that must be transported into a nearby mountainous area, either to be distributed or to reach its final destination, may be stored in that area to take advantage of the lower temperature.

3.5.4 Underground Storage

Underground storage is good for keeping produce cool but not for removing field heat. Soils have a low thermal conductivity, which limits the rate of transfer away from the storage. This problem can be solved partially by installing a network of pipes in the soil through which air can be pumped to transfer heat away from the soil. A greater soil volume can then be used to cool the produce.

3.5.5 Thermoelectric Cooling (Peltier Effect)

The Peltier effect is the heat absorbed or evolved when a current is passed across a junction between two dissimilar metals (32). Dissimilar thermoelectric metals are those that have different available electron energy levels. When electrons flow from one such material to another, the electrons must change energy levels and thus absorb or release heat. The direction of the flow of electrons determines whether the heat is absorbed or evolved. The electrons are forced to flow by applying a voltage difference across the two dissimilar materials. A circuit that consists of two dissimilar thermoelectric materials and a voltage source is known as a *Peltier thermoelectric couple*.

The principles described can be used to produce a refrigeration system that acts much as a heat pump. Thermoelectric cooling has advantages over mechanical refrigeration since no moving parts are involved. However, a series of many Peltier thermoelectric couples is needed to produce sufficient heat exchange for horticultural needs, since each couple produces a very small amount of heat exchange. The thermoelectric coolers need a direct current source, and systems that operate on 120-V source or from 12-V car batteries have been built (32). Some systems have capacities of up to 35.2 kW of air conditioning (32). To make the system as efficient as possible, heat transfer surfaces that operate in either gas or liquids are used. For gas systems, the heat transfer component consists of fins to increase the surface area and fans to increase the convective heat transfer rate. Liquid systems use pumps to circulate a fluid over the heat transfer surface. Apart from the fans or pumps that are used to increase the heat exchange, the system does not have parts that can wear out or generate noise. There is no refrigerant; thus there is no concern about toxicity, corrosiveness, chemical stability, and flammability that may occur in conventional vapor recompression systems.

The main disadvantage of thermoelectric cooling is that it has a low coefficient of performance (32): the ratio of heat extracted divided by the necessary energy input to achieve the heat removal. Although thermoelectric cooling has not been extensively used as a source of refrigeration for cooling or storage of fruits and vegetables, it offers many appealing advantages that may incite further development.

3.6 Calculating Refrigeration Requirements

The amount of refrigeration that is required for cooling or storage is often expressed in kilowatts or tons of refrigeration. By definition, a *ton* of refrigeration is the amount of heat absorbed in 24 h by a ton of ice melting at 0°C. One ton of refrigeration is equivalent to 12 660 kJ $\cdot$ h^{-1} (12,000 Btu $\cdot$ h^{-1}).

The refrigeration load for a storage can be calculated after accounting for all the sources of heat to be removed. There are three main sources of heat: (a) the produce itself, whether the heat is produced or contained within; (b) the walls by conduction or leaks; and (c) equipment needed in the storage area. The size and type of equipment chosen to refrigerate the storage room are based on the refrigeration load required, as well as the desired temperature and relative humidity.

Heat from the produce consists of field heat and heat of respiration. Field heat is the amount of sensible heat that must be removed when cooling produce to the desired storage temperature. With increased use of bulk handling, heat loads may reach unusually high levels and thus may sometimes exceed the cooling capacity of the storage. The heat of respiration is the amount that is produced by the metabolic reactions occurring within the produce. The amount of respiration heat given off by stored produce varies with the type of commodity, its age, and its initial temperature. As the temperature increases, the rate of respiration increases, and thus the amount of heat evolved also increases. Heat generated by products with high respiration rates may contribute substantially to the cooling load.

The second source is heat from outside sources, such as the storage walls. This type of heat may be in two forms, conductive or convective. Conductive heat is gained or lost through the building floor, walls, and ceiling. The greater the temperature difference between the storage room and the environment, the greater the heat transfer rate. This type of heat transfer can be limited by proper insulation. Convective heat arises from infiltration of outside air, such as through open storage room doors or poor airtight walls and ceiling.

The final source of heat is that produced by continuously or periodically used equipment, such as forklifts, lights, motors, pumps, scrubbers, and human occupancy. The sum of the heat gained from all three main sources can be totaled then multiplied by a service factor of 1.1 to 1.2 and a defrost factor of 1.1 to 1.2 to give the total refrigeration load (24).

4 MODIFICATION OF ATMOSPHERIC COMPOSITION

4.1 Structural Design of the Storage Room

The size and location of the storage area are important. The room should be sized to handle the peak amounts of produce (2). The advantages of square room construction are lower heat loss and lower construction costs, which are due to higher volume to surface area ratio (2). However, the most economical roof span for framing systems is 12 to 19 m; therefore, buildings often have this range for the width and then are simply extended in length (33). The height of the refrigerated spaces is often in the range of 6 to 8 m, and pallets may be stacked to a height of 4.5 to 5.5 m.

The availability of proper utilities must be considered when choosing the site. Three-phase power and an adequate water supply must be present (24). Possible expansion must also be taken into account, including areas for parking, shipping and receiving, cold storage, and packing facilities (24).

The type and amount of insulation to be used for the storage rooms are dependent on the time of year when the storage room is to be used. Rooms designed to operate throughout the year require more insulation than rooms that are to be used only in the fall or summer (29). Ceilings require the most insulation, followed by walls and then floors (24). The minimal insulation requirements are 4.6 $m^2 \cdot °C \cdot W^{-1}$ in the ceiling, 4.2 $m^2 \cdot °C \cdot W^{-1}$ in the walls, and 2.3 $m^2 \cdot °C \cdot W^{-1}$ in the floors (34). Vapor barriers are needed to prevent moisture from forming within the insulated spaces in the walls and ceilings (24). Rooms that are to be used for CA or MA storage should be well sealed, because infiltration rates fluctuate with external atmospheric pressure changes and wind speed. Such changes make it difficult to maintain steady-state conditions when the passive methods of MA storage are chosen. Although conditions can be maintained in a CA storage (active control of air composition), high infiltration rates result in higher operating costs (9). The seals of controlled and MA storage should be checked for air leaks. A standard test is to pressurize the storage room to 25 mm (1 in) of water and to determine how long the pressure takes to decrease to 12 mm (1/2 in) of water (24). A 20-minute room, or a room that takes 20 minutes to drop half of its pressure, is acceptable for 3.0% O_2 levels (35). A 30-minute room is adequate for all low-O_2 storage (35). Many sources of information can be found that suggest the proper construction of storage rooms, including the amount and type of insulation, vapor barriers, and type of structures (21, 31, 36, 37).

CA and MA storage rooms must be equipped with a pressure relief system, since the gas seals or structure may otherwise be damaged from changes in pressure, as a result of the addition and removal of gases or temperature changes (2). A simple method to control the pressure range is a water trap, which may be constructed by building a trough

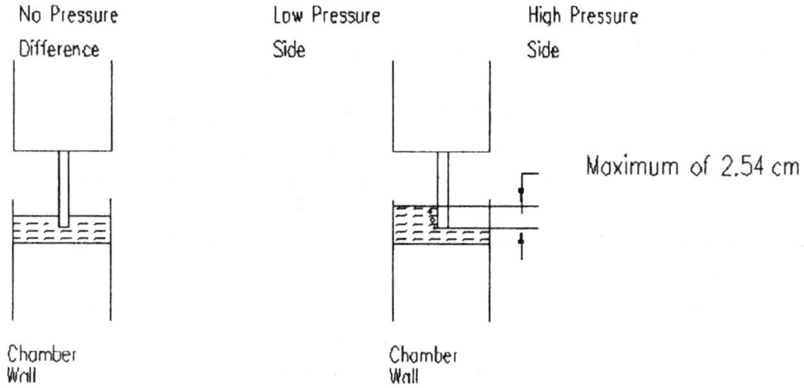

Fig. 8 Water trap used as a pressure relief system.

in the wall and filling it with water (Fig. 8). A thin wall is built from the main wall and extends down into the trough. The trough is filled with water until it extends up above the lower limit of the thin wall. When the pressure in the room changes, the water level in the inside rises or drops depending on the direction of the change. When the pressure reaches a certain degree, the water level will have dropped low enough on each side to allow air to enter or exit the room and relieve the pressure. The system should be designed so that the water height differential is only 2.54 cm (24), resulting in a maximal pressure difference of approximately 250 Pa. Glycol is often used instead of water since it does not evaporate readily at storage room temperatures. Another method for controlling the pressure limits in a room is to use spring-loaded or weight-loaded check valves. These valves are more expensive than the water traps (2). For small pressure changes a breather bag may also be used. Breather bags are installed by cutting a hole in the wall of the storage room and sealing the open end of a larger bag around the perimeter of the hole. The increase or decrease of pressure in the room allows the bag to fill with or empty of gas. Essentially, it allows the effective volume of the room to increase or decrease, minimizing the change in pressure. Bags should have 0.35 to 0.4 m^3 of capacity for every 100 m^3 of room volume (2).

4.2 Controlled Atmosphere Storage

CA storage implies precise control of the gas concentrations inside the storage room. Modification of atmospheric gas levels may reduce the respiration rate of fresh produce, as well as control the level of ethylene (C_2H_4) and thus retard ripening. The gas concentrations of ambient air are 78.08% N_2, 20.95% O_2, and 0.03% CO_2 (2). In most CA storage systems, the O_2 level is decreased and/or the CO_2 level is increased. Either generally causes a decrease in product respiration rate. Different types of produce respond differently to these two gases, and thus the proper atmosphere for a given commodity should be predetermined experimentally. In some cases, ideal concentrations of these gases for long-term storage of one commodity may prove harmful to another. For example, cauliflower stored in 10% CO_2 at 5°C is injured after a week in storage, whereas broccoli in the very same environment remains in excellent condition (23). Some recommendations of the proper storage requirements are given in Table 1.

The choice of CA system to use depends primarily on the gas composition that is desired and the rate at which it is to be achieved. The standard free volume (SFV) is the ratio of the volume of air to the volume of commodity. The SFV in typical warehouses ranges from 1.5 to 3.0 and is a function of the stacking arrangement, room geometrical characteristics, commodity shape and density, and method of packing, either in bulk or in crates (3). CA rooms with a higher SFV have more stable gas composition with time but also require more intervention to modify the gas composition. For example, a storage room with an SFV of 3.0 requires approximately twice as much time to reach its equilibrium gas composition as a room with an SFV of 1.5, since twice as much O_2 must be removed by the respiration of the produce or by the control system.

Generally, the greatest benefits of CA storage result from the rate of decrease in the O_2 level. Rooms that require very low O_2 levels, 1% to 2%, or that require a very fast pull-down should have a very efficient CA system and a low SFV. The system must also take into consideration the level of CO_2. High levels of CO_2 can cause damage to some types of fruits and vegetables. For example, it has been shown that some apple cultivars may be damaged by 3% CO_2 (38), whereas lettuce cannot tolerate any CO_2 (39). Methods of CO_2 scrubbing therefore need to be considered when designing CA storage for produce that is susceptible to CO_2 injury. Finally, attention should be given to the control of C_2H_4 levels in the storage room if the produce is susceptible to hastened ripening from increased C_2H_4 concentrations. Thus, there are three main control systems that may be used to obtain the desired gas concentrations: (a) O_2 control systems, (b) CO_2 control systems, and (c) C_2H_4 control systems.

4.2.1 Oxygen Control Systems

If product respiration does not decrease the O_2 level quickly enough, one of four active methods may be used: (a) external burners, (b) liquid or gaseous nitrogen (N_2), (c) gas separator systems, or (d) hypobaric storage. External burners use the combustion of propane or natural gas to remove O_2 from air that enters the storage room (Fig. 9). The combustion process produces a mixture of CO_2 and water vapor and usually requires a CO_2 scrubber to prevent high accumulations of the gas. The air should be cooled after combustion and before it is injected into the room. Open flame or catalytic burners may be used. The open flame burner has the disadvantage of not permitting air recirculation because of the risk of extinguishing the flame by the O_2-depleted recirculating airstream. Catalytic burners are therefore preferred. Furthermore, catalytic burners provide more

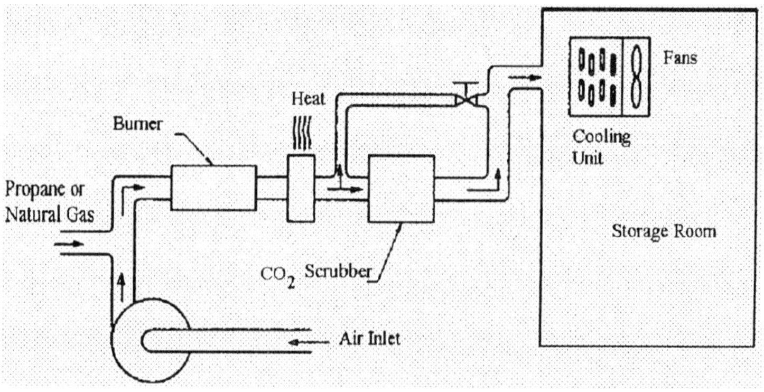

Fig. 9 Schematic of an external gas generator for an O_2 control system.

complete combustion and can reduce O_2 levels to 3%. Catalytic burners are more expensive to install if used in a recirculating system, but this expense is compensated for quickly by lower operating costs. Overall, external burners are inexpensive but may be safety hazards if they use highly flammable fuels.

Flushing with liquid or gaseous N_2 is an effective rapid O_2 pull-down technique. The amount of N_2 needed to reduce the O_2 level is a function of the desired O_2 level and the SFV (Fig. 10). Usually, liquid N_2 is injected into the room through spray headers that atomize the N_2 into a fine mist. It is best to have the spray headers placed in front of the evaporator fans. This method also assists in refrigerating the room because of the latent heat of evaporation of the liquid N_2; however, this advantage may be offset by the cost of insulating the N_2 supply lines. This method should not be used on unpacked produce since freeze burning may occur, and adequate venting should be used to prevent overpressurization of the room (40). Nevertheless, liquid N_2 can create appropriate CA conditions very rapidly with a purge rate of up to 35 $m^3 \cdot h^{-1}$ (40).

Gas separator systems may also be used for O_2 control. The different gas separator systems are (a) pressure swing absorption (PSA), (b) hollow fiber membrane separators (HFMSs), and (c) high-temperature ammonia cracking (HTAC). PSA systems are used to generate a stream of air that is very high in N_2 and low in O_2. This system compresses dry air and forces it through a bed of pelletized carbon material (molecular sieve), which absorbs O_2 and yields an N_2-enriched stream (40) (Fig. 11). The air that enters the molecular sieve should be dry and free of contaminants (40); it is best to have the air filtered before it passes through the sieve. The bed pressure is typically 830 kPa (120 psi) (40). The purity of the N_2 stream can range from 90% to 99.9%, depending on factors such as the pressure, temperature, and airflow rate. After a few minutes of operation the molecular sieve becomes saturated with O_2. The O_2 may be removed by decreasing the pressure and venting the sieve (40). Two molecular sieves are usually connected in parallel so that the operation can be continuous: one is regenerated while the other is depleting O_2 from the supply stream. PSA systems can provide rates of 105 to 385 $m^3 \cdot h^{-1}$ at 98% purity with

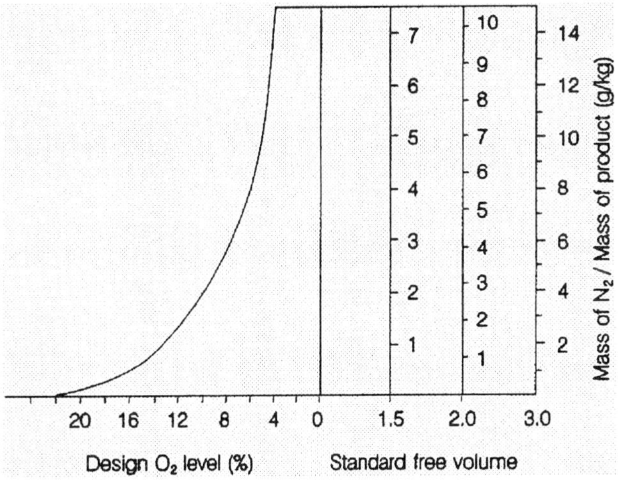

Fig. 10 Amount of N_2 needed for the flushing of O_2 as a function of the desired O_2 concentration and the standard free volume.

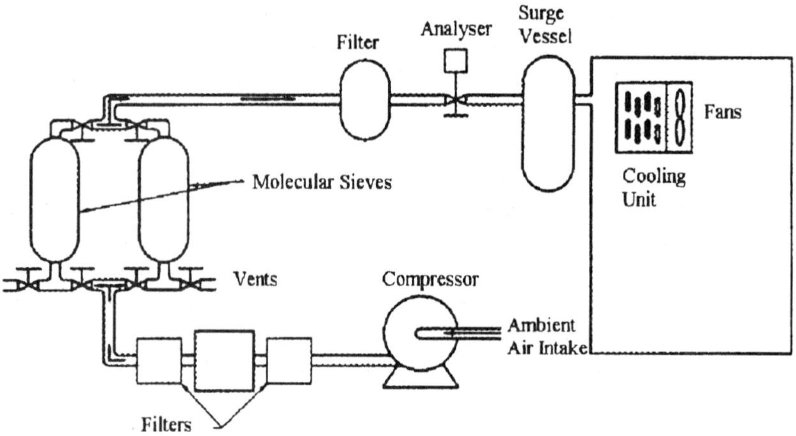

Fig. 11 Schematic of a pressure swing absorption system.

compressors ranging from 30 to 112 kW, respectively (40). The initial cost of the system is fairly high, and the system requires regular inspections.

HFMS works on the principle that some gases can diffuse through membranes at higher rates than others. In the case of air, CO_2 and O_2 have much higher permeation rates than N_2. In HFMS, compressed hot air is forced into a hollow fiber membrane chamber. The O_2 and CO_2 quickly pass through the membrane and are vented to the ambient air (Fig. 12). The concentration of N_2 increases, since N_2 does not pass through the membrane as quickly. The stream leaving the chamber is nearly pure N_2 and is fed to the storage room. The purity of the output may be changed by modifying the rate at which the N_2 stream leaves the chamber (40). For CA storage, the aim is to produce an airstream of 97% to 99% N_2 (40). The temperature of the air has an effect on the output of the system, which increases with higher temperatures (40); however, it should be kept in mind that the membrane can be damaged if the temperature is too high (40). The air that is

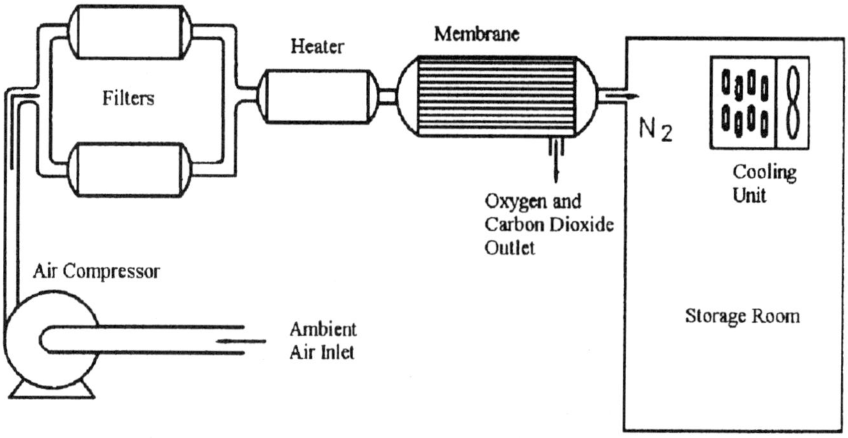

Fig. 12 Schematic of a hollow fiber membrane system.

used in the system should be filtered to prevent contamination of the membranes. The initial cost and the cost of replacement of an HFMS are very high, but maintenance costs are lower than those for a PSA system. This method may also be used for scrubbing by recirculating air from the CA rooms (40) since water vapor and CO_2 permeate the membrane walls.

Ammonia cracking is another type of O_2 control system. High-temperature anhydrous ammonia (NH_3) gas is reacted with air from the CA room. The reaction involves the splitting of the NH_3 into hydrogen and inert N_2 gas. The hydrogen then reacts with the O_2 present in the air to form water vapor. The air returning to the room consists of N_2 and water vapor but is free of CO_2, carbon monoxide (CO), and hydrocarbons. The returning airstream is cooled before it returns to the room. The efficiency of the combustion process is determined by the combustion temperature and the ammonia/oxygen ratio (24). The operating costs are high and the use of ammonia gas can be extremely dangerous since the gas is toxic at concentrations of 0.5% by volume in air (37). In addition to these disadvantages, processes of ammonia manufacture require a lot of fossil fuel energy simply to combine the inert atmospheric N_2 with hydrogen.

Hypobaric storage involves storing the produce at reduced pressure, usually between 10 and 80 mm of mercury. The air in the reinforced, airtight refrigerated room is continually removed by a vacuum pump (41). When the pressure in the room reaches the desired level, air is allowed to enter the room at a rate that creates one to four air changes per hour (41). This system provides relatively easy manipulation of O_2 concentration and relative humidity. CO_2, C_2H_4, and other volatile gases of metabolism are also removed, making it possible to store together commodities that are otherwise not normally compatible in storage (41). The system can also be used as a vacuum cooler. However, there are some major difficulties with hypobaric storage (40). The cost associated with providing a room with acceptable structural strength for the required vacuum is very high. It is also difficult to permit accumulation of CO_2 to levels that are beneficial to product quality. Finally, hypobaric storage can affect the flavor of the produce and can cause unsatisfactory ripening after storage.

4.2.2 Carbon Dioxide Control Systems

There are five commercially available scrubbing systems for the removal of excess CO_2. These are based on (a) caustic soda, (b) hydrated lime, (c) water, (d) activated charcoal, and (e) molecular sieves. The level of CO_2 in the room is controlled by adjusting the gas or liquid flow rate through the scrubber. The atmospheric composition of the room should be measured to determine the flow rate required.

Caustic soda (NaOH) dissolved in water can be used to remove CO_2 from the air of the storage room by circulating it in open tubes. The amount of CO_2 removed can be adjusted by changing the duration of caustic soda solution exposure to the room's atmosphere. The use of caustic soda has been largely discontinued because of its corrosiveness and the potential danger in handling. Systems using a dry caustic material in a closed container that have been introduced may provide noncorrosive alternatives to aqueous solutions (24).

The hydrated lime ($CA(OH)_2$) scrubber is one of the simplest and most effective systems for controlling the level of CO_2. Hydrated lime is placed in an insulated and airtight box outside the storage room to which it is connected by two pipes (Fig. 13). One pipe is an inlet that allows air from the storage room to enter the box containing the lime.

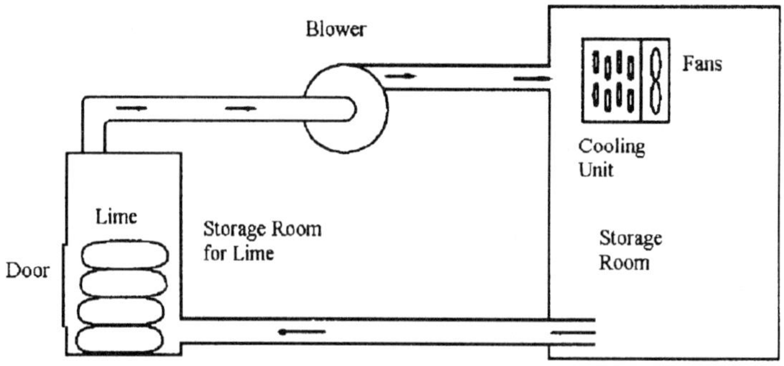

Fig. 13 Schematic of a hydrated lime system for the removal of CO_2.

The other pipe is a return through which CO_2-depleted air is returned to the CA room. The return pipe may be equipped with blowers or dampers to regulate the airflow passing through the scrubber.

The box is usually large enough to contain sufficient lime for the entire storage period. The lime should be placed on a pallet with a 10-cm space between layers of 25-kg bags to allow for good air circulation. The effectiveness may be further increased if the bags are only partially filled because less than 2% of the lime in a 25-kg bag is consumed as a result of the hardening of the outer layer. The amount of lime required depends on the amount of produce in the storage room, the respiration rate, and the rate of CO_2 addition to the room. Some O_2 scrubbers produce CO_2 as a by-product, which must also be removed by the lime. A 25-kg bag of lime has a volume of approximately 0.1 m^3 (24), and roughly 10 bags are needed for every 19 tonnes (1000 bushels) of fruit, whereas about 40 bags are needed for 100 tonnes of vegetables (24). The hydrated lime and the CO_2 react in a 1:1 ratio to form limestone ($CaCO_3$) and water. The lime that is to be used can be agricultural or chemical hydrated lime as long as it is fresh, high in calcium, and fine enough to pass through a 100 mesh sieve (24). One should choose limes that have less than 2% magnesium hydrate (MgO) since the calcium hydrate (CaO) is more reactive than the MgO (24). Hydrated lime with an assay of 75% CaO contains 99% $Ca(OH)_2$ (24).

It is possible to add lime to the storage room directly, but only if the lime is not responsible for scrubbing all the CO_2. This is most often used during the initial O_2 pull-down period (24). The lime can be placed directly under the evaporator, in front of the door, or on top of the stacks, as long as it does not disturb the airflow in the room. Some heat is given off during the reaction between CO_2 and the lime, so thermostats or thermometers should not be placed close to the lime to prevent the room temperature from being biased upward (24).

Water CO_2 scrubbers use the inside to outside differential pressures of CO_2 to remove it from the CA room. A brine solution is pumped or sprayed over the evaporator coils in the CA room (Fig. 14). As the brine is exposed to the atmosphere in the CA room it absorbs CO_2, since the CO_2 partial pressure is very low in the incoming water and very high in the atmosphere of the CA room. The brine solution is collected in the room and allowed to travel into a reservoir located outside the room. Here the brine solution is pumped to aerators where the CO_2 is dissipated to the air because of a differential pressure between the CO_2 in water and that in air. There is corrosion due to the brine solution, but

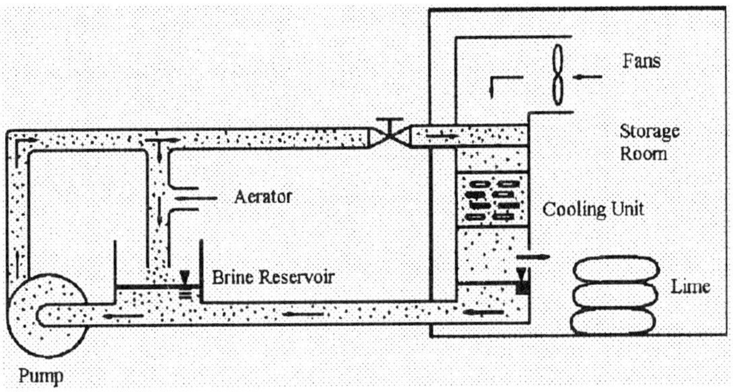

Fig. 14 Schematic of a water scrubber to remove CO_2.

this may be prevented if a dry cooling unit is used. A modified system (42) uses two aerators, one inside the room and one outside the room. The size of the water CO_2 scrubber is based on the quantity of produce stored, its CO_2 production rate, and the desired concentration level. The rate of removal can be controlled through the water flow rate. Peak CO_2 removal rate generally occurs in the early stages of storage when respiration rates are high and atmosphere generators producing CO_2 may be in use (24). If hydrated lime scrubbing is used to assist the water scrubber for the peak CO_2 production period, then a smaller water scrubber can be used (i.e., one that can handle the steady-state CO_2-evolution rate rather than the peak rate). It has been recommended that the flow rate should be 100 L per hour per ton for apples stored at 1°C with 3% O_2 and 5% CO_2 (43).

Water scrubbers have some disadvantages that have made their popularity decline. Small amounts of O_2 are added to the room since there is a higher partial pressure of O_2 in the incoming water solution than there is in the CA room (24). This addition of O_2 makes the control of the atmosphere more difficult, especially if the room is not already perfectly airtight. Finally, since the water scrubber is only controlling CO_2, another system must be used in parallel for O_2 pull-down and control.

Activated charcoal and molecular sieve scrubbers are based on the adsorption of CO_2. They consist of a unit filled with an adsorbent material, two blowers, and four solenoid valves dedicated to a control unit (Fig. 15). Air from the CA room is blown through the absorbing unit, and the CO_2-depleted stream is then directed back to the CA room. The adsorbent loses its ability to trap CO_2 as a result saturation of active sites and thus after some time must be reactivated. This is done by purging the adsorbent with ambient air. The control unit, generally based on timers, shuts off the valves that are connected to the CA room and opens other valves to force outside air through the absorbent material to remove the CO_2 (reactivation process). Molecular sieves should be heated during the reactivation phase to enhance the rate of CO_2 removal (24). Since purging with ambient air can add O_2 back into the CA room, the absorbing unit is often purged with pure N_2.

It is possible to use one large charcoal scrubber to scrub two or more CA rooms or to blend atmospheres from different rooms to obtain desired atmospheric compositions (24). CA rooms should be equipped with pressure relief systems to prevent damage to gas seals or structural damage due to malfunction of the automatic valve controls or improperly set manual valves (24). Activated charcoal and molecular sieve systems have

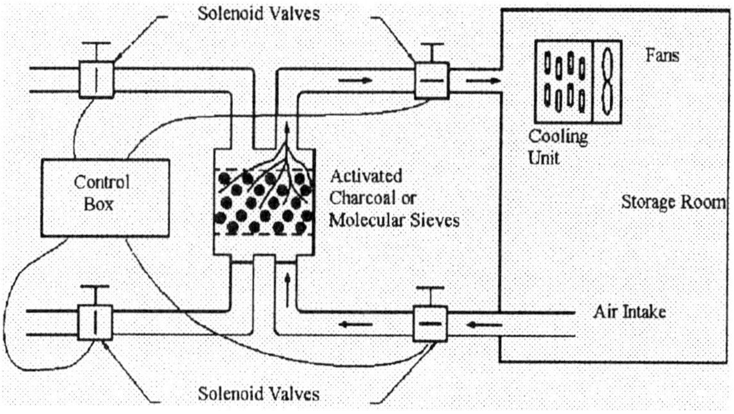

Fig. 15 Schematic of an activated charcoal/molecular sieve system for the removal of CO_2.

very low operating costs, and the absorbent material usually lasts for about 5 years before it must be replaced. The molecular sieve system requires more energy to operate than the activated charcoal system because of the heat requirements of the reactivation phase (24).

4.2.3 Ethylene Control Systems

Ethylene (C_2H_4) induces ripening in many fruits and can also cause some physiological disorders in vegetables (23). The amount of C_2H_4 produced by the commodity can be reduced by decreasing the surrounding O_2 level and increasing the CO_2 level (23). Low temperature levels, 0°C to 4.4°C, can prevent the production or inhibit the action of C_2H_4 (23). Nonetheless, C_2H_4 levels less than 1 ppm can produce physiological responses in many fruits and vegetables (23). For example, kiwifruit may be affected by C_2H_4 concentrations as low as 0.1 ppm (35). Therefore, removal of C_2H_4 from the storage environment is important. It should be mentioned that C_2H_4 sensitive commodities should not be stored near or with commodities that produce high levels of it, such as climacteric fruit (e.g., apple). Most of the C_2H_4 may be removed if N_2 generators or gas flushing methods are used. This may be acceptable for types of produce that have low C_2H_4 production rates, but other control systems are often required for commodities that have high C_2H_4 production rates. Commercial C_2H_4 scrubbers include (a) the heated catalyst scrubber, (b) C_2H_4-absorbing beads, and (c) ozone.

A heated catalyst can be used to maintain relatively low levels of C_2H_4. Air from the CA room is forced through ceramic packings, which are used as heat exchangers, and contacts a heated catalyst. The catalyst is electrically heated and the high temperature promotes the oxidation of C_2H_4 to CO_2 and water vapor. Up to 87% of the C_2H_4 can be removed from the air in one pass through the catalyst. The system is set up so that flow reversal through the ceramic packings occurs at timed intervals. The energy requirements for heating the air and then cooling it after breakdown of C_2H_4 are the main disadvantages of the heated catalyst approach.

C_2H_4-absorbing beads are small spherical particles of aluminium silicate impregnated with potassium permanganate ($KMnO_4$). The beads are placed in a sealed unit and air from the CA room is circulated through the unit. The C_2H_4 reacts with the $KMnO_4$, changing the color of the beads from purple to brown as they become saturated with C_2H_4.

Once saturated, the beads must be replaced. The cost of the bead system is similar to that of the heated catalyst (44) and is a function of the required rate of C_2H_4 removal. For long-term storage of a product that has a high C_2H_4 production rate, it is less costly to use a heated catalyst. On the other hand, if the production rate of C_2H_4 is low, it is cheaper to use a $KMnO_4$ scrubber (44).

Some research has been done on the use of ultraviolet (UV) radiation as a means to remove C_2H_4. UV light reacts with O_2 to form ozone (O_3), which has the ability to destroy C_2H_4 (44). The reaction rate between O_3 and C_2H_4 is very slow, therefore, a fairly large reactor is needed for efficient removal (44).

4.3 Modified Atmosphere Storage

MA storage differs from CA storage in that the atmospheric composition is not actively controlled. Most MA systems use semipermeable membranes to regulate gas exchange between the MA and the ambient air. The composition of the air changes as a result of the respiratory action of the produce and the permeability characteristics of the membrane used. For example, silicone membranes allow the gases to diffuse at different rates, which are determined by the chemical and physical characteristics of the gases (45). Other important factors in the amount of gas diffusing through the membrane include temperature, membrane surface area, permeability of the membrane, and the gas partial pressure difference across the membrane. Generally the membranes are more permeable to CO_2 than they are to O_2. Therefore, the concentration of CO_2 increases and the concentration of the O_2 decreases, as O_2 is consumed and CO_2 is released as a by-product during respiration of the produce. Although MA storage does not achieve the same degree of atmospheric control as does the CA approach, it is less expensive (46). The membranes can be made of polymeric films or wax or may be edible coatings of individual fruits. MA storage is often a better approach for short-term storage of small quantities of produce than is CA storage (46), and it is often used in association with packaging.

The time required for the gas composition to stabilize to quasi steady state is an important factor in MA storage. The stabilization period is a function of the designed gas composition, commodity respiration, relative stacking volume, and airtightness of the system. The time required to reach the stable gas levels may be in the order of days to weeks. If the stabilization period is too long, the product may deteriorate more than desired. In this situation, it may be advantageous to use a rapid O_2 pull-down method for the initial modification of the atmosphere.

MA storage systems are widely used in European countries, whereas they are only beginning to be popular in North America (45). Several silicone membrane–based systems have been conceived since the early 1970s, the most popular of which is the Marcellin system. This system consists of a series of silicone rubber bags that are connected in parallel (Fig. 16). Air from the storage room is circulated through the exchanger, resulting in an exchange of gases as they diffuse through the silicone rubber bags. The amount of diffusion taking place can be changed by changing the number of bags that are being used. The maintenance of 5% CO_2 and 3% O_2 requires 50 m^2 of silicone membrane for every 100 tons of fruit at a bulk density of 200 to 250 $kg \cdot m^{-3}$ (41). A modified version of the Marcellin system uses a diffusion unit that functions similarly to a plate heat exchanger. Within an airtight box, a number of gas diffusion panels made by using square framing hold the semipermeable silicone membrane in place. Air from the storage room is passed through the diffusion unit by a blower, and in the same manner outside air is

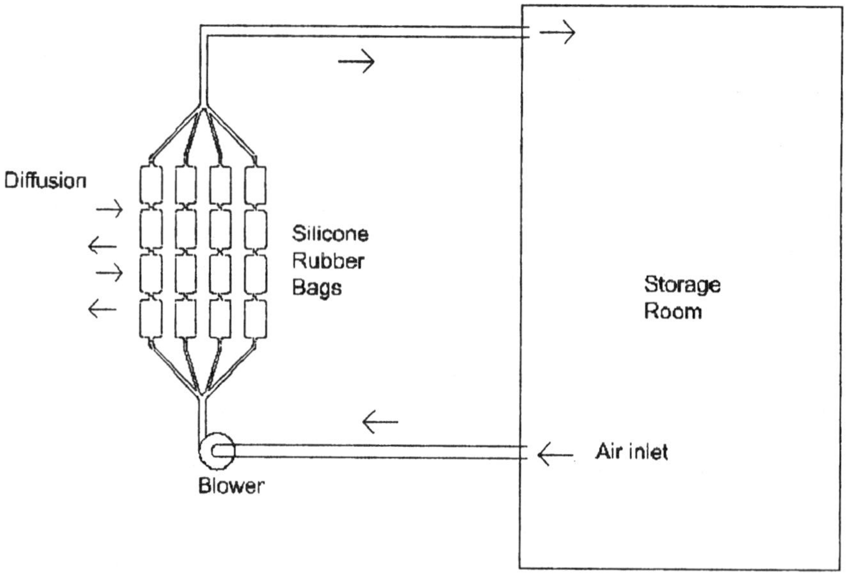

Fig. 16 The Marcellin system.

passed through the diffusion unit. A silicone membrane separates the two types of air flowing through the system and allows the diffusion of gas from one side of the membrane to the other (45). The air from the storage room picks up some O_2 and releases some CO_2 as it contacts the silicone membrane. The outside air passing through the diffusion unit absorbs the excess CO_2 and releases O_2.

A pallet MA package system has been used successfully for fruits and vegetables during transportation and storage. It requires little initial investment and involves wrapping individual pallets of produce with a polyethylene wrap of a thickness of 80 to 150 μm. An area of the polyethylene wrap is replaced by a silicone membrane window, which allows for gas exchange. Calibrated orifices are also used to allow for pressure regulation within the pallet package. This system has a lot of flexibility, as the produce may be moved from place to place while still under modified conditions, though extra care must be taken to prevent damaging the packages. The produce can be stored under optimal atmospheric composition without any modifications to preexisting buildings or investment in expensive equipment to change the air composition. This system allows for easy management of produce; small quantities of produce may be marketed without disturbing the atmospheric composition of all the pallets, as occurs when a sealed room is opened. However, there are a couple of disadvantages to the system: the space required to store the produce is greater because of the need to provide adequate ventilation around all the pallets, and the time and material required to wrap the pallets and install the correct size of membrane add to the storage cost of the produce.

Some produce may be packaged in permeable film wraps or bags. These wraps or films allow the passage of gases at different rates, much in the same manner as other MA storage techniques. These films have been tested on many commodities, such as apples, avocados, kiwifruits, peaches, and tomatoes (45). However, there are situations in which damage to the produce can occur as a result of a very high CO_2 or very low O_2 concentration (47). A solution to this problem is the addition of microperforations to the polyethylene bags. In some experiments it has been shown that microperforations are more useful at higher temperatures and short storage duration. The size and number of microperforations have an effect on the efficiency of the microperforations (48). Presence of too many holes takes away the effect of the semipermeable nature of the bags, whereas presence of too few holes does not provide the necessary ventilation (48).

5 CONCLUDING REMARKS

Product temperature is the critical factor to be controlled in the postharvest storage of fruits and vegetables. Temperature has the single greatest effect on respiration rate and thus deterioration rate of the product. Rapid product cooling after harvest is essential in the preservation of most fresh commodities. Though there are a number of precooling methods, care should be taken when selecting a method for a specific commodity. Product properties differ from commodity to commodity. Factors that must be taken into account when selecting a precooling method include product deterioration rate, sensitivity to water contact, temperature of freezing, temperature of chilling, sensitivity to water loss, economic considerations, and the expected storage duration. Because of the combination of these factors, some precooling methods are not suited for certain products.

Optimal storage of commodities is necessary to prolong their fresh quality. Storage temperature vary from product to product, depending on chilling sensitivity. Other factors that can be controlled to maximize storage duration include relative humidity and gas concentrations. The recommended relative humidities are commodity specific, depending on the nature of the product. Proper relative humidity levels decrease the moisture loss of the product, preventing shriveling or wilting. Controlling or modifying the gas compositions of O_2, CO_2, and C_2H_4 can also be useful in prolonging the storage life. Though these changes can increase the storage life of a commodity, they are best used in conjunction with low-temperature management as this factor remains the most critical. When modifying the gas composition, care must be taken not to allow the O_2 levels to reach anaerobic conditions or CO_2 levels to become harmful to the product. Reducing C_2H_4 levels can retard the ripening process in many fruits. The recommended levels of these gases vary from product to product, and some produce may not benefit from such changes. The decision to use modified or controlled gas compositions also depends on the economic factors involved. These systems require well sealed storage rooms and the necessary equipment to produce the desired levels. Hence, for short-term storage they may not be economical.

REFERENCES

1. AA Kader (ed.). 1992. Postharvest Technology of Horticultural Crops, 2nd ed. Cooperative Extension University of California, Davis, CA, Publ. No 3311.
2. R Wills, B McGlasson, D Graham, and D Joyce. 1998. Postharvest: An Introduction to the Physiology and Handling of Fruits, Vegetables, and Ornamentals, 4th ed. CAB International, New York.
3. GSV Raghavan, P Alvo, Y Gariepy, and C Vigneault. 1996. Refrigerated and Controlled Atmosphere Storage. In: Processing Fruits: Science and Technology. Vol. 1. Biology, Principles, and Applications. (ed. LP Somogyi, HS Ramaswamy, and YH Hui). Technomic, Lancaster, PA.

4. C Dennis. 1984. Effect of storage and distribution conditions on the quality of vegetables. Acta Hortic. 85–104.
5. LL Morris. 1982. Chilling injury of horticultural crops: An overview. HortScience 161–162.
6. AL Ryall and WJ Lipton. 1972. Handling, Transportation, and Storage of Fruits, and Vegetables. Vol. 1. Vegetables and Melons. Avi, Westport, CT.
7. KLB Gast and R Flores. 1991. Precooling Produce—Fruits and Vegetables. Cooperative Extension Service, Kansas State University, Manhatten, Kansas.
8. W Boa and RT Lindsay. 1976. Vegetable preparation, cooling and storage. ARC Res. Rev. 2(3): 86–87.
9. American Society of Heating, Refrigeration and Air Conditioning Engineers. 1986. Methods of precooling of fruits, vegetables and ornamentals. In: Refrigeration Systems and Applications Handbook. ASHRAE, Atlanta.
10. JF Thompson. 1995. Hydrocooling fresh market commodities. Perishables Handling Newsletter. University of California, Issue 84: 2–10.
11. SA Sargent, MT Talbot, and JK Brecht. 1991. Evaluating Precooling Methods for Vegetable Packinghouse Operations. Florida Cooperative Extension Service. Institute of Food and Agricultural Sciences. University of Florida, Gainesville. Document No. SSVEC-47.
12. FG Mitchell, R Guillou, and RA Parsons. 1972. Commercial Cooling of Fruits and Vegetables. Manual 43, University of California, Division of Agricultural Sciences.
13. EH Hardenburg, AE Watada, and CY Wang. 1986. The Commercial Storage of Fruits, Vegetables and, Florist and Nursery Stocks. USDA, Washington, D.C. Agricultural Handbook 66.
14. SE Prussia and RL Shewfelt. 1984. Ice Distribution for Improved Quality of Leafy Greens. ASAE Paper No. 84-6014, St. Joseph, MI.
15. MD Boyette and EA Estes. 1992. Postharvest Technology Series: Crushed and Liquid Ice Cooling. North Carolina Cooperative Extension Service, North Carolina State University, AG-414-5.
16. C Vigneault, B Goyette, and GSV Raghavan. 1995. Continuous flow liquid-ice system tested on broccoli. Can. Agric. Eng. 37(3): 225–230.
17. JP Émond and C Vigneault. 1998. Reusable Containers for the Preservation of Fresh Fruits and Vegetables. U.S. Patent No. 5,727,711.
18. AP Longmore. 1973. The pros and cons of vacuum cooling. Food Industries of South Africa, No. 26: 6–7, 9, 11.
19. CC Cheyney, RF Kasmire, and LL Morris. 1979. Vacuum cooling wrapped lettuce. California Agric. 33:10, 18–19.
20. MD Boyette. 1996. Forced-air cooling packaged blueberries. Appl. Eng. Agric. 12(3): 213–217.
21. HW Fraser. 1991. Forced-Air Cooling of Fresh Ontario Fruits and Vegetables. Ministry of Agricultural and Food, Toronto, Ontario, AGDEX 202–736.
22. HW Fraser and I MacKinnon. 1992. Experiences with forced-air precooling of horticultural crops in Canada. Presented at the 1992 International Summer Meeting, June 21–24. Paper No. 92-6017. American Society of Agricultural Engineers, St. Joseph, MI.
23. AL Ryall, and WT Pentzer. 1974. Handling, Transportation, and Storage of Fruits and Vegetables. Vol. 2. Fruits and Tree Nuts. Avi, Westport, CT.
24. JA Bartsch and GD Blanpied. 1984. Refrigeration and Controlled Atmosphere Storage for Horticultural Crops. Northeast Regional Agricultural Engineering Service, Cornell University, Ithaca, NY, NRAES J. Paper No. 22.
25. American Society of Heating, Refrigeration and Air-Conditioning Engineers. 1983. Forced-circulation air cooling and defrosting. In: Equipment Handbook. ASHRAE, Atlanta.
26. American Society of Heating, Refrigeration and Air Conditioning Engineers. 1984. Evaporative air cooling. In: Systems Handbook. ASHRAE, Atlanta.

27. F Gamache and D Désilets. 1987. Cooling of Strawberries in Liquid Carbon Dioxide. CSAE Paper No. 87-501.
28. RP Rohrbach, R Ferrell, EO Beasley, and JR Fowler. 1984. Precooling blueberries and Muscadine grapes with liquid carbon dioxide. Transactions of the ASAE 27(9): 1950–1955.
29. MJ Ceponis and RA Cappelini. 1979. Control of postharvest decays of blueberry fruits by precooling, fungicide and modified atmospheres. Plant Dis. Reporter 63(12): 1049–1053.
30. HJ Braud. 1979. Water Source Heat Pumps for Agricultural Applications. ASAE, St. Joseph, Michigan. ASAE paper No. 79-4562.
31. Public Works Canada 1985. ENERSTOCK 85: Proceedings of the III International Conference on Energy Storage for Building Heating and Cooling. September 22–26, Toronto.
32. American Society of Heating, Refrigeration and Air Conditioning Engineers. 1981. Thermodynamics and refrigeration cycles. In: Fundamentals Handbook. ASHRAE, Atlanta.
33. JA Lindley and JH Whitaker. 1996. Agricultural Buildings and Structures, rev. ed. American Society of Agricultural Engineers, St. Joseph, MI.
34. American Society of Heating, Refrigeration and Air Conditioning Engineers. 1986. Refrigerated warehouse design. In: Refrigeration Systems and Applications Handbook. ASHRAE, Atlanta.
35. D Bishop. 1990. Controlled Atmosphere Storage. In: Cold and Chilled Storage Technology (ed. CKJ Dellino). Van Nostrand Reinhold, New York, pp. 66–98.
36. JW Layer. 1971. Refrigerated Farm Storage. Cornell Univ. Info. Bull. 16.
37. PE Brecht, AA Kader, and LL Morris. 1973. The effect of composition of the atmosphere and duration of exposure on brown stain of lettuce. J. Am. Soc. Hortic. Sci 98: 536–538.
38. PD Lidster, GD Blanpied, and RK Prange. 1990. Controlled-Atmosphere Disorders of Commercial Fruits and Vegetables. Agricultural Canada Publication 1847/E.
39. H Waelti, and RP Cavalieri. 1990. Matching nitrogen equipment to your needs. Washington State University Tree Fruit Postharvest Journal 1(3): 3–13.
40. RP Singh and DR Heldman 1993. Introduction to Food Engineering, 2nd ed. Academic Press, San Diego.
41. GSV Raghavan and Y Gariépy. 1984. Structure and instrumentation aspects of storage systems. Acta Hortic. 157: 5–30.
42. RM Smock, LL Creasy, and GD Blanpied. 1960. Water scrubbing in CA Rooms. Cornell University, Ithaca, NY, Paper No. s-508.
43. IJ Pflug. 1960. Oxygen reduction in CA storage: A comparison of water versus caustic soda absorbers. Mich. Agric. Exp. Stat. Q. Bull. 43(3): 455–466.
44. JF Thompson and MS Reid. 1989. Economical ethylene control. Perishables Handling Newsletter. University of California, Issue 67: 6.
45. Y Gariépy, GSV Raghavan, and JA Munroe. 1986. CO_2 and O_2 relation in the design of the silicone membrane system for long-term CA storage of fruits and vegetables. 1986 Annual AIC Conference, Canadian Society of Agricultural Engineering, Paper No. 86–407.
46. JD Mannapperuma, D Zagory, RP Singh, and AA Kader. 1989. Design of polymeric packages for modified atmosphere storage of fresh produce. Proceedings of the Fifth International Conference. Vol. 2: 225–233.
47. CB Watkins and CJ Thompson 1992. An evaluation of microperforated polyethylene film bags for storage of ''Cox's Orange Pippin'' apples. Postharvest Biol. Technol. 2: 89–100.
48. EW Hewett and CJ Thompson. 1989. Modified atmosphere storage and bitter pit reduction in ''Cox's Orange Pippin'' Apples. Sci. Hortic. 39: 117–129.
49. JA Leshuk and ME Saltveit. 1990. Controlled Atmosphere Storage Requirements and Recommendations for Vegetables. In: Food Preservation by Modified Atmospheres. (ed. M Calderon and R Barkai-Golan) CRC Press, Boca Raton, FL.

50. J Robbins, and PP Moore. 1992. Fruit quality of stored, fresh red raspberries after a delay in precooling. Hortic. Technol. 2(7): 468–470.
51. CH Crisosto. 1991. Sweet cherry harvesting, postharvest handling and storage. In: Perishables Handling. University of California 71: 2–6.
52. MT Talbot, SA Sargent, and JK Brecht. 1991. Cooling Florida Sweet Corn. Florida Extension Service, University of Florida, Circular 941.

19

Packaging of Fruits and Vegetables

JAMES P. SMITH, HOSAHALLI S. RAMASWAMY, and G. S. VIJAYA RAGHAVAN

McGill University, Sainte-Anne-de-Bellevue, Quebec, Canada

BYRAPPA RANGANNA

University of Agricultural Sciences, Bangalore, India

1 INTRODUCTION

The postharvest losses of important commercial fruits and vegetables vary from 20% to 50% before they reach consumers. Fruits and vegetables are high in moisture, ranging from 70% to 95%. Their equilibrium humidities are as high as 98%. Under normal atmospheric conditions they dry rapidly, which causes wilting and shriveling as a result of loss of rigidity and shrinkage of cells. The primary objective of packaging of fruits and vegetables is to protect the contents during storage, transportation and distribution against deterioration, which may be physical, chemical, or biological. Packaging is hence provided at the point of production or processing or at distribution centers. Though packaging forms the last link in the chain of production, storage, marketing, and distribution, it still plays an important role in delivering the contents safe from the "farm gate to the consumer plate." Increase in production can have an impact on the consumer only when the food is wholesome, unadulterated, and available under hygienic conditions at an economical price. As mentioned, about 25% to 40% of fruits and vegetables are spoiled or become substandard during storage and distribution. This enormous wastage, which results in product scarcity and higher prices, is attributed mainly to poor packaging, improper handling methods, and inadequate transportation facilities.

Robertson (1992) defines *packaging* as "the enclosure of products, items or packages in a wrapped pouch, bag, box, cup, tray, can, tube, bottle or other container to perform the following functions: containment; protection; and/or preservation; communication;

and utility or performance.'' Since the 1970s there has been tremendous growth in new food processing/packaging technologies. The growth of these new packaging/processing technologies, for both short- and long-term preservation of food, is due to interrelated factors: (a) developments in new polymeric barrier packaging materials, (b) increased urbanization, (c) market needs and consumer demands for convenience, and (d) increasing energy costs. As a result of these interrelated factors, food packaging technology has gone through a tremendous transformation. Packaging now provides increased consumer information, is used very effectively as a marketing tool, and has clearly evolved from its primary and previously single role of protection to be a more multifaceted tool. There are a multitude of packaging materials in today's marketplace, each designed with specific properties. The correct choice of packaging is dependent not only on a knowledge of the physical, chemical, and microbiological characteristics of fruits and vegetables, but also on the functional properties of the packaging materials available for a particular product or preservation technology.

This chapter gives a brief overview of the properties of the materials most commonly used for packaging of fresh and processed fruits and vegetables and the packaging technologies that can be applied for shelf life extension of products.

2 PACKAGE REQUIREMENTS/FUNCTIONS

The two main functions of packaging are (a) to assemble the produce into convenient units for handling and (b) to protect the produce during distribution, storage, and marketing. Modern packages for fresh fruits are expected to meet a wide range of requirements, which may be summarized as follows: (a) The packages must have sufficient mechanical strength to protect the contents during handling and transport and while stacked. (b) The construction material must not contain chemicals that can transfer to the produce and cause it to become toxic to humans. (c) The package must meet handling and marketing requirements in terms of weight, size, and shape. The current trend is to reduce many sizes and shapes of packages by standardization. Palletizing and mechanical handling make standardization essential for economical operation. (d) The packages should allow rapid cooling of the contents. Furthermore, the permeability of plastic films to respiratory gases may also be an important requirement. (e) The security of the package or its ease of opening and closing may be important in some marketing situations. (f) The package should identify its contents. (g) The package may be required either to exclude light or to be transparent. (h) The package may be required to aid retail presentation. (i) The package may need to be designed for ease of disposal, reuse, or recycling. (j) The cost of the package should be as low as possible.

Packaging may or may not delay or prevent spoiling of fresh fruit and vegetables; however, incorrect packaging can accelerate spoilage. Packaging should serve to protect against contamination, damage, and excess moisture loss. An excessive moisture barrier causes excessively high relative humidity in the package and results in accelerated spoilage due to microorganisms or skin splitting of some fruits.

3 TYPES OF CONTAINERS

After harvest, fruit and vegetables are handled in different containers from the field up to the retail stores. Approaches followed by both the developed and tropical countries are similar in many of the situations.

3.1 Field Containers

Picking or harvesting containers are of many types, depending on the crop, region, and availability of materials. Picking bags of canvas or burlap, mesh hampers, and baskets of woven veneer or bamboo are widely used.

3.2 Shipping Containers

A shipping container is a handling unit used to facilitate moving horticultural produce from one location to another. Packaging for shipping and handling requires suitable containers to protect produce from bruising, vibration, and the weight of other stacked containers. The container should be sturdy enough to permit reasonable stacking without collapse or pressure damage to the produce. It should not affect exchange of O_2 and CO_2 and at the same time should be permeable to heat of respiration and transpiration of fresh fruits. The ideal pack consists of a tight-fill without a bulge in a lidded container having sufficient stacking strength to protect the contents under all handling conditions. In many of the developed countries the shipping containers are used only once and are not returned to the shipper. In India and other less developed countries, baskets and boxes are often returned or sold and receive multiple use. Common types include nailed wooden boxes and crates, wirebound boxes and crates, plywood boxes, and baskets.

It is difficult to name specific containers for different fruits, since several types may be satisfactory, depending on the region, distance to market, method of precooling, quantity or weight shipped, and availability and cost of materials. Fiberboard (corrugated) cartons are becoming popular for shipping both tropical and subtropical fruits. Their light weight and low cost are advantages.

3.3 Consumer Packages

Use of small consumer-sized packages for produce has grown with the increase in large self-service markets for retailing. It may consist of a paper or a plastic bag made available for customers to select, package, and weigh their purchases. Consumer packages are of the following types: (a) bags made of paper, film, or cotton or plastic mesh; (b) trays of molded pulp, paperboard, plastic, or foamed plastic; (c) folding paperboard cartons, sometimes with a clear plastic window or with dividers for individual fruits; and (d) small rectangular or round baskets made of coated or waxed paper board or other material.

3.3.1 Bags

The most widely used consumer-unit package is the bag. It is inexpensive, is easily filled and closed, and is available in many sizes and many materials. It provides less protection from physical damage than most other packages.

3.3.2 Plastic-Film Bags

Numerous transparent and translucent films of various compositions are available commercially, and some of them at lower cost than kraft paper, cotton cloth, or burlap. Prefabricated bags are available from many manufacturers and may also be fabricated at the user's premises by machines that will form and heat-seal bags from rolls of flat film. The advantages of plastic-film bags are (a) good visibility of the packaged product, (b) limited permeability to water vapor and reduced moisture loss from the product, and (c) strength and tear resistance of 1.0 to 1.5 mil film.

3.3.3 Mesh Bags

The ultimate in ventilation of contents is achieved in the mesh bag. The netting with openings between strands of 3 to 6 mm allows free movement of air to and from the interior of the bag. Mesh bags are fabricated from several materials, the most common of which are fine plastic strands, cotton thread, and twisted strands of processed paper. The important advantages of mesh bags are (a) excellent ventilation for heat exchange during precooling, (b) avoidance of high relative humidity in the bag for commodities and situations in which high relative humidity is undesirable, (c) good visibility of product, and (d) easy closure. Mesh bags are widely used for origin prepackaging of oranges, because the fruit cools faster in mesh bags than in plastic film, and the hazard of fruit decay during a prolonged marketing period is somewhat lower in mesh than in film. Apples are less commonly packed in mesh bags because they are usually cooled before prepackaging and the possibility of decay during marketing is lower for apples than for oranges.

3.3.4 Shrink-Film Wraps

Films of a number of types can be given heat-shrink characteristics by stretching under controlled temperatures and tensions to form molecular orientation, after which the film is cooled in the stretched condition to maintain its form. Films such as polypropylene, polystyrene, polyethylene, and rubber hydrochloride can be converted to shrink-films by the molecular orientation method. After the shrink-film is applied to the filled trays in tubular or heat-sealed wrap form, the packages are passed through a heat tunnel to shrink the film cover. This immobilizes the fruits to reduce the possibility of physical damage during handling. Studies at the retail level showed that apples in overwrapped trays were better protected from mechanical injury than those in open trays (Wills et al., 1989). Labor costs for packing film covering the trays were not high, and often were lower than those for preparing standard packs in film box liners.

3.3.5 Consumer Trays

A favorite type of consumer package for preparation at the wholesale level or in the retail store is the molded tray. Trays are made of chipboard, molded foam plastic, or clear plastic. Trays are made in many sizes; the most commonly used trays are designed to hold four to six apples or oranges in a single layer. Larger and deeper trays hold a dozen or more medium-sized fruits in two layers or in a pyramid.

4 TYPES OF PACKAGING MATERIALS

A variety of packaging materials, with specific functional properties, are commercially available for packaging fresh fruits and vegetables. These include wood, cloth, paper, and plastics.

4.1 Wood and Textiles

Wooden containers were traditionally used for the bulk transportation of fruits and vegetables to the marketplace. Wood offers good mechanical protection, good stacking characteristics, and a high weight-to-strength ratio. However, it is not a great moisture or gas barrier and can be a source of microbial contamination, particularly by molds. With the advent of plastics, wooden containers are gradually being replaced by polystyrene, polypropylene, and polyethylene containers, which are lighter in weight and have lower transportation costs. Textile containers, e.g., jute sacks, are also used sparingly for the bulk transportation of fruits and vegetables to market. Although jute sacks are durable and have high tear resistance, they have low extensibility, they are very poor barriers to moisture and gases, and they are subject to mold spoilage. They are also being replaced by multiwall paper sacks.

4.2 Paper and Board

Paper and board are still very popular packaging materials in North America. Kelsey (1989) estimated that paper and paperboard packaging constitutes about 31% of the approximately 70 million tons of paper products produced annually with a market value of approximately $16 billion.

Paper pulp is produced from wood chips by acid or alkaline hydrolysis. The pulp is suspended in water and beaten with rotating impellers and knives to split the cellulose fibers longitudinally. The fibers are then refined and passed through heated rollers to reduce the moisture content, and then through finishing rollers to give the final surface properties to the paper. Alkaline hydrolysis produces sulfate pulp, and acid hydrolysis produces sulfite pulp. The various types of paper and paperboard containers used as food packaging are shown in Table 1.

Kraft paper is made from at least 80% sulfate wood pulp. It is a very strong paper, which is used to make grocery bags, multiwall bags, shipping sacks, and specialty bags that require both economy and strength for bulk packaging of powders, flour, sugar, fruits, and vegetables. Bleached papers are more expensive and weaker than unbleached and have excellent printability. Vegetable parchment is produced from sulfate pulp that is passed through a bath of sulfuric acid. It has a more intact surface than kraft paper and therefore has greater grease resistance and wet strength properties than kraft paper.

Table 1 Types of Paper Commonly Used as Packaging Material

Product	Characteristics	Example
Kraft paper	Brown, unbleached paper. Good strength and resistant to bursting when dry	Heavy duty bags and sacks
Bleached Kraft paper	White paper, may be glossy. Less strength than unbleached paper	White bags, wrapping paper
Parchment paper	Translucent paper treated with H_2SO_4 to gelatinize surface layers	Butter and margarine wrap
Greaseproof paper	High-density paper, very smooth surface	Wrapping paper requiring high resistance to grease
Glassine	High-density greaseproof paper. Transparent, brittle	Overwraps on candy
Tissue	Lightweight paper produced from most pulps	Lightweight and used to protect soft products from dust and bruising
Paperboard/cardboard	Compacted paper pulp	Cartons, boxes, trays, separators
Corrugated cardboard	Paperboard sheets interspersed with paper corrugations	Secondary boxes of many kinds

Source: Adapted from Jelen (1985) and Brown (1992).

Because of its high grease resistance and wet strength, it is used for packaging butter and shortening (Fellows, 1988; Brown, 1992).

Sulfite paper is lighter and weaker than sulfate paper. Greaseproof paper is made from sulfite pulp in which the fibers are more thoroughly beaten to produce a closer structure. It is resistant to oils and fats when dry, but these properties are lost when the paper becomes wet. Packaging applications for greaseproof papers include margarine wraps, french-fry bags, inner liners for multiwall sacks, and liners in composite cans for packaging frozen juices. Glassine is a greaseproof sulfite paper that is given a high gloss finish by the finishing rollers. It is used as wrapping material for candy products and certain bakery products. Chocolate-coated glassine acts as a barrier to ultraviolet (UV) light to prevent rancidity problems in chocolates and potato chips. Tissue paper is a soft, nonresilient paper used to protect fruits against dust and bruising (Brown, 1992).

A major disadvantage of paper as a packaging material is its poor barrier properties against moisture, gases, grease, and odors. Furthermore, it cannot be heat sealed. To improve its barrier and heat sealability properties, paper is often combined with wax, plastic film, metal foil, or a combination of foil and plastic film.

Paperboard is made in a similar way to paper but is thicker to protect foods from mechanical damage. The main characteristics of board are thickness, stiffness, ability to crease without cracking, degree of whiteness, surface properties, and suitability for printing (Brown, 1992). White board is suitable for contact with food and is often coated with polyethylene, polyvinyl chloride, or wax for heat sealability. It is commonly used to prevent freezer burn in stored frozen products. Pulp containers are made from paper pulp compressed in molds to remove moisture. Pulp containers are used for egg cartons, low-cost food trays, and cushioning for food products.

Corrugated board is the most common form of secondary food packaging and is used by virtually every industry. According to Kelsey (1989), 280 billion square feet of corrugated board, with a market value of $11.8 billion, was produced in 1986.

Corrugated board has an outer and an inner lining of kraft paper with a central corrugating (or fluting) material. This is made by softening kraft paperboard with steam and passing it over corrugating rollers. The linear are then applied to each side, using a suitable adhesive. The board is formed into ''cut-outs'' that are then assembled into cases at the filling line. There are four different flute sizes, A, B, C, and E flutes, which vary in height and the number of flutes per unit length of board. They can be used alone or in combination with one another to produce single-face, single-wall, double-wall, and triple-wall corrugated board constructions as shown in Fig. 1. Corrugated board has good impact abrasion and compression strength and is mainly used in secondary packaging containers. The most standard type of secondary packaging material is single-wall C flute. High storage humidity that causes delamination of the corrugated material is prevented by lining with polyethylene or greaseproof paper to coating with microcrystalline wax and polyethylene (Brown, 1992).

4.3 Plastics

Since the 1970s there has been a tremendous increase in the use of plastics, replacing traditional packaging materials such as glass, metal, and paper. The raw materials for plastics are petroleum, natural gas, and coal. They are formed by a polymerization method that creates linkages between many small repeating chemical units (monomers) to form large molecules or polymers. Examples of common plastic materials and their monomer

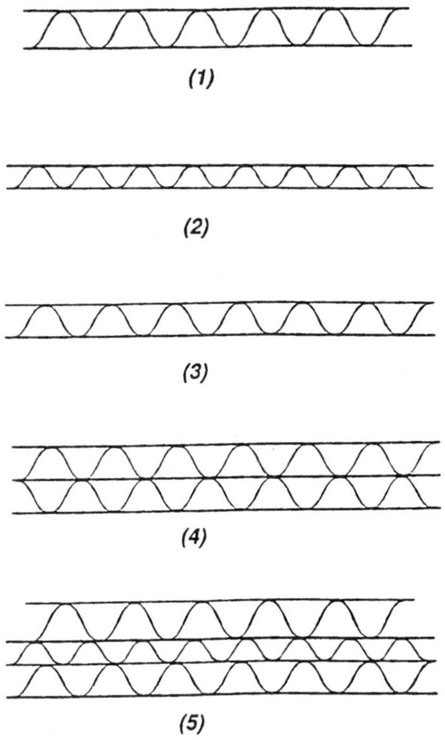

Fig. 1 Various types of corrugated board construction: (1) "A" flute–single wall, (2) "B" flute–single wall, (3) "C" flute–single wall, (4) "C" and "C"–double wall, and (5) "A," "B," and "C"–triple wall. (Courtesy of Smurfit-MBI, Montreal, Quebec, Canada.)

building blocks are listed in Table 2. Many plastics contain very small amounts of additives, such as plasticizers, antioxidant, lubricants, antistatic agents, heat stabilizers, and UV stabilizers. These are added to facilitate processing of plastics or to impart desirable properties to them. For example, plasticizers are added to soften plastics, thus making them more flexible and less brittle for use in cold climates or with frozen stored products. According to Fellows (1988), the advantages of plastics as packaging materials are their relatively low cost, good barrier properties against moisture and gases, heat sealablity to prevent leakage of contents, suitability for high-speed filling, wet and dry strength, suitability for printing, ease in handling and convenience for the manufacturer, retailer, and consumer; addition of little weight to the product and close fit to the shape of the food, thereby wasting little space during storage and distribution.

Plastics may be made as flexible films or as semirigid and rigid containers to meet the varied packaging and processing requirements of food. Plastic films are made with a wide range of mechanical, optical, heat sealable, and barrier properties. Furthermore, they can be coated with another polymer or metallized to produce a laminated structure with superior properties. Examples of common flexible films are cellulose, polyethylene, polyester, polyamide, polypropylene, polystyrene, polyvinyl chloride, polyvinylidene chloride, ethylene vinyl acetate, ethylene vinyl alcohol, and Surlyn. A summary of the important characteristics of these films is included in Table 2.

Table 2 Examples of Plastics Used as Packaging Materials

Materials	Structural unit	Important properties
Cellulose	Glucose	Good strength, poor H_2O and gas barrier, good printability, no heat sealability
Polyethylene (PE)	Ethylene	Good strength, flexibility, extensibility, high H_2O barrier, poor gas barrier, low melting point, good heat sealability
Polyester (PET)	Ethylene glycol + terephthalic acid	Stiff, strong, inert, excellent mechanical properties, poor heat sealability, moderate H_2O and gas barrier
Polyamide (PA)	Diamine + various acids	Stiff, strong, inert, clear, excellent machinability, heat sealability, poor H_2O barrier, high gas barrier when dry
Polypropylene (PP)	Propylene	Tough, inert, clear, low melting point, high H_2O barrier, poor gas barrier
Polystyrene (PS)	Styrene	Stiff, strong, brittle, low H_2O and gas barriers
Polyvinyl chloride (PVC)	Vinyl chloride	Soft, inert, clear, extensibility, good H_2O barrier, moderate moisture barrier
Polyvinylidene chloride (PVDC, Saran®)	Vinyl alcohol + vinylidene chloride	Inert, clear, not very strong, high melting point, heat sealability at high temperature, excellent H_2O and gas barriers
Ethlyene vinyl acetate (EVA)	Vinyl acetate + ethylene	Tough, clear, inert, high extensibility, low melting point, heat sealability, intermediate H_2O barrier, poor gas barrier
Ethylene vinyl alcohol (EVOH)	Vinyl alcohol + ethylene	Strong, stiff, inert, heat sealability at low temperature, low H_2O barrier, high gas barrier
Ionomer (Surlyn)	Ethylene + methacrylic acid	Tough, inert, clear, heat sealability at low temperature, intermediate H_2O barrier, low gas barrier

Source: Adapted from Jelen (1985) and Brown (1992).

4.4 Coated Films

Individual films are often coated with other polymers or aluminum to improve their barrier properties or to impart heat sealability. For example, nitrocellulose is coated on one side of cellulose film to provide a moisture barrier but to retain oxygen permeability. A coating of vinyl chloride or vinyl acetate gives a stiffer film, which has intermediate permeability. Sleeves of this material are tough, stretchable, and permeable to air and moisture. A thin coating of aluminum (termed *metallization*) produces a very good barrier to oils, gases, moisture, odors, and light. Metallized film is less expensive and more flexible than foil laminates that have similar barrier properties and is therefore suitable for high-speed filling on form-fill-seal equipment. Cellulose, polypropylene, or polyester is metallized by depositing vaporized aluminum onto the surface of a film under vacuum. Metallized polyester has higher barrier properties than metallized polypropylene, but polypropylene is finding more widespread use as it is currently less expensive.

4.5 Laminated Films

Lamination of two or more films improves the appearance, barrier properties, or mechanical strength of a package. Materials that can be laminated to each other include plastic to plastics, paper to plastic, paper to aluminum foil, and paper to aluminum foil to plastic. Several methods can be used to laminate materials, including dry and wet bonding, thermal extrusion, and coextrusion.

Laminated materials are used when high gas and moisture characteristics are required for a long shelf life. Laminated structures usually consist of an outer protective tougher layer, e.g., nylon or polypropylene, a middle high gas barrier layer, e.g., EVOH or PVDC; and an inner heat sealant layer. Low density polyethylene (LDPE) is commonly used as a heat sealant layer because of its low melting temperature. However, it sometimes does not produce a good seal with starchy or greasy food products. The choice of sealant layer for these food products is either EVA or Surlyn.

5 BARRIER PROPERTIES OF PACKAGING MATERIALS

Many materials can be selected for packaging fruit and vegetable products. When choosing the appropriate packaging material, the following factors should be considered: gas barrier properties, moisture barrier properties, antifog properties, machinability, mechanical strength, sealability, and performance versus cost

Among the most important characteristics are their barrier properties to both oxygen and moisture vapor, which vary greatly from material to material. High barrier materials usually are high barriers to both moisture and oxygen, e.g., glass, tinplate, and aluminum foil. However, the barrier properties to both oxygen and moisture may be different and may also vary as a function of the relative humidity and temperature of the storage conditions. A good example is EVOH, hygroscopic film that is an excellent oxygen barrier at low relative humidities. However, at higher relative humidities it absorbs moisture, which has a plasticizing effect and reduces the barrier characteristics to oxygen. Some films have mixed barrier properties, i.e., low oxygen barrier characteristics and high moisture vapor barriers. A good example is LDPE; for this reason this film is selected for packaging fresh meat and produce and for frozen stored products to prevent freezer burn.

6 PACKAGING OF FRUITS AND VEGETABLES

The shelf life of packaged fruits and vegetables is controlled by the properties of the product (including water activity, pH, susceptibility to enzymic or microbiological deterioration, mechanism of spoilage, and requirement for or sensitivity to oxygen, light, carbon dioxide, and moisture) and the properties of the package.

Moisture loss or uptake is one of the most important factors that control the shelf life of fruits and vegetables. Fruits and vegetables are high-moisture products with moisture content ranging from 75% to 95%. Loss of moisture under normal storage conditions causes wilting and shriveling of the product. However, proper packaging is able to extend storage life of fresh products by maintaining moisture loss during storage at 10% or less, thereby preventing wilting. The rate of moisture loss varies with each product's respiration rate and the water vapor permeability of the packaging film. The use of small perforations in some film to ensure a constant supply of oxygen has no appreciable effect on moisture loss.

Fruits and vegetables are living organisms, and even after harvest they continue to respire and transpire. Respiration involves the uptake of oxygen and breakdown of organic matter into water and carbon dioxide. If there is not enough oxygen, fermentation occurs and small amounts of alcohol are produced. This results in the production of off-flavors and off-odors and spoilage of the commodity. Therefore, packaging materials for fruits and vegetables should not produce an excessively high barrier to oxygen.

The thermal properties of the packaging material should also be taken into consideration to minimize temperature fluctuations. Maturation can be slowed by storage at refrigeration temperatures, as this reduces the respiration and the synthesis of ethylene, which causes maturation. However, an excessively low temperature may cause chilling damage to the products. Therefore, proper packaging can also ensure temperature distribution within the package and prevent chilling injury. Some packages are required to withstand processing conditions (for example, hot filling, heat sterilization, boil-in-the-bag).

Packaging should retain desirable odors, e.g., of strawberries, or prevent odor pickup in dried products. There should also be negligible odor pickup from the plasticizers, printing inks, adhesives, or solvents used in the manufacture of the packaging material.

Packaging protects food from mechanical damage caused by transportation or handling, e.g., vibration and compression damage. Compression damage during storage may arise as a result of overstacking. Metal, wood, or fiberboard shipping cases prevent mechanical damage, and products are held tightly within retail containers by molded trays or fin seals to prevent movement. Groups of retail containers are similarly immobilized by shrink- or stretch-wrapping.

Molds and bacteria are a major cause of spoilage of fruits and vegetables. They may either grow on the surface of the products or spread inside the product as a result of surface bruising or cuts and cause internal decay. Careful handling and proper packaging can minimize physical damage and delay microbial spoilage of fruits and vegetables. However, an excessive moisture barrier can cause high internal relative humidity in the package that is conducive to microbial growth.

Packaging cannot prevent fresh fruits and vegetables from spoilage and decay. On the other hand, incorrect packaging can accelerate spoilage. However, packaging can serve to protect against contamination, physical damage, and excessive moisture loss.

6.1 Choice of Package

As fruits and vegetables vary in their physical attributes (shape, size, toughness) as well as their moisture content and respiration rate, a specific type of packaging has to be selected to minimize the perishability of the specific product.

Anaerobic spoilage and rotting are the major spoilage problems of soft fruits as they bruise and squash readily. Typical soft fruits are grapes, blueberries, strawberries, raspberries, and plums. Usually these are packaged in semirigid plastic containers, e.g., high-density polyethylene (HDPE) or polypropylene (PP), with a plastic overwrap of polystyrene (PS), which is vented to prevent fogging. Under ideal storage conditions, berries remain fresh for only 2 or 3 days.

Hard fruits, such as avocados, peaches, pears, bananas, citrus fruits, and tomatoes, have lower respiration rates and are less sensitive to handling. Commonly used packaging systems are open PS or paperboard trays with a plastic overwap of low-density polyethylene (LDPE) or polyvinyl chloride (PVC). Hard fruits, e.g., apples, may be bagged in LDPE films or in LDPE/HDPE nets, e.g., oranges. In certain cases, hard fruits may cut the packages, e.g., pineapples. In this case, heat shrink PVC is used. This film is tear resistant and can withstand the sharp contours of the fruit.

Stem products, e.g., celery, rhubarb, and asparagus, are highly perishable as they lose moisture rapidly. They are usually packaged in high-barrier moisture films such as LDPE with ventilation. They can also be banded or sleeved with shrink films.

Root vegetables usually have a long shelf life. Typical root vegetables are carrots, radish, onions, beets, and potatoes. These are usually packed in LDPE bags to prevent moisture loss during prolonged storage at ambient temperature. In the case of potatoes that are light sensitive, printing the film or tinting it an amber color can be used to prevent greening of potatoes.

Green vegetables, such as brussel sprouts, cabbages, lettuce, brocoli, and cauliflower, tend to lose moisture rapidly, resulting in wilting. Furthermore, they have high respiration rates and so anaerobic conditions must be prevented within the packaged product. Packaging materials that prevent moisture loss and are also low barriers to oxygen include LDPE and PVC. PP is also commonly used as it is a high moisture barrier. However, it must be perforated as it is also a high oxygen barrier.

Ready prepared cut vegetables and salads have a high surface area. Therefore, they lose moisture rapidly and respire faster. The packaging film of choice is again LDPE or LDPE/EVA, which ensures the desired shelf life of these products.

6.2 Modified Atmosphere Packaging

Modified atmosphere packaging (MAP) can be defined as "the enclosure of food products in a barrier film in which the gaseous environment has been changed or modified to slow respiration rates, reduce microbiological growth and retard enzymatic spoilage with the intent of extending shelf life" (Young et al., 1988).

MAP is becoming an increasingly popular method of shelf life extension of food products when an extended shelf life at refrigerated temperatures is required. Several methods can be used to modify the atmosphere within the packaged products. These employ (a) passive modification or (b) active modification.

In passive modification, the atmosphere is modified as a result of a commodity's respiration, i.e., O_2 consumption and CO_2 generation. In active modification, the package headspace is flushed with a known concentration of O_2, CO_2, and N_2 (Smith et al., 1990). Products differ in their tolerance to O_2 and CO_2. For example, the optimal atmosphere for peas is 10% O_2 and 7% CO_2; for avocados it is 1% O_2 and 10% CO_2. Therefore, the optimal film for peas may not be suitable for avocados. The recommended MA conditions for selected fruits and vegetables are shown in Table 3.

Selection of a film of the correct permeability to oxygen, carbon dioxide, and water vapor is critical to the success of MAP of fruits and vegetables. If the film is too permeable to oxygen, the product respires, produces ethylene, and ripens. If the permeability is too

Table 3 Recommended Modified Atmosphere Conditions for Fruits and Vegetables

Commodity	Storage Temperature (°C)	Atmosphere O_2	Atmosphere CO_2
Apple	0–5	2–3	1–2
Apricot	0–5	2–3	1–2
Avocado	5–13	2–5	3–10
Banana	12–15	2–5	2–5
Cherry (sweet)	0–5	3–10	10–12
Grapefruit	10–15	3–10	5–10
Kiwifruit	0–5	2	5
Mango	10–15	5	5
Papaya	10–15	5	10
Peach	0–5	1–2	5
Pear	0–5	2–3	0–1
Pineapple	10–15	5	10
Strawberry	0–5	10	15–20
Asparagus	0–5	20	5–10
Beans, snap	5–10	2–3	5–10
Broccoli	0–5	1–2	5–10
Brussels sprouts	0–5	1–2	5–7
Cabbage	0–5	3–5	5–7
Cantaloupe	3–7	3–5	10–15
Cauliflower	0–5	2–5	2–5
Corn, sweet	0–5	2–4	10–20
Cucumber	8–12	3–5	0
Honeydew melon	10–12	3–5	0
Lettuce	0–5	2–5	0
Mushrooms	0–5	Air	10–15
Bell peppers	8–12	3–5	0
Spinach	0–5	Air	10–20
Tomatoes (mature)	12–20	3–5	0
Tomatoes (partly ripe)	8–12	3–5	0

Source: From Kader (1986).

low, anaerobic conditions are soon reached and the product ferments. Examples of permeabilities of films used for MAP of fruits and vegetables are listed in Table 4. Generally, films with a CO_2:O_2 ratio of ~3:1 are most suitable. Films with this ratio include LDPE and PVC and laminates of EVA/LDPE. These films ensure that a desirable equilibrium modified atmosphere (EMA) is established: i.e., the rates of permeation of O_2 and CO_2 through the packaging material equal the product's respiration rate. In addition, these films are excellent barriers to moisture and minimize moisture/weight loss of the product during prolonged storage under the MAP atmosphere. Packaging materials used for MAP of fruits and vegetables must also have sufficient strength to resist puncture, withstand repeated flexing, and endure mechanical stresses during handling and distribution. Options include flexible pillow packs, semirigid trays and closing systems, and bag-in-box containers.

Table 4 Permeabilities of Packaging Films Used in Modified Atmosphere Packaging of Fruits and Vegetables

	Permeability to[a]	
Film type	Carbon dioxide	Oxygen
Polyethylene, low density (LDPE)	7,700–77,000	3,900–13,000
Polyvinyl chloride (PVC)	4,263–8,138	620–2,248
Polypropylene (PP)	7,700–21,000	1,300–6,400
Polystyrene (PS)	10,000–26,000	2,600–7,700
Saran (PVDC)	52–150	8–26
Polyester (PET)	180–390	52–130

[a] Permeability expressed as cubic centimeters per square meter per mil per day per atmosphere ($cm^3/m^2/mil/day/atm$).
Source: From Zagory and Kader (1988).

Examples of commodities packaged under modified atmospheres and extension in shelf life possible are listed in Table 5.

6.2.1 Individual Seal Packaging

Developed since the 1980s, individual seal packaging (ISP) involves the use of heat shrinkable polymeric film (usually 0.5–0.75 mil HDPE) that is wrapped around individual units of fruits or vegetables and shrunk by blowing hot air over the package. ISP products have many advantages. Ripening is delayed by the microatmosphere created around the product as a result of the product's metabolic activities. Furthermore, the film acts as a good barrier to water so that no moisture loss occurs. ISP also prevents the spread of disease from one product to another, improves handling and sanitation of the product, and facilitates pricing and labeling of individual products. A marked reduction in shrinkage and weight loss without deleterious effects on flavor has been shown to result from ISP (Ben-Yehoshua, 1989). ISP has also been shown to result in a two- to threefold extension of shelf life in terms of appearance, firmness, weight loss, and other quality attributes. However, off-odors may occur as a result of poor gas exchange, and the high relative humidity (RH) in the microatmosphere of the packaged product may enhance fungal spoilage despite the low-O_2–high-CO_2 atmosphere within the packaged product. In an attempt to overcome these limitations, ISP with perforations can be used. These perforations cover ~10% of the films' surface and perforations vary in diameter from 0.7 to 0.16 mm (Ben-Yehoshua, 1989).

6.2.2 Edible Films

Coating of fruits and vegetables with edible materials to preserve their quality and extend their shelf life has been in practice for centuries. The most common form of coating of fruits and vegetables is wax coating to retard respiration, dehydration, and senescence. Hot-melt waxes and carnauba-oil-in-water emulsions have been used effectively for citrus fruits, apples, tomatoes, and eggplants (Kester and Fennema, 1986).

Interest in edible films has intensified over the past few years as a result of increased consumer demand for fresh, frozen, and convenience foods and consumer concerns about the environment. The most important characteristics of edible films are that they are good

Table 5 Shelf Life of Selected Fruits and Vegetables Stored Under Modified Atmosphere Packaging Conditions

		Modified atmosphere environment		
Commodity	Packaging film	$\%O_2$	$\%CO_2$	Shelf life
Apples + Pears	Sealed PE bags	10–15	0.5–2.5	6 Months
Apples	Sealed PE film tubes	2–5	5–7	4 Months
Blueberries	PE pallet covers	1–2	3–5	~6 Weeks
Peaches	Cryovac PE film	10–15	15–25	NA
Avocados	Sealed PE bags	3–5	7–9	8–10 Days
Kiwifruit	Sealed PE bags	N/A	3–4	6 Months
Banana	PVC overwrap	3	3	15 Days
Cabbage	PVC overwrap	2–3	3–4	2–3 Weeks
Brussels sprouts	PVC overwrap	2–3	3–4	2–3 Weeks
Lettuce	Sealed PE bags	5	10	~12 Days
Beans	Cellophane film	0.5	~30	7 Days
Peppers	PVC film	14	3	NA
	PE bags	6–11	4–6	NA
Sweet corn	PE bags	2–5	5–10	NA
Artichokes	PE bags	3–4	3–6	8 Weeks
Broccoli	PE bags + 4.5% EVA	1–2	8	3 Weeks
Celery	PE bags/liners	5	9	5 Weeks
Carrots	PE bags	17	3	15 Months
Mushrooms	PVC overwrap	2	10–12	5 Days
Mixed salad greens	PVC overwrap	2	10	6–7 Days

Source: Adapted from Prince (1989).

gas and moisture barriers and may be suitable for coating with additives such as antimicrobials, antioxidants, nutrients, and coloring agents.

Edible films most commonly used are derived from polysaccharides, protein, and lipid. Polysaccharide films may be made from starch, dextrins, and cellulose derivatives. Protein films are made from collagen, gelatin, wheat and corn gluten, and zein. Lipid films may be made from natural waxes and surfactants (Kester and Fennema, 1986). Fruits and vegetables coated with lipid (hydrophobic) films have good moisture barriers. Fungicides can be applied to these films to retard yeast and mold spoilage. However, the coating must not cause an excessively high gas barrier because it must reduce anaerobic respiration, which can cause physiological disorders (Kester and Fennema, 1986). Other materials that have been used as edible film coatings have films made from alginate, gelatin, acetylated monoglyceride, and chitosan.

7 CONCLUSIONS

Food packaging is an essential unit operation for preserving food quality, minimizing food wastage, and reducing the use of chemicals, additives, and stabilizers. The food package serves the important functions of containing the food, providing protection against chemical and physical damage, providing convenience in using the product, and conveying consumer information to the consumers. It protects the food by acting as a barrier to oxygen, moisture, chemical compounds, and microorganisms that are detrimental to the quality of the food product. It provides consumers with convenient features such as microwavability, resealability, and ease of opening. It conveys useful information such as a description of food contents, weight-to-volume ratio, manufacturer's name and address, directions for preparing foods, and nutrition values. It also serves as an effective marketing tool for promoting product identification and selling the product.

REFERENCES

Ben-Yehoshua, S. (1989). Individual seal packaging of fruit and vegetable in plastic film. In Controlled/Modified Atmosphere/VacuumPackaging of Foods (ed., A. L. Brody). Food and Nutrition Press, Trumbell, CT, pp. 101–118.

Brown, W. E. (1992). Properties of plastics used in food packaging. In: Plastics in Food Packaging: Properties, Design and Fabrication. Marcel Dekker, New York, pp. 103–139.

Fellows, P. (1988). Packaging. In: Food Processing Technology. Ellis Norwood Ltd., Chichester, England, U.K., pp. 421–447.

Jelen, P. (1985). Food packaging technology. In: Introduction to Food Processing. Reston, Reston, VA, pp. 249–266.

Kader, A. A. (1986). Biochemical and physiological basis for effects of controlled and modified atmospheres on fruits and vegetables. Food Technol. 40:99–104.

Kelsey, R. J. (1989). Packaging in Today's Society. Technomic, Lancaster, PA.

Kester, J. J., and Fennema, O. R. (1986). Edible films and coatings: A review. Food Technol. 40: 47–59.

Prince, T. (1989). Modified atmosphere packaging of horticultural products. In: Controlled/Modified Atmosphere/Vacuum Packaging of Foods (ed., A. L. Brody). Food and Nutrition Press, Trumbull, CT, pp. 67–100.

Robertson, G. L. (1992). Food Packaging: Principles and Practice. Marcel Dekker, New York.

Smith, J. P., Ramaswamy, H. S., and Simpson, B. K. (1990). Developments in food packaging technology. Part II. Storage aspects. Trends Food Sci. Technol. Today 1:112–119.

Wills, R. B. H., McGlasson, W. B., Graham, D., Lee, T. H., and Hall, E. G. (1989). Handling, packaging and distribution. In: Postharvest: An Introduction to the Physiology and Handling of Fruits and Vegetables. New South Wales University Press, Sydney, NSW, Australia, pp. 132–144.

Young, L. L., Reviere, R. D., and Cole, A. B. (1988). Fresh red meats: A place to apply modified atmospheres. Food Technol. 42:65–69.

Zagory, D., and Kader, A. A. (1988). Modified atmosphere packaging of fresh produce. Food Technol. 42:70–77.

20

Transportation and Handling of Fresh Fruits and Vegetables

CATHERINE K. P. HUI and CLÉMENT VIGNEAULT

Agriculture and Agri-Food Canada, Saint-Jean-sur-Richelieu, Quebec, Canada

DENYSE I. LEBLANC

Agriculture and Agri-Food Canada, Kentville, Nova Scotia, Canada

JENNIFER R. DeELL

Ontario Ministry of Agriculture and Food, Vineland Station, Ontario, Canada

SAMSON A. SOTOCINAL

McGill University, Sainte-Anne-de-Bellevue, Quebec, Canada

1 INTRODUCTION

Fruits and vegetables are usually transported several times before reaching the point of sale. Locally grown produce may be transported directly from farms to retail outlets. However, imported fruits and vegetables may initially be transported by road or rail to a central shipping facility in one country, then shipped by sea or air to a broker's facility in another country, where it is transported again by road or rail to a wholesaler's warehouse, and finally by road to a retail store. Between these two situations, many other scenarios exist. This shows that the produce is often handled many times before reaching the consumers.

The quality of fresh fruits and vegetables at the point of sale depends on several factors, which include the produce quality at harvest, its physiological stage, the amount of precooling, the packaging materials, and the temperature at which it was maintained.

To prevent physical damage and slow down the natural deterioration processes, fresh fruits and vegetables must be properly selected, packaged, handled, and kept at optimal temperature at all stages of distribution and handling, including each step in transportation.

In 1992, it was estimated that 25 million metric tons of fruits (approximately 12% of the annual production) were exported worldwide (International Institute of Refrigeration, 1995). These exports were often transported over vast distances. Approximately 74% were sent by sea, a small percentage were sent by air, and the remainder were transported by either rail or road. Intracontinental transport of perishable foods is estimated to be as great in volume as intercontinental transport (International Institute of Refrigeration, 1995). However, intracontinental shipments are usually delivered by road, with a smaller proportion going by rail (International Institute of Refrigeration, 1995). Maintaining the optimal temperature of such a large quantity and variety of produce on all modes of transport is a considerable challenge.

The worldwide fleet of transportation vehicles capable of carrying perishable foods is extensive. The International Institute of Refrigeration (1995) estimates that in 1993 there were approximately 800 conventional refrigerated ships of over 3000-m^3 capacity in service worldwide; 370 container ships with 215,000 container-adapted spaces, 250,000 refrigerated containers, and 80,000 insulated containers; 75,000 mechanically refrigerated railway wagons; and between 850,000 and 1,250,000 mechanically refrigerated trucks or transport trailers.

If these transport vehicles are well designed and constructed, properly loaded with well precooled produce, and maintained at optimal temperature, quality losses of fruits and vegetables are minimized throughout the journey. Detailed information on the factors affecting produce quality, the methods for maintaining the desired environmental conditions, the handling and loading techniques, as well as the particularities for the four modes of transport is presented in this chapter.

2 FACTORS AFFECTING PRODUCE QUALITY DURING TRANSPORT

Produce quality at the point of sale is dependent on a series of physiological and environmental factors. Deterioration of fruits and vegetables is an inevitable and irreversible process. The quality of produce can only be preserved, not improved (McGregor, 1989). During transport and handling, produce may be exposed to temperature fluctuations, water loss, noncompatible gases, physical injuries, and mixed loads. To ensure the delivery of good-quality fruits and vegetables at the point of sale, high-quality produce must be shipped and this quality must be maintained throughout the entire distribution system.

2.1 Initial Quality

Generally, produce handlers focus their attention on methods of preserving produce quality during transport. Nevertheless, the initial quality of fruits and vegetables prior to shipment should also be considered since it has a significant effect on the quality of produce at arrival. Healthy produce can resist disease, mechanical damage, and physiological disorder better than produce that has defects. When transported under identical conditions, produce of poor initial quality would incur more losses than produce of good quality. With the high cost of transportation, it becomes important to ship only top-quality produce (McGregor, 1989).

Physiological factors such as cultivar and harvest maturity can influence the initial quality of produce. These factors may alter produce's resistance to pathogens and to adverse environmental conditions often found during transport. For example, it has been shown that cultivar and fruit firmness at harvest affect blackberry shelf life, as soft fruit are more susceptible to infection and damage (Perkins-Veazie et al., 1997). Moreover, fruits and vegetables are often subjected to fluctuating atmospheric conditions during transport, and the produce may not be held at its optimal storage temperature throughout the journey. Problems become more complex and severe when horticultural produce is shipped as mixed loads. In order to ensure the marketability of produce, it is essential to select varieties (harvested at the optimal maturity) that are more resistant to pathogens, physical injuries, and atmospheric variations for export.

2.2 Temperature

Wills and associates (1998) describe temperature as ''the single most important factor'' that influences the quality of horticultural products. Produce storage and shelf life is a function of both temperature and time (Thompson et al., 1998). The pulp temperature of the produce influences its metabolism and consequently determines the deterioration rate. Ambient temperature favors the growth and development of rot organisms (Thompson et al., 1998). Generally, produce becomes more perishable as its pulp temperature increases. It is therefore essential to lower the temperature of produce after harvest and maintain this low temperature during transport and handling. Nevertheless, overcooling the produce may cause some adverse effects. Freezing and chilling injuries can result from indiscriminate use of refrigeration. In general, non-chilling-sensitive produce should be held slightly above its freezing temperature, whereas chilling-sensitive produce should be kept slightly above its threshold temperature for chilling injury (Wills et al., 1998).

The effect of temperature on fruits and vegetables is also cumulative. Produce quality is eventually affected even if produce is exposed to high temperatures for short, intermittent periods (Thompson et al., 1998). Consequently, it is essential to maintain fruits and vegetables at lower temperatures even for relatively short periods of transport and handling. Problems of poor temperature management are magnified in cases of long-distance and export shipments, in which produce may be subjected to constantly changing or adverse climatic conditions. For example, in Canada, horticultural produce is imported from the southern parts of the United States during winter. Transport trailers have to move perishable produce from a climatic zone where ambient temperatures are generally above 30°C to a zone where ambient temperatures can reach as low as −40°C. In such circumstances, it is extremely important to consider the wide variation in ambient temperature. It becomes a challenge to maintain the ideal environmental conditions required by the produce throughout the journey.

2.3 Humidity and Water Loss

Fruits and vegetables lose water after harvest through transpiration. In the absence of water supply from the mother plant they may wilt, shrivel, become tough, and lack flavor (Nicholas, 1985). Excessive water loss directly affects the appearance and marketable weight of the produce.

Transpiration is a function of the produce's moisture content and the ambient relative humidity. Most fruits and vegetables contain around 85% to 90% water by weight (Nicholas, 1985). Unless the relative humidity of the surrounding air is close to 100%, produce continues to lose water to its environment. The transpiration rate is also affected by certain properties of the fruits and vegetables (i.e., morphological and anatomical characteristics, surface-to-volume ratio, surface injuries, and maturity stage) and other environmental conditions (i.e., temperature, air movement, and atmospheric pressure) (Kader, 1992).

During transport, produce water loss can be reduced by

1. Maintaining a high relative humidity around the produce
   - Maintain a small temperature gradient between the evaporator coils and the circulating air (if a large temperature gradient exists, water vapor in the air condenses on the coils and the air is dry).
   - Install a humidity control system.
   - Top-ice produce that is water-resistant or add ice within the produce packaging.
   - Cover the produce or its container with a semipermeable wrap.
2. Reducing the water permeability of produce's surface
   - Apply wax to the surface of the produce.
3. Reducing the ambient air movement
   - Provide only enough air movement to remove the heat of respiration.
4. Reducing the produce's moisture loss through respiration
   - Precool the produce thoroughly, prior to shipping, to reduce the temperature differential between the produce and the surrounding air.
5. Reducing the water holding capacity of the surrounding air
   - Maintain low temperatures within the transport vehicle (warm air can absorb more moisture).

2.4 Gas Composition

The gas composition in transport vehicles can affect the shelf life of fresh produce. Changes in concentrations of gases such as oxygen (O_2) and carbon dioxide (CO_2) may prolong storage life. However, extremely high or low levels of any gas may lead to physiological disorders on some fruits and vegetables. In addition, the accumulation of volatile gases released by the produce (such as ethylene [C_2H_4]) or other sources may be harmful to the produce. A modified atmosphere inside the vehicle can thus be beneficial or detrimental to horticultural produce.

To prevent the accumulation of any undesirable gases or volatiles in produce shipments, adequate ventilation is required. This can normally be accomplished by adjusting vents on trailers, containers, and railcars.

During transport, controlled or modified atmosphere (CA or MA) can be used as a preservation method to supplement refrigeration. The terms *CA* and *MA* imply the addition or removal of gases to alter the normal atmospheric composition surrounding the produce (Wills et al., 1998). The levels of CO_2, O_2, N_2, and C_2H_4 within the vehicle can thus be manipulated. CA is attained when the concentrations of gases are actively maintained at predetermined levels. In MA, the gas composition is modified initially by injecting a mixture of gases into the sealed environment. However, the gas composition is not consistently replenished afterward. The benefit or hazard of such atmospheric modifications varies greatly with commodity, cultivar, physiological stage, atmospheric composition, temperature, and duration of the application (Kader, 1992). Recommended CA/MA conditions during transport and/or storage for selected produce are listed in Kader (1992).

MA can be used in all modes of transport. Nevertheless, it is more commonly used in railcars, trailers, and intermodal containers (Kader, 1992). MA storage of strawberries is probably the best example of the application of such technology in transport. Strawberries are commonly shipped in a low-O_2 and high-CO_2 environment. High levels of CO_2 are commonly used in berry and cherry shipments as they retard the growth of molds (Ashby, 1995). Nitrogen (N_2) is generally used for leafy vegetables since CO_2 creates discoloration on such produce (Ashby, 1995). Reduced levels of O_2 can normally retard ripening and decay on fruits (Ashby, 1995).

Various methods were developed to create MA within transport vehicles. For semitrailers, a plastic curtain is used to seal around the inside of the rear doorway before the air is evacuated, and the desired gas composition is injected into the trailer (Ashby, 1995). Another alternative is to insert into the trailer, prior to loading, a large plastic bag that can enclose the entire cargo (Ashby, 1995). After loading of the produce into the bag and sealing of the bag, the required gas composition can then be injected into the bag (Ashby, 1995). CO_2 absorbers (generally lime) in sealed polyethylene pallet covers can also be used to create and maintain the desired atmospheric composition (Kader, 1992). For MA/CA shipments, vehicles or palletized loads are usually prepared by specialized companies (Ashby, 1995). These companies also market the gases and in some cases also provide monitoring and recharging services in transit.

2.5 Mixed Load

Optimal transport conditions vary with produce and cultivar. When several types of horticultural produce are loaded in the same transport vehicle, shippers must consider their compatibility. Produce must be compatible in terms of the following:

1. Recommended storage temperature
2. Recommended relative humidity
3. Sensitivity to chilling or freezing injury
4. Production and sensitivity to gases and volatiles
5. Production and absorption of odors

Welby and McGregor (1993) have divided fruits and vegetables into 10 compatibility groups, mainly on the basis of their recommended temperature and relative humidity. Tables of produce sensitive to chilling, freezing, moisture loss, C_2H_4, and odors are also included. TransFRESH Corporation (1988) has also developed a slide chart that provides a quick reference on produce compatibility.

Extra attention should be given when loading mixed loads in transport vehicles. Fruits and vegetables should never be loaded with nonfood cargo as this involves risks of contamination through transfer of odor or toxic chemical residues (Welby and McGregor, 1993). Similar size containers should be loaded on the same pallet to increase stability. Produce weight should be well distributed, with heavier produce across the floor and lighter produce placed above. Lock bars, load gates, and vertically placed pallets can be used to separate and secure stacks of different sized packages (Welby and McGregor, 1993). A representative sample of each type of produce should also be available near the door to facilitate inspection at entry ports so that cargo will not have to be unloaded for inspection (Welby and McGregor, 1993).

2.6 Physical Injury

During transport and handling horticultural produce is constantly subjected to rubbing, compression, impact, and vibration, which result in bruised, crushed, or punctured pro-

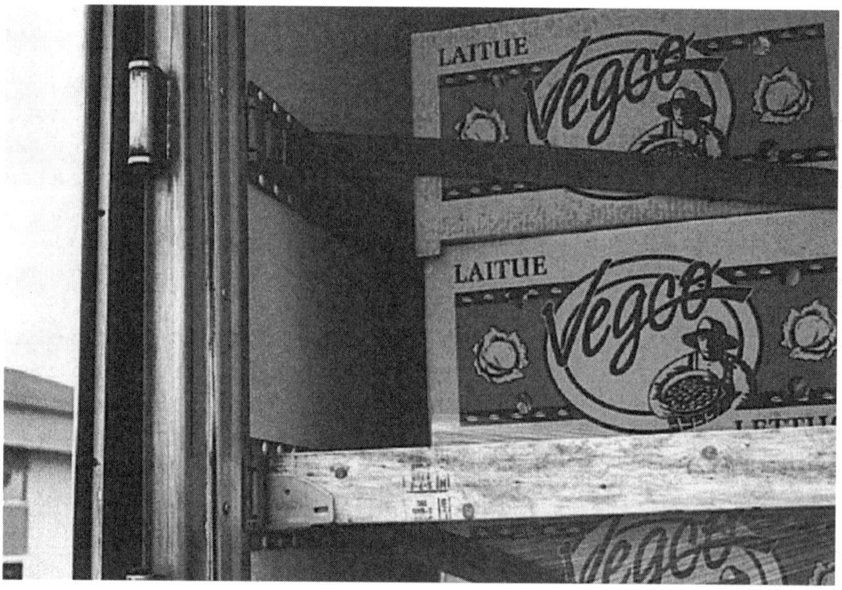

Fig. 1 Bracing system using straps and wooden bars to hold produce boxes.

duce. Physically damaged produce tends to have a higher respiration rate; therefore, it generates more heat and C_2H_4 gas. Also, damaged tissue decays more rapidly and this decay may spread and ruin other produce within the load.

Physical injuries can be reduced by selecting resistant cultivars, harvesting produce at the correct maturity, using proper packaging materials, stacking or loading produce carefully in vehicles, and securing the cargo within the vehicle. Loads within a vehicle can be secured by using aluminum or wood load locks, fiberboard honeycomb fillers, wood blocking and nailing strips, inflatable kraft paper air bags, cargo nets and straps, or wood load gates (Figs. 1 and 2) (McGregor, 1989).

2.7 Transport Conditions

Transport conditions for horticultural produce are usually the same as the storage requirements. Transport/storage conditions vary significantly for each type of produce. The following is a list of references that provide valuable information on transport requirements and other related topics:

1. Ashby (1995): information on recommended transport temperature, relative humidity, highest freezing points, and ice cooling compatibility of major produce
2. ASHRAE (1994): information on recommended storage temperatures, relative humidity, approximate storage life, water content, highest freezing point, specific heat, and latent heat of many perishables
3. Guide to Food Transport (1989): information on produce shelf life, ideal transport temperature and humidity, C_2H_4 and chilling sensitivity, heat production rates for a wide range of temperatures, specific heat, specific weight, color photos, and descriptions of 100 types of produce
4. McGregor (1989): information on recommended transport temperature, relative

Fig. 2 Bracing systems used to prevent produce from shifting backward to the rear doors.

humidity, and approximate storage life of over 150 types of fruits and vegetables, including many types of tropical produce

5. Hardenburg and colleagues (1986): information on recommended storage temperatures, relative humidity, approximate storage life, highest freezing point, water content, specific heat, and respiration rate of various horticultural commodities

3 CONTROL SYSTEMS FOR PRESERVING QUALITY DURING TRANSPORT

Once perishable goods are loaded for transport, it is important to protect them from variations in environmental conditions. The transport vehicle must therefore be capable of maintaining fruits and vegetables at their required temperature and humidity throughout the journey. Systems are available to maintain optimal temperatures in transport vehicles;

selection of the refrigeration system must take into account the cooling capacity required and the characteristics of the produce being transported.

3.1 Calculation of the Required Cooling Capacity

The refrigeration system used in a transport vehicle must be capable of removing all heat entering the vehicle from the outside, all heat generated within the vehicle, and any heat contained in the vehicle itself. One must remember that the cooling capacity varies significantly, depending on the outside ambient conditions. Therefore, it is important to select the extreme conditions that can be encountered during transport to calculate the cooling capacity required. One must also remember that when transporting produce in a region where outside ambient temperatures are below 0°C, a heating system is required to prevent produce from freezing.

3.1.1 Internal Heat Loads

The internal heat loads include respiratory heat generated by the produce as well as any field heat that remains within the produce. Fruits and vegetables are usually harvested at ambient temperature and then precooled to optimal storage or transport temperature. However, if produce is not adequately precooled, its internal heat load is larger. In addition, loading areas are often at ambient temperature, thus, produce transferred from cold storage to a transport vehicle may gain heat from the surroundings. This heat is also considered as field heat. In some cases, the amount of heat gain is significant especially when produce is left for long periods in the loading area as a result of delays during the loading process.

Field heat is a function of produce mass, specific heat, and the difference between the initial and desired transport temperatures. Specific heat is the amount of heat gained or lost by a unit weight of produce per unit change in temperature. Both specific heat and optimal transport temperature vary among types of produce.

Horticultural produce continues to respire even after harvest. As produce respires, CO_2, moisture, and heat are released. Some types of produce, such as asparagus, corns, and strawberries, respire more than others, such as apples, oranges, and potatoes (Ashby, 1995). Highly perishable produce tends to generate much more respiratory heat. Since respiration rate is higher when produce pulp temperature is higher, more heat is released as temperature increases.

Values of specific heat, recommended transport temperature, respiration rates, and heat generation of different commodities can be found in references listed in Sec. 2.7. ASHRAE (1986) also provides heat evolution rates for additional produce.

3.1.2 External Heat Loads

The external heat loads include all sources of heat that enter the transport vehicle from the outside. The usual sources of external heat include the heat of conduction, the air infiltration, as well as the heat of radiation.

Heat is conducted through the floor, walls, and ceiling of a transport vehicle whenever a temperature gradient exists between the outside and inside of the vehicle. The amount of heat transferred through conduction is a function of the surface area where heat conduction occurs, the type and thickness of insulation within those surfaces, and the temperature difference between the outside and inside of the vehicle. Insulating materials for use in transport vehicles should be of moderate cost, easy to apply, and light in weight and should have low thermal conductivity, low moisture permeability, and water retention (ASHRAE, 1998). In addition, the insulation should be resistant to fire, to breakdown at extreme temperatures, cracking, crumbling, shifting, and vibration or any other types of mechanical abrasion (ASHRAE, 1998). Insulation used for the vehicle floor should have enough strength to support a forklift or other loading equipment. Any moisture penetration or physical damage decreases the insulation value of the vehicle, and since this normally occurs when transport vehicles age, the amount of conduction heat load is generally higher for older vehicles.

Infiltration is one of the largest sources of heat gain in transport vehicles. Warm air enters the transport vehicle through small holes, cracks, drainage holes, and broken door seals. The opening of the vehicle's doors also significantly increases the refrigeration requirements. For example, in a transport trailer involved in a multiple-stop delivery, the heat gained through door openings can be five times the conduction heat load (ASHRAE, 1998).

Solar radiation also increases the refrigeration load in a transport vehicle. It was found that the cooling requirement of stationary vehicles increased by 20% after exposure to sunlight for several hours (ASHRAE, 1998). Heat gain through radiation can be minimized with the use of highly polished steel, aluminum plates, or reflective paints on the outside surfaces of transport vehicles (Ashby, 1995). Frequent cleaning is required to maintain the reflective properties of the exterior of the vehicle.

3.1.3 Residual Heat Loads

The residual heat loads include any heat initially contained in the transport vehicle or any heat load not included as internal or external loads. The usual sources of residual heat include the heat initially contained in the air and the inner surfaces inside the transport vehicle (Ashby, 1995), and the heat contained in the packaging boxes and the devices used to secure the load. Depending on the mode of transport, one should also consider the heat generated by loading equipment.

3.1.4 Total Cooling Capacity

In general, the total cooling capacity of the refrigeration system should be equal to the sum of the internal, external, and residual heat loads, multiplied by a safety factor. However, the actual calculation of cooling capacity is more complex, since other factors have to be considered in the design of the cooling system. Additional heat gain may also have to be included in the calculations to account for specific requirements related to the mode of transport. A few examples of cooling load calculations, for various cooling systems and for various modes of transport, are included in ASHRAE (1998). In the design of mechanical refrigeration systems for trucks and trailers, manufacturers normally use computer programs or charts to match the refrigeration unit to the vehicle used for specific operations.

3.2 Temperature Control Systems

Different systems are available to maintain optimal temperatures in transport vehicles. Mechanical refrigeration, ice cooling, and cryogenic cooling are the three most commonly used methods. Other methods are also available, depending on the mode of transport. Basic concepts of the cooling methods are presented here, other cooling methods are presented in Sec. 5.

3.2.1 Mechanical Refrigeration

Mechanical refrigeration is the most widely used cooling method in road and rail transport. It has also gained popularity in marine and intermodal transport. Similar to household refrigerators, mechanical refrigeration systems used in transport vehicles operate on the vapor-compression-refrigeration cycle (Wark, 1988). In such systems, a refrigerant changes phase from liquid to vapor and back to liquid in a closed cycle. As a result of these phase changes, the refrigerant is able to absorb heat in the evaporator located within the vehicle and release it through the condenser into the air outside the vehicle.

The main components of a mechanical refrigeration system are the evaporator, compressor, condenser, and expansion valve. In the evaporator, the liquid refrigerant absorbs heat from the warm air surrounding the produce and changes into vapor. The vaporized refrigerant is then compressed by the compressor to a pressure high enough to allow the refrigerant to release heat and condense at ambient temperature inside the condenser. Depending on the cargo temperature, a variable amount of liquefied refrigerant is released through the expansion valve into the evaporator, where it continues the cycle and maintains the desired temperature inside the vehicle.

The refrigeration unit may be driven by a diesel engine, an electric motor, or a combination of both. Refrigeration systems on trucks and trailers are typically driven by diesel engines; some models have an optional electric motor (ASHRAE, 1998). Intermodal containers are usually equipped with compressors run by electric motors (usually hermetic), with or without standby engine-driven generators (ASHRAE, 1998). The refrigeration unit is plugged into an electrical power outlet when the intermodal container is stored on a dock. Once the container is loaded on a trailer, a flatcar, or a ship, then the engine-driven generator provides electricity to run the refrigeration unit.

Thermostats are used to control the cooling rate supplied by the mechanical refrigeration units. Modern refrigerated trailers and containers are equipped with microprocessors to regulate the temperature. These systems are capable of controlling temperature to within $\pm 1°C$ of the setpoint. To prevent damage to the produce in transport, ASHRAE (1998) suggests that thermostats be set 1°C to 2°C higher than the freezing point for cool season vegetables. Recommended thermostat settings for most perishables are listed in ASHRAE (1998).

Frost accumulation on the evaporator coils reduces the cooling capacity of refrigeration units. Defrosting is automatic in mechanical refrigeration systems. The process is controlled by an electric timer or a differential pressure switch (ASHRAE, 1994). Defrosting is accomplished by using electric heaters or hot refrigerant gas. During the defrost cycle, airflow is stopped to prevent heat produced in the coils from reaching the produce.

All of the components of a mechanical refrigeration system are normally built into a self-contained unit attached to the front end of the vehicle. The unit can either be nose-mounted or recessed in the front wall of the vehicle. Most highway trailers use nose-mounted systems. The engine, condenser, and other accessories are located on the outside of the front wall, whereas the evaporator and air fans are found inside the vehicle (Fig. 3) (Ashby, 1995). Intermodal containers use recessed-type systems, in which the entire refrigeration unit is fused to the front section of the container so that the containers are the same length as nonrefrigerated containers and can therefore be stacked on flatcars or container ships (Fig. 4) (Ashby, 1995).

Fig. 3 Nose-mounted refrigeration system used in a semitrailer.

3.2.2 Ice Cooling

Ice cooling is the oldest and simplest cooling method. Ice is a good heat absorber, having a latent heat of fusion of 335 $kJ \cdot kg^{-1}$ (144 $Btu \cdot lb^{-1}$). In addition to providing a cool environment for horticultural produce, it maintains high humidity levels within the transport vehicle. It can be used as the main cooling source or as a supplement to mechanical refrigeration systems. In the latter case, ice cooling provides protection for the produce if the mechanical refrigeration unit fails or its capacity is not sufficient to cool the produce to the desired temperature. Ice cooling is used in land and air transport.

Finely crushed ice, flaked-ice, or liquid ice (slurry of ice and water) can be blown onto the top of the produce. This process is known as *top-icing*. Crushed ice or liquid ice can also be applied on the top of individual pallets or injected into individual boxes (package icing) before loading. Package icing can also be used as a precooling method. Highly perishable produce such as broccoli and sweet corn are commonly precooled with liquid ice. The remaining ice particles within packages serve to protect the produce from rewarming and dehydration during transport.

Compared to other cooling methods, ice cooling is the only method that provides thermal and humidity protection. Nevertheless, it can only be used for produce that can tolerate high levels of humidity because fruits and vegetables may decay as a result of excess moisture. Welby and McGregor (1993) and Ashby (1995) have prepared a list of produce that can be top- or package-iced. Packaging materials for produce and interior surfaces of transport vehicles have to be water resistant since excessive moisture can weaken the strength of packaging boxes and corrode metal body structures. To prevent expensive damage to produce and the vehicle, a mechanism to evacuate water from melting ice must be available. Trailers and containers usually have drainage holes located at the four corners of the floors. Airlines that accept the use of liquid ice normally require that

Fig. 4 Recessed refrigeration system fused with the front wall of an intermodal container.

the ice be sealed in polyethylene bags and put into a leakproof container that has a moisture absorbent pad (McGregor, 1989). The use of ice cooling requires more time and labor for handling than other cooling methods. Finally, the weight of ice reduces the quantity of produce that can be loaded on trucks and trailers, since there are limits on the weights of vehicles circulating on highways.

The freezing of melted ice is one of the problems that arise when ice cooling is used along with mechanical refrigeration. For vehicles equipped with air circulation ducts on the ceiling, airflow in top-iced cargoes may be severely restricted when melted ice is frozen by the refrigeration unit on top of the pallets (ASHRAE, 1998).

In package icing, a layer of air appears between the ice particles and the produce when the ice melts. This air gap slows the rate of heat removal since it acts as an insulator.

3.2.3 Cryogenic Cooling

In some low-temperature systems, liquid N_2, liquid CO_2, or solid CO_2 (dry ice) is used to keep fruits and vegetables at their optimal temperatures. These systems were developed

Table 1 Heat Absorption Characteristics of Liquid Ice and Various Cryogenic Refrigerants

Cooling medium	Initial temperature (°C)	Amount of heat absorbed to reach 0°C or 1°C (kJ/kg)
Liquid ice	0	335
Liquid ice	−13	372
Liquid N_2	−196	396
Liquid CO_2 (at 5.7×10^6 Pa)	+20	215
Liquid CO_2 (at 2.1×10^6 Pa)	−18	326
Solid CO_2 (dry ice)	−79	633

Source: From International Institute of Refrigeration (1995).

before the extensive application of mechanical refrigeration. Because of the difficulties involved in controlling atmospheric conditions in cryogenic cooling and the increased efficiency of mechanical refrigerating units, the use of cryogenic cooling systems is generally limited to local delivery vehicles and frozen product delivery (Ryall and Pentzer, 1982). Liquid N_2 and dry ice are the most common cryogenic cooling methods used in air transport (Ryall and Pentzer, 1982).

When a cryogenic refrigerant is in contact with produce, it extracts heat from the produce and vaporizes. A problem associated with the use of liquid and solid cryogenic substances is the disposal of gas after evaporation or sublimation (solid to gas). Different cryogenic refrigerants vaporize at different temperatures. The heat absorption characteristics of the various cryogenic refrigerants are compared to those of liquid ice in Table 1. Other physical and thermophysical properties of these refrigerants are presented in ASHRAE (1997).

3.2.3.1 Liquid Nitrogen and Liquid Carbon Dioxide

During transport, liquid cryogenic refrigerants are stored in pressurized tanks. The storage vessels are located either inside or outside the vehicle. Liquid cryogenic refrigerants have contact with warm air through direct injection systems or heat exchangers within cargo. For direct injection systems, liquid refrigerants are released from spray nozzles mounted on the ceiling of the vehicle. Once the desire temperature is reached, the thermostatically controlled release valve shuts off the supply of refrigerant to the nozzles. For heat exchanger systems, liquid refrigerants are circulated within coils or plates. After contact with warm air, the refrigerants absorb heat, vaporize, and are vented away. This type of design is said to provide better temperature distribution than direct injection systems (Ryall and Pentzer, 1982).

The direct injection systems are easy to use; have a high cooling capacity, with no defrosting problems; are less costly; and are lighter in weight than mechanical refrigeration systems (Ryall and Pentzer, 1982). Nevertheless, the advantages of these systems are offset by some severe problems in their use. Excessive concentrations of N_2 or CO_2 gases create risks of freezing and/or suffocating the workers (ASHRAE, 1998). High concentrations of such gases may also be harmful to certain types of horticultural produce. It is known that an atmosphere filled with 100% N_2 causes physiological disorders on most fruits and vegetables, even though many of them can withstand high levels of N_2 for several days with no injury (Ashby, 1995). In spite of the fact that moderate levels of CO_2 can retard produce decay and ripening, high concentrations may cause off-flavors, off-colors, and physiological disorders (Ashby, 1995). Leafy green vegetables are found to be incompatible with CO_2 refrigeration systems (McGregor, 1989). Furthermore, liquid CO_2 must be stored at pressures higher than 416 kPa (gauge) and below 31°C (ASHARE, 1998). The possibility of very low cargo temperatures as a result of an excessive accumulation of gases and the problem of availability of recharging gases in transport points make liquid cryogenic cooling less acceptable for use in fruit and vegetable transport (Ryall and Pentzer, 1982).

3.2.3.2 Dry Ice

Solid CO_2 or dry ice may be used in the form of blocks, snow, or pellets in transport vehicles. It is mostly used in frozen food and ice cream delivery trucks (Ashby, 1995) but is also used during the air transport of strawberries (ASHRAE, 1998). Direct contact with dry ice should be avoided to prevent damage to the produce (McGregor, 1989). Dry ice is sometimes used in air cargo along with iced water, to slow the melting of iced water and prolong the cooling period during transport (ASHRAE, 1998).

3.2.4 Heating During Winter

During winter, when outside temperatures are lower than the temperature required inside the vehicle, the transport vehicles must be heated to prevent the chilling or freezing of fresh produce. Mechanically refrigerated trucks, trailers, railcars, and containers are usually heated either by electric heating or by reverse-cycle running of the refrigeration unit. Once the cargo temperature is set, the built-in thermostatic controller automatically switches from cooling to the heating mode whenever it is needed. However, the heating capacity of the mechanical refrigeration system operating in reverse cycle is only about a third of its refrigeration capacity when set at 0°C in a 30°C ambient temperature (International Institute of Refrigeration, 1995). Therefore, supplemental electric heating may be required when transporting produce in very cold regions.

When the temperature control system used in a transport vehicle is ice cooling or cryogenic cooling, a heating device must be incorporated in the vehicle. Heating devices commonly used in transport vehicles are burners (with an air-to-air heat exchanger) and electric heaters.

3.3 Air Circulation Systems

Air circulation plays a dominant role in maintaining ideal transport temperature within transport vehicles. In fact, the refrigeration system becomes useless if cooled air is not circulated throughout the cargo. The air circulation system is used to distribute cool air around and through the cargo to absorb any heat that enters the vehicle or to absorb the heat generated by the produce. The warm air returns to the refrigeration unit, where it is cooled and cycled back around the cargo. If the air circulation system is poorly designed and the cooled air is not evenly distributed, overwarming or overcooling occurs in different parts of the vehicle. During winter, the air circulation system is required to circulate warm air around the produce, to prevent chilling or freezing injuries. Two different types of air delivery systems are commonly used in mechanically refrigerated semitrailers, intermodal containers, and railcars: top-air and bottom-air.

3.3.1 Top-Air Delivery System

The top-air delivery system is the most widely used air circulation method in mechanically refrigerated semitrailers and railcars. In this system, the refrigeration unit in the front of the vehicle blows cold air along the ceiling. The cold air circulates above the cargo from the front to the rear of the vehicle. Some of the cold air also flows downward along the side walls. At the rear doors, the cold air flows downward and returns along the floor, underneath the cargo, as well as through the load from rear to front. As it reaches the front, the air flows upward and returns to the refrigeration unit. The cold air, as it circulates along all surfaces of the vehicle and through the load, picks up any heat that enters the vehicle and any heat that is generated by the produce.

A typical top-air delivery system is composed of an air delivery duct along the ceiling, horizontal floor ducts, and a return-air bulkhead (Fig. 5). The actual materials used to construct these components differ from vehicle to vehicle. Because semitrailers are the predominant intracontinental transport vehicles, the arrangement of their top-air delivery system is discussed in detail in the following.

For semitrailers, the air delivery duct is usually a wide strip of canvas or vinyl material hanging along the ceiling from the front to the rear of the vehicle. The air delivery

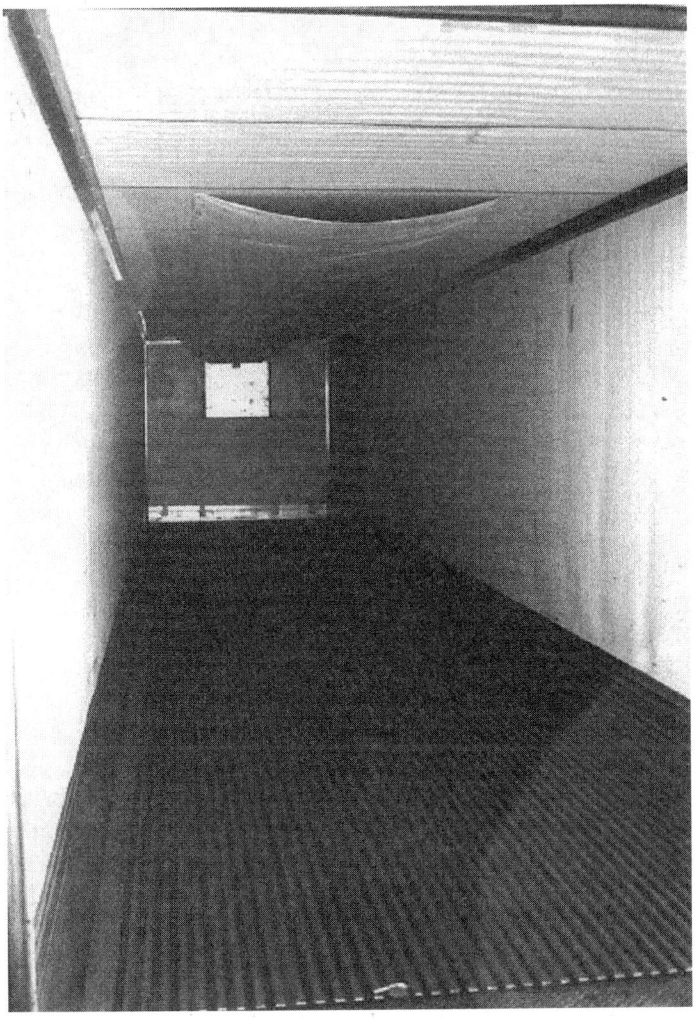

Fig. 5 Typical arrangement of the air circulation system in a semitrailer.

duct is used to ensure that the air moves from the blower to the rear of the vehicle. The National Perishable Logistics Association/Refrigerated Transportation Foundation (NPLA/RTF) recommends a minimal cross-sectional area of 0.1548 m^2 (240 in^2) for the air duct (Ashby, 1995). The air delivery duct should also extend from the front of the vehicle to 3 to 5 m from the rear (Ashby, 1995). Openings on both sides of the air duct should also be provided to allow air ''spill'' down the side walls. The International Institute of Refrigeration recommends a different configuration for the air delivery duct (1995). Rather than having air spill down the side walls along the full length of the air delivery duct, it recommends discharging the air at three different positions along the ceiling: 20% should be discharged near the front of the vehicle, 50% at approximately one-third (1/3) of its length (from the front to the rear), and 30% at three-quarters (3/4) of its length. This duct configuration also ensures that air will spill down the side walls along the full length of the vehicle.

Different floor duct designs are used in semitrailers. The floor configuration affects the amount of air that circulates underneath the cargo and therefore affects the capacity to absorb the heat conducted through the floor. The most common types of floors found in transport trailers and railcars are the duct board floor and the T-beam floor. Ashby (1995) states that about 0.1548 m^2 (240 in^2) of return air passage is required for the average trailer blower to operate at full capacity. However, the average duct board floor offers only approximately 0.0303 m^2 of return air passage. The average T-beam floor offers approximately 0.1290 m^2 of return air passage, which is much better than that of the duct board floor. Nevertheless, the T-beam floor is found to be less resistant to physical damage during loading and unloading operations. Placing produce on pallets significantly increases the area available for air circulation and should therefore be considered for trailers with duct board floors. For trailers that are not equipped with a duct floor, produce must be palletized to ensure sufficient air passage.

The use of a return air bulkhead is highly recommended in semitrailers because it helps to isolate the load from the front wall and prevents the load from blocking the return airports of the refrigeration unit (Kasmire et al., 1996). Both a frame bulkhead and a solid bulkhead are commonly used. The frame bulkhead prevents the blockage of the return airports but does not prevent the short-circuiting of air from the refrigeration unit's air supply to the return airports. The use of a solid bulkhead ensures that the air circulates around the cargo rather than short-circuiting the cargo.

Vertical channels on the rear doors and side walls have been found to improve air movement and reduce heat conduction through the rear doors and side walls (Kasmire et al., 1996). Nevertheless, ribbed or fluted walls are more likely to be damaged by forklifts during loading and unloading. Modern semitrailers are normally equipped with smooth doors and flat side walls.

3.3.2 Bottom-Air Delivery System

The bottom-air delivery system is used extensively in intermodal containers. The use of this system in highway trailers is limited, since additional structural strength is required for the pressurized air (Ashby, 1995). The bottom-air delivery system used in sea containers generally consists of a horizontal T-beam floor, vertically ribbed rear doors, vertically ribbed side walls, and a solid bulkhead in the front (Fig. 6). In this type of air delivery system, most air movement is vertical. The refrigeration system blows cold air in the T-beam floor of the container. This air flows from the front to rear of the vehicle and is forced upward through the cargo. When the air reaches the ceiling of the container, it

Fig. 6 T-beam floor and ribbed side wall used in an intermodal container.

flows from the rear to the front and returns to the refrigeration unit through the bulkhead openings.

The loading pattern used to fill a transport vehicle with a bottom-air delivery system is very important (International Institute of Refrigeration, 1995). The entire floor should be covered in order to force air up and through the cargo. Any floor spaces that are not covered by the cargo (e.g., between pallets, the last pallets and the rear doors) must be filled with empty boxes, blocks, or insulation to prevent cold air from short-circuiting.

4 LOADING OF TRANSPORT VEHICLES

4.1 Preparation before Loading

As presented in previous sections, to maintain produce at recommended temperature during transport, produce should already be at this temperature prior to loading. In addition, the refrigeration system used must be able to remove external, internal, and residual heat contained in the transport vehicle. To ensure that these conditions are met, three important steps must be taken prior to loading of produce: (a) the produce must be well precooled, (b) the transport vehicle must be cleaned, and (c) the transport vehicle must be precooled or prewarmed.

4.1.1 Precooling of Produce

Precooling implies the rapid removal of produce field heat after harvest. Room cooling, forced-air cooling, hydrocooling, vacuum cooling, and package icing are commonly used to precool horticultural produce. Before shipment, produce should be precooled as near

Fig. 7 Accumulation of debris obstructing the airflow pathways for the air circulation system.

as possible to the recommended transport temperature. Improperly precooled produce increases the cooling requirement for the refrigeration system used in transport. Most cooling systems are designed only to maintain the low temperature of precooled produce and thus normally do not have the extra refrigeration capacity to achieve precooling.

4.1.2 Cleaning of Transport Vehicle

Cleanliness ensures proper air circulation around the transported load. It also prevents the occurrence of bacterial, chemical, and odor contamination within transport vehicles. Floors should be washed or swept thoroughly before loading. Floor ducts should be free of debris to provide clear pathways for air circulation (Fig. 7). To ensure proper drainage and prevent disease development, floor drain holes should be free of obstructions. Semitrailers, containers, or railcars used for transporting fruits and vegetables may also be used to deliver seafood or frozen products. In such cases, thorough cleaning and airing of the vehicle become necessary for produce that is odor absorbing (such as apples or bananas). Freshly opened cans of ground coffee may be helpful in absorbing odors if they are left for 8 hours or more in a closed vehicle (Ashby, 1995). Ground coffee can also be spread directly on the floor and swept away before loading.

4.1.3 Precooling or Prewarming of the Transport Vehicle

Transport vehicles should be precooled or prewarmed before loading to reduce the initial cooling or heating load on the vehicle's refrigeration or heating system. During summer, precooling the vehicle reduces the chance of the produce's rewarming after loading, causing a larger demand on the refrigeration system. During winter and especially during extreme subfreezing weather, prewarming of the vehicle reduces the chance of chilling or freezing injuries of produce.

Ashby (1995) recommends setting the thermostat at the desired transport temperature, closing the vehicle doors, and running the cooling or heating system until the vehicle body reaches the setpoint temperature. Modern mechanically refrigerated vehicles are equipped with microprocessors that automatically run the precooling cycle. In all cases, sufficient time has to be scheduled ahead of loading, as the precooling process may be long.

For all types of vehicles, it is important to turn off the mechanical refrigeration system once the doors are opened for loading. When the refrigeration system is on while the doors are open, the system's fan draws in outside air. Since warm outside air has a higher moisture content than the air in the vehicle, this additional moisture condenses on the refrigeration system's coils. The ice formed on the evaporator coils limits or blocks off the air circulation during transport (Ashby, 1995). Similar conditions occur when the vehicle's doors are left open longer than required, even when the refrigeration system is turned off.

4.2 Considerations during Loading

The loading pattern used for arranging produce in a vehicle depends on the type of transport vehicle (semitrailer, container, railcar, ship, or airplane), the type of load (palletized or unitized), and the type of air delivery system (top or bottom). Special attention has to be given to mixed loads (see Sec. 2.5). Horticultural produce can be handled as either palletized or unitized loads. The palletized load is the most commonly used loading arrangement because of its ease of handling and of its reduced labor requirement. Nevertheless, a larger quantity of produce can be transported if produce is individually placed in transport vehicles. The unitized loads also allow better air circulation through the cargo when boxes are stacked evenly and air passages are left between rows of boxes.

Numerous loading patterns are available for both types of loads. No matter what type of loading pattern is used, the basic guideline is to load produce such that conduction heat transfer is minimized, air circulation is maximized, and the quantity of produce transported is maximized. Conduction heat transfer can be reduced by leaving an air space between the produce and the walls and rear doors. Loading boxes on pallets prevents heat conduction between floor and produce. When loading a top-air delivery vehicle, open pathways should be made available for air circulation through the load. For bottom-air delivery vehicles, it is important to cover the entire floor with cargo or fillers, as a high static pressure has to be maintained to force air upward. For mechanically refrigerated semitrailers or railcars using a top-air delivery system, the centerline loading pattern is recommended for palletized cargo. Details on other loading patterns are presented in McGregor (1989) and Ashby (1995).

5 PARTICULARITIES FOR EACH TRANSPORTATION MODE

Fresh fruits and vegetables can be transported by road, rail, ship, or air. Selection of the mode of transport and the type of vehicle should be based on the following factors:

1. Destination
2. Produce value
3. Produce perishability
4. Quantity of produce to be transported
5. Recommended transport conditions

6. Climatic conditions at origin and destination
7. Time required to reach destination
8. Freight rates
9. Quality of transport service

5.1 Highway Transport

Highway transport is the predominant method used to haul large volumes of produce across the nation. In regions where infrastructures are well developed, highway transport is normally selected for intracontinental shipments because of the presence of a well-developed road network, the reliability of transport equipment, the moderate cost of transportation, and the flexibility to deliver any type of produce almost everywhere. Highway transport can also be combined with rail transport to deliver produce over long distances on land. Two types of vehicles are used in highway transport: refrigerated semitrailers and intermodal containers. The latter are discussed in Sec. 5.4.

Refrigerated semitrailers, often referred to as *trailers*, can be considered as intermodal transport vehicles. These consist of an insulated rectangular structure with wheels in the rear and support at the front. Uncoupled from the tractors, they can be transported on railroad flatcars (piggyback or trailer-on-flatcar [TOFC]). Trailers can also be driven right onto sea vessels (roll-on–roll-off [RORO]) or simply hauled by a tractor on the highway (tractor-trailer). Refrigerated trailers are available in lengths of 12 m (40 ft), 13.7 m (45 ft), 14.6 m (48 ft), or 16.2 m (53 ft) (McGregor, 1989).

Most modern trailers are mechanically refrigerated with a nose-mounted system. Liquid ice, liquid N_2, and liquid CO_2 can also be used as sources of cooling. Most trailers are equipped with top-air delivery systems, as discussed in Sec. 3.3.1.

5.2 Rail Transport

Rail transport used to be the predominant method for land transport. With the increased use of highway transport, rail transport has steadily declined. This mode of transport is still used in regions where roads or highways are not well developed. Four different vehicles can be used for rail transport: the ice-refrigerated railcar, mechanically refrigerated railcar, refrigerated semitrailer on flatcar ([TOFC]), and intermodal container on flatcar (COFC). The latter two types of vehicles are discussed in Sec. 5.4.

5.2.1 Ice-Refrigerated Railcar

The ice-refrigerated railcar is rapidly being replaced by mechanically refrigerated systems in Europe and the United States. Originally, ice-refrigerated railcars consisted of an insulated load compartment with one or several ice compartments. Later on, more sophisticated systems were developed, and ice bunkers were placed in the end of the cars with raised racks for air circulation (Ryall and Pentzer, 1982). Underload floor racks, basket-type bunkers, air-circulation fans, adjustable ice racks, and additional insulation were also added to improve the cooling capacity of railcars (Ryall and Pentzer, 1982).

5.2.2 Mechanically Refrigerated Railcar

The mechanically refrigerated railcar was initially designed to replace the ice-refrigerated railcar. Nevertheless, the use of rail transport has gradually declined with the increased use of refrigerated trailers and intermodal containers. The refrigeration system used in railcars is similar to the one used in trailers and containers. With the refrigeration system located at one end of the car, air is distributed vertically downward through the load.

A centrifugal blower is normally used to draw cool air from the evaporator coil and air is discharged through the duct or plenum placed on the ceiling. Air then flows downward either through slots or holes in the plenum or through wall flues on side walls and the far end of the car. In both systems, warm air returns to the refrigeration unit through the floor ducts (Ryall and Pentzer, 1982).

5.3 Marine Transport

Marine transport is generally used to carry produce that has a relatively long shelf life (e.g., bananas, kiwis, citrus fruits) over long distances. The majority of exported fruits are shipped by sea: 98% of bananas, 75% of exotic fruits, 62% of citrus fruits, and 48% of temperate climate fruits as of 1992 (International Institute of Refrigeration, 1995). Marine transport is normally selected for intercontinental shipments since it is the most economical mode of transportation over long distances. Produce being shipped overseas may remain on board for 5 to 6 weeks; therefore, good temperature and humidity control is essential.

Marine transport is carried out in conventional refrigerated ships or in container ships. In the early 1990s, it was estimated that 70% of the refrigerated goods transported by sea were carried in conventional refrigerated ships, and 30% was carried in container ships (International Institute of Refrigeration, 1995). In 1993, there were approximately 800 conventional refrigerated ships of over 3,000-m^3 capacity in service worldwide, with an overall capacity of 80 million m^3, and 40% of the fleet was made up of pallet-friendly ships (International Institute of Refrigeration, 1995). Also in 1993, there were approximately 370 container ships in service with 215,000 container-adapted spaces (i.e., a volume-equivalent unit for a 20-ft container). Of these spaces, 165,000 were for containers equipped with refrigerating units and 50,000 for insulated porthole containers (International Institute of Refrigeration, 1995).

5.3.1 Conventional Refrigerated Ships

Conventional refrigerated ships are completely insulated and have a series of holds divided into rooms. The rooms are vertically aligned and have a common hatch for loading and unloading. These ships can carry refrigerated goods, frozen foods, and nonperishable products in different holds or rooms at the same time.

Two cargo handling techniques are used in refrigerated ships: manual handling (breakbulk system) and palletized handling (Ryall and Pentzer, 1982). The breakbulk ship loading system involves lowering the shipping boxes into the hold with a net and manually stacking the boxes in the hold. Palletized handling involves using pallet slings to lower pallet loads of boxes into the ship's hold. A larger number of ships now have side doors (or vertical hatches) so that pallets of produce can be loaded onto ships with forklifts or other equipment (International Institute of Refrigeration, 1995). Although the breakbulk system makes better use of the ship's hold, labor costs are very high and more produce or container damage occurs because the boxes are handled several or many times. The effective volume used in a palletized load is smaller than in a manually stacked load. However, handling is much faster and labor costs are lower.

The method of stacking cargo in the hold is critical if specific cargo temperatures must be maintained (Ryall and Lipton, 1979). Spaces must be left between rows, stacks, or layers, so that circulating air can reach one or more surfaces of each shipping container. Such spacing can be provided by placing wood strips (dunnage) vertically between rows or stacks as the cargo is loaded. Sometimes it is necessary to place dunnage between layers, but this is not desirable when the produce is packed in corrugated shipping containers, which tend to sag or crease as they absorb moisture. This process may close the air channels between layers or damage the containers and produce (Ryall and Lipton, 1979). The new generation of pallet-friendly ships includes a special layout of the hold so that complete palletization is possible (International Institute of Refrigeration, 1995). In such ships, vertical gratings or bins are used to mark off geometrical spaces for the pallets. These spaces are aligned so that air circulation between pallets is possible to remove the heat contained in the boxes of produce.

Because of the handling method used to load conventional refrigerated ships and the lack of precooling of some produce before shipping, the load often has to be cooled initially. Therefore, the ship must have sufficient refrigeration capacity and air circulation to lower the temperature of the cargo. To speed up the initial cooling of the cargo, the refrigeration system can be set at a lower temperature at the start. However, the produce temperature must never fall below the lowest allowable temperature. The air circulation rate may also have to be increased to two to four times the normal rate during the initial cooling down phase (International Institute of Refrigeration, 1995). Thus, when the air circulation rate is high (e.g., between 60 and 90 air changes per hour), it may be necessary to adjust the delivery air temperature a few degrees above the required temperature. Otherwise, evaporative cooling can lower the temperature of the produce well below the setpoint of the refrigeration system.

In addition to proper air circulation, fresh air must be added to remove gases (e.g., CO_2, C_2H_4) produced during the respiration process. The addition of fresh air, especially in tropical conditions, inevitably includes the introduction of large amounts of water vapor. This poses an additional load on the cooling coils and may have an effect on humidity levels in the cargo space. Conversely, in very cold ambience the introduction of fresh air adds to heating requirements and decreases humidity in the cargo space (International Institute of Refrigeration, 1995).

Ships' holds are generally refrigerated with either a direct expansion refrigeration system or an indirect refrigeration system using low-temperature brine (International Institute of Refrigeration, 1995). In addition, winter shipments of produce to Western Europe can sometimes be adequately protected in holds of ships by judicious use of cool outside air (Ryall and Lipton, 1979). This service is generally limited to vessels using the North Atlantic route during the months of December through March. The service is not suitable for highly perishable items such as leafy vegetables but when properly operated should provide adequate protection for potatoes, most melons, peppers, and tomatoes.

5.3.2 Container Ships

Container ships have specially designed holds that have vertical guides to stack containers below the deck. Typically, containers can be stacked six to nine high below deck, and three to four high above deck. In the latter case, the containers are clamped and lashed to hatch covers as well as to the decks (International Institute of Refrigeration, 1995). Container ships can generally accommodate all types of containers, including insulated porthole containers, mechanically refrigerated containers, and dry cargo containers.

Since loading and unloading of containers are much more rapid than the breakbulk handling of a ship's cargo, savings in port time is an important advantage of container ships. Also, the container's cargo is handled fewer times and is subjected to fewer interruptions in refrigeration than when it is handled by the breakbulk system. Therefore, the quality of the produce once it reaches the destination should be better than when it is handled by the breakbulk system in conventional refrigerated ships. However, container ships can be only loaded and unloaded in ports equipped with special cranes and handling equipment that can lift and lower containers.

Refrigerated containers for marine transport have essentially the same construction, insulation, and temperature control systems as those for use exclusively on truck chassis and railroad flatcars. The principal additional requirement for sea containers is sufficient structural strength to permit stacking six to nine units high and to withstand the strains of accelerations and decelerations experienced aboard ships and road vehicles (Ryall and Lipton, 1979; International Institute of Refrigeration, 1995). The bottoms of marine containers can withstand these strains only with the help of corner castings. The upper face also has a low resistance to compression. Therefore, the handling and loading of containers require specially designed devices, at the port, on the dock, and on board. These include straddle carriers and spreader cranes (International Institute of Refrigeration, 1995).

Perishable foods may be carried in insulated porthole containers or in mechanically refrigerated containers. A clip-on refrigeration unit is attached to the insulated porthole containers during transportation on land, during waiting periods on the loading or unloading docks, and on board when stacked on the deck. However, porthole containers are generally carried below deck, where they are connected, via their portholes, to the ship's refrigeration system. Refrigerated air is then circulated, via the bottom porthole, through the container and returned to the ship's refrigeration system via the top porthole. In one of the most common arrangements of porthole containers they are stacked and connected to a vertical duct equipped with a fan and an air cooler capable of cooling the entire stack. The stack must be composed of produce with the same temperature requirement and compatible odor and respiration (International Institute of Refrigeration, 1995). The air coolers are generally fed by a multitemperature brine system.

Mechanically refrigerated containers are usually equipped with a recessed refrigeration unit. The unit composes the front of the container to permit on-deck stowage compatible with dry cargo containers. Many of the refrigeration units operate with a diesel engine during highway or rail travel and then are plugged on shipboard to operate with an electric motor (Ryall and Lipton, 1979). Mechanically refrigerated containers are generally carried on deck to ensure that there is sufficient air to remove the heat rejected by the refrigeration unit's condenser. Some recently built ships also provide space below the deck for mechanically refrigerated containers. These ships have additional air renewal below deck to take away the heat of the condensers (International Institute of Refrigeration, 1995). Mechanically refrigerated containers should preferably be carried on the center part of the deck, which is the part that is least sensitive to vertical acceleration (International Institute of Refrigeration, 1995).

Despite the precautions that are normally taken during loading (e.g., placement at midlength or at the stern of the ship and placement of mechanically refrigerated containers on the side to protect them best from sea spray), experience has shown that containers are, in some very exceptional cases, susceptible to damage (International Institute of Refrigeration, 1995). Failure of the defrosting system, bursting of pipes, and damage to the internal electric installations are the most frequently encountered problems. Preventative measures can be taken to save the produce if damage is limited. However, equipment is not always well maintained on board because of a lack of qualified personnel or lack of spare parts. Therefore, some ship owners prefer porthole containers with centralized refrigerating installations under the sole responsibility of the personnel on board. This arrangement makes transport more reliable and in the event of damage simplifies any litigation that might arise. Mechanically refrigerated containers are nevertheless suitable for marine transport and are in fact the most popular type of sea containers used because of their flexibility of use (International Institute of Refrigeration, 1995).

Proper air circulation through and around the load is as important in a container as in a conventional refrigerated ship or a transport trailer.

5.4 Intermodal Transport

Intermodal equipment allows produce to be transported on land, on sea, and again on land, with little handling between each mode of transportation. Since the goods are switched from one mode of transportation to the other, without breaking up the cargo, there is continuity in the cold chain and less required handling of the individual goods. This should result in reduced shipping costs and less damage to the produce.

Sea containers are the most common type of equipment used in intermodal transport (International Institute of Refrigeration, 1995). They can be loaded or unloaded outside port areas. On land, they can be transported on flatbed trailers or on railroad flatcars. However, the large capital cost of marine containers requires that their usage rate be relatively high. Other limitations of marine containers are the difficulty of gathering at a single location of enough cargo to fill a container or of enough containers requiring a similar final destination. Nevertheless, the degree of containerization is increasing on intercontinental maritime routes. For example, more than 90% of perishable foodstuffs carried on the North America–Western Europe routes are containerized (International Institute of Refrigeration, 1995).

Vehicles used for land transport, such as trucks, semitrailers, trailers, and railroad cars, can also travel on board RORO ships as long as they are equipped with a self-contained refrigeration system. These vehicles are either rolled on ramps or lifted vertically onto the ship. Immobilizing these vehicles on the ship is an expensive operation, and thus this form of intermodal transport is more suitable for short crossings between longer journeys on land (International Institute of Refrigeration, 1995).

Refrigerated semitrailers or trailers can be transported on railroad flatcars. However, only one trailer can be piggybacked on a flatcar at a time. Road-rail containers, also called *swap bodies*, can be carried on trailer chassis or on railroad flatcars. However, these containers must not be mistaken for sea containers. They are only used for bimodal transport that is on roads and railroads. These containers are not as sturdy as marine containers, therefore, they cannot be stacked when loaded. Road-rail containers are not suitable for transport in container ships. However, sea containers can be single- or double-stacked on flatcars when carried on railroads.

Intermodal transfers require standardization of container dimensions, fastenings and fittings for lifting on and off truck chassis, railroad flatcars, and ships' holds or decks. Uniformity of dimensions is particularly important in international movement of containers. Highway regulations, bridge capacities, underpass heights, and street widths vary considerably in different countries. If containers are not designed to meet these varying conditions, true intermodal transportation on an international basis is impossible (Ryall and Lipton, 1979). Containers belonging to one shipping company can often be used by another company. Therefore, equipment has to be interchangeable. To be truly substitutable, the mechanical and thermal specifications of the containers, as well as the dimensions, must be standardized (International Institute of Refrigeration, 1995).

Marine containers are standardized by the International Organization for Standardization (ISO) under reference 1496-2 of 1988. Containers are 2.438 m (8 ft) wide and vary in height from 2.438 m (8 ft) to 2.890 m (9 ft 6 in). The most common container lengths are 6.096 m (20 ft) and 12.192 m (40 ft). The mechanical resistance, airtightness, insulation, and, in certain cases, features of the refrigeration unit and electrical connections are all specified in the ISO standard (International Institute of Refrigeration, 1995).

5.5 Air Transport

Air transport is generally used to carry, over long distances, produce that has a short shelf life and therefore cannot easily be transported by road, rail, or sea. Airfreight can also be used when there is a shortage of a particular fruit or vegetable in the market or when exceptionally high prices are available for out-of-season produce.

Air transport is more expensive than road, rail, or marine transport; however, it gives a quicker return on capital that is tied up during transportation. Since the 1970s, the use of larger, more efficient aircraft has resulted in continually decreasing costs (per metric ton-kilometer) for airfreight of produce (Peleg, 1985). Nowadays, specially modified wide-bodied jets are dedicated specifically to the airfreight of fresh produce and general cargo. For example, a Boeing 747F freighter aircraft may have a payload of about 120 metric tons.

5.5.1 Cargo Flights versus Passenger Flights

Produce can be carried on either passenger flights or cargo flights. Generally, passenger flights are more regularly scheduled than cargo flights and, therefore, may be more convenient for produce with a short shelf life. However, on passenger flights, passengers and baggage are given a higher priority than cargo. Therefore, there may not always be room for perishable produce on these flights. Some airlines offer special services at a premium to guarantee the transport of perishable freight. The higher rate is necessary to compensate for the loss of valuable payload to accommodate the heavier thermal units and the redistribution of these units (International Institute of Refrigeration, 1995).

On passenger flights, freight can be carried on the same deck as passengers and on the lower deck. When both freight and passengers are on the main deck, the passenger section is in the front or rear of the aircraft. On cargo flights, freight is carried on both the main and lower decks.

5.5.2 Loading Devices

Loading and unloading of an aircraft's hold must be done rapidly since the time between the flight's arrival and its departure is short. To shorten the time required for loading and unloading, cargo is generally stacked in unit load devices (ULDs) well ahead of the time of loading.

Many sizes of ULD are available to fit the internal dimensions of various aircraft cargo holds. The most common sizes of the base of these devices are 2235 mm × 3175 mm (88 in × 125 in) and 2438 mm × 3175 mm (96 in × 125 in) (International Institute of Refrigeration, 1995). Each model of ULD has a specific gross weight it can hold. Half-size models generally hold 1500 kg of cargo; full size models can hold up to 6800 kg (International Institute of Refrigeration, 1995). ULDs are constructed to be lightweight, because weight is a major concern of aircraft. ULDs are therefore readily damaged and require special handling devices. Consequently, their use is generally restricted to the airfreight stage of the journey.

The two main categories of ULDs used are aircraft pallets and aircraft containers. However, in the newer wide-bodied cargo aircraft, intermodal containers can also be used.

5.5.2.1 Aircraft Pallets

Aircraft pallets are lightweight aluminum sheets about 5 mm thick, with special edge rail sections about 15 mm thick, to which the restraining hardware secures the pallet to the aircraft. The edge rails also incorporate slots for engaging the end hooks of the netting used to secure the cargo to the pallet (Peleg, 1985). Besides the netting, aircraft pallets are generally only covered with polyethylene film to protect against dust and rain.

Cargo must be stacked on aircraft pallets so that it does not exceed the contour of the hold. In the lower deck, cargo can extend out over the pallet itself to maximize space utilization. A pallet wing or pallet extension can be used in this arrangement to support the overhanging boxes (International Institute of Refrigeration, 1995).

Produce shipping containers can be stacked directly on the aircraft pallets or can be unitized on regular pallets or slipsheets beforehand (Peleg, 1985). In the first case, there is better volume utilization of the aircraft; however, the individual produce shipping containers must be handled more frequently. In the latter case, unit loads of produce can be loaded onto the aircraft pallet with a forklift, but some space is lost because of the presence of the regular pallets and gaps between them. The whole stack is also less stable.

Aircraft pallets are not necessarily handled by forklifts. They are generally transferred on either roller systems, ball mat, or inverted casters.

5.5.2.2 Aircraft Containers

Aircraft containers are built to fit the standard contours of aircraft holds and, therefore, maximize the utilization of the cargo holds. Containers used for the lower deck of an aircraft usually have an overhang, whereas those for the main deck are often shaped like igloos. Like aircraft pallets, containers are of lightweight construction and therefore must be supported from below. They must not be stacked nor moved with forklifts. Aircraft containers are mainly transferred on roller systems and are generally made of aluminum alloy.

5.5.2.3 Intermodal Containers

For door-to-door, superfast transport, intermodal 20- and 40-ft refrigerated containers may also be used (Peleg, 1985). Their heavier structure adds considerable weight to the payload of an aircraft. However, they have the advantage of reducing the number of times the individual produce shipping containers need to be handled and ensure that the load is held under refrigerated conditions before and after the flight. The refrigeration unit on the containers must not be operated in flight, since the heat generated by the unit's condenser cannot be easily disposed of in the pressurized cargo compartment. During relatively short flying times, the container insulation and the heat capacity of the produce usually prevent noticeable produce temperature rise (Peleg, 1985). Regular nonrefrigerated intermodal containers can be used if the transport logistics are well organized, that is, if the trucking and airplane loading times can be precisely coordinated (Peleg, 1985).

A Boeing 747F freighter aircraft can carry 5 40-ft containers (2438 mm × 12,192 mm) and 10 10-ft containers (2438 mm × 3048 mm), or 13 20-ft containers (2438 mm ×

6096 mm) and 4 10-ft containers (2438 mm × 3048 mm), or 28 aircraft pallets (2438 mm × 3175 mm) on its main deck.

5.5.3 Temperature Control on Aircraft

Aircraft cargo holds are pressurized similarly to passenger cabins. At high altitudes, outside air temperatures can be as low as −55°C. Once compressed to cabin pressure, the inside cargo air ranges in temperature from 0°C to 10°C. The air is then used to remove heat from the electronic instrument bay. Further heat can be supplied by the aircraft's heating system so that the temperature in the cargo hold is generally held between 15°C and 25°C (International Institute of Refrigeration, 1995). Certain wide-bodied aircraft now have temperature controlled holds that are suitable for perishable foods that require lower temperatures during the flight. At a normal cruising altitude of 10,000 m, there is sufficient capacity to cool the main cargo deck to approximately 7°C on a hot day (e.g., 38°C at sea level). Conversely, there is sufficient heating capability to maintain the compartment temperature at 25°C on an extremely cold day (−50°C at sea level) (Peleg, 1985).

The air taken into the aircraft is initially at a low temperature, and thus has a very low relative humidity (around 5%) (International Institute of Refrigeration, 1995). Also, at altitudes of about 10,000 m, the cabin pressure is around 600 to 650 mm Hg rather than 760 mm Hg at sea level. This may cause an increase of about 20% in the rate of water loss from the produce, compared to the loss on the ground at the same temperature and relative humidity (Peleg, 1985). Produce that is susceptible to moisture loss should therefore be protected by suitable packaging. Overwrapping pallet loads with a plastic film or tarpaulin cover generally takes care of this problem (Peleg, 1985).

Airlines generally segregate a load to ensure that produce with similar temperature requirements are loaded together (International Institute of Refrigeration, 1995). This is done in accordance with the standards set by the International Air Transport Association (IATA).

Mechanical refrigeration systems are not generally used on aircraft containers because they add weight and reduce the effective volume that can be used for cargo. Also, it is difficult to remove the heat rejected by the condensing unit of the system. Instead, some aircraft containers are built with insulated fiber-glass panels and doors to ensure that the perishable produce within is maintained at a lower temperature. However, these insulated or thermal containers are heavier, have smaller internal volume, are more expensive to repair, and are not used as often as conventional aircraft containers. Therefore, they are frequently returned empty.

General purpose aircraft containers can always be used for perishable fruits and vegetables. To maintain produce temperatures more efficiently, the containers can be lined with expanded polystyrene panels. However, the lining process can be time-consuming, since aircraft containers have an irregular shape. Another alternative is to cover the produce with flexible insulating material composed of gas-filled bubbles or polyethylene foam sandwiched between layers of metallized polyethylene. The insulating material used to line the container or to cover the produce in the container can be used for one or more trips. When produce is loaded on an aircraft pallet, covering the pallet with flexible insulating material slows the warming of produce.

Another alternative for shippers is to pack shipping containers of produce in fiberboard pallet boxes made of multiflute fiberboard, often with a moisture resistant coating. The walls of these boxes, which are generally 15 to 30 mm thick, provide some insulation and are laminated with metallized foil to reduce heat transfer. The main disadvantages of fiberboard pallet boxes are their cost and the problem of disposal once at the destination. There are two sizes of pallet boxes: small pallet boxes (e.g., 1056 mm × 1005 mm × 634 mm), which fit 8 to 12 per aircraft pallet and can carry up to 500 kg of produce, and larger pallet bins (e.g., 1960 mm × 1450 mm × 1530 mm), which fit 2 per aircraft pallet and carry up to 2000 kg of produce (International Institute of Refrigeration, 1995).

Dry ice can also be used to maintain cold temperatures in aircraft containers. To prevent produce from freezing and to ensure that the refrigeration provided by the dry ice is not localized, dry ice must be packaged and spaced above the cargo in the container. Also, the cargo must be spaced away from the walls of the aircraft container to permit the gas/air mixture to circulate by natural convection. This allows the heat entering through the walls of the container to be removed. Sublimating dry ice generates CO_2 gas, which at high concentrations is hazardous to the occupants of the aircraft. The airline must therefore be notified if dry ice is included in any consignment. Finally, frozen containers of chemical eutectic gel can also be used within shipping containers to maintain low temperature (McGregor, 1989).

Sharp (1989) suggests that for journeys lasting 24 hours or less, an insulated, nonrefrigerated aircraft container is adequate for maintaining recommended produce temperatures. The insulated container provides protection from freezing in subzero environments and retards heat gain from warm surroundings. However, for journeys of 48 hours or more, a source of refrigeration is required, such as dry ice. Sharp (1989) also found that container insulation significantly delays the rate of temperature rise in precooled produce, and without insulation the benefits of precooling are short-lived. Also, dry ice is of little benefit when used without insulation. Finally, insulation is of greater value to a load precooled to 0°C than to one less cooled. This is because a higher temperature load of produce has a higher initial respiration rate than that of a 0°C load. Therefore, the higher temperature load warms up more rapidly because of the higher respiration rate. Once the load has reached the surrounding temperature, the insulation actually slows the escape of the heat of respiration produced by the load.

5.5.4 General Recommendations

Although fresh produce cannot be mechanically refrigerated during air transport, temperature abuse often occurs before or after the airfreight segment of the journey. Ground handling can amount to over 70% of the total elapsed time from shipper to receiver (ASHRAE, 1998). To prevent, or at least limit, temperature abuse during air transport, the International Institute of Refrigeration (1995) has prepared a list of recommendations:

1. Precool produce to the desired temperature before shipping.
2. Use refrigerated transport to the airport.
3. Transfer boxes of produce onto aircraft pallets or containers under controlled temperatures.
4. Pack boxes of produce as densely as possible in the ULD.
5. If refrigerated storage is not available at the airport, protect the produce and hold it under cover from direct sun, rain, and all other climatic conditions that could damage it.
6. Use insulated aircraft containers whenever available, or cover pallets with flexible insulating material that has a weather-protecting surface, or use fiberboard pallet boxes.
7. Use, whenever possible, a temperature recorder that can be read at arrival.

8. Choose a direct flight to prevent transfers between aircraft.
9. If possible, choose a flight that arrives at a suitable time on a suitable day of the week for a direct transfer to the market.
10. Where possible, arrange customs and quarantine clearance in advance, to minimize delay for whoever is to collect the cargo at the destination.
11. Make sure that the shipment is collected as soon as possible from the destination airport in a temperature-controlled vehicle and transferred without delay to the market. If direct transport to the final destination is impossible, send the shipment to an intermediate cold storage.

6 CONCLUSIONS

The last section of the chapter presented the particularities of the four modes of transport. Equipment used for transporting fresh fruits and vegetables varies greatly among the various mode of transport. Nevertheless, no matter which mode of transport is selected, or which type of vehicle and control system is used, the key factors in preserving produce quality remain the same. To ensure optimal quality on arrival, only high-quality produce should be selected for transport. Then the produce has to be protected against temperature variations, water loss, harmful gases or volatiles, and physical injuries through the journey. These considerations are applicable throughout the entire distribution system for fresh fruits and vegetables.

REFERENCES

Ashby, B. H. 1995. Protecting Perishable Foods During Transport by Truck. Handbook No. 669. Washington D.C.: Transporation and Marketing Division, Agricultural Marketing Service, U.S. Department of Agriculture.

American Society of Heating, Refrigerating and Air-Conditioning Engineers. 1986. ASHRAE Refrigeration Handbook: Systems and Applications (SI). Atlanta: ASHRAE.

American Society of Heating, Refrigerating and Air-Conditioning Engineers. 1994. ASHRAE Refrigeration Handbook: Systems and Applications (SI). Atlanta: ASHRAE.

American Society of Heating, Refrigerating and Air-Conditioning Engineers. 1997. ASHRAE Fundamentals Handbook (SI). Atlanta: ASHRAE.

American Society of Heating, Refrigerating and Air-Conditioning Engineers. 1998. ASHRAE Refrigeration Handbook (SI). Atlanta: ASHRAE.

Guide to Food Transport—Fruit and Vegetables. 1989. Copenhagen: Mercantila.

Hardenburg, E. H., A. E. Watada, and C. Y. Wang. 1986. The Commercial Storage of Fruits, Vegetables, and Florist and Nursery Stocks. Agriculture Handbook 66. Washington D.C.: Agricultural Research Service, U.S. Department of Agriculture.

International Institute of Refrigeration. 1995. Guide to Refrigerated Transport. Paris: IIR.

Kadar, A. A. 1992. Postharvest technology of horticultural crops, 2nd ed. Publication 3311. Oakland: Division of Agriculture and Natural Resources, University of California.

Kasmire, R. F., R. T. Hinsch, and J. F. Thompson. 1996. Maintaining Optimum Perishable Product Temperatures in Truck Shipments. Postharvest Horticulture Series No. 12. Davis: University of California.

McGregor, B. M. 1989. Tropical Products Transport Handbook. Agriculture Handbook No. 668, rev. ed. Washington D.C.: Office of Transportation, U.S. Department of Agriculture.

Nicholas, C. J. 1985. Export Handbook for U.S. Agricultural Products. Agricultural Handbook No. 593. Washington D.C.: Office of Transportation, U.S. Department of Agriculture.

Peleg, K. 1985. Produce Handling, Packaging and Distribution. Westport, CT: AVI.

Perkins-Veazie, P., J. K. Collins, J. R. Clark, and L. Risse. 1997. Air shipment of ''Navaho'' blackberry fruit to Europe is feasible. HortScience 32(1): 132.

Ryall, A. L. and W. J. Lipton. 1979. Handling, Transportation and Storage of Fruits and Vegetables. Vol. 1. Vegetables and Melons. 2nd ed. Westport, CT: AVI.

Ryall, A. L. and W. T. Pentzer. 1982. Handling, Transportation and Storage of Fruits and Vegetables. Vol. 2. Fruits and Tree Nuts, 2nd ed. Westport, CT: AVI.

Sharp, A. K. 1989. Air transport of strawberries from Australia: requirements and possibilities. Food Australia 41: 755–760.

Thompson, J. F., F. G. Mitchell, T. R. Rumsey, R. F. Kasmire, and C. H. Crisosto. 1998. Commercial Cooling of Fruits, Vegetables, and Flowers. Publication 21567. California: Division of Agriculture and Natural Resources, University of California.

TransFRESH Corporation. 1988. Fresh Produce Mixer and Loading Guide. Salinas: TransFRESH.

Wark, K. 1988. Thermodynamics, 5th ed. Montreal: McGraw-Hill.

Welby, E. A. and B. M. McGregor. 1993. Agricultural Export Transportation Workbook. Agricultural Handbook 700. Washington D.C.: Transportation and Marketing Division, Agricultural Marketing Service, U.S. Department of Agriculture.

Wills, R., B. McGlasson, D. Graham, and D. Joyce. 1998. Postharvest: An Introduction to the Physiology and Handling of Fruit, Vegetables and Ornamentals, 4th ed. New York: CAB International.

21

Potential Applications of Volatile Monitoring in Storage

PETER ALVO, GEORGES DODDS, G. S. VIJAYA RAGHAVAN, and AJJAMADA C. KUSHALAPPA

McGill University, Sainte-Anne-de-Bellevue, Quebec, Canada

CRISTINA RATTI

Université Laval, Sainte-Foy, Quebec, Canada

1 INTRODUCTION

There are few technologies available to help storage managers detect disease in fruit, vegetable, and grain storage facilities. The only widespread technique is regular inspection by the manager, who uses visual, olfactory, and tactile observations to assess the condition of the stored produce (1). The principal indicators used by the human observer are: unpleasant odors emanating from decaying produce, condensation on the crop or the ceiling of the storage, excessive moisture on the floor or on the sides of pallet boxes, visible molds or diseased tissues, and, in the case of bulk storage such as for potatoes, collapse of the pile. However, significant losses have usually been incurred by the time the problem has been detected (2–4).

Temperature monitoring is essential to environment control systems, and the possibility of using temperature changes as indicators of the incidence of disease has been investigated (2,3,5,6). Efforts in this direction were based on the property that the heat of respiration of infected produce with active pathogens should be higher than that of uninfected produce, primarily because activity of the pathogen results in more rapid conversion of substrates than occurs in the metabolic activity of the stored, healthy produce. The detection of ''hotspots'' in bulk storage nevertheless requires a large number of sensing locations, particularly in a well-ventilated facility where the heat of respiration is rapidly dissipated (7). Temperature monitoring for environmental control purposes requires a sensing point for every 50–100 tonnes of potatoes (8); however, this is not considered adequate for early disease detection.

Thermal infrared scanning can be used to obtain a higher resolution of the temperature distribution at the surface of a pile (6), and this technique has been used successfully to detect disease hotspots in unventilated piles of sugar beets (9). Although the method has been commercialized in the United States with satisfactory results, disease detection methods based on temperature rise are not considered more effective than visual and olfactory signs since the disease is usually advanced before the thermal change is apparent at the surface of the pile (2,4,10).

An alternative to human observation and temperature monitoring, and that which is the focus of this chapter, are the trapping and quantification of the volatile compounds in the storage headspace by using modern analytical equipment with low detection thresholds. The hypothesis of this detection technique is that the volatiles that emanate from diseased or decaying produce are qualitatively and/or quantitatively different from those of the healthy produce. Many volatile compounds are associated with the normal metabolism of stored produce (e.g., production of aromas). The background includes substances released by the storage structure itself or infiltrating into the storage enclosure from the surroundings.

When a disease organism is present, the volatile profile includes normal metabolites of active pathogens, by-products of the interaction of the pathogen and its host, and the background. Thus, the headspace profile should be different when there are no disorders or infections than when such problems exist. Even dried or processed commodities are subject to attack by microorganisms, to chemical changes such as oxidation, or to residual enzymatic action, which can contribute to differences in the normal volatile profiles. Thus, the technology of monitoring volatiles is not limited to early disease detection in fresh produce; it also has potential in quality assessment at many points of agrifood production and marketing.

Although a headspace analysis system could also be subject to the problem of rapid localization of infection sites, as is the temperature approach, it should have the advantage of providing an indication of a specific problem, be it the type of pathogen or a particular storage disorder. A great deal of research needs to be done before volatile monitoring systems for application to the storage of any given type of produce can be commercialized. Although the predominant volatiles of many commodities have been identified, there may be differences due to natural factors such as stage of maturity, cultivar, and production and handling methods, including cultural practices and prestorage environment (11).

New storage technologies such as controlled atmosphere (CA) and hypobaric storage can alter the metabolism of the produce, e.g., ethylene metabolism in apples (12), and lead to quantitatively different volatile profiles than under normal atmosphere storage conditions. Qualitative differences may also occur (12); however, these may actually be quantitative (3). For example, compounds that are present in very small amounts may not be detectable at one temperature or for a given volume of airflow through an adsorbent trap, whereas at a higher temperature or filtered volume, they may appear as peaks in a chromatographic trace. There are also instances in which natural compounds are degraded if the gas chromatography (GC) apparatus is operated at high temperatures. These factors, and the nature of the adsorbent trap, must be taken into account in assessing the published literature.

This chapter reviews some relevant fundamental research on volatiles of agricultural commodities and of their storage pathogens and then assesses the overall potential of volatile monitoring for early disease detection on the basis of research more specifically directed toward that goal.

2 RESEARCH OBJECTIVES

The research needed for the eventual development of an early disease detection system for stored commodities requires two basic sets of information: (a) the volatile profiles of the commodity over time under normal circumstances and (b) changes in these profiles attributable to the proliferation of pathogens or to the onset of physiological disorders (e.g., chilling injury, anaerobiosis). Since these problems may be caused, to a certain extent, by breakdown of the environmental control system, disease detection and environmental control should be integrated.

The literature indicates that the volatile profile can depend on factors such as: variety (and even cultivar), characteristics of the production site, cultural practices, and type of chemicals used in production and handling. Those that determine the expected (or normal) profiles of a particular commodity include physiological age at harvest; prestorage treatments or processing; storage conditions (temperature, pressure, gas composition); evolution of the profile over the storage time due to processes such as ripening, change of substrate for respiration, and change of mode of respiration from aerobic to anaerobic (fermentation); and background profiles for the particular site.

One would expect particular changes in the normal profiles if a physiological disorder occurs or if a pathogen infects the produce. Physiological disorders cause different chemical activities within the produce from that in a ''normal'' situation and the volatiles resulting from the biochemical activity of a pathogen become superimposed. At advanced stages, the pathogen also interferes with the normal metabolism of the commodity. For example, it is well known that the development of anaerobic storage conditions leads to fermentation in commodities such as apples, in which case the anaerobic profile includes high concentrations of ethyl alcohol (11). The storage headspace of commodities that are subject to invasion by pathogenic microorganisms may exhibit pathogen-specific volatiles (2,3,13) or volatile profiles associated with the principal substrates attacked. If particular disorders or pathogens can be associated with characteristic changes in the profiles, the information can be used in risk assessment and can suggest appropriate remedial actions.

Assuming that differences in volatile profiles can be related to activity of pathogens and physiological state of the stored product, as is indicated in the literature, quantitative relationships among key factors must be analyzed to determine whether the chosen monitoring technology can provide a real benefit. First, the detection limits and magnitudes of change of the discriminant volatiles should be known relative to the extent of infection. Second, the relationship between the detectable extent of infection and the epidemiological characteristics of the disease must be assessed since it is possible that at detectable levels, the pathogen has already spread through or proliferated to the point at which extensive damage has already occurred such that investment in and operation of the system become a financial burden rather than a method of reducing losses. Since the symptoms of disease are likely to be initially localized in one or a few individuals, storage facilities with several tons of produce likely require multiple sampling points; the precise strategy implemented is dependent on quantities stored, epidemiological features of the pathogen, and air circula-

Table 1 Studies on Volatiles of Selected Aromatic Herbs and Spices

Herb or spice	References
Angelica	14
Apple flowers	15–17
Rooibos tea	18
Basil	19–23
Cardamom	24
Caucus	25
Ceratolia ericoides	26
Cloves	27
Coriander	28,29
Dill	30
Garlic	31,32
Ginger	33–37
Honeysuckle	35,38,39
Maté (*Ilex paraguayensis*)	40,41
Juniper berries	42–46
Lovage	35
Pepper (*Piper nigrum*)	47
Rice foliage	48
Rosemary	49
Tabasco peppers	50
Vetiver	51

tion needs. Thus, the implementation problem concerns the relationships among rates of release of volatiles, infectivity of the pathogen, airflow rates and circulation patterns, and detection limits for specific compounds.

Research concerning the characterization of volatiles of various commodities is extensive and dates back to the turn of the 20th century. Although the important role that the earlier works have had in laying the foundations for the many applications of volatile monitoring is recognized, a comprehensive review is far beyond the scope of this chapter. We have therefore chosen to provide lists of earlier references pertaining to specific groups of commodities in Tables 1 to 6 (14–325). Many of the listed references relate to identification of volatiles from fresh or dry material, in an effort either to elucidate biochemical pathways in general or to determine the factors involved in the flavor and aroma characteristics of commodities in particular and the reasons for their deterioration. Studies concerning differences among varieties and cultivars of major fruit and vegetable crops are to be found in the tables. This fundamental work has led to research aimed at improving the performance of the agrifood industry in various respects.

Sec. 3 reviews some recent relevant work concerning the study of volatiles in various contexts, with the aim of giving an idea of the scope of volatile monitoring as a technique and of the issues that one must consider in developing a disease detection technology. Sec. 4 focuses on disease detection in particular, and Sec. 5 evaluates the quantitative potential of headspace analysis for disease detection in commercial storage facilities.

Table 2 Volatiles of Temperate Fruit

Fruit	References
Apple	52–64
Apricot	65–73
Blackberry	74,75
Sour cherry	76
Cranberries	77
Grapes	78–85
Honeydew melon	86
Lemon/lime	87
Muskmelon/cantaloupe	88–98
Nectarine	99–101
Orange	102–112
Peach	67,100,113–117
Pears	58,118–127
Plum	67
Quince	128–131
Raspberry	132–148
Strawberry	56,149–160
Tangerine	161
Watermelon	93,162

Table 3 Volatiles of Tropical Fruit

Fruit	References
Babaco	163
Banana	164–174
Ceriman (*Monstera deliciosa*)	175
Cocoa	164
Guava	176
Kiwifruit	177
Mango	154,178–189
Papaya	57,190–205
Passion fruit	206–213
Pineapple	37,154,214–224

Table 4 Volatiles of Wild Fruit

Fruit	References
Arctic bramble	225–228
Bilberry	229
Black chokeberry	230
Cloudberry	228
Nectarberry	139,231
Strawberry	155

Table 5 Volatiles of Selected Vegetables

Vegetables	References
Broccoli	232
Bunching onions	32
Carrots	1,233–242
Caucus (*Allium victorialis*)	25
Celery	35
Chickpea	243–245
Cucumbers	246,247
Green pepper	248
Leeks	32
Onions	32,249–259
Pea	260–264
Peanuts	265–270
Popcorn (popped)	271,272
Potato	2,3,4,5,10,13,273–277
Sweet potato	273
Tomato	56,150,278–310

Table 6 Volatiles of Grain Crops

Grains	References
Review for cereals	311
Barley	312,313
Maize (*Zea mays*)	314–316
Oats	312
Pearl millet	317
Rice	318,319
Wheat	312,320–325

3 RECENT RESEARCH ON VOLATILES OF COMMODITIES

Color, flavor, and aroma development in fruit has been the object of many studies since the advantages of some harvest and postharvest strategies are sometimes accompanied by reductions in consumer acceptance. For example, strawberries picked at underripe stages, white or pink, have a longer shelf life and are less susceptible to damage during transportation than ripe fruit. However, once exposed to ripening temperature and light, they produce noticeably less aroma than ripe-picked berries and do not have the same taste, even though color may fully develop (326). These same authors found several quantitative and qualitative differences in the evolution of volatile profiles of Kent strawberries over a 10-day ripening period at 15°C.

A comprehensive study of the 30 major tomato volatiles indicated significant differences in profiles due to cultivar (4 tested), degree of ripeness when picked, and storage temperature (327). However, it was concluded that final volatile production was mainly related to final ripening temperature after storage, regardless of storage temperature. Nevertheless, final volatile production at full ripeness was lower in all instances than in those picked table-ripe.

Controlled atmospheres (CAs) of various gas compositions and temperatures are used in the long-term storage of various commodities. Since the main influence of CA storage is to reduce metabolic activity to a minimum by reducing the availability of O_2, production of volatiles, including respiratory CO_2, is usually lower than for the same commodity stored at the same temperature in regular atmospheres. The same is true for hypobaric storage. On the other hand, a study of McIntosh and Cortland apples (328) indicated increases in the production of certain volatiles under low-ethylene controlled atmosphere (LCA) storage (3% O_2, 3% CO_2) compared to storage in air at the same temperature (3.3°C), and that Cortland apples produce the same amount of total volatiles under LCA as in air.

Although storage and shelf lives of many fruits were substantially increased by technologies such as CA, LCA, and hypobaric storage, the observation that the fruits had less aroma and flavor spurred research into possible causes and remedies. Examination of volatile profiles has shown that although changes related to reduced aroma production are strikingly quantitative, some qualitative changes may also be induced (12). Other work (329) indicated that apple cultivars whose aroma precursors are derived from fatty acids (ester-type cultivars) are more susceptible to aroma suppression in CA storage than those that derive aroma precursors from amino acids (alcohol-type cultivars). However, suppression of aroma development is not necessarily due to the CA conditions per se, but to aging of the fruit, albeit at a slower rate than in normal atmosphere storage (328). Essentially, it is quite plausible that substrates necessary for aroma and flavor development are more depleted after prolonged CA storage than they would be in a short period of normal atmosphere storage, since the amount of substrate available at the end of the storage period is the initial concentration minus the time integral of depletion rate, D_r. Although D_r is smaller in CA conditions than in normal atmosphere, the time is longer. This could also explain the observations of many that even treatments meant to stimulate aroma and flavor production after CA do not succeed in creating the same levels of volatiles as those manifested by the air-ripened fruit.

Comparisons of volatiles generated under CA and other storage-extending techniques are nevertheless more complex than these summaries indicate. The outstanding difference between CA and normal atmosphere is in the rate of evolution of volatile profiles as a result of the much shorter life span of the commodity stored in normal air. Total volatile production tends to peak within a few days of the beginning of storage, in association with ripening and aroma production, whereas it drops rapidly as substrates are depleted and final senescence sets in. The CA-stored commodity, in general, exhibits a low production for a long time unless an O_2 deficit occurs, in which case one expects to see rises in anaerobic metabolites such as ethanol. If oxygen levels rise for some reason, then one should observe a rise in all aerobic metabolites, compared to those of the normal CA profile.

The CA concept is also applied to marketing of fresh produce with short shelf lives. Modified atmosphere (MA) packaging is a method of passive induction of low-oxygen conditions in the immediate environment of the produce. The permeability of the packaging material, whether integral or perforated, in conjunction with the respiration activity of the packaged commodity determines the O_2 levels in the package. Changes in respiration rate due to temperature changes can result in anaerobic conditions and the development of off-odors in species such as broccoli (232). These odors have been attributed to specific compounds (232), and efforts have been made to elaborate the mechanisms of formation (330) in order to improve product quality. Nevertheless, it is still unclear as to whether the compounds involved are the result of metabolic changes in the product or to the activity of microorganisms, even though the latter study was based on surface-sterilized material.

CA has also been used for storage of some dry goods, and quality has been assessed by analysis of volatile profiles. Volatile profiles of Assam black tea leaves packaged in different conditions have been reported (331). Samples were placed in standard tea pouches and lots of 40 were either vacuum packed at 20 mbar in polyethylene pouches laminated with 0.24-μm metallized polyester, packed in the same kind of pouches nitrogen-flushed to <1.0% O_2, or held in a standard atmosphere in 30-μm metallized polypropylene pouches. There were significant differences in volatile profiles after 16 and 48 weeks of storage. It was determined that 6 of 16 volatiles identified could serve as indicators of tea quality, and that the changes in these volatiles were consistent with oxidative deterioration, which could be reduced significantly by excluding oxygen from the storage pouches. It was also suggested that profile analysis could serve to distinguish the age of tea lots for the different packaging conditions. However, efforts to use headspace analysis as an indicator of storage time of roasted ground coffee led to the conclusion that such methods are only dependable if the temperature history of the product is known (332).

Along similar lines, it has been suggested that propanethiol S-oxide (PSO) can be used as a marker for freshness of stored scallions (333) on the basis of linear approximations of PSO production as a function of time by scallions stored at 0°C or 5°C. However, further work is needed to determine relationships between PSO levels and remaining fresh-quality lifetime. There have also been several studies showing the promise of hexanal, a principal volatile from the oxidative breakdown of fatty acids, as an indicator of stability of oils and oil-bearing grains and grain products, e.g., oats (334) and soybeans (335). Many commodities are subjected to chemical or heat treatments to improve their storability, to reduce the possibility of invasion by bacterial and fungal pathogens, and to prevent damage by insects. Pretreatments involving short-term storage in anaerobic conditions have recently been proposed as alternative measures of insect and pathogen control, and as techniques to enhance flavor and aroma production of fresh commodities. Since such pretreatments alter the volatile profiles, the observed trends in profile during storage may also be different from those in the untreated cases.

Various chemicals are currently used in the potato industry to inhibit sprouting during storage. Maleic hydrazide is a preharvest foliage spray used for this purpose. Isopropyl-N-(3-chlorophenyl)carbamate (CIPC) is an inhibitor that is applied as a thermal fog through the storage ventilation system after volatilization of the product and is normally suspended in a methanol or propylene glycol and HiSol 10 carrier which is a mixture of aromatic hydrocarbons. The volatiles of potatoes treated with each of these inhibitors were investigated in an attempt to determine why significant quantities of french fries made from Russet Burbank potatoes grown and stored in Manitoba had developed off-flavors and were unmarketable (336). Sensory evaluations showed that off-flavors were associated with potatoes that had been treated with the CIPC in propylene glycol and HiSol 10 carrier. Headspace GC profiles from the off-flavored and pleasantly flavored potatoes did not contain volatiles usually associated with earthy, musty flavors. Rather, it was determined that six known potato volatiles were present in concentrations 2 to 10 times higher than normal. All six were benzene derivatives and three of them are known to be the major components of the HiSol 10 formulation. The authors speculate as to the possible mechanisms of flavor changes associated with differences in concentrations of these compounds and not that aromatic compounds from fuel, pesticides, and decaying plant material may be accumulated by crops by adsorption.

Treatment of strawberries with acetaldehyde (AA) vapor has been shown to suppress the growth of *Botrytis cinerea* and *Rhizopus stolonifer* (337). Dipping in AA can improve color (338) and inhibits mold development (338,339). However, AA treatments also alter the volatile profiles of strawberries and many other fruits, primarily because this compound is metabolized when it diffuses into the tissues. The changes in the volatile profile depend on the concentration of AA used and the treatment duration (339). Moreover, the nature of the changes seems to depend on the biochemical pathways involved for each aroma or flavor constituent. The increase of certain volatiles may be linear with treatment time or concentration, whereas others may level off or even decrease after the initial rise over the same range of conditions. The authors reported the development of off-flavors in strawberries when AA treatment involved a high concentration (5000 $\mu L\ L^{-1}$) over a long duration (3–4 h). Although browning of the calyx also occurred under these conditions, there was a significant reduction in the percentage of decayed berries.

In a study of aroma enhancement in feijoa fruit by anaerobic treatment (340), it was found that the major aroma constituent, ethyl butyrate, was present in much higher concentrations in fruit kept under anaerobic conditions for 24 h prior to storage at 20°C for 13 days than in the control fruits. An atmosphere of 98:2 N_2:O_2 was the most effective gas combination in enhancing ethyl butyrate concentration. Treatment with AA also increased the ethyl butyrate concentration significantly over the first week of storage; however, the difference declined to nonsignificant levels after 13 days. Concentrations of ethanol (EtOH), AA, and ethyl acetate in the fruit varied in time and were usually significantly higher in the anaerobic conditions than in the controls. Anaerobic pretreatment has also been shown to increase volatiles associated with sensory quality of citrus fruits, their juices and oil extracts, e.g., oranges (341), with important implications for the citrus industry.

The possibility of using high-CO_2–low-O_2 treatments to eradicate insects from Thomson Seedless table grapes was studied (342). The treatments were conducted at 5°C and 20°C for up to 6 days, followed by storage for 10 days. Although a 100% mortality rate was achieved in 4.4 days at 20°C, anaerobically induced AA and EtOH accumulation was high, negatively affecting sensory evaluations. The CA treatment at 5°C took 6 days to achieve a 100% mortality rate but resulted in substantially lower accumulations of AA and EtOH, resulting in acceptability to the taste panel. Although accumulated EtOH and acetaldehyde should reduce to normal levels with time after restoring aerobic conditions, the authors attributed the sustained level of EtOH and slight increase in AA level to the low temperature (0°C) used for the rest of the storage period. Similar results had been obtained in an experiment to determine appropriate quarantine conditions for Bartlett pears (343). Although lethality studies on insects were not conducted, it was determined that the pears could tolerate anaerobic conditions for up to 10 days, with low O_2 concentrations varying from 0.25% to 1%.

Some studies provide evidence of the influence of the mode of thermal treatment on the volatile profiles of certain commodities. For example, it has been found that microwave heating of sweet potatoes significantly reduces the quantity of volatiles produced during subsequent baking, as compared to simply baking in a convection oven (344). The rapid initial heating with microwaves deactivates amylases before they can convert starch to maltose, the sugar that is precursor to many of the compounds in the profile. The storage stability of soybeans after thermal inactivation of enzymes by using microwaves has been studied (335). The heating time significantly influences the storage stability, as represented by the concentration of hexanal in the headspace of oil or meal. Microwave thermal inactivation for 4 or 6 minutes resulted in the best stability over an 8-week period, whereas 8-minute inactivation resulted in heat damage. Quality changes in stored raw and steam-treated oat products from three varieties have been assessed (334). Some flavor and aroma characteristics have been correlated with the headspace volatiles and found to depend on variety in some instances (334). This work indicated that increased storage stability resulted from heat treatment but also suggested that the conditions of heat treatment could influence the sensory characteristics by altering the volatile profile. The levels of hexanal in the heat-treated samples remained low over 42 weeks of storage, whereas hexanal concentrations associated with the raw samples increased rapidly after the 18th week. Both of the studies described support earlier conclusions by several workers that hexanal is a good indicator of the oxidative stability of lipids. This suggests that headspace analysis for this compound may be a useful quality control method for commodities containing significant oil fractions.

The pigments and main flavor volatiles of saffron that had been dried by seven different methods at several temperatures were studied (345). It was concluded that drying in a solar or oven dryer at 40°C ± 5°C was optimal, yielding a better product in terms of pigment and volatile retention than traditional sun or shade drying. The major volatile was 4-hydroxysafranal under vacuum and cross-flow drying; however, this component dehydrated to safranal and related flavor volatiles after several weeks of storage, resulting in a profile similar to that of the standard product.

In an effort to develop a better long-term storage strategy for preservation of aroma in fragrant rice, the volatile profiles of white rice, brown rice, and paddy rice were studied after 3-month storage followed by boiling (346). Storage temperature was 30°C with relative humidity at 84%, as is representative of conditions in the tropics. Pressure was atmospheric or 150-Pa vacuum pressure. Comparisons with freshly prepared white rice showed that the air-stored samples all developed aldehyde and ketone off-odors as a result of lipid oxidation. The authors attributed the higher aldehyde and ketone concentrations of the white rice to inhibition of oxidation by the additional layers present in the unmilled brown and unmilled, undehulled paddy rice. However, the volatile profile of the vacuum-packed white rice was the most similar to that of the fresh white rice. None of the storage strategies was able to reduce the 40%–60% loss of the main fragrance volatile, 2-acetyl-1-pyrroline. Not surprisingly, a number of volatiles were present in far greater concentrations in the paddy and/or brown rice than in the stored or fresh white rices.

The use of volatile insect pheromones to attract stored-product insects into traps and away from food products has been extensively studied (347). The association of stored-product insects with areas of fungal infestation are well known. Some volatiles from stored-product fungi appear to serve as insect attractants (348,349); however, the response of mites to various fungal extracts (ethanol, benzene, chloroform, or hexane) appears to be less than the response to the living fungus (349). Thus, a complex interaction exists between mite and fungus within the stored commodity, probably also involving moisture and CO_2 levels.

The volatiles produced by foreign grain beetles *Ahasverus advena* (Waltl) feeding on oats have been analyzed (350). Volatiles from headspace were captured on a trap (Porapak Q), eluted in pentane, and concentrated. Separation was performed on a GC (SP-1000) or a gas chromatography–mass spectrometry (GC-MS) (Carbowax 20M) coated open tubular glass column. The primary volatile compound detected was (*R*)-1-octen-3-ol (approximately 90%), along with (*Z*)-2-octenol, 1-nonanol, and (*Z*)-3-nonenol (approximately 10%). Male beetles produced roughly three times as much 1-octen-3-ol as female beetles. In a mixed population 1-octen-3-ol production decreased approximately

10-fold as density (beetles/gram oats) increased 10-fold. By using a pitfall olfactometer, it was shown that 1-octen-3-ol was a beetle attractant, whereas the other volatiles had no such effect, whether alone or in combination, and did not enhance attraction by 1-octen-3-ol. This suggested that 1-octen-3-ol is an aggregation pheromone for *A. advena*. Since 1-octen-3-ol is also produced by storage fungi (322), it may have a role in attracting beetles to deteriorating grain stores. Similar tests showed that 1-octen-3-ol elicits attraction responses from other cucujid beetles—the sawtoothed grain beetle, *Oryzaephilus surinamensis* (L.); merchant grain beetle, *O. mercator* (Fauvel); and rusty grain beetle, *Cryptolestes ferrugineus* (Stephens) (351–354). The compounds 3-octanol, 3-octanone, and 3-methyl-1-butanol identified from fungally infested wheat (321,323) were also attractive to the grain beetles (354). However, all these compounds, except 3-methyl-1-butanol, were repulsive to the squarenecked grain beetle, *Cathartus quadricolis* (Guér.).

Three fungal volatiles (1-octanol, 1-octen-3-ol, and 3-methyl-1-butanol) and one mite-produced volatile (tridecane) were tested on 1- to 4-week-old mixed sex adults of five species of stored-product beetles: *Tribolium confusum* J. du Val, *T. castaneum* Herbst., *Rhyzopertha dominica* Fabr, *Ahasverus advena* Waltl., and *Sitophilus granarius* Linn. (355). Tridecane had no effect on the development from the egg stage and subsequent multiplication of malathion-susceptible and resistant strains of *T. castaneum*. The fungal volatile 3-methyl-1-butanol was highly inhibitory to both processes, when applied at high doses. This compound also reduced oviposition by *T. castaneum* in the malathion-susceptible strain but increased oviposition in malathion-resistant strains.

Finally, there is some evidence that volatiles may be emitted by certain pathogens at particular stages of their life cycle (356). The sexual stage of *Puccina punctiformis* (Strauss), a rust of the Canada thistle, emits fragrance volatiles presumably to attract cross-fertilizing insects. Although it is not evident that storage fungi would necessarily emit different volatiles at various life stages, such activity would certainly help in determining the urgency of remedial measures.

3.1 Summary

The cited references represent but a small portion of the literature in this area. However, it is hoped that this short review has given a broad picture of the actual and potential applications of volatile monitoring (VM), with the exception of VM for disease detection, which is discussed in Sec. 4. There are clearly many factors involved in the production of volatiles by various commodities. The presence of some volatiles may be natural (e.g., aroma production only at ripening), whereas others are due to human input (application of chemicals) or particular storage conditions, and still others may be due to the activity of microorganisms in packaged commodities. Because there may be several hundred naturally occurring volatiles for certain commodities, and usually 30 or more major volatiles involved in aromas, these were not tabulated. Suffice it to say that naturally occurring volatiles of plant products do overlap to a certain extent with those related to microbial and fungal metabolisms.

4 DETECTION OF DISEASES BY VOLATILE MONITORING

4.1 Microorganisms

Characterization of bacteria by chromatography began by separation of cellular components by two-dimensional (2-D) paper chromatography (357,358), but these methods were reported to be poor in differentiating between genera and species. Some early GC work was done to distinguish bacterial strains on the basis of their pyrolysis products (359,360) or on the basis of their fatty acid methyl esters (361,362).

The first attempts to distinguish bacteria on the basis of their native volatile metabolites were made in the 1960s (363–369). Volatiles from ether extracts of several strains of seven *Bacillus* species and from *Escherichia coli*, *Aerobacter aerogenes*, and *Pseudomonas aeruginosa* grown in the same volatile-poor nutrient solution were analyzed by using flame ionization detection (FID) and electron capture detection (ECD) (363). Average peak areas for four replicates were ordered from largest to smallest, and differences in size tested statistically. These "signatures" were tabulated against genus and strain for both detection systems and led the identification of specific characters to distinguish between strains. The strains of the different *Bacillus* species were easy to distinguish because a large number of peaks were available for comparison, and in general, the genera that produced acetoin (*A. aerogenes* and *Bacillus* spp.) produced a greater number and total quantity of volatiles than the others. *E. coli* and *P. aeruginosa* produced few or no peaks, despite good growth. Low volatile production by *E. coli*, *Pseudomonas* spp., and *Alkaligenes* spp. was corroborated in other studies (363).

Different species and strains within a species of *Aerobacter aerogenes*, *Escherichia coli*, *Pseudomonas aeruginosa*, *Streptococcus faecalis*, and *Staphylococcus aureus* cultured on glucose-containing medium have been differentiated by comparison of signatures from gas-liquid chromatographic separation (364). Volatile metabolites from exponential phase cultures were also found to be different from those obtained from stationary phase cultures. In a study of dextrose broth cultures of several Enterobacteriaceae (*Escherichia coli*, *Escherichia freundii*, *Enterobacter aerogenes*, *Alkaligenes faecalis*, *Proteus vulgaris*, *Pseudomonas aeruginosa*, and eight species of *Salmonella*), *E. freundii* strains were distinguished from all other cultures by a distinct (uncharacterized) peak at 1.3 min, 10-fold greater than for any other organism tested (365). The chromatographic pattern from headspace could serve as a general test for the presence of Enterobacteriaceae in noninoculated foodstuffs that have very low initial bacterial loads (365).

Headspace vapors of sterile deodorized milk inoculated with either *Lactobacillus casei*, *L. acidophilus*, *Escherichia coli*, *Streptococcus lactis*, *S. diacetilactis*, *S. liquifaciens*, *S. thermophilus*, *Achromobacter lipolyticum*, *Aerobacter aerogenes*, *Pseudomonas fragi*, or *Escherichia freundii* were sampled and separated by GC (366,367). Headspace gases from deodorized milk cultures, obtained from 16 to 32 h after inoculation, were analyzed. The principal volatiles were acetaldehyde, diacetyl, ethanol, and methyl sulfide. Presence of coliform bacteria (*A. aerogenes*, *E. coli*, *E. freundii*) was characterized by high levels of ethanol, some acetaldehyde and methyl sulfide, and no diacetyl. On the other hand, *Ach. lipolyticum* and *S. diacetilactis* produced significant levels of diacetyl, little or no ethanol, and similar amounts of acetaldehyde to the coliforms. *L. casei* produced only acetaldehyde and diacetyl. Timing of volatile production also permitted differentiation of some of the other genera/species studied.

It has been shown that methyl sulfide is the volatile responsible for "cowy" or "feedy" milk and that this is produced by *Aerobacter aerogenes* (368). This organism was the only one common to several samples of defective milk obtained from different milk processing and distribution sites. The GC detection threshold and flavor threshold for methyl sulfide in milk were 50 ppb and 115 ppb, respectively. The methyl sulfide and titratable acidity both began rising when the culture entered the exponential growth phase (3 days). Organoleptic tests were only able to detect methyl sulfide after 4 days. Methyl sulfide levels decreased in the later stationary phase samples (days 5–7), at which point the other odor components, acetaldehyde, acetone, and ethanol, became more important, leading to a "doughlike" odor.

Volatiles from *E. coli*, *E. freundii*, and *E. intermedia*, *Aerobacter aerogenes*, and *Paracolobactrum aerogenoides* growing in milk have also been compared by headspace GC analysis (369). Peaks at 14.5 and 48 min were present in all but the *E. coli* inoculated cultures; these were tentatively identified as isobutyl alcohol and a carbonyl compound. Differences between the other organisms were based on peak height.

These and other techniques are still used to differentiate between microorganisms in culture or in food products (370–373). The pyrolysis-gas chromatography technique remains of questionable utility in distinguishing microorganisms (373). Nevertheless, gas chromatographic characterization of bacterial fatty acid methyl esters is still commonly used in the differentiation of microorganisms (374–376).

Although the preceding studies are not particularly encouraging with respect to detection of specific bacteria in a headspace analysis context, they do indicate that microorganic activity in general can be diagnosed.

4.2 Stored Cereals

The volatiles of cereals were reviewed by Maga (311). Here we review, more specifically, work connected with volatiles emitted from cereal storage pathogens and storage insects.

The headspace of coarse wheat meal infested with different mold organisms was sampled (318). Five *Aspergillus* species, four *Penicillium* species, and single *Alternaria*, *Cephalosporium*, *Fusarium*, and *Rhizopus* species all produced 3-methyl butanol, 3-octanone, and 1-octen-3-ol, whereas *Syncephalastrum nigricans* produced the former two, but not the latter compound.

Aspergillus flavus isolated from wheat grain was grown on sterilized wheat meal in order to investigate the volatiles involved in producing the musty odor characteristic of this infection (315,317). The main volatiles were 1-octen-3-ol and *cis*-2-octen-1-ol, as identified by infrared and mass spectrometry. These compounds were also found for *A. flavus*–infected corn, barley, oats, rice, soybean, and rape. Other *A. flavus*–specific volatiles on wheat included 3-methylbutanol, 3-octanone, 3-octanol, and 1-octanol. This work was repeated with four other *Aspergillus* species, five *Penicillium* species, and one species each of *Alternaria*, *Cephalosporium*, and *Fusarium* species, all isolated from wheat grain (316). In all but one *Aspergillus* species, 1-octen-3-ol accounted for over 60% of the total volatiles. The production of this compound (as percentage of total volatiles) by *Aspergillus ochraeus* was dependent on the medium. Growth on wheat meal, gluten, starch, and media containing plant oils gave 84%, 27%, 9%, and <1% of 1-octen-3-ol, respectively. This compound was also detected as being the major volatile produced by the mushrooms *Boletus edulis* and *Agaricus bisporus*.

Odor formation in moist wheat, barley, and oats has been investigated (312,313, 318). It was found that a complex population of fungi, including *Alternaria alternata* (Fr.) Keissler, *Penicillium* spp., *Eurotium* spp., and *Aspergillus* spp., developed over a 20-week period (312). Samples of intergranular air (400 ml) trapped using molecular sieve (Chromosorb 105), thermally desorbed, and analyzed showed that the major odor compounds from these fungi were 1-octanol, 3-octanone, and 3-methyl-1-butanol (330). Maximal levels were observed after 7 weeks, octanol was the main component, 3-octanone accounted for 5%–8% of total volatiles, and 3-methyl-1-butanol was either absent or present in trace (<1%) amounts. Volatile levels decreased substantially after 105 days of incubation. The total fungal propagule count rose initially, peaked at 2–4 weeks, then decreased before rising slowly through the remainder of the incubation. Thus the maximum in volatiles appears to have occurred in the lull between the initial growth burst of the fungi and the later gradual rise in fungal population. Similar volatile production patterns have been reported in stored barley (312), with greater emission levels at 20% moisture content (m.c.) than at 16% m.c. barley.

The fungal odor compounds 3-methyl-1-butanol, 1-octen-3-ol, and 3-octanone were monitored in experimental wheat storage bins (377). Wheats at roughly 15.6% and 18% moisture were stored in 3.66-m-high, 0.61-m-diameter cylinders, with or without ventilation through a perforated floor. Samples (355 g) were placed under a N_2 stream and volatiles trapped on 60/80 mesh (Chromosorb 105). Volatile levels were higher in the high-moisture wheat (18%) than in the lower-moisture wheat (15.6%). Whereas all three volatiles were detected in freshly harvested wheat (both moisture levels) from nonventilated bins, only the former was detected in ventilated bins. In general, volatile production was lower in ventilated bins of stored wheat (378). The main fungi present included *Alternaria alternata*, *Aspergillus repens*, and *Penicillium* spp. Correlations between biotic and abiotic factors were studied through simple correlation, principal component analysis, and stepwise regression analyses. Levels of 3-methyl-1-butanol, the principal volatile in 25% moisture content wheat, were correlated with bacteria, *Penicillium* spp., and *Fusarium* spp. In 20% moisture content this volatile was best correlated with infection by *Asp. glaucus* and bacteria. Levels of 1-octen-3-ol were correlated with populations of *Penicillium* sp. at 20% and 35% m.c.

Maize was stored for 20 days at 5°C or 22°C, at various humidities from 20% to 75%, resulting in seed moisture contents of 33%, 20%, and 15% (316). The high-humidity treatment was quickly colonized by natural mycoflora *Cladosporium* sp., *Verticillium* sp., and *Cephalosporium* sp. By day 3, storage mycoflora (*Penicillium* sp.) proliferated. Yeasts multiplied actively until day 3 and then remained constant. Samples of 100 g dry weight (d.w.) were taken at 0, 1, 3, 5, 9, 12, and 20 days; were ground; and volatiles extracted under vacuum and cold trapped at −78°C. Trapped material was extracted into methylene chloride, reduced in volume under N_2, and separated on a column (Carbowax 20M). After 3 days, volatile levels increased significantly, such that at 12 days' incubation at 22°C the following volatile compounds could be identified: ethylanisol, hexanal, 1-hexanol, 3-methyl-butanol, 1-pentanol, 3-octanol, 1-octen-3-ol, octanone, and 3-hydroxy-2-butanone. Volatile levels in the seed stored at 5°C were much lower; 3-methyl butanol, pentanol, hexanol, and 1-octen-3-ol were the major components. The fungal microflora also developed much more slowly. In particular, *Penicillium* sp. grew poorly. The compound 3-hydroxy-2-butanone appeared early in samples from 5°C and 22°C and was present throughout the incubation.

When 20% m.c. seed was stored, bacterial and yeast growth was almost nil, and natural mycoflora were greatly reduced; however, *Penicillium* sp. grew rapidly. Levels of the volatile 1-octen-3-ol rose sharply compared to those in storage of higher-moisture seeds; levels of other volatiles remained constant or rose slowly. Again, storage at 5°C retarded mycofloral growth and volatile production. Seed at 15% moisture showed no change in volatile levels from day 0 through 50, though some strains of *Aspergillus glaucus* were able to grow, albeit slowly. However, these inoculated seeds did show 1-octen-3-ol levels higher than those in noninoculated seeds.

The volatiles involved in the development of "mousy" odors in wetted/dried pearl millet were determined by GC-MS and infrared spectroscopy (317). The compound responsible for the off-odor in millet was determined to be 2-acetyl-1-pyrroline, an important "popcorn like" aroma in aromatic rice (379).

The same experimental setup was used to study volatiles associated with stored-product mite pests (380). Wheat stored at 15.2% moisture was infested with either *Acarus siro* (L.), *Aeroglyplus robustus* Banks, or *Lepidoglyphus destructor* (Schrank). Three volatiles were associated with mite-infested wheat: tridecane, neral, and geranial. The latter two are components of the mite pheromone citral and occurred at concentrations three- to fivefold lower than those of tridecane. In laboratory scale experiments in addition to lower level of tridecane, lower levels of perillen were also detected in mite-infested wheat. Tridecane did not occur in any *Acarus* sp. mite–free fungally infected wheat; however, tridecane levels were higher in coinfected than in uninfected wheat. Tridecane levels were significantly higher for a *Fusarium semitectum–Acarus* sp. association than those with *Alternaria alternata*, *Aspergillus repens*, *Fusarium moniliform*, and *Arthrobotrys* spp., although all did exhibit some tridecane production.

The presence of 2-methyl propanal and pentanal in the headspace vapors of ground kernels of triticale (*Triticum hexaploide* Lart.) and its parents, rye (*Secale cereale* L.) and durum wheat (*Triticum durum* Desf.), were determined after heating to 120°C for 2 h (320). Peaks were identified by comparison with known standards. It was found that these compounds were at intermediate levels in triticale as compared to its parents. A comparison of high-quality wheat, wheat heavily damaged by weevils (*Sitophilus* spp.) with weevils removed, and weevils themselves did not show any volatiles specific to weevil infestation, though quantitative differences in some volatiles were found between the clean and the damaged wheat.

The studies described indicate that the volatile profiles may depend qualitatively and/or quantitatively on the pathogen involved, on the types of substrates available to a given organism, and on interactions of pathogens and other factors.

4.3 Volatiles of Storage Pathogens of Fruits and Vegetables

Volatiles of lemons infected or uninfected by the common green lemon mold (*Penicillium digitatum*) were trapped by passing water-saturated air over samples and then through activated carbon traps (381). The volatiles were then heat desorbed under vacuum and cold trapped in ice salt, dry ice methylcellosolve, and liquid nitrogen for analysis by infrared gas and mass spectroscopy. Results showed a much greater diversity of compounds in the mold infected fractions.

Several workers have studied carrot volatiles through destructive means (1,239,242). In one study (242), carrots were sliced longitudinally or transversely, blended, or grated. Volatiles were trapped (Tenax GC) and separated on a (Carbowax 20M or Igepal CO-880) GC column. There were significant differences in volatile levels due to direction of slice (longitudinal or transverse) and levels of volatiles increased (particularly terpinolene, terpinen-4-ol, and myrcene) from tip to crown, and decreased (particularly terpinolene and caryophyllene) from the outside (phloem) to the inside (xylem). Profiles were also different according to whether the carrots had been sliced, blended, or grated. This finding emphasizes the importance of the method of preparation for volatile monitoring. Although prone to other influences, volatile monitoring of intact commodities is not subject to these pitfalls.

Direct headspace sampling was used to obtain volatile profiles from six cultivars of carrots that were stored in separate chambers for 1, 7, 16, or 26 weeks at 7.2°C, 95%

relative humidity (RH), and then diced (239). Twelve different peaks (A–L) were obtained. There were differences in peaks observed and magnitudes associated with storage time and cultivar. There were also interactions between cultivar and storage time. Although the peaks were unidentified, the ratio (A + B + 0.1E)/(0.1H + 0.1I + K) was found to be best correlated with storage duration.

The volatiles present in 7.7 L containers of healthy potatoes and potatoes infected with *Sclerotinia sclerotiorum* (Lib.) de Bary or *Botrytis cinerea* Pers. were collected in traps (Chromosorb 105) and analyzed after heat desorption, cryofocusing on a precolumn, 5% OV-101 on 60/80 mesh (Chromosorb-W), and separating in a column (Supelcowax 20M) (1). The *Sclerotinia* sp.–infected carrots produced somewhat higher levels of methyl(3-methylethenyl)-benzene than the uninfected carrots. This compound is associated with some essential oils and may be a natural compound associated with carrots. It was interesting to note that dichlorobenzene, a fumigant, was detected 16–20 days after the incubation began. The *Botrytis* sp.–infected carrots produced an (uncharacterized) volatile not present in the other samples, which decreased progressively in concentration as the incubation continued. These results indicate that the release of certain compounds may be assisted by pathogenic action, if it is not a direct result of conversion of substrates by the microorganisms.

The bulb volatiles of onion (*Allium cepa* L.), bunching onion (*A. fistulosum* L.), garlic (*A. sativum* L.), and leeks (*A. porrum*) may be involved in stimulating germination of sclerotia of the soil-borne pathogenic fungi *Sclerotium cepivorum* Berk (32). Headspace from chopped material was captured in a dry ice trap and subsequently separated on a column (Carbowax 1540). Leeks, which are significantly less affected by this disease, showed lower overall amounts of flavor (pyruvate) and germination-inducing aroma precursors, but higher relative levels of S-1-methyl-L-cysteine sulfoxide, a compound inactive in sclerotial germination.

A study of volatiles from shelled peanuts adjusted to 25% moisture, noninoculated or inoculated with either *Aspergillus parasiticus* or *A. flavus*, suggested that the presence of acetone is contingent on actively metabolizing *Aspergillus* sp. (266). Although acetone would not be expected to accumulate under ideal storage conditions, wet spots in bulk storage units were detected. In moldy peanuts the volatile acetone was present in concentrations 30 times the concentration in fresh peanuts; other volatiles remained at similar levels. Reducing moisture content to 6% removed this difference.

Sclerotial germination of *Sclerotium rolfsii* has been studied in the presence of healthy disks of sweet potato *Ipomoea batatas* (L.) or in the presence of disks infected with different storage rots: *Erwinia chrysanthemi* Burkholder, McFadden, and Dimock; *Ceratocystis fimbriata* Ell. & Halst.; or *Diplodia gossypina* Cooke (273). In the presence of decaying sweet potatoes, sclerotia underwent eruptive germination to a much greater extent than in the presence of healthy roots or of butanol, a known stimulant of eruptive germination. Furthermore, directional hyphal growth toward healthy tissues was enhanced by the presence of volatiles from decaying sweet potatoes.

In efforts to find the growth inhibiting volatiles of stored potatoes (cv. King Edward), compounds obtained from an ether extract of freeze-dried potato peels were separated (Carbowax 20M) and identified by mass spectrometry (276). These compounds were then tested for their ability to suppress sprouting in stored potatoes. The compounds 1,4- and 1,6-dimethylnaphthalene were the most potent sprouting inhibitors.

Cryogenic air liquefaction and subsequent concentration (TENAX) have been used to study the volatiles associated with soft rot (*Erwinia carotovora* var. *atroseptica*) of potato (cv. Chieftain) (4). Volatiles were separated (Porapak P). The infection-specific volatiles included ethanol, acetone, 2-butanone, and 3-hydroxy-2-butanone. In commercial storage sites, ethanol, acetone, and particularly 2-butanone were at much higher concentrations in bins of diseased potatoes than in bins of healthy potatoes, for both cv. Kennebec and cv. Russet Burbank. The compound 3-hydroxy-2-butanone was not detected in the commercial storage units (4).

The volatiles produced by the interaction of potatoes (cv. Russet Burbank) and the soft rot pathogen *Erwinia carotovora* var. *carotovora* stored in sealed plastic bags were trapped (Chromosorb 105) and separated by GC on a column (Porapak P) (3). The infection-specific compounds were methanol, propanal, 1-butanol, 2-butanol, and those reported in Ref. 4. Ethanol, methanol, and an unidentified compound were most characteristic of infection. It was demonstrated that the rate of aeration and temperature influenced the rate of volatile production and the nature of the compounds produced.

Volatiles of potatoes (cv. Russet Burbank) infected with either *Erwinia carotovora* (soft rot) or *Corynebacterium sepedonicum* (bacterial ring rot) were compared (2). It was found that soft rot developed much more quickly than ring rot and caused a 200-fold increase in total volatile levels over the first 5 days of storage. The volatiles were normal metabolites of potatoes but were present in very high concentrations. Levels of volatiles from ring rot infected potatoes only began to increase significantly after more than 5 days of storage. A single unidentified volatile peak was found to be specific to ring rot, and ring rot–associated volatiles did not include acetaldehyde, propionaldehyde, acetone, or three other compounds that were present in the headspace of soft rot infected potatoes. Besides those with ethanol, the dominant volatile peaks associated with soft rot were completely different from those reported in Ref. 4, suggesting that differences in test conditions and sampling/detection procedures must be taken into account. A comparison of two varieties of *Erwinia carotovora* (*E.c. carotovora* and *E.c. atroseptica*) showed that there were few qualitative differences in profiles, but several small quantitative differences in pathogen-host volatile signatures (5).

Volatile profiles from potatoes (cv. Atlantic) infected with either *Erwinia carotovora* var. *carotovora* or *Fusarium roseum* var. *sambucinum* (fusarium dry rot) were studied (13). Potatoes were inoculated and stored in plexiglass containers and volatiles were trapped (Chromosorb 105). After thermal desorption, the volatiles were cryogenically focused on a precolumn (OV-101) and then separated (Supelcowax 10). Many of the compounds reported earlier (3–5) were present to varying degrees as background. The compounds pentane and dimethyl disulfide were specific to both infections, and one additional unidentified compound was unique to fusarium dry rot infection.

5 ASSESSMENT OF VOLATILE MONITORING

The preceding literature indicates that volatile profiles of diseased commodities are different from those of the intact commodities. In some cases, specific compounds associated with pathogenic activity have apparently been identified; however, pathogen specificity is not clear in general since many types of pathogen produce similar volatile metabolites, and the method of collection and analysis can influence the results (382). Furthermore, none of the work expressly aimed at disease monitoring in storage has managed to quantify either detected levels or rates of emanation. Thus, the question remains as to whether or not techniques such as headspace analysis can be used in practice. This section attempts to shed light on this question. Since no quantitative data are available, certain assumptions are made and some operating conditions are discussed in terms of minimal detectable levels of low-molecular-weight organic compounds.

5.1 Assumptions

The first assumptions are that the monitoring system is based on trapping headspace volatiles with an adsorbent, and that the trapping ratio of the adsorbent is 1 for one or more volatile species (i.e., 100% of the volatile molecules entering the trap are adsorbed). In other words, the adsorbent is not saturated during the collection period. On the basis of discussions with industry specialists, a reasonable assumption for the minimal detectable quantity of a low-molecular-weight volatile trapped in a suitable adsorbent is of the order of 1 ng (10^{-12} kg). A reasonably distinguishable chromatographic peak should be obtained if 100 ng (10^{-10} kg) is trapped, thermally desorbed, and identified with a flame ion detector (FID).

Several questions regarding concentrations of volatiles now arise. First, if there is a pathogen-specific compound emanated, what percentage of the total volatile spectrum would it account for? It is assumed that the compound to be detected accounts for 1%, by mass, of the volatile emanations from diseased individuals, and that its molecular weight is close to the average of all the molecular species emanated, not including CO_2. Second, the question arises as to whether the rate of emanation (in kilograms per kilogram per day [kg kg^{-1} day^{-1}]) changes as the infection progresses. Since the number of infected individuals should increase in time, it is assumed that at some time after initial infection, there are a sufficient number infected that various stages of infection are represented and a constant emission rate may be assumed.

Suction rate through traps can be assumed to be about 50 L h^{-1}, per trap, on the basis of equipment tested in a commercial potato storage (383). This is roughly equivalent to 1 kg day^{-1} airflow, per trap. For a compound of molecular weight roughly 100 to be collected in sufficient amounts to be clearly detected (100 ng) in one trap, it would have to be present in air in a concentration of roughly 5×10^{-7} ppm, if collection were done over a full day in a sealed room. In a ventilated storage, the detectable concentration would have to be several times higher, depending on the number of air changes per day as mitigated by the proximity of the trap to the site of emanations. In a large storage room, a grid of traps would be necessary to permit determination of the position of the infection locus.

The size of the storage facility, the mass of produce stored, the level of infection, the emanation rate of volatiles, the free air space, and the ventilation requirements all come into play in evaluating the feasibility of volatile monitoring as a disease detection technique. The less free volume there is, the greater the concentration of a volatile emitted by a given mass of infected produce. Furthermore, since there is more produce when the free air space is small, this greater concentration is representative of a lower percentage of infected produce even though the infected mass is the same. The unknown at this point is the rate of emanation of volatiles.

5.2 Estimation of Volatile Emanation Rates of Potatoes

The respiration rate of potatoes stored at 5°C–10°C is in the range of about 108 to 132 mg CO_2 kg^{-1} day^{-1} (384). Although volatile emanation rates have never been quantified for stored potatoes, whether healthy or diseased, they can be estimated on the basis of data on apples and pears obtained by cerate oxidimetry (58). Respiration rates of apples stored at −0.5°C were about 80 mg CO_2 kg^{-1} day^{-1}, or slightly lower than potatoes at 5°C–10°C. Nonethylene volatile emanation rates for apples were reported in terms of milligrams $Ce(SO_4)_2$-reduced kg^{-1} day^{-1}; rates ranged from 1 at the beginning of storage to 5 for McIntosh apples and 25 for Golden Delicious by 175 days. Assuming an average molecular weight of volatiles of 100 and a stoichiometric ratio of 2.5:1 in the reduction reactions, the volatile production rates ranged from 0.1 to about 3 mg kg^{-1} day^{-1}, or of the order of 3% of the CO_2 emanations. When fruits were removed from storage and allowed to ripen at 18°C, respiration increased by 5 to 6 times and production of nonethylene volatiles increased 10- to 40-fold, depending on variety, by the time decay was visible. Given that these rates were obtained at 18°C, the increases would have been perhaps only 5- to 10-fold at 5°C to 10°C, assuming a Q_{10} of 2. It may therefore be reasonable to assume that the volatile emanation rates from stored healthy potatoes are about 3–4 mg kg^{-1} day^{-1} and that with the onset of disease at low temperatures they rise to the order of 15 to 40 mg kg^{-1} day^{-1}.

5.3 Case Study

With this background, one can assess the possibility of disease monitoring by headspace analysis using adsorbent traps. The preceding estimates are used as a starting point and applied to a commercial potato storage facility in Quebec where some preliminary work has been conducted (383). The storage facility consisted of several enclosures, one of which contained stacked crates of potatoes. The enclosure held 900 crates, each containing the contents of 30 22.73-kg bags. Volume and mass characteristics are presented in Table 7. The sample calculation given in the table show that, assuming an emission rate of 15 mg kg^{-1} day^{-1} for diseased potatoes, it is possible to collect a detectable quantity of a disease-specific volatile in at least 1 of 20 traps set out in a grid, if the infection level is 1% and the disease specific volatile accounts for 1% of the total volatiles, excluding CO_2. Assuming that the volatiles are instantaneously and homogeneously distributed about the

Table 7 Actual and Assumed Values Used in Estimating Detectability of Volatiles in a Potato Storage Warehouse

Volume of warehouse	1,728 m^3
Mass of potatoes	613,636 kg
Density of potatoes	1,100 kg m^{-3}
Volume occupied by potatoes	558 m^3
Free air volume	1,170 m^3
Emission rate (assumed)	0.000015 kg kg^{-1} day^{-1}
Suction rate through trap	1 kg day^{-1}
Sampling time	1 h
Density of air	1 kg m^{-3}
Trapping efficiency (ratio)	1.0
Proportion infected potatoes	1%
Number of traps	20
Total volatiles emitted	0.092 kg day^{-1}
Proportion of volatiles in free air	1%
Number of air changes per day	5
Trapped emissions per day	1.3×10^{-6} kg
Volatiles adsorbed per trap	6.6×10^{-8} kg
Specific volatile (proportion of total)	1%
Specific volatile adsorbed per trap	6.6×10^{-10} kg
Detection limits of instrument	1.0×10^{-12} kg

room, each of the 20 traps should adsorb about 0.6 μg of the volatile in an hour of sampling. This is two orders of magnitude greater than the level required for an easily observable peak (100 ng) and four orders of magnitude greater than the detection limit. As can be seen in the table, five air changes per day are assumed, giving a dilution factor of 0.2 due to ventilation. That is, only one-fifth of the volatiles emanated over a given period are assumed to be in the room at the time of sampling; the rest are pushed out by ventilation air or diffusion through leaks in the room. A proportion of 90% has been allowed for adsorption/absorption of emitted volatiles by other potatoes or materials in the room (wood from the crates, walls, etc.).

Even with such reasonably conservative estimates, it is clear that headspace analysis has the potential to provide earlier warning of disease presence than present procedures. Had the emission rate been overestimated by a factor of 10, such a system should still be able to detect the disease-specific volatile or set of volatiles accounting for 1% of the total emissions. The warehouse manager estimated that in years when there are disease problems losses range from 5% to 20%. Since this particular warehouse buys potatoes from producers in both Quebec and Ontario, the manager was interested in knowing whether this type of system could give information as to which of several lots in the room at a given time could be experiencing difficulty. If the varieties are stacked in sections, then analysis of the volatiles over the 20-trap grid should provide location information, which the manager would then use to check a specific lot.

Since potatoes are usually stored in piles rather than crates, with ventilation provided through slots in the floor, positioning the infection loci will be more difficult unless provisions for positioning sampling tubes within the pile are made. Measurements and analysis can be simplified by sampling during the periods between ventilation cycles. In such a situation, the analysis of Table 1 could be adjusted to accommodate negligible dilution with a lower quantity of volatiles present. For example, if one assumes that the room has been completely replenished with fresh air at the end of a cycle, 4 hours later, the quantity of volatiles present is 1/6 of the quantity in the table. However, since air changes due to leakage during the 4-h period should amount to only a fraction of a complete air change, the dilution factor should be small enough to cancel out the effect of having a lower quantity of volatiles produced in the shorter period. Thus, it seems likely that a specific volatile or volatiles should still be fairly easily detected with a network of sampling tubes within the pile.

These rather straightforward analyses nevertheless provide a good indication as to the viability of the volatile monitoring technique for detecting disease in storage.

ACKNOWLEDGMENTS

The authors wish to acknowledge the editorial suggestions of one of the pioneers in volatile monitoring for detection of disease in storage, Dr. M. K. Pritchard of the University of Manitoba. The authors also gratefully acknowledge financial support from the Natural Sciences and Engineering Research Council (NSERC) and the Conseil des recherches en pêche et en agroalimentaire du Québec (CORPAQ) for past and ongoing research conducted at Macdonald Campus of McGill University. Finally, the efforts of the many researchers who have contributed directly and indirectly to this area of research over the century are sincerely acknowledged.

REFERENCES

1. Ouellette, E., Raghavan, G. S. V., Reeleder, R. D. Volatile profiles for disease detection in stored carrots. *Can. Agric. Eng.* 1990, *32*, 255–261.
2. Waterer, D. R., Pritchard, M. K. Monitoring of volatiles: A technique for detection of soft rot (*Erwinia carotovora*) in potato tubers. *Can. J. Plant Pathol.* 1984, *6*, 165–171.
3. Waterer, D. R., Pritchard, M. K. Volatile monitoring as a technique for differentiating between *E. carotovora* and *C. sepedonium* infections in stored potatoes. *Am. Potato J.* 1984, *61*, 345–353.
4. Varns, J. L., Glynn, M. T. Detection of disease in stored potatoes by volatile monitoring. *Am. Potato J.* 1979, *56*, 185–197.
5. Waterer, D. R., Pritchard, M. K. Production of volatile metabolites in potatoes infected by *Erwinia carotovora* var. *carotovora* and *E. carotovora* var. *atroseptica. Can. J. Plant Pathol.* 1985, *7*, 47–51.
6. Hyder, L., Cacka, J., Brackett, S. Detecting rot with infrared. Spudman 1984, *4*, 14–15.
7. Rastovski, A., van Es, A., et al. *Storage of Potataoes: Post-Harvest Behaviour, Store Design, Storage Practice, Handling*. Centre for Agricultural Publishing and Documentation, Waganingen, Netherlands, 1981.
8. Statham, O. J. H. Better storage design. Agric. Eng. 1983, *2*, 46–48.
9. Remote sensing: A new twist sends old techniques soaring. Co-Op County News, April 17, 1978.
10. Schapper, L. A., Varns, J. L., Glynn, M. T., Medal, D., Hilley, J. D. A computerized gas sampling and analysis system for potato storages. *ASAE paper* 84-5532, 1984.
11. Mattheis, J. P., Buchanan, D. A., Fellman, J. K. Change in apple fruit volatiles after storage in atmospheres inducing anaerobic metabolism. J. Agric. Food Chem. 1991, *39*, 1602–1605.
12. Bangerth, F., Streif, J. Effect of aminoethoxyvinylglycine and low-pressure storage on the post-storage production of aroma volatiles by Golden Delicious Apples. J. Sci. Food Agric. 1987, *41*, 351–360.
13. Ouellette, E., Raghavan, G. S. V., Reeleder, R. D., Greenhalgh, R. Volatile monitoring technique for disease detection in stored potatoes. *J. Food Process. Preservation* 1990, *14*, 279–300.
14. Nykänen, I., Nykänen, L., Alkio, M. Composition of angelica root oils obtained by supercritical CO_2 extraction and steam distillation. *J. Essent. Oils Res.* 1991, *3*, 229–236.
15. Bicchi, C., Joulain, D. 1990. Headspace-gas chromatographic analysis of aromatic plants and flowers. *Flavour Fragrance J.* 1990, *5*, 131–145.
16. Buchbauer, G., Jirovetz, L., Wasicky, M., Nikoforov, A. Headspace and essential oil analysis of apple flowers. *J. Agric. Food Chem.* 1993, *41*, 116–118.
17. Loughrin, J. H., Hamilton-Kemp, T. R., Anderson, R. A., Hildebrand, D. F. Volatiles from flowers of *Nicotiana sylvestris, N. Otophora*, and *Malus* × *Domestica* headspace components and day/night changes in their relative concentrations. *Phytochemistry* 1990, *29*, 2473–2477.
18. Habu, T., Flath, R. A., Mon, T. R., Morton, J. F. Volatile components of Rooibos tea (*Aspalathus linearis*). *J. Agric. Food Chem.* 1985, *33*, 249–254.
19. Gaydon, E. M., Faure, R., Bianchi, J.-P., Lamaty, G., Rakonotonirainy, O., Randriamiharisoa, R. Sesquiterpene composition of basil oil: Assignment of the ^{1}H and ^{13}C NMR spectra of β-elemene with two dimensional NMR. *J. Agric. Food Chem.* 1989, *37*, 1032–1037.
20. Nykänen, I. High resolution gas chromatographic mass spectrometric determination of the flavour composition of basil (*Ocimum basilicum* L.) cultivated in Finland. *Z. Lebensm. Unters. Forsch.* 1986, *182*, 205–211.
21. Sharma, A., Tewari, R., Virmani, O. P. French basil (*Ocimum basilicum* L.), a review. *Curr. Res. Med. Aromat. Plants* 1987, *9*, 136–151.
22. Sheen, L.-Y., Ou, Y.-H. T., Tsai, S.-J. Flavor characteristic compounds found in the essential oil of *Ocimum basilicum* L. with sensory evaluation and statistical analyses. *J. Agric. Food Chem.* 1991, *39*, 939–943.

23. Vernin, G., Metzger, J., Vernin, G., Fraisse, D., Suon, K. N., Scharff, C. Analysis of basil oils by GC-MS data bank. *Perfum. Flavor.* 1984, *9*, 71–86.
24. Gopalakrishnan, N., Narayanan, C. S. Supercritical carbon dioxide extraction of cardamom. *J. Agric. Food Chem.* 1991, *39*, 1976–1978.
25. Nishimura, H., Fujiwara, K., Mizutani, J., Obata, Y. Volatile flavor components of Caucus. *J. Agric. Food Chem.* 1971, *19*, 992–994.
26. Jordan, E. D., Hsieh, T. C.-Y., Fischer, N. H. Volatile compounds from *Ceratolia ericoides* by dynamic headspace sampling. *Phytochemistry* 1992, *31*, 1203–1208.
27. Gopalakrishnan, N., Shanti, P. P. V., Narayanan, C. S. Composition of clove (*Syzygium aromaticum*) bud oil extracted using carbon dioxide. *J. Sci. Food Agric.* 1990, *50*, 111–117.
28. Hirvi, T., Salovaara I., Oksanen, H., Honkanen, E. Volatile constituents of coriander fruit cultivated at different localities and isolated by different methods. In Brunke, E.-J., ed., *Progress in Essential Oil Research.* deGruyter: Berlin, 1986, pp. 111–116.
29. Kerrola, K., Kallio, H. Volatile compounds and odor characteristics of carbon dioxide extracts of coriander (*Coriandrum sativum* L.) fruits. *J. Agric. Food Chem.* 1993, *41*, 785–790.
30. Brunke, E.-J., Hammerschmidt, F.-J., Koester, F.-H., Mair, P. Constituents of dill (*Anethum graveolens* L.) with sensory importance. *J. Essent. Oils Res.* 1991, *3*, 257–267.
31. Brodnitz, M. H., Pascale, J. V., Van Derslice, L. Flavor components of garlic extract. *J. Agric. Food Chem.* 1971, *19*, 273–275.
32. Coley-Smith, J. R. A comparison of flavour and odour components of onion, leek, garlic and *Allium fistulosum* in relation to germination of sclerotia of *Sclerotium cepivorum. Plant Pathol.* 1986, *35*, 370–376.
33. Chen, C.-C., Kuo, M.-C., Wu, C.-M., Ho, C.-T. Pungent compounds of ginger (*Zingiber officinale* Roscoe) extracted by liquid carbon dioxide. *J. Agric. Food Chem.* 1986, *34*, 477–480.
34. Chen, C.-C., Ho, C.-T. Gas chromatographic analysis of volatile components of ginger oil (*Zingiber officinale* Roscoe) extracted with liquid carbon dioxide. *J. Agric. Food Chem.* 1988, *36*, 322–328.
35. DePooter, H. L., Coolsaet, B. A., Dirinck, P. J., Schamp, N. M. GLC of the headspace after concentration on Tenax GC and of the essential oils of apples, fresh celery, fresh lovage, honeysuckle and ginger powder. In Baerheim Svendsen, A., Scheffer, J. J. C., eds., *Essential Oils and Aromatic Plants.* Martinus Nijhoff/Dr. W. Junk: Dordrecht, Netherlands, 1985, pp. 67–77.
36. Sakamura, F. Changes in volatile constituents of *Zingiber officinale* rhizomes during storage and cultivation. *Phytochemistry* 1987, *26*, 2207–2212.
37. Wu, P., Kuo, M.-C., Ho, C.-T. Glycosidically bound aroma compounds of ginger (*Zingiber officinale* Roscoe). *J. Agric. Food Chem.* 1990, *38*, 1553–1555.
38. Hedin, P. A., Phillips, J. A., Dysart, R. J. Volatile constituents from honeysuckle aphids, *Hyadaphis tataricae*, and the honeysuckle, *Lonicera* spp.: Search for assembling pheromones. *J. Agric. Food Chem.* 1991, *39*, 1304–1306.
39. Wu, Y. L., Fang, H. J. Constituents of the essential oil from the flowers of *Lonicera japonica* (Thunb.). *Hua Hsieh Hsieh Pao* 1980, *38*, 573–580.
40. Kawakami, M., Kobayashi, A. Volatile constituents of green mate and roasted mate. *J. Agric. Food Chem.* 1991, *39*, 1275–1279.
41. Kubo, I., Muroi, H., Himejima, M. Antibacterial activity against *Streptococcus mutans* of Mate tea flavor components. *J. Agric. Food Chem.* 1993, *41*, 107–111.
42. Gebomini, N., Vidrich, V., Fusi, P., Michelozzi, M. Capillary gas chromatography of the terpenic fraction of *Juniperus communis* L. black, green, berry and leaf extracts. *J. High Resolution Chromatogr. Chromatogr. Commun.* 1988, *11*, 218–220.
43. Hörster, H., Csedo, C., Racz, G. Gas chromatographic analysis of volatile oil from ripe and

unripe juniper fruit (*Juniperus communis* L.) harvested in Romania. *Rev. Med.* 1975, *20*, 215.

44. Kallio, H., Junger-Mannermaa, K. Maritime influence on the volatile terpenes in the berries of different ecotypes of juniper (*Juniperus communis* L.) in Finland. *J. Agric. Food Chem.* 1989, *37*, 1013–1016.
45. Nykänen, I., Nykänen, L., Alkio, M. The composition of the flavour of juniper berries isolated by supercritical CO_2 extraction and steam distillation. In Bessiere, Y., Thomas, A. F., eds., *Flavour Science and Technology*. Wiley: Chichester, England, 1990, pp. 217–220.
46. Taskinen, J., Nykänen, L. Volatile constituents of an alcoholic extract of juniper berry. *Int. Flavours Food Addit.* 1976, *7*, 288, 233.
47. Sankar, K. U. Studies on the physicochemical characteristics of volatile oil from pepper (*Piper nigrum*) extracted by supercritical carbon dioxide. *J. Sci. Food Agric.* 1989, *48*, 483–493.
48. Hernandez, H., Hsieh, T. C.-Y., Smith, M., Fischer, N. Foliage volatile of two rice cultivars. *Phytochemistry* 1989, *28*, 2959–2962.
49. Reverchon, E., Senatore, F. Isolation of rosemary oil: Comparison between hydrodistillation and supercritical CO_2 extraction. *Flavour Fragrance J.* 1992, *7*, 227–230.
50. Ingham, B. H., Hsieh, T. C.-Y., Sundstrom, F. J., Cohn, M. A. Volatile compounds released during dry afterripening of Tabasco pepper seeds. *J. Agric. Food Chem.* 1993, *41*, 951–954.
51. Nikoforov, A., Buckbauer, G., Jirovetz, L., Remberg, B., Remberg, G. Headspace constituents of vetiver oil. *Z. Naturforsch.* 1991, *47B*, 439–440.
52. Carelli, A., Lozano, J. E. Apple aroma from Argentina: Quality evaluation by capillary gas chromatography. *J. High Res. Chromatogr.* 1989, *12*, 488–490.
53. Cunningham, D. G., Acree, T. E., Barnard, J., Butts, R. M., Braell, P. A. Charm analysis of apple volatiles. *Food Chem.* 1986, *19*, 137–147.
54. DePooter, H. L., Montens, J. P., Willaert, G. A., Dirinck, P. J., Schamp, N. M. Treatment of Golden Delicious apples with aldehydes and carboxylic acids: Effects on the headspace composition. *J. Agric. Food Chem.* 1983, *31*, 813–818.
55. DePooter, H. L., Schamp, N. M. The study of aroma formation and ripening of apples cv. Golden Delicious by headspace analysis. *J. Essent. Oils Res.* 1989, *2*, 47–56.
56. Dirinck, P., Schreyen, L., Schamp, N. Aroma quality evaluation of tomatoes, apples, and strawberries. *J. Agric. Food Chem.* 1977, *25*, 759–762.
57. Flath, R. A., Black, D. R., Guadagni, D. G., McFadden, W. H., Schultz, T. H. Identification and organoleptic evaluation of compounds in Delicious apple essence. *J. Agric. Food Chem.* 1967, *15*, 29–35.
58. Gerhardt, F. Rate of emanation of volatiles from pears and apples. *Proc. Am. Soc. Hortic. Sci.* 1954, *64*, 248–253.
59. Grevers, G., Doesburg, J. J. Gas chromatographic determination of some volatiles emanated by stored apples. *Sci. Technol Commun. Int. Fed. Fruit Juice Prod. Rep.* 1962, *4*, 319–321.
60. Grevers, G., Doesburg, J. J. Volatiles of apples during storage and ripening. *J. Food Sci.* 1965, *30*, 412–415.
61. Iwamoto, M., Takagi, Y., Kogami, K., Hayashi, K. Synthesis of characteristic flavor constituents of Kogyoku apple (Jonathan) essential oil. *Agric. Biol. Chem.* 1983, *47*, 117–119.
62. Olias, J. M., Sanz, L. C., Rios, J. J., Perez, A. G. Inhibitory effect of methyl jasmonate on the volatile ester-forming enzyme system in Golden Delicious apples. *J. Agric. Food Chem.* 1992, *40*, 266–270.
63. Petro-Turza, M., Szarfildi-Szalma, I., Maclarassy-Mersich, E., Teleky-Vamossy, G., Füzesi-Kardos, K. Correlation between chemical composition and sensory quality of natural apple aroma condensates. *Nahrung* 1986, *30*, 765–774.
64. Stanley, G., Algie, J. E., Brophy, J. J. 1,3,3-Trimethyl-2,7-dioxabicyclo[2.2.1] heptane: A volatile oxidation product of α-farnesene in Granny Smith apples. *Chem. Ind.* 1986, 556.

65. Chairote, G., Rodriguez, F., Crouzet, J. Characterization of additional volatile flavor components of apricot. *J. Food Sci.* 1981, *46*, 1898–1901, 1906.
66. Guichard, E., Souty, M. Comparison of the relative quantities of aroma compounds found in fresh apricot (*Prunus armeniaca*) from six different varieties. *Z. Lebensm. Unters. Forsch.* 1988, *186*, 301–307.
67. Krammer, G., Winterhalter, P., Schwab, M., Schreier, P. Glycosidically bound aroma compounds in the fruits of *Prunus* species, Apricot (*P. armeniaca*, L.), Peach (*P. persica*), yellow plum (*P. domestica* L. ssp, *Syriaca*). *J. Agric. Food Chem.* 1991, *39*, 778–781.
68. Rhoades, J. W., Millar, J. D. Gas chromatographic method for comparative analysis of fruit flavors. *J. Agric. Food Chem.* 1965, *13*, 5–9.
69. Rodriguez, F., Seck, S., Crouzet, J. Constituants volatils de l'abricot variété Rouge du Roussillon. *Lebensm. Wiss. Technol.* 1980, *13*, 152–155.
70. Salles, C., Jallageas, J.-C., Fournier, F., Tabet, J.-C., Crouzet, J. C. Apricot glycosidically bound volatile components. *J. Agric. Food Chem.* 1991, *39*, 1979–1983.
71. Takeoka, G. R., Flath, R. A., Mon, T. R., Teranishi, R., Guentert, M. Volatile constituents of apricot (*Prunus armeniaca*). *J. Agric. Food. Chem.* 1990, *38*, 471–477.
72. Tang, C. S., Jennings, W. G. Volatile components of apricot. *J. Agric. Food Chem.* 1967, *15*, 24–28.
73. Tang, C. S., Jennings, W. G. Lactonic compounds of apricot. *J. Agric. Food Chem.* 1968, *16*, 252–254.
74. Houchen, M., Scanlan, R. A., Libbey, L. M., Bills, D. D. Possible precursor for 1-methyl-4-isopropenylbenzene in commercial blackberry flavor essence. *J. Agric. Food Chem.* 1972, *20*, 170–175.
75. Scanlan, R. A., Bills, D., Libbey, L. M. Blackberry flavor components of natural essence. *J. Agric. Food Chem.* 1970, *18*, 744.
76. Schwab, W., Scheller, G., Schreier, P. Glycosidically bound aroma compounds from sour cherry. *Phytochemistry* 1990, *29*, 607–612.
77. Anjou, K., von Sydow, E. 1967. The aroma of cranberries. I. *Vaccinium vitis-idaea* L. *Acta Chem. Scand.* 1967, *21*, 945–952.
78. Gunata, Y. Z., Bayonove, C. L., Baumes, R. L., Cordonnier, R. E. The aroma of grapes. I. Extraction and determination of free and glycosidically bound fractions of some grape aroma constituents. *J. Chromatogr.* 1985, *331*, 83–90.
79. Stevens, K. L., Lee, A., McFadden, W. H., Teranishi, R. Volatiles from grapes. I. Some volatiles from Concord essence. *J. Food Sci.* 1965, *30*, 1006–1010.
80. Strauss, C. R., Dimitriadis, E., Wilson, B., Williams, P. J. Studies on the hydrolysis of two megastigma-3,6,9-triols rationalizing the origins of some volatile C_{13} norisoprenoids of *Vitis vinifera* grapes. *J. Agric. Food Chem.* 1986, *34*, 145–149.
81. Voirin, S. G., Baumes, R. L., Bitteur, S. M., Gunata, Z., Bayonone, C. L. Novel monoterpene disaccharide glycosides of *Vitis vinifera* grapes. *J. Agric. Food Chem.* 1990, *38*, 1373–1378.
82. Williams, P. J., Strauss, C. R., Wilson, B., Massy-Westropp, R. A. Novel monoterpene disaccharide glycosides of *Vitis vinifera* grapes and wines. *Phytochemistry* 1982, *8*, 2013–2020.
83. Williams, P. J., Strauss, C. R., Wilson, B., Massy-Westropp, R. A. Use of C_{18} reversed-phase liquid chromatography for the isolation of monoterpene glycosides and non-isoprenoid precursors from grape juice and wines. *J. Chromatogr.* 1982, *135*, 471–480.
84. Williams, P. J., Strauss, C. R., Wilson, B., Massy-Westropp, R. A. Glycosides of 2-phenylethanol and benzyl alcohol in *Vitis vinifera* grapes. *Phytochemistry* 1983, *22*, 2039–2041.
85. Wilson, B., Strauss, C. R., Williams, P. J. Changes in free and glycosidically bound monoterpenes in developing muscat grapes. *J. Agric. Food Chem.* 1984, *32*, 919–924.
86. Buttery, R. G., Seifert, R. M., Ling, L. C., Soderstrom, E. L., Ogawa, J. M., Turnbaugh, J. G. Additional aroma components of honeydew melon. *J. Agric. Food Chem.* 1982, *30*, 1208–1211.

87. Moshonas, M. G., Shaw, P. E. Analysis of volatile flavor constituents from lemon and lime essence. *J. Agric. Food Chem.* 1972, *20*, 1029–1030.
88. Homatidou, V., Karvouni, S., Dourtoglou, V. V. Determination of characteristic aroma components of "Cantaloupe" *Cucumis melo* using multidimensional gas chromatography (MDGC). In Charlambous, G., ed., *Flavors and Off-Flavors*, Proc. 6th International Flavor Conference, Crete, Greece. Elsevier: Amsterdam, 1989, pp. 1011–1023.
89. Horvat, R. J., Senter, S. D. Identification of additional volatile compounds from cantaloupe. *J. Food Sci.* 1987, *52*, 1097–1098.
90. Kemp, T. R. Volatile *Cucumis melo* components, identification of additional compounds and effects of storage conditions. *Phytochemistry* 1973, *12*, 2921–2924.
91. Kemp, T. R., Knavel, D. E., Stoltz, L. P. Cis-6-nonenal, a flavor component of muskmelon fruit. *Phytochemistry* 1972, *11*, 3321–3322.
92. Kemp, T. R., Stoltz, L. P., Knavel, D. E. Volatile components of muskmelon fruit. *J. Agric. Food Chem.* 1972, *20*, 196–198.
93. Kemp, T. R., Lundin, R. E. 3,6-Nonadien-1-ol from *Citrullus vulgaris* and *Cucumis melo. Phytochemistry* 1974, *13*, 1167–1170.
94. Schieberle, P., Ofner, S., Grosch, W. Evaluation of potent odorants in cucumber (*Cucumis sativus*) and muskmelons (*Cucumis melo*) by aroma extract dilution analysis. *J. Food Sci.* 1990, *55*, 193–195.
95. Wyllie, S. G., Leach, D. N. Sulfur-containing compounds in the aroma volatiles of melons (*Cucumis melo*). *J. Agric. Food Chem.* 1992, *40*, 253–256.
96. Yabumoto, K., Jennings, W. G., Yamaguchi, M. Volatile constituents of cantaloupe, *Cucumis melo*, and their biogenesis. *J. Food Sci.* 1977, *42*, 32–37.
97. Yabumoto, K., Yamaguchi, M., Jennings, W. G. Production of volatile compounds by muskmelon, *Cucumis melo. Food Chem.* 1978, *3*, 7–16.
98. Yamaguchi, M., Hughes, D. K., Yabumoto, Y., Jennings, W. G. Quality of cantaloupe muskmelons: Variability and attributes. *Sci. Hortic.* 1975, *6*, 59–70.
99. Engel, K.-H., Flath, R. A., Buttery, R. G., Mon, T. R., Ramming, D. W., Teranishi, R. Investigation of volatile constituents of aroma components in some nectarine cultivars. *J. Agric. Food Chem.* 1988, *36*, 549–553.
100. Lim, L., Romani, R. J. Volatiles and the harvest maturity of peaches and nectarines. *J. Food Sci.* 1964, *29*, 246–253.
101. Takeoka, G. R., Flath, R. A., Guentert, M., Jennings, W. Nectarine volatiles: Vacuum steam distillation versus headspace sampling. *J. Agric. Food Chem.* 1988, *36*, 553–560.
102. Ahmed, E. M., Dennison, R. A., Dougherty, R. H., Shaw, P. E. Flavor and odor thresholds in water of selected orange juice components. *J. Agric. Food Chem.* 1978, *26*, 187–191.
103. Attaway, J. A., Oberbacher, M. F. Studies on the aroma of intact Hamlin oranges. *J. Food Sci.* 1968, *33*, 287–289.
104. Marsili, R. Measuring volatiles and limonene oxidation products in orange juice by capillary GC. *LC-GC Mag.* 1986, *4*, 358–362.
105. Moshonas, M. G., Shaw, P. E. Quantitative analysis of orange juice flavor volatiles by direct-injection gas chromatography. *J. Agric. Food Chem.* 1987, *35*, 161–165.
106. Moshonas, M. G., Shaw, P. E. Changes in composition of volatile components in aseptically packaged orange juice during storage. *J. Agric. Food Chem.* 1989, *37*, 157–161.
107. Moshonas, M. G., Shaw, P. E. Flavor evaluation of volatile flavor constituents of stored aseptically packaged orange juice. *J. Food Sci.* 1989, *54*, 82–85.
108. Nisperos-Carriedo, M. O., Shaw, P. E. Comparison of volatile flavor components in fresh and processed orange juices. *J. Agric. Food Chem.* 1990, *38*, 1048–1052.
109. Rodriguez, P. A., Culbertson, C. R. Quantitative headspace analysis of selected compounds in equilibrium with orange juice. In Charalambous, G., Inglett, G., eds., *Instrumental Analysis of Foods: Recent Progress*, Vol. 2. Academic Press: New York, 1983, pp. 187–195.

110. Schreier, P. Changes in flavour compounds during the processing of fruit juices. *Proc. Long Ashton Symp.* 1981, *7*, 355–371.
111. Schultz, T. H., Flath, R. A., Mon, T. R. Analysis of orange volatiles with vapor sampling. *J. Agric. Food Chem.* 1971, *19*, 1060–1065.
112. Wolford, R. W., Attaway, J. A., Alberding, G. E., Atkins, C. D. Analysis of flavor and aroma constituents of Florida orange juices by gas chromatography. *J. Food Sci.* 1963, *28*, 320–323.
113. Do, J. Y., Salunkhe, D. K., Olson, L. E. Isolation, identification, and comparison of the volatiles of peach fruit as related to harvest maturity and artificial ripening. *J. Food Sci.* 1969, *34*, 618–621.
114. Lim, L. S. Studies on the relationship between the production of volatiles and the maturity of peaches and pears. Ph.D. Thesis, University of California, Davis, 1963.
115. Power, F. B., Chesnut, V. K. Odorous constituents of peaches. *J. Am. Chem. Soc.* 1921, *43*, 1725–1727.
116. Sevenants, M. R., Jennings, W. G. Volatile components of peach. Part II. *J. Food Sci.* 1966, *31*, 81–86.
117. Spencer, M. D., Pangborn, R. M., Jennings, W. G. Gas chromatographic and sensory analysis of volatiles from cling peaches. *J. Agric. Food Chem.* 1978, *26*, 725–732.
118. Heinz, D. E., Creveling, R. K., Jennings, W. G. Direct determination of aroma compounds as an index of pear maturity. *J. Food Sci.* 1965, *30*, 641–643.
119. Heinz, D. E., Jennings, W. G. Volatile esters of Bartlett pears, Part V. *J. Food Sci.* 1966, *31*, 69–80.
120. Heinz, D. E., Pangborn, R. M., Jennings, W. G. Pear aroma, relation of instrumental and sensory techniques. *J. Food Sci.* 1964, *29*, 756–763.
121. Heinz, D. E., Sevenants, M. R., Jennings, W. G. Preparation of fruit essences for gas chromatography. *J. Food Sci.* 1966, *31*, 63–68.
122. Jennings, W. G. Volatile esters of Bartlett pears. *J. Food Sci.* 1961, *26*, 564–568.
123. Jennings, W. G., Leonard, S., Pangborn, R. M. Volatiles contributing to the flavor of Bartlett pears. *J. Food Technol.* 1960, *14*, 587–563.
124. Jennings, W. G., Creveling, R. K., Heinz, D. E. Volatile esters of Bartlett pears. IV. Esters of *trans*, 2-*cis*, 4-decadienoic acid. *J. Food Sci.* 1964, *29*, 730–734.
125. Jennings, W. G., Creveling, R. K. Volatile esters of Bartlett pears, Part II. *J. Food Sci.* 1963, *28*, 91–94.
126. Jennings, W. G., Sevenants, M. R. Volatile esters of Bartlett pears, Part III. *J. Food Sci.* 1964, *29*, 158–162.
127. Romani, R. J., Ku, L. Direct gas chromatographic analysis of volatiles produced by ripening pears. *J. Food Sci.* 1966, *31*, 558–560.
128. Tsuyena, T., Ishihara, M., Shiota, H., Shiga, M. Volatile components of quince fruit (*Cydonia oblongata* Mill.). *Agric. Biol. Chem.* 1983, *47*, 2495–2502.
129. Umano, K., Shoji, A., Hagi, Y., Shibamoto, T. Volatile constituents of peel of quince fruit, *Cydonia oblonga* Miller. *J. Agric. Food Chem.* 1986, *34*, 593–596.
130. Winterhalter, P., Schreier, P. 4-Hydroxy-7,8-dihydro-β-ionol: Natural precursor of theaspiranes in quince fruits (*Cydonia oblonga*, Mill.). *J. Agric. Food Chem.* 1988, *36*, 560–562.
131. Winterhalter, P., Schreier, P. Free and bound C_{13} norisoprenoids in quince (*Cydonia oblonga* Mill.) fruit. *J. Agric. Food Chem.* 1988, *36*, 1251–1256.
132. Bohnsach, H. Beitrag zur Kenntnis der ätherischen Ö, Riechund Geschmackstoffe. 19. Mitteilung, Über Untersuchungsergebnisse des natürlichen Himbeerfruchtöles. 2. Teil. A. mit Wasserdampf Aromen-Anteile, B. mit Wasserdampf nicht flüchtige Anteile. *Riechstoffe, Aromen, Körperpflegem.* 1967, *17*, 358–361.
133. Bohnsach, H. Beitrag zur Kenntnis der ätherischen Ö, Riechund Geschmackstoffe. 20. Mitteilung, Über Untersuchungsergebnisse des natürlichen Himbeerfruchtöles. 3. Teil. Die Inhalt-

stoffe des himbeertrester-Extractöles. *Riechstoffe, Aromen, Körperpflegem.* 1967, *17*, 514–517.

134. Deifel, A. 4-(4-Hydroxyphenyl)-2-butanon-himbeerketon. *Z. Lebensm. Unters.-Forsch.* 1989, *188*, 330–332.
135. Duclos, J., Latrasse, A. Comparaison des produits fixes et volatiles de variétés de framboises à différent stades de maturité. *Ann. Technol. Agric.* 1971, *20*, 141–146.
136. Hiirsalmi, H., Kallio, H., Pyysalo, T., Linko, R. R., Koponen, P. The ionone content of raspberries, nectarberries and nectar raspberries and its influence on their flavour. *Ann. Agric. Fenn.* 1974, *13*, 23–28.
137. Honkanen, E. Tutkimuksia eräiden Rubus-suvun marjojen aromiainekoostumuksesta 1. *Tutkimus Tekniikka* 1972, *3*, 48–53.
138. Honkanen, E., Kallio, H., Pyysalo, T. Tutkimuksia eräiden Rubus-suvun marjojen aromiainekoostumuksesta. *Tutkimus Tekniikka* 1973, *10*, 51–54.
139. Obretenov, T., Lazarov, K., Genov, N. Study of raspberry aroma. 2. Characteristics of aromatic substances of wild raspberries (*Rubus idaeus* L.) and Willamette and Malling Promise cultivars. *Nauch. Tri. Vyssh. Inst. Khran. Vkus. Prom.* 1972, *19*, 275–279.
140. Pabst, A., Barron, D., Etiévant, P., Schreier, P. Studies on the enzymatic hydrolysis of bound aroma constituents from raspberry fruit pulp. *J. Agric. Food Chem.* 1991, *39*, 173–175.
141. Palluy, E., Sundt, E., Winter, M. Recherche sur les arômes. Analyse de l'arôme des framboises. 3. Les acides et esters inférieurs. *Helv. Chim. Acta* 1963, *46*, 2297–2301.
142. Pisarnitskii, A. F., Vereshchagin, G. I., Macgaroshvili, G. I., Bogatova, E. G. Carbonyl compounds and their effect on the aroma of fruit, including berries. *Prikl. Biokhim. Mikrobiol.* 1970, *6*, 13–17.
143. Schmidlin-Meszaros, J. Probleme im Gebiet der Lebensmittelzusatzstoffe, Ermittlung des *p*-Hydroxyphenylbutanon-(3)s, des Himbeerketons. *Alimenta* 1971, *10*, 39–44.
144. Sundt, E., Winter, M. Untersuchungen über Aromastoffe. 3. Mitteilung. Die Isolierung von Geraniol aus Himbeeren. *Helv. Chim. Acta* 1960, *43*, 1120–1123.
145. Sundt, E., Winter, M. Recherches sur les arômes. 6^e^ communication. Analyse de l'arome volatils des framboises. 2. Les alcools. *Helv. Chim. Acta* 1962, *45*, 2212–2215.
146. Weurman, C. GLC studies on the enzymatic formation of volatile compounds in raspberries. *Food Technol.* 1961, *15*, 531–534.
147. Winter, M., Enggist, E. Recherche sur les arômes. 17^e^ communication. Sur l'arôme de framboise. 4. *Helv. Chim. Acta* 1971, *54*, 1891.
148. Winter, M., Sundt, E. Recherche sur les arômes. 5^e^ communication. Analyse de l'arôme des framboises. 1. Les constituants carbonyles volatils. *Helv. Chim. Acta* 1962, *45*, 2195–2197.
149. Dimick, K. P., Makower, B. Volatile flavor of strawberry essence. I. Identification of the carbonyls and certain low boiling point substances. *Food Technol.* 1956, *10*, 73–76.
150. Drawert, F., Heiman, W., Emberger, R. Enzymatische bildung von hexen-(2)-al-(1), hexanal and deren vorstufen. *Annalen Chem.* 1966, *694*, 200–204.
151. McFadden, W. H., Teranishi, R., Corse, J., Black, E. R., Mon, T. R. Volatiles from strawberries. II. Combined gas chromatography and mass spectrometry on complex mixtures. *J. Chromatogr* 1965, *18*, 10–19.
152. Mosandl, A., Hener, U., Hagenauer-Hener, U., Kustermann, A. Stereoisomeric flavor compounds. 33. Multidimensional gas chromatography direct enantiomer separation of γ-lactones from fruits, foods, and beverages. *J. Agric. Food Chem.* 1990, *38*, 767–771.
153. Mussinan, C. J., Walradt, J. P. Organic acids from fresh California strawberries. *J. Agric. Food Chem.* 1975, *23*, 482–484.
154. Pickenhangen, W., Velluz, A., Passerat, J.-P., Ohloff, G. Estimation of 2,5-dimethyl-4-hydroxy-3(*2H*)-furanone (FURANEOL) in cultivated and wild strawberries, pineapples and mangoes. *J. Sci. Food Agric.* 1981, *32*, 1132–1134.
155. Pyysalo, T., Honkanen, E., Hirvi, T. Volatiles of wild strawberries, *Fragaria vesca* L., com-

pared to those of cultivated berries, *Fragaria ananassa* cv. Senga Senganna. *J. Agric. Food Chem.* 1979, *27*, 19–22.
156. Schreier, P. Quantitative composition of volatile constituents in cultivated strawberries, *Fragaria ananassa* cv. Senga Sengana, Senga Litessa, and Senga Gourmella. *J. Sci. Food Agric.* 1980, *31*, 487–494.
157. Staudt, G., Drawert, F., Tressl, R. Gaschromatographischmassenspektrometrische Differenzierung von Erdbeerarten. II. *Fragaria nilgerrensis. Z. Pflanzenzüchtg.* 1975, *75*, 36–42.
158. Tressl, R., Drawert, F., Heimann, W. Gaschromatographischmassenspektrometrische Bestandsaufnahme von Erdbeer-Aromastoffen. *Z. Naturforsch.* 1969, *24b*, 1201–1202.
159. Willhalm, B., Palluy, E., Winter, M. Recherches sur les arômes. XII. Sur l'arôme des fraises fraîches. Identification des acides volatils et de quelques autres composés. *Helv. Chim. Acta* 1966, *49*, 65–67.
160. Winter, M., Willhalm, B. Recherches sur les arômes. XI. Sur l'arôme des fraises fraîches. Analyse des composés carbonylés, esters et alcools volatils. *Helv. Chim. Acta* 1964, *47*, 1215–1227.
161. Moshonas, M. G., Shaw, P. E. Analysis of volatile flavor constituents from tangerine essence. *J. Agric. Food Chem.* 1972, *20*, 70–71.
162. Kemp, T. R. Identification of some volatile compounds from *Citrullus vulgaris. Phytochemistry* 1975, *14*, 2637–2638.
163. Shaw, G. J., Allen, J. M., Visser, F. R. Volatile flavor components of babaco fruit (*Carica pentagona* Heilborn). *J. Agric. Food Chem.* 1985, *33*, 795–797.
164. Bailey, S. D., Mitchell, D. G., Bazinet M. L., Weurman, C. Studies on the volatile components of different varieties of cocoa beans. *J. Food Sci.* 1962, *27*, 165–170.
165. Hultin, H. O., Proctor, B. E. Changes in some volatile constituents of the banana during ripening, storage and processing. *Food Technol.* 1961, *15*, 440–445.
166. Issenberg, P., Wick, E. L. Volatile components of banana. 1. Isolation. 2. Separation and Identification. *J. Agric. Food Chem.* 1963, *11*, 2–6.
167. Kleber, C. The occurrence of amyl acetate in bananas. *Am. Perfumer* 1912, *7*, 235–238.
168. McCarthy, A. I., Palmer, J. K. Production of volatile compounds by the banana fruit during ripening. In Proc. 1st Intern. Congress of Food Science and Technology. Gordon and Breach: New York, 1964, pp. 483–488.
169. McCarthy, A. I., Palmer, J. K., Shaw, C. P., Anderson, E. E. Correlation of gas chromatographic data with flavour profiles of fresh banana fruit. *J. Food Sci.* 1963, *28*, 379–384.
170. Myers, M. J., Issenberg, P., Wick, E. A. Vapor analysis of the production by banana fruit of certain volatile constituents. *J. Food Sci.* 1969, *34*, 504–509.
171. Rotherbach, F., Eberlein, L. Über das vorkommen von estern in der Fruchten der Bananen. *Deutsch. Essigind.* 1905, *9*, 81–87.
172. Tressl, R., Jennings, W. G. Production of volatile compounds in ripening banana. *J. Agric. Food Chem.* 1972, *20*, 189–192.
173. Wick, E. L., McCarthy, A. L., Myers, M., Murray, E., Nursten, H., Issenberg, P. I. Flavor and biochemistry of volatile banana components. *Adv. Chem.* 1966, *56*, 241–260.
174. Wick, E. L., Yamanishi, T., Kobayashi, A., Valenzuela, S., Issenberg, P. Volatile constituents of banana (*M. cavendishii*, variety Valery). *J. Agric. Food Chem.* 1969, *17*, 751–759.
175. Peppard, T. L. Volatile flavor constituents of *Monstera deliciosa. J. Agric. Food Chem.* 1992, *40*, 257–262.
176. Idstein, H., Schreier, P. Volatile constituents from guava (*Psidium guajava* L.) fruit. *J. Agric. Food. Chem.* 1985, *33*, 138–143.
177. Bartley, J. P., Schwede, A. Production of volatile compounds in ripening kiwi fruit (*Actinidia chinensis*). *J. Agric. Food Chem.* 1989, *37*, 1023–1025.
178. Abd-El-Baki, M. M., Askar, A., El-Samahy, S. K., Abd-El-Fadeel, M. G. Studies on mango flavour. *Deutsch. Lebensm. Rundsch.* 1981, *77*, 139–142.

179. Ackerman, L. G. J., Torline, P. A. Volatile components in the headspace of eight mango cultivars. *Lebensm. Wiss. Technol.* 1984, *17*, 339–341.
180. Bartley, J. P., Schwede, A. Volatile flavor components in the headspace of the Australian or ''Bowen'' mango. *J. Food Sci.* 1987, *52*, 353–355.
181. Engel, K.-H., Tressl, R. Studies on the volatile components of two mango varieties. *J. Agric. Food Chem.* 1983, *31*, 798–801.
182. Gholap, A. S., Bandyopadhyay, C. Characterization of green aroma of raw mango (*Mangifera indica* L.). *J. Sci. Food Agric.* 1980, *28*, 885–889.
183. Hunter, G. L. K., Bucek, W. A., Radford, T. Volatile components of canned Alphonso mango. *J. Food Sci.* 1974, *39*, 900–903.
184. Idstein, H., Schreier, P. Volatile constituents of Alphonso mango (*Mangifera indica*). *Phytochemistry* 1985, *24*, 2313–2316.
185. MacLeod, A. J., DeTroconis, N. G. Volatile flavour components of some mango fruit. *Phytochemistry* 1982, *21*, 2523–2526.
186. MacLeod, A. J., Pieris, N. M. Comparison of the volatile components of some mango cultivars. *Phytochemistry* 1984, *23*, 361–366.
187. MacLeod, A. J., Snyder, C. H. Volatile components of two cultivars of mango from Florida. *J. Agric. Food Chem.* 1985, *33*, 380–384.
188. Sakho, M., Crouset, J., Seck, S. Volatile components of African mango. *J. Food Sci.* 1985, *50*, 548–550.
189. Wilson, C. W., III, Shaw, P. E., Knight, R. J., Jr. Importance of some lactones and 2,5-dimethyl-4-hydroxy-3(*2H*)-furanone to mango (*Mangifera indica* L.) aroma. *J. Agric. Food Chem.* 1990, *38*, 1556–1559.
190. Ettlinger, M. G., Hodgkins, J. E. The mustard oil of papaya seed. *J. Org. Chem.* 1956, *21*, 204–207.
191. Flath, R. A., Forrey, R. R. Volatile components of papaya (*Carica papaya* L., Solo variety). *J. Agric. Food Chem.* 1977, *25*, 103–109.
192. Flath, R. A., Light, D. M., Jang, E. B., Mon, T. R., John, J. O. Headspace examination of volatile emissions from ripening papaya (*Carica papaya* L., Solo variety). *J. Agric. Food. Chem.* 1980, *38*, 1060–1063.
193. Gmelin, R., Kjaer, A. Glucosinolates in some New World species of Capparidaceae. *Phytochemistry* 1970, *9*, 601–603.
194. Heidlas, J., Lehr, M., Idstein, H., Schreier, P. Free and bound terpene compounds in papaya (*Carica papaya* L.) fruit pulp. *J. Agric. Food Chem.* 1984, *32*, 1020–1021.
195. Idstein, H., Schreier, P. Volatile constituents from papaya fruit (*Carica papaya* L., var. Solo). *Lebensm. Wiss. Technol.* 1985, *18*, 164–169.
196. Ismail, H. H., Tucknott, O. G., Williams, A. A. The collection and concentration of aroma components of soft fruit using Porapak. *J. Sci. Food Agric.* 1980, *31*, 262–266.
197. Katague, D. B., Kirch, E. R. Chromatographic analysis of the volatile components of papaya fruit. *J. Pharm. Sci.* 1965, *54*, 891–894.
198. MacLeod, A. J., Pieris, N. M. Volatile components of papaya (*Carica papaya* L.) with particular reference to glucosinolate products. *J. Agric. Food Chem.* 1983, *31*, 1005–1008.
199. Schreier, P., Lehr, M., Heidlas, J., Idstein, H. Über das aroma der papaya-frucht (*Carica papaya* L.), Hinweise auf vorstufen flüchtiger terpenverbindungen. *Z. Lebensm. Unters. Forsch.* 1985, *180*, 297–302.
200. Schreier, P., Lehr, M., Heidlas, J., Idstein, H. Volatiles from papaya (*Carica papaya* L.) fruit: Indication of precursors of terpene compounds. *Z. Lebensm. Unters. Forsch.* 1985, *180*, 297–302.
201. Schwab, W., Mahr, C., Schreier, P. Studies on the enzymic hydrolysis of bound aroma components from *Carica papaya* fruit. *J. Agric. Food Chem.* 1989, *37*, 1009–1012.
202. Schwab, W., Schreier, P. Simultaneous enzyme catalysis extraction: A versatile technique for the study of flavor precursors. *J. Agric. Food Chem.* 1988, *36*, 1238–1242.

203. Tang, C. S. Benzyl isothiocyanate of papaya fruit. *Phytochemistry* 1971, *10*, 117–119.
204. Tang, C. S. Localization of benzyl glucosinolate and thioglucosidase in *Carica papaya* fruit. *Phytochemistry* 1973, *12*, 769–772.
205. Winterhalter, P., Katzenberger, D., Schreier, P. 6,7-Epoxylinalool and related oxygenated terpenoids from *Carica papaya* fruit. *Phytochemistry* 1986, *25*, 1347–1350.
206. Casimir, D. J., Kelford, J. F., Whitfield, F. B. Technology and flavor chemistry of passion fruit juices and concentrates. *Adv. Food Res.* 1981, *27*, 243–295.
207. Chen, C.-C., Kuo, M.-C., Hwang, L. S., Wu, J. S.-B., Wu, C.-M. Headspace components of passion fruit juice. *J. Agric. Food Chem.* 1982, *30*, 1211–1215.
208. Engel, K.-H., Tressl, R. Formation of aroma components from nonvolatile precursors in passion fruit. *J. Agric. Food Chem.* 1983, *31*, 998–1002.
209. Heiderich, M., Winterhalter, P. 3-Hydroxy-*retro*-α-ionol, a natural precursor of isomeric edulans in purple passion fruit (*Passiflora edulis* Sims.). *J. Agric. Food Chem.* 1991, *39*, 1270–1274.
210. Parliment, T. H. Some volatile constituents of passion fruit. *J. Agric. Food Chem.* 1972, *20*, 1043–1045.
211. Whitfield, F. B., Sugowdz, G. The 6-(but-2′-enylidene)-1,5,5-trimethyl-cyclohex-1-enes: Important volatile constituents of the juice of the purple passion fruit *Passiflora edulis. Aust. J. Chem.* 1979, *32*, 891–903.
212. Winter, M., Kloti, R. Über das aroma der gelben passionsfrucht (*Passiflora edulis* f. *flavicarpa*). *Helv. Chim. Acta* 1972, *55*, 1916–1921.
213. Winterhalter, P. Bound terpenoids in the juice of the purple passion fruit (*Passiflora edulis* Sims). *J. Agric. Food Chem.* 1990, *38*, 452–455.
214. Berger, R. G., Drawert, F., Nitz, S. Sesquiterpene hydrocarbons in pineapple fruit. *J. Agric. Food Chem.* 1983, *31*, 1237–1239.
215. Berger, R. G., Drawert, F., Kollmannsberger, H., Nitz, S., Schaufstetter, B. Novel volatiles in pineapple fruit and their sensory properties. *J. Agric. Food Chem.* 1985, *33*, 232–235.
216. Connell, D. W. Volatile flavoring constituents of the pineapple. *Aust. J. Chem.* 1964, *17*, 130–136.
217. Creveling, R. K., Silverstein, R. M., Jennings, W. G. Volatile components of pineapple. *J. Food Sci.* 1968, *33*, 284–287.
218. Gawler, J. H. Constituents of canned Malayan pineapple juice. *J. Sci. Food Agric.* 1962, *13*, 57–63.
219. Rodin, J. O., Coulson, D. M., Silverstein, R. M., Leeper, R. W. Volatile flavor and aroma components of pineapple. III. The sulfur-containing components. *J. Food Sci.* 1966, *31*, 721–725.
220. Rodin, J. O., Himel, C. M., Silverstein, R. M., Leeper, R. W., Gortner, W. A. Volatile flavor and aroma components of pineapple. 1. Isolation and tentative identification of 2,5-dimethyl-4-hydroxy-3(*2H*)-furanone. *J. Food Sci.* 1965, *30*, 280–285.
221. Silverstein, R. M., Rodin, J. O., Himel, C. M., Leeper, R. W. Volatile flavor and aroma components of pineapple. II. Isolation and identification of chavicol and γ-caprolactone. *J. Food Sci.* 1965, *30*, 668–672.
222. Takeoka, G., Buttery, R. G., Flath, R. A., Teranishi, R., Wheeler, E. L., Wieszorek, R. L., Guentert, M. Volatile constituents of pineapple [*Ananas comosus* (L.) Merr.]. In Teranishi, R., Buttery, R. G., Shahidi, F., eds., *Flavor Chemistry, Trends and Developments*. American Chemical Society: Washington, D.C., 1989, pp. 223–237.
223. Wu, P., Kuo, M.-C., Zhang, K. Q., Hartmann, T. G., Rogen, R. T., Ho, C.-T. Analysis of glycosidically bound 2,5-dimethyl-4-hydroxy-3(*2H*)-furanone in pineapple. *Perfum. Flavor.* 1990, *15*, 51–53.
224. Wu, P., Kuo, M.-C., Hartmann, T. G., Rogen, R. T., Ho, C.-T. Free and glycosidically bound aroma compounds in pineapple [*Ananas comosus* (L.) Merr.]. *J. Agric. Food Chem.* 1991, *39*, 1984–1989.

225. Kallio, H. Development of volatile aroma compounds in Arctic bramble, *Rubus arcticus* L. *J. Food Sci.* 1976, *41*, 563–566.
226. Kallio, H. Identification of vacuum steam-distilled aroma compounds in the press juice of arctic bramble, *Rubus arcticus* L. *J. Food Sci.* 1976, *41*, 555–562.
227. Kallio, H., Linko, R. R. Volatile monocarbonyl compounds of arctic bramble (*Rubus arcticus* L.) at various stages of ripeness. *Z. Lebensm. Unters. Forsch.* 1973, *153*, 23–26.
228. Pyysalo, T., Suinko, M., Honkanen, E. Odor thresholds of the major volatiles identified in cloudberry (*Rubus chamaemorus* L.) and arctic bramble (*Rubus arcticus* L.). *Z. Lebensm. Wiss. Technol.* 1977, *10*, 36–39.
229. Sydow, E., Andersson, J., Anjou, K., Karlsson, G., Land, D., Griffiths, N. The aroma of bilberries (*Vaccinium myrtilus* L.). II. Evaluation of the press juice by sensory methods and by gas chromatography and mass spectroscopy. *Lebens. Wiss. Technol.* 1970, *3*, 11–17.
230. Hirvi, T., Honkanen, E. Analysis of the volatile constituents of black chokeberry (*Aronia melanocarpa* Ell.). *J. Agric. Food Chem.* 1985, *36*, 808–810.
231. Honkanen, E., Pyysalo, T., Hirvi, T. The aroma of Finnish wild raspberries, *Rubus idaeus* L. *Z. Lebensm. Unters. Forsch.* 1980, *171*, 180–182.
232. Forney, C. F., Mattheis, J. P., Austin, R. K. Volatile compounds produced by broccoli under anaerobic conditions. *J. Agric. Food Chem.* 1991, *39*, 2257–2259.
233. Buttery, R. G., Seifert, R. M., Guadagni, D. R., Black, D. R., Ling, L. C. Characteristics of some volatile constituents of carrots. *J. Agric. Food Chem.* 1968, *16*, 1009–1015.
234. Heatherbell, D. A., Wrolstad, R. E., Libbey, L. M. Isolation, concentration, and analysis of carrot volatiles using on-column trapping and gas-liquid chromatography–mass spectrometry. *J. Agric. Food Chem.* 1971, *19*, 1069–1073.
235. Heatherbell, D. A., Wrolstad, R. E., Libbey, L. M. Carrot volatiles. 1. Characterization and effects of canning and freeze-drying. *J. Food Sci.* 1971, *36*, 219–230.
236. Heatherbell, D. A., Wrolstad, R. E., Libbey, L. M. Carrot volatiles. 2. Influence of variety, maturity and storage. *J. Food Sci.* 1971, *36*, 225–230.
237. Heatherbell, D. A., Wrolstad, R. E. The enzymatic regeneration of volatile flavor components in carrots. *J. Agric. Food Chem.* 1971, *19*, 281–284.
238. Longan, B. J., Hruzek, G. A., Burns, E. E. Effect of processing variables on volatile retention of freeze-dried carrots. *J. Food Sci.* 1974, *39*, 1191–1194.
239. Rasekh, J., Kramer, A. Gas chromatographic profiles of stored carrots. *J. Am. Soc. Hortic. Sci.* 1971, *96*, 572–575.
240. Seifert, R. M., Buttery, R. G. Characterization of some previously unidentified sesquiterpenes in carrot roots. *J. Agric. Food Chem.* 1978, *26*, 181–182.
241. Senalik, D., Simon, P. W. Quantifying intra-plant variation of volatile terpenoids in carrot. *Phytochemistry* 1987, *26*, 1975–1979.
242. Simon, P. W., Lindsay, R. C., Peterson, C. E. Analysis of carrot volatiles collected on porous polymer traps. *J. Agric. Food Chem.* 1980, *28*, 549–552.
243. Rembold, H., Wallner, P., Nitz, S., Kollmannsburger, H., Drawert, F. Volatile components of chickpea (*Cicer arietinum* L.) seed. *J. Agric. Food Chem.* 1989, *37*, 659–662.
244. Rembold, H., Wallner, P., Singh, A. K. Attractiveness of volatile chickpea (*Cicer arietinum* L.) seed components to *Heliothis armigera* larvae. *Z. Angew. Entomol.* 1989, *107*, 65–70.
245. Saxena, K. N., Rembold, H. Attraction of *Heliothis armigera* (Hübner) larvae by chickpea seed powder constituents. *Z. Angew. Entomol.* 1984, *97*, 145–153.
246. Fleming, H. P., Cobb, W. Y., Etchells, J. L., Bell, T. A. The formation of carbonyl compounds in cucumbers. *J. Food Sci.* 1968, *33*, 572–576.
247. Forss, D. A., Dunstone, E. A., Ramshaw, E. H., Stark, W. The flavor of cucumbers. *J. Food Sci.* 1962, *27*, 90–93.
248. Buttery, R. G., Seifert, R. M., Guadagni, D. G., Ling, L. C. Characterization of some volatile constituents of bell peppers. *J. Agric. Food Chem.* 1969, *17*, 1322–1327.

249. Bernhard, R. A. Comparative distribution of volatile aliphatic disulfides derived from fresh and dehydrated onions. *J. Food Sci.* 1968, *33*, 298–304.
250. Boelens, M., de Valois, P. J., Wobben, H. J., van der Gen, A. Volatile flavor compounds of onion. *J. Agric. Food Chem.* 1971, *19*, 984–991.
251. Brodnitz, M. H., Pollock, C. L., Vallon, P. P. Flavor components of onion oil. *J. Agric. Food Chem.* 1969, *17*, 760–763.
252. Freeman, G. G., Mossdeghi, N. Effect of sulphate nutrition on flavour components of onion (*Allium cepa*). *J. Sci. Food Agric.* 1970, *21*, 610–615.
253. Freeman, G. G., Whenham, R. J. Changes in onion (*Allium cepa* L.) flavour components resulting from some post-harvest processes. *J. Sci. Food Agric.* 1974, *25*, 499–515.
254. Freeman, G. G., Whenham, R. J. A survey of volatile components of some *Allium* species in terms of S-alk(en)yl-L-cysteine sulphoxides present as flavout precursors. *J. Sci. Food Agric.* 1975, *26*, 1869–1886.
255. Kallio, H., Salorinne, L. Comparison of onion varieties by headspace gas chromatography-mass spectrometry. *J. Agric. Food Chem.* 1990, *38*, 1560–1564.
256. Mazza, G. Relative volatilities of some onion flavour compounds. *J. Food Technol.* 1980, *15*, 35–41.
257. Mazza, G., LeMaguer, M., Hadziyev, D. Headspace sampling procedures for onion (*Allium cepa* L.) aroma assessment. *Can. Inst. Food Sci. Technol. J.* 1980, *13*, 87–96.
258. Mazza, G., Lemaguer, M. Volatiles retention during the dehydration of onion (*Allium cepa* L.). *Lebensm. Wiss. Technol.* 1979, *12*, 333–337.
259. Saghir, A. R., Mann, L. K., Bernhard, R. A., Jacobsen, J. V. Determination of aliphatic mono- and disulfides in *Allium* by gas chromatography and their distribution in the common food species. *Proc. Am. Soc. Hortic. Sci.* 1964, *84*, 386–389.
260. Bengtsson, B., Bosund, I. Gas chromatographic evaluation of volatile substances in stored peas. *Food Technol.* 1964, *18*, 773–776.
261. Murray, K. E., Shipton, J., Whitfield, F. B., Kennett, B. H., Stanley, G. Volatile flavor components from green peas (*Pisum sativum*). 1. Alcohols in unblanched frozen peas. *J. Food Sci.* 1968, *33*, 290–294.
262. Ralls, J. W., McFadden, W. H., Seifert, R. M., Black, D. R., Kilpatrick, P. W. Volatiles from a commercial pea blancher: Mass spectral identifications. *J. Food Sci.* 1965, *30*, 228–237.
263. Shipton, J., Whitfield, F. B., Last, J. H. Extraction of volatile compounds from green peas (*Pisum sativum*) *J. Agric. Food Chem.* 1969, *17*, 1113–1118.
264. Whitfield, F. B., Shipton, J. Volatile carbonyls in stored unblanched frozen peas. *J. Food Sci.* 1966, *31*, 328–331.
265. Dickens, J. W., Slate, A. B., Pattee, H. E. Equipment and procedures to measure peanut headspace volatiles. *Peanut Sci.* 1987, *14*, 97–100.
266. Lee, L. S., Cucullu, A. F., Pons, W. A., Jr. Gas-liquid chromatographic detection of actively metabolizing *Aspergillus parasiticus* in peanut stocks. *J. Agric. Food Chem.* 1973, *21*, 470–473.
267. Pattee, H. E., Beasley, E. O., Singleton, J. A. Isolation and identification of volatile components from high-temperature cured off-flavor peanut. *J. Food Sci.* 1965, *30*, 388–392.
268. Pattee, H. E., Rogister, E. W., Giesbrecht, F. G. Interrelationships between headspace volatile concentration, selected seed-size categories and flavor in large-seeded Virginia-type peanuts. *Peanut Sci.* 1989, *16*, 38–42.
269. Pattee, H. E., Yokohama, W. H., Collins, M. F., Giesbrecht, F. G. Interrelationships between headspace volatile concentrations, marketing grades, and flavor in runner-type peanuts. *J. Agric. Food Chem.* 1990, *38*, 1055–1060.
270. Singleton, J. A., Pattee, H. E. Effects of induced low temperature stress on raw peanuts. *J. Food Sci.* 1987, *52*, 242–244.

271. Schieberle, P. Primary odorants in popcorn. *J. Agric. Food Chem.* 1991, *39*, 1141–1144.
272. Walradt, J. P., Lindsay, R. C., Libbey, L. M. Popcorn flavor, identification of volatile compounds. *J. Agric. Food Chem.* 1970, *18*, 926–928.
273. Clark, C. A. Influence of volatiles from healthy and decaying sweet potato storage roots on sclerotial germination and hyphal growth of *Sclerotium rolfsii. Can. J. Bot.* 1989, *67*, 53–57.
274. Gumbmann, M. R., Burr, H. K. Volatile sulfur compounds in potatoes. *J. Agric. Food Chem.* 1964, *12*, 404–408.
275. Khan, I., Müller, K., Warmbier, H. Emfluss von sorte und düngung auf das spectrum flüchtiger aromastoffe in kartoffeln. *Potato Res.* 1977, *20*, 235–242.
276. Meigh, D. F., Filmer, A. A. E., Self, R. Growth-inhibitory volatile compounds produced by *Solanum tuberosum* tubers. *Phytochemistry* 1973, *12*, 987–993.
277. Varns, J. L., Schaper, L. A. Volatile monitoring for disease detection in storage. *Valley Potato Grower* 1981, *47*, 36.
278. Baldwin, E. A., Nisperos-Carriedo, M. O., Baker, R., Scott, J. W. Quantitative analysis of flavor parameters in six Florida tomato cultivars (*Lycopersicon esculentum* Mill). *J. Agric. Food Chem.* 1991, *39*, 1135–1140.
279. Baldwin, E. A., Nisperos-Carriedo, M. O., Moshonas, M. G. Quantitative analysis of flavor and other volatiles and for certain constituents of two tomato cultivars during ripening. *J. Am. Soc. Hortic. Sci.* 1991, *116*, 265–269.
280. Buttery, R. G., Seifert, R. M. Volatile tomato constituents: Identification of 2,6-dimethylundeca-2,6-dien-10-one. *J. Agric. Food Chem.* 1968, *16*, 1053.
281. Buttery, R. G., Seifert, R. M., Ling, L. C. Volatile tomato components, characterisation of 6,10,14-trimethylpentadec-5,9,13-trien-2-one. *Chem. Ind.* (London) 1969, *8*, 238.
282. Buttery, R. G., Teranishi, R., Ling, L. C., Flath, R. A., Stern, D. J. Quantitative studies on origins of fresh tomato aroma volatiles. *J. Agric. Food Chem.* 1988, *36*, 1247–1250.
283. Buttery, R. G., Takeoka, G., Teranishi, R., Ling, L. C. Tomato aroma components: Identification of glycoside hydrolysis volatiles. *J. Agric. Food Chem.* 1990, *38*, 2050–2053.
284. Buttery, R. G., Teranishi, R., Flath, R. A., Ling, L. C. Identification of additional tomato paste volatiles. *J. Agric. Food Chem.* 1990, *38*, 792–795.
285. Buttery, R. G., Teranishi, R., Ling, L. C., Turnbaugh, J. G. Quantitative and sensory studies on tomato paste volatiles. *J. Agric. Food Chem.* 1990, *38*, 336–340.
286. Dalal, K. B., Salunkhe, D. K., Olson, L. E., Do, J. Y., Yu, M. H. Volatile components of developing tomato fruit grown under field and greenhouse conditions. *Plant Cell Physiol.* 1968, *9*, 389–393.
287. Dirinck, P., Schreyen, L., van Wassenhove, F., Schamp, N. Flavor quality of tomatoes, *J. Sci. Food Agric.* 1976, *27*, 499–508.
288. Giannone, L., Baldrati, G. Valutazione quantitativa dei constituenti volatili dell'aroma delgi alimenti vegetali mediante analisi gas-chromatografica del gas dello spazio de testa. I. L'aroma del pomodoro. *Ind. Conserve* (Parma) 1967, *42*, 176–180.
289. Gustafson, F. G. Production of alcohol and acetaldehyde by tomatoes. *Plant Physiol.* 1934, *9*, 359.
290. Hein, R. E., Fuller, G. W. Gas chromatographic studies on volatile compounds from processed tomatoes. Conf. on Advances in Flavor Research, 1963, Southern Utiliz. R&D Division, USDA, New Orleans.
291. Johnson, J. H., Gould, W. A., Badenhop, A. F., Johnson, R. M., Jr. Quantitative comparison of isoamylol, pentanol and 3-hexenol-1 in tomato juice: Varietal and harvest differences and processing effects. *J. Agric. Food Chem.* 1968, *16*, 255–259.
292. Katayama, O., Tubata, K., Yamato, I. The aroma components in fruits and vegetables. II. Volatile components of tomato. *Nippon Shokuhin Kogyo Gakkaishi.* 1967, *14*, 44–46.

293. Kazeniac, S. J., Hall, R. M. Flavor chemistry of tomato volatiles. *J. Food Sci.* 1970, *35*, 519–530.
294. Marlatt, C., Ho, C.-T., Chien, M. Studies of aroma constituents bound as glycosides in tomato. *J. Agric. Food Chem.* 1992, *40*, 249–252.
295. Meigh, D. F., Pratt, H. K., Cole, C. Production of carbonyl compounds by tomato fruit tissue in the presence of aliphatic alcohols. *Nature* (London) 1966, *211*, 419–420.
296. Miers, J. C. Formation of volatile sulfur compounds in processed tomato products. *J. Agric. Food Chem.* 1966, *14*, 419–423.
297. Nelson, P. E., Hoff, J. E. Tomato volatiles, effect of variety, processing and storage time. *J. Food Sci.* 1969, *34*, 53–57.
298. Petro-Turza, M. Flavor of tomato and tomato products. *Food Rev. Int.* 1987, *2*, 309–351.
299. Pyne, A. W., Wick, E. L. Volatile components of tomatoes. *J. Food Sci.* 1965, *30*, 192–196.
300. Schormüller, J., Grosch, W. Untersuchungen über aromastoffe von lebensmitteln. I. Ein beitrag zur analytik neutraler, in tomaten vorkommender carbonylverbindungen. *Z. Lebensm. Untersuch.-Forsch.* 1962, *118*, 385–387.
301. Schormüller, J., Grosch, W. Untersuchungen über aromastoffe von lebensmitteln. II. Über das vorkommen weiterer carbonylverbindungen in der tomate. *Z. Lebensm. Untersuch.-Forsch.* 1965, *126*, 38–42.
302. Schormüller, J., Grosch, W. Untersuchungen über aromastoffe von lebensmitteln. III. Ein beitrag zur analytik in tomaten vorkommender flüchtiger alkohole. *Z. Lebensm. Untersuch.-Forsch.* 1965, *126*, 188–193.
303. Shah, B. M., Salunkhe, D. K., Olson, L. E. Effects of ripening processes on chemistry of tomato volatiles. *J. Am. Soc. Hortic. Sci.* 1969, *94*, 171–175.
304. Siesso, V., Crouzet, J. Tomato volatile components, effects of processing. *Food Chem.* 1977, *2*, 241–244.
305. Spencer, M. S., Stanley, W. L. Flavor and odor components in the tomato. *J. Agric. Food Chem.* 1954, *2*, 1113–1115.
306. Stevens, M. A. Inheritance and flavor contribution of 2-isobutylthiazole, methyl salicylate and eugenol in tomatoes. *J. Am. Soc. Hortic. Sci.* 1970, *95*, 9–13.
307. Stevens, M. A., Kader, A. A., Albright-Holton, M., Algazi, M. Genotypic variation for flavor and composition in fresh market tomatoes. *J. Am. Soc. Hortic. Sci.* 1977, *102*, 680–689.
308. Viani, R., Bricout, J., Marion, J. P., Müggler-Chavan, F., Reymond, D., Egli, R. H. Sur la composition de l'arôme de tomate. *Helv. Chim. Acta* 1969, *52*, 887–890.
309. Yu, M. H., Olson, L. E., Salunkhe, D. K. Precursors of volatile components in tomato fruit. II. Enzymatic production of carbonyl compounds. *Phytochemistry* 1968, *7*, 555–557.
310. Yu, M. H., Salunkhe, D. K., Olson, L. E. Production of 3-methyl-butanal from L-leucine by tomato extract. *Plant Cell Physiol.* 1968, *9*, 633–637.
311. Maga, J. A. Cereal volatiles, a review. *J. Agric. Food Chem.* 1978, *26*, 175–178.
312. Abramson, D., Sinha, R. N., Mills, J. T. Mycotoxin and odor formation in moist cereal grain during granary storage. *Cereal Chem.* 1980, *57*, 346–351.
313. Abramson, D., Sinha, R. N., Mills, J. T. Mycotoxin and odor formation in barley stored at 16 and 20% moisture in Manitoba, Canada. *Cereal Chem.* 1983, *60*, 350–355.
314. Dravnieks, A., Watson, C. A. Corn odor classification from low-resolution gas-chromatographic profiles of headspace volatiles. *J. Food Sci.* 1973, *38*, 1024–1027.
315. Dravnieks, A., Reilich, H. G., Whitfield, J., Watson, C. A. Classification of corn odor by statistical analysis of gas chromatographic patterns of headspace volatiles. *J. Food Sci.* 1973, *38*, 34–39.
316. Richard-Moulard, D., Cahagnier, B., Poisson, J., Drapron, R. Evolution comparées des constituants volatils et de la microflore de maïs stockées sous différentes conditions de température et d'humidité. *Ann. Technol. Agric.* 1976, *25*, 29–44.
317. Seitz, L. M., Wright, R. L., Waruska, R. D., Rooney, L. W. Contribution of 2-acetyl-1-

pyrroline to odors from wetted ground pearl millet. *J. Agric. Food Chem.* 1993, *41*, 955–958.
318. Bullard, R. W., Holguin, G. Volatile components of unprocessed rice (*Oryza sativa* L.) *J. Agric. Food Chem.* 1977, *25*, 99–103.
319. Yasumatsu, K., Moritaka, S., Wada, S. Stale flavor of stored rice. *Agric. Biol. Chem.* 1966, *30*, 483–486.
320. Hougen, F. J., Quilliam, M. A., Curran, W. A. Headspace vapors from cereal grains. *J. Agric. Food Chem.* 1971, *19*, 182–187.
321. Kaminski, E., Libbey, L. M., Stawicka, S., Wasowicz, E. Identification of the predominant volatile compounds produced by *Aspergillus flavus*. *Appl. Microbiol.* 1972, *24*, 721–726.
322. Kaminski, E., Stawicki, S., Wasowicz, E. Volatile flavor compound produced by molds of *Aspergillus, Penicillium* and other *Fungi Imperfecti*. *Appl. Microbiol.* 1974, *27*, 1001–1004.
323. Kaminski, E., Stawicki, S., Wasowicz, E. Volatile flavor substances produced by molds on wheat grain. *Acta Aliment. Polon.* 1975, *1*, 153–157.
324. Kaminski, E., Stawicki, S., Wasowicz, E., Przybylski, R. Detection of deterioration of grain by gas chromatography. *Ann. Technol. Agric.* 1973, *22*, 401–407.
325. McWilliams, M., Mackey, A. C. Wheat Flavor Components. *J. Food Sci.* 1969, *34*, 493–496.
326. Miszczak, A., Forney, C. F., Prange, R. K. Development of aroma volatiles and color during postharvest ripening of 'Kent' strawberries. *J. Am. Soc. Hortic. Sci.* 1995, *120*(4), 650–655.
327. Stern, D. J., Buttery, R. G., Teranishi, R., Ling, L. Effect of storage and ripening on fresh tomato quality, Part I. *Food Chem.* 1994, *49*, 225–231.
328. Yahia, E. M., Liu, F. W., Acree, T. E. Changes of some odor-active volatiles in low-ethylene controlled atmosphere stored apples. *Food Sci. Technol.* 1991, *24*, 145–151.
329. Brackmann, A., Streif, J., Bangerth, F. Relationship between a reduced aroma production and lipid metabolism of apples after long-term controlled atmosphere storage. *J. Am. Soc. Hortic. Sci.* 1993, *118*(2), 243–247.
330. Hansen, M., Buttery, R. G., Stern, D. J., Cantwell, M. I., Ling, L. C. Broccoli storage under low-oxygen atmosphere: Identification of higher boiling volatiles. *J. Agric. Food Chem.* 1992, *40*, 850–852.
331. Springett, M. B., Williams, B. M., Barnes, R. J. The effect of packaging conditions and storage time on the volatile composition of Assam black tea leaf. *Food Chem.* 1994, *49*, 393–398.
332. Kallio, H., Leino, M., Koullias, K., Kallio, S., Kaitaranta, J. Headspace of roasted ground coffee as an indicator of storage time. *Food Chem.* 1990, *36*, 135–148.
333. Yamane, A., Yamane, A., Shibamoto, T. Propanethiol S-oxide content in scallions (*Allium fistulosum* L. variety Caespitosum) as a possible marker for freshness during cold storage. *J. Agric. Food Chem.* 1994, *40*, 1010–1012.
334. Moltenberg, E. L., Magnus, E. M., Bjorge, J. M., Nilsson, A. Sensory and chemical studies of lipid oxidation in raw and heat-treated oat flours. *Cereal Chem.* 1996, *73*(5), 579–587.
335. Snyder, J. M., Mounts, T. L., Holloway, R. K. Volatiles from microwave-treated, stored soybeans. *J. Am. Oil Chemist Soc.* 1991, *68*, 744–747.
336. Mazza, G., Pietrzak, E. M. 1990. Headspace volatiles and sensory characteristics of earthy, musty flavoured potatoes. *Food Chem.* 1990, *36*, 97–112.
337. Prasad, K., Stadelbacher, G. J. Effect of acetaldehyde on postharvest decay and market quality of fresh strawberries. *Phytopathology* 1974, *64*, 948–951.
338. Morris, J. R., Cawthon, D. L., Buescher, R. W. Effect of acetaldehyde on postharvest quality of mechanically harvested strawberries for processing. *J. Am. Hortic. Soc.* 1979, *104*, 262–264.
339. Pesis, E., Avissar, I. Effect of postharvest application of acetaldehyde vapour on strawberry decay, taste and certain volatiles. *J. Sci. Food Agric.* 1990, *52*, 377–385.
340. Pesis, E., Zauberman, G., Avissar, I. Induction of certain aroma volatiles in feijoa fruit by

postharvest application of acetaldehyde or anaerobic conditions. *J. Sci. Food Agric.* 1991, *54*, 329–337.

341. Shaw, P. E., Moshonas, M. G., Pesis, E. Changes during storage of oranges pretreated with nitrogen, carbon dioxide and acetaldehyde in air. *J. Food Sci.* 1991, *56*, 469–474.
342. Ahumada, H. M., Mitcham, E. J., Moore, D. G. Postharvest quality of ''Thomson Seedless'' grapes after insecticidal controlled-atmosphere treatments. *HortScience* 1996, *31*(5), 833–836.
343. Ke, D., van Gorsel, H., Kader, A. A. Physiological and quality responses of ''Bartlett'' pears to reduced O_2 and enhanced CO_2 levels and storage temperature. *J. Am. Soc. Hortic. Sci.* 1990, *115*(3), 435–439.
344. Sun, J.-B., Severson, R. F., Kays, S. J. Effect of heating temperature and microwave pretreatment on the formation of sugars in Jewel sweet potato. *J. Food Quality* 1994, *17*, 447–456.
345. Raina, B. L., Agarwal, S. G., Bhatia, A. K., Gaur, G. S. Changes in pigments and volatiles of saffron (*Crocus sativus* L.) during processing and storage. *J. Sci. Food Agric.* 1996, *71*, 27–32.
346. Widjaja, R., Craske, J. D., Wootton, M. Changes in volatile components of paddy, brown and white fragrant rice during storage. *J. Sci. Food Agric.* 1996, *71*, 218–224.
347. Burkholder, W. E., Ma, M. Pheromones for monitoring and control of stored-product insects. *Annu. Rev. Entomol.* 1985, *30*, 257–272.
348. Thomas, C. M., Dicke, R. J. Attraction of the grain mite, *Acarus siro* (Acarina, Acaridae), to solvent extracts of fungi associated with stored-food commodities. *Ann. Entomol. Soc. Am.* 1972, *65*, 1069–1073.
349. Thomas, C. M., Dicke, R. J. Response of the grain mite *Acarus siro* (Acarina, Acaridae) to fungi associated with stored-food commodities. *Ann. Entomol. Soc. Am.* 1971, *64*, 63–68.
350. Pierce, A. M., Pierce, H. D., Jr., Borden, J. H., Oehlschlager, A. C. Fungal volatiles, semiochemicals for stored-product beetles (Coleoptera, Cucujidae). *J. Chem. Ecol.* 1991, *17*, 581–597.
351. Dolinski, M. G., Loschiavo, S. R. The effect of fungi and moisture on the locomotory behavior of the rusty grain beetle, *Cryptolestes ferrugineus* (Coleoptera, Cucujidae). *Can. Entomol.* 1973, *105*, 485–490.
352. Pierce, H. D., Jr., Pierce, A. M., Johnston, B. D., Oehlschlager, A. C., Borden, J. H. Aggregation pheromone of square-necked grain beetle, *Cathartus quadricolis* (Guér.). *J. Chem. Ecol.* 1988, *14*, 2169–2184.
353. Pierce, A. M., Pierce, H. D., Jr., Borden, J. H., Oehlschlager, A. C. Production dynamics of cucujolide pheromones and identification of 1-octen-3-ol as a new aggregation pheromone for *Oryzaephilus surinamensis* and *O. mercator* (Coleoptera, Cucujidae). *Environ. Entomol.* 1989, *18*, 747–755.
354. Pierce, A. M., Pierce, H. D., Jr., Oehlschlager, A. C., Borden, J. H. 1-Octen-3-ol, attractive semiochemical for foreign grain beetle, *Ahasverus advena* (Waltl) (Coleoptera, Cucujidae). *J. Chem. Ecol.* 1991, *17*, 567–580.
355. Sinha, R. N., White, N. D. G., Demianyk, C. J., Kawamoto, H. Effects of fungal and acarine volatile chemicals on some stored-product beetles. *J. Appl. Zool. Res.* 1992, *3*, 106–117.
356. Connick, W. J., French, R. C. Volatiles emitted during the sexual stage of the Canada Thistle Rust fungus and by thistle flowers. *J. Agric. Food Chem.* 1991, *39*, 185–188.
357. Mattick, A. T. R., Cheeseman, G. C., Berridge, N. J., Bottazzi, V. The differentiation of species of *Lactobacilli* and *Streptococci* by means of paper partition chromatography. *J. Appl. Bacteriol.* 1956, *19*, 310–313.
358. Cheeseman, G. C., Berridge, N. J., Mattick, A. T. R., Bottazzi, V. The differentiation of bacterial species by paper chromatography. III. An explanation of the *Lactobacillus casi—plantarum* group. *J. Appl. Bacteriol.* 1957, *20*, 205–209.
359. Reiner, E. Identification of bacterial strains by pyrolysis-gas liquid chromatography. *Nature (London)* 1965, *206*, 1272–1274.

360. Garner W., Gennaro, R. M. Related bacteria differentiated by gas chromatography. *Chem. Eng. News* 1965, *43*, 69–70.
361. Brown, J. P., Cosenza, B. J. Fatty acid composition of lipids extracted from three spherical bacteria. *Nature (London)* 1964, *204*, 802–803.
362. Cattaneo, C., Lucchesi, M., De Ritis, G. C., Pietropaolo, C., Rossi, P., Frisani, G. Gas chromatographic separation of the C_{12}-C_{20} fatty acids in the lipids of mycobacteria. *Chem. Abstr.* 1965, *62*, 16646F.
363. Henis, U., Gould, J. R., Alexander, M. Detection and identification of bacteria by gas chromatography. *Appl. Microbiol.* 1966, *14*, 513–524.
364. O'Brien, R. T. Identification of micro-organisms by gas chromatographic analysis of metabolic products. 26th Annual Meeting Inst. Food Technologists, Portland, Oregon, Abstract 120, 1966.
365. Guarino, P. A., Kramer, A. Gas chromatographic analysis of head-space vapors to identify micro-organisms in foods. *J. Food Sci.* 1969, *34*, 31–37.
366. Bassette, R., Claydon, T. J. Characterization of some bacteria by gas chromatographic analysis of head space vapors from milk cultures. *J. Dairy Sci.* 1965, *48*, 775.
367. Bassette, R., Bawdon, R. E., Claydon, T. J. Production of volatile materials in milk by some species of bacteria. *J. Dairy Sci.* 1966, *50*, 167–171.
368. Toan, T. T., Bassette, R., Claydon, T. J. Methyl sulfide production by *Aerobacter aerogenes* in milk. *J. Dairy Sci.* 1965, *48*, 1174–1178.
369. Bawdon, R. E., Bassette, R. Differentiation of *Escherichia coli* and *Aerobacter aerogenes* by gas-liquid chromatography. *J. Dairy Sci.* 1966, *49*, 624–627.
370. Detection and identification of microorganisms, enterotoxins, and mycotoxins. *Food Technol.* 1987, *41*, 59–72.
371. Fox, A., ed. *Analytical Microbiology Methods, Chromatography and Mass Spectrometry*. Plenum Press: New York. 1990, p. viii.
372. Morgan, S. L., Fox, A. *Chemotaxonomic Characterization of Microorganisms by Capillary Gas Chromatography-Mass Spectrometry*. Columbia: University of South Carolina, Dept. of Chemistry, 1988.
373. Stern, N. J. The inability of pyrolysis gas-liquid chromatography to differentiate selected foodborne bacteria. *J. Food Protect.* 1982, *45*, 229–233, 237.
374. Decallonne, J., Delmee, M., Wauthoz, P., El-Lioui, M., Lambert, R. A rapid procedure for the identification of lactic acid bacteria based on the gas chromatographic analysis of the cellular fatty acids. *J. Food Protect.* 1991, *54*, 217–224.
375. Gitaitis, R. D., Beaver, R. W. Characterization of the fatty acid methyl ester content of *Clavibacter michiganensis* subsp. *michiganensis. Phytopathology* 1990, *80*, 318–321.
376. Roy, M. A. Use of fatty acids for the identification of phytopathogenic bacteria. *Plant Dis.* 1988, *72*, 460.
377. Sinha, R. N., Tuma, D., Abramson, D., Muir, W. E. Fungal volatiles associated with moldy grain in ventilated and non-ventilated bin-stored wheat. *Mycopathologia* 1988, *101*, 53–60.
378. Tuma, D., Sinha, R. N., Muir, W. E., Abramson, D. Odor volatiles associated with microflora in damp ventilated and non-ventilated bin-stored bulk wheat. *International J. Food Microbiol.* 1989, *8*, 103–119.
379. Tanchotikul, U., Hsieh, T. C.-Y. An improved method of quantification of 2-acetyl-1-pyrroline, a ''popcorn'' like aroma, in aromatic rice by high-resolution gas chromatography/mass spectrometry/selected ion monitoring. *J. Agric. Food Chem.* 1991, *39*, 944–947.
380. Tuma, D., Sinha, R. N., Muir, W. E., Abramson, D. Odor volatiles associated with mite-infested bin-stored wheat. *J. Chem. Ecol.* 1990, *16*, 713–724.
381. Turk, A., Messer, P. J. Green lemon mould gaseous emanation products. *J. Agric. Food Chem.* 1953, *1*, 264–268.
382. Wyllie, S. G., Allie, S., Filsoot, M., Jennings, W. G. Headspace sampling: Use and abuse.

In Charalambous, G., ed., *Analysis of Foods and Beverages, Headspace Techniques*. Academic Press: New York, 1978, pp. 1–79.

383. Lyew, D., Gariépy, Y., Ratti, C., Raghavan, G. S. V., Kushalappa, A. C. An apparatus to sample volatiles in a commercial potato storage facility. Applied Eng. in Agric. 1999, 15, 243–247.

384. van Es, A., Hartmans, K. J. Respiration. In: *Storage of Potatoes: Post-Harvest Behaviour, Store Design, Storage Practice, Handling*. Pudoc: Wageningen, 1987, pp. 133–140.

22

Irradiation of Fruits, Vegetables, Nuts and Spices

MONIQUE LACROIX

Canadian Irradiation Centre and University of Quebec, Laval, Quebec, Canada

MICHÈLE MARCOTTE

Agriculture and Agri-Food Canada, Saint-Hyacinthe, Quebec, Canada

HOSAHALLI S. RAMASWAMY

McGill University, Sainte-Anne-de-Bellevue, Quebec, Canada

1 INTRODUCTION

The world population approaches 7 billion people. This level of population adds to the pressure on our limited resources and significantly affects our ability to feed ourselves (Satin, 1996). Worldwide alimentary self-sufficiency and security are the main objectives that we are trying to reach to suppress hunger and malnutrition, to reduce alimentary losses/waste, and to protect human health. Unfavorable climatic conditions, such as very hot and humid temperatures and the absence of a continuous cold chain for protection during storage, transportation, and marketing of foods, contribute to the increase of alimentary losses. In developed countries, losses are usually small. Nevertheless, these losses may reach 40% for some products (e.g., strawberries and mushrooms). In developing countries, losses tend to be rather high. Storage losses alone have been estimated as low as 10% for cereal grains to as high as 75% for more vulnerable crops. Significant insect damage can occur even before storage, resulting in losses that may account up to 60% (Satin, 1996). In addition to these losses consumers have to cope with diseases resulting from the development of pathogenic organisms that may be found naturally in foods and result from the production of toxic substances. Microbial contamination is responsible for a variety of alimentary diseases, such as toxoplasmosis, salmonellosis, campilobacteriosis, listeriosis, trichinellosis, and cholera. According to the United Nations, over 30% of the mortality rate worldwide is caused by alimentary diseases. In the 1990s ingestion of raw vegetables, fruits, and fruit juices was linked to outbreaks of food-borne illnesses in the United States (Thayer and Rajkowski, 1999). In contrast to the extensive studies on irradiation to control pathogens on meat and poultry products, very few studies on the value of ionizing radiation for the elimination of pathogens on or in fruit juices and fresh fruits and vegetables are available.

Irradiation technology is one solution to these problems. If used in combination with other technologies such as refrigeration, irradiation technology reinforces the global efficacy of the treatment through synergy. Moreover, by adding another type of treatment, it is possible to reduce radiation doses, therefore retaining/improving product quality. It is now well recognized that the most effective combined disinfestation treatment is the use of variations in temperatures before, during, or after irradiation (Tilton and Brower, 1987). In the control of food-borne microorganisms, without adversely affecting organoleptic qualities of foods, one of the most promising combined treatments is irradiation with mild heat (Farkas, 1990). However, some spices or seasoning agents are also known to possess powerful antimicrobial and antioxydant activities (Conner and Beuchat, 1984; Lattaoui and Tantaoui-Elaraki, 1994; Lacroix et al., 1997; Mahrour et al., 1998a, 1998b; Ouattara et al., 2001). The utilization of these plants has a synergistic effect with irradiation and contributes a flavoring agent to foods (Aboultab et al., 1986; Mahrour et al., 1998a, 1998b).

Today there is a general consensus in the scientific community about the wholesomeness of foods treated by gamma irradiation (CAST, 1986; WHO, 1988; Diehl, 1990). Several agricultural products are important commodities in international trade, often the trade of these products is seriously hampered by the infestation of several species of insects and mites. The presence of parasites, some microorganisms, yeasts, and molds is also the source of problems. Irradiation alone or combined with other processes can contribute to ensuring food safety of healthy and compromised consumers (pregnant mothers, immunocompromised acquired immunodeficiency syndrome [AIDS] patients, medicated patients, and aging persons), satisfying quarantine requirements, and controlling severe losses during transportation and commercialization. Several investigations and transfer technology studies have demonstrated the usefulness of mild heat treatment prior to low-dose irradiation in extending the shelf life of certain fresh exotic fruits without affecting their normal quality attributes. By using combined treatments, a minimal level of irradiation that is less detrimental to quality attributes of the commodity and may be effective in inactivating toxigenic microorganisms can be used.

2 TECHNICAL ASPECTS OF FOOD IRRADIATION

2.1 Characteristics of Ionizing Radiation

The ionizing radiation that is being discussed is part of the electromagnetic spectrum with the radio waves on one end and the high-energy x-rays and gamma rays at the other end. In the middle there are the visible light rays, with infrared and ultraviolet rays on each side. Between the radio and infrared rays are the microwaves, which are becoming common in households. Since all these are part of the electromagnetic spectrum, they also have several characteristics in common. They are all waves with a characteristic wavelength and frequency and a certain amount of associated energy. The higher the wavelength, the smaller is the associated energy.

The radio waves with long wavelengths of 30 cm to 3 km have very limited energy associated with them. Microwaves, although relatively low-energy waves, can cause molecular vibrations in materials like food that contain moisture and fat and can result in very rapid heating. The x- and gamma rays at the other end of the spectrum emit very-short-wavelength radiation and have very high associated energy levels. When made to bombard against materials, they can knock off an electron from an atom or molecule, causing ionization. For this reason, these are often called *ionizing radiations*. When food irradiation is discussed, it is mainly with respect to ionizing radiations.

The energy associated with electromagnetic radiation is related to wavelength and frequency, as follows:

Frequency = velocity of light/wavelength of radiation
Energy = Planck's constant × frequency

2.2 Sources of Ionizing Radiations

The main source of gamma radiation is cobalt 60, which is a radioactive isotope produced from cobalt 59. A second source is cesium 137, which is a spent fuel of nuclear reactors and also produces gamma rays. A third source is β-rays, which are a stream of electrons. Since the associated energy levels of these rays are too low to have any practical value in terms of radiation preservation, they need to be accelerated (in cyclotrons, linear accelerators, etc.) to make them acquire the required energy. Careful precautions should be taken to ensure that all electrons have enough energy. If the acquired energy is too high, induced radioactivity in foods can occur with irradiation.

2.3 Units of Radiation

2.3.1 Intensity

There are two types of units for measuring radiation: intensity and dose. The intensity refers to the source and the dose to the product. The intensity of a radioactive source is measured in number of disintegrations per second; the unit of intensity is the curie, which represents 3.7×10^{10} disintegrations per second.

The radioactive source is different from conventional energy sources in that it is continuously emitting radiation. As it does, it decays, in the process of radioactive decay. The radioactive decay is assumed to follow first-order kinetics, hence, changes in intensity with respect to time can be represented by

$$I = I_0 \exp(-dt)$$

where I is the intestity at time t, I_0 is the initial intensity, and d is the radioactive decay constant. Another term that is frequently used with respect to radioactive decay is *half-life* (λ) of the radioactive isotope, which is defined as the time interval in which the intensity of a radioactive source is reduced by one-half. It can be shown that $\lambda = 0.693/d$. Half-life of cobalt 60 is 5.27 years, of cesium 137, 30.2 years, and of radium, 1600 years.

2.3.2 Dose

The treatment received by the food product is characterized by the radiation dose, the quantity of energy absorbed by the food while it is exposed to the radiation field. The international unit of measurement is the gray (Gy); 1 Gy represents 1 J of energy absorbed per kilogram of irradiated product. One gray is equivalent to 100 rad. The desired dose is achieved by the time of exposition and by the location of the product relative to the source. The amount of energy absorbed by the food also depends on the mass, bulk density, and thickness of the food. For each kind of food, a specific dose has to be delivered to achieve a desired result. If the dose is less than appropriate, the intended preservation effect may not be achieved. If the dose is excessive, the food may be damaged and unacceptable for consumption. In 1980, an international expert committee (Food Agriculture Organization [FAO], International Atomic Energy Agency [IAEA], and the World Health Organization [WHO]) concluded, on the basis of the data available in the literature, that irradiation of any food product up to a total dose of 10 kGy (10,000 Gy) does not present any toxicological danger and does not introduce any particular nutritional or microbiological problem. Doses are classified into three categories: low, medium, and high doses. For fruits and vegetables, low doses of up to 1 kGy are used to inhibit sprouting of certain crops such as potatoes and onions, to disinfect fruits and vegetables from insects and parasites, and to delay physiological processes (e.g., ripening) of fruits. A delay in postharvest ripening can occur only in a climacteric fruit that ripens normally after harvest. To eliminate spoilage microorganisms and to extend the shelf life of fruits and vegetables, medium doses of 1 to 10 kGy are necessary. Fungi and bacteria are known to have serious pathological effects on fruits. Senescence and physiological breakdown occur in both climacteric and nonclimacteric fruits as they age in postharvest storage. High doses of 10 to 50 kGy are used to decontaminate herbs, spices, and food ingredients. At dose levels at which the growth of spoilage microorganisms such as fungi on a fruit can be controlled, a number of problems may be encountered because the dose levels used are always higher than those used for delaying ripening of fruits. The problems are usually related to the physiological response of the commodity being irradiated. Thus, an effective treatment is dependent on the relative sensitivity of the host (e.g., fruit) and the spoilage microorganism (e.g., fungi). The host must be able to tolerate the treatment without any apparent injury or other undesirable side effect. It is clear that ionizing energy has some potential applications to fruit commodities but has some limitations. In most cases for fruits, doses are determined by the host rather than by the pathogen.

3 IRRADIATION FACILITIES

A variety of food irradiation facilities are available on the market. They vary with respect to design and physical arrangement. An essential requirement for the industrial use of food irradiation is an economical source of radiation energy that can deliver the treatment. Today, electron accelerators and synthesized radioisotopes are the two types that satisfy this requirement. The systems have comparable efficiency but differ in the constraints implied in their use. These systems must be installed in a shielded room to allow the treatment to take place without compromising the health of humans in the vicinity of the irradiator.

Electron accelerators (Fig. 1) are simpler to operate because they can be turned on and off easily. There are available at a relatively low cost. The limitations of these machines are related to the low penetrating power of accelerated electrons. This technology is suitable only for processing of thin layers or surface treatment of food materials. The high voltage of electron beams must be lower than or equal to 10 MeV in order to prevent induction of radioactivity in food materials being irradiated. The power of the machine

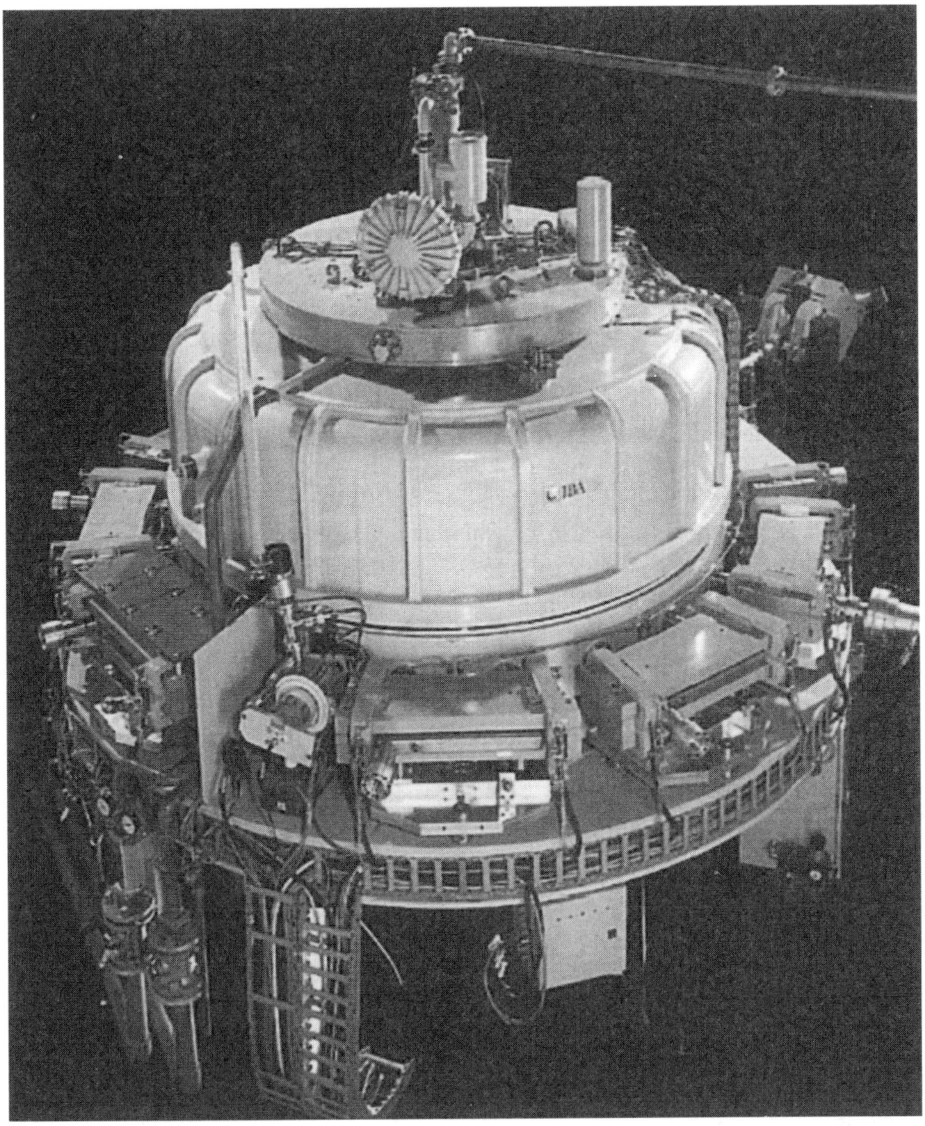

Fig. 1 View of an electron accelerator.

can vary from 5 to 20 kW. Accelerated electrons penetrate to a depth of about 4 cm in water. There are many types of electron accelerators; the main differences among them are in the methods used for creating the high voltage required. From one side, a hot filament emits electrons into an evacuated chamber. These electrons are attracted by a high positive electric potential and are focused into a narrow beam. The electron beam passes between the poles of an electron magnet with a changing magnetic field. The beam sweeps from side to side. Food products are usually loaded with a conveyor to the irradiation room. It is very important to survey the energy level of accelerated electrons continuously and to ensure that it does not surpass the value that would render the food radioactive.

With synthesized radionuclides, gamma radiation is produced continuously at a specific energy level. Selected radioisotopes such ^{60}Co and ^{137}Cs are used. These radioisotopes do not produce gamma radiation with enough energy to induce radioactivity in the food material and ^{60}Co can be produced commercially in large quantities and in a short time. A separate shielded storage space, usually a pool of water, in which the ^{60}Co source can be withdrawn while there is no irradiation treatment must be provided. Very careful procedures must be performed during the transportation and installation of the source. An advantage of gamma rays over a machine source is their ability to penetrate bulk items. Gamma rays with an energy of 0.15 to 4 MeV penetrate deeply into about 30 cm of water. This capacity allows the processing of large quantities of foods. Isotope sources are also inherently reliable. However, exposure time can be fairly long as compared to that of electron accelerators. During a food irradiation treatment, products are loaded in the irradiation room and the source is then raised to an irradiation position. The vast majority of irradiation plants operate with a radionuclide or ^{60}Co source of energy because of its ease of operation.

Industrial processes using gamma rays and accelerated electron sources can also be divided into two main categories: batch and continuous processes. In batch facilities, a given quantity of food is irradiated for a precise period. In continuous irradiation plants, food is passed through the irradiation chamber at a controlled rate calculated to ensure that all the food receives exactly the intended dose. Batch facilities are flexible and simpler in operation. A wider range of doses is available, and the spatial relationship between the source and the target can be arranged. These facilities are well adapted to experimentation

Fig. 2 Batch type irradiator at CRDA.

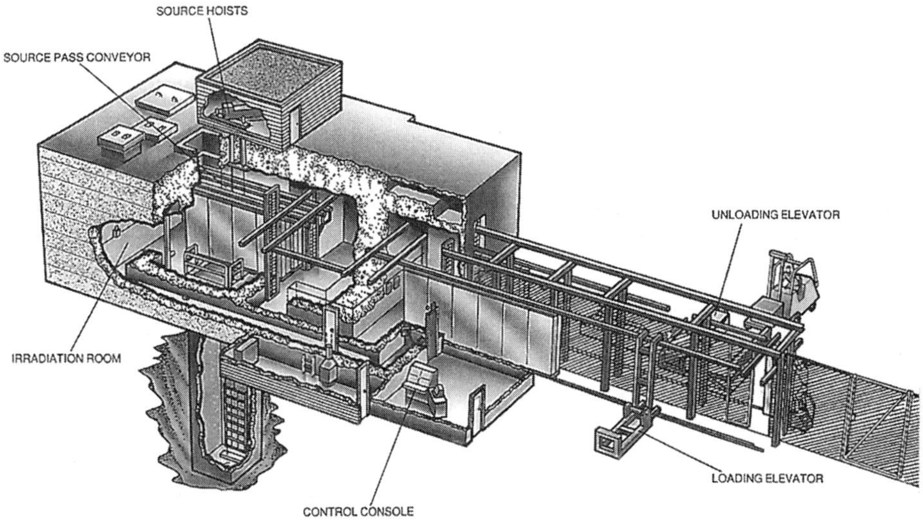

Fig. 3 View of a carrier irradiator at CIC.

and research. Figure 2 shows a batch-type system installed at Agriculture and Agri-Food Canada's Food Research and Development Centre (CRDA).

Continuous plants are suitable for large volumes, leading to a significant economy of scale and a lower labor cost. In continuous facilities, the exposure time is controlled by regulating the speed of the transport mechanism. Figure 3 shows a carrier-type irradiator installed at the Canadian Irradiation Centre (CIC).

4 MEASUREMENT OF THE IRRADIATION TREATMENT

Measuring the irradiation dose is known as *dosimetry*. The three main purposes of dosimetry in food irradiation are (a) to develop the proper dose for the food commodity under research, (b) to obtain data for commissioning the food product through the regulatory agency, and (c) to establish the quality control procedure in the food production plant. It is also used to determine the configuration of the irradiation field after the installation of the irradiator or any changes that occurred in the irradiation facility. McLaughlin and associates (1982) have reviewed in detail the technical aspects of dosimetry.

In general, experimental dosimetry is preferred to calculation methods. Dosimeters are placed within the food product being irradiated to measure the distribution of the absorbed energy and to determine the maximal and minimal doses absorbed by the food. The measurement of the absorbed dose must be as accurate as possible to establish correct procedures for food preservation and quality control in irradiation processing. The choice of a dosimeter is mainly dependent upon (a) its reliability for calibration and standardization; (b) its reproducibility at all specified dose levels; (c) its limited dependence on process conditions (e.g., radiation spectra, dose rate), (d) its capacity for correction for systematic errors caused by environmental factors (e.g., humidity, temperature, light effects), (e) its degree of equivalence to the food product, (f) its commercial availability at low cost, and (g) its ease of handling. Primary or reference dosimeters (e.g., calorimeters) are used to calculate the amount of energy absorbed or the dose directly from the measurements. Secondary or routine dosimeters give measurements that vary with the amount of energy absorbed. The dose is estimated from a calibration curve in which measurements obtained from a secondary dosimeter are plotted against the measurements made under similar conditions by the primary dosimeter. Once secondary dosimeters are properly calibrated, they can be used to estimate the dose absorbed by the food. Of course, a secondary dosimeter that has absorption properties that match those of the food has to be selected. The main sources of inaccuracy are variations of dosimeter response with the dose rate, spectral energy, dosimeter batch variations, readout anomalies, instabilities, and environmental fluctuations during and after the treatment. Therefore, it is extremely important to have a quality control procedure to test the dosimeters. The upper and lower limits of the dose of the food product for a given process depend on two factors, (a) the dose gradients and (b) the dose extremes of a low-density item in a process run.

Ceric-cerous (MDS Nordion International Inc.) is an aqueous chemical dosimeter. Ceric-cerous dosimeters are used mainly for research purposes. On irradiation, ceric ions are reduced to cerous ions. The readout system is simple, composed of an electrochemical cell and a digital multimeter. It is relatively inexpensive. The range covers most food requirements (low dose, 0.6 to 10 kGy; high dose, 5 to 50 kGy). Aqueous solutions are stored in glass ampoules that are breakable and are relatively large. It is usually used in the development of the technology.

Fricke is a dosimeter generally prepared in-house. The Fricke dosimeter is widely used for calibration and for comparison between laboratories. It is one of the most useful reference dosimeters for intercomparisons by national laboratories. Aqueous solutions of Fe^{+2} ions are converted into Fe^{+3} ions. An ultraviolet spectrophotometer is used as a readout system to determine the concentration of Fe^{+3} at 305 nm; the range is 0.02 to 0.4 kGy. Aqueous solutions are kept in glass ampoules that are breakable and are relatively large.

For routine dosimetry during processing of fruits, *FWT 60 radiochromic film* (Far West Technology) is the most commonly used dosimeter. Its ruggedness is excellent; its size is small, and it is relatively inexpensive. The dosimeter is usable throughout the anticipated dose range (0.1–100 kGy) in food processing. Other commercial dosimeters available for routine measurement of the absorbed dose are the Harwell YR (Didcot, United Kingdom) and the Farwest Optichromic (Goleta, California).

4.1 Detection of Radiation

Irradiation detection methods do not require any special approval from regulatory agencies even though irradiation doses for treating various food products have been accepted in over 30 countries. Nevertheless, the development of methods to detect irradiation is important for a number of reasons, such as differences in national legislation, control of international trade, quality control, and consumer reassurance. Several attempts to detect irradiation on an experimental basis have proved most difficult because the changes induced by irradiation are small and are not radiation specific. It is not likely that a universal method will be developed for all foods, but methods can be devised for specific groups. Willemot and colleagues (1996) have reviewed detection methods for fruits in detail. At present, the most promising irradiation detection method are electron spin resonance (ESR) and thermoluminescence. Further studies in the development of instrumentation and standardization of methods will eventually lead to general practical applications.

5 LEGISLATION

National governments enforce regulations for irradiated foods and irradiation facilities. Regulatory agencies determine which food may be treated by irradiation under which conditions and for which purpose. They also prescribe the type of information that must be included in the label. Food irradiation plants are inspected as all food processing operations are and have to comply with regulations by authorities responsible for the safety of the application of irradiation.

5.1 World

In 1983, the Codex Alimentarius Commission accepted that foods irradiated up to 10 kGy were safe and wholesome and therefore toxicological testing was no longer necessary. In 1988, the FAO, WHO, IAEA and the International Trade Centre–United Nations Conference on Trade and Development (UNCTAD)–General Agreement on Tariffs and Trade (GATT) jointly organized the international conference on the acceptance, control of, and trade in irradiated food in Geneva, Switzerland (IAEA, 1989). The use of food irradiation was endorsed by government-designated experts of 57 countries. Since then, countries around the world have aligned their regulations with the Codex General Standard of Irradiated Foods and have cleared many foods for irradiation. It is in South Africa that the greatest variety of fruits are processed annually with ionizing energy, including mangoes, papayas, bananas, litchis, tomatoes, and strawberries. Although many attempts have been made to internationalize regulation of this technology in the world, most countries continue to approve its use on a case-by-case basis. Today, at least 36 countries have collectively approved irradiation of more than 50 different foods.

5.2 United States

In 1986 the U.S. Food and Drug Administration (FDA) issued a regulation (CFR title 21, part 179) that permits irradiation of fruits at a maximal dose up to 1 kGy for certain benefits such as insect disinfestation and ripening, growth, and maturation inhibition. Not only specific applications to foods are specified in this regulation; type and sources of radiation—gamma rays from cobalt 60 or cesium 137, accelerated electrons from a machine source not to exceed 10 MeV, or X rays from a machine source not to exceed 5 MeV—also are. As well, labeling is required. The internationally recognized logo (Fig. 4) must appear on the package or the display as well as either of the following statements: ''Treated with radiation'' or ''Treated by irradiation.'' If irradiated fruits are shipped to a manufacturer for further processing, a ''Do not irradiate again'' label must be displayed. Anyone may irradiate fruits in compliance with this regulation without further permission from FDA. This regulation is a full authorization; there are no FDA licensing procedures for plant facilities. Plants must comply with current FDA regulations for good manufacturing practices for the production, handling, and storage of foods (CFR title 21, part 110). FDA monitors food processing by periodic unannounced inspections of the facilities under its jurisdiction. Moreover, industrial activity must conform to laws designed to protect workers' safety and the environment. The use of a radioisotope such as cobalt 60 or cesium 137 as a source of radiation requires a license issued by the United States Nuclear Regulation Commission (USNRC). In the case of electron accelerators and x-ray machines, safety concerns arise only when the machine is in use. Manufacturers of machine sources of radiation are not required to be licensed by the USNRC, but they must submit a report

Fig. 4 Internationally recognized logo for irradiated foods.

to FDA's Center for Devices and Radiological Control for Health and Safety Act. Although this legislation does not apply to food processors, they should ensure that the equipment has been reported by the manufacturer. It has to be pointed out that some states require registration or licensing for facilities that use machine generated radiation.

5.3 Canada

In Canada, products must be irradiated in plants specially constructed for this purpose. These plants must possess a license issued by the Atomic Energy Control Board (AECB), which inspects regularly to verify safety and compliance. As well, irradiated food products must be approved in advance by Health Canada. In the current legislation, a dose of 0.15 kGy may be used to inhibit germination of potatoes and onions. As well, wheat, flour, and whole wheat flour may be irradiated with a dose of 0.75 kGy to prevent insect infestation during storage. Whole or ground spices and dehydrated seasonings may be irradiated with a maximal dose of 10 kGy to reduce the initial microbial level (Jategaonkar and

Marcotte, 1993). Today, irradiation of fruits is not permitted by the law. To get permission to irradiate fruits, processors must prepare a submission to Health Canada according to the preclearance requirements for proposed irradiated food. Details of the purpose of the proposed irradiation; the irradiation dose required; the chemical, physical, microbial, and nutritional effects; other processes to be applied before and after irradiation; and recommended conditions of storage and shipment must be specified. Then, the petition is evaluated in terms of whether it addresses and satisfies the preclearance requirements. If accepted, the petition follows due regulatory processes and the fruit is added to the clearance list for irradiation in the Canadian Food and Drug Regulations. If the irradiated food is sold in the marketplace, the processor must also comply with labeling requirements. Identification of the irradiated food product or food ingredient, using the international symbol, is required if it makes up more than 10% of product content. As well, a written statement that the food has been irradiated must be either written on the package or displayed beside the irradiated food.

6 ECONOMIC ASPECTS

The beneficial effects of irradiation of fruits such as disinfestation, reduction of spoilage, and extension of shelf life offer potential for significant cost savings. As a result of the limited application of irradiation, the economical feasibility of food irradiation has to be examined on a case-by-case basis. One concern is whether the process is economically viable. According to many authors (Vasseur, 1991), it is well known that the cost of irradiating with gamma rays from an isotope source such as ^{60}Co is similar to the cost of irradiating with electrons from an accelerator such as a 10-MeV linear accelerator. Both the gamma ray apparatus and electron accelerators are costly.

The cost of irradiating various foods has been estimated at between U.S. $0.02 and U.S. $0.40 per kilogram (WHO, 1988). According to Mills (1987), a study carried out by Fundacion Chili in 1985 demonstrated that the cost of the irradiation treatment for fruit disinfestation would be $0.025 per kilogram on a $1 million capital investment in a cobalt 60 batch-type irradiator. Kunstadt and Steeves (1993) discussed the economics of food irradiation application and the effects of various parameters on the cost of the treatment or of unit processing. They presented a detailed analysis of the costs for differents types of cobalt 60 irradiators designed for fruit disinfestation. A number of calculations were performed for a variety of doses, packing densities, and mass of production. Unit processing cost decreases rapidly with the increase of throughputs. Costs stabilize once the minimal economic throughput is reached and surpassed. Thus, a successful irradiator operates at a level exceeding the minimal economic mass of production. The effect of increasing the dose on unit processing costs is linear. As the dose is increased, the unit processing cost is also increased. It is dependent only on processing time. The effect of densities is also linear and related to the irradiator's cobalt utilization efficiency. As density is lowered, the cost increases. Similar costs were derived for different types of cobalt 60 irradiators. According to this design (i.e., 60-Mkg/year, 0.15 kGy, and 0.4 gcm^3), unit processing costs vary from below $0.01/kg or 250 Mkg/year to $0.107/kg or 10 Mkg/year.

It has to be pointed out that the calculated cost of the irradiation treatment is often offset by the increased market price differential because produce that was not previously accessible is marketed, with the result that waste and food losses are reduced at every level (Corrigan, 1993).

7 IRRADIATION TREATMENTS FOR FRUITS AND VEGETABLES

7.1 Dried Fruits and Vegetables

Dried fruits and vegetables have played a prominent role in the feeding of humans and animals in many civilizations and in different parts of the world. However, spoilage of dried fruits and vegetables as a result of insect infestation, microbial growth, deterioration, and chemical changes is a serious problem, especially in hot and humid climates (Wahid et al., 1989, 1987). The low moisture content of dried fruits and vegetables ordinarily provides the needed preservative action; in cases in which the moisture content is not sufficiently low, the fruit, irradiation treatments may be useful. Most commonly, the principal use for irradiation of dried fruits and vegetables has been to secure insect disinfestation. Dried vegetables are commonly used as soup ingredients and contain a large number of microorganisms capable of causing spoilage (Wilkinson and Gould, 1996). A radiation dose of 5–10 kGy can reduce contamination, depending on the organisms present. Yeasts and bacterial spores are more radiation resistant than Enterobacteriaceae. Insect pests are easily controlled by radiation treatment. Losses, caused by weevils, of dry green beans used for bean sprouts can be reduced by irradiation with no change in germination or rootlet length (Lan et al., 1991). Changes in flavor and color are minimal in some products, even at doses of 10 kGy (Kiss et al., 1974). However, certain vegetable products (e.g., asparagus, mushrooms, and onions) undergo browning. Irradiation at above 5 kGy decreases required cooking time of dried vegetables (Farkas, 1988; Kiss et al., 1974). Irradiation causes partial breakdown of cellulose, pectin, and starch and mobilization of the calcium in the plant tissue. This results in a decrease in product hardness and an increase in absorption capacity of the dried product. Gamma irradiation also slightly increases solubility of red kidney beans and has been effective in reducing mold contamination (Dogbevi et al., 1999).

Dried fruits of interest are apples, apricots, currants, dates, figs, pears, prunes, quinces, and raisins (Urbain, 1986). For raisins, zante currants, prunes, and dried apricots, typical insects that infest dried fruits include the saw-toothed grain beetle, *Oryzaephilus surinamensis* (Brower and Tilton, 1970); the fig moth, *Ephesia cautella*; and the Indian meal moth, *Plodia interpunctella*. The moth *Nemapogon granellus* (L) has been identified in dried mushrooms. Other insects may be present in various dried fruits. As one would expect, destruction of other insect forms (e.g., eggs or larvae) requires smaller irradiation doses. For example, eggs of *P. interpunctella* treated with 0.2 kGy do not hatch. If only eggs or larvae are present, a dose of 0.2 kGy is sufficient. In dried apricots and figs infested with *Corcyra cephalonica* and *Cadra cautella* or dried dates and raisins infested with *Tribolium castaneum* a radiation dose of 0.25 kGy to be combined with storage at low temperatures (10°C–20°C) is needed to ensure disinfestation for 1 year (Wahid et al., 1989). Adult forms may maintain activity for several days after irradiation with doses as large as 1 kGy and present problems for quarantine requirements (Urbain, 1986). Several researchers have tried to reduce the irradiation dose for disinfestation by adding another treatment that predisposes insects to be damaged by irradiation. One of the more promising areas of research to increase the efficacy of radiation disinfestation is the use of temperature variations before, during, or after the irradiation treatment (Tilton and Brower, 1983).

Changes in temperature apparently modify the effects of radiation primarily by affecting the metabolic state of the insect (Tilton and Brower, 1983). Moreover, the response to radiation of adult granary weevils was modified by the temperature before, during, or after the treatment. Pendlebury (1966) reported that high temperatures (30°C) before

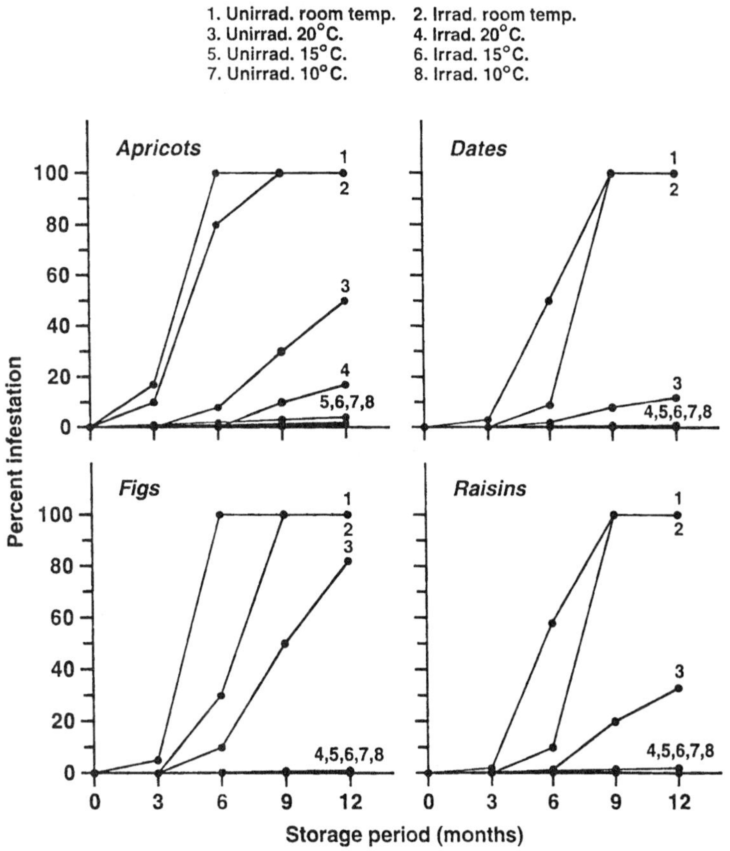

Fig. 5 Effect of irradiation and storage temperatures on insect disinfestations of dried fruits. (From Wahid et al., 1989.)

irradiation sensitized the insects to radiation, but the same high temperature during treatment did not increase the rate of adult mortality. Prolonged exposure to high temperature after irradiation also accelerated death at all doses investigated. A 100% mortality rate was reached in 14 days and 56 days at 30°C and 15°C, respectively. Apparently, the temperature modified the lethal response to radiation through its effects on metabolic rate, speed of the cell cycle, and repair of radiation damage, but the degree of sterilization induced by irradiation was not affected by the temperature. According to a report from Pakistan, insect control can also be achieved by a combination of irradiation and cold storage (Wahid et al., 1989). It has been observed that the infestation rate was higher in apricots and figs than in dates and raisins (Fig. 5). Initially, all samples were apparently free of any insect infestation. However, insect attacks increased consistently in irradiated and unirradiated samples, especially during room temperature storage.

Ionizing radiation at a dose of 0.15–0.75 kGy has been proposed for insect disinfestation. Most insects are sterilized at these doses. However, it was found that 1 kGy may not sterilize some moth species (Tilton and Burditt, 1983). Doses of 0.25 and 0.5 kGy decreased insect infestation, but in all cases after 12 months of storage at room temperature, insect damage reached 100%. The radiation dose of 0.25 kGy coupled with lower

Table 1 Estimated Mean Period of Time in Days Required to Produce 50% (LT_{50}) or 95% (LT_{95}) Mortality when Gamma-Irradiated *Oryzaephilus surinamensis* Adults Were Kept at 25° or 40°C (6–10 Replicates of 5 Vials with 50–70 Insects)

Dose	25°C		40°C	
(krad)	LT_{50}	LT_{95}	LT_{50}	LT_{95}
0	34.0[a]	87.0[a]	10.5	28.3
35	7.6	17.0	3.4	9.6
70	6.9	16.5	3.5	9.0
105	5.5	16.1	1.5	8.0

[a] Extrapolation of 29 days only.
Source: From Ahmed et al. (1981).

temperature completely blocked the development of all types of insects during a storage period of 1 year. Irradiation treatment had no adverse effects on the appearance or taste of any sample. It was observed that packaging of dried fruits in amber bottles or black/silver polyethylene protected the color of dried fruits better. These results indicated that the packaging of dry fruits in colored packages provided better protection for the quality of fruits.

Dates are important commodities in international commercial trading. Among the main factors affecting the quality of dates are insects. These pests damage the date fruit externally and internally and limit the export capability of these nutritive fruits. Before shipping, dry dates are usually disinfested by using methyl bromide, a fumigant that is not easy to handle and requires certain precautions. Ahmed and coworkers (1981) carried out an experiment with low radiation doses (0.35, 0.7, and 1.05 kGy) combined with heat (40°C) or cold (25°C) treatment in comparison with fumigation and irradiation alone (0.7 kGy) for insect disinfestation of dates. This study demonstrated that a treatment with irradiation alone (0.7 kGy) resulted in complete killing of insects only after a storage period that exceeded 1 month. However, a high degree of killing was achieved within a short period when unirradiated products were held at 40°C and irradiated at the lowest dose of irradiation (0.35 kGy) (Table 1). Therefore, a combination treatment of low doses of gamma irradiation and heat would be advantageous if a short storage period of time is required to cause complete kill of insects in dates, instead of using somewhat high doses which might render radiation disinfestation of dried dates prohibitive. The use of a dose in excess of 4kGy (dose for decontamination) can modify the texture due to the conversion of pectin substances. If there is a need, improvement of sensory characteristics can be obtained by irradiating at very low temperatures (Urbain, 1986).

7.2 Fresh Fruits and Vegetables

All fruits and vegetables are perishable as a result of physiological changes, postharvest fungal diseases, other pathological breakdown, and insect infestation. A good preservation technique for fresh fruits and vegetables must be efficient as a postharvest treatment. It should retain the qualities and nutrients of the commodity in a freshlike state. It should have a proven capacity for controlling insect larvae and eggs if insect infestation is a problem. It should have a synergistic effect on the commodity if and when combined with other preservation techniques (Moy and Nagai, 1985). Disinfestation and shelf life extension have been extensively studied and have a great deal of potential and promise, especially for tropical fruits. The use of irradiation at a pasteurization dose (2–5 kGy) may control postharvest spoilage and diseases that affect fruits and vegetables without impairing their sensory qualities (Moy and Nagai, 1985).

The use of irradiation at a pasteurization dose can produce quality changes such as softening of the products. Therefore, this technology should be used in combination with other treatments. The more promising application is the use of irradiation with heat. Through synergy, a lower dose can be used to achieve pasteurization. According to Buckley and associates (1969), nongerminating spores were markedly resistant to single applications of heat (hot water at 39°C–46°C) or irradiation (0.5–2 kGy); a strong inactivation effect (<1% survival rate) was obtained when irradiation plus heat (1.25 kGy + 46°C, 5 minutes) was applied in sequence. The interaction was less pronounced with treatments in reverse sequence. With germinating spores (6-h incubation) an even greater synergistic effect (0.1% survival rate) resulted from heating (39°C, 5 min) followed by irradiation (125 kGy). Thus maximal synergy was obtained in nongerminating spores with a radiation-heat sequence whereas in germinating spores maximal synergy was obtained with a heat-irradiation sequence. For both nongerminating and germinating spores, heating and irradiation produced greater interaction than heating and chilling (Moy, 1983).

7.2.1 Tropical Fruits

Tropical fruits are frequently wasted as a result of deterioration during handling, transport, and storage caused by the senescence of the fruit combined with fungal decay and infestation with pests such as fruit flies and weevils (Grandison et al., 1995). Extension of storage or shelf life through control of postharvest rots is the most difficult goal of irradiation of fruits and vegetables. In most instances, doses of irradiation required to inactivate pathogens result in tissue damage (Heather, 1986). A widely reported exception is the strawberry. This fruit has been shown to be tolerant to irradiation and can be treated at doses that eliminate infections of fruit rot.

The minimal effective absorbed dose required is in the range of 1.5 to 2 kGy, sometimes as high as 3 kGy (Dennison and Ahmed, 1971). Provided that fruits are subsequently kept cool, high quality can be maintained for more than 14 days. The result of 1.5 kGy applied to strawberry coated with a cross-linked edible film (Lacroix and Vachon, 1999) indicated that combined treatments were effective in reducing water losses and mold growth. The treatment was effective in extending the shelf life by more than 15 days when the fruit was stored at 4°C. Coating with edible film can also be efficient in delaying browning by acting as an oxygen barrier or scavenger (Le tien et al., 2001). Another promising way of increasing the effectiveness of radiation in the control of food-borne microorganisms without adversely affecting organoleptic qualities of foods is to combine it with heat treatment. However, the sequence of the application of the combination may also play an important role (Farkas, 1990). In the case of fungi, the heat treatment preceding irradiation usually results in greater antimicrobial effect of all combinations (Padwal-Desai, 1974). With bacterial spores, the reverse order of treatment (i.e., irradiation followed by heating) was found to be more synergistic; the effect of heating followed by irradiation seemed to be additive or only slightly more than additive (Farkas and Roberts, 1976). Unfortunately, few other fruits can tolerate treatments at such high doses.

Several investigations have demonstrated the usefulness of mild heat treatment prior

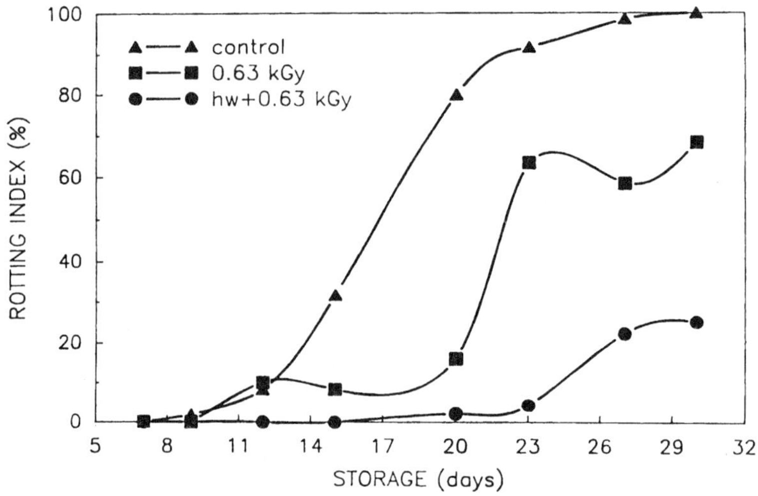

Fig. 6 Rotating index of irradiated mangoes with or without hot water treatment (hw).

to low-dose irradiation in extending the shelf life of certain fresh fruits without affecting their normal quality. The heat treatment used during these experiments was generally a hot water dip. For fresh fruits and vegetables, it is important to establish exact parameters such as proper ripening stage and proper pretreatment, posttreatment, and transport conditions to yield optimal results (Langerak, 1982)

In South African studies, the combination of hot water dipping (55°C, 5 min) for mangoes and 50°C for 10 min in papayas waxing and irradiation (0.75 kGy), and low-temperature storage and shipment (7°C–11°C) has been shown to be particularly effective in controlling fungus and insect attack and in delaying senescence (Thomas, 1977). Transportation trials from South Africa to Europe demonstrated that combination-treated mangoes and papayas may be transported long distances with much lower losses in quality than untreated lots (Brodick and Thomas, 1978). The combined treatment of mild heat and low-dose irradiation also offers possibilities for delayed ripening and reduction of microbial spoilage of tomatoes (Langerak, 1979) and mangoes (Gagnon, et al., 1993; Lacroix et al., 1993)

In 1988, a project was initiated by MDS-Nordion International Inc., the Canadian Irradiation Centre (CIC), the Research Centre of Sciences Applied to Food (CRESALA), and the Office of Atomic Energy for Peace (OAEP) in Thaïland to investigate the possibility of using ionizing radiation in combination with a hot water dip treatment to achieve longer fresh storage life, while ensuring the quality of mangoes on an industrial scale.

The percentage of rotted mangoes results (Fig. 6) showed that rotting took place rapidly in the control samples. Irradiated samples showed significantly ($p \leq 0.05$) slower rot development; the hot water dip plus irradiation treatment slowed the rate of rotting further. The causal fungus of anthracnose was sensitive to hot water, and dipping for 5 minutes at 55°C (Thomas, 1986a) was sufficient to control this disease. On the other hand, radiation processing targets nuclear deoxyribonucleic acid (DNA) and inactivates the microorganisms (Josephson and Peterson, 1983). The percentage of rotted fruits increased more slowly for hot water dipped plus irradiated mangoes than for the control and irradiated samples. A 50% level of rejection was reached between days 20 to 23 for irradiated

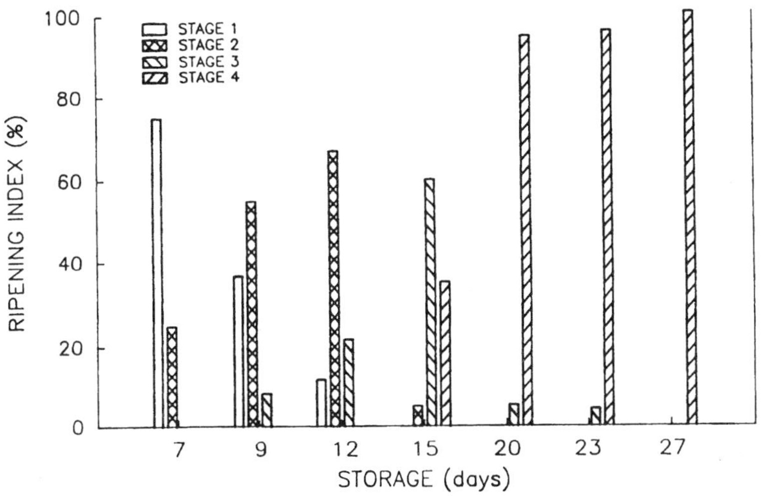

Fig. 7 Ripening index of control mangoes stored at 15°C and 60–65% R.H.

mangoes. The mangoes treated with hot water plus irradiation did not reach this level even after 30 days of storage at 20°C. The development of ripe skin color (stage 4) was delayed in irradiated and hot water dip plus irradiated mangoes. On day 15, 60% of the control mangoes were in stage 3 and 35% in stage 4 (Fig. 7) as compared to irradiated and hot water dip plus irradiated mangoes, which were essentially in stages 2 and 3. On day 27, 100% of the control were in stage 4 whereas only 66% of irradiated (Fig. 8) and 47% of hot water plus irradiated mangoes were in that stage (Fig. 9) (Gagnon et al., 1993; Lacroix et al., 1993).

In this study, the hot water treatment and/or the gamma irradiation of mangoes had no adverse effects on nutritional and chemical qualities. The combined treatment of mild

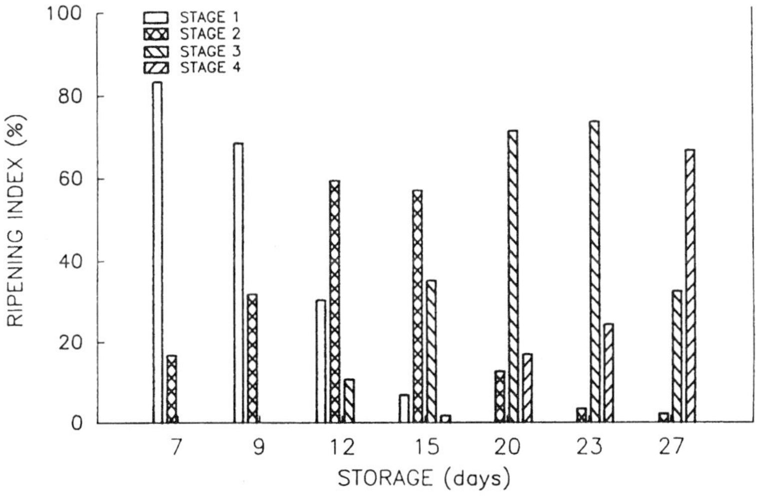

Fig. 8 Ripening index of irradiated mangoes stored at 15°C and 60–65% R.H.

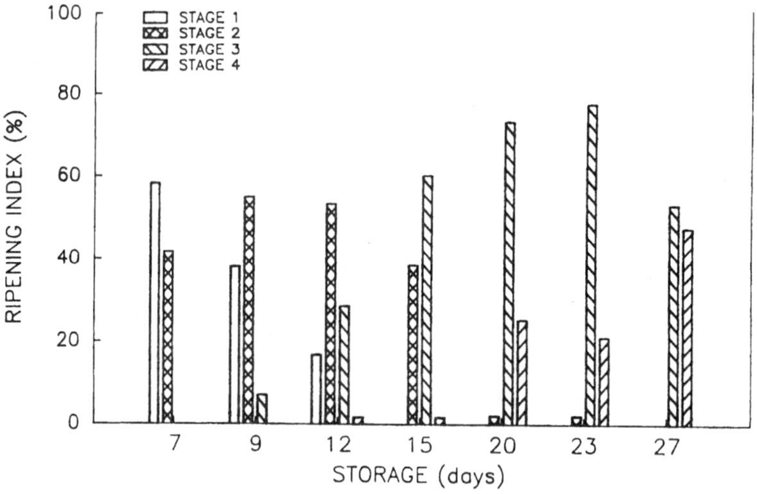

Fig. 9 Ripening index of hot water dipped and irradiated mangoes stored at 15°C and 60–65% R.H.

heat and low-dose irradiation also offered possibilities for delayed ripening and reduction of microbial spoilage during storage at controlled temperatures (El-Sayed, 1978; Langerak, 1979).

7.2.2 Citrus Fruits

Studies on Moroccan tangerines and clementines were also undertaken at the Canadian Irradiation Centre. Results demonstrated that washing with warm water and waxing were treatments that caused significant losses in fruits during the storing period. However, clementines irradiated at 0.3 kGy and stored at 3°C for a period of 8 weeks had losses of less than 11% (Mahrouz et al., 1999; Oufedjikh et al., 2000).

It was also observed that throughout the entire preservation period, the content of vitamin C in irradiated clementines was significantly higher. Moreover, results of the sensory evaluation showed that organoleptic qualities were maintained throughout the entire storage period, especially for irradiated fruit. According to the panelists, the taste of irradiated fruits was sweeter than that of nonirradiated fruits (Mahrouz et al., 2002; Jobin et al., 1992; Abdellaoui et al., 1995).

The effect of waxing and irradiation on the phenol content and the color of oranges was also studied. Experiments demonstrated that irradiation and storage at 20°C favored the synthesis of phenolic compounds, such as flavonods. This phenomenon was found to be related to a decrease in intensity of the red coloring in oranges, favoring more yellowish color. Waxing, however, maintains the color and the concentration of phenolic compounds in oranges (Moussaïd et al., 2000a).

Among essential oils in oranges, D-limonene is the major compound in the orange skin. Irradiation at a dose of 1 kGy did not show a significant effect on essential oils. However, a dose of 2 kGy stimulated the D-limonene synthesis and favored the degradation of other essential oils such as methyl anthranilate. It was also noted that the D-limonene was significantly inferior in waxed oranges. Storage did not affect in any way the D-limonene content of oranges (Moussaïd et al., 2000b).

The control of the quantity of phenolic compounds is important in oranges because the appearance of brown spots in fruits is related to an important concentration of these in tissues. These studies allowed the researchers to observe that waxing has a protective effect on the quality of the oranges during irradiation. Moreover, a dose of irradiation up to 1 kGy did not in any way affect the content of essential oils while ensuring the safety of the product.

A combination of mild heat treatment followed by subsequent irradiation for the preservation of fruit juice was also studied as a promising treatment. Hoang and Julien (1975) investigated this combined treatment for the preservation of apple juice. Irradiation at 4 kGy subsequent to heating at 50°C for 10 min completely controlled the spoilage of apple and pear juice concentrates by osmophylic yeasts (Kaupert et al., 1981). Organoleptic tests have shown that juice heated at 70°C for 8 s and irradiated at 3.5 kGy had an excellent score after 4 weeks of storage at 30°C. It is also interesting to note that irradiation of fresh fruits generally increases the rate of juice extraction in treated fruits in part because of the hydrolysis of pectins by irradiation.

7.2.3 Banana

According to Thomas (1986a), a dose of 0.2–0.5 kGy causes a delay of ripening in bananas and can increase the shelf life of the fruits two- to three fold. However, the maximal delay in ripening occurs with preclimateric bananas; as fruit maturity progresses, the treatment is less effective (Wilkinson and Gould, 1996). Mathur (1968) concluded that an application of a dose of 0.2 kGy could extend the shelf life of Cavendish banana when stored at 19°C. Irradiated bananas require 45 days to ripen without ethylene treatment as against 25 days for nonirradiated fruits. According to Padwal-Desai and colleagues (1973), a combination of heat treatment (50°C, 5 min) and 0.25–0.35 kGy increased the storage life of bananas by 3–5 days. Hot water treatment is used as a supplementary process to control stem-end rot.

In bananas, the combined effect of mild heat and low-dose irradiation offered the possibility of delay of ripening from 0 to 7 days at doses of 0.2–0.5 kGy (Moy et al., 1971).

7.2.4 Papayas

Delaying ripening of papaya is possible when the fruits are irradiated at a dose up to 0.75 kGy (Thomas, 1986a). Combining irradiation with hot water dip treatment (50°C, 10 min) can reduce decay by molds. This combined treatment can increase the shelf life by a minimum of 3 days if papayas are refrigerated at 10°C for 3 weeks and maintained at 21°C–23°C for 6 days (Moy and Nagai, 1985). Akamine and Goo (1969) have demonstrated that subsequent control atmosphere storage further extended the shelf life by 2 days.

7.2.5 Apples

A radiation dose up to 1 kGy was effective for disinfestation and delayed ripening of apples. Initial softening occurred but hydrolysis of starch was retarded (Olsen et al., 1989); however, irradiated apples are firmer than nonirradiated fruit after storage (Thomas, 1986b). All apples surpassed the export standard for firmness after 11 months (Narvaiz et al., 1988). At present results of combination treatment with modified atmospheres, surface coating, chemicals, and heat appear to be inconclusive (Wilkinson and Gould, 1996). However, there is a potential for irradiation treatment insect disinfestation of apples, with a possible improvement in shelf life. Irradiation at a dose of 0.4 kGy is permitted in China. The economic benefits of this treatment have been demonstrated in this country (Zu and Sha, 1993). Marketing of irradiated apples began in the United States in 1988 (Terry and Tabor, 1988).

7.2.6 Sweet Cherries

The shelf life of sweet cherries was extended by a combined treatment of 2–4 kGy with storage at 4°C (Salunkhe, 1961).

7.2.7 Figs

The shelf life of fresh figs is limited by fungal spoilage (Wilkinson and Gould, 1996). A combined treatment of hot dip (50°C, 5 min) followed by a dose of 1.5 kGy is reported by Padwal-Desai and coworkers (1973) to be an effective treatment for figs. An extension of shelf life of 3–4 days or 8–10 days, at room temperature and at 15°C, respectively, was obtained by using these treatments.

7.3 Vegetables

7.3.1 Tomatoes

According to Abdel-Kader and associates (1968), irradiation at a mean dose of 2.8 kGy combined with storage at a temperature of 12°C–15°C increased the shelf life of tomatoes by 4–12 days as a result of decay control. However, Mathur (1968) considered a dose of 0.75 kGy optimal for tomatoes when the storage temperature was 9°C–10°C and 85%–90% relative humidity. Suceptibility to decay increased with irradiation dose. Mathur (1968) found that decay could be controlled with postirradiation treatment with 500 ppm methyl ester of indolyl-3-acetic acid, which also reduced losses due to transpiration and respiration. The storage life of irradiated tomatoes treated with the hormone and stored at 9°C–10°C was 80 days as against 47 days of control fruits.

7.3.2 Onions

Onion is one of the more economically important vegetables grown and exported in many countries of the world. Compared to many other fresh vegetable crops, matured and cured onion bulbs can be kept for relatively longer periods because of their natural dormancy mechanisms and low respiration and metabolic activity. However, considerable losses still occur during long-term storage especially under unfavorable environmental conditions. The interest in the use of irradiation for sprout inhibition of onions and other bulb crops of the allium family was stimulated by the report of Dallyn and colleagues (1955) that complete inhibition of sprouting in onions could be obtained after gamma irradiation with doses as low as 0.04 to 0.74 kGy. Subsequently, several investigations have been carried out throughout the world of the application of ionizing radiation for sprout inhibition of different cultivars and strains of onions grown under varying agroclimatic conditions. Most of the results seemed to agree that for maximal sprout inhibition, irradiation should be carried out very soon after harvesting of onions, when they are in the dormant period, with doses in the range of 0.02 to 0.12 kGy. Moreover, Ogata and Chackin (1972) found that a higher dose rate of 8 kGy/h produced slightly better sprout inhibition than the same dose at a dose rate of 0.5 kGy/h (Horis and Kawasaki, 1965).

7.3.3 Mushrooms

Mushrooms are popular food products. The level of loss during marketing can be as high as 40%. The loss of mushroom quality is marked by brown discoloration of the surface

(Lutz and Hardenberg, 1968), cap opening, stipe elongation, cap diameter increase, weight loss, and texture changes (Gormley, 1975). The loss of whiteness in storage is a complex process (Mac Canna and Gormley, 1968). Polyphenol oxidase enzymes present in the mushroom cells cause browning when they are subjected to forces that can disrupt cellular integrity, such as vibrations, rough handling, and aging (Guthrie, 1984; Hughes, 1958). Polyphenol oxidase enzyme, oxidize phenolic compounds present in the mushroom such as tyrosine to quinones, which condense to form the brown melanin pigments (Long and Alben, 1969). The presence of bacterial infection can also cause a brown discoloration. Bacterial disorders are the main postharvest factor affecting the shelf life of mushroom, especially after removal from refrigerated storage, resulting in slimy texture. *Pseudomonas tolaasii*, regarded as a normal constituent of the microflora of the mushroom bed, can produce a metabolic substance toxic to mushrooms under certain conditions. The infection appears as a brown injury. Wong and coworkers (1982) described another mushroom disease, ginger blotch (GBO), caused by *Pseudomonas*, which produces superficial discoloration (ginger). Molds can also affect the quality of fresh mushrooms. *Mycogone perniciosa magnus* can also grow on mushrooms. Low-dose irradiation in combination with controlled storage is a very effective method for controlling the deterioration of fresh mushrooms caused by senescence and microbial activity. Irradiation treatment at 1 kGy can extend the shelf life from 2 to 4 days at 10°C by reduction of the microflora, slowing the respiration and the morphological transformation as well as lowering polyphenol enzyme activity and the browning reaction. Pathogenic bacteria such as *Pseudomonas tolaasii* and *Mycogone perniciosa* can be eliminated at 2 kGy and shelf life can be increased from 2 to 8 days with storage at 10°C (Skou et al., 1974). However, at such a dose, the irradiation can affect some physical characteristics (hyphes contraction and discoloration). Achieving the same dose at different dose rates appears to have some significant effects (hyphes contraction and discoloration).

Irradiation treatment (2 kGy) at a dose rates of 4.5 kGy/h (I^-) and 32 kGy/h (I^+) has been studied to evaluate their effects on the overall quality of the mushrooms (Beaulieu et al., 1992a; Beaulieu et al., 1992b). The irradiation effect of the two dose rate was effective in lowering the number of yeasts, molds, and bacteria and extending the shelf life by 6 days (Beaulieu et al., 1992). However, on the basis of mushroom whiteness, I^+ and I^- showed, respectively, a shelf life increase of 2 and 4 days. After storage of 9 days, the I^+ treated mushrooms were significantly ($p \leq 0.05$) darker than I^- and highest L whiteness color was observed in I^-. In treated mushrooms, polyphenol oxidase activity was significantly lower ($p \leq 0.05$) than in the control. The total content of phenolic compounds throughout storage was significantly higher ($p \leq 0.05$) only in I^-, as compared to I^+ and control samples (Beaulieu et al., 1999). The greater phenol accumulation was in accordance with the better preserved coloration observed in I^- indicating a diminished oxidation rate in the sample. In I^-, the higher phenol concentration and the better preservation of whiteness throughout the storage period would be explained by diminished activity of polyphenol oxidases. But this explanation was not valid for I^+. The coloration and browning expected to be less in I^+ than in I^- were more pronounced and were similar in I^+ and in the control. The cellular membranes observed by electron microscopy revealed better preserved integrity in I^- than in I^+ samples. The thickening of the membranes observed in I^+, which caused stress to the cell, altered the cell permeability. This alteration allowed the release of vacuolar phenolic compounds. In contact with oxygen in the cell cytoplasm, an increase enzymatic browing was observed. This could explain the more pronounced browning in I^+ samples even though the phenolic compounds were in smaller concentration. This study suggested that the correct dose rate should be applied for each product. A lower dose rate (2 kGy/h) should be applied to fresh mushrooms (Beaulieu et al., 1992b, 1999).

7.3.4 Groundnuts (Peanuts) and Cocoa Beans

In groundnuts a combination of heat plus irradiation (65°C, 35 min, and 0.5 kGy) was found to inactivate toxigenic fungi like *Aspergillus flavus* (Padwal-Desai, 1974). Irradiated nuts were shelf stable for several months when vacuum sealing as packaging under nitrogen was employed. Encouraging results have also been obtained with cashew nuts with combined treatments (Farkas, 1990). Treatment of dried cocoa beans and maize with a combination of moist hot air (30 min at 60°C and 85% relative humidity [RH]) and irradiation (4 kGy) proved to be effective in inactivating fungi including *Aspergillus flavus* (Odamtten et al., 1985).

7.4 Spices

The rate of microbiological deterioration of a composite food usually depends on the degree of contamination of its constituents. In many cases, an important source of microbial contamination may be one of the minor components of the food product. Under the prevailing production and handling practices, many food ingredients such as spices contain a large amount of microorganisms (Farkas, 1988). Microbial counts in dry ingredients can vary from one storage period to another, depending on the moisture content and storage conditions. According to Farkas (1988), count frequently decreases during storage. Spices contain an important level of microorganisms, which cause real problems for the food industry, which uses spices for canned meat and sausage products. Spices contain microorganisms indigenous to the soil and the plants that grow in it that survive drying and heat processes. Spices are also a prolific sources of heat resistant spores of bacteria (Krishnaswamy et al., 1971). The source of contamination may be dust, insects, fecal materials from birds and rodents, and possibly the water used in some processes. Also, fungi may grow in spices before drying or during drying, storing, and shipping (Farkas, 1988). A survey of bacteriological contamination of untreated spices demonstrated that black pepper, turmeric, paprika, allspice, and marjoram are the spices most highly contaminated with bacteria. The aerobic plate count of these spices may sometimes reach a level of 80–100 million per gram (Glas, 1965; Soedarman et al., 1984). The use of only 1% of a spice containing 10^7 microorganisms per gram corresponds to an infection of 10^5 bacteria per gram of the composite food products, which can constitute a significant contamination problem and contribute to the spoilage of the product (Farkas, 1988). According to deBoer and Janssen (1983) and Volkova (1970), *Bacillus cereus* is frequently present in spices. A relatively high incidence of *Clostridium perfringens* is also found (deBoer and Boot, 1983); *C. perfringens* was found in 80% of 54 different kinds of spices. Occasionally, *Salmonella* spp. contamination is observed in spices. This bacterium has been found in samples of pepper, coriander, peppermint, and paprika (Laidley et al., 1973; WHO, 1974). Fecal streptococci occur in spice samples, usually in low number (deBoer and Janssen, 1983; Baxter and Holzapfel, 1982; Masson, 1978). Spices play a major role in mold contamination of meat products. White pepper, black pepper, chili, and coriander seem to be most heavily contaminated with molds (International Commission on Microbiological Specifications for Foods, Spices, 1980). A relatively high incidence of toxigenic molds has been detected in black pepper, ginger, turmeric, celery seed, and nutmeg (Pal and Kundu, 1972; Christensen, 1972; Shank et al., 1972; Farkas, 1992). Whereas certain spices and herbs, especially cinnamon, cloves, and possibly oregano and mustard, inhibit mycelial growth and subsequent toxin production, others, particularly sesame seed, ginger, and rosemary leaves, appear to be conductive to aflatoxin occurrence (Llewellyn et al., 1971). According to Jacquet and Teherani (1974), pepper can be an important source of toxigenic *A. flavus* and a high level of aflatoxin contamination in sausages and pepper cheese. The radiation dose requirements for spice treatments depend on the number and types of microorganisms present and the chemical composition of spices. Spices and herbs are dry and resistant to irradiation treatment. In general, they can tolerate a dose up to 10 kGy (Wilkinson and Gould, 1996). According to Farkas (1988), changes in odor and flavor may occur at a dose above 15 kGy. The dose range of 3–10 kGy is approximately equal to the microbicidal effect of established commercial fumigation processes (Farkas, 1988). Kiss and Farkas (1981) have followed the kinetics of the growth of microflora in paprika and spices (pepper, ginger, chili, mixed spices) under the effects of irradiation and heat treatment. According to their study, mesophilic aerobic bacteria of spices had higher thermoresistance than the bacteria of paprika. Thermal resistance of surviving microflora, of irradiated spices decreased with heat treatment, and with an increase in the dose of irradiation, thermal sensitivity also increased.

8 THE FUTURE OF IRRADIATION FOR PRESERVATION

Irradiation applied to fruits, vegetables, nuts, and spices is increasingly recognized as an effective method for reducing postharvest losses, ensuring hygienic quality of produce, and faciliting international trade of particular exotic fruits. Loaharanu (1989) reported on market trials that was performed successfully in the 1980s mainly for fruits. In 1986, 2 metric tons of mangoes was irradiated at a dose up to 1 kGy in Puerto Rico and flown to Miami. In 1987, Hawaiian papayas were shipped to be irradiated at low doses in Los Angeles. In both trials, products were sold out quickly. Another market trial was performed in Lyon, France, in 1988. Strawberries irradiated up to 2 kGy were sold at a slightly higher cost and consumers preferred their quality. Since the opening of the 6.8M$ Vindicator Inc. in January 1992, the industrial facility has irradiated many fruits, namely, strawberries, tomatoes, and citrus, which have been sold in Florida and Northbrook, near Chicago, Illinois. In the United States, markets that have offered irradiated produce have had good results.

Progress in the commercial use of food irradiation has been slow, but there have been positive signs along the way (Pszczola, 1997). In 1992, irradiated strawberries were sold at an independent produce and grocery store in Florida (Marcotte, 1992) and at one in the Chicago area (Pszczola, 1992). The latter store also sold a variety of other irradiated produce, including grapefruits, oranges, onions, tomatoes, mushrooms, and blackberries (Pszczola, 1997). In 1993, selected stores in the United States sold irradiated poultry (Pszczola, 1993). On occasion, stores in the United States have been selling irradiated exotic fruit such as papaya, mango, and rambuttan. Plans are also currently under way for the building of a facility in the state of Washington for irradiation of fresh fruits and vegetables as well as poultry, meat, and seafood. The quantity of spices treated by irradiation has continued to increase each year. In California, irradiation of dried vegetable seasonings increased by 20% in 1993 and 20% in 1994. In 1996, Sterigenics Int, in Fremont, California, started a project to allow irradiation of fresh produce that is in demand in the food service industry (Pszczola, 1997).

For this technology to be successful, the application should be fulfilling a real need.

As well, the irradiation process should either be the only solution to a specific problem or possess real advantages over existing technologies. The cost should be comparable to that of other food processes. The trend in the practical application of irradiation for fruits is likely to increase in the coming years in view of the prohibition or restriction of fumigants used for insect disinfestation. Since the U.S. ban on ethylene dibromide, many producers in several countries have employed hot water dips to disinfest fruits such as mangoes and papayas, but this technology destroyed the quality of these fruits. Methyl bromide, another fumigant used to prevent the migration of microbes and pests around the world, is still used in the United States to disinfest grapes, pineapples, and dried fruits. This toxic gas has been related to the depletion of the ozone layer. Because of the U.S. Clean Air Act, it was anticipated that the phasing out of the utilization of methyl bromide as a fumigant would take place by the year 2005.

Because irradiation can be applied to fruits in a ''freshlike'' state; because it can kill microbial contaminants; because it can sterilize or kill adult insects as well larvae and eggs, it is an alternative process of considerable interest. The increase in consumption of exotic fruits originating from developing countries and the demand for safe, nutritious, and convenient foods in industrialized countries will likely contribute to the application of this technology on a broader scale. In developing countries, postharvest food losses can be enormous. A low-lose application of irradiation (up to 1 kGy) to fruits at postharvest offers a unique opportunity to eliminate insect infestation. Developing countries will likely benefit from this technology by reducing their losses and offering an increase in the supply of certain produce. Consumers will benefit from greater price stability as a result of the availability of many commodities including tropical fruits throughout the year.

REFERENCES

Abdellaoui, S., Lacroix, M., Jobin, M., Boubekri, C., and Gagnon, M. 1995. Effects de l'irradiation gamma avec et sans traitement à l'eau chaude sur les propriétés physicochimiques, la teneur en vitamine C et les propriétés organoleptiques des clémentines. Sciences des aliments 15(3): 217–235.

Adbel-Kader, A. S., Morris, L. L., and Maxie, E. C. 1968. Physiological studies of gamma irradiated tomato fruits. II. Effects on deterioration and shelf life. Proc. Am. Soc. Hortic. Sci., 93:831.

Aboultab, E. A., Soliman, F. M., El-Zalabami, S. M., Brunke, E. J., and El Kersh, T. A. 1986. Essential oils of *Thymus bovei* Benth. Egypt. J. Pharm. Sci. 27(1–4):209–214.

Ahmed, M. S. H., Al-Hakkak, Z. S., Al-Maliky, S. K., Kadhum, A. A., and Lamooza, S. B. 1981. Irradiation disinfestation of dry dates and the possibility of using combination treatments. In Combination Processes in Food Irradiation. IAEA, Vienna, pp. 217–228.

Akamine, E. K., and Goo, T. 1969. Controlled atmosphere storage of irradiated papayas (*Carica papaya var.* SOLO). In Dosimetry, Tolerance and Shelf Life Extension Related to Disinfestation of Fruits and Vegetables by Gamma Irradiation Report 1967 to 1968. Agreement No. 5. Division of Isotopes Development, U.S. Atomic Energy Commission, Washington, D.C., 63. (Contract No. AT (04-3)-235, Project.

Baxter, R., and Holzapfel, W. H. 1982. A microbial investigation of selected spices, herbs and additives in South Africa. J. Food. Sci. 47:570.

Beaulieu, M., Lacroix, M., Charbonneau, R., Laberge, I., and Gagnon, M. 1992a. Effects of gamma irradiation dose rate on microbiological and physical quality of mushrooms (*Agaricus bisporus*). Sciences des Aliments 12:289–303.

Beaulieu, M., Lacroix, M., Béliveau, M., Athanasious, R., Hegedus, P., and Gagnon, M. 1992b. Effect of gamma irradiation dose rate on phenolic compounds, polyphenol oxidase and browning of mushrooms *Agaricus bisporus*. Proceedings of the XVI International Conference in Association with the Royal Society of Chemistry, Lisbon, July 13–16, 1992, Tome II, 16: 78–83.

Beaulieu, M., Béliveau, M., D'Aprano, G., and Lacroix, M. 1999. Dose rate effect of gamma irradiation on phenolic compounds, polyphenol oxidase and browning of mushrooms (*Agaricus bisporus*). J. Agric. Food Chem. 47:2537–2543.

Brodick, H. T., and Thomas, A. C. 1978. Radiation preservation of subtropical fruits in South Africa. In Food Preservation by Irradiation, Vol. 1. IAEA, Vienna, pp. 167–178.

Brower, J. H., and Tilton, E. W. 1970. Insect disinfestation of dried fruit by using gamma radiation. Food Irradiation 11:1–2, 10–14.

Buckely, P. M., Sommer, N. F., Coon, D. A., Dolly, M., and Maxie, F. C. 1969. Inactivation of *Rhizopus stolonifer* sporangiospores by single and combined treatments of heating, chilling and gamma irradiation. Radiat. Bot. 40:26.

Christensen, C. M. 1972. Pure spices: How pure? Am. Soc. Microbiol. News 38:165.

Code of Federal Regulations (CFR). 1992. FDA, Washington, D.C.

Codex Alimentarius Commission. 1984. Codex General Standard for Irradiated Foods and Recommended International Code of Practice for Operation of Radiation Facilities Used for the Treatment of Foods, Vol XV, WHO, Geneva.

Conner, D. E., and Beuchat, L. R. 1984. Effects of essential oils from plants on growth of food spoilage yeasts. J. Food. Sci. 49:429–434.

Corrigan, J. 1993. Presentation at the International Symposium on Cost/Benefit Aspects of Food Irradiation Processing, Aix-en-Provence, France, March 10–13.

Council for Agricultural Science and Technology (CAST). 1986. Ionizing Energy in Food Processing and Pest Control. 1. Wholesomeness of food treated with ionizing energy. Report No. 109.

Council for Agricultural Science and Technology (CAST). 1989. Ionizing Energy in Food Processing and Pest Control. 11. Applications. Report No. 115.

Dallyn, S. L., Sawyer, R. L., and Sparrow, A. H. 1955. Extending onion storage life by gamma irradiation. Nucleonics 13:48.

de Boer, E., and Boot, E. M. 1983. Comparison of methods for isolation and confirmation of *Clostridium perfringens* from spices and herbs. J. Food. Prot. 46:533.

de Boer, E., and Janssen, F. W. 1983. Microbiology of spices and herbs. Proceedings of Dutch Symposium on Food Microbiology, Delft, November 17.

Dennisson, R. M., and Ahmed, E. M. 1971. Effects of low level irradiation upon the preservation of food products. Final Summary Report to U.S. Atomic Energy Commission (USAEC), May 1963–March 1970. USAEC, Isot. Rad. Technol. Oak Ridge Operations Office, Contract Report ORO-680.

Diehl, J. F. 1990. Safety of irradiated foods. Marcel Dekker, New York.

Dogbevi, M. K., Vachon, C., and Lacroix, M. 1999. Effect of gamma irradiation on the physicochemical properties and natural microflora of dry red kidney beans (*Phaseolus vulgaris*). J. Food. Sci. 64:3, 540–542.

El-Sayed, S. A. 1978. Changes in keeping quality of tomato fruits after post-harvest treatment with gamma irradiation combined with heat. Egypt. J. Hortic. 5:167.

Farkas, J. 1988. Microbial contamination of dry ingredients and its significance to the food industry and public health. In Irradiation of Dry Food Ingredients (J. Farkas, ed.). CRC Press, Boca Raton, Fla, pp. 1–11.

Farkas, J. 1990. Combination of irradiation with mild treatment: Review. Food Control 1:4, 223–229.

Farkas, 1992. Radiation treatment of spices. Prehramberotechnol. Biotechnol. Rev. 30:159–163.

Farkas, J., and Roberts, T. A. 1976. The effect of sodium chloride, gamma irradiation and/or heating on germination and development of spores of *Bacillus cereus* T. in single germinants and complex media. Acta Aliment. 5:289–302.

Gagnon, M., Lacroix, M., Pringsulaka, V., Latreille, B., Jobin, M., Nouchpramool, K., Prachasitthisak, Y., Charoen, S., Abdulyatham, P., Lettre, J., and Grad, B. 1993. Effect of gamma irradiation combined with hot water dip and transportation from Thailand to Canada on biochemical and physical characteristics of their mangoes (Nahng Glahng Wahn variety). Radiat. Phys. Chem. 42(1–3):283–287.

Glas, A. 1965. Praktisches Handbuch der Lebensmittel. Bayerischer Landwirtschaftsverlag, Munich.

Gormley, T. R. 1975. Chill storage of mushrooms. J. Sci. Food. Agric. 26:401–411.

Grandisson, A. S., Brennan, J. G., and Campbell-Platt, G. 1995. Combination Treatments Including Electron Beam Irradiation for Extending the Shelf Life of Fresh Fruit. IAEA, Vienna.

Guthrie, B. D. 1984. Studies on the control of bacterial deterioration of fresh, washed mushrooms (*Agaricus bisporus/brunescens*). Master's thesis, Pennsylvania State University.

Heather, N. W. 1986. Irradiation of fruit and vegetables. Queensland Agricultural Journal, March–April, pp. 85–87.

Hoang, N. H., and Julien, J. P. 1975. Effets du chauffage et de l'irradiation (post-irradiation) sur la saveur et la couleur de jus de pomme. Can. Inst. Food. Sci. Technol. J. 8:12–15.

Horis, S., and Kawasaki, T. 1965. Gamma irradiation of onion bulbs to inhibit sprouting. II. Browning phenomenon of innerbuds of irradiated onion: Bulbs and method for early detection for inhibition of sprouting. Annu. Rep. Radiation Center Osaka Prefect 5:90.

Hughes, D. M. 1958. Discoloration findings noted longer shelf life is the goal. Mush-room News 4:9–13.

International Atomic Energy Agency (IAEA). 1989. Acceptance, Control of and Trade in Irradiated Food. Conference Proceedings, Geneva, December 12–16, 1988, Jointly Organized by FAO, WHO, ITC-UNCTAD GATT, Vienna.

International Commission on Microbiological Specifications for Foods, Spices. 1980. In Microbial Ecology of Foods, vol. 2, Silliker, J. H., Elliott R. P., Baird-Parker, A. C., Bryan, F. L., Christian, J. H. B., Clark, D. S., Olson, J. C. Jr., and Roberts, T. A., Eds. Academic Press, New York, p. 731.

Jacquet, J., and Teherani, M. 1974. An unusual presence of aflatoxin in certain animal products: Possible role of pepper. Bull. Acad. Vet. Fr. 47:313.

Jategaonkar, L., and Marcotte, M. 1993. Effects of irradiation on spices, herbs and seasoning: A review of selected literature. Internal Research Report, MDS Nordion, p. 3.

Jobin, M., Lacroix, M., Abdellaoui, S., Bergeron, G., Boubekri, C., and Gagnon, M. 1992. Effet de l'irradiation gamma avec et sans traitment à l'eau chaude sur les propriétés physiques, chimiques et sensorielles des tangerines. Microbiol. Aliment. Nutr. 10:115–128.

Josephson, E. S., and Peterson, M. S. 1983. Preservation of food by ionizing radiation, Vol. I–III. CRC Press, Boca Raton, Fla.

Kaupert, N. L., Lescano, H. G., and Kotliar, N. 1981. Conservation of apple juice and pear juice concentrates: Synergic effect of heat and radiation. In Combination Processes in Food Irradiation. IAEA, Vienna, pp. 205–216.

Kiss, I., and Farkas, J. 1981. Combined effect of gamma irradiation and heat treatment on microflora of spices. In Combination Processes in Food Irradiation. IAEA, Vienna, pp. 107–115.

Kiss, I., Farkas, J., Farkas-Csentes, P., and Gasztonyi, K. 1974. Enhancement of the keeping quality of sliced packaged bread by ionizing radiation. In Proceedings of the 4th International Congress of Food Science and Technology (IUFOST), Madrid, 4:155–162.

Krishnaswamy, M. A., Patel, J. D., and Parthasarathy, N. 1971. Enumeration of microorganisms in spices and spices mixtures. J. Food. Sci. Technol. 8:191.

Kundstadt, P., and Steeves, C. 1993. Economics of food irradiation. Proceedings of the International Symposium on Cost/Benefit Aspects of Food Irradiation Processing, Aix-en-Provence, March 10–13, 1993, IAEA/FAO/WHO, pp. 395–416.

Lacroix, M., and Vachon, C. 1999. Utilization of irradiation in combination with other processes for preserving food products. Rec. Res. Dev. Agric. Food Chem. 3:313–328.

Lacroix, M., Gagnon, M., Pringsulaka, V., Jobin, M., Latreille, B., Nouchpramool, K., Prachasitthisak, V., Charoen, S., Adulyatham, P., Lettre, J., and Grad, B. 1993. Effect of gamma irradiation with or without water dip and transportation from Thailand to Canada on nutritional qualities, ripening index and sensorial characteristics of their mangoes (Nahng Glahng Wahn Variety). Radiat. Phys. Chem. 42(1–3):273–277.

Lacroix, M., Smoragiewicz, L., Pazdernik, M. I., Koné, M. I., and Krzystyniak, K. 1997. Prevention of lipid radiolysis by natural antioxydants from rosemary (Rosmarinus officinalis L.) and thyme (Thymus vulgaris L.) Food Res. Int. 30(6):457–462.

Laidley, R., Handzel, S., Severs, D., and Butler, R. 1974. *Salmonella* Weltevreden Outbreak Associated with Contaminated Pepper. Epidemiol. Bull. (Dept. Natl. Health Welfare, Ottawa) 18:4, 62.

Lan, D. N., Hien, B. C., Tu, N. T., Quan, V. H., Dung, P. T., Thao, D. P., and Mai, H. H. 1991. Preservation of dry green beans by irradiation. In Insect Disinfestation of Food and Agricultural Products by Irradiation, Proceedings of the Final Research Co-Ordination Meeting, Beijing, 1987. International Atomic Energy Agency, Vienna, pp. 89–92.

Langerak, D. I. S. 1979. The preservation of fruits and vegetables by ionizing radiation. In Course Book of the IFFIT Inter-Regional Training Course on Food Irradiation, International Facility for Food Irradiation Technology. Wageningen, Netherlands, Lecture No. L91.

Langerak, D. I. S. 1982. Combined heat and irradiation treatments to control mould contamination in fruits and vegetables. In Course Book of the IFFIT Inter-Regional Training Course on Food Irradiation, International Facility for Food Irradiation Technology. Wageningen, Netherlands, Lecture No. L98.

Lattaoui, N., and Tantaoui-Elaraki, A. 1994. Comparative kinetics of microbial destruction by the essential oils of *Thymus broussonetti*, *T. zygis* and *T. satureioides*. J. Essent. Oil. Res. 6:165–171.

Le tien, C., Vachon, C., Mateescu, M. A., and Lacroix, M. 2001. Milk protein coatings prevent browning of apples and potatoes. J. Food Sci. 66(4):512–516.

Llewellyn, G. C., Burkett, M. L., and Eadie, T. 1971. Potential molds growth, aflatoxin production and antimycotic activity of selected natural spices and herbs. J. Assoc. Off. Anal. Chem. 64: 955.

Loaharanu, P. 1989. International trade in irradiated foods: Regional status and outlook. Food Technol. 43(7):77–83.

Long, T. J., and Alben, J. O. 1969. Preliminary studies of mushroom tyrosinase (polyphenol oxidase). Mushroom Sci. 5:281–299.

Lutz, J. M., and Hardenberg, R. E. 1968. The commercial storage of fruits, vegetables and florist and nursery stock. Agric. Handbook 66.

Mac Canna, C., and Gormley, T. R. 1968. Quality assessment of mushrooms: Relationship between moisture loss, color, and toughness of harvested cultivated mushrooms. Mushroom Sci. 7:485–492.

Mahrouz, M., Lacroix, M., Jobin, M., Oufedjikh, H., Boubekri, C., and Gagnon, M. 1996. Étude de la conservation des clémentines par irradiation gamma: Impact sur certaines propriétés. (unpublished data).

Mahrouz, A., Lacroix, M., Nketsia-Tabiri, J., Calderon, N., and Gagnon, M. 1998a. Antimicrobial properties of natural substances in irradiated fresh poultry. Radiat. Phys. Chem. 52:81–84.

Mahrouz, A., Lacroix, M., Nketsia-Tabiri, J., Calderon, N., and Gagnon, M. 1998b. Antioiydant properties of natural substances in irradiated fresh poultry. Radiat. Phys. Chem. 52:77–80.

Mahrouz, M., Lacroix, M., D'Aprano, G., Oufedjikh, H., Boubekri, G., Gagnon, M. 2002. Effect of γ-irradiation combined with washing and waxing treatment on physicochemical properties, vitamin C and organoleptic quality of citrus clementina Hort. Ex. Taneka. J. Agric. Food. Chem. (revised version under revision).

Marcotte, M. 1992. Irradiated strawberries enter the U.S. market. Food Technol. 46(5):80–86.

Masson, A. 1978. Hygienic quality of spices. Mitt. Geb. Lebensmittelunters. Hyg. 69:544.

Mathur, P. B. 1968. Application of atomic energy in the preservation of fresh fruits and vegetables. Indian Food Packer 22:1.

McLaughlin, W. L., Jarret, R. D., and Olejnik, T. A. 1982. Dosimetry. In: Preservation of Food by Ionizing Radiation, Vol. 1 (E. S. Josephson and M. S. Peterson, eds.). CRC Press, Boca Raton, Fla.

Mills, S. 1987. Discussion Paper: Issues in Food Irradiation. Science Council of Canada, Ottawa.

Moussaïd, M., Lacroix, M., Nketsia-Tabiri, J., and Boubekri, C. 2000a. Phenolic compounds and the colour oranges subjected to a combination treatment of waxing and irradiation. Radiat. Phys. Chem. 57:273–275.

Moussaid, M., Lacroix, M., Nketsia-Tabiri, J., and Boubekri, C. 2000b. Effects of irradiation in combination with waxing of the essential oils in orange peel. Radiat. Phys. Chem. 57:269–271.

Moy, J. H., Akamine, E. K., Brewbaker, J. L., Buddenhagen, I. W., Ross, E., and Spielmann, H. 1971. Dosimetry, tolerance and shelf-life extension related to disinfestation of tropical fruits by gamma irradiation. In Disinfestation of Fruit by Irradiation. Proceedings of the FAO/IAEA Panel, Honolulu, 1970. IAEA, Vienna, pp. 43–57.

Moy, J. H. 1983. Radurization and radicidation: Fruits and vegetables. In: Preservation of Food by Ionizing Radiation, Vol. 3 (E. S. Josephson and M. S. Peterson, eds.). CRC Press, Boca Raton, Fla.

Moy, J. H., and Nagai, N. Y. 1985. Quality of fresh fruits irradiated at disinfestation doses. In: radiation disinfestation of food and agricultural products, Proceedings of an International Conference (J. H. Moy, ed.), Honolulu, Hawaii, 1983, pp. 135–147.

Narvaiz, P., Lescano, H. G., and Kaupert, N. L. 1988. Preservation of apples by irradiation. Food Chem. 27:273–281.

Odamtten, G. T., Appiah, V., and Langerak, D. I. 1985. Microbiological quality and production of aflatoxin B_1 by *Aspergillus flavus* link NRRL 5906 during storage of artificially inoculated maize grain treated by a combination of heat and gamma radiation. In Food Irradiation Processing. IAEA, Vienna, pp. 245–246.

Ogata, K., and Chackin, K. 1972. Irradiation of fruits and vegetables: Sprout inhibition and fruit ripening. Kagaku to Seibutsu 10:234.

Olsen, K. L., Hungate, F. P., Drake, S. R., and Eakin, D. E. 1989. 'Red Delicious' apples' response to low dose radiation. J. Food. Quality 12:107–113.

Ouattara, B., Sabato, S. F., and Lacroix, M. 2001. Combined effect of antimicrobial coating and gamma irradiation on the shelf life extension of precooked shrimp (Penaeus spp.) J. of Food Microbiology 68(1–2):1–9.

Oufedjikh, H., Mahrour, M., Amiot, M. J., and Lacroix, M. 2000. Effect of gamma irradiation on phenolic compounds and phenylalanine ammonia-lyase activity during storage in relation to peel injury from peel of citrus clementine (*Horticultura Extanaka*). J. Agric. Food Chem. 48: 559–565.

Padwal-Desai, S. R. 1974. Studies on control of foodborne moulds by gamma-irradiation. Ph.D. Thesis, Bhabha Atomic Research Centre, Bombay, University of Bombay, India.

Padwal-Desai, S. R., Ghanekar, A. S., Thomas, P., and Sreenivasan, A. 1973. Heat-radiation combination for control of mold infection in harvested fruits and processed cereal foods. Acta Alimentaria, 2(2):189–207.

Pal, N., and Kundu, A. K. 1972. Studies of *Aspergillus* ssp. from Indian spices in relation to aflatoxin production. Sci. Cult. 30:252.

Pendlebury, J. B. 1966. The influence of temperature upon the radiation susceptibility of *Sitophilos granarius* (L.). In Entomology of radiation disinfestation of grain (P. B. Cornwell, ed). Pergamon Press, London, chap. 3.

Pszczola, D. E. 1992. Irradiated produce reaches Midwest market. Food Technol. 46(5):89–92.

Pszczola, D. E. 1993. Irradiated poultry makes U.S. debut in Midwest and Florida markets. Food Technol. 47(11):89–96.

Pszczola, D. E. 1997. 20 Ways to market the concept of food irradiation. Food Technol. 51(2):46–49.

Salunkhe, D. K. 1961. Gamma irradiation effects on fruits and vegetables. Econ. Bot. 15:28.

Satin, M. 1996. The prevention of Food Losses after Harvesting. In Food Irradiation, 2nd ed. Technomic, Lancaster PA, pp. 81–94.

Shank, R. C., Wogan, G. N., and Gibson, J. F. 1972. Dietary aflatoxins and human liver cancer. I. Toxigenic moulds in foods and foodstuffs of tropical South East Asia. Food Cosmet. Toxicol. 10(1):51.

Skou, J. P., Bech, K., Lundsten, K. 1974. Effects of ionizing irradiation on mushrooms as influenced by physiological and environmental conditions. Radiat. Bot. 14:287–299.

Soedarman, H., Stegeman, H., Farkas, J., and Mossel, D. A. A. 1984. Decontamination of black pepper by gamma radiation. In Microbial Associations and Interactions in Food I. Kiss, T. Deak and K. Incze, eds. Akademiai Kiado, Budapest, p. 401.

Terry, D., and Tabor, R. 1988. Consumer acceptance of irradiated produce: A value added approach. Department of Agriculture, Central Missouri State University.

Thayer, D. W., and Rajkowski, K. T. 1999. Developments in irradiation of fresh fruits and vegetables. Food Technol. 53(11):62–65.

Thomas, A. C. 1977. Radiation preservation of sub-tropical fruits. Food. Irrad. Newslett. 1(2):19.

Thomas, P. 1984. Radiation preservation of foods of plant origin. Part III. Tropical fruits: Bananas, mangoes and papayas. CRC. Crit. Rev. Food Sci. Nutr. 23:147–206.

Thomas, P. 1986a. Radiation preservation of foods of plant origin. Part III. Tropical fruits: Bananas, mangoes and papayas. CRC Crit. Rev. Food Sci. Nutr. 23(2):147–205.

Thomas, P. 1986b. Radiation preservation of foods of plant origin. Part V. Temperate fruits: Pome fruits, stone fruits and berries. CRC Crit. Rev. Food Sci. Nutr. 24(4):357–400.

Tilton, E. W., and Brower, J. H. 1983. Radiation effects on arthropods. In Preservation of Food by Ionizing Radiation. Vol. II. (E. S. Josephson and M. S. Peterson, eds.). CRC Press, Boca Raton, Fla., p. 269.

Tilton, E. W., and Brower, J. H. 1987. Ionizing radiation for insect control in grain and grain products: Technical review. Am. Assoc. Cereal Chem. 32(4):330–335.

Tilton, E. W., and Burditt, A. K. Jr. 1983. Insect disinfestation of grain and fruit. In Preservation of Food by Ionizing Radiation, Vol. III. CRC Press, Boca Raton, Fla., p. 215.

Urbain, W. M. 1986. Food Irradiation. Academic Press, New York.

Vasseur, J. P. 1991. Ionisation des produits alimentaires:Technique et documentation. Lavoisier, Paris.

Volkova, R. S. 1970. *Bacillus cereus* contamination of foods and environment at institutional feeding points. Gig. Sanit. 36:108.

Wahid, M., Sattar, A., and Khan, I. 1989. Effect of combination methods on insect disinfestation and quality of dry fruits. J. Food Processing Preservation 13(1):79–85.

Wahid, M., Sattar, A., Neelofar, S., Atta, I., Khan, and Ehlermann, D. A. E. 1987. Radiation disinfestation and quality of dried fruits. Acta Alimentaria 16(2):159–166.

WHO. 1974. *Salmonella* surveillance. Weekly Epidemol. Rec. WHO Geneva 42:351.

WHO. 1988. Food Irradiation: A Technique for Preserving and Improving the Safety of Food. WHO, Geneva.

Willemot, C., Marcotte, M., and Deschênes, L. 1996. Ionizing radiation for preservation of fruits. In Processing Fruits: Science and Technology, Vol. 1. Biology, Principles, and Applications. (L. S. Somogyi, H. S. Ramaswamy, and Y. H. Hui, eds.), Technomic, Lancaster, Pa., pp. 221–260.

Wilkinson, V. M., and Gould, G. W. 1996. Food Irradiation: A Reference Guide. Butterworth Heinemann, UK, p. 158.

Wong, W. C., Fletcher, J. T., Unsworth, B. A., and Preece, T. F. 1982. A note on ginger blotch, a new bacterial disease of the cultivated mushroom, *Agaricus bisporus*. J. Appl. Bact. 52:43–48.

World Health Organization. 1988. Food Irradiation: A Technique for Preserving and Improving the Safety of Food. WHO, Geneva.

Zu, Z., and Sha, Z. 1993. Cost-benefit analysis of irradiation of vegetables and fruits at the Shanghai Irradiation Centre. In: Cost-benefit aspects of food irradiation processing. Proceedings of a IAEA/FAO/WHO symposium, Aix-en-Provence, March 1993, International Atomic Energy Agency, Vienna, pp. 201–211.

23

Drying of Fruits, Vegetables, and Spices

STEFAN GRABOWSKI and MICHÈLE MARCOTTE

Agriculture and Agri-Food Canada, Saint-Hyacinthe, Quebec, Canada

HOSAHALLI S. RAMASWAMY

McGill University, Sainte-Anne-de-Bellevue, Quebec, Canada

1 INTRODUCTION

For many agricultural products, such as fruits, vegetables, spices, and herbs, dehydration, or drying, is still the most typical postharvest operation throughout the world. The term *drying* is usually reserved for those processes that accomplish removal of water by evaporation, rather than by pressure or other physical means (1). Production data of dry fruits and vegetables suffer from poor statistics and are not available in many countries. According to an approximate estimation, about 20% of continuously growing world fruit and vegetable production (Table 1) is subjected to drying, more than 50% is consumed as fresh, 20% as frozen, 5% as canned, and 5% as pickled. The technological developments of the drying process vary from the simplest method of field sun drying to relatively much more technologically developed methods, for example, explosion puffing. This of course does not necessarily mean that the product dried in the modern technology scheme has better quality or value than that dried traditionally in the sun. The quality depends on several other parameters, such as properties of the wet material, climatic and environmental conditions, and scale of production.

In this chapter, basic principles of the drying process are omitted as they are described in Chapter 5 and in several other books and periodicals. The following are useful sources of information: ''Dehydration of foods'' by Barbosa-Canovas and Vega-Mercado

Table 1 World Fruit and Vegetable Production 1996–2001

Year	Fruits excluding melons (MT[a])	Vegetables including melons (MT[a])
1996	413,932,200	565,523,300
1997	442,459,900	608,123,500
1998	433,767,200	631,037,300
1999	457,797,300	667,632,800
2000	466,414,200	691,894,500
2001	466,340,300	698,127,200

[a] MT, metric ton
Source: From Ref. 103.

(2), ''Handbook of Industrial Drying'' by Mujumdar (3), ''Drying Principles, Applications, and Design'' by Strumillo and Kudra (4), ''Drying of Solids'' by Mujumdar (5), ''Drying Principles and Practice'' by Keey (6), ''Advances in Drying,'' Vol. 1–5, by Mujumdar (7–11), and ''Dictionary of Drying'' by Hall (12). The content of this chapter is devoted to practical rather than theoretical aspects of fruit, vegetable, and spice drying. Torrey (13), Hartwooth (14), and Kudra and Mujumdar (15) provide details on patents, new ideas, and tendencies in this area. Several other books and publications are important sources of drying data for foods generally, and fruits and vegetables specifically: Van Arsdel and coworkers (1), Crapiste and Rotstein (16), Arhey and Ashurst (17), Ratti and Mujumdar (18), Jayaraman and Das Gupta (19), Sokhansanj and Jayas (20), Grabowski and Mujumdar (21), Arthey and Dennis (22), Brennan (23), Somogyi and Luh (24), and Karel (25). Additional information, especially of commercial importance, is available through the Internet.

2 FOOD AS A DRYING OBJECT

A key parameter in the food dehydration processes is *water activity* (a_w), defined as the ratio of water vapor pressure (p) of the food system to the vapor pressure of pure water (p_0) at the same temperature.

$$a_w = p/p_0 \tag{1}$$

The microbial stability of dehydrated foods, or more precisely of foods stabilized by a reduced water activity, results from interruption of vital processes essential to microbial growth or spore germination, which is mediated by a depressed availability of water in the food. The number and types of microorganisms that may be associated with foods are extremely large and may not remain constant during the life of a food. These microorganisms may originate from the raw material or from contamination (by people, animals and insects, water or air, contact surfaces, etc.). Every microorganism has an optimal and a minimal water activity for growth. A reduction of a_w below the optimum delays spore germination and decreases the growth rate, and reduction of a_w below the minimum, presented in Table 2, totally inhibits growth or spore germination. However, the reduction of water activity is not sufficient to destroy all microorganisms. During air drying the increased temperature of the food can affect the living forms of the microorganisms except the spores of species such as *Bacillus* and *Clostridium* species. For example, dried onion

Table 2 Minimal Water Activity for the Growth of Some Microorganisms in Food

Group of microorganisms	Examples	Minimal water activity
Normal bacteria, viruses	*Clostridium botulinum, Salmonella, Bacillus, Pseudomonas, Escherichia, Lactobacillus, Vibrio* spp.	0.91
Normal yeasts	*Candida, Torulopsis,* most *Saccharomyces* spp. (also *Staphylococcus aureus*)	0.88 0.80–0.87
Normal molds	Mycotoxigenic penicillia	0.80
Halophilic bacteria	Most halophilic bacteria	0.75
Some molds	Mycotoxigenic aspergilli	0.75
Xerophilic molds	*Aspergillus chevalieri, A. candidus, Wallemia sebi* (also yeast *S. bisporus*)	0.65
Osmophilic yeasts	*Saccharomyces rouxi* (also molds: *A. echinulatus, Monoascus bisporus*)	0.60

Source: From Refs. 26 and 93.

may contain a heavy load of *Clostridium perfringens* spores, which can cause food poisoning when used in a poultry stuffing (26). It should be noted that the drying process does not necessarily destroy the food toxins occurring as contaminants before or during drying. In the case of the food intoxicated by toxins produced by some food poisoning bacteria (*Clostridium botulinum, Staphylococcus aureus, Bacillus cereus*) this food should be excluded from the drying process. Other typical microorganisms as viruses, protozoa, algae, and prions do not grow on food, so the only aspects of importance are their pathogenicity or toxigenicity and their resistance to thermal drying procedures as they can be transferred to a human body and cause some diseases (26). Almost all of these are more sensitive than the average vegetative bacterium.

The effect of water activity on microbial growth is influenced by many other factors, such as temperature, pH, nutrients, preservatives and other components of food, and oxygen supply. With the same water content in the food, several different water activity values can be achieved by changing the composition of food. Table 3 presents examples of maximal water activity values that are permissible for certain types of dehydrated foods. The practical relation between water activity and water content in the food is indicated by food stability maps (27). At present, the measurement of water activity is easy with the use of several types of commercially available single-channel or multichannel instruments; however, moisture content, wet (w.b.) or dry basis (d.b.), remains the most practical criterion in the food drying technology. The water activity of fresh fruits and vegetables falls in the range 0.97–0.99. Dried fruits with a moisture content of 18%–28% have water activity of 0.7–0.8, and dried vegetables with 14%–24% moisture have water activities between 0.7 and 0.77. The water activity of most dehydrated foods is below the minimal water activity values of food pathogens, which are inhibited at a water activity of 0.9 (only *Staphylococcus aureus* is capable of growing at a_w values down to 0.85). Most other microorganisms such as food spoilage–type bacteria and fungi, which can grow at lower water activities, multiply very slowly in low a_w conditions or require a special environment

Table 3 Maximal Permissible Water Activity for Some Dry Fruits, Vegetables, and Additives (20°C)

Dry product	Maximal a_w	Dry product	Maximal a_w
Apples	0.70	Potato flakes	0.11
Apricots	0.65	Dry soup mix	0.60
Dates	0.60	Soluble coffee	0.45
Peaches	0.70	Sucrose	0.85
Pears	0.65	Fructose	0.63
Plums	0.60	Dextrose	0.83
Orange powder	0.10	Maltose	0.92
Peas	0.25–0.45	Sorbitol	0.55–0.65
Beans	0.12–0.80	Honey	0.75

Source: From Refs. 23, 26, and 93.

for growth. Fungi (yeasts and molds) tend to grow more slowly than bacteria and are often outcompeted in many types of food spoilage, unless bacterial growth is limited, e.g., by drying below their minimal water activity or presence of preservatives. In general, yeasts and molds tend to be more resistant than bacteria to harsh environmental conditions (low a_w) and so cause spoilage under such conditions. The nature of this type of spoilage may be similar to that caused by bacteria except that the growth may be much more pronounced. Some molds may produce toxins (especially mycotoxins) that can result in a variety of acute or chronic toxicity syndromes in humans and animals. The foods that may be affected by molds associated with mycotoxin production are, for example, nuts, figs, cocoa, coffee, and grains. Production of mycotoxin is generally associated with poor practices during harvesting, drying, or storage of these foods.

A dehydrated product remains stable only as long as it is protected from water, air, sunlight, and contaminants. For small packages of dehydrated fruits and vegetables, the use of an in-package desiccant (to change the moisture content of the storage environment to 1% or lower) permits storage for 6 months or more at room temperatures without significant losses of vitamins. Some useful information on proper storage condition of dried foods is presented by Labuza (27).

Extension shelf life of foods is not the only objective of the drying process, however, it is the most important. Other rationales for controlled fruit and vegetable drying are (28) to reduce the product seasonality, to improve transportability and reduce the costs of transport, to improve storage capability, and to reduce nutritional fluctuations.

These advantages of fruit and vegetable drying are compensated by some negative changes that occur during drying, for example (1,29), some ''heat damage'' of heat-sensitive constituents (vitamins, enzymes, etc.); browning, shrinkage, and ''case hardening''; irreversible loss of ability to rehydrate; loss of volatile constituents; and changes in moisture distribution within the product. Generally, these negative aspects of drying are of at least the same order of magnitude as those encountered in other preservation methods (canning, aseptic processing or freezing), and drying of fruits, vegetables, and spices remains a popular method of food preservation.

3 PRACTICAL DRYING TECHNIQUES

Drying of agricultural products can be natural, based on a natural action of sun and wind (field sun drying); artificial (air drying, contact drying, osmotic dehydration, etc.), or mixed (solar-assisted air/contact drying), using both natural and artificial energy sources, for example, in a hybrid solar dryer (30). All these drying methods are playing an important role in fruit, vegetable, spice, and herb dehydration. The heat necessary for the drying is transferred from the energy source to the material by means of conduction, convection, internal generation (microwaves, radio frequency), surface radiation (sun, light), or a combination of all these. According to the latest trends, in many existing conventional drying systems, solar, wind, and other natural energy is increasingly providing all or part of the energy required for the drying process.

Fig. 1 presents a general scheme of fruit, vegetable, and spice drying processes. As the variety of these products is extremely large, in individual case some of the processing steps may be omitted or others added. In this chapter, only the drying operation that is the main part of the full processing scheme is considered.

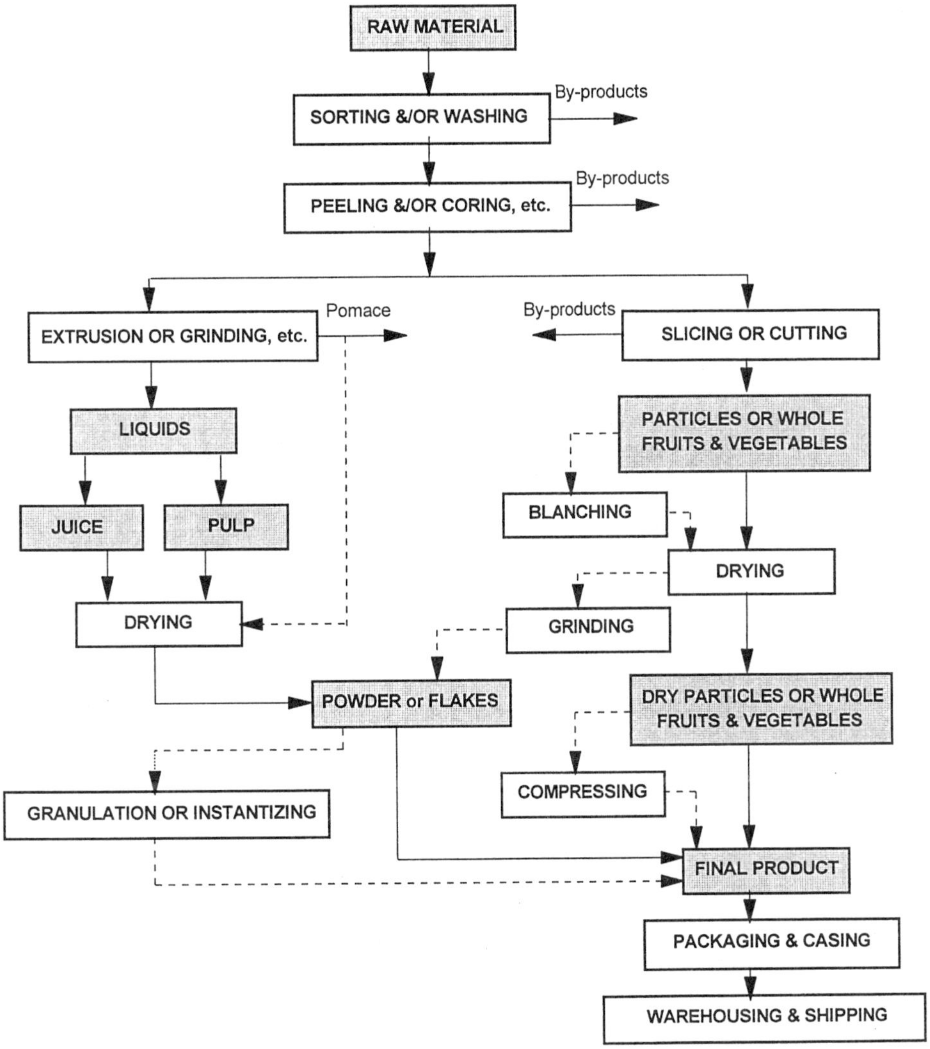

Fig. 1 A general scheme for fruit and vegetable drying.

Table 4 Total Horizontal Solar Insolation and Sunshine Hours for Some Locations Known from Fruit, Vegetable, and Herb Sun/Solar Drying

Location	Average insolation (kWh/[m^2 day])	Sunshine hours (h/day)
California (Santa Barbara)	5.6	—
Cameroon	3.8–5.5	4.5–8
Egypt (Cairo)	6	9.6
Guatemala	5–5.3	—
India	5.8	8–10
Indonesia	4.24	—
Kenya	5.25–5.6	6–7
Malaysia	4.41	—
Mali	4.34	8.4
Mexico (Jalapa, Veracruz)	4.65	—
Nicaragua	5.43	—
Nigeria	3.8–7.15	5–7
Papua New Guinea	4.6–9.6	4.5–8
Philippines (Metro Manila)	4.55	—
Sierra Leone	3.4–5.3	3–7.5
Thailand	4.25–5.66	—
Trinidad Tobago	5.4–6	8–10
Wisconsin (Madison)	3.94	—

Source: From Refs. 31 and 33.

3.1 Sun and Solar Drying

Field sun drying, used for large tonnages of fruits and vegetables, can be very inexpensive in areas where climatic conditions are adequate and additionally (but not necessarily, as, for example, in California) labor costs are low. No costs are incurred to heat or to circulate the air. Additionally, the solar energy is natural, abundant, and environment-friendly. Typically, fruits, herbs, spices, and to a lesser extent vegetables are spread out in the sun and wind on the field, special mats, concretes, etc. The radiant energy of the sun provides the heat to evaporate the water while the wind helps to move the moisture and accelerates the process. This type of drying performs well in warm and dry conditions in fields or other locations (e.g., in shadow for light-sensitive products like herbs and spices). At night and during the rainy periods, this type of drying cannot be used. The temperature of the food during sun drying is usually 5°C–15°C above ambient temperature (20). Sun drying techniques are based mainly on experience and tradition. Thickness of the wet material layer, drying time, and material handling before, during, and after drying are the conditions normally standardized on the basis of experience and can differ from one location to another. Very approximate predictions of sun drying effectiveness can be based on the energy balance from available data on the distribution of solar energy radiation throughout the world (Table 4) (31–33), absorbency/reflectance characteristic of the wet and the dry material (20), and local climatic conditions such as ambient air temperature, wind speed and relative humidity, and rainfall frequency, and intensity (34). For example, the thermal efficiency of sun drying of Colombian coffee beans, calculated as a ratio of heat necessary to remove the moisture to the total solar radiation heat, has been reported to be in the range of 13.7%–22.6% (35).

The time of sun drying depends on the product characteristics and drying conditions and typically ranges from 3 to 4 days but can be longer, for example, 3 to 4 weeks for raisins and apricots (20). During that time, the product has to be protected from rain, insects, birds, other animals, etc. Several practical methods have been developed to reduce the length of drying time. One of the most important, especially for larger production scale, is the application of different product pretreatments before sun drying. Table 5 presents examples of pretreatment techniques applied for some fruits prior to sun drying in California (36). Several other pretreatment methods and materials are used for this purpose, for example, steaming, immersion in boiling water or sugar and/or salt solutions, and use of ethyl-, methyl- or sodium oleate.

Another way to reduce sun drying time is to use solar energy concentrators with or without natural or forced airflow inside the dryer. This technique is normally called *solar drying*. More detailed description of solar drying principles, equipment, and conditions may be found in selected publications (28,30,34,35,37). Some practical values of solar collector efficiencies as applied to drying are given by Hall (12). Some of the solar drying setups for fruits, vegetables, herbs, and spices are detailed in the following section.

3.1.1 Solar Natural Dryers (Cabinet, Tent, Greenhouse, and Others)

The solar cabinet is the simplest dryer and very popular in many locations. It is used on a small production (family) scale. The wet product is placed in an enclosure, and the solar heat, generated through a conversion of solar radiation into low-grade heat, accelerates the evaporation of moisture from the product. The airflow inside the cabinet is driven by natural convection. The dryer throughput and the drying time are estimated from practice (Table 6); however, considerable effort has been made to model solar collector performance (37). The thermal efficiencies of solar drying of some fruits were reported to be 11%, 11%, and 13% for apricots, peaches, and apples, respectively (38), with highest values at the beginning of the drying run. For peaches, for instance, values were 25%, 8%, and 3% for the first, second, and the third drying days, respectively. Very similar values of the total thermal efficiency of solar drying were estimated by Bansal and Garg (28), who calculated the evaporation heat (Q) from the equation

$$Q = \eta_1 \eta_2 A I \tag{2}$$

where η_1 and η_2 are the thermal efficiency coefficients of the solar collector (η_1) and conventional cabinet dryer (η_2) with approximate values of 1/3 for both, A is the collector surface, and I is the solar insolation intensity. By multiplying both thermal efficiencies η_1 and η_2, a value of 0.11 or 11% efficiency as reported earlier (38) can be realized. This is not a general value for all solar drying applications. An individual level of this parameter should be estimated in each specific drying process.

The quality of the dried product is the most important parameter in the drying technology. Patil (39) compared the quality of peppers after sun drying, polyethylene solar drying, and solar cabinet drying. He found that solar cabinet drying offers some advantages over direct sun drying in terms of better quality and faster rate of drying. Jayaraman and associates (40) compared ascorbic acid retention in cauliflower, cabbage, and bitter gourd dried by three different methods: (a) directly under the sun, (b) under the sun in a black polyethylene tent, and (c) in a drying cabinet with three solar collectors. They reported (Table 7) that the direct exposure of these products to the sun decreased quality factors such as vitamin retention. The indirect use of solar energy may improve the quality of the final product. In some cases, this was not practical because of high costs of solar

Table 5 Outline of California Practice in Sun Drying Fruits

Fruit	Typical pretreatment	Drying practice	Treatment during drying	Drying time (days)	Yield
Prunes	Dipping in hot lye solution: fresh prunes are immersed in 0.25%–1% for 5–30 s Imperial prunes are checked in boiling water or 0.25% lye	Spread on 3- × 8-ft wooden or other tray, 35–40 kg/tray	Imperial prunes turned to prevent molding; all prunes stacked when 2/3 to 3/4 dry, usually after 4–5 days	7–14	2.4:1
Silver prunes	Silver plums are lye dipped and than sulfured for 4 h	As above	None	7–14	—
Raisins natural	None	Picked and dried in vineyard on paper or 2- × 3-ft wooden trays, 10 kg/tray	Turned on trays after partially dry or rolled into sausages on short pieces of heavy wrapping paper; stacked 5–6 days after turning	10–25	4:1
Soda-dip raisins	Dipping in 0.5% hot lye solution for 3–6 s	As above	As above	10–20	—
Sulfur-bleached raisins	Short lye dip as above then sulfured for 4 h	As above	As above	1–25	—
Apricots	Cut, pitted, trayed, and sulfured for 3 h	Dried on 3- × 8-ft wooden trays	Trays spread in sun for 1–4 days, then stacked	2–8	5:1
Peaches	Cut, pitted, trayed, and sulfured for 4–6 h	Dried on 3- × 8-ft wooden trays in drying yard	As above	4–8	4.5:1
Pears	Ripened, sorted, washed to remove spray residue; halved; stemmed; sulfured up to 24 h	Dried on wooden trays in drying yard	Stacked when partly dry, usually after 1–2 days	14–25	—
Nectarines	Cut, pitted, sulfured for 3–4 h	As for apricots	As for apricots	4–8	—

Source: From Ref. 36.

Table 6 Solar Cabinet Drying Throughput

Product	Amount of fresh matter dried per unit time[a]	Maximal allowable temperature (°C)
Apricots	4.0 kg/2 days	66
Garlic	2.6 kg/2 days	60
Grapes	5.7 kg/4 days	88
Okra	3.0 kg/2 days	66
Onions	3.0 kg/2 days	71

[a] Cabinet dimensions, 1.93 × 0.6 m; location, Syria; approximate cost of the dryer, $ U.S. 21.3 (in 1980).
Source: From Ref. 42.

Table 7 Ascorbic Acid Retention (Percentage) in Some Sun/Solar Dried Vegetables

Drying mode	Cauliflower[a]	Cabbage[a]	Bitter gourd[a]	Potato[b]
Direct sun	2.9	0.005	2.5	45.5
Polyethylene black tent	18.1	24.4	20.2	—
Solar cabinet (with three collectors)	27.1	32.9	49.9	61.3[c]

Sources: [a] From Ref. 40.
[b] From Ref. 87.
[c] Osmosed + Solar from Ref. 87.

collectors. Less expensive solar tents or covers are often used on a larger production scale (e.g., dates in Tunisia, raisins in Greece). Also solar greenhouses and terrace or room dryers are very practical for larger production as they are simple in construction and relatively low in investment cost (41). Fig. 2 presents two examples of low-cost solar dryers for a relatively larger production scale.

3.1.2 Dryers with Solar Collectors and Natural or Artificial Airflow

Some fruit, vegetable, and spice dryers utilize both direct and indirect solar radiation. In these dryers, radiant energy from the sun falls directly onto the product being dried. In addition, an airpreheater (solar collector) is used to raise the drying air temperature. The circulation of air in the solar preheater is through either free convection or use of a fan. In both situations, this airstream significantly accelerates the drying rate. Fig. 3 presents two examples of very simple and inexpensive dryers of this type. According to published data (37,42), a family-scale dryer (Fig. 3a) works well for drying apricots, apples, mushrooms, etc., usually in 1 day. Similarly, by adding a wind-driven fan for more intensive airflow, a higher throughput of the dryer can be achieved (Fig. 3b). The fan provides air circulation in the dryer and pulls the air up through the product. Dampers can be installed in the stack to control airflow rate, thus also controlling the amount of heat buildup. Several other modifications of this concept are in use for drying purposes in tropical and hot-climate areas. Further extension of the dryer throughput can be achieved by applying a forced convection airflow. For example, as reported by Bolin and Salunkhe (43), a forced convection–type dryer was successfully used for grape drying from 77% to 54% moisture

(a)

(b)

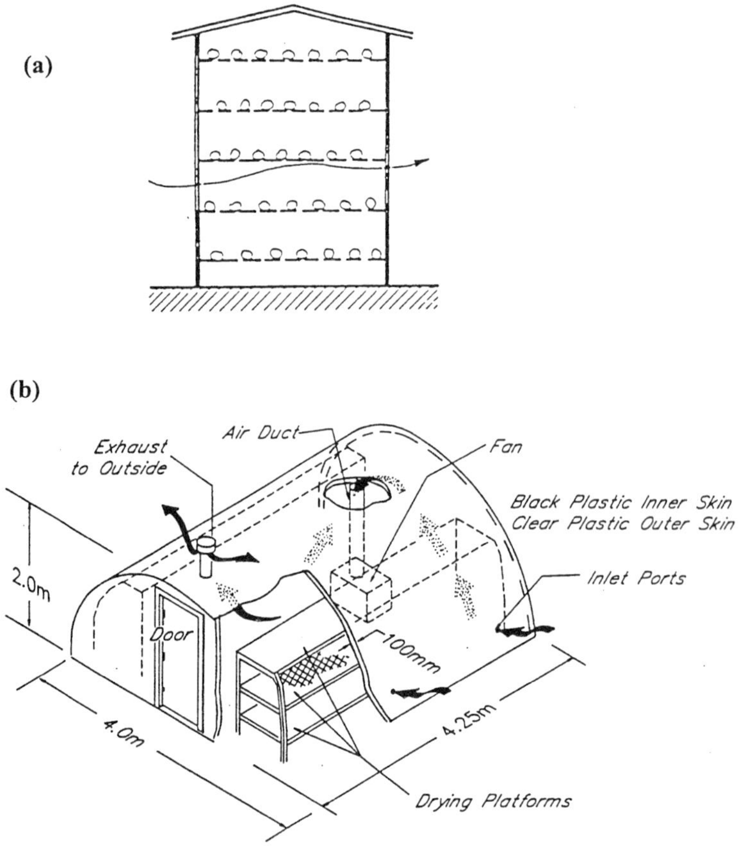

Fig. 2 Examples of large-scale solar dryers: (a) Australian type dryer. (From Ref. 33.) (b) Plastic-tent solar dryer. (From Ref. 37.)

content (wet basis) within 2 days, and to less than 10% in 6 days. This is a considerable improvement when compared to 8 days for regular sun drying of oleate-sprayed grapes, or more than 2 weeks regular sun drying of untreated grapes. The authors indicated that this type of dryer has an added advantage over sun drying for raisins, i.e., protecting the fruit from inclement rains as well as insects. Another solution in this type of installations is the use of very long metal air preheater pipes (44) or large-surface solar collectors (45). The flow of air is then artificial with the use of powerful fans. In the construction design presented in Ref. 44, the length of the solar air preheating pipe was about 7 miles. The heating rate of this preheater was reported to be sufficient for proper operation of a solar-assisted sonic dryer for several fruits and vegetables.

3.1.3 Solar-Assisted Artificial Dryers

For commercial applications, in which a large tonnage of product needs to be dried, some source of auxiliary energy is needed to initiate forced air movement and/or to provide the supplemental energy. In this scenario, the solar drying equipment is part of a whole drying system and generally assists operation of a typical air dryer, supplying it some amount of heat. Construction of a solar-assisted artificial drying system can be simple or

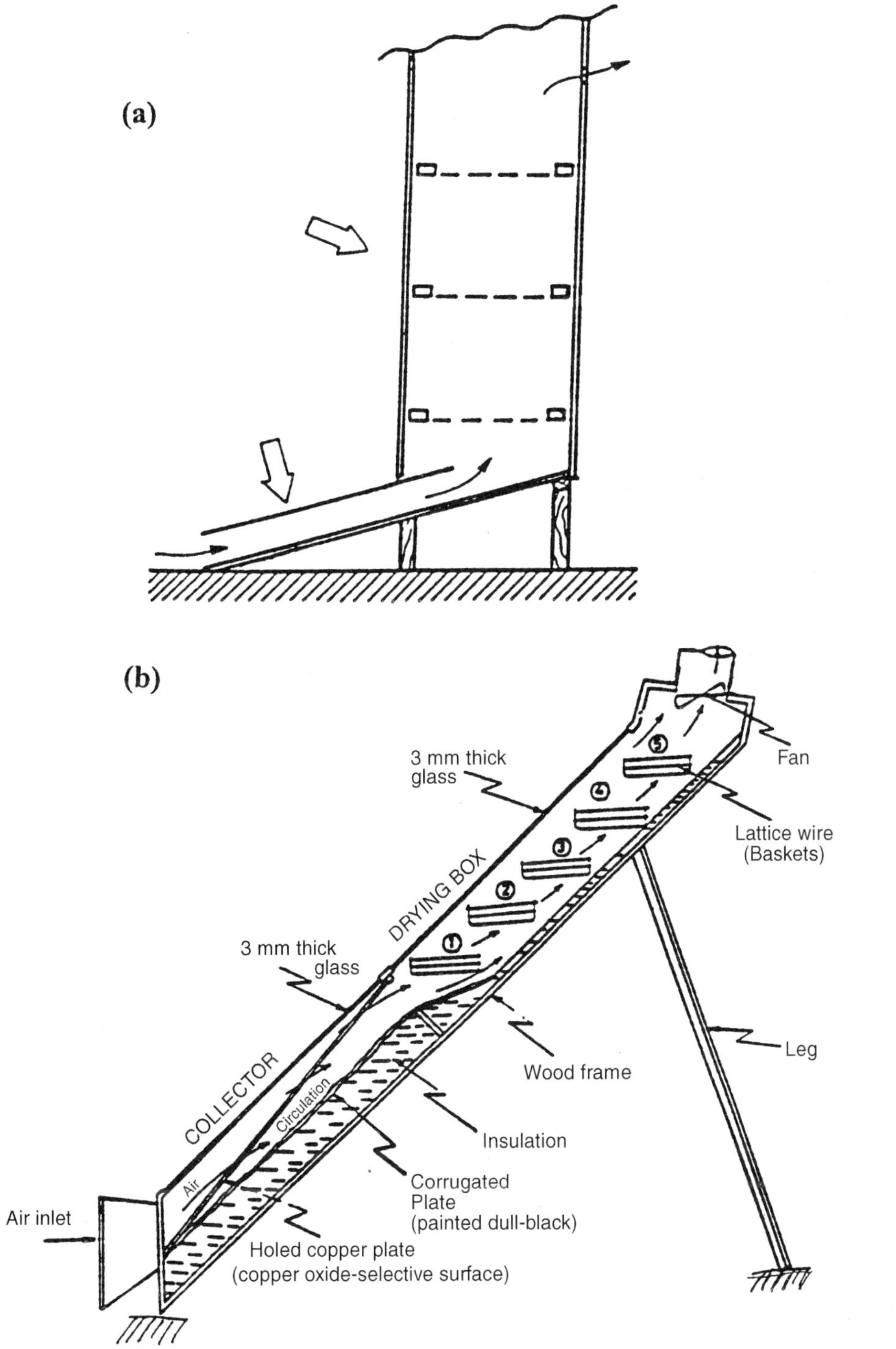

Fig. 3 Low-cost fruit and vegetable solar dryers. (Courtesy of Brace Research Institute, McGill University, Montreal, Canada.)

combined with other elements, such as a source of stored heat or heat pumps. In the simplest system the solar energy is used directly to provide a part of the heat required for moisture evaporation in a conventional air dryer. The drying of the food is normally intensive during sunshine hours; at night or low-radiation periods an auxiliary heat source supplements the heat energy. Several conventional types of dryers can be combined with solar energy collectors. For example, Smith and colleagues (46) describe a commercial convective air drying installation for the final dehydration of potato cubes from 11% down to 4% moisture (wet basis). This system utilizes solar double-glazed collectors and an auxiliary gas burner to heat the drying air. The solar collectors operate at about 40°C above the ambient temperature. In another example, considerable reduction of the fuel gas was reported by West Growers and Packers (43) when a tunnel dryer was modified for utilization of solar radiation and exhaust air was used to preheat inlet air to the dryer. Another drying procedure, described in Ref. 47, incorporated combined osmotic-solar-vacuum drying for papaya chunks. The solar energy was used for heating the osmotic syrup, which was pumped over stainless steel racks in a vacuum acrylic plastic (Plexiglas) tank. A very similar concept was proposed as a solar-assisted osmotic dehydration technique (48) for some tropical fruits. Another technique, which has not been fully explored, is the use of a desiccant to remove moisture from the drying agent (used for dehydration of heat-sensitive foods), followed by solar energy to regenerate the desiccant.

To reduce the effect of periodicity of solar radiation several physical heat storage systems have been developed for drying applications. Water-type and rock bed–type heat storage systems are the most typical (30). Carnegie (49) described one of the largest experimental installations built for commercial scale raisin drying (Fig. 4). This unit consisted of 1885 m^2 of a single-glazed solar flat-plate collector, a 700,000 kg (354 m^3) rock heat storage system, and a heat recovery wheel. All these were connected to one commercial dehydration tunnel. A supplementary energy source was a gas burner used when necessary. According to the author, the rock heat storage had limited success because of the pressure drop across the bed and because of temperature stratification. During the system's operation between 1978 and 1981, it was reported that the solar collector supplied from 26%

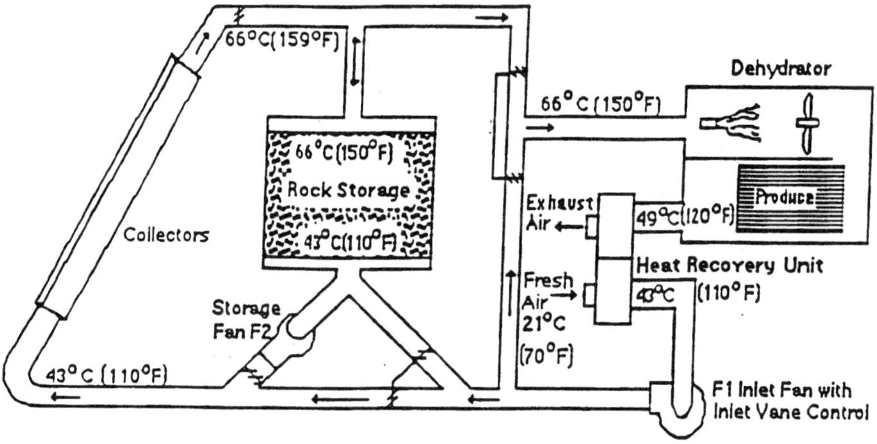

Fig. 4 Solar raisin dryer with rock-bed heat storage. (From Ref. 49.)

to 33% of the required heat energy, the heat recovery unit supplied approximately 58%, and the burners supplied the remaining 15%.

Proposed schemes of solar energy application should be integrated with other drying systems. More generally, solar energy should be applied as much as possible in energy consuming processes such as drying. Active research and development of various methods and components of solar drying systems have been taking place internationally, especially for the reduction of investment cost associated with solar collectors and adequate control systems (30,49).

3.2 Hot Air Dryers

Artificial hot air drying of fruits, vegetables, and spices employs a very wide range of methods and installations of different production scales. The main mechanism of drying in this case is moisture migration from inside the wet material to the surface and then evaporation to the surrounding air. The drying rate is a function of several parameters, such as temperature, humidity and velocity of drying air, moisture content of the moist and dry products, rheological characteristics, and specific surface and geometrical form of the initial product. Some general types of constructions of the hot air dryers suitable for drying of fruits, vegetables, and spices are presented in the sections that follow.

3.2.1 Tray, Cabinet, Kiln, and Bin Dryers

A tray or cabinet dryer is the simplest type of drying equipment. The operation of the dryer is periodic and the moist product is subjected to non-steady-state process. The dryer allows processing of different products, from liquid slurries to solid piece-form materials. A typical cabinet dryer consists of an insulated chamber into which tray loads of prepared food are placed. A fan pushes or pulls air through a heater and then either horizontally or vertically between the trays. By means of a damper system, part of the air may be recycled and part discharged to the atmosphere. Such dryers vary in size from very small units, containing one or few trays, to very large units used singly or in groups with a throughput of fresh material up to 20,000 kg per day. The material to be dried is placed in relatively thin layers (1 to 6 cm thick). Good control of drying conditions is possible. Among main disadvantages of these dryers are the large amount of manual work needed to operate the dryer and the relatively low intensity of the drying. However, a wide variety of materials can be dried in them. Because of its versatility, their use is widespread in the food industry. An example of modeling of this type of drying process is given in Ref. 16.

For relatively bigger production loads, a kiln dryer is the solution. It consists of a two-story building with a slotted floor that separates the drying section (upper part of the building) from the air-heating section in the lower floor. The heated air is forced from the lower to the upper section through the floor and the bed of the product. Drying times are quite long because of the relatively large thickness of the product layer and the low air speed. At the end of a drying cycle, dry product is removed and a new batch of a wet product is loaded. A combination system that uses solar energy produces some reduction in gas, oil, or steam consumption.

Bin-type finishing dryers are sometimes used for the final drying and equilibration of some dried fruits and vegetables, although often with large modern conveyor dryers they are no longer needed. They are essentially large vertical cylindrical bins with perforated bottoms through which a constant stream of warm air is blown (14).

3.2.2 Tunnel Dryers

Tunnel dryers possess all advantages of tray dryers. In addition, they have semicontinuous operation. This type of dryer is very popular for drying fruits and vegetables. As compared to that of a tray dryer, the investment cost is higher. A typical dryer may be more than 20 m long with a square or rectangular cross section of 2 m × 2 m. The wet material is loaded into trays that are stacked on trolleys (Fig. 5). The trolleys are introduced periodically into one end of the tunnel (the ''wet'' end), advance through the tunnel, and are removed at the other end (the ''dry'' end). A typical tunnel dryer can be operated in a cocurrent or countercurrent flow of air and trolleys. Two-stage dryers that feature a short cocurrent stage followed by a longer countercurrent stage are also used. In cross-flow designs the drying air moves at right angles to the path of the trays of food. Basic modeling principles for this type of drying are similar to that of a cabinet dryer. The drying times are similar to those in cabinet tray drying, and the airflow rate is linked to the total number of trays. The dimensions of the tunnel are calculated on the basis of the necessary drying throughput, necessary drying time, and dimensions and capacity of a single trolley. Information on the design, constructional details, and operation of these dryers has been published (1,20,36).

3.2.3 Conveyor-Type Dryers

Fully continuous drying operation is achieved in conveyor, belt, or band dryers. This type of dryer is very popular in the fruit and vegetable processing industry. The wet product formed or placed in a bed of different thicknesses is carried through a tunnel on perforated (mesh, slotted, or louvered) conveyors. Heated air is directed up or down through the conveyor and the layer of the product, usually up in the early drying stages and down toward the dry product exit, or directed across the material surface for products in thin layers on a nonperforated band. Some models consist of two or more conveyors in a series. Also, either cocurrent or countercurrent configurations can be used. Such dryers are limited to foods that form a porous bed (cut, granulated, pelleted, or naturally particulate foods). For vegetable or fruit drying, multiple conveyors (up to five), one above the other, (Fig. 6), can be used. The wet product is introduced onto the top conveyor and progresses downward from one conveyor to the next. Air circulation is usually a combination of cross-flow and through-flow. A final drying step for some vegetables and fruits is often provided in this type of dryer. Gradually, the conveyor dryers are replacing drying trays

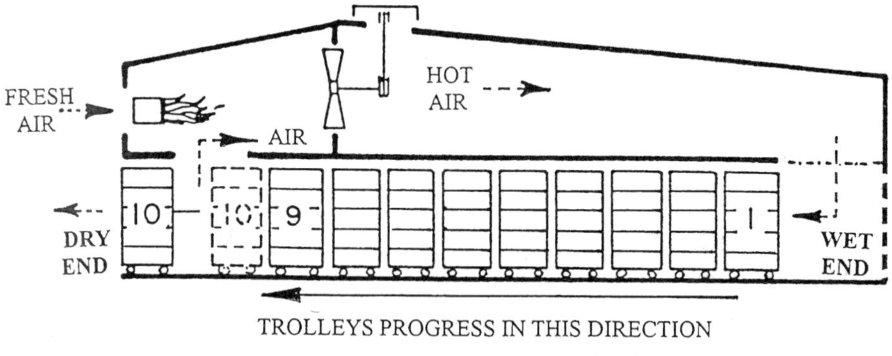

Fig. 5 Simple cocurrent tunnel dryer with partial air recirculation.

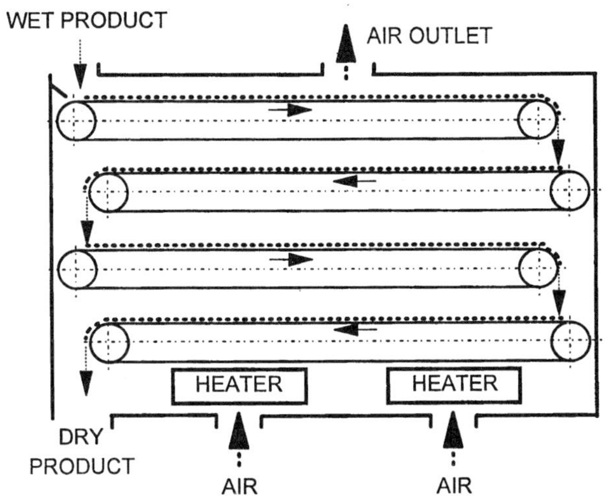

Fig. 6 Diagram of multiple-conveyor dryer.

in tunnels for fruit and vegetable pieces such as apples, carrots, onions, and potatoes (12). Infrared, microwave, or radio frequency energy is sometimes additionally supplied to the product conveyed through the tunnel. The drying unit may operate under vacuum or atmospheric conditions.

3.2.4 Fluidized Bed Dryers

Particles of fruits and vegetables (whole or diced) may be dried in fluidized bed dryers (Fig. 7) or several of their modifications: vibrofluidized, pulse fluidized, or spouted bed dryers (4,50,51). The main advantage of this type of drying is its brevity as a result of high-intensity heat and mass transfer achieved through enhanced air turbulence in the fluidized bed. Batch or continuous systems are typically used for the drying of food particles in the range of 20 μm to 10 mm. Peas, beans, diced carrots, onions, and potatoes are typical vegetables dried in this type of dryer. If the particle size of the food covers a wide range, uniform fluidization of the product may be difficult. In such cases, mechanically or pneumatically (pulse fluidization) induced vibration may overcome the problem. The frequency of the vibration is usually in the range 5–25 Hz with an amplitude of few millimeters (51). This type of dryer is often used as a second-stage powder dryer or as a granulator after, for example, spray drying in a first stage (Fig. 8).

Another modification of the fluidized bed dryer, a spouted bed dryer, may be used in some practical applications, for example, for paprika breaks, carrot cubes, mushrooms, and tomato seed drying (52). The spouted bed technique can be also used for fruit and vegetable concentrate or paste drying, when it is spread into a bed of inert particles (synthetic resin [Teflon], alumina, etc.) circulating as a fountain inside a drying chamber (Fig. 9). The pulse-fluidized bed dryer, adopting regular pulsations of the drying air, as well as other fluidized bed modifications, the swirl fluidizer and the centrifugal fluidized bed dryer (50), are still in a development stage. However, they many offer promising technology for wide range of diced or flaked vegetables: bell pepper, beet, carrot, cabbage, onion, and mushroom (19,53). Much more detail on the fluidized bed drying technique can be found in published sources (3,4,12,50).

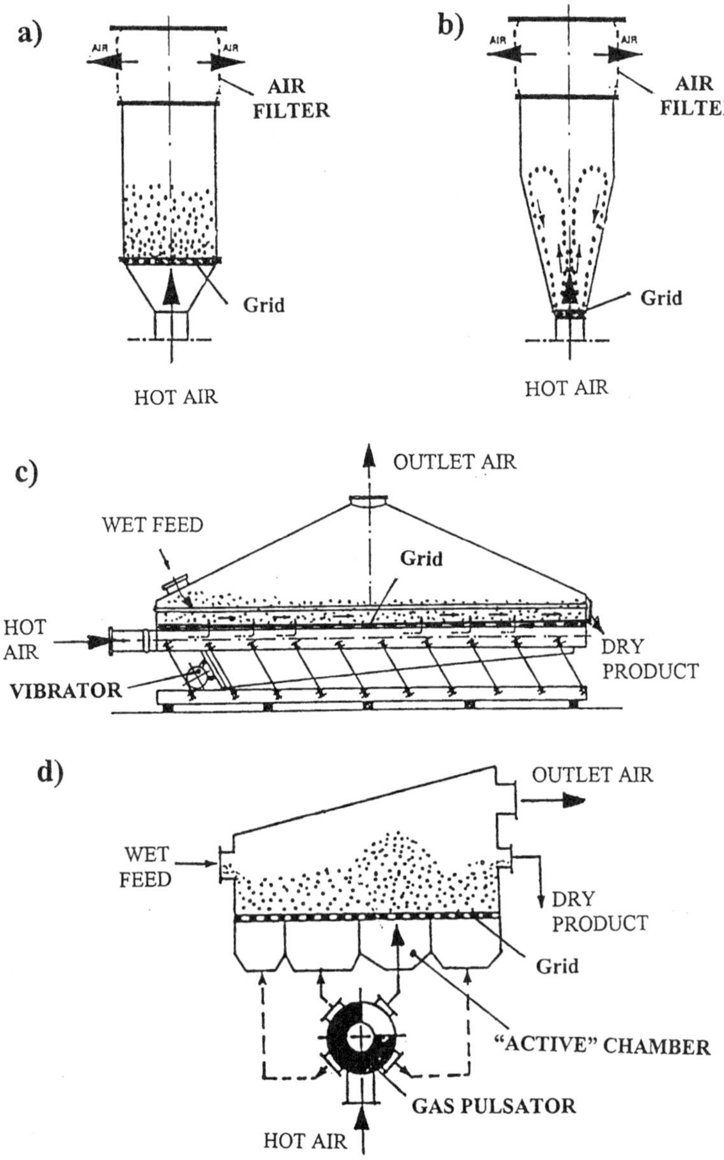

Fig. 7 Basic schemes of fluidized bed dryers: (a) standard, (b) spouted bed, (c) vibrofluidized bed, and (d) pulse fluidized bed. (Modified from Refs. 3 and 53.)

3.2.5 Spray Dryers

Some fruit or vegetable powders are produced from juices, concentrates, or pulps by using a spray drying technique. Dry powders can be directly used as important constituents of dry soups, yogurt, etc. Principles of spray drying are well documented in the literature (54,55). However, its application to fruit and vegetable pastes poses several problems. For example, stickiness, caking, and hygroscopicity of the final products must be overcome. Drying of tomato pulp is a typical difficult process as the dried tomato powder

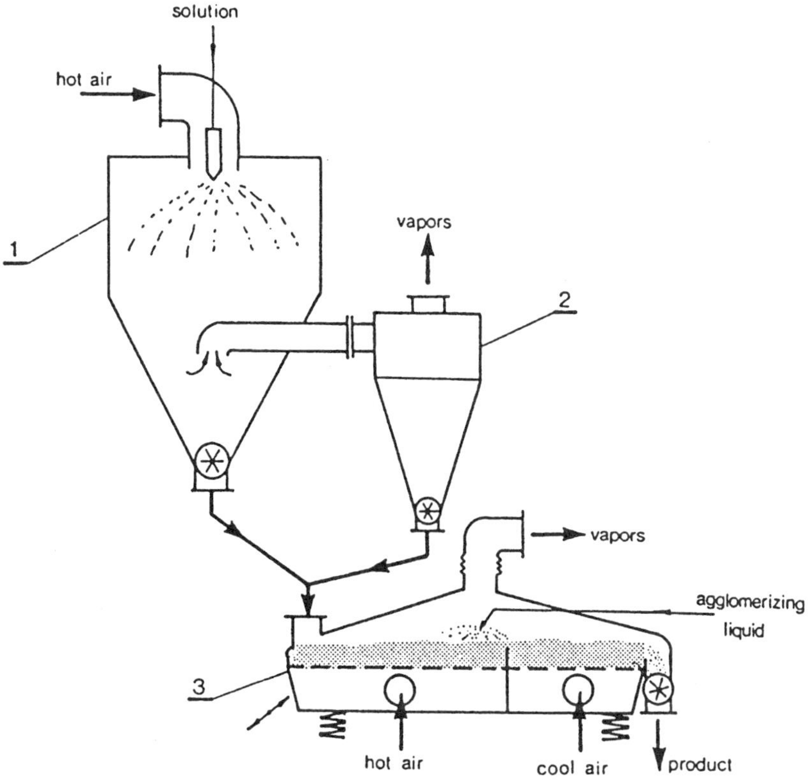

Fig. 8 An example of a combined spray drying–agglomerating system: (1) spray dryer, (2) cyclone, (3) vibrofluidized bed dryer-granulator. (From Ref. 3.)

tends to become soft and sticky while still warm. The drying is achieved by spraying of the slurry into an airstream at a temperature of 138°C to 150°C and introducing cold dry air either into the outlet end of the dryer or to the dryer walls to cool them to 38°C–50°C (19). A wide range of fruit and vegetable powders can be dried, agglomerated, and instantized in spray drying units specially equipped with an internal static fluidized bed, integral filter, or external vibrofluidizer (56). Bananas, peaches, apricots, and to a lesser extent citrus powders are examples of products dried by such techniques.

A typical spray drying unit consists of a large-volume drying chamber, feed atomizer, hot air supply, and dry product recovery/dust collection systems. Three consecutive steps of the process can be distinguished: (a) atomization of the feed, (b) spray air mixing and moisture evaporation, and (c) separation of the dry product from the exit air.

Atomization is the most critical operation in the spray drying operation. Three general types of atomizers are available with many other individual developments. The most commonly used are rotary wheel atomizers and single-fluid pressure nozzle atomizers. Pneumatic two-fluid (compressed air/liquid feed) nozzles are used rarely in very special applications (slurries, pastelike materials, etc.). Cocurrent, countercurrent, or mixed flow of the droplets and drying air is typical in spray drying chambers. The geometrical properties of the drying chambers depend mainly on the type of atomizer used. They are generally short and wide (conical type) for rotary wheel atomizers and long and narrow (tall type)

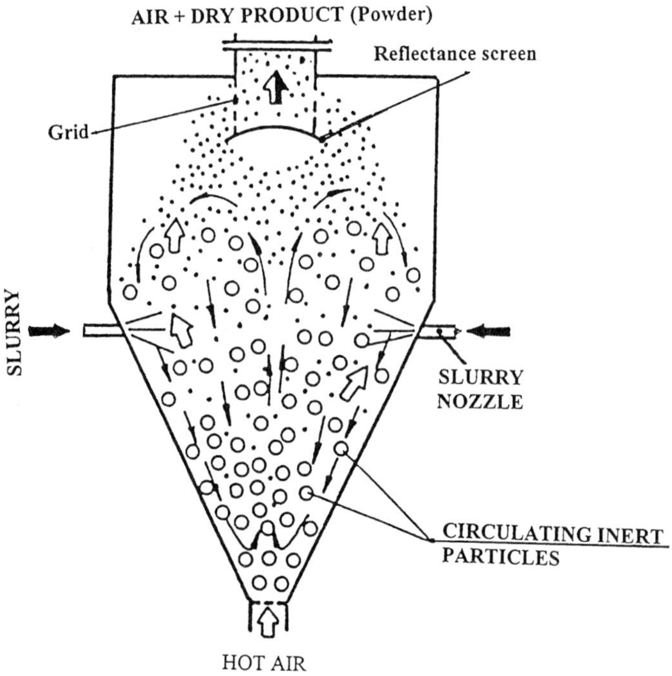

Fig. 9 Principles of drying of the slurry in a spouted bed of inert particles. (Modified from Ref. 3.)

for nozzle atomizers. The dry product is collected at the bottom of the drying chamber and/or in cyclones, bag filters, or electrostatic precipitators. The final wet separation of the powder from the exhaust air is provided in wet scrubbers, wet cyclones, or irrigated fans (54,55). Some spray drying systems, equipped with special atomizing devices, are used for drying pastelike materials such as fruit and vegetable concentrates.

3.2.6 Other Hot Air Dryers

Pneumatic conveying dryers are generally used for finish drying of powders and granulates of fruits and vegetable products and are extensively used for making potato granules (Fig. 10). As the material entering the dryer must be conveyable in an airstream, the incoming

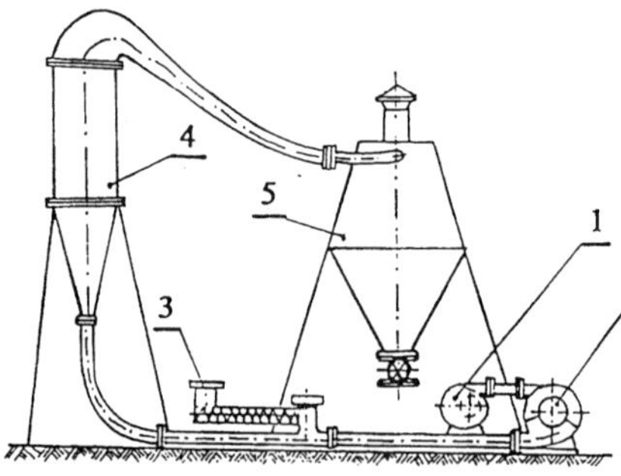

Fig. 10 Modified pneumatic dryer for potato granulates (Aradi-Rybanski dryer): (1) fan, (2) air heater, (3) feeder, (4) upper drying chamber, (5) cyclone. (From Ref. 21.)

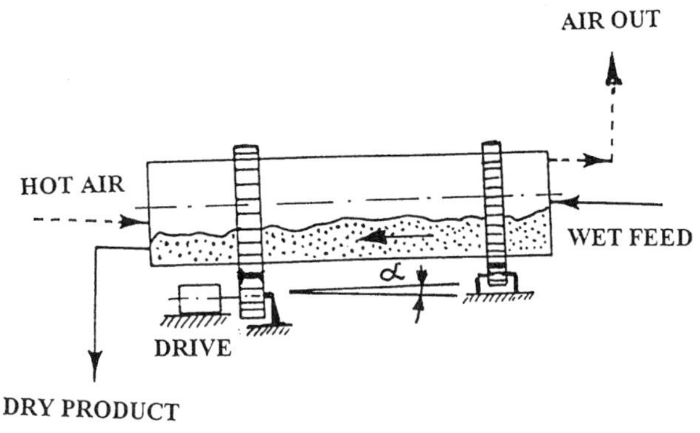

Fig. 11 Simplified schematic of a rotary dryer.

material usually is predried in another way (spray, fluidized, spouted bed dryers, etc.) to a moisture level below 40% (w.b.).

Rotary dryers may sometimes be used for drying of particulate solid foods. In these dryers, (Fig. 11) the wet solids (particles, crystals, etc.) enter one end of a rotating drum, move down the drum length, cascading from peripherally mounted lifting flights, and exit at the other end suitably dried. The drying medium is either hot air or combustion gases that flow cocurrently or countercurrently to the direction of the solids through the drum length. The size of these dryers varies from approximately 0.3 m in diameter by 2 m in length to 5 m in diameter by 90 m in length. Many details of these last techniques are discussed in the literature (1,12,57).

3.3 Contact Dryers

The hot surface of the contact dryer is the source of heat necessary to evaporation of moisture from the fruit or vegetable concentrates, pulps, pastes, or slurries. This type of drying is more economical than, for example, spray drying, but the dry product usually has different characteristics from those of a spray dryer. A typical example of the contact dryer is a drum dryer with a single, double, or twin drums. The single-drum comprises only one roll. A double-drum dryer comprises two rolls, which rotate toward each other at the top. This arrangement makes it possible to adjust the feed thickness layer by changing the space between them. The twin-drum dryer is similar in appearance to the double-drum dryer but is quite different in its operation. The two drums occupy the same general position as in a double-drum dryer but rotate away from each other at the top and are not spaced closely together. The fruit or vegetable pulp is fed onto the metal surface of the drums by means of various types of feeders. The dry product is removed by doctor blades. Depending on the material properties, the product is removed in the form of powders, flakes, or webs. The drum system can be entirely enclosed in a hood for vacuum operation or for supply of the necessary amount of drying air. Thus, organic solvents or volatiles can be recovered in this process. A typical example of a single-drum drying operation is the drying of mashed potato or apricot pulp (to produce a sheet of apricot; see Fig. 12). Double-drum or twin-drum dryers are in commercial use for drying apple sauce, mashed white potatoes, and purees of tomato, pumpkin, and banana. Success has been reported in drum-drying mixtures of apple sauce with strawberry, cranberry, blueberry, black currant, loganberry, and banana (1).

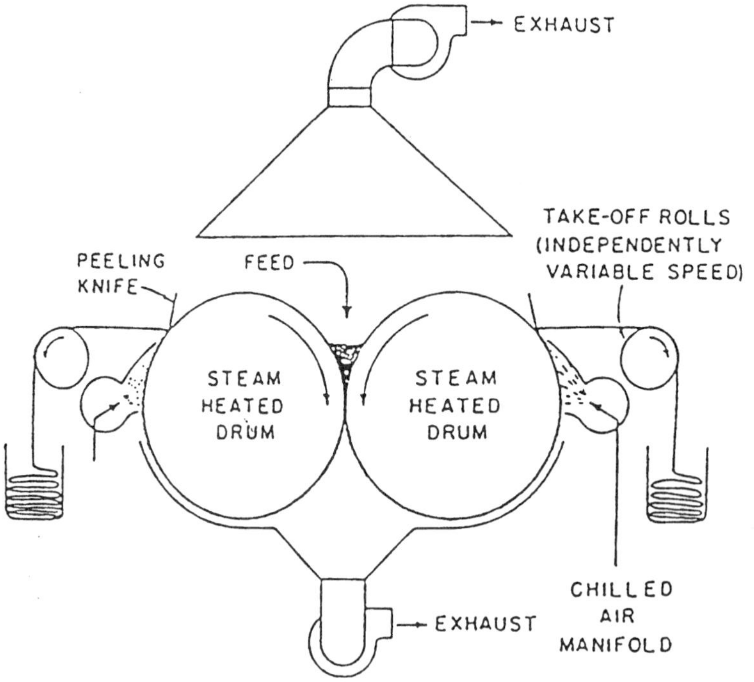

Fig. 12 Double-drum dryer. (From Ref. 19.)

Special types of contact dryers—plate dryers with rotating radial arms—are dedicated to drying of solid particulate materials (53,58). All contact drying units may operate under a vacuum, at atmospheric pressure, or in gas-tight chambers.

3.4 Foam Drying

Two foam drying methods are of practical importance: foam mat and foam spray drying. Foam mat dried fruit or vegetable powders have fewer heat-induced changes in color and flavor than conventional spray dried or drum dried products. A product with density less than that in a conventional dryer is obtained. The product density is about equal to the density of instantized or agglomerated powder.

A stable gas-liquid foam is the primary condition for successful foam drying. Glycerol monostearate, solubilized soya protein, and propylene glycol monostearate are the typical additives for the fruit and vegetable foam formulation from juice or pulp. Foam mat drying involves drying a thin layer (0.1–0.5 mm) of the stabilized foam in air at 65°C–70°C for only a few minutes, as the foam structure decreases drying time to about one-third. The foam is spread on perforated floor craters as the airstream is forced through the bed. A continuous belt tray dryer as well as slightly modified spray dryer (12) can also be used to this process. Good quality tomato, apple, grape, orange, and pineapple powders can be produced by using this technique. Optimal initial concentration of feed solids is in the range of 30% for tomato and 55% for orange. The cost of such a drying process may be higher than that of spray or drum drying, although lower than that of vacuum and freeze drying. More technical details on foam mat drying are given in van Arsdel and coworkers (1) and Arthey and Ashurst (17).

3.5 Explosion Puffing

Overheating of a wet food product in a pressurized chamber, called a *gun*, followed by a quick release of the pressure to atmospheric condition is the technique, called *explosion puffing*, that allows the production of a good-quality porous dry material. The amount of water typically contained in fruits and vegetables is usually too large to evaporate entirely during the explosion puffing process. An initial predrying of the fresh product is therefore necessary when using this technology. Table 8 (19) presents some examples of practical values of this initial moisture content for the explosion puffing of selected fruits and vegetables along with some other necessary processing conditions. The average moisture content of fruits and vegetables in order to start the process should be lowered to 15%–35%. Such partly dried food pieces are put into a puffing gun. The gun is sealed and pressurized by heating directly (superheated steam) or indirectly through the walls (gas flame, electric heater, etc.). As the pressure and temperature inside the chamber reach a certain level (Table 8), the door of the gun is suddenly opened and explosion of the moisture from the food occurs. The advantages of this technique include shorter drying time (which may be lower by a factor of 2 or 3, as compared with conventional air drying time), good reconstitution characteristics that result from the porous structure, and good organoleptic properties. A final-stage drying is sometimes necessary after the explosion puffing.

A simple batch explosion puffing gun is essentially a cylindrical pressure chamber with a quick-release lid, heated internally or externally. A more sophisticated continuous explosion puffing system also has been developed (Fig. 13) (59); this system allows up to 44% reduction in steam consumption as compared with conventional dehydration.

3.6 Vacuum and Freeze Drying

To intensify moisture removal and lower drying temperature to protect heat-sensitive food components, a vacuum drying technique is employed. Some fruits, vegetables, herbs, and spices are often dried by this technique. Typical batch or continuous drying equipment

Table 8 Explosion Puffing Processing Conditions of Some Fruits and Vegetables

Commodity	Moisture content before puffing (% w.b.)	Steam pressure (kPa)	Temperature of process (°C)	Dwell time (s)	Rehydration time (s)
Apples	15	117	121	35	5
Blueberries	18	138	204	39	4
Cranberries	17–26	138	163	64	3
Pears	18	228	154	60	5
Pineapples	18	83	166	60	1
Strawberries	25	90	177	—	3
Beets	20–26	276	163	120	5
Carrots	25	275	149	49	5
Celery	25	275	149	39	5
Mushrooms	20	193	121	39	5
Onions	15	414	154	30	5
Peppers	19	207	149	45	2
Potatoes	25	414	176	60	5

Source: From Ref. 19.

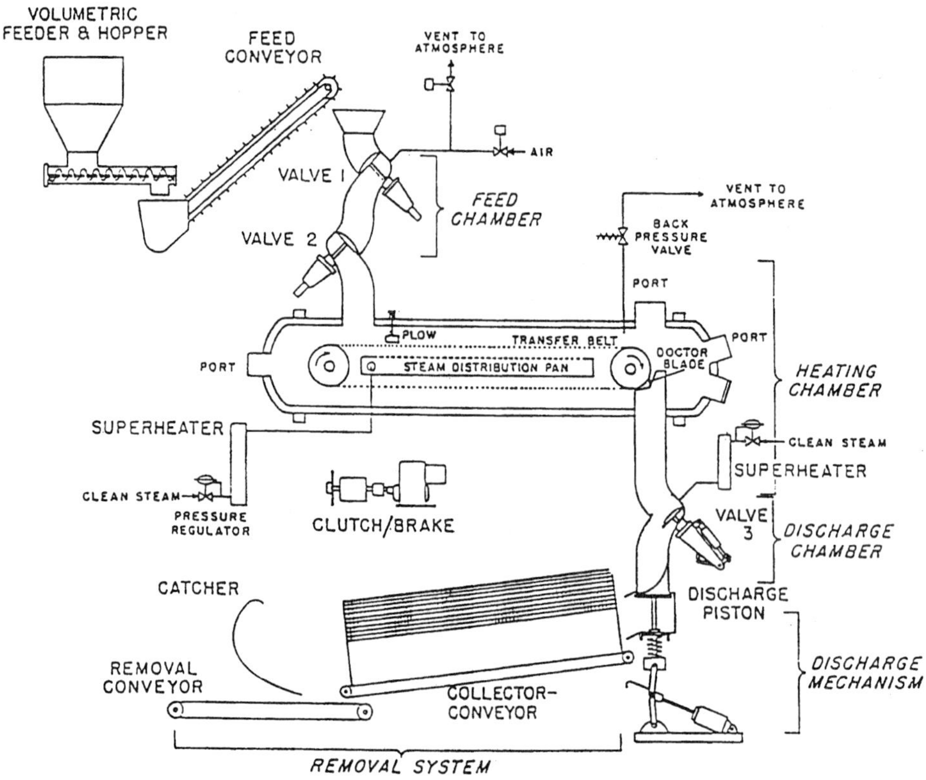

Fig. 13 Basic concept of continuous explosion puffing dryer. (From Ref. 59.)

is used with some modifications to sustain the vacuum created by a pump connected through a moisture condenser.

One form of vacuum drying, i.e., freeze drying, is very important in food drying operations. Removal of the moisture is a result of sublimation, without a phase change from solid (ice crystals) to liquid. Fig. 14 presents a phase diagram of the water with an indication of conditions for normal pressure air drying, vacuum drying, and freeze drying processes. The advantage of freeze drying over other methods of drying is the superior quality of the product obtained. Little or no shrinkage occurs. The dry product has a porous structure and a color almost as fresh as that of the raw material. The only disadvantage of this process is the high equipment and operational cost—the highest of all typical drying methods. This explains why freeze drying is only used for more expensive products, such as coffee extracts, instant dry soups, and powders of exotic fruits, spices, and herbs. Bananas, oranges, strawberries, peaches, plums, tomato, fruit juices and flavors, asparagus, beans, cabbage, cauliflower, celery, mushrooms, onions, peas, parsley, chives, and many herbs are processed for freeze dried products.

There are two main stages in the freeze drying process: (a) freezing of the food, when most of the water is converted into ice, and (b) sublimation, when the bulk or all of the ice is transferred into vapor under very low pressure or high vacuum, and this vapor is removed from the dryer. In some cases, additional final drying, in the same or other equipment, is necessary. Cabinet or tunnel batch-type dryers are typically used with pressures in the range 13.5–270 Pa. Commercial high-performance freeze drying cabinets are

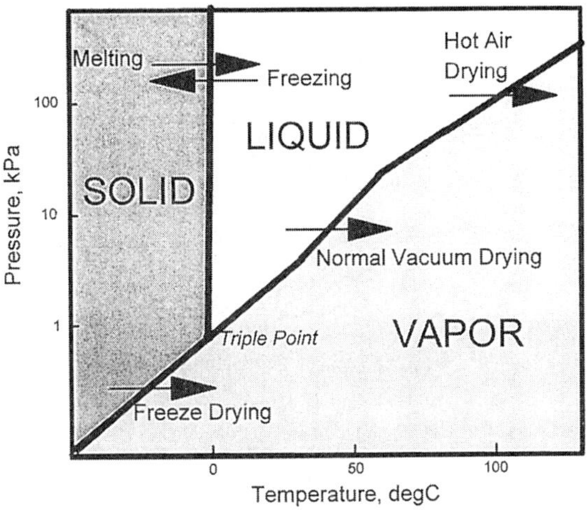

Fig. 14 Phase diagram of water in application to drying process.

fitted with heated shelves on which trays of frozen food are placed. Heat necessary to sublimation is transferred from the heated shelves by conduction, while additional heat may sometimes be supplied by microwave or infrared radiation (60). In a tunnel freeze dryer the food is placed on specially designed trays that are assembled on tracks. The tracks enter the tunnel through a vacuum lock. In the main tunnel, the trays move between stationary heated plates and exit through another lock at the dry end. Fully continuous freeze dryers are developed for specific applications. Particulate solids are conveyed by a screw, belt, or vibrating plates through vacuum chambers. Spray (23) and fluidized bed atmospheric freeze drying (61) are examples of modifications of the typical freeze drying process. Details of the freeze drying technique can be found in several publications (2,60,62).

3.7 Radiative and Acoustic Drying

Microwave drying is becoming steadily more popular in drying technology. However, research is needed, especially for larger production scales. Drying of potato chips, peas, onions, beans, and juice concentrates is an example of such applications (23).

Infrared drying is used for thin layer drying of some foods and especially spices (23). These products are conveyed beneath infrared radiators on belts or vibrating plates. Some research work on infrared drying of potatoes and carrots has been done by Karuri (63).

Acoustic drying with waves of 12–19 kHz and intensities between 128 and 132 dB was applied to drying of green rice in a fluidized bed at 20°C and 40°C (64). A successful combination of solar and acoustic energy for drying of different agricultural products has also been reported (44).

3.8 Osmotic Dehydration

An example of a minimal processing dehydration method for foods, especially fruits and vegetables, is osmotic dehydration. The process is based on a tendency to reach an equilibrium between osmotic pressure inside the biological cells (fruit, vegetable, etc.) and the surrounding osmotic solution, which has an increased osmotic pressure caused by high concentration of a soluble osmotic agent. Molecular diffusion of water through semipermeable cell membranes takes place and the product loses its water. The osmotic dehydration process is stopped when the osmotic pressure inside the tissue reaches the osmotic pressure of the surrounding syrup. Theoretically, molecules of the solute, being large, do not migrate en masse from the osmotic syrup into the product. Practically, however, some migration (infusion) of the solute occurs. A net mass loss is therefore the difference between water removal and solids gain. Osmotic dehydration can reduce water activity, which is limited mainly by the water activity of the osmotic syrup itself, to some extent. For example, a typical osmotic syrup, such as a sucrose solution (60° Brix), has a water activity of about 0.85. This means that food osmotically dehydrated in the solution is equilibrated to a water activity of 0.85 and therefore is not necessarily safe from a microbial point of view, therefore, a second stage of drying is often necessary. This second stage may be typical hot air drying, solar drying, or vacuum, freeze, microwave, or any other type of drying. For example, Kim and Toledo (65) combined osmotic dehydration of blueberries with high-temperature short-period fluidized bed drying. The process reduces the dehydration time compared to that of conventional air dehydration and simultaneously puffs the blueberries during the process. The dry product was reported to compare favorably with the commercial product in the market. Several other two-step drying processes, with osmotic dehydration in the first step, have been developed for many fruits and vegetables. For example, high-quality dry apple, carrot, and potato products have been produced; however, dehydrated celery has not been satisfactory (66).

Typical osmotic dehydration equipment consists of a mixer tank for the syrup with heating elements to maintain temperature around 50°C; however, higher or lower temperatures are possible, depending on the type of dehydrated product. For example, Nsonzi and Ramaswamy (67) found the optimal dehydration conditions for blueberries at 50°C for 4.5 h in 55°Brix sucrose syrup. Continuous osmotic dehydration units have also been developed. Fig. 15 shows a pilot scale continuous osmotic dehydration unit applied to

Fig. 15 Blueberry dehydration in continuous pilot-scale osmotic dehydration unit. (Courtesy of Food Research and Development Center, Saint-Hyacinthe, Canada.)

blueberries as well as other fruits and vegetables. More detailed data on practical and theoretical aspects of fruit and vegetable osmotic dehydration have been published (2,48,67–70). Osmotic infusion of some additives, for example, caffeine and vitamins, before regular drying is a technique used for new fortified dry foods (71,72).

4 DRYING OF FRUITS AND VEGETABLES

A large number of fruits and vegetables have been successfully dried by one or more techniques described in the preceding section. It is beyond the scope of this book to detail drying aspects for each fruit and vegetable. Further, more than one technique can be successfully applied for a given fruit or vegetable. This section is limited to those fruits and vegetables of greater commercial importance. Also, the discussion is limited to the most typical application; therefore, other methods and conditions used in the fruit and vegetable processing industry are not detailed. Raisin, apple, apricot, fig, peach, pear, and prune are important dried fruits; raisins account for the greatest volume. Hence, these are considered. Similarly potato, carrot, pea, beet, onion, and garlic occupy spots of commercial significance and are discussed.

4.1 Drying of Fruits

4.1.1 Grape (Raisin)

Grapes of certain varieties (Thompson Seedless, European Muscat, Muscat of Alexandria, etc.) are processed for raisins. California is the biggest producer, contributing about 50% to the total world production. Turkey, Australia, Greece, Iran, the Republic of South Africa, and Spain are other important raisin-producing countries.

As previously mentioned, sun drying of grapes is the most typical drying procedure. The grapes are spread on wooden or paper trays or strips of plastic in the vineyards. Drying is typically preceded by dipping in special solutions to remove the waxy layer. Grapes may also be sulfur treated (36,73); however, large quantities are dried without any pretreatment (74). The recommended (24) dipping treatments are (a) soda dip: 0.2% to 0.3% NaOH at 93°C for 2 s, followed by a water rinse; (b) Australian-type mix dip: 0.3% NaOH + 0.5% K_2CO_3 + 0.4% olive oil at 82°C for 2–3 s without rinsing; (c) Australian cold dip: 5% K_2CO_3 emulsified with 0.4% olive oil for 1–4 min at 35°C–38°C; (d) California soda-oil dip: 0.4% solution of soda ash and Na_2CO_3 (Wyandotte powder) for 30–60 s at 35°C–38°C. Pretreatment techniques are applied with different solutions, temperatures, and dipping times. The sun drying process normally takes 2–3 weeks and produces dark raisins—if no sulfur pretreatment is applied.

Artificial air drying of grapes is also applied, but only to a limited extent. The process starts with dipping of grapes in 0.25% hot lye, washing, spreading on trays, and sulfuring (exposing to burning sulfur). An average 1.8–2.3 kg of sulfur per metric ton of fresh fruits is burned in a sulfur house for about 4 h. A tunnel dryer is typically used for the grape drying up to 9%–14% w.b., the process takes 24–45 h (12). The sulfur dioxide concentration of the final product ('Golden' variety, bleached raisins) ranges between 1500 and 2000 ppm. A small amount of grapes is puffed in vacuum shelf dryers, giving a high-quality, low-density porous material of less than 4% moisture, which is used as an ingredient of instant cereals, soups, desserts, cakes, etc. Additional information about raisin production can be found (73–75). Some new techniques, employing, for example, microwave energy, have also been developed (24).

4.1.2 Pome Fruits (Apple and Pear)

4.1.2.1 Apple

Apples are often preserved by drying on a large scale. The largest quantities of apples are dried in the United States; significant, but smaller, quantities are also dried in Canada, Australia, the Republic of South Africa, and Italy. Only artificial dryers are used by commercial apple drying plants. For drying, the peeled and sliced apples are typically treated with a weak solution of citric acid and potassium meta/bisulfite dip. The latter provides SO_2 in solution, which inhibits the enzymatic browning. Normally, apples of size bigger than 6.35 cm (2.5 in) of certain cultivars, for example, 'Red Delicious' or 'Golden Delicious,' 'Jonathan,' 'Gravenstein,' 'Yellow Newton Pippin,' 'Winesap,' and 'Rome Beauty,' are peeled, trimmed, and cut to pieces 9.5–12.5 mm (3/8–1/2 in) in thickness. The treated apple slices may be then cold stored for about 24 h to allow the fungicide to penetrate the slices. Then, the slices are spread on slotted floors of natural draft or loft-type kilns, where the heated air rises through slices and dehydrates them. Drying in the kiln requires 14–18 h, depending on the rate of air circulation. Air temperatures of about 65°C–75°C are maintained, and the fruit is turned several times during drying. A typical ratio of fresh to dry mass of apples is approximately 7:1 for a typical high final moisture content of 10%–25% w.b. This type of dry product is sometimes called an *evaporated apple*. Low-moisture apple slices, with less than 3% of moisture, are another dehydrated apple product. Final products may contain up to 2500–3000 ppm of SO_2.

Tunnel and conveyor dryers are also often used to dry apple slices. In the tunnel dryer, the apple slices are spread onto trays at about 10 kg/m^2. Typical examples of inlet and outlet air temperatures and approximate drying time for apple slices, as well as for other fruits, are presented in Table 9. Table 10 presents examples of drying conditions for apple slices, apricot, and plums in a four-band-conveyor dryer (Fig. 6). Continuous belt, vacuum, or other types of dryers are used for finish drying of apple pieces from evaporated apples to low-moisture apples. Other methods, such as explosion puffing, freeze drying, foam mat, and drum drying, have all been used in experimental or relatively low-production-scale test runs. In addition to slices, several other forms of dried apples are available. Small-size apples, dried as whole fruits, are also used for apple chop production. Apple powder (flour) consists of dehydrated apples prepared mainly from dry slices or chops by a grinding process. A dry applesauce is produced from apple pulp. All dried

Table 9 Examples of the Drying Process Parameters in the Tunnel Dryer

	Temperature of the air in the		
Commodity	Inlet (°C)	Outlet (°C)	Drying time (h)
Apples	74	54	12–14
Apricots	80–82	58	29.5
Blueberries	45–55	—	8–12
Cherries	78–82	58–62	12
Grapes	78	50–54	18–24
Pears	75–78	52–58	48
Plums	78–80	45–55	24

Source: From Refs. 1, 92, and 93.

Table 10 Examples of the Drying process Parameters in the Four-Band Dryer

Parameter	Apples	Apricots, (large)	Apricots, (small)	Plums
Unit load on a first band (kg/m^2)	7.6	12	16	14
Temperature of air (°C)				
Under 1st band	63	80	80	80
Under 2nd band	69	73	73	78
Under 3rd band	62	65	65	70
Under 4th band	41	60	60	64
Linear velocity (m/s)				
of 1st band	0.163	0.072	0.073	0.065
of 2nd band	0.105	0.048	0.057	0.047
of 3rd band	0.078	0.034	0.037	0.033
of 4th band	0.066	0.029	0.025	0.025
Typical drying time (min)	200	900	420	960
Relative humidity of air on the dryer outlet (%)	38–40	45–50	45–50	60–67

Source: From Ref. 92.

apple products are packaged directly or after compression in moistureproof packages or containers.

4.1.2.2 Pear

Dry pears are less popular than dry apples: less than 1% of pears produced are dried (76). Sun drying is the most typical method used for pear dehydration. Typically a two-step drying procedure is applied: first, direct sun drying on trays for 1 to 2 days; second, 3 to 6-week drying in trays stacked beneath sheds, open at the sides to allow good air circulation. Air drying in trays is also applied sometimes (1,76). Pears are halved, peeled, and sliced, or peeled, halved, and cored, before sulfuring and drying. Prepared pieces are loaded onto trays with an average unit load of 10 kg/m^2. Maximal temperature in the dryer (hot end) is typically 65°C. The drying time varies from 6 to 48 h, depending on what size the fruit pieces are and whether they are peeled or not.

4.1.3 Drupe Fruits

Prunes, apricots, peaches, and cherries are examples of drupe fruits commercially available in dried form. Prunes contribute most, followed by apricots, peaches, and cherries in decreasing order. By comparison, in U.S. dried fruit production only raisins are produced in larger quantities than prunes: 391,000 tons of raisins and 184,000 tons of prunes in 1991 (77).

4.1.3.1 Prunes

Dehydrated to a moisture content of 19%–35%, plums are usually called *prunes*. Sun drying of plums/prunes is the most common drying technique, there is a growing tendency to use artificial drying, especially in tunnel and belt dryers. The 'French,' 'Imperial,' and 'Sugar' cultivars are the most used varieties of prunes in North America, especially in California, which contribute approximately 75% of the total world production (78). Good quality prunes are initially dipped in cold or hot water and spread in a single layer on the drying trays. Drying time ranges from 24 to 36 h with hot air temperature in the tunnel around 75°C. The final product has a moisture content of about 18% (w.b.). According to Raoult (79) countercurrent tunnel drying of prunes takes 18–26 h when temperature of the air at the inlet to the tunnel is 50°C–60°C and at the exit around 74°C. Linear velocity of hot air is maintained at about 3 m/s at the tunnel inlet and 6–7 m/s at the exit from the tunnel. Dry good quality prunes of 22%–23% moisture (w.b.) are produced.

Other dry prune products are also commercially available. Low-moisture prunes have 4% moisture (w.b.) after vacuum second-step drying. Dry prune flakes, nuggets, granules, and dices are ingredients used in the bakery and confectionary industries.

4.1.3.2 Apricot

Apricot is primarily sun dried to 15%–20% moisture (w.b.); however, a small part is dried artificially. Iran, the United States, Australia, Spain, and the Republic of South Africa are the major producers of dry apricot. As with the apple, two different types of dry fruits are produced: evaporated apricot with water content about 20% (w.b.) and low-moisture apricot with 5% moisture (w.b.).

For evaporated apricot, fruits fully ripened on the tree are harvested, pitted, and cut into halves; loaded into wooden trays; and exposed to sulfur fumes for a minimum of 3 h. An average 4 days of open sun drying is subsequently performed. After this period, the trays are stacked and exposed to the wind for long-term final drying. Presorted dry fruits are stored in wooden boxes for sweating for several weeks. Final dry product may contain from 2000 to 5000 ppm of sulfur dioxide.

For low-moisture dry apricot a second-stage drying is applied. Vacuum shelf dryers are commonly used for this purpose, starting with evaporated fruits. Different shades of the final product, from lemon yellow to deep orange, may be achieved, depending on the drying conditions (1).

Tunnel dryers are also used for apricot drying. The process consists of two steps, both in tunnel dryers. The first step is drying to produce 50% weight reduction. The second consists of steam blanching and returning to the tunnel for final drying.

Other dry apricot products, flakes and powder, are formed from the low-moisture product; nuggets and granules are formed from dried extruded apricot paste (80). Puff drying of apricot juice provides a spongelike final product (81).

4.1.3.3 Peach and Cherry

Certain varieties of freestone peaches are sun or artificially dried to about 25% to 28% (w.b.) moisture content (1). Pitted and lye peeled fruits are halved and sulfured for about 4–6 h. The sun drying process takes longer than it does for apricots as the fruit pieces are larger. Artificial drying is normally provided in countercurrent tunnel dryers with hot air temperature of 68°C. The drying time varies from 24 to 30 h with unit tray loads of about 10 kg/m^2.

Sweet or sour cherries are also dried as whole fruits with or without stems or as pitted fruits. For drying of whole fruits a special skin pretreatment is applied: dipping the fruits into a boiling 0.5% solution of NaOH, followed by rinsing in water. Sulfuring of the white or pink cherries requires about 30 min, less for black cherries (80). Countercurrent tunnel drying with hot air at a temperature of about 77°C for 8–12 h is the typical procedure for tunnel drying of cherries. Additionally, vacuum drying of cherries in a sugar solution up to a final moisture content of 5% produces special types of decorative cherry products.

4.1.4 Berries

For most berry fruits such as strawberries, raspberries, blueberries, cranberries, gooseberries, and currants, sun drying is not very practical. The structure of the fruits, especially the skin, poses problems for drying. But this difficulty can be easy overcome by using, for example, freeze drying or its modifications. For example, freeze drying of strawberries consists of cooling the fruit to just above the freezing point of water, then slowly freezing the water to develop larger crystals that expand and rupture the structure of fruit, and finally freeze drying the fruit to below 3% moisture content (82). Perforation of the skin of frozen blueberries (83) increases the freeze drying rate significantly and improves quality.

The high cost of freeze drying is a strongly limiting factor; less expensive methods are continuously being developed. Cranberries, for example, may be tunnel dried after some prehandling, which consists of slicing the berries or perforating them. The drying of such fruits for 6–10 h at a temperature of 77°C yields a product of about 10% moisture content (w.b.). The unit load of the trays in a tunnel dryer is about 5 kg/m^2. Another final product, with moisture content of 5% (w.b.), is normally achieved after the second drying step of the tunnel predried berries, which normally takes place in a bin dryer (1). The dry product may be additionally pressed and then packed into cans, cartons, etc.

Explosion puffing is also sometimes used for the drying of whole berries (Table 8) as well as other methods described previously. More recently, a two-step drying process, with osmotic dehydration for the first stage and air, vacuum, or freeze drying for the second stage, has been developed (67–72). The main advantages of this process are significant energy saving and potential for creating of new products by infusion of specific additives into the berry structure through osmotic action (71).

Dry strawberry and raspberry puree, granules, and powders are usually processed in drum and spray dryers (77). High-temperature short-period spray drying of fruit paste is carried out after mixing with some carrier products such as maltodextrin, starch, or corn syrup solids. The added carrier materials prevent clumping of the product during drying and storage. Foam mat drying is also a potential method.

4.1.5 Tropical and Subtropical Fruits

Dry tropical and subtropical fruit products are becoming more popular as it is more convenient to handle them in the dry form than as fresh products. Further, they are more microbiologically safe because some thermal inactivation of microorganisms occurs during drying (84,85). A wide variety of fruits are dried as whole fruit or as slices, chunks, granules, or powder. Some of these are described in the sections that follow.

4.1.5.1 Fig

Depending on the cultivar, dry figs are processed in slightly different ways, both by natural sun and by artificial means. The difference is a sulfuring pretreatment, which is practical for some cultivars, such as 'Adriatic' and 'Kadota,' but never used for 'Calimyrna' and 'Mission.' The process starts with defuzzing (removal of surface hair) in a 0.2% solution of salt and hydrated lime. This is followed by sulfuring and sun drying, or alternatively drying is carried out directly in the sun without sulfur treatment. For a better skin appearance, drying in shadows and stacks is the most typical procedure. For artificial drying, the hair-free figs are halved, sometimes sulfured, and dried on trays in a countercurrent tunnel dryer. The final moisture of 5% (w.b.) is obtained after 8–12 h of drying with hot air at a temperature about 74°C. In some operations, the fig is dried to less than 3.5% and cut into slices of about 16 × 8 mm or powdered. More details of fig drying are presented by Desai and Kotecha (86).

4.1.5.2 Dates

Typically, freshly harvested dates are dried for commercial applications, with a moisture content less than 20%–24% (w.b.), as a result of sun drying directly on the palm. Small quantities of the fruit are additionally dried for a low-moisture product, which contains less than 4% moisture. Vacuum shelf dryers are used for such processes with sun dried dates as starting material. Grinding, cutting, and powdering of the low-moisture product are the final procedures before packaging.

4.1.5.3 Banana

Banana may be dried in different forms by different methods (77,86). Sun or solar dried bananas are used for flour or powder. To prevent darkening of the product, SO_2 pretreatment is normally used. Diced or sliced and peeled bananas are dried on trays in heated air chambers or tunnels. Hot air of less than 93°C initially, and about 65°C finally, is used to dry bananas to 20% moisture content (w.b.). Freeze drying of banana slices is still not very practical, although high-quality banana granules can be produced with this method (77). Banana pulp is dried into flakes by drum drying, typically in double-drum atmospheric dryers. The flakes may be pulverized to make a powder. Spray dried banana powder is produced after special preparation of the feed, mainly dilution with water. Several other procedures have been developed for banana drying. One example is the osmovac process, which starts with an osmotic treatment of banana pieces in a sugar syrup of 70° Brix for 8–10 h. After the surface syrup is strained from the fruit, the pieces are dried in a vacuum dryer at 66°C–71°C and 10 mm Hg to a final moisture content of 2.5%.

4.1.5.4 Pineapple and Mango

Peeled and diced fruits are dried in a tray dryer after appropriate sulfuring. Further drying to low-moisture products may be achieved in a vacuum dryer. Dried mango puree is produced from a pulp by several different methods, such as drum and spray drying and foam mat, puff, and freeze drying. More details of these procedures are given in Refs. 24, 80, and 84.

4.1.5.5 Papaya

Papaya slices 2.5 cm thick are sulfured for about 2 h and dried in a cross-flow dryer in a two-step operation: first at 74°C–76°C for 2 h and then for approximately 4 h in air at 60°C. Papaya chunks (1.5 × 1.5 × 1 cm) are also occasionally freeze dried for the best quality of product (84). Papaya powder is ground from dry (8%–10% w.b) papaya slices of 5-mm thickness. Drying on the trays with hot air at 65.5°C is the typical procedure. Other dry products such as papaya leather (rolls) are dried up to 12%–13% (w.b.) in different types of dryers from papaya puree enriched with 10% (w/w) sucrose.

4.1.5.6 Guava

Cross-flow cabinet air drying of guava pieces is the simplest and most typical drying method. Guavas are quartered, deseeded, steam blanched, sulfured for 20 min in a sulfur house, and dried at 65°C for 13.5–15 h with an airflow rate of about 3 m/s to a final moisture content of 6%–7% (w.b.). Drying to a final moisture content of 4% (w.b.) requires about 11.5 h at 55°C or 8 h at 77°C (84). Osmotic dehydration followed by convective drying on glycerine coated trays as well as osmovac and foam mat drying are practical methods for guava pieces and powder drying. Various foaming agents such as egg albumin, glycerol monostearate, peanut protein, guar gum, and carboxymethyl cellulose have been tested; egg albumin proved to be the best (84).

4.1.5.7 Citrus Fruit Powders

Citrus products are mostly dehydrated from juice or pulp to make powder, granules, or instant products. Almost every type of dehydration process has been evaluated for citrus juices. Drum, spray, and freeze drying as well as foam mat and vacuum puff drying are commonly applied. All aspects of the initial and final product characteristics (moisture content, granulation, rheological properties, particle shape, etc.) and production scale should be considered for selection of the best method.

4.2 Drying of Selected Vegetables

4.2.1 Potato and Potato Products

With many different final products, potatoes contribute most to the vegetable drying industry. They can be dehydrated by several methods to give dices, flakes, granules, and flour. Crisps, chips, and other snack foods are also typical potato products. Detailed descriptions of potato drying techniques can be found in the literature (1,21,87,88,89). All processes preceding drying, i.e., washing, peeling, slicing, blanching, sulfiting, calcium salt treatment and/or precooking, and mashing, are well described by Willard and associates (90).

4.2.1.1 Potato Dices

Diced potatoes are prepared after cut from good-quality peeled potatoes to regular shape cube or cubicoid with normalized dimensions presented in Table 11. At this stage, an average of 25% of fresh material is wasted while after drying, inspection, and screening, and the final product yield is about 14% (91). Blanching in steam or hot water precedes the drying. Several types of dryers suitable for particulate solid materials have been used. Cabinet, tunnel, or conveyor dryers are the most common. For a typical cabinet dryer, Burits and Berki (92) specify the following characteristics of the drying process:

Table 11 Typical Sizes and Shapes of Diced Potatoes Before Drying

No.	Dimensions (United States) (in)	Dimensions (Europe) (mm)
1	3/8 × 3/8 × 3/8	10 × 10 × 10
2	1/4 × 1/4 × 1/4	8 × 8 × 8
3	1/4 × 1/4 × 3/8	6 × 6 × 6
4	3/8 × 3/8 × 3/16	16 × 16 × 5
5	3/16 × 3/16 × 1/4	10 × 10 × 3
6	1/2 × 1/2 × 1/8	10 × 10 × 3
7	3/4 × 3/4 × 1/8	10 × 10 × 2
8	3/4 × 1 × 3/8	8 × 8 × 1.5

Source: From Refs. 88 and 93.

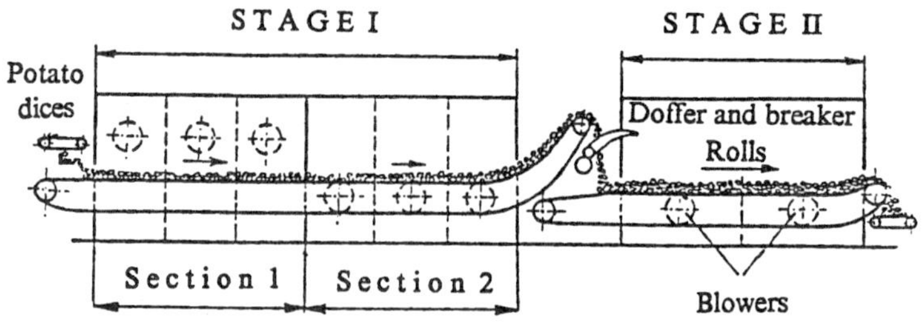

Fig. 16 Two-stage conveyor dryer of potato dices. (From Ref. 92.)

Fresh potato load	
For dices 5 × 5 × 5 mm	75–90 kg/m^2
For dices 10 × 10 × 10 mm	Up to 75 kg/m^2
For dices greater than 10 × 10 × 10 mm	Up to 65 kg/m^2
Air temperature during the first-stage drying	65°C–75°C
Air temperature during the second-stage drying	55°C–60°C
Average throughput of dry matter of 8%–9% (w.b.)	7–12 kg/(m^{2}h)
Drying time	4–8 h

Cabinet and tunnel dryers have been used for diced potatoes in the older plants, but now these are used mainly for other vegetables and fruits. Conveyor potato dice dryers are the most popular. The cost of fresh and dry product handling is reduced significantly by the automatic loading and unloading in these dryers. Also, the product quality is improved through better process control. Although conveyor or belt dryers are expensive to install, they are more economical in contemporary potato processing.

The most common dryers for diced potatoes are the two-stage conveyor dryers shown in Fig. 16. The first stage of the dryer is divided into two sections with three parts in each section. Each section has a fan, which allows for individual programming of proper drying parameters. The second stage of the dryer consists of two parts, again with two separate fans. Typical potato bed heights are in the range of 7.5–15 cm. The bed is normally higher in the second stage while the conveyor is slowly driven. Part of the drying air passing through each section can be recirculated to reduce the energy demand. The temperature of the drying air in the first and second sections of the first stage are in the range of 93°C–127°C and 71°C–105°C, respectively, and in the second stage, 55°C–82°C.

In Europe, vertical multiple (two to six) conveyor dryers are more commonly used for potato dice drying. For instance, the following are typical drying conditions in a five-conveyor dryer: unit surface load on the first conveyor, 15–22 kg/m^2; air temperature and linear (through bed) velocity, 50°C–80°C and 0.12–0.35 m/s, respectively; and total drying time, 180–210 min (92). A final moisture content of about 11%–12% (w.b.) can be achieved.

To extend the shelf life of dry (11%–12%) diced potato, an additional second step of drying is usually performed. Dry product of 6%–7% (w.b.) can be stored approximately three times longer than a product with 11%–12% moisture content (93). This final drying process may be performed in a multiconveyor dryer. For example, a four-stack-conveyor dryer with temperatures of drying air of 55°C, 60°C, 60°C, and 40°C in the first to the fourth conveyor, respectively, produces the appropriate moisture content within 80 min.

Successful drying of diced potatoes in a belt-through dryer has been reported (36,94). A standard belt through-flow dryer, which has a bed 1.2 m wide and 3 m long, has been reported to evaporate 450 kg of moisture per hour, giving a product of nearly uniform moisture in each piece. Other types of drying such as freeze and microwave drying are still expensive for industrial-scale applications. However, a lot of research is continuously being pursued.

4.2.1.2 Potato Strips

A considerable portion of the total potato production is utilized as livestock feed. The most efficient method of preserving potato as livestock feed is drying. Potato tubers for such processing undergo either a wet or a dry cleaning and are sliced. Potato strips thus prepared are usually dried in cocurrent rotary dryers, 1–3 m in diameter. The ratio of the diameter to the length of the dryer is typically 1:5. The inlet air temperature is about 500°C and the outlet is about 100°C. Dry potato strips contain about 12% (w.b.) moisture (89).

4.2.1.3 Potato Flakes

Potato flakes are the leading form of dried mashed potatoes. Part of the total potato flakes production is finely ground to produce potato flour after milling and screening of the final flakes. The process starts with washing, peeling, precooking, cooling, and then mashing of the potatoes. Mashed raw material before drying contains potato pulp and several additives. Degree of broken potato cells (for example, 15%–20%) in raw material before drying is one of the parameters controlling the drying kinetics. Typically mashed potato pulp contains 78.5%–81.5% water (w.b.) and dry potato flakes contain 6%–7% (w.b.).

Generally, single- and double-drum dryers are used to produce potato flakes. The most frequently used cylinders are 600–1250 mm in diameter and 2200–3200 mm in length (single drum) or 900–2200 mm in length (double drums). The temperature of the cylinder surface, heated by saturated steam, typically reaches 140°C and above. For temperatures in the range of 250°C, mineral oil or gas is commonly used as a heating agent. Such high temperatures allow higher drum speeds, which are typically in the range 2–8 rpm.

Drum dryers are the most economical and energy efficient dryers—with thermal efficiency in the range of 76%–90%, including heat losses and radiation. They require between 1.1 and 1.3 kg of steam/kg of evaporated water.

Some modifications of the typical drum dryers for specific application in potato flake production have been developed. A Foerster double-drum dryer is an example of such a modification (89). A modified pneumatic dryer, the so-called Aradi-Rybanski dryer (21), has been used for granulated mashed potato drying (Fig. 10). More details of the drying techniques in granulated mashed potato production (in the so-called add-back process or direct process) can be found in Refs. 21 and 88.

4.2.1.4 Potato Starch

Drying is the end-stage operation of great importance for the final product quality of potato starch. Raw to-be-dried starch is typically rotary vacuum dried (centrifugally in earlier technologies) to form a filter cake of about 36%–40% moisture (w.b.). The finished potato starch product has approximately 17%–20% moisture (w.b.) and consists of about 98%–98.5% starch on a dry basis.

Potato starch is very heat-sensitive material, which when heated moist at above 45°C swells and gelatinizes. Practically moist starch should not be heated to temperatures higher than 35°C. Several different drying methods have been employed for filter-cake potato starch drying, from the simplest open-air drying on racks in stove-heated drying houses to the adoption of rotary-turbo dryers and, more recently, to continuous-belt or flash dryers. Flash drying and its modification—cyclone flash drying—are the most common methods of potato starch drying in modern technologies. Because of the very high moisture evaporation intensity in flash dryers, the entire drying process takes only 2–5 s, and the product temperature is kept below 40°C. The linear velocity of hot (160°C–165°C) air in the drying duct is in the range 10–20 m/s. Dry starch at about 40°C is cooled during pneumatic transport and mixing to storage temperatures of 10°C–15°C.

4.2.1.5 French Fries

The surface moisture of blanched potato strips prepared for frozen french fry production is usually removed before frying by drying in hot air. As reported by Lisinska and Leszczynski (89), in large processing plants the production lines are equipped with continuous conveyor dryers. The following advantages were observed for predrying french fries before frying (89):

- Uniform product color
- Optimal product stability and quality
- Prevention of weak and soggy product formation
- Decreased oil content in final product

4.2.1.6 Potato Chips

In potato chip technology, raw sliced potatoes are sometimes partially dried before frying to reduce the frying time. As for french fries, predrying results in lowering of the oil content in the final product (low-oil chips). Several practical drying methods have been employed. Drying in perforated drums or belts is typical. The aim of this operation is surface water removal, as predrying decreases the water content of raw slices by only about 4% (89).

4.2.2 Carrot

Carrots can be cross-cut or crinkle-cut as rings and dried as such. Starch coating may be used to extend the shelf life of the dry product because oxidation of carotene together with off-flavor and off-odor can be significant problems during storage. Typically, steam blanched and sulfited or steam blanched and starched dices are spread onto trays at a loading of about 6 kg/m^2. Carrots are then dried in the tunnel or on continuous single-conveyor or multiconveyor dryers to a final moisture content of about 4% (w.b.). Hot air temperature in such a typical counterflow tunnel dryer is 71°C, and about 7 h of drying is necessary to achieve an optimal moisture content of 8% (w.b.) in the product. Additional bin drying at 60°C through 7 h yields final product moisture content of 4% (w.b.). Relatively fast drying (50 min) of diced carrot in continuous belt-through dryers was reported by Bolin and colleagues (80). Freeze drying (temperature −33°C and absolute pressure 0.2 mm Hg) as well as explosion puffing are other drying techniques employed for carrots on a limited scale. Low-oxygen packages (vacuum, nitrogen, or carbon dioxide atmosphere) with O_2 levels lower than 2% are necessary to prevent carotenoid deterioration during storage. Details of carrot drying technology have been published (80,95).

4.2.3 Beets

Peeled beets are sliced, diced, or strip-cut to specific sizes. Steam blanching for about 6 min is used in many cases, but not always. Sulfiting of the raw beets is not practical. Tunnel and conveyor dryers of different configurations are typically used. Relatively high drying air temperatures (95°C–100°C) may be used in the first drying stage; however, a lower temperature of air (71°C) is desirable in the second stage. Final moisture content of beets in such a dryer is 11% (w.b.), and normally additional drying to 5% (w.b.) in stationary bins is necessary.

Explosion puffing of dices is commercially used for a good-quality dry product. Air predried diced beets of 45% moisture (w.b.) are processed in a puffing gun. They are heated to the pressure of 0.3 MPa in steam; then the pressure is quickly released to the atmospheric level. The exploded product, which has lost about 5% moisture, is then finish dried on trays, continuous belt, etc. The porous structure of the final product results in much lower dehydration time than is required in conventional drying techniques.

4.2.4 Onion and Garlic

Dried root vegetables, potatoes, and onions have found extensive use in food manufacturing, both in dry product formulation as well as in ingredients for wet soups, baby foods, and ready to eat or canned meals.

4.2.4.1 Onion

Sliced, flaked, or kibbled onions are mainly dried on continuous belt conveyors or tunnels to a target moisture content of 5% (w.b.) or less. Onions are normally not blanched, to prevent the flavor loss that would otherwise occur during this process. When dried in tunnel-type dryers, the sliced onions are automatically spread on wooden trays with a unit load of about 5 kg/m^2. The trays are automatically stacked on cars and conveyed through a typical two-stage tunnel dryer. In the first (cocurrent) stage, hot air at a temperature of 70°C–88°C is normally used; in the second (countercurrent) it is 55°C–60°C. After 10–15 h the dry material has 5%–7% moisture, it may then be dried in bins to final moisture less than 5%.

More practical equipment for onion drying is a multistage continuous belt conveyor. A stainless steel perforated belt is normally used with hot air flowing through the belt and a bed of onion slices of 10–15 cm in thickness. The temperature of the drying air is gradually reduced from about 80°C to 55°C as the sliced onions move through the dryer. The average drying time is 6 h with a final moisture content of 6% (w.b.). Additional drying (12–30 h) in stationary bins, using partially dehumidified air of 50°C, gives the final product 4% moisture or less. A small proportion of dry onion is produced by freeze drying.

For commercial use, sliced, chopped, minced, granulated, or powdered dry onions are produced. High-hygroscopic dry onion products require a special low air-humidity atmosphere in the final processing stages such as screening, grinding, and packaging. A variety of moisture resistant containers, such as plastic or aluminum bags, jars, clipboard boxes, and metal cans, are used for the retail market.

4.2.4.2 Garlic

Drying technology of garlic is almost the same as that of onion (1). The garlic bulbs are broken into individual cloves, washed, sliced, and then dried up to 6.5% (w.b.). Garlic is widely used, in the formulation of spice mixtures, garlic salt, etc. All the same forms and types of dry garlic products as of onion are commercially available.

4.2.5 Other Vegetables

Many other vegetables are being dried, however, to a much lower extent. Generally, similar drying methods and conditions can be used for most vegetables. Gentle drying conditions are recommended as vegetable tissues are very heat sensitive.

4.2.5.1 Green Beans

Green beans are dried on the trays of tunnel dryers or in belt dryers, either while still frozen or in a thawed state. Two-step tunnel drying is recommended (96): first at 90°C and then at 62°C. A final moisture content of 5% (w.b.) is typically a target value.

4.2.5.2 Cabbage

Cabbage typically is cored, trimmed, washed, and shredded by kraut cutters before blanching and sulfiting. The pretreated material is dried in a two-stage tunnel dryer with an inlet air temperature of about 80°C in the first (parallel airflow) stage and 60°C in the second (countercurrent airflow) stage. A finish drying step to lower moisture content of about 7% to 4% (w.b.) is provided in stationary bin dryers, supplied with air at 50°C. Strong compression of the dry product before packing was the standard procedure for cabbage packed in cans during World War II, but this procedure is not in use anymore.

Practical information about other vegetable dehydration procedures can be found in specialized literature and recent publications (1,2,22,94).

5 SPICES

The quality of dry spices, culinary herbs, and spice blends is strongly affected by the drying method and drying conditions that are applied. The main attribute of spices is the improvement of food flavor and taste by some characteristic components, especially volatiles. These components are very heat-sensitive and should be preserved (to prevent evaporation or decomposition) to the maximal level during all processing steps. The drying process involves some thermal treatment, so the drying method and the conditions should be very delicate and individually chosen for each type of spice or herb. Low-temperature natural or artificial drying is the common practice.

The measurement of moisture content in spices presents some unusual problems. Moisture is usually measured in food products by measuring the weight loss of a sample during a drying test at elevated (around 100°C) temperature or in a vacuum. The volatile oils that typically exist in spices are also lost during drying, and this weight loss is measured as moisture. To resolve this problem, the trade has adopted a codistillation method for most spices. In this test, the spice is covered with toluene and the toluene raised to its boiling temperature. The moisture in the spice codistills with the toluene, and as the toluene is condensed, the moisture separates from the toluene and is measured. Karl Fisher titration can be also employed for some spices, whereas for other spice products (paprika and other capsicum, etc.) a standard drying oven test is sufficient. Vacuum drying techniques are also frequently used to evaporate moisture at lower temperatures.

The final moisture level in dry spices is of practical importance, as for all other perishable foods. It should be low enough to prevent microbiological growth but not too low, to prevent unfavorable changes in overdry tissue such as color losses and increased

Table 12 Characteristics of Drying of Selected Spices and Herbs

				Drying characteristics					
Spice	Part of plant used	Pretreatment before drying	Form of dry product	In sun	In shadow	Artificial	Drying time (h)	Remarks	Major sources
Black pepper (*Piper nigrum*)	Berries	Fermentation	Whole corns, powder	Yes	Rather not	Possible	20	On mats or concrete floors	Indonesia, India, Brazil, Malaysia
White pepper (*Piper nigrum*)	Berries	Soaking (8 days)	Whole corns, powder	Yes	Rather not	Possible	Several days	On mats or concrete floors	Indonesia, Brazil
Chili pepper (*Capsicum frutescens*)	Pods	Curing in shelters	Whole or ground	Yes	Possible	Tunnels, steel belts	Sun, 5–15 days; artif 30 h	85% of water has to be removed	United States, Mexico
Red pepper (*Capsicum frutescens*)	Pods	Curing in shelters	Whole or ground	Yes	Possible	Tunnels, steel belts	Sun, 5–15 days; artificial, 30 h		Pakistan, China, Mexico, India, Turkey
Cloves (*Syzygium aromaticum*)	Flower buds	—	Whole buds	Yes	Open air	Not practical	Several days	Reduction of mass by 3:1	Brazil, Tanzania, Indonesia, Sri Lanka
Cinnamon/cassia (*Cinnamomum zeylanicum* or *C. cassia*)	Coppiced shoots	Fermentation	Quills, powder	As a second stage	In a first stage	Not practical	3–4 Days	Drying until smooth, pale brown color	Cinnamon: Sri Lanka, India; cassia: Taiwan, Indonesia, China
Coriander (*Coriandrum sativum*)	Seeds or leaves	—	Whole seeds or leaves	Seeds	Leaves	Not practical	Several days	Cut, whole plants drying on field, then threshed and drying	Morocco, Egypt, Argentina, United States, Mexico
Dill (*Anethum s.*)	Seeds or leaves	—	Whole seeds or whole and chopped leaves	Seeds	Leaves	Yes	Sun, several days	Drying on trays	India, Pakistan, United States, Egypt

Table 12 Continued

Spice	Part of plant used	Pretreatment before drying	Form of dry product	Drying characteristics: In sun	In shadow	Artificial	Drying time (h)	Remarks	Major sources
Ginger (*Zingiber o.*)	Rhizomes	Peeling, boiling	Rhizomes (hands)	Yes	Not practical	Not practical	8 Days		China, Nigeria, Jamaica, India
Mustard (*Brassica hirta* or *B. juncea*)	Seeds	—	Mustard flour	Yes	Yes	Possible	—	Powdered dry mustard has no aroma and requires moisture to develop flavor	Canada, United States
Chives (*Allium schoenoprasum*)	Leaves, shoots	—	Tiny lengths, shoots, or powder	—	—	Freeze drying	—	Increased demand	United States
Vanilla (*Vanilla planifolia*)	Pods	Hot water	Pods, powder	Yes	No	New methods	Traditional, 5–6 months	In sun daily, in blankets and room nights	Mexico, Brazil, Paraguay
Parsley (*Petroselinum c.*)	Stems	—	Flakes, granules	To avoid	Yes	Yes	Variable	Low-temperature drying	United States, Israel, Mexico
Bay (*Laurus nobilis*)	Leaves	—	Leaves	To avoid	Yes	On trays	15 Days	Thin layer of leaves in moderate warm shelters	Turkey
Marjoram (*Majorana hortensis*)	Whole herb	—	Leaves, granules	No	Yes	On trays	Variable	Immediate storage in closed packages	Egypt
Mint (*Mentha piperita* or *M. spicata*)	Leaves	—	Leaves, flakes	No	Yes	Yes	Variable	Prevent heat damage	United States, Egypt
Rosemary (*Rosmarinus officinalis*)	Leaves, whole herbs	—	Leaves, flakes	No	Yes	Possible	Variable	Contain antioxidants	Portugal, former Yugoslavia

Source: From Refs. 97–102.

hardness. An average of 10% moisture content (w.b.) is desirable for most spices. Detailed data on several typical dry spice chemical and physical specifications and applications are presented by Charalambous (97), Tainter and Grenis (98), Richard (99), Farrell (100), Rosengarten (101), and Purseglove and coworkers (102).

Sun and field drying of spices and herbs is the oldest and still the most typical drying method for the majority of these products. In many cases direct exposure to the sun can create discoloration of the spice or herb tissue so shadow-open-air or cabinet drying is applied. Larger-scale drying of spices and herbs is performed as for drying of typical fruits and vegetables (e.g., tunnel and conveyor dryers) (99). Multistep processes, with hot air temperatures decreasing with the advance of drying (for example, 110°C initially, then 60°C and 40°C at the end of drying) is very typical. Table 12 presents examples of drying procedures and drying conditions for some popular spices and culinary herbs. More details of preparation, characteristics, and application of the spices can be found in other references (97–102).

6 CONCLUDING REMARKS

Fruits and vegetables as well as spices and herbs play an extremely important role in human nutrition. Additionally, the recognition that the intake of nutrients, minerals, and antioxidants (vitamin C, E, etc.) may have a protective role against illnesses and tiredness will mean that fruits and vegetables—a potent source of relevant nutrients, minerals, and vitamins—will continue to be in demand. However, they have to be preserved in a form that protects these constituents. It is likely that the market for fresh, frozen, and properly dried fruits and vegetables will outgrow that for other forms of processing in the short term. For drying technology, this means that only the methods that yield appropriate quality of the final products may have good practical potential. Research should be dedicated to the development of proper drying processes. Improvement of existing as well as introduction of new drying techniques are equally important aspects of research. It is well known that spices have not only a flavor effect on food, but some antioxidant and microorganism inhibition activity too. Proper drying techniques have to be applied to each individual product to assure its highest quality. Energy and environmental aspects of this process should also be considered.

REFERENCES

1. W. B. Van Arsdel, M. J. Copley, and A. I. Morgan, *Food dehydration*, Vols. 1 and 2, AVI Publ. Co., Westport, Connecticut, 1973.
2. G. V. Barbosa-Canovas, and H. Vega-Mercado, *Dehydration of foods*, Chapman & Hall, New York, 1996.
3. A. S. Mujumdar, *Handbook of industrial drying*, 2nd ed. Marcel Dekker, New York, 1995.
4. C. Strumillo, and T. Kudra, *Drying: Principles, applications and design*, Gordon and Breach, London, 1986.
5. A. S. Mujumdar, *Drying of solids*, International Science and Oxford & IBH, New York and New Delhi, 1992.
6. R. B. Keey, *Drying: Principles and practice*, Pergamon Press, Oxford, 1972.
7. A. S. Mujumdar, *Advances in drying*, Vol. 1, Hemisphere, Washington, D.C., 1980.
8. A. S. Mujumdar, *Advances in drying*, Vol. 2, Hemisphere, Washington, D.C., 1983.
9. A. S. Mujumdar, *Advances in drying*, Vol. 3, Hemisphere, Washington, D.C., 1984.
10. A. S. Mujumdar, *Advances in drying*, Vol. 4, Hemisphere, Washington, D.C., 1987.

11. A. S. Mujumdar, *Advances in drying*, Vol. 5, Hemisphere, Washington, D.C., 1991.
12. C. W. Hall, *Dictionary of drying*, Marcel Dekker, New York, 1979.
13. M. Torrey, *Dehydration of fruits and vegetables*, Noyes Data Corporation, Park Ridge, New Jersey, 1974.
14. S. D. Holdworth, Advances in the dehydration of fruits and vegetables. In: *Concentration and drying of foods*, ed. D. McCarthy, Elsevier Applied Science, London, pp. 293–303, 1986.
15. T. Kudra, and A. S. Mujumdar, Advanced Drying Technologies, Marcel Dekker, New York, 2001.
16. G. H. Crapiste, and E. Rotstein, Design and performance evaluation of dryers. In: *Handbook of food engineering practice*, ed. K. J. Valentas, E. Rotstein, and R. P. Singh, CRC Press, Boca Raton, Florida, pp. 125–166, 1997.
17. D. Arthey, and P. R. Ashurst, *Fruit processing*, Blackie Academic and Professional, London, 1996.
18. C. Ratti, and A. S. Mujumdar, Drying of fruits. In: *Processing fruits: Science and technology*, Vol. 1, ed. L. P. Somogyi, H. Ramaswamy, and Y. H. Hui, Technomic, Lancaster, Pennsylvania, 1996.
19. K. S. Jayaraman, and D. K. Das Gupta, Drying of fruits and vegetables. In: *Handbook of industrial drying*, ed. A. S. Mujumdar, Marcel Dekker, New York, pp. 643–690, 1995.
20. S. Sokhansanj, and D. S. Jayas, Drying of foodstuffs. In: *Handbook of industrial drying*, ed. A. S. Mujumdar, Marcel Dekker, New York, pp. 589–625, 1995.
21. S. Grabowski, and A. S. Mujumdar, Drying technologies in potato processing. In: *Drying of solids*, ed. A. S. Mujumdar, International Science and Oxford & IBH, New York and New Delhi, pp. 303–325, 1992.
22. D. Arthey, and C. Dennis, *Vegetable processing*, Blackie, Glasgow and London, 1991.
23. J. G. Brennan, Dehydration of foodstuffs. In: *Water and food quality*, ed. T. M. Hardman, Elsevier Applied Science, London, 1989.
24. L. P. Somogyi, and B. S. Luh, Dehydration of fruits. In: *Commercial fruit processing*, 2nd ed., ed. J. G. Woodroof and B. S. Luh, AVI, Westport, Connecticut, pp. 353–405, 1986.
25. M. Karel, Physical structure and quality of dehydrated foods. In: *Drying '91*, ed. A. S. Mujumdar, Elsevier Science, Amsterdam, pp. 26–35, 1991.
26. W. F. Harrigan, and R. W. Park, *Making safe food: A management guide for microbiological quality*, Academic Press, London, 1991.
27. T. D. Labuza, *Shelf-life dating of food*, Food & Nutrition Press, Westport, Connecticut, 1982.
28. N. K. Bansal, and H. P. Garg, Solar crop drying. In: *Advances in drying*, Vol. 4, ed. A. S. Mujumdar, Hemisphere, Washington, D.C., pp. 279–299, 1987.
29. P. E. Hubble, Consider microwave drying, *Chem. Eng.* Vol. 89, pp. 125–132, 1982.
30. L. Imre, Solar drying. In: *Handbook of industrial drying*, ed. A. S. Mujumdar, Marcel Dekker, New York, pp. 373–452, 1995.
31. B. de Jong, *Net radiation received by a horizontal surface at the Earth*, Delft University Press, Delft, Netherlands, 1973.
32. B. Coppolino, Validation of very simple model for computing global solar radiation in the European, African, Asian and North American areas, *Solar and Wind Technology*, Vol. 7, pp. 489–494, 1990.
33. D. Collivar, Techniques of estimating incident solar radiation. In: *Solar energy in agriculture*, ed. B. F. Parker, Elsevier, Amsterdam, pp. 1–66, 1991.
34. T. A. Lawand, and J. LeNormand, *Methodological approach for determining potentials for solar dehydration and other applications*, Brace Research Institute, Macdonald Campus of McGill University, Report No. R-140, 1980.
35. W. Szulmayer, *Drying principles and thermodynamics of sun drying*, Brace Research Institute, Macdonald Campus of McGill University, Technical Report No. T-124, 1976.

36. J. L. Heid, and M. A. Joslyn, *Fundamentals of food processing operations*, AVI, Westport, Connecticut, 1967.
37. B. F. Parker, *Solar energy in agriculture*, Elsevier, Amsterdam, 1991.
38. J. R. Puigalli, and A. Tiguert, *Drying Technology*, Vol. 4, pp. 555–581, 1986.
39. R. T. Patil, Drying studies on black pepper, *J. Food Sci. Technol.*, Vol. 26, pp. 230–231, 1989.
40. K. S. Jayaraman, D. K. Das Gupta, and N. Babu Rao, Solar drying of vegetables: Quality improvement using cabinet with multiple flat plate collectors and pretreatment. In: *Drying of solids*, ed. A. S. Mujumdar, International Science, and Oxford & IBH, New York and New Delhi, pp. 405–432, 1992.
41. A. Kamaruddin, and Mursalim, Drying of vanilla pods using a greenhouse effect solar dryer. *Drying Technology*, Vol. 15, pp. 685–698, 1997.
42. T. A. Lawand, *A surrey of solar agricultural dryers*, Brace Research Institute, Macdonald Campus of McGill University, Technical Report No. T-99, 1975.
43. H. R. Bolin, and D. K. Salunkhe, Food dehydration by solar energy, *Crit. Rev. Food Sci. Nutr.* Vol. 16, pp. 327–353, 1982.
44. Westeco Drying Inc., First commercial sonic-assisted drying plant starts up in California. *Chilton's Food Engineering*, Vol. 59. No. 12, p. 120, 1987.
45. G. Roa, and I. C. Macedo, Stationary bin and solar collector module, *Solar Energy*, Vol. 18, p. 445, 1976.
46. C. C. Smith, J. A. Maga, and J. C. Chapman, *Solar process drying of potato products*, Paper No. 77-6521, ASAE Meeting, Chicago, 1977.
47. J. H. Moy, *Solar energy application in food processing: Direct radiant drying, air drying and osmovac—dehydration of foods with solar energy*, Final Report CRIS, University of Hawaii, Honolulu, June 1977 (after Bolin and Salunkhe, 1982).
48. S. Grabowski, and A. S. Mujumdar, Solar-assisted osmotic dehydration. In: *Drying of solids*, ed. A. S. Mujumdar, International Science Oxford & IBH, New York and New Delhi, pp. 367–404, 1992.
49. E. J. Carnegie, Solar fruit drying, In: *Solar energy in agriculture*, ed. B. F. Parker, Elsevier, Amsterdam, pp. 335–349, 1991.
50. S. Hovmand, Fluidized bed drying, In: *Handbook of industrial drying*, ed. A. S. Mujumdar, Marcel Dekker, New York, pp. 195–248, 1995.
51. Z. Pakowski, A. S. Mujumdar, and C. Strumillo, Theory and application of vibrated beds and vibrated fluid beds for drying processes In: *Advances in drying*, Vol. 3, ed. A. S. Mujumdar, Hemisphere, Washington, D.C., pp. 245–306, 1984.
52. E. Pallai, T. Szentmarjay, and A. S. Mujumdar, Spouted bed drying. In: *Handbook of industrial drying*, ed. A. S. Mujumdar, Marcel Dekker, New York, pp. 453–488, 1995.
53. T. Kudra, and A. S. Mujumdar, Special drying techniques and novel dryers. In: *Handbook of industrial drying*, ed. A. S. Mujumdar, Marcel Dekker, New York, pp. 1087–1150, 1995.
54. K. Masters, *Spray drying handbook*, George Godwin, London, 1979.
55. I. Filkova, and A. S. Mujumdar, Industrial spray drying systems. In: *Handbook of industrial drying*, ed. A. S. Mujumdar, Marcel Dekker, New York, pp. 263–308, 1995.
56. Integral filter revolutionize spray drying design, *Food Technol.*, Vol. 51, No. 5, p. 95, 1997.
57. J. J. Kelly, Rotary drying. In: *Handbook of industrial drying*, ed. A. S. Mujumdar, Marcel Dekker, New York, pp. 161–184, 1995.
58. M. Hasan, and A. S. Mujumdar, Drying of polymers. In: *Handbook of industrial drying*, ed. A. S. Mujumdar, Marcel Dekker, New York, pp. 1039–1070, 1995.
59. W. K. Heiland, J. F. Sullivan, R. P. Konstance, J. C. Craig, Jr., J. Cording, Jr., and N. C. Aceto, A continuous explosion puffing system, *Food Technol.*, Vol. 54, pp. 772–779, 1977.
60. J. M. N. Dalgleish, *Freeze-drying for the food industries*, Elsevier Applied Science, London, 1990.

61. G. J. Malecki, P. Shinde, A. I. Morgan, Jr., and D. F. Farkas, Atmospheric fluidized bed freeze drying, *Food Technol.*, Vol. 24, pp. 93–99, 1970.
62. A. I. Liapis, and R. Bruttini, Freeze drying. In: *Handbook of industrial drying*, ed. A. S. Mujumdar, Marcel Dekker, New York, pp. 309–344, 1995.
63. F. G. Karuri, *The application of radiant energy to the dehydration of foods*, Ph.D. Thesis, University of Reading, Reading, England, 1984.
64. H. S. Muralidhara, and D. Ensminger, Acoustic drying of green rice, *Drying Technol.*, Vol. 4, pp. 137–143, 1986.
65. M. H. Kim, and R. T. Toledo, Effect of osmotic dehydration and high temperature-fluidised bed drying on properties of dehydrated Rabbiteye blueberries, *J. Food Sci.*, Vol. 52., pp. 980–988, 1987.
66. J. J. Jen, Postharvest handling and processing of selected fruits and vegetables. In: *Trends in food processing. II. Proceedings of the 7th World Congress of Food Science and Technology*, Singapore, October 1987, ed. Ghee, A. H., Singapore Inst. Food Sci. and Technol., Singapore, pp. 261–263, 1989.
67. F. Nsonzi, and H. S. Ramaswamy, Quality evaluation of osmo-convective dried blueberries, *Drying Technol.*, Vol. 16, pp. 705–723, 1998.
68. P. P. Lewicki, and A. Lenart, Osmotic dehydration of fruits and vegetables. In: *Handbook of industrial drying*, ed. A. S. Mujumdar, Marcel Dekker, New York, pp. 691–714, 1995.
69. H. N. Lazarides, Osmotic preconcentration: Developments and prospects. In: *Processing of foods and process optimization: An interface*, ed. R. P. Singh, and F. A. R. Oliveira, CRC Press, Boca Raton, Florida, pp. 73–121, 1994.
70. A. L. Raoult-Wack, A. Lenart, and S. Guilbert, Recent advances in dewatering through immersion in concentrated solutions. In: *Drying of solids*, ed. A. S. Mujumdar, International Science, and Oxford & IBH, New York and New Delhi, pp. 21–51, 1992.
71. J. Cohen, and T. C. S. Yang, *Osmotic dehydration and its application in nutrient infusion of various foods*, U.S. Army Research, Development and Engineering Center, Natick, Massachusetts, Report No. TR-95/034, 1995.
72. H. L. Mantius, and P. R. Peterson, *Fruit extraction and infusion*, US Patent No. 5,320,861, 1994.
73. M. D. Ranken, *Food industries manual*, Blackie, Glasgow, p. 403, 1988.
74. V. K. Patil, V. R. Chakrawar, B. Narwadkar, and G. S. Shinde, Grape. In: *Handbook of fruit science and technology*, ed. D. K. Salunkhe, and S. S. Kadam, Marcel Dekker, New York, pp. 7–38, 1995.
75. R. K. Strigler, G. T. Berg, and J. R. Morris, Raisin production and processing, In: *Processing fruits: Science and technology, Vol. 2, Major processed products*, ed. L. P. Somogyi, D. M. Barrett, and Y. H. Hui, Technomic, Lancaster and Basel, pp. 235–263, 1996.
76. P. Y. Kadam, S. A. Dhumal, and N. N. Shinde, Pear. In: *Handbook of fruit science and technology*, ed. D. K. Salunkhe, and S. S. Kadam, Marcel Dekker, New York, pp. 183–202, 1995.
77. L. P. Somogyi, H. S. Ramaswamy, and Y. H. Hui, *Processing of fruits: Science and technology, Vol. 1, Biology, principles, and applications*, Technomic, Lancaster and Basel, 1996.
78. V. P. Bhutani, and V. K. Joshi, Plum. In: *Handbook of fruit science and technology*, ed. D. K. Salunkhe, and S. S. Kadam, Marcel Dekker, New York, pp. 203–241, 1995.
79. M.-P. Raoult, *Transformation des fruits: Jus, confitures, fruits secs*, Altersil, ENSIA, Cirad Ceemat, France (in French), 1984.
80. H. R. Bolin, G. Fuller, and J. Powers, Product development and applications for a dry apricot concentrate, *Food Product Dev.*, Vol. 7, pp. 30–41, 1973.
81. V. M. Ghorpade, M. A. Hanna, and S. S. Kadam, Apricot. In: *Handbook of fruit science and technology*, ed. D. K. Salunkhe, and S. S. Kadam, Marcel Dekker, New York, pp. 335–361, 1995.

82. W. I. Vollink, R. E. Kenyon, S. Barnett, and H. Bowden, Freeze dried strawberries for incorporation into breakfast cereal, US Patent No. 3,395,022, 1971.
83. R. K. Scharschmidt, and R. E. Kenyon, *Freeze drying of blueberries*, US Patent No. 3,467,530, 1971.
84. J. Jagtiani, H. T. Chan, and W. S. Sakai, *Tropical fruit processing*, Academic Press, San Diego, 1988.
85. S. Nagy, P. E. Shaw, and W. F. Wardowski, *Fruits of tropical and subtropical origin*, Florida Science Source, Inc., Lake Alfred, Florida, 1990.
86. U. T. Desai, and P. M. Kotecha, Fig. In: *Handbook of fruit science and technology*, ed. D. K. Salunkhe, and S. S. Kadam, Marcel Dekker, New York, pp. 407–417, 1995.
87. M. N. Islam, and J. M. Flink, Dehydration of potato. II. Osmotic concentration and its effect on air drying behavior, *J. Food Technol.*, Vol. 17, pp. 373–385, 1982.
88. W. F. Talburt, and O. Smith, *Potato processing*, Van Nostrand Reinhold, New York, 1987.
89. G. Lisinska, and W. Leszczynski, *Potato science and technology*, Elsevier Applied Science, London, 1989.
90. M. J. Willard, V. M. Hix, and G. Kluge, Dehydrated mashed potatoes-potato flakes in potato processing. In: *Potato processing*, ed. W. F. Talburt, and O. Smith, Van Nostrand Reinhold, New York, pp. 123–222, 1987.
91. H. K. Burr, and R. M. Reeve, Potatoes. In: *Food dehydration*, Vol. 2, ed. W. B. Van Arsdel, M. J. Copley, and A. I. Morgan, AVI, Westport, Connecticut, pp. 83–157, 1973.
92. O. Burits, and F. Berki, *Drying of fruits and vegetables*, Ed. Izd. Pistchevaya Prom., Moscow (in Russian), 1978.
93. W. A. Woskobnikov, W. N. Gulaeev, Z. A. Kac, and O. A. Popov, *Drying of vegetables and fruits*, Ed. Izd. Pistchevaya Prom., Moscow (in Russian), 1983.
94. B. S. Luh, and J. G. Woodroof, *Commercial vegetable processing*, 2nd. ed., AVI, Westport, Connecticut, 1988.
95. G. Mazza, Carrot. In: *Quality and preservation of vegetables*, ed. N.A.M. Eskin, CRC Press, Boca Raton, Florida, 1989.
96. D. K. Tressler, New developments in the dehydration of fruits and vegetables, *Food Technol.*, Vol. 10, pp. 119–124, 1956.
97. G. Charalambous, *Spices, herbs and edible fungi*, Elsevier, Amsterdam, 1994.
98. D. R. Tainter, and A. T. Grenis, *Spices and seasonings*, VCH, New York, 1993.
99. H. Richard, Epices et aromates, TEC&DCC-Lavoisier, Paris (in French), 1992.
100. K. T. Farrell, *Spices, condiments and seasonings*, AVI, Daytona Beach, Florida, 1985.
101. F. Rosengarten, Jr., *The book of spices*, Jove, New York, 1981.
102. J. W. Purselove, E. G. Brown, C. L. Green, and S. R. J. Robins, *Spices*, Tropical Agriculture Series, Longman Science and Technology, London, 1981.
103. FAO Internet Homepage, http://apps.fao.org, 2001.

24

Coffee: A Perspective on Processing and Products

KULATHOORAN RAMALAKSHMI and BASHYAM RAGHAVAN

Central Food Technological Research Institute, Mysore, India

1 INTRODUCTION

Over the centuries numerous legends have accumulated about the discovery of coffee. Possibly the earliest references to the use of coffee are to be seen in the Old Testament. Although cultivation may have begun as early as the 6th century C.E., the first written mention of coffee as such is by Razes, a 10th-century Arabian physician. The most well known story of the discovery of the coffee plant is concerned with a goatherd tending his flock in the hills around a monastery on the banks of the Red Sea. He noticed that his goats, after chewing berries from the bushes growing there, started prancing excitedly. A monk from the monastery observed this behavior, took some of the berries, roasted them, and brewed them. When served, the brew kept his people more alert during the long prayers at night. And this saw the birth of the world's most stimulating beverage (1).

The word *coffee* is derived from the Arabic word *quahweh*, which is a poetic term for "wine." Since wine is forbidden to devout Muslims, the name was changed to *coffee*. The wild coffee plant is indigenous to Ethiopia, from which it spread to Arabia and nearby countries. The transport of coffee from the countries near Arabia to other parts of the world was limited; the raw beans were not allowed out of the country without steeping in boiling water or heating to destroy their germinating power. Strangers were not allowed to visit the plantations; it was Baba Budan, a pilgrim from India, who smuggled out a few seeds capable of germination. He planted the seeds in western ghats of Coorg in South India around 1600 C.E. The cultivation was expanded during British rule. In Brazil, coffee entered through a Brazilian officer who, while on a visit to French Guyana in 1727, received a plant hidden in a bouquet of flowers as a token of affection from the governor's

Table 1 Coffee Producing Countries

Arabica		Robusta
Brazil	Kenya	Angola
Colombia	Mexico	Cameroon
Honduras	Nicaragua	Côte d'Ivoire
India	Papua New Guinea	Indonesia
Costa Rica	Peru	Madagascar
Dominican Republic	Ethiopia	Philippines
United Republic of Tanzania	Uganda	Vietnam
		Zaire
Ecuador		
El Salvador		
Guatemela		

Source: From Ref. 13.

wife. This was the start of coffee plantations in Brazil, which now holds supremacy in the world.

The coffee plant belongs to the Rubiaceae family, which has over 70 species of coffee. But only seven of them have significant economic importance. The commercially cultivated species are arabica (*Coffea arabica*) and robusta (*Coffea canephora*). *Coffea liberica*, another species, was devastated during the 1940s by an epidemic of trache-mycosis due to infection by *Fusarium xylaroides*, and the commercial growth of this species has effectively ceased since then. *Coffea robusta*, which is noted for its resistance to disease, contains more caffeine than *C. arabica* and is thus more economical in the manufacture of instant coffee (2).

Coffee is one of the most important agricultural products traded worldwide. It is grown and exported by over 70 developing countries in the tropical and subtropical belt (Table 1), but industrialized countries import and consume most of it. Of these 70, 51 countries, including Brazil, Colombia, Guatemala, India, and Mexico, are responsible for more than 99% of world output and are exporting members of the International Coffee Agreement. The world coffee production and its export in recent years are given in Table 2 (3).

Table 2 World Coffee Balance (1000 Bags Green Coffee)

Crop year	96–97	95–96	94–95
Beginning stock	36,470.3	46,122.5	40,973.2
Production	98,992.0	87,884.0	95,172.2
Robusta	33,470.0	29,737.0	29,087.0
Arabica	65,522.0	58,147.0	66,085.0
Domestic use	25,478.0	24,681.0	24,482.0
Exports	77,786.5	73,358.0	65,884.0
Green	73,676.9	68,983.7	62,317.7
Roasted	560.8	95.3	474.9
Solubles	3,548.8	3,779.3	3,091.5
Ending stock	33,546.9	36,470.3	46,122.5

Source: From Ref. 3.

In more recent times, a hybrid between *C. arabica* and *C. robusta* was developed in Ivory Coast, namely, *C. arabusta*, which is reported to have better flavor quality than *C. robusta* and more disease resistance than *C. arabica*. The beans also have lower caffeine content. Cultivation of this species is currently restricted to limited areas because of poor yield and diminished size. Attempts have been made to develop a coffee that is free of or has a lower level of caffeine but with little success.

2 BOTANY

The coffee plant is a small tree, which in the case of *C. arabica* is maintained at 5 ft (150 cm) and in the case of *C. robusta* at 5.5 to 6.0 ft (170 to 185 cm). This can grow up to 25 ft (7.6 m) in the wild state but is pruned for two reasons: (a) to facilitate harvesting and (b) to maintain optimal tree shape. The primary branches are opposed and horizontally drooping, and the leaves grow in pairs on short stalks. They are about 15 cm in *Coffea arabica* and longer in *Coffea canephora*, oval, and fairly dark green in color (4).

Various methods of propagation may be used, including cuttings, grafting, and layering. The use of cuttings is the normal commercial practice. The coffee plants yield within 3–5 years and last up to 30–40 years. An altitude of 2000–4000 ft (600–1200 m) is ideal for coffee, but it can also be grown at up to 6000 ft (1800 m). As in the case of tea, the higher the altitude, the better the quality. But the limiting factor is that coffee cannot withstand frost. The ideal climatic condition for coffee is 75°C–80°C and rainfall 60–80 in (150–200 cm). The yield of coffee is totally dependent on the flowers produced by the plants and, more importantly, on the percentage of fruit set from the flowers. The blossoming is largely dependent on timely rainfall, which induces the flower buds to open within 7 to 10 days. In recent years, sprinkler spraying is widely used for opening flower buds. Pollination of the flowers is done mainly by insects. The normal development from flower to fruit requires 5 to 8 months in *C. arabica* and 9 to 10 months in *C. robusta.* The fruits are borne in clusters at each leaf axile.

In the case of *C. arabica* each cluster carries 15 to 20 fruits per node, whereas there are 40 to 80 fruits in *C. robusta* (5) (see Fig. 1). The fresh fruit consists of an outer skin over a fleshy pulp in which are embedded two seeds, each flat on one side and convex on the other. Occasionally only one seed, rounded on both sides (peaberry), may be found. The seeds are covered by a thin silver skin and a thick layer called *parchment.*

3 PROCESSING

Quality is a summative index of many characteristics of coffee, such as its appearance in the raw, roasted, and liquid states and qualities comprising factors such as aroma, body, and acidity. Quality of coffee depends on the variety, environmental factors (soil, altitude), insect or fungal attack, nutritional factors, method of processing, drying, hulling, and grading. Although it is possible to overcome the influence of these factors by adopting improved cultural practices, correct processing technique is necessary to prevent deterioration in quality. Faulty processing can bring about deterioration of even the best quality coffee. Proper processing in the plantations can go a long way to preserve and enhance the inherent qualities of good coffee.

The two main species of coffee, *C. arabica* and *C. robusta*, begin to yield from the age of 4 to 5 years. The produce, that is, the coffee fruits picked at the estate, has to be processed there before dispatch to the curing works. There are two methods of coffee

Fig. 1 Coffee arabica shrub with ripe berries. (Courtesy of Coffee Research Sub-Station, Chettally, Kodagu, India.)

processing, the *wet method* and the *dry method* (Fig. 2). In the wet method, only ripe fruits having a reddish brown color are picked, graded, and fed to a pulper to remove the outer skin and mucilage. The parchment is washed thoroughly and dried to prescribed standard test weight. The coffee prepared by this method is called *parchment coffee*. This method is practiced in Colombia, Kenya, and most of the South and Central American countries; India also processes *C. arabica* by this method. A major problem associated with this is its requirement of a large volume of water. In the dry method, ripe, green, and underripe fruits are sorted out and then dried separately to prescribed standard test weight. The coffee obtained by this method is called *cherry coffee*. It may be noted that coffee can be considered to be dry when a fistful of it produces a rattling sound on shaking.

Coffee Board, India, has published a list of "do"s and "don't"s in the processing of parchment and cherry coffee, which is reproduced in Table 3.

Parchment coffee prepared by the wet method is generally favored by the market. Cherry coffee, because of its very nature of preparation and its longer contact with the mucilage and fruit skin, is usually associated with the characteristic ''fruity'' flavor. Hence, it is desirable to process the largest quantity possible by the wet method (6).

3.1 Harvesting

For the preparation of both parchment and cherry coffee, picking of the right type of fruit forms an essential part of processing. Coffee fruits should be picked as they become ripe

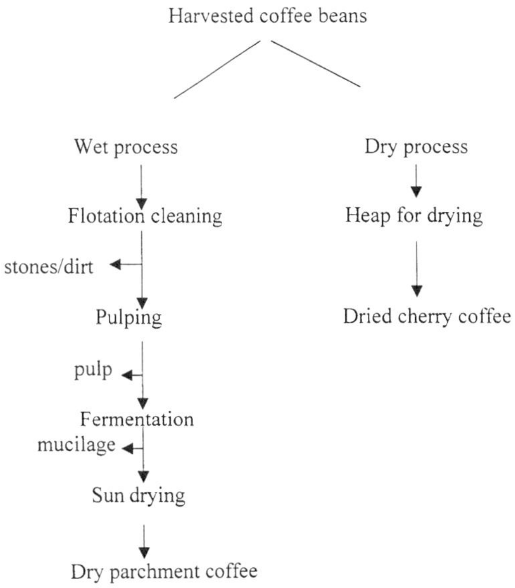

Fig. 2 Processing of green coffee.

Table 3 "Do"s and "Don't"s in the Processing of Coffee

"Do"s for parchment coffee

- Pick just ripe fruits.
- Overripe or green cherries, if harvested, should be sorted out on a clean surface.
- Clean bags should be used for collection of harvested fruits.
- Where fertilizer or cattle feed bags are used, they should be washed thoroughly in running water, many times, before use.
- Maintain cleanliness in the pulper house (a house where the pulper is installed). The pulper, washing machine, tank, vats, trays, etc., should be kept clean.
- Grade the fruits into uniform sizes before pulping.
- Adjust the pulper to suit the size of the graded fruits.
- Pulp the fruits on the same day as the harvest.
- Separate the fruit skin and unpulped fruits from the pulped fruits before fermentation.
- Pulped fruits, left for fermentation in the vats, should be covered with clean bags.
- Clean water should be used for washing coffee.
- Use correct concentration of alkali for alkali washing (1% of sodium hydroxide in 10 liters of water).
- Raoeng or Aquapulper machinery used for demucilaging should be thoroughly cleaned before use.
- Wash the parchment with clean water as many times as required to have pebble clean parchment.
- Soak the parchment either under water or under 1% sodium metabisulfite solution overnight for enhancing the quality of coffee.
- Clean the pulper machine, vats, etc., after the day's work.
- Greens should be processed separately.
- Baskets lined with polythene sheet may be used for carrying wet parchment to the drying yard.
- Drain off the excess water rapidly.

Table 3 Continued

- Dry the coffee on wire mesh trays initially or on coir mat spread on a table.
- Dry the coffee on clean, tiled, or concrete drying yards.
- Sort out all pulper cuts, naked beans, blacks, and other defective beans during drying of parchment.
- Dry the coffee to constant weight.
- Provide adequate drainage to the drying yards.
- Heap the coffee and keep it covered during the night to prevent any type of direct contact with moisture (mist or rain).
- Dry the coffee to the prescribed test weight standards.
- Store the coffee in clean gunny socks, in well ventilated and clean godowns, free of dampness. Line the floor with wooden planks.
- Dispatch the coffee to the curing works at the earliest opportunity.
- Coffee obtained from greens and floats should be pooled separately.

The "do"s enumerated are of paramount importance in the processing of coffee at the estate to ensure good quality coffee.

"Don't"s for parchment coffee

- Overripe and green cherries should not be mixed with ripe fruits during processing.
- Gunny sacks used for fertilizer or cattle feed should not be used for harvesting of coffee.
- Prevent contamination with earth and grid at fruit or parchment stage.
- Prolonged heaping of fruits and delayed pulping should be prevented.
- Prevent underfermentation or overfermentation.
- Recycling of used water for pulping and washing should be prevented.
- Do not carry wet parchment to the drying yard in cow-dung-plastered baskets.
- Do not dry coffee on mud or cow-dung-plastered drying yards.
- Prevent overdrying or underdrying.
- Avoid hard tools for raking coffee.
- Don't store coffee along with fertilizers, pesticides, pepper, cardamom, cloves, and other such materials that may contaminate the bean.

Observing the "don't"s is as important as observing the "do"s since well-processed coffee can be spoiled even if any one of the "don't"s is disregarded.

The cherry coffee, which is obtained by the dry processing of coffee, needs great attention in preparation.

"Do"s for cherry coffee

- Fruits should be picked as and when they ripen.
- Greens and underripe fruits should be sorted out and dried separately.
- When resorting to stripping whole crop, it is preferable to do so when 50% fruits are ripe, but green and ripe fruits need to be dried separately.
- The fruits should be spread evenly to a thickness of about 8 cm (3 in) on clean drying ground. It is desirable that drying be carried out on a tiled or concrete floor.
- Stir and ridge the fruits at least once every hour.
- As in the case of parchment, heap the coffee and cover every day early in the evening, and spread again the next morning after the mist clears up.
- Dry the cherry to the prescribed standard test weights.
- Each lot of cherry should be bagged separately in clean, dry gunny sacks in a clean, dry room, preventing storage of pesticides and fertilizers in the same room.

Source: From Coffee Board, India.

Fig. 3 Sorting of greens and leaves/twigs from coffee fruits. (Courtesy of Coffee Research Sub-Station, Chettally, Kodagu, India.)

and when, on gently squeezing the fruit, the bean inside pops out easily. Underripe and overripe fruits have a deterioration in quality; the former tend to produce ''immature beans'' and the latter ''foxy'' coffee. If, for any reason, it is not possible to pick coffee as it ripens, the over- and underripe fruits should be scrupulously sorted out before using them for pulping (Fig. 3). They may be dried separately as cherry beans. It is advisable to wash and dry the bags that are used for collecting the harvested fruits frequently. Bags in which fertilizers, pesticides, and fungicides are stored should never be used for this purpose. Indefinite delay in pulping and keeping of the fruits in a heaped condition for more than 10 hours cause a fruity/winelike taste. Provisional storage of fruits in water before pulping prevents overheating (7).

Harvesting is usually a manual process, employing seasonal labor. Wages are usually paid according to the weight harvested. Control is needed to ensure that the amount of unripe cherry harvested is kept to a minimum and damage to the coffee tree is limited. A number of approaches have been tried for mechanization. Tree and branch shakers, which shake ripe berries off the tree and onto mesh nets, have been tried, but a significant amount of unripe berries and leaves is also removed in this process.

Coffee beans readily absorb foreign taints and odors. Pulper yards, pulper vats, syphons, and channels as well as sieves and gorumanes (wooden ladles with long handles) should be checked daily and kept clean. No fruit, fruit skin, or beans from the previous day's harvest should be allowed to remain and mix with fresh coffee fruits or pulped mass. Fermented beans of the previous day's lots when mixed with fresh and clean parchment cause deterioration of quality of the entire lot. Clean water should be used for pulping and washing, and all extraneous matter such as leaves and twigs should be removed.

3.2 Preparation of Parchment Coffee

3.2.1 Classification

Selectively picked coffee is transported to a central processing unit on the day of harvest, when moisture may be reduced from 65%–70% to 45%. Ripe or soft cherries are dumped into a funnel-shaped dry tank partly filled with water. Floats and sinks are removed; storing under running water helps to reduce heat and microbiological activity caused by temperature (8).

3.2.2 Pulping

Preparation of coffee by the wet method requires pulping equipment and an adequate supply of clean water. Fruits should be pulped on the harvest day to prevent fermentation before pulping. Pulping is carried out in two steps. In the first step the fruit is squeezed between the roughened surface of a rotating cylinder or a disk and a stationary part that is smooth. The distance between the two surfaces is carefully adjusted so that the space narrows as the fruit is carried through the breast. The breast is often held by springs that permit tension. The passage produces a squeezing action that detaches skin and flesh from the fruit. The rotating drum has small projections that assist the dragging of the pulp. In the second step, the seeds are separated from the pulp. This is accomplished by a plate with a sharply ground edge. It is fixed at a right angle to the drum with a precisely regulated clearance of about 1/16 in (0.16 cm). The rough surface of the moving drum or disk forces most of the flexible fibrous pulp through the narrow gap, which is too small for the seeds to pass through. Seeds are separated onto a conveyor and passed to the next step of operation. The pulper should be properly adjusted and checked every day to ensure satisfactory pulping and to prevent cuts. Pulper-nipped beans and other deformed beans cause defective parchment. Fruits may be fed to the pulper through a syphon arrangement to ensure uniform feeding and to separate lights and floats from sound fruits. Uniform feeding ensures proper removal of skin and prevents cuts. The pulped parchment should be sieved to eliminate any unpulped fruits and fruit skin (9).

In plantations where there is little water available, recirculation of water in the pulping section is recommended. Adoption of this measure has become necessary even in places where water is available in abundance because of water pollution problems. Recirculated water may be used no longer than 1 working day. If it is left overnight, the water is soon contaminated, and use of such water affects the quality of coffee adversely.

It is desirable to separate lights from well-filled, mature, heavy beans at every stage. At the prefermentation stage this can be done by passing the coffee in a channel having a pronounced slope (at least 4%) and an ample cross section (minimum of 30×30 cm). The channel should be cemented and smoothed so that no residue passes through surfaces and cracks. Where facilities are available, grading at the prefermentation stage can also be done by passing the parchment through an Aagard pregrader or a grading sieve to separate heavy beans from light. The skins separated by pulping should be led away from the vats into collection pits so that microbial decomposition of the skin does not affect the bean quality when it is mixed with the bean. The approximate chemical composition is given in Table 4.

In the market three types of pulpers, viz., disk, drum, and vertical, are available. The disk type is not 100% efficient because of variation in size of berries and the fixed gap between the moving and stationary parts of the pulper (Fig. 4). The important drum-type pulper (Fig. 5) is capable of rejecting the unpulpable fruits and has facility for green cherry separation. The unpulped fruits are pulped by use of a repasser pulping system. The vertical-type pulper is capable of pulping any size of berries, yielding better turnout and using less energy and water (10).

Table 4 Chemical Composition of Pulp

Component	Composition (%)
Moisture	41.0–43.0
Raw fiber	26.0–28.0
Sugars	8.0–10.0
Tannins	7.0–9.0
Minerals	2.0–4.0
Waxes, fats, resins	1.0–2.0
Volatile oil	0.1
Others	6.0–7.0

Source: From Ref. 11.

3.2.3 Demucilaging and Washing

The freshly pulped coffee beans are covered with a slippery mucilagenous layer approximately 0.80 mm thick. When freshly pulped, it is translucent and colorless; on exposure to air it turns brown, probably through enzyme oxidation, resembling that of freshly cut apples. It consists chemically of protopectin, pectin, pectinase, and small amounts of sugars along with the naturally occurring enzymes pectase, pectinase, pectiesterase, and proto-

Fig. 4 Pulper cum washer. (Courtesy of Coffee Research Sub-Station, Chettally, Kodagu, India.)

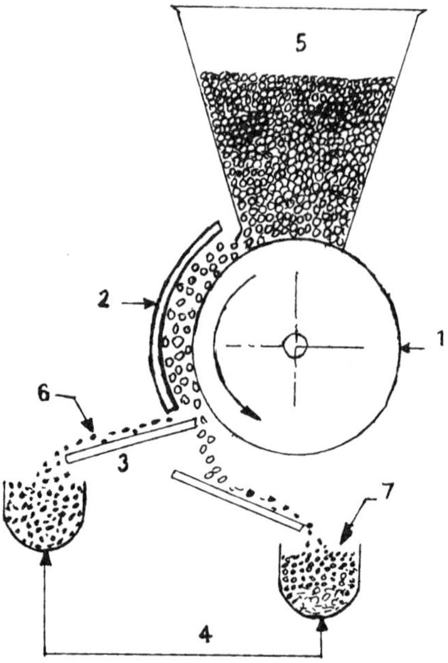

Fig. 5 Diagrammatic details of a drum pulper in side view showing (1) rotary drum, (2) breast-plate, (3) separating plate, (4) receiving troughs, (5) cherries, (6) beans, (7) pulp. (From Ref. 10. Reproduced with permission.)

pectinase. The mucilage is insoluble in water and appears as an amorphous gel. The chemical composition of mucilage is given in Table 5.

3.2.3.1 Chemical Characteristics of Fermentation

The pectinic acids are composed of linear polymerized chains of acids formed mainly from hexose (hydroxy sugar), chiefly galactose with little arabinose and other sugars. Esters of these acids are also formed. Pectinic acids are classified into protopectins, pectinase, and pectic acids (with decreasing molecular weight from protopectin, at 70,000). In addition, the enzymes present break down the long polymer chains into small monomers of acids and esters. Yeasts generally produce alcohols from sugars; bacteria use this alcohol as food and form acetic acid, lactic acid, and other carboxylic acids by oxidation. The mucilage on the parchment skin can be removed by any one of the following methods:

Table 5 Chemical Composition of Mucilage

Component	Composition (%)
Moisture	84.0–86.0
Pectin	0.8–1.0
Sugar	3.0–5.0
Ash	0.5–1.0
Protein	8.0–10.0

Source: From Ref. 11.

natural fermentation, treatment with alkali, the enzymatic method, and attrition (frictional removal) in machines (Raoeng or Aquapulper).

3.2.3.2 Natural Fermentation

The most common method for mucilage removal is natural fermentation. The hydrolysis of pectins is brought about by enzymes that are naturally present in fruits. Fermentation is a critical stage in processing and has profound influence on quality. Fermentation should be controlled so that it is wholly alcoholic and not acetic. The latter type of fermentation can occur under dry, high-temperature, and high-pH conditions, which usually occur in the top 10-cm layer of the fermenting mass of wet parchment. This should be prevented as it is likely to cause stinkers and beans producing unpleasant odors and taints. The fermenting mass should not be allowed to dry up and should be kept covered.

The mucilage breaks down in the process of fermentation. In the case of *C. arabica* it is complete in about 24 to 36 h. Fermentation takes longer in cool weather than in warmer conditions. Great care is to be exercised at this stage, as otherwise overfermented, sticky mucilage is left on the parchment. This subsequently leads to absorption of moisture by the beans and ''mustiness'' in the final product. When correctly fermented, the mucilage is removed easily and the parchment does not stick to the hand after washing. The beans feel rough and gritty when squeezed by hand, a feeling similar to squeezing pebbles (6). When the mucilage breakdown is complete, clean water is let in and the parchment washed with three to four changes of water. Fermentation is often carried out in concrete tanks (Fig. 6).

Coffea robusta has a thicker and more sticky mucilage and hence fermentation is not complete even after 72 h. Quite often, the mucilage breakdown is not complete even after a very long period. It is, therefore, desirable to resort to warm water treatment, alkali treatment, or frictional removal (attrition) of mucilage. A simple and quick method of removal of mucilage is to mix equal weights of pulped coffee and water and heat to 50°C ± 3°C as quickly as possible when the structure of pectic materials of the mucilage gel is ruptured. The coffee may be washed afterward by contact with warm water for about 3 minutes. The only constraint on this method is the cost of the fuel.

3.2.3.3 Treament with Alkali

Removal of mucilage by treatment with alkali takes 1.0 h for *C. arabica* and 1.5–2.0 h for *C. robusta*. The beans obtained after pulping are drained of excess water and spread out in the vats uniformly and furrowed with *gorumanes*. A 10% solution of caustic soda (sodium hydroxide) is evenly applied into the furrows by using a can. About 1 kg sodium hydroxide dissolved in 10 L water is sufficient to treat 1000 to 1200 L of wet parchment. The parchment is agitated thoroughly by means of gorumanes to produce contact of the alkali with the parchment and trampled by feet for about half an hour. When the parchment is no longer slimy and makes a rattling noise, clean water is let in and the parchment is washed clean with three or four changes of water.

3.2.3.4 Enzymatic Removal of Mucilage

Commercial formulations of pectinolytic enzymes can be used to hasten the fermentation process. The length of fermentation can be reduced by use of pectinolytic enzymes; both concentrations of enzymes and temperature of the ambient air determine the fermentation time. Generally the pulped coffee is taken to digestion tanks every evening and fermenta-

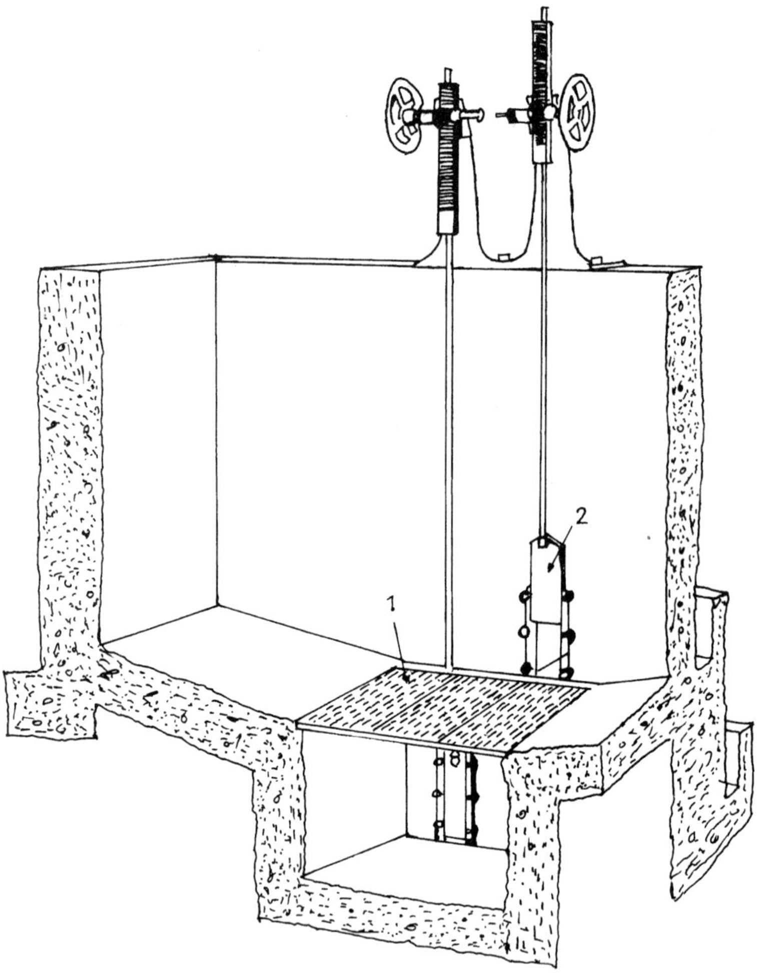

Fig. 6 Construction details of a typical fermentation tank, showing (1) grid and (2) sluicegate. (From Ref. 10. Reproduced with permission.)

tion is carried out overnight. Alkali treatment requires strict control of the time of contact with alkali so that the quality of the coffee is not affected.

3.2.3.5 Attrition Method

Some pulpers (Raoeng and Aquapulper) pulp and demucilage the beans in one operation. These machines are especially suitable for demucilaging *C. robusta* parchment. Rubbing of the mucilage from pulped coffee by frictional force or scrubbing is known as *attrition*. Several machines have been designed to demucilage the coffee beans. In one such machine, the pulped coffee beans are pressed against each other and against a rough surface while being forcibly fed through the machine by a screw against resistance generated by a partially throttled discharge. Clearances are carefully adjusted so that the parchment layer is not damaged, and the corners of the ribs or projections are rounded to prevent cutting the beans. However, a number of naked and bruised beans may result in the parchment. It is, therefore, necessary to adjust the machines carefully to obtain uniform pulping and demucilaging. Sorting of fruits into different sizes and uniform feeding by using a siphon arrangement may also rectify this defect to a considerable extent. If naked and bruised beans occur in the parchment, in spite of careful manipulation, they may be garbled out on the drying trays (11).

Prefermentation of pulped coffee for 12 h in the case of *C. arabica* and 24 to 36 h in the case of *C. robusta* is desirable before use of an aquawasher for effective removal of mucilage. These machines are also often used only for demucilaging after removing the fruit skin in the traditional pulpers.

Opinion is divided on the merits and demerits of processing coffee by methods involving natural fermentation and those without prolonged fermentation. Cup-test results have indicated that there is no difference in cup quality between coffees processed by different methods. It may, however, be pointed out that coffee processed without prolonged natural fermentation dries faster (6, 9, 11).

3.2.3.6 Postfermentation Soaking

Wherever water supply is abundant and additional vats are available, the washed parchment may be soaked under water for about 12–24 h (overnight) and then given a final wash. This method seems to improve the quality of substandard coffees both in appearance and in the cup. The washed coffee can be graded in channels to separate heavy beans from ''lights.''

3.2.4 Speed of Processing

Picking of fruits from early morning to late afternoon, prompt transportation to a central processing unit so that moisture can decrease from 65%–70% to 45%, pulping and removal of mucilage overnight, and washing cannot take more than 36 h before the beans are subjected to drying (10).

3.3 Preparation of Cherry Coffee

The harvested cherries are spread out evenly on concrete or paved patios on trays or simply on the ground. Drying on the bare ground is not advisable as cherries become stained by dust and earth and often acquire an earthy smell and taste. Drying in the sun takes 3 to 5 weeks depending on the thickness of the layers of cherries, temperature, and amount of daily sunshine. During the dry process the cherries have to be turned over at regular intervals to ensure even drying to prevent fermentation of the lower layers when the temperature drops or ultimately rains occur. The coffee must be covered or placed under shelter during night timings. Drying is complete when the moisture content of the dried cherries is around 12%. At this stage the outer shell is dark brown and brittle. The bean rattles inside the husk. Dried beans are bagged and stored for several weeks. Hulling of dried cherries is not usually practiced after drying. During this time the green beans inside the dried shell continue to lose some of their moisture content and spread out evenly by osmosis. The ratio of dry cherry to green beans is usually 2:1.

3.4 Drying

The next stage in processing is drying the parchment in the sun until the moisture content is sufficiently reduced to permit storage of beans till they are dispatched to curing works. It is necessary to emphasize that proper drying contributes to the healthy color and quality of the bean. Underdried parchment turns ''moldy'' and is ''bleached'' during storage and subsequent curing operations. The wet parchment coffee has a water content of around 50%–55%, which has to be reduced to 10%. In preparation of cherry coffee, the moisture content of 65% in fruits is reduced to 11% for safe storage.

3.4.1 Sun Drying

Most of the world's coffee is dried in sun, although the present tendency is to dry more and more in mechanical dryers because of rising labor costs, unfavorable weather, and the goal of obtaining uniformly dried beans. Freshly pulped and washed coffee usually contains 52% to 54% moisture. After draining of excess water, beans are spread on the drying terrace. The thickness of the layer is usually 2–4 in (5–10 cm) and coffee is raked at frequent intervals during daytime. To prevent uneven drying and cracking of the parchment during intense heat or sunshine, the coffee beans should be covered or dried in shade; in the case of cherry coffee, the fruits are heaped for 1 or 2 days to allow softening and then dried in open yards. Fruits are dried up to a moisture level of 12%; drying may take about 1 week for washed coffee and about 3 weeks for natural coffee (12).

3.4.2 Machine Drying

Generally, machine drying consists of passing hot air through a bed of coffee. Two methods used are through-flow and cross-flow drying; through-flow drying is much more efficient. It is a good practice to use drying air at a temperature of 60°C in the first stages of drying and slowly decrease the temperature.

In mechanical drying, heated air is passed through a bed of coffee, which is stirred constantly by either rotating the container or using rakes or conveyers. One of the simplest instruments consists of a wire screen or perforated metal on which coffee is loaded and heated by passing hot air through it. Loading, unloading, and stirring are carried out manually. There have been several improvements in the design and operation of dryers, such as automatic loading and unloading, rotating power-driven rakes, recirculation of hot air, and furnaces that accomplish smokeless combustion. Several commercial dryers (Wilken, Guardiola, Torres, Moreira, and American vertical grain) are marketed. Some of these are specially designed for washed coffee, some for cherry, and others for both types, namely, *C. arabica* and *C. robusta.*

The processing operations up to drying are carried out at the estate plantation level. The coffee beans with about 12% moisture level (with husk or parchment, as the case may be) are packed in gunny sacks and sent to curing works for further processing, including hulling, grading, bulking, packing, and storage (Fig. 7).

3.5 Hulling

Hulling is a general term used to describe the separation of parchment hulls and silver skin from parchment coffee and husk from dried cherry coffee (Fig. 8). Polishing is an additional action to remove any silver skin that may still cling to the bean after hulling. The hulling of dried cherries is often preceded by mechanical grading to prevent passage of small cherries through the unhulled beans, necessitating their separation from the clean beans (13).

Most hullers are similar in type; some are best suited for dried cherry and others are more appropriate for parchment coffee. Hullers can be separated into two main groups, friction hullers and impact hullers.

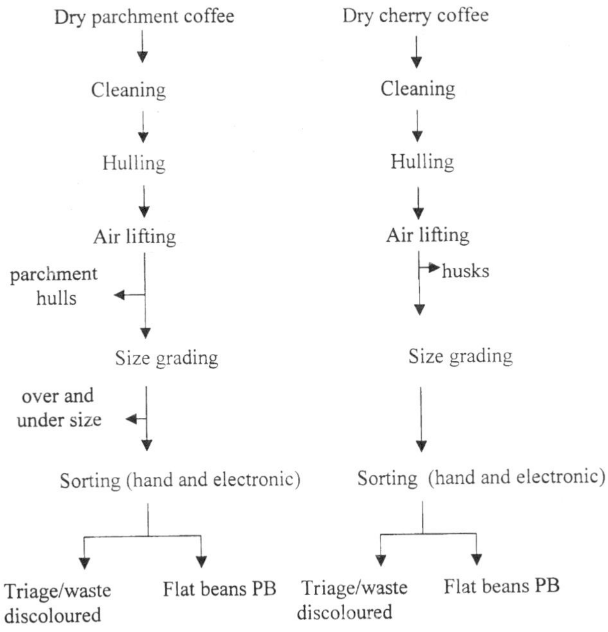

Fig. 7 Processing steps in curing works.

Friction hullers accept both cherry and parchment. These types of hullers have a horizontal, cylindrical casing, the upper part made of steel and the lower part made of strong wire mesh or perforated sheeting. A cylinder of slightly smaller diameter revolves inside this casing. Its surface is fitted with interchangeable steel ribs that form a spiral. A horizontal knife projects from outside into the casing. The distance between this hulling knife and the hulling bars or ribs on the cylinder is set to be less than the diameter of the

Fig. 8 Coffee Huller. (Courtesy of Chamundi Coffee Curing Works, Mysore, India.)

product that is to be hulled. As coffee is fed into the machine, the coffee is squeezed between the ribs and the knife; the shells shatter and separate from the beans. Types of frictional hullers are smout-type hullers, roll hullers, crossbar hullers, and compact hullers. The important smout-type hullers consist of a rotor with a helical pitch that increases toward the discharge end. The rotor turns inside a horizontal casing comprising two matching concave halves with spiral grooves. The spirals on the rotor turn in the opposite direction from those on the casing. The flow of coffee is controlled by a balanced gate at the exit, which can be set to increase or reduce the flow out of the huller, thereby regulating the pressure buildup inside the casing.

Impact hullers rely on impact rather than friction. The advantage is that virtually no heat is generated and less power is used than in the traditional hullers. Most impact hullers cannot handle cherry coffee and are mainly used for parchment coffee. The impact huller consists of a horizontal disk spinning rapidly on a vertical axle inside a circular chamber. The edge of the disk is ringed by hardened steel pins or round bars. Coffee is fed onto the center of the spinning disk and is thrown out toward the ring steel pins by centrifugal force. The parchment hulls shatter when they hit the steel pins, leaving the beans free. Hulls and beans are separated pneumatically. The disc speed can be adjusted to suit the degree of brittleness of the product.

3.6 Polishing, Grading, and Sorting

Green coffee beans undergo further processing, such as polishing, grading, and sorting.

3.6.1 Polishing

Polishing involves removal of the silver skin except that retained in the center cut of the beans. The purpose is purely cosmetic, to improve the appearance of the bean. Polishing may not be required for wet processed coffee. *C. robusta* has a tough and tenacious silver skin, which requires a wet polishing operation to remove it. Polishers (e.g., the Okrassa huller-polisher; see Fig. 9) resemble these of smout friction hullers but use phosphor bronze bars, which are softer than steel and do not harm the beans.

3.6.2 Grading

Grading is the separation of beans according to size and density. The coffee bean has three dimensions: length, width, and thickness. It is usually oblong, fairly flat, and boat shaped. However, peaberry has an elongated oval shape. Although round perforations are sufficient to separate most beans, peaberry requires oblong slots. A flat bed grader and a drum grader separate the beans according to size, and pneumatic separators, often called *catadors*, grade the beans by density (Figs. 10, 11).

It must be noted that in grading of coffee, a lot can be labeled only as "Graded and Garbled" or as "Bulk." It must be ensured that standards (of percentage range of "Blacks") specified for grading a lot as "Graded and Garbled," "Bulk," or "Blacks" are strictly observed. The sieving standards for different types and grades of coffee should also be strictly enforced. Screen (sieve) sizes are expressed as numbers, e.g., Robusta Grade One Screen 16; or by letters, e.g., Arabica Grade AA for bold bean, or by description, e.g., bold, medium, or small. Grading depends on trade custom in any given country. However, all coffee for export is graded to exclude the largest and smallest beans and particles (14).

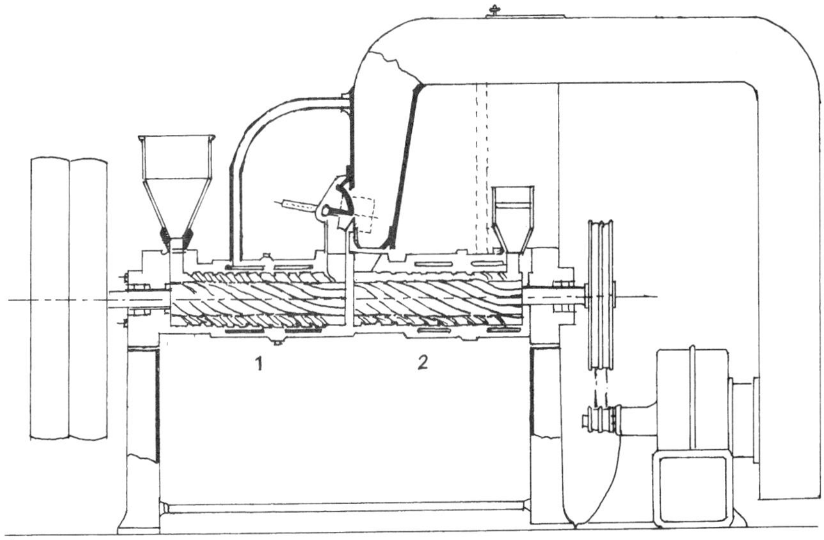

Fig. 9 Construction details of the Okrassa huller-polisher, showing the two separate compartments: (1) huller, (2) polisher. (From Ref. 10. Reproduced with permission.)

Fig. 10 Grading of coffee beans. (Courtesy of Chamundi Coffee Curing Works, Mysore, India.)

Fig. 11 Peaberry separator. (Courtesy of Chamundi Coffee Curing Works, Mysore, India.)

The Indian grade designations and standards for *Coffea arabica* and *C. robusta* are listed in Table 6.

3.6.3 Sorting

Sorting is usually the final stage in the preparation of coffee for export. It is required to remove any defective bean remaining after processing. A certain amount of extraneous material, including whole berries, twigs, stones, and fragments of husks and parchment, may be present and require removal from the beans. Sorting may be carried out physically by blasting air upward through the beans; the process of air lifting removes the defective beans. In some cases, beans removed by air lifting are exposed as a low-grade triage coffee. Hand sorting is a traditional process and is highly labor-intensive, but the cost can be justified by the high quality of coffee hand sorted by expert sorters. Although discolored sour beans are difficult to recognize, complete removal is essential since presence of only a small number adversely affects the flavor of the brew.

Electronic color sorting machines of the monochromatic type use the reflection of white light to identify beans whose surface brightness differs from that of the rest (Fig. 12). Bichromatic sorting machines apply a combination of colors and can eliminate a considerable range of offending beans such as black beans, foxy beans, yellow sours, marble beans, and water-damaged beans. Ultraviolet sorting machines, which operate on the same mechanical principles as bichromatic machines, identify defective beans by exposing them to ultraviolet light.

Table 6 Indian Grade Designations and Standards for *Coffea Arabica* and *Coffea Robusta*

Type	Grade designation	Sieving requirements: Aperture size, in mm	Sieving requirements: Percentage by weight retained (minimum)	Sieving requirements: Garbled standard permissible percentage of Triage/BBB[a]
Arabica plantation (washed)	Plantation A	6.65	90	2 of PB 2 of T
	B	6.00	75	2 of PB 3 of T
	C	5.50	75	—
	PB	—	—	3 of T
	Bulk	—	—	3 of T
Arabica cherry (unwashed)	Arabica Cherry AB	6.00	90	2 of PB 3 of T
	C	5.50	75	2 of BBB
	PB	—	—	2 of AB (flats) 3 of T
	Bulk	ungraded & ungarbled		18 of T
Robusta parchment (washed) and Cherry (unwashed)	Robusta Parchment AB & Robusta Cherry AB	6.00	90	2 of PB 3 of T
	Robusta Parchment C & Robusta Cherry C	5.50	75	2 of BBB
	Robusta Parchment PB & Robusta Cherry PB	—	—	2 of AB (flats) 3 of T
	Robusta Parchment Bulk & Robusta Cherry Bulk	ungraded & ungarbled		10 of T

[a] Triage: broken, withered, spotted, elephant, small, discoloured, malformed beans; pales; and pulper cuts. BBB, blacks, bits, browns.
Source: From Ref. 14.

3.7 Packing and Storage

Green coffee, after curing, has to be stored under suitable conditions to ensure proper appearance and saleability of the beans. The problems of appearance and color become more important, especially in plantation coffee. The main factors influencing color and appearance in green coffee are moisture content and time and temperature of storage. The color of the bean can be maintained satisfactorily at moisture level below 10%. The bluish green color of plantation coffee becomes bleached in coffee with moisture level above

Fig. 12 Electronic color sorter (Sortex). (Courtesy of Chamundi Coffee Curing Works, Mysore, India.)

10% after storage. The changes are accelerated in high-humidity conditions during monsoon season.

Coffee can absorb moisture from the humid atmosphere and become soft, lose its color, and become susceptible to microbial growth and insect infestation during storage. Although a large number of insects are potential pests of stored coffee, the most important is the beetle *Araecerus fasciculatus*. It is generally accepted that *Coffea arabica* with less than 13.5% and *C. robusta* with less than 12.0% moisture content are not likely to be attacked by this beetle (15). In Brazil, a dose of 3–5 to 15 aluminum phosphide tablets (Celphos) per ton with an exposure period of 24 h has been tried to control all stages of this insect (16, 17). A mixture of ethylene dibromide and methyl bromide has also been used successfully for fumigation of monsooned coffee. Other control measures adapted are spraying of vacant premises with pyrethrin and fumigating of empty gunny sacks.

Coffee should be stored inside a well constructed godown. Godowns should be concrete-floored, and uncured coffee and cured coffee should be stored separately. The storage temperature and humidity in the godowns are the major factors influencing the quality of the stored coffee beans. The temperature inside the godown should be maintained at less than 20°C and dry environmental humidity at 50%–70% without much fluctuation. During storage of green coffee, suitable dunnage (wooden platform) should be provided to all stalks of raw coffee, both parchment and cherry coffee, to prevent dampness of the floor from affecting the quality of coffee. Sufficient space should be left between the walls of the godown and the coffee stalks, between the roof and the top of the stalk; the required

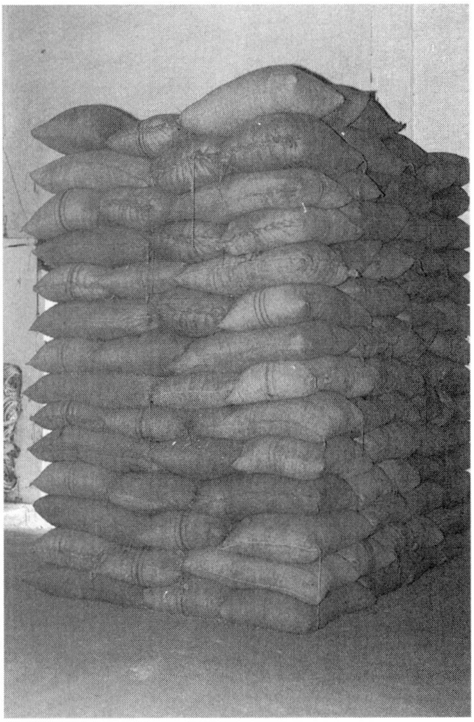

Fig. 13 Stacking of coffee bags. (Courtesy of Chamundi Coffee Curing Works, Mysore, India.)

moving space between the stalks should also be maintained. The godown should have proper ventilation and drainage and lighting facilities and should be fitted with fire extinguishers. Optimal stack height for storing of uncured coffee is 20–25 bags and for clean coffee 14 bags, a standard bag containing 60 kg of beans (18) (Fig. 13).

4 PRODUCTS

4.1 Roasted and Ground Coffee

The first step in the preparation of any consumable product from green coffee is roasting. Green coffee possesses very little or no aroma. The characteristic aroma is developed only through roasting. Coffee received at the roasting plant should be free of extraneous matter, but in practice heavy foreign materials, such as nails, coins, buttons, stones, and pebbles, and light contaminants, such as wood splinters, strings, chaff, and beans other than coffee, often appear and therefore must be cleaned well. Large bodies may be removed by screens, light bodies may be blown off by air blasting, and nails and other materials may be screened by magnetic separators.

Roasting of coffee is a process of exposing the coffee beans to a warming process that is sufficient to drive off the free and bound moisture; dry beans are heated to a temperature of 200°C–250°C. The degree of roast is critical to flavor development in the bean and determines many of the flavor characteristics of the brewed coffee. The relationships

Table 7 Different Types of Roasting

Temperature °C	Roast	Weight loss %	Cup quality
200	Very light	11–12	Acid taste, less aroma
220	Light	14–15	Better aroma, less astringent, acidic cup
230	Medium	16–20	Optimal quality
240	Dark	21–23	Preferred in Europe
250	Italian	24–25	Dark color, good taste

Source: From Ref. 19.

among final bean temperature, weight loss, and cup quality are summarized in Table 7. Many types of chemical and physical changes occur during roasting, including changes in color, size, and shape of the bean. Chemical reactions, such as oxidation, reduction, hydrolysis, polymerization, and decarboxylation, take place during roasting (19, 20).

4.1.1 Roasting Equipment

The conventional roasting equipment consists of a metal container in which green coffee is heated while it is continuously rotated. Heat may be supplied by conduction from hot metal surfaces or convection from hot air or more generally a mixture of both methods of heat transfer together. Earlier roasters made use of various types of frying pans and hand rotated cylinders. It is necessary that during the roasting process heat be applied quickly and uniformly and the beans be continuously stirred. Continuous roasters are used in large-scale processing plants because they have greater efficiency and ensure more uniformity than batch-type roasters. The horizontal rotating drum has either a solid or a perforated wall in which hot air from a furnace is passed through the tumbling green beans. Continuous roasters consist of either a perforated drum or a cylinder for roasting and subsequent cooling of the beans. Coffee beans are compartmentalized within the cylinder and separated from the cooling section by a central disk. The rotating drum principle is used in commercial roasters (Thermalo and Jubilee). The Gothot roaster (Fig. 14) uses a fixed vertical vessel with rotating paddles to assist heat transfer from hot air to green beans, providing short roasting time. Probat has marketed a batch-type roaster (Fig. 15) in which a rotating horizontal bowl is used for intimate contact of hot gases and beans. Hot air blast assisted by centrifugal force carries the seeds to the periphery; they then fall back to the center. After a short roasting time discharge occurs from the periphery of the bowl. Fluidized bed roasters are used for large-scale roasting of coffee beans. Both heating and cooling are achieved in the same vessel by a fluidized-solid contact technique. Fluidized roasters have better control of process parameters and deliver the product with uniform roasting. The spouted bed roaster (Fig. 16) is a variant of the fluidized bed roaster that has an advantage in large-scale roasting and tends to develop unstable fluidization. Pressure roasting is considered to be potentially more efficient and is generally considered to increase the acidity of the brew. Nitrogen is used as the pressurizing agent (9, 21, 22).

4.1.2 Grinding

After developing the coffee flavor by roasting, efficient extraction of the roasted coffee solubles and volatiles that contribute to coffee flavor and aroma is desirable. The solubles could be extracted from the whole roasted beans, but the yield would be low and flavor

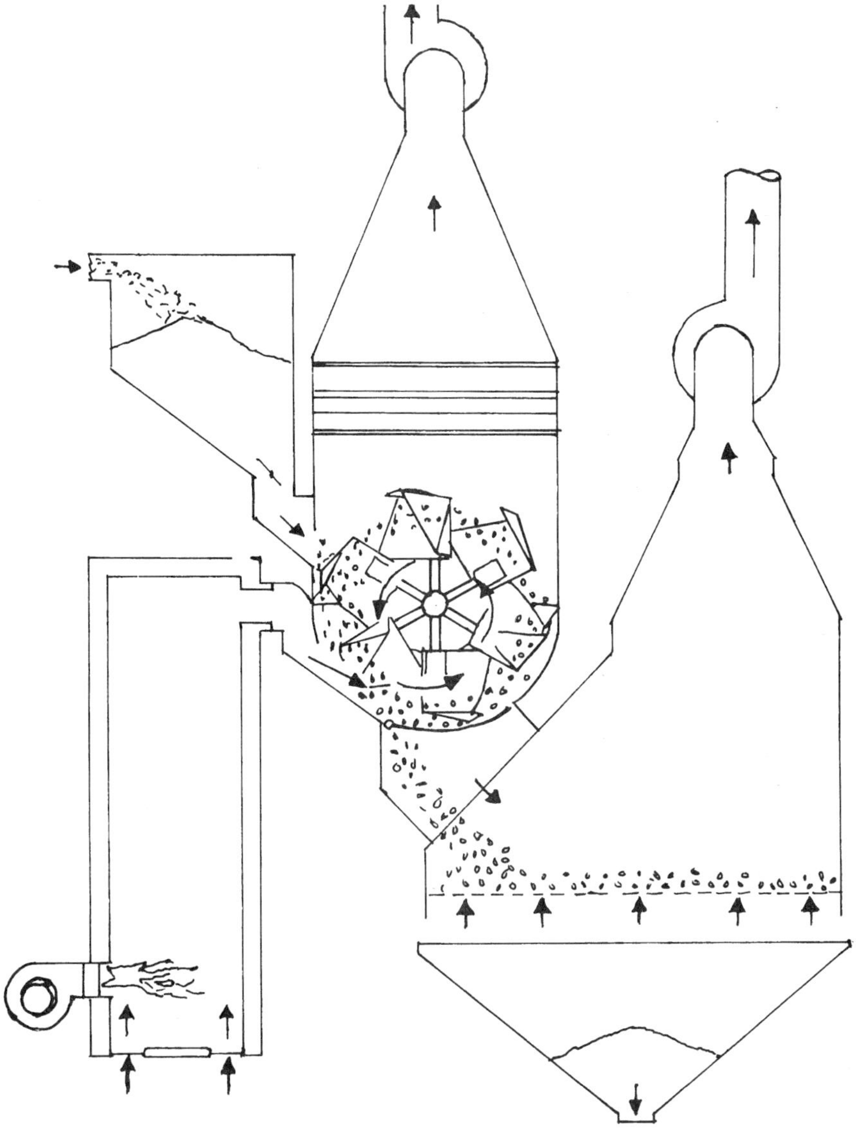

Fig. 14 The Gothot Rapido-Nova standard-type batch roaster, showing fixed roasting drum with paddles, furnace, and cooling car. (From Ref. 10. Reproduced with permission.)

would be poor. Extraction may be made to give a higher yield of solubles by breaking down the whole bean to smaller pieces. The roasted beans cannot be ground directly after roasting as they are too soft and would be crushed, flattened, and scarred. Therefore, water quenching is an essential step before grinding. When the beans are cooled, they become hard and brittle and can be ground. Various types of grinders are available for large-scale grinding of coffee beans. The extraction and keeping quality of the coffee powder is influenced by the size of the particle. There are three standards: regular grind, drip grind, and fine grind. Fine grind is suitable for steeping methods and vacuum coffee makers;

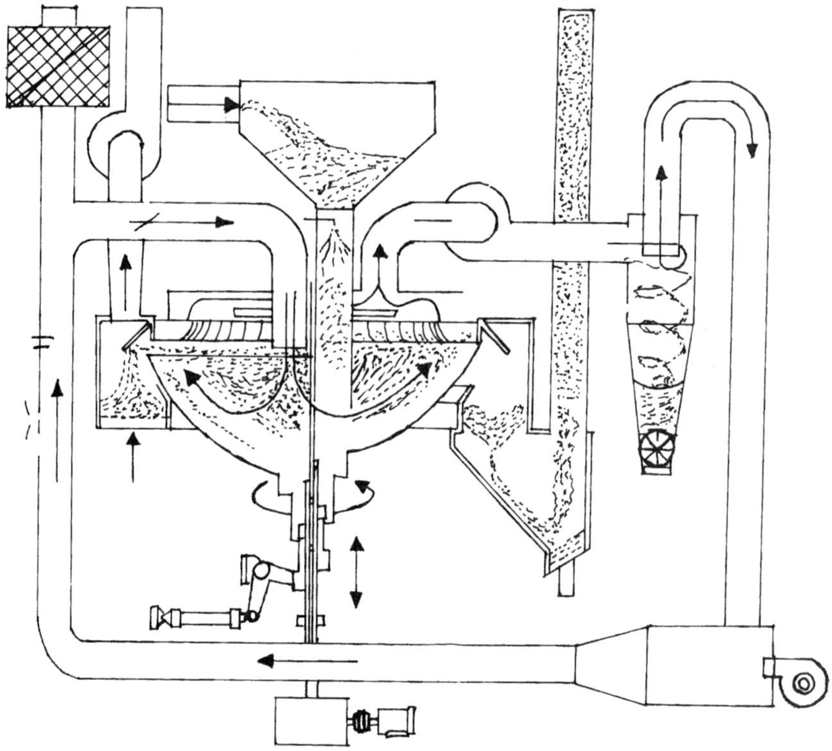

Fig. 15 Probat batch-type RZ, showing the rotating bowl. (Courtesy of Probat-Werke GmbH.)

coarse grinds are better for filter methods. Light roasted coffees are tenacious, pliable, and tough and do not break down as readily as hard, brittle dark roasted beans. Dark roasts, however, always yield more ''fines'' than lighter roasts (23).

4.1.3 Chemical Characteristics of Coffee

The chemical composition of green coffee depends mainly on the variety of coffee, though slight variations due to climatic conditions, agricultural practices, and processing and storage conditions are possible. The average approximate composition of green coffee and the compositional data for *C. arabica*, *C. robusta*, and instant coffee are given in Tables 8 and 9, respectively.

4.1.3.1 Moisture

Moisture is an important parameter, which affects the quality and storage behavior of coffee. It is generally recognized that green coffee should be allowed to reach a moisture content in excess of 12%, corresponding to a relative humidity of 70%. A high moisture content results in loss of green color and favors mold growth, flavor deterioration, and potential for mold toxin formation. A moisture content of 10.5% for plantation coffee and 11.0% for cherry coffee is generally recommended. The moisture content in green, roast, and instant coffees is determined by the air oven, vacuum oven, or Karl Fischer method. The simplest and most straightforward method of determining the moisture content in green coffee is the air oven method. A two-stage process involving heating at 130°C for

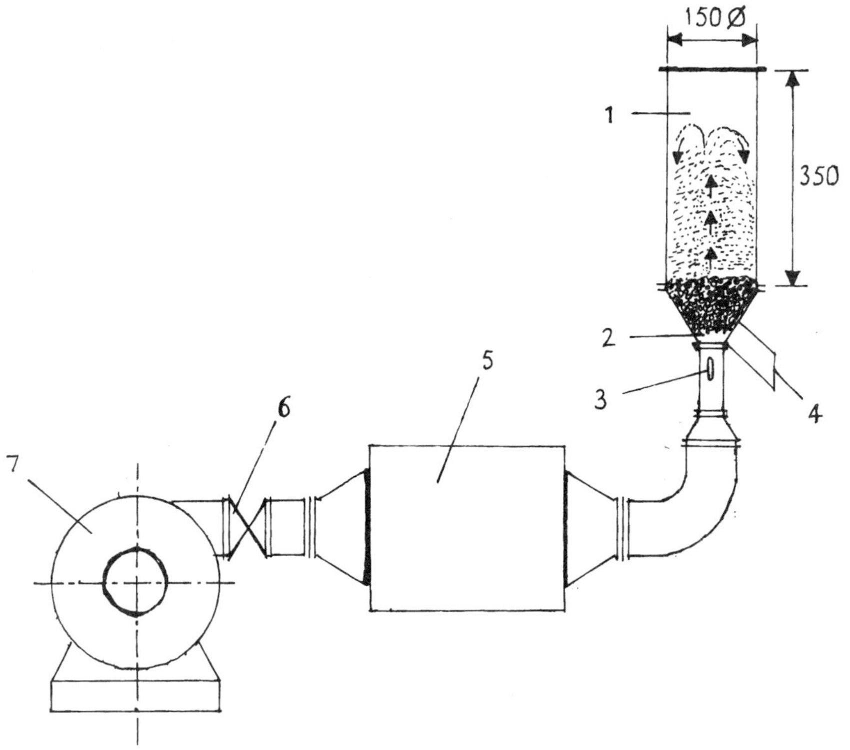

Fig. 16 Schematic diagram of a spouted bed roaster: (1) Glass column, (2) screen (25.4 mm), (3) temperature recorder, (4) discharge chute, (5) heater, (6) inlet air control valve, (7) blower. All dimensions are in millimeters. (From Ref. 21. Reproduced with permission from Elsevier Science.)

Table 8 Chemical Composition of Green Coffee

Component	Composition (%)
Reducing sugars	1.0
Sucrose	7.0
Pectin	3.0
Starch	10.0
Pentosan	5.0
Hemicellulose	15.0
Holocellulose (fiber)	18.0
Lignin	2.0
Oils	13.0
Protein	13.0
Ash	4.0
Chlorogenic acid	7.0
Other acids	1.0
Trigonelline	1.0
Caffeine	1.0

Source: From Ref. 38.

Table 9 Compositional Data (%) for Arabica, Robusta, and Instant Coffee

Component	Arabica		Robusta		Instant coffee powder
	Green	Roasted	Green	Roasted	
Mineral	3.0–4.2	3.5–4.5	4.0–4.5	4.6–5.0	9.0–10.0
Caffeine	0.9–1.2	~1.0	1.6–2.4	~2.0	4.5–5.1
Trigonelline	1.0–1.2	0.5–1.0	0.6–0.7	0.3–0.6	—
Lipids	12.0–18.0	14.5–20.0	9.0–13.0	11.0–16.0	1.5–1.6
Total chlorogenic acid	5.5–8.0	1.2–2.3	7.0–10.0	3.9–4.6	5.2–7.4
Aliphatic acid	1.5–2.0	1.0–1.5	1.5–2.0	1.0–1.5	—
Oligosaccharides	6.0–8.0	0.0–3.5	5.0–7.0	0.0–3.5	0.7–5.2
Total polysaccharides	50.0–55.0	24.0–39.0	37.0–47.0	—	~6.5
Amino acids	2.0	0.0	2.0	0.0	0.0
Proteins	11.0–13.0	13.0–15.0	11.0–13.0	13.0–15.0	16.0–21.0

6 h followed by a rest period in a desiccator for 15 h and then a second drying period of 4 h is employed. Normally green coffee contains about 10%–13% (w/w) water content. Since roasted and ground and soluble coffees have lower water content (~2%–5%), the air oven method cannot be employed because of its lack of accuracy. Alternatively, the Karl Fischer method, in which water in the sample is extracted with methanol prior to filtration, is often employed (24). Moisture meters are now available for quick measurement of water content in green coffee. The most commonly used moisture meter (Kappa) works on the principle of the dielectric constant. The instrument requires calibration using a reference method such as the air oven method. The moisture meters, once calibrated, are very handy and can be used to determine water content in coffee either in the plantations or in the curing works.

4.1.3.2 Nitrogenous Compounds

Coffee contains some nitrogenous compounds, such as caffeine, trigonelline, betaine, choline, and ammonia. Of these, caffeine is the most important with respect to its concentration as well as its effect on human physiological characteristics.

4.1.3.2.1 CAFFEINE. Caffeine is an alkaloid with a substituted purine ring system. The caffeine content of green coffee beans varies according to the species: *C. robusta* contains about 2.2%, *C. arabica* about 1.2%, and the hybrid *C. arabusta* about 1.72%. Environmental and agricultural factors appear to have minimal effect on caffeine control. Reports on the biosynthesis and degradation of caffeine in coffee are limited. The main biosynthesis route utilized the purine nucleotide for the formation of caffeine as shown in Fig. 17. The caffeine is degraded relatively slowly; demethylation yields theobromine, theophylline, and xanthine, which are metabolized to urea, as shown in Fig. 18 (25). The fatal dose of caffeine is 10 g. On roasting, caffeine is unchanged though some loss occurs through sublimation.

There are several analytical methods for the determination of caffeine content in coffee. Earlier methods were based on the extraction of caffeine into organic solvents such as chloroform, followed by purification on a column of celite or alumina and quantification by absorption measurement at 272 nm. At present the preferred methodology is

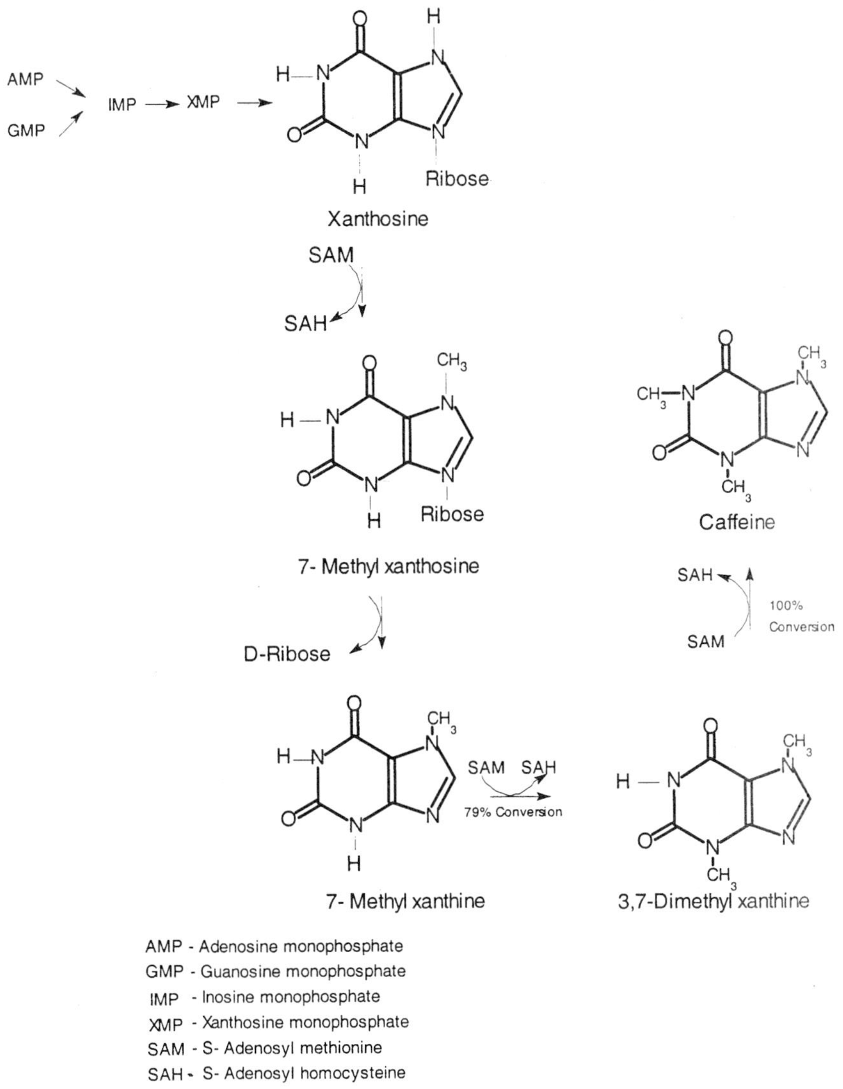

Fig. 17 Formation of caffeine. (From Ref. 25.)

the high-performance liquid chromatography (HPLC) method, which consists of preparation of an aqueous extract of coffee followed by precipitation or solid-phase extraction before injection into HPLC. Methods using reversed phase columns (C_{18}) with aqueous methanol as the mobile phase are also available. Detection is achieved by ultraviolet (UV) absorption at 272 nm (26, 27).

4.1.3.2.2 OTHER NITROGENOUS COMPOUNDS. The other nitrogenous compounds present in coffee can be classified into two groups: (a) those that are inherently stable at roasting temperature and (b) those that readily decompose, giving rise to volatile compounds. The components in the first group are ammonia, betaine, and choline present in trace amounts. The second group of compounds consists mainly of trigonelline and seroto-

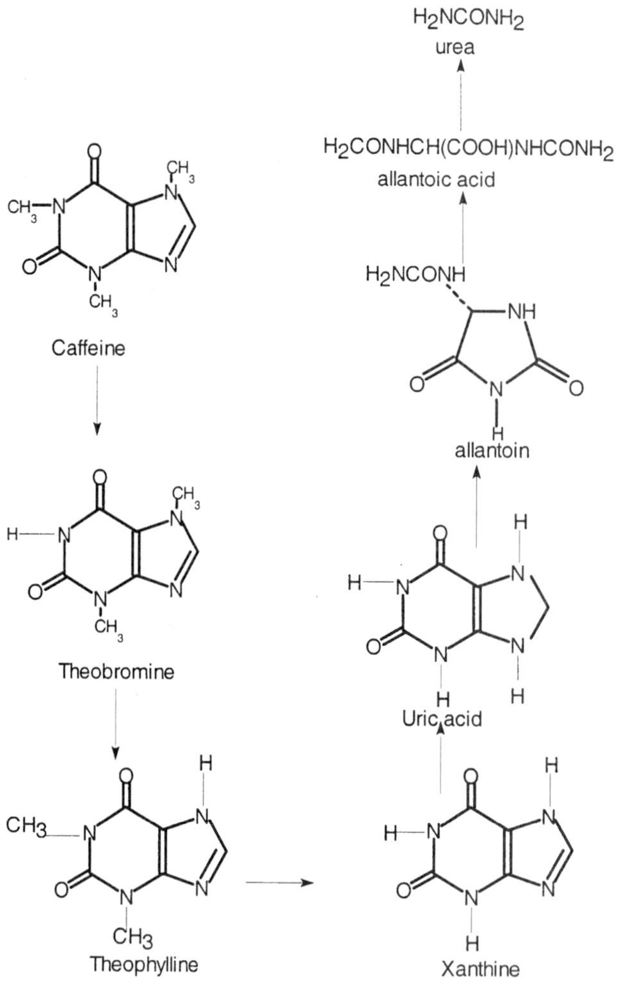

Fig. 18 Degradation of caffeine. (From Ref. 25.)

nin amides. The level of trigonelline (Fig. 19) present in coffee depends on the species: *C. arabica* coffee contains about 1% and *C. robusta* about 0.7%.

4.1.3.3 Chlorogenic Acid

Chlorogenic acid is another compound present in coffee; it is considered the second most important component and its level is generally analyzed. Levels of chlorogenic acid appear to be dependent on species and are unaffected by differences in agronomic practice or

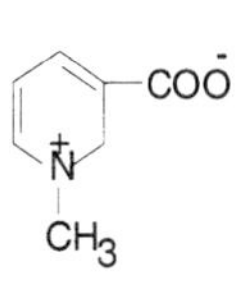

Fig. 19 Trigonelline. (From Ref. 28.)

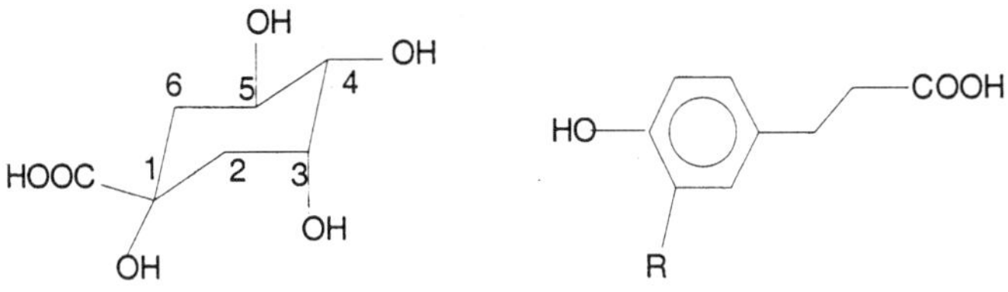

A: Quinic Acid　　B: E- Cinnamic Acids

R -H p-Coumaric Acid
R -OH Caffeic Acid
R -OCH_3 Ferulic Acid

Fig. 20 Chlorogenic acids. (From Ref. 28.)

method of processing. The level of chlorogenic acid in *Coffea arabica* varies from 5% to 7.5% and in *C. robusta* from 7.0% to 10.5%. Chlorogenic acid (Fig. 20) comprises a group of compounds: (a) caffeoyl quinic acid, (b) dicaffeoyl quinic acids, (c) coumaroyl quinic acid, (d) feruloyl quinic acids, (e) caffeoyl feruloyl quinic acids, (f) feruloyl caffeoyl quinic acids. The astringency of coffee is contributed by chlorogenic acid. It is generally agreed that high chlorogenic acid level indicates lower quality. These compounds suffer heavy losses during roasting; the degree of loss depends on the type of roasting (28).

Analytical methods can broadly be grouped into two kinds: spectrometric (colorimetric) methods, which determine total chlorogenic acid content, and chromatographic methods, which allow separation and quantification of individual isomers. A typical colorimetric method is based on the formation of a molybdenum-dihydroxy phenol complex (29), but quantification is complicated by the different molar absorbances of the complexes formed from various chlorogenic acids. In another method, chlorogenic acid is estimated by UV spectrometry before and after lead acetate treatment of coffee extract, followed by measurement of the absorbance at 325 nm (30). The preferred chromatographic method is based on HPLC (31), wherein a good degree of selectivity of detection is achieved by using UV absorption at 320 nm.

4.1.3.4 Coffee Oil

Lipids constitute a major component of green coffee, varying from 14% to 16% in *C. arabica* and from 9% to 13% in *C. robusta*. The lipid component of green coffee bean comprises coffee oil, which is primarily present in endosperm, and coffee wax, which is present on the outer layer. Coffee oil contains over 5% of unsaponifiable material and 46% linoleic acid. Saturated and unsaturated acids of the C_{20}, C_{22}, C_{24} series are also present in small concentrations. Coffee oil is a liquid at room temperature but settles out fatty acid crystals during storage. It contains an excessive amount of unusual unsaponifiable matter, the presence of which makes the oil unfit for most uses. The physical properties of coffee oil are similar to those of other vegetable oils (32). Coffee oil has a specific gravity of 0.9440–0.9450, a refractive index of 1.468–1.469, and viscosity with temperature slightly higher than soya bean or linseed oil. Coffee beans are coated with a polar wax that consists of fatty esters of 5-hydroxytryptamine that are known to be mucosal irritants.

Table 10 Formation of Coffee Volatiles from Nonvolatiles in Green Coffee During Roasting

Green beans	Roast coffee
Lipids, Fatty acid	Alphatic hydrocarbons
Higher terpenoids	Monoterpenoids
Lignin	Phenolic compounds
Starch, Sugars	Acids, Aldehydes, Ketones, Sulfurous compounds
Peptides, Amino acids	Ketones
Trigonelline	Nitrogenous compounds

4.1.3.5 Coffee Aroma and Volatile Compounds

The volatile compounds of coffee are largely responsible for the aroma. The green bean does not possess any appealing flavor and its water infusion is unpalatable. The desirable taste and aroma are mainly formed during roasting. Roasted coffee contains more than 800 volatile compounds. The important precursors are amino acids, sugars, and chlorogenic acids (Table 10). There can also be significant differences in the volatile composition depending on coffee species, conditions during cultivation and harvesting, and method of processing. The search for individual key flavor compounds responsible for overall flavor of coffee has not yielded good results. However, only a relatively small number may make a significant contribution to flavor (19, 33, 34). Some of the volatile compounds responsible for the coffee aroma are listed in Table 11.

Furfuryl mercaptan is considered the most important aroma compound in coffee; at high dilution it emits a smell resembling that of coffee. Kahweofuran also has an aroma associated with roasted coffee that becomes sulfurous at higher concentrations; dimethyl sulfide is reported to have an effect on mild coffee.

4.1.4 Changes During Roasting

The important changes that take place during roasting are loss of moisture, loss of organic matter and production of CO_2, swelling of bean and consequent changes in density of the bean, decrease in the breaking strength of the bean, caramelization of sugar and other constituents with consequent changes in color, formation of typical aroma compounds, decrease in the tanninlike constituents and sugars, increase in water-soluble matter, and formation of niacin and increase in its content during roasting. Roasting results in loss of weight of the bean. Roasting produces a large amount of CO_2, which under high pressure inside the bean bloats it, with a corresponding reduction in the specific gravity from 1.2–

Table 11 Aroma Volatiles of Coffee

Class of compound		No. of compounds	Class of compound	No. of compounds
Hydrocarbon:	Aliphatic	35	Pyrazine	70
	Aromatic	32	Quinoxalines	11
Alcohols:	Aliphatic	16	Furans	84
	Aromatic	2	Pyrones	3
Aldehydes:	Aliphatic	16	Oxazole	28
	Aromatic	9	Thiols	6
Ketones:	Aliphatic	57	Sulfides	15
	Aromatic	5	Disulfides	10
Acids		19	Trisulfides	3
Esters		29	Thioethers	2
Lactones		9	Thiophenes	27
Anhydrides		3	Dithiolanes	2
Amines		11	Thiozoles	27
Imide		1	Acetals	1
Pyrroles		66	Nitriles	2
Indole		4	Oximes	1
Pyridines		11	Phenols	39
Quinoline		4		

Source: From Ref. 19.

1.3 to 0.6–0.8. This results in porous structure and reduction in the breaking strength of the roasted bean. The color of the bean changes from gray green to light brown, dark brown, or almost black, depending on the type of roast. The pH of the roasted coffee brew falls down to 5.5–5.0 from pH 6.0 of the green beans. This reaction is mainly due to the formation of volatile organic acids. In general light roasts produce more acid per cup than dark roasts. Medium roasts have a cup pH of 5.0 and darker roasts have pH up to 5.3 (35).

Sugars and proteins break down to aldehydes, alcohols, and acids. Sucrose is the major sugar that suffers heavy loss during roasting. Proteins are denatured and are broken down to amino acids. The most significant change occurring during roasting is the formation of aroma compounds. Roasted whole beans retain the characteristic aroma for about 1 week under normal atmospheric conditions. This is mainly due to carbon dioxide buildup inside the bean, which provides an inert atmosphere.

4.1.5 Brewing of Coffee

The extraction of soluble solids from roasted and ground (R&G) coffee by using hot water is called the *brewing process*. The quality of the beverage depends, to a great extent, on the brewing process, viz., the type of equipment used, quality and quantity of water used, yield of soluble solids, and concentration of the brew. There are several brewing practices: steeping, drip filter, vacuum coffee maker, and percolator. In steeping, coffee powder is mixed with boiling water and kept covered with a lid for 5–10 min and the brew is decanted. The drip filter is the most common method employed for preparing coffee beverages, R&G coffee (coarse powder) is placed in a steel vessel with a perforated bottom, boiling water is added, and the brew is collected in the vessel below. The efficiency of

Table 12 Chemical Composition per Cup

Total soluble solids in 6.7 fl oz	Roasted and ground coffee (1.379 g)
Caffeine	0.085
Chlorogenic acid	0.190
Reducing sugars	0.015
Other carbohydrates	0.205
Peptides	0.062
Potassium	0.103
Other minerals	0.140
Acids	0.178
Trigonelline	0.053
Volatiles	<0.0001

Source: From Ref. 36.

extraction is improved by wetting the powder with half its volume of water. A vacuum coffee maker consists of two glass bowls, one above the other. A cloth or glass filter is provided in the top bowl. Coffee powder is added to the top bowl and water to the bottom, which is heated. The water is forced by steam to the top bowl. Once the source of heat is removed, a vacuum created in the lower bowl drains the brew. A percolator consists of a basket, which holds the powder, supported by a hollow tube, through which hot water rises and steeps through the powder. The ground coffee is subjected to continuous contact with hot water (36, 37).

The yield of solubles in the brew depends on the brewing conditions and the powder-to-water ratio. Generally, a ratio of 1:20 is used for domestic and restaurant brewing. This results in a brew of 1.1% to 1.3% concentration, with a yield of solubles of 20%–21%. The composition of the soluble solids is given in Table 12.

4.1.6 Cup Tasting of Coffee

Coffee products are evaluated by trained and experienced tasters who have an extensive vocabulary to describe the desirable and undesirable attributes of the beverages—*sweet, salty, acidic, sour, bitter, balanced, flat, stale, rancid, astringent, metallic, burnt*, etc. Off-flavors are described by several less well-defined terms: *bricky, cereal, chemical, earthy, grassy, green, harsh, onion, oxidized, papery, unclean, wood*, etc. Some of the important terms are discussed in the following sections (38).

4.1.6.1 Acidity and Sourness

Acidity is a desirable attribute, whereas sourness is undesirable, although the layperson considers the terms synonymous. The sour note is associated with a mixture of acids, alcohols, and esters produced by microbial fermentation. Acidity is associated with protons. Wet processed high grown *C. arabica* produces the most acid beverage; dry processed *C. robusta* the least, and dry processed *C. arabica* intermediate, for a given roast and color. Acidity is considered an important characteristic of medium roasts: high acidity provides better quality and more intense aroma to the beverage. It is generally agreed that pH 4.9 to 5.2 is the ideal range for coffee beverages. Under ideal roasting conditions, roasted *C. arabica* yields a brew of the beverage in that pH range, and *C. robusta* yields a less acid brew of pH 5.0–5.8.

4.1.6.2 Bitterness

An element of bitterness is desirable in coffee. Bitterness has been attributed to caffeine, but even decaffeinated coffees have been found to possess profound bitterness. It appears that other heterocyclics also contribute to bitterness.

4.1.6.3 Astringency

Astringency is not a primary taste. Many astringent molecules are bitter, and these sensations may be confused; they are distinguished by expert tasters. Caffeoyl quinic acids, dicaffeoyl quinic acid, and caffeic acid are the likely astringent components of coffee.

4.1.6.4 Staling

Staling refers to the deterioration of taste and odor on storage of coffee powder. The roasted and ground coffee preserves its aroma for up to 6 months under inert conditions at low temperatures. The entrapped CO_2 is instrumental in achieving this storage life. During storage, many volatiles decline in concentration as a result of volatility and oxidation. Similarly many new compounds are formed as a result of oxidation and other interactions.

4.2 Soluble Coffee

Soluble coffee is an important product, second only to R&G coffee. It is prepared by dehydrating the coffee brew and is instantly soluble in water (Fig. 21). The product is inferior in taste and aroma to the freshly prepared brew. As a coffee product, this has gained popularity in most countries.

The processing for soluble coffee consists mainly of two steps, viz., extraction of

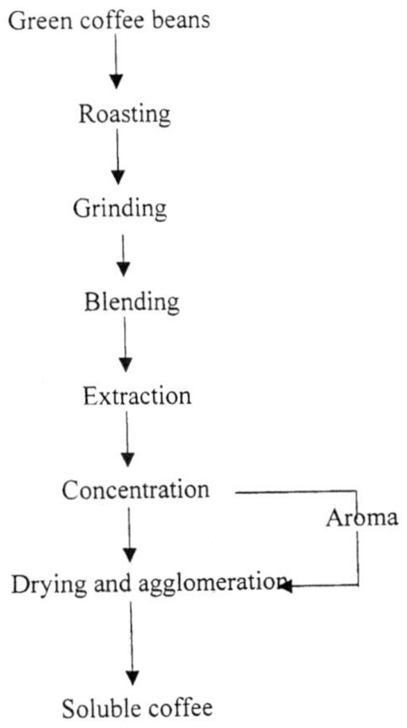

Fig. 21 Conversion of green beans into soluble coffee.

the brew and drying. The extraction of R&G coffee by using domestic brewing devices yields very dilute solutions with about 1% solids concentration. The solubles represent 15%–20% of the roasted powder, depending on factors such as extracting device, powder-to-water ratio, and time and temperature of extraction. In instant coffee manufacture, it is necessary to get a brew of much higher concentration and also a higher yield of solubles. On an industrial scale extraction is carried out by slurry and percolation methods (39, 40).

4.2.1 Slurry Method

In slurry extraction, coffee powder is taken in large tanks and heated with water under agitation. The solids are separated from the brew by using a centrifuge. In this type of extraction, it is possible to use finely ground coffee. However, large-scale slurry processing is more expensive than percolator processing.

4.2.2 Percolation Method

The extraction is carried out by passing hot water through a bed of coffee powder packed in cylindrical vessels, called *percolators*. The size of the grind is important in large-scale extraction. Finer grinds give better extraction but may result in high back pressure. Excessive coarseness of powder leads to decreased efficiency. The R&G coffee yields approximately 25% solubles under normal conditions. Higher extractives up to 50% can be achieved by use of higher temperature and pressure, when starch, proteins, and other insoluble carbohydrates are solubilized by hydrolysis. Percolation consists of three distinct processes, viz., wetting, extraction, and hydrolysis.

4.2.3 Wetting

The R&G coffee is practically free of moisture. It can hold twice its weight of water. When coffee powder is treated with hot water, it absorbs water, the particles swell, occluded gases like CO_2 and air are displaced, and the powder is ready for extraction. Generally, wetting is carried out by coffee brew, instead of water. Lower concentration in the brew and higher temperature effect better wetting. Some heat is liberated during wetting.

4.2.4 Extraction

Extraction of solubles occurs rapidly, after the particles are wetted and saturated with the extract. The rate of extraction is influenced by factors such as grind size and uniformity, column change, and temperature. A solubles yield of about 25% is obtained under normal conditions.

4.2.5 Hydrolysis

Hydrolysis is a process of solubilizing part of the carbohydrates and proteins that are insoluble under normal conditions. Solubles yield of up to 50% is achieved by the use of higher temperatures (160°C–170°C). Solubilization of the insoluble solids by acid hydrolysis is also reported. The coffee powder, after extraction of about 25% solubles, is treated with phosphoric acid to lower the pH to about 1.5 to 2.0. Hydrolysis is continued for 1 h with agitation at 100°C. The grounds are then centrifuged and the infusion neutralized with calcium oxide. The extract obtained by hydrolysis imparts poor taste and flavor characteristics to the brew and the instant coffee.

4.2.6 Extraction Equipment

The simplest method of extraction is the batch process, wherein ground coffee is mixed with boiling water, drained off the extract, and more water added to leach out the extractives. This results in a yield of about 25% solubles, but the extract thus obtained is dilute. The solubles concentration in the extract is improved by a countercurrent procedure wherein a series of extractors are used. The extract leaving the first percolator is passed through the second. This is continued until the required concentration is achieved. The hot water enters the percolator with the most spent coffee grounds (with least solubles left) and moves through the percolators of less spent grounds. The extract leaves the freshest percolator with a solubles concentration of about 35%. As each percolator is exhausted, it is isolated from the battery, and the spent grounds are discharged, refilled with fresh coffee, and replaced ''on stream'' (24).

The commercial percolation equipment consists of a battery of five to eight columns, usually made of stainless steel. The design of the columns—size, shape, and number—is based on the production rate, solubles yield, and concentration required. The columns are generally of 1 to 4 ft (0.3 to 1.2 m) diameter and 6 to 20 ft (1.8 to 6.1 m) tall. The extractor is equipped with line refractometers to check solubles concentration with a provision for intercolumn heating. A feed water temperature of 160°C–180°C and pressure ranging from 10.7 to 14.4 bar are used. Particle size of powder should be neither too fine nor too coarse. Approximate size specifications are 20% to be retained on British sieve mesh size 8 and 5% to pass through 20 mesh. After leaving the extractor, the coffee extract is clarified while hot. This is achieved by passing through strainers, but more commonly by centrifuging. The clarified extract is then weighed and kept in storage tanks.

4.2.7 Drying of Coffee Extract

The drying of coffee extract is achieved by one of two methods, spray drying or freeze drying. The extract is concentrated to 35%–45% solids content by employing film/plate/centritherm. Generally, the extract of the fresh powder that has most of the aroma is kept separate and the extracts from the comparatively spent grounds are concentrated and added back to the aroma-rich extract.

4.2.7.1 Spray Drying

In the spray drying process, the atomized coffee extract (~35% solubles) is subjected to flash evaporation by using air at a relatively higher temperature (200°C–260°C). Spray drying involves three steps: dispersion of the extract as a fine spray, exposure of the atomized particles to a current of hot air, and separation of air from the dried product. The atomization can be effected by either of two methods: flowing of extract through a rotating disk or passing through a pressure nozzle. Two types of spray dryers are available, cocurrent and countercurrent, depending on the air and the feed flow. However, cocurrent dryers are by far more popular. Similarly, spray dryers of the vertical type are more common than the horizontal type. The atomized particles on exposure to air at high temperature lose water almost instantaneously. The exposure of hot air is short, a residence time of 10–30 s, and little heat damage is caused. During drying, a protective film is formed around the droplets. This film is more permeable to water vapor than volatiles, and hence there is little loss of aroma during spray drying. Under carefully controlled conditions, flavor retention up to 85%–90% is possible. The dehydrated powder carried by a cyclone separator falls into the receiver (41, 42).

The desirable properties of instant coffee are low moisture content, instant solubility, dark reddish color, beady structure, and free flow nature. These are achieved by the careful control of spray drying conditions and coffee brew properties. For example, increased solubles concentration, lower inlet temperature, and higher viscosity of the brew result in thick walled particles, dark color, better solubility, and higher bulk density. On the other hand, the presence of gas (CO_2) in the extract reduces the powder bulk density, and therefore also the increased inlet air temperature.

4.2.7.2 Freeze Drying of Coffee

Freeze drying is achieved by freezing of the coffee extract and sublimation of ice under reduced pressure. The major advantage is that the brew is not subjected to higher temperatures during drying, thus causing minimal heat damage to the aroma volatiles. The limitations of this process are the poor capacities of freeze dryers and the high cost of power in some countries. Freeze drying is generally more time consuming and essentially a batch process (43).

4.2.8 Aromatization of Soluble Coffee Powder

The aroma volatiles are lost during different stages of instant coffee manufacture such as roasting, grinding, extraction, concentration, and drying. Therefore, attempts have been made to recover the aroma and add it back to the dry powder. The aroma recovery is effected by condensing or absorbing of the volatiles from the roaster/grinder gases, gas stripping of the powder, steam distillation, solvent extraction of powder/CO_2 extraction of the powder, and condensation of percolator vent gases. The isolated aroma components can be swept through the powder by means of an inert gas or can be sprayed against the inside of a falling powder curtain.

4.2.9 Agglomeration Technique

The physical characteristics of the dried powder generally affect its functional properties. Fine particles tend to lump while dissolving in water. Bigger particles with beady structure possess dark attractive color and better solubility. New and improved spray dryers that yield powders of desired properties such as bulk density, appearance, and shape have been developed, but the production costs in such cases may be prohibitive. As an alternative, the method of agglomeration is generally employed. The process consists of drawing the fine particles together to form bigger aggregates. This is achieved by partially solubilizing the surface of the particles by water or steam. The wet particles adhere to each other, forming bigger granules. It is necessary that agglomeration not result in lumping or caking of the powder. Water or steam is introduced as fine spray onto the air-conveyed or tumbled powder, thereby subjecting it to the wetting action. Some spray dryers are equipped with multiple nozzles; at different stages of drying the particles are in contact for adhesion. Sometimes it is necessary to redry the particles after agglomeration for removal of the excess moisture (44, 45).

4.3 Other Convenience Products

The most important convenience product of coffee is the soluble coffee. Soluble coffee is also prepared from coffee-chicory mixtures.

4.3.1 Liquid Coffee

Liquid coffee is made of coffee extract. It is a concentrated, sweet, and ready to drink coffee sold in plastic bottles or paper cartons. Iced coffee made from coffee syrup is a popular beverage in Japan; canned coffee was commercialized by UCC Ueshima Coffee in 1970 in Osaka. The development of more sophisticated vending machines with provision for heating in winter and refrigeration in summer has stimulated the sales of canned coffee. The remarkable success of canned coffee in Japan prompted the manufacturer (Tadao Ueshima) to introduce the product in glass bottles (46–48).

4.3.2 Coffee Brew Concentrate

Yet another convenience product prepared from coffee brew is the coffee concentrate, wherein the aroma-rich extract is bottled under inert atmosphere with or without preservatives. The major expensive step of drying the brew is eliminated in this product. Since the brew is not subjected to any further heat treatment, the product is much superior in aroma quality to the soluble coffee. The product retains its fresh flavor for periods up to 6 months.

4.3.3 Additives Used in Coffee

Several additives or substitutes are used in coffee in order to increase the brew strength. Cyclodextrin used as an additive couples with undesirable taste components and imparts smoothness to the flavor of the beverage. Cyclodextrins are heat stable, and their addition to the concentrated extract does not alter the characteristics of coffee. Chelating agents such as phytic acid and its alkali metal salts are used to prevent foam and scum formation in the reconstituted instant coffee beverage. Substitutes such as chicory, barley, malt, and rye are used to increase the brew strength (49–51).

4.3.4 Ready Mix Coffee Beverage

Ready mix coffee contains all the ingredients of a coffee beverage in a dry form. Thus soluble coffee, milk powder, and sugar are mixed in the proportion 1:5:10. About 25 g of the mix is needed for a cup (8 oz/240 ml) of coffee. It is necessary that the ingredients be thoroughly mixed to get a homogenous blend. To achieve this, the particle size should be the same for all the ingredients. The product needs special packaging considerations to protect it from both air and moisture. This can be disposed of in vending machines and the shelf life of the product (in the machine) is about 10 days. For consumers vacuum packaging or gas packaging may be adopted, to extend the shelf life to reasonable periods.

The ready mix coffee, in spite of careful blending, may tend to disperse individually, indicating a definite lack of homogeneity, resulting in a beverage of poorer taste, aroma, and mouth feel. This defect is overcome by agglomeration of coffee with sugar and milk solids. The process is meant to yield a product wherein the composite particles have a discrete, porous, spongelike texture, with low bulk density. The composition of the blend is not altered under normal handling and shipping conditions, thus ensuring uniform quality of the product. Soluble coffee powder is thoroughly mixed with powdered sugar and milk powder. The mixture is finely ground, moistened, and agglomerated to obtain a product that has half the bulk density of the ground blend. The product has instant solubility and no tendency to separation of the individual ingredients (52).

4.3.5 Monsooned Coffee

A specialty Indian coffee is the monsooned coffee, which is in demand in Europe, especially in the Scandinavian countries. Monsooned coffee is prepared on the west coast of India during monsoon season. This process uses only A grades of whole crop cherry or AB grades of *C. arabica* and *C. robusta* coffee that have been exposed to a humid atmosphere that causes them to absorb moisture up to 15%–16%. The process of monsooning is as follows: green coffee is spread in well ventilated brick floored godowns to a thickness of 4–6 in (10–15 cm) during the monsoon season. The coffee beans are raked periodically and exposed to humid atmosphere for periods ranging from 6 to 8 weeks. Beans are then packed in loosely woven gunny sacks and stacked, bulked, and rebagged. This results in swelling to one and half times the normal size of cherry beans and a change in color to pale white or golden/light brown. These swollen beans are then polished through hullers, graded, and garbled by sorter. Through monsooning the dry processed coffee acquires a special natural mellow flavor. The grades that are specially sought in the international market are Monsooned Arabica/Monsooned AA and Monsooned Robusta AA (53, 54).

4.3.6 Coffee Paste

Coffee paste is a ready-to-serve coffee beverage in highly concentrated form. The liquid ingredients, milk and coffee brew, are separately concentrated and mixed with sugar thoroughly to obtain the product in a paste form. During extraction of coffee powder, the aroma-rich initial extract is collected separately and mixed with the concentrated milk and brew in the end. This product is best canned and stored under refrigerated conditions. Consumer (unit) packs can be made in aluminum foil laminated pouches.

4.3.7 Fortified Coffee Beverages

Fortified coffee beverages are prepared by fortifying the soluble coffee with sugars, hydrolyzed cereal solids, proteins, vitamins, minerals, and other ingredients, which are normally added to the extract before freeze drying. Iron-fortified soluble coffee is prepared by precipitating polyhydroxyphenols and polyhydroxyphenol polysaccharides from the coffee extract by holding at 35° to 70°F (2°C to 20°C) for sufficient time and adding an assimilable elemental iron at the rate of 0.01% to 1% by weight of coffee solids (55).

4.3.8 Coffee Brewing Bag

Filter paper infusion bags have been developed for brewing coffee. A disposable, porous, sealable paper pouch is made and precharged with coffee grounds and sealed. Coffee brewed from these bags has better aroma and flavor. The roasted and ground coffee is given a treatment with surfactants such as dimethyl polysiloxane (DMPS) to increase the extract yield. DMPS is sprayed to a level of 20 to 1600 ppm on coffee prior to packing in the infusion bag (56).

4.3.9 Decaffeinated Coffee

It has already been mentioned that caffeine present in coffee is responsible for most of the physiological effects of the beverage. Caffeine is not tolerated by some people. It can cause excessive influence on the central nervous system and respiratory system. The symptoms are restlessness, excitement, and insomnia. Roasted and ground coffee contains caffeine ranging from 1% to 2%, depending on the variety. The caffeine content in a cup of coffee ranges from 50 to 150 mg. The removal of caffeine from coffee minimizes the health hazards (adverse physiological effects), at the same time retaining the desirable attributes of coffee beverage. With the increased emphasis on health and the known ill effects of caffeine, more people are switching over to beverages with less or no caffeine. The share of decaffeinated coffee in the U.S. (coffee) market has risen above 20% and is still on the increase. Kaffee HAG in Germany was the first to commercialize the process of decaffeination at the beginning of the 20th century. Since then, there have been several patents on decaffeination of coffee. Decaffeination is carried out on green coffee beans before roasting. This minimizes the loss or alteration of flavor that develops only during roasting.

Green coffee beans are hard and nonporous in nature, and it is difficult to remove caffeine by extraction. Hence, the beans are first steamed to cause swelling and softening of the tissue. The general procedure of decaffeination consists of steaming the green beans and extracting with a solvent trichloro ethylene, chloroform, ethylacetate, water, or liquid carbon dioxide (Fig. 22). Dichloromethane is preferred as a solvent in recent times by virtue of its low boiling point and higher caffeine solubility. However, in view of the suspected carcinogenic properties of chlorinated hydrocarbons, the trend now is for use of safer solvents such as ethyl acetate, water, and supercritical carbon dioxide.

4.3.9.1 Decaffeination Using Organic Solvents

The extraction of caffeine is carried out by using a battery of columns, by the countercurrent method. Green coffee beans are steamed for about half an hour, thereby increasing the moisture to 40% by a prewetting step. The wetted beans are then extracted by using solvents such as dichloromethane and ethyl acetate (57, 58) at temperatures between 50°C and 120°C. After removal of caffeine (95°C–98°C) the residual solvent in the beans is removed by steam stripping and the wet beans are dried as quickly as possible. The crude caffeine is recovered by desolventizing the extract and purified by charcoal treatment

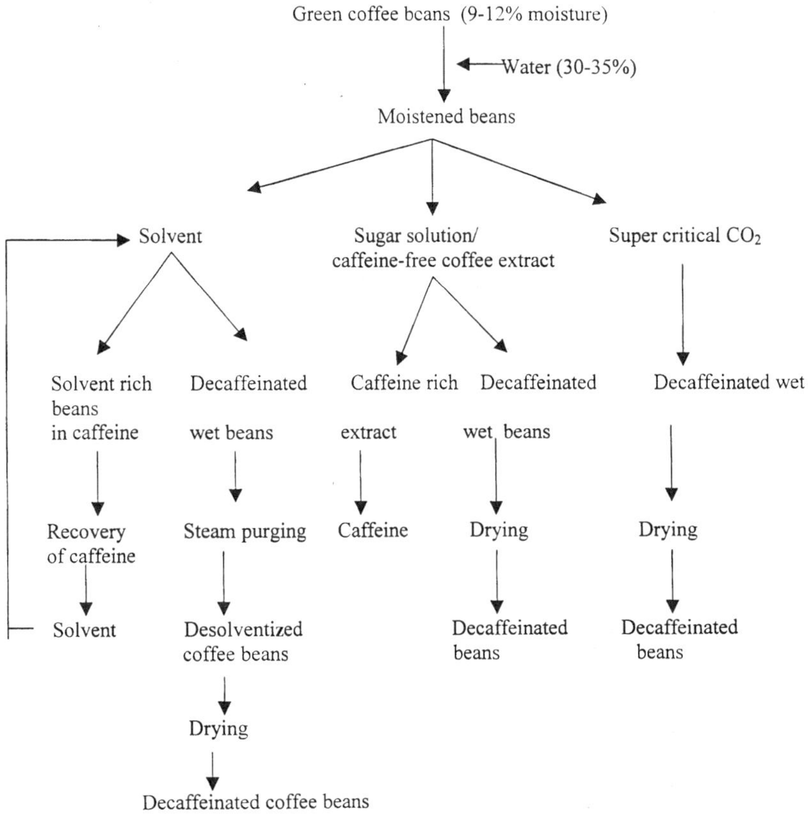

Fig. 22 Decaffeination of coffee beans.

followed by crystallization. Caffeine is a valuable by-product of the decaffeination process used in the food and pharmaceutical industries.

4.3.9.2 Water Decaffeination

In the water decaffeination process, water is used as the extracting solvent and extraction is carried out at about 90°C–95°C. The water used for extraction contains equilibrium quantities (~15%) of coffee solubles (other than caffeine) to prevent extraction of more solubles from green coffee. Decaffeinated wet coffee beans with about 53% moisture are dried. Caffeine is recovered from the aqueous solution by extraction by solvents such as dichloromethane. The aqueous extract, after caffeine recovery and complete removal of solvent, is recycled for extraction purposes. About 98% of the original caffeine is removed in 8 h of extraction. Water decaffeination is claimed to be more economical than the organic solvent method. Another advantage of the process is that the beans are not exposed to organic solvents and there is no problem of solvent residue in the product (59).

4.3.9.3 Decaffeination Employing Supercritical Carbon Dioxide

Another method of extraction of caffeine from coffee beans uses supercritical carbon dioxide. Under high pressure, the gas acts as a fluid and can be used as a solvent. A process has been commercialized by HAG-GF in Germany, using carbon dioxide as the extractant. The steamed and wetted coffee beans are subjected to carbon dioxide extraction at high pressures (120–180 bars) and temperatures ranging from 40°C to 80°C. The major advantage of this process is that it yields decaffeinated coffee of high quality. Moreover, supercritical carbon dioxide has a high selectivity for caffeine and the noncaffeine solids are not extracted. The caffeine extracted is also of high purity. This is a batch process involving the use of a high-pressure system and expensive equipment (60).

4.3.9.4 Decaffeinated Soluble Coffee

There have been several patents in recent years for decaffeination of roast coffee extracts. The major problem is removing caffeine without removing aroma and flavor components. In most of the processes, therefore, the initial steps are stripping and collection of aroma volatiles. The brew after aroma removal is subjected to solvent extraction. In some processes, first, the aroma volatiles are removed by steam distillation of the coffee powder then water extraction of the brew follows. Decaffeination can also be achieved by the use of ion exchange columns. However, the ion exchange resin reduces the pH of the brew to such an extent as to develop off-taste and induce sedimentation of solids on standing. These can be eliminated by neutralizing the decaffeinated extract to its original pH of 5.0–5.2. The extract is then spray dried in the usual way to yield the instant coffee powder. The major drawback of all these processes is that during the aroma recovery step there is a tendency to sedimentation of coffee solids, creating problems for the extraction of caffeine. The heat treatment may also affect the taste and flavor of the product. The solvent extraction of the coffee brew generally results in emulsification of solvent and extract, and elimination of emulsion is necessary for a successful process.

4.3.10 Health Coffees

Coffee brew has been reported to cause gastrointestinal irritation and other symptoms including emetic activity. This "indigestibility" of coffee has been an issue of interest, especially in countries such as Germany. This problem has resulted in the so-called health coffees, both roasted and ground and instant, wherein the suspected harmful components (mainly phenolic) are removed from the green coffee. Two processes are reported, dewaxing of green coffee and steaming. Dewaxing is achieved by solvent washing of the green coffee beans. The efficiency of the solvent washing process is generally monitored by analyzing the carboxy-5-hydroxytryptamides content. In the other process, the green coffee is subjected to a steaming operation that results in an overall reduction of the chlorogenic acid complex. Part of the chlorogenic acid is hydrolyzed and steam distilled, and the roasted coffee contains much lower amounts of phenolics than untreated coffee.

4.3.11 Carbonated Coffee

Carbonated coffee is a coffee flavored novel beverage containing coffee brew, sugar, acid, CO_2, preservatives, and other components. It is an amber colored, sparklingly clear carbonated beverage with the refreshing aroma of coffee. The general process consists of preparation of sugar syrup; mixing of coffee brew, acids, preservatives, etc.; dilution; and carbonation. It is best enjoyed as a chilled beverage. It is possible to replace the coffee brew with soluble coffee powder. Similarly, other compatible flavors, including chocolate, cardamom, and vanilla, may be added along with the coffee solubles. The product has a shelf life comparable to that of any other soft drink.

4.3.12 Espresso Coffee

Italian espresso coffee, a traditional Italian drink consumed in Europe, America, Japan, and other countries, was discovered after the Second World War. It contains 30–40 ml of thick concentrated brew prepared by using an espresso machine. Milk and a little cream and sugar are added to taste. Espresso can also include flavors such as chocolate, cardamom, and clove (61).

5 SUMMARY

Coffee is one of the most fascinating subjects for researchers throughout the world to dwell on and to generate new knowledge, novel products, and improved processes. Innumerable patents and voluminous literature on various aspects of coffee processing bear testimony to the sustained interest in the subject. The chemical composition of coffee has been the subject of investigation since the 1980s, Data on the chemical composition of different varieties of coffee produced in different parts of the world are well documented. Several methods have been standardized for the determination of two important constituents of coffee, namely, caffeine and chlorogenic acid.

A major problem faced by the coffee plantation is effluent released by the wet processing of coffee. Use of enzymes may substantially reduce the consumption of water and the processing time to produce parchment coffee.

Although coffee flavored with nut and berry flavors is popular in the West, Asian flavors should be tried. Some of the most compatible among spice flavors and citrus flavors may have a deeper impact on coffee consumers. Decaffeinated coffees are also catching up in the market, in the form of health drinks.

The aroma of coffee has undergone thorough investigation with modern instrumental methods that have identified about 800 compounds, perhaps the largest number reported in any food. Although certain compounds have been reported to have some effect on the coffee aroma, the real "character impact" compound(s) is yet to be established.

REFERENCES

1. M. N. Clifford and K. C. Willson, Coffee: Botany, Biochemistry and Production of Bean and Beverages, Croom Helm, London, 1985.
2. A. H. Varnam and J. P. Sutherland, Beverages: Technology, Chemistry and Microbiology, Chapman and Hall, London, 1994.
3. Indian Coffee, 61(7): 6 (1997).
4. Indian Coffee, 61(10): 26 (1997).
5. D. Schoenholt, Tea Coffee Trade J., 164(4): 42 (1992).
6. Indian Coffee, 61(1): 11 (1997).
7. C. H. J. Brando, Tea Coffee Trade J., 168(3): 180 (1996).
8. Wealth of India, Raw Materials: A Dictionary of Indian Raw Materials and Industrial Products, Vol. II, 288 (1950).
9. M. Sivetz and H. E. Foote, Coffee Processing Technology, Vol. I, AVI, Westport, Connecticut, 1963.
10. R. J. Clarke and R. Macrae, Coffee: Technology, Vol. II, Elsevier Applied Science, London and New York, 1985.
11. M. Sivetz and N. Desrosier, Coffee Technology, AVI, Westport, Connecticut, 1979.
12. Indian Coffee, 61(12): 3 (1997).
13. Commodity Handbook, Coffee: An Exporters Guide, International Trade Centre, UNCTAD/GATT, Geneva, 1992.
14. Indian Coffee, 46(1, 2): 3 (1982).
15. J. R. Rangaswamy, Indian Coffee, 53(6): 5 (1989).
16. D. Puzzi and A. Orlando, Biologico Brazil, 29: 127 (1963).
17. J. M. Countinho, D. Puzzi, and A. Orlando, Biologico Brazil, 27: 271 (1961).
18. N. Balasubramanyam, A. R. Indiramma, T. Rajasekaran, K. O. Abraham, and M. L. Shankaranarayana, Indian Coffee, 53(1): 5 (1989).
19. K. O. Abraham and M. L. Shankaranarayana, J. Coffee Res., 20(2): 93 (1990).
20. V. Subramanyan, D. S. Bhatia, C. P. Natarajan, R. Balakrishnan Nair, C. S. Viraktamath, and A. Balachandran. Indian Coffee, 24(1): 11 (1960).
21. V. D. Nagaraju, C. T. Murthy, K. Ramalakshmi, and P. N. Srinivasa Rao, J. Food Eng., 31: 263–270 (1997).
22. M. Sivetz, 14th International Scientific Colloquium on Coffee, ASIC, San Francisco, 313 (1991).
23. S. Sturdivant, Tea Coffee Trade J., 162(3): 32 (1990).
24. R. Macrae, R. K. Robinson, and M. J. Sadler, Encyclopaedia of Food Science, Food Technology and Nutrition, Vol. II, Academic Press, London, 1993.
25. G. R. Waler and T. Suzuki, 13th International Colloquium on Coffee, ASIC, Paipa, 351 (1989).
26. Official Methods of Analysis of AOAC International, 16th ed., II, 1998.
27. T. Kazi, 11th International Scientific Colloquium on Coffee, ASIC, Lome, 227 (1985).
28. B. Raghavan and K. Ramalakshmi, Indian Coffee, 52(11): 3 (1998).
29. P. Ribereau-Gayon, Plant Phenolics, Hafner, New York, 1972.
30. L. C. Weiss, Assoc. Offic. Agric. Chem. 40(2): 350 (1957).
31. K. J. Balayaya and M. N. Clifford, J. Food Sci. Technol., 32(2): 104 (1995).
32. N. A. Khan and J. B. Brown, J. Am. Oil. Chem. Soc., 30(10): 606 (1953).
33. C. P. Natarajan, R. Balakrishnan Nair, N. Gopalakrishna Rao, C. S. Viraktamath, D. S. Bhatia, and A. N. Sankaran, J. Sci. Ind. Res., 19(A): 32 (1960).
34. T. Shibamoto, 14th International Scientific Colloquium on Coffee, ASIC, San Francisco, 107 (1991).
35. K. O. Abraham and M. L. Shankaranarayana, Indian Coffee, 51(8): 6 (1987).
36. W. P. Clinton, 11th International Colloquium on Coffee, ASIC, Lome, 87 (1991).
37. K. Basavaraj and S. Pusphalatha, Indian Coffee, 51(10): 13 (1998).
38. K. O. Abraham, Guide on Food Products, Vol. II, Spelt Trade Publication, Bombay, 1992.

39. K. F. Sylla, 13th International Colloquium on Coffee, ASIC, Paipa, 219 (1989).
40. M. Sivetz, Tea Coffee Trade J., 145(3): 37 (1973).
41. K. Masters, Spray Drying, 2nd ed., George Godwin, London, 1976.
42. J. H. Perry, Chemical Engineers' Handbook, 4th ed., McGraw-Hill, New York, 1976.
43. M. Sivetz, Tea Coffee Trade J., 146(1): 74 (1974).
44. W. F. Purves, W. F. Lee, P. H. Davies, and W. J. Jeffrey, U.S. Patent 3 740 232 (1973).
45. K. E. Hansen and O. Hansen, U.S. Patent 3 966 979 (1976).
46. Indian Coffee 55(2): 15 (1991).
47. R. L. Colton, Tea Coffee Trade J., 155(7): 18 (1983).
48. Kazuo Karaswa, Tea Coffee Trade J., 161(5): 34 (1989).
49. Indian Coffee 55(10): 15 (1991).
50. N. Gainaris, U.S. Patent 3 908 033 (1976).
51. R. M. Hamilton and R. E. Heady, U.S. Patent 3 528 819 (1970).
52. N. Gopalakrishna Rao, C. P. Natarajan, and A. Balachandran, Indian Coffee, 34(1): 12 (1970).
53. L. Nagabhushana Rao, Indian Coffee, 53(3): 7 (1989).
54. K. J. Balayaya and M. N. Clifford, 16th International Colloquium on Coffee, ASIC, Kyoto, 316 (1995).
55. L. R. Monrea, F. J. Patrizo, and W. J. Einstman, U.S. Patent 3 933 838 (1976).
56. R. A. Kaplan, U.S. Patent 3 846 569 (1974).
57. L. R. Morrison and J. H. Phillips, U.S. Patent 4 486 453 (1984).
58. L. R. Morrison and J. H. Phillips, European Patent EP 0 114 426 B1 (1987).
59. S. Brown, Tea Coffee Trade J., 160(2): 12 (1988).
60. N. Gopalakrishna, Indian Coffee 56(9): 7 (1990).
61. K. Basavaraj, Indian Coffee, 61(7): 3 (1997).

25

Tea: An Appraisal of Processing Methods and Products

SRIKANTAYYA NAGALAKSHMI

Central Food Technological Research Institute, Mysore, India

1 INTRODUCTION

The tea plant is cultivated in tropical and subtropical regions of varying climatic conditions: with a temperature range of 13°C–29°C, altitude 2460 m above sea level, and acidic soil rich in iron and manganese with a pH range of 3.3–6.0, preferably 4.5–5.5 (1).

Tea is the most widely consumed of ancient beverages. The world production of tea in 1996 was 2,622,000 metric tons (2). India is the largest producer, with a total production of tea approaching 820,000 metric tons in 1998 and exports of 200,000 metric tons (3). The other major tea producing and exporting countries are China, Sri Lanka, Turkey, Kenya, and Indonesia. Among 30–40 countries, the United Kingdom, Russia, Japan, Pakistan, the United States, and Egypt are the top importers and consumers of large quantities of tea and tea products. According to a recent U.K. market survey, tea was assessed far ahead of coffee as a beverage, considered the ideal choice in the morning by 81%, at breakfast by 75%, at midmorning by 67%, and in the afternoon by 77%.

The tea plant belongs to the genus *Camellia* (*Camellia sinensis*), a member of the family Theaceae. Taxonomically, two basic varieties of the tea plant are recognized: viz., the Chinese variety *C. s. sinensis*, and the Assamese variety *C. s. assamica*. The plant originated in East Asian countries, China, Burma, Laos, and Vietnam. Its use as a beverage in China dates back to 350 C.E.. Tea reached Europe in the 16th century. Tea was discovered in 1823 in the Indian province of Assam, where it was first cultivated, and gradually spread to South India, Indonesia, and Sri Lanka. Large-scale cultivation spread to many other countries in Asia, the former Soviet Union, Africa, and South America (4).

Conventional teas are (a) totally fermented black tea, (b) raw or unfermented green tea, and (c) partially fermented Oolong (red and yellow) tea, nonconventional tea products are instant tea (cold- and hot-soluble), flavored tea, and decaffeinated tea. Canned or bottled teas, soluble tea mixes, tea beverages, frozen tea liquid, and tea tablets are convenience products. Liquid tea concentrates, tea mixes, iced tea mixes, and fruit tea mixes have shown tremendous increase in the U.S. market. Decaffeinated tea, both plain and flavored, is finding expanding markets in the United States, United Kingdom, Germany, and other countries.

2 CONVENTIONAL TEAS

2.1 Black Tea

Black tea is made from young leaves and unopened buds of the tea plant. The major steps involved in the manufacture of black tea are plucking, withering, leaf distortion, fermentation, firing, grading, packing, and storage.

2.1.1 Harvesting

The fresh green tea leaves are usually harvested by hand, at intervals of 7–14 days, throughout the year. Generally, only the rapidly growing shoot tips down to about the second or third unfolded leaf are plucked and used.

Long plucking interval or use of clones that deviate from the desirable leaf standard results in varying chemical parameters, such as lowering of levels of theaflavin, caffeine, and volatile flavor compounds and imparting of poor sensory properties, such as sweet, flowery, and grassy aromas. The chemical composition of tea flush as percentage of dry weight is as follows: flavanols 25%, caffeine 4%, amino acids 4%, organic acids 0.5%, monosaccharides 4%, polysaccharides 13%, protein 15%, cellulose 7%, lignin 6%, lipids 3%, chlorophyll and other pigments 0.5%, ash 5%, and volatiles 0.1% (5).

2.1.2 Withering

The withering step makes freshly plucked tea leaves undergo certain biochemical and physiological changes that assist the further processing steps, rolling and fermentation. The physiological and biochemical changes that occur in the living tea leaf continue, but withering alters the pattern and rate of these changes. The following chemical and biochemical changes during withering ensure the quality of black tea: (a) increase in amino acid, simple carbohydrate, and caffeine levels; (b) maximal activity of polyphenol oxidase; (c) loss of pectinase activity, and (d) breakdown of chlorophyll.

Withering is carried out with leaves spread in thin layers (0.3–0.7 kg/m^2) in trays/tats/open loft system/rooms on the upper story of the factory in traditional processing of tea. The withering period varies from 16 to 20 h depending upon the condition of the leaf and the requirement of made tea. Hot air is blown from the bottom of the withering tray/trough system. The moisture in the leaves is evaporated by air, which causes drying. Low-altitude factories use ambient air and factories at high elevation use hot air for drying. It is observed that short withering periods (12 h) and low temperature (10°C–15°C) result in good flavor quality in made tea, whereas longer withering (20–30 h), forced withering, and high temperatures (25°C–30°C) have a good effect on color, but an adverse effect on the chemical and flavor qualities.

The various withering techniques use drum, tunnel, trough, and continuous withering systems. Trough withering is most popular because of its low cost of construction and maintenance (5).

2.1.3 Leaf Distortion/Rolling for Manufacture of Orthodox Tea

After the withering process, the leaf is distorted by rolling or cutting. Conventional processing of leaf requires rolling in order to produce orthodox teas, viz., black tea and green tea. Leaf distortion immediately after plucking is not advisable. The rolling technique wrings out the juice from the leaf and twists the leaf. A roller consists of a circular table, a cylindrical box or jacket, and a cap to apply pressure. The leaf is bruised and twisted, then broken into small pieces by increased pressure and sifted by green leaf sifters. The remaining bulk is rolled. The duration of each roll varies from 15 to 60 min, and the number of rolls varies from two to five in normal practice, depending on such factors as degree of wither, type of tea required, roller charge and speed, rolling conditions, and temperature (5).

2.1.4 Leaf Distortion for Nonwither Teas

In the case of nonwither teas, a number of leaf distorting machines are used. Versatile modern machines (Legg-cut, CTC [crushing, tearing, curling], and Rotorvane) are used singly or in combination. The object is intensive maceration of the tea leaf to ensure rapid and complete fermentation. The CTC machine consists of two engraved metal rollers running close together and works as a mangle, one at 70 rpm and the other at 700 rpm. The soft withered leaf is cut, torn, or rolled in the small gap between the serrated surfaces of the rollers. The made tea is discriminated in the trade. The liquor qualities are enhanced.

Machines used for leaf distortion include the Triturator, Ceylon continuous tea processing machine, Tocklai continuous roller (TCR), Barbora leaf conditioner (BLC), and USSR continuous rolling expresser (5).

2.1.5 Fermentation

Fermentation is the step in black tea processing most important for the necessary chemical and biochemical changes. The process starts at the onset of leaf maceration and is allowed to continue under ambient conditions. The green leaf after rolling and sifting in the case of orthodox tea or macerated leaf (CTC type) is spread in thin layers 5–8 cm deep on the factory floor or on racked trays in a fermentation room. Temperature control and air diffusion are facilitated by using humidifiers or cool air. Time of fermentation varies between 45 min and 3 h, depending on the nature of leaf, maceration techniques, ambient temperatures, and requirement of made tea. Temperature varies between 24°C and 27°C. Low temperature (15°C–25°C) improves flavor.

At the end of fermentation leaf color changes from green to coppery red along with development of a pleasant characteristic aroma. The termination point is determined by the skill of the tea maker or by instrumental techniques. Fermentation is terminated by the firing step. Fermentation can be assessed by measuring the theaflavin and thearubigin content, which are formed in the ratio of 1:10, under ideal conditions of fermentation. Estimation of tannin is another useful method. Tannin decreases during this period, from 20% in green tea leaf to 10%–12% in fermented tea. Modern developments in fermentation technology such as skip, trough, and drum continuous fermenting systems have merits such as controlled optimal temperatures, reduced cost, lowered floor space requirement, and improved briskness in tea liquor (5).

2.1.6 Pigments

Flavanols of tea in the group of catechins constitute 25%–30% of the dry matter of the tea leaf. Tea quality is mainly related to the flavanol concentration of the fresh leaf.

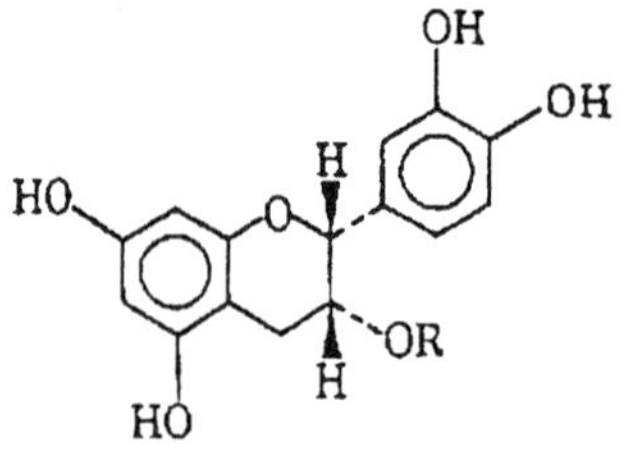

(a) (-)-epicatechin;
R = H; 1-3% of dry wt

(b) (-) -epicatechin gallate;
R = 3,4,5-trihydroxybenzoyl;
3-6% of dry wt

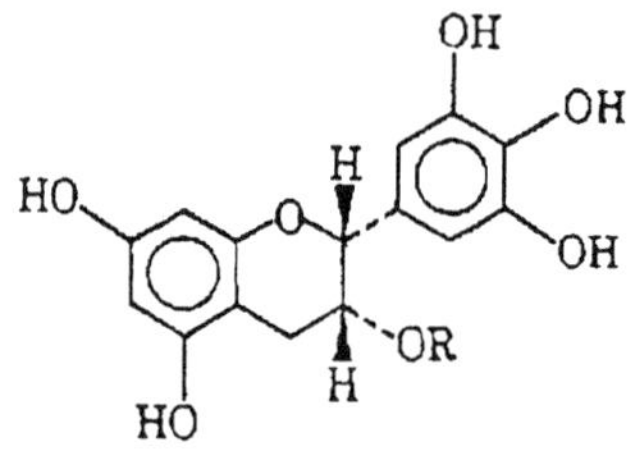

(c) (-) -epigallocatechin;
R = H; 3-6% of dry wt

(d) (-) -epigallocatechin gallate;
R = 3,4,5-trihydroxybenzoyl;
9-13% of dry wt

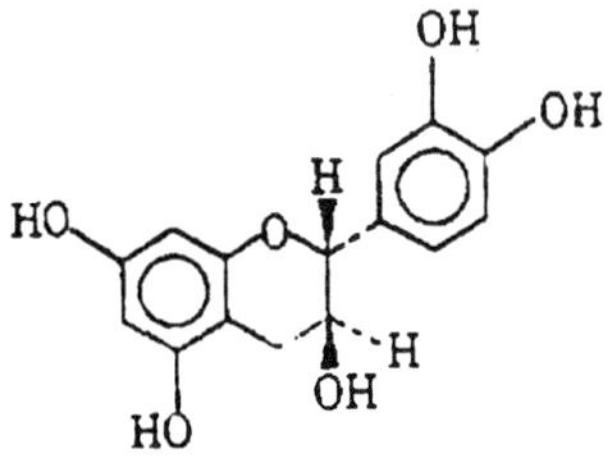

(e) (+) -Catechin; 1-2% of dry wt

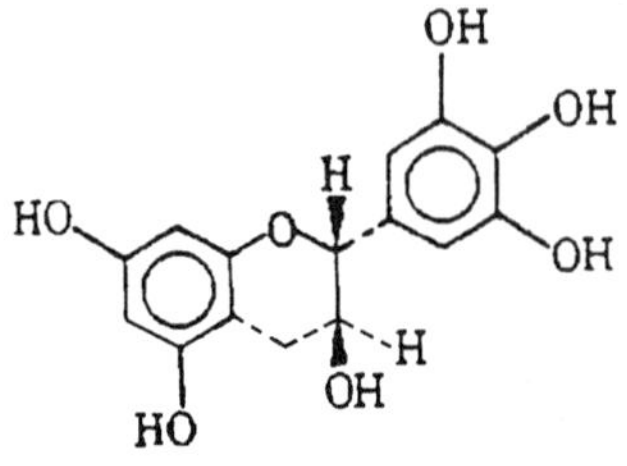

(f) (+) -Gallocatechin; 3- 4% of dry wt

Fig. 1 The flavanols that occur in tea.

(−) Epigallocatechin gallate constitutes a major portion of the flavanols in the fresh leaf and levels of gallated flavanols decrease as the leaf ages (6, 7). The major catechins and the reactions in the formation of tea pigments are shown in Fig. 1.

All catechins undergo oxidation. (−) Epigallocatechin and (−) epigallocatechin gallate are readily oxidized. Quinone formation (Fig. 2) is the primary driving force in fermentation; it results in the formation of the coloring pigments of black tea.

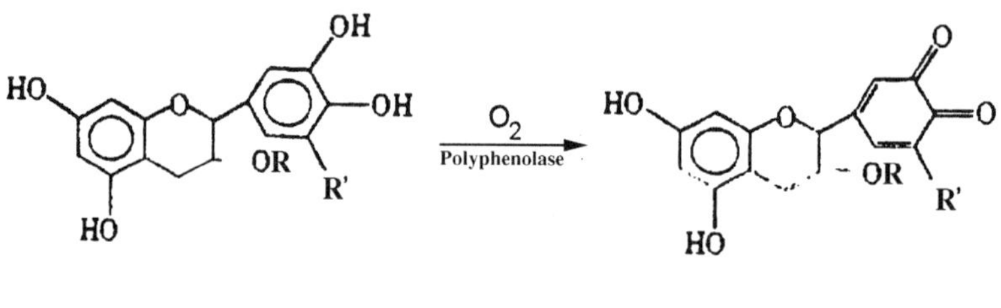

R = H or 3,4,5 trihydroxybenzoyl
R' = H or OH

Fig. 2 Quinone formation.

2.1.7 Other Polyphenols

Other polyphenols include gallic acid; flavanols, such as quercetin, kaempherol, myricetin, and their glycosides; and depsides, such as chlorogenic acid and *p*-coumaryl-quinic acid. Theogallin (3-galloylquinic acid) is unique to tea.

2.1.8 Theaflavins

A quinone derived from a simple catechin (a), (b), or (e) reacts with a quinone derived from a gallacatechin, (c), (d), or (f), to form a compound with a benztropolone ring known as *theaflavin* (Fig. 3).

The four theaflavins theoretically derivable from compounds (a), (b), (c), and (d) are isolated and characterized; one other form, known as *isotheaflavin*, is also isolated and characterized.

Theaflavins exhibit a bright orange-red color in solution. They are important in the determination of brightness, which is a desirable attribute of tea beverage and are distinctive enough in color to be determined spectrophotometrically in tea brew (6, 7).

The total theaflavin concentration in black tea varies from 0.3% to 2.0%. Only 10% of the original catechin content of tea flush is accounted for as theaflavins in black tea. Prolongation of fermentation decreases theaflavin content.

2.1.9 Theaflavic Acids

Although gallic acid is not directly oxidized by tea polyphenolase, it is converted to gallic acid quinone by catechin oxidation products. Further reaction with the oxidized form of (a), (b), or (e) yields theaflavic acids (Fig. 4).

The theaflavic acids are bright red, acidic substances. They are present only in very small quantities in black tea. Oxidized products of epigallocatechin coupled with gallic acid form a series of colorless substances known as *bisflavanols* (Fig. 5), which occur in very small quantities in black tea (7).

Catechin quinone + gallacatechin quinone ⟶

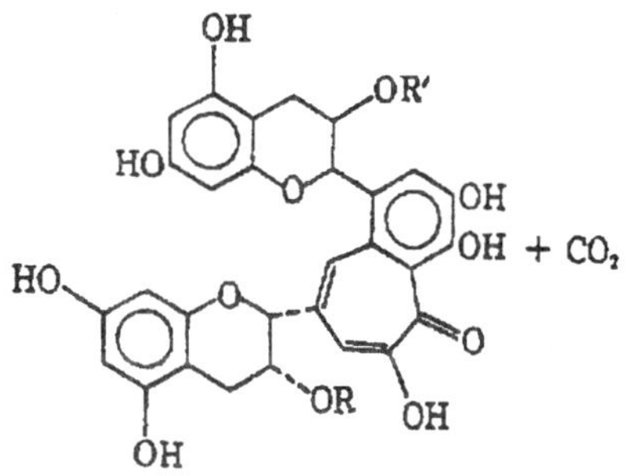

theaflavin; R = R' = H
theaflavin gallate A; R = 3,4,5-trihydroxybenzoyl, R' = H
theaflavin gallate G; R = H_1
R' = 3,4,5-trihydroxybenzoyl
theaflavin digallate; R = R' = 3,4,5-trihydroxybenzoyl

Fig. 3 Theaflavin formation.

gallic acid quinone + epicatechin quinone →

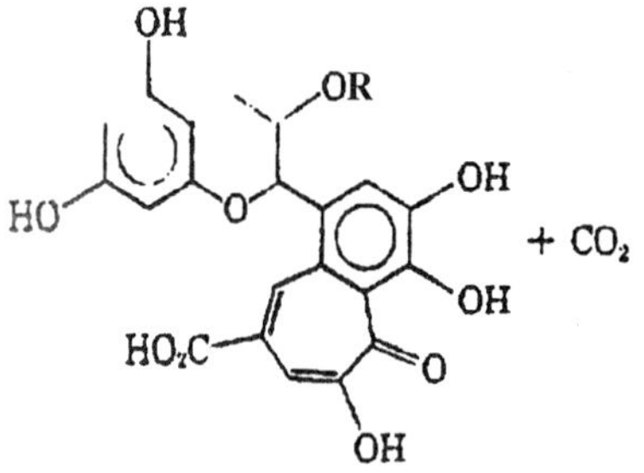

epitheaflavic acid, R = H
epitheaflavic acid 3'-gallate;
R = 3,4,5-trihydroxybenzoyl

Fig. 4 Theaflavic acids formation.

2.1.10 Thearubigins

During the formation of black tea, about 15% of tea catechins remain unchanged and 10% are accounted for by theaflavin, theaflavic acid, and bisflavanol formation. About 75% of the catechins are converted to a complex, poorly defined, and incompletely separated group of substances, known as *thearubigins* because of their red-brown color. Molecular weight determinations indicate a range of 700–40,000. Reductive hydrolysis of thearubigins yields small quantities of all of the catechins found in tea. Evidence exists for the presence of polymeric proanthocyanidine, which accounts for only a small fraction.

Thearubigins constitute the largest group of compounds in black tea (up to 20% of dry weight) and contribute significantly to color, strength, and mouth feel of the beverage. The oxidized polyphenol complex of black tea is generally referred to as *tannin* but has no relationship to tannic acid (pentadigalloyl ester of glucose), which does not occur in tea. Thearubigins impart a specific bright and vivid red color to the liquor. The ratio of

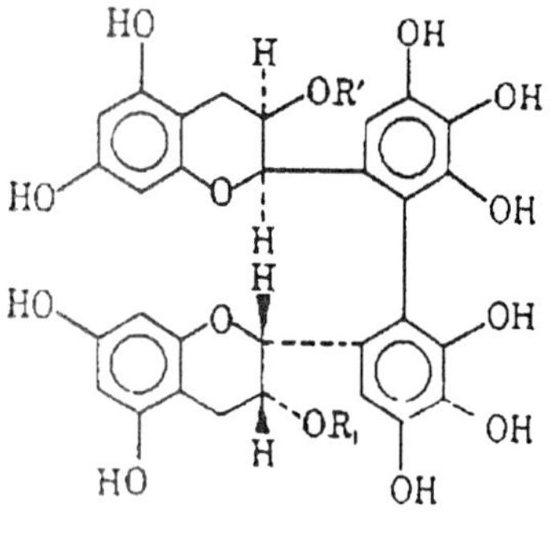

bisflavanol A; R = R' = 3,4,5-trihydroxybenzoyl
bisflavanol B; R = 3,4,5-trihydroxybenzoyl; 4' = H
bisflavanol C; R = R' = H

Fig. 5 Bisflavanols formation.

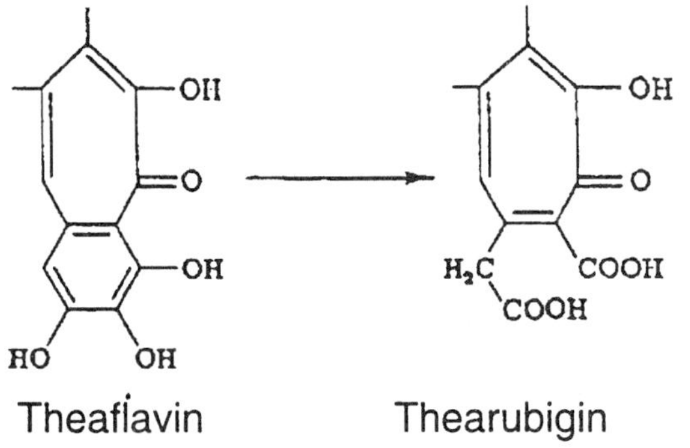

Fig. 6 Theaflavin and thearubigin.

theaflavin to thearubigins is taken as an index of strength of the liquor. The control of these two factors depends upon fermentation time and temperature (6, 7). The structure of thearubigin is illustrated in Fig. 6.

2.1.11 Aroma Formation

The oxidized catechin reacts with the precursor molecules present in green tea and produces the volatile compounds in black tea, which are mainly dependent on genetic, cultural, and manufacturing variables. Oxidation of amino acids, carotenes, and unsaturated lipids leads to the formation of aroma compounds during the fermentation period. Aldehydes of amino acids formed as a result of Strecker's degradation reaction are of significant importance. Ionones, terpene alcohols, terpene aldehydes, and their oxidation products such as theaspirone and dihydroactinidiole result from oxidation of the carotenoids present in the flush by quinones. More than 638 aroma compounds in tea have been identified. The most important of aroma components are terpenes, terpene alcohols, lactones, ketones, esters, and spiro compounds. Aroma compounds of some of the world renowned teas with unique flavor characteristics have been identified: India's Darjeeling tea has linalool, linalool oxides I and II, geraniol; China's Keemun tea contains 2-phenylethanol and geraniol; Sri Lanka's Uva tea contains methylsalicylate, linalool, and linalool oxides (8, 9). The concentrations of aroma constituents and volatile oil content are high in orthodox tea compared to those of CTC teas. CTC teas give strong brew/body, more cuppage, and malty aroma, whereas orthodox teas produce a light brew with fine, flowery, and rich aromas due to high concentrations of linalool, linalool oxides, methylsalicylate, genaniol phenyl ethanol, and cis̄jasmone^{+}, β-ionone with other components (10, 11). The formation of pigments and aroma compounds in black tea is represented in Fig. 7.

2.1.12 Firing

Firing is done in a special dryer at a high temperature, 90°C–95°C, immediately after fermentation. The time taken for firing is 20 min to reduce the moisture by 3%–4% and to inactivate the enzymes. Hot air causes evaporation, with the result that the tea shoots lose their coppery red and brown color and are transformed into black tea. Firing decreases aroma constituents with an increase in some aliphatic carboxylic acids, which suggests oxidative reactions (5).

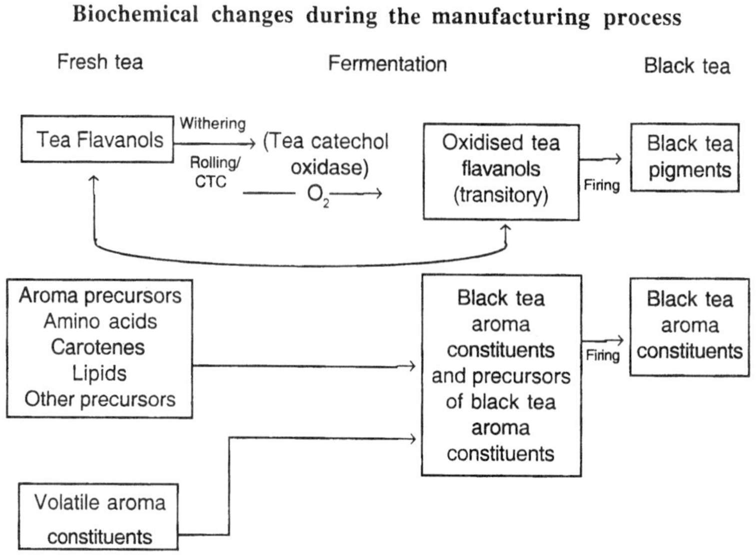

Fig. 7 Formation of coloring pigments and aroma compounds in black tea.

2.1.13 Grading and Storage

Tea is often winnowed to remove stalky material and sieved to obtain different grades, which are based on particle size. Chemical changes take place during the storage of finished tea products, which tend to lose all residual greenness and harshness within a few weeks' time. Tea remains sound and full of flavor for more than 1 year, if kept in a cool place and protected from moisture and oxygen (5).

2.1.14 New Manufacturing Technology

The traditional technology of tea manufacture has certain disadvantages: (a) undamaged tea leaf tissues (20%–25%) during rolling or CTC method, which in turn produce a nonuniform rate of oxidation that results in high losses of polyphenols; (b) loss of 70% to 80% of essential oil in the firing stage, which weakens the aroma; and (c) rapid aging of tea, which results in loss of its high quality. Hence, thermal treatment of underfermented teas in a factory or storehouse by ambient temperatures that reach 40°C is practiced in many parts of the world, particularly in China and India. Thermal treatment eliminates the grassy odor and coarse taste in the unfermented teas that are due to polyphenols, catechins, and other constituents present in high proportions. Thermal treatment results in isomerization and epimerization of catechins, degradation of chlorophyll, and synthesis of aldehydes and essential oils and improves the flavor quality of manufactured tea (12).

2.2 Green Tea

Around 21% of total tea production is consumed as green tea, which contains larger amounts of catechins and vitamins. Green tea has a pleasant taste, flowery aroma, and light green color with an olive shade. Green tea is produced from a bud and two to three lead flushes of tea shrubs. Because green tea is unfermented, the development of the oxidative process is regarded as an adverse factor. In the Japanese method, the enzyme is inactivated by steaming, whereas in the Chinese method the enzyme is inactivated by roasting leaves in a pan. The steps of manufacturing of green tea are plucking, steaming/roasting, primary heating and rolling, rolling, secondary rolling, drying, refining, firing, sorting, and packing. There are 18 different types of green tea, classified as pan fixed (viz., pan dried, basket dried, cured roasted, sun dried) and steam fixed (13).

2.3 Yellow Tea

Yellow tea is an unfermented tea that occupies an intermediate position between black and green tea but is closer to green tea. The yellow tea infusion is of brighter color than that of green tea. Yellow tea, which has a milder taste and a stronger aroma than green tea, is a pleasant and refreshing beverage. It is characterized by a higher content of catechins, vitamins, and extractives than black tea and therefore is physiologically more effective.

Yellow tea is produced from a bud and two or three young leaves and tender shoots of the tea plant. The chemical transformation occurring in raw tea during withering in essence distinguishes the yellow tea from green tea. The tender parts of the shoots, the bud, and first leaf should be withered to the same extent as the third leaf and stalk. The steps in manufacturing yellow tea are withering, roasting, rolling, sorting, and firing (13).

2.4 Red Tea

Red tea is a fermented tea; its production is based on a controlled combination of enzymic and thermochemical processes in which the rate of the enzymic process is slower than that in black tea manufacture. Red tea has a pleasant, mild, and astringent taste and a strong, stable aroma. The steps in manufacture are withering, rolling, roasting, firing, sorting, and final firing (13).

2.5 Dark Tea

Dark tea is an unfermented type; about 40,000 to 60,000 tons are produced only in China, mainly for the internal market. It is brownish yellow or brownish red in infusion, stale in aroma, and piny, smokelike (of microbial fermentation), and mellow in taste. Fresh leaves, one bud, and four to six leaves with some stalks are withered, rolled, piled, dried first time/level, dried finally, steamed, and compressed to obtain the finished product. Piling is an important process, in which temperature and humidity are kept high for a long time so that microorganisms such as *Aspergillus glaucus*, *Saccharomyces* spp.s and others, are produced naturally. During piling, polyphenols are oxidized as a result of dampness, heat, and the action of microorganisms. Dark tea contains no chlorophyll, and the levels of catechins and free amino acid concentration are lower than in yellow tea. These chemical changes mellow the flavor of dark tea. There are many dark tea varieties in three major groups according to piling, i.e., piled before first drying, piled after first drying, and piled after final drying (14).

2.6 White Tea

White tea is a fermented type; about 1000 tons are produced in China only for export. Fresh leaves, one bud, and one to two leaves with profuse hairs or only buds from the first plucking of spring tea with moderate content of polyphenols are withered, fired to obtain raw tea, sorted, and packed as made tea. During withering for 1–3 days, levels of amino acids increase, and sugar and polyphenol levels decrease. The infusion of white tea is light orange-yellow, the infused leaf is open and mixed, the aroma is fresh and pure, and the taste is plain (14).

2.7 Green Brick Tea

Green brick tea is both a flavor product and a foodstuff, which is consumed as a soup cooked with water or milk and supplemented with butter, mutton fat, and salt. It is in high demand in Mongolia, China, and Russia. The two stages of its production are (a) preparation of half-finished *Lao-cha* (meaning ''old tea'') and (b) pressing into green brick tea. Lao-cha is produced in factories from November to February or March, after the production of regular black and green teas is completed. Lao-cha is made from old leaves of the tea plant. Two kinds of raw materials are used, one for the coating and the other for the inner portion of brick tea. The coating raw material, usually gathered in autumn, is of high quality, composed of coarse leaves without lignified brown stalks, but tender enough to be subjected to rolling. The inner portion is made of coarse shoots with up to 12 leaves and green/brown stalks (70% leaves and not more than 30% lignified stalks).

The Lao-cha material is processed immediately after harvest either by an ancient traditional Chinese method requiring 16–20 days or by a new method requiring only 10–20 h. In the Chinese method, raw tea is roasted, allowed to stand for 2–3 h in small stacks, cooled, rolled, dried to 18%–20% moisture, and subjected to fermentation in large stacks of 3–11 tons of tamped tea bulk. The fermentation process ends when the temperature inside the stack increases to 55–65°C by self-maturation. The inner material at that temperature is placed on the outside of the stack and again allowed to ferment for 5 days to reach the temperature. Then, the stack is taken apart, loosened, cooled, and exposed to final firing at 85°C–90°C with 8%–9% moisture content. After taste testing, the Lao-cha tea produced is stored in separate batches for pressing. In the new method, the process consists of roasting, high-temperature rolling, thermal treatment, and firing. Half the coating material covers the lower layer of the brick, the other half the upper layer, and the inner material is placed between them. It is then wrapped up in a cloth napkin, exposed to water vapor at 6–7 atm for 2 min at 95°C–110°C so that the Lao-cha bulk loosens; then it is transferred to a metal press machine preheated to 70°C and pressed by a hydraulic press at 100–110 atm pressure. The press machine is slowly transferred onto the conveyor line and allowed to stand for 60 min to cool. The bricks are removed, trimmed, and dried in a chamber at 34°C–36°C, 50%–55% relative humidity, for 15–20 days, to yield approximately 11% moisture. Then 16 bricks of size 350 × 160 × 32 mm wrapped in paper are packed in a standard plywood box (13).

3 NONCONVENTIONAL AND NEW TEA PRODUCTS

3.1 Flavored Teas

Flavored teas are produced by incorporating various natural and nature-identical flavors in various processed teas. Flavored black tea, green tea, Oolong tea, instant tea, decaffeinated black tea, and decaffeinated instant tea are available in the market and are highly profitable in generating foreign exchange earnings.

Medium- and poor-equality teas that are highly unremunerative are incorporated with the choice blend of the good-flavor notes, masking the undesirable characteristics but retaining the essential background flavor of tea. Incorporation of natural flowers or flower petals or skin peels of citrus fruits is done after the firing step in factories. Spiced teas with ginger, cardamon, clove, and cinnamon in certain parts of India and scented teas with lemon, orange, mint, bergamot, and rose in the Middle Eastern countries, Japan, and the United States are popular (15, 16).

The quality of low-grade teas is improved by addition of chemicals: alcohols, aldehydes, ketones, esters, geraniol, citronellol, phenylethylalcohol, cinnamaldehyde, *p*-hydroxy-benzaldehyde, acetophenone, benzyl acetate and methylsalicylate, 1,3-dihydroxy-2-propane, oxygenated cyclohexanes, 8-oxo-1,3,6,6-tetramethyl, 5,6,7,8-tetrahydroisoquinolene, 2,6,6-trimethyl-cyclohex-2-ene-4-ol-l-one, and others (15).

Addition of different amino acids to hot tea brew produces different aromas: e.g., phenyl alanine produces a roselike odor; glutamic acid and alanine, a flowery odor; norleucine, a spicy odor; and threonine, a winelike odor (8). The flavor enhancers are developed to impart fresh tea flavor and aroma into cold- and hot-water-soluble teas. Some of the flavoring agents are a combination of 1-penten-3-ol (10.55%), hexanol (1.25%), *cis*-3-hexen-1-ol (15.1%), linalool oxide (11.92%), linalool (55%–66%), geraniol (0.52%), isobutanol (5%), and essence of apricot, banana, apple, raisin, currant, date, prune, and fig (15, 16).

3.2 Methods of Incorporation of Flavors

3.2.1 Incorporation of Natural Fragrant Flowers

Chloranthus inconspicuus, *Jasminium sambee*, *Gardenia florida*, *Murraya exotica*, and *Aglaia odorata* are commonly employed for scenting black teas. Scented green tea is prepared by using fully fired tea. When tea is still warm, it is poured into the tea chest and spread to form a 2-in-thick layer. A layer of the desired scenting flowers is spread on this layer. The process of spreading of alternate layers of tea leaves and flowers is repeated till the chest is completely filled. A proportion of about 3 parts by weight of flowers to 100 parts of tea is generally employed. Then it is heated by the *poey* (bamboo basket) process for about 2 h, by which time the flowers are quite crisp. The flowers are then sifted out, and the tea is packed. The tea thus prepared is later mixed with unscented teas in the ratio of 1:20. All the material is then slightly heated in an iron pan and packed. This tea was known as "Cowslip Hysons" in England in earlier days (17).

3.2.2 Flavor Incorporation Through Tea Bags

Flavor incorporation into tea is done through tea bags. The tea bag is initially flavored; subsequently it is filled with tea. In this method 0.2 mg of theaspirone and 0.1 mg of wax of tea leaf are dissolved in 100 ml of 20% ethanol. The solution is applied in a continuous narrow line on a rolled paper of 10-cm width; a glass pen holds the solution, to apply 0.1 ml of the solution per 15-cm length of the paper. This treated paper is cut to a desired size and 2 g of black tea leaf is wrapped in the paper to prepare tea bags (15).

3.2.3 Flavor Incorporation into Soluble Tea

The liquid tea extract, or instant tea, is flavored by addition of 1 to 2 parts of Δ-decalactone in a concentration of 0.5–9 ppm, in terms of tea-soluble solids, and 1 part of geranyl acetate. The flavor is added to an infusion of tea leaves or to solid instant tea, as a solution or emulsion, in a diluent such as alcohol.

Liquid spiced teas with ginger, cardamom, clove, cinnamon, coriander, and cumin are prepared by adding them at the brewing stage and concentrating the tea extract (15).

3.2.4 Flavor Incorporation by Diffusion

There are two standardized methods of diffusion. The first method involves mixing a small quantity of the desired grade of tea directly with the flavor component and then remixing this blend with the bulk. Here, sufficient time has to be allowed for stabilization of the flavors with tea. In the second method, the tea sample has indirect contact with the flavor blend, which is sprayed on a wad of cotton or paper. Here, only vapors are in contact with tea, by means of thorough mixing. In another method, scenting is done in specially constructed scenting drums. The flavor is made to flow from a glass container onto the chromatographic blotting paper; when the drum filled with tea is rotated, it contacts the moving pad and flavoring is done for 30–40 min per batch. This process ensures optimal flavor and uniform flavoring and is used by Sri Lankan exporters (16).

3.2.5 Flavor Incorporation by Spray

Black tea is spread in uniform layers in trays, which are kept in specially constructed rooms. Flavor is sprayed on them by automatic spraying devices. Mixing is done at regular intervals to ensure uniform distribution of flavor. This method has the advantage of rapidly flavoring tea in small quantities, although some flavor loss results from volatilization of flavor into the atmosphere (16, 18).

3.2.6 Use of Encapsulated Flavors

Proper control of quantity of flavor to be incorporated is possible when encapsulated flavors (spray dried or molten extruded) are used in the preparation of flavored teas and flavor loss can be minimized during storage. One disadvantage is that flavor distribution cannot be uniform. Therefore, unit packing, or small packs for servings of 5 cups to 25 cups, are useful, and the entire quantity should be utilized in order to effect uniform distribution of flavor in the brew. The moisture level in these products should preferably be below 6%. Packaging is done immediately after processing, in triple-laminated aluminum foil, and in tea bags with laminated sachets in order to retain the flavor (16, 18, 19).

3.2.7 Herbal Teas or Tisanes or Tea Substitutes

The herbal teas, tisanes, and substitutes are fragrant infusions prepared by dried leaves, flowers, roots, or combinations of aromatic/culinary herbs. They do not contain tea leaves; however, they are popularly called herbal teas. The infusions are prepared in the same way as tea. They vary widely in color and flavor from pale liquids, delicate in taste, to those that are intense in color and robust in flavor. The most popular ones are leaves of boldo (*Peumus boldus*), camomile herbs and flowers (*Anthemis nobilis*), elder flower (*Sambuscus nigro*), hibiscus flower (*Hibuscus sabdariffa*), leaves of lemon verbena (*Lippia citriodora*), lime blossoms and leaves (*Tilia cordota and Tilia silvestris*), orange flowers (*Citrus aurantium*), peppermint leaf (*Mentha piperita*), rose hip fruits (*Rosa canina*), sage leaves (*Salvia officinalis*), and thyme leaves (*Thymus vulgaris*). Some are good as a hot beverage with sugar; others, such as rose hips, lemon verbena, and hibiscus, are good when cold. Lemon juice enhances the flavor of tisanes. These are sold by herbalists and chemists and in health food shops and delicatessens (20).

In Germany, a breakfast tea beverage is made by using a blend of heather leaf, blackberry, strawberry, raspberry, cornflower, and spice. Fennel herb, peppermint, teliac, cloves, and rose are used in another drink. Hibiscus, in large quantities, blended with apple and blackberry, is used in yet another herbal drink.

Tea substitutes such as Doojoong tea, ginseng tea, and hibiscus tea are popular in the Orient. Ginseng tea is made by extracting ginseng root, rich in mineral content, yields a popular drink in Korea.

A tea substitute manufacturing process involves comminuting of bilberry leaves, curling, fermentation, and drying (21). To improve the quality characteristics and therapeutic properties of the tea substitute, the leaves are withered before processing (60%–62% moisture in leaves); additionally, comminuted leaves of Cornelian cherry and eucalyptus are added to the bilberry leaves (in defined proportions) at the curling stage and subsequently double curled and graded (16, 21).

Rooibosch red bush tea or Rooibos tea (*Aspalathus lincaris*) is another tea substitute. Young leaves and branches are processed in a way similar to that used for black tea. Convenience products such as tea concentrate and bottled tea have been produced from this native of South Africa since the early 20th century (16).

3.3 Decaffeinated Teas

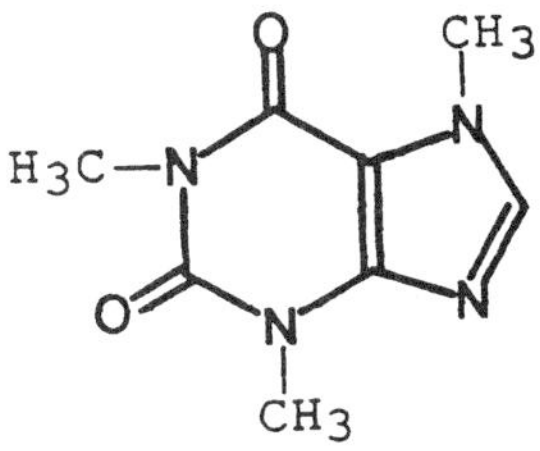

Caffeine

The effect of caffeine is most commonly linked to stimulating action on the central nervous system. Caffeine is useful for several purposes, including increase in muscular work, allergy relief, obesity control, diuresis, and treatment of the common cold. However, excessive consumption can lead to caffeism, which causes anxiety and neurosis. Doses of 100 mg/kg body weight are a serious poisoning hazard in children; the acute human fatal dose is 170 mg/kg body weight. Increase in caffeine intake may also lead to increased heartbeat, promote secretion of stomach acids, step up production of urine, and cause skin disorders (22). Hence, decaffeination of tea is desirable to regulate caffeine intake. Decaffeinated tea is produced by removing 60% to 90% of the caffeine from black tea by solvent extraction as prescribed by local legislation. The decaffeinated product generally contains 0.2% to 0.8% caffeine. The raw material contains bound caffeine, about 1% to 1.5% which cannot be removed by simple extraction, unless some chemical/enzyme treatment is given. Hence decaffeination means the percentage of caffeine removal from black tea, depending on soluble caffeine content of the raw material. In decaffeinated teas, total solubles vary 34.8%–38.5%, tannins 7.23%–7.84%, TR 3.16%–5.90%, and TF 0.11%–0.20% (16).

Decaffeinated tea is always found to be inferior to normal tea in appearance and flavor quality because of the various processing steps involved in its manufacture. Ethyl acetate and supercritical liquid carbon dioxide are the only two solvents permitted for decaffeinating black tea. Tea requires high moisture content for decaffeination. A minimal water content of 18% is required and as much as 55% moisture is helpful for solvent extraction to free caffeine from the tea matrix. Water is also used as a solvent for decaffeination purposes. The following steps are involved in the process: (a) the moisture content is increased from 10% to 40%; (b) the tea is treated with an organic solvent for 12 to 18 h to extract 97% of the caffeine; (c) the tea is steamed with live steam to remove all residual solvents; and (d) excess moisture is removed by drying (16, 23).

3.3.1 Solvent Decaffeination Technology

3.3.1.1 Sample Screening

Raw material with low caffeine content, 2%–3%, is preferred. Particle size, bulk density, moisture content, and soluble solids are also important. Leafy grades, fannings (broken grade), are good since they do not form lumps in both the extraction and steaming steps.

3.3.1.2 Sample Preparation/Conditioning of the Leaf

Prewetting of the leaf is done by adding extra moisture to the leaf and allowing it to stand for 1 h. Normally 18% to 40% extra moisture is added.

3.3.1.3 Extraction

The optimal ratio of purified solvent to material has to be fixed in column extraction, which is a batch operation and requires 6- to 8-h contact time. Both cocurrent and countercurrent extractions are carried out in order to achieve higher yield.

3.3.1.4 Steaming

After the solvent is drained out completely, the leaf is steamed for nearly 30 min to ensure the recovery of solvent that is held in the leaf.

3.3.1.5 Drying

Drying is done at low temperature (75°C to 80°C) for a period of 1 h.

3.3.1.6 Residual Solvent in Manufactured Decaffeinated Tea

The general procedure of residual solvent methylene chloride or ethyl acetate determination consists of distilling of the solvent and its estimation by the gas-chromatographic technique.

The tea sample, with added methylene bromide (MB) as an internal standard, is equilibrated for 1.5 h at 100°C in aqueous sodium sulfate, before manual head space sampling. Methylene chloride (MC) and MB are separated on a column (porosil A) at 160°C and detected by an electrolytic conductivity detector (Coulson). MC is a solvent permitted in the decaffeination of black tea in European countries, at a permissible residual level of 30 ppm (24).

3.3.1.7 Caffeine Estimation by High-Performance Liquid Chromatography

Various methods such as the spectrophotometric, gas chromatographic, gravimetric, AOAC, and high-performance liquid chromatographic (HPLC), are available for estimation of caffeine in tea samples. The HPLC method is quick and estimates only that amount of caffeine that is dissolved in the brew within 6 min. This method gives an accurate estimate of caffeine intake per cup of brew. The method involves brewing 2 g of the tea sample using 177 ml water for 6 min, cooling the filtered brew, and injecting into HPLC. The peak area represents the concentration of caffeine in the sample (25, 26).

3.3.1.8 Decaffeination Using Liquid/Supercritical Carbon Dioxide

Liquid carbon dioxide exists within the temperature range of −55°C to +31°C and a pressure range of 3 to 74 bar, can be handled as an ordinary solvent provided it is kept under constant pressure, and is miscible with all organic solvents. It is a nonpolar solvent, its polarity approximating that of hexane or pentane, and its solvent power is not high when compared with that of other solvents. There are some standardized methods for the decaffeination of black tea by liquid carbon dioxide. Black tea, 400 g, is moistened with 226 g water and 97% of the caffeine extracted with carbon dioxide at 300 bar, at 20°C in 4 h. The sensory characteristics of the decaffeinated tea are good. In another method, 10 kg tea (3.3% caffeine, 5.9% moisture) is moistened with 3.5 liters of water and extracted with supercritical carbon dioxide at 290 bar and 63°C. The carbon dioxide is then passed through an activated carbon absorbent at 280 bar and 61°C, then passed over the tea again in cycle for 2–5 h, to produce decaffeinated tea (after drying at 70°C to 4.5% moisture content) containing 0.8% caffeine (16).

In another method, the flavor of moistened black tea (25%–50% water) is extracted with carbon dioxide at 60–150 bar at 20°C–70°C. In the second step, deflavored tea is decaffeinated with carbon dioxide at 150–400 bar at 10°C–100°C. Subsequently, flavoring materials removed in the first step are reintroduced into decaffeinated tea and dried. Carbon dioxide is separated from flavor by pressure reduction. The process reduces the loss of flavor during decaffeination (16).

3.3.2 Decaffeination of Soluble Tea

3.3.2.1 Soluble Tea Process

In the soluble tea process, tea extract is cooled to form an aqueous soluble phase, including a small portion of caffeine, and an insoluble phase, containing a major portion of caffeine as a complex with tannins. The insoluble phase is decaffeinated and then added to the aqueous soluble phase, which is concentrated and spray dried to obtain decaffeinated soluble tea.

3.3.2.2 Liquid-Liquid Extraction

Caffeine is selectively extracted from the aqueous phase to trichloroethylene/methylene chloride by liquid-liquid extraction. An extract rich in caffeine is placed in contact with solvent methylene chloride or trichloroethane, to transfer caffeine. Recently rotating disk contactors have proved to be more efficient and versatile. The extraction of caffeine in methylene chloride increases with temperature. The solvent-to-extract ratio is about 4:8 to ensure 98% caffeine removal. This procedure is applicable to the manufacture of decaffeinated soluble tea (16).

3.3.2.3 Use of Adsorbents

An aqueous extract of tea is decaffeinated by mixing it with a solid adsorbent (e.g., activated carbon, clay, hydrated silicates, zeolites, and ion exchange resins). Caffeine is recovered from the adsorbent by extraction with water-immiscible solvent specific to caffeine (16).

3.3.2.4 Complexing of Caffeine with Cyclodextrin

The complexing method consists of mixing aqueous caffeine-containing solution with monomeric and polymeric cyclodextrins. Cyclodextrin is mixed with caffeine solution and incubated at 50°C for various durations (30 min to 18 h). The mixture is then incubated at 4°C for 20–90 min and the precipitate is removed by either centrifugation or filtration. The polymerized β-cyclodextrin is found to be the best decaffeinating agent; it causes approximately 80% removal of caffeine from the aqueous solution (16).

3.3.2.5 Microbial Decaffeination

Aqueous tea extracts are decaffeinated by fermentation with a *Pseudomonas* sp. organism. Thus, 50 ml of 1% caffeine is inoculated with exponentially growing cells of *Pseudomonas* sp. NRRL B-8051, B-8052, and B-8060 and held at 30°C for 48 h with shaking. Analysis showed the following caffeine removal percentages by different methods: solvent extraction, 97%; liquid-liquid extraction, 98%; liquid/supercritical carbon dioxide, 75%; cyclodextrin, 80%; and microbial, 99% (16).

3.4 Instant Teas

Instant tea obtained by dehydration of tea infusion of black tea or green tea is a dry powder that is valued for its convenience. The product is highly hygroscopic and hence requires suitable packaging. Different types of instant tea, viz., hot-soluble and cold-soluble, are available, depending on end usage. Both types are largely used in vending machines. Other forms marketed are decaffeinated soluble tea, either plain or with added flavors, and flavored instant teas.

3.4.1 Technology

Instant tea is prepared by using different raw materials—green tea leaves after fermentation, black tea, low-grade teas, crude tea leaves collected after pruning of tea plantations, and tea waste. Russians use specially prepared dried raw tea.

Instant tea preparation requires quality control of raw material, withering, leaf distortion to a size suitable for extraction, and optimal fermentation time to produce the best body of the liquor. Flavor losses, however, cannot be completely eliminated during processing. The combination of a rotorvane coupled with a CTC machine is the best choice for leaf distortion, and generally coarse cutting is adopted to prevent clogging in the extractors. This step is followed by optimal fermentation to achieve the required body of the liquor (27).

3.4.1.1 Extraction

Water is used for extraction. The quality of water plays an important role. Demineralized or distilled water must be used for preparation of infusion. Usually, cocurrent or countercurrent extraction is carried out, followed by a dewatering step. Vibroscreens, sieves, hydraulic presses, screw presses, and gravity and forced filtration techniques are generally adopted to separate the extract. The leaves are mixed with hot water in different material-to-water ratios. In a multistage cocurrent system, the exhausted leaves are successively extracted in different stages, and the extracts obtained are mixed together before the final concentration. In countercurrent operation, the material and the enriched extracts move in opposite directions so that there is a gradual buildup of solids concentration in the successive extractions. Generally, the solids buildup is at a 7%–10% level. In each operation, a portion of the first extract is preserved and added finally before spray drying to make up for the flavor loss (28).

3.4.1.2 Decreaming and Cream Solubilization

The haze formation in normal tea brew, when it cools to room temperature, technically termed *cream*, is due to a caffeine-polyphenol-protein complex. This affects the appearance, quantity, and quality of cold-soluble tea beverages and iced teas. Hence, chilling tea extract to remove the cream is known as *decreaming*. To prevent cream formation, (a) addition of various carbohydrates during fermentation; (b) solvent decaffeination of black tea and solubilization by the addition of microbial tannase enzyme to tea infusions; (c) decreaming using centrifuges, filters, and membranes; and (d) solubilization of the tea cream by chemical and enzymatic methods are adopted (27).

3.4.1.3 Aroma Stripping and Aromatization

Aroma stripping is carried out immediately after extraction, and before concentration. Stripping is done with either steam or inert gas. The volatiles are concentrated under low temperature and pressure conditions, so that they can be added back to the final product (27).

3.4.1.4 Concentration

Concentration of the extract from 7% to 10% level of solids up to 30% level is done using a vacuum concentrator or a falling film, rising film, or thin film evaporator. This is a crucial step for aroma retention. Hence, low temperature and short retention times of the extract in the concentrators are employed. Use of permitted additive as a filler helps increase the solids concentration to the 30% level, which is desirable for the drying operation.

3.4.1.5 Drying

The drying of the infusion is achieved by spray drying, freeze drying, or vacuum shelf drying. These methods are not commonly adopted, because of high energy costs and lower throughputs. Spray drying is generally employed. Tall tower-type spray dyers provided with pressure- or nozzle-type atomizers are preferred, depending on whether a puffed or a powdery product is required. Flavor losses are high because of the high temperatures used. Bulk density of the product is increased by bubbling of carbon dioxide or ammonium carbonate followed by spray drying. The agglomeration technique is employed to increase the bulk density. Addition of anticaking agents is practiced since the product is highly hygroscopic (27).

3.4.2 Chemical Composition

Chemical composition (percentage) of hot-water-soluble instant tea varies: tannins, 40.50%–49.00%; total catechins, 55.20%–89.40%; caffeine, 3.50%–5.10%; total amino acids, 7.00%–8.00%; total nitrogen, 2.90%–4.00%; and total protein nitrogen, 0.14%–0.40%. In cold-soluble instant tea percentages are much lower (14, 16).

3.4.3 Blending

Instant tea can be used to fortify other teas. Addition of instant tea to regular tea increases the content of extractives, including polyphenols, caffeine, and other constituents, up to 60%–70% from 35%–40%, and thereby enhances the nutritive and biological value of the product. An addition of 15% instant tea to a regular tea of any grade does not affect its taste or flavor and is very profitable because instant teas are usually produced from low-cost raw materials. Indian instant teas with superior flavor are generally used for upgrading the instant teas produced in Europe (14, 16).

3.5 Tea Concentrates

Tea leaves are extracted with hot water at 80°C–90°C. A secondary extraction ensures complete recovery of soluble solids, using water at 95°C–100°C. Additives such as citric acid and sodium bicarbonate facilitate better extraction. The extract is clarified and concentrated by using either evaporation or the freeze-concentration method. To the concentrated brew, sugar, stabilizers (xanthan, CMC, or carrageen), emulsifiers (mono- and diglycerides, lecithin, or ethoxy fatty alcohols), preservatives (sodium benzoate, potassium sorbate), and flavorings (lemon oil or orange oil) are added. The product is storable (60°F) for 6 months without deterioration and is used for the preparation of liquid tea–based beverages (16).

3.6 Tea Beverages

3.6.1 Ready-to-Serve Beverages

Fresh tea brew (brown tea) is consumed hot over most of the world; in India and Britain it is consumed with milk and sugar. Nowadays ready-to-serve (RTS) products based on tea, e.g., Teh Botole and related products, popular in Indonesia, represent a significant share of the soft drink market. This is a noncarbonated bottled tea beverage prepared from Teh Wangi, a processed green tea with jasmine flavor. Clarity of the beverage and its shelf life are well maintained. Tea beverages in various forms that can be stored for long periods are common (29).

Vinification of tea extracts is possible. Clarified and unclarified aqueous tea extract is fermented by *Saccharomyces cerevisiae* var. *Montachet* and by *S. bayanus*, after sweetening with sugar to 20°Brix. Theaflavin, caffeine, and theobromine levels in the fermenting process must decrease progressively over a period of 15 weeks. The soluble amino acids are reduced by more than 90% by *S. bayanus* in 1 week in the presence of added ammonium phosphate. Thearubigins remain unchanged. Marked astringency development is attributed to changes in the proportions of the soluble tea constituents. Interactions between oxidized polyphenols and proteins/polysaccharides are observed (30).

3.6.2 Carbonated Tea Beverages

Black tea or decaffeinated black tea is extracted with water to introduce the required solubles into the water medium and then filtered through fine filters. Sugar syrup prepared separately, a preservative such as sodium benzoate, and an acidulant such as citric acid are incorporated into the beverage. Turbidity is a major problem in the case of bottled cold tea beverage. Chemicals are generally used in order to solubilize the cream, which causes turbidity and some undesirable taste in the beverage; however, use of decaffeinated tea in the preparation of beverage has helped to eliminate the turbidity to some extent. The finished product is cooled and carbonated. The composition of the carbonated tea beverage in general is as follows: tea solids, 0.3% to 0.4%; sugar, 10%; citric acid, 0.1% to 0.15%; and sodium benzoate, 100 to 120 ppm (16).

3.6.3 Green Tea Extracts

Green tea extracts are gaining markets in the international trade because of their potential health benefits. Green tea polyphenols/extracts are prepared from green tea after extraction with water and drying. The dried powder is dissolved in hot water and extracted with chloroform to make it free of caffeine. The caffeine-free aqueous layer is extracted with ethylacetate solvent and further concentrated, dried, and preserved. The catechins are subjected to HPLC analysis to estimate percentage of each catechin. The extract must contain no less than 35% of (−) epigallo catechin gallate according to trade demand standards (31).

4 QUALITY STANDARDS AND SPECIFICATIONS

The United States Department of Agriculture (USDA) defines *tea* as "the tender leaves, buds and tender internodes of different varieties of *Thea sinensis* L., prepared and cured by recognized methods of manufacture."

Quality parameters for black tea, decaffeinated black tea, and flavored tea are physical, chemical, sensory, and microbiological aspects. Physical characteristics are grades, particle size, bulk density, color, and appearance that reflect the visual quality. The chemical parameters include polyphenols (tannins), total solubles, and extraneous matter. ISI and PFA have specified the limits for ash, water solubles, and fiber content to safeguard the interest of the consumer. In the case of instant tea, the liquor characteristics and free flowability of the product are considered. Microbiological parameters are important in liquid tea products such as carbonated beverage and canned/frozen tea concentrates.

4.1 Physical Characteristics

The processed black/green teas are separated from stalks, fiber, and foreign matter. According to the market requirements of particle size, they are sorted into different grades by sorting machines fitted with meshes of different sizes. They broadly fall into four categories, leaf grade, brokens, fannings, and dust, based on the appearance, size, and liquor qualities. Specifications for intermediate grades in the categories of black tea, CTC tea, and green tea are available in the literature.

4.2 Chemical Quality of Tea

Chemical constituents such as polyphenols (tannins); caffeine, pectic substances, pigments (theaflavins [TFs] and thearubigins [TRs]), noncaffeine nitrogenous compounds, sugars, minerals, and lipids contribute to the quality of tea (32). The processed black tea contains 10% of tannins, compared to green tea leaves, which contain 18%–20% of tannins. TF and TR, which are formed in the ratio of 1:10 or 1:12 during the fermentation process, give a correct quality assessment of the made tea, indicating efficient fermentation. Ethylacetate extracts of tea contain the whole of TF which give the bright color and brisk sensation on the palate. High-quality tea, with good colour, strength, briskness, and flavor, has a low TR/TF ratio. Since TR content is roughly 10 times that of TF, a tea with a low TR/TF ratio, say, 10:12, has the best quality.

The quality of raw material can also be tested before processing to determine the quality of the final product. It is established that the four measurable characteristics—viz., crop yield per bush (C), hairiness of leaf (H), phloem index (Pi), and ratio of xylem of phloem vessels on vascular bundles (Vs)—in the following multiple regression equations can predict tea flavor quality and strength:

$$\text{Quality} = -27 + 0.06\,C + 5.5\,H + 0.16\,Pi + 50\,Vb$$
$$\text{Strength} = +22 + 0.11\,C + 4.5\,H + 0.05\,Pi - 22\,Vb$$

Phloem index is the frequency of occurrence of calcium oxalate crystals in the phloem tissue of the petiole.

The estimation of TF and TR consists of extracting them into the ethyl acetate layer from the aqueous layer; further separation is achieved by shaking with sodium bicarbonate solution. The optical densities of TF and TR are measured at 380 and 460 nm, respectively.

Tannins are estimated by Lowenthal's method, which is based on extraction of tannins into an aqueous medium and precipitation by addition of gelatin. Estimation of tannins is carried out before and after precipitation with gelatin by titrating with standard permanganate solution. This step yields the oxidizable tannins present in the sample.

Total caffeine is estimated by Baily-Andrew's method. Caffeine is extracted into aqueous medium by digesting tea leaves with heavy magnesium oxides. Then it is extracted into a chloroform layer and desolventized to produce caffeine in solid form. Nitrogen is estimated in the crude caffeine by the micro-Kjeldhal method and multiplied by a conversion factor. It is reported as percentage of caffeine in the sample. The analysis of ash, water-soluble ash, alkalinity of ash, acid-insoluble ash, water extract, and crude fiber uses standard procedures.

Limits of specifications by the Bureau of Indian Standards (percentage by mass) are total ash, 4.0%–8.0%; water-soluble ash (expressed as percentage of total ash), 40.0% (minimum); acid-insoluble ash, 1.0% (maximum); alkalinity of water-soluble ash (as K_2O), 1.0%–2.2%; water extract, 32.0% (minimum); and crude fiber, 16.5% (maximum).

4.3 Foreign Substances and Adulteration

Contamination by foreign substances such as copper, lead, and iron is caused by equipment used in processing. Spent tea as such or mixed with coaltar dyes, catechu, caramel, and curcuma after drying is used as an adulterant. Blackgram husk and sawdust may also be mixed after coloring. Foreign leaves that are astringent are used after drying for adulterating tea. Limits are fixed for such foreign substances in several countries.

4.3.1 Pesticide Residues

Insecticides and fungicides used for the protection of crops may impart residues in made tea. It is observed that sulfur contributes an undesirable odor, and benzene hexachloride (BHC) a musty taint, whereas dichlorodiphenyl trichloroethane (DDT) does not. So far no tolerance limits for pesticide residues in tea have been legalized. Efforts are made to gain biological control over tea plant insects.

4.4 Blending of Teas

The graded teas, at 3% moisture level, are packed in tea chests, preferably with aluminum foil or pliofilm lining in humidity controlled rooms. The graded teas, whose quality and value are known to brokers, simplify selling and buying in specific world marketing centers according to the demand. The buyers, through expert tasters, blend the grades in different proportions according to the consumer needs of different areas for retail sale. Blending is mixing of teas with different characteristics, is to create products that best satisfy the requirements of the consumer market. Retail tea is invariably a blend of different grades that contains 20–40 different components in various proportions on the basis of quality, flavor, strength, body, size and style of leaf, and price of the selected teas. In order to balance the cost, relatively cheap neutral teas or ''fillers'' are normally used. The grades of black and green teas are listed in Appendix 1.

Since no tea grade has consistent quality, because of variations in processing and seasons, regional blending is always done only by expert tasters, as there are no alternative objective methods available.

4.5 Tea Tasting and Sensory Quality of Tea

The quality of tea leaf is determined by the combined effects of nonvolatile solids (extractable from tea leaf to about 0.30% to 0.45% solids concentration under normal brewing conditions) as well as the very complex volatile fraction (which constitutes less than 0.1% of tea leaf by dry weight basis). Caffeine is the important component of the nonvolatile fraction that gives a modified pleasant taste to the infusion in the presence of polyphenols. Tea leaves contain 2.5% to 4.0% caffeine and traces of theophylline and theobromine. The refreshing property of tea is due to the presence of caffeine. The other nonvolatiles that contribute to taste include carbohydrates, proteins, and pectins. Caffeine determines the level of tangy astringency in black tea infusion (28).

Infusion is made by pouring boiling water into a mug (earthenware, china, or plastic, preferably the former) of 0.142-L capacity in which a graded/blended tea sample (2.83 g) is placed (33). After 5–6 min the brew (of 2% solids) is evaluated by an expert for color (depth, brightness), quality (good, coarse), strength (strong, weak), briskness (pungent, soft), flavor (fine, poor), manuring faults, taints, cream (bright, dull), and other characteristics. The evaluation also includes (a) the quality of the infused leaf, viz., color, together with appearance (bright copper, bright greenish, mixed, dull, dark); (b) the smell of the wet, hot infused leaf; and (c) the quality of dry leaf, viz., the grade (uniformity of size and form), color (black, grayish), make and style (general appearance as desirable to the specified grade), aroma (smell, warmth, and dampness), and feel (spongy, lack of density).

Tea tasting terms, terms relating to tea manufacture, and those used by producers, buyers, blenders, and brokers in the trade are well defined. The most important and comprehensive glossary of tasters' terms for dry leaf, infused leaf, liquour, and defects is available in the literature (33).

Aroma and appearance, which cannot be determined chemically, play a vital role in tea quality and price. To satisfy the changing market requirements and variation in quality arising from processing, seasonal, and regional influences, sensory quality assessment by expert tasters is essential for routine evaluation of both commercial samples and estate varieties. The analytical results are also useful when tasting reports are contradictory and help to rectify processing variations. Tasting is more economical than chemical analysis but the two are complementary. The chemical composition of tea liquors in relation to their cup characteristics is compared as color versus theaflavins and thearubigins; strength versus total solids, theaflavins, and thearubigins; briskness versus theaflavins; and quality versus caffeine, theaflavins, and thearubigins.

4.6 Particulars

4.6.1 Flavored Teas

The quality parameters described also hold good for flavored teas. A standard quality flavor should be incorporated. Moisture content of the sample is another vital factor that alters the added flavor profile during storage. Hence, moisture should be kept at well below 6%.

4.6.2 Decaffeinated Teas

Apart from quality factors described, residual solvent is another important parameter for decaffeinated tea. Methylene chloride is a solvent permitted for use in decaffeination in European countries. The permissible level is 30 ppm. Color and appearance of the decaffeinated product are usually dull.

4.6.3 Instant Teas

The quality of instant tea is reflected by moisture, bulk density, ash, caffeine, and teaflavin, and thearubigin. Normally, bulk density of the product varies between 0.06 and 0.17

g/cm^3, but density less than 0.07 g/cm^3, is not desirable. A particle size of 200–500 μm gives a free-flowing powder and good solubility. Moisture content should be kept well below 5% in order to prevent caking and darkening of color. The cold-water-soluble instant tea is valued for its clarity and appearance. The clarity of the brew is increased by addition of chemicals at the decreaming step of the process. This increases the ash content in the final product and causes adverse taste (27).

5 STORAGE AND PACKAGING ASPECTS

5.1 Tea

Manufactured black teas are subject to different conditions of storage for a period of 3–6 months before they reach the consumer. Chemical changes generally observed during storage are loss of flavor and astringency and development of undesirable taints due to lipid hydrolysis; loss of theaflavins, amino acids, sugar, and pigments; and increase in moisture content. The changes are accelerated by high moisture and elevated temperature. Storage at low temperature, low moisture, and low oxygen availability and treatment with acid during the fermentation stage of manufacture lead to positive reduction in residual enzyme activity, thereby maintaining theaflavin content during storage and resulting in good cup characteristics. Black tea can store well for 300 days, retaining its flavor, at 32% relative humidity, 20°C, and light-absent conditions.

Storage of green tea results in reduction in vitamin C content, changes in tea color from bright green to olive green to dull brownish green, change in color of tea liquor, and dullness in aroma. These changes are accelerated by moisture, oxygen, elevated temperatures, and light sensitivity (34–36).

Suitable packing materials, both unit and bulk packing, are essential to protect tea from moisture and oxygen. Packet teas of various sizes from 5 to 500 g are packed in single-use paper pouches and foil-laminate 500-g packs made up of paperboard cartons or polycoated foil-lined paperboard cartons. For tea bags single-dip-type unit packing is used. Polyethylene unit bags are used for shorter storage periods and foil laminates for longer storage.

Bulk packing is done in plywood chests with an inner foil lining. Cellulose film or metallized polyester is also used in place of aluminum foil, for export purposes. Recently, multiwall sacks have been introduced for bulk packing. Multiwall paper sacks made from extensible kraft paper are found suitable for bulk packaging in place of conventional plywood chests. In the course of time plywood boxes may be replaced by corrugated board boxes (36).

5.2 Tea Products

The same type of packing described for tea is used for tea products such as flavored and decaffeinated teas. Instant tea is usually packed in glass jars, but tagger top tin and refill packs made with folding cartons lined with aluminum foil laminate or paper/foil/poly-laminate pouches are also used. Carbonated teas/liquid teas are packed in plastic bottles, nonreturnable glass, refillable glass, and metal cans. PET bottles of all sizes are used for the carbonated beverage. Steel cans are replacing aluminum cans (36).

6 MEDICINAL AND NUTRITIVE PROPERTIES OF TEA

Some of the valuable compounds in tea, such as caffeine, theobromine, theophylline, tannin, essential oils, minerals, and vitamins, through their pharmacological and physiological properties have beneficial effects on digestion, the nervous system, blood vessels, cardiovascular function, blood pressure, and vital energy. Tea, an excellent diaphoretic, stimulates metabolic processes and plays a certain prophylactic role. Tea alkaloids are very good for mental fatigue and contribute to cerebral vasodilatation. Tea's beneficial effects on cold induced fever are commonly known. Recent findings show that the tannin-catechin complex has high vitamin P activity, superior to that of capillary-strengthening drugs, e.g., citrin, rutin, esculin, and activity increases as a result of accumulation of ascorbic acid. Clinical tests proved that scurvy cannot be cured with the aid of crystalline ascorbic acid alone, but in combination with vitamin P it can, through their synergistic effect, increase total resistance to infectious diseases. Tea flavonols also have vitamin P activity.

The antioxidative property of tea catechins has an antiradiation effect. In irradiation experiments, addition of catechins to oxygen-containing tissues decreased their radiation induced injury as a result of rapid development of oxidative processes. Tea catechins are used as therapeutic and prophylactic drugs, and catechin-rich tea can provide some protection against radiation. The tannin-catechin complex of green tea infusion almost completely eliminates adverse effects of strontium-90 on the human body, absorbing the radioactive isotope and excreting it before it can reach bone marrow. Tea phenolic compounds, as important biological antioxidants, can delay arteriosclerotic development in the human body, which involves peroxidation of membrane lipids that is due to lack of biological antioxidants such as tocopherols. Tea polyphenols have bactericidal and bacteriostatic activity. Because of this antimicrobial action, green tea brew is effective in the control of dysentery. The strong protistocidal effect of tea extract is used against certain protozoan diseases of humans and farm animals.

Hypertension, diabetes, poliomyelitis, scarlet fever, other infectious diseases, and heavy muscle work, especially in hot workshops, result in a phenomenal increase in the requirement of vitamin P. Green tea, taken in the form of a hot liquor, can replenish this deficiency. Catechin-rich teas favorably affect the metabolism of healthy people. Green tea brew alleviates headache and beneficially influences the cardiovascular system, fluid electrolyte balance, hemopoiesis, and renal function. Green tea brew regulates cholesterol metabolism in atherosclerotic patients. Green tea, when incorporated into combined antirheumatic therapy at an active stage of rheumatism, produces favorable effects on the general condition of patients and their capillary resistance. The effect of green tea on chronic hepatitis of mainly viral origin is found to alleviate inflammatory processes and regulate vital components of human metabolism. Green tea is claimed to protect against the incidence of cancer, especially lung, esophageal, skin, and gastrointestinal cancers.

Another important nutritional function of tea is that of maintaining fluid balance, which is due to its versatility, absence of harm, and potential health benefits. In addition, tea is apparently a good source of fluoride and manganese, which prevents dental cavities and aids protein and energy metabolism, respectively (38, 39).

7 BY-PRODUCTS OF TEA

Tea waste is a rich source of a number of by-products, such as caffeine, polyphenols, pigments, polymer extenders, vitamins, minerals, growth regulators, and foaming agents. Tea seed, another waste product, is utilized for production of oil, saponins, and furfural from its husk. Spent tea and sludge obtained during manufacture of cold-soluble instant tea are used for making *n*-triacontanol, caffeine, and other minor constituents (16, 41).

7.1 Caffeine

Although caffeine is produced synthetically there is always a significant preference for the natural caffeine extracted from tea waste. Tea waste is first denatured with lime or alkali; caffeine is extracted with solvents such as benzene, chloroform, methylene chloride, and other halogenated hydrocarbons; and the solvent is evaporated and the residue extracted with hot water, purified, and crystallized. In another method, the denatured tea waste is extracted with hot water, followed by transfer of caffeine to any of the water-immiscible solvents, concentration of the organic phase, purification, and crystallization (16, 41).

7.2 Polyphenols and Pigments

Polyphenols and pigments from tea waste mainly comprise theaflavins, thearubigins, and catechins, which constitute about 10%–15% of the tea extractives. Tea polyphenols have considerable potential as food antioxidants and in the production of food color. When tea extracts are coupled with beet root pigments, a red dye is obtained; an orange dye is produced by mixing with the ethylacetate-soluble portion. By treating tea extracts with iron lactate solution, a black dye can be produced. Tea polyphenols impart stability to these colors, making them heat-stable and resistant to enzymic action and microbial growth (16, 41).

7.3 Vitamins and Minerals

Dried tea leaves and tea waste are comparable to dried alfalfa leaves in their vitamin and mineral content, and a vitamin flan can be produced from vegetative waste of tea bushes.

7.4 Extenders in Polymers

In the production of phenolformaldehyde resin, a partial replacement of phenol with tea waste polyphenols increases the flowability of the resultant resin, without adversely affecting the polymer characteristics, and reduces the cost. Adhesives for wood that have a high water content are prepared by using tea powder as filler, producing high flow values. The replacement of the existing resins in plywood adhesive with tannin polymers from tea waste appears to be a possibility.

7.5 Animal Feed and Composting

Spent tea waste, which is rich in crude protein and amino acids, can be a good source for animal feed formulations.

7.6 Tobacco Substitutes and Foaming Agents

Tea waste tannins are used in filters for tobacco smoke and as tobacco substitute sheets by grinding to a paste, mixing with plasticizing substances, and rolling into a sheet. A foaming agent produced from tea extract can be used as a substitute for egg white.

7.7 Triacontanol, a New Plant Growth Regulator from Tea

Instant tea waste, obtained from green tea leaves, contains a maximal amount of triacontanol. Triacontanol in quantities as small as 50 mg/Ha enhances the yield of some crops, such as tomato, cucumber, lettuce, maize, corn, and rice. Trials on tea plants showed an increase in harvestable yield of 25%–30% with reduction in the dormant shoots. Tests of these materials on other crops also showed good and encouraging results.

7.8 Tea Seed Oil

Tea seed contains 20%–45% oil, which resembles groundnut and olive oils. It is rich in unsaturated fatty acids, especially linoleic acid and vitamin E. It can be hydrogenated in the presence of a nickel catalyst to yield a hydrogenated fat. It is used in the cosmetic and pharmaceutical industries as a substitute for olive oil.

7.9 Saponins

Saponins constitute 10%–14% in tea seeds and can be removed by solvent extraction. Saponins are plant glycosides with good emulsifying properties. They are used as stabilizers for photographic film emulsions, as a means of increasing the solubility of carbon dioxide in carbonated beverages, and for control of predaceous fish in ponds and earthworms in golf courses. Treatment of pruned tea bushes with pure tea seed saponins increased yield of flush and nursery plants, root weight, and shoot growth (16, 41).

8 MARKETING AND SALES

The marketing of tea in bulk auction and the retail sale of tea fluctuate widely, as a result of many factors. The domestic demand for tea is steadily increasing in some of the producing countries, including India and Malaysia, curtailing the quantity for export. For the past few years, India has restricted export of CTC tea. Sometimes the price of tea in foreign markets falls suddenly, as a result of the presence of substandard tea or other causes affecting further export. The pattern of consumer purchases changes with price and availability of tea. Tea has to compete with a large number of soft drinks that are widely promoted through mass media. Changes of lifestyle and likes and dislikes have changed the pattern of tea consumption among different age groups: for example, in Canada, hot beverage sales have declined, and in Europe bagged teas are generally in high demand. In some markets, for instance, Malaysia, demand for herbal tea and ginger tea has suddenly increased. Bottled tea is a popular beverage in Indonesia. Proper and continuous market research is essential to forecast trends, meet demand, and yield maximal profit to the producer. The Food Service Committee of the Tea Association of the United States strives to educate people on the preparation of good tea and induce them to supply ''hot tea hot,'' to promote the tea drinking habit. China hosts tea weeks, e.g., Tea and Chinese Culture. The Tea Council of London Tea Pot—2000, Singapore's High Tea, and India's Brand Promotion are geared to establishing a steady market for tea.

9 CURRENT AND FUTURE RESEARCH AND DEVELOPMENT WORK

Research and development for processing methods continue. Development of biosensors to test the completion of chemical oxidation at different stages of processing is being attempted to allow the production of consistently high-quality black tea. Simulation of processing variables may reduce different operations, time, energy, labor, and therefore cost. There is continued research in new product development, value-added products, convenience products, and development of compatible flavors to satisfy customer requirements. Low-grade teas and surplus teas can be profitably diverted to manufacturing useful tea products like brick tea, which has longer shelf life.

Well-defined, rigorous quality control procedures will help exports, preventing financial loss. Development of quality standards for all major tea products and convenience forms will be useful. Analytical quality records on traditional tea samples, if made available, will enhance cup tasting and blending of tea for different consumer markets.

Intensified applied research to establish healthful properties of tea, health hazards due to chemical fertilizers and pesticides, and usefulness of organic fertilizers and biological control is being pursued. By application of genetic engineering techniques attempts are being made to develop high-yielding, high-quality tea clones.

ACKNOWLEDGMENTS

Grateful acknowledgments are due to Dr. N. Krishnamurthy, Head, Plantation Products, Spices and Flavour Technology Department, for his keen interest and encouragement during the preparation of the manuscript, and Dr. V. Prakash, Director, Central Food Technological Research Institute, Mysore, India, for permission to publish this chapter.

APPENDICES

Appendix 1 Catechin Content and Quality of Various Types of Tea

Tea type	Catechin (mg/g)	Appearance	Color of infusion	Aroma	Taste
Nonfermented type					
Green tea	105.30	Leaf shape and brokens	Bright green/ olive green	Fresh Brisk Raw Greenish	Brisk Bitter
Yellow tea	53.40	Leafy	Yellow Bright	Fresh characteristic	Mellowed Less bitter
Dark tea	35.78	Pressed leaf	Brown with yellowish tinge	Smoky Piny Stale Flowery Fruity	Mellowed
Semifermented type					
Oolong tea	35.64	Leaf shape	Golden yellow	Rich flowery odor	Strong Mellow
Fermented type					
White tea	50.00	Fresh Leafy with hairs	Light orange Yellow	Fresh Leafy Raw	Plain
Black tea	6.90	Leafy Brokens Even	Reddish orange color	Malty Flowery	Strong Brisk

Appendix 2 Specifications for Green Tea and Black Tea

Constituents (%)	Green tea (nonfermented)	Black tea (fermented)
Water extract (m/m)		
minimum	32.0	32.0
Total ash (m/m)		
minimum	4.0	4.0
Acid-insoluble ash (m/m)		
maximum	1.0	1.0
Crude fiber (m/m)		
maximum	16.5	16.5
minimum	9.0	—
Total catechins (m/m)		
maximum	19.0	—

Appendix 3 Concentrations of Catechins and Caffeine in Tea Infusions

Constituents (mg/100ml)	Green tea (nonfermented)	Oolong tea (semifermented)	Black tea (fermented)
Catechins	19.65	16.70	6.90
Caffeine	24.51	23.80	18.70

Appendix 4 Vitamin C Content in Different Types of Tea

Tea type	Vitamin C (mg/100g tea)
Nonfermented type	
Japanese green teas	
Sencha	245 (500 maximum)
Gyokuro	106 (190 maximum)
Kamaricha	205
Chinese green tea	
Maofeng Cha	140
Semifermented type	
Oolong tea	6
Fermented type	
Black tea	0

Appendix 5 Black Tea Grades—Orthodox

SI no.	Kind of tea	Grade name	Nomenclature
1.	Whole leaf	FP	Flowery pekoe
		FTGFOP	Fine tippy golden flowery orange pekoe
		TGFOP	Tippy golden flowery orange pekoe
		TGFOP1	Tippy golden flowery orange pekoe one
		GFOP	Golden flowery orange pekoe
		FOP	Flowery orange pekoe
		OP	Orange pekoe
2.	Broken	BOP1	Broken orange pekoe one
		GFBOP	Golden flowery broken orange pekoe
		BPS	Broken pekoe souchong
		GBOP	Golden broken orange pekoe
		FBOP	Flowery broken orange pekoe
		BOP	Broken orange pekoe
3.	Fannings	GOF	Golden orange fannings
		FOF	Flowery orange fannings
		BOPF	Broken orange pekoe fannings
4.	Dust	OPD	Orthodox pekoe dust
		OCD	Orthodox churamani dust
		BOPD	Broken orange pekoe dust
		BOPFD	Broken orange pekoe fine dust
		FD	Fine dust
		D-A	Dust A
		Spl. D	Special dust
		GD	Golden dust
		OD	Orthodox dust

Appendix 6 Black Tea Grades—Crush, Tear, Curl

SI no.	Kind of tea	Grade name	Nomenclature
1.	Broken	PEK	Pekoe
		BP	Broken pekoe
		BOP	Broken orange pekoe
		BPS	Broken pekoe souchong
		BP1	Broken pekoe one
		FP	Flowery pekoe
2.	Fannings	OF	Orange fannings
		PF	Pekoe fannings
		PF1	Pekoe fannings one
		BOPF	Broken orange pekoe fannings
3.	Dust	PD	Pekoe dust
		D	Dust
		CD	Churamani dust
		PD1	Pekoe dust one
		D1	Dust one
		CD1	Churamani dust one
		RD	Red dust
		FD	Fine dust
		SFD	Super fine dust
		RD1	Red dust one
		GD	Golden dust
		SRD	Super red dust

Appendix 7 Composition of Fresh Tea Shoots

Crude fiber, cellulose, lignin, etc.	22.0%
Proteins	16.0%
Fats	8.0%
Chlorophyll and other pigment	1.5%
Pectin	4.0%
Starches	0.5%
(i) Insoluble in water (total)	52.0%
Fermentable polyphenols	20.0%
Other polyphenols	10.0%
Caffeine	4.0%
Sugar and gummy matter	3.0%
Amino acids	7.0%
Minerals (ash)	4.0%
(ii) Soluble in water (total)	48.0%

Appendix 8 (top photo) Withering
(bottom photo) CTC machine

Appendix 9 (top) Fluid bed dryer
(bottom) Cup tasting

Appendix 10 (top) Grades of tea
(bottom) Black tea brew

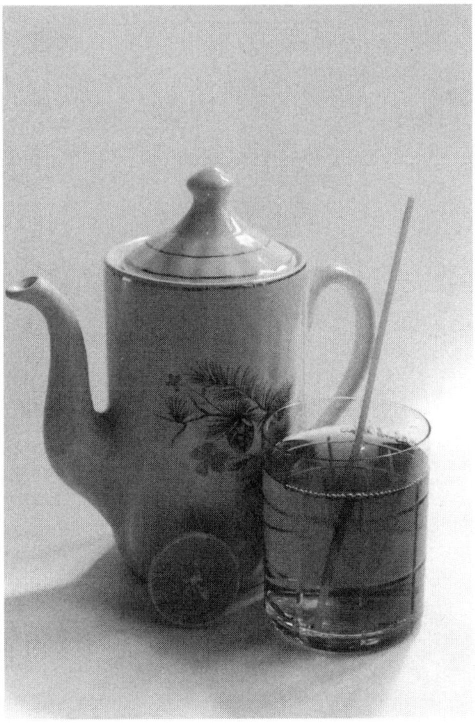

Appendix 11 (top) Semifermented tea brew (bottom) Lemon tea

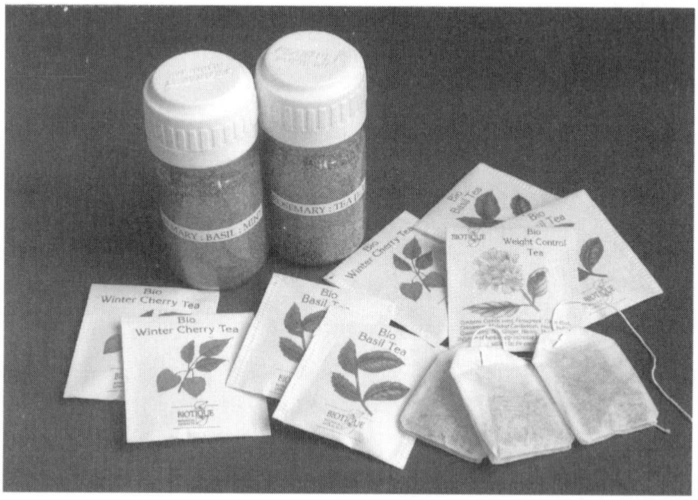

Appendix 12 (top) Tea products
(bottom) Herbal teas

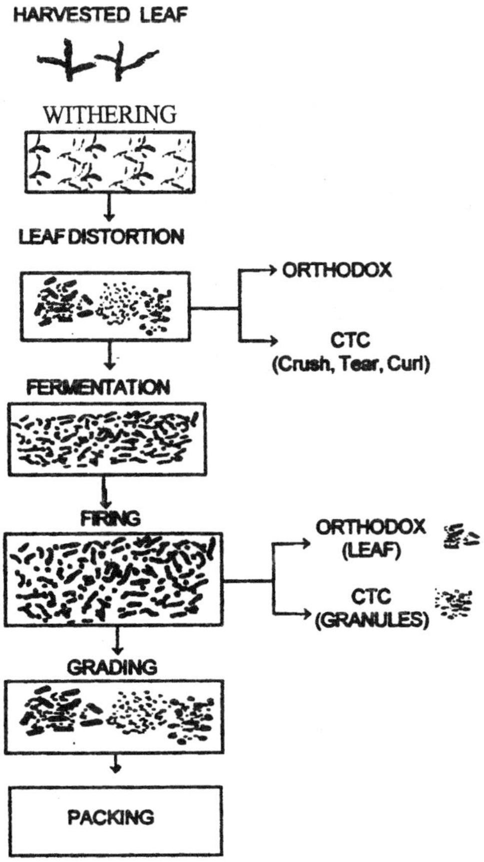

Appendix 13 Flow chart for processing tea

Appendix 14 Two leaves and a bud

Appendix 15 Tea estate

REFERENCES

1. Balasubramanyam, Tea in India, Publication and Information Directorate/Wiley Eastern, New Delhi, India (1995).
2. FAO, Production Year, Vol. 50 (1996).
3. Food Digest CFTRI, Mysore 21(3), p. 134, 135 (1998).
4. T. Eden, Tea, 2nd ed., Western Printing Services, Bristol, p. 148 (1965).
5. J. Workhoven, FAO Bulletin, Rome, No. 26 (1974).
6. G. W. Sanderson, Structural and functional aspects of phytochemistry, Rec. Adv. Phytochem., 5, p. 251 (1972).
7. H. Graham, Tea. In Kirkothmer Encyclopedia of Chemical Technology, 3rd ed., Wiley Interscience, 22, 628 (1978).
8. G. W. Sanderson. In Geruch and Gesmack Stoffe (Drawert, D. F., Ed.), Verlag Hans Carl, Nurenberg, p. 65 (1975).
9. G. W. Sanderson and H. N. Graham, J. Agric. Food Chem. 21, p. 576 (1973).
10. I. Flement, Coffee, Cocoa, Tea, 5(3), p. 317 (1989).
11. T. Takeo and P. K. Mahantha, J. Sci. Food. Agric., 34, p. 307 (1983).
12. T. Yamanishi. In Flavour of Foods and Beverages, Chemistry and Tech (George Charalambous and George E. Inglett, eds.), p. 305 (1978).
13. M. A. Bokuchava and N. I. Skobeleva, CRC Crit. Rev. Food Sci. Nutr., 12, p. 303 (July 1980).
14. Food Rev. Int., Special issue on Tea, 11(3), p. 371 (1995).
15. S. Kanthamani, United Planters' Association of South India, 47 (1971).
16. S. Nagalakshmi, Guide on Food Products, Year Book, Vol. III(3), Trade India Publications, Bombay, p. 1 (1993).
17. C. R. Harler, The Culture and Marketing of Tea, Oxford University Press, London (1964).
18. T. Samnasivam, Tea Coffee Trade J., 162(1), p. 36 (1990).
19. D. Nagalakshmi, C. P. Natarajan, and R. Seshadri, J. Sci. Food Agric., 37, p. 185 (1986).
20. Beverages. In The Good Cook, Time Life Books, Amsterdam, p. 14 (1982).
21. N. I. Orgavelidze, I. D. Bochorishvili, G. S. Dzhgerenaya, Z. A. Rizhamadze, and E. R. Kvanchinani, USSR Patent SU 12025 40A (1986) (RU).
22. P. S. Elias, Association Scientifique International du Cafe, 93 (1985).
23. M. S. Sivetz, Coffee Processing Technology, AVI, Westport, Connecticut, Vol. 2, p. 207, (1963).
24. B. D. Page and C. F. Charbonneu, J. Assoc Official Analytical Chem, 67(4), p. 757 (1984).
25. J. Puranaik and S. Nagalakshmi, J. Agric. Food Chem., 45, p. 3973 (1997).
26. J. L. Blauch and S. M. Tarkajr, J. Food Sci., 48, p. 745 (1983).
27. N. D. Pintauro, Noyes Data Corporation, New Jersey (1977).
28. S. Nagalakshmi and R. Seshadri, Indian Food Ind. 5, p. 13 (1986).
29. K. K. Mitra, Int. Tea J., No. 3, p. 9 (1982).
30. A. Ekanayake and P. B. Chandradasa, J. Sci. Food. Agric., 41(1), p. 87 (1987).
31. T. Matsuzaki and Y. Hara, Nippon Nogeikagaku Kaishi, 59, p. 129 (1985).
32. G. W. Sanderson, S. R. Arvind, S. E. Larry, J. F. Francis, S. Robert, H. M. Charles, and C. Philip, ACS Symposium Series No. 26, p. 17 (1975)
33. Glossary of Tea Tasting Terms. In Preparation of Tea Infusion—IS 6400 (1971), Indian Standards Institution, New Delhi, IS 4541 (1986).
34. G. V. Stagg, J. Sci. Food Agric., 25, p. 1015 (1974).
35. S. Jayaratnam and D. Kirtisinghe, Tea Q., 44, p. 164 (1974).
36. J. Dougan, E. J. Glossop, G. E. Howard, and B. D. Jones, A Study of Changes Occurring in Black Tea during Storage, Tropical Products Institute, London (1978).
37. H. K. Das, Two and a Bud, 35(1/2), 34 (1988) FSTA 23(6) 6H 103 (1991).
38. Critical Rev. Food Sci. Nutri., Special issue, Tea and Health (F. M. Clydesdale, ed.), 37(8) (1997).

39. Chemistry and Applications of Green Tea (T. Yamanoto, L. R. Juneja, D. Chi Chu, and M. Kim, eds.), CRC Press, Boca Raton, Florida, pp. 37, 61, 87, 109, 137 (1997).
40. T. Yamanishi, Handbook of Food and Beverage Stability: Chemical Biochemical, Microbiological and Nutritional Aspects, Academic Press, New York, p. 665 (1986).
41. R. Seshadri, S. Nagalakshmi, J. Madhusudhana Rao, and C. P. Natarajan, Trop. Agric., 63(1), p. 2 (1986).

26

Postharvest Technology of Cocoa

KAMARUDDIN ABDULLAH

Institut Pertanian Bogor, Bogor, Indonesia

1 INTRODUCTION

Cocoa was known as *chocolatl*, the royal beverage of the king Montezuma during the Mayan period. It also has a medicinal effect and was so famous and high priced that it was once used as money and for tax payment in Mexico.

Dried cocoa beans are usually processed into cocoa butter, which contains fat between 56% and 58%, made up of a mixture of triglycerides. A typical cocoa butter from West Africa is composed of the following fatty acids: palmitic, 25.3%; stearic, 36.6%; oleic, 33.3%; linoleic, 2.8%; and other fatty acids 2.0%.

Cocoa butter with this composition is good for making chocolate and has favorable hardness and a melting point of body temperature (Wood and Lass, 1985). Most chocolate manufacturers require that cocoa beans can be processed into good chocolate with good flavor. Such quality depends very much on the variety of tree and the method of postharvest handling, such as fermentation and drying. Indonesia began to diversify its cocoa production in the mid-1970s not only to produce dried cocoa beans (CCCN No. 1801000), but also to process the beans into cocoa butter (CCCN No. 1801000), cocoa paste (CCCN No. 1803000), cocoa powder (CCCN No. 1805000), chocolate milk crumbs (CCCN No. 1806120), food preparation containing cocoa (CCCN No. 1806290), confectionery containing cocoa (CCCN No. 1806390), and Choco (CCCN No. 1806000) (Spillane, 1995).

Theobroma cacao originated from the jungle of the Amazon in Brazil—the area between 18° N and 15° S. *Theobroma cacao* consists of two sub-species: the *Cacao*, a subspecies known as the *Criollo*, which can be found in Central and South America, and *spharocarpum*, which is also called the *Forastero* (Cuatrecasas, 1964). The *Criollo* has superior quality. The pods are red and the beans have no color. *Forastero* has a lower quality and green pods. The *Forastero* has one hybrid called the *Trinitario*, which is widely adopted. The color of the pods is yellow, the beans are violet, and the pods can have spherical or enlongated forms (Spillane, 1995). The *Forestero* grows well at the elevation of 400 m above sea level, and has larger beans; it is faster in bearing pods but has less aroma.

The world producers of cocoa (*Theobroma cacao L.*) are on the African continent (Ivory Coast, Ghana, Nigeria, and Cameroon), in Latin America (Brazil and Equador), and in Asia (Malaysia and Indonesia). In 1987, the world production of cocoa in net output was 2.2 million tons and increased to 2.3 million tons in 1993. In the Asia-Pacific region, Indonesia was the largest cocoa producer in 1993 with 280,000 tons, followed by Malaysia with 220,000 tons, and Papua New Guinea with 38,000 tons. In 1998, total production from Indonesia reached 306,000 tons (BPS, 1998).

Cocoa was first introduced to Indonesia in the 16th century from the Philippines. It was first grown in the Sulawesi area and later in Java. The main area of production in the early stage of its introduction into the country was Java and Sumatra as a substitute for coffee destroyed by a coffee leaf rust. On Java, the *Criollo* was grown in significant amounts in the 1960s when the production of *Forastero* cocoa in East Java reached 5000 tons. In Indonesia, major cocoa plantations can be found in West Java, East Java, Bali, Sulawesi, and North Sumatra. The average yield per hectare is estimated to be between 1.7 and 2.1 tons. Cocoa in Indonesia can be harvested almost all year, except the months of August and September. The total export between January and September 1996 was about 166,000 tons, worth U.S. $192 million. Table 1 shows the main producers of cocoa in the Asia and Oceania region.

2 POSTHARVEST ACTIVITIES

Postharvest activities commonly practiced in cocoa producing countries are shown in Fig. 1 and will be described in the following sections. These include (a) harvesting, (b) curing/fementation, (c) drying, (d) sorting and quality determination, and (e) packaging.

2.1 Harvesting

There are three types of cocoa tree that are widely grown: the *Criollo*, the *Forastero*, and the *Trinitario*. The basic differences among these trees are shown in Table 2.

Cacao pods have several variations in form, length, hardness of the skin, and color. The forms can be spherical to cylindrical, while the skin may have spikes and holes. The following are common forms of cocoa pods (Spillane, 1995):

1. *Angoleta*: deeply ridged, warty, with square form at the stalk end
2. *Cundeamor*: more or less the same as *Angoleta*, but the neck resembles a bottle
3. *Amelonado*: smooth, shallow furrows; resembles a melon with one end blunt; resembles a bottle
4. *Calabacillo*: small and spherical

Harvest time of mature cocoa varies from place to place from 115–150 days after flowering in Nigeria to 165–200 days after flowering in Papua New Guinea. According to Alvim et al. (1972) maturity can be estimated using the following empirical formula:

$$N = 2500/(T - 9) \qquad (1)$$

Table 1 Production of Cocoa from the Asian and Oceania Region in Thousands of Tons

Country/Year	1984 /1985	1985 /1986	1986 /1987	1987 /1988	1988 /1989	1989 /1990	1990 /1991	1991 /1992	1992 /1993	1993 /1994
1. Fiji	0.1	0.3	0.3	0.3	0.2	0.4	0.3	0.4	0.3	0.3
2. India	4	6	6	6	6	6	6	6	6	6
3. Indonesia	31	40	48	58	93	115	150	180	245	280
4. Malaysia	9	131.2	167	227	225	243	221	220	225	220
5. Papua New Guinea	32.4	32.3	30.6	33.8	48	10.8	33.4	40.9	38.8	38
6. Philippines	5	5	6	5	8	7	5	5	6	6
7. Samoa	0.7	0.8	0.5	0.7	0.5	0.5	—	—	—	—
8. Solomon Islands	1.2	1.4	1.2	2.5	2.7	3.6	4.1	3.5	3	3
9. Sri Lanka	2.5	2.5	2.5	2	2	2	1.4	1.4	1.4	1.4
10. Thailand	0.2	0.2	0.3	0.4	0.4	0.4	0.4	0.4	0.4	0.4
11. Vanuatu	0.6	1.2	1.1	0.8	1.4	2.2	2.2	1.5	1.6	1.8
12. Other countries	0.1	0.1	0.1	0.1	0.1	0.1	0.1	0.1	0.1	0.1

Source: BPS, 1995.

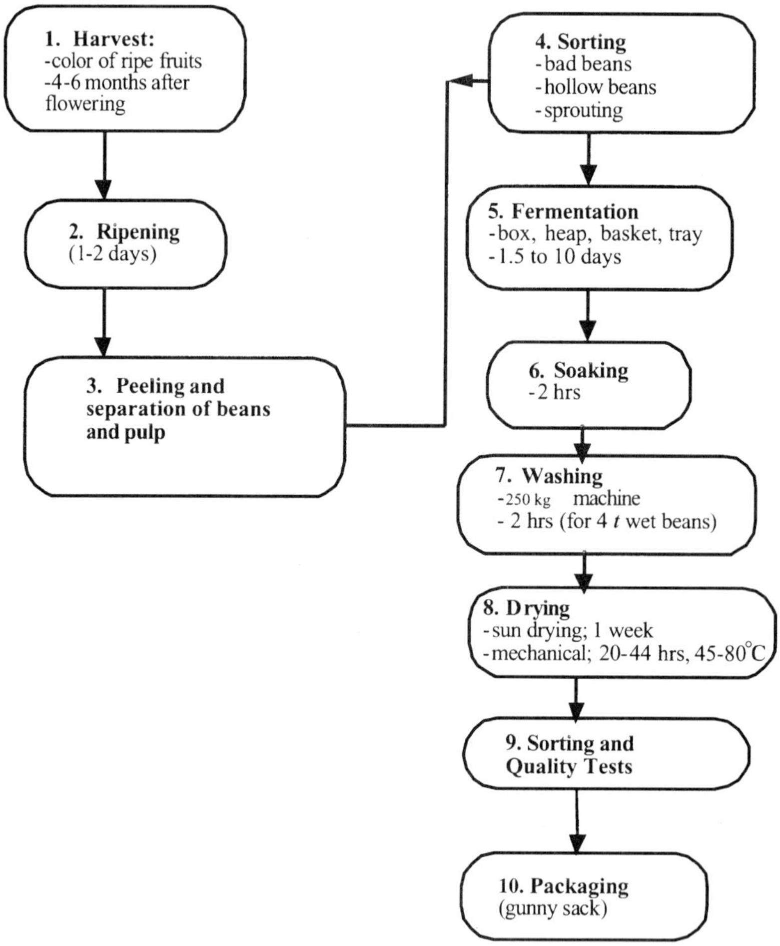

Fig. 1 Typical postharvest activities for cocoa processing technology.

Table 2 Characteristics of the *Criollo*, *Forastero*, and *Trinitario*

	Criollo	*Forastero*	*Trinitario*
1. Pod husk/cocoa tree:			
a. Texture	Soft	Hard	Mostly hard
b. Color	Some reds	Green	Several colors
2. Beans:			
a. Average per pod	20–30	>30	>30
b. Color of cotyledon	White, ivory or pink	Pink to violet	Several colors

Source: Oxopeus, 1985.

where N equals number of days to reach maturity and T equals average temperature of the location (°C).

Cocoa can be harvested almost year round, but in a monsoon climate cocoa beans are harvested during a 5 to 6 month period. With *Amelonado* cocoa in Nigeria, for example, 75% of the crop is harvested between September and January. In Ghana, 25% of the crop is harvested during the peak harvest season of November, while in Sabah, Malaysia, only 12% is harvested in the peak seasons in April and October (Wood and Lass, 1985). Mature cocoa can be harvested by observing the color of the fruits: green pods change from green to orange-yellow, and red pods turn to orange, especially in the furrow section. The change of color of cocoa with red pods is not as identifiable as it is for the green pods.

Most cocoa trees bear pods on the branches and sometimes on the trunk, and pods are easily reached within the arm's length of an adult. Pods are removed with gentle care from the tree by using knives of different forms without leaving damage or wounds that may induce fungus infestation. Sometimes it is necessary to use a bamboo pole with knives attached to the top to cut the stalk of the pods when the pods are high on the branches. Harvest time is usually conducted from early morning through noon.

2.2 Curing and Fermentation

Harvested pods are usually collected in one or more convenient places in the field where the cutting of pods to obtain cocoa beans takes place. Sometimes mature fruits are collected in gunny sacks and are ripened for 2 days. After ripening, the fruits are cut into two sections in order to collect the beans (seeds). Pods can also be opened by using bamboo sticks to hit the pods until they split into two pieces. The beans are then collected and piled on plastic sheets or placed in wooden baskets. The skin of the pods may be buried in the soil or used as animal feed. The practice of providing an interval of 3 to 4 days between harvesting and opening is recommended for the benefit of fermentation. The fermentation process is mainly aimed at killing the beans so that changes within the beans can take place, such as changes in color, improvement in aroma and flavor, as well as enhancement of the 'beans' keeping quality. During the fermentation process, the temperature of cocoa beans increases to about 40°C as a result of the exothermic chemical reactions in the pulp that result from the activity of a succession of microorganisms. The pH of the cotyledons drops from 6 to about 5. At 36 to 72 hours after fermentation starts, the beans are ''killed'' and thereafter many chemical reactions take place inside the beans (Wood and Lass, 1985). Some microorganisms active during fermentation processes are the *Saccharomyces cerevisiae*, *S. theobromae*, *S. ellipsoideus*, *S. apiculatus*, *S. mumalus*, and *Eutrolupsis theobromae*. During fermentation, microorganisms are added into the fermentation box at 0.5 g/kg of fresh beans in the form of yeasts to accelerate the fermentation process (Tryono, 1996). In practice, there are two types of fermentation; box fermentation and heap fermentation, where piles of cocoa beans are covered with banana leaves to attain the required fermentation process. The latter method is common among smallholders.

Modern fermentation processes used perforated boxes with 2 m × 1 m × 1 m dimensions and are loaded with 400 to 500 kg fresh beans. Fermentation takes place in 4 to 6 days; every 48 hours, the beans in the boxes are mixed for better fermentation. The boxes can be arranged so that a continuous process can be achieved. For example, at Chikasungka Plantation in Bogor, Indonesia, fermentation processes are conducted with two boxes: beans are kept 2 days in the first box and the remaining 3 days in the second box. Fermentation should be terminated at the right time, since it influences the quality of the beans. The end of fermentation can be detected by observing the color of the beans, which usually turn brown; the pulp can be easily removed and the strong smell of acetic acid is removed from the beans. Usually the *Criollo* beans can be fermented faster than the *Forastero* because of the difference in pigment content. Good fermentation is achieved if the pH of the cotyledons has reached 4.8, the amount of pulp is reduced to 0.16%, and when the beans develop a good aroma. After the fermentation process is completed, the pulp can be cleaned by soaking the beans in running water for about 2 to 3 hours. During this process, the beans lose about 4.5% of their weight. After cleaning, the seeds or beans are ready for drying (Tryono, 1996).

2.3 Drying

The main objectives of cocoa drying besides removing moisture are to reduce bitterness and astringency and to develop a chocolate brown color. Another objective is to prevent off-flavors that arise from faulty drying or excessive acidity, which may result from rapid drying. The rate of drying is an important factor affecting the flavor and quality of dried beans. If the drying process is too slow, there is a danger that molds will develop, particularly at a low drying temperature, and penetrate the pulp and shell, which may lead to off-flavor taste in cocoa. On the other hand, rapid drying may prevent the oxidative changes that are needed and may cause excessive acidity (Wood and Lass, 1985).

The drying characteristics of cocoa beans indicate the existence of a constant rate and two falling rate periods. The critical moisture content (m.c.) lies at 40% w.b., and the first falling rate lies between 40% m.c., w.b. and 23% m.c., w.b. The last period occurs at lower moisture levels where the moisture moves by diffusion from the nib to shell and is finally dispersed to the surrounding air (Bravo and McGaw, 1974). Cocoa drying can be accomplished either by sun drying or by artificial drying.

2.3.1 Sun Drying

Direct sun drying of cocoa beans has been used widely in Indonesia as well as in many other cocoa producing countries. Besides its low cost, drying using direct sunshine on a concrete floor or on bamboo mats raised above the ground produces an adequate quality of dried beans when conducted properly. The common practice in sun drying is to provide a space of at least 2.8 m^2 for 50 kg beans, with thickness or depth of spread of about 5 cm. In Trinidad, the normal size of a concrete floor for drying is 18 m $\times$ 6 m for 2,250 kg of dry beans. To provide even drying, the beans are raked with a wooden palette so that they are in ridges with the drying floor exposed between them. The length of drying time for sun drying depends on the weather. In Indonesia the time required is about 1 week, but during cloudy weather it may be extended to 2 weeks.

2.3.2 Artificial Drying

2.3.2.1 Artificial Drying Using Fossil Fuel and Biomass

In artificial drying methods, hot air is produced by flowing ambient air passing through kerosene or biomass fuel fired stoves on to the drying chamber. The drying chamber can take the form of a rectangular or cylindrical bin for batch drying, with or without a stirrer or in the form of a drum dryer. As shown in Table 3, the drying temperature varies between 60°C and 80°C in a natural convection dryer and between 45°C and 80°C in a forced convection dryer. The time required to reduce the moisture from about 65% w.b. to the final moisture content of 7% w.b. is 36 to 44 hrs for natural convection and 20 to 44 hrs

Table 3 Operating Characteristics and Efficiencies of Some Cocoa Dryers

Type of dryer	Platform Area (m^2)	Spec. dry weight loading (kg/m^2)	Temperature (°C)	Airflow (m/s)	Drying time (hrs)	Overall Drying Efficiency (%)
A. Natural convection:						
1. Samoan	11.1	23.8	60–80	n.a.	40–44	9
2. Secador	36	37.5	60	0–0.15	36.4	15.8
B. Forced convection:						
1. Platform	11.8	42.7	70–80	0.046–0.102	20–28	18.5–29.9
2. Platform	18	46–49.5	63–75	0.03	36–43	27
3. Platform circular		171.2	55–65	n.a.	40–44	32.8–52.5
C. Forced convection with recirculation platform	20.8	27.5–51.7	45–75	0.123	20–23	25.4–46.3
D. Greenhouse Effect solar dryer:						
1. Vibrating racks	13	7.5	45.2	0.15–0.5	40	17.8
2. Bin with stirrer	13	14.5	49.5	0.5	33	46

Source: Modified from Wood and Lass, 1985.

for forced convection. The drying efficiency of the artificial dryer varies from 9% to 16% for the natural convection dryer and from 18.0% to 53% for the forced convection dryer (see Table 3). These values were calculated using the following formula (McDonald et al., 1981).

$$\eta = \frac{\Delta W_w\,(\Delta H_{fg})\,100}{W_f\,C_f} \tag{2}$$

where ΔW_w denotes the weight of water evaporated in kilograms and ΔH_{fg} is the latent heat of evaporation (2256 kJ/kg), W_f is the amount of fuel consumed in kilograms, and C_f is the calorific value of fuel at 25°C (kJ/kg).

An experiment conducted by Shelton (1967) on a small scale showed that a deep layer of 25 cm, a low airflow of 0.05 m/s, and a moderate temperature of 60°C–65°C constituted the most economical conditions. By manipulating the rate of hot airflow, a high flow rate during predrying and a slower rate during the falling rate period can reduce the consumption of fuel.

In Indonesia, the drying process uses the following procedure. First, the beans are conditioned for about 72–80 hours in wooden boxes in bed depths of 15 cm. The beans are then aerated at an air velocity of about 0.3 m/s. The drying air temperature for cocoa is between 60°C to 65°C, but using 55°C to 60°C is recommended to prevent shrinkage.

The Rajamandala estate plantation in Indonesia uses 5 cylindrical bins to dry about 12 tons of wet cocoa beans. Each bin is provided with a mechanical mixer to overcome the sticky seed cover by pulp and is loaded with cocoa beans to a depth of 25 cm. The mixer is powered by a 2.2 kW electric motor, 15 KW centrifugal blowers are used to deliver the drying air at 3.3. m/s. Fuel wood is the common fuel for heating. However, because of the recent scarcity of wood, the use of solar energy as a heating source has been tested on several cocoa plantations (Kamaruddin, 1993; Sri Mulato, et al., 1996; Mühlbauer et al., 1997; Halawa and Arjuno, 1996). The drying process usually takes place for about 27 hours continuously, day and night, and the maximum time allowed is usually 38 hours. Under this system, the total energy required is about 9 MJ/kg water evaporated (see also Table 3).

2.3.2.2 *Solar Drying*

Esper and Mühlbauer (1996) at the University of Hohenheim, Germany, produced about 250 solar tunnel dryers that are being used in 30 countries including Indonesia. The standard version of the tunnel dryer has 2 m width and 10 m length and consists of two sections—a solar air heater and the drying tunnel—and has a capacity of 100 to 300 kg. The whole structure is covered with transparent UV-stabilized PE plastic foil 0.2 mm thick, with transmissivity of 92% for visible radiation. Some designs of this tunnel dryer are provided with a PV electric generator to drive a blower, in order to improve the drying process. Leis et al. (1999) reported that using a solar processing center located in East Java, Indonesia, consisting of a 216 m^2 solar collector unit, drying bins for cocoa, and a biomass furnace could dry cocoa beans from 60% m.c.,w.b. to its final moisture for 1640 kg of dried cocoa beans within 40 hrs. The consumption of fuel wood was 193 kg per ton of dried cocoa, producing a drying air temperature of 60°C. It was reported that the amount of wood conserved by the new furnace design was 80%, compared to the conventional heating method.

In order to reduce the construction cost of drying using solar energy as the main source of heating, optimization techniques can be used. Research results using the Lagrange multiplier have shown that the initial cost of a solar drying system can be further reduced by using a greenhouse effect mechanism (Kamaruddin, 1993). The latter system, called a Greenhouse Effect (GHE) solar dryer and shown in Fig. 2, consists of a transparent structure made of polycarbonate sheet, several vibrating rocks with perforated aluminum trays, and a 240 W blower unit. The floor size was 3.27 m by 3.27 m, and the height of the structure was 1.98 m on one side and 2.73 m on the other. In addition, a blackened steel sheet was installed at the upper position inside the structure to enhance the thermal performance.

Test results of the GHE solar dryer showed that using this system, an initial loading of 228 kg (60.4% m.c.,w.b.) of wet cocoa beans could be dried to a final moisture content of 6.7 w.b. within 40 hours. Under this condition, the average temperature and RH were 45.2°C and 35% respectively. The drying efficiency of the system was 18.4%, as calculated using the following formula:

$$\eta_d = \frac{[m_p\, Cp_p\, (Tp_f - Tp_i) + m_w\, \Delta H_{fg}]\, 100}{[P_w + I(\theta)\, A_f]\, 3.6\, \theta_d + \mathbf{m}_f C_f} \tag{3}$$

where m_p is the mass of the product dried (in kilograms), Cp_p is the heat capacity of the product (kJ/kg °K), Tp_f is the final temperature of the product (°K), Tp_i is the initial

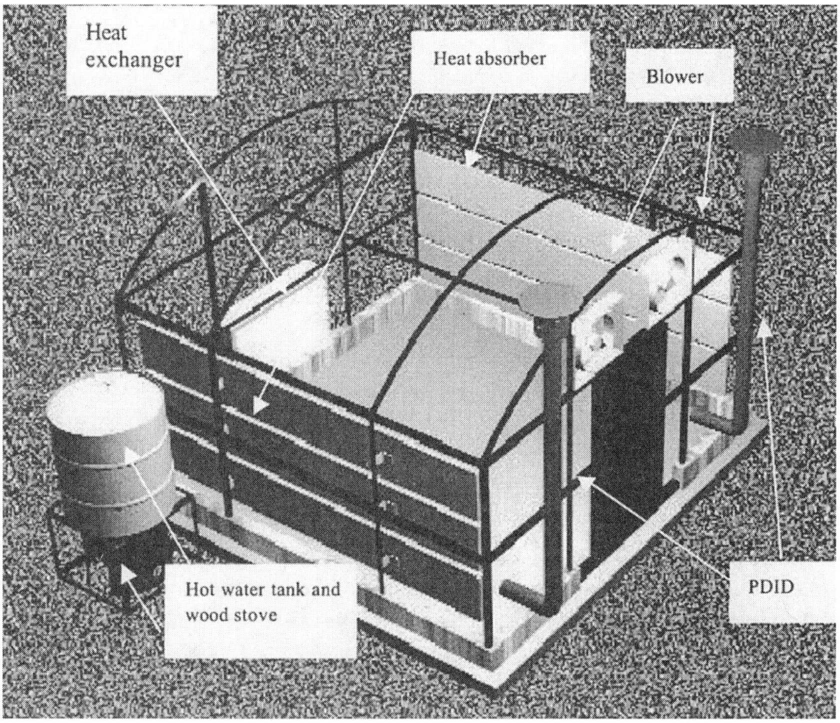

Fig. 2a Schematic diagram of a standard GHE solar dryer.

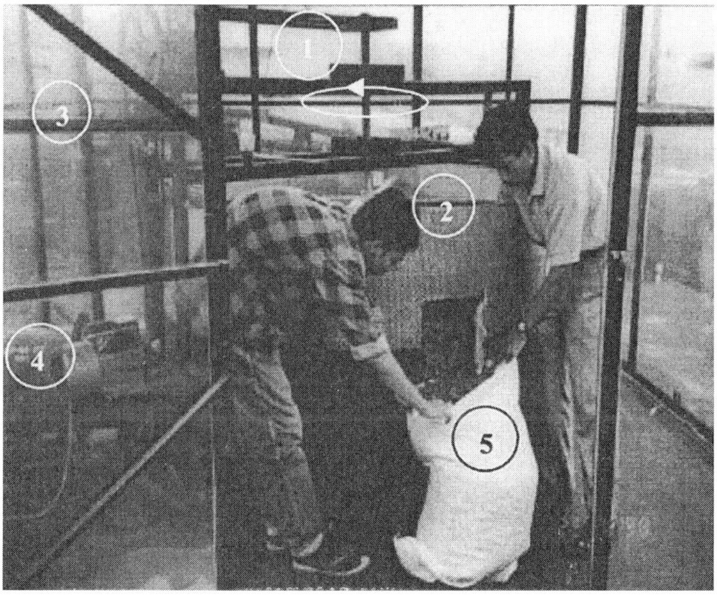

Fig. 2b GHE solar dryer for cocoa bin with mechanical stirrer. Shown in the picture are (1) the mechanical stirrer, (2) polycarbonate wall, (3) perforated drying bin, (4) exhaust blower, and (5) the dried cocoa beans.

temperature of the product (°K), m_w is the total amount of evaporated water from the product (kg), ΔH_{fg} is the latent heat of evaporation (kJ/kg), P_w is the electric power input (kW), $I(\theta)$ is solar irradiation (kW/m^2), θ_d is drying time (h), $\mathbf{m}_f$ is the combustion rate (kg/h), C_f is the calorific value of fuel (kJ/kg), and A_f is the effective drying floor area (m^2).

The total specific energy for drying, E_s—a parameter defined as the ratio between the total input energy both from solar and commercial energy and the total amount of evaporated moisture from the beans—was also used to compare the drying performance.

$$E_s = \frac{\{[P_w + I(\theta)\, A_f]\, 3.6\, \theta_d + \mathbf{m}_f C_f]\}\ 100}{m_w} \tag{4}$$

Using this parameter, it was found that for the above GHE solar dryer test the specific energy required to evaporate moisture from the beans was 12.90 MJ/kg (Kamaruddin, 1998 and Nelwan, 1997). Further study in developing new designs using a drying bin equipped with a mechanical stirrer and a larger loading capacity of 400 kg had improved the specific total energy for drying to 6.2 MJ/kg water evaporated (Manalu, 1999). For comparison, a commercial grain dryer using only commercial energy input, for example, the value of specific energy for drying lies in the range of 3 to 10 MJ/kg water evaporated (Bakker-Arkema and Saleh, 1985). The drying time of the latter GHE solar dryer design was between 32 and 33 hours. More information on test results can be seen from Table 4 and Figs. 4 and 5. Judging from the resulting quality obtained using the GHE solar dryers it is recommended that such systems can be utilized for cocoa drying. The operating temperature, as shown in Table 4, could also be lowered than those currently used in most cocoa plantations.

2.3.2.2.1 SYSTEM SIMULATION. In order to improve the drying performances of the GHE solar dryer system, simulation techniques can be used. The basic governing equations for heat and mass transfer within a transparent structure were developed by

Table 4 Test Performances of Greenhouse Effect (GHE) Solar Dryer

Parameter/drying mode	Fermented cocoa (using vibrating racks)	Fermented cocoa (using mechanical stirrer)
1. Initial load (kg)	228	400
2. Initial m.c. (% w.b.)	60.4	55.9
3. Average drying temperature (°C)	45.2	49.2
4. Average RH (%)	35	30.4
5. Drying time (h)	40	32
6. Specific energy (MJ/kg)	12.9	6.17
7. Drying efficiency (%)	19	55
8. Floor area (M^2)	10.2	10.2
9. Share of solar energy (%)	16	17
10. Additional heating (MJ)	1.41	1.05
11. Transparent material	Polycarbonate sheet	Polycarbonate sheet

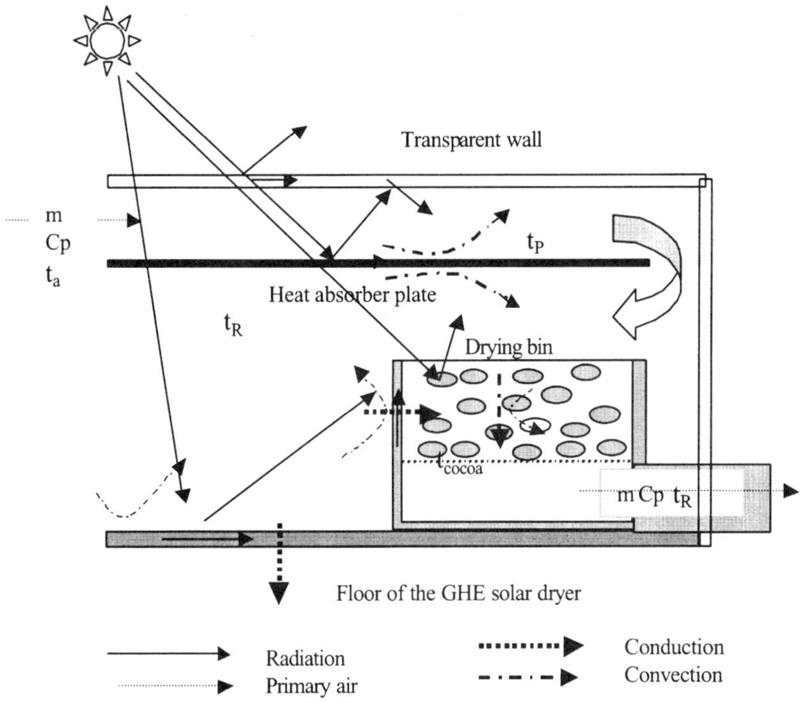

Fig. 3 Transfer of solar energy to generate drying air in a GHE solar drying system.

studying the heating effect of the incoming solar energy to the drying air within the structure through simultaneous radiative, convective, and conductive heat transfer mechanisms. As shown in Fig. 3, some of the heat is finally consumed to increase the temperature of cocoa beans and evaporate moisture from the beans.

A typical result of the simulation study is shown in Figs. 4 and 5 to estimate the change in operating parameters as well as the resulting moisture content of the cocoa beans. Other results of the simulation have been reported elsewhere (Kamaruddin, 1998).

When the GHE solar dryer is used in combination with drying racks, direct application of Henderson's solution for thin layer drying may be applied, provided that the surrounding RH and temperature at each rack are known. Accordingly, the change in average moisture content can be simplified into the following relation (Henderson and Perry, 1976):

$$\frac{M - M_e}{M_o - M_e} = A\ e^{-k\theta} \tag{5}$$

Such estimates for moisture change within cocoa beans can also be extended for batch drying using a mechanical stirrer, since the moisture content in each layer is almost uniform due to the mixing action of the stirrer. Data inputs in the drying simulation usually consist of solar radiation, airflow rate, inlet air temperature, and relative humidity (RH), dimensions of the GHE structure, transmissivity of the transparent wall, absorbtivity and emissivity of components such as racks and steel frames within the structure, floor pavement, and the products. The magnitude of related heat transfer coefficients within and

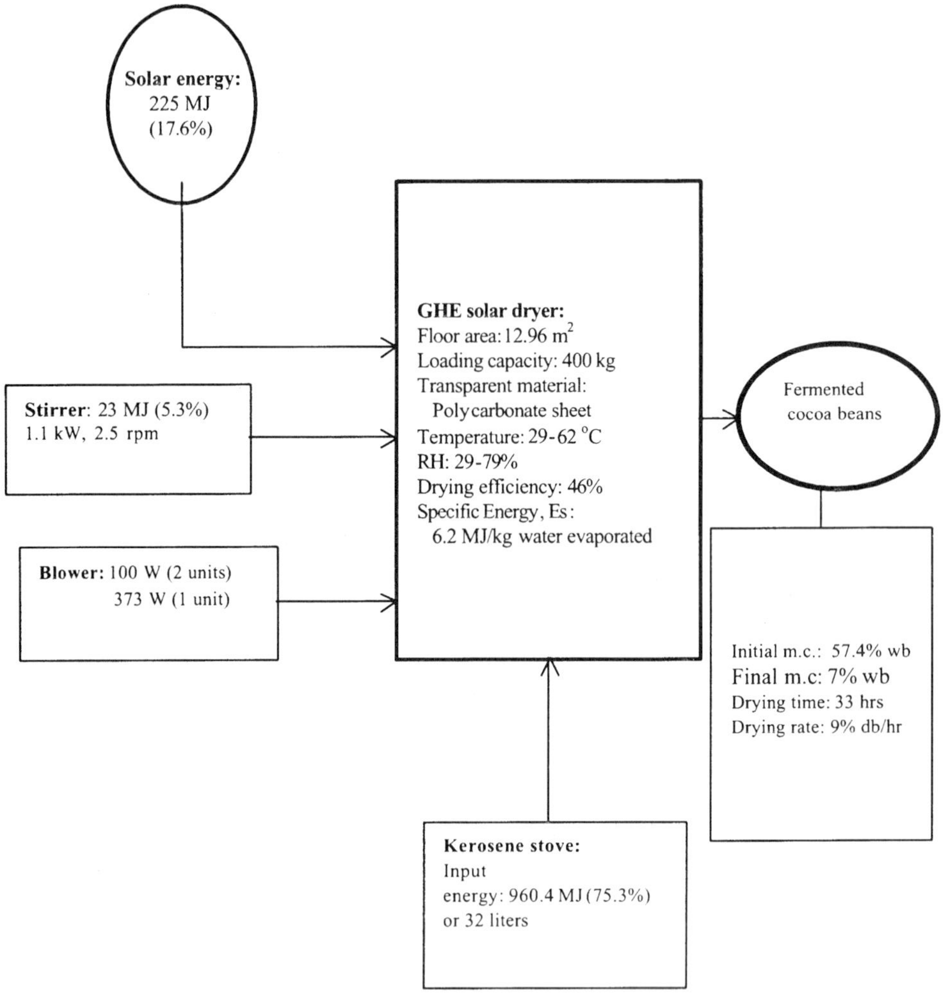

Fig. 4 Typical performance of GHE solar dryer with stirrer.

outside the structure can be estimated from the already established correlation functions in standard heat transfer handbooks or textbooks. In addition, basic data on the thermophysical properties of cocoa beans are also of key parameters, particularly in our effort to make further optimization of the drying process.

Some of the thermal properties were measured and correlated as described below (Nelwan, 1997).

For the equilibrium moisture content, M_e in % m.c., d.b. (dry basis) for cocoa beans, was derived from data of Talib et al. (1995) as shown in Table 5.

$$M_e = -0.036\,(t_d - t_w)^2 + 1.1583\,(t_d - t_w) + 2.0106 \quad (6)$$

The drying constant k (1/min) is expressed by

$$k = \exp\,[5.376 - (3707.65/T_a)] \quad (7)$$

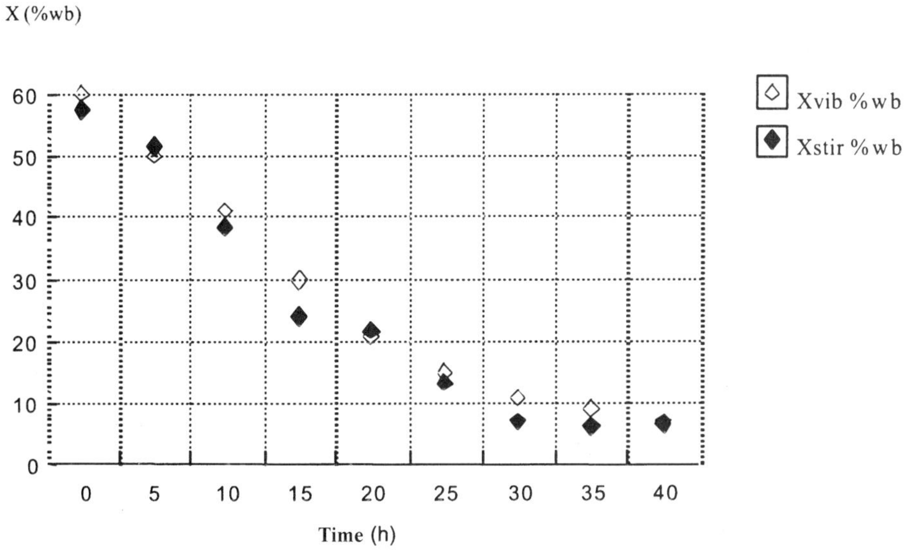

Fig. 5 Change in moisture content of cocoa beans when dried using a GHE solar dryer. Xvib and Xstir are the drying curves for drying using vibrating racks and mechanical stirrer, respectively.

using a slab model (Henderson and Perry, 1976).

The latent heat of vaporization ΔH_{fg} was determined using the Clausius-Clapeyron equation and later was correlated with the equilibrium moisture content. The resulting formula includes the heat of vaporization of pure water as represented in the first term in the bracket. The M_e values were obtained from data of Daud et al (1994).

$$\Delta H_{fg} = [2502 - 2.3775\ t]\ [1 + 0.7297 \exp(-0.1361\ M_e\ T_a)] \tag{8}$$

Figure 6 shows an example of predicted results of the simulation on the drying process of cocoa beans (Nelwan, 1997).

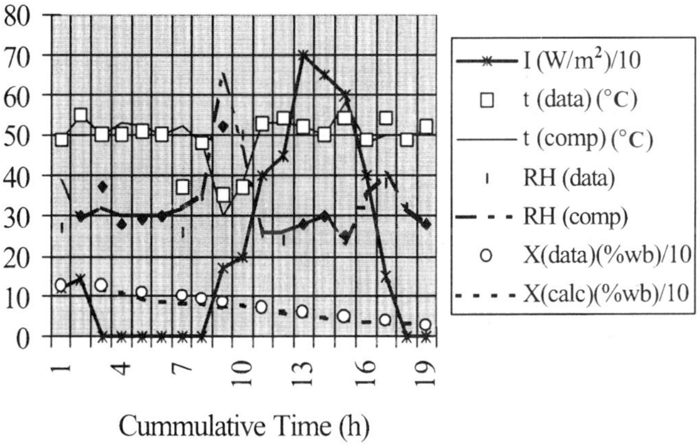

Fig. 6 Comparison between computer results and data for the first two consecutive days of drying of fermented cocoa beans. (From Kamaruddin, 1998.)

Table 5 Equilibrium Moisture Content (M_e in % d.b.) of Cocoa Beans

	Temperature (°C)					
RH (%)	20	30	40	50	60	70
30	—	—	—	7.79	7.72	5.39
40	—	—	14.91	10.89	7.84	6.77
50	38.0	22.84	15.96	10.65	8.33	6.46
60	38.99	24.06	17.53	11.8	8.61	7.85
70	39.83	25.54	16.72	12.95	10.67	9.11
80	41.97	30.09	18.95	15.9	14.44	11.84
90	49.29	40.28	28.11	25.9	22.84	16.05

Source: After Daud et al., 1994.

2.3.3 Moisture Content and Testing

The final moisture for drying is set between 6% to 7% w.b. in order to maintain its quality during long storage. Moisture content greater than those values makes the beans susceptible to the growth of mold. On the other hand, moisture content less than 6% w.b. will make the beans very brittle. When the beans are dried artificially, the final moisture content can be determined easily by a manual method by taking some samples and cutting them into two pieces after they have cooled. The half-beans should then be broken, and should snap into two parts rather than bend or shatter. Final moisture content can also be measured using the standard test procedure such as the conventional vacuum oven, and various moisture meters. These meters can be used by inserting the probes into bags containing cocoa beans and the moisture content can be read instantly. The Marconi moisture meter requires grinding of the beans before they are put in a small cell (Wood and Lass, 1985).

2.3.4 Recovery

The productivity of a plantation or farm can be determined by calculating the recovery factor expressed as (Rohan, 1963);

$$R = \frac{100\ W_b}{W_{uf}} \tag{9}$$

where R is recovery factor (percentage), W_b is the weight of dried fermented beans (kg), and W_{uf} is the weight of unfermented beans (kg). The common value for the recovery factor is 40%, but it may vary from 31% in Ecuador to 46% in Zaire.

2.3.5 Cleaning and Bagging

After the drying process is complete, the dried beans, at about 7% m.c.,w.b., are bagged in jute sacks or other types of gunny sacks. During bagging, initial cleaning and sorting of damaged and inferior beans is conducted and flat and broken beans can be removed. The International Cocoa Standards (ICS) require that exported cocoa be bagged in jute sacks, with each sack containing 62.5 kg net weight of dried cocoa beans.

2.3.6 Storage

In order to store cocoa beans in sacks for 3 months or more, a good storage facility should be provided. The storage room should have a cement floor and good ventilation. Care should be taken to make sure that mold cannot grow and that the products are free from pest attacks. For this purpose, the relative humidity within cocoa stores should not exceed 80% and the temperature variation within the store should be kept to a minimum. The sacks can then be arranged by placing them directly on the dry floor, or stacked on pallets or stillages which raise the sacks 5-10 cm above the floor. When cocoa beans have to be stored for longer periods of time while the RH is relatively high, polyethylene liners inside the normal jute bag can be used. In this way, the adsorption of moisture by the highly hygroscopic beans can be prevented.

2.3.7 Sorting and Quality Determination

In some cocoa producing countries, such as Indonesia, manual sorting is still practiced. Sorting is based on the weight of beans, purity, color, foreign materials, and mold infestation. For quality evaluation, the cut test is used where 300 beans taken from random samples are cut lengthwise using the MAGRA cut test device. The cut beans are then laid out on a board for quality assessment. According to the International Cocoa Standards known as the Model Ordinance, "cocoa merchantable quality" is defined as follows (Wood and Lass, 1985):

1. Cocoa of merchantable quality must be fermented, thoroughly dry, free from smoky beans, free of abnormal or foreign odors, and free of evidence of adulteration.
2. It must be reasonably free of living insects.
3. It must be reasonably uniform in size; reasonably free of broken beans, fragments, and pieces of shell; and virtually free of foreign matter.

The quality requirements and other aspects relevant to quality determinations are further defined as follows (Wood and Lass, 1985):

1. *Thoroughly dry* is defined as cocoa beans that have been dried throughout. The moisture content must not exceed 7.5% w.b. as determined at first port of destination or subsequent points of delivery.
2. *Smoky beans* are defined as cocoa beans that have a smoky smell or taste or show signs of contamination by smoke.
3. *Uniform size* is defined as cocoa beans that have size uniformity and not more than 12% of the beans should be outside the range of 1/3 average weight.
4. *Moldy beans* are cocoa beans on which internal parts have mold visible to the naked eye.
5. *Slaty beans* are beans that show a slaty color over half or more of the surface exposed by a cut made lengthwise through the center.
6. *Insect-damaged* beans have internal parts that are found to contain insects at any stage of development, or show signs of damage caused thereby that are visible to the naked eye.
7. *Germinated beans* are cocoa beans whose shells have been pierced, slit, or broken by the growth of the seed germ.
8. *Flat beans* are beans whose cotyledons are too thin to be cut to give a surface of cotyledon.

Table 6 describes in details the standard quality of dried cocoa beans imposed by the Indonesian Cacao Association (Askindo), (Spillane, 1995). Quality assurance for exported cocoa beans is of utmost importance. Cocoa beans that do not comply with standard quality can suffer significant price reduction or be rejected by the receiving country. Therefore, good quality management in cocoa processing before shipment should be enforced, since it will prevent great loss to cocoa producing countries and ultimately to their farmers. On the other hand, cocoa with prime quality will be awarded with premium prices. In 1991, the amount of premium for prime quality cocoa was £300/ton.

LIST OF SYMBOLS

A	geometric coefficient in Eq. (5)
A_f	floor area of GHE solar dryer (m^2)
C_f	calorific value of fuel (kJ/kg)
C_p	heat capacity (kJ/kg°K)
E_s	specific energy for drying (MJ/kg water evaporated)
ΔH_{fg}	latent heat of evaporation (kJ/kg)
I	solar radiation (kW/m^2)
K	drying constant (1/min)
$\mathbf{m}_f$	combustion rate of fuel (kg/h)
m_w	mass of water evaporated from product (kg)
M	moisture content (% d.b.)
M_e	equilibrium moisture content (% d.b.)
M_o	initial moisture content (% d.b.)
N	number of days to reach maturity (days)
P	electrical power of blower, pumps etc. (kW)
R	recovery factor (%)
t_a	temperature of the air (°C)
t_d	dry bulb temperature (°C)
t_w	wet bulb temperature (°C)
T	absolute temperature (°K)
T_L	average temperature of location (°C)
T_{pf}	final drying temperature of cacao (°K)
T_{pi}	initial temperature of cocoa during drying (°K)
W_b	weight of dried fermented beans (kg)
W_{uf}	weight of unfermented beans (kg)
ΔW_w	amount of water evaporated (kg)
X	moisture content (% w.b.)

Greek Letters

η	drying efficiently as defined by Eq. (2) (%)
ηd	drying efficiency (%)
θ	time (h)
θd	drying time (h)

REFERENCES

Alvim P de T, Machado AD, and Vello F, 1972. Physiological responses of cocoa to environmental factors, Proc. 4th International Cocoa Research Conference, Trinidad. pp. 210–225.

Bakker-Arkema FW and Saleh HM, 1985. In-store drying of grain: the state of the art. Preserving grain quality by aeration and in-store drying. ACIAR Proceedings, No. 15, pp. 24–30.

BPS, 1999. Central Statistic Bureau of Indonesia.

Bravo A and McGraw DR, 1974. Experimental artificial drying characteristics of cocoa beans, Tropical Agriculture, Trin. 51:395–406.

Bird RB, Stewart WE and Lightfoot EN, 1976. Transport Phenomena, Wiley International Edition, John Wiley & Sons, New York.

Cuatrecasas, J, 1964. Cacao and its Allies: A taxonomic Revision of the Genus Theobroma. U.S. National Herb., Vol. 35, No. 6, pp. 379–614.

Daud WRW, Talib MZM, and Ibrahim MH, 1994. Drying of Cocoa Beans. Proc. Drying '94. Edited by V. Rudolph, R.B. Keey and A.S. Mujumdar pp. 1129–1135.

Esper A and Mühlbauer W, 1996. Solar tunnel dryer for fruits, Plant Research and Development, vol. 44, pp. 61–80.

Halawa EEH and Arjuno B, 1996. Proceedings, Workshop on Monitoring, Evaluation and Adoption Strategy. ASEAN-Canada Project on Solar Energy in Drying Process.

Henderson SM and Perry RL, 1976. Agricultural Process Engineering, 3rd ed. The AVI Publishing Co. Inc., Westport, CT.

Huang F, Li Z, and Li L, 1990. Review of Decadal Solar Drying in China. Energy and the environment into the 1990s. Sayigh AAM, ed. Pergamon Press, Oxford, Vol. 2.

Kamaruddin A, 1993. System Optimization in Solar Drying. Paper No. 30-1. Proceedings of the 5th International Energy Conference, Energex '93. Seoul, Korea. Vol. III, pp. 86–102.

Kamaruddin A, 1995. Optimization in the design of solar drying system for agricultural products. Final Report of a Competitive Research Grant No. 064/P4M/DPPM/PHB I/3/1994 of the Directorate General of Higher Education. Indonesia. 83 pages.

Kamaruddin A, 1998. Greenhouse Effect Solar Dryer for Coffee and Cocoa Beans. Final Report. University Research for Graduate Education. Contract No. 032/HTPP-II/URGE/1996. Directorate General of Higher Education, Indonesia.

Kamaruddin A, Armansyah H, Tamrin T, Wenur F, and Dyah W, 1994. Heat and Mass Transfer Within a Fiber-glass House Solar Dryer. Proceedings of the International Conference on Fluid and Thermal Energy Conversion, FTEC '94. Bali, Indonesia. Vol. II, pp. 179–191.

Kamaruddin A, Tamrin T, Wenur F, and Dyah W, 1994. Drying of black pepper using solar energy, Proceed. International Drying Symposium, IDS'94, Gold Coast, Australia.

Manalu LP, 1999. MSc thesis. IPB Graduate Program, Bogor, Indonesia.

McDonald CR, Lass RA, and Lopez ASF, 1981. Cocoa Drying—A Review. Cocoa Growers Bulletin, Vol. 31. pp. 5–41.

Müller J, 1992. Prediction of Drying Rate for Solar Drying., Paper No. 926040. ASAE 1992 International Summer Meeting.

Mühlbauer W, Muller J, and Esper A, 1997. Agricultural Crop Drying and Storage, F.WW. Bakker-Arkema and D.E. Maier, Ed. Marcel Dekker Inc.

Nelwan LO, 1997. MSc thesis. IPB Graduate Program, Bogor, Indonesia.

Rohan TA, 1963. Processing of raw cocoa for the market. FAO, Agric. Studies, No. 5, Rome. p. 96.

Shelton B, 1967. Artificial Drying of Cocoa Beans. Trop. Agric., Trin. 44:125–32.

Spillane JJ, 1995. Cocoa commodity—its role in the Indonesian economy (In Indonesian language). Kanisius Publisher, Yogyakarta, Indonesia.

Siegel R and Howell JR, 1972. Thermal Radiation Heat Transfer. McGraw Hill, Kogakusha, Ltd. Tokyo.

Sri Mulato T, Atmawinata O, Purwadaria HK, and Muhlbauer W, 1999. Performance of solar collector dryer combined with biomass furnace for coffee processing. In Proc. The First Asian-Australian Drying Conference. ADC99, Denpassar, Bali, Ed. Kamaruddin A., A.H. Tambunan and A.S. Mujumdar. pp. 479–491.

Sri Mulato T, Atmawinata O, Yusianto and Handaka, 1996. Development of a solar cocoa processing centre for cooperative in Indonesia. Paper presented in Industrial Drying Technology Workshop, Bogor, August 20–22.

Sri Mulato T, Pass A, Atmawinata O, Yusianto, Handaka and Mühlbauer W, 1997. Development of a solar cocoa processing center for cooperative in Indonesia, (in press).

Talib MZM, Daud WR, and Ibrahim MH, 1995. Moisture Desorption Isotherms of Cocoa Beans. Transaction of the ASAE, Vol. 38 (4). pp. 1153–1155.

Tryono, 1996. Graduation thesis. (in Indonesian Language) Faculty of Agricultural Technology, IPB, Bogor, Indonesia.

Tryono, 1996. Field Practice Report. (in Indonesian Language) Faculty of Agricultural Technology, IPB, Indonesia.

Wood GAR, and Lass RA, 1985. Cocoa, Longman Scientific & Technical, co-published in the US with John Wiley & Sons, Inc. New York.

27

Conversion and Utilization of Biomass

AMALENDU CHAKRAVERTY

Indian Institute of Technology, Kharagpur, India

1 INTRODUCTION

The biomass is the renewable and biodegradable organic matter generated through life processes. A large quantity of biomass is produced annually by land and aquatic plants. There is also a vast resource of lignocellulosic residues in the world that sometimes poses serious disposal problems (1).

The organic matter produced by plants has been in use since primitive times. Even in this modern age, firewood and wood charcoal are considered common commercial sources of energy all over the world, particularly in the third world. Wood charcoal used in the metallurgical and other industries cannot be produced by any chemical method as yet (1–2). Alcohol, a versatile organic base compound, is produced from biomass of high carbohydrate content. The paper industry is also based on fibrous cellulosic biomass. Hence there is scope for the introduction of improved technologies of the conversion of the huge biomass resources into food, feed, fuel, energy, chemicals, and other value-added materials. Development of viable technologies, recycling and efficient management of wastes, control of environmental pollution, and improvement of the quality of rural life should be the aims of these projects.

The following relevant points have to be taken into account during the feasibility studies of the biomass conversion technologies:

- Collection and handling cost
- Pretreatment and enrichment cost
- Lack of commercially viable technology
- Integration of biomass conversion processes for complete utilization of products and coproducts
- Lack of reliable biomass statistics
- Ecological balance

Table 1 Bulk Density, True Density, and Angle of Repose of Rice Husk and Bran

Material	Moisture content, percentage (w.b.)	Bulk density, kg/m^3	True density, kg/m^3	Angle of repose, degrees
Rice husk	11.38	102.0	1022.0	50.7
	15.29	103.0	1031.0	52.2
Rice bran	7.66	255.0	—	51.8
	10.23	243.0	—	—

Source: From Refs. 3, 4, and 5.

It is apparent that the improved biomass conversion technologies have great potentialities. Hence all aspects of biomass and its conversion technologies should be considered thoroughly for commercial exploitation.

2 BIOMASS

2.1 Classification

The biomass sources including solid wastes can be classified under the following categories:

1. Crop residues and farm wastes
2. Forest products
3. Agroindustrial wastes
4. Energy farm products
5. Aquatic biomass
6. Marine products
7. Animal wastes
8. Municipal solid wastes
9. Municipal sewage sludge

Of these biomass sources, conversion and utilization of the first three and a part of the seventh items are discussed in this chapter.

2.2 Characteristics

Some of the physical and thermal properties of common biomass are presented in Tables 1, 2, and 3.

Table 2 Thermal Conductivity of Rice Husk and Bran

Material	Moisture content, percentage (w.b.)	Thermal conductivity	
		kcal/(hm°C)	kW/(mK)
Rice husk	11.51	0.035	0.041
	14.72	0.046	0.054
Rice bran	7.66	0.075	0.087
	10.23	0.084	0.098

Source: From Refs. 3, 4, and 5.

Table 3 Specific Heat (C_p) of Rice Husk and Bran

Material	Moisture content, percentage (w.b.)	Specific heat (C_p) cal/(g°C)	kJ/(kgK)
Rice husk	9.97	0.288	1.21
	13.86	0.368	1.54
Rice bran	7.66	0.399	1.67
	10.23	0.437	1.83

Source: Refs. 3, 4, and 5.

2.3 Biomass Conversion

Generally the conversion technologies of biomass or wastes include the following basic processes:

1. Thermal and thermochemical processing: combustion or incineration, pyrolysis or destructive distillation, gasification, liquefaction, and others
2. Chemical processing: production of pulp and paper, furfural, silica, silicon and silicates, and others
3. Biochemical processing: fermentation, biogasification, composting, and others

In fact, both thermal and thermochemical technologies involve high temperature. These biomass conversion technologies are covered in this chapter. However, the common methods of waste disposal are land filling or sanitary land filling and disposal in rivers and the sea.

3 COMBUSTION

Combustion is the process of liberation of heat by the oxidation of the combustible constituents of the fuel. Generally, fuels are compounds of carbon and hydrogen; in addition, variable percentages of oxygen and small percentages of sulfur and nitrogen are present. Biomass fuels are thermally degradable organic materials. The combustion of biomass fuels not only generates carbon dioxide and water vapor but also produces carbonaceous residues, smoke, tar, and some harmful gases, such as carbon monoxide. Combustion and incineration are the same in principle. Incineration is used as a common method of disposing of solid wastes, such as municipal solid waste by high-temperature oxidation, though its primary objective is to reduce the volume of solid waste and its secondary purpose is to produce energy or to destroy putrescibles or hazardous wastes (6,7).

3.1 Principles

3.1.1 Combustion Parameters

The fuels generated from plants are mainly composed of moisture and different organic and inorganic compounds. The effects of these on combustion are described in the following sections.

3.1.2 Moisture

The raw wood plant residues, bagasse, etc., may contain high moisture, and these should be dried to a lower moisture level. Some agricultural residues, namely, straw and husks, have 8% to 12% storage moisture contents. The water content not only acts as a heat sink and lowers the combustion efficiency but also affects the economics of the fuel utilization.

3.1.3 Organic Compounds

Cellulose, hemicellulose, and lignin are the main organic constituents of biomass fuels. Biomass also contains some lipids, carbohydrates, and proteins. The variation in composition of these organics affects combustion.

3.1.4 Minerals

Mineral (ash) contents of biomass vary widely with the species, locality, and soil. Some agricultural waste, such as rice husk, contains silica content as high as 15%–20%. Silica acts as a heat insulator, whereas some soluble electrovalent compounds may act as catalysts in gasification and combustion of biomass.

3.1.5 Definitions

Some important terms related to combustion are defined here for ready reference.

The *gross calorific value* or *higher heating value* is the amount of heat energy released when a unit mass of solid or liquid fuel or a unit volume of gaseous fuel is completely burned and the products of combustion are cooled to 15°C, thereby condensing the water vapor present therein.

The *net calorific value* is defined as the gross calorific value minus the latent heat of condensation (at 15°C) of the water vapor present in the products of combustion.

The thermochemical equations (8) in burning fuel containing carbon and hydrogen are as follows:

$$C + O_2 = CO_2 + 408816 \text{ kJ/kg mole} \tag{1}$$

and

$$H_2 + \tfrac{1}{2}O_2 = H_2O + 288889 \text{ kJ/kg mole} \tag{2}$$

Therefore, the gross calorific value, C_G, (kilojoules per kilogram [kJ/kg]), of the fuel can be calculated from the following mathematical expression:

$$C_G = \frac{\text{C} \times 34068 + \text{H} \times 144{,}445}{100} \tag{3}$$

where C and H are percentages of carbon and hydrogen, respectively.

The equation for C_G of a fuel containing carbon, hydrogen, and oxygen in percentages is as follows:

$$C_G = \frac{\text{C} \times 34068 + \left(\text{H} - \dfrac{\text{O}}{8}\right) \times 144{,}445}{100} \tag{4}$$

3.1.6 Combustion Process

The combustion of biomass occurs through two pathways. At higher temperatures, pyrolysis of biomass produces a mixture of combustible gas, which leads to flaming combustion. But at lower temperatures, pyrolysis mainly generates carbonaceous char and a mixture of water vapor and carbon dioxide. Oxidation of the resulting active char then provides glowing or smoldering combustion.

3.1.7 Efficient Combustion

Usually combustion reactions take place at high temperature and high speed. The same amount of heat is released independently of the conditions of the reactions, provided the end products are the same and the combustion reactions are complete.

In order to avail oneself of the heat of combustion of the fuel to the maximal extent, reactions should be as complete as possible; that requirement implies that no less than the stoichiometric (theoretical) amount of air must be supplied in the proper manner. In actual practice more than the theoretical amount of air (i.e., excess air) is supplied to ensure the completion of combustion reactions.

Important conditions required for efficient combustion of any fuel in a furnace are as follows:

- A higher than stoichiometric amount of air has to be supplied for complete combustion.
- During burning of fuels, the volatiles leaving the fuel bed must be intimately mixed with the secondary air for flaming combustion.
- The volatiles cannot be allowed to cool below the ignition point until the combustion reactions are complete.
- The fuel and the air must be in free and intimate contact.
- A provision for expansion of the gases during combustion at high temperature is to be made.

3.1.8 Air Requirement

With the help of the chemical equations the quantity of theoretical (stoichiometric) air required for complete combustion of a fuel can be determined if the ultimate analysis of the fuel is known. The volumes of air actually required for different percentages of excess air, the flue gas produced, and the enthalpy of the flue gas can also be determined from the empirical equations and statistical tables available for combustion calculations when the net calorific values of the fuels are known.

3.2 Furnaces

The common furnaces for combustion of biomass fuels can be divided into two groups, (a) fixed grate furnaces and (b) cyclone furnaces. The fixed grate furnaces can be further divided into (a) horizontal grate furnaces and (b) inclined grate furnaces. Other furnaces are also used in industries (8). Some of these furnaces are described in this chapter.

3.2.1 Horizontal Grate Furnace

The horizontal grate furnace consists of a furnace chamber, a precipitation chamber, and an air mixing chamber. The furnace walls and roof are made of refractory bricks to minimize heat losses and to withstand high temperature.

The air required for combustion of the fuel on the horizontal fire grate is admitted through the draft opening. The secondary air for combustion of volatile gases is allowed to pass through some additional holes. Cast-iron bars are employed for the fire grates because of their stability under high heating and resistance to corrosion. Ambient air is

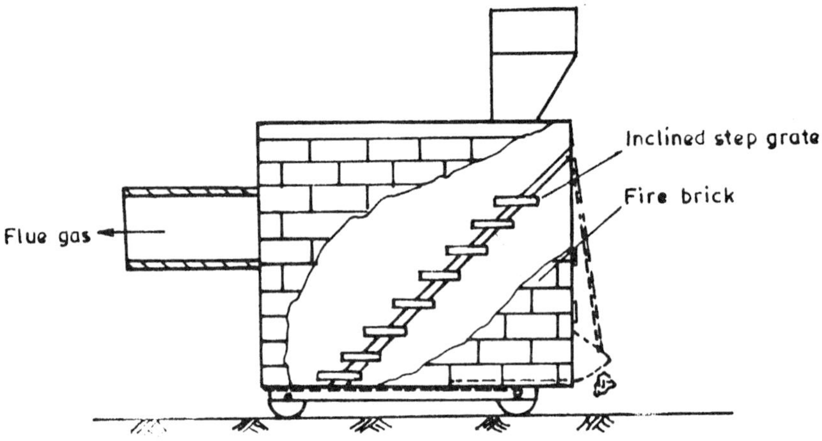

Fig. 1 Schematic of an inclined step grate furnace.

allowed to mix with the flue gas in the mixing chamber through some ports to reduce the temperature of the flue.

3.2.2 Inclined Grate Furnace

An inclined step grate furnace consists of a hopper, an inclined grate fixed at an angle of 45° with the horizontal, three side walls, and a roof. The grates are made of cast-iron bars, which are arranged in a staircase fashion. The furnace is lined with refractory bricks (Fig. 1).

Thin layers of husk are fed through the inlet at the top of the inclined grate, and husk flows down the inclined grate by gravity during its combustion. The efficiency of this furnace is low (6).

3.2.3 Cyclone Furnaces

The cyclone principle of combustion has introduced a new era in the development of boiler furnaces. The cyclone furnaces are found to be efficient for combustion of pulverized coal. These may also be suitable for firing biomass fuels. Both horizontal and vertical cyclone furnaces can be designed according to need. A horizontal cyclone furnace consists of a slightly inclined cylinder, lined with firebricks, into which air is injected tangentially at high velocity. The fuel introduced at one end is entrained by the revolving mass and is thrown against the cyclone walls, where it mixes intimately with air and burns. As a result, in a cyclone furnace combustion occurs in a comparatively short space, where the flow conditions can be controlled and adjusted from outside. The flue gas, which leaves through the aperture at the other end of the cyclone, is quite free of fly ash (8).

In a horizontal cyclone furnace, a length-to-diameter ratio of 1:1.3 is normally maintained. Cyclones are designed for positive-pressure operation. Generally cyclones require an air blower with high static pressure. The schematics of some cyclone furnaces are shown in Fig. 2.

3.2.4 Furnace Design

The values of thermal load of the furnace grate area and thermal load of the furnace volume are important for the design of a furnace.

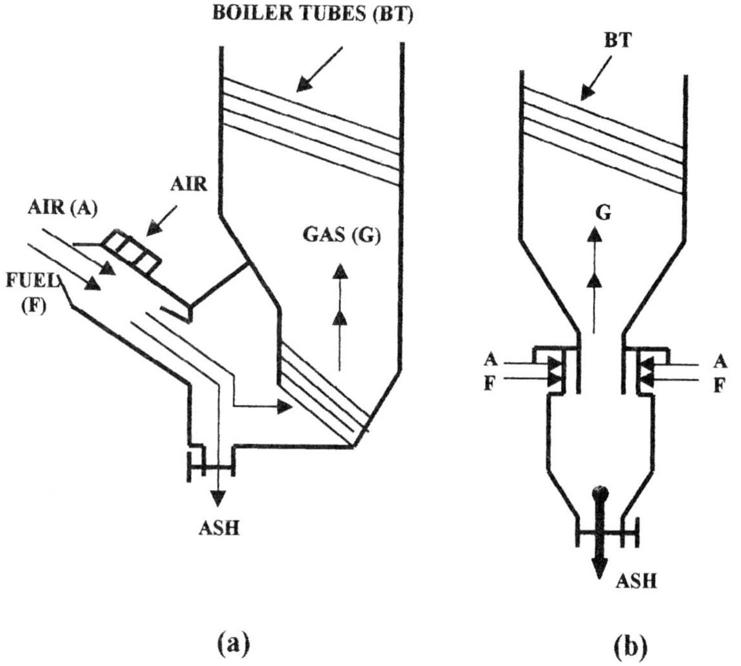

Fig. 2 Schematics of different cyclone furnaces.

The thermal load of the furnace grate area is the amount of heat energy generated by complete combustion of a solid fuel on unit area of a fire grate in unit time.

The thermal load of the furnace volume is the amount of heat energy generated by complete combustion of a solid fuel in unit volume of the furnace in unit time.

Accordingly, the following mathematical expressions can be obtained directly.

$$q_A = \frac{W_F \cdot C_N}{A} \tag{5}$$

and

$$q_V = \frac{W_F \cdot C_N}{V} \tag{6}$$

where q_A = thermal load of furnace grate area, kW/m^2, q_V = thermal load of furnace volume (kW/m^3), W_F = rate of fuel burned (kg/s), C_N = net calorific value of the fuel (kJ/kg), A = furnace grate area (m^2), and V = volume of a furnace (m^3).

The thermal load of the furnace grate area and the thermal load of the furnace volume for rice husk fuels are about 232.6 kW/m^2 and 174.5 kW/m^3, respectively.

As a result of various kinds of losses the amount of heat contained in the fuel gas leaving the furnace is less than that in the fuel charged.

Therefore, the efficiency of a furnace (η_f) can be expressed as follows:

$$\eta_f = \frac{\text{Heat output in the flue gas}}{\text{Heat input from the fuel}} = \frac{m \cdot h}{W_F \cdot C_N} \tag{7}$$

where m = flow rate of flue gas (nm^3/s), and h = enthalpy of flue gas (kJ/nm^3).

4 GASIFICATION

Wood gasifiers had been operated successfully to generate low-calorie gas at the beginning of the 20th century. Then a good number of commercial gasifiers were installed in the United States to provide electricity and street lighting to many cities. Within a few years a large number of automotive gas producers were operating all over the world. Surprisingly, these technologies had been almost abandoned after the 1950s (9). At present a wide range of raw materials from agricultural residues to municipal solid wastes have been suitably used for gasification.

It is important to note the following:

A gasifier can be used to generate low- or medium-calorie gas for direct combustion or to produce a fuel gas for an internal combustion engine.

A gasifier has to be equipped with an effective gas cleaning system if the gas is to be used for an internal combustion engine.

Some of the products of gasification are not only damaging to engines and burners, but also harmful to human beings. That is why these gases cannot be used as cooking gases.

In general, the gasifiers may produce some condensable volatile product such as tar that is not desirable. But its amount can be reduced significantly by introducing the following three tar conversion mechanisms: (a) thermal cracking at high temperature, (b) combustion of the tar with air, and (c) catalytic thermochemical conversion.

4.1 Principles

A few important terms used in gasification are defined in the subsequent paragraphs. Generally *gasification* refers to the process of thermochemical conversion of organic matter into the maximal usable chemical energy generated in the gaseous products (9).

The *equivalence ratio*, Ψ, is defined as the ratio of actual air supplied to the stoichiometric air required for complete combustion. It can be expressed as follows:

$$\Psi = \frac{\text{Actual air required for gasification}}{\text{Stoichiometric air required for combustion}} \tag{8}$$

The *efficiency*, η_g, of a gasifier can be defined as the ratio of the chemical energy output in the dry producer gas at 15°C to the energy input from the biomass feed.

$$\eta_g = \frac{\text{Chemical energy output in the dry producer gas at 15°C}}{\text{Energy input from the biomass}} \tag{9}$$

During gasification the total energy as well as the sensible heat energy in the gas increase with the increase of equivalence ratio, Ψ. But the chemical energy in the gas increases up to a certain value of Ψ and then decreases sharply with a further increase of Ψ (10). That is why a substoichiometric quantity of air is always supplied to the gasifier for gasification.

4.1.1 Parameters Affecting Gasification

Some important parameters affecting the fixed bed gasification are (a) moisture content, (b) volatile matter content, (c) ash content, (d) calorific value of the fuel, (e) composition of the biomass fuel, and (f) orientation of the fuel bed structure.

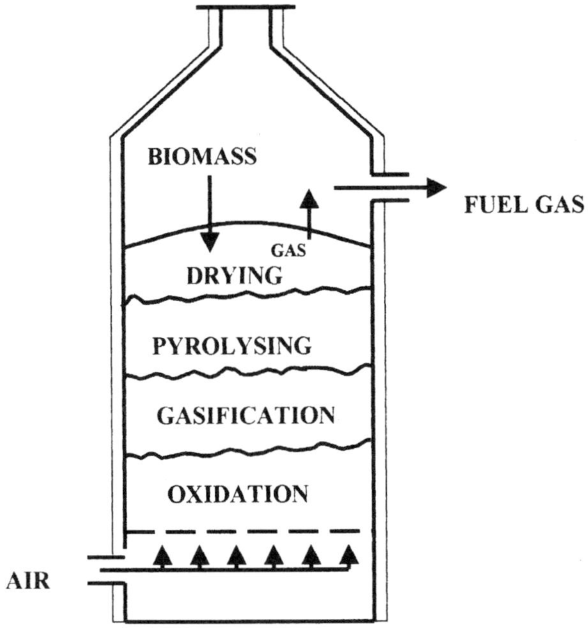

Fig. 3 Schematic of different zones of an updraft gasifier.

In the different stages of gasification, a number of thermochemical reactions take place. As discussed earlier, in gasification a substoichiometric amount of air is always supplied. Shortly after ignition, four distinct reaction zones are set up in the gasifier unit (Fig. 3). These are oxidation, reduction, pyrolysis, and drying zones. Some of the important thermochemical reactions corresponding to the different reaction zones are shown in the table below.

Zone	Reaction	
Drying	Wet organic matter + heat	→ Dry organic matter
Devolatilization	Dry organic matter + heat	→ Condensable hydrocarbon + CH_4 + Char
Pyrolysis	$CO_2 + H_2$ + heat	→ $CO_2 + H_2O$ (gas)
	$CH_4 + \frac{1}{2} O_2$ + heat	→ $CO + 2H_2$
Gasification	$C + H_2O$ + heat	→ $CO + H_2$
	$C + 2H_2O$ + heat	→ $CO_2 + 2H_2$
	$C + CO_2$ + heat	→ $2CO$
Oxidation	$C + O_2$	→ CO_2 + heat

In the oxidation zone, heat is generated as a result of an exothermic reaction between the carbon of the char and the oxygen of the combustion air to produce CO_2. On the other hand, the heat of the exothermic reaction supplies heat energy required for completion of the endothermic reactions in the reduction, pyrolysis, and drying zones.

In general, a gasification process produces CO, H_2, CO_2, CH_4, C_2H_4, C_3H_6, NH_3, H_2S, N_2, H_2O, tar vapors, and low-molecular-weight organic liquids (8).

4.2 Gasifiers

The gasifiers can be mainly grouped into (a) fixed bed and (b) fluidized bed gasifiers. Other gasifiers are (c) stirred bed, (d) tumbling bed, and (e) entrained bed gasifiers.

Either air or oxygen can be supplied to the gasifier. In an air blown gasifier a low-calorie gas is produced, whereas in an oxygen blown gasifier a medium-calorie gas is generated. Most of the fixed bed gasifiers operate under atmospheric pressure. Different types of gasifiers ranging from a low to a high capacity of fuel gas output have been developed (11).

4.2.1 Fixed Bed Gasifier

The fixed bed gasifiers can further be subdivided into (a) updraft, (b) downdraft, and (c) cross-draft vertical reactors.

4.2.2 Updraft Gasifier

In an updraft (or a countercurrent flow) gasifier, the biomass feed moves downward from the top, whereas the air moves upward from the bottom (Fig. 4). Hence the oxidation zone resides at the bottom and gasification takes place through zones of decreasing temperature as the gas rises through the fuel bed of the gasifier.

The gas flows opposite to the direction of the relatively cold feed material and leaves the reactor at a lower temperature. Thus the fuel gas generated by an updraft gasifier has high tar content. The height-to-diameter ratio is usually kept at 3:1.

4.2.3 Downdraft Gasifier

In a downdraft (or a concurrent flow) gasifier, both the biomass and the air move downward toward the bottom of the gasifier (Fig. 5). But the gases pass through the higher-temperature zones at the bottom, where a certain amount of tars, etc., are being thermally

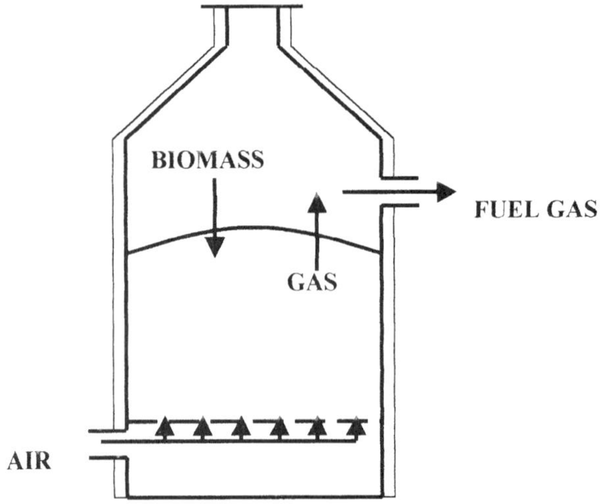

Fig. 4 Updraft gasifier.

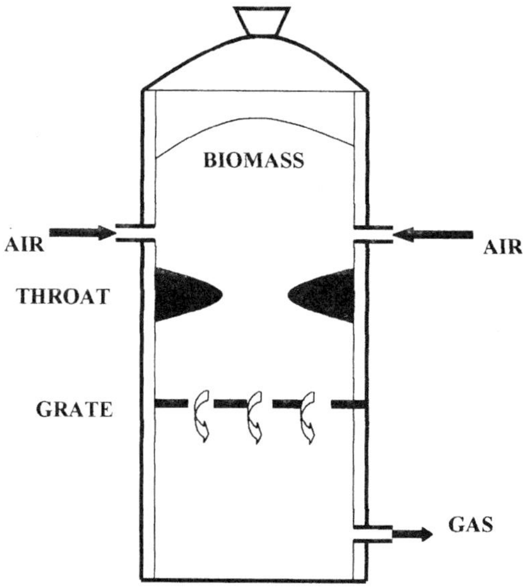

Fig. 5 Downdraft gasifier.

cracked. Hence the downdraft gasifier produces cleaner gas with a relatively lower amount of tar, when compared to the updraft and cross-draft gasifiers.

4.2.4 Cross-Draft Gasifier

The air is fed to a cross-draft (or a cross-flow) gasifier through a horizontal nozzle. As and when the fuel gas is produced, it is discharged through a vertical grate on the opposite side of the air nozzles.

4.2.5 Fluidized Bed Gasifier

In the refractory lined fluidized bed gasifier (Fig. 6), sometimes inert materials are used as a fluidizing medium. The desired fluidization can be achieved by supplying air or oxygen at a fluidization velocity through the perforated supporting plate. As a result the entire bed is kept under a state of suspension. The temperature is uniformly maintained throughout the bed. Simultaneous oxidation and gasification occur rapidly at the constant fluidized bed medium. Thorough mixing of the fluidized bed results from high turbulence. The bed temperature should be lower than 1100°C, in order to prevent slagging of the ash. Generally the height-to-diameter ratio of a fluidized bed reactor is 10:1.

4.2.6 Design Parameters

The design parameters to be considered for rice husk gasification are (a) equivalence ratio, (b) diameter of the gasifier, (c) rice husk feed rate, (d) time of operation, (e) superficial gas velocity within the fuel bed, (f) ultimate analysis of rice husk, (g) gas composition, (h) specific gasification rate, (i) degree of rice husk conversion, (j) volume reduction of rice

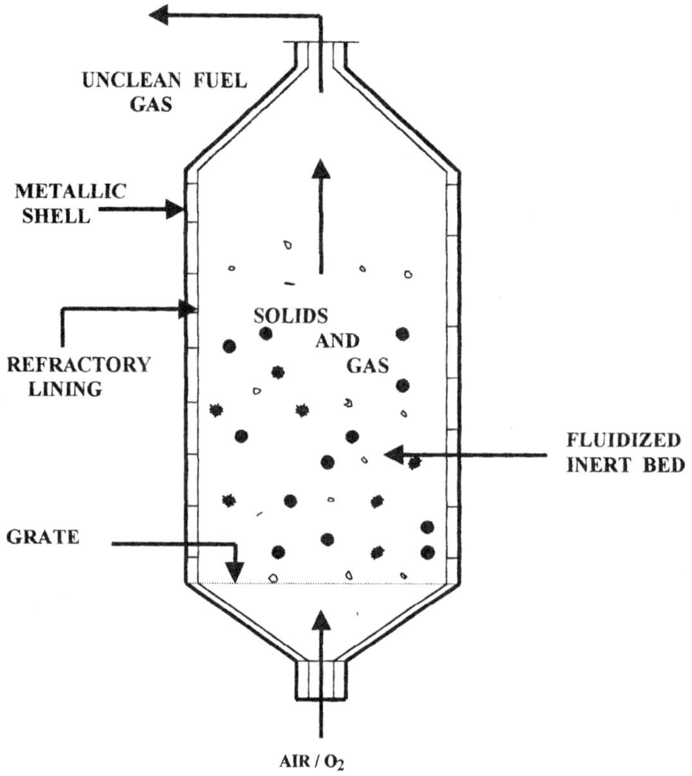

Fig. 6 Fluidized bed gasifier.

husks, (k) carbon conversion percentage, (l) higher heating value of the dry gas produced, (m) efficiency of the process, and (n) ash removal rate for a continuous system (9).

5 PYROLYSIS

Before the First World War charcoal, acetic acid, methanol, acetone, and wood oil were produced by the wood distillation industry only. Today, there is no synthetic substitute for charcoal (12). Charcoal is a versatile material, which is used in the metallurgical processes as well as in the chemical industries. It is a raw material for the production of activated carbon, which is utilized in various industries. Moreover, firewood and charcoal are noncommercial sources of energy for cooking and heating in many developing countries. By utilizing the vast agricultural residues and other biomass wastes of the world, the charcoal industry can make a significant contribution to humankind.

The destructive distillation of biomass is the thermal decomposition of biomass in the absence of air, whereas the thermal decomposition of the biomass in an inert atmosphere is known as *pyrolysis*. However, the major objectives of both are the same. Pyrolysis or destructive distillation produces a mixture of gases, tar, oils, acetic acid, methanol, and other organic compounds and a solid residue containing the inerts of the waste and a char. The yields of pyrolysis vary with the heating rate and the final exposure temperature of the solid wastes. In general, the higher the heating rate and the higher the final temperature, the greater is the conversion of wastes into gaseous and liquid products. The major objectives of pyrolysis of biomass are production of char/charcoal with a higher calorific value than that of the raw biomass and production of a less smoky, clean-burning char/charcoal and of a more reactive fuel.

However, there are disadvantages of pyrolysis:

A charred product may be so brittle that it is difficult to handle.
A sizable fraction of the energy in the raw feed is lost in the gas and liquid products.
If the valuable by-products are not recovered, not only loss but pollution results.

5.1 Principles

The following two definitions of charcoal are used:

Charcoal refers to a black, solid, nonlustrous residue, or amorphous carbon, from vegetable or animal substances; or a coal made by charring wood in a kiln or retort from which air is excluded (2).

Charcoal is the residue of solid nonagglomerating organic matter, of vegetable or animal origin, that results from carbonization by heat in the absence of air at a temperature above 300°C (2).

The second definition is more specific. The efficiency of charcoal conversion (η_c) can be expressed as follows:

$$\eta_c = \frac{\text{Charcoal output, kg}}{\text{Biomass input, kg}} \tag{10}$$

5.1.1 Different Stages of Thermal Decomposition

The pyrolysis or carbonization of wood takes place in the following stages when it is heated to temperatures above 300°C in the absence of oxygen or under controlled limited air intake:

Up to a temperature of 170°C mainly water is evaporated from the wood.
At temperatures ranging from 170°C to 270°C carbon monoxide (CO), carbon dioxide (CO_2), and some condensable vapors evolve.
At temperatures between 270°C and 280°C an exothermic reaction starts.
At temperatures ranging from 270°C to 400°C evolution of methanol and other organics takes place.
At temperatures between 400°C and 500°C maximal charring or carbonization occurs.

When the pyrolysis process enters the exothermic state no further external heating is necessary as the temperature in the retort reaches a level between 400°C and 500°C. The thermal analyses (TGA, DTG, and DTA) of rice straw are presented in Fig. 7 to show different stages of thermal decomposition (13).

5.2 Practice

5.2.1 Traditional Methods

Two methods—namely, the charcoal pit and the earth mound kiln—are in practice. In the first system a pit is dug out. Wood logs are put into it and it is covered with earth to

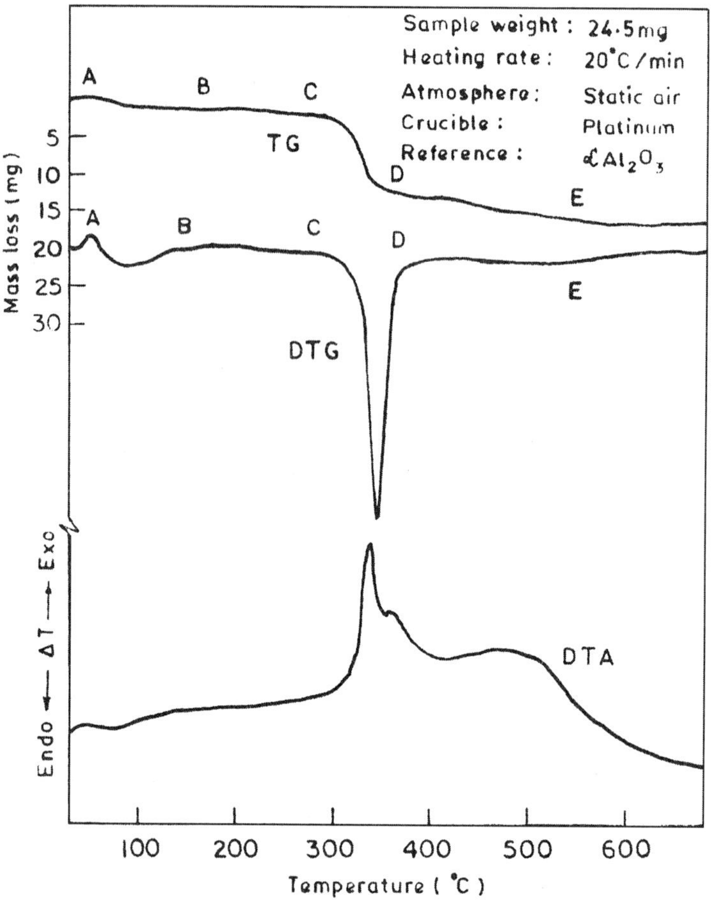

Fig. 7 Thermo gravimetric (TG) analysis, differential thermo gravimetric (DTG) analysis and differential thermal analysis (DTA) for rice straw (13).

seal and insulate the chamber. The carbonization of wood is accomplished. In another system a pile of wood logs on the ground covered with earth, sand, and leaves is carbonized. Carbonization takes place in both methods in the absence of air or with limited air. But these methods are wasteful, as the valuable by-products are not recovered. Recently an earth mound kiln with the addition of a chimney and an earth mound chimney kiln with a tar oil recovery system have been developed. In operating a charcoal pit or an earth mound kiln, care must be taken to control and seal the air inlets; otherwise, the charcoal will burn to ashes.

One cycle of operation requires about 3 to 3½ days for carbonization and a subsequent 3 days for cooling. On average, the yield is about 15% (2).

5.2.2 Large-Scale Production

In large-scale production of charcoal a battery of six to seven Beehive brick, Argentine, or other kilns (2) are generally employed. Any kiln that has no by-product recovery system will not only cause pollution but will also make the whole process uneconomical. The main features of large-scale charcoal production are the following:

Capital investment is high.
Limited recovery of some by-products is possible.
Different raw materials may be used.
Labor-saving devices can also be implemented.

5.2.3 Argentine Kiln

The hemispherical Argentine kiln is also known as a *half-orange kiln* (Fig. 8). The inner volume of a kiln is about 15 m^3. The kilns are made of bricks with a door, and large kilns should have two doors. The operational procedures for a half-orange kiln are as follows:

The size of the fairly dry wood log is reduced to about 1 m in length. These wood logs are placed vertically inside the kiln. When the kiln is fully charged, the door is sealed, with all small inlet ports kept open. A small amount of glowing charcoal is thrown inside the kiln through an ignition eye. Generally the complete carbonization process may take about 10 to 12 hours. After the kiln is made airtight by sealing all ports, it is allowed to cool down sufficiently. Then the door is opened, the fire is completely extinguished with water, and the charcoal is discharged. A complete cycle of operation takes about 2 days, including a day for cooling, for the production of charcoal in this kiln. Then the charcoal has to be cured on the storage space for 8 to 10 days. On average the yield is as high as 25% to 26%. Some other types of kilns for charcoal making are also available (2). However, an established chemical process, namely, the Othmer (or Azeotropic) distillation process, is superior to the preceding methods as far as recovery of acetic acid, methanol, and other wood chemicals is concerned (12).

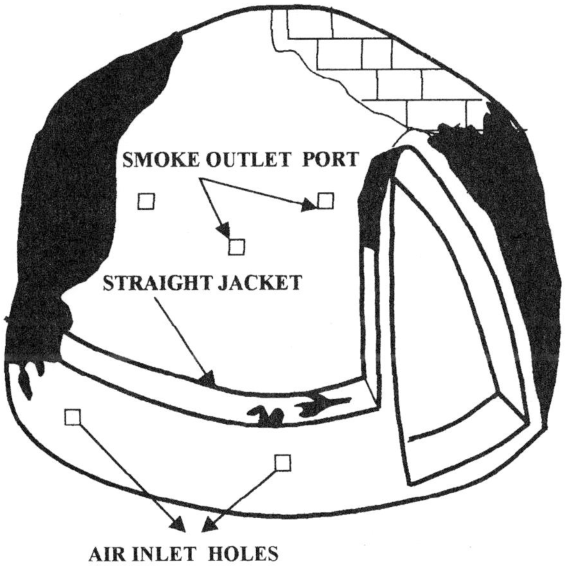

Fig. 8 Half-orange kiln.

6 CHEMICAL AND BIOCHEMICAL PROCESSING

6.1 Amorphous Silica and Combustible Gas

The rice husk is composed of approximately 38% cellulose, 18% petosan, 22% lignin, 20% ash (mainly silica), and 2% other organics. The biogenic silica present in rice husk and rice straw is inherently amorphous. The carbon-free white ash (silica) obtained from rice straw and husk has opened a new dimension in its utilization concept. This silica has been found to be an attractive source for manufacturing pure silicon and other chemical products (14–16).

The straws (1 cm in length) or husks are thoroughly cleaned with water and sun dried. The dried materials are leached with mineral acids at 75°C for 1 h and washed with distilled water several times to remove the acids and then dried in the sun to lower the moisture content (m.c.) to 6% to 7% wet basis (w.b.).

The temperature of a vertical electric furnace is raised to the desired constant temperature and then straw or husk is fed into the furnace to expose it to the stable furnace temperature with a natural draft. Combustion continues till the entire mass is converted into white ash. The structure of silica in the white ash has been examined by the x-ray diffraction technique to determine the formation of any crystallinity during combustion of straw or husk at high temperatures.

When the bed thickness of straw and husk increases in the furnace, the flame temperature and the duration of flame also increase. If the furnace temperature increases, the flame temperature also increases for both straw and husk. For the production of low-calorie combustible gas along with the amorphous silica, both raw and acid-treated straw and husk containing 6% to 7% (w.b.) moisture should be subjected to a furnace temperature of 450°C (14–15).

At a furnace temperature of 500°C the raw rice straw and husk are not converted to milky white ash even after their long exposure. It is necessary to pretreat them with mineral acids at the elevated temperature for 1 h to remove their metallic impurities for the production of pure white ash by complete combustion.

The acid treated straw and husk samples of 20-cm bed thickness subjected to 450°C to 500°C furnace temperatures correspond to the complete combustion times of about 3 h and 1.5 h, for the production of milky white ash.

The diffractograms of both untreated and acid-treated straw and husk at 500°C for a bed depth of 20 cm show no sharp peak, thus indicating that the silica in the white ash samples is amorphous in nature. As the bed temperature may exceed the critical temperature of 720°C for crystallinity of amorphous silica at higher bed depth and higher furnace temperature, a furnace temperature of 450°C and a bed depth of 20 cm are suitable for the production of pure amorphous silica from acid treated straw and husk (14–15).

6.2 Pure Silicon from Rice Husk and Straw

The amorphous silica available from rice husk and straw finds varied uses, depending on its form and purity. It is used as an antisticking agent, as a reinforcing material in rubber, as a filler in paper, and as a base material in silicate industries. Because amorphous silica is active, it can be reduced to silicon at a comparatively lower temperature by metallothermic reduction and other processes. Solar grade and semiconductor grade silicon have to be very pure.

The processes developed for conversion of ricc husk silica to silicon is described in the subsequent paragraphs.

6.2.1 Calcium Reduction Process

The pure rice husk silica can be reduced to silicon of reasonable purity by using calcium as a strong reducing agent. The white ash obtained by burning rice husk at about 500°C is leached with moderately hot concentrated mineral acids to remove most of the soluble impurities present in it. The material is then thoroughly washed with warm distilled water and dried. The purified rice husk silica is thoroughly mixed with granular calcium, and the mixture is then reduced inside a furnace at a moderately high temperature. The reduced mass, after being cooled to room temperature, is taken out. The main reaction product can be purified to a level of 99.9% silicon with the simple treatment of various acids only. x-ray diffraction, emission spectrographic, and other studies have indicated that the silicon product is of high purity (17).

6.2.2 Halide Reduction Process

In another process $SiCl_4$ is synthesized by chlorinating pyrolized rice husk, which is then used for production of silicon. Polycrystalline and amorphous silicon are produced by Zn reduction of $SiCl_4$ in a vapor-phase reaction. Metallic Zn is kept inside the reaction chamber in a special crucible at an elevated temperature at one end of the reactor, and the $SiCl_4$ stream passes over it. Silicon is deposited on the wall of the reactor (18).

6.3 Furfural from Plants Rich in Pentosan

The name *furfural* was derived from the words *furfur* (i.e., ''bran'') and *oleum* (i.e., ''oil''). Fresh furfural is a colorless, inflammable organic liquid. On exposure to light and air, it turns to a reddish-brown liquid.

Furfural is commonly used as a selective solvent for refining lubricating oils and vegetable oils. Furfuryl alcohol is produced by catalytic hydrogenation of furfural at 200°C. The furfuryl alcohol can be used to produce resins. Furfuryl alcohol is also used in the areas of pharmaceuticals, insecticides, and solvents. Furfural is a base material for tetrahydrofuran, which can be converted to Nylon 6.6. Various industrially important organic chemicals can be produced from furfural (19).

6.3.1 Principle

Furfural is a versatile derivative of furan with the aldehydic group CHO in the 2-position as shown in the following chemical structure:

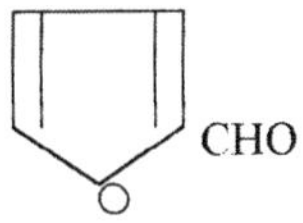

The furfural can be synthesized from any plant material containing pentosans. On acid hydrolysis the pentosan forms xylose, which subsequently loses three molecules of water to form furfural:

$$\underset{\text{Pentosan}}{C_5H_8O_4} + H_2O \xrightarrow[\text{Heat}]{HCl/H_2SO_4} \underset{\text{Pentose}}{C_5H_{10}O_5} \longrightarrow \underset{\text{Furfural}}{C_5H_4O_2} + 3H_2O$$

The chemical reaction can also be presented as follows:

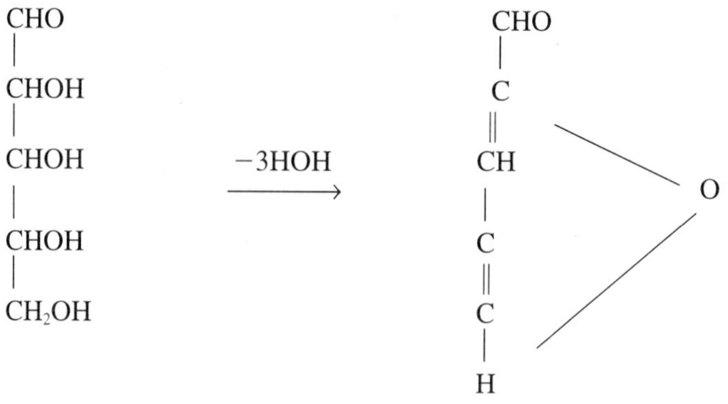

Pentose Furfural

6.3.2 Production

Although corn oats are the main raw materials, bagasse, oat hulls, cottonseed hulls, and rice hulls are also used (19).

6.3.3 Process

In the past, crushed corncobs were treated with sulfuric acid at a temperature of 100°C–120°C for 50–55 minutes. When corncobs were hydrolyzed with dilute sulfuric acid of 1.9% concentration at a temperature of about 121°C for 50 minutes, using a solid/liquid ratio of 30:100, the yield of furfural was 9%. Subsequently a number of methods have been developed for the production of furfural (20–22).

6.4 Biodegradable Starch-Based Plastics Films

The polysaccharides are abundantly available renewable sources, which can serve polymer needs in addition to their conventional applications. Starch is such a natural polymer available commercially at a comparatively low price. Attempts have been made to use unmodified starch directly in polymer systems.

A mixture of cornstarch, low-density polyethylene (LDPE), a coupling agent, and urea is prepared. A small amount of water is added to it so that starch is moistened enough to smear itself uniformly around the granules of LDPE and the coupling agent. The compounded material is discharged in the form of a continuous strand, which is cut with rotary knives in a granulator. The granules can be blown into a smooth film on the film-blowing extruder.

The tensile strength of these starch-based plastics films is comparable to that of unmodified LDPE, whereas the elongation at break is reduced. The values for density, water absorption, and water-soluble matter are relatively higher in comparison with those of LDPE. Some of these films are found to be biodegradable under the soil. Hence these films can be used as carrier bags, agricultural mulch, and nursery bags (23).

6.5 Anaerobic Blogasification

The anaerobic decomposition of organic matter to methane (CH_4) and carbon dioxide (CO_2) is called *biogasification*. Though the process of biogasification has been known for a long time, its potential for converting various substrates ranging from animal secretory products to municipal wastes into biogas has been realized recently. Now different designs of biogasification plants are in use. Biogas can be utilized for operating diesel engines, lighting, cooking, and other heating purposes as well.

The main advantages of biogasification are as follows:

The biogas contains a higher percentage of methane (about 60%).
The digested material containing significant amounts of nitrogen and other soil nutrients can be utilized as an organic fertilizer.
The overall thermal efficiency of a burner using biogas is fairly high.
The biogas can be purified to methane enriched fuel.
A simple design for a biogas plant is available, and it can be manufactured locally.

6.5.1 Principles

The conversion of biomass to CH_4 and CO_2 during anaerobic fermentation occurs by the concerted action of three main groups of bacteria—namely, fermentative, acetogenic, and methanogenic bacteria. The fermentative bacteria hydrolyze the raw organic substrates mainly to acetate, fatty acids, CO_2, and H_2. The H_2 producing acetogenic bacteria produce H_2, acetate, and sometimes CO_2. The methanogenic bacteria mainly catabolize acetate, CO_2, and H_2 to CH_4 and other compounds (24).

6.5.2 Factors Affecting Anaerobic Biogasification

The main factors that affect the formation of biogas from an organic substrate are: temperature; pH; percentages of C and N, solid organic matter, and water in the slurry; retention time; loading rate; and nutrients.

For improving the efficiency of the biomethanation process it is important to find a suitable mixture of substrates—namely, algae, water hyacinth, cow dung, etc. Single substrate provides a suboptimal carbon-to-nitrogen ratio, thus restricting the activity of several enzymes that are responsible for decomposition of various components present in the residue. For example, the population of the cellulolytic bacteria can be improved by using a mixture of water hyacinth and cow dung in comparison with cow dung alone. Moreover, higher hemicellulose content of water hyacinth also causes a high rate of formation of biogas. The methane content can be further increased with the incorporation of algae, which contain vitamin B_{12}. The lower yield of methane from cow dung alone is due to the presence of a considerable proportion of lignin and less of hemicellulose. If acetic acid and propionic acid are produced in the system in appreciable quantity, a maximal quantity of methane is formed, thereby indicating that acid composition is probably one of the most important factors for higher methane content in the biogas (25–26).

CONCLUSIONS

Combustion of biomass has been commercially exploited all over the world. Wood gasifiers were successfully operated at the beginning of the 20th century. Gasification of many other types of biomass has been studied with success in the recent past. Before the First World War charcoal, wood oils, etc., were solely produced by the wood distillation industry; subsequently methanol, acetic acid, etc., were synthetically produced by the chemical industries, but charcoal is still being produced by wood pyrolysis only. Biogasification is an old technology whose full potential has been realized recently. The various chemical technologies for conversion of biomass have yet to be developed for commercial exploitation. The different conversion technologies employed for any particular biomass for the synthesis of various products and by-products should be integrated for economy. Hence research and development (R&D) activities have to be concentrated on these potential areas in order to meet the future demand for food, feed, chemicals, and energy from the vast renewable biomass resources.

ACKNOWLEDGMENTS

The author wishes to express his gratitude to his colleagues, PG Scholars at PHTC, Indian Institute of Technology, Kharagpur, India, for their cooperation and encouragement. He is also grateful to his wife and son for their untiring assistance in the preparation of the manuscript.

REFERENCES

1. Pal, W., Coombs, J., and Hall, D. O. (Editors). Energy from Biomass, 3rd European Committee Conference. Elsevier Applied Science, 1985.
2. Emrich, W. Energy from Biomass. Series E, Vol. 7, D. Reidil, Barking, Essex, Holland, 1985.
3. Mishra, P. Investigation on physico-thermal properties and thermal decomposition of rice husk, production of pure amorphous white ash and its conversion to pure silicon. Ph.D. Thesis, IIT, Kharagpur, India, 1986.
4. Devedattam, D. S. K. Studies on some properties of rice bran and its stabilization by dry heat and wet heat treatments using a conduction heating system. Ph.D. Thesis, IIT, Kharagpur, India, 1983.
5. Chakraverty, A and Paul Singh, R. Post Harvest Technology of Cereals, Pulses, Fruits and Vegetables. Science Pub. Inc., Enfield, NH, 2001.
6. Beagle, E. C. Rice-husk conversion to energy. FAO Services Bulletin 31, Rome, 1978.
7. Wilson, D. G. (Editor). Handbook of Solid Waste Management. Van Nostrand Reinhold, New York, 1977.
8. Shvets, I., Tolubinsky, V., Kirakovsky, N., Neduzhy, I., and Sheludko, I. Heat Engineering, 2nd ed., Mir, Moscow, 1975.
9. Kaupp, A. Gasification of Rice Hulls. Friedr. Vieweg and Sohn, Braunschweig, 1984.
10. Tillman, D. A. Wood as an Energy Source. McGraw-Hill, New York, 1982.
11. Kaupp, A., and Goss, J. R. State of the art for small scale (to 50 kW) gas producer–engine systems. U.S. Agency for International Development, Washington, D.C., 1981.
12. Shreve, N. R. The Chemical Process Industries. McGraw-Hill, New York, 1956.
13. Chakraverty, A., Banerjee, H. D., and Pandey, S. K. Studies on thermal decomposition of rice straw. Thermochimica Acta, 120, 241–255, 1987.
14. Chakraverty, A., and Kaleemullah, S. Conversion of rice husk into amorphous silica and combustible gas, Energy Convers. Mgmt. 32(6), 565–770, 1991.
15. Chakraverty, A., and Kaleemullah, S. Production of amorphous silica and combustible gas from rice straw. J. Materials Sci. 26, 4554–4560, 1991.
16. Chakraverty, A., Mishra, P., and Banerjee, H. D. Investigation of thermal decomposition of rice husk. Thermochimica Acta, 94, 267–275, 1985.
17. Mishra, P., Chakraverty, A., and Banerjee, H. D. Production and purification of silicon by calcium reduction of rice-husk white ash. J. Materials Sci. 20, 4387–4391, 1985.

18. Acharya, H. N., Datta, S. K., Banerjee, H. D., and Basu, S. Low temperature preparation of polycrystalline silicon from silicon tetrachloride. Mater. Lett. (2), 1982.
19. Paturau, J. M. By-Products of the Cane Sugar Industry, 2nd ed., Sugar Series 3. Elsevier Scientific, New York, 1982.
20. Nanda, S. K. Laboratory and pilot plant investigation on the production of furfural from rice husk. Unpub. M. Tech. Thesis, PHTC, IIT, Kharagpur, India, 1979.
21. Riclis, S. G., and Kolotilo, D. M. Production of furfural from corn cobs under atmospheric pressure in a medium of superheated steam. Chem. Abstr., 59, 15487, 1963.
22. Zakoshchikov, A. P., and Abramyants, S. V. New method for obtaining furfural from different types of plants. Chem. Abstr., 58, 2419, 1963.
23. Nanda, S. K., Chakraverty, A., and Maiti, S. Starch-based plastic films. Part 1. Preparation of the film. J. Polym. Matr., 7, 331–333, 1990.
24. Bryant, M. P. In Schlegel, H. G., and Barnea, J. Eds. Microbial Energy Conversion, Verlag Enrich Gotze K. G., Gottingen, pp. 107–118, 1976.
25. Das, D., Ghose, T. K., and Gopalkrishnan, K. S. Improved conversion efficiencies of the biomethanation process by using mixed agricultural residues in a batch system, Proceedings of National Symposium on Biotechnology, Chandigarh, India, March 13–15, 1982.
26. Chakraverty, A. Biotechnology and Other Alternative Technologies for Utilization of Biomass/Agricultural Wastes. Oxford, IBH, New Delhi, 1995.

28

Utilization of By-Products of Fruit and Vegetable Processing

WALIAVEETIL E. EIPESON and RAMESH S. RAMTEKE

Central Food Technological Research Institute, Mysore, India

1 INTRODUCTION

The current world production of fruits and vegetables is placed at 429 and 596 million tonnes, respectively (1). Among the fruits and vegetables, citrus, apple, pineapple, grape, tomato, and potato are processed in substantial quantities.

The waste generated in the fruit and vegetable processing industries includes both solids and liquids. Proper management of the waste is both a regulatory requirement as well as an economic necessity. Treatment and reuse of liquid wastewater from processing industries is a highly developed area and is beyond the scope of the present review.

The fruit and vegetable processing industries generate 10%–60% of the raw materials as solid waste. The composition of these wastes suggests enormous potential for producing value-added products. The typical example is the citrus processing industry, which produces a number of value-added by-products from waste, some of them even more valuable than the main product, viz., citrus juice concentrates. In fact, the waste from one processing plant could become the raw material of another plant, and utilization of waste rather than disposal should be the goal of the industry.

Bioremediation of fruit and vegetable processing wastes to produce value-added products has also opened up enormous product opportunities. The future of the food processing industry and in particular the fruit and vegetable processing industry lies in achieving zero-waste processing systems.

Table 1 Major Apple Producing Countries

Country	Production (1997) (1000 MT)
China	18,410
United States	4,639
Iran	1,925
France	1,918
India	1,200
Chile	940
World	56,087

Source: From Ref. 1.

2 BY-PRODUCTS FROM FRUIT PROCESSING WASTE

2.1 Apple

Apple (*Mallus domestica Borkh*) is a deciduous fruit grown mostly in temperate regions in the world. The total world production of apples is estimated at 56 million tonnes; China, the United States, and France are the leading producers (Table 1).

Apple pomace or apple press cake is the major waste of the apple processing industry, although some quantity of industrially discarded apples and belt rejections also constitutes the waste.

2.1.1 Composition of Pomace

The proximate composition of apple pomace (2) shows high potential for effective utilization for the production of various products of commercial importance. The pomace is rich in carbohydrates (9.5%–22%) and pectin (1.5%–2.5%) and also contains 1% to 1.8% protein.

2.1.2 By-Products of Apple Waste

The various value-added products that can be prepared from apple waste are summarized in Table 2.

Table 2 Possible By-Products of Apple Waste

Apple waste	By-products	Refs.
Discarded apples, belt rejections	Pulp, ready-to-serve beverages, squashes	3
Pomace	Alcohol	4, 5
	Flavor components	6, 7
	Pectin	8
	Nutrition fiber	9
	Biogas	10
	Beer, soft drinks, vinegar, cider	3
	Citric acid	11
Peel, core, offcuts, stem, and seeds	Pectin methyl esterase	12, 13

2.1.3 Alcohol

In recent years there has been considerable interest in the production of ethanol from fruit processing waste. Although traditionally alcohol is produced by submerged fermentation, a solid-state fermentation process with a yield of about 42 liters ethanol per metric ton of wet pomace has been reported by Hang and coworkers (4). This represents an energy recovery of approximately 20% of the total energy of the pomace. Gupta and associates (5) studied the solid-state fermentation of apple pomace, using *Saccharomyces cerevisiae*, *S. diastaticus*, *Pitchia fermentans*, *Candida utilis*, and *C. tropicalis*. Among the organisms, *S. diastaticus* was found to be the best, giving maximal yield of 2.8% with a fermentation efficiency of 43.8%, followed by *S. cerevisiae*, with net yield of 2.6% and a fermentation efficiency of 40.6%. The fermentation efficiency of these strains can be improved by supplementing the substrate with nitrogen, phosphorus, and trace elements. Nitrogen supplementation was found to improve the alcohol production more than either phosphate or trace element supplementation and is related to the specific requirements of each strain.

2.1.4 Flavor

Apple pomace is used for producing natural flavoring compounds. Almosnino and Belin (6,7) have quantitatively generated hexanal and 2,4-decadienal in an apple pomace–linoleic acid mixture, thereby confirming the presence of enzyme activities intrinsic to apple pomace in the presence of precursor. The production of aroma components can be enhanced by micronization of the pomace and by addition of ascorbic acid. It is also reported that factors such as pH, temperature, and oxygen greatly influence the production of aroma components.

2.1.5 Pectin

Though the yield and quality of pectin from apple are slightly inferior to those of citrus, apple pomace could be a potential source for pectin. Sulc and colleagues (8) have also described methods for the production of pectin from apple pomace. Apple pomace is high in pectin content and can be converted into thickeners and modifiers for use in apple products.

2.1.6 Nutrition Fiber

Hang (9) has outlined a systematic approach for the production of edible fibers, citric acid, and fuels from apple pomace.

2.1.7 Other Products

Discarded apples and belt rejections in large apple processing plants are converted into pulp and used for the preparation of beverages and squashes (3). Fresh apple pomace used as a partial replacement of fodder to a herd of milk cows of Holstein Friesian breed significantly reduced fodder expense with no detrimental effect on health of the animal or milk yield and quality.

A solid-state fermentation process using apple pomace as a substrate and strains of *Aspergillus niger* for the production of citric acid has been reported (11). The yield of citric acid was found to be dependent on (a) methanol content, (b) strains of *Aspergillus niger* used, (c) fermentation time and temperature, and (d) apple pomace variety fermented. The yield was 88% on the basis of the amount of sugar consumed. King (12,13) obtained a crude enzyme preparation of pectic esterase from Bramley apple waste material such as peel, cores, and offcuts. Enzyme with activities up to 60 units/g of dry material was obtained. Various characteristics of enzyme such as pH, temperature, and activation energy were also reported.

2.2 Banana

Banana (*Musa paradisica*), though marketed mainly as a table fruit, is also processed primarily into puree, which is used in the manufacture of baby foods. The main waste material from banana processing is the peel, which constitutes about 20%–40% of the fruit weight. Depending upon the marketing conditions 10%–30% of banana harvest is also discarded because of over- and underripeness, blemishes, and other characteristics that make the fruit unacceptable for the export market.

2.2.1 Bioconversion of Waste

El-Shimi and associates (14) carried out studies on bioconversion of banana by *Aspergillus foetidus* for production of proteins and glucoamylase. α-amylase production from banana waste has been investigated by Krishna and Chandrashekharan (15). In the process developed, the wastes (banana fruit stalks) were cooked, dried, and ground prior to use in solid-state fermentation. Effects of cooking time and temperature used for processing banana wastes, particle size, moisture, and fermentation parameters (pH, temperatures, nutrients, inoculation size, and incubation period) on yield of α-amylase using *Bacillus subtilis* CBTK 106 were studied. A maximal enzyme activity of 5,345,000 U mg min^{-1} was achieved. Maximal yields were obtained by using banana waste that had been cooked at 121°C for 20 min and with a particle size of 400 μm and an initial moisture content of 70%. Yields of α-amylase from the banana waste medium were 2.65-folds higher than those of wheat bran medium. Tiwari and coworkers (16) undertook comparative studies on saccharification of banana peels by acid, enzyme, and stem and subsequent production of ethanol using *Saccharomyces cerevisiae* var. *ellipsoidus*

Detailed studies of production of bioprotein from banana waste have been carried out at Louisiana State University (17). It has been shown that maximal biomass yields with *Pichia spartinae* under optimal conditions were 53% on ripe banana pulp liquid and 58% on ripe banana skin liquid. Amino acid analysis of the protein revealed a high lysine content and an essential amino acid pattern comparing favorably with those of the FAO reference protein (17).

Fermentation of whole banana waste liquor for lactic acid production has been investigated (18). Fermentation of sugar in whole banana waste using mixed heterogenous inoculum was better than that of the axenic or mixed inoculum because lactic acid is the main product of fermentation.

2.2.2 Other Products

A number of products have been developed from banana waste at the Central Food Technological Research Institute (CFTRI) in Mysore, India (19). About 1000 banana plants is estimated to yield 20–25 tonnes pseudostem. The inner core, which is tender, is used as a vegetable. The pseudostem contains about 5% starch, which is used for eating purposes and for sizing in the textile industry. The pseudostem is crushed and stirred after mixing with water. The extract is sieved to remove the fiber, and the starch is allowed to settle. The settled starch is washed several times and dried.

The banana bunch, after removal of fruits, yields 8%–10% of the stem portion, which has very little commercial value. After steam processing under pressure it can be used as a deworming feed for pigs. The stem as well as the pseudostem fibers can be used for making paper (20). Hygienically collected peels of ripe bananas can be processed into *murraba*, a type of sweet product. Cooked banana skins are dipped in 40°Brix syrup, the concentration of which is gradually increased to 70°Brix or more. The product is cooked and packed into bottles (19).

2.3 Citrus

Oranges, mandarins, limes, lemons, pomelos, and grapefruit belong to the citrus family. The world production of oranges in 1997 was 63.8 million tonnes (Table 3). Brazil is the largest producer of oranges in the world, followed by the United States and Mexico.

Citrus processing is one of the most important food industries of the world. The main processed products of the industry are juice and juice concentrate. During the production of citrus juice, a large amount of processing residue, consisting of peel and rag, pulp wash, and seed, which amounts to about 50%–60%, is generated (21). Citrus waste is a rich source of essential oils, pectin, and a variety of by-products, including cattle feed. Kesterson and Braddock (22) have reviewed the processing details of by-products and specialty products of Florida citrus.

2.3.1 Dried Citrus Pulp and Molasses

The majority of citrus processing waste is converted into dried citrus pulp and used as a cattle feed. It is low in crude proteins, fiber, and fat but rich in carbohydrates. The low bulk density of dried citrus pulp results in high storage and handling costs, which have led to the development of the palletizing industry.

Citrus molasses is made from pressed liquor obtained during the manufacture of dried citrus pulp. The liquor contains 9%–15% soluble solids, of which 60%–75% are sugars. Citrus molasses is similar to cane molasses and is widely used as an animal feed and can be converted into alcohol by fermentation and for production of yeast, vinegar, 2,3-butyl glycol, riboflavin, and citric acid (22).

Table 3 Major Countries Producing Oranges

Country	Production (1997) (1000 MT)
Brazil	22,999
United States	10,538
Mexico	4,052
India	2,080
Italy	2,079
China	2,308
Iran	1,600
Pakistan	1,410
World	63,838

Source: From Ref. 1.

2.3.2 Citrus Peel Oil

The flavedo portion of the peel contains numerous oil cells, which on crushing or puncturing eject oil. Fresh orange peel yields about 0.54% oil by the cold press method (23). The citrus peel oil has attained a great commercial importance in the flavoring of several kinds of beverages, confectionery, and bakery products. It is also used in cosmetics and perfumery. The purity of citrus oils is determined by measuring their specific gravity, refractive index, optical rotation, and content of aldehydes and esters. The physicochemical characteristics of mandarin and lime oils have been investigated (23); a process for the recovery of essential oil from citrus waste has been described (24). Citrus stripper oil is peel oil that has been subjected to lime treatment and recovered by distillation from citrus peel liquor. The condensate contains approximately 60%–80% oil, and limonene accounts for about 90% of the oil.

2.3.3 Flavonoids

Hesperidin and naringin are the two important flavonone glycosides found in citrus. Hesperidin is found in the peel of the mature sweet orange. It is recovered by extracting chopped citrus peel with hot methanol and then allowing it to crystallize (22). Naringin is the predominant flavonoid in grapefruit, having an extremely bitter taste, which is extracted by a method similar to the one used for hesperidin. Park and associates (25) reported the identification and quantification of flavonoids in orange and lemon peels, Wilson (26) showed that grapefruit skin contains a considerable amount of neohesperidin and naringin. They reported the levels of these compounds in various samples and their use for conversion to the sweet dehydrochalcone derivatives.

The important citrus bioflavonoids in human nutrition and medicine have been reviewed. Bioflavonoids understood and applied in modern nutrition and medicine would appear to confer advantages in increased resistance to a variety of diseases against which we do not have effective measures at this stage (27).

2.3.4 Carotenoids

Orange peel is a rich source of a number of carbonyl carotenoids, and many of them are of epoxy character. The role of carotenoids in the treatment of skin cancer (28) and as an antioxidant (29) is well known. Molnar and Szabolcs (30) reported the identification of β-citraurin epoxide (3-hydroxy-5,6-dihydro-8′-apo-β-carotenal) and several isomers of violaxanthin (5,6,5′, 6-diepoxy-5,6,5′,6-tetrahydro-β-β caroten-3,3′-diol) in Valencia orange peel. Rosenberg and colleagues (31) have shown a two-stage countercurrent process using 1:2 peel-to-solvent ratio to be more efficient than stirring in a vessel. Limonene was found to be a more efficient solvent for the extraction of citrus peel carotenoids with a recovery of 4.5 g crude pigment concentrate/kg of citrus peel. Yen and Chen (32) isolated various xanthophylls, such as violeoxanthin, violaxanthin, violaxanthin epimer, neoxanthin, sinesia-xanthin, β-cryptoxanthin, luteinepoxide, and lutein in Taiwanese orange peel. The antioxidant effect of each pigment on soybean oil was shown to be dependent on the concentration of each pigment.

2.3.5 Citrus Seed Oil and Meal

Citrus seeds have long been recognized as a source of edible oil with characteristics suitable for human consumption. It has been an industrial practice to include seeds from processed citrus fruit with the peel and pulp portion during the production of dried citrus feed. Since the seed contains valuable oil and protein suitable for human consumption, it would be more profitable if they were processed separately and not included in feed for livestock.

The dried seeds with or without flaking are pressed at high pressure in an oil expeller to recover the turbid oil, which can be clarified in a plate and frame filter press (22). Citrus oil contains a considerable quantity of polyunsaturated fatty acid and linoleic and linolenic acids.

Extensive work has been done on citrus seed meal as a raw material for feed purposes (22). The dry oil extracted meal obtained during the oil manufacturing operation makes a valuable poultry feed.

2.3.6 Pectin

The citrus fruit components such as flavedo, albedo, membrane, juice vesicles, and core contain varying quantities of pectin, which is mainly in the form of protopectin. Pectin is widely used in the manufacture of foods as gelling agents, thickeners, and stabilizers of dispersed systems. Normally lemon, lime, grapefruit, and orange waste is used for pectin production. The general procedure for the commercial production of pectin from citrus fruit is described by La Fuente (24). Kratchanova and coworkers (33) reported that preliminary microwave heating of crushed fruit material of orange and lemon gave a higher pectin yield. Poonia and associates (34), while evaluating the pectin content of four species of citrus waste using various extractants, such as sodium hexametaphosphate, ammonium oxalate, ethylenediaminetetrachloroacetic acid, and water, showed that crude pectin and jelly units were significantly higher when ammonium oxalate was used as extractant. Pectin quality in terms of methoxyl content, anhydrouronic acid, and degree of esterification was better when sodium hexametaphosphate was used. The rag portion of the fruit yielded higher crude pectin and jelly units than peel. Zafiris and Oreopouloiu (35) studied the effect of temperature, time, and pH on the extraction of pectin yield and jelly units. According to ash and methoxyl content determination, the product was classified as low-ash and high-methoxyl pectin. Its purity expressed as anhydrogalacturonic acid content varied from 68.5% to 75.0%.

2.3.7 Beverage Clouding Agent

Citrus by-products and waste contain large amounts of coloring material in addition to their complex polysaccharide content. Citrus waste products, including citrus seeds, juice vesicles, peel, and waste liquid from essential oil and pectin manufacturing, have been used to make natural beverage clouding agents (36). Citrus by-products and waste such as molasses, concentrated citrus peel juice, and pulp wash after fermentation, proteolytic treatment, and extraction with alcohol yielded a natural beverage cloud (37).

2.3.8 Products by Bioconversion

Bahar and Azuaje (38) reported that *Aspergillus* and *Fusarium* spp. grown on the clear filtrate of orange peel gave 58% protein, 26% carbohydrates, 4.75% ribonucleic acid (RNA), 0.76% deoxyribonucleic acid (DNA), and 0.73% phosphorus of the final biomass. Nwabueze and Oguntimein (39) reported the production of biomass containing 57% protein in a sweet orange waste medium (peel, pulp, and seed). Clementi and colleagues (40) studied the growth behavior of *Memnoniella echinta* and *Fusarium roseum* in slurry fermentation systems using untreated orange peel as a substrate. The more concentrated the peel slurry, the greater the substrate inhibitory effect on microbial growth. Essilfie and coworkers (41) studied the protein upgradation of orange peel waste by using *Aspergillus niger*, containing potassium dihydrogen orthophosphate, ammonium sulfate, and molasses. The crude protein level increased from 5% to 13% after incubation at 30°C for 72 h. Utilization of citrus molasses as a substrate for the production of citric acid (42), pyruvic acid (43), xanthan (44,45), and acetate and ethanol (46) has been reported.

2.3.9 Specialty Products

Various specialty products, including gelled citrus peel, frozen citrus juice sacs, dried juice sacs, citrus puree, beverage bases of frozen fruit desserts (Popsicles™), brined and sulfured peels, candied and glycerated peels, marmalades, bland syrup, and peel seasoning, have been extensively reviewed (22,42).

2.4 Grapes

Grape is an important fruit crop in many countries; the present production is about 58 million tonnes (1). Though grapes are mainly used as table fruit and converted into raisins, a major proportion of the production is used in juice and wine manufacture. The primary waste generated in the process is grape pomace, which consists of the skin, seeds, and stem and represents as much as 20% of the processed fruit (47). In recent years, there has been increased interest in the recovery and reuse of by-products from grape pomace.

Ethanol can be produced from grape pomace by solid-state or submerged fermentation. For submerged fermentation, water is added to leach sugar out from the pomace and then fermented. After fermentation the alcohol is recovered by distillation. The concentration of ethanol in the fermented dilute medium is usually too low to be recovered economically. Besides, this method poses serious disposal problems for the resulting waste effluents. The solid-state fermentation process described by Hang and associates (48) overcomes this problem. This process yields more than 53 g ethanol/kg pomace.

2.4.1 Tartrate

Grape pomace as a source of tartrates has been investigated. Amerine and colleagues (49) have reviewed methods of tartrate extraction from grape pomace. In a batch process described, pomace is mixed with three parts of water, heated to 60°C–100°C, drained, and heated again with two parts of water, and the pomace is pressed to recover extract. The combined extract is cooled to crystallize tartrates.

2.4.2 Pigments

Blue grapes have been extensively studied for extraction of natural anthocyanin food colorant. The production of natural food colorants from grape pomace and several other plant materials, such as annato (*Bixa orellana*), turmeric (*Curcuma longa*), red beet (*Beta vulgaris*), paprika (*Capsicum annum*), safflower (*Carthamus tinctoris*), and kokum (*Garcinia indica*), has been reported. The Bangalore blue variety of purple grapes, which is probably a hybrid between the American *Vitis labrusca* and *V. vinifera*, is cultivated extensively in India. Because of its high acid content, it is not a choice table fruit and hence is processed into juice. The pomace is used for the production of color concentrate and powder. The process essentially involves extraction of the color under certain optimized conditions, clarification of the extract, pasteurization, and either vacuum concentration or vacuum drying to obtain the concentrate or powder (50).

2.4.3 Seed Oil and Protein

Grape seed represents about 2%–3% of the whole grape berry, and protein and oil content of the seed ranges from 10% to 13% and 12% to 16%, respectively (51). Grape seed oil can be recovered by mechanical pressing or by solvent extraction of ground seeds. The oil has low saturated fatty acid content and a high level of linoleic acid. This oil has a pleasant flavor and is stable when used as a frying oil (52).

Grape seed is a potential source of protein. However, it is important to remove polyphenols from the protein to improve its digestibility. Fantozzi (51) described a process for production of protein concentrate having improved protein digestibility. The process involves soaking whole seeds in an alkaline solution to extract polyphenols, drying and milling the seeds, than extracting the protein with 20% sodium chloride. The extract is acidified to pH 4 to precipitate the protein, which is recovered by centrifugation. The precipitated protein is desalted by dialysis and dried.

2.5 Mango

Mango (*Mangifera indica L.*) is one of the most important tropical fruits. The world production of mangoes during 1997 was 23 million tonnes; India had a share of 51% (1). The other major mango producing countries are China and Mexico (Table 4).

Even though mango is highly valued as a table fruit, substantial quantities of fruits are processed. Ripe mangoes are processed into (a) frozen mango products, (b) canned products, (c) dehydrated products, and (d) ready-to-serve beverages (53). During the manufacture of mango products large quantities of waste are generated; they account for 35%–55% of the fruit (54), depending on the variety.

Actual figures on the quantity of mango waste generated commercially are not readily available. However, a rough estimate of the waste available annually in India alone based on the quantities of mango processed (0.2 million tonne) is of the order of 90,000 tonnes.

Mango pulp, which is the most important semifinished product of mangoes, is extracted commercially by feeding the mangoes after washing and cutting into a pulper fitted with a fine cylindrical sieve with approximately 0.8-mm-diameter opening. During the

Table 4 Major Mango Producing Countries

Country	Production (1997) (1000 MT)
India	12,000
China	2,150
Mexico	1,444
Indonesia	1,128
Pakistan	914
Nigeria	500
Philippines	480
Chile	456
Haiti	210
World	23,428

Source: From Ref. 1.

Table 5 Different Components Obtained During Mango Pulp Extraction

Component	Percentage
Mango pulp	45–65
Peels	15–20
Pulper waste	10–15
Stones (seeds)	10–20

Source: From Ref. 55.

operation, coarse mango pulp flows out and the peel along with the stones (seeds) are ejected. The coarse pulp is again passed through a pulper fitted with a finer sieve (0.5-mm-diameter opening) to recover finished pulp. The fiber portion along with some quantity of adhering pulp is rejected as pulper waste. The approximate proportions of the different components are given in Table 5.

Extensive work has been done on characterization of the waste components and preparation of value-added products from them. Some of the aspects have been reviewed (56,57).

2.5.1 Peel Waste

Mango peel along with pulper waste sometimes referred to as *peel waste* constitutes 20%–30%. The peel waste is a good source of sugars and pectin and also contains substantial quantities of protein, tannins, and crude fiber.

2.5.2 Mango Peel Pectin

Srirangarajan and Shrikhande (58) extracted and characterized pectin from three varieties of mango, viz., Alphonso, Langra, and Dashehari. The quality of pectin was comparable to that of orange peel pectin.

Tandon and coworkers (59) showed that the yield of pectin from Dashehari varieties of mango was dependent upon the method of extraction. Whereas hydrochloric acid extraction yielded 9.6%, the yield was 12.97% by sodium hexameta phosphate. They also showed that the yield and quality of pectin were better from ripe than immature mangoes. Yu and Delvalle (60), however, have shown that pectin yield from peels of partially ripe mangoes was higher (14.7%) than that from fully ripe mangoes (10.9%). The data on the yield and characteristics of pectin extracted from the peels of different varieties of mangoes show varietal variations (58); in general, the quality is suitable for food and pharmaceutical purposes. On a dry weight basis, the pectin content ranges from 13% to 19% and the jelly grade ranges from 155 to 200. The methoxyl content (8.1% to 9%) shows that the pectin is of high-methoxyl type.

Beerh and associates (61) have shown the possibility of preparing juice, wine, and vinegar from peel and pulper waste of Totapuri mangoes. The juice yield was 51% from peels and 75%–78% from pulper waste on enzyme liquefaction and pressing. The characteristics of the juice showed that the juice from the peels was astringent, however, it could be removed by gelatin treatment. When the recovered juices were added to mango pulp for mango nectar preparation, the quality of the beverage was not affected.

Beerh and colleagues (61) and Ethiraj and Suresh (62) have shown the possibility of producing wine and vinegar from mango peel waste. Ethiraj and Suresh (62) found that an average yield of 2.5 L absolute alcohol could be obtained from 100 kg mango peel waste. Since the alcohol content in the fermented peel extract was only 2.5% to 3.0%, to obtain vinegar of 5.0% acetic acid content, it was necessary to raise the alcohol content either by fortification or by secondary fermentation with the addition of cane sugar.

Sethi and coworkers (63) used mango peel as a substrate for the production of fungal proteins, carboxymethyl-cellulase, and polygalacturonase by fermentation. Rasheed and associates (64) used mango waste for the production of single-cell protein and other metabolites by yeast. There appears to be good potential for the extraction of bioactive components from mango peel. Cojocaru and coworkers (65) isolated a mixture of 5-(12-cis-heptadecenyl) and 5-(pentadecenyl) resorcinol from mango peel having antifungal activity against *Alternaria alterata*, which is responsible for black spot disease in mango fruit. The potential for the production of dietary fiber (66) and wax (67) from mango peel waste has been demonstrated.

Mango aroma concentrate has been prepared from mango peel waste (68). Aroma concentrate was obtained by steaming the peel waste in a vapor generator under inert atmosphere, condensing the vapors, and concentrating the aroma by successive evaporation. The aroma concentrate obtained from peel waste of the 'Totapuri' variety of mango has similar composition and sensory quality to those obtained from mango pulp. The aroma concentrate can be used for flavoring beverages and dairy products. The potential of utilizing mango peel waste as cattle feed has been investigated (69). It was shown that poultry diets containing mango peel waste significantly improved the performance of layers and could be a substitute for rice polishing.

2.5.3 Mango Seed (Stone)

Mango seed or stone consists of a tenacious coat enclosing the kernel. The seed content of different varieties of mangoes ranges from 9% to 23% of the fruit weight (70) and the kernel content of the seed ranges from 45.7% to 72.8% (71).

A number of workers have investigated the composition of mango kernel (72–74). The reports indicate that the mango kernel is comparable to most of the cereals in respect to carbohydrates, protein, fat, and minerals. The average content of the components comprises carbohydrate (69.22%–78%), fat (8.35%–16.13%), protein (5.6%–9.5%), and a fair amount of fiber (0.14%–2.95%) and ash (0.35%–3.66%) (75).

Reduction of the tannin content is a prerequisite for direct utilization of mango kernel flour. Dhingra and Kapoor (75) showed that tannin can be effectively removed if the kernels are soaked two to three times in water for 20 min each at 80°C. Traditionally mango kernels are soaked in water, deskinned, ground, washed in water to remove astringency, and dried (76).

In traditional food items, 20%–30% of kernel flour can be used without adversely affecting their acceptability (75). However, Zia-Ur-Rehman and colleagues (69) showed that egg production and egg size were adversely affected when mango stone was added to the layer's diet at 8% level. They opined that tannin and saponin contents in mango stone might have made the diet unpalatable with subsequent reduction in feed consumption, thereby adversely affecting the egg size and yield.

Lasztity and coworkers (73) have investigated the kernel for protein of four Egyptian varieties of mangoes. They found that the protein content varied from 5.0% to 7.2% and the amino acids ranged from 31.0% to 39.4%, showing a relatively good total amino

Table 6 Fatty Acid Composition of Mango Seed Kernel Fat

Fatty acid		Percentage	Reference
Palmitic	(16:0)	6.80–9.70	67, 78
Stearic	(18:0)	32.70–44.00	67, 78
Oleic	(18:1)	43.70–53.40	67, 78
Linoleic	(18:2)	3.60–6.90	67, 78
Linolenic	(18:3)	0.30–1.00	78
Arachidonic	(20:0)	1.10–2.50	78

acid proportion (essential amino acid/total amino acid ratio). However, the proteins were deficient in sulfur amino acids, lysine, threonine, phenylalanin, and tyrosine. Attempts have been made partially to substitute kernel flour in bakery products. Though the absence of gluten is a drawback for use in bread, it can be substituted in up to 10% of wheat flour for cakes (77).

Carbohydrates constitute the major component (69.22% to 79.78%) of mango kernels, and starch constitutes about 92% (74,75).

2.5.4 Mango Kernel Fat

Extensive work has been carried out on mango kernel fat. The reported values on kernel fat content vary from 8.85% to 16.13% (74). The lipids of mango kernel are composed of neutral lipids, phospholipids, and unsaponifiable matter. Van Pee (78) reported that mango kernel fat contains 94.8%–97% of neutral lipids, 1.1%–2.8% of phospholipids, and 1.3%–2.2% of unsaponifiable matter. The unsaponifiable matter (USM) on chromatographic separation did not show any unusual constituents, other than those present in the USM of other oils, such as palm, sesame, groundnut, and cotton seed oils (79).

Narsimhachar and associates (80) processed mango kernel for recovery of fat and refined the fat. Mango stones were decorticated by using a specially adapted decorticator. Fat was extracted both by hexane as well as by a combined method of expeller followed by hexane extractions. Blandness, plasticity, and absence of toxic substances render it a potential edible fat in sweetmeat and in pastries in which saturated fats are required. It can be used as a substitute for tallow in the preparation of high-quality soaps and as an extender to cocoa butter. The fat extracted meal can be used in animal feed rations and as a direct sizing agent in textiles (81).

The fatty acid composition of kernel fat has been investigated by many workers (Table 6). Stearic and oleic acids are the major constituents of mango kernel fat, accounting for over 85%. Palmer and Sharma (82) showed that mango kernel fat added to ghee from buffalo milk at the rate of 1% acted as an antioxidant.

2.6 Pineapple

Pineapple (*Ananas comosus Lin*) is one the most popular noncitrus fruits processed. Some of the leading pineapple producing countries with their production are listed in Table 7.

Because of the unique structure of the fruit and the presence of very active protease, pineapple does not find extensive use as a table fruit and hence is mostly processed into a number of products. The major products are canned slices and tidbits, juice, concentrate, beverages, and jam. Depending on the variety and the shape of the fruits, the total waste

Table 7 Production of Pineapple in Some Countries

Countries	Production (1997) (1000 MT)
Thailand	2,000
Brazil	1,937
Philippines	1,452
India	1,100
China	899
Nigeria	800
Indonesia	538
Colombia	320
Mexico	301
Vietnam	185
Malaysia	163
World	12,794

Source: From Ref. 1.

generated is 30%–60% of the fruit. The important waste components include crown, stem, core, leaves, and mill juice (Table 8).

2.6.1 Mill Juice

Most of the pineapple waste components can be milled and pressed to extract milled juice, and the compact residue can be either easily disposed of or further dried to produce pineapple bran. A number of value-added products can be prepared from the milled juice. A typical composition of the soluble solids of milled juice includes 75%–85% sugar, 7%–9% citric acid, 2% malic acid, 2.5%–4% protein, 3.5% gum, and several mineral constituents (83).

Properly prepared high-quality sugar syrup from mill juice is at present the most valuable by-product. The high acidity of the juice is removed by treating with either lime or a suitable ion exchange resin. The deacidified and clarified juice syrup can be used successfully in canned pineapples (84).

Table 8 Pineapple Waste and Its Products

Waste component	By-products	Refs.
Mill juice	Sugar syrup	84
	Bromelin	85, 86, 87
	Wine	88, 89
Centrifuged juice	Oleoresin	93
Underflow	Hemicellulose-β	93
	α-Cellulose	93
Core	Syrup	94
	Vinegar	89
Crown, peels, and leaves	Fiber	95

2.6.2 Bromelin

Among all the pineapple by-products, extraction and characterization of bromelin, a proteolytic enzyme, has been the most extensively studied. A process for the production of bromelin from pineapple has been patented in India (85). The enzyme is extracted into dibasic potassium or sodium phosphate solution or potassium phosphate buffer (pH 5.5–7.5). The extracted bromelin is precipitated with ethyl alcohol, acetone, or *n*-butanol and dried at reduced pressure. In the Chinese process (86), the mill juice is clarified and the enzyme is precipitated with alcohol. The settled bromelin slurry is spray dried to obtain the crude enzyme. Tisseau (87) has developed a method for bromelin extraction based on milled juice clarification followed by ammonium sulfate precipitation.

2.6.3 Wine

Attempts have been made by several workers to produce wine from pineapple waste juice. Alien and Musenge (88) prepared wine from pineapple pulp and waste juice. The results of the studies revealed that boiled waste juice produced the best wine. Production of still and sparkling wines from Philippines pineapple juice has been reported. General acceptability, flavor, appearance, and aroma were good in both cases (89).

2.6.4 Vinegar

Pineapple canary waste juice, a good source of sugar (7%–10%), can provide a good substrate for vinegar production. Satyavati and colleagues (89) have developed a method for the production of good-quality vinegar from pineapple waste juice. The process consists of fermenting the extracted juice with *Saccharomyces cerevisiae* after adjusting the final Brix to 10° and then passing the alcoholic fermented juice through a generator for acetic fermentation with *Acetobacter* sp. culture.

Submerged acidification of waste pineapple juice was studied by using pure culture inoculum. Vinegar containing up to 7% acetic acid could be produced in less than 24 h with a conversion efficiency greater than 90% in both laboratory and pilot equipment (90).

2.6.5 Feed

Extensive work has been reported on the feed value of the pineapple waste (90). Ismail and Abdulla (91) have reported the preparation of pineapple bran, which is the dried product obtained after extracting mill juice from the waste. The bran so prepared had 6%–8% crude protein and 22%–28% crude fiber. The analytical data of the product indicated that it could be a potential source of feed for ruminants.

In a series of trials in Taiwan, wet or partially dried pineapple bran was ensiled with rice straw, maize meal, dried sweet potato chips, molasses, and urea in various combinations and proportions. Conservation methods and feeding value of these pineapple bran mixtures for ruminants were studied. When adequately supplemented, the bran was found to be a valuable and cheap feeding stuff well suited to use in a variety of canneries (90).

2.6.6 Other Products

Chicaya (92) described a method for the preparation of sucrose and molasses containing glucose and levulose from the juice by hot precipitation of extract with lime filtering to produce clarified juice, which after demineralization with ion exchange resins is concentrated to produce syrup containing crystals of sucrose, which is recovered by centrifugation.

Chan and Moy (93) utilized the underflow from centrifugation of pineapple juice. The underflow, which is viscous fluid, was used to prepare potentially useful by-products, including oleoresin having a strong pineapple flavor, a colorless pineapple syrup after ion exchange resin removal of phenolic substances, a water-soluble hemicellulose-β, a food thickener, and purified α-cellulose fraction.

The process of obtaining the fiber consists of removal of green tissue by scraping, soaking of the fiber, and sun drying. Pina, the delicate and expensive fabric of the Philippines, is made from this fiber. The waste material left during extraction of fiber is suitable for paper making. Physical and other characteristics of pineapple leaf fiber have been studied to ascertain its textile value (95). This fiber was found to be as thin as the finest quality of jute but about 10 times as coarse as cotton.

3 BY-PRODUCTS OF VEGETABLE PROCESSING WASTE

3.1 Potato

Potato (*Solanum tuberosum*) is the vegetable most produced in the world with a current estimated production of 295 million tonnes (1). China is the largest producer of potatoes, followed by The Russian Federation (Table 9).

Major waste material in the potato processing plants is in the form of peel and juice containing solids. Potatoes are processed for different products and for starch. The waste effluent from the potato starch manufacturing is similar to that of other effluents from cutting and chipping operations of other processes. The other source of potato waste are damaged produce from harvest; sprouted, green, and damaged potato from storage; and rejected potato. A number of products can be prepared from potato waste (Table 10).

Sagar and coworkers (96) discussed the preparation of a variety of products, such as chips, flour, burfee, halwa, and gulab jamun from sprouted and green and damaged potatoes.

Table 9 Production of Potato by Major Countries

Country	Production (1997) (1000 MT)
China	45,534
Russian Federation	40,000
Poland	20,776
India	19,240
Ukraine	19,000
Germany	12,438
Belarus	11,500
Netherlands	8,081
United Kingdom	7,154
France	6,500
World	295,407

Source: From Ref. 1.

Table 10 Products from Potato Processing Waste

Waste Component	By-Products	Refs.
Damaged produce from harvest, sprouted green and damaged potato from storage, and rejected potato	Chips, flour, tikki, burfee, halva, gulab jamun	96
Potato pulp and liquor	Starch	98
	Ethanol	98, 99
	Proteins	103
	Enzymes	98, 103
Peel	Dietary Fiber	105, 106

3.1.1 Starch

Various workers have carried out studies on the recovery of starch from potato waste. Ritcher and Grosser (97) discussed the application of enzymatic decomposition of potato components to reduce waste and upgrading of the waste.

3.1.2 Alcohol

Klingspohn and associates (98) fractionated pectin, starch, cellulose, and hemicellulose from potato pulp. Cellulose and hemicellulose enzymatically degraded into glucose and xylose can be converted into alcohol. Grobben and colleagues (99) reported the production of acetone, butanol, and ethanol from potato waste inoculated with *Clostridium acetobutylicum*. Using an integrated membrane system, production of 20 g/L of solvent was obtained. With a polypropylene perstraction system and oleyl alcohol decane mixture used as an extractant, the product yield increased from 0.13 g/g to 0.23 g/g. By incorporating a microfiltration step having a hydrophilic membrane and using fatty acid methyl esters from sunflower oil as an extractant, an increase in the production of acetone, butanol, and ethanol was observed.

3.1.3 Proteins

Wilhem and Kempf (100) carried out studies on the preparation of protein from potato juice, using a centrifugation, ultrafiltration, and thermal coagulation method. They also compared the functional properties and industrial importance of protein manufactured by these methods. Luther and coworkers (101) carried out studies on protein recovery from waste potato juice by a coagulation method. It is also possible to extrude the potato protein to get a food supplement for athletes (102).

3.1.4 Enzymes

Klingspohn and associates (98) used pulp residue (PR) consisting of hemicellulose and cellulose, which was enzymatically converted to glucose and xylose and fermented by *Pachysolen tannophils* to produce ethanol. The pulp residue and the residue of the enzymatic decomposition were used as a substrate for enzyme production by *Trichoderma reesei* Rut C-30 and *Trichoderma reesei* MGC 77 in batch and continuous cultures. The highest enzyme specific activities (6.1 U/mg avicelase, 23.4 U/mg CMCase [carboxy methyl cellulase], and 66.8 U/mg xylanase) and productivities (8.9 U/L/h avicelase, 34.3 U/L/h CMCase, and 97.8 U/L/h xylanase) were obtained with the residues of enzymatic decomposition. The enzyme mixture in the cultivation medium was concentrated by ultrafiltration and diafiltration. Trojanowaski and colleagues (103) utilized potato processing waste as a carbon and nitrogen source for the production of the enzyme laccase (polyphenol oxidase) using strains of *Trametes villosa, T. versicolour*, and *Polyporous bromalis*. In addition, *Trametes villosa* can synthesize α-amylase

3.1.5 Biomass

Several microbial processes for treatment of potato wastes have been described by different workers mainly to reduce biological oxygen demand. Senez (104) described the procedure for protein enrichment of starchy raw material of potato waste by solid-state fermentation with strains of *Aspergillus niger*. The protein quality was found to be comparable to that of soyabean meal.

3.1.6 Dietary Fiber

Toma and coworkers (105) recommended the use of potato peel as a source of dietary fiber and a better substitute for wheat bran in the preparation of bread because of its mineral content, total dietary fiber, water holding capacity, and lack of phytate. Nebesny (106) reported that aqueous potato pulp after enzyme liquefaction and saccharification could be dried to obtain fiber containing 79% dry solids (d.s.) of dietary fiber, which can be used in the preparation of bread.

3.2 Tomato

Tomato (*Lycopersicum esculentum L.*) is one of the most important vegetable crops in the world with an estimated production of 88 million tonnes (1). Unlike that of many other vegetables, a major percentage of tomatoes is used for processing into juice, paste, puree, soup, ketchup, sauce, and canned tomatoes.

Tomato pomace left over as a waste after extracting tomato juice constitutes about 20% to 30% of the raw material. The pomace essentially consists of seeds and skin. The seed component of the waste, which accounts for about 55%, has received considerable scientific investigation

3.2.1 Seed Oil

A number of workers have studied the extraction and characterization of oil from tomato seeds (107,108). Tomato seed oil is characterized by its high content of monounsaturated fatty acids, namely, oleic acid (C 18:1, 81.34%). Although this edible fatty acid is of low nutritional value compared to polyunsaturated fatty acids, especially linoleic acid (C 18:2), which is predominant in such crops as safflower, sunflower, and corn, nevertheless oleic acid can be used as a soap base in the manufacture of oleate ointments, cosmetics, polishing compounds, lubricants, and food grade additives, and other products (109).

3.2.2 Seed Protein

Studies on the amino acid composition of the protein indicated the presence of all essential amino acids. The most limiting amino acids are cysteines, methonine sulfur amino acids, isoleucine, valine, leucine, and phenyl alanine; threonine, hystidine, and lysine were found to be the most abundant essential amino acids of tomato seeds (109). The high level of lysine in tomato seed protein has promoted supplementation of the seed meal in wheat flour, which is deficient in the amino acid. Supplementation of wheat flour bread at 10% and 20% levels increased lysine by 40.2% and 6.9% (110). Wheat breads with the addition of 1% or 2% of tomato seed flour to the wheat flour were equal in quality to the control wheat bread, but greater addition resulted in crumb darkening. Tests with rye flour showed that adding 7% tomato seed flour resulted in bread of very good quality with greater volume and porosity than control bread (107).

3.2.3 Coloring Matter

The possibility of extracting coloring matter from tomato skin has been investigated (111). The major pigment in tomato skin is lycopene (11.98 mg/100 g), which corresponds to about 71% of the lycopene found in tomato paste (109). The increasing safety awareness of and restrictions imposed on artificial food colors have resulted in considerable attention to natural colors as substitutes. Tomato skin has the potential for extraction of lycopene red color for food use.

Kramer and Kwiss (112) have investigated the possibility of recovering valuable nutrient components from tomato cannery waste and from the entire tomato plant, including not only undersized, green, and overripe fruits but vines as well. They have shown that tomato vines contain over 14% protein, though cannery waste has higher levels (21%). In the processing scheme developed, the waste material is milled at pH 8 and the pulp is pressed. The press cake, containing approximately 60% of the total solids with the protein content reduced to about 10% dry weight basis, can be utilized as such for cattle feed. The extract may be acidified directly to pH 3.5 and a precipitate containing 48%–58% protein recovered; it can be fractionally precipitated by stepwise acidification to pH 4.8, 4, and 3.5. All four precipitates may be combined at this point and utilized as tomato protein concentrate, which includes tomato pigments, flavonones, and other substances, or they may be further precipitated by solvent extraction with acetone. The acetone residue consists of a highly concentrated, bland, protein isolate (70%–85%). Upon recovery of the acetone a residue having a strong concentrated tomato color and flavor is obtained. Tomatine and perhaps other substances, including sugars and acids, may be recovered from the spent liquor at several stages in the process.

The available literature on tomato waste only indicates the possibility of preparing a few by-products. Since the quantity of processed tomato products is increasing steadily, there is a need to develop commercially viable technologies for the production of other by-products.

4 BIOGAS PRODUCTION FROM FRUIT AND VEGETABLE PROCESSING WASTE

A promising method for economic utilization of fruit and vegetable processing wastes that otherwise are unsuitable or uneconomical for production of by-products is biomethanation. This method has certain advantages because it provides fuel in the form of methane, digested slurry can be used as manure, and enriched effluent can be used for irrigation purposes. Some of the processing wastes that contain antinutritional or toxic factors can also be utilized for biogas production after suitable modifications.

Kranzier and Davis (10) have studied the potential for recovery of energy from apple pomace by means of biogas conversion technology. It has been reported that citrus peel can be used only up to 4% level for biogas generation because the citrus oil becomes inhibiting for anaerobic digestion (113). This problem has been greatly overcome by fungal pretreatment of the waste (114). Solid-state fermentation of wastes using selected strains of *Sporotrichum, Aspergillus, Fusarium, and Penicillium* spp. reduced the level of antimicrobial substances, allowed the use of a loading rate of 8%–10% (dry weight), and improved the overall productivity of biogas and methane.

Wastes from fruit and vegetable processing lack certain essential nutrients required for efficient biogas production. Carbon, nitrogen, and phosphorus (C:N:P) ratio is important for anaerobic digestion. Mango peel waste supplemented with rice bran could generate biogas of 0.6 m^3/kg volatile solid (VS) added (115). Trace elements including cobalt, nickel, and iron also showed a beneficial effect on biogas yield and its methane content (116).

Lenschner and associates (117) reported the engineering and economic analysis of methane generation from potato, fruits, tomato, and dairy wastes. They were of the opinion that the seasonality of the waste was the major drawback to commercialization of the process. Krishnanand (118) has shown that mango peel, citrus peel, and green pea pods, can be ensilaged and subsequently used for biogas production, thus overcoming the seasonal availability problem.

The future of anaerobic digestion of fruit and vegetable processing wastes depends on the cost competitiveness of the biogas produced. With the ever-increasing price of petroleum products, a favorable solution is likely to be realized in the immediate future.

5 COMPOSTING

In the context of growing concern over environmental pollution caused by indiscriminate disposal of organic wastes such as fruit and vegetable processing wastes, composting is an environmentally friendly process. The end product obtained by composting can be added back into the soil as an organic fertilizer. During the process of composting, the volume of the waste is reduced by 25%–40% and the thermophilic temperatures (40°C–70°C) attained destroy most of the pathogenic organisms as well as weed seeds that may be present in the waste material.

Composting is a natural aerobic biochemical process in which thermophilic microorganisms transform organic material into stable soil-like product (119). Several books and reviews have exhaustively documented the process of composting (119–122).

A large spectrum of microorganisms are involved in the composting process. Though the main classes are bacteria, fungi, actinomycetes, and protozoa, within each class there are many strains and at least 2000 strains of bacteria are present. The number of microbes in compost is phenomenal, approximately 2×10^9/g (122).

For an efficient composting process, the requirements for the microorganisms to survive and grow are very specific. The key control parameters are moisture content (40%–60%) (119), carbon-to-nitrogen ratio (25:1) (120), and pH (6–8) (119–120) of the waste material; temperature; and aeration (oxygen concentration). Therefore, the initial part of preparing the waste material for composting should achieve these parameters.

Moisture content and aeration are related. High moisture content impairs oxygen availability by blocking the pore space of the waste material. Low oxygen concentration leads to anaerobic decomposition, resulting in the production of by-products having offensive odor. Finely-ground waste material may have to be mixed with bulking agents such as wood chips to maintain the desired porosity for proper aeration.

The C:N ratio is important because low C:N ratio results in production of ammonia, and excessively high C:N ratio causes the available nitrogen to be used up before the carbonaceous material is completely decomposed. Therefore, proper blending of different waste materials to achieve the desired C:N ratio is necessary before composting (120).

It is also necessary to adjust the pH of the waste materials within the range of 6–8; outside this range microbial activity is compromised and decompositions are slowed or even stopped (119–120).

During the composting process, initially mesophilic bacteria are responsible for decomposition. As a result of the bacterial activity, the temperature increases, and above 40°C, thermophilic bacteria take over and the process continues for several days with a temperature range of 50°C–65°C. This is a very important stage because here the levels of pathogenic bacteria and other undesirable organisms are reduced to acceptable levels. Subsequently as the temperature falls, mesophilic organisms continue the composting process at a slower rate. The active composting period is followed by the curing period, when the microbial population is stabilized. The curing time, which varies from a few months to 1 or 2 years, depends on the nature of the waste material used. The cured compost should meet certain quality standards (123–124).

There are three major types of composting methods (125), viz., passive piles or windrows, aerated windrows, and in-vessel systems. Among the methods passive windrows require the least capital investment and in-vessel systems the highest. However, sophisticated systems provide better process control and higher throughputs.

5.1 Composting of Fruit and Vegetable Processing Wastes

The major limitations of composting of fruit and vegetable processing wastes are their high moisture content (80%–90%) and acidic pH (126–128). Therefore, it is necessary to modify the material to reduce the moisture content to 50%–60% and neutralize the acidity to obtain the desired pH range. Moisture content can be partially reduced by draining and pressing. Bulking agents are also used to reduce moisture content and to increase the porosity. They include sawdust (128,129), rice hulls (126), paper (129,130), and coffee ground residues (125). While mixing the bulking agents with the waste materials, care should be taken to maintain the C:N ratio. To reduce high C:N ratio, sometimes nitrogen in the form of urea is added. Wood ash and lime are added to neutralize excess acidity.

Even though aerated static piles and passive piles with no aeration have been used (126,130), the most widely used method for composting fruit and vegetable processing wastes is turned windrows (126,129,130).

A composting machine with a rotary turning device that processes a large quantity of waste in a relatively short time from the processing of Satsuma mandarins has been developed (131). Aerobic and thermophilic bacteria, including *Bacillus sterothermophilus* and *Thermus* species isolated from sewage, were used as the starting material for the treatment. Batches are turned over once a day by the rotary device and composting can be completed in approximately 2 weeks.

The several advantages offered by composting—which include environmental friendliness, waste volume reduction, and value additions—as well as new technological developments, are likely to encourage more and more fruit and vegetable processing industries to adopt composting as a waste management option.

REFERENCES

1. FAO Production Year Book, Vol. 51, Food and Agriculture Organization, Rome 1997.
2. Smock, R. N., and A. M. Neubert, Apples and Apple Products, Interscience, New York, 1950.

3. Shah, G. H., and F. A. Masoodi, Studies on the utilization of waste from apple processing plants, Indian Food Packer 5:47 (1994).
4. Hang, Y. D., C. Y. Lee, and E. E. Woodams, A solid state fermentation system for the production of ethanol from apple pomace, J. Food Sci. 47:1851 (1982).
5. Gupta, L. K., G. Pathak, and R. P. Tiwari, Effect of nutrition variables on solid state alcoholic fermentation of apple pomace by yeasts, J. Sci. Food. Agric. 50:55 (1990).
6. Almosnino, A. M., and J. M. Belin, Apple pomace: An enzyme system for producing aroma compounds from polyunsaturated fatty acids, Biotechnology Lett. 13(12):893 (1991).
7. Almosnino, A. M., M. Bensoussan, and J. M. Belin, Unsaturated fatty acids bioconversion by apple pomace enzyme system: Factors influencing the production of aroma compounds, Food Chem. 55(4):327 (1996).
8. Sulc, D., R. Tamandiuk, J. Terkulov, Z. Baskar, and S. Petrov, Utilization of secondary raw materials in food production, Hrana I Ishrana 257(10):199 (1982).
9. Hang, Y. D., Production of fuels and chemicals from apple pomace, Food Technol. 41(3): 115 (1987).
10. Kranzier, G. A., and D. C. Davis, Energy potential of fruit juice processing residue, ASAE Pap. 81:6006 (1981).
11. Hang, Y. D., and E. E. Woodams, Apple pomace: A potential substrate for citric acid production by *Asp. Niger*, Biotechnology Lett. 6(11):763 (1994).
12. King K., Partial characterization of the *in situ* activity of pectin esterase in Bramley apple, Int. J. Food Sci. Technol. 25:188 (1990).
13. King K., Characteristics of pectin esterase isolated from Bramley apple waste, J. Sci. Food Agric. 57:43 (1991).
14. El-Shimi, M. L.; S. M. Mohsin, and Abd El-Magid, Bioconversion of some agricultural wastes by *Aspergillus foetidus*, Egypt. J. Food Sci. 15(1):121 (1987).
15. Krishna, C., and M. Chandrashekharan, Banana waste as a substrate for α-amylase production by *Bacillus subtilis* (CBTK 106) under solid state fermentation, Appl. Microbiol. Biotechnol. 46(2):106 (1996).
16. Tiwari, H. K., S. S. Marwaha, and K. Rupal, Ethanol from banana peel, Agric. Wastes 16(2): 135 (1986).
17. Chung, S. L., and S. P. Meyers, Bioprotein from banana wastes, Dev. Ind. Microbiol. 20: 723 (1979).
18. Lopez-Baca, A., and J. Gomez, Fermentation pattern of whole banana waste liquor with four inocula, J. Sci. Food Agr. 60(1):85 (1992).
19. Banana in India. Production, Preservation and Processing, Central Food Technological Research Institute, Mysore, India, 1989.
20. Chellappa, S., and Srinivasan, Wealth from agricultural waste, Indian Chem. J. 15(6):27 (1980).
21. Anand, J. C., and S. B. Maini, Utilisation of fruit and vegetable waste, Indian Food Packer 51(2):45 (1997).
22. Kesterson, J. W., and R. J. Braddock, By-products and speciality products, Florida Citrus Technical Bulletin, Agric. Expt. Station, University of Florida, 1976, pp. 1.
23. Ranganna, S., V. S. Govindrajan, and K.V.R. Ramana, Citrus fruits II, varieties, chemistry, technology and quality evaluation. Part II. Chemistry, technology and quality evaluation. A. Chemistry, CRC Crit. Rev. Food Sci. Nutr. 18:313 (1983).
24. La Fuente, F. S., New prospects for utilizing by products from citrus juice manufacturing, Revista Agroqu. Technol. Alimentos 20(1):13 (1980).
25. Park, G. L., J. M. Avery, J. L. Byers, and D. B. Nelson, Identification of bioflavonoids from citrus, Food Technol. 37(12):98 (1983).
26. Wilson, R. D., New Zealand grapefruit skin: Waste product or potential gold mine, Food Technol. N. Z. 17(10):71 (1982).
27. Robbins, R. C., Medical and Nutritional aspects of citrus bioflavonoids. In Citrus Nutrition and Quality (Steven Nagy and J. A. Attaway, eds.), ACS, Washington, D.C., 1980, pp. 43.

28. Mathews-Roth, M. M., Carotenoids and cancer prevention and experimental and epidemiological studies, Pure Appl. Chem. 57(5):717 (1985).
29. Burton, G. W., and K. U. Ingold, β-Carotene: As unusual type of lipid antioxidant, Science 224:569 (1984).
30. Molnar, P. and J. Szabolcs, β-citraurin epoxide, a new carotenoid from Valencia orange peel, Phytochemistry 19:633 (1980).
31. Rosenberg M., C. H. Mannheim, and I. J. Kopelman, Carotenoid based food colourant extracted from orange peel by d-limonene-extraction process and use, Lebensm. Wiss.-u-Technol. 17:270 (1983).
32. Yen, W. J., and C. H. Chen, Isolation of xanthophylls from Taiwanese orange peel and their effects on oxidation stability of Soybean oil, Food Chem. 53:417 (1995).
33. Kratchanova, M., I. Panchev, E. Pavlova, and L. Shtereva, Extraction of pectin from fruit materials pretreated in an electromagnetic field of super-high frequency, Carbohydrate Polym. 25:141 (1994).
34. Poonia, S., R. Yamadagni, and S. S. Dhawan, Studies on the utilization of citrus waste for pectin extraction, Haryana J. Hortic. Sci. 23(1):28 (1994).
35. Zafiris, A. G., and V. Oreopoulou, The effect of nitric acid extraction variables on orange pectin, J. Sci. Food Agric. 60(1):127 (1992).
36. Wiener, C., and C. J. Hass, Dried albedo clouding agents, U.S. Patent application 192,262 (1980).
37. Sreenath, S. K., P. G. Crandall, and R. A. Baker, Utilisation of citrus by-products and wastes as a beverage clouding agent, J. Fermen. Biotechnol. 80(2):190 (1995).
38. Bahar, S. and T. J. Azuaje, Studies on the growth of *Fusarium sp.* on citrus waste for the production of single cell protein, J. Food Sci. Technol. 21(2):63 (1984).
39. Nwabueze, T. U., and Oguntimein, Sweet orange (*Citrus sinensis*) residue as a substrate for a single cell protein production, Biol. Wastes 20(1):71 (1987).
40. Clementi, F., M. Moresi, and J. Rossi, Effect of medium composition on microbial utilization of citrus waste by mixed fungal culture, Appl. Microbiol. Biotechnol. 22:26 (1985).
41. Essilfie, R. J., J. H. Bavour, and G. Scurray, Protein upgrading of orange peel waste for potential stock feed by solid state fermentation, ASEAN Food J. 2(1):25 (1986).
42. Kesterson, J. W., R. J. Braddock, and P. G. Crandall, Recovery of citrus by products and speciality products from Florida citrus. Transaction of Citrus Engineering Conference 24:34 (1978).
43. Mariguchi, M., Fermentative production of pyruvic acid from citrus peel extract by *Debsryomyces clouderii*, Agric. Biochem. 46:955 (1982).
44. Green, M., G. Shelef, and D. Bilanovic, The effect of various citrus waste fractions on xanthan fermentation, Chem. Eng. J. 56(1):B37–B41 (1994).
45. Bilanovic, D., G. Shelef, and M. Green, Xanthan fermentation of citrus waste, Bioresource Technol. 48(2):169 (1994).
46. Grohmann, K., E. A. Baldwin, B. S. Buslig, L. O' Neal Ingramm, Fermentation of galacturonic acid and other sugars in orange peel hydrolysates by ethanologenic strain of *Escherichia coli*, Biotechnol. Lett. 16(3):281 (1994).
47. Hang, Y. D., Recovery of ingredients from grape pomace, Process Biochem. 2:2 (1988).
48. Hang, Y. D., C. Y. Lee, J. M. Lucia, and E. E. Woodams, A solid state fermentation process for alcohol production from grape pomace. Paper presented at 37th Annual meeting of American Society of Enology and Viticulture at Anaheim, CA, June 26–28, 1986.
49. Amerine, M. A., H. W. Baig, and W. B. Cruess, The Technology of Wine Making, AVI, Westport, CT, 1972.
50. Sampathu, S. R., N. Krishnamurthy, S. Sivashankar, R. Shankaranarayana, P. N. Srinivasarao, and Y. S. Lewis, Natural food colours, Indian Food Packer 35(2):97 (1981).
51. Fantozzi, R., Grape seed, a potential source of protein, J. Am. Oil Chem. Soc. 58(2):1027 (1981).

52. Kinsella, J. E., Grape seed oil: A rich source of linolic acid, Food Technol. 28(5): 58 (1974).
53. Ramteke, R. S. and W. E. Eipeson (1997). Mango products and mango based ingredients. Proceed. 1st International Convention, Food Ingredients: New Technologies, Fruit and Veg., Cuneo, Italy. Allione Ricerca Agroalimentare S.p.a., p. 90.
54. Bhalerao, S. D., G. V. Mulmuley, S. M. Anathakrishna, and V. H. Potty, Wash and waste water management in food industry. I. Fruit and vegetable processing, Indian Food Packer 43(2): 5 (1989).
55. CFTRI, Mysore (1985–86). Mango pulp concentration. Annual report.
56. Tandon, D. K. and S. K. Kalra, Utilization of mango waste, Beverage Food World 16(1): 21 (1989).
57. Maini, S. B. and J. C. Anand, Utilization of fruit waste, In: Advances in Horticulture, Vol. 4, (Chadha, K. L. and Parekh, O. P. eds.), Malhotra, New Delhi, 1993, pp. 1967.
58. Srirangarajan, A. N. and A. J. Shrikhande, Comparative aspects of pectin extracted from the peel of different varieties of mango, J. Food Technol. 14: 539 (1979).
59. Tandon, D. K., S. K. Kalra, B. P. Singh, and N. Garg, Characterization of pectin from mango fruit waste, Indian Food Packer 45(4): 9 (1991).
60. Yu, A. N. and M. Delvalle, Preparation and characterization of mango (*Mangifera indica L*) peel pectin, U. P. Home Econ. J. 7(1/1): 17 (1979).
61. Beerh, O. P., B. Raghuramaiah, G. V. Krishnamurthy, and N. Giridhar, Utilization of mango waste: Recovery of juice from waste pulp and peel, J. Food Sci. Technol. 13(3): 138 (1976).
62. Ethiraj, S., and E. R. Suresh, Studies on the utilisation of mango processing wastes for production of vinegar, J. Food Sci. Technol. 29(1): 48 (1992).
63. Sethi, R. P., H. K. Tiwari, and S. M. Sood, Mango peel as a carbon source for fungal protein, carboxymethyl cellulase and polygalacturonase production, Indian J. Agric. Res. 11(1): 46 (1977).
64. Rasheed, M. M., S. A. Moharib, and E. W. Jwanny, Yeast conversion of mango waste or methanol to single cell protein and other metabolites, Biol. Wastes, 32(4): 277 (1990).
65. Cojocaru, M., S. Droby, E. Glotter, A. Goldman, H. E. Gotlieb, B. Jacoby, and D. Prusky, 5-(12-Heptadecenyl)-resorcinol, the major component of the antifungal activity in the peel of mango fruit, Phytochemistry 25(5): 1093 (1986).
66. All India Coordinated Research Project on Post Harvest Technology of Fruits and Vegetables Annual Report, IARI, New Delhi, 1991.
67. Narsimhachar, B. L., and G. A. Azeaomoddin, Extraction and analysis of mango peel wax, Acta Hortic. 231(2): 749 (1989).
68. Aroma recovery from mango peel, Annual report, CFTRI, Mysore, India, 1991, pp. 58.
69. Zia-Ur-Rehman, Sekhawat-Ali, A. D. Khan, and F. H. Shah, Utilization of fruit and vegetable wastes in layer's diet, J. Sci. Food Agric. 65:381 (1994).
70. Palaniswamy, K. P., C. R. Muthukrishna, and K. G. Shanmugavelu, Physicochemical characteristics of some varieties of mango, Indian Food Packer 28(5): 12 (1974).
71. Hemavathy, J., J. V. Prabhakar, and D. P. Sen, Drying and storage behaviour of mango (Mangifera indica) seeds and composition of kernel fat, Asean Food J. 4(2): 59 (1988).
72. Laxhminarayana, G., T. Chandrasekhar Rao, and P. A. Ramlingaswamy, Varietal variations in content, characteristics and composition of mango seeds and fat. J. Am. Oil Chem. Soc. 60:88 (1983).
73. Lasztity, R., M. A. El- Shafei, M. B. Abdil Samai, F. S. Hotour, and M. Labib, Biochemical studies of some nonconventional sources of protein. Part 4. The proteins of mango waste stone kernel, Nahrung 30: 867 (1986).
74. Dhingra, S., and A. C. Kapoor, Nutritive value of mango seed kernel, J. Sci. Food Agric. 36(8): 752 (1985).
75. Dhingra, S., and A. C. Kapoor, Acceptability of mango seed kernel flour in conventional food items, Indian J. Agric. Sci. 55(8): 550 (1985).

76. Parmar, S. S. and R. S. Sharma, Mango seed kernel: A review on chemical composition and edible uses, Indian Food Packer 38(5): 40 (1984).
77. Subramaoyam, H., I. S. Shantha and H. A. B. Parpia, Physiology and biochemistry of mango fruit, Advances in Food Research 21:223 (1976).
78. Van Pee, W. M., L. E. Boni, M. N. Foma, and A. Hendrix, Fatty acid composition of kernel fat of different mango (*Mangifera indica*) varieties, J. Sci. Food. Agric. 32: 485 (1981).
79. Gopalkrishna, A. G., and D. P. Sen, Unsaponifiable matter of mango kernel fat, J. Oil Techol. Assoc. India 12(1):14 (1980).
80. Narsimhachar, B. L., B. R. Reddy, S. D. Tirumalarao, Processing of mango stone for fat, J. Am. Oil Chem. Soc. 54:494 (1977).
81. Baliga, B. P., and A. D. Shitbole, Cocoa butter substitute from mango fat, J. Am. Oil Chem. Soc. 58:110 (1977).
82. Parmar, S. S., and S. Sharma, Use of mango seed kernel in enhancing the oxidative stability of ghee, Indian J. Dairy Res. 5(2):91 (1986).
83. Collins, J. L., The Pineapple, Wiley Interscience, New York, 1960.
84. Felton, G. E., Use of ion exchangers in by-product recovery from pineapple waste. Food Technol. 3:40 (1949).
85. Indian Patent 135303 (1974).
86. Nutritional design, Food Eng. Int. 2(6):38 (1977).
87. Tisseau, R., Utilization of pineapple-canning waste: potential for extraction of bromelin. Fruits II(12):703 (1986).
88. Alien, A., and H. M. Musenge, Utilization of pineapple waste for wine making, Zambia J. Sci. Technol. 1(1):29 (1976).
89. Satyavati, V. K., A. V. Bhat, G. Verkey, and K. K. Mukherjee, Preparation of vinegar from pineapple waste, Indian Food Packer 26(3):50 (1972).
90. Dev, D. K., and U. M. Ingle, Utilization of pineapple by-products and wastes: A review, Indian Food Packer 5:15 (1982).
91. Ismail. K. B., J. Abdulla, and Y. S. Yin, Bulletin Food Technology Division, Ministry of Agriculture and Fisheries, Malayasia No. 135, 1974, pp. 11.
92. Chicaya, P., French Patent application 2273068 (1975).
93. Chan, C., and J. H. Moy, Hemicellulose-β from commercial pineapple juice underflow, J. Food Sci. 42(6):1451 (1977).
94. Murti, I. A. S., H. N. Malathi, D. Rajalaxmi, and S. C. Bhattacharya, Industrial waste and by-products of food processing industry in India: A survey, Indian Food Packer 30(3):49 (1976).
95. Ghosh, S. K., and M. K. Sinha, Indian Text. J. 88(2):111 (1977).
96. Sagar, V. R., S. B. Maini, and Sahani, C. Kaur, Potato flour from potato waste, Beverage Food World 23(1):59 (1996).
97. Ritcher, G., and D. Grosser, Enzymatic processes of potato processing, Starke 35(4):113 (1983).
98. Klingspohn, U., J. Bader, B. Kruse, P. V. Kishor, K. Schugerl, H. A. Kracke-Helm, and Z. Likidis, Utilization of potato pulp from potato starch processing, Process Biochem. 28:91 (1993).
99. Grobben, N. G., G. Eggink, F. P. Cuperous, and H. J. Huizing, Production of acetone, butanol and ethanol (ABE) from potato waste: Fermentation with integrated membrane extraction, Appl. Microbiol. Biotechnol. 39:494 (1993).
100. Wilhem, E., and W. Kempf, High quality protein from potato water: Advances in separation and purification, 8th Conf. Assoc. of Potaotes Res. Abst., 1981, p. 265.
101. Luther, H., E. Weber, and E. Chuster, Processing of potato for potato press juice, Lebens. Ind. 30(4):167 (1983).
102. Potato waste: Hiden profits in those mountains of peels? Food Eng. 57(1):135 (1985).
103. Trojanowski, J., A. Kharazipour, F. Mayer, and A. Hutterman, Verwendung von kartoffel-

pulpe und kartoffelfruchtwassser also Nahrmedium Zur Annzucht von laccase produzierended pilzu, Starke 47:116 (1995).
104. Senez, J. C., Solid state fermentation of starchy substrate, Food Nutr. Bull. 1(2):18 (1979).
105. Toma, R. B., P. H. Orr, B. D' Appolonia, F. R., Dintiz, and M. M. Tabekhia, Physical and chemical properties of potato peel as a source of dietary fiber in bread, J. Food Sci. 44:1403 (1979).
106. Nebesny, E., Utilization of potato pulp for baking of bread, Starke 47(1):36 (1995).
107. Hussain Al-Wandawi, Maha Abdul Rahman, and Kaib Al-Shaikhly, Tomato processing wastes as essential raw material source, J. Agric. Food Chem. 33:804 (1985).
108. Yassen, A. A. E., Mohamed H. A. Shams El-din, and A. Ramy Abd El-Latif, Fortification of Balady bread with tomato seed meal, Cereal Chem. 68(2):159 (1991).
109. Maharram, Y. G., and A.S.F. Messallan, Utilization of tomato seeds as a source of oil and protein, Alexandria J. Agric. Res. 28(1):147 (1980).
110. Carlson, B. L., D. Knnor, and T. R. Watkins, Influence of tomato seed addition on the quality of wheat flour breads, J. Food Sci. 46:1029 (1981).
111. Zhvalevskii, I. S., Z. P. Kamnera, and G. V. Nesterenko, Integrated programme of reuse of wastes from canning industry, Pishchevaya Promyshlennost 3:20 (1977).
112. Kramer, A., and W. H. Kwiss, Utilization of tomato processing wastes, J. Food Sci. 42:212 (1977).
113. Lane, A. G., Anaerobic digestion of orange peel, Food Technol. Aus. 36(3):125 (1984).
114. Srilatha, H. R. Krisha Nand, K. Sudhakar babu, and K. Madhukara, Fungal pretreatment of orange processing waste by solid—state fermentation for improved production of methane, Process Biochem. 30(4):327 (1995).
115. Krishna Nand, Consolidated report of the project entitled: Pilot scale studies on the methane generation from fruit and vegetable processing waste (No. 5/2/29/88- BE) for the period March 1989 to September submitted to MNES, Govt. of India, New Delhi, 1991.
116. Raju, N. R., S. Sumitra devi, and K. Krishna Nand, Influence of trace element on biogas production from mango processing waste in 1.5 m^{-3} KVIC digester, Biotechnol. Lett. 13(6): 461 (1991).
117. Lenschner, A. P., C. E. Werk and E. Ashare, Assessment of secondary agricultural residues. Part II. Engineering and economic analysis for conversion of methane and or alcohol fuels: Fuel gas system (D.L. Wise, ed.), CRC Series in Biotechnology Systems, CRC Press, Boca Raton, FL, USA.
118. Krishananand, Biogas from food waste, Indian Food Ind. 13(3):22 (1994).
119. Miller, F. C., Composting as a process based on the control of ecologically selective factors. In soil microbial ecology (F. B. Metting, ed.), Marcel Dekker, New York, 1993, p. 515.
120. Rynk, R, (ed.), Composting methods in on-farm composting handbook, Northeast Regional Agricultural Engineering Service, Cooperative Extension, Ithaca, NY, 1992, p. 24
121. Haug, R. T., The Practical Handbook of Compost Engineering, Lewis, Boca Raton, FL., 1993.
122. Anderson, D., Disposal of sludge solids from food industry waste treatment, CRC Crit. Rev. Food Technol. 3(1):27 (1972).
123. Canadian Council of Ministers of Environment, Guidelines for compost quality, Report CCME 106 E, Environment Canada, Ottawa, Ontario, Canada, 1996.
124. Mathur, S. P., G. Owen, H. Dinel and M. Schnitzes, Determination of compost biomaturity. I. liturature review, Biol. Agric. Hortic. 10:65 (1993).
125. Mathur, S. P., Composting processing in biocon version of waste materials to industrial products (A. M. Martin, ed.), 1991, p. 147.
126. Riggle, D., Revival time for composting food industry wastes, BioCycle 29(5):35 (1989).
127. Stapleton, J., E. J. Morreale, and E. Kiviat, No land fill space for apple wastes, BioCycle 34(4): 46 (1994).

128. Grobe, K., Compostes links up with food processor, BioCycle 34(7):40, 42 (1994).
129. Verville, R. and B. Seeking, New use for blueberry residuals, BioCycle 33(4):71 (1993).
130. Lowe, E. D. and D. R. Buckmaster, Dewatering makes big difference in compost strategies, BioCycle 35(1):78 (1995).
131. Nobuya Inaba, Kazuyuki Maruyama, Masaki Miyake and Yusaku Fujio, Composting of the processing waste of Satsuma Mandarin (Citrus Unshiu Marc) with a developed composting machine Nippon Shokuhin Kagaku Kogaku. 43(1):1205 (1996).

29

Control Aspects of Postharvest Technologies

ISTVÁN FARKAS

Szent István University, Gödöllő, Hungary

1 INTRODUCTION

Control of postharvest operations, especially the drying process, for different agricultural materials plays an important role in influencing energy savings, quality of end products, and environmental issues, as discussed by Farkas (10).

The control of drying, therefore, can be formulated as a multiobjective optimization problem. Most of the commercially available dryers have a control unit; however, in some special cases individually designed controllers must be developed. In most cases the control algorithms developed for given processes can be derived from models, taking into account the necessary physical requirements. The physically based models are mainly used for design and performance analysis of the process. For control purposes various black-box models are more suitable and are widely used because of their simplicity. Along with the classic control strategies the intelligent approaches (expert systems, neural net, and fuzzy controllers) are also becoming the focus of interest (Farkas et al., 11; Hashimoto, 17; Siettos et al., 32).

In an agricultural farm there is great demand to collect all the available energy resources and distribute them optimally among the different consumers; among those resources are dryers, which require a fairly great portion of the total energy. This task requires hierarchical control, as suggested by Farkas (3).

This chapter considers the dryer control system; the off-line grain dryer operation; the special aspects of grain, fruit, vegetable, grape, and wood drying processes; and an integrated farm energy system.

2 DRYER CONTROL SYSTEM BACKGROUND

The most important requirements and characteristics of agricultural dryer control systems include the main objectives of a dryer control and the arrangements of typical control systems. The similarities of the control characteristics of agricultural and industrial dryers have been described by Jumah and Mujumdar (18).

2.1 Scope of the Dryer Control System

The aim of postharvest technologies is to produce a solid end product of a certain percentage of moisture content for immediate use or for further safe storage. The use of automatic control for achieving the final target of the postharvest technologies provides a good opportunity to improve the efficiency in dryer operation.

The control of dryers has probably not progressed concurrently with improvements in dryer design, which may be attributed to various factors summarized and listed in Refs. 18 and 28. The factors are the following:

The lack of emphasis on product quality in the past.
An apparent lack of knowledge of the important role dryer control plays in product quality and drying efficiency.
The lack of direct, on-line, reliable methods for sensing product moisture content.
The complex and highly nonlinear dynamics of drying processes, leading to difficulties in modeling the process adequately.

The main objectives of a dryer control system (18) include the following:

Maintenance of desired dried product quality, irrespective of disturbances in the drying operation and variations in feed supply.
Maximization of throughput at optimal energy efficiency and minimal cost.
Prevention of overdrying and underdrying: Underdrying may result in spoilage, in the case of grains and foodstuffs. Overdrying of the product results in increased energy costs and reduced yields since the price of some products is based on a specific moisture content; it may also cause thermal damage to heat-sensitive products.
Reduction of fire hazard, defective product, and particle emission.
Suppression of the influence of external disturbances.
Stable drying process.
Optimization of the performance of the drying process.

2.2 Dryer Control System Characteristics

There are several parameters affecting the drying process so the control system requires consideration of those factors. The requirements of an agricultural dryer control system (18,22) are as follows:

1. *Accuracy*: The end product moisture content must be close to the desired value.
2. *Stability*: The system must not oscillate; if it does, large fluctuations in output moisture content occur.
3. *Speed of response*: Any disturbances (e.g., changes in input moisture content) should be quickly offset by the controller in order to provide acceptable upset recovery time and system stability.

4. *Robustness*: The control system should be able to operate successfully over a wide range of process conditions.

In relation to process control, the variables (temperatures, moisture contents, flow rates, pressures, etc.) associated with an agricultural drying process can be generally divided into two groups (18,35):

1. *Input variables*, which denote the effect of the surroundings on the drying process
2. *Output or controlled variables*, which denote the effect of the drying process on the surroundings

The input variables can be classified as *manipulated variables*, which can be adjusted either manually or automatically, and *disturbances or load variables*, which cannot be adjusted by a control system.

The most important manipulated inputs to a dryer are (a) heating rate (e.g., inlet air temperature for direct dryers), (b) airflow rate (for direct dryers), and (c) solids feed rate.

The most common dryer disturbances are

1. Ambient air humidity
2. Ambient air temperature
3. Feed moisture content
4. Feed temperature
5. Feed composition

The dryer output or controlled variables may be classified as *measured output variables* and *unmeasured or difficult-to-measure output variables*. For instance, the product moisture content can be a measured or unmeasured variable, depending on the capability of the moisture sensor applied.

The most typical dryer output variables are

1. Exhaust air humidity
2. Exhaust air temperature
3. Dried product moisture content
4. Dried product temperature
5. Product quality (color, flavor, activity, etc.)

The relationships among output, manipulated, and load variables, which also apply to the control system of the drying process, can be seen in Fig. 1.

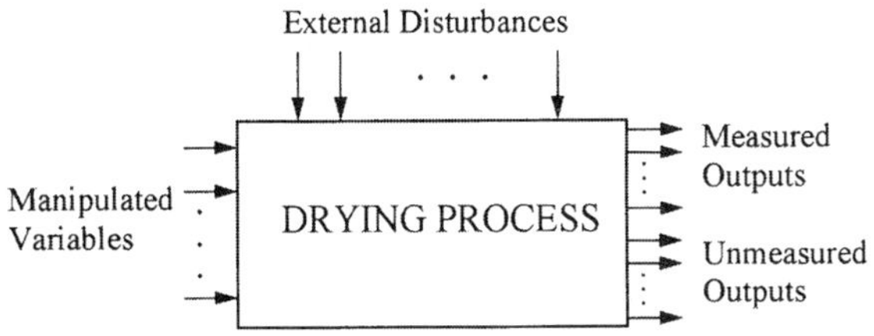

Fig. 1 Scheme of drying process variables.

The purpose of a dryer control system is to produce a desired output by adjusting the manipulated variables to compensate for changes caused by disturbances. In a drying process the most important controlled variable is the product moisture content.

2.3 Dryer Control System Principles

There are several principles that can be applied successfully to set up a dryer control system for drying processes.

Manual, feedback, feedforward, model-based, and computer-based direct digital control are considered.

2.3.1 Manual Control

In practice in some cases there is no way to apply automatic control solutions to maintain the drying process. In such cases manual control is necessary. In manual control, expert judgment of the operator is relied on for judging the end point and setting the initial throughput of the drying process (Fig. 2).

The dryer manual control schedule may be described by the following sequence (18,22,28):

1. Turn on the dryer.
2. Set the initial throughput.
3. Measure the output moisture content and compare the measurement with the desired value.
4. On the basis of the difference between the desired and the measured moisture content values, make adjustments to the manipulated variable(s) (e.g., energy input, feed rate) to maintain the desired moisture content.

Manual control is simpler and less expensive and requires less expertise than automatic control. It can be applied to small-scale plants (mainly batch systems) and to easy-to-dry materials. It is also labor-intensive.

2.3.2 Feedback Control

The principle of feedback control is one of the most commonly used control strategies in drying process control. The main function of a dryer feedback controller is to hold the controlled variable at a desired value or set point.

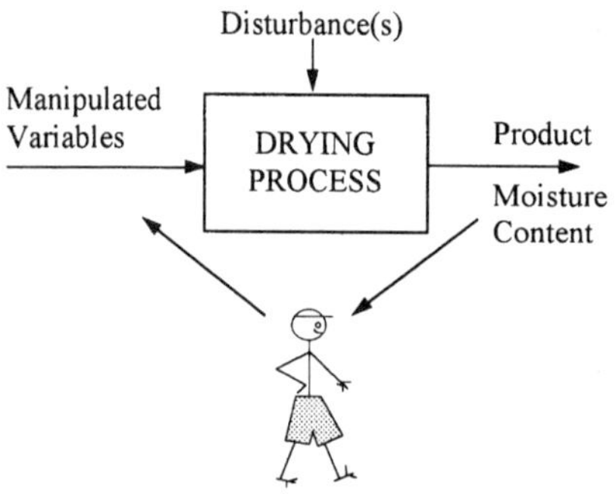

Fig. 2 Scheme of manual control of the drying process.

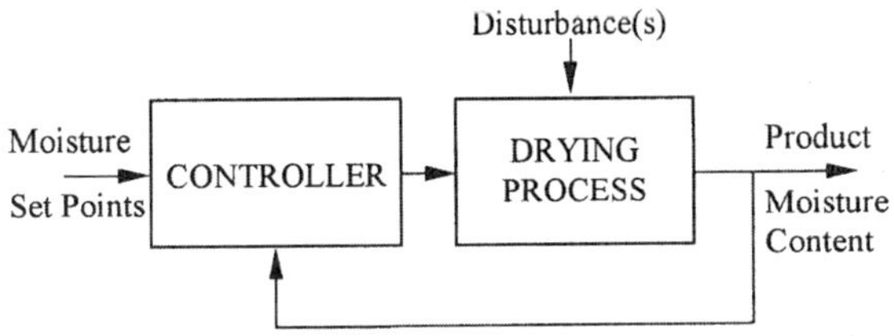

Fig. 3 Scheme of feedback control of the drying process.

The control system receives a measured signal of the controlled output variable (i.e., moisture content) and compares it with the set point value, which generates an error signal. On the basis of the error signal, the controller changes the value of the manipulated variable in order to reduce the magnitude of the error.

Usually, the controller does not affect the manipulated variable directly but through another device, known as the *final control element*, such as a control valve, motor, fan, or heater, depending on the application. Ideally, control results in an exact correction in the process output variable, forcing it back to the desired value (set point). A typical feedback control loop is shown in Fig. 3.

Three basic types of feedback controller actions are proportional, integral, and derivative actions. Proportional action actuates the manipulated variable in direct proportion to the error signal. Integral action eliminates any steady-state residual errors or offsets. Integral action moves the manipulated variable on the basis of the time integral of the error. The purpose of the derivative action is to forecast fast changes in the error signal by using a control mode proportional to the time rate of change of the error signal.

In industrial dryer control applications the three control actions described can be used individually or in combination as proportional (P), proportional-integral (PI), or proportional-integral-derivative (PID) controllers. In most of the drying applications the general three-action (PID) controllers, investigated and reviewed by Robinson (28) and Marchant (22), are used.

2.3.3 Feedforward Control

The residence time of the solid material may be relatively long in certain types of industrial dryers. Also, in certain dryers the thermal inertia of the heat transfer mode may be long relative to the drying time. In such cases, there is a significant time lag between a change made to an input and perception of its effect on the outputs. If this dead time is large, it may result in inadequate performance of the feedback controller. To overcome this problem, a predictive type of control, known as *feedforward control*, is used.

In the case of the feedforward controller, process disturbances are measured and compensation is made for them without waiting for a change in the output variable to indicate that a disturbance has occurred. The control scheme is implemented by measuring the disturbance or load input (e.g., inlet moisture content, solid flow); the controller uses a system model to determine the relationship between disturbances and the manipulated and controlled variables.

The objective is to keep the value of the controlled output variable at the desired level by eliminating the impact that the disturbances would have on the output. It is clear that the effectiveness of the feedforward control system depends on the accuracy of the

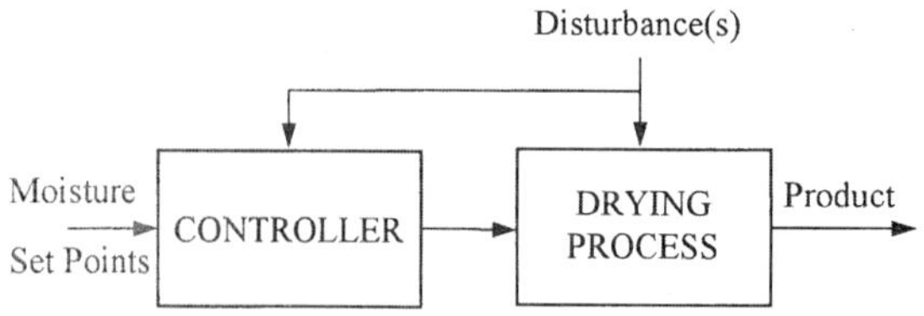

Fig. 4 Scheme of feedforward control of the drying process.

system's mathematical model in predicting the response of the process to input and disturbance changes. The principle of feedforward control is illustrated in Fig. 4.

2.3.4 Model-Based Control

Model-based control (MBC) has recently become a process control approach of much interest. The basic idea is to use a dynamic model of the process in the control system (1,12,18,21,35). In practice, several different types of model-based controllers are applied. The most frequently used strategies are (a) inferential control and (b) internal model control (IMC).

The *inferential control* is an early model-based approach for process control (1,18,35). This control strategy is useful when the main dryer-controlled variable (i.e., product moisture content) cannot be measured directly as a result of technical difficulties or insufficient economic justification for its measurement. Inferential control uses the values of measured outputs (e.g., product/air temperature and/or humidity) together with the process model to infer the value of the unmeasured control variable. The estimate is used by the controller to adjust the values of the manipulated variables in order to keep the moisture content at the desired levels (Fig. 5).

The basic process of *internal model control* (IMC) uses a process model and relates the controller settings to the model parameters in such a way that the selection of the specified closed-loop response yields a physically realizable feedback controller (1,18,21,28). IMC is advantageous because it can be adjusted to balance controller performance with control system robustness when either modeling errors or changes in process dynamics occur. Clearly, the effectiveness of IMC depends on availability of a reliable dryer model. The structure of an IMC system is depicted in the scheme represented in Fig. 6.

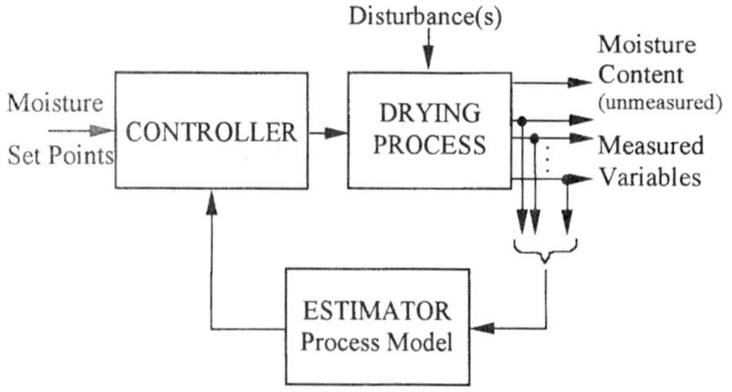

Fig. 5 Scheme of inferential control of the drying process.

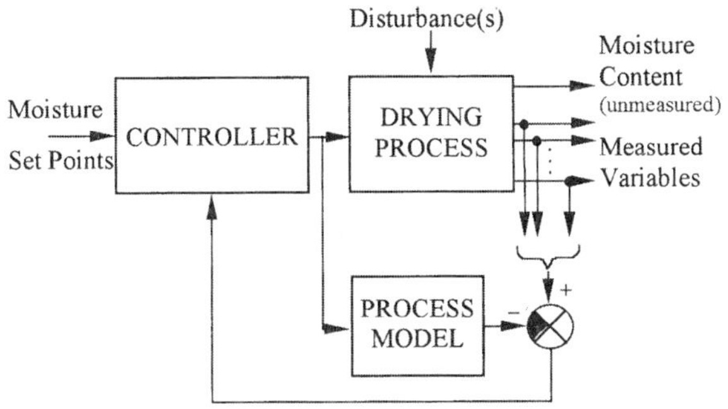

Fig. 6 Scheme of internal model control of the drying process.

2.3.5 Direct Digital Control

Traditionally, analog instrumentation was used, and in some cases is still being used for process control. Following up on the fast development in computer technology, coupled with significant reduction in cost in the early 1970s, sophisticated computer-based control systems and controllers have evolved.

Although the primary task of a computer-based controller is implementation of a control algorithm (PID or more sophisticated algorithms), the presence of a computer makes it possible to achieve more than basic control and to assign a number of tasks that are useful in process control. The following computer control system characteristics illustrate some of these tasks (18,21,35):

1. Implementation of classic and advanced control algorithms.
2. A single digital computer (or microprocessor) that services a number of control loops (time-shared basis).
3. Distributed data processing by which data can be collected from different process instruments and processed for monitoring and control purposes. The computer system can also be connected to local analytical instruments (e.g., moisture and humidity meters), which usually have their own microcomputer.
4. Static and dynamic displays on monitors or other visual display units.
5. Mathematical functions.
6. Data acquisition and storage for different process measurements, such as temperature, flow rate, pressure, humidity, and moisture content.
7. Planning, supervision, optimization, quality control, and control of operation mode.

Figure 7 shows the structure of a computer-based direct digital control (DDC) system. The computer system collects data from the process measurements and on the basis of the control algorithm already programmed and stored in the memory of the computer calculates the values of the manipulated variable and implements the control action on the process.

The signals are converted via analog-to-digital (A/D) and digital-to-analog (D/A) converters. Besides the converters, multiplexers, amplifiers, transducers, and other signal processing units are also used in the computer-to-process and process-to-computer interfaces. The operator communicates with the control system with a keyboard, a monitor, and a printer or plotter.

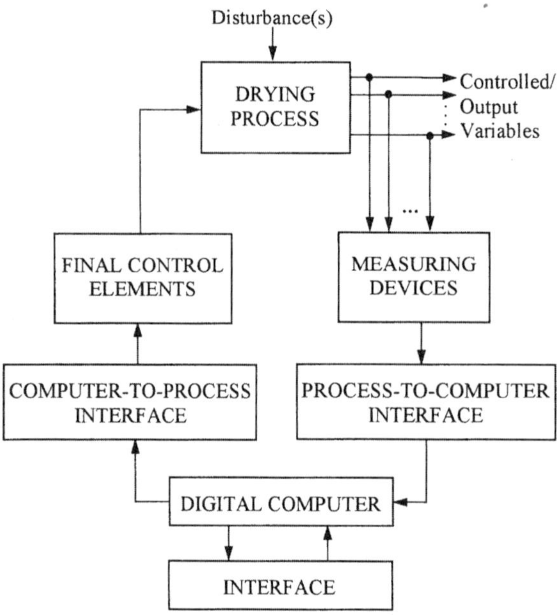

Fig. 7 Scheme of computer-based direct digital control of the drying process.

3 CONTROL SYSTEMS FOR POSTHARVEST PROCESSES

Many attempts have been made to design high-level control systems to ensure quality of the end product. Recent examples—grain, corn, food, brown rice, fruits, apple, vegetable, potato, grape, wood, bean, and tea—are briefly described in this section.

To solve the control problem of the grain drying process, first, the black-box models involved in the control algorithm should be identified. Determination of the optimal operational parameters of the system can be carried out by performing measurements. In some special cases even physically based models can provide useful information, as discussed by Farkas (6,7,9). Drying of food products, especially corn, is often performed in a batch dryer. Such processes are rather difficult to control by classic controllers. The process model can be used by predictive control techniques, as discussed by Patwadian and associates (30). Trelea and colleagues (43) have developed a multivariable predictive optimal control algorithm for on-line control of a batch drying process. The control problem is stated in time domain along with technological and product constraints.

Germinated brown rice is being developed as a valuable and novel food product. Unpolished brown rice contains useful elements, such as dietary fiber, vitamins E and B, iron, magnesium, phosphorus, and zinc. Suzuki and Maekava (37) have developed a method of environmental control (called *sprouting control*) for the root emergence. Most of the fruits are matured and color-changed while maintaining freshness; their commercial prices are mainly determined during the storage process, as indicated by De Baerdemaeker and Hashimoto (2). The optimization of the storage process is very important for fruit production. For optimization of such a complex system intelligent control techniques are more suitable than conventional approaches. To optimize the cold storage process for apples, Morimoto and coworkers (25) apply a new intelligent control technique based on neural networks and genetic algorithms. They use relative humidity as the manipulated variable and the quality parameters as the control variables. Mizrach and associates (23) have made a study of the effects of storage time and temperature on the softening process of avocado fruit by means of nondestructive ultrasonic measurements.

It is known that the available oxygen content is fairly low during controlled atmosphere (CA) storage of fruits and vegetables as it is really applied in order to prevent fermentation. Application of lower amounts of oxygen demands a dynamic control system, as developed by Schouten (31). Ethanol production is used to indicate fermentation.

In order to improve the traditional control of the raw vegetable drying process a new contour of the control system was introduced by Stefanovic and Stakic (34). The computer-based control system allows reduction of energy consumption and provides higher productivity during dehydration.

Since the 1990s, computerized environmental control systems have been applied in potato storage. The control systems can be analyzed in terms of their purpose, environmental factors, control functions, management functions, and additional advantages, as discussed by Lamber (20). In large warehouses the bulk of potatoes are air-conditioned by forcing through outdoor air. The inlet air may be mixed with the outlet air by dampers. Gottschalk (14) has discussed control equipment to set the position of dampers and the ventilation rate. He has developed a heat and mass transfer model to be included in the control algorithm using fuzzy-logic methods. A general algorithm was also developed to vary the relevant parameters to secure high quality of potatoes and to reduce the ventilation energy and therefore the cost of storage.

The dehydration of grapes is very important commercially. Kiranoudis and associates (19) have offered a solution for the optimization of a tunnel type of grape dryer. The dryer has semibatch-type operation with trucks and trays. The control objective was to reduce fuel consumption along with maintaining quality under the constraints on the production rate of the dryer and the maximal permissible drying air temperature. The optimization parameters were the temperature and the humidity of the drying stream.

The control aspects of wood, especially lumber, drying have been extensively studied by Tarasiewicz and colleagues (38,39). In addition to the classic feedback controller they have discussed the internal model control and model predictive control systems. They have developed a sophisticated new control system for wood drying that integrates the principles of minimal energy consumption and minimal drying time with the constraint of wood quality. For optimal design and operation of the heat pump dehumidifier wood drying kiln, a comprehensive model has been developed by Sun and Carrington (36). The model is suitable for analyzing the influence of design and control variables on the system.

Vasconcelos and coworkers (45) have proposed alternative designs and suitable definitions of operation policy through optimization procedures for typical beans. The proposed dryer is a column type with multisection structure.

In India alone the largest tea producer reaches an annual production of about 7×10^8 kg. During production of black tea, shoots of three leaves and a bud are withered to approximately 72% moisture content w.b. Tea is further processed and dried, and the moisture content is reduced to about 3%. A continuous fluid bed dryer can achieve that moisture content within 20 minutes. Manual control of such an operation is difficult because of the time required to reach stability. Temple and van Boxtel (40,41) have developed a simple feedback control system from the exhaust air temperature, that can also reduce energy consumption.

4 OFF-LINE GRAIN DRYER OPERATION

For optimal operation of agricultural grain dryers computer-based control can be used. There are, however, large-scale cross-flow grain dryers that are poorly instrumented, e.g., having no capacity to perform on-line measurements of process variables. In this situation we cannot use on-line control, but the control strategy developed on the basis of a physically based model of the drying process can be used off-line to adjust the manipulating variables. This is practically manual control; however, the process model is used directly for making decisions on the adjustments.

4.1 Dryer Description

The scheme of a classic high-capacity widely used cross-flow grain dryer is presented in Fig. 8. The length of the dryer is over 10 m, which is divided into two drying zones and one cooling zone in order to take advantage of air recirculation to improve energy efficiency. Natural gas or liquid propane is used as the fuel for preheating the ambient air. The holding capacity is almost 10^5 kg. The main features of the dryer are the following:

The exhausted air from the cooling zone can be recycled to the second drying zone.
The airflow direction is reversed in the second drying zone.
The exhausted air from the second drying zone can be recycled to the first drying zone.

4.2 Dryer Control Strategy

The control strategy of the dryer operation described is as follows:

Keep the average outlet grain moisture content as a constant at a desired value of $X_{out} = 12\%–13\%$, w.b.
Minimize the energy supplied by proper heater use of the recirculated air.

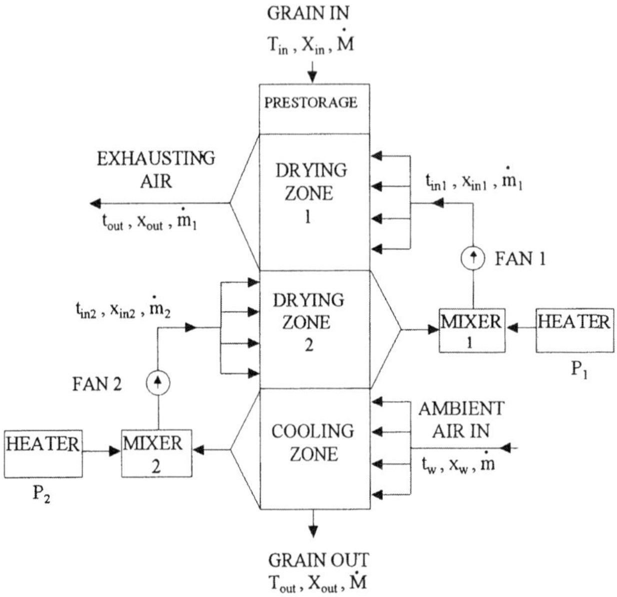

Fig. 8 Scheme of a high-performance cross-flow dryer.

In practice, the grain flow rate (e.g., discharge rate) can be most easily regulated. In principle, however, the following input variables can be manipulated:

Product flow rate, $\dot{M}$
Heating capacity of burners, P_1, P_2
Airflow rate of fans, $\dot{m}$, $\dot{m}_1$, $\dot{m}_2$

One way to carry out the control problem is to install an on-line microcomputer-based data acquisition and control system. In the case of newly designed grain dryers, this principle should be considered. The drying process model should be inserted directly into the control algorithm, giving the application potential for the most appropriate control. Unfortunately, in the case of badly instrumented existing dryers we cannot follow this approach. Therefore, the results of the simulation model can be used only off-line.

4.3 Simulation Model for Control

For the control of a cross-flow grain dryer the block oriented simulation system (BOSS), including the physically based model for the drying process, can be applied successfully (6). In this specific case the inlet variables to the simulation model are the following:

Inlet product temperature, T_{in}
Inlet product moisture content, X_{in}
Airflow rates, $\dot{m}$, $\dot{m}_1$, $\dot{m}_2$
Heating capacity of burners, P_1, P_2
Product feed rate, $\dot{M}$
Ambient air temperature and humidity, t_W, x_W

The variation of the values of so-called manipulating parameters, $\dot{m}$, $\dot{m}_1$, $\dot{m}_2$, P_1, P_2, and $\dot{M}$, as input to the simulation model allows the determination of their optimal values to reach the desired operational conditions for the entire drying process.

5 CONTROL OF CROSS-FLOW GRAIN DRYERS

One of the main reasons to control dryers automatically is that drying is a slow process that requires manual supervision and control time, which is demanding and expensive. Another reason is that dryers exhibit relatively complicated dynamics, as a result, an automatic controller often effects more accurate control than the manual alternative.

5.1 Dryer Description

In the following, a continuous cross-flow dryer described in detail in Nybrant (26,27) and Farkas and Nybrant (8) is considered for the control study. A schematic view of such a dryer is given in Fig. 9.

The wet grain is normally kept in a container on the top of the dryer. While it continuously flows through the dryer, it first passes the drying section, where it is dried by preheated air of typically 40°C–60°C. In the cooling section the grain is cooled by ambient air before it leaves the dryer. The speed of the discharge motor determines the grain flow through the dryer and thereby the average amount of moisture removed from the grain.

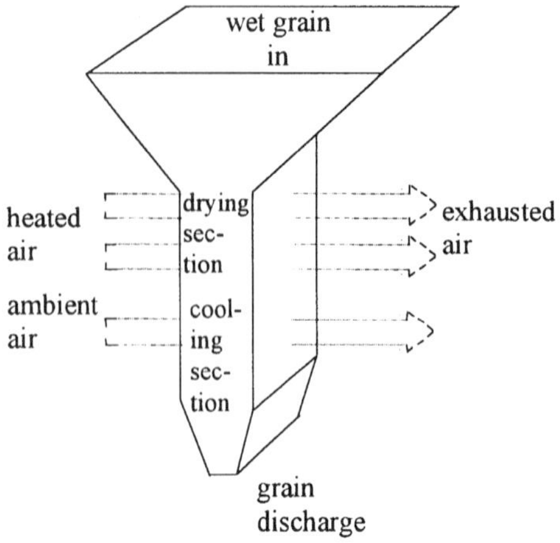

Fig. 9 Schematic view of a continuous cross-flow dryer.

5.2 Dryer Control Strategy

A control objective for a continuous cross-flow dryer can be formulated to determine the discharge rate so that the outlet moisture content is as close as possible to a reference value, regardless of variations in inlet moisture content and grain and dryer characteristics.

According to the control objective, an obvious control strategy would be to use feedback from the outlet moisture content and feedforward from the inlet moisture content, which is the most dominant measurable disturbance. Such feedback-feedforward controllers are described, for example, by Toftdal and Olesen (42), Nybrant (26,27), and Moreira and Bakker-Arkema (24). In a physically based model for dryer control, the grain moisture distribution can then be calculated and used for control purposes, as discussed by Farkas (5).

Since it is expensive to measure moisture content accurately, an alternative, especially for smaller dryers, is to measure and control the exhaust air temperature. This temperature is thermodynamically related to the moisture content of outlet grain; therefore, good control of this temperature can be expected to lead to reasonably good control of moisture content.

An adaptation of an integrating pole placement controller, originally described by Tuffs and Clarke (44) and modified by Nybrant (26) to include feedforward potential, can be used successfully. As discussed in Refs. 26 and 27, when the wetter and cooler grain reaches the sensor at the bottom of the drying section, the controller is able to recover the desired temperature quickly and accurate control is maintained. The convergence is, as expected, very fast, and the process takes just a few samples to reach good control. This finding indicates that an adaptive control system that includes only one parameter to be estimated can also indirectly control the outlet moisture content fairly well.

6 CONTROL OF SOLAR CROP DRYERS

To investigate the physical behavior of a solar grain dryer on the basis of the physical modeling discussed earlier, a microcomputer-based data acquisition system was designed

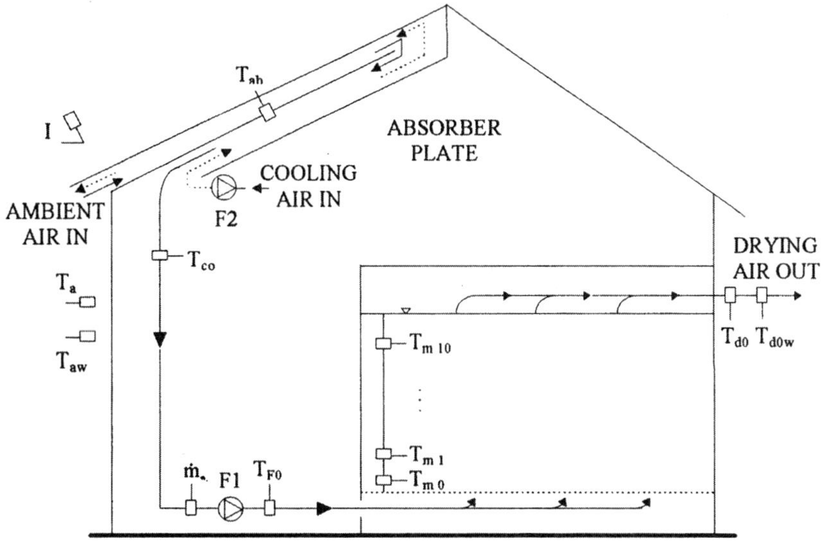

Fig. 10 Setup of a solar crop dryer.

to make energy calculations and to address the control problems determined by the desired drying technology and the optimal use of all available solar and/or ambient energy.

6.1 Solar Dryer Description

Small-scale solar equipment can be used effectively worldwide for drying of various agricultural crops. The design used at Colorado State University, Fort Collins, is shown in Fig. 10, which illustrates its design, which was introduced by Farkas and Smith (4) and by Farkas (7). The dryer is designed to produce 2,000 kg dried material, starting at 20%–40% w.b. initial moisture content. The following are the key specifications for this unit:

1. *Collector*
   The structure is integrated into the roof
   The area is 36 m^2
   The absorber plates are 0.6-mm-thick aluminum, coated with optically selective silicon paint
   The glass is 3 mm thick, tempered, and with low iron content
   The insulation is 25-mm-thick fiberglass
2. *Dryer*
   The size is 3 m long by 2.5 m wide by 1 m high
   The wall insulation is about 38 mm thick
   A 3 by 3-mm wire mesh serves to hold the grain
3. *Air system*
   The cross section of the air channel is 0.15 m^2
   The airflow rate is 500 L/s at a pressure of 10 kPa
   The fan motor capacity required is about 1 kW

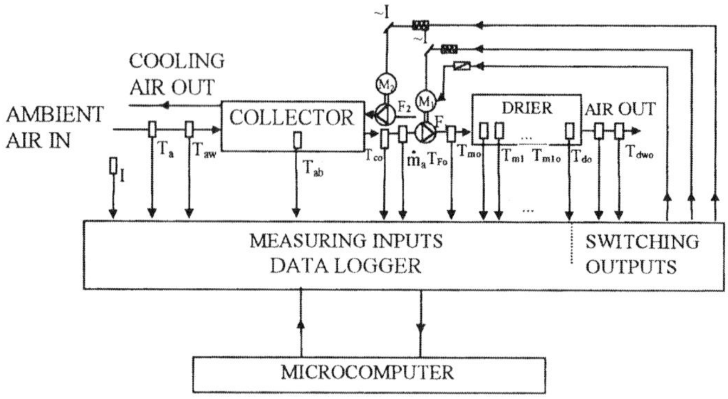

Fig. 11 Control scheme of a solar crop dryer.

6.2 Data Acquisition and Control System

The scheme of the data acquisition and control system of the solar crop dryer can be seen in Fig. 11.

The control program running on the microcomputer checks the data logging and provides the following functions:

- Evaluation and printout of momentary values
- Setting of necessary control signals
- Data saving for future processing

The locations of the sensors are shown in Fig. 11. The temperature and radiation sensors are connected directly; the airflow rate sensor is indirectly connected to the computer, through a signal transformer. The scanning time is 300 s, except the measurement of layer temperatures, which is hourly. First, the control system switches off the fan, *F1*, while taking the real temperatures, and after the measurement, it switches it on again.

6.3 Dryer Control Strategy

The strategy of optimal control involves satisfying the drying requirements. The maximal thermal efficiency for the whole solar system must be sought, as follows:

1. Quality requirements

   For barley drying the material temperature cannot rise above a permissible value, which was limited as 60°C. This critical state may mainly occur intermittently at night as a result of biological heat development. In such a situation use of an air blower is required.

   The goal of drying is to reach a desired final moisture content of 12% w.b.
2. Optimal use of total available energy

   The precondition for starting the drying process is that the solar radiation must exceed a given value. On the basis of experience this value has been identified as 15 MJ/h.

   To stop the drying process the drying rate can be used. Its value was determined at 0.5 kg/h. This condition allows drying to continue by ambient energy without solar assistance.

To decrease losses it is advisable to maintain collector outlet air temperature at below 60°C.

3. Additional conditions

In the absence of drying, the collector absorber temperature can rise to 150°C; then the fan, *F2*, is applied for cooling.

4. Support of the material layer temperature measurement procedure

7 CONTROL OF FRUIT STORAGE

In developing control strategy for agricultural processes difficulties always occur as a result of biological elements. The model-based controllers should take into account all those mechanisms that are rather difficult to describe by mathematical equations (16). For that reason, biological elements of fruit storage systems can be measured but are not included in the control of the process. So far, a simple environmental feedback control has been applied without accurate responses from the real biological mechanisms. That gives somehow an explanation for those efforts for using artificial intelligence (AI) in the control of fruit storage as well (2,15). A scheme of the ''speaking fruit approach'' (2) is shown in Fig. 12.

The basic control is an environmental feedback control used to maintain the air temperature and humidity corresponding to the set points. The set points, however, can be adjusted by the system operator. The information for the set points is taken from the fruit measurements. In this loop the fruit responses are identified by an artificial neural network (ANN) or by a genetic algorithm (GA), which are more effective in this case. The final decisions on optimal operation can be made by the operator of a computer. So this loop can be practically operated off-line or on-line.

The artificial neural network technique is a good tool to make a dynamic model of an unknown system, especially to identify nonlinear behavior. For training of the ANN, model measurements must be performed.

The genetic algorithm (GA) technique serves initially for optimization of a multivariable objective function. The optimization is carried out by simulating the evolutionary process with the use of *crossover* and *mutation* as genetic operators. Because of the binary digit implementation of the operators, all the decision variables are also coded to a finite-length binary string.

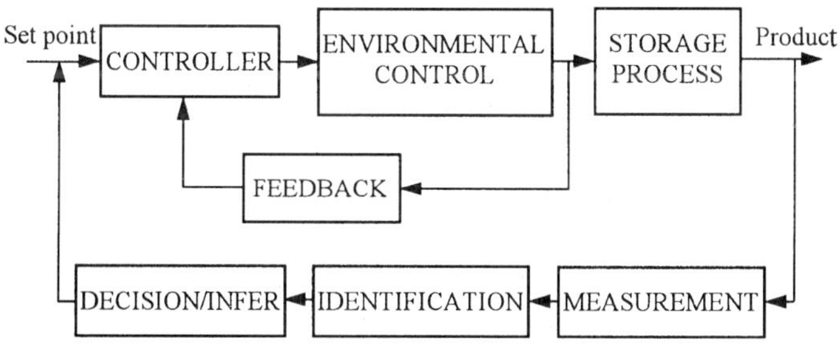

Fig. 12 Intelligent control scheme of a fruit storage system.

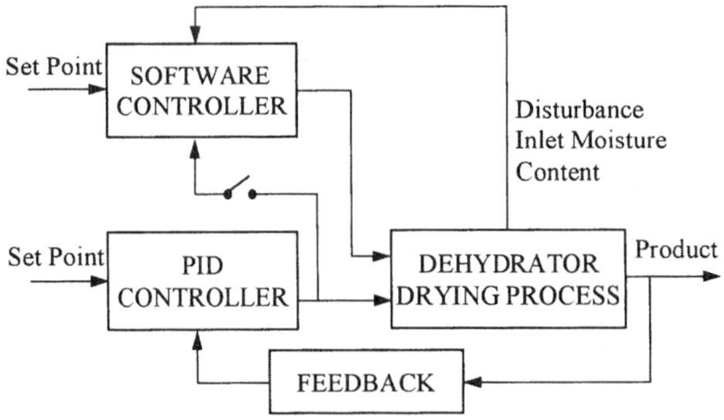

Fig. 13 Control scheme of a vegetable dehydrator.

8 CONTROL OF RAW VEGETABLE DEHYDRATION

Raw vegetables can be dried in a pneumatic drum dryer (dehydrator). The drying agent is a mixture of hot gases obtained by fuel combustion with ambient air that is forced through the dryer by a fan. The vegetable mass is carried into the dehydrator by means of a belt conveyor. After dehydration, the dried product goes to a cyclone to separate the solids, then to a mill for chopping, and finally to a packing device (34). The conveyor speed can be varied in a range by changing the revolutions per minute (rpm) of the driving engine. The temperature of drying gas is controlled by the fuel flow rate by means of a valve.

Traditionally, a feedback PID temperature controller has been used for maintaining the proper dehydration process to ensure optimal energy consumption during the dehydration process. In the case of a significant change in the inlet moisture of the raw material caused by a disturbance, extension of a feedforward feed rate control has to be used for achieving maximal capacity of the dryer. To handle the feedforward part of the control an additional computer-based discrete control system can be applied (34). The scheme of the control system is shown in Fig. 13. The role of the computer-based discrete control system beyond the pure control function is also to measure the system variables and operate the data acquisition and monitoring system.

9 CONTROL OF GRAPE DRYING

The grape is traditionally dried on open-air trays by direct exposure to solar radiation. In order to achieve a high quality of dried product the drying should be carried out under controlled conditions. In one method drying takes place in storehouses; the process uses a semibatch tunnel dryer operation with trucks and trays.

The air is blown through the trays containing the grapes. In order to provide uniform moisture distribution for the dried product the air movement is reversed for each truck. The trucks enter and move along the dryer at a certain interval opposite the drying air. When the grape on the first truck reaches the desired final moisture content, it is removed from the dryer and simultaneously the rest of the trucks proceed one step forward to the exit (19).

The recirculated air to the dryer is heated by combustion gases from a burner. The fraction of recirculation is determined on the basis of the gas amount from the burner.

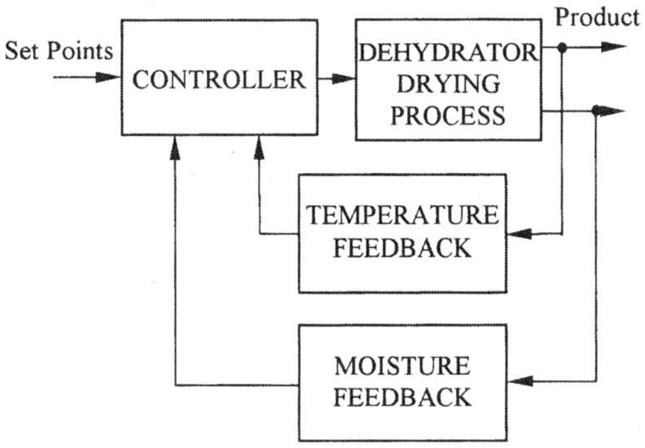

Fig. 14 Control scheme of a tunnel grape dryer.

Thus the temperature and the humidity of the airstream can be controlled by a classic feedback controller. The scheme of the control system can be seen in Fig. 14.

To accomplish real control of the dryer, a model-based approach that takes into account the heat and mass transfer balances of the airstream and product feed and the additional nonlinear operational constraints is required. The fuel consumption per unit mass of dried grape is introduced as the objective function (19).

10 CONTROL OF WOOD DRYING

The major wood drying problem is drying of lumber pieces to a desired final moisture content. Traditionally, the main parameters affecting the wood drying process in a kiln are the temperature, humidity, and velocity of the airstream exposed to the lumber board. Another important factor to be taken into account is, however, the geometric arrangement of the lumber board, which mainly depends on the raw material itself (38). The objective of wood drying control is to decrease the energy consumption used in the process along with providing a high quality of the dried product in minimal possible time (39).

The classic control scheme for the wood drying process is shown in Fig. 15. The set points of the control systems are the drying air temperature and humidity. These manipulated variables are dependent on the temperature and moisture content of the lumber. These output variables are used as feedback signals in the control system. The outer control loop is based on the off-line measurement of the wood specimen.

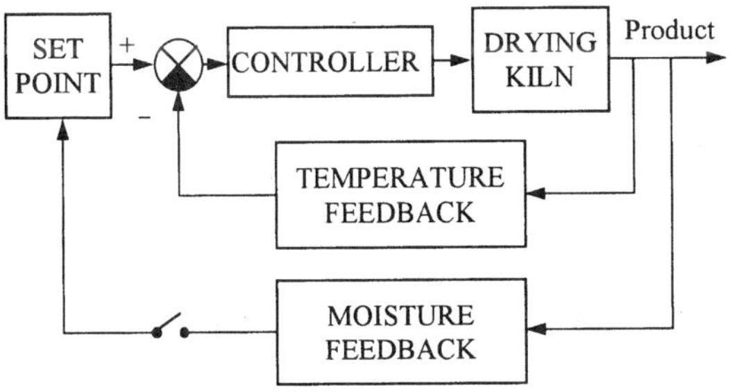

Fig. 15 Scheme of classic control of wood drying.

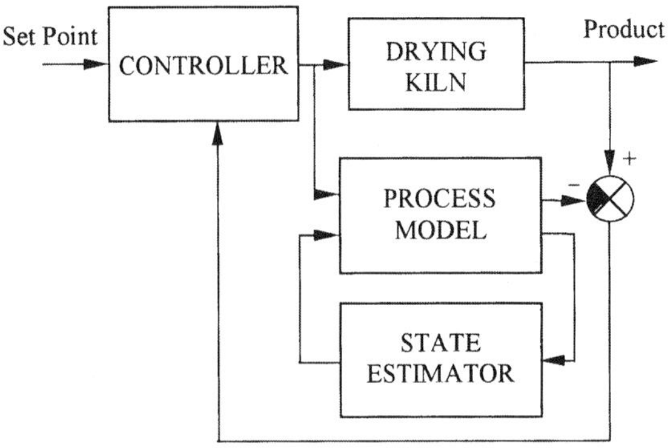

Fig. 16 Scheme of model predictive control of wood drying.

Recently, model-based control approaches have been used. Tarasiewitcz and Leger (38) have described an internal process model for system control. Application of a pure internal model structure in the control system is ideal if the disturbances cannot be estimated empirically and the time delay is negligible. Otherwise a model predictive control can be used more effectively. Such a system permits a global solution of the control problem through its capacity to include the specific constraints of the dryer unit. The scheme of a model predictive control system with the model state estimator (39) is shown in Fig. 16.

11 CONTROL OF INTEGRATED FARM ENERGY AND TECHNOLOGY SYSTEMS

Optimal consumption of energy is a key question on a farm. It is necessary to integrate all available energy sources including renewable ones and to distribute them optimally among the different technological consumers, such as greenhouses, animal buildings, crop dryers, service water heating, and hot water storage.

Setting up a solar preheating system can be economically justified if the solar energy gathered is used all year. In present conditions integration of solar energy into farm energy systems seems to have potential for economical application. The storage of the collected solar energy is a key question in such cases, as discussed by Gordon and Rabl (13) and Sriramulu and associates (33).

It is necessary to solve the control problem so that a coupled energy system can provide optimal operation under any operating condition. The system must provide control of energy distribution by selection of the operation mode for the actual source of energy and self-control of each technological process. Recently the former problem, i.e., selection of the operating mode, has been studied; selection can be based on a hierarchy that takes into consideration the relevant energy conditions and the demands of the consumer subsystems. A priority principle can be determined for a given type of energy source by the consumers.

11.1 Farm Energy Technology System Description

A farm energy/technology system includes all available energy sources and those technological processes that can be directly connected to the integrated energy system. On a farm the factors to be considered are electrical heating, use of a thermogenerator with oil or gas burners, and solar energy by water collectors, which provide a very promising and clean way of applying renewable energy. However, the economic aspects of such systems have to be considered.

The chief energy consumers on a farm are usually the crop dryers, which require preheated air during the drying process. In spite of this fact, in terms of energy integration, it is advisable to use water flow solar collectors because they can be easily connected to the existing layout, for example, linking a hot water storage tank, which can serve simultaneously as an energy storage. In such a way the dryer is connected indirectly to the energy system through an air/water heat exchanger. Hot water can be produced through a water/water heat exchanger, and heat energy can be used by different heat consumers with the aid of additional heat exchangers to preheat the biogas containers, for instance.

A simplified layout of such a coupled energy/technology system is shown in Fig. 17, where the necessary fan, pumps, and valves are also indicated. The solar collector can be connected by means of the *Pump(1)* to the dryer, to the hot storage tank, to the hot water producing system, and to the other technological heat consumers. This arrangement has the advantage of multipurpose use of solar energy.

The *Fan* forces the air through the drying bed, which can also be preheated by a thermogenerator if necessary. At night, the storage tank takes over. In this case the dryer or other technological heat consumer can be operated from the storage tank by *Pump(1)* or *Pump(2)*.

In unfavorable weather conditions additional energy is needed. Equipment includes thermogenerators and an electrical heating system built into the storage tank.

In the scheme shown in Fig. 12 in conjunction with the operation modes, the main categories (3) are the following:

- Drying with ambient air
- Drying energy from collector
- Collector works for the storage tank
- Drying energy from storage tank
- Other service from storage
- Loading of the storage tank
- Heating by thermogenerator

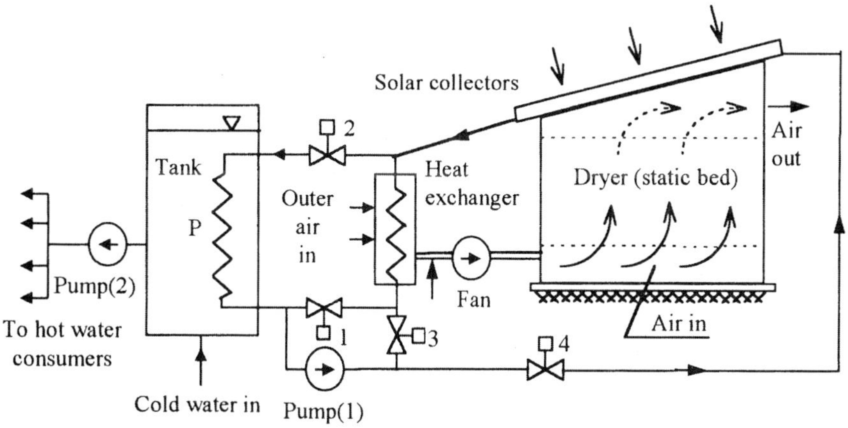

Fig. 17 Layout of an integrated energy/technology system.

11.2 Control Strategy of Operation Modes

Selection of the desired operation mode entails a coupled control problem that takes into account the collection and distribution of all sorts of energy sources available in the farm, taking economic aspects into consideration. This is a hierarchical control task that can be derived from the energy/technology circumstances on the farm. A possible solution for this problem is as follows:

Because the solar system uses a hot water storage tank as an energy storage, ability to fulfill hot water demand at any time is required. It can be done directly from solar energy or indirectly from a storage tank. This implies a given loading level in the storage to fulfill this requirement.

The second important aim is to provide the desired temperature of drying air. It can be provided first from direct solar energy. If the required amount of direct solar energy is not available, drying can be continued with stored solar energy from the tank. If stored energy is not sufficient, oil or gas thermogenerators can preheat the drying air. Electrical energy can be used only as a last resort because it is the most expensive source.

In consuming direct solar energy, the dryer, always the main energy consumer, has priority. The storage tank has next priority, assuming that the stratified tank can usually be loaded because of continuous water usage on the farm.

After the storage tank is loaded, water is heated; additional technological heat consumers use direct solar energy.

In drying, if solar energy is not sufficient to meet the demand of all subsystems, then it is necessary to remove subsystems from the direct solar energy resource in the following sequence: first, additional technological heat consumers; second, water heating; third, the storage tank. Of course, priority order can be reversed.

To perform the control tasks a block-oriented simulation model of the integrated system can include the following: (a) a submodel of technological processes, (b) an organizing submodel that allows construction of an appropriate layout from the available subsystems, and (c) a control subsystem that has potential for study of different control algorithms.

The structure of the simulation model permits flexible extension feasibility, i.e., including new types of blocks for additional technological processes on the farm and/or new control units.

In the selection of the appropriate operating mode the easily measured system variables are the ambient air temperature, the outlet and inlet water temperatures of the solar collector, the inlet and outlet air temperatures of the dryer, the layer temperatures of the storage tank, and the temperature of the service hot water.

REFERENCES

1. Boseley, J. R., T. F. Edgar, A. A. Patwardhan, and G. T. Wright, Model-based control: A survey, In Advanced Control of Chemical Processes (ed. Najim, K. and E. Dufour), IFAC Symposia Series, No. 8, Pergamon Press, Oxford, New York, 1992.
2. De Baerdemaeker, J., and Y. Hashimoto, Speaking fruit approach to the intelligent control of the storage system, Proceedings of 12th CIGR World Congress, Vol. 1, 1994, pp. 190–197.
3. Farkas, I., Control and computer simulation of a complex solar drying system, Ph.D. Thesis, Budapest, 1985.
4. Farkas, I., and C. C. Smith, Computer simulation and control of a solar crop dryer, Third

Technical Meeting of the CNRE-FAO on Solar Drying, Stuttgart, Germany, September 9–11, 1987. (Also in FAO CNRE Bulletin No. 19, 1988, pp. 11–14.)
5. Farkas, I., Computer aided off-line grain dryer operation, Problems of Agriculture and Forestry Engineering, 2nd National Scientific Conference, Warsaw, June 20–21, 1989, pp. 221–224.
6. Farkas, I., Modelling and identification of agricultural dryers, 11th IFAC World Congress, Tallinn, Estonia, Vol. 12, 1990, pp. 4–8.
7. Farkas, I., Modeling and identification for control of solar and connected technological systems, Unpublished D.Sc. Thesis, Hungarian Academy of Sciences, Budapest, 1992.
8. Farkas, I., and T. Nybrant, Control of Post-Harvesting Processes: Drying 1992, Part B (ed. A. S. Mujumdar), Elsevier Science, 1992, pp. 1379–1388.
9. Farkas, I. (ed.), Modelling, Control and Optimization: Greenhouse, Drying and Farm Energy System, Gödöllõ University of Agricultural Science, Gödöllõ, Hungary, 1998.
10. Farkas, I., Control aspects of post-harvest and related processing technologies, Computers and Electronics in Agriculture, Vol. 26, No. 2, 2000, pp. 81–82.
11. Farkas, I., Reményi, P., and Biró, A., A neural network topology for modelling grain drying, Computers and Electronics in Agriculture, Vol. 26, No. 2, 2000, pp. 147–158.
12. Garcia, C. E., and M. Morari, Internal model control. 1. A unifying review and some new results, Ind. Eng. Chem. Proc. Des. Dev., Vol. 21, 1982, p. 308.
13. Gordon, J. M., and A. Rabl, Design analysis and optimisation of solar industrial process heat plants without storage, Solar Energy, Vol. 28, No. 6, 1982, pp. 519–530.
14. Gottschalk, K., Optimization of the climate for potato storehouses using adaptive fuzzy-control methods, CAPP 195, 1st IFAC/CIGR/EURAGENG/ISHS Workshop on Control Applications in Post-Harvest and Processing Technology Ostend, Belgium, 1–2 June, 1995, pp. 125–129.
15. Hashimoto, Y., Recent strategies of optimal growth regulation by ''speaking plant concept,'' Acta Horticulturae, Vol. 260, 1989, pp. 115–121.
16. Hashimoto, Y., Computer integrated plant growth factory for agriculture and horticulture, Proceedings of IFAC/ISHS 1st Workshop on Mathematical and Control Applications in Agriculture and Horticulture, 1991, pp. 105–110.
17. Hashimoto, Y., Application of artificial neural networks and genetic algorithms to agricultural systems, Computers and Electronics in Agriculture, Vol. 18, No. 2, 1997, pp. 71–72.
18. Jumah, R. Y., and A. S. Mujumdar, Control of industrial dryers, In Handbook of industrial drying, 2nd Edition (ed. A. S. Mujumdar), Marcel Dekker, New York, 1995, pp. 1343–1368.
19. Kiranoudis, C. T., Z. B. Maroulis, and D. Marinos-Kouris, Modelling and optimization of a tunnel grape dryer, Drying Technology, Vol. 14, No. 7–8, 1996, pp. 1695–1718.
20. Lamber, F., The development of computer controlled environment in potato storage, CAPP 195, 1st IFAC/CIGR/EURAGENG/ISHS Workshop on Control Applications in Post-Harvest and Processing Technology Ostend, Belgium, 1–2 June, 1995, pp. 117–118.
21. Luyben, W. L., Process Modelling, Simulation and Control for Chemical Engineers, 2nd Edition, McGraw-Hill, New York, 1990.
22. Marchant, J. A., Control of high temperature continuous dryers, Agricultural Engineering, Vol. 40, 1995, pp. 145–149.
23. Mizrach, A., Flitsanov, U., Akerman, M., and Zauberman, G., Monitoring avocado softening in low-temperature storage using ultrasonic measurements, Computers and Electronics in Agriculture, Vol. 26, No. 2, 2000, pp. 199–207.
24. Moreira, R. G., and F. W. Bakker-Arkema, A feedforward/feedback adaptive controller for commercial crossflow grain driers, Journal of Agricultural Engineering Research, Vol. 45, 1990, pp. 107–116.
25. Morimoto, T., W. Purwanto, J. De Baerdemaeker, and Y. Hashimoto, Optimization for fruit quality during a storage process, CAPP 195, 1st IFAC/CIGR/EURAGENG/ISHS Workshop on Control Applications in Post-Harvest and Processing Technology Ostend, Belgium, 1–2 June, 1995, pp. 19–24.

26. Nybrant, T. G., Modelling and control of grain driers, Report UPTEC 8625R, Institute of Technology, Uppsala University, Uppsala, Sweden, 1986.
27. Nybrant, T. G., Modelling and adaptive control of continuous grain driers, Journal of Agricultural Engineering Research, Vol. 40, pp. 165–173, 1988.
28. Robinson, J., Improve dryer control, Chemical Engineering Progress, Vol. 88(12), 1992, pp. 28–33.
29. Panda, R. C., Dynamics and control of Continuous fluidized bed dryer, Ph.D. Thesis, Indian Institute of Technology, Madras, India, 1993.
30. Patwadian, A. A., J. B. Rawlings, and T. F. Edgar, Non-linear model predictive control, Chemical Eng. Commun., Vol. 87, 1990, pp. 123–141.
31. Schouten, S. P., Dynamic control of the oxygen content during CA storage of fruits and vegetables, CAPP 195, 1st IFAC/CIGR/EURAGENG/ISHS Workshop on Control Applications in Post-Harvest and Processing Technology, Ostend, Belgium, 1–2 June, 1995, pp. 131–136.
32. Siettos, C. I., Kiranoudis, C. T., and Bafas, G. V., Advanced control strategies for fluidized bed dryers, Drying Technology, Vol. 17, No. 10, 1999, pp. 2271–2291.
33. Sriramulu, V., S. B. Ahmed, and M. C. Gupta, Investigation of thermal storage unit for solar power generation, Solar Energy and Conservation (ed. T. N. Veziroglu), Vol. 3, Pergamon Press, New York, 1979, pp. 202–226.
34. Stefanovic, M. M., and M. B. Stakic, Computer system for the control of a pneumatic drum dryer, Drying 1998, Proceedings of the 11th International Drying Symposium (IDS 198) (ed. C. B. Akritidis, D. Marinos-Kouris, and G. D. Saravacos), Halkidiki, Greece, August 19–22, 1998, Vol. A, pp. 597–604.
35. Stephanopoulis, Chemical Process Control, Prentice-Hall, Englewood Cliffs, NJ, 1984.
36. Sun, Z. F., and Carrington, C. G., Dynamic modelling of a dehumidifier wood drying kiln, Drying Technology, Vol. 17, No. 4–5, 1999, pp. 711–729.
37. Susuki K., and T. Maekawa, Studies on the germinated brown rice for cooking foods homogeneous sprouting control in the liquid culture, Proceedings of 2nd IFAC/ISHS/CIGR/EURAGENG Workshop on Control Applications in Post-Harvest and Processing Technology (ed. I. Farkas), Pergamon Press, Elmsford NY, 1998, pp. 31–35.
38. Tarasiewicz, S., and F. Leger, Modelling simulation and control of the wood drying process. Part 1. A set of PDE's as an internal model, Drying Technology, Vol. 16, No. 6, 1998, pp. 1075–1084.
39. Tarasiewicz, S., F. Ding, and F. Leger, Modelling simulation and control of the wood drying process. Part 2. Variable control system, Drying Technology, Vol. 16, No. 6, 1998, pp. 1085–1100.
40. Temple, S. J., and A. J. B. van Boxtel, Control of fluid bed drying of tea, Proceedings of 2nd IFAC/ISHS/CIGR/EURAGENG Workshop on Control Applications in Post-Harvest and Processing Technology (ed. I. Farkas), Pergamon Press, Elmsford, NY, 1998, pp. 37–41.
41. Temple, S. J., and van Boxtel, A. J. B., Control of fluid bed tea dryers: controller design and tuning, Computers and Electronics in Agriculture, Vol. 26, No. 2, 2000, pp. 159–170.
42. Toftdal, A., and H. Olesen, Grain drying, In Innovation Development Engineering, Thisted, Denmark, 1982.
43. Trelea, I. C., F. Courtois, and G. Trystram, Non-linear optimal control of batch corn dryer, Drying 1996, Proceedings of the 10th International Drying Symposium (IDS 196), Kraków, Poland, 30 July–2 August, Vol. B, 1996, pp. 1417–1424.
44. Tuffs, P. S., and D. W. Clarke, Self-tuning control of offset: a unified approach, IEEE Proceedings, No. 132, 1985, pp. 100–110.
45. Vasconcelos L. G. S., R. Maciel Filho, and S. C. S. Rocha, Optimal design of industrial grain dryer: Alternative configurations, Drying 1998, Proceedings of the 11th International Drying Symposium (IDS 198), Halkidiki, Greece, August 19–22, 1998, Vol. B, pp. 1432–1439.

Index